Zoology

Zoology

Eleventh Edition

Stephen A. Miller
College of the Ozarks–Professor Emeritus
zoology.miller@gmail.com

Todd A. Tupper
Northern Virginia Community College
zoology.tupper@gmail.com

ZOOLOGY, ELEVENTH EDITION

Published by McGraw-Hill Education, 2 Penn Plaza, New York, NY 10121.

This book is printed on acid-free paper.

3 4 5 6 7 8 9 LKV 21 20

ISBN 978-1-259-88002-5 (bound edition)
MHID 1-259-88002-8 (bound edition)

ISBN 978-1-260-16204-2 (loose-leaf edition)
MHID 1-260-16204-4 (loose-leaf edition)

Portfolio Manager: *Michael Ivanov*
Product Developer: *Elizabeth Sievers*
Marketing Manager: *Kelly Brown*
Content Project Managers: *Lisa Bruflodt, Amber Bettcher*
Buyer: *Laura Fuller*
Design: *Matt Backhaus*
Content Licensing Specialist: *Lorraine Buczek*
Cover Image: ©*Kiran Bahra*
Compositor: *SPi Global*

Library of Congress Cataloging-in-Publication Data

Names: Miller, Stephen A., author.
Title: Zoology / Stephen A. Miller, College of the Ozarks-Professor Emeritus,
Todd A. Tupper, Northern Virginia Community College.
Description: Eleventh edition. | New York, NY: McGraw-Hill Education, [2019]
| Includes bibliographical references and index.
Identifiers: LCCN 2018011462| ISBN 9781259880025 (bound edition: alk. paper)
| ISBN 9781260162042 (loose-leaf edition)
Subjects: LCSH: Zoology.
Classification: LCC QL47.2 .M55 2019 | DDC 590–dc23 LC record available at https://lccn.loc.gov/2018011462

BRIEF CONTENTS

Preface xi

1 Zoology: An Evolutionary and Ecological Perspective 1
2 The Structure and Function of Animal Cells 11
3 Cell Division and Inheritance 31
4 Evolution: History and Evidence 53
5 Evolution and Gene Frequencies 72
6 Ecology: Preserving the Animal Kingdom 87
7 Animal Taxonomy, Phylogeny, and Organization 106
8 Animal Origins and Phylogenetic Highlights 121
9 The Basal Animal Phyla 134
10 The Smaller Lophotrochozoan Phyla 158
11 Molluscan Success 183
12 Annelida: The Metameric Body Form 206
13 The Smaller Ecdysozoan Phyla 226
14 The Arthropods: Blueprint for Success 239
15 The Pancrustacea: Crustacea and Hexapoda 257
16 Ambulacraria: Echinoderms and Hemichordates 284
17 Chordata: Urochordata and Cephalochordata 303
18 The Fishes: Vertebrate Success in Water 314
19 Amphibians: The First Terrestrial Vertebrates 336
20 Nonavian Reptiles: Diapsid Amniotes 355
21 Birds: The Avian Reptiles 373
22 Mammals: Synapsid Amniotes 392
23 Protection, Support, and Movement 418
24 Communication I: Nervous and Sensory Systems 435
25 Communication II: The Endocrine System and Chemical Messengers 462
26 Circulation and Gas Exchange 481
27 Nutrition and Digestion 501
28 Temperature and Body Fluid Regulation 521
29 Reproduction and Development 542

Appendix A: Animal Phylogeny 564
Appendix B: The History of the Earth: Geological Eons, Eras, Periods, and Major Biological Events 566
Appendix C: Animal-Like Protists: The Protozoa 568
Glossary 573
Index 599

CONTENTS

Preface xi

CHAPTER 1

ZOOLOGY: AN EVOLUTIONARY AND ECOLOGICAL PERSPECTIVE 1

Chapter Outline 1
Introduction to Zoology 1
Zoology: An Evolutionary Perspective 2
Zoology: An Ecological Perspective 4
WILDLIFE ALERT 8
Summary 9
Concept Review Questions 9
Analysis and Application Questions 10

CHAPTER 2

THE STRUCTURE AND FUNCTION OF ANIMAL CELLS 11

Chapter Outline 11
Cells: The Common Unit of All Life 11
Cellular Membranes and Membrane Transport 13
Energy Processing 17
The Nucleus, Ribosomes, and Vaults 23
The Endomembrane System 24
Peroxisomes 26
The Cytoskeleton and Cellular Movement 27
Levels of Organization in an Animal 28
Summary 29
Concept Review Questions 30
Analysis and Application Questions 30

CHAPTER 3

CELL DIVISION AND INHERITANCE 31

Chapter Outline 31
Eukaryotic Chromosomes 31
The Cell Cycle and Mitotic Cell Division 34
Meiosis: The Basis of Sexual Reproduction 36
DNA: The Genetic Material 38
Inheritance Patterns in Animals 44
WILDLIFE ALERT 50
Summary 51
Concept Review Questions 51
Analysis and Application Questions 52

CHAPTER 4

EVOLUTION: HISTORY AND EVIDENCE 53

Chapter Outline 53
Organic Evolution and Pre-Darwinian Theories of Change 53
Darwin's Early Years and His Journey 54
Early Development of Darwin's Ideas of Evolution 55
The Theory of Evolution by Natural Selection 57
Geological Time and Mass Extinctions 60
Microevolution, Macroevolution, and Evidence of Macroevolutionary Change 62
Summary 70
Concept Review Questions 71
Analysis and Application Questions 71

CHAPTER 5

EVOLUTION AND GENE FREQUENCIES 72

Chapter Outline 72
Populations and Gene Pools 72
Must Evolution Happen? 73
Evolutionary Mechanisms 73
Species and Speciation 80
Rates of Evolution 81
Molecular Evolution 84
Mosaic Evolution 84
Summary 85
Concept Review Questions 85
Analysis and Application Questions 86

CHAPTER 6

ECOLOGY: PRESERVING THE ANIMAL KINGDOM 87

Chapter Outline 87
Animals and Their Abiotic Environment 87
Biotic Factors: Populations 89
Biotic Factors: Interspecific Interactions 91
Communities 94
Trophic Structure of Ecosystems 95
Cycling within Ecosystems 97
Ecological Problems 99
Summary 104
Concept Review Questions 105
Analysis and Application Questions 105

CHAPTER 7

Animal Taxonomy, Phylogeny, and Organization 106

Chapter Outline 106
Taxonomy and Phylogeny 106
Patterns of Organization 115
Summary 119
Concept Review Questions 120
Analysis and Application Questions 120

CHAPTER 8

Animal Origins and Phylogenetic Highlights 121

Chapter Outline 121
Earth's Beginning and Evidence of Early Life 121
Life's Beginning and the First 3 Billion Years 122
Multicellularity and Animal Origins 126
Phylogenetic Highlights of Animalia 129
Summary 132
Concept Review Questions 132
Analysis and Application Questions 133

CHAPTER 9

The Basal Animal Phyla 134

Chapter Outline 134
Evolutionary Perspective 134
Phylum Porifera 136
Phylum Cnidaria 141
Phylum Ctenophora 151
WILDLIFE ALERT 153
Further Phylogenetic Considerations 154
Summary 156
Concept Review Questions 157
Analysis and Application Questions 157

CHAPTER 10

The Smaller Lophotrochozoan Phyla 158

Chapter Outline 158
Evolutionary Perspective 158
Platyzoa: Phylum Platyhelminthes 160
Platyzoa: Smaller Phyla 171
Other Lophotrochozoans 176
Further Phylogenetic Considerations 179
Summary 181
Concept Review Questions 181
Analysis and Application Questions 182

CHAPTER 11

Molluscan Success 183

Chapter Outline 183
Evolutionary Perspective 183
Molluscan Characteristics 184
Class Gastropoda 186
Class Bivalvia 190
Class Cephalopoda 194
Class Polyplacophora 199
Class Scaphopoda 200
Class Monoplacophora 200
Class Solenogastres 201
Class Caudofoveata 201
Further Phylogenetic Considerations 202
WILDLIFE ALERT 202
Summary 204
Concept Review Questions 205
Analysis and Application Questions 205

CHAPTER 12

Annelida: The Metameric Body Form 206

Chapter Outline 206
Evolutionary Perspective 206
Annelid Structure and Function 209
Clade (Class) Errantia 214
Clade (Class) Sedentaria 215
Basal Annelid Groups 221
Further Phylogenetic Considerations 222
Summary 224
Concept Review Questions 225
Analysis and Application Questions 225

CHAPTER 13

The Smaller Ecdysozoan Phyla 226

Chapter Outline 226
Evolutionary Perspective 226
Phylum Nematoda (Roundworms) 227
Other Ecdysozoan Phyla 234
Further Phylogenetic Considerations 236
Summary 238
Concept Review Questions 238
Analysis and Application Questions 238

CHAPTER 14

The Arthropods: Blueprint for Success 239

Chapter Outline 239
Evolutionary Perspective 239

Metamerism and Tagmatization 241
The Exoskeleton 241
The Hemocoel 242
Metamorphosis 243
Subphylum Trilobitomorpha 244
Subphylum Chelicerata 244
Subphylum Myriapoda 253
Further Phylogenetic Considerations 254
Summary 255
Concept Review Questions 255
Analysis and Application Questions 256

CHAPTER 15

THE PANCRUSTACEA: CRUSTACEA AND HEXAPODA 257

Chapter Outline 257
Evolutionary Perspective 257
Subphylum Crustacea 258
WILDLIFE ALERT 266
Subphylum Hexapoda 268
Further Phylogenetic Considerations 279
Summary 282
Concept Review Questions 283
Analysis and Application Questions 283

CHAPTER 16

AMBULACRARIA: ECHINODERMS AND HEMICHORDATES 284

Chapter Outline 284
Evolutionary Perspective 284
Phylum Echinodermata 285
Phylum Hemichordata 295
WILDLIFE ALERT 296
Further Phylogenetic Considerations 299
Summary 301
Concept Review Questions 302
Analysis and Application Questions 302

CHAPTER 17

CHORDATA: UROCHORDATA AND CEPHALOCHORDATA 303

Chapter Outline 303
Evolutionary Perspective 303
Phylum Chordata 304
Further Phylogenetic Considerations 310
Summary 312
Concept Review Questions 312
Analysis and Application Questions 313

CHAPTER 18

THE FISHES: VERTEBRATE SUCCESS IN WATER 314

Chapter Outline 314
Evolutionary Perspective 314
Survey of Fishes 317
Evolutionary Pressures 324
WILDLIFE ALERT 330
Further Phylogenetic Considerations 331
Summary 334
Concept Review Questions 334
Analysis and Application Questions 335

CHAPTER 19

AMPHIBIANS: THE FIRST TERRESTRIAL VERTEBRATES 336

Chapter Outline 336
Evolutionary Perspective 336
Survey of Amphibians 338
Evolutionary Pressures 341
Amphibians in Peril 350
WILDLIFE ALERT 351
Further Phylogenetic Considerations 352
Summary 353
Concept Review Questions 354
Analysis and Application Questions 354

CHAPTER 20

NONAVIAN REPTILES: DIAPSID AMNIOTES 355

Chapter Outline 355
Evolutionary Perspective 355
Survey of the Nonavian Reptiles 357
WILDLIFE ALERT 362
Evolutionary Pressures 364
Further Phylogenetic Considerations 371
Summary 371
Concept Review Questions 372
Analysis and Application Questions 372

CHAPTER 21

BIRDS: THE AVIAN REPTILES 373

Chapter Outline 373
Evolutionary Perspective 373
Evolutionary Pressures 376
WILDLIFE ALERT 390
Summary 391
Concept Review Questions 391
Analysis and Application Questions 391

CHAPTER 22

Mammals: Synapsid Amniotes 392

Chapter Outline 392
Evolutionary Perspective 392
Diversity of Mammals 394
Evolutionary Pressures 397
WILDLIFE ALERT 408
Human Evolution 409
Summary 416
Concept Review Questions 416
Analysis and Application Questions 417

CHAPTER 23

Protection, Support, and Movement 418

Chapter Outline 418
Integumentary Systems 418
Skeletal Systems 422
Nonmuscular Movement and Muscular Systems 427
Summary 433
Concept Review Questions 434
Analysis and Application Questions 434

CHAPTER 24

Communication I: Nervous and Sensory Systems 435

Chapter Outline 435
Neurons: The Basic Functional Units of the Nervous System 435
Neuron Communication 437
Invertebrate Nervous Systems 439
Vertebrate Nervous Systems 442
Sensory Reception 446
Invertebrate Sensory Receptors 448
Vertebrate Sensory Receptors 451
Summary 460
Concept Review Questions 461
Analysis and Application Questions 461

CHAPTER 25

Communication II: The Endocrine System and Chemical Messengers 462

Chapter Outline 462
The Evolution and Diversity of Chemical Messengers 462
Hormones and Their Feedback Systems 463
Mechanisms of Hormone Action 465
Endocrine Glands 466
Invertebrate Endocrine Control 466
Endocrine Systems of Fishes and Amphibians 469
Endocrine Systems of Amniotes 471
Some Hormones Are Not Produced by Endocrine Glands 478
Evolution of Endocrine Systems 478
Summary 479
Concept Review Questions 479
Analysis and Application Questions 480

CHAPTER 26

Circulation and Gas Exchange 481

Chapter Outline 481
Transport Systems in Invertebrates 481
Transport Systems in Vertebrates 483
Vertebrate Circulatory Systems 485
The Mammalian Circulatory System 487
Lymphatic Systems 489
Gas Exchange 489
Vertebrate Respiratory Systems 492
The Mammalian Respiratory System 495
Summary 499
Concept Review Questions 500
Analysis and Application Questions 500

CHAPTER 27

Nutrition and Digestion 501

Chapter Outline 501
Evolution of Heterotrophy 501
The Metabolic Fates of Nutrients in Heterotrophs 502
Digestion 503
Animal Strategies for Getting and Using Food 504
Diversity in Digestive Structures: Invertebrates 506
Diversity in Digestive Structures: Vertebrates 507
The Mammalian Digestive System 512
Summary 519
Concept Review Questions 520
Analysis and Application Questions 520

CHAPTER 28

Temperature and Body Fluid Regulation 521

Chapter Outline 521
Homeostasis and Temperature Regulation 521
Control of Water and Electrolytes 528
Invertebrate Excretory Systems 530
Vertebrate Excretory Systems 532
Summary 540
Concept Review Questions 541
Analysis and Application Questions 541

CHAPTER 29

REPRODUCTION AND DEVELOPMENT 542

Chapter Outline 542
Asexual Reproduction in Invertebrates 542
Sexual Reproduction in Invertebrates 544
Sexual Reproduction in Vertebrates 546
The Human Male Reproductive System 549
The Human Female Reproductive System 551
Prenatal Development and Birth in a Human 556
Summary 561
Concept Review Questions 561
Analysis and Application Questions 562

Appendix A: Animal Phylogeny 564
Appendix B: The History of the Earth: Geological Eons, Eras, Periods, and Major Biological Events 566
Appendix C: Animal-Like Protists: The Protozoa 568
Evolutionary Perspective 568
Protozoans: Major Ecological Roles 568
Supergroup Excavata 568
Supergroup Amoebozoa 570
Supergroup Chromalveolata 571
Glossary 573
Index 599

PREFACE

As the senior author of *Zoology*, I (Stephen Miller) am very pleased to present the eleventh edition of this textbook. I have been associated with this book from its inception. The initial planning, writing, and revision of *Zoology* through preparing this latest edition involved tasks that have spanned more than 30 years of my life. I am honored to have played a small part in the training of zoologists and other biologists, most of whom I will never know personally. Similarly, it has been an honor to feel a connection to the professors who use *Zoology* in their courses and who deserve the credit for mentoring future scientists. The processes of writing and revision have involved many joys, and some headaches, for me personally. I hope these efforts have helped inspire bright young minds to strive to make our planet a better place than Earth would have otherwise been.

I believe the comments of student users and reviewers of this textbook when they say that the strengths of this book lie in presentation of relevant, up-to-date zoological concepts in a friendly writing style. A book that is not read cannot inform and inspire. These strengths are carried on into the eleventh edition. Our goal has been to present zoology as a dynamic field that is critical to understanding and preserving of our planet and to present zoology in a format that is adaptable to a variety of one- or two-semester course formats.

As important as it has been for us to maintain the focus and goals of *Zoology*, change is inevitable. Two important changes faced us in this revision. The first change was the retirement of Dr. John Harley as coauthor. He has been with the book, nearly from the beginning. I wish him the best in his retirement. The second, and very welcome, change was the addition of Dr. Todd Tupper to the book team. Todd teaches general zoology at Northern Virginia Community College. He has earned both teaching and professional achievement awards. Todd involves students in his research and writing, has been a reviewer of the book, has served as an invited contributor to the tenth edition, and has used the book in his classes for many years. I have come to know Todd as someone who lives and breathes zoology and has a hunger for passing his devotion on to students. His expertise is in herpetological conservation, and you will see his ecologically minded fingerprints throughout this edition. Thank you, Todd.

Todd's involvement with zoology began with brief emailed feedback with John and me. His comments were honest and constructive, and they made the book better. We have had similar feedback from other users and students, and we welcome more of the same from you. We take all of your comments very seriously. As always, as soon as we close the cover on one revision, we begin thinking about the next. That planning begins now. Please feel free to contact us at zoology.miller@gmail.com or zoology.tupper@gmail.com.

CONTENT AND ORGANIZATION

Revision for the eleventh edition of *Zoology* has been a bigger than usual undertaking. Every chapter has been carefully scrutinized, and most chapters have had significant changes. The evolutionary and ecological perspectives that characterize this book have been retained. Both are extremely important for preserving the integrity of our discipline and the health of our planet. These perspectives are at the forefront of every chapter. A detailed explanation of the changes made to each chapter is presented in "New to the Eleventh Edition" later in this preface. The basic organization of the book is described in the following paragraphs.

Chapters 1 through 6 present cellular, evolutionary, and ecological concepts that unite zoology to biology as a whole. Chapter 2 (The Structure and Function of Animal Cells) has been completely rewritten, and the other chapters are updated with new population statistics, fresh examples, and new illustrations and photographs.

Chapters 7 through 22 cover animal taxonomy and phylogeny (chapter 7), animal origins and phylogenetic highlights (chapter 8), and survey the animal phyla (chapters 9 through 22). For those familiar with previous editions, you will note that chapter 8 is new. We believe that the addition of this chapter makes sense. It should help students move from understanding evolutionary principles (chapters 4 and 5), to knowing how zoologists organize life's diversity (chapter 7), to grasping the "big picture" of animal phylogeny (chapter 8), and to fitting animal phyla into the "big picture" (chapters 9 through 22). The trade-off that was made in presenting animal origins and diversification material in chapter 8 was to condense the coverage of the protists (chapter 8 in earlier editions) by moving it to a newly created appendix C. Instructors who want to cover these important nonanimal eukaryotes can still do so using appendix C. All of the chapters in this group, 7 through 22, include new explanations, photographs, and illustrations.

Chapters 23 through 29 have seen big changes. Although many of the excellent illustrations and photographs from the tenth edition have been retained, the text in these chapters has been largely rewritten, and in some cases reorganized. The goals of this reorganization and rewriting were to create a stronger evolutionary focus, to illustrate important unifying principles of structure and function without presenting excessive detail, and to enhance the readability of these chapters. We believe that we have accomplished these goals.

PEDAGOGY

Integrated Learning Outcomes and Critical Thinking

We have retained pedagogical elements useful to science faculty in identifying measurable learning outcomes. **Learning Outcomes** have been retained and enhanced in the eleventh edition for each major section of each chapter. Each major section ends with a **Thinking Beyond the Facts** question, which requires students to apply the information presented in the section. Possible answers to these questions are available to instructors using Connect Zoology. These elements allow students to self-test and instructors to document student learning. In addition, instructors and students using Connect Zoology can access auto-gradable and interactive assessment material tied to learning outcomes from the text. These Connect features include the new LearnSmart and SmartBook adaptive learning tools and are described under "Teaching and Learning Resources."

Each chapter ends with a **Summary** of key concepts organized by section, a set of **Concept Review Questions,** and **Analysis and Application Questions**. These questions have been carefully reviewed and revised as needed. They allow students to test their understanding of chapter concepts and to apply concepts they have learned in each chapter. Suggested answers to these questions are available to instructors through Connect. In addition to being printed in the back of the book, the glossary is available electronically through Connect and in SmartBook.

An Evolutionary and Ecological Focus

Zoology emphasizes ecological and evolutionary concepts and helps students understand the process of science through elements of chapter organization and boxed readings. Each chapter in chapters 9 through 22 begins with a section entitled **Evolutionary Perspective**. This section discusses the relationship of the phylum or phyla covered in the current chapter to the animal kingdom as a whole and to animals discussed in previous chapters. Students are frequently reminded to consult appendices A and B to reorient themselves to phylogenetic relationships and geological time frames of evolutionary events. Similarly, each survey chapter ends with a section entitled **Further Phylogenetic Considerations**. This section discusses phylogenetic relationships of groups (subphyla or classes) within the phylum or phyla being studied and is a point of transition between chapters. The discussion in this section is usually supported by a cladogram illustrating important phylogenetic relationships.

To further explain and support evolutionary concepts, a set of themed boxed readings entitled **Evolutionary Insights** is present throughout the book. These boxes provide detailed examples of principles covered in a chapter and provide insight into how evolutionary biology works. For example, chapter 4 includes a reading on cat biogeography that illustrates how a variety of sources of evidence are used to paint a picture of the history of one group of animals. Chapter 5 has a reading on speciation of Darwin's finches that illustrates how speciation can occur. Chapter 18 has a reading on the evolution of the vertebrate limb, and chapter 25 has a reading on the evolution of hormone receptors.

The ecological perspective of *Zoology* is stressed throughout chapters 1 to 22. Human population and endangered species statistics have been updated. Ecological problems are discussed including a newly written section entitled "Earth's Resources and Global Inequality" in chapter 6. The ecological perspective is reinforced by boxed readings entitled **Wildlife Alerts**. Wildlife Alerts first appeared in the fourth edition and have been very well received by students and professors. Each boxed reading depicts the plight of selected animal species or broader ecosystem issues relating to preserving animal species. These readings have been revised and updated. Apart from these boxed readings, numerous examples of threatened and endangered species are woven into chapters 4 through 22 that remind students of the delicate status of natural ecosystems of our planet. Students who read and study this book should have an enhanced understanding of ecological principles and how human ignorance and misplaced values have had detrimental effects on our environment in general and on specific animal groups in particular.

The Process of Science

To help students understand that science is a process, not just a body of facts, **How Do We Know** boxed readings are retained in this edition and they highlight research results that provide insight into biological processes. Chapter 9 has a boxed reading entitled "How Do We Know about Sponge Defenses?" This reading describes how zoologists investigated sponge defense mechanisms. Chapter 19 has a boxed reading entitled "How Do We Know about Amphibian Skin Toxins?" This reading describes how scientists are studying antibacterial and anticancer effects of amphibian skin toxins. Students learn that these studies have implications for studying naturally occurring compounds that may aid in the development of novel pharmaceutical drugs.

Digital Assets and Media Integration

Beginning with the ninth edition of *Zoology,* digital resources were integrated into the book through the Connect Zoology site. Many of the sections within most chapters are linked to animations from McGraw Hill's library of animations. These animations will enhance students' understanding of material within the chapter and are available through Connect.

NEW TO THE ELEVENTH EDITION

As with earlier revisions of *Zoology*, the focus for this revision has been on presenting evolutionary and ecological concepts clearly and accurately using examples from current literature as convincingly as possible. The revisions highlighted below should impress students with the excitement experienced in zoology as new information clarifies zoological concepts and informs our understanding of phylogenetic relationships.

- **Chapter 1 (Zoology: An Evolutionary and Ecological Perspective)**

  A new introduction emphasizes the importance of zoological studies in our modern world. Population, world resources, and threatened and endangered species statistics have been updated with 2018 data. Table 1.4 is new and presents information on the major contributors of anthropogenic greenhouse gas emissions.

- **Chapter 2 (The Structure and Function of Animal Cells)**

  This chapter has been completely rewritten. It begins with the properties and common origin of cells. Coverage of cells is organized by common functions. It begins with cellular membranes and membrane transport, followed by energy processing. Five new figures (2.11–2.13 and 2.15–2.16) and new text art support an introduction to enzymes, ATP, glycolysis, the citric acid cycle, electron transport and chemiosmosis. This coverage provides important background for material that comes later, on energy flow in ecosystems, animal temperature regulation, nutrition, and related topics. The descriptions of the nucleus, ribosomes, and vaults are followed by coverage of functions of the endomembrane system, which are supported by the newly rendered figure 2.19. Peroxisomes, the cytoskeleton, and an introduction to tissues, organs, and organ systems follow. Detailed descriptions of tissue types have been moved to the appropriate system in chapters 23 through 29.

- **Chapter 3 (Cell Division and Inheritance)**

  A new section on cell-cycle control has been added to this chapter.

- **Chapter 4 (Evolution: History and Evidence)**

  The introduction to chapter 4 has been rewritten. In an era in which the validity of science is under attack, students need to be reminded of what a scientific theory is and the work involved with establishing scientific theories. We believe that the beginning of this chapter is the right place to present such a reminder. The revival of the epigenetics movement is briefly addressed. A new section entitled "Geological Time and Mass Extinctions" has been added to this chapter. This content is referenced throughout the book, and it is presented visually in appendix B, where it can be easily accessed. The "Evolutionary Insights" reading on cat phylogeny has been updated and illustrates how data from multiple disciples contribute to our understanding of evolutionary pathways. The cat phylogeny in box figure 4.1 has also been updated.

- **Chapter 5 (Evolution and Gene Frequencies)**

  In chapter 5, we have included an expanded discussion of neutral theory and genetic drift. A new example of founder effect (the brown anole lizard) is presented. The discussion of the effects of limiting gene flow through human influences is expanded with new examples that describe the loss of genetic diversity in bighorn sheep through highway construction and in polar bear populations as a result of sea-ice melting.

- **Chapter 6 (Ecology: Preserving the Animal Kingdom)**

  In chapter 6, the discussions of community stability and ecosystem trophic structure have been revised, including a discussion of biomagnification in the latter. Statistics on human population growth have been updated. We have added a new section entitled "Earth's Resources and Global Inequality," where students are challenged to consider overconsumption in western cultures. The material on biodiversity is revised with additional discussion on its preservation and five reasons it is threatened.

- **Chapter 7 (Animal Taxonomy, Phylogeny, and Organization)**

  Chapter 7 is updated with a newly written section on taxonomic methods. It compares the usefulness of analyses of genomic DNA, mitochondrial DNA, ribosomal RNA, and noncoding DNA sequences in phylogenetic studies. It emphasizes the importance of both molecular and traditional taxonomic data in these studies. There is a new discussion, and supporting figures, illustrating the construction of simple cladograms from phenotypic characters. A new "How Do We Know" boxed reading does the same using data from DNA base sequences.

- **Chapter 8 (Animal Origins and Phylogenetic Highlights)**

  Chapter 8 is new in the eleventh edition of *Zoology*. It begins with a discussion of Earth's beginning and evidence of early life. It continues by presenting what is known of events occurring within the first 3 billion years of Earth's history. It surveys hypotheses regarding life's origin and the evolution of the three domains, including the importance of horizontal gene transfer in early evolutionary events. The role of endosymbiosis in the origin of eukaryons is discussed. This presentation is followed by a discussion of the origins of multicellularity and animal origins within the eukaryotic clade Opisthokonta. The name "Apoikozoa" is introduced to designate the monophyletic choanoflagellate/animal lineage within Opisthokonta. A discussion of the Cambrian explosion follows. Chapter 8 ends with phylogenetic highlights of Animalia, including an introduction to the basal phyla, Protostomia, and Deuterostomia. We believe that this new chapter establishes a firm connection between the discussion of evolutionary theory in chapters 4 and 5, phylogenetics in chapter 7, and the survey of the animal phyla in chapters 9 through 22.

- **Chapters 9 through 17**

  Chapters 9 through 17 survey the animal phyla through the invertebrate chordates. They have been carefully revised and include smaller changes that clarify phylogenetics and natural history. We have included additional examples, often focusing on anthropogenic changes that threaten members of each group of animals. Wildlife Alerts have been rewritten in chapter 11 (freshwater bivalves) and chapter 16 (sea cucumbers). Boxed readings on sponge defenses (chapter 9) and early deuterostome evolution (chapter 17) have also been significantly revised.

- **Chapter 18 (The Fishes: Vertebrate Success in Water)**

  Chapter 18's coverage of early craniate evolution has been rewritten. The evolutionary history of early bony fishes has been revised, including new information on the evolution of paired appendages and jaws within the placoderms.

- **Chapter 19 (Amphibians: The First Terrestrial Vertebrates)**

  Revision of chapter 19 includes updated information on amphibian origins and origins of the Lissamphibia. Descriptions of all the extant amphibian orders have been rewritten with expanded coverage of the natural history of members of each order. The "Evolutionary Pressures" section has been updated and enhanced with expanded coverage and additional examples of amphibian adaptations. The section describing the endangered status of many amphibians has been rewritten and expanded. This section includes both anthropogenic threats like habitat destruction and "enigmatic" threats, including the spread of the chytrid fungus and climate change.

- **Chapter 20 (Nonavian Reptiles: Diapsid Amniotes)**

  Chapter 20 includes a new introduction to the two extant amniote lineages. We have consistently used the designations "nonavian reptiles" to refer to the traditional vertebrate class "Reptilia" and "avian reptiles" to refer to the traditional vertebrate class "Aves." These designations are intended as constant reminders of the accepted makeup of the reptilian lineage. Information on the evolution of Testudines and shell adaptations has been added. There are expanded descriptions of the natural history of the nonavian reptilian orders and squamate suborders. The "Evolutionary Pressures" section has been updated and enhanced with new examples.

- **Chapter 21 (Birds: The Avian Reptiles)**

  In chapter 21, the theropod origin of avian reptiles has been rewritten and expanded, as has the coverage of the evolution of the early birds. The discussion of the appearance of feathers and the evolution of flight has been enhanced. Figures 21.4 and 21.5 on feather structure and development are new, and the "Evolutionary Pressures" section is updated and includes additional examples to illustrate important concepts.

- **Chapter 22 (Mammals: Synapsid Amniotes)**

  Mammalian evolution within the therapsid lineage has been updated, as has mammalian classification. The "Evolutionary Pressures" section has been enhanced, and new information on human evolution has been added.

- **Chapters 23 through 29**

  Chapters 23 through 29 have been largely rewritten. The organization of all subjects in this block of chapters has been carefully examined to streamline their presentations, to ensure that they accurately portray evolutionary changes in animal systems, and to make them as reader friendly as possible. The organ-system histology formerly presented in chapter 2 is now appropriately placed in these chapters. In addition to these organizational revisions, the following content changes are present.

- **Chapter 23 (Protection, Support, and Movement)**

  The discussion of the mechanism of skeletal muscle contraction has been expanded, including new and revised figures. The comparative histology and physiology of fast and slow twitch skeletal muscle fibers, insect flight muscle, cardiac muscle, and vertebrate and invertebrate smooth muscles are explained.

- **Chapter 24 (Communication I: Nervous and Sensory Systems)**

  The presentation of concepts concerning resting membrane potentials, graded potentials, and action potentials has been clarified and enhanced. The presentation of invertebrate nervous systems has been revised to clearly reflect trends toward centralization and cephalization. Similarly, coverage of the evolution of the vertebrate brain has been rewritten with the same goal. A new "Evolutionary Insights" box describes surprising new roles for bitter taste receptors.

- **Chapter 25 (Communication II: The Endocrine System and Chemical Messengers)**

  Chapter 25 has been rewritten to provide clearer descriptions of the evolution and diversity of chemical messengers and the principles of negative and positive feedback control of hormonal functions. Both invertebrate and vertebrate endocrine systems are discussed in a phylogenetically meaningful organization.

- **Chapter 26 (Circulation and Gas Exchange)**

  In addition to clearer and more accurate descriptions of the evolution of vertebrate hearts, chapter 26 has a new discussion of oxygen and carbon dioxide transport, including discussions and illustrations of oxyhemoglobin dissociation curves.

- **Chapter 27 (Nutrition and Digestion)**

  Chapter 27 has a new discussion of the evolution of heterotrophy and a new "Evolutionary Insights" boxed reading on gut microbiomes.

- **Chapter 28 (Temperature and Body Fluid Regulation)**

  Chapter 28 includes a new presentation of temperature regulation, including a new figure 28.3 depicting strategies for animal thermoregulation. The sections on osmoregulation and excretion have been updated. The descriptions of invertebrate nephridia are presented with an evolutionary focus. More detail is provided on the evolution of vertebrate kidneys and the physiology of the metanephric kidney.

- **Chapter 29 (Reproduction and Development)**

  This chapter has been streamlined throughout, and the section on vertebrate reproduction has been enhanced; it now includes a stronger evolutionary focus and a clearer description of hormonal regulation of the human reproductive function.

- **Appendices**

  The eleventh edition of *Zoology* has moved some material into three appendices. Appendix A presents one interpretation of animal phylogeny that is developed and used in the textbook. Appendix B provides an expanded view of Earth's geological history and some major biological events occurring in each eon, era, and period. Frequent references are made to both of these appendices throughout the textbook. Placing this information in appendices provides students easy access to this important information. We recommend that students tab these appendices for quick referral. Appendix C covers the Protozoa. The content of appendix C has been revised and condensed from its appearance in chapter 8 of the tenth edition. While protists are not animals, they are still very important subjects in many general zoology courses. Appendix C will allow instructors who wish to cover these organisms to do so.

ACKNOWLEDGMENTS

We wish to thank reviewers who provided feedback and analysis of the revision plan for the eleventh edition. In the midst of their busy teaching and research schedules, they took time to consider our proposed revisions and offered constructive advice that greatly improved the eleventh edition. We also appreciate the help of individuals who reviewed portions of the eleventh edition manuscript. Dr. Chris Barnhart from Missouri State University provided valuable feedback on the content of the Wildlife Alert on endangered freshwater mussels (chapter 11). His remarkable photographs greatly enhance our understanding of the plight of these animals. Lastly, we'd like to sincerely thank the students and biology faculty at NOVA, Alexandria, for helping to prompt meaningful inquiry and discussion that ultimately led to many of the revisions present in the eleventh edition.

REVIEWERS

Calhoun Bond, *Greensboro College*
Levi Castle, *South Piedmont Community College*
Dawn Cummings, *Community College of Denver*
Quentin R. Hays, *Eastern New Mexico University*
Megan Keith, *South Plains College*
Mark Schlueter, *Georgia Gwinnett College*
Cristina Summers, *Central Texas College*
Evan W. Thomas, *MacMurray College*
Travis Vail, *Golden West College*
Daniece H. Williams, *Hinds Community College, Rankin Campus*

SPECIAL THANKS AND DEDICATIONS

The publication of a textbook requires the efforts of many people. We are grateful for the work of our colleagues at McGraw-Hill Education who have shown extraordinary patience, skill, and commitment to this textbook. Justin Wyatt and Michael Ivanov served as our Portfolio Mangers during this revision. They have guided *Zoology* through the author transition process, and they have worked very hard as advocates of the extensive revisions present in the eleventh edition. Elizabeth Sievers, Senior Product Developer, coordinated all of the tasks involved with publishing this edition. We learned to expect her emails at all hours of the day, and we are still amazed at her ability to guide reviews, manuscript, figure and table revisions, and new photographs into their proper places in the final version you have in front of you. Thank you for your patience with us on the many occasions that we submitted revised material and then resubmitted the same with additional changes. We know that we must have caused you moments of frustration beyond words. Lisa A. Bruflodt and Amber Bettcher served as Content Project Managers for this edition. We appreciate their efficiency and organization.

We wish to extend special appreciation to our families and loved ones for their patience and encouragement. Janice A. Miller lived through many months of planning and writing of the first edition of *Zoology*. She died suddenly two months before it was released. Steve's wife, Carol A. Miller, has been especially supportive throughout the lengthy revisions of the eleventh edition. Carol, an accomplished musician, spent many hours proofreading *Zoology* for grammatical errors. Over the past 25 years, she has become a much better zoologist than her husband has become a musician—something about practicing got in his way. Todd's parents, Pamela and Edward Tupper, and his long-time mentor, Dr. Paul Gurn, have been steadfastly supportive and consistently encouraged his commitment to studying and protecting wildlife. We dedicate this book to our families. We send a special memorial dedication to Kasia Wilczak and Dr. Noble S. Proctor—may you rest in peace.

McGraw-Hill Connect® is a highly reliable, easy-to-use homework and learning management solution that utilizes learning science and award-winning adaptive tools to improve student results.

Homework and Adaptive Learning

- Connect's assignments help students contextualize what they've learned through application, so they can better understand the material and think critically.
- Connect will create a personalized study path customized to individual student needs through SmartBook®.
- SmartBook helps students study more efficiently by delivering an interactive reading experience through adaptive highlighting and review.

Over **7 billion questions** have been answered, making McGraw-Hill Education products more intelligent, reliable, and precise.

Connect's Impact on Retention Rates, Pass Rates, and Average Exam Scores

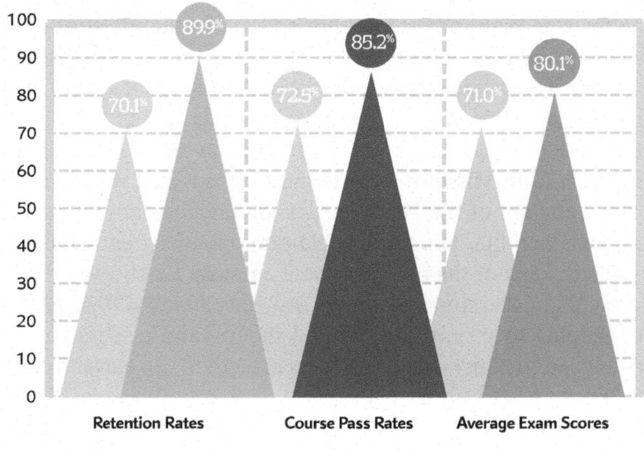

Using **Connect** improves retention rates by **19.8** percentage points, passing rates by **12.7** percentage points, and exam scores by **9.1** percentage points.

73% of instructors who use **Connect** require it; instructor satisfaction **increases** by 28% when **Connect** is required.

Quality Content and Learning Resources

- Connect content is authored by the world's best subject matter experts, and is available to your class through a simple and intuitive interface.
- The Connect eBook makes it easy for students to access their reading material on smartphones and tablets. They can study on the go and don't need internet access to use the eBook as a reference, with full functionality.
- Multimedia content such as videos, simulations, and games drive student engagement and critical thinking skills.

Robust Analytics and Reporting

©Hero Images/Getty Images

- Connect Insight® generates easy-to-read reports on individual students, the class as a whole, and on specific assignments.
- The Connect Insight dashboard delivers data on performance, study behavior, and effort. Instructors can quickly identify students who struggle and focus on material that the class has yet to master.
- Connect automatically grades assignments and quizzes, providing easy-to-read reports on individual and class performance.

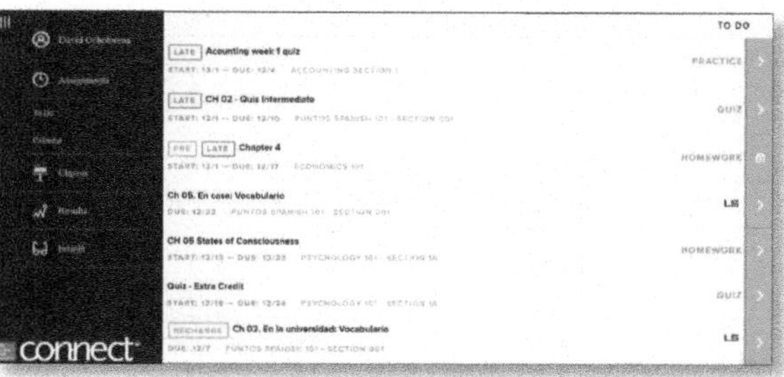

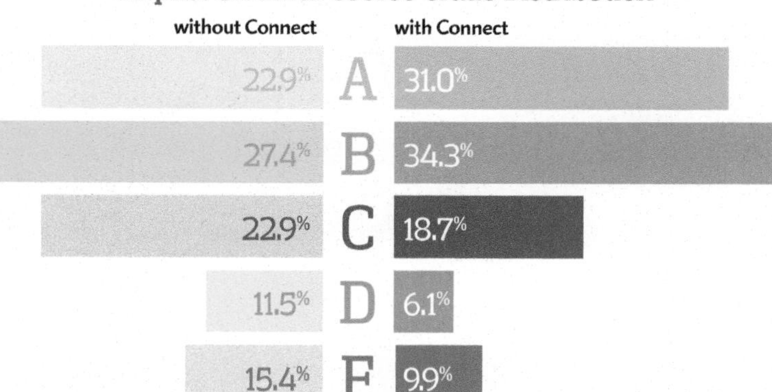

More students earn **As** and **Bs** when they use **Connect**.

Trusted Service and Support

- Connect integrates with your LMS to provide single sign-on and automatic syncing of grades. Integration with Blackboard®, D2L®, and Canvas also provides automatic syncing of the course calendar and assignment-level linking.
- Connect offers comprehensive service, support, and training throughout every phase of your implementation.
- If you're looking for some guidance on how to use Connect, or want to learn tips and tricks from super users, you can find tutorials as you work. Our Digital Faculty Consultants and Student Ambassadors offer insight into how to achieve the results you want with Connect.

Generations of Luo fishermen on Lake Victoria, Africa, have caught cichlid fish, including tilapia, as a mainstay of their economy. Recent introductions of the Nile perch (*Lates niloticus*) has changed the Lake Victoria ecosystem and the fishing economy of the lake.
©Nigel Pavitt/Getty Images

1

Zoology: An Evolutionary and Ecological Perspective

Chapter Outline

1.1 Introduction to Zoology
1.2 Zoology: An Evolutionary Perspective
Evolutionary Processes
Animal Classification and Evolutionary Relationships
1.3 Zoology: An Ecological Perspective
World Resources and Endangered Animals

You are about to begin a journey into the study of animals—a journey that the authors hope informs a deeper appreciation for the diversity of animal life and the evolutionary processes that produced this diversity. As you read and study, we also hope that you will become more aware of events and practices that threaten animal diversity. You will encounter ecological and zoological principles that help us understand how delicate ecological balances are being challenged and why these challenges impact animals and their environments in specific ways. Welcome to zoology. We hope your journey is one that enhances your life and promotes the welfare of life on our planet.

1.1 INTRODUCTION TO ZOOLOGY

LEARNING OUTCOME

1. Differentiate various approaches to the science of zoology.

Zoology (Gr. *zoon,* animal + *logos,* to study) is the study of animals. It is one of the broadest fields in all of science because of the immense variety of animals and the complexity of the processes occurring within animals. There are, for example, more than 28,000 described species of bony fishes and more than 400,000 described (and many more undescribed) species of beetles! It is no wonder that zoologists usually specialize in one or more of the subdisciplines of zoology. They may study particular functional, structural, or ecological aspects of one or more animal groups (table 1.1), or they may choose to specialize in a particular group of animals (table 1.2).

Ichthyology, for example, is the study of fishes, and ichthyologists work to understand the structure, function, ecology, and evolution of fishes. These studies have uncovered an amazing diversity of fishes. One large family of bony fish, Cichlidae, contains 2,000 to 3,000 species. Members of this family include the familiar *Tilapia* species that grace our dinner plates and a host fish that hobbyists maintain in freshwater aquaria. Cichlid species range in length from 2.5 cm to 1 m and have an enormous variety of color patterns (figure 1.1), habitats, and body forms. Ichthyologists have described a wide variety of feeding habits in cichlids. These fish include algae scrapers like *Eretmodus* that nip algae with chisel-like teeth; insect pickers like *Tanganicodus;* and scale eaters like *Perissodus.*

TABLE 1.1
EXAMPLES OF SPECIALIZATIONS IN ZOOLOGY

SUBDISCIPLINE	DESCRIPTION
Anatomy	Study of the structure of entire organisms and their parts
Cytology	Study of the structure and function of cells
Comparative Genomics and Bioinformatics	Study of the structure, function, and evolution of the genetic composition of groups of animals using computer-based computational methods
Ecology	Study of the interaction of organisms with their environment
Embryology	Study of the development of an animal from the fertilized egg to birth or hatching
Genetics	Study of the mechanisms of transmission of traits from parents to offspring
Histology	Study of tissues
Molecular biology	Study of subcellular details of structure and function
Parasitology	Study of animals that live in or on other organisms at the expense of the host
Physiology	Study of the function of organisms and their parts
Systematics	Study of the classification of, and the evolutionary interrelationships among, animal groups

TABLE 1.2
EXAMPLES OF SPECIALIZATIONS IN ZOOLOGY BY TAXONOMIC CATEGORIES

SUBDISCIPLINE	DESCRIPTION
Entomology	Study of insects
Herpetology	Study of amphibians and reptiles
Ichthyology	Study of fishes
Mammalogy	Study of mammals
Ornithology	Study of birds
Protozoology	Study of protozoa

(a)

(b)

FIGURE 1.1
Cichlids. Cichlids of Africa exist in an amazing variety of color patterns, habitats, and body forms. (*a*) This dogtooth cichlid (*Cynotilapia afra*) is native to Lake Malawi in Africa. The female of the species broods developing eggs in her mouth to protect them from predators. (*b*) The fontosa (*Cyphontilapia fontosa*) is native to Lake Tanganyika in Africa.
(a) ©Blickwinkle/Alamy (b) ©ella1977/Shutterstock

All cichlids have two pairs of jaws. The mouth jaws are used for scraping or nipping food, and the throat jaws are used for crushing or macerating food before it is swallowed.

Many cichlids mouth brood their young. A female takes eggs into her mouth after the eggs are spawned. She then inhales sperm released by the male, and fertilization and development take place within the female's mouth! Even after the eggs hatch, young are taken back into the mouth of the female if danger threatens (figure 1.2). Hundreds of variations in color pattern, body form, and behavior in this family of fishes illustrate the remarkable diversity present in one relatively small branch of the animal kingdom. Zoologists are working around the world to understand and preserve this enormous diversity.

SECTION 1.1 THINKING BEYOND THE FACTS

Why is it often necessary for Zoologists to specialize in a subdiscipline within zoology?

1.2 ZOOLOGY: AN EVOLUTIONARY PERSPECTIVE

LEARNING OUTCOMES

1. Appraise the importance of evolution as a unifying concept in zoology.
2. Explain how our taxonomic system is hierarchical.

Animals share a common evolutionary past and evolutionary forces that influenced their history. Evolutionary processes are

FIGURE 1.2

A Scale-Eating Cichlid. Scale-eaters (*Perissodus microlepis*) attack from behind as they feed on scales of prey fish. Two body forms are maintained in the population. In one form, the mouth is asymmetrically curved to the right and attacks the prey's left side. The second form has the mouth curved to the left and attacks the prey's right side. Both right- and left-jawed forms are maintained in the population and prey do not become wary of being attacked from one side. *Perissodus microlepis* is endemic (found only in) to Lake Tanganyika. A male with its brood of young is shown here.
Courtesy of Dr. Kazutaka Ota, University of Kyoto, Laboratory of Animal Ecology, Japan

remarkable for their relative simplicity, yet they have had awesome effects on life-forms. These processes have resulted in an estimated 4 to 10 million species of animals living today. (Over 1.5 million animal species have been described, and about 10,000 new species are added each year.) Many more, about 90%, existed in the past and have become extinct. Zoologists must understand evolutionary processes if they are to understand what an animal is and how it originated.

Evolutionary Processes

Organic evolution (L. *evolutus,* unroll) is change in the genetic makeup of populations of organisms over time. Charles Darwin published convincing evidence of evolution in 1859 and proposed a mechanism that could explain evolutionary change. Since that time, biologists have become convinced that evolution occurs. The mechanism proposed by Darwin has been confirmed and now serves as the nucleus of our broader understanding of evolutionary change (*see chapters 4 and 5*). Our modern understanding of evolution has become one of the two most important unifying concepts within zoology and the most important concept that links zoology to other biological sciences. Evolutionary principles help us understand the origin and evolution of early life-forms (*see chapter 8*) and why animals look and behave as they do.

Understanding how the diversity of animal structure and function arose is one of the many challenges faced by zoologists. For example, the cichlid scale eaters of Africa feed on the scales of other cichlids. They approach a prey cichlid from behind and bite a mouthful of scales from the body. The scales are then stacked and crushed by the second set of jaws and sent to the stomach and intestine for protein digestion. Michio Hori of Kyoto University found that there were two body forms within the species *Perissodus microlepis.* One form had a mouth that was asymmetrically curved to the right, and the other form had a mouth that was asymmetrically curved to the left. The asymmetry results in right-jawed fish approaching and biting scales from the left side of their prey and the left-jawed fish approaching and biting scales from the right side of their prey. Both right- and left-jawed fish have been maintained in the population; otherwise, the prey would eventually become wary of being attacked from one side. The variety of color patterns within the species *Topheus duboisi* has also been explained in an evolutionary context. Different color patterns arose as a result of the isolation of populations among sheltering rock piles separated by expanses of sandy bottom. Breeding is more likely to occur within their isolated populations because fish that venture over the sand are exposed to predators.

Animal Classification and Evolutionary Relationships

Evolutionary principles also help us understand the origin of the diversity of life on our planet, and to organize that diversity in ways that make sense (*see chapters 7 and 8*). This evolutionary perspective helps us understand animal relationships on the grand scale of the animal kingdom, but it also helps us understand relationships on smaller scales—for example, among the 2,000 to 3,000 cichlid species in lakes around the world.

Groups of individuals are more closely related if they share more of their genetic material (DNA) with each other than with individuals in other groups. (You are more closely related to your brother or sister than to your cousin for the same reason.) Genetic studies suggest that the oldest populations of African cichlids are found in Lakes Tanganyika and Kivu, and from these the fish invaded African rivers and Lakes Malawi, Victoria, and other smaller lakes (figure 1.3). The history of these events is beginning to be understood and represents the most rapid known origin of species of any animal group. For example, the origin of Lake Victoria's cichlid species has been traced to an invasion of ancestral cichlids, probably from Lake Kivu approximately 100,000 years ago. Today, Lake Kivu has only 15 species of cichlids. This invasion continued up to about 40,000 years ago when volcanic eruptions isolated the fauna of Lakes Kivu and Victoria. That time period is long from the perspective of a human lifetime, but it is a blink of the eye from the perspective of evolutionary time. There is firm geological evidence that Lake Victoria nearly dried out and then refilled 14,700 years ago. This event probably did not result in the extinction of all cichlids in the lake because the lake basin may have retained smaller bodies of water, and thus refuges for some cichlid species. After Lake Victoria refilled, these refuge populations provided the stock for recolonizing the lake. More than 500 species of cichlids inhabited Lake Victoria by the beginning of the twentieth century. Many of these species evolved in fewer than 15,000 years. This very rapid evolution is a phenomenon referred to as evolutionary plasticity (*see chapter 5*).

How Do We Know about Genetic Relationships among Animals?

As shown by the example of Lake Victorian cichlids, zoologists often ask questions about genetic relationships among groups of animals. These family relationships are depicted in tree diagrams throughout this book. Early studies of genetic relationships involved the analysis of inherited morphological characteristics like jaw and fin structure that can be readily measured. With the advent of molecular biological techniques, zoologists have added to their repertoire of tools the analysis of variation in a series of enzymes, called allozymes, and DNA structure. These techniques allow zoologists to directly observe genetic relationships because the more DNA that two individuals, or groups of individuals, share, the more closely they are related. Because proteins, like enzymes, are encoded by DNA, variations in the structure of a protein also reflect genetic relationships. The genetic relationships of cichlids described in this chapter were investigated using a combination of morphological characteristics and molecular techniques. These topics are discussed in more detail in chapters 3, 4, and 5.

Like all organisms, animals are named and classified into a hierarchy of relatedness. Although Carl von Linne (1707–1778) is primarily remembered for collecting and classifying plants, his system of naming—**binomial nomenclature**—has also been adopted for animals. A two-part name describes each kind of organism. The first part is the genus name, and the second part is the species epithet. Each kind of organism (a species)—for example, the cichlid scale-eater *Perissodus microlepis*—is recognized throughout the world by its two-part species name. Verbal or written reference to a species refers to an organism identified by this two-part name. The species epithet is generally not used without the accompanying genus name or its abbreviation (*see chapter* 7). Above the genus level, organisms are grouped into families, orders, classes, phyla, kingdoms, and domains, based on a hierarchy of relatedness (figure 1.4). Organisms in the same species are more closely related than organisms in the same genus, and organisms in the same genus are more closely related than organisms in the same family, and so on. When zoologists classify animals into taxonomic groupings they are making hypotheses about the extent to which groups of animals share DNA, even when they study variations in traits like jaw structure, color patterns, and behavior, because these kinds of traits ultimately are based on the genes that they share.

FIGURE 1.3

Lakes Victoria, Kivu, Tanganyika, and Malawi. These lakes have cichlid populations that have been traced by zoologists to an ancestry that is approximately 200,000 years old. Cichlid populations originated in Lake Kivu and Lake Tanganyika and then spread to the other lakes.

Evolutionary theory has affected zoology like no other single theory. It has impressed scientists with the fundamental unity of all of life. As the cichlids of Africa illustrate, evolutionary concepts hold the key to understanding why animals look and act in their unique ways, live in their particular geographical regions and habitats, and share characteristics with other related animals.

SECTION 1.2 THINKING BEYOND THE FACTS

Why can taxonomists use similarities in DNA, similarities in morphological characteristics, or both when investigating taxonomic (evolutionary) relationships among animals?

1.3 ZOOLOGY: AN ECOLOGICAL PERSPECTIVE

LEARNING OUTCOMES

1. Use an example to generate an explanation for the importance of ecology as a unifying concept in zoology.
2. Analyze the relationships between human population growth and threats to world resources.

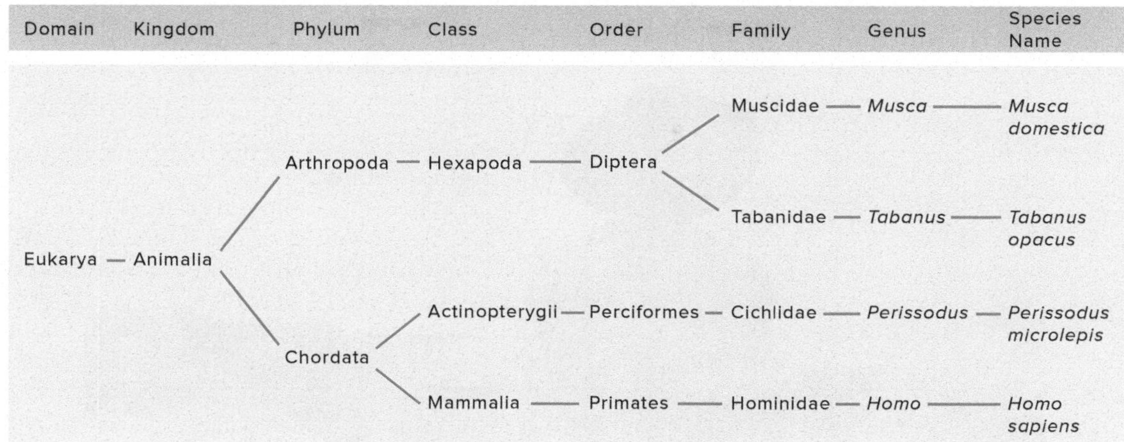

FIGURE 1.4

Hierarchy of Relatedness. The classification of a housefly, horsefly, cichlid fish, and human illustrates how the classification system depicts degrees of relatedness.

Modern zoology requires an ecological perspective. It is the second major unifying theme in zoology. An ecological perspective recognizes that animals can never be understood apart from other organisms and the nonliving components of their environment. **Ecology** (Gr. *okios,* house + *logos,* to study) is the study of the relationships between organisms and their environment (*see chapter 6*). Throughout our history, humans have depended on animals, and that dependence too often has led to exploitation. We depend on animals for food, medicines, and clothing. We also depend on animals in other, more subtle ways. This dependence may not be noticed until human activities upset the delicate ecological balances that have evolved over hundreds of thousands of years.

In the 1950s, the giant Nile perch *(Lates niloticus)* was introduced into Lake Victoria in an attempt to increase the lake's fishery (figure 1.5). This voracious predator reduced the cichlid population from 80% to less than 1% of the total fish biomass (total mass of all fish in the lake). Predation by the Nile perch has also resulted in the extinction of 65% of the cichlid species. Because many of the cichlids fed on algae, the algae in the lake grew uncontrolled. When algae died and decayed, much of the lake became depleted of its oxygen. The introduction of nonnative water hyacinth, which has overgrown portions of the lake, has resulted in further habitat loss. To make matters worse, when Nile perch are caught, their excessively oily flesh must be dried. Fishermen cut local forests for the wood needed to smoke the fish. This practice has resulted in severe deforestation around Lake Victoria. The resulting runoff of soil into the lake has caused further degradation. Decreased water quality not only presented problems for the survival of individual cichlids, but also increased turbidity that interfered with critical behavioral functions. Many of these species rely on their bright colors as visual cues during mating. Mouth-brooding species rely on vision to pick up developing eggs. The loss of Lake Victorian cichlids may be the largest extinction event of vertebrate species in modern human history.

There are some hopeful signs in this story. Although many Lake Victorian species have been lost forever, some cichlids are recovering. Heavy fishing pressure on the Nile perch has reduced its population density. (It still comprises more than 50% of catch weight—down from about 90% in the 1980s.) This decline has promoted the recovery of some cichlids that feed on small animals in the upper portions of open-water areas. (The Nile perch is predominately a bottom-dwelling predator.) One cichlid (*Haplochromis pyrrhocephalus*) is faring better than most other cichlid species. Over a 20-year period, scientists have observed rapid evolution of increased gill surface area and associated changes in head morphology, which have allowed this species to survive the lowered oxygen concentrations now present in Lake Victoria.

The Lake Victoria example also illustrates how ecological decisions made for economic reasons can have far-ranging economic and ecological consequences. Nile perch are marketed to Nairobi, the Middle East, and Europe to restaurants and fish markets. The hide is used in belts and purses, and the urinary bladder is used in oriental soup stock and as filter material by European alcohol producers. Catching, processing, and marketing such large fish to diverse foreign markets have resulted in the fishing and processing industries being taken from the hands of local fishermen and processors. These functions are primarily the work of large-boat fishing fleets and large fish-processing corporations. Changes in the local economy to agriculture have resulted in deforestation of the surrounding landscapes, and untreated sewage and agricultural and industrial runoff have further polluted Lake Victoria.

World Resources and Endangered Animals

There is grave concern for the ecology of the entire world, not just Africa's greatest lakes. The problems, however, are most acute in developing countries, which are striving to attain the same wealth as industrialized nations. Two problems, global overpopulation and the exploitation of world resources, are the focus of our ecological concerns.

FIGURE 1.5

Introduction of the Nile perch (*Lates niloticus*) in an attempt to improve Lake Victoria's fishery has resulted in the extinction of many cichlid species and has indirectly contributed to decreased water quality and deforestation. Commercial fishing is the primary means for control of Nile perch, and it is reducing perch populations and aiding in the recovery of native species.

Population

Global overpopulation is at the root of virtually all other environmental problems. Human population growth is expected to continue in the twenty-first century. Virtually all of this growth is in less developed countries, where 6.3 billion out of a total of 7.6 billion humans now live. Since a high proportion of the population is of childbearing age, the growth rate will increase in the twenty-first century. By the year 2050, the total population of India (1.71 billion) is expected to surpass that of China (1.38 billion) and the total world population will reach 9.7 billion. The 2018 U.S. population will reach 329 million. In 2050, it is projected to increase to 389 million. Even though Africa does not have the highest human population, its population is increasing more rapidly than other major regions of the world (table 1.3). As the human population grows, the disparity between the wealthiest and poorest nations is likely to increase.

TABLE 1.3

WORLD POPULATION PROJECTIONS FOR MAJOR WORLD REGIONS: 2018 AND 2050 (PROJECTED)

WORLD REGION	2018	2050 (PROJECTED)
World	7.6	9.7
Africa	1.03	2.39
Asia	4.16	5.16
Europe	0.74	0.71
Latin America and Caribbean	0.60	0.78
North America	0.35	0.45

Population sizes are based on figures from the United Nations Department of Economics and Social Affairs (2015) and expressed in billions of people.

World Resources

Human overpopulation is stressing world resources. Although new technologies continue to increase food production, most food is produced in industrialized countries that already have a high per-capita food consumption. Maximum oil production is expected to continue in this millennium. Deforestation of large areas of the world results from continued demand for forest products, fuel, and agricultural land. This trend contributes to climate change by increasing atmospheric carbon dioxide from burning forests and impairing the ability of the earth to return carbon to organic matter through photosynthesis. Deforestation also causes severe regional water shortages and results in the extinction of many plant and animal species, especially in tropical forests. Forest preservation would result in the identification of new species of plants and animals that could be important human resources: new foods, drugs, building materials, and predators of pests (figure 1.6). Nature also has intrinsic value that is just as important as its provision of resources for humans. Recognition of this intrinsic worth provides important aesthetic and moral impetus for preservation.

The stress being placed on world resources is a worldwide problem—not just one caused by, or occurring in, less developed countries. Major causes of resource problems are directly associated with activities in, and demands created by, industrialized nations. This fact is illustrated by looking at the anthropogenic (human-made) contribution of greenhouse gases to the atmosphere. The major sources of greenhouse gases are tied to the overuse of fossil fuels, and the greatest contributors to anthropogenic greenhouse gas

(a)

(b)

FIGURE 1.6

Tropical Rain Forests: A Threatened World Resource. (*a*) A Brazilian tropical rain forest. (*b*) A bulldozer clear-cutting a rain forest in the Solomon Islands. Clear-cutting for agriculture causes rain forest soils to quickly become depleted, and then the land is often abandoned for richer soils. Cutting for roads breaks continuous forest coverage and allows for easy access to remote areas for exploitation. Loss of tropical forests results in the extinction of many valuable forest species.

WILDLIFE ALERT
An Overview of the Problems

Extinction has been the fate of most plant and animal species. It is a natural process that will continue. In recent years, however, the threat to the welfare of wild plants and animals has increased dramatically—mostly as a result of habitat destruction. Tropical rain forests are one of the most threatened areas on the earth. It is estimated that rain forests once occupied 14% of the earth's land surface. Today this has been reduced to approximately 6%. Each year we lose about 150,000 km^2 of rain forest. This is an area of the size of England and Wales combined. This decrease in habitat has resulted in tens of thousands of extinctions. Accurately estimating the number of extinctions is impossible in areas like rain forests, where taxonomists have not even described most species. We are losing species that we do not know exist, and we are losing resources that could lead to new medicines, foods, and textiles. Other causes of extinction include climate change, pollution, and invasions from foreign species. Habitats other than rain forests—grasslands, marshes, deserts, and coral reefs—are also being seriously threatened.

No one knows how many species living today are close to extinction. As of 2017, the U.S. Fish and Wildlife Service lists 2,408 species in the United States as endangered or threatened. The IUCN has assessed over 91,000 species worldwide and of these nearly 26,000 species are listed as endangered or threatened. (Recall that it is estimated that there are between 4 and 100 million species of animals living today.) An **endangered species** is in imminent danger of extinction throughout its range (where it lives). A **threatened species** is likely to become endangered in the near future. Box figure 1.1 shows the number of endangered and threatened species in different regions of the United States. Clearly, much work is needed to improve these alarming statistics.

In the chapters that follow, you will learn that saving species requires more than preserving a few remnant individuals. It requires a large diversity of genes within species groups to promote species survival in changing environments. This genetic diversity requires large populations of plants and animals.

Preservation of endangered species depends on a multifaceted conservation plan that includes the following components:

1. A global system of national parks to protect large tracts of land and wildlife corridors that allow movement between natural areas
2. Protected landscapes and multiple-use areas that allow controlled private activity and also retain value as a wildlife habitat
3. Zoos and botanical gardens to save species whose extinction is imminent

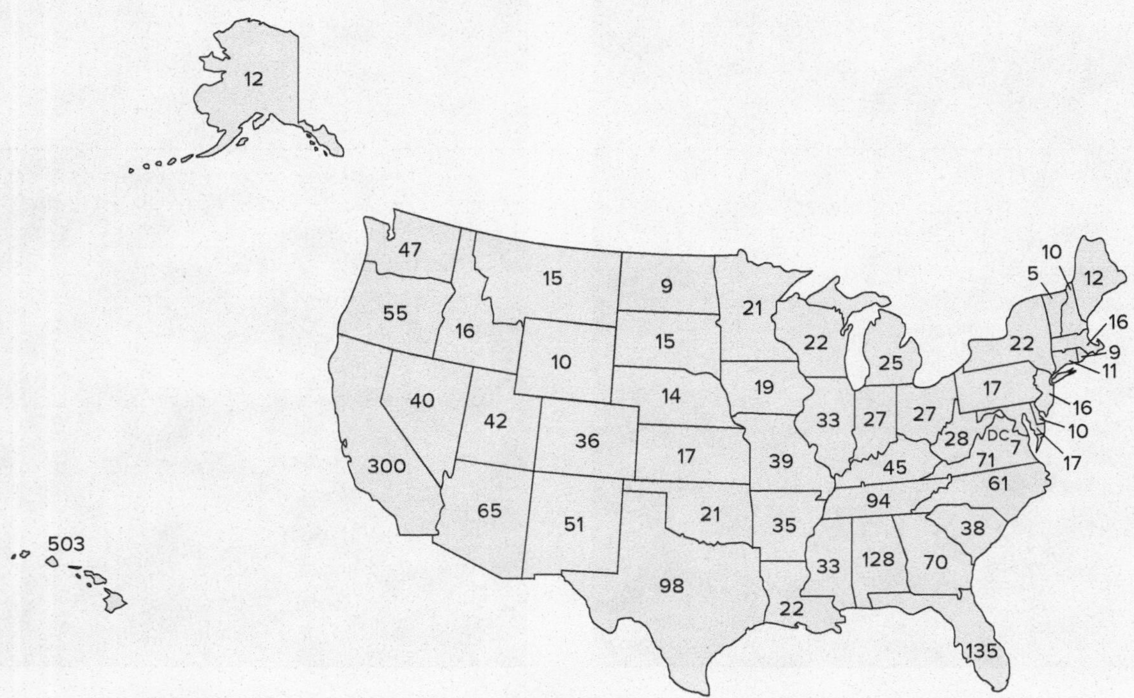

BOX FIGURE 1.1 Map Showing Approximate Numbers of Endangered and Threatened Species in the United States. The number for each state includes all endangered or threatened species believed or known to occur in the state. Because the ranges of some organisms overlap two or more states, the sum of all numbers is greater than the sum of all endangered and threatened species. The total number of endangered and threatened species in all listing categories in the United States is 2,408. The total number of listed animals is 1,458, with fish having the greatest number of listed species.

TABLE 1.4
MAJOR CONTRIBUTORS OF ANTHROPOGENIC GREENHOUSE GAS EMISSIONS

COUNTRY	$MTCO_2EQ$*	PERCENT OF WORLD TOTAL
China	11,735	26.9
United States	6,280	14.4
European Union (28 countries)	4,225	10.0
India	2,909	6.7
Russia	2,199	5.0
Canada	738	1.7

*Greenhouse gas emissions are given in metric ton carbon dioxide equivalents. This number takes into account all greenhouse gases including CO_2, methane, nitrous oxide, and others and is a calculation of their global warming potential. These 2013 data exclude land-use change and forestry and are from the World Resources Institute.

emissions are China, the United States, and the European Union (table 1.4). These three industrialized regions account for one-half of the greenhouse gas emissions.

Solutions

An understanding of basic ecological principles can help prevent ecological disasters like those we have described. Understanding how matter is cycled and recycled in nature, how populations grow, and how organisms in our lakes and forests use energy is fundamental to preserving the environment. There are no easy solutions to our ecological problems. Unless we deal with the problem of human overpopulation, however, solving the other problems will be impossible. We must work as a world community to prevent the spread of disease, famine, and other forms of suffering that accompany overpopulation. Bold and imaginative steps toward improved social and economic conditions and better resource management are needed.

"Wildlife Alerts" that appear within selected chapters of this text remind us of the peril that an unprecedented number of species face around the world. Endangered or threatened species from a diverse group of animal phyla are highlighted.

Section 1.3 Thinking Beyond the Facts

What is another example of how the careless disregard of ecological relationships has resulted in detrimental environmental consequences? (If you cannot think of an example on your own, see the "Wildlife Alert" boxes in subsequent chapters.)

Summary

1.1 **Introduction to Zoology**

- Zoology is the study of animals. It is a broad field that requires zoologists to specialize in one or more subdisciplines.

1.2 **Zoology: An Evolutionary Perspective**

- Animals share a common evolutionary past and evolutionary forces that influenced their history.
- Evolution explains how the diversity of animals arose.
- Evolutionary relationships are the basis for the classification of animals into a hierarchical system. This classification system uses a two-part name for every kind of animal. Higher levels of classification denote more distant evolutionary relationships.

1.3 **Zoology: An Ecological Perspective**

- Animals share common environments, and ecological principles help us understand how animals interact within those environments.
- Human overpopulation is at the root of virtually all other environmental problems. It stresses world resources and results in pollution, climate change, deforestation, and the extinction of many plant and animal species. Overuse of world resources by industrialized nations is a major contribution to environmental degradation.

Concept Review Questions

1. At least three of the following are examples of specialization in zoology. Select the one choice that is not a specialization in zoology or select choice "e."
   a. Ichthyology
   b. Mammalogy
   c. Ornithology
   d. Histology
   e. All of the above are examples of specializations in zoology.
2. A change in the genetic makeup of populations of organisms over time is a definition of
   a. binomial nomenclature.
   b. organic evolution.
   c. evolution.
   d. ecology.
3. Which of the following do zoologists use to study the genetic relationships among animals?
   a. Inherited morphological characteristics
   b. Enzyme structure
   c. DNA structure
   d. All of the above are used by zoologists to study genetic relationships.

4. Which one of the following statements is true?
   a. A species name is the same as the species epithet.
   b. The class-level ranking in the binomial system is less inclusive than the order-level ranking.
   c. The binomial nomenclature system reflects hypotheses on degrees of relatedness among animals.
   d. Members of different classes are likely to have gene combinations that are more similar than the gene combinations of members of the the same genus.
5. All of the following may result from deforestation except one. Select the exception.
   a. Climate change is promoted.
   b. Extinction of many plant and animal species occurs.
   c. Regional water shortages occur.
   d. Long-term improvement in the standard of living in less developed countries occurs.
   e. Loss of important human resources such as new drugs and food occurs.
6. By the year 2050, most human population growth will occur in ______________ and result in a world population of about ______________.
   a. less developed countries; 7 billion
   b. less developed countries; 9.7 billion
   c. less developed countries; 20.5 billion
   d. developed countries; 5.5 billion
   e. developed countries; 10.2 billion

Analysis and Application Questions

1. How is zoology related to biology? What major biological concepts, in addition to evolution and ecology, are unifying principles shared between the two disciplines?
2. What are some current issues that involve both zoology and questions of ethics or public policy? What should be the role of zoologists in helping resolve these issues?
3. Many of the ecological problems facing our world concern events and practices that occur in less developed countries. Many of these practices are the result of centuries of cultural evolution. What approach should people and institutions of developed countries take in helping encourage ecologically minded resource use?
4. Why should people in all parts of the world be concerned with the extinction of cichlids in Lake Victoria?

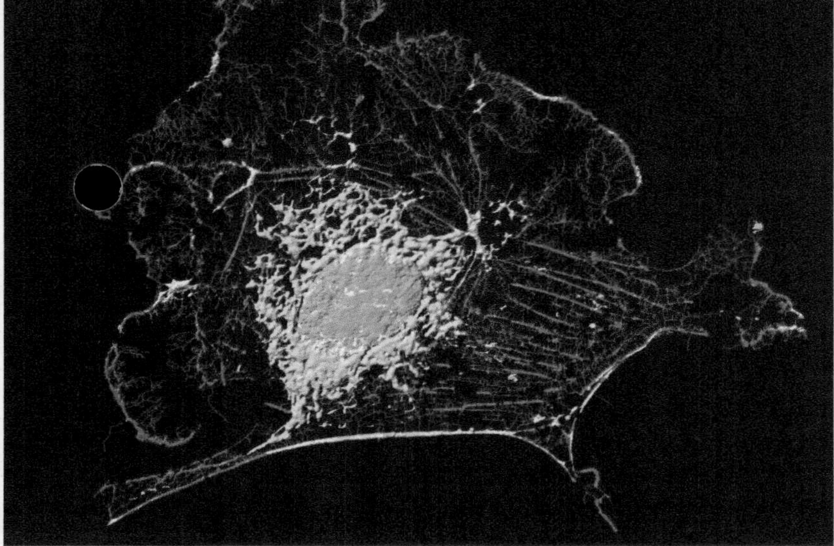

Animal cells are present in a variety of shapes and sizes and often specialized in function. Fibroblasts, like the one shown here, are 10 to 15 micrometers ($10–15x10^{-6}$M) in size. Fibroblasts are responsible for producing the extracellular matrix and collagen present in animal connective tissues and essential for supporting an animal's multicellular existance.

2

The Structure and Function of Animal Cells

Chapter Outline

2.1 Cells: The Common Unit of All Life
Properties and Varieties of Cells
2.2 Cellular Membranes and Membrane Transport
The Plasma Membrane
Membrane Transport
2.3 Energy Processing
An Overview of Energy Metabolism
Glycolysis and Fermentation
The Mitochondrion and Aerobic Respiration
Alternative Pathways
2.4 The Nucleus, Ribosomes, and Vaults
2.5 The Endomembrane System
The Endoplasmic Reticulum and the Golgi Apparatus
Vesicles and Cellular Transport
Endocytosis and Exocytosis
2.6 Peroxisomes
2.7 The Cytoskeleton and Cellular Movement
Microtubules, Intermediate Filaments, and Microfilaments
Cilia and Flagella
2.8 Levels of Organization in an Animal

What does it mean to be an animal? For most of this book we will be imagining the organism moving through its wooded habitat, swimming through the ocean, hiding from the waves at the base of a coral reef, or burrowing through rich garden soil. We will see animals interacting with their physical environment and with other organisms. We will study the history of animal life on our planet. We will see animal life. We will see animals doing what all life must do for their species to survive: use energy, grow, reproduce and adapt to their environment. Failing in any of these essential processes results extinction–an end common to 99% of species that have ever existed. In this chapter, our focus is not on the animal. Our focus is on an entity that comprises an animal and shares all of life's functions with an animal, but at the most basic level. Before we can truly understand an animal, or any organism, we must understand the functions of the most basic unit of life—the cell.

2.1 CELLS: THE COMMON UNIT OF ALL LIFE

LEARNING OUTCOMES

1. Analyze the concept depicted by cell theory that states (in part) "cells are the basic unit of structure and function of life."
2. Describe three elements of cell structure common to all cells.

There are many common cellular properties that unite all living forms. There is also an immense variety in life that has its roots in the specialization of cells for various functions. In this section we will examine some basic properties of cells that help us understand this unity and diversity.

Properties and Varieties of Cells

The unity of life stems, in part, from the characteristics of the cells that comprise living organisms. Cells have a fundamentally similar organization, and most cells are very small. The diversity we see in cell structure and function is real. Given that all cells had a common origin and 3.9 billion years of evolutionary history, one could argue that this diversity is quite small.

Common Organization

There are three major components of cells (figure 2.1). All cells have a **plasma membrane**. It is the outer boundary of the cell that separates internal events from the environment. All cells have a nuclear region. In organisms other than bacteria, the nuclear region is surrounded by

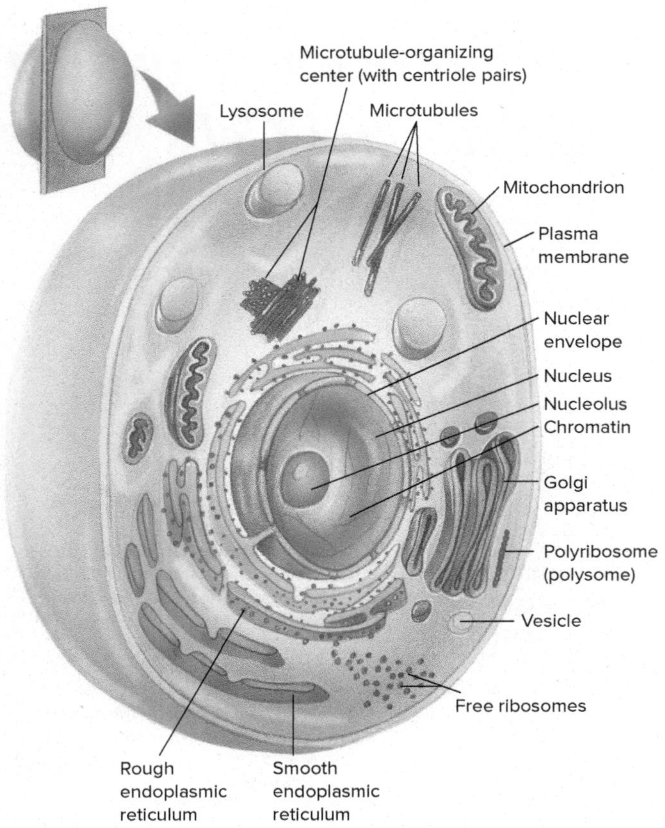

FIGURE 2.1

A Generalized Animal Cell. This representation is based mainly on electron microscopy. The sizes of some organelles are exaggerated to show detail.

its own membrane system called the nuclear envelope. The **nucleus** (or nuclear region) is the genetic control center of the cell. The portion of the cell outside the nucleus but inside the plasma membrane is the cell's **cytoplasm** (Gr. *kytos*, hollow vessel + *plasm*, fluid). The cytoplasm consists of a semifluid cytosol and small structures called **organelles** (L. *organum*, organ + *elle*, little) that perform specific functions. In animals, cells function in cooperation with other cells. They are bound together in functional entities by an extracellular protein matrix. These functional entities will be described at the end of this chapter.

Limits on Size

Most cells are microscopic in size (see *How Do Zoologists Investigate the Inner Workings of Cells?*). Exceptions include the eggs of some vertebrates, like reptiles and birds, and some long nerve cells. There are good reasons for this small size.

To sustain life, cells must function by exchanging nutrients, wastes, and other chemical components with the environment. In the process, organelles must concentrate raw materials and move products around within the cell. These exchanges depend on transport functions discussed later in this chapter. Cells must be kept small because as the radius of a cell increases, cell volume increases more rapidly than surface area (figure 2.2). The surface of a cell is where all exchanges with the environment occur. If cell volume becomes too large, the surface area of the cell becomes too small to support the required exchanges of nutrients and wastes.

As the volume of a cell increases, materials within the cell can become relatively scattered. For this reason, organelles compartmentalize materials to provide membranous surfaces that allow reactants and enzymes to attach, collide, and form products essential for life. The small size of cells promotes the concentration of reactants and enzymes, increasing the opportunities for cellular reactions and reducing the distances that materials must traverse within a cell.

Common Origin

Cells come in many shapes and sizes. In animals, cells vary in size and function from red blood cells that carry blood gases in vertebrates to flagellated cells that create water currents and filter food within the bodies of sponges. Even cells studied in introductory

How Do Zoologists Investigate the Inner Workings of Cells?

The small size of cells is the greatest obstacle to discovering their nature and the anatomy of the tiny structures within cells. The evolution of science often parallels the invention of instruments that extend human senses to new limits. Cells were discovered after microscopes were invented, and high-magnification microscopes are needed to see the smallest structures within a cell. Most commonly used are the light microscope, the transmission electron microscope, the scanning electron microscope, the fluorescence microscope, the scanning tunneling microscope, and the atomic force microscope.

Microscopes are the most important tools of cytology, the study of cell structure. But simply describing the diverse structures within a cell reveals little about their function. Today's modern cell biology developed from an integration of cytology with biochemistry, the study of molecules and the chemical processes of metabolism. Throughout this book, many photographs are presented using various microscopes to show different types of cells and the various tiny structures within. From these photographs, it will become apparent that similarities among cells reveal the evolutionary unity of life.

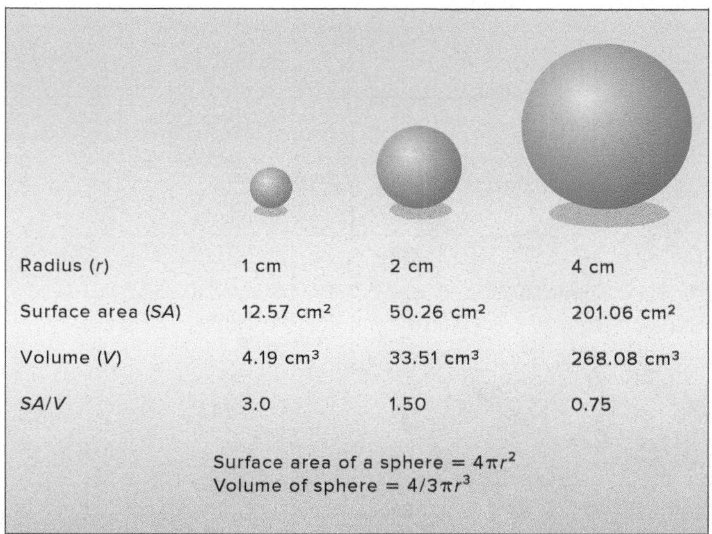

FIGURE 2.2

The Relationship between Surface Area and Volume. As the radius of a sphere increases, its volume increases more rapidly than its surface area. If cell volume becomes too large, the surface area of a cell becomes too small to support the required exchanges of nutrients and wastes with the environment. (SA/V = surface-area-to-volume ratio.)

botany and microbiology courses have remarkable similarities to those you will encounter in the study of zoology. How can cells in all living organisms be so similar? Even though evolution has acted on cells through their 3.9 billion years of existence to produce great variety, the clues to the single common origin of all cells are not hidden from scientists.

What are these clues? First, all cells share a common genetic makeup. **Genes** (Gr. *genos*, race) are heritable units that that comprise all organisms (*see chapter 3*). They are made of **deoxyribonucleic acid (DNA)**, a double helix of nitrogenous bases that codes for a sequence of amino acids that makes up the proteins of organisms. These bases and the genetic code, with a few minor exceptions, are universal to all of life. Second, other chemical components of cells are also universal. These include **ribonucleic acid (RNA)**, which is a single-stranded chain of nitrogenous bases that participates with DNA in the synthesis of proteins (*see chapter 3*). Additionally, the energy-carrying molecule, adenosine triphosphate (ATP), is universal to all life. (The role of ATP in cells is discussed later in this chapter.)

Three Types of Cells

Evolution has been acting on cells for the last 3.9 billion years. It should be no surprise that cells have adapted to a great variety of environments and carry out a tremendous variety of functions. Evolutionarily, there are three types of cells. In chapter 8, these three cell types will define the three largest subdivisions of life on earth. Cells present in two (Bacteria and Archaea) of these three subdivisions lack membrane-bound nuclei and other membranous organelles. These cell types have been referred to as being prokaryotic (Gr. *pro*, before + *karyon*, kernel). These two groups are distantly related and should not be lumped together based on characteristics that are absent. Rather, they are more meaningfully divided into two groups based on characteristics that define each group.

Bacteria are commonly encountered organisms such as the gut inhabiting *Escherichia coli* and others. **Archaea** inhabit extreme habitats. They are found in acidic hot springs and the acidic, oxygen-free environment of a cow's stomach where they help digest cellulose. Superficially these two groups have cell types that resemble each other in many ways such as size (0.5 to 5 μm), type of cell division, and reactions that sustain their lives. On the other hand, at the molecular level their cells are as different from each other as either is different from the third type of cell.

The third type of cell is found in the Eukarya and is said to be a eukaryotic (Gr. *eu*, true) cell. The **Eukarya** include animals, plants, fungi, and protists (e.g., *Amoeba* and *Paramecium*). The **eukaryotic cell** is usually 10 to 30 μm in size. It is characterized by a membrane-bound nucleus, mitochondria, chloroplasts (plants and some protists), and other membranous organelles. It is also characterized by having DNA complexed with protein, unique types of cell division, and unique metabolic and molecular features. The structure and function of eukaryotic animal cells are considered in the following pages.

SECTION 2.1 THINKING BEYOND THE FACTS

Both evolutionary theory and cell theory are unifying themes for all of biology. How do both theories unify biology when considering life at a cellular level?

2.2 CELLULAR MEMBRANES AND MEMBRANE TRANSPORT

LEARNING OUTCOMES

1. Discuss how our understanding of the structure of the plasma membrane informs our understanding of the following membrane functions: restricting passage of some polar molecules but promoting transport of other polar molecules, promoting the passage of most nonpolar molecules, and recognition of specific types of cells by other cells (e.g., an egg by a sperm cell).
2. Differentiate non-transporter-mediated membrane exchanges from carrier-mediated exchanges, explaining why each type of exchange is important for a cell.

Membrane systems play very important functions within animal cells. Membranes surround many organelles where they compartmentalize specific functions carried out by the cell. For example, mitochondria rely on an extensive membrane system that is used in energy processing. The endomembrane system includes the endoplasmic reticulum and the Golgi apparatus that work together packaging and transporting proteins within the cell. And of course the nuclear envelope, which helps to define the eukaryotic cell, shares many properties of cellular membranes in general. These organelles

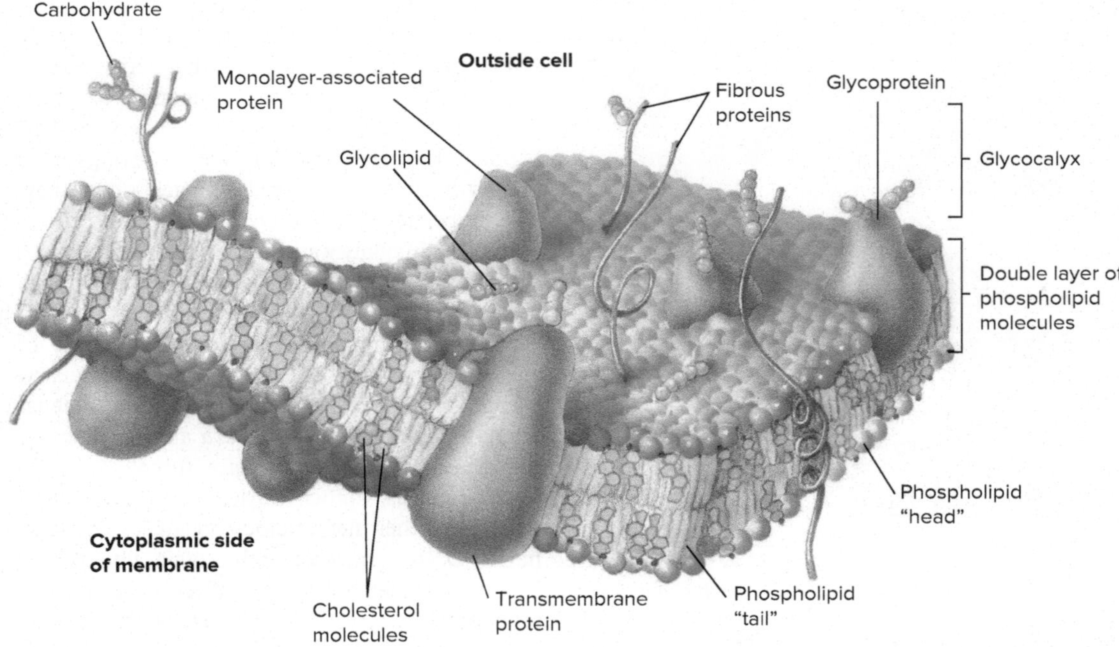

FIGURE 2.3

Structure of the Plasma Membrane. The plasma membrane is a very fluid phospholipid bilayer that acts as a two-dimensional fluid where phospholipid molecules can move within the membrane. Cholesterol molecules and transmembrane proteins are embedded in the bilayer. Monolayer-associated proteins attach to the inner and outer surfaces of the membrane. The glycocalyx protects the cell and functions in cell-to-cell recognition and adhesion.

will be described in subsequent sections. The membrane that confines the contents and functions of the cell, and separates life from nonlife, is the plasma membrane. The structure and function of the plasma membrane is considered next.

The Plasma Membrane

The plasma membrane is a phospholipid bilayer with interspersed proteins and other macromolecules (figure 2.3). Each phospholipid molecule usually consists of two hydrophobic (Gr. *hydro*, water + *phobia*, fear) "tails" and a hydrophilic (Gr. *philos*, loving) phosphate "head." The tails of these molecules are comprised of uncharged atoms that orient toward the interior of the two lipid layers. The charged hydrophilic "head" of molecules on one side of the bilayer is attracted to the watery medium on the outside of the cell, and the "head" of molecules on the other side of the bilayer is attracted to the watery cytoplasm on the inside of the cell. The hydrophilic/hydrophobic properties of phospholipids cause them to spontaneously assemble in a watery medium. These membranes are very fluid. Water limits the movement of phospholipids out of the bilayer, but nothing prevents phospholipids from moving about within the membrane. Cellular membranes are two-dimensional fluids—a characteristic essential to the functions that will be described in following sections of this chapter.

The plasma membrane, and membranes in general, contain sterols—principally cholesterol. A small charged end of the cholesterol molecule associates with the hydrophilic "head" of a phospholipid and the remainder of the cholesterol associates with the phospholipid "tails" (figure 2.4). Cholesterol modulates membrane fluidity. It reduces the tendency of the membrane to become more fluid with increasing temperature and to become less fluid with decreasing temperature. It also fills in spaces between hydrocarbon chains making the membrane less permeable to very small ions and molecules.

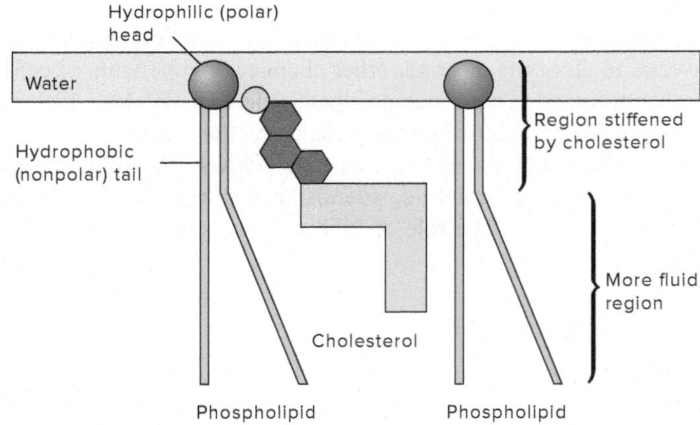

FIGURE 2.4

The Arrangement of Cholesterol between Lipid Molecules of a Lipid Bilayer. The charged end of a cholesterol molecule associates with the hydrophilic head of a phospholipid, and the remainder of the cholesterol molecule associates with the hydrophobic tail of a phospholipid. Cholesterol stabilizes membrane fluidity with varying temperatures. One half of the lipid bilayer is shown; the other half is a mirror image.

Membrane proteins attach to the inner and outer membrane surfaces (monolayer-associated proteins) and are embedded in the membrane (transmembrane proteins) (*see figure 2.3*). Membrane proteins transport ions and various molecules across the membrane, are points of attachment for cellular structures, form junctions between cells, serve as hormone receptors, and function as enzymes.

The plasma membrane of eukaryotic cells has a carbohydrate layer on its outer surface. Carbohydrate chains attach to proteins to form glycoproteins and to lipids to form glycolipids. In many animal cells, these carbohydrate groups protrude from the cell surface to form a coat called the **glycocalyx** (Gr. *glyco*, sugar + *kalyx*, coat) (*see figure 2.3*). Surface carbohydrate coats protect the cell from mechanical and chemical damage and have important roles in cell-to-cell recognition and adhesion. Like the jersey of your school's athletic team identifies your school, the glycocalyx of an egg cell identifies it to the sperm cells trying to fertilize the egg.

Membrane Transport

It is very wrong to simply think of cellular membranes as barriers to movement of cellular materials. Life depends upon keeping very precise balances of materials on both sides of a cellular membrane. The primary barrier to movement across a membrane is the lipid bilayer. Hydrophilic molecules cannot permeate these fatty layers. On the other hand, small nonpolar molecules (e.g., molecular oxygen [O_2] and carbon dioxide [CO_2]) and larger hydrophobic molecules (like lipid soluble hormones) move through the lipid bilayer easily.

Other exchanges across a membrane involve passage of a substance either through a membrane channel or with the assistance of protein transporters. For example, in order to conduct a nerve impulse (action potential, *see chapter 24*) nerve cells keep sodium ions 10 to 30 times more concentrated on the outside of a nerve cell membrane than on the inside. Potassium ions are 28 times more concentrated on the inside of a nerve cell membrane than on the outside. Large negatively charged proteins confined by the plasma membrane within a cell maintain a "sink" of negative charges within the cell. A nerve impulse occurs when membrane channels open and close allowing sodium and potassium ions to move in specific directions. Protein transporters in the plasma membrane use energy to pump these ions back to their original locations inside or outside the cell. The property of membranes that allows some substances to move across the membrane while other substances are prevented from crossing is called **selective permeability**.

Non-Transporter Gradient Exchanges

Molecules are in constant random motion because of heat energy. Molecules and ions inside or outside a cell "explore" their environment through these random motions, colliding with and bouncing off one another. These collisions occur more frequently where molecules are more crowded (concentrated), and the bouncing is more frequent and vigorous. The result is that molecules and ions in regions where they are more concentrated tend to bounce into regions where they are less concentrated. The difference in concentration of a substance between two points of reference is called a **concentration gradient**, and the movement of a substance along (down) a concentration gradient (from an area of higher concentration to an area of lower concentration) is called **simple diffusion** (L. *diffundere*, to spread). Oxygen is often delivered to cells through the blood of animals. Oxygen is in a higher concentration in the blood, and cells typically have a lower concentration of oxygen because oxygen is continually used in cellular reactions. This sets up a concentration (diffusion) gradient, and oxygen moves easily through the lipid bilayer into the cell "down this gradient." Other substances, like various ions, diffuse through membrane channels (figure 2.5).

Osmosis is the diffusion of water across a selectively permeable membrane. Because water is moving from an area of higher concentration to an area of lower concentration across the membrane, osmosis is a special case of diffusion. Water's concentration difference is a result of the membrane's selective permeability; one side of the membrane has a solute (e.g., a salt or sugar) that is not permeable to the membrane. In figure 2.6, water will move from compartment 2 to compartment 1 of the beaker due to the higher concentration of water molecules in compartment 2. Sugar molecules are too large to pass through the membrane.

In animal cells, water moves through membranes very easily—more so than predicted by water's solubility in the lipid bilayer. This movement is facilitated by water channels called **aquaporins** (L. *aqua*, river + *porus*, tiny opening). Water movements across animal plasma membranes are usually through osmosis. Osmosis occurs when water is absorbed from the gut of an animal into the bloodstream or reabsorbed in the kidneys during urine processing.

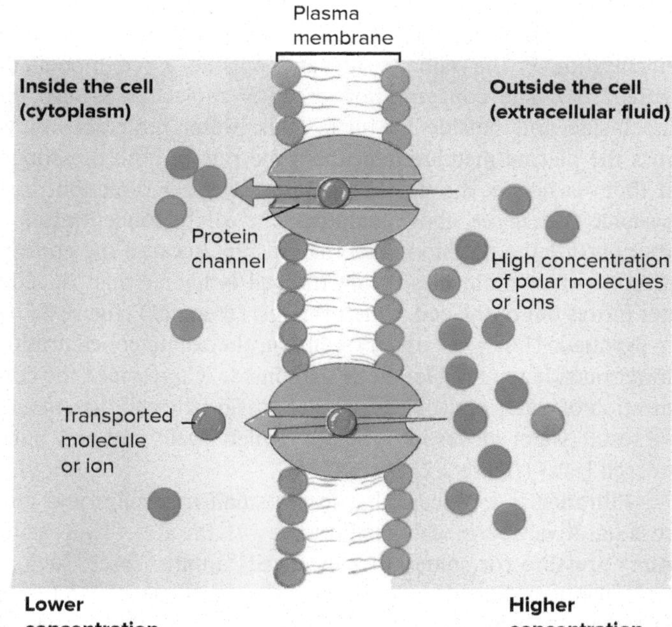

FIGURE 2.5

Diffusion through Membrane Channels. Diffusion is the movement of a substance from an area of higher concentration to an area of lower concentration (along or down a concentration gradient). Small polar molecules and various ions diffuse through membrane channels. Nonpolar molecules, like steroid hormones, do not require membrane channels but can diffuse through the phospholipid bilayer.

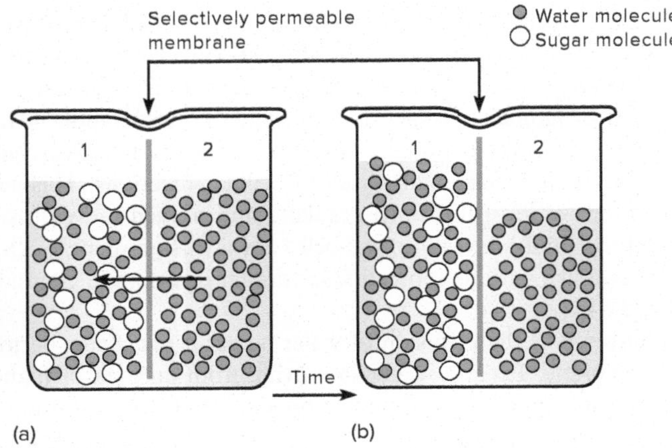

FIGURE 2.6

Osmosis. (*a*) A selectively permeable membrane separates the beaker into two compartments. The membrane is permeable to water but impermeable to large sugar molecules. Compartment 1 contains sugar and water molecules, and compartment 2 contains only water molecules. Because compartment 1 contains dissolved sugar molecules, which cannot move across the membrane, water in compartment 1 diffuses down its concentration gradient from compartment 2 into compartment 1. (*b*) Over time, the net diffusion into compartment 1 slows and stops because the increased volume in compartment 1 creates osmotic pressure that counteracts water's movement into compartment 1. At this point, osmotic equilibrium has been achieved.

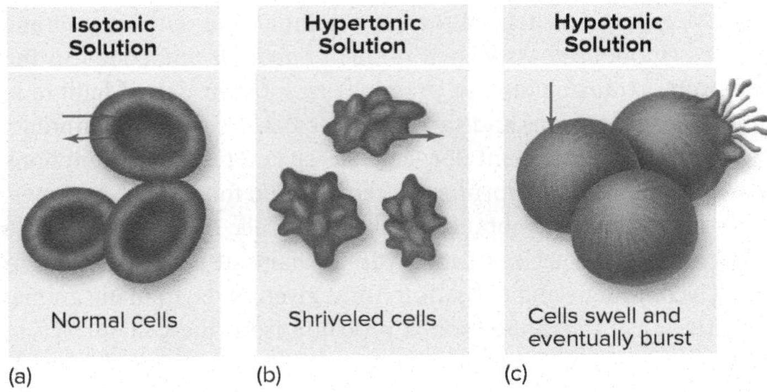

FIGURE 2.7

Tonicity. In this example, red blood cells are placed in varying concentrations of a salt solution. Red blood cell membranes are permeable to water but not to salt. (*a*) Isotonic solutions have solute concentrations that are the same as the solute concentrations inside cells. Red blood cells placed in an isotonic solution exchange water across their membranes equally in both directions. Blood cell volume does not change. (*b*) Hypertonic solutions have solute concentrations that are higher than the solute concentrations inside cells. Red blood cells placed in a hypertonic solution have a higher concentration of water inside the cells than outside the cells. Water diffuses from the cells, and cell volume decreases. The red blood cells shrivel in a process called crenation. (*c*) Hypotonic solutions have solute concentrations that are lower than the solute concentrations inside cells. Red blood cells placed in a hypotonic solution have a lower concentration of water inside the cells than outside the cells. Water diffuses into the cells and cell volume increases. The red blood cells swell and may undergo lysis.

The term **tonicity** (Gr. *tonus*, tension) refers to the relative concentration of solutes in the water inside and outside the cell. For example, in an **isotonic** (Gr. *isos*, equal) solution, the solute concentration is the same inside and outside a red blood cell (figure 2.7*a*). The concentration of water molecules is also the same inside and outside the cell. Thus, water molecules move across the plasma membrane at the same rate in both directions, and there is no net movement of water in either direction. In a **hypertonic** (Gr. *hyper*, above) solution, the solute concentration is higher outside the red blood cell than inside. Because the concentration of water molecules inside the cell is higher than outside, water moves out of the cell, which shrivels (crenates) (figure 2.7*b*). In a **hypotonic** (Gr. *hypo*, under) solution, the solute concentration is lower outside the red blood cell than inside. Conversely, the concentration of water molecules is higher outside the cell than inside. As a result, water moves into the cell, which swells and may burst (undergo lysis) (figure 2.7*c*).

Filtration is a process that forces small molecules and ions across selectively permeable membranes with the aid of hydrostatic (water) pressure (or some other externally applied force, such as blood pressure). For example, filtration occurs when blood pressure forces water and dissolved molecules through the permeable walls of small blood vessels called capillaries (figure 2.8). In filtration, large molecules, such as proteins, do not pass through the smaller membrane pores or between cellular junctions. Filtration also takes place in the kidneys when blood pressure forces water, wastes, and other molecules out of the blood vessels and into the kidney tubules during urine formation.

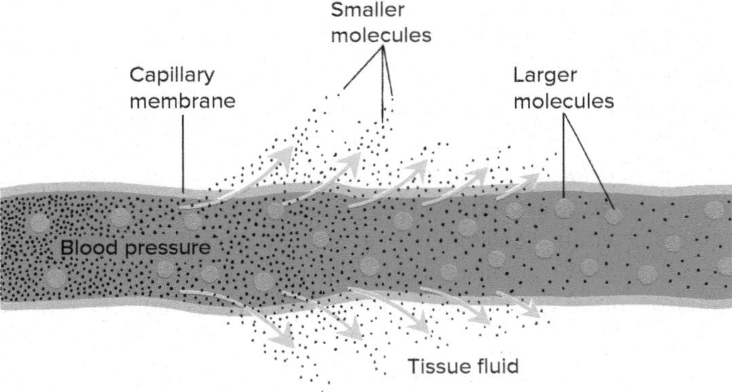

FIGURE 2.8

Filtration. Filtration occurs when hydrostatic pressure forces water and dissolved molecules and ions through the permeable walls of a membrane. In this example, the high blood pressure in the capillary forces small molecules through pores or between cell junctions of the capillary membrane. Larger molecules cannot pass through the small openings in the capillary membrane and remain in the capillary. Arrows indicate the direction of small molecule movement.

Carrier-Mediated Transport

Most molecules and other substances are large and/or polar and cannot cross a membrane by simple diffusion. These molecules require the aid of transport proteins to cross the plasma membrane. Because transport proteins are finite in their abundance in a plasma

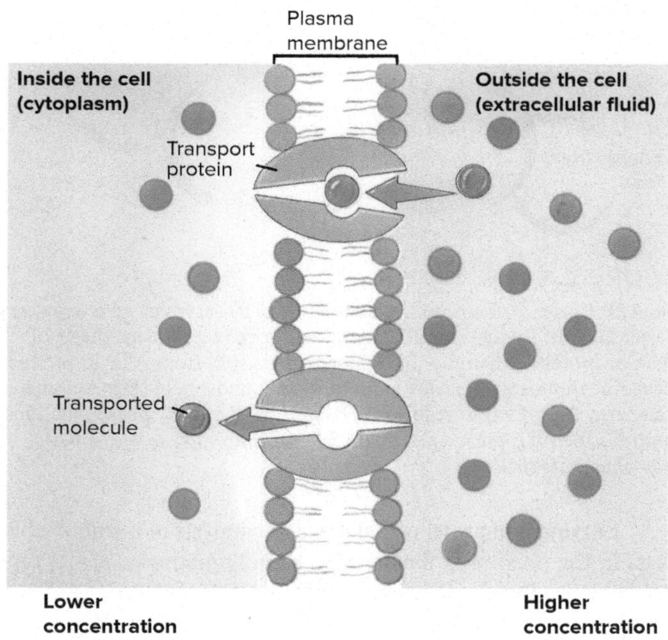

FIGURE 2.9

Facilitated Diffusion. Facilitated diffusion is a form of carrier-mediated transport. Large or polar molecules are in a high concentration on one side of the plasma membrane. A concentration gradient favors the movement of the molecules through the membrane, but they are too large and polar to move unaided through the membrane. In this illustration, a transport protein is shown changing configuration, which allows the protein to pick up a molecule on one side of the membrane and, using the concentration gradient that exists for the transported molecule, deposit the molecule on the other side of the membrane. The concentration gradient provides the energy for facilitated diffusion. No cellular energy input is required.

membrane, they can become saturated when the concentration of transported molecules is high.

Facilitated diffusion occurs when large or polar molecules are in a high concentration on one side of the plasma membrane, and the concentration gradient provides the energy required for the molecules to move to the other side of the membrane. A transport protein is required to "facilitate" the movement of the molecules through the membrane. Glucose is a large, simple sugar. It is often transported by facilitated diffusion, for example during the absorption of glucose from your gut tract into your blood stream after a meal (figure 2.9).

Active transport moves molecules across a selectively permeable membrane against a concentration gradient—that is, from an area of lower concentration to one of higher concentration (figure 2.10). This movement against a concentration gradient requires ATP energy (*see figure 2.11*). These carrier proteins are called uniporters if they transport a single type of molecule or ion, symporters if they transport two molecules or ions in the same direction, and antiporters if they transport two molecules or ions in the opposite direction. One antiporter mechanism, the sodium-potassium pump, helps maintain the high concentrations of potassium ions and low concentrations of sodium ions inside nerve cells that are necessary for the transmission of electrical impulses. The calcium pump is a uniporter that keeps the calcium concentration hundreds of times lower within the cytosol of the cell as compared to the outside of the cell or within a membranous structure called smooth endoplasmic reticulum (*see figure 2.19*). Muscle contraction depends on this very steep calcium gradient.

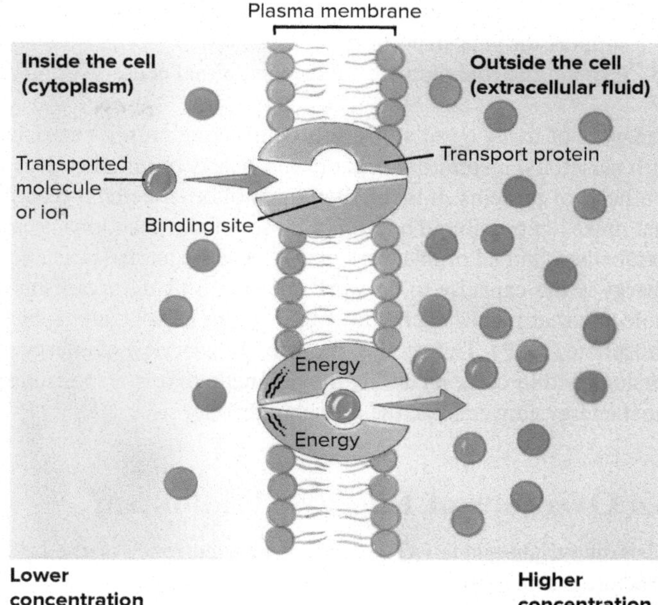

FIGURE 2.10

Active Transport. Active transport is a form of carrier-mediated transport in which cellular energy is used to transport a molecule or ion against a concentration gradient (from an area of lower concentration to an area of higher concentration). In this illustration, a transport protein is shown combining with the transported molecule on one side of the membrane, changing configuration with the input of cellular energy (in the form of ATP), and depositing the molecule on the opposite side of the membrane where the molecule is accumulating.

SECTION 2.2 THINKING BEYOND THE FACTS

Think about what happens to lipids (fats) as they cool. Next, think about the function of sterols, including cholesterol, in the plasma membrane. Finally, think about a snake or lizard on a cold day versus a hot day. How would the functions of these animals' plasma membranes be affected in the absence of membrane cholesterol in temperature extremes?

2.3 ENERGY PROCESSING

LEARNING OUTCOMES

1. Cellular functions are usually carried out in multistep metabolic pathways. Use examples from the reactions of cellular respiration to explain why these multistep pathways are advantageous, and to explain the roles of enzymes and energy in these pathways.
2. Hypothesize on possible reasons that animals are aerobic organisms rather than anaerobic like some bacteria.

Explain how the reactions of cellular respiration provide support for your hypothesis.
3. Explain the role of mitochondria in animal cells.

The work of living never stops. It is contracting muscles associated with skeletons, respiratory organs, and hearts of animals. It is the synthesis of proteins. It is the thousands of other cellular reactions that make life possible. The sum of these cellular reactions is called **metabolism**, and all of this work occurs because animals use energy. **Energy** is the capacity to do work, and in all living organisms the molecule that makes energy available within cells is adenosine triphosphate (ATP). This section provides an overview of energy conversions within cells, including the organelle that is responsible for most energy conversions, the mitochondrion.

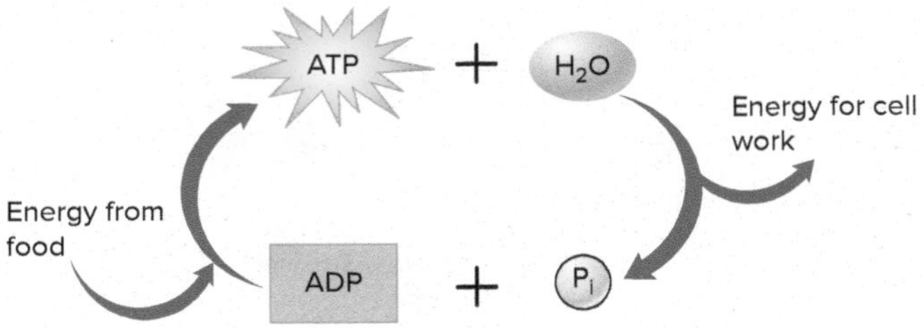

FIGURE 2.11

The ATP Cycle. Adenosine triphosphate (ATP) serves as an energy carrier in cells. Energy is available for cell work (e.g., the synthesis of DNA or protein) when the third phosphate is split from ATP to produce adenosine diphosphate (ADP) and inorganic phosphate (P_i). Animals use energy from food to re-form ATP from ADP and P_i. These reactions comprise the ATP cycle. (*See figure 3.8 for an example of nucleotide monophosphate structure.*)

An Overview of Energy Metabolism

Adenosine triphosphate (ATP) is the energy currency of the cell. Its structure is similar to that of other nucleotides that make up DNA and RNA (*see figure 3.8*). ATP is comprised of a nitrogen-containing base (adenine), a sugar (ribose), and three phosphate groups. The bonds between phosphates are high energy bonds–it takes energy to form the bonds, and energy is released when the bonds are broken. In a cell, the bond between the second and third phosphates is continually broken and reformed in the ATP cycle (figure 2.11). The energy given off when the bond is broken is used to drive cellular reactions, and the energy used to reform the bond is derived from foods animals eat.

Metabolic reactions that require energy input are coupled to the release of the third phosphate from ATP. The released phosphate may combine with a reactant, changing its shape, or provide energy for the formation of a new chemical bond. Either way, the reactant is raised to a higher energy state to become the product as follows:

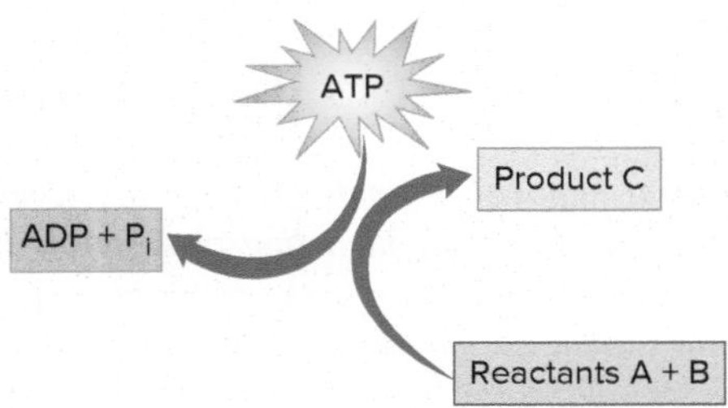

Cellular reactions usually do not occur in single-reaction steps. Instead, cellular reactions occur in multistep, linked metabolic pathways. Multistep pathways are advantageous because they provide multiple intermediate compounds that can serve as branch points for alternative pathways and provide multiple points for control of the pathway. The following is an example of a metabolic pathway.

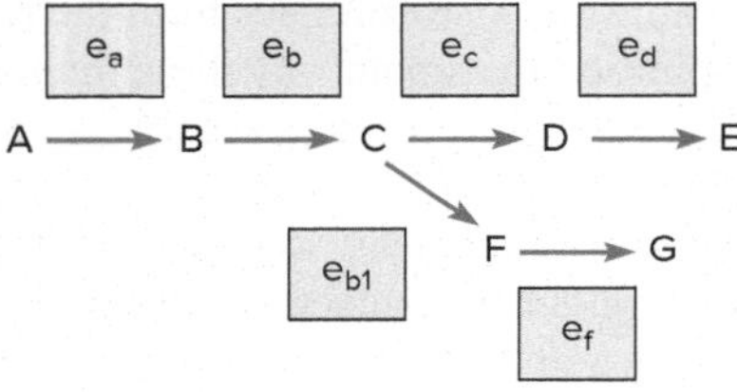

Enzymes are proteins that serve as catalysts in metabolic pathways. In the previously illustrated metabolic pathway, e_a - e_f represent enzymes that catalyze each reaction. As catalysts, enzymes do not get incorporated into the products of their reaction, but bring reactants, called **substrates**, into proximity in order to promote their reaction. When enzymes combine with their substrates in ways that stress the substrates, they lower the energy required for the reaction to occur. The energy input required for a reaction to occur is called the **energy of activation** and is reduced in an enzymatic reaction. The lowering of the energy of activation allows reactions to occur in cells at temperatures that are compatible with life (figure 2.12).

Glycolysis and Fermentation

Ultimately, the energy that animals require to support life comes from the sun in the form of solar radiation. **Autotrophs** (Gr. *autos*, self + *tropho*, feeder), like green plants and algae, use solar energy to reduce carbon dioxide (CO_2) to sugars through the process of photosynthesis. Photosynthetic products are consumed by herbivorous animals, and this energy is passed through food webs (*see figure 6.10*). Animals obtain energy by ingesting other organisms and are referred to as **heterotrophs** (Gr. *hetero*, other). This energy supports **anabolic pathways** (Gr. *ana*, up + *metaballein*, to change) that build up larger molecules of life–like proteins and DNA (*see chapter 3*). In this section, we will discuss the pathways that harvest energy from food and convert it into the energy that is temporarily caught in ATP molecules. These energy-harvesting metabolic pathways are called **catabolic pathways** (Gr. *kata*, down) because, as they proceed, they break down large organic molecules.

The first set of reactions that animals use to harvest energy occurs in the cytoplasm of cells, and the second set of reactions occurs in organelles called the mitochondria. These reactions are collectively called **cellular respiration** because cells need oxygen (O_2) to bring the second set of reactions to completion. The first set of reactions occurs in all organisms (Bacteria, Archaea, and Eukarya), can occur in the absence of oxygen, and is called **glycolysis** (Gr. *glykos*, sweet + *lysis*, splitting).

Glycolysis is a 10-step metabolic sequence that begins with a six-carbon sugar, glucose, the primary sugar used in animal

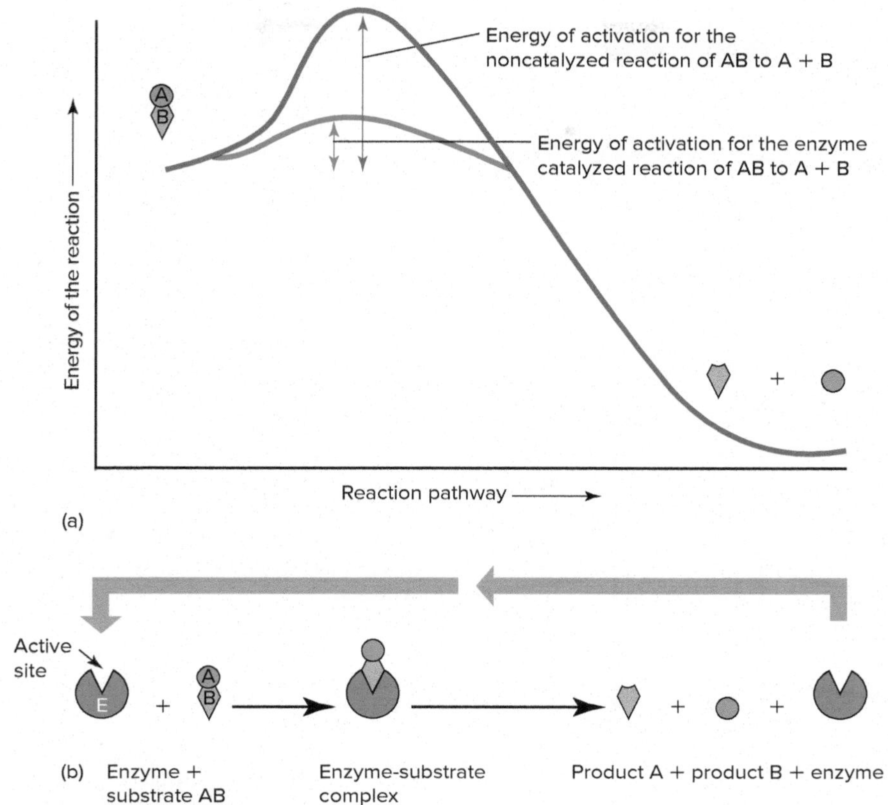

FIGURE 2.12

Energy of Activation and Enzyme Function. (*a*) AB is a relatively stable compound that contains energy that can be released during its reaction to A + B. Because it is stable, the reaction of AB requires energy input to get it started. This energy input is called the energy of activation. The energy of activation is supplied in cells in the form of ATP. The amount of energy of activation needed can be lowered through the use of an enzyme. (*b*) An enzyme lowers the energy of activation by combining with its substrate at the enzyme's active site to form an enzyme-substrate complex. The enzyme thus stresses the substrate, or brings two substrates together in a very tight association, to make the reaction possible under conditions compatible with life. The enzyme can bind to other AB substrate molecules to repeat the reaction many times.

metabolism. Each step in glycolysis is mediated by a specific enzyme, and glucose is eventually broken down into two, three-carbon molecules called pyruvate. The initial steps require an expenditure of two ATPs, but by the end of the set of reactions four ATPs are produced for a net gain of two ATPs. In addition, two reduced coenzymes (NADH) are produced (figure 2.13). The reaction (reduction) of NAD to form NADH converts that molecule into an energy-carrying molecule that may be used within the mitochondrion to produce more ATP.

Under most circumstances, the pyruvate and NADH produced in glycolysis move from the cytosol of the cell into mitochondria for further processing and more ATP production. Occasionally, oxygen may not be present, and tissues are temporarily anaerobic (Gr. *an*, without + *aer*, air + *bios*, life). This can occur in muscle cells during strenuous exercise when the cells use oxygen more rapidly than oxygen is delivered to a muscle by the respiratory and circulatory systems. Anaerobic conditions also occur in tissues of diving animals (*see page 497*). Under anaerobic conditions, pyruvate is metabolized one step further using the hydrogens from NADH to produce lactic acid. This is called **lactate fermentation**. Lactate fermentation supplies a very small amount of ATP to withstand brief anaerobic conditions, but animal tissues (especially nervous tissues) cannot survive long anaerobically. Lactic acid must be removed from tissues and recycled, usually in the liver. This removal and recycling occurs when aerobic conditions are restored.

The Mitochondrion and Aerobic Respiration

Animals require oxygen for their survival, and under aerobic conditions pyruvate and NADH enter the mitochondria for further processing. Mitochondrial processing involves two subset reaction sequences The first subset is called the Krebs cycle, which oxidizes pyruvate to CO_2 and results in additional ATP and reduced coenzymes. The second subset is called the electron transport system, which harvests the energy within reduced coenzymes produced in glycolysis and the Krebs cycle.

Mitochondria (sing., mitochondrion) are double-membrane bound organelles that are spherical to elongate in shape. An inner membrane folds to form incomplete partitions called cristae

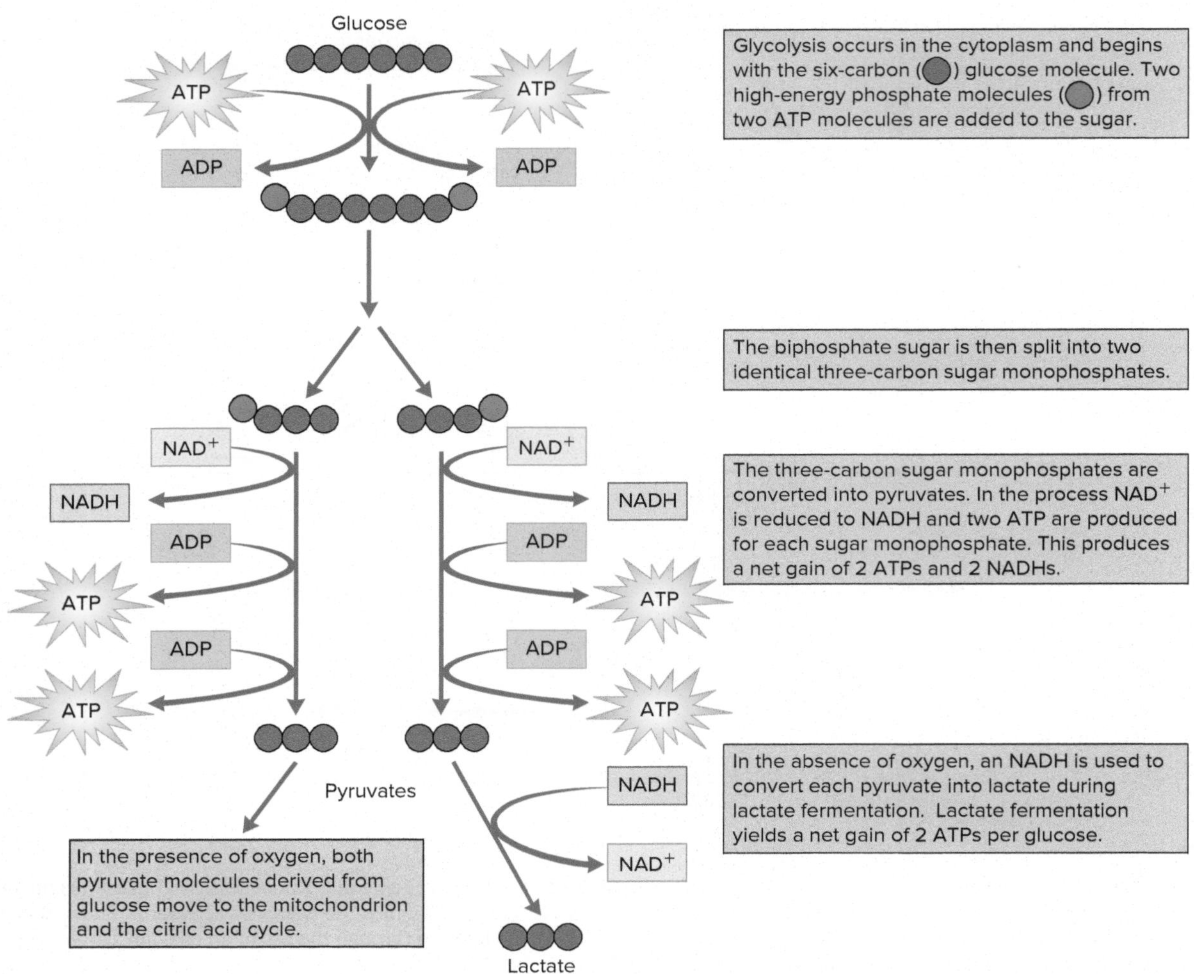

FIGURE 2.13

Glycolysis and Lactate Fermentation. Glycolysis occurs in the cytoplasm of a cell. Under aerobic conditions the pyruvate that is produced enters the mitochondrion for further processing. Under anaerobic conditions in animals lactate fermentation occurs.

(sing., crista; figure 2.14). The chemical "soup" within the inner membrane is the matrix. The matrix contains ribosomes, circular DNA, and a host of enzymes and other molecules involved with pyruvate processing and producing more mitochondria (*see page 125*).

The first reactions that process pyruvate occur within the mitochondrial matrix (*see figure 2.14*). These reactions include a preparatory step and a series of reactions that begin and end with the same molecule; thus these reactions form a cycle. Because an early identified component of this cycle is citric acid, the cycle is called the **citric acid cycle**. The preparatory step for the citric acid cycle occurs when pyruvate (a C_3) loses CO_2, and the remaining two carbons join a molecule called coenzyme A (C_2CoA). NAD is reduced to NADH in the process. The two carbons from C_2CoA then enter the citric acid cycle (figure 2.15). During the cycle each of these two carbons is split off an intermediate molecule as CO_2, GTP is produced, NAD is reduced to NADH, and FAD is reduced to form $FADH_2$. GTP is guanosine triphosphate (*see guanine structure in figure 3.8*) and is quickly converted into ATP. FAD is a coenzyme similar to NAD. Recall that two pyruvates were produced in glycolysis from each glucose; thus, it takes two turns of the citric acid cycle to account for all of the carbons in each glucose. All of the carbons of glucose are now in the form of CO_2, they diffuse out of the mitochondrion and cell, and they are transported to a respiratory surface and released to the environment.

The purpose of cellular respiration is to produce ATP for cellular work. At this point in the catabolism of glucose, there is a net gain of four ATP—two from glycolysis and two from two turns of the citric acid cycle. These four ATP represent less than 10% of the total energy within the original glucose molecule. Where is the 90% of the energy that remains unaccounted for by the ATP produced to this point? It is within the NADH and $FADH_2$. These reduced coenzymes are high energy molecules, and the energy they contain will be harvested in a series of electron transfers occurring on the inner mitochondrial membranes we have identified as cristae. Proteins of these membranes accept electrons from NADH and $FADH_2$ and pass them on through a series of other membrane proteins. In the process, the electrons that originated with glucose eventually combine with molecular oxygen (O_2)—the oxygen that animals breathe

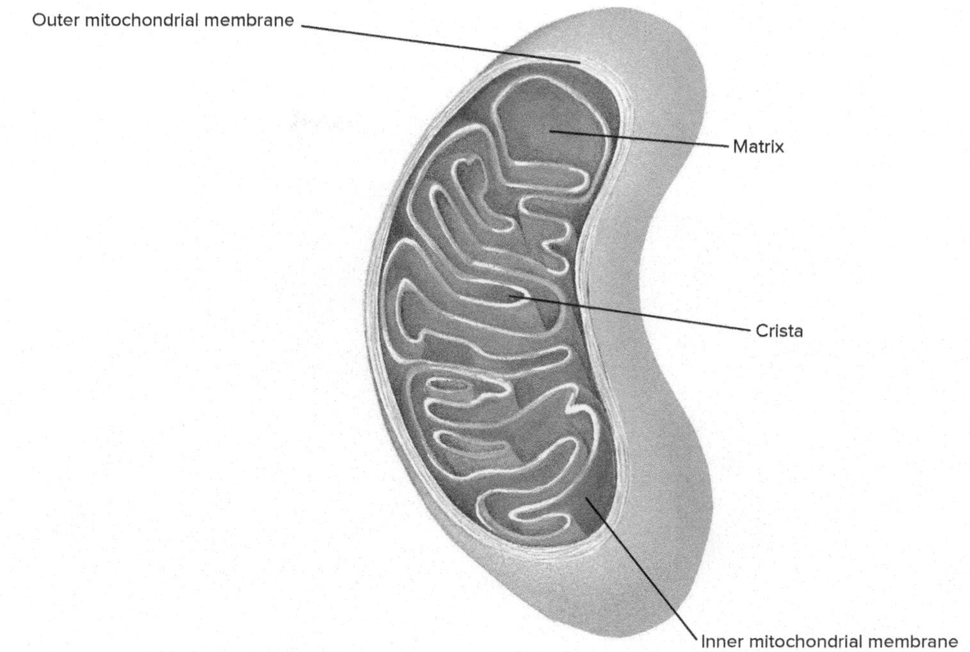

FIGURE 2.14

Mitochondrion. The mitochondrion is thought of as the powerhouse of the cell. Pyruvate from glycolysis enters the mitochondrion, and the first stages of its processing occur within the bounds of the inner mitochondrial membrane, called the matrix. These reactions are called the citric acid cycle. ATP-producing reactions are completed on the surfaces of the folded membranes, called cristae, and within the spaces between inner and outer mitochondrial membranes. These reactions are called the electron transport system and chemiosmosis.

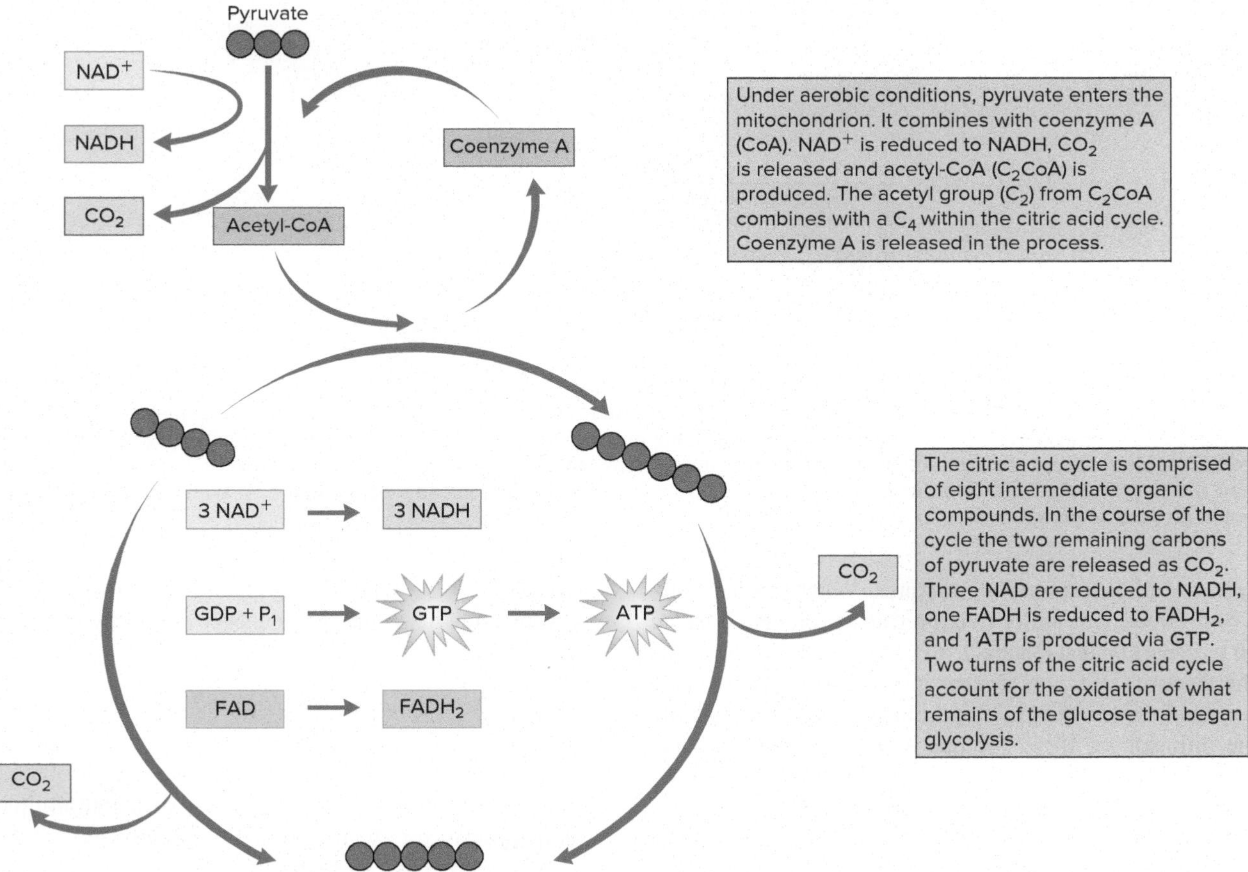

FIGURE 2.15

The Citric Acid Cycle. The citric acid cycle occurs in the matrix of the mitochondrion. Under aerobic conditions, pyruvate from glycolysis enters the matrix of the mitochondrion. Carbons are split off as CO_2, and NADH, $FADH_2$, and ATP are produced.

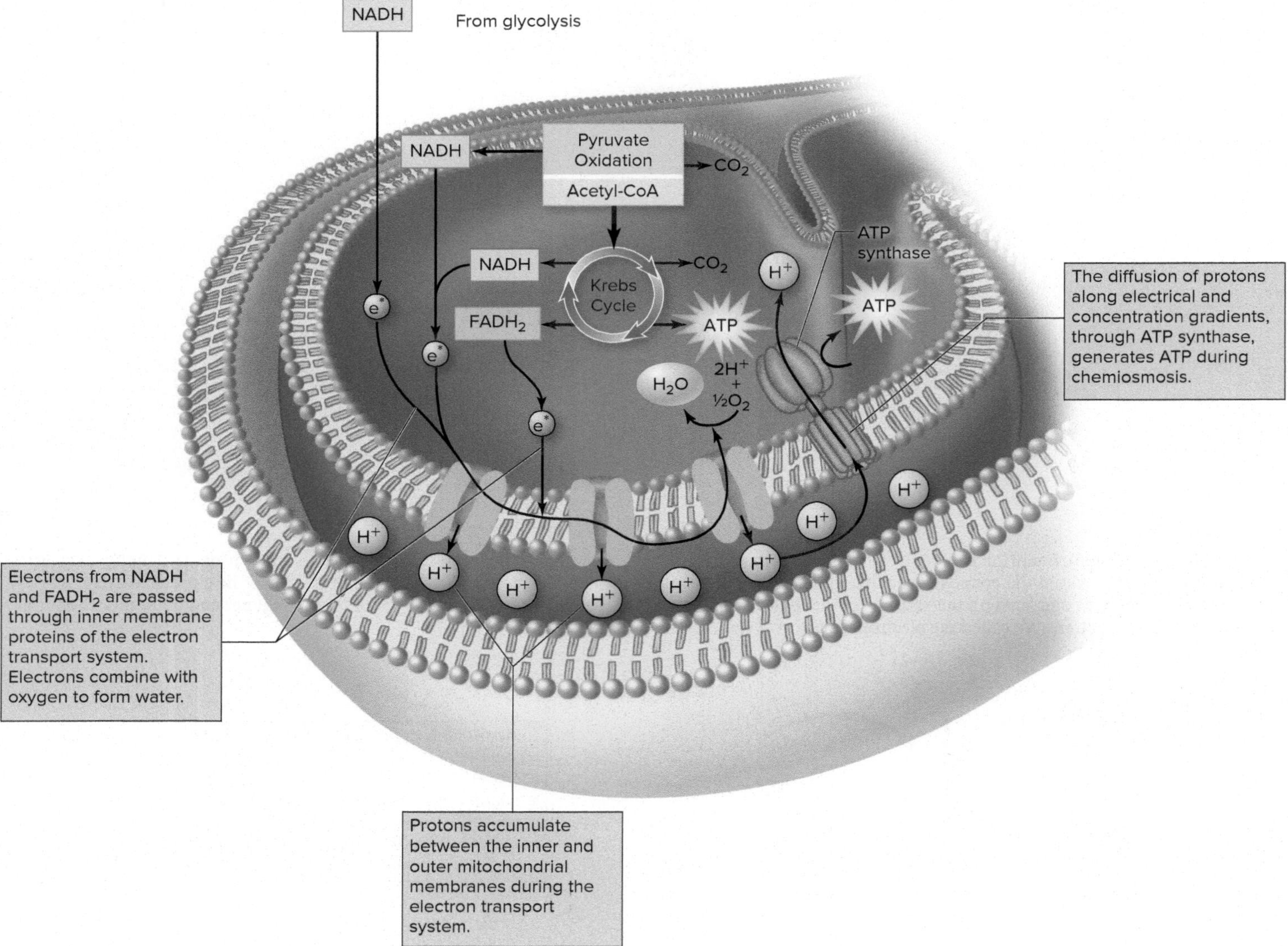

FIGURE 2.16

The Electron Transport System and Chemiosmosis. The electron transport system and chemiosmosis harvest the energy in high energy electrons of NADH and $FADH_2$. Electrons are transported through membrane proteins of the inner mitochondrial membranes and eventually combine with molecular oxygen to produce water. As electrons are passed through membrane proteins, protons accumulate between the inner and outer mitochondrial membranes. Concentration and electrical gradients favor the movement of these protons across the inner membrane through ATP synthase. ATP synthase uses energy from this proton gradient to phosphorylate ADP to form ATP.

via lungs, gills, or other respiratory structures. Water (H_2O) is produced as a by-product. This entire electron transfer process is called the **electron transport system** (figure 2.16).

In the electron transport system, protons (H^+) accumulate within the space between the inner and outer mitochondrial membranes. The diffusion of these protons along concentration and charge gradients through ATP-generating enzymes (ATP synthase) produces ATP. This process is called **chemiosmosis**.

The total maximum theoretical yield for cellular respiration is 36 to 38 ATP depending on the tissue involved. The actual yield of ATP is usually about 30 per glucose because varying quantities of ATP are used to complete glucose metabolism depending on conditions within the cell (table 2.1). This represents an energy-harvesting efficiency of 52 to 55%, far greater than the efficiency of most human-made machines. (The energy efficiency of an automobile is about 15%.)

Alternative Pathways

Cellular respiration is a versatile set of metabolic pathways—the pathways can do much more than metabolize glucose. The pathways of glycolysis can be reversed by liver cells to synthesize glucose from pyruvate and lactic acid. Sugars other than glucose can be catabolized directly (e.g., fructose) or after being first converted to glucose (e.g., lactose, maltose, and galactose), and

TABLE 2.1

ENERGY YIELD FROM GLUCOSE METABOLISM IN CELLULAR RESPIRATION*

STAGE OF METABOLISM	PRODUCT OF STAGE	ATP YIELD PER GLUCOSE
Glycolysis	2 ATP	2
	2 NADH	3**
Pyruvate Oxidation to C_2CoA (two per glucose)	2 NADH	5
Citric Acid Cycle (two turns per glucose)	6 NADH	15
	2 $FADH_2$	3
	2 GTP	2
Total		30

*The maximum theoretical yield of ATP produced during the aerobic metabolism of glucose is 36 to 38 ATP. ATP expenditures required for the process vary depending on the concentrations of reagents in the cytosol and mitochondrial matrix and the specific tissue being considered, and these ATP expenditures must be subtracted from the theoretical yield. This table presents a reasonable "typical" yield.

**Each NADH produced in the cytosol produces fewer ATP molecules than are produced by an NADH from the citric acid cycle. Transporting electrons from cytosolic NADH into the mitochondrion requires an expenditure of ATP energy.

a storage carbohydrate called glycogen can be synthesized in liver and muscle cells. The citric acid cycle plays a key role in the catabolism of fats and protein. Fats, comprised of glycerol and long fatty acid chains, are important long-term energy storage molecules. The metabolism of glycerol utilizes the reactions of glycolysis, and two-carbon units of fatty acids combine with coenzyme A and enter into the citric acid cycle in the same way that the carbons of pyruvate are metabolized. Catabolism of a fat molecule yields huge energy benefits. The citric acid cycle is the location where proteins are catabolized for the production of ATP. Very importantly, these same pathways can be reversed to produce fat when animals consume more food than is required to meet their immediate energy requirements, and other reactions based on the citric acid cycle can convert one amino acid to another for the production of proteins (*see chapter 3*).

Section 2.3 Thinking Beyond the Facts

How are the reactions of cellular respiration tied to the structure of mitochondria?

2.4 THE NUCLEUS, RIBOSOMES, AND VAULTS

LEARNING OUTCOME

1. Assess the related functions of chromatin, the nuclear envelope, ribosomes, and vaults.

Three related cellular structures are involved with genetic control of the cell. These structures are involved with coding for and producing protein. Their functions are discussed in detail in chapter 3. The organelles are briefly described in this section.

The nucleus (L. *nucleus*, kernel or nut) contains the DNA and is the control and information center for the eukaryotic cell. As discussed in chapter 3, eukaryotic DNA exists in combination with protein in the form of darkly staining chromatin, and at certain times in the life of a cell chromatin is condensed into chromosomes. The nucleus is the location where genetic information from DNA is transcribed into RNA. RNA is then translated at ribosomes into proteins (e.g., enzymes) that determine a cell's activities. The nucleus is also the vehicle for transferring genetic information between generations of cells and organisms.

The **nuclear envelope** is a membrane that separates the nucleus from the cytoplasm and is continuous with the endoplasmic reticulum at a number of points. More than 3,000 nuclear pores penetrate the surface of the nuclear envelope (figure 2.17). These pores allow materials to enter and leave the nucleus, and they give the nucleus direct contact with the endoplasmic reticulum (*see figures 2.1 and 2.19*). Nuclear pores are not simply holes in the nuclear envelope; each pore is composed of an ordered array of globular and filamentous proteins. The size of the pores prevents DNA from leaving the nucleus but permits RNA to be moved out. Cellular organelles, called **vaults**, are cytoplasmic ribonucleoproteins shaped like octagonal barrels (figure 2.18). They are located in large numbers outside the nucleus and are believed to dock at nuclear pores and aid in transport of materials, including RNA, between the nucleus and the cytoplasm. Other vault functions are under investigation.

The nucleus of nondividing cells also contains one or more nonmembranous structures called nucleoli. A **nucleolus** is comprised of RNA and protein and is the preassembly point for ribosomes. After preassembly, ribosome subunits leave the nucleus through nuclear pores. In the cytoplasm, **ribosomes** are sites where the genetic message transcribed from DNA is translated into protein (*see chapter 3*). Ribosomes contain almost equal amounts of protein and a special kind of RNA called ribosomal RNA (rRNA). Some ribosomes attach to the endoplasmic reticulum (*see section 2.5*), and some ribosomes float freely in the cytoplasm.

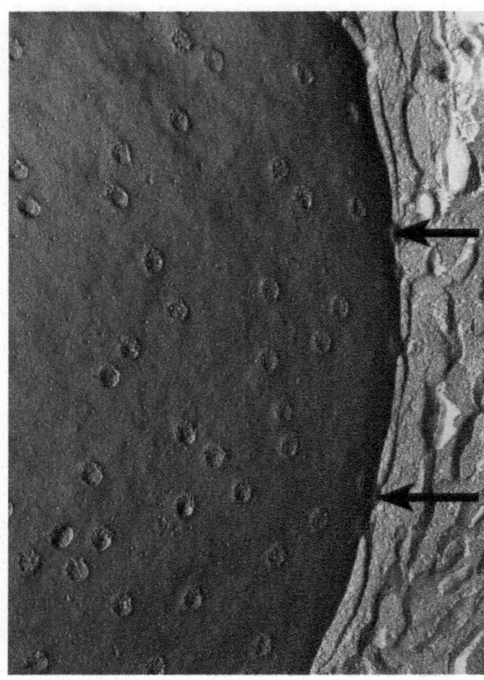

FIGURE 2.17

The Nuclear Envelope. This photograph is a color-enhanced electron micrograph of the nuclear envelope, showing nuclear pores (arrows).
©Don W. Fawcett/Science Source

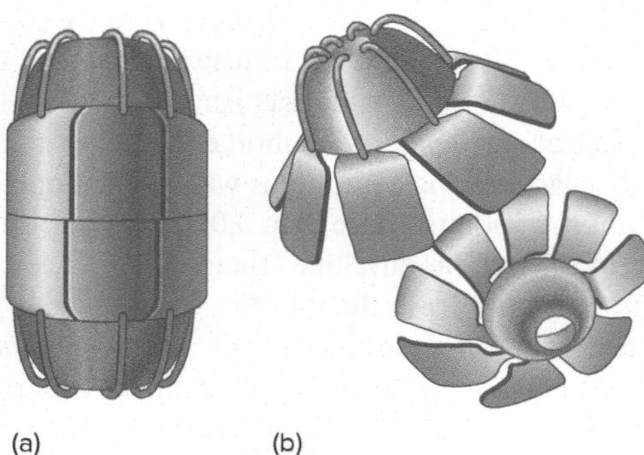

FIGURE 2.18

Vaults. (*a*) A three-dimensional drawing of the octagonal barrel-shaped organelle believed to transport mRNA from the nucleus to the ribosomes. (*b*) A vault opened to show its octagonal structure.

Whether ribosomes are free or attached, they usually cluster in groups connected by a strand of another kind of RNA called messenger RNA (mRNA). These clusters are called polyribosomes or polysomes (*see figure 2.1*).

SECTION 2.4 THINKING BEYOND THE FACTS

Eukarya have a nucleus. Archaea and Bacteria have their DNA collected into a nuclear area, but it is not set off by a nuclear envelope. Would you expect archaeal and bacterial cells to have RNA? Vaults? Ribosomes? Explain your answers.

2.5 THE ENDOMEMBRANE SYSTEM

LEARNING OUTCOMES

1. Explain why the endoplasmic reticulum, Golgi apparatus, endosomes, and lysosomes are functionally related and comprise the endomembrane system.
2. Contrast the roles of exocytosis in the functions of the cells of the pancreas (which produces and secretes digestive enzymes), and endocytosis in the function of certain white blood cells (which engulf and destroy bacteria intracellularly).

The **endomembrane system** consists of an interconnected system of membranes that includes the endoplasmic reticulum, the Golgi apparatus, various types of vesicles, and the nuclear envelope. The membranous connections between these organelles reflect their roles in promoting and regulating the flow of materials within cells.

The Endoplasmic Reticulum and the Golgi Apparatus

The **endoplasmic reticulum** (ER) is a complex, membrane-bound labyrinth of flattened sheets, sacs, and tubules that branches and spreads throughout the cytoplasm. The ER is continuous from the nuclear envelope to the plasma membrane (*see figure 2.1*), and it is a series of channels that helps various materials move throughout the cytoplasm. It also is a storage unit for enzymes and other proteins that are synthesized by the attached ribosomes (*see chapter 3*). ER with attached ribosomes is rough ER. ER without attached ribosomes is smooth ER (figure 2.19) and is the site for lipid production, detoxification of a wide variety of organic molecules, and storage of calcium ions (e.g., in muscle cells). Most cells contain both types of ER, although the relative proportion varies among cells.

The **Golgi apparatus** (named for Camillo Golgi, who discovered it in 1898) is a collection of membranes associated physically and functionally with the ER in the cytoplasm (*see figure 2.19*). It is composed of flattened stacks of membrane-bound cisternae (sing., *cisterna*, L.; closed spaces serving as fluid reservoirs). The Golgi apparatus sorts, packages, and secretes proteins and lipids.

Proteins that ribosomes synthesize are passed into the ER and sealed off in little packets called transfer vesicles. Transfer vesicles pass from the ER to the Golgi apparatus and fuse with it (*see figure 2.19*). In the Golgi apparatus, the proteins are concentrated and chemically modified. One function of this chemical modification is to mark and sort the proteins into different batches for different destinations. Eventually, the proteins are packaged into membrane-enclosed vesicles, which are released into the cytoplasm to be transported to various locations. Many vesicles

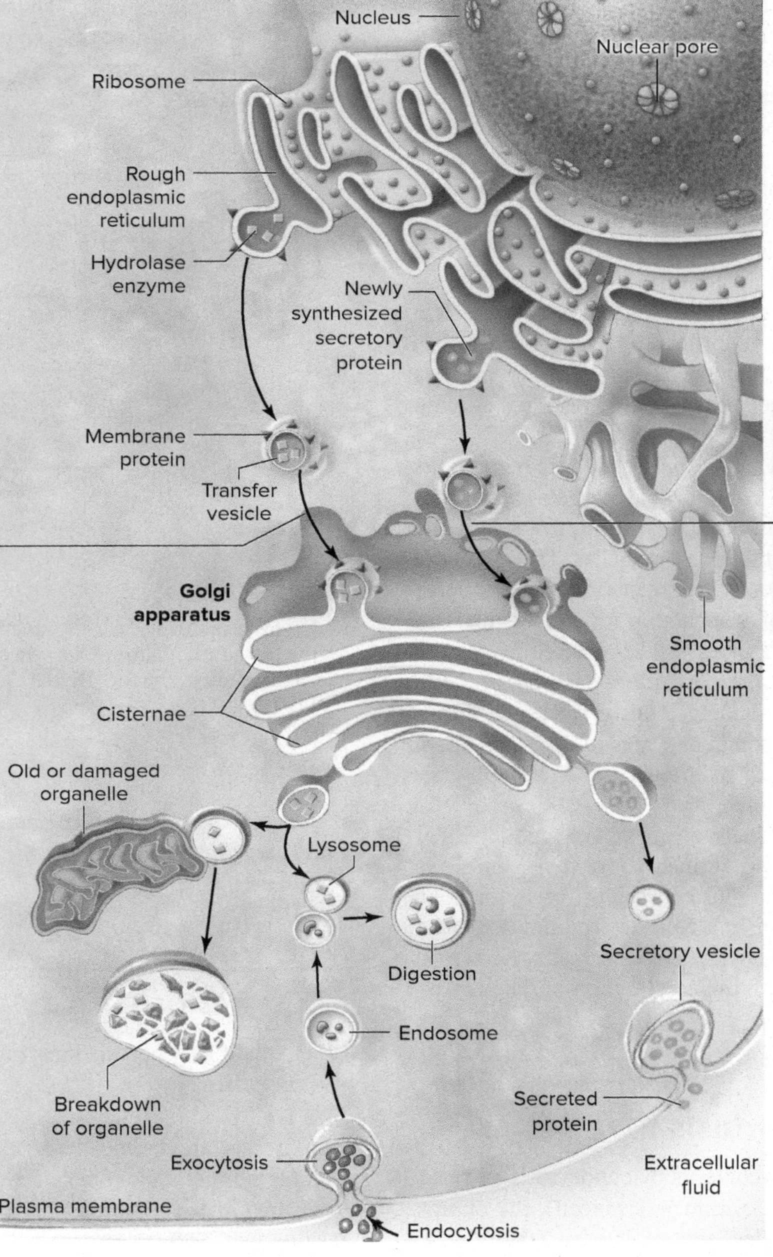

FIGURE 2.19

The Endomembrane System. The endomembrane system consists of the nuclear membrane, smooth and rough endoplasmic reticulum, Golgi apparatus, and vesicles derived from these membranes. This illustration shows two lysosomal pathways on the left and a secretory pathway on the right.

contain secretory products. Golgi apparatuses are most abundant in cells that secrete chemical substances to the outside of the cell (e.g., pancreatic cells secreting digestive enzymes and nerve cells secreting neurotransmitters) and hormone-producing cells and tissues.

Vesicles and Cellular Transport

Some materials diffuse through the cell dissolved in the cytosol. Other large molecules may move directly from one organelle to another. For example, proteins move from ribosomes into the ER as the proteins are being produced. The movement of still other materials may be aided by an organelle. The possible role of vaults in aiding the movement of mRNA from the nucleus to ribosome was described earlier. Many cellular transport processes involve the small membrane-enclosed structures described in the previous section, called **vesicles.** Vesicles are a part of the endomembrane system because they form from, or receive materials from, the ER or the Golgi apparatus.

Vesicles take on many different forms that you will encounter as you study further. The transfer vesicles that carry protein from the ER to the Golgi apparatus for processing, and secretory vesicles that move materials to the plasma membrane for release to the outside of the cell were described previously. **Vacuoles** are a type of vesicle used for temporary storage and transport. For example, some freshwater sponges have contractile vacuoles that collect and remove excess water from within cells to maintain solute concentrations appropriate for homeostasis. Sponges also incorporate food filtered from water into food vacuoles for intracellular digestion (*see figure 9.3*). Food vacuoles are an example of an **endosome**—a vesicle created when the plasma membrane invaginates to engulf materials from the outside of the cell (*see figures 2.19 and 2.20*) and pinches off and will eventually fuse with a lysosome, which is a vesicle considered next.

Lysosomes (Gr. *lyso*, dissolving + *soma*, body) are membrane-bound spherical organelles that contain acid hydrolase enzymes that break down extracellular material brought into the cell, waste intracellular organic molecules, and worn organelles. The hydrolase enzymes are synthesized by ribosomes, transported in transport vesicles from the ER to the Golgi apparatus for processing, and then secreted by the Golgi apparatus in vesicles that either mature into lysosomes or fuse with existing lysosomes (*see figure 2.19*). Lysosomes fuse with vesicles containing food or wastes or engulf intracellular debris, allowing lysosomal enzymes to digest the food, wastes, or debris. Breakdown products are used in energy processing or recycled for anabolic functions.

Scientists are discovering that lysosomes have functions that go beyond the disposal processes described above. Lysosomes are implicated in helping to control the shift between primarily anabolic or primarily catabolic metabolism in a cell, determining longevity of a cell, and influencing neurodegenerative diseases. A lysosomal gene malfunction is associated with the acceleration of the progression of Alzheimer's disease.

Endocytosis and Exocytosis

Vesicles form and discharge their contents by endocytosis and exocytosis. In **endocytosis** (Gr. *endon*, within + *cyto*, cell), the plasma membrane envelops large particles and molecules (figure 2.20 and *see figure 2.19*) and moves them in bulk across the membrane. The three forms of endocytosis are pinocytosis, phagocytosis, and receptor-mediated endocytosis.

Pinocytosis (Gr. *pinein*, to drink) is the nonspecific uptake of small droplets of extracellular fluid. **Phagocytosis** (Gr. *phagein*, to eat) is similar to pinocytosis except that the cell takes in solid material rather than liquid. **Receptor-mediated endocytosis** involves a specific receptor protein on the plasma membrane that "recognizes" an extracellular molecule and binds with it. The reaction stimulates the membrane to indent and create a vesicle containing the selected molecule.

In the process of **exocytosis** (Gr. *exo*, outside), a secretory vesicle fuses with the plasma membrane and releases its contents into the extracellular environment (*see figures 2.19 and 2.20*). This process adds new membrane material, which replaces the portion of the plasma membrane lost during endocytosis.

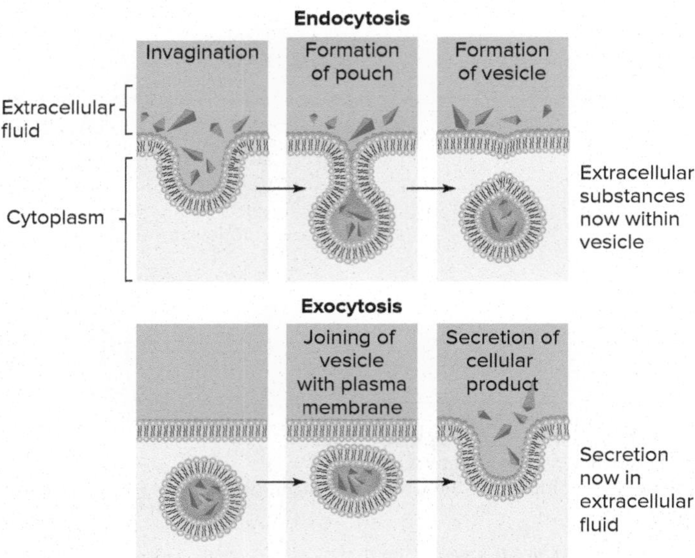

FIGURE 2.20

Endocytosis and Exocytosis. Endocytosis and exocytosis are responsible for the transport of large molecules into and out of cells and for packaging proteins by the ER and Golgi apparatus.

Section 2.5 Thinking Beyond the Facts

Why is the Golgi apparatus central to our definition of what organelles comprise the endomembrane system?

2.6 PEROXISOMES

LEARNING OUTCOME

1. Explain how peroxisomes protect animals from degradative processes.

Peroxisomes are vesicle-like structures. They are not a part of the endomembrane system because they do not receive their contents from the Golgi apparatus, and they form from the division of preexisting peroxisomes rather than by budding from the ER or Golgi apparatus. They contain enzymes that catalyze the removal of electrons and associated hydrogen atoms from acids, including amino acids and fatty acids. They detoxify alcohol. They break down the very reactive hydrogen peroxide molecule to water and oxygen. A small amount of hydrogen peroxide is generated in the mitochondria within the electron transport chain. This highly oxidizing molecule, if not broken down, can induce DNA damage and cause cell death. These degradative processes occur in inflammation, aging, and the metabolism of some cancers.

Section 2.6 Thinking Beyond the Facts

What would be the result of a disease that prevented the formation of peroxisomes?

2.7 THE CYTOSKELETON AND CELLULAR MOVEMENT

LEARNING OUTCOMES

1. Contrast the structure and function of microtubules, intermediate filaments, and microfilaments.
2. Compare and contrast the structure and function of cilia and flagella.

The cytosol of a cell is often misrepresented as a fluid interior with floating organelles. Nothing could be further from the truth. The cytosol is highly structured and dynamic. It contains a complex network of filaments and tubules that connect structures and give shape and texture to a cell. This network of interconnected filaments and tubules is called the **cytoskeleton**. Some of these same tubules project from the cell, still surrounded by the plasma membrane, and provide locomotion for a cell (e.g., a sperm cell) or move materials across outer cell surfaces.

Microtubules, Intermediate Filaments, and Microfilaments

Microtubules, intermediate filaments, and microfilaments form the cytoskeleton (figure 2.21). **Microtubules** are hollow, slender, cylindrical structures in animal cells. Each microtubule is made of spiraling subunits of paired globular proteins called tubulin subunits (figure 2.22*a*). Microtubules function in the movement of organelles, such as secretory vesicles, and in chromosome movement during division of the nucleus of a cell. They are also part of a transport system within the cell. For example, in nerve cells, they help move materials through the long nerve processes. Microtubules are an important part of the cytoskeleton in the cytoplasm, and they are involved in the overall shape changes that cells undergo during periods of specialization.

Microtubules are usually assembled at a structure in the cell called a **microtubule-organizing center (MTOC)** (*see figure 2.1*). The MTOC also anchors one end of a microtubule. Many cells have a MTOC that is called the **centrosome** that is located near the nucleus. It is associated with two structures called centrioles. **Centrioles** help organize the microtubules involved with cell division (*see chapter 3*), and they also contribute to the formation of a portion of cilia and flagella called the basal body (discussed later in this section).

Intermediate filaments are a chemically heterogeneous group of protein fibers, the specific proteins of which can vary with cell type (figure 2.22*b*). These filaments help maintain cell shape and the spatial organization of organelles, as well as promote mechanical activities within the cytoplasm.

Microfilaments (actin filaments) are solid strings of protein (actin) molecules (figure 2.22*c*). Actin microfilaments are present in muscle cells as myofibrils, which help muscle cells to shorten or contract. Actin microfilaments in nonmuscle cells provide mechanical support for various cellular structures and help maintain cell shape in most animal cells.

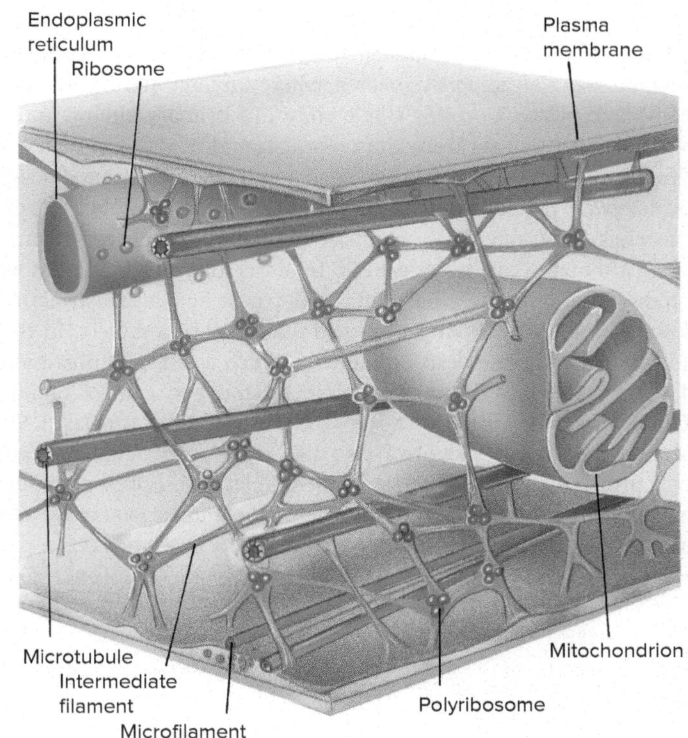

FIGURE 2.21

The Cytoskeleton. This illustration shows the three-dimensional arrangement of microtubules, intermediate filaments, and microfilaments of the cytoskeleton.

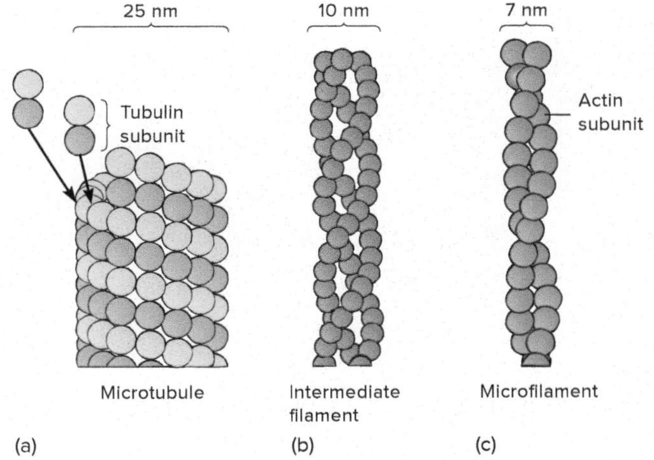

FIGURE 2.22

Three Classes of Protein Fibers Making Up the Cytoskeleton of Eukaryotic Cells. (*a*) Microtubules consist of paired globular proteins, called tubulin subunits, linked in parallel rows. (*b*) Intermediate filaments in different cell types are composed of different protein subunits. (*c*) The protein actin is the primary subunit in microfilaments.

Cilia and Flagella

Cilia and flagella arise from cytoskeletal elements. **Cilia** (sing., cilium; L. *cilium*, eyelash) and **flagella** (sing., flagellum; L. *flagrum*, whip) are elongated appendages on the surface of some cells by which the cells, including many unicellular organisms, propel themselves. In stationary cells, cilia move material over the cells' surfaces. A cilium may also act as a signal-receiving "antenna" for the cell. The membrane proteins on this single cilium (a primary cilium) transmit molecular signals from the cell's external environment to the cytoplasm. When the molecular signal gets to the cytoplasm, it leads to changes in the cell's activities. Cilia-based signaling appears to be a necessity for proper brain function and embryonic development.

Although flagella are 5 to 20 times as long as cilia and move somewhat differently, cilia and flagella have a similar structure. Both are membrane-bound cylinders that enclose a matrix. In this matrix is an axoneme or axial filament, which consists of nine pairs of microtubules arranged in a circle around two central tubules (figure 2.23). This is called a 9 + 2 pattern of microtubules. Each microtubule pair (a doublet) also has pairs of dynein (protein) arms projecting toward a neighboring doublet and spokes extending toward the central pair of microtubules. Cilia and flagella move as a result of the microtubule doublets sliding along one another.

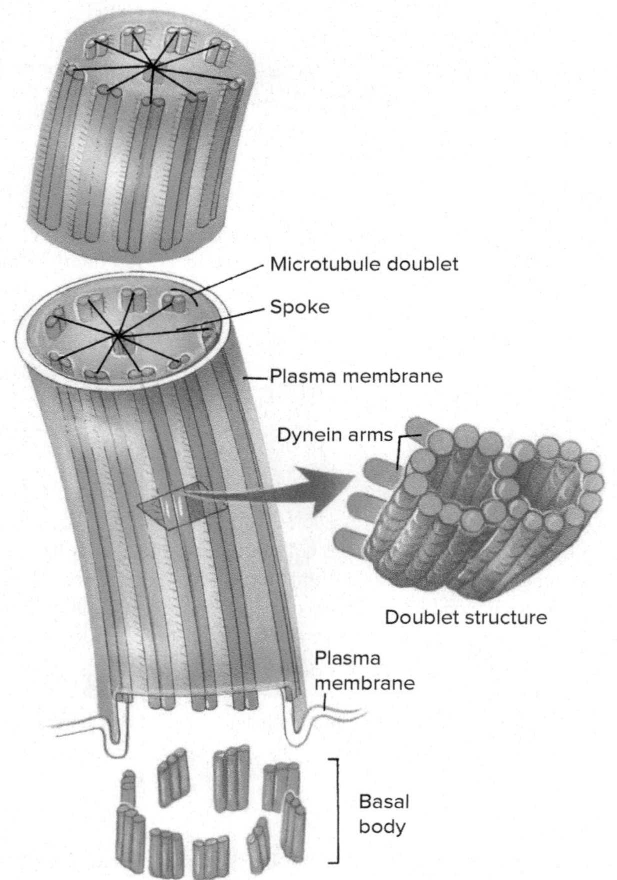

FIGURE 2.23

The Structure of Cilia and Flagella. The axoneme of a cilium or flagellum consists of nine pairs of microtubule doublets arranged around two central microtubules (9 + 2 pattern). Each doublet has dynein arms that extend toward a neighboring doublet. Spokes extend toward the central pair of microtubules. The dynein arms push against the adjacent microtubule doublet to bring about movement. The basal body forms the base of each cilium or flagellum. It is within the cytoplasm and has a 9 + 0 arrangement of microtubules.

In the cytoplasm at the base of each cilium or flagellum lies a short, cylindrical **basal body** that is also comprised of microtubules. The basal body is formed at, and structurally identical to, the centriole. The basal body controls the growth of microtubules in cilia or flagella. The microtubules in the basal body form a 9 + 0 pattern: nine sets of three with none in the middle.

Section 2.7 Thinking Beyond the Facts

What would a cell be like without its cytoskeleton?

2.8 Levels of Organization in an Animal

Learning Outcomes

1. Describe the relationships of tissues to organs and organs to organ systems.
2. Contrast the functions of the four types of tissues found in animals.

Animals are multicellular organisms; that is, they consist of groups of cells that function together with varying levels of interdependence. Chapter 7 describes patterns of organization that characterize multicellular organisms, from those that are loose associations of cells to those that have highly organized and interdependent groups of cells that are related in function. The latter groups of animals exhibit up to three major levels of cellular organization—tissue, organ, and organ-system levels.

Tissues are groups of cells that have similar structure and embryonic origin that perform a specialized function. The study of tissues is called **histology** (Gr. *histos*, tissue + *logos*, discourse). Animals may have up to four tissue types—some have fewer.

Epithelial tissues cover or line a body surface or cavity. Epithelial tissues rest on a **basement membrane** that separates epithelial tissues from underlying tissues. The basement membrane is a thin fibrous matrix that includes collagen proteins—a protein diagnostic of animal tissues. Epithelial tissues are specialized for many different kinds of functions. In the small intestine, epithelial tissues secrete enzymes into the digestive tract and absorb the products of digestion. The surface of the skin is an epithelial layer that protects an animal from external assaults.

Connective tissues support and bind an animal's body parts. They are comprised of a fiber matrix embedded in a ground substance that gives connective tissues textures that vary from fluid (e.g., blood) to solid (e.g., cartilage and bone). The nature of this matrix determines the functional properties of the different kinds of connective tissues.

Nervous tissues are involved with communication within animals. They are comprised of nerve cells, called neurons, which conduct impulses. Other cells, called neuroglia, protect, support, and provide nourishment for the entire nervous tissue. Nervous tissues make up nerves, ganglia, and brains of animals.

Muscle tissues allow movement. Some muscle attaches to skeletal elements of an animal (e.g., your skeletal muscle and the flight muscle that attaches to an insect's exoskeleton). Other muscle propels food through digestive tracts of animals. Still other muscle pumps blood. You will see many examples of all of these tissues in action in the chapters that follow.

Tissues are often (depending on the animal group) organized into two higher levels of organization. **Organs** are composed of one or more tissues that function together. Organs are present in flatworms, where reproductive organs, eyespots, and feeding structures are present. The organs of your body are additional examples. **Organ systems** are groups of organs that work together to perform a common function. Organ systems such as the circulatory, digestive, reproductive, and respiratory systems are examples. Organ systems are present in diverse groups of animals from worms to vertebrates.

Section 2.8 Thinking Beyond the Facts

All animals are multicellular, and most animals have tissues. Why do you think tissues and multicellularity go hand-in-hand?

Summary

2.1 **Cells: The Common Unit of All Life**

- Cells have a common organization consisting of a plasma membrane, nucleus, and cytoplasm.
- Cells must be small in order to support required exchanges of nutrients and wastes and to promote the concentration of the reactants and enzymes required for life's chemical reactions.
- Cells have a common origin as evidenced by a shared genetic makeup and a common energy-carrying molecule, ATP.
- There are three basic types of cells. Each type is characteristic of one of the three main groups of organisms: Bacteria, Archaea, and Eukarya.

2.2 **Cellular Membranes and Membrane Transport**

- The plasma membrane is a phospholipid bilayer that contains interspersed proteins and other macromolecules that form a two-dimensional fluid. Membrane proteins transport ions and molecules, form junctions between cells, and serve as receptors. A glycocalyx protects the cell and functions in cell identification.
- The plasma membrane is selectively permeable. Exchanges across membranes may occur through non-transporter gradient exchanges: simple diffusion, osmosis, and filtration. Exchanges also occur through carrier-mediated transport: facilitated diffusion and active transport. Active transport involves transport against a concentration gradient and requires an input of energy.

2.3 **Energy Processing**

- The sum of cellular reactions in cells is called metabolism, and these reactions require energy to sustain them.
- Adenosine triphosphate (ATP) is the energy currency of the cell. The release of the third phosphate from ATP to form ADP and P_i releases energy for metabolic reactions. Enzymes promote reactions in metabolic pathways.
- Animals are heterotrophs and acquire energy by ingesting other organisms. The energy in food is harvested in the reactions of cellular respiration. Glycolysis occurs in the cytoplasm of cells and results in the breakdown of glucose to pyruvate. NAD is also reduced to NADH. There is a small energy yield from glycolysis. If oxygen is not present, lactate fermentation occurs. If oxygen is present, the pyruvates move into the mitochondrial matrix where the reactions of the citric acid cycle occur. In this cycle, carbons of pyruvate are released as CO_2, ATP is produced, NAD is reduced to NADH, and FAD is reduced to $FADH_2$. NADH and $FADH_2$ carry electrons from glucose into the electron transport system occurring on mitochondrial cristae. Electrons are passed through membrane proteins and eventually combine with O_2 to produce water. In the process, chemiosmosis occurs, and ATP synthase enzymes produce more ATP. The total maximum theoretical yield for aerobic cellular respiration is 36 to 38 ATP per glucose, although the actual yield is usually about 30 ATP.
- Alternative metabolic pathways are able to metabolize sugars other than glucose, fats, and proteins. These alternative pathways can be reversed to synthesize fats and convert one amino acid to another amino acid.

2.4 **The Nucleus, Ribosomes, and Vaults**

- The nucleus of the cell contains DNA in combination with proteins (chromatin). The nucleus is the location where DNA is transcribed into RNA. The nuclear envelope separates the nucleus from the cytoplasm. RNA and other materials move between the nucleus and the cytoplasm through pores in the nuclear envelope. These movements are aided by vaults. Ribosomal subunits are preassembled in the nucleolus. They exit the nucleus to the cytoplasm where they function in translating RNA into protein.

2.5 **The Endomembrane System**

- The rough endoplasmic reticulum (ER) is a labyrinth of membranes that transports and stores proteins that are synthesized by the attached ribosomes. Smooth endoplasmic reticulum is the site for lipid production, detoxification of organic molecules, and the storage of calcium ions. The Golgi apparatus concentrates, modifies, and packages proteins synthesized by ribosomes at the ER.
- Proteins are packaged into membrane-enclosed vesicles of various types. Transfer vesicles carry protein from the ER to the Golgi apparatus. Secretory vesicles move materials to the plasma membrane for release to the outside of the cell. Vacuoles are used for temporary storage and transport. Lysosomes break down extracellular materials brought into the cell and wastes generated within the cell.
- Vesicles form and discharge their contents by endocytosis and exocytosis.

2.6 **Peroxisomes**

- Peroxisomes are vesicle-like structures that contain enzymes that detoxify acids and alcohol. They also break down hydrogen peroxide, which can induce DNA damage and cell death.

2.7 **The Cytoskeleton and Cellular Movement**

- The cytoskeleton is a network of interconnected filaments and tubules that gives shape and texture to the cell and participates in cellular movement. Microtubules form from, and are anchored at, the microtubule-organizing center (centrosome). The centrosome is associated with centrioles that help organize the microtubules involved with cell division. Intermediate filaments help maintain cell shape and the spatial organization of organelles. Microfilaments (actin filaments) function in muscle contraction and provide mechanical support for various cellular structures.

- Cilia and flagella are on the surface of some cells. They are the means by which many unicellular organisms propel themselves and stationary cells move materials over cell surfaces. Cilia and flagella are comprised of a 9 + 2 pattern of microtubules.

2.8 **Levels of Organization in an Animal**

- Animals are multicellular, and their cells function together with varying levels of interdependence. Tissues are groups of cells that have similar structure and embryonic origin that perform a specialized function. Animals have up to four tissue types. Epithelial tissues cover or line a body surface or cavity. Connective tissues support and bind an animal's body parts. Nervous tissues conduct impulses or support a nervous system. Muscle tissues allow movement. Organs are composed of one or more tissues that function together. Organ systems are groups of organs that work together.

Concept Review Questions

1. All of the following statements regarding a cell's plasma membrane are true except one. Select the statement that is NOT true.
   a. The hydrophilic/hydrophobic properties of membrane phospholipids cause them to spontaneously assemble into a phospholipid bilayer in the watery medium of life. This property is responsible for our visualization of cellular membranes as two-dimensional fluids.
   b. Membrane proteins are always, and only, found attached to the inner surface of a membrane where they serve as attachment points for secretory vesicles.
   c. Cholesterol within the membrane stabilizes membrane fluidity with temperature extremes.
   d. Animal cells have a surface coat of glycolipids and glycoproteins called the glycocalyx. This coat protects the cell, and it plays roles in cell-to-cell recognition and adhesion.
2. The reactions of cellular respiration are examples of metabolic pathways. Which one of the following statements is true of cellular respiration?
   a. Glucose is the only substrate that can be broken down to derive energy in the form of ATP.
   b. Glycolysis occurs in the cytoplasm, ends with pyruvate, and results in the major ATP energy yield in cellular respiration.
   c. The CO_2 that animals produce as a waste product of cellular respiration is produced as carbons from pyruvate are released in the matrix of mitochondria.
   d. The electron transport system occurs in the matrix of mitochondria and results in the release of CO_2 as well as the production of a second by-product, water.
3. Which one of the following statements is true regarding the endomembrane system of a cell?
   a. It is an interconnected system of membranes, and it includes the endoplasmic reticulum (ER), the Golgi apparatus, lysosomes, and peroxisomes.
   b. The Golgi apparatus and ER are related functionally because one job of the Golgi apparatus is to modify, sort, and package proteins produced by the ribosomes that are attached to the ER. These proteins were transported to the Golgi apparatus through transfer vesicles produced by the ER.
   c. Lysosomes are not part of the endomembrane system because they do not form from membranes of the ER or other endomembrane structure.
   d. Transfer vesicles pinch off from the plasma membrane and eventually fuse with lysosomes.
4. Which one of the following combinations of cellular structures includes only structures that have something to do with the production or function of microtubules or are structures comprised of microtubules?
   a. Basal bodies, cilia, actin filaments
   b. Flagella, 9 + 2, tubulins, centrosomes, basal bodies
   c. Muscle contraction, intermediate filaments, microfilaments
   d. 9 + 2, MTOC, microfilaments, tubulins
5. Which of the following tissue types would make up the inner lining of the digestive tract of a snake?
   a. Muscular tissue
   b. Connective tissue
   c. Nervous tissue
   d. Epithelial tissue

Analysis and Application Questions

1. Cells come in many sizes and shapes, and they perform many diverse functions. In spite of these observations, cell theory is a concept that unifies all of biology. Why is the latter statement true?
2. Why it is accurate to visualize the plasma membrane as a two-dimensional fluid?
3. How do animal cells transport materials against concentration gradients? What functions can you think of that could not occur if animal cells lacked this ability? (You will become aware of many more of these functions as you study chapters that follow.)
4. Why are reactions involving ATP thought of as coupled reactions?
5. Evolutionarily, which do you think came first—aerobic or anaerobic cellular respiration? Why?
6. Why are the ER, the Golgi apparatus, and various vesicles (e.g., endosomes, lysosomes, and vacuoles) considered part of the endomembrane system of a cell, but peroxisomes are not?

Humans have known for centuries that traits are inherited by offspring from their parents. Through trial and error we have manipulated breeding to achieve desired characteristics in our animals. We now understand more about the molecules controlling inheritance and the positive and negative effects of humans interacting with animal reproduction (*see pages 49 and 50*).

3

Cell Division and Inheritance

Chapter Outline

3.1 Eukaryotic Chromosomes
Sex Chromosomes and Autosomes
Number of Chromosomes
3.2 The Cell Cycle and Mitotic Cell Division
Interphase: Replicating the Hereditary Material
M-Phase: Mitosis
M-Phase: Cytokinesis
Cell-Cycle Control
3.3 Meiosis: The Basis of Sexual Reproduction
The First Meiotic Division
The Second Meiotic Division
Spermatogenesis and Oogenesis
3.4 DNA: The Genetic Material
The Double Helix Model
DNA Replication in Eukaryotes
Genes in Action
Changes in DNA and Chromosomes
3.5 Inheritance Patterns in Animals
Segregation
Independent Assortment
Other Inheritance Patterns
The Molecular Basis of Inheritance Patterns

Reproduction is essential for life. Each organism exists solely because its ancestors succeeded in producing progeny that could develop, survive, and reach reproductive age. At its most basic level, reproduction involves a single cell reproducing itself. Unicellular organisms undergo asexual reproduction through cell division. For multicellular organisms, cellular reproduction is involved in growth, repair, and the formation of sperm and egg cells that enable the organism to reproduce.

At the molecular level, reproduction involves the cell's unique capacity to manipulate large amounts of DNA, DNA's ability to replicate, and DNA's ability to carry information that will determine the characteristics of cells in the next generation. Information carried in DNA is manifested in the kinds of proteins that exist in each individual. Proteins contribute to observable traits, such as coat color of mammals and plumage characteristics of birds, and they function as enzymes that regulate the rates of chemical reactions in organisms. Within certain environmental limits, animals exist as they do because of the proteins that they synthesize.

At the level of the organism, reproduction involves passing DNA from individuals of one generation to the next generation. The classical approach to studying inheritance involves experimental manipulation of reproduction and observing patterns of inheritance between generations. This work began with Gregor Mendel (1822–1884), and it continues today.

Gregor Mendel began a revolution that has had a tremendous effect on biology and our society. Genetic mechanisms explain how traits are passed between generations. They also help explain how species change over time. Genetic and evolutionary themes are interdependent in biology, and biology without either would be unrecognizable from its present form. Genetic precepts have tremendous potential to help us understand how human activities have affected animal populations. This understanding may help us preserve wildlife and improve the conditions of our own lives. This chapter introduces principles of cell division and genetics that are essential to understanding the zoological and evolutionary foundations presented throughout this textbook.

3.1 EUKARYOTIC CHROMOSOMES

LEARNING OUTCOMES

1. Critique the statement that one cannot understand genetics without understanding how DNA is packaged within cells.
2. Differentiate between sex chromosomes and autosomes in a diploid animal.
3. Explain how chromosome numbers can vary in animal cells.

Genetics (Gr. *gennan*, to produce) is the study of how biological information is transmitted from one generation to the next. The starting point for understanding how genetic information is transmitted between generations is learning how genetic information in the form of DNA is packaged within cells. In eukaryotic cells, long strands of DNA are associated with proteins. These associations are called **chromosomes**. During most of the life of a cell, chromosomes are in a highly dispersed state called chromatin. During these times, units of inheritance called genes (Gr. *genos,* race) may actively participate in the formation of protein. When a cell is dividing, however, chromosomes exist in a highly folded and condensed state that allows them to be distributed between new cells being produced. The structure of these chromosomes will be described in more detail in the discussion of cell division that follows.

Chromatin consists of DNA and histone proteins. This association of DNA and protein helps with the complex jobs of packing DNA into chromosomes (chromosome condensation) and regulating DNA activity.

There are five different histone proteins. The amino acid composition of these proteins creates positive charges that attract the negative charges of DNA's phosphate groups. Some of these proteins form a core particle. DNA wraps in a coil around the proteins, a combination called a **nucleosome** (figure 3.1). The fifth histone, sometimes called the linker protein, is not needed to form the nucleosome but may help anchor the DNA to the core and promote the winding of the chain of nucleosomes into a solenoid. Higher-order folding forms chromatin loops, rosettes, and the final chromosome. The details of this higher-order folding are still under investigation.

Not all chromatin is equally active. Some chromosomal regions (e.g., regions around centromeres, *see figure 3.4*) are always inactive, meaning that the base sequences in DNA in these regions is not transcribed (processed) to produce protein. In other cases entire chromosomes may be inactivated during embryonic development. (*See the discussion of X-chromosome inactivation in the section that follows.*) Inactive portions of chromosomes produce dark banding patterns with certain staining procedures and thus are called **heterochromatic regions,** whereas active portions of chromosomes are called **euchromatic regions.** Alterations of chromatin structure including the addition of chemical groups to histone proteins and DNA, removal or repositioning nucleosomes, and interactions with a host of proteins control the activity of chromatin. The importance of these kinds of genetic controls is easy to understand when one considers the changing genetic states in various animal tissues as embryonic tissues differentiate to perform specific functions in adults or the changes that occur in gene activity during the metamorphosis of a caterpillar to butterfly.

Sex Chromosomes and Autosomes

In the early 1900s, work with the insect *Protenor* provided our first insights into the chromosomal basis for maleness or femaleness. One

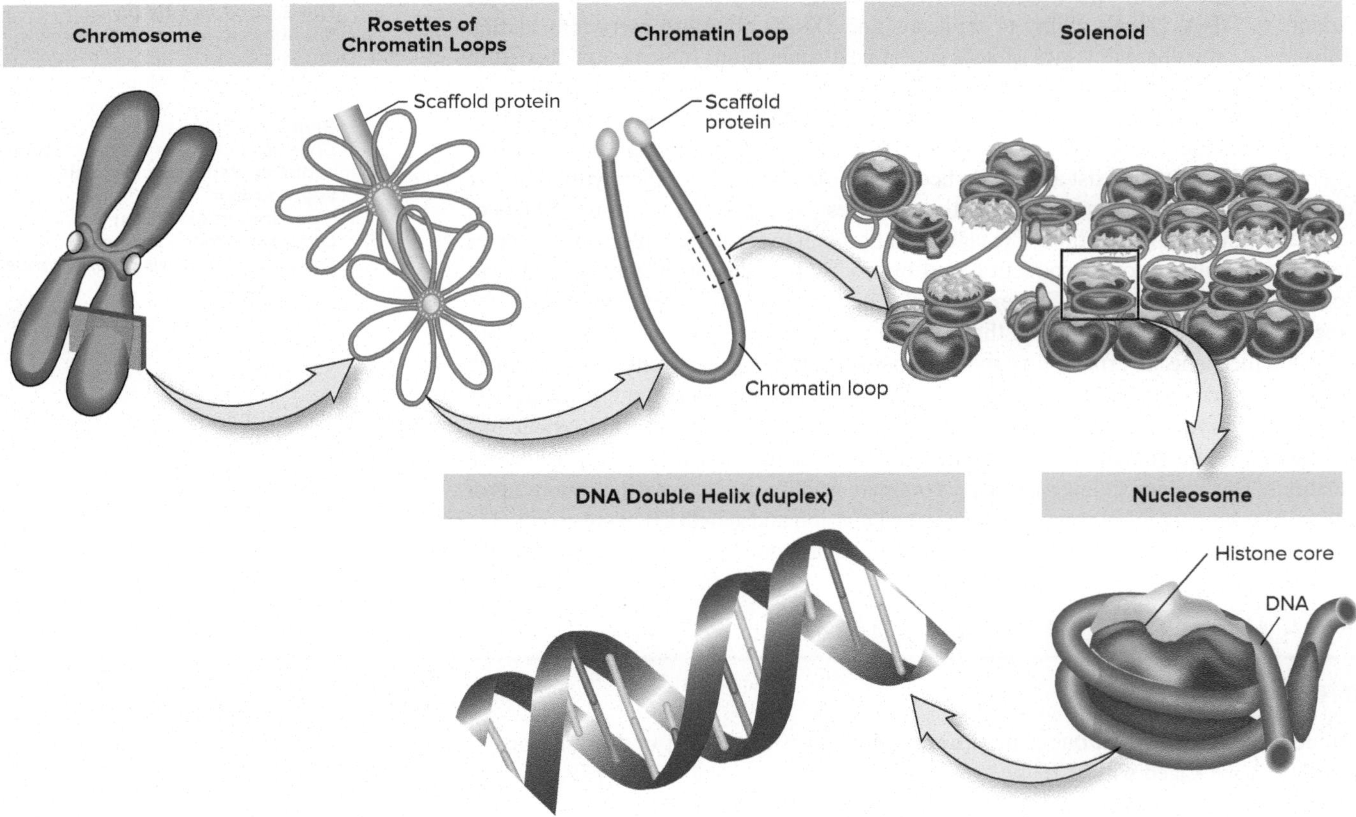

FIGURE 3.1

Organization of Eukaryotic Chromosomes. Chromosomes consist of long DNA molecules that wrap histone proteins. The DNA and histone complex is called a nucleosome, and the chain of nucleosomes is coiled into a solenoid. The solenoid is then looped into rosettes around a scaffold protein. Further compaction results in the eukaryotic chromosome.

darkly staining chromosome of *Protenor,* called the X chromosome, is represented differently in males and females. All somatic (body) cells of males have one X chromosome (XO), and all somatic cells of females have two X chromosomes (XX). Similarly, half of all sperm contain a single X, and half contain no X, whereas all female gametes contain a single X. Fertilization involving an X-bearing sperm will result in a female offspring and fertilization involving a sperm with no X chromosome will result in a male offspring. This sex determination system explains the approximately 50:50 ratio of females to males in this insect species (figure 3.2). Chromosomes that are represented differently in females than in males and function in sex determination are **sex chromosomes.** Chromosomes that are alike in both sexes are **autosomes** (Gr. *autos,* self + *soma,* body).

The system of sex determination described for *Protenor* is called the X-O system. It is the simplest system for determining sex because it involves only one kind of chromosome. Many other

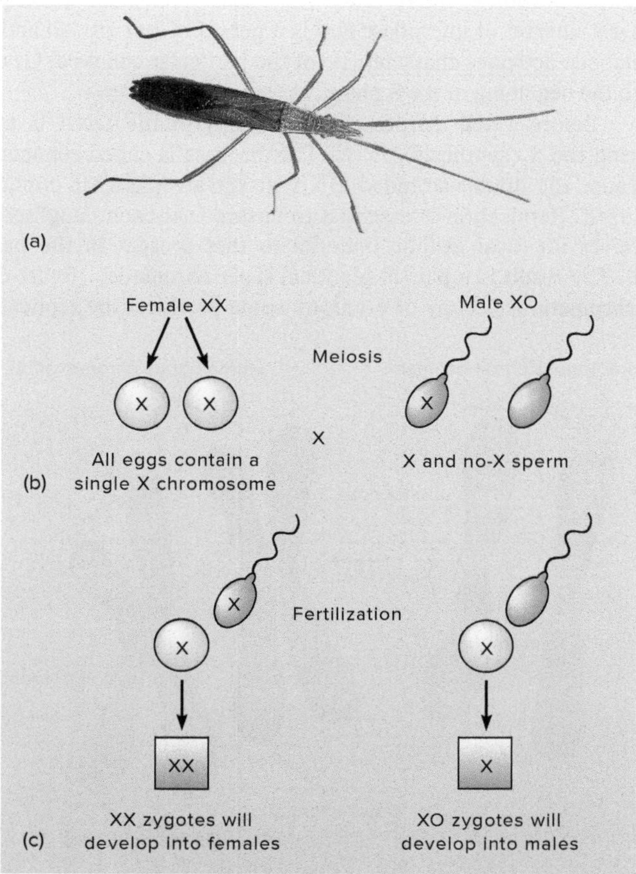

FIGURE 3.2

X-O System of Sex Determination for the Insect *Protenor.*
(*a*) *Protenor belfragei* belongs to a family of plant bugs (order Hemiptera) that feed on plant juices of grasses and sedges. (*b*) In females, all cells except gametes possess two X chromosomes. During meiosis, homologous X chromosomes segregate, and all eggs contain one X chromosome. Males possess one X chromosome per cell. Meiosis results in half of all sperm cells having one X, and half of all sperm cells having no X. (*c*) Fertilization results in half of all offspring having one X chromosome (males), and half of all offspring having two X chromosomes (females).
©Brandon Woo

animals (e.g., humans and fruit flies) have an X-Y system of sex determination. In the X-Y system, males and females have an equal number of chromosomes, but the male is usually XY, and the female is XX. (In birds, the sex chromosomes are designated Z and W, and the female is ZW.) Even though the X and Y chromosomes are called "sex chromosomes," they also help determine non-sex-related traits. This is especially true for the X chromosome of most animals. It is very large and has genes that code for many traits. Similarly, autosomal chromosomes frequently carry genes that influence sexual characteristics. This mode of sex determination also results in approximately equal numbers of male and female offspring:

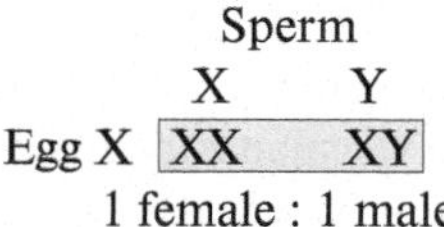

The fact that female mammals have two large X chromosomes and males have only one X chromosome suggests that females will have a "double dose" of proteins encoded by that chromosome. This is not the case. Early in mammalian development, one of the two X chromosomes of female embryonic cells is randomly converted to heterochromatin by the addition of methyl groups ($-CH_3$) to histone proteins. For example, the calico cat has a patterning of three coat colors (white, black, and orange) that originates by the random inactivation of X chromosomes in different embryonic cells. A cell that has an active X chromosome carrying an "orange gene" will result in a population of skin cells that produces orange fur, and a cell that has an active X chromosome carrying a "non-orange gene" will result in a population of skin cells that produces black fur. White patches are produced by the interaction of spotting genes with coat color genes. Calico cats are usually female because females normally have two X chromosomes. Rare male calicos occur when a chromosomal event called nondisjunction occurs and results in a male with two X chromosomes along with the Y chromosome (*see figure 3.15*).

Number of Chromosomes

Even though the number of chromosomes is constant within a species, chromosome number varies greatly among species. The chromosome number of animals usually varies between 10 and 50.

Chromosomes are present in sets, with the number in a set being characteristic of each kind of animal and expressed as "N." N identifies the number of different kinds of chromosomes. Most animals have two sets, or $2N$ chromosomes. This is the **diploid** (Gr. *di,* two + *eoides,* doubled) condition. Some animals have only one set, or N chromosomes (like gametes) and are **haploid** (Gr. *hapl,* single) (e.g., male honeybees and some rotifers).

Some animals (e.g., brine shrimp, snout beetles, some flatworms, and some sow bugs) have more than the diploid number of chromosomes, a condition called **polyploidy** (Gr. *polys,* more). The upset in numbers of sex chromosomes often interferes with reproductive success. Asexual reproduction often accompanies polyploidy. Other animals function normally with polyploid chromosomes. Many fish species (e.g., salmon) are polyploid and grow faster, reach larger sizes, and live longer than diploid species.

SECTION 3.1 THINKING BEYOND THE FACTS

Bacteria have a single circular chromosome that remains in a dispersed state throughout their life cycles. Bacteria lack histone proteins. Do you think that bacteria regulate gene function by the formation of hetrochromatic chromosomal regions? Explain.

3.2 THE CELL CYCLE AND MITOTIC CELL DIVISION

LEARNING OUTCOMES

1. Contrast the cell-cycle activities of an embryonic cell and a mature bone cell in a way that explains the importance of these activities for each type of cell.
2. Explain why the events of mitotic cell division result in daughter cells being identical to parental cells.
3. Hypothesize on the outcome of the cell cycle when cell-cycle control mechanisms fail.

The life of a cell begins when a parent cell divides to produce the new cell. The new cell then goes through maintenance and growth processes until it matures and ultimately divides to produce another generation of two cells. The life of a cell, from its beginning until it divides to produce the new generation of cells, is called the **cell cycle** (figure 3.3).

Mitosis (Gr. *mitos,* thread) is the distribution of chromosomes between two daughter cells, and **cytokinesis** (Gr. *kytos,* hollow vessel + *kinesis,* motion) is the partitioning of the cytoplasm between the two daughter cells. **Interphase** (L. *inter,* between) is the time between the end of cytokinesis and the beginning of the next mitotic division. It is a time of cell growth, DNA synthesis, and preparation for the next mitotic division.

The G_1 (first growth or gap) phase represents the early growth phase of the cell. During the S (DNA synthesis) phase, growth continues, but this phase also involves DNA replication. The G_2 (second growth or gap) phase prepares the cell for division. It includes replication of the mitochondria and other organelles, synthesis of microtubules and protein that will make up the mitotic spindle fibers, and chromosome condensation. The **M (mitotic) phase** includes events associated with partitioning chromosomes between two daughter cells and the division of the cytoplasm (cytokinesis).

Interphase: Replicating the Hereditary Material

The first portion of interphase is gap phase 1 (G_1). It is usually the longest interval of interphase and is a period of cell growth and the metabolic activities characteristic of the particular cell type. G_1 ends with the beginning of the S phase.

Before a cell divides, an exact copy of the DNA is made during the S (synthesis) phase. This process is called replication, because the double-stranded DNA makes a replica, or duplicate, of itself. Replication is essential to ensure that each daughter cell receives identical genetic material to that present in the parent cell. The result is a pair of identical **sister chromatids** (figure 3.4). A **chromatid** is a copy of a chromosome produced by replication.

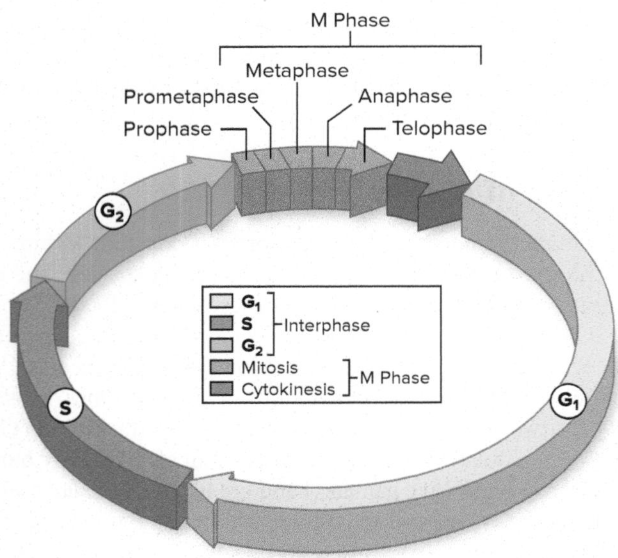

FIGURE 3.3

Life Cycle of a Eukaryotic Cell. During the G_1 phase, cell components are synthesized and metabolism occurs, often resulting in cell growth. During the S (synthesis) phase, the chromosomes replicate, resulting in two identical copies called sister chromatids. During the G_2 phase, metabolism and growth continue until the mitotic phase is reached. This drawing is generalized, and the length of different stages varies greatly from one cell to the next.

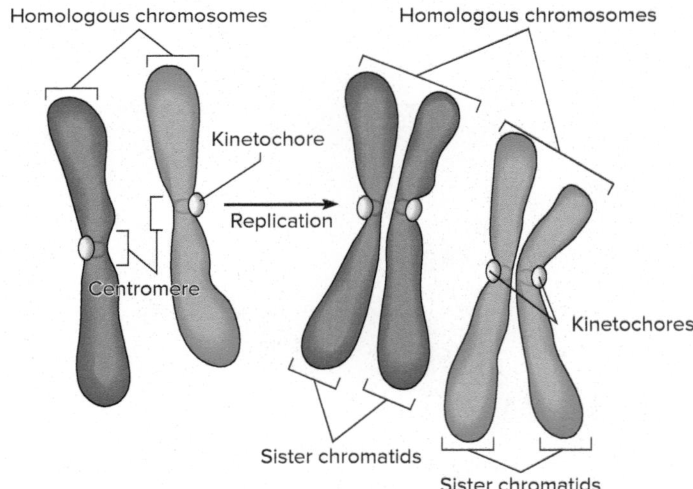

FIGURE 3.4

Chromosome Replication and Homologous Chromosomes. Chromosome replication occurs during interphase of the cell cycle. Before replication (S phase of the cell cycle), chromosomes consist of a single chromatid. Nonreplicated chromosomes are shown diagrammatically in a condensed state for comparative purposes. They would actually be in the form of uncondensed chromatin during replication. Following replication, chromosomes consist of two identical chromatids held together at the centromere. Homologous chromosomes (described later in this chapter) are represented by red and blue colors. These chromosomes carry genes for the same traits; one homolog was received from the maternal parent and the other from the paternal parent.

Each chromatid attaches to its other copy, or sister, at a point of constriction called a centromere. The **centromere** is a specific DNA sequence of about 220 nucleotides and has a specific location on any given chromosome. Bound to each centromere is a disk of protein called a **kinetochore,** which eventually is an attachment site for the microtubules of the mitotic spindle.

The final stage of interphase is gap phase 2 (G_2). As the cell cycle moves into the G_2 phase, the chromosomes begin condensation. During the G_2 phase, the cell also begins to assemble the structures that it will later use to move the chromosomes to opposite poles (ends) of the cell. For example, centrioles replicate, and there is extensive synthesis of the proteins that make up the microtubules.

The time spent by a cell in interphase varies greatly depending on the cell. Rapidly dividing embryonic cells move very quickly through G_1 to S, and again quickly through G_2 to M. The entire cell cycle may occur within a few minutes time. Rapidly dividing cells produce a many-celled embryo from a single fertilized egg within hours. On the other hand, the cell cycle of mature cells is often arrested in G_1, and these cells are said to have entered the G_0 phase. They may remain in G_0 for an extended period of time. For example, mature liver cells remain in G_0 for a year or two before dividing again. Mature bone, muscle, and nerve cells remain in G_0 indefinitely, or they can reenter G_1 quickly when cell division is required (e.g., in response to an injury).

M-Phase: Mitosis

Mitosis is divided into five phases: prophase, prometaphase, metaphase, anaphase, and telophase. In a dividing cell, however, the process is actually continuous, with each phase smoothly flowing into the next (figure 3.5).

The first phase of mitosis, **prophase** (Gr. *pro,* before + phase), begins when chromosomes become visible with the light microscope as threadlike structures. The nucleoli and nuclear envelope begin to break up, and the two centriole pairs move apart. By the end of prophase, the centriole pairs are at opposite poles of the cell. The centrioles radiate an array of microtubules called **asters** (L. *aster,* little star), which brace each centriole against the plasma membrane. Between the centrioles, the microtubules form a spindle of fibers that extends from pole to pole. The asters, spindle, centrioles, and microtubules are collectively called the **mitotic spindle** (or mitotic apparatus).

Prometaphase follows the break-up of the nuclear envelope. A second group of microtubles attach at one end to the kinetochore of each chromatid and to one of the poles of the cell at the other end of the microtuble. This bipolar attachment of spindle fibers to chromatids is critical to the movement of the chromatids of each chromosome to opposite poles of the cell in subsequent phases of mitosis.

As the dividing cell moves into **metaphase** (Gr. *meta,* after + phase), the chromosomes (consisting of two replicated chromatids) begin to align in the center of the cell, along the spindle equator.

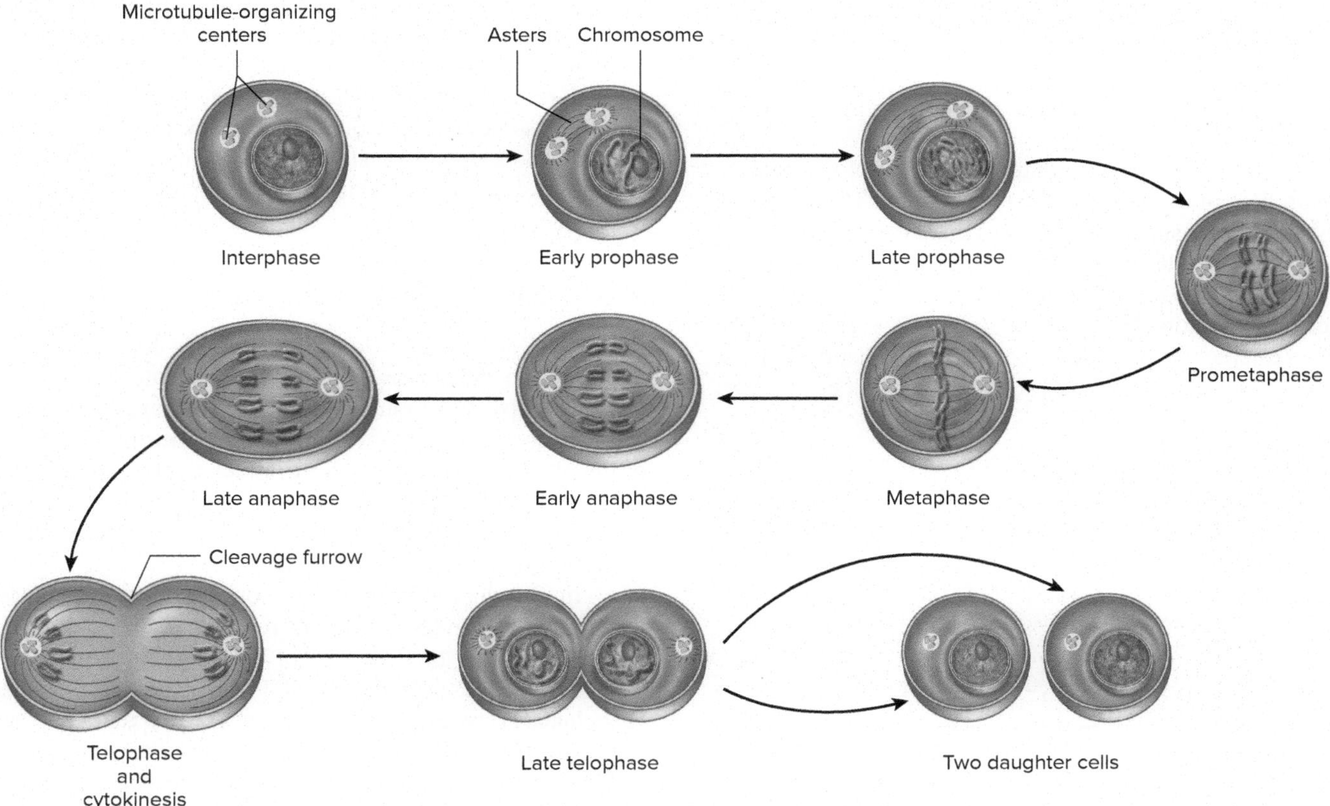

FIGURE 3.5

Continuum of Mitosis and Cytokinesis. Mitosis is a continuous process during which the nuclear parts of a cell divide into two equal portions. Cytokinesis is the division of the cytoplasm of a cell.

Toward the end of metaphase, the centromeres divide and detach the two sister chromatids from each other, although the chromatids remain aligned next to each other. After the centromeres divide, the sister chromatids are considered full-fledged chromosomes (called daughter chromosomes).

During **anaphase** (Gr. *ana,* back again + phase), the shortening of the microtubules in the mitotic spindle, and perhaps the activity of motor proteins of the kinetochore, pulls each daughter chromosome apart from its copy and moves it toward its respective pole. Anaphase ends when all the daughter chromosomes have moved to the poles of the cell. Each pole now has a complete, identical set of chromosomes.

Telophase (Gr. *telos,* end + phase) begins once the daughter chromosomes arrive at the opposite poles of the cell. During telophase, the mitotic spindle disassembles. A nuclear envelope re-forms around each set of chromosomes, which begin to uncoil for gene expression, and the nucleolus is resynthesized. The cell also begins to pinch in the middle. Mitosis is over, but cell division is not.

M-Phase: Cytokinesis

The final phase of cell division is cytokinesis, in which the cytoplasm divides. Cytokinesis usually starts sometime during late anaphase or early telophase. A contracting belt of microfilaments called the contractile ring pinches the plasma membrane to form the cleavage furrow. The furrow deepens, and two new, genetically identical, daughter cells form.

Cell-Cycle Control

The cell cycle must be precisely controlled. If control mechanisms fail, uncontrolled growth (e.g., cancer) can occur. These controls involve cyclin proteins and kinase enzymes that interact with cyclins. There are three cell-cycle checkpoints (G_1, G_2, and M-phase) where the the cell cycle can be halted by DNA damage, starvation, or the absence of growth factors. The cell cycle is also influenced by cellular responses to external factors such as contact with other cells. These controls are the subject of active research in the field of cell biology.

Section 3.2 Thinking Beyond the Facts

Why is mitotic cell division of a diploid cell useful for growth and repair processes but not useful in the production of egg and sperm cells?

3.3 Meiosis: The Basis of Sexual Reproduction

Learning Outcomes

1. Contrast the importance of meiotic cell division and mitotic cell division in animals.
2. Explain why meiotic cell division produces haploid cells after the first and second divisions.
3. Explain the importance of prophase I to the outcome of meiosis.
4. Contrast spermatogenesis and oogenesis.

Sexual reproduction requires a genetic contribution from two different sex cells. Egg and sperm cells are specialized sex cells called **gametes** (Gr. *gamete,* wife; *gametes,* husband). In animals, a male gamete (sperm) unites with a female gamete (egg) during fertilization to form a single cell called a **zygote** (Gr. *zygotos,* yoked together). The zygote is the first cell of the new animal. The fusion of nuclei within the zygote brings together genetic information from the two parents, and each parent contributes half of the genetic information to the zygote.

To maintain a constant number of chromosomes in the next generation, animals that reproduce sexually must produce gametes with half the chromosome number of their ordinary body cells (called **somatic cells**). All of the cells in the bodies of most animals, except for the egg and sperm cells, have the diploid ($2N$) number of chromosomes. Gametes are produced by germ-line cells. **Germ-line cells** comprise a population of cells that develop the ability to pass genes on to progeny, usually through sexual reproduction. These cells undergo a type of cell division called **meiosis** (Gr. *meiosis,* diminution). Meiosis occurs in the ovaries and testes and reduces the number of chromosomes in gametes to the haploid ($1N$) number. The nuclei of the two gametes combine during fertilization and restore the diploid number.

Meiosis begins after the G_2 phase in the cell cycle–after DNA replication. Two successive nuclear divisions, designated meiosis I and meiosis II, take place. The two nuclear divisions of meiosis result in four daughter cells, each with half the number of chromosomes of the parent cell. Moreover, these daughter cells are not genetically identical. Like mitosis, meiosis is a continuous process, and biologists divide it into the phases that follow only for convenience.

The First Meiotic Division

Prophase I of meiosis is long and complex. Its events are vitally important to the successful completion of meiosis. Cell biologists have divided prophase I into five substages that begin with the condensation of chromatin and end with the fragmentation of the nuclear membrane and the attachment of spindle fibers to the kinetochore of each chromosome. These latter events parallel those of prometaphase of mitosis–but are also different in very important ways.

In prophase I, chromatin condenses and chromosomes become visible under a light microscope (figure 3.6*a*). Because a cell has a copy of each type of chromosome from each original parent cell, it contains the diploid number of chromosomes. **Homologous chromosomes** (homologues) carry genes for the same traits, are the same length, and have a similar staining pattern, making them identifiable as matching pairs (*see figure 3.4*). During prophase I, homologous chromosomes line up side-by-side in a process called **synapsis** (Gr. *synapsis,* conjunction), forming a **tetrad** of chromatids (also called a bivalent). The tetrad thus contains the two homologous chromosomes, one is maternal in origin and one

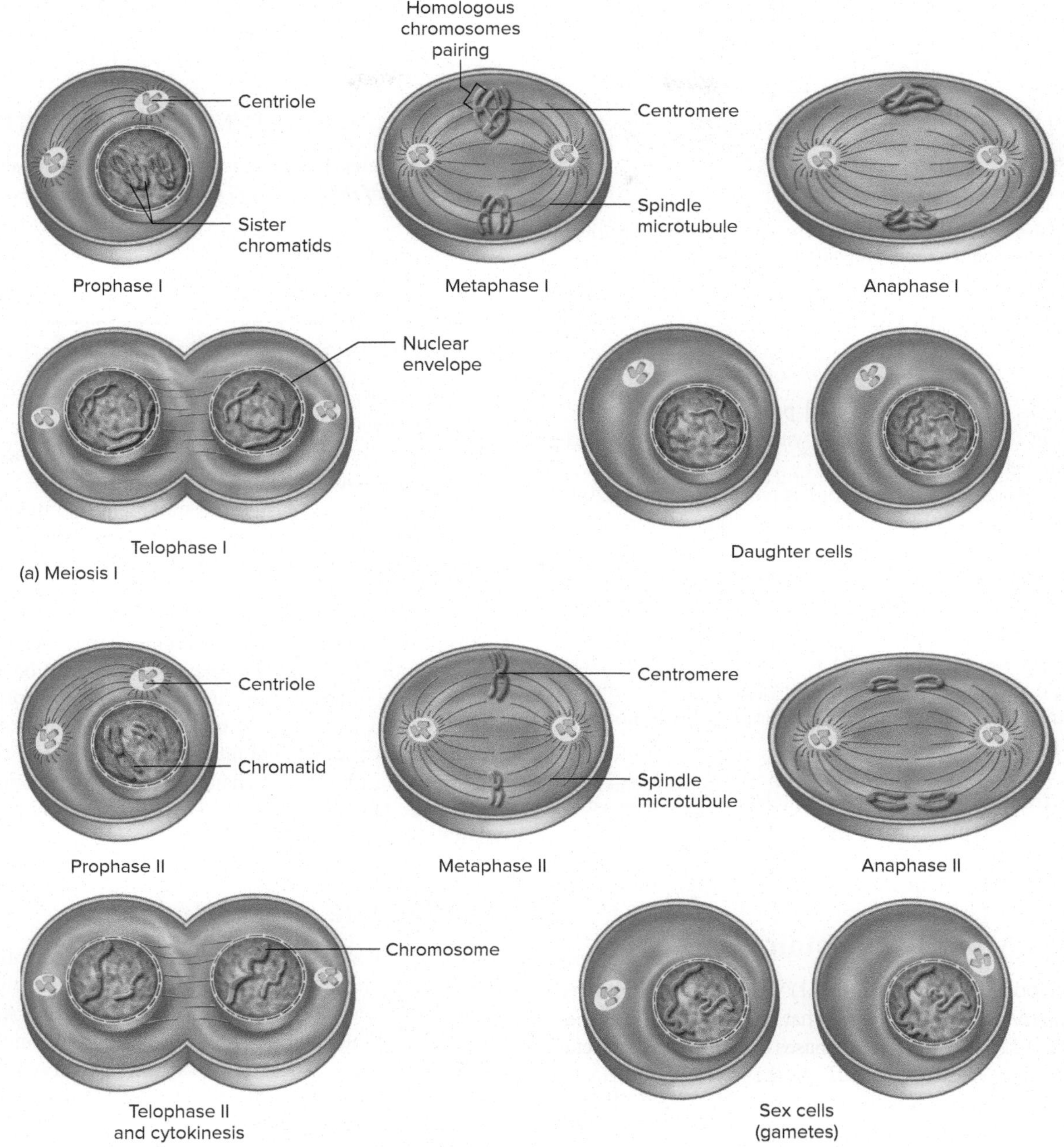

FIGURE 3.6

Meiosis and Cytokinesis. (*a*) Stages in the first meiotic division. Chromosomes of maternal origin are shown in red. Chromosomes of paternal origin are shown in blue. Homologous pairs of chromosomes are indicated by differences in size. (*b*) Stages in the second meiotic division. One of the two daughter cells from the first division is followed through the second division.

is paternal in origin (figure 3.7). An elaborate network of protein is laid down between the two homologous chromosomes. This network holds the homologous chromosomes in a precise union so that corresponding genetic regions of the homologous chromosomes are exactly aligned.

Synapsis also initiates a series of events called **crossing-over,** whereby the nonsister chromatids of the two homologous chromosomes in a tetrad exchange DNA segments (*see figure 3.7*). This process effectively redistributes genetic information among the paired homologous chromosomes and produces new combinations of genes on the various chromatids in homologous pairs. Thus, each chromatid ends up with new combinations of instructions for a variety of traits. Crossing-over is a form of **genetic recombination** and is a major source of genetic variation in a population of a given species.

In metaphase I, the microtubules form a spindle apparatus just as in mitosis (*see figures 3.4 and 3.5*). However, unlike mitosis,

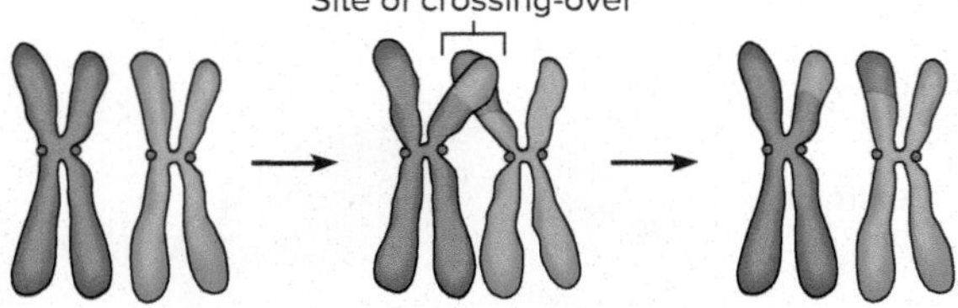

FIGURE 3.7

Synapsis and Crossing-Over. Synapsis is the very tight gene-to-gene pairing of homologous chromosomes during prophase I of meiosis. Molecular interactions between homologous chromosomes result in the snipping and rejoining of nonsister chromatids and the exchange of regions of nonsister arms. This exchange of chromatid arms is called crossing-over.

where homologous chromosomes do not pair, each pair of homologues lines up in the center of the cell, with centromeres on each side of the spindle equator.

Anaphase I begins when homologous chromosomes separate and begin to move toward each pole. Because the orientation of each pair of homologous chromosomes in the center of the cell is random, the specific chromosomes that each pole receives from each pair of homologues are also random. This random distribution of members of each homologous pair to the poles of the cell, along with the genetic recombination between homologous chromosomes that occurs during crossing-over (prophase I), means that no two daughter cells produced by meiotic cell division will be identical.

Meiotic telophase I is similar to mitotic telophase. The transition to the second nuclear division is called interkinesis. Cells proceeding through interkinesis do not replicate their DNA. After a varying time period, meiosis II occurs.

The Second Meiotic Division

The second meiotic division (meiosis II) resembles an ordinary mitotic division (*see figure* 3.6*b*), except that the number of chromosomes has been reduced by half. The phases are prophase II, metaphase II, anaphase II, and telophase II. At the end of telophase II and cytokinesis, the final products of these two divisions of meiosis are four new "division products." In most animals, each of these "division products" is haploid and may function directly as a gamete (sex cell).

Spermatogenesis and Oogenesis

The result of meiosis in most animals is the formation of sperm and egg cells. **Spermatogenesis** produces mature sperm cells and follows the sequence previously described. All four products of meiosis often acquire a flagellum for locomotion and a caplike structure that aids in the penetration of the egg. **Oogenesis** produces a mature ovum or egg. It differs from spermatogenesis in that only one of the four meiotic products develops into the functional gamete. The other products of meiosis are called polar bodies and eventually disintegrate. In some animals the mature egg is the product of the first meiotic division and only completes meiosis if it is fertilized by a sperm cell.

SECTION 3.3 THINKING BEYOND THE FACTS

Why are the events of the first meiotic division so very important in the outcome of the entire meiotic cell division process?

3.4 DNA: THE GENETIC MATERIAL

LEARNING OUTCOMES

1. Explain the structure of DNA and how that structure allows the molecule to undergo replication.
2. Explain how the genetic code in DNA is transcribed into messenger RNA and then translated into protein.
3. Use examples to justify the statement that most changes in DNA are detrimental for an organism but some are vitally important for the evolution of populations.

Twentieth-century biologists realized that a molecule that serves as the genetic material must have certain characteristics to explain the properties of life: First, the genetic material must be able to code for the sequence of amino acids in proteins and control protein synthesis. Second, it must be able to replicate itself prior to cell division. Third, the genetic material must be in the nuclei of eukaryotic cells. Fourth, it must be able to change over time to account for evolutionary change. Only one molecule, DNA (deoxyribonucleic acid), fulfills all of these requirements.

The Double Helix Model

Two kinds of molecules participate in protein synthesis. Both are based on a similar building block, the nucleotide, giving them their name–nucleic acids. One of these molecules, DNA, is the genetic material, and the other, RNA, is produced in the nucleus and moves to the cytoplasm, where it participates in protein synthesis. The study of how the information stored in DNA codes for RNA and protein is **molecular genetics.**

DNA and RNA are large molecules made up of subunits called **nucleotides** (figure 3.8). A nucleotide consists of a nitrogen-containing organic base in the form of either a double ring **(purine)** or a single ring **(pyrimidine).** Nucleotides also contain a pentose (five-carbon) sugar and a phosphate ($-PO_4$) group. DNA and RNA molecules, however, differ in several ways. Both DNA and RNA contain the purine bases adenine and guanine, and the pyrimidine base cytosine. The second pyrimidine in DNA, however, is thymine, whereas in RNA it is uracil. A second difference between DNA and RNA involves the sugar present in the nucleotides. The pentose of DNA is deoxyribose, and in RNA it is ribose. A third important difference between DNA and RNA is that DNA is a double-stranded molecule and RNA is single stranded, although it may fold back on itself and coil.

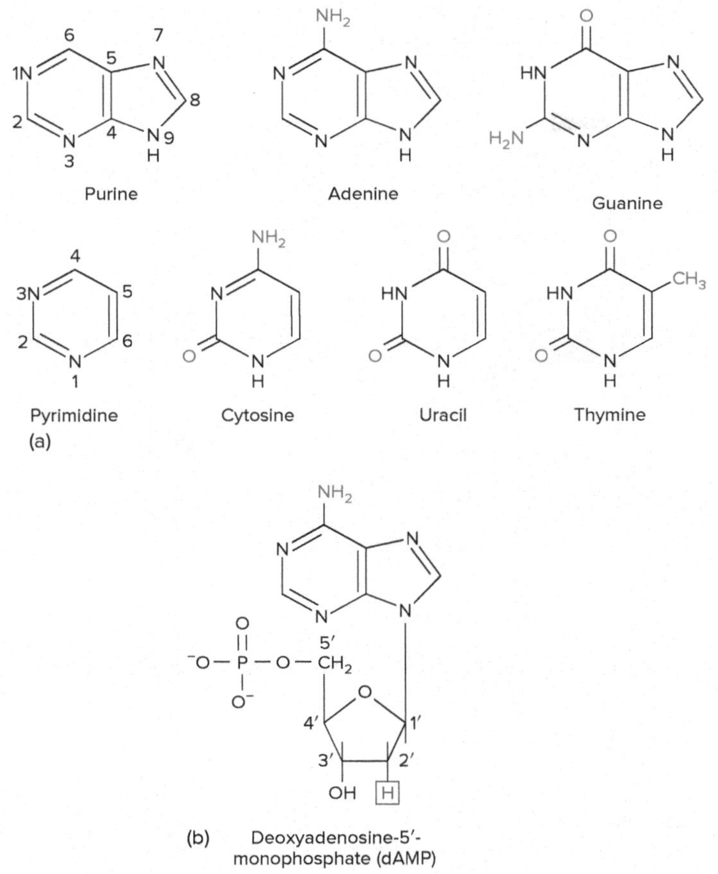

FIGURE 3.8

Components of Nucleic Acids. (*a*) The nitrogenous bases in DNA and RNA. (*b*) Nucleotides form by attaching a nitrogenous base to the 1′ carbon of a pentose sugar and attaching a phosphoric acid to the 5′ carbon of the sugar. (Carbons of the sugar are numbered with primes to distinguish them from the carbons of the nitrogenous base.) The sugar in DNA is deoxyribose, and the sugar in RNA is ribose. In ribose, a hydroxyl group (—OH) would replace the hydrogen enclosed by the purple box.

The key to understanding the function of DNA is knowing how nucleotides link into a three-dimensional structure. The DNA molecule is ladderlike, with the rails of the ladder consisting of alternating sugar-phosphate groups (figure 3.9*a*). The phosphate of a nucleotide attaches at the fifth (5′) carbon of deoxyribose. Adjacent nucleotides attach to one another by a covalent bond between the phosphate of one nucleotide and the third (3′) carbon of deoxyribose. The pairing of nitrogenous bases between strands holds the two strands together. Adenine (a purine) is hydrogen bonded to its complement, thymine (a pyrimidine), and guanine (a purine) is hydrogen bonded to its complement, cytosine (a pyrimidine) (*see figure 3.9*a). Each strand of DNA is oriented such that the 3′ carbons of deoxyribose in one strand are oriented in the opposite directions from the 3′ carbons in the other strand. Thus, the two strands of DNA have opposite polarity and the DNA molecule is said to be **antiparallel** (Gr. *anti,* against + *para,* beside + *allelon,* of one another). The entire molecule is twisted into a right-handed helix, with one complete spiral every 10 base pairs (figure 3.9*b*).

DNA Replication in Eukaryotes

In order for cells to pass genes to a subsequent generation of cells during mitotic cell division, or between generations of individuals through meiotic cell division and sexual reproduction, DNA must be faithfully copied in the S-phase of interphase. During DNA replication, each DNA strand is a template for a new strand. Replication begins simultaneously at many initiation sites along the length of a chromosome and involves a suite of enzymes. Replication begins in both directions from an initiation site, and each original strand serves as a template for the synthesis of a new strand (figure 3.10). Enzymes unwind the double helix, synthesize RNA primers needed for polymerase enzymes, relieve torque in the untwisting parent molecule, link DNA nucleotides (DNA polymerase), cut out primers and fill gaps (another DNA polymerase), and join ends of synthesized DNA segments. Because half of the old molecule is conserved in each of the two new molecules, DNA replication is said to be seimiconservative.

Animation
DNA Replication

Genes in Action

A gene can be defined as a sequence of bases in DNA that codes for the synthesis of one polypeptide, and genes must somehow transmit their information from the nucleus to the cytoplasm, where protein synthesis occurs. The synthesis of an RNA molecule from DNA is called **transcription** (L. *trans,* across + *scriba,* to write), and the formation of a protein from RNA at the ribosome is called **translation** (L. *trans,* across + *latere,* to remain hidden).

Three Major Kinds of RNA

Each of the three major kinds of RNA has a specific role in protein synthesis and is produced in the nucleus from DNA. **Messenger RNA (mRNA)** is a linear strand that carries a set of genetic instructions for synthesizing proteins to the cytoplasm. **Transfer RNA (tRNA)** picks up amino acids in the cytoplasm, carries them to ribosomes, and helps position them for incorporation into a polypeptide. **Ribosomal RNA (rRNA),** along with proteins, makes up ribosomes.

The Genetic Code

DNA must code for the 20 different amino acids found in all organisms. The information-carrying capabilities of DNA reside in the sequence of nitrogenous bases. The genetic code is a sequence of three bases—a triplet code. Figure 3.11 shows the genetic code as reflected in the mRNA that will be produced from DNA. Each three-base combination is a **codon.** More than one codon can specify the same amino acid because there are 64 possible codons, but only 20 amino acids. This characteristic of the code is referred to as

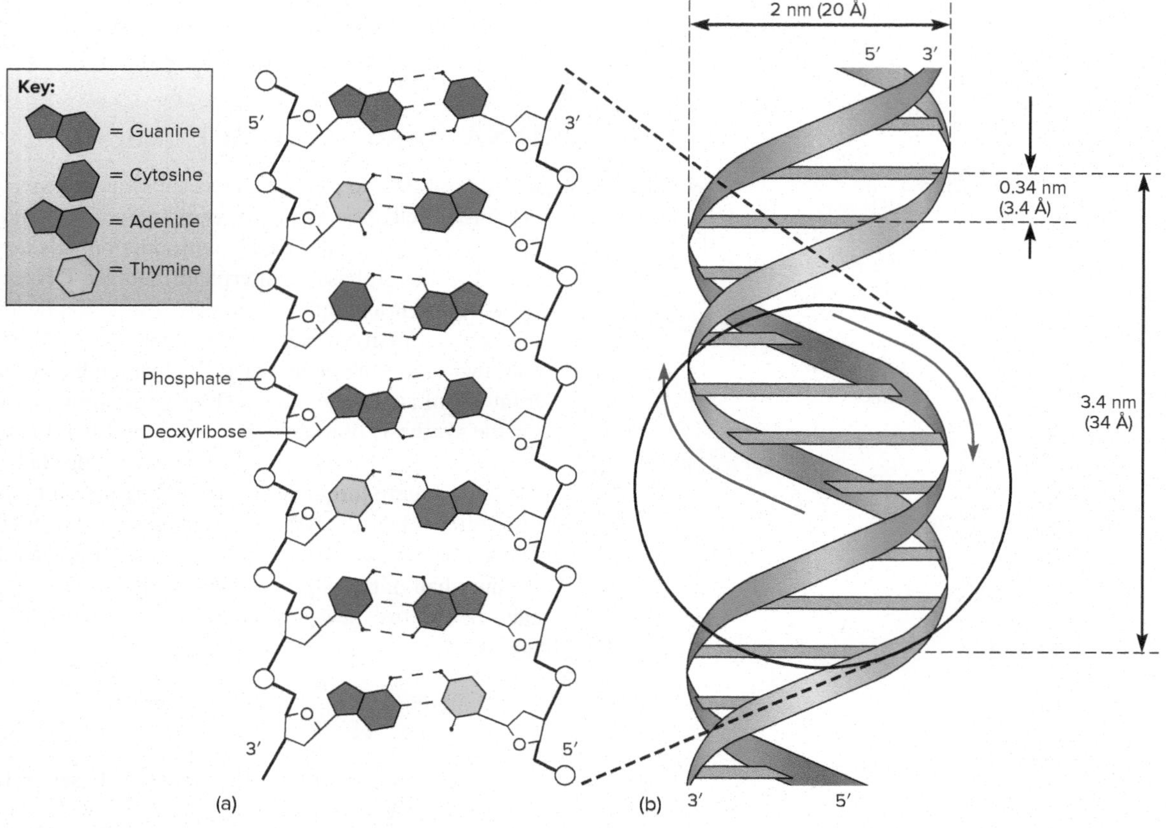

FIGURE 3.9

Structure of DNA. (*a*) Nucleotides of one strand of nucleic acid join by linking the phosphate of one nucleotide to the 3′ carbon of an adjacent nucleotide. Dashed lines between the nitrogenous bases indicate hydrogen bonds. Three hydrogen bonds are between cytosine and guanine, and two are between thymine and adenine. The antiparallel orientation of the two strands is indicated by using the 3′ and 5′ carbons at the ends of each strand. (*b*) Three-dimensional representation of DNA. The antiparallel nature of the strands is indicated by the curved arrows.

degeneracy. Note that not all codons code for an amino acid. The base sequences UAA, UAG, and UGA are all stop signals that indicate where polypeptide synthesis should end. The base sequence AUG codes for the amino acid methionine, which is a start signal.

Transcription

The genetic information in DNA is not translated directly into proteins, but is first transcribed into mRNA. Transcription involves numerous enzymes that unwind a region of a DNA molecule, initiate and end mRNA synthesis, and modify the mRNA after transcription is complete. Unlike DNA replication, only one or a few genes are exposed, and only one of the two DNA strands is transcribed (figure 3.12).

One of the important enzymes of this process is RNA polymerase. After a section of DNA is unwound, RNA polymerase recognizes a specific sequence of DNA nucleotides. RNA polymerase attaches and begins joining ribose nucleotides, which are complementary to the 3′ end of the DNA strand. In RNA, the same complementary bases in DNA are paired, except that in RNA, the base uracil replaces the base thymine as a complement to adenine.

Newly transcribed mRNA, called the primary transcript, must be modified before leaving the nucleus to carry out protein synthesis. Some base sequences in newly transcribed mRNA do not code for proteins. RNA splicing involves cutting out noncoding regions so that the mRNA coding region can be read continuously at the ribosome.

Translation

Translation is protein synthesis at the ribosomes in the cytoplasm, based on the genetic information in the transcribed mRNA. Another type of RNA, called transfer RNA (tRNA), is important in the translation process (figure 3.13). It brings the different amino acids coded for by the mRNA into alignment so that a polypeptide can be made. Complementary pairing of bases across the molecule maintains tRNA's configuration. The presence of some unusual bases (i.e., other than adenine, cytosine, guanine, or uracil) disrupts the normal base pairing and forms loops in the molecule. The center loop (the "anticodon loop") has a sequence of three unpaired bases called the **anticodon.** During translation, pairing of the mRNA codon with its complementary anticodon of tRNA appropriately positions the amino acid that tRNA carries.

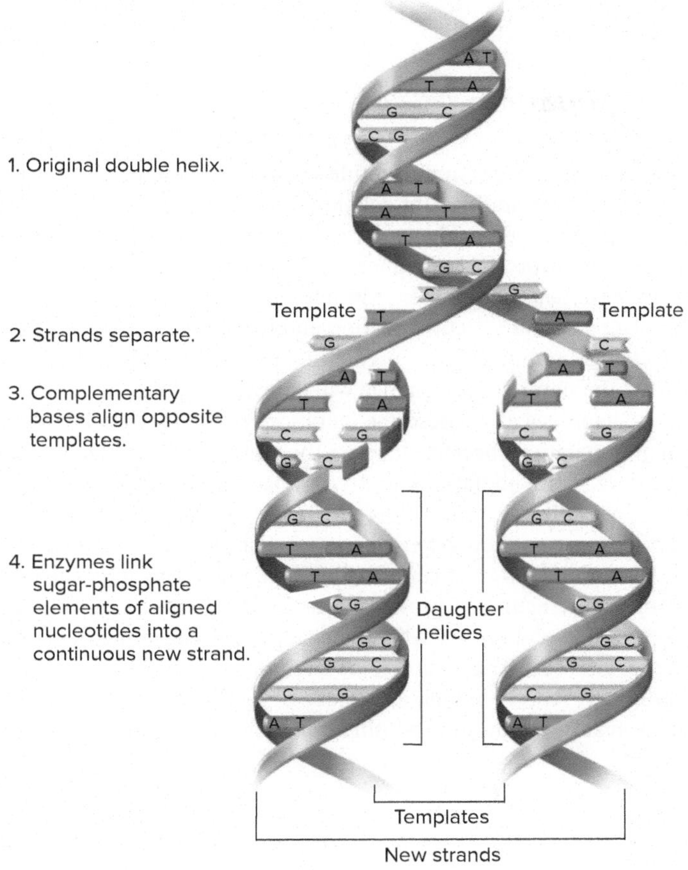

1. Original double helix.

2. Strands separate.

3. Complementary bases align opposite templates.

4. Enzymes link sugar-phosphate elements of aligned nucleotides into a continuous new strand.

Template

Template

Daughter helices

Templates

New strands

FIGURE 3.10

DNA Replication. (*1*) Replication begins. Only one direction is shown here for simplicity. (*2*) An enzyme causes the double helix to unwind and the two strands to separate. Each original strand serves as a template for the synthesis of a new strand. (*3*) Other enzymes help align nucleotides with exposed, unpaired bases on the unwound portions of the original DNA and (*4*) link nucleotides into continuous new strands.

Second position

First position	U	C	A	G	Third position
U	UUU Phe	UCU Ser	UAU Tyr	UGU Cys	U
	UUC Phe	UCC Ser	UAC Tyr	UGC Cys	C
	UUA Leu	UCA Ser	UAA STOP	UGA STOP	A
	UUG Leu	UCG Ser	UAG STOP	UGG Trp	G
C	CUU Leu	CCU Pro	CAU His	CGU Arg	U
	CUC Leu	CCC Pro	CAC His	CGC Arg	C
	CUA Leu	CCA Pro	CAA Gln	CGA Arg	A
	CUG Leu	CCG Pro	CAG Gln	CGG Arg	G
A	AUU Ile	ACU Thr	AAU Asn	AGU Ser	U
	AUC Ile	ACC Thr	AAC Asn	AGC Ser	C
	AUA Ile	ACA Thr	AAA Lys	AGA Arg	A
	AUG Met	ACG Thr	AAG Lys	AGG Arg	G
G	GUU Val	GCU Ala	GAU Asp	GGU Gly	U
	GUC Val	GCC Ala	GAC Asp	GGC Gly	C
	GUA Val	GCA Ala	GAA Glu	GGA Gly	A
	GUG Val	GCG Ala	GAG Glu	GGG Gly	G

Abbreviation	Amino acid	Abbreviation	Amino acid
Ala	Alanine	Leu	Leucine
Arg	Arginine	Lys	Lysine
Asn	Aparagine	Met	Methionine
Asp	Aspartic acid	Phe	Phenylalanine
Cys	Cysteine	Pro	Proline
Gln	Glutamine	Ser	Serine
Glu	Glutamic acid	Thr	Threonine
Gly	Glycine	Trp	Tryptophan
His	Histidine	Tyr	Tyrosine
Ile	Isoleucine	Val	Valine

FIGURE 3.11

Genetic Code. Sixty-four mRNA codons are shown here. The first base of the triplet is on the left side of the figure, the second base is at the top, and the third base is on the right side. The abbreviations for amino acids are also shown. In addition to coding for the amino acid methionine, the AUG codon is the initiator codon. Three codons–UAA, UAG, and UGA–do not code for an amino acid but act as a signal to stop protein synthesis.

Ribosomes, the sites of protein synthesis, consist of large and small subunits that organize the pairing between the codon and the anticodon. Several sites on the ribosome are binding sites for mRNA and tRNA. At the initiation of translation, mRNA binds to a small, separate ribosomal subunit. Attachment of the mRNA requires that the initiation codon (AUG) of mRNA be aligned with the P (peptidyl) site of the ribosome. A tRNA with a complementary anticodon for methionine binds to the mRNA, and a large subunit joins, forming a complete ribosome.

Polypeptide formation can now begin. Another site, the A (aminoacyl) site, is next to the P site. A second tRNA, whose anticodon is complementary to the codon in the A site, is positioned. Two tRNA molecules with their attached amino acids are now side-by-side in the P and A sites (figure 3.14). This step requires enzyme aid and energy, in the form of guanine triphosphate (GTP). An enzyme (peptidyl transferase), which is actually a part of the larger ribosomal subunit, breaks the bond between the amino acid and tRNA in the P site, and catalyzes the formation of a peptide bond between that amino acid and the amino acid in the A site.

The mRNA strand then moves along the ribosome a distance of one codon. The tRNA with two amino acids attached to it that was in the A site is now in the P site. A third tRNA can now enter the exposed A site. This process continues until the entire mRNA has been translated, and a polypeptide chain has been synthesized. Translation ends when a termination codon (e.g., UAA) is encountered.

Protein synthesis often occurs on ribosomes on the surface of the rough endoplasmic reticulum (*see figure 2.19*). The positioning of ribosomes on the ER allows proteins to move into the ER as the protein is being synthesized. The protein can then be moved to the Golgi apparatus for packaging into a secretory vesicle or into a lysosome.

Changes in DNA and Chromosomes

The genetic material of a cell can change, and these changes increase genetic variability and help increase the likelihood of survival in changing environments. These changes include alterations in the base sequence of DNA and changes that alter the structure or number of chromosomes.

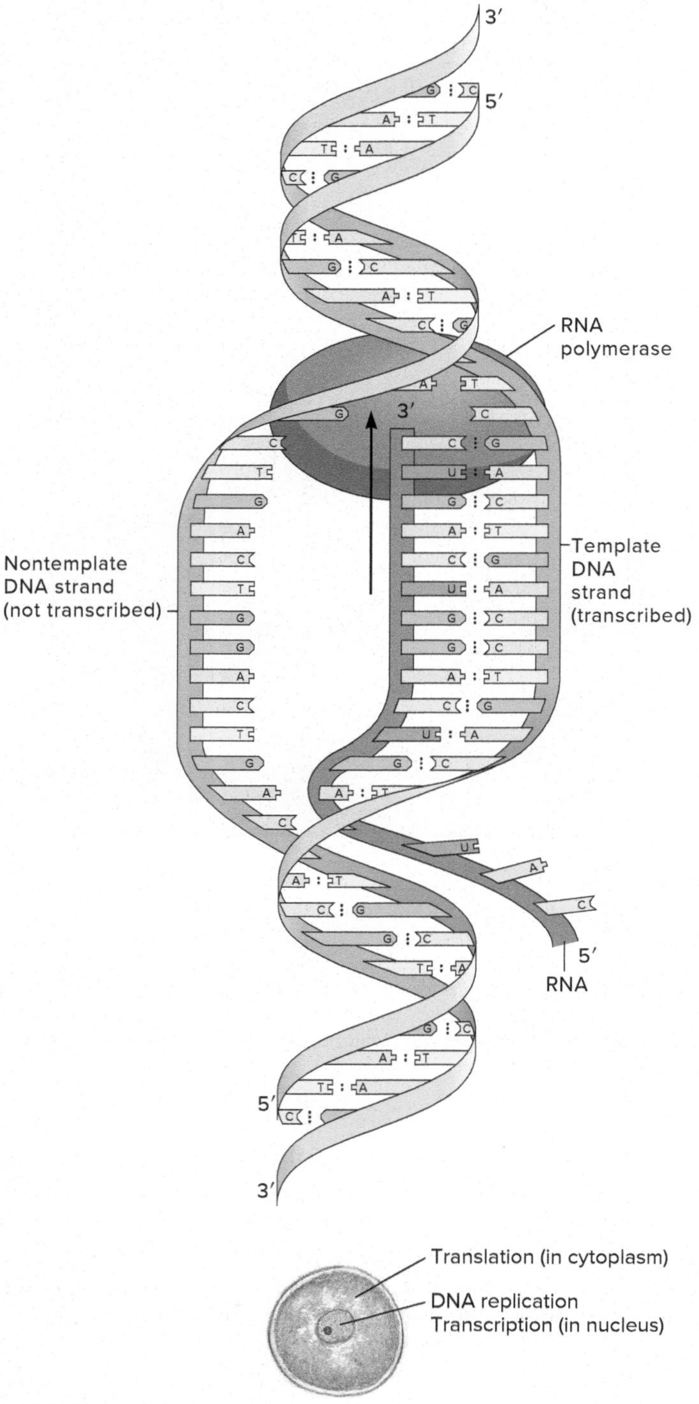

FIGURE 3.12

Transcription. Transcription involves the production of an mRNA molecule from the DNA segment. Note that transcription is similar to DNA replication in that the molecule is synthesized in the 5′ to 3′ direction.

Point Mutations

Genetic material must account for evolutionary change. **Point mutations** are changes in nucleotide sequences and may result from the replacement, addition, or deletion of nucleotides. Mutations are always random events. They may occur spontaneously as a result of base-pairing errors during replication, which result in a substitution of one base pair for another. Although certain environmental factors (e.g., electromagnetic radiation and many chemical mutagens) may change mutation rates, predicting what genes will be affected or what the nature of the change will be is impossible.

Chapters 4 and 5 describe mutations as the fuel for the evolution of populations because they are the only source for new genetic variations. Point mutations and crossing-over are two sources of genetic variations covered thus far in this chapter. Mutations are the only source of new genetic material. For individuals, mutations can be a source of great suffering because mutations in genes that disturb the structure of proteins that are the products of millions of years of evolution are usually negative and cause many of our genetic diseases. The majority of mutations arise in body cells. These often remain hidden and cause no problems for the individual because either they are within a gene that is not being expressed

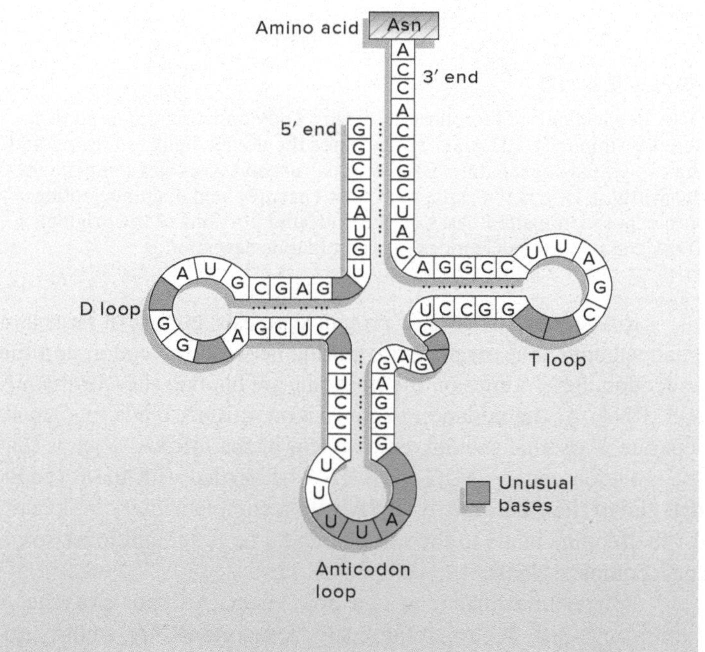

FIGURE 3.13

Structure of Transfer RNA. Diagrammatic representation of the secondary structure of transfer RNA (tRNA). An amino acid attaches to the 3′ end of the molecule. The anticodon is the sequence of three bases that pairs with the codon in mRNA, thus positioning the amino acid that tRNA carries. Other aspects of tRNA structure position the tRNA at the ribosome and in the enzyme that attaches the correct amino acid to the tRNA.

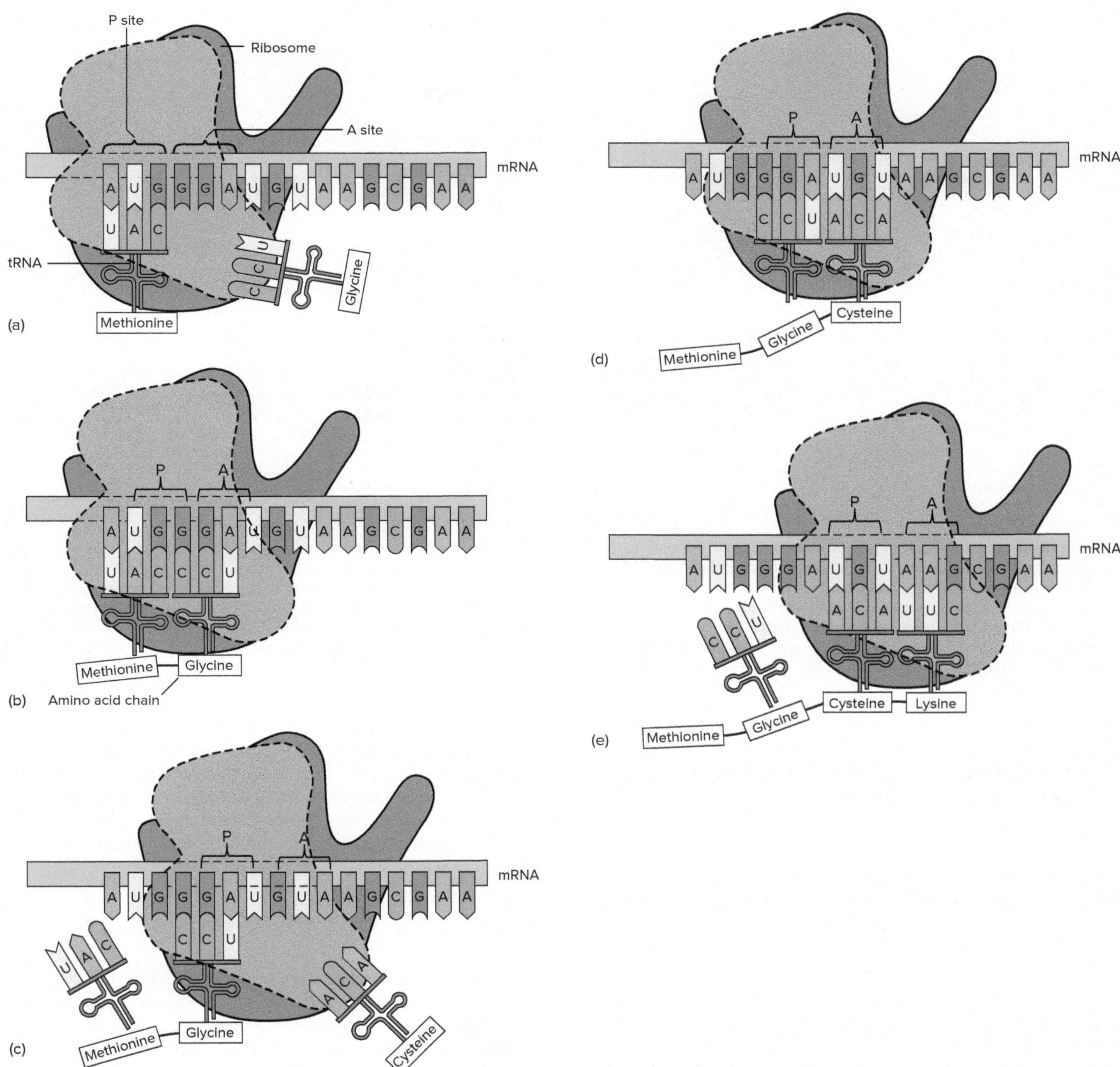

FIGURE 3.14

Events of Translation. (*a*) Translation begins when a methionine tRNA associates with the P site of the smaller ribosomal subunit and the initiation codon of mRNA associated with that subunit. The larger ribosomal subunit attaches to the small subunit/tRNA complex. (*b*) A second tRNA carrying the next amino acid enters the A site. A peptide bond forms between the two amino acids, freeing the first tRNA in the P site. (*c*) The mRNA, along with the second tRNA and its attached dipeptide, moves the distance of one codon. The first tRNA is discharged, leaving its amino acid behind. The second tRNA is now in the P site, and the A site is exposed and ready to receive another tRNA-amino acid. (*d*) A second peptide bond forms. (*e*) This process continues until an mRNA stop signal is encountered.

Animation Translation of mRNA

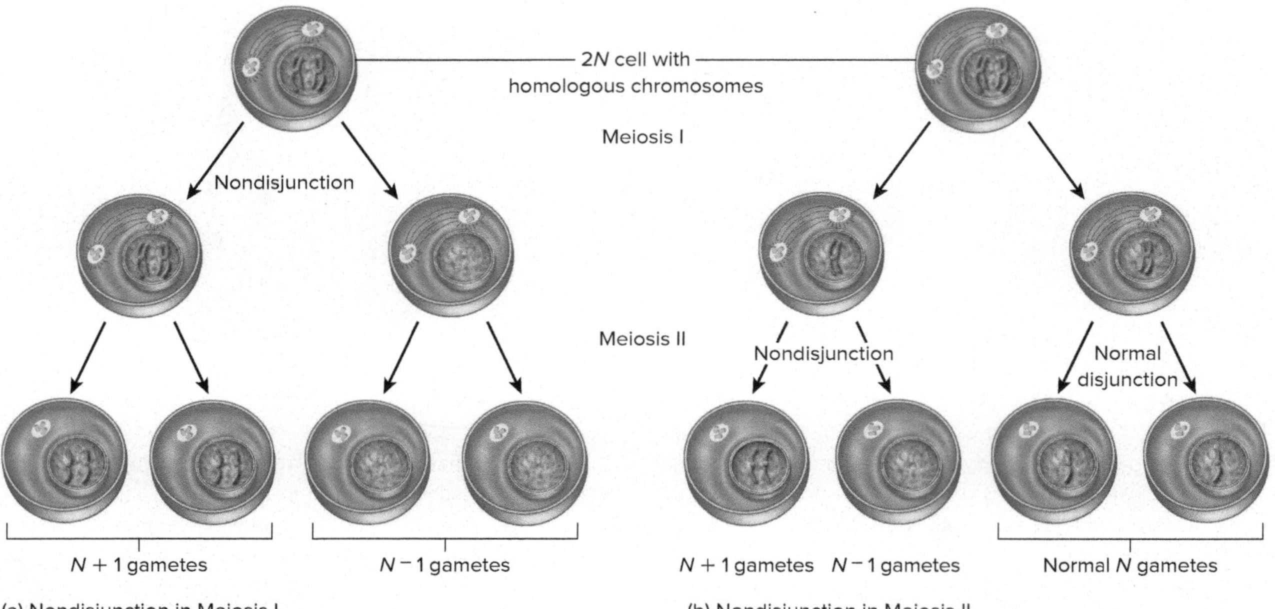

FIGURE 3.15

Results of Primary and Secondary Nondisjunction in Gamete Formation. (*a*) Primary nondisjunction occurs in meiosis I and results from the failure of homologous pairs to separate normally. Both members of the homologous pair of chromosomes end up in one cell. A normal second meiotic division results in half of all gametes having both members of the homologous pair of chromosomes ($N + 1$). The other half of all gametes lacks members of this pair of homologous chromosomes ($N - 1$). (*b*) Secondary nondisjunction occurs after a normal first meiotic division. The failure of chromatids of one chromosome to separate in the second division means that a fourth of the gametes will be missing a member of one homologous pair ($N - 1$), and a fourth of the gametes will have an extra member of that homologous pair ($N + 1$). This illustration assumes that the second cell that resulted from meiosis I undergoes a normal second meiotic division.

in the cell or they may be altering the structure of DNA that is not coding for a protein. We all harbor hundreds of millions of these somatic mutations. The only mutations that affect future generations are those that arise in germ cells of the testes or ovaries.

Variation in Chromosome Number

Changes in chromosome number may involve entire sets of chromosomes, as in polyploidy, which was discussed earlier. **Aneuploidy** (Gr. *a,* without), on the other hand, involves the addition or deletion of one or more chromosomes, not entire sets. The addition of one chromosome to the normal $2N$ chromosome number ($2N + 1$) is a trisomy (Gr. *tri,* three + ME *some,* a group of), and the deletion of a chromosome from the normal $2N$ chromosome number ($2N - 1$) is a monosomy (Gr. *monos,* single).

Errors during meiosis usually cause aneuploidy. **Nondisjunction** occurs when a homologous pair fails to segregate during meiosis I or when chromatids fail to separate at meiosis II (figure 3.15). Gametes produced either lack one chromosome or have an extra chromosome. If one of these gametes is involved in fertilization with a normal gamete, the monosomic or trisomic condition results (*see male calico cats, page 33*). Aneuploid variations usually result in severe consequences involving mental retardation and sterility.

Variation in Chromosome Structure

Some changes may involve breaks in chromosomes. After breaking, pieces of chromosomes may be lost, or they may reattach, but not necessarily in their original position. The result is a chromosome that may have a different sequence of genes, multiple copies of genes, or missing genes. All of these changes can occur spontaneously. Various environmental agents, such as ionizing radiation and certain chemicals, can also induce these changes. The effects of changes in chromosome structure may be mild or severe, depending on the amount of genetic material duplicated or lost.

SECTION 3.4 THINKING BEYOND THE FACTS

One strand of DNA has the base sequence 3′AGTGCATTC5′. Write the sequence of bases in the second strand. Using the strand provided here as a template, show the mRNA produced in transcription and the sequence of amino acids produced in translation.

3.5 INHERITANCE PATTERNS IN ANIMALS

LEARNING OUTCOMES

1. Explain and apply Mendelian principles.
2. Predict the results of crosses involving incompletely dominant and codominant alleles.
3. Relate dominance concepts to the molecular basis of inheritance.

Classical genetics began with the work of Gregor Mendel and remains an important basis for understanding gene transfer between generations of animals. Understanding these genetics principles helps us predict how traits will be expressed in offspring before these offspring are produced, something that has had profound implications in wildlife conservation, agriculture, and medicine. One of the challenges of modern genetics is to understand the molecular basis for these inheritance patterns.

The fruit fly, *Drosophila melanogaster,* is a classic tool for studying inheritance patterns. Its utility stems from its ease of handling, short life cycle, and easily recognized characteristics.

Studies of any fruit-fly trait always make comparisons to a wild-type fly. If a fly has a characteristic similar to that found in wild flies, it is said to have the wild-type expression of that trait. (In the examples that follow, wild-type wings lay over the back at rest and extend past the posterior tip of the body, and wild-type eyes are red.) Numerous mutations from the wild-type body form, such as vestigial wings (reduced, shriveled wings) and sepia (dark brown) eyes, have been described (figure 3.16).

Segregation

During gamete formation, genes in each parent are incorporated into separate gametes. During anaphase I of meiosis, homologous chromosomes move toward opposite poles of the cell, and the resulting gametes have only one member of each chromosome pair. Genes carried on one member of a pair of homologous chromosomes end up in one gamete, and genes carried on the other member are segregated into a different gamete. The **principle of segregation** states that pairs of genes are distributed between gametes during gamete formation. Fertilization results in the random combination of gametes and brings homologous chromosomes together again.

A cross of wild-type fruit flies with flies having vestigial wings illustrates the principle of segregation. (The flies come from stocks that have been inbred for generations to ensure that they breed true for wild-type wings or vestigial wings.) The offspring (progeny) of this cross have wild-type wings and are the first generation of offspring, or the first filial (F_1) generation (figure 3.17). If these flies are allowed to mate with each other, their progeny are the second filial (F_2) generation. Approximately a fourth of these F_2 generation of flies have vestigial wings, and three-fourths have wild-type wings (*see figure 3.17*). Note that the vestigial characteristic, although present in the parental generation, disappears in the F_1 generation and reappears in the F_2 generation. In addition, the ratio of wild-type flies to vestigial-winged flies in the F_2 generation is approximately 3:1. Reciprocal crosses, which involve the same characteristics but a reversal of the sexes of the individuals introducing a particular expression of the trait into the cross, yield similar results.

Genes that determine the expression of a particular trait can exist in alternative forms called **alleles** (Gr. *allelos,* each other). In the fruit-fly cross, the vestigial allele is present in the F_1 generation, and even though it is masked by the wild-type allele for wing shape, it retains its uniqueness because it is expressed again in some members of the F_2 generation. **Dominant** alleles hide the expression of another allele; **recessive** alleles are those whose expression can be masked. In the fruit-fly example, the wild-type allele is dominant

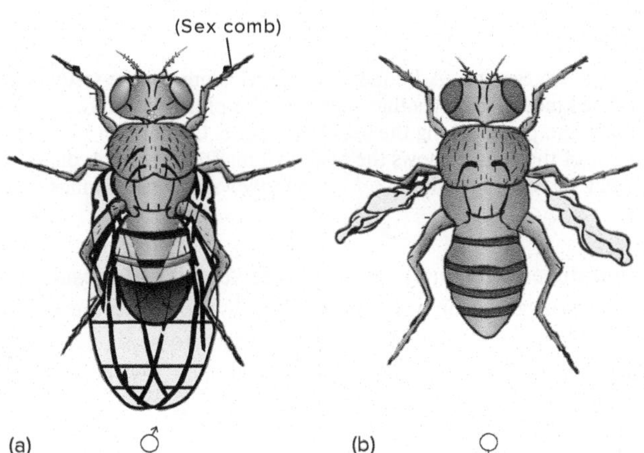

FIGURE 3.16

Distinguishing Sexes and Phenotypes of *Drosophila melanogaster.* (*a*) Male with wild-type wings and wild-type eyes. (*b*) Female with vestigial wings and sepia eyes. In contrast to the female, the posterior aspect of the male's abdomen has a wide, dark band and a rounded tip.

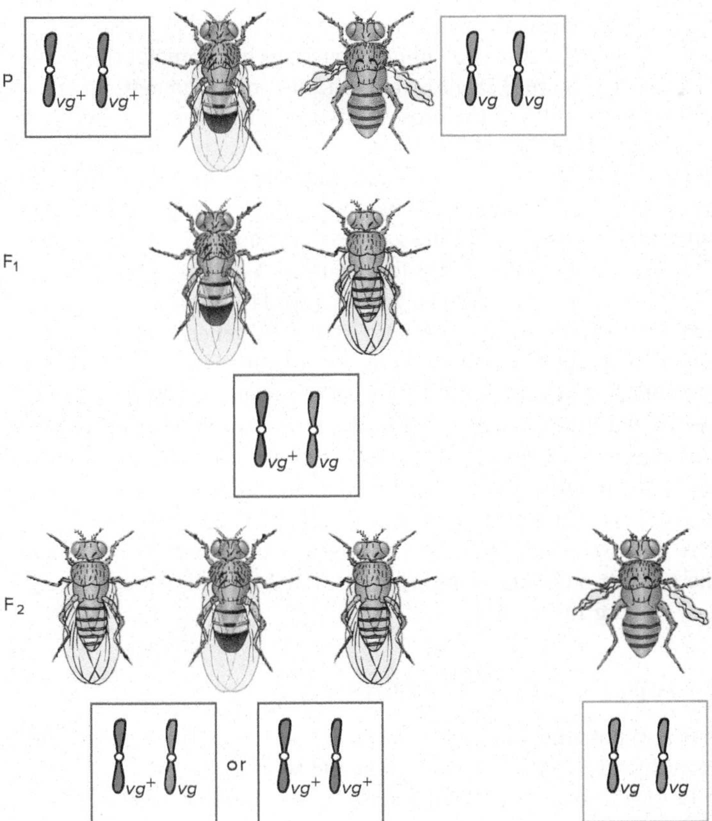

FIGURE 3.17

Cross Involving a Single Trait. Cross between parental flies (P) with wild-type (vg^+) wings and vestigial (vg) wings, carried through two generations (F_1 and F_2).

because it can mask the expression of the vestigial allele, which is therefore recessive.

The visual expression of alleles may not always indicate the underlying genetic makeup of an organism. This visual expression is the **phenotype,** and the genetic makeup is the **genotype.** In the example, the flies of the F_1 generation have the same phenotype as one of the parents, but they differ genotypically because they carry both a dominant and recessive allele. They are hybrids, and because this cross concerns only one pair of genes and a single trait, it is a **monohybrid cross** (Gr. *monos,* one + L. *hybrida,* offspring of two kinds of parents).

An organism is **homozygous** (L. *homo,* same + Gr. *zygon,* paired) if it carries two identical genes for a given trait and **heterozygous** (Gr. *heteros,* other) if the genes are different (alleles of each other). Thus, in the example, all members of the parental generation are homozygous because only truebreeding flies are crossed. All members of the F_1 generation are heterozygous.

Crosses are often diagrammed using a letter or letters descriptive of the trait in question. The first letter of the description of the dominant allele commonly is used. In fruit flies, and other organisms where all mutants are compared with a wild-type, the symbol is taken from the allele that was derived by a mutation from the wild condition. A superscript "+" next to the symbol represents the wild-type allele. A capital letter means that the mutant allele being represented is dominant, and a lowercase letter means that the mutant allele being represented is recessive.

Geneticists use the Punnett square to help predict the results of crosses. Figure 3.18 illustrates the use of a **Punnett square** to predict the results of the cross of two F_1 flies. The first step is to determine the kinds of gametes that each parent produces. One of the two axes of a square is designated for each parent, and the different kinds of gametes each parent produces are listed along the appropriate axis. Combining gametes in the interior of the square shows the results of random fertilization. As figure 3.18 indicates, the F_1 flies are heterozygous, with one wild-type allele and one vestigial allele. The two phenotypes of the F_2 generation are shown inside the Punnett square and are in a 3:1 ratio.

The **phenotypic ratio** expresses the results of a cross according to the relative numbers of progeny in each visually distinct class (e.g., 3 wild-type:1 vestigial). The Punnett square has thus explained in another way the F_2 results in figure 3.17. It also shows that F_2 individuals may have one of three different genotypes. The **genotypic ratio** expresses the results of a cross according to the relative numbers of progeny in each genotypic category (e.g., 1 vg^+vg^+:2 vg^+vg:1 $vgvg$).

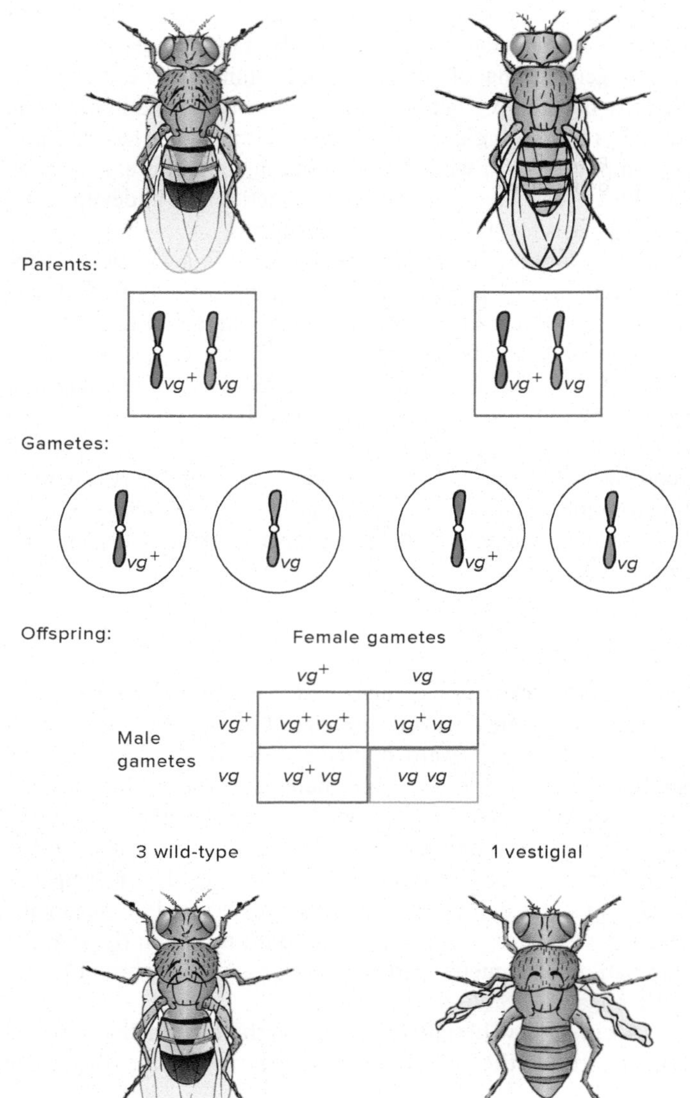

FIGURE 3.18

Use of a Punnett Square. A Punnett square helps predict the results of a cross. The kinds of gametes that each member of a cross produces are determined and placed along the axes of a square. Combining gametes in the interior of the square shows the results of mating: a phenotypic ratio of three flies with wild-type wings (vg^+) to one fly with vestigial wings (vg).

Independent Assortment

It is also possible to make crosses using flies with two pairs of characteristics: flies with vestigial wings and sepia eyes, and flies that are wild for these characteristics. Sepia eyes are dark brown, and wild-type eyes are red. Figure 3.19 shows the results of crosses carried through two generations.

Note that flies in the parental generation are homozygous for the traits in question and that each parent produces only one kind of gamete. Gametes have one allele for each trait. Because each parent produces only one kind of gamete, fertilization results in offspring heterozygous for both traits. The F_1 flies have the wild-type phenotype; thus, wild-type eyes are dominant to sepia eyes. The F_1 flies are hybrids, and because the cross involves two pairs of genes and two traits, it is a **dihybrid cross** (Gr. *di,* two + L. *hybrida,* offspring of two kinds of parents).

The 9:3:3:1 ratio is typical of a dihybrid cross. During gamete formation, the distribution of genes determining one trait does not influence how genes determining the other trait are distributed. In the example, this means that an F_1 gamete with a vg^+ gene for wing condition may also have either the *se* gene or the

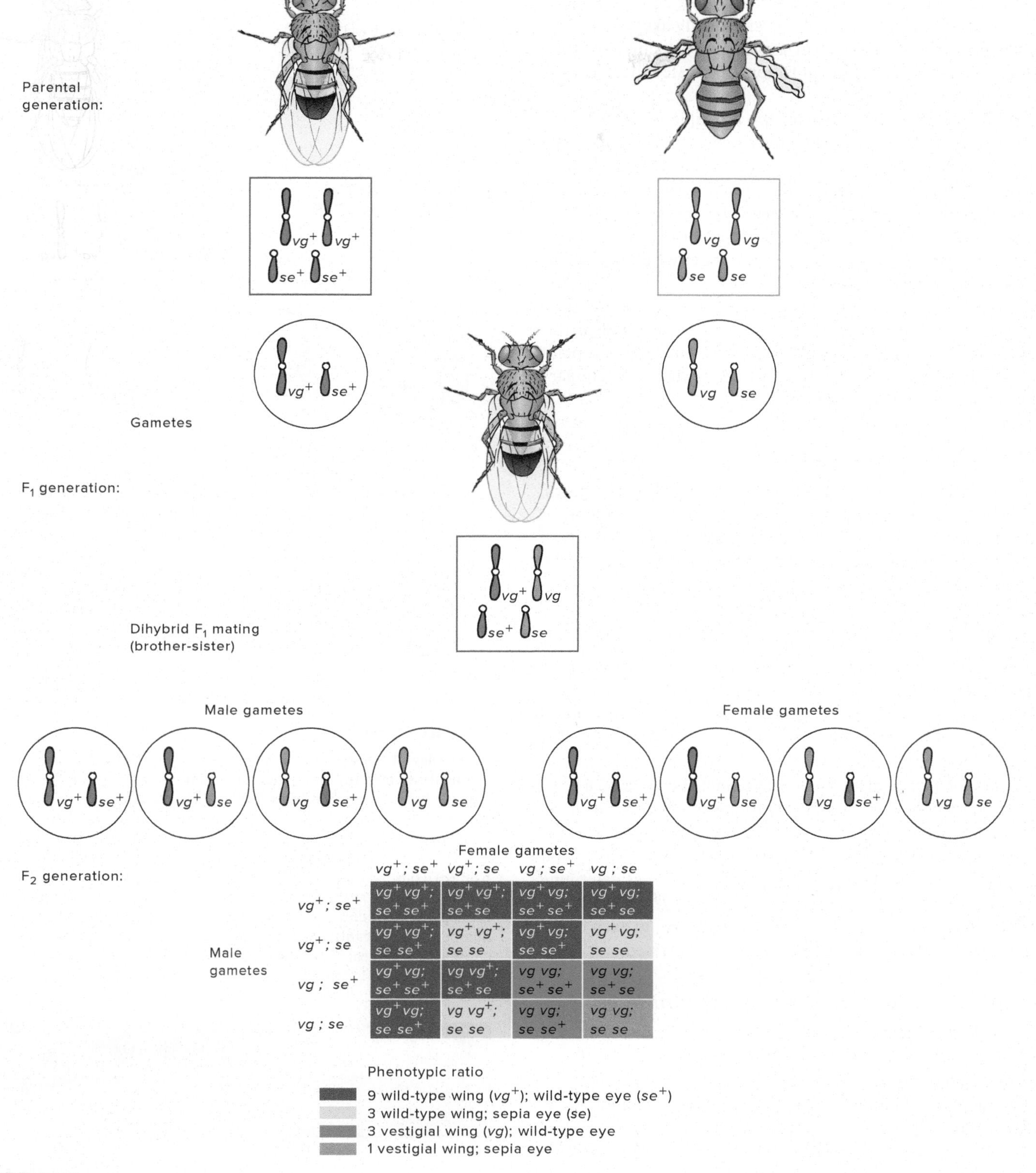

	vg+; se+	vg+; se	vg ; se+	vg ; se
vg+; se+	vg+ vg+; se+ se+	vg+ vg+; se+ se	vg+ vg; se+ se+	vg+ vg; se+ se
vg+; se	vg+ vg+; se se+	vg+ vg+; se se	vg+ vg; se se+	vg+ vg; se se
vg ; se+	vg+ vg; se+ se+	vg vg+; se+ se	vg vg; se+ se+	vg vg; se+ se
vg ; se	vg+ vg; se se+	vg vg+; se se	vg vg; se se+	vg vg; se se

FIGURE 3.19

Constructing a Punnett Square for a Cross Involving Two Characteristics. Note that every gamete has one allele for each trait and that all combinations of alleles for each trait are represented.

se^+ gene for eye color, as the F_1 gametes of figure 3.19 show. Note that all combinations of the eye color and wing condition genes are present, and that all combinations are equally likely. This illustrates the **principle of independent assortment,** which states that, during gamete formation, pairs of factors segregate independently of one another.

The events of meiosis explain the principle of independent assortment (*see figure 3.6*). Cells produced during meiosis have one member of each homologous pair of chromosomes. Independent assortment simply means that when homologous chromosomes line up at metaphase I and then segregate, the behavior of one pair of chromosomes does not influence the behavior of any other pair (figure 3.20). After meiosis, maternal and paternal chromosomes are distributed randomly among cells.

This independent assortment of maternal and paternal chromosomes is the third source of genetic variation covered in this chapter. Independent assortment as well as crossing-over and point mutations provide the genetic variation upon which evolutionary processes act (*see chapters 4 and 5*).

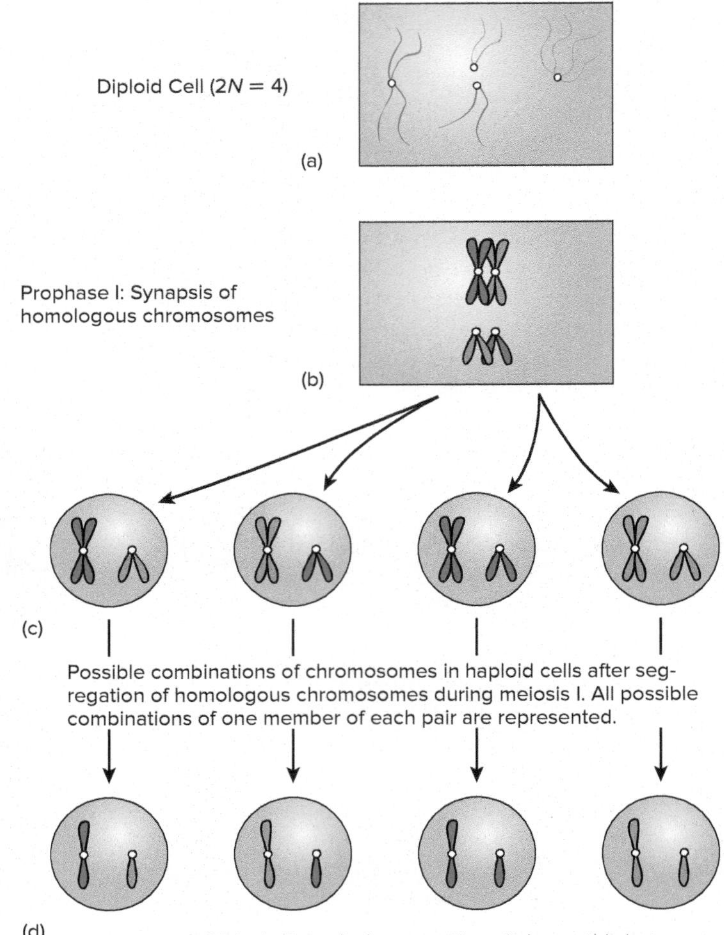

FIGURE 3.20

Independent Assortment of Chromosomes during Meiosis. Color distinguishes maternal and paternal chromosomes. Similar size and shape indicate homologous chromosomes. (*a*) This cell has a diploid (2*N*) chromosome number of four. (*b*) During the first meiotic division, one homologous pair of chromosomes (and hence, the genes this pair carries) is segregated without regard to the movements of any other homologous pair. (*c*) Thus, all combinations of large and small chromosomes in the cells are possible at the end of meiosis I. (*d*) Meiosis II simply results in the separation of chromatids without further reduction in chromosome number. Most organisms have more than two pairs of homologous chromosomes in each cell. As the number of homologous pairs increases, the number of different kinds of gametes also increases.

Other Inheritance Patterns

The traits considered thus far have been determined by two genes, where one allele is dominant to a second. In this section, you learn that there are often many alleles in a population and that not all traits are determined by an interaction between a single pair of dominant or recessive genes.

Multiple Alleles

Two genes, one carried on each chromosome of a homologous pair, determine traits in one individual. A population, on the other hand, may have many different alleles with the potential to contribute to the phenotype of any member of the population. These are called **multiple alleles.**

Genes for a particular trait are at the same position on a chromosome. The gene's position on the chromosome is called its **locus** (L. *loca,* place). Numerous human loci have multiple alleles. Three alleles, symbolized I^A, I^B, and *i,* determine the familiar ABO blood types. Table 3.1 shows the combinations of alleles that determine a person's phenotype. Note that *i* is recessive to I^A and to I^B. I^A and I^B, however, are neither dominant nor recessive to each other. When I^A and I^B are present together, both are expressed.

Incomplete Dominance and Codominance

Incomplete dominance is an interaction between two alleles that are expressed more or less equally, and the heterozygote is different from either homozygote. For example, in cattle, the alleles for red coat color and for white coat color interact to produce an intermediate coat color called roan. Because neither the red nor the white allele is dominant, uppercase letters and a prime or a superscript are used to represent genes. Thus, red cattle are symbolized *RR*, white cattle are symbolized *R′R′*, and roan cattle are symbolized *RR′*.

Codominance occurs when the heterozygote expresses the phenotypes of both homozygotes. Thus, in the ABO blood types, the I^AI^B heterozygote expresses both alleles.

TABLE 3.1
GENOTYPES AND PHENOTYPES IN THE ABO BLOOD GROUPS

GENOTYPE(S)	PHENOTYPE
I^AI^A, I^Ai	*A*
I^BI^B, I^Bi	*B*
I^AI^B	*A* and *B*
ii	*O*

The Molecular Basis of Inheritance Patterns

Just as the principles of segregation and independent assortment can be explained based on our knowledge of the events of meiosis, concepts related to dominance can be explained in molecular terms. When we say that one allele is dominant to another, we do not mean that the recessive allele is somehow "turned off" when the dominant allele is present. Instead, the product of a gene's function is the result of a sequence of metabolic steps mediated by enzymes, which are encoded by the gene(s) in question. A functional enzyme is usually encoded by a dominant gene, and when that enzyme is present a particular product is produced. A recessive allele usually arises by a mutation of the dominant gene, and the enzyme necessary for the production of the product is altered and does not function. In the homozygous dominant state, both dominant genes code for the enzyme that produces the product (figure 3.21*a*). In the heterozygous state, the activity of the single dominant allele is sufficient to produce enough enzyme to form the product and the dominant phenotype (figure 3.21*b*). In the homozygous recessive state, no product can be formed and the recessive phenotype results (figure 3.21*c*).

In the same way, one can explain incomplete dominance and codominance. In these cases both alleles of a heterozygous individual produce approximately equal quantities of two enzymes and products, and the phenotype that results would either be intermediate or show the products of both alleles.

Section 3.5 Thinking Beyond the Facts

What events of meiotic cell division are reflected in the principles of segregation and independent assortment?

Homozygous Dominant

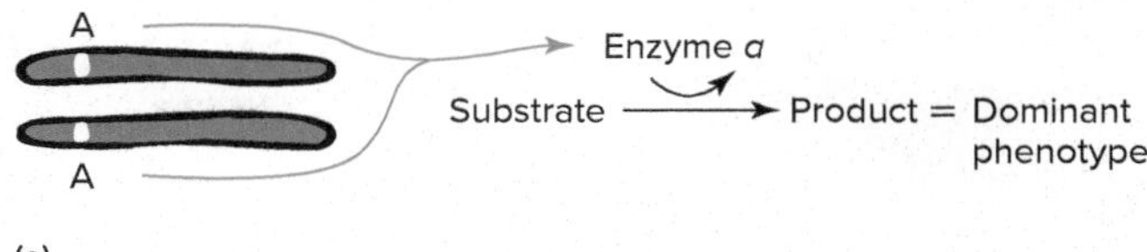

(a)

Heterozygous

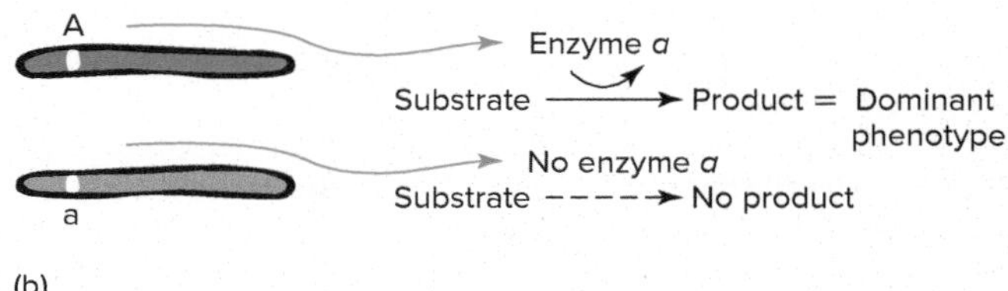

(b)

Homozygous Recessive

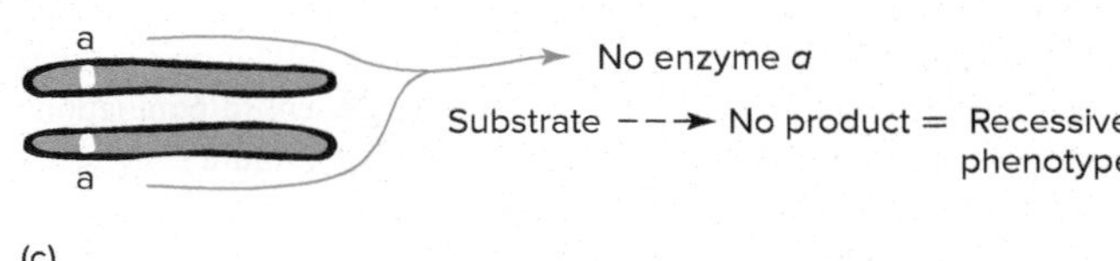

(c)

FIGURE 3.21

The Molecular Basis of Dominance. (*a*) In a homozygous dominant individual, both dominant genes code for enzymes that produce the product and the dominant phenotype. (*b*) In the heterozygous state, the single dominant allele is sufficient to produce enough enzyme to form the product and the dominant phenotype. (*c*) In the homozygous recessive state, no product can be formed and the recessive phenotype results.

How Do We Know—Inbreeding for Speed Hurts

California Chrome, Northern Dancer, Seattle Slew, Secretariat, and all the other horses in the Thoroughbred breed have been bred for speed. They all trace their genetic makeup back to three stallions in England in the early 1700s. They are fast, they command stud fees that have soared up to $1 million, and their foals sell for as much as $13 million. Inbreeding for speed may be killing the breed. Recent genetic analysis has revealed that Thoroughbreds are so genetically similar that they are almost clones of one another—a potentially dangerous situation.

Breeding for speed is not breeding for health. Large powerful muscles acting on slim legs and small hooves create legs that are more likely to break when hooves strike the track (Barbaro, 2006 Preakness Stakes and Eight Belles, 2008 Kentucky Derby) and leg bones that chip at the joints. The lungs of Thoroughbreds bleed and wheeze in an effort to get adequate air into the lungs. Inbreeding is also suspected as the cause of a higher rate of infertility and miscarriage in this breed and possibly a greater susceptibility to disease as compared to other breeds of horses. Apparently breeding for speed is not breeding for durability.

There has been an explosion of genetic research searching for the genetic basis of traits that have accumulated over centuries of inbreeding. The horse genome has been sequenced, and gene chips are being developed to screen horses for genetic defects. It is yet to be seen whether this information will be used to breed for health or faster finishes.

WILDLIFE ALERT

Preserving Genetic Diversity

One of the ways in which scientists evaluate the environmental health of a region is to assess the variety of organisms present in an area. Environments that have a great variety or diversity in species (biodiversity, *see chapter 6*) are usually considered healthier than environments with less diversity. Diversity can be reduced through habitat loss, the exploitation of animals or plants through hunting or harvesting, and the introduction of foreign species.

Another criterion used to evaluate environmental health is genetic diversity. Genetic diversity is the variety of alleles within a species. When a species on the brink of extinction is preserved, reduced genetic diversity within the species threatens the health of the species. Near-extinction events, in which many individuals die, eradicate many alleles from populations (*see figure 5.2*). Lowered numbers of individuals result in inbreeding, which also reduces genetic diversity. The result is that populations that survive near-extinction events tend to be genetically uniform. The effect of genetic uniformity on populations is nearly always detrimental because when environmental conditions change, entire populations can be adversely affected. For example, if one individual in a genetically uniform population is susceptible to a particular disease, all individuals will be susceptible, and the disease will spread very quickly. High genetic diversity improves the likelihood that some individuals will survive the disease outbreak, and the species will be less likely to face extinction. Since mutation is the ultimate source of new variation within species, lost genetic diversity can only be replaced over evolutionary timescales. For all practical purposes, when genetic diversity is lost, it is gone forever.

Conservation geneticists evaluate the genetic health of populations of organisms and try to preserve the genetic variation that exists within species. These efforts involve the use of virtually every genetic tool available to modern science, including the molecular techniques for studying DNA and the proteins of endangered organisms. Conservation geneticists search native populations for individuals that could be used to enhance the genetic makeup of endangered organisms. They recommend breeding programs to preserve alleles that could easily be lost. Many zoos throughout the world cooperate in breeding programs that exchange threatened animals, or gametes from threatened animals, to preserve alleles.

One example of a conservation program attempting to preserve an endangered species is focused on the snow leopard (*Panthera uncia*). Little is known of the historic population size of this species. Today there are between 4,000 and 6,500 snow leopards distributed throughout the mountains of central Asia. Snow leopards live at the snow line of mountains at altitudes between 2,000 and 5,000 meters. Climate change is causing the snow line to recede to higher altitudes, reducing snow leopard habitat. Poaching the snow leopard to supply its coat for black-market trade and bones and other body parts for use in traditional Asian medicine are serious threats to the cats that remain. Hunting wild prey and habitat destruction for farming and grazing livestock also threaten these cats (box figure 3.1). Unfortunately, when the blue sheep (*Pseudois nayaur*) and ibex (*Capra sibirica*) prey become scarce snow leopards prey on livestock, which often results in retribution killing of snow leopards by farmers.

There are encouraging steps being taken to preserve snow leopards. Conservation organizations are helping farmers and herders understand how to live with these cats by securing barns and livestock holding areas from these cats. Most countries that share the snow leopard's range have made hunting and trafficking of snow leopards illegal. National parks have been established throughout its range to protect snow leopards and their habitat, and zoos around the world are cooperating in helping to maintain genetic diversity of this majestic cat species.

BOX FIGURE 3.1 Snow leopard (*Panthera uncia*).

SUMMARY

3.1 **Eukaryotic Chromosomes**

- Eukaryotic chromosomes are complexly coiled associations of DNA and histone proteins.
- The presence or absence of certain chromosomes that are represented differently in males and females determines the sex of an animal. The X-Y system of sex determination is most common.

3.2 **The Cell Cycle and Mitotic Cell Division**

- The replication of DNA and its subsequent allocation to daughter cells during mitotic cell division involves a number of phases collectively called the cell cycle. The cell cycle is that period from the time a cell is produced until it completes mitosis.
  Mitosis maintains the parental number of chromosome sets in each daughter nucleus. It separates the sister chromatids of each (replicated) chromosome for distribution to daughter nuclei.
- Interphase represents about 90% of the total cell cycle. It includes periods of cell growth and normal cell function. It also includes the time when DNA is replicated.
- Mitosis is divided into five phases. During prophase, the mitotic spindle forms and the nuclear envelope disintegrates. During prometaphase the microtubules attach at one end to the kinetochore of a chromatid and at the opposite end to one pole of the cell. During metaphase, the replicated chromosomes align along the spindle equator. During anaphase, the centromeres joining sister chromatids divide and microtubules pull sister chromatids to opposite poles of the cell. During telophase, the mitotic spindle disassembles, the nuclear envelope re-forms, and chromosomes unfold.
- Cytokinesis, the division of the cytoplasm, begins in late anaphase and is completed in telophase.
- The cell cycle must be precisely controlled in response to DNA damage or environmental factors. These controls involve cyclin proteins and kinase enzymes that interact with cyclins. There are three cell-cycle checkpoints (G_1, G_2, and M-phase) where the cell cycle can be halted.

3.3 **Meiosis: The Basis of Sexual Reproduction**

- Meiosis is a special form of nuclear division that results in the formation of haploid (1*N*) gametes. Gamete formation involves two meiotic and cytoplasmic divisions. In the first division, homologous pairs of chromosomes undergo synapsis, including crossing-over, followed by the separation of members of each chromosome pair into gametes. This reduces the chromosome number to the haploid condition in two daughter cells.
- The second division is similar to mitotic cell division. Fertilization restores the diploid (2*N*) chromosome number in the zygote.
- In the life cycle of most animals, germ-line cells undergo gametogenesis to form haploid gametes (sperm in males and eggs in females). Fusion of a sperm and an egg nucleus at fertilization produces a new diploid cell (zygote).

3.4 **DNA: The Genetic Material**

- DNA is the hereditary material of the cell. RNA participates in protein synthesis.
  Nucleotides are nucleic acid building blocks. Nucleotides consist of a nitrogenous (purine or pyrimidine) base, a phosphate, and a pentose sugar.
- DNA replication is semiconservative. During replication, the DNA strands separate, and each strand is a template for a new strand. New strands are assembled according to a complementary relationship between DNA nucleotides.
- Protein synthesis is a result of two processes. Transcription occurs in the nucleus and involves the production of a messenger RNA (mRNA) molecule from a DNA molecule. A section of DNA is unwound, and RNA polymerase links ribose nucleotides in a sequence that is complementary to one DNA strand. Translation involves the movement of mRNA to the cytoplasm where it associates with ribosomes. Transfer RNA (tRNA) carries amino acids to the ribosomes and positions amino acids according to the pairing relationships between the mRNA codon and the tRNA anticodon. Peptide bonds are formed between two amino acids at the ribosome. Translation of the entire mRNA molecule produces a polypeptide.
- Changes in DNA and chromosomes include point mutations, which alter the bases in DNA, and changes in chromosome number and structure. These changes are usually deleterious for the organism.

3.5 **Inheritance Patterns in Animals**

- The principle of segregation states that pairs of genes are distributed between gametes during gamete formation when homologous chromosomes are distributed to different gametes during meiosis.
- The principle of independent assortment states that, during gamete formation, pairs of genes segregate independently of one another. This is a result of meiotic processes in which members of one homologous pair of chromosomes are not influenced by the movements of any other pair of chromosomes.
- Populations may have many alternative expressions of a gene at any locus. Human traits, like the ABO blood group, are traits determined by multiple alleles.
  Incomplete dominance is an interaction between two alleles in which the alleles contribute more or less equally to the phenotype. Codominance is an interaction between two alleles in which both alleles are expressed in the heterozygote.
- Patterns of inheritance observed at an organismal level are explained at a molecular level by the presence or absence of functional enzymes. A dominant allele usually encodes a functional enzyme, and a recessive allele usually encodes a nonfunctional enzyme.

CONCEPT REVIEW QUESTIONS

1. These are represented differently in males and females of the same species.
   a. Autosomes
   b. Nucleosomes
   c. Sex chromosomes
   d. Histones
2. Which of the following would be more nearly identical?
   a. Homologous chromosomes
   b. Nonhomologous chromosomes
   c. Sister chromatids before meiotic prophase I
   d. Sister chromatids after meiotic prophase I
   e. Chromosomes at metaphase II

3. Chromatids move toward opposite poles of the cell during
   a. prophase I of meiosis.
   b. metaphase of mitosis.
   c. anaphase of mitosis.
   d. anaphase I of meiosis.
   e. anaphase II of meiosis.
   f. Both c and d are correct.
   g. Both c and e are correct.
4. A student carried out a cross between two fruit flies. One fly is heterozygous for the vestigial-wing trait and one is homozygous for the vestigial-wing trait. The offspring expected from this cross would
   a. all be vestigial winged.
   b. all be wild winged.
   c. include flies with vestigial wings and wild wings in a ratio of 3:1.
   d. include flies with vestigial wings and wild wings in a ratio of 1:1.
5. A student carried out a cross between two fruit flies. One fly was homozygous for vestigial wings and also homozygous for sepia eyes. The second fly was heterozygous for vestigial wings and homozygous for wild eyes. The offspring expected from this cross would
   a. all be vestigial winged, but one-half of the flies would have sepia eyes and one-half would have wild eyes.
   b. all have wild eyes, but one-half of the flies would have vestigial wings and one-half would have wild wings.
   c. show the following phenotypes in equal numbers: wild wings, wild eyes; wild wings, sepia eyes; vestigial wings, wild eyes; and vestigial wings, sepia eyes.
   d. show the following phenotypes in a 9:3:3:1 ratio: wild wings, wild eyes; wild wings, sepia eyes; vestigial wings, wild eyes; and vestigial wings, sepia eyes.

Analysis and Application Questions

1. Which do you think evolved first—meiotic cell division or mitotic cell division? Why? What do you think may have been some of the stages in the evolution of one from the other?
2. Assume that a cell containing a 2*N* chromosome number of 6 has just completed prophase of mitosis. Assume mitosis will be completed. What would result if prometaphase kinetochore microtubules of both chromatids of one chromosome were attached to the same pole of the cell?
3. Why is it important that all regions of chromosomes are not continually active?
4. Do you think that Mendel's conclusions regarding the assortment of genes for two traits would have been any different if he had used traits encoded by genes carried on the same chromosome? Explain.
5. Nondisjunction followed by normal fertilization results in a trisomy that often has detrimental consequences. Male calico cats are trisomic, but they function normally. How would you explain this observation?

Charles Darwin described organic evolution as "descent with modification." Many of his ideas on how organic evolution happens were formulated during and after his visit to the Galápagos Islands–the home of this land iguana (*Conolophus subcristatus*).

4 Evolution: History and Evidence

Chapter Outline

4.1 Organic Evolution and Pre-Darwinian Theories of Change
4.2 Darwin's Early Years and His Journey
Voyage of the HMS Beagle
4.3 Early Development of Darwin's Ideas of Evolution
Geology
Fossil Evidence
Galápagos Islands
4.4 The Theory of Evolution by Natural Selection
Natural Selection
Adaptation
Alfred Russel Wallace
4.5 Geological Time and Mass Extinctions
4.6 Microevolution, Macroevolution, and Evidence of Macroevolutionary Change
Biogeography
Paleontology
Analogy and Homology
Interpreting the Evidence: Phylogeny and Common Descent

Chapters 4 and 5 are devoted to an introduction to evolutionary theory. In science, the use of the word "theory" is very different from its use in casual conversations among nonscientists. A scientific theory is a concept that has explanatory power. It is a concept that is supported by many related observations and experiments, debated by many scientists, and refined over decades of study. The uncertainty suggested in the word "theory" is only in recognition that science never has a complete and final answer. While the essence of any theory is true and unlikely to change substantially, the details of all theories can be subject to new observations, experiments, and refinement. Such is the case with evolutionary theory. It is a prime example of how scientific theories come into being, change through years of study, and become cornerstones of scientific disciplines.

4.1 ORGANIC EVOLUTION AND PRE-DARWINIAN THEORIES OF CHANGE

LEARNING OUTCOMES

1. Evaluate how scientific thought on evolutionary change prior to the work of Charles Darwin influenced Darwin's thinking.
2. Compare Lamarckian ideas of evolutionary change to modern epigenetic ideas of change.

Charles Darwin is the author of the theory of evolution by natural selection that has been supported and refined by the work of scientists over the last 150 years. **Organic evolution**, according to Charles Darwin, is "descent with modification." This simply means that populations change over time. Populations consist of individuals of the same species that occupy a given area at the same time (*see chapter 5*). They share a unique set of genes. Even though Darwin originated the theory of evolution by natural selection, his ideas were influenced by scientists who preceded him.

Some of the earliest references to evolutionary change are from the ancient Greeks. The philosophers Empedocles (495–435 B.C.) and Aristotle (384–322 B.C.) described concepts of change in living organisms over time. Georges-Louis Buffon (1707–1788) spent many years studying comparative anatomy. His observations of structural variations in particular organs of related animals convinced him that change must have occurred during the history of life on earth. Buffon attributed change in organisms to the action of the environment.

He believed in a special creation of species and considered change as being degenerate—for example, he described apes as degenerate humans. Erasmus Darwin (1731–1802), a physician and the grandfather of Charles Darwin, was intensely interested in questions of origin and change. He accepted the idea of a common ancestry of all organisms.

Jean Baptiste Lamarck (1744–1829) was a distinguished French zoologist. His contributions to zoology include important studies of animal classification. Lamarck published a set of invertebrate zoology books. His theory was based on a widely accepted theory of inheritance that organisms develop new organs, or modify existing organs, as needs arise. (Charles Darwin also accepted this idea of inheritance.) Similarly, he hypothesized that disuse resulted in the degeneration of organs. Lamarck thought that "need" was dictated by environmental change and that change involved movement toward perfection. The idea that change in a species is directed by need logically led Lamarck to the conclusion that species could not become extinct—they simply evolved into different species.

Lamarck illustrated his ideas of change with the often-quoted example of the giraffe. He contended that ancestral giraffes had short necks, much like those of any other mammal. Straining to reach higher branches during browsing resulted in their acquiring higher shoulders and longer necks. These modifications, produced in one generation, were passed on to the next generation. Lamarck published his theory in 1802 and included it in one of his invertebrate zoology books, *Philosophie Zoologique* (1809). He defended his ideas in spite of intense social criticism.

Lamarck's acceptance of a theory of inheritance that we now know is largely incorrect led him to erroneous conclusions about how evolution occurs. We now know that random changes in the structure of DNA (mutation) and chance processes involved in the assortment of genes into gametes (e.g., independent assortment, crossing-over, and random fertilization—*see chapter 3*) result in variation among offspring. The environment then plays a role in determining the survival of these variations in subsequent generations.

Within the last few years, a stir has developed within evolutionary theory with the revival of Lamarckian ideas. A significant body of evidence is demonstrating that the environment can influence gene activity, and these altered activity states can be passed on to offspring. Altered activity states do not result from changes in DNA base sequences, as occurs with mutations. The altered activity occurs as a result of DNA methylation and other gene control mechanisms (*see chapter 3*). Concepts of gene control are not new, but the idea that gene control can be environmentally induced and that these control states can be passed to offspring is revolutionary. The inheritance of environmentally induced change in a species is called epigenetics (Gr. *epi*, upon + *gennan*, to produce), and its overall contribution to evolutionary change is uncertain.

Section 4.1 Thinking Beyond the Facts

Extinction is one possible outcome of evolutionary change. If Lamarck had been correct regarding the mechanism of evolutionary change, would extinction be more a more likely or a less likely outcome of evolutionary change? Explain.

4.2 DARWIN'S EARLY YEARS AND HIS JOURNEY

LEARNING OUTCOMES

1. Describe the circumstances that led to Charles Darwin becoming a naturalist on HMS *Beagle*.
2. Describe the dates of the voyage of HMS *Beagle* and the path taken by the ship during its voyage.

Charles Robert Darwin (1809–1882) was born on February 12, 1809. His father, like his grandfather, was a physician. During Darwin's youth in Shrewsbury, England, his interests centered around dogs, collecting, and hunting birds—all popular pastimes in wealthy families of nineteenth-century England. These activities captivated him far more than the traditional education he received at boarding school. At the age of 16 (1825), he entered medical school in Edinburgh, Scotland. For two years, he enjoyed the company of the school's well-established scientists. Darwin, however, was not interested in a career in medicine because he could not bear the sight of people experiencing pain. This prompted his father to suggest that he train for the clergy in the Church of England. With this in mind, Charles enrolled at Christ's College in Cambridge and graduated with honors in 1831. This training, like the medical training he received, was disappointing for Darwin. Again, his most memorable experiences were those with Cambridge scientists. During his stay at Cambridge, Darwin developed a keen interest in collecting beetles and made valuable contributions to beetle taxonomy.

Voyage of the HMS *Beagle*

One of his Cambridge mentors, a botanist by the name of John S. Henslow, nominated Darwin to serve as a naturalist on a mapping expedition that was to travel around the world. Darwin was commissioned as a naturalist on the HMS *Beagle,* which set sail on December 27, 1831, on a five-year voyage (figure 4.1). Darwin helped with routine seafaring tasks and made numerous collections, which he sent to Cambridge. The voyage gave him ample opportunity to explore tropical rain forests, fossil beds, the volcanic peaks of South America, and the coral atolls of the South Pacific. Most important, Darwin spent five weeks on the **Galápagos Islands,** a group of volcanic islands 900 km off the coast of Ecuador. Some of his most revolutionary ideas came from his observations of plant and animal life on these islands. At the end of the voyage, Darwin was just 27 years old.

By 1842, Darwin had developed the essence of his conclusions but delayed their publication because of uncertainty about how they would be received. His ideas were eventually presented before the Linnean Society in London in 1858, and *On the Origin of Species by Means of Natural Selection* was published in 1859 and revolutionized biology. In the years after his voyage, Darwin was an extremely prolific scientist. He published five volumes on *Zoology of the Beagle Voyage* (1843), *Fertilisation of Orchids* (1862), *The Variation of Plants and Animals under Domestication* (1873), *The Descent of Man* (1871), and numerous other works.

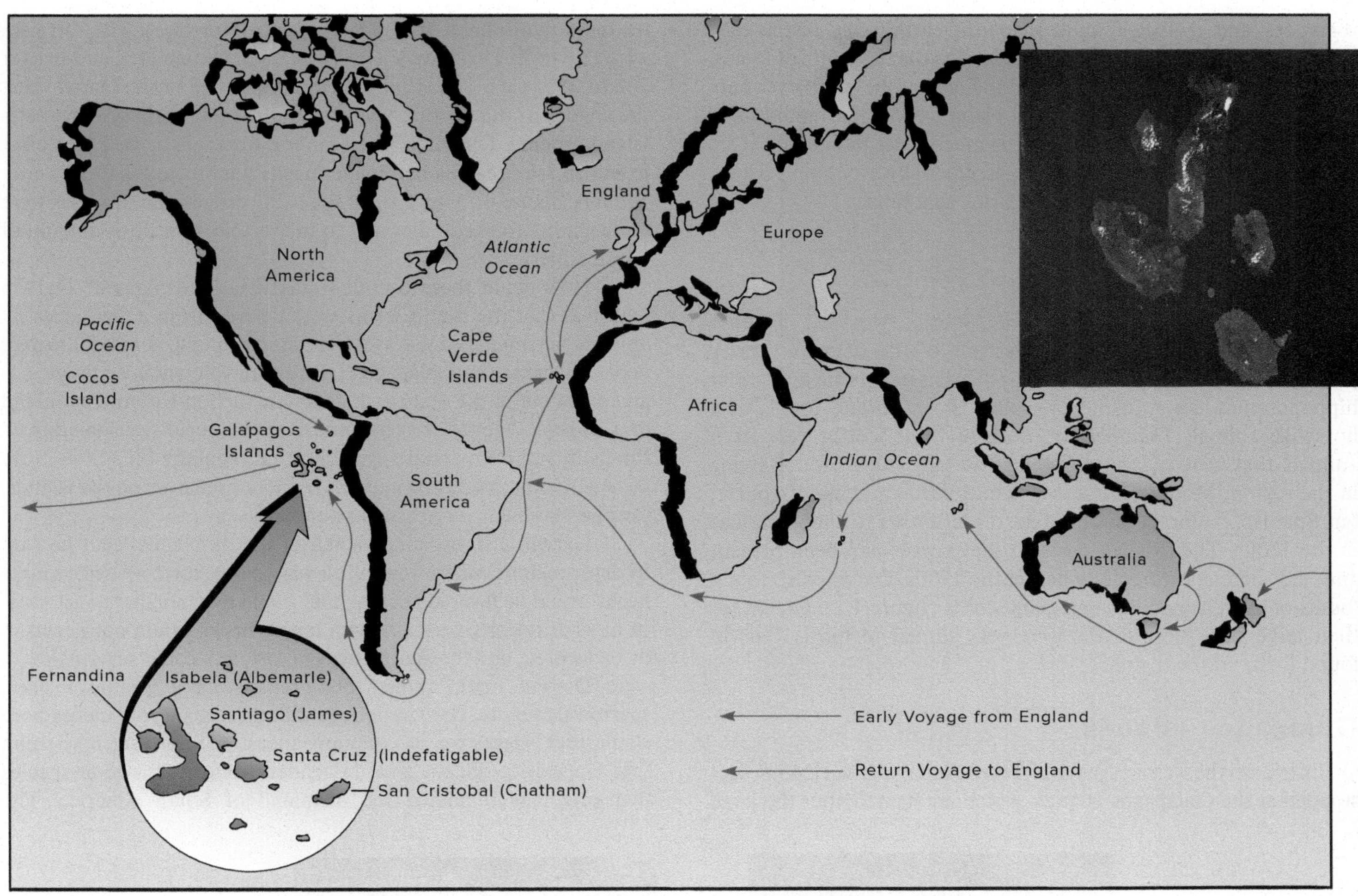

FIGURE 4.1

Voyage of the HMS *Beagle*. The inset shows four of the major islands as photographed from outer space: Fernandina, Isabella (formed from three volcanic peaks), Santiago, and Santa Cruz.

Source: JSC/NASA

SECTION 4.2 THINKING BEYOND THE FACTS

How does the life of Charles Darwin illustrate the fact that major scientific advancements are usually careful and painstaking enterprises, not ideas that come through inspirational flashes?

4.3 EARLY DEVELOPMENT OF DARWIN'S IDEAS OF EVOLUTION

LEARNING OUTCOMES

1. List the sources of evidence that convinced Charles Darwin that evolutionary change occurs.
2. Formulate a hypothetical scenario that illustrates the concept of adaptive radiation.

The development of Darwin's theory of evolution by natural selection was a long, painstaking process. Darwin had to become convinced that change occurs over time. Before leaving on his voyage, Darwin accepted the prevailing opinion that the earth and its inhabitants had been created 6,000 years ago and had not changed since. Because his observations during his voyage suggested that change does occur, he realized that 6,000 years could not account for the diversity of modern species if they arose through gradual change. Once ideas of change were established in Darwin's thinking, it took about 20 years of study to conceive, and thoroughly document, the mechanism by which change occurs. Darwin died without knowing the genetic principles that support his theory.

Geology

During his voyage, Darwin read Charles Lyell's (1779–1875) *Principles of Geology*. In this book, Lyell developed the ideas of another geologist, James Hutton, into the theory of **uniformitarianism.** His theory was based on the idea that the forces of wind, rain, rivers, volcanoes, and geological uplift shape the earth today, just as they have in the past. Lyell and Hutton contended that it was these forces, not catastrophic events, that shaped the face of the earth over hundreds of millions of years. This book planted two important ideas in Darwin's mind: (1) the earth could be much older than 6,000 years and (2) if the face of the earth changed gradually over long periods, could not living forms also change during that time?

Fossil Evidence

Once the HMS *Beagle* reached South America, Darwin spent time digging in the dry riverbeds of the pampas (grassy plains) of Argentina. He found the fossil remains of an extinct hippopotamus-like animal, now called *Toxodon,* and fossils of a horselike animal, *Thoatherium*. Both of these fossils were from animals that were clearly different from any other animal living in the region. Modern horses were in South America, of course, but Spanish explorers had brought these horses to the Americas in the 1500s. The fossils suggested that horses had been present and had become extinct long before the 1500s. Darwin also found fossils of giant armadillos and giant sloths (figure 4.2). Except for their large size, these fossils were very similar to forms Darwin found living in the region.

Galápagos Islands

On its trip up the western shore of South America, the HMS *Beagle* stopped at the Galápagos Islands, which are named after the large tortoises that inhabit them (Sp. *galápago,* tortoise). The tortoises weigh up to 250 kg, have shells up to 1.8 m in diameter, and live for 200 to 250 years. The islands' governor pointed out to Darwin that the shapes of the tortoise shells from different parts of Albemarle Island differed. Darwin noticed other differences as well. Tortoises from the drier regions had longer necks than tortoises from wetter habitats (figure 4.3). In spite of their differences, the tortoises were quite similar to each other and to the tortoises on the mainland of South America.

How could these overall similarities be explained? Darwin reasoned that the island forms were derived from a few ancestral animals that managed to travel from the mainland, across 900 km of ocean. Because the Galápagos Islands are volcanic (*see figure 4.1*) and arose out of the seabed, no land connection with the mainland ever existed. One modern hypothesis is that tortoises floated from the mainland on mats of vegetation that regularly break free from coastal riverbanks during storms. Without predators on the islands, tortoises gradually increased in number.

Darwin also explained some of the differences that he saw. In dryer regions, where vegetation was sparse, tortoises with longer necks would be favored because they could reach higher to get food. In moister regions, tortoises with longer necks would not necessarily be favored, and the shorter-necked tortoises could survive.

Darwin made similar observations of a group of dark, sparrow-like birds. Darwin noticed that the Galápagos finches bore similarities suggestive of common ancestry. Scientists now think that Galápagos finches also descended from an ancestral species that originally inhabited the mainland of South America. The

(a)

(b)

FIGURE 4.2

The Giant Sloth. (*a*) Charles Darwin found evidence of the existence of giant sloths in South America similar to this *Megatherium*. Giant sloths lived about 10,000 years ago and weighed in excess of 1,000 kg. They certainly did not move through branches like this living relative, the brown-throated three-toed sloth (*Bradypus variegatus*), 3–6 kg. (*b*). Instead, they probably fed on leaves of lower tree branches that they could reach from the ground. The similarity of giant sloths and modern-day sloths impressed Darwin with the fact that species change over time. Many species have become extinct. As in this case, they often leave descendants that provide evidence of evolutionary change.

(a)

(b)

FIGURE 4.3

Galápagos Tortoises. (*a*) Shorter-necked subspecies of *Chelonoidis nigra** live in moister regions and feed on low-growing vegetation. (*b*) Longer-necked subspecies live in drier regions and feed on high-growing vegetation. This tortoise, known as Lonesome George, was the last of the *G. nigra abingdonii* subspecies. He died in 2012. *This species name replaces the older name (*Geochelone elephantopus*) based on recent phylogenetic evidence.

chance arrival of a few finches, in either single or multiple colonization events, probably set up the first bird populations on the islands. Early finches encountered many different habitats containing few other birds and predators. Ancestral finches, probably seed eaters, multiplied rapidly and filled the seed-bearing habitats most attractive to them. Fourteen species of finches arose from this ancestral group, including one species found on small Cocos Island northeast of the Galápagos Islands. Each species is adapted to a specific habitat on the islands. The most obvious difference between these finches relates to dietary adaptations and is reflected in the size and shape of their bills. The finches of the Galápagos Islands provide an example of **adaptive radiation**—the formation of new forms from an ancestral species, usually in response to the opening of new habitats (figure 4.4).

Darwin's experiences in South America and the Galápagos Islands convinced him that animals change over time. It took the remaining years of his life for Darwin to formulate and document his ideas, and to publish a description of the mechanism of evolutionary change.

Section 4.3 Thinking Beyond the Facts

How does the fact that the Galápagos Islands are volcanic inform our understanding of the origin and evolution of animal populations living on the islands?

4.4 THE THEORY OF EVOLUTION BY NATURAL SELECTION

LEARNING OUTCOMES

1. Describe the four requirements for evolution to occur by natural selection.
2. Explain how reproductive success, phenotype, and environment are related to evolutionary adaptation.
3. Describe the contributions made by Thomas Malthus and Alfred Russel Wallace to the development of evolutionary theory.

On his return to England in 1836, Darwin worked diligently on the notes and specimens he had collected and made new observations. He was familiar with the obvious success of breeders in developing desired variations in plant and animal stocks (figure 4.5). He wondered if this artificial selection of traits could have a parallel in the natural world.

Ideas of how change occurred began to develop on his voyage. They took on their final form after 1838 when he read an essay by Thomas Malthus (1766–1834) entitled *Essay on the Principle of Population*. Malthus hypothesized that the human population has the potential to increase geometrically. (Geometric growth involves increasing by doubling or by some other multiple rather than by adding a fixed number of individuals with each new generation.) However, because resources cannot keep pace with the increased demands of a

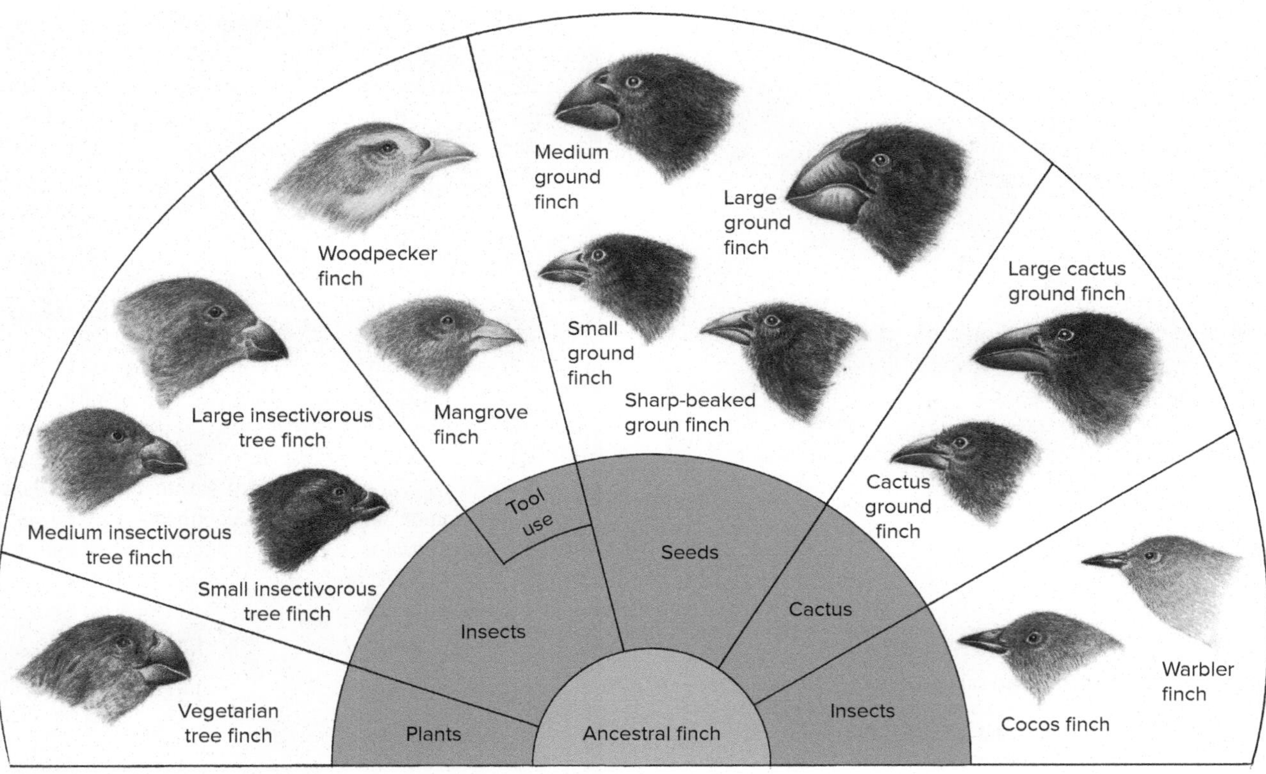

FIGURE 4.4

Adaptive Radiation of the Galápagos Finches. Ancestral finches from the South American mainland colonized the Galápagos Islands. Open habitats and few predators promoted the radiation of finches into 14 different species.

FIGURE 4.5

Artificial Selection. Dogs (*Canis familiaris*) were domesticated between 30,000 and 20,000 years ago. Although 99.9% genetically similar to *Canis lupis,* the grey wolf (*a*), their ancestral wolf species is unknown. Since then, humans have been selectively breeding dogs for many purposes. Some toy, or tea cup, dogs have primarily been bred for the enjoyment and status of the rich. Other dogs were bred for working. The Shetland Sheep Dog (*b*) was bred for herding sheep in England. Still other dogs were bred for hunting. The Irish Wolfhound (*c*) originated in Ireland and was used for hunting deer and wolves. Ancient Romans trained this breed to pull enemies from their horses during battle.

burgeoning population, population-restraining factors, such as poverty, wars, plagues, and famine, begin to have an influence. Darwin realized that a similar struggle to survive occurs in nature. When viewed over generations, this struggle could be a means of **natural selection.** Traits that were detrimental for an animal would be eliminated by the failure of the animal containing them to reproduce.

Natural Selection

Charles Darwin had no knowledge of modern genetic concepts, and therefore had no knowledge of the genetic principles that are the basis of evolutionary theory as it exists today. The modern version of his theory can be summarized as follows:

1. All organisms have a far greater reproductive potential than is ever realized. For example, a female oyster releases about 100,000 eggs with each spawning, a female sea star releases about 1 million eggs each season, and a female robin may lay four fertile eggs each season. What if all of these eggs were fertilized and developed to reproductive adults by the following year? A half million female sea stars (half of the million eggs would produce females and half would produce males), each producing another million eggs, repeated over just a few generations, would soon fill up the oceans! Even the adult female robins, each producing four more robins, would result in unimaginable resource problems in just a few years.
2. Inherited variations exist. They arise from a variety of sources, including mutation, genetic recombination (*see chapter 3*), and random fertilization. Seldom are any two individuals exactly alike. Some of these genetic variations may confer an advantage to the individual possessing them. In other instances, variations may be harmful to an individual. In still other instances, particular variations may be neither helpful nor harmful. (These are said to be neutral.) These variations can be passed on to offspring.
3. Because resources are limited, existence is a constant struggle. Many more offspring are produced than resources can support; therefore, many individuals die. Darwin reasoned that the individuals that die are those with the traits (variations) that make survival and successful reproduction less likely. Traits that promote successful reproduction are said to be adaptive.
4. Adaptive traits become more common in subsequent generations. Because organisms with maladaptive traits are less likely to reproduce, the maladaptive traits become less frequent in a population.

With these ideas, Darwin formulated a theory that explained how the tortoises and finches of the Galápagos Islands changed over time. In addition, Darwin's theory explained how some animals, such as the ancient South American horses, could become extinct. What if a group of animals is faced with a new environment to which it is ill-adapted? Climatic changes, food shortages, and other environmental stressors could lead to extinction.

Adaptation

Adaptation occurs when a heritable change in a phenotype increases an animal's chances of successful reproduction in a specified environment. Adaptations must be heritable changes to be passed to subsequent generations. Adaptations are defined in the context of enhancing reproductive success because survival of a species occurs through successful reproduction, and survival of the species is the ultimate measure of success. Adaptations are defined in the context of a specified environment because a change that promotes successful reproduction in one environment may be detrimental to reproductive success in a different environment.

Even though adaptations are defined in the context of reproductive success, they can manifest themselves in a variety of ways. Adaptations may be behavioral, physiological, or morphological. For example, snowshoe hares show adaptations to their environment in the coniferous and boreal forests of North America from the Pacific Northwest to New England. Their common name is derived from their larger hind limbs (as compared to other hares) with more hair and larger toes, which provide more surface area for moving across snowy ground. Their reddish brown summer coat (pelage) provides camouflage against predators (lynx, bobcats, raptors, and foxes) in their spring and summer forest-floor habitat. In response to day-length changes, they begin a month-long October molt and take on a white coat with their ears tipped in black. They are now camouflaged in a snowy forest. In April, another month-long molt returns them to their summer pelage. These adaptations promote survival and make successful reproduction more likely. (figure 4.6).

New adaptations arise as a result of mutations, and they are perpetuated by natural selection. Mutations are chance events, and most mutations are either harmful or neutral (*see chapter 3*). Adaptive mutations never occur as a result of a need, and there is

FIGURE 4.6

Adaptations of the Snowshoe Hare (*Lepus americanus*) The snowshoe hare is shown here in its winter pelage. Notice the enlarged hind limbs. Scott Mills (North Carolina State University and the University of Montana) has documented the effects of climate change on this hare's pelage adaptation. Climate change is decreasing the snow-cover season throughout temperate North America. Coat-color molts in the snowshoe hare now precede the snow-cover season in the winter and extend past the time of melting snow in the spring. During these in-between times, a white winter coat color against a brown forest floor becomes a white flag for predators and makes this hare more vulnerable to predators. Dr. Mills and his team are studying the effects of this predation on hare populations. Will genetic variations within the hare populations that promote later molts in the fall and earlier molts in the spring become selectively advantageous and more common?

no guarantee that a species will change in order to meet the challenges of a changing environment. If adaptive changes did occur in response to need, extinction would not occur–and extinction is a fact of life for the majority of species. Most genetic variations exist as neutral alleles, having arisen by mutation years earlier, and are expressed as adaptive traits only when a population encounters a new environment and natural selection acts on the population. Adaptation may result in the evolution of multiple new groups if the environment can be exploited in different ways. When the evolution of multiple groups occurs, adaptive radiation results (*see figure 4.4*).

Not every characteristic is an adaptation to some kind of environmental situation. An allele that provided some adaptive trait in one environment may be neutral when the environment changes, but persist in the population because the trait is not detrimental. Other alleles may result in traits that were never adaptive. Because these alleles result in neutral traits, they are not selected against by natural selection and persist in the population.

Alfred Russel Wallace

Alfred Russel Wallace (1823–1913) was an explorer of the Amazon Valley and led a zoological expedition to the Malay Archipelago, which is an area of great biogeographical importance. Wallace, like Darwin, was impressed with evolutionary change and had read the writings of Thomas Malthus on human populations. He synthesized a theory of evolution similar to Darwin's theory of evolution by natural selection. After writing the details of his theory, Wallace sent his paper to Darwin for criticism. Darwin recognized the similarity of Wallace's ideas and prepared a short summary of his own theory. Both Wallace's and Darwin's papers were published in the *Journal of the Proceedings of the Linnean Society* in 1858. Darwin's insistence on having Wallace's ideas presented along with his own shows Darwin's integrity. Darwin then shortened a manuscript he had been working on since 1856 and published it as *On the Origin of Species by Means of Natural Selection* in November 1859. The 1,250 copies prepared in the first printing sold out the day the book was released.

In spite of the similarities in the theories of Wallace and Darwin, there were also important differences. Wallace, for example, accepted that every evolutionary modification was a product of selection and, therefore, had to be adaptive for the organism. Darwin, on the other hand, admitted that natural selection may not explain all evolutionary changes. He did not insist on finding adaptive significance for every modification. Further, unlike Darwin, Wallace stopped short of attributing human intellectual functions and the ability to make moral judgments to evolution. On both of these matters, Darwin's ideas are closer to the views of most modern scientists.

Wallace's work motivated Darwin to publish his own ideas. The theory of natural selection, however, is usually credited to Charles Darwin. Darwin's years of work and massive accumulations of evidence led even Wallace to attribute the theory to Darwin. Wallace wrote to Darwin in 1864:

> I shall always maintain [the theory of evolution by natural selection] to be actually yours and yours only. You had worked it out in details I had never thought of years before I had a ray of light on the subject.

Section 4.4 Thinking Beyond the Facts

Is there a difference in thinking about natural selection as weeding out less fit variations versus selecting for adaptive variations? Explain.

4.5 Geological Time and Mass Extinctions

LEARNING OUTCOMES

1. Compare the methods and information provided by relative and absolute dating techniques.
2. Explain two hypotheses regarding the occurrence of mass extinction events.

The geological timescale is a system of dating events and relationships among organisms during the earth's history. Evidence for time labeling within the geological timescale is based on relative dating techniques and absolute dating techniques. Relative dating techniques (stratigraphy) estimate time relationships between events based on the position of one event in a rock stratum (layer) relative to surrounding strata. Different strata of rock result from differing rates of sedimentation. Climatic and geological events influence rates of sedimentation. When rates of sedimentation change, a break in deposition occurs, leaving a distinct layer or stratum. Successive strata are piled on top of each other, with younger strata on top of older strata. Fossils in younger strata are of animals that lived more recently than fossils in older strata. Relative dating does not assign absolute dates to events, but geologists can use it to correlate strata around the world. Radiometric and molecular dating techniques, on the other hand, can be used to assign dates to rock strata and events that occurred in the past (*see How Do We Know Evolutionary Timescales?*). Absolute and relative dating techniques are used together to help provide a timescale for evolutionary events.

The geological timescale is presented in appendix B at the back of this book. The timescale is divided into blocks of time that are progressively subdivided into smaller units. The largest divisions of Earth's history are eons, followed by eras, and then periods. Although not shown in appendix B, periods are further subdivided into epochs, and epochs are subdivided into ages. The divisions of time were first established by early geologists and paleontologists based on geological events and life-forms that characterized particular rock strata. The time intervals were named by European scientists based on the locations and characteristics of particular geological sites (e.g., the Devonian period was named after the county of Devon in England). Later geologists and paleontologists have revised the time intervals and the geological and biological characteristics of the intervals.

With the exception of discussions of the origin of life and the origin of animals in chapter 8, we will focus our attention on events occurring in the Phanerozoic eon. It began 541 million years ago (mya). Examination of the timescale in appendix B reveals that geological events have had profound influences on animal life. In particular, five major extinction events occurred during Phanerozoic eon. These extinction events are marked in red on the timescale.

How Do We Know Evolutionary Timescales?

Evolutionary timescales have been studied for many years using relative dating techniques that place geological events in sequential order that is determined by the geological record. Absolute dating provides absolute dates for events and is determined by radiometric and molecular techniques. These techniques have been used to provide the timescales shown in appendix B of this textbook.

Igneous rocks form when lava cools. These rocks contain radioactive isotopes of elements. For example, uranium-238 is the radioactive isotope that undergoes atomic nuclear decay through a series of isotopes to produce lead-206. These decay processes occur at a constant rate for a particular isotope. The rate of decay is described in terms of the isotope's half-life—the time required for one-half of the unstable atomic nuclei to decay. The half-life of uranium 238 is 4.5 billion years. Since no radioactive isotope is added to a rock once it is formed, the ratio of radioactive isotope to its decay product can be used to date a rock, and a fossil contained within that rock. Because different isotopes have different rates of decay, varying time frames can be measured. The very long half-life of uranium-238 makes it useful for dating rock formations that are hundreds of millions of years old. Carbon-14 has a half-life of 5,730 years and is used to date fossils thousands of years old.

Molecular techniques have provided another absolute dating technique, the molecular clock. Molecular biologists have found that within each kind of molecule, the rate of change is relatively constant. Different molecules have different rates of change. Changes in some molecules are detrimental and selected against, so the "clock" for changes in these molecules runs very slowly. In other molecules, changes are less detrimental to function and are tolerated. The "clock" for change in these molecules runs less slowly. This dating principle also works in regions of DNA that do not code for functional proteins. If the rate of change in a region of DNA is relatively constant over time, the amount of change can be used to date evolutionary events. This concept is called the **molecular clock.** The accuracy of the molecular clock is controversial, but when dates can be substantiated by radiometric and relative dating techniques, it is a very effective tool for evolutionary biologists. The timescales depicted in the Evolutionary Insights reading in this chapter are an example of how molecular and radiometric techniques can be used to establish when evolutionary events occurred.

The first of these events occurred at the Ordovician/Silurian boundary (444 mya) and the last occurred at the Cretaceous/Tertiary boundary (66 mya). The largest mass extinction was the Permian extinction (252 mya). The Permian extinction resulted in the elimination of 95% of marine species and 70% of terrestrial species.

The causes of mass extinctions are being investigated, and multiple hypotheses are currently debated. In 1980, Luis Walter Alvarez proposed that the Cretaceous/Tertiary mass extinction was caused by the impact of a huge asteroid with Earth in the vicinity of the Yucatán Peninsula in Mexico. This impact would have hurled debris into the atmosphere, creating an "impact winter": global cooling, the death of photosynthetic organisms, and possibly acid rain. Another hypothesis suggests volcanic events may have caused mass extinctions. Four of the five extinction events (all but the Ordovician extinction) correspond to the timing of cataclysmic volcanic eruptions of major volcanic fields called traps. These were not the eruptions we see today. They were huge and resulted in deadly sulfurous and CO_2 emissions that spread across the globe. The fact that an asteroid impact in the Yucatán corresponded with volcanic eruptions 66 mya may mean that these combined forces are responsible for that mass extinction.

The nature of the recovery of organisms after mass extinction events is investigated using the fossil record. Surviving species usually undergo a burst of diversification following a mass extinction event as they fill habitats left open by species that perished. The burst of diversification levels off fairly quickly. For example, although there were diverse groups of small mammals during the Cretaceous period, the largest and dominant land animals were the dinosaurs. The Cretaceous/Tertiary mass extinction led to the demise of most dinosaurs, and the groups of mammals that survived rapidly diversified and filled habitats and ecological roles formerly occupied by the dinosaurs.

In the same way that the work of Lyell and Hutton provided a timescale for Darwin to work with in the development of the theory of evolution by natural selection, geological time periods spanning hundreds of millions of years allow us to understand how small genetic changes can accumulate to produce major evolutionary transitions. In the next section, we present evidence for these major transitions, and in chapter 5 we describe how small genetic changes occur within groups of animals and may eventually result in changes at the species level.

Section 4.5 Thinking Beyond the Facts

How do you think the world would be different today had the Cretaceous/Tertiary mass extinction not occurred?

4.6 MICROEVOLUTION, MACROEVOLUTION, AND EVIDENCE OF MACROEVOLUTIONARY CHANGE

LEARNING OUTCOMES

1. Compare microevolution and macroevolution.
2. Describe the sources of evidence for macroevolution.
3. Describe the kind of information contained in a phylogenetic tree and how that information is represented.

Organic evolution was defined earlier as a change in populations over time, or simply "descent with modification." The change must

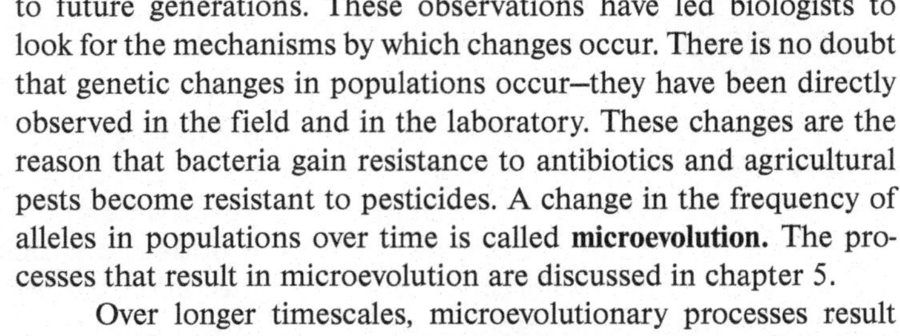

involve the genetic makeup of the population in order to be passed to future generations. These observations have led biologists to look for the mechanisms by which changes occur. There is no doubt that genetic changes in populations occur—they have been directly observed in the field and in the laboratory. These changes are the reason that bacteria gain resistance to antibiotics and agricultural pests become resistant to pesticides. A change in the frequency of alleles in populations over time is called **microevolution.** The processes that result in microevolution are discussed in chapter 5.

Over longer timescales, microevolutionary processes result in large-scale changes. Large-scale changes that result in extinction and the formation of new species are called **macroevolution.** Macroevolutionary changes are difficult to observe in progress because of the geological timescales that are usually involved. Evidence that

(a)

(b)

Bering Strait

Leopard *(Panthera pardus)*

Jaguar *(Panthera onca)*

(c)

FIGURE 4.7

Biogeography as Evidence of Evolutionary Change. (*a*) The leopard (*Panthera pardus*) of Africa and Asia has a similar ecological role to that of the (*b*) jaguar (*Panthera onca*) of Central and South America. Their similar form suggests common ancestry, even though they are separated by apparently insurmountable oceanic barriers (*c*). Spotted varieties of these species are distinguished by the presence (jaguar) or absence (leopard) of small spots within dark rosette markings of their coats. Biogeographers have provided probable explanations for these observations. (*See Evolutionary Insights, pages 69–70.*)

(a) ©Digital Vision./Getty Images (b) ©AlexanderDavid/Getty Images

macroevolution occurs, however, is compelling. This evidence is in the form of patterns of plant and animal distribution, fossils, biochemical molecules, anatomical structures, and developmental processes. Organisms leave evidence of what they looked like and how they lived. Evolutionary investigators piece this evidence together and provide detailed accounts of the lives of extinct organisms and their relationships to modern forms. The sources of evidence for macroevolution are described in the next section.

Biogeography

Biogeography is the study of the geographic distribution of plants and animals. Biogeographers try to explain why organisms are distributed as they are. Biogeographic studies show that life-forms in different parts of the world have distinctive evolutionary histories.

One of the distribution patterns that biogeographers try to explain is how similar groups of organisms have dispersed to places separated by seemingly impenetrable barriers. For example, native cats are inhabitants of most continents of the earth, yet they cannot cross expanses of open oceans. Obvious similarities suggest a common ancestry, but similarly obvious differences result from millions of years of independent evolution (figure 4.7 and Evolutionary Insights, pages 69–70). Biogeographers also try to explain why plants and animals, separated by geographical barriers, are often very different in spite of similar environments. For example, why are so many of the animals that inhabit Australia and Tasmania so very different from animals in any other part of the world? The major native herbivores of Australia and Tasmania are the many species of kangaroos. In other parts of the world, members of the deer and cattle groups fill these roles. Similarly, the Tasmanian wolf/tiger (*Thylacinus cynocephalus*), now believed to be extinct, was a predatory marsupial that was unlike any other large predator. Finally, biogeographers try to explain why oceanic islands often have relatively few, but unique, resident species. They try to document island colonization and subsequent evolutionary events, which may be very different from the evolutionary events in ancestral, mainland groups. The discussion that follows will illustrate some of Charles Darwin's conclusions about the island biogeography of the Galápagos Islands.

Modern evolutionary biologists recognize the importance of geological events, such as volcanic activity, the movement of great landmasses, climatic changes, and geological uplift, in creating or removing barriers to the movements of plants and animals. Biogeographers divide the world into six major biogeographic regions (figure 4.8). As they observe the characteristic plants and animals in each of these regions and learn about the earth's geological history, we understand more about animal distribution patterns and factors that played important roles in animal evolution. Only in understanding how the surface of the earth came to its present form can we understand its inhabitants.

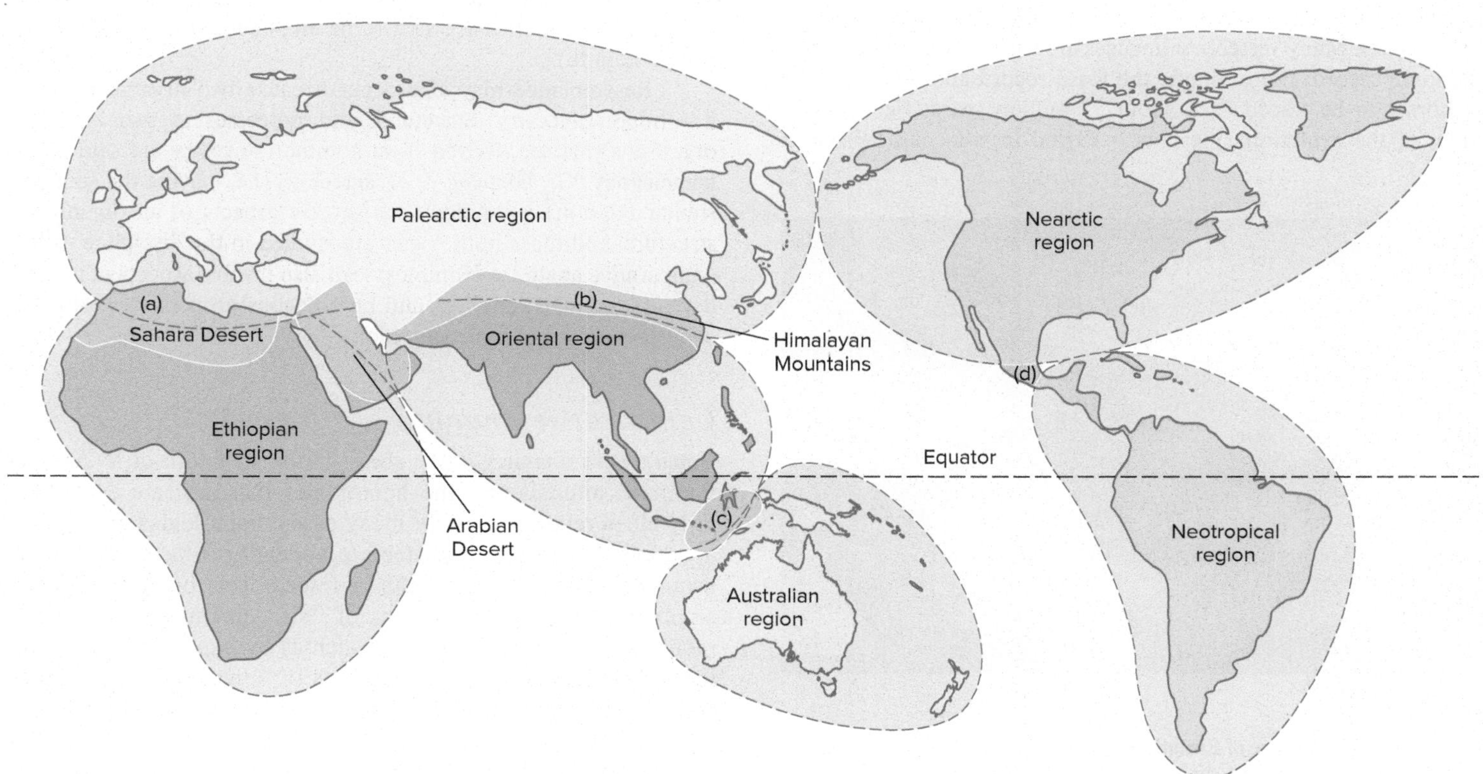

FIGURE 4.8

Biogeographic Regions of the World. Barriers, such as oceans, mountain ranges, and deserts, separate biogeographic regions of the world. (*a*) The Sahara and Arabian Deserts separate the Ethiopian and Palearctic regions, (*b*) the Himalayan Mountains separate the Palearctic and Oriental regions, (*c*) deep ocean channels separate the Oriental and Australian regions, and (*d*) the mountains of southern Mexico and Mexico's tropical lowlands separate the Nearctic and Neotropical regions.

Paleontology

Paleontology (Gr. *palaios*, old + *on*, existing + *logos*, to study), which is the study of the fossil record, provides some of the most direct evidence for evolution. **Fossils** (L. *fossilis*, to dig) are evidence of plants and animals that existed in the past and have become incorporated into the earth's crust (e.g., as rock or mineral) (figure 4.9). Fossils are formed in sedimentary rock by a variety of methods. Most commonly, fossilization occurs when sediments (e.g., silt, sand, or volcanic ash) quickly cover an organism to prevent scavenging and in a way that seals out oxygen and slows decomposition. As sediments continue to be piled on top of the dead organism, pressures build. Water infiltrates the remains and inorganic compounds and ions replace organic components. Hard parts of the organism are most likely to be fossilized, but delicate structures are sometimes fossilized when silica is involved with replacement. These pressure and chemical changes transform the organism into a stony replica. Other fossilization processes form molds, casts of organisms, or carbon skeletons. Tracks and burrows, and even mummified remains, are sometimes found. Fossilization is most likely to occur in aquatic or semiaquatic environments. The fossil record is therefore more complete for those groups of organisms living in or around water and for organisms with hard parts.

The fossil record provides information regarding sequences in the appearance and disappearance of organisms. Paleontologists use this information to provide an understanding of many evolutionary lineages. Many vertebrate lineages are very well documented in the fossil record. For example, the fossil record allows the history of horses to be traced back about 55 million years (figure 4.10). Most of the evolutionary events occurred in what is now North America. *Hyracotherium* was a dog-sized animal (0.2 m in height at the shoulder). Fossils reveal the presence of four prominent toes on each foot and a tooth structure indicative of a browsing lifestyle. As the habitat became more grassland-like, natural selection favored animals with longer legs used for outrunning predators and larger more durable teeth (molars) used for grazing. A loss of some foot bones, and a reduction in others, was accompanied by an elongation of the middle digit. The shift from browsing to grazing was also accompanied by an elongation of the face. Paleontologists have also used the fossil record to describe the history of life on earth (*see appendix B*). Evidence from paleontology is clearly some of the most convincing evidence of macroevolution.

FIGURE 4.9

Paleontological Evidence of Evolutionary Change. Fossils, such as these trilobites (*Paradoxides*), are direct evidence of evolutionary change. Trilobites existed about 500 mya and became extinct about 250 mya. Fossils may form when an animal dies and is covered with sediments. Water dissolves calcium from hard body parts and replaces it with another mineral, forming a hard replica of the original animal. This process is called mineralization.

©Alan Morgan

Analogy and Homology

Structures and processes of organisms may be alike. There are two reasons for similarities, and both cases provide evidence of evolution. Resemblance may occur when two unrelated organisms adapt to similar conditions. For example, adaptations for flight have produced flat, gliding surfaces in the wings of birds and insects. These similarities indicate that independent evolution in these two groups produced superficially similar structures and has allowed these two groups of animals to exploit a common aerial environment. The evolution of superficially similar structures in unrelated organisms is called **convergent evolution,** and the similar structures are said to be **analogous** (Gr. *analog* + *os,* proportionate).

Resemblance may also occur because two organisms share a common ancestry. Structures and processes in two kinds of organisms that are derived from common ancestry are said to be **homologous** (Gr. *homolog* + *os,* agreeing) (i.e., having the same or similar relation). Homology can involve aspects of an organism's structure, and these homologies are studied in the discipline called comparative anatomy. Homology can also involve aspects of animal development and function, and homologous processes are studied using techniques of molecular biology.

Comparative Anatomy

Comparative anatomy is the study of the structure of living and fossilized animals and the homologies that indicate evolutionarily close relationships. In many cases, homologies are readily apparent. For example, vertebrate appendages have a common arrangement of bones that can be traced back through primitive amphibians and certain groups of fish. Appendages have been modified for different functions such as swimming, running, flying, and grasping, but the basic sets of bones and the relationships of bones to each other have been retained (figure 4.11). The similarities in structure of these bones reflect their common ancestry and the fact that vertebrate appendages, although modified in their details of structure, have retained their primary functions of locomotion.

Other structures may be homologous even though they differ in appearance and function. The origin of the middle ear bones of terrestrial vertebrates provides an example. Fish do not have

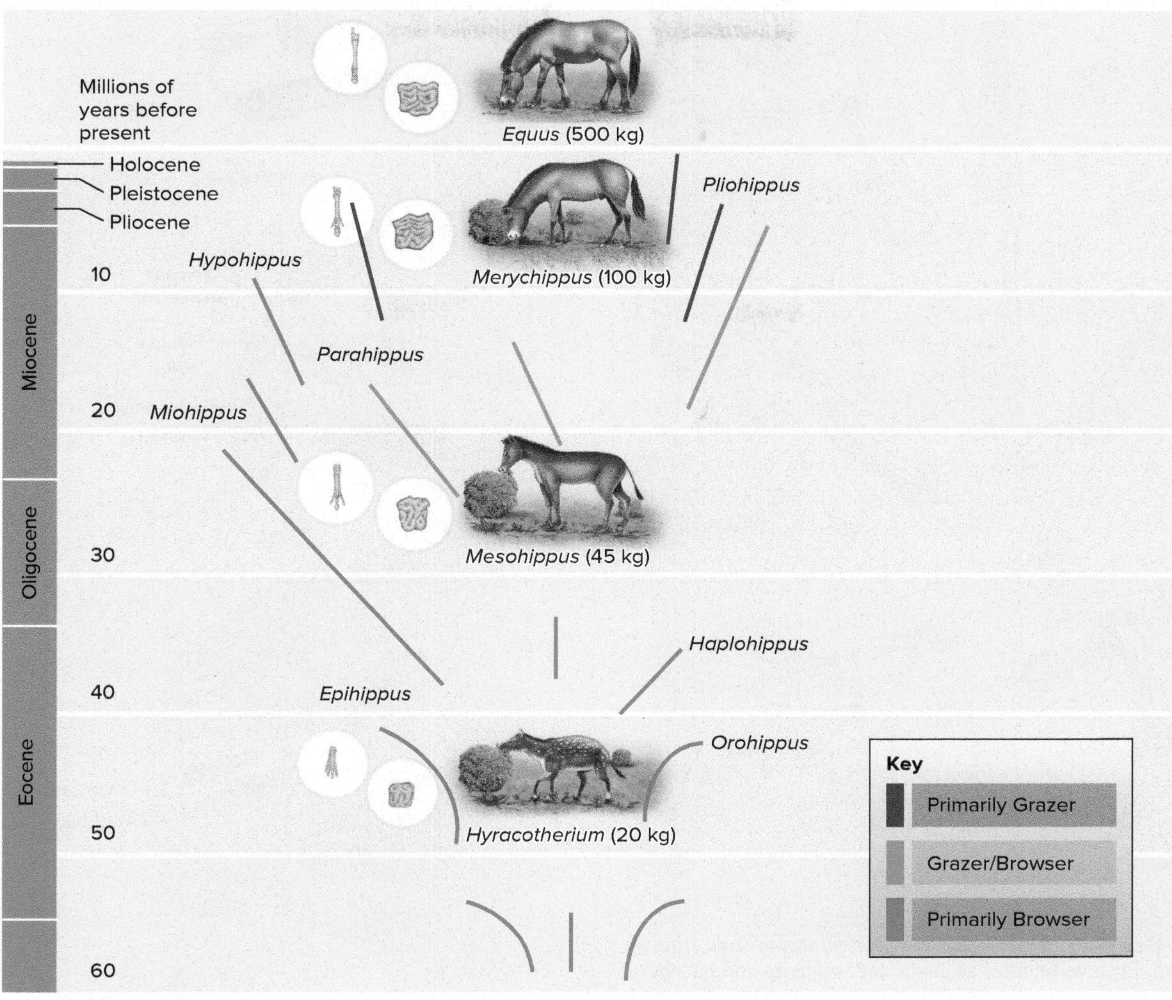

FIGURE 4.10

Reconstruction of an Evolutionary Lineage from Evidence in the Fossil Record. The fossil record allows horse evolution to be traced back about 55 million years. The horse ancestors illustrated were not direct ancestor/descendant sequences. The illustrations depict anatomical changes that occurred during horse evolution. Horse ancestors were small, primarily browsing animals that walked on the tips of three or four toes. Evolution resulted in larger animals adapted to a grazing lifestyle and that walked or ran on the tips of their middle toe digits. Note that evolutionary lineages are seldom simple ladders of change. Instead, numerous evolutionary side branches often meet with extinction.

middle and outer ears. Their inner ear provides for the senses of equilibrium, balance, and hearing, with sound waves being transmitted through bones of the skull. Terrestrial vertebrates evolved from primitive fish, and the evolution of life on land resulted in an ear that could detect airborne vibrations. These vibrations are transmitted to receptors of the inner ear through a middle ear and, in some cases, an outer ear. One (amphibians, reptiles, and birds) or more (mammals) small bones of the middle ear transmit vibrations from the eardrum (tympanic membrane) to the inner ear. Studies of the fossil record reveal the origin of these middle ear bones (figure 4.12). Small bones involved in jaw suspension in primitive fish are incorporated into the remnants of a pharyngeal (gill) slit to form the middle ear. In amphibians, reptiles, and birds, a single bone of fish (the hyomandibular bone) forms the middle ear bone (the columella or stapes). In the evolution of mammals, two additional bones that contributed to jaw support in ancient fish (the quadrate and articular bones) are used in the middle ear (the incus and malleus, respectively).

There are many other examples of structures that have changed from an ancestral form. Sometimes these changes result in a structure that can be detrimental to the organism. The human vermiform appendix evolved from a large fermentation pouch, and it is still used in this manner in animals like rabbits and many other herbivores. In humans, the vermiform appendix may have functions related to the lymphatic system, but these functions pale in comparison to the problems that result when it becomes infected. Boa constrictors have minute remnants of hindlimb (pelvic) bones that are remnants of the appendages of their reptilian ancestors. Baleen whales, like all whales, evolved from land mammals and also retain remnants of the pelvic appendages their ancestors used for walking on land. In both of these cases the remnant pelvic appendages serve no apparent function. Structures that have no apparent function in

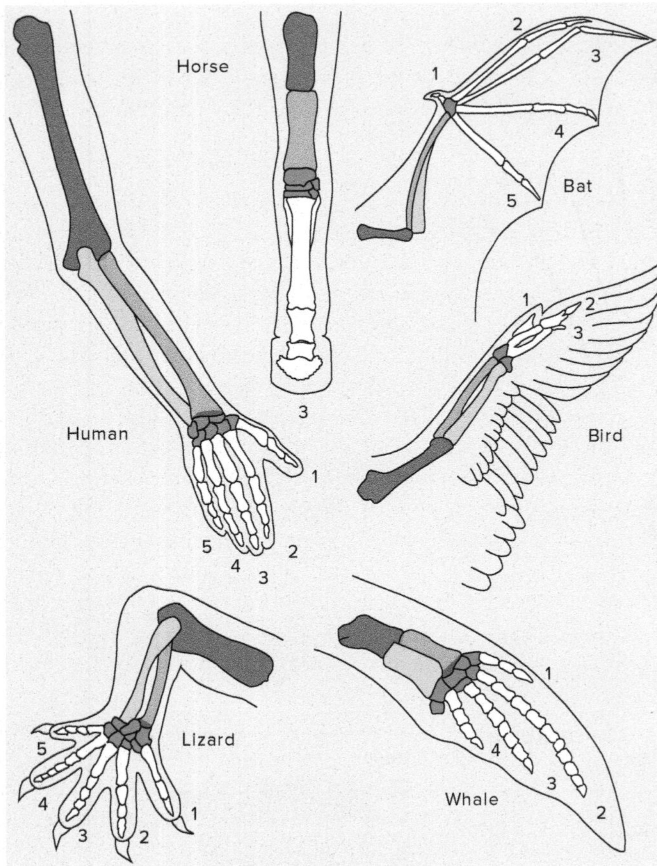

FIGURE 4.11

The Concept of Homology. The forelimbs of vertebrates evolved from an ancestral pattern. Even vertebrates as dissimilar as whales and bats have the same basic arrangement of bones. The digits (fingers) are numbered 1 (thumb) to 5 (little finger). Color coding indicates homologous bones.

modern animals but clearly evolved from functioning structures in ancestors are called **vestigial structures** and provide another source of evidence of evolutionary change.

Developmental Patterns

Evidence of evolution also comes from observing the developmental patterns of organisms. The developmental stages of related animals often retain common features because changes in the genes that control the development of animals are usually harmful and are eliminated by natural selection. For example, early embryonic stages of vertebrates are remarkably similar (figure 4.13*a*). Many organ systems of vertebrates also show similar developmental patterns (figure 4.13*b*). These similarities are compelling evidence of evolutionary relationships within animal groups.

Adults in vertebrate groups are obviously different from one another. If developmental patterns are so similar, how did differences in adult stages arise? Evolution is again the answer. These differences are a result of evolutionary changes in the genes that control the onset of developmental stages and the rate at which development occurs. These changes result in differences in the size and proportions of

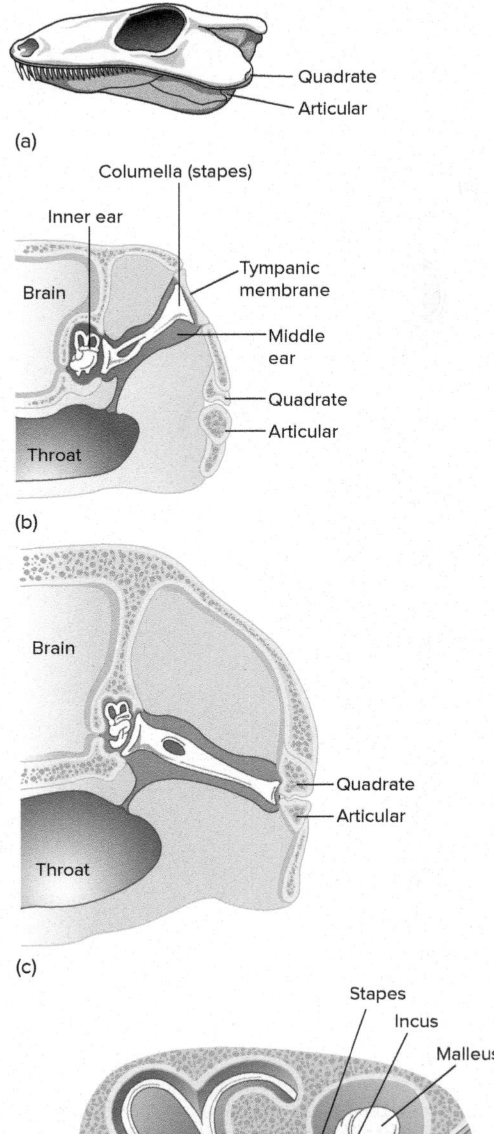

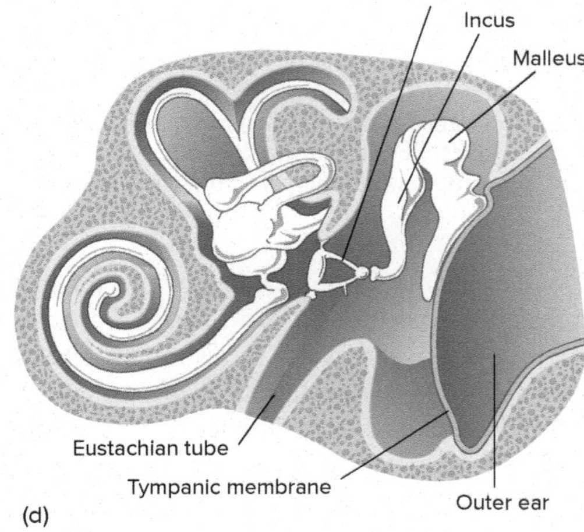

FIGURE 4.12

Evolution of the Vertebrate Ear Ossicles. (*a*) Lateral view of the skull of a primitive amphibian showing the two bones (quadrate and articular) that function in jaw support and contribute to the middle ear bones of mammals. Diagrammatic sections of the heads of (*b*) a primitive amphibian, (*c*) a primitive reptile, and (*d*) a mammal showing the fate of three bones of primitive fish. The columella (stapes) is derived from a bone called the hyomandibular, the incus is derived from the quadrate bone, and the malleus is derived from the articular bone.

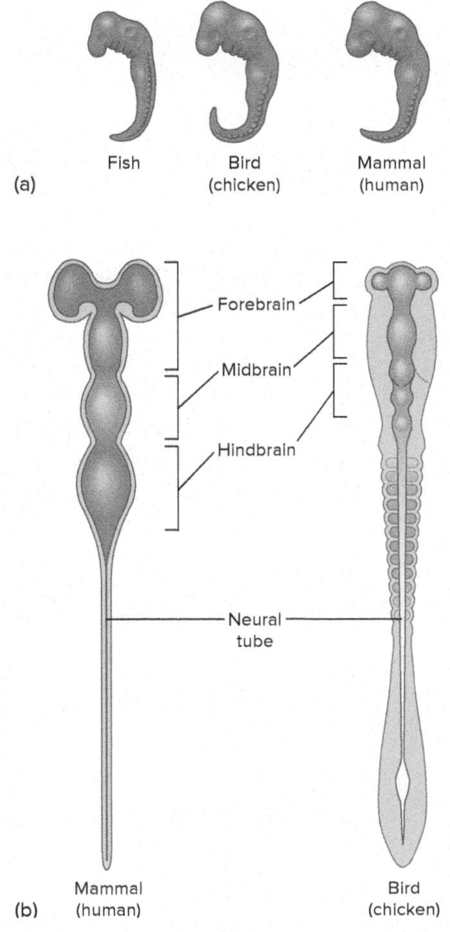

FIGURE 4.13

Developmental Patterns. (*a*) The early embryonic stages of various vertebrates are remarkably similar. These similarities result in the preservation of developmental sequences that evolved in early common ancestors of vertebrates. (*b*) Organ systems, like the nervous system, also show similar developmental patterns. Later developmental differences may result from evolutionary changes in the timing of developmental events.

organs. Differing growth rates of the bones of the skull, for example, can explain the proportional differences between bones of the human and ape skulls. Modern developmental biology is providing a growing appreciation of how evolution has conserved many genes that control the developmental similarities of animal groups. At the same time, it is helping to explain developmental changes that result in the great diversity of animal life.

Molecular Biology

Within the last 25 years, molecular biology has yielded a wealth of information on evolutionary relationships. Studying changes in anatomical structures and physiological processes reflects genetic change and evolution. Unfortunately, it is often difficult to sort out the relationship between genes and the structures and functions they control. Studying nuclear DNA, mitochondrial DNA, ribosomal RNA, and proteins is particularly useful in evolutionary biology because these molecules can provide direct evidence of changes in genes and thus evolution. Just as animals can have homologous structures, animals also have homologous biochemical processes that can be studied using molecular biological methods.

Molecular methods have several advantages: they are useful with all organisms, the data are quantifiable with readily available computer software, and databases of molecular information for many organisms are very large and growing. The use of molecular data allows biologists to investigate the causes of the genetic variation and molecular processes that influence evolution. These data also provide information for the construction of evolutionary trees (phylogenies).

The principle behind molecular analysis is that closely related organisms will be genetically more similar than distantly related organisms. Genetic similarity or degree of relatedness is reflected in the variation (or lack of it) in the amino acids that comprise a protein or in the bases that comprise DNA. This genetic variation can be quantified in a number of ways. Genetic variation is often measured by the **proportion of polymorphic loci** in a population. A polymorphic locus is one where two or more alleles exist. For example, imagine a researcher examined 20 loci from representatives of two populations. In the first population, the researcher found that five of these loci had more than one allele. The proportion of polymorphic loci would be 5⁄20 or 0.25. In the second population, 10 of these loci had more than one allele, and the proportion of polymorphic loci was 0.5. Genetic variation could be greater in the second population for a number of reasons. For example, it could indicate greater time since divergence from an ancestor and thus more time for variations to accumulate, or it could indicate genetic mixing with more than one ancestral group. Documenting genetic variation is important in evolutionary studies because this variation is the fuel for natural selection. (Recall that genetic variation was the second of four points in the earlier description of natural selection.)

Techniques for isolating and manipulating DNA have provided very powerful tools for the analysis of genetic variation among groups of animals. The polymerase chain reaction (PCR) and automated DNA sequencers allow researchers to begin with very small amounts of sample DNA and quickly and inexpensively determine the base sequences of DNA and other genetic fingerprinting patterns. Variation in DNA in homologous genes and regions suggests relationships between genes and groups of organisms. The tree diagram in appendix A is based largely on the study of variation in the base sequence of ribosomal RNA.

Interpreting the Evidence: Phylogeny and Common Descent

Scientists use the data gathered from studies just described to understand how organisms are related to each other. **Phylogeny** refers to the evolutionary relationships among species. It includes the depiction of ancestral species and the relationships of modern descendants of a common ancestor. These depictions involve the use of **phylogenetic trees** showing lines of descent. In the past, biologists relied mainly on the fossil record and anatomical traits in the construction of phylogenetic trees. The addition of molecular data is revolutionizing phylogenetic studies.

How Do We Know about Evolution—"Evo-Devo"?

The study of development has revealed that animals in groups as diverse as insects and humans share genes that direct certain stages of development. **Homeobox (*Hox*) genes** determine the identity of body regions in early embryos. They identify, for example, where a limb of a fly or a fish will be located. Mutations in these genes may cause body parts to appear in the wrong place, to be duplicated, or to be lost. Developmental biologists are finding that the same *Hox* genes direct the development of structures in diverse groups of animals. Body segmentation is present in both arthropods (insects and their relatives) and vertebrates. In the past, this observation has been explained as a case of convergent evolution. It now appears that the same gene appears to regulate the development of segmentation in both groups, which means that body segmentation was probably present in a common ancestor of both groups. The study of evolution through the analysis of development is sometimes called "evo-devo" and is revealing that a relatively small set of common genes underlies basic developmental processes in many organisms. Evo-devo is helping to explain how small changes in these development-directing genes can have far-reaching evolutionary consequences. (*See chapters 8 and 18 for other examples.*)

Phylogenies are reconstructed by examining variations in homologous structures, proteins, and genes. The process begins by gathering data from different organisms being compared. In molecular studies it usually involves sampling DNA or proteins from present-day organisms. Proteins or DNA from a particular locus or set of loci are compared using computer programs that include assumptions about the types of changes that are more likely to occur. (Changes of purine to pyrimidine, or vice versa, are less likely than changes of one purine to another purine or one pyrimidine to another pyrimidine.) The computer program examines all possible relationships between the different organisms and groups the organisms based on the fewest number of evolutionary changes that must have taken place since they shared a common ancestor.

Figure 4.14 shows a phylogenetic tree for the hemoglobin molecule. Hemoglobin genes are grouped into alpha and beta families. The products of these genes are incorporated into the hemoglobin molecule that transports oxygen in red blood cells. Another related gene codes for the oxygen storage pigment in muscle, myoglobin. Genes in the beta family have branch points or **nodes** that represent ancestral molecules. In other trees (*see Evolutionary Insights box figure 4.1*), nodes could represent individuals, populations, or species. A **branch** represents an evolutionary connection between molecules (individuals, populations, or species). The longer the branch, the greater the variation and the more distant the evolutionary relationship between molecules (individuals, populations, or species). The relatively short branch distance between nodes in the beta family reflects close evolutionary ties. Similar conclusions are drawn for genes in the alpha family. The conclusions drawn from this tree are that all of the modern hemoglobin genes are derived from a single gene that existed between 600 and 800 mya. Phylogenetic trees like this one, and the many that will follow in chapters 7 through 22, reinforce our ideas of common descent. All life is related, and the evidence of this relationship is overwhelming.

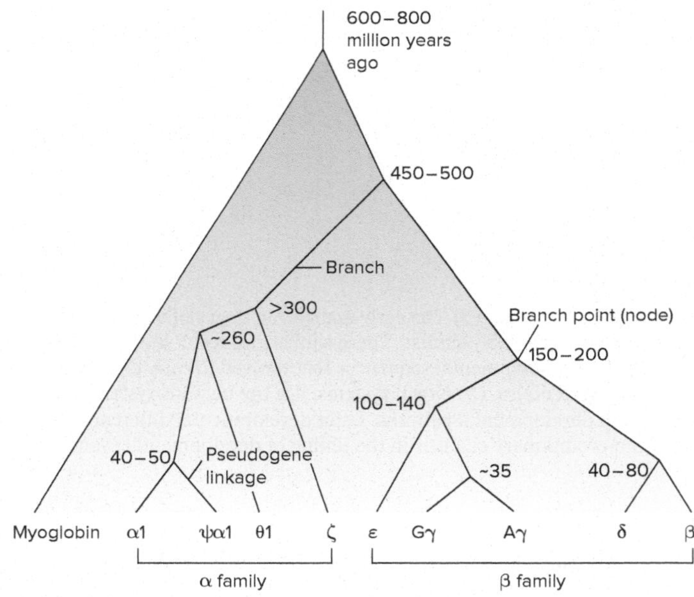

FIGURE 4.14

A Phylogenetic Tree of Hemoglobin. This phylogenetic tree is derived from molecular studies of the hemoglobin molecule. The numbers associated with each branch point indicate the approximate time in millions of years. The pseudogene in the alpha family is a gene that is apparently nonfunctional.

Evolution is the major unifying theme in biology because it helps explain both the similarities and the diversity of life. There is no doubt that it has occurred in the past and continues to occur today. Chapter 5 examines how the principles of population genetics have been combined with Darwinian evolutionary theory into what is often called the **modern synthesis.**

EVOLUTIONARY INSIGHTS

An Example from Cat Phylogeny

All sources of evidence are used in piecing together the evolutionary relationships of animals. Evidence from paleontology, biogeography, molecular biology, and ecology is being used to understand the evolutionary past of cat species.

Cats are members of the mammalian family Felidae. Paleontologists have established that all cats arose from a common ancestor approximately 34 mya. Fossils from France reveal that a bobcat-sized animal (*Proailurus*) lived in Europe about 25 mya. Early cats gave rise to multiple cat lineages—all but one met with extinction. The surviving lineage includes modern cats, members of the subfamily Felinae (box figure 4.1).

Biogeographers have helped to explain the evolution and distribution of wild cats around the globe. These scientists have used evidence from the fossil record, characters observed in living animals, molecular biology, and a knowledge of current distribution patterns to develop their hypotheses. Molecular phylogenetic studies and the fossil record reveal the Felinae lineage arose approximately 10 mya. This lineage diverged into the eight smaller lineages shown in box figure 4.1. Each of these lineages has its own evolutionary story, but one of these, the *Panthera* lineage, is particularly interesting.

The *Panthera* lineage includes two genera, *Neofelis* (clouded leopards) and *Panthera* (lions, tigers, jaguars, and leopards). Fossils show that a tiger/jaguar-like cat existed in northern China between 3 and 6 mya and may be the ancestor of the *Panthera* species. This date is based on studies of base sequence differences in certain ribosomal RNA genes and mitochondrial genes among *Panthera* species. One hypothesis is that this common ancestor spread into Europe, the rest of Asia, Africa, and also eastward. About 1 mya some ancestral panthers crossed the Bering land bridge that connected Asia and North America. The breakup of the Bering land bridge then isolated these groups. The jaguar's current distribution in Central and South America may be explained by the presence of other large North American cats that are now extinct. Competition may have driven jaguars southward.

The leopard (*Panthera pardus*) and the jaguar (*Panthera onca*) are remarkably similar in appearance (*see figure 4.7*) and ecological roles. In the wild, their similar appearance presents no problem in the identification of these species because of their distribution. The leopard is found in Africa, the Middle East, and Asia. The jaguar is found only in Central America and northern and central South America.

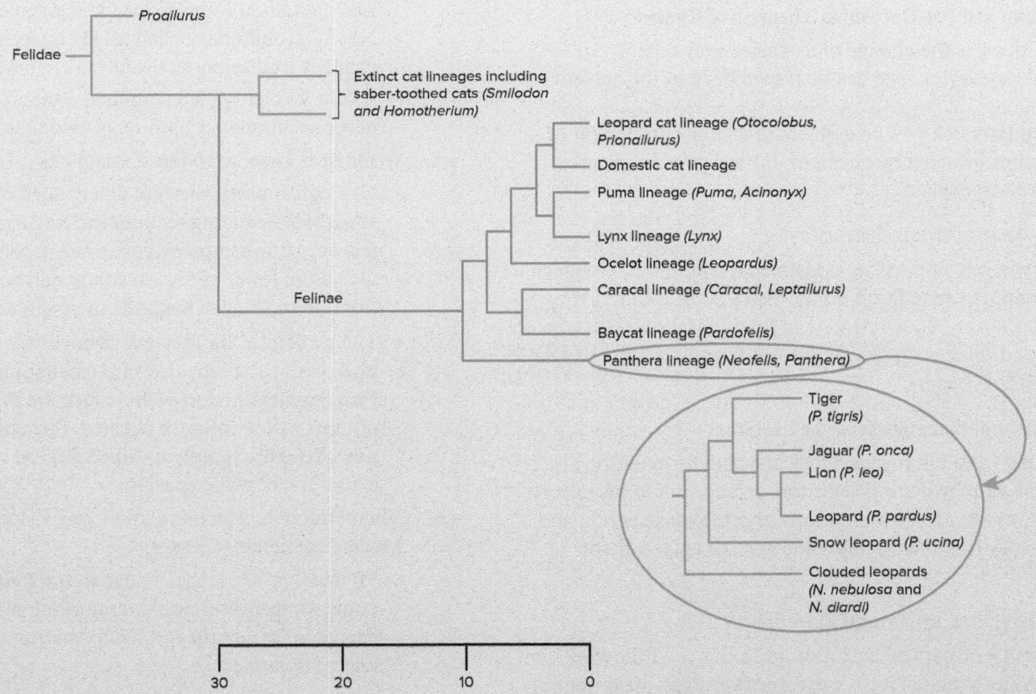

BOX FIGURE 4.1 Phylogeny of the Family Felidae. Felid cats were derived from a single ancestor about 30 mya. A single lineage (subfamily Felinae) gave rise to eight modern cat groups that originated within the last 10 million years.

(*Continued*)

EVOLUTIONARY INSIGHTS *Continued*

The morphological (structural) and behavioral differences that do exist between leopards and jaguars may be explained by habitat differences. Most jaguars are found in the densely forested areas of the Amazon Basin. Their smaller size is thought to be an adaptation to climatic and vegetational changes encountered as the cats moved south. Leopards, on the other hand, have evolved into a complex group of subspecies as they adapted to diverse environments across their range. One habit of African and Asian leopards is to use powerful neck and limb muscles to cache prey high in the boughs of trees. Antelope and other prey may be three times the body weight of the leopard. This behavior reduces competition from scavenging hyenas and opportunistic lions that may happen upon a leopard's kill.

Both of these species are threatened by habitat loss and hunting—both have been prized for their fur. Although the leopard has a very large range and diverse prey base, a number of subspecies are gone from many parts of their original range. Jaguars are severely threatened by deforestation. It is estimated that there are 15,000 individuals left in the wild.

SECTION 4.6 THINKING BEYOND THE FACTS

Some opponents of evolutionary theory contend that evolutionary theory is not valid science because it concerns events of the past that cannot be observed or re-created in the laboratory. How would you respond to this criticism?

SUMMARY

4.1 **Organic Evolution and Pre-Darwinian Theories of Change**

- Organic evolution is the change of a species over time.
  Ideas of evolutionary change can be traced back to the ancient Greeks.
  Jean Baptiste Lamarck was an eighteenth-century proponent of evolution and proposed a mechanism—inheritance of acquired characteristics—to explain it.

4.2 **Darwin's Early Years and His Journey**

- Charles Darwin saw impressive evidence for evolutionary change while on a mapping expedition on the HMS *Beagle*. The theory of uniformitarianism, South American fossils, and observations of tortoises and finches on the Galápagos Islands convinced Darwin that evolution occurs.

4.3 **Early Development of Darwin's Ideas of Evolution**

- After returning from his voyage, Darwin began formulating his theory of evolution by natural selection. In addition to his experiences on his voyage, later observations of artificial selection and Malthus's theory of human population growth helped shape his theory.

4.4 **The Theory of Evolution by Natural Selection**

- Darwin's theory of natural selection includes the following elements: (*1*) All organisms have a greater reproductive potential than is ever attained; (*2*) inherited variations arise by mutation; (*3*) in a constant struggle for existence, those organisms that are least suited to their environment die; and (*4*) the adaptive traits present in the survivors tend to be passed on to subsequent generations, and the nonadaptive traits tend to be lost.
- Adaptation may refer to a process of change or a result of change. An adaptation is a characteristic that increases an organism's potential to reproduce in a given environment.
  Not all evolutionary changes are adaptive, nor do all evolutionary changes lead to perfect solutions to environmental problems.
- Alfred Russel Wallace outlined a theory similar to Darwin's but never accumulated significant evidence documenting his theory.

4.5 **Geological Time and Mass Extinctions**

- Geological timescales are determined by geologists using relative and absolute dating techniques. Relative dating compares geological events based on evidence left in rock strata, older events are located in lower strata. Absolute dating techniques use radioactive and molecular methods to assign dates to geological events.
- The geological timescale is divided into blocks of time that are progressively subdivided into smaller units.
  Five mass extinction events mark the Phanerozoic eon. Causes of mass extinctions are debated. Species that survive mass extinctions diversify rapidly to fill ecological roles left empty by the extinction of other species.

4.6 **Microevolution, Macroevolution, and Evidence of Macroevolutionary Change**

- Microevolution is the change in the frequency of alleles in populations over time. Macroevolution is large-scale change that results in extinction and the formation of new species over geological timescales.
- Evidence of macroevolutionary change comes from the study of biogeography, paleontology, comparative anatomy, molecular biology, and developmental biology.
- All sources of evidence are used in studying the phylogeny of animals. These studies have resulted in the wealth of information on animal lineages that will be presented in chapters that follow.

Concept Review Questions

1. Charles Lyell's *Principles of Geology* influenced Charles Darwin's ideas of evolutionary change in which of the following ways?
   a. His description of fossils of South America convinced Darwin that species that were present in the past became extinct, thus reinforcing ideas of change.
   b. Lyell developed the ideas of another geologist, James Hutton, who advocated that uniformitarianism, not catastrophic change, was responsible for shaping the face of the earth.
   c. Lyell's work convinced Darwin that the earth must be much older than 6,000 years.
   d. Lyell described the volcanic origin of the Galápagos Islands, providing Darwin with ideas of how the islands might have been colonized by tortoises and finches.
   e. Both b and c are correct.
   f. All of the above (a–d) are correct.
2. The evolution of a number of new characteristics from an ancestral form is called ____________ and results from ____________.
   a. natural selection; the opening of new habitats
   b. adaptive radiation; the opening of new habitats
   c. punctuated equilibrium; loss of genetic variation in a species
   d. microevolution; extinction of one of two closely related species
3. A heritable change that increases an animal's chance of successful reproduction
   a. is an adaptation.
   b. prevents evolutionary change.
   c. always results in the evolution of a new species.
   d. is called genetic drift.
4. Resemblance between two organisms can occur because of shared ancestry. Structures that are similar because of shared ancestry are said to be
   a. homologous.
   b. analogous.
   c. convergent.
   d. adaptive.
5. Millions of years ago two populations of a species were separated as a result of two continents drifting apart. Members of these two populations adapted to different environments and today are very different from one another and classified as different species. Accounts like this that explain how geographical barriers help account for the evolution of groups of animals is the work of
   a. paleontologists.
   b. molecular biologists.
   c. comparative anatomists.
   d. biogeographers.

Analysis and Application Questions

1. Outline a hypothesis and design a test of "inheritance of acquired characteristics," and define what is meant by the word "theory" in the theory of evolution by natural selection.
2. Describe the implications of inheritance of acquired characteristics for our modern concepts of how genes function.
3. How would you explain the presence of gaps in the fossil record? Would you be more likely to see gaps in the fossil record of rodents, fish, molluscs, or segmented worms? Explain your answer.
4. Why is the stipulation of "a specified environment" included in the definition of adaptation?
5. Imagine that you could go back in time and meet simultaneously with Charles Darwin and Gregor Mendel. Construct a dialogue in which you explain to both the effect of their ideas on each other's theories and their theories on modern biology. Include their responses and questions throughout the dialogue.

5

Evolution and Gene Frequencies

Inherited variations help determine whether or not this African white-fronted bee-eater (*Merops bullockoides*) can catch enough food to enable it to reproduce. The evolution of this species will be reflected in how common, or how rare, specific alleles are in future generations.

Chapter Outline

5.1 Populations and Gene Pools
5.2 Must Evolution Happen?
The Hardy–Weinberg Theorem
5.3 Evolutionary Mechanisms
Neutral Theory and Genetic Drift
Gene Flow
Mutation
Natural Selection Reexamined
Balanced Polymorphism and Heterozygote Superiority
5.4 Species and Speciation
Allopatric Speciation
Parapatric Speciation
Sympatric Speciation
5.5 Rates of Evolution
5.6 Molecular Evolution
Gene Duplication
5.7 Mosaic Evolution

Natural selection can be envisioned as operating in two ways, and both are important perspectives on evolution. One way (e.g., the focus of chapter 4) looks at characteristics of individual animals. When a population of birds acquires an adaptation through natural selection that permits its members to feed more efficiently on butterflies, the trait is described in terms of physical characteristics (e.g., bill shape) or inherited behaviors. This description of natural selection recognizes that natural selection must act in the context of living organisms.

The organism, however, must be viewed as a vehicle that permits the phenotypic expression of genes. This chapter examines the second way that natural selection operates—on genes. Birds and butterflies are not permanent—they die. The genes they carry, however, are potentially immortal. The result of natural selection (and evolution in general) is reflected in how common, or how rare, specific alleles are in a group of animals that are interbreeding—and therefore sharing genes.

5.1 POPULATIONS AND GENE POOLS

LEARNING OUTCOMES

1. Relate the concept of a gene pool to a population of animals.
2. Explain why different individuals within a population are genetically different from each other.

Individuals do not evolve. Evolution requires that genetic changes are passed from one generation to another within larger groups called populations. **Populations** are groups of individuals of the same species that occupy a given area at the same time and share a common set of genes. With the possible exception of male and female differences, individuals of a population have the same number of genes and the same kinds of genes. A "kind" of gene would be a gene that codes for a given trait, such as hair length or the color of a mammal's coat. Differences within a population are based on variety within each trait, such as red or white hair in a mammal's coat. As described in chapter 3, this variety results from varying expressions of genes at each of the loci of an animal's chromosomes. Recall that these varying expressions of genes at each locus are called alleles. A population can be characterized by the frequency of alleles for a given trait, that is, the abundance of a particular allele in relation to the sum of all alleles at that locus.

The sum of all the alleles for all traits in a sexually reproducing population is a pool of hereditary resources for the entire population and is called the **gene pool.**

Variety within individuals of a population results from having various combinations of alleles at each locus. Some of the sources of this variety have been discussed in chapter 3. These sources of variation include (1) the independent assortment of chromosomes that results in the random distribution of chromosomes into gametes, (2) the crossing-over that results in a shuffling of alleles between homologous chromosomes, and (3) the chance fertilization of an egg by a sperm cell. Variations also arise from (4) rearrangements in the number and structure of chromosomes and (5) mutations of existing alleles. Mutations are the only source of new alleles and will be discussed in greater detail later in this chapter. This chapter also describes how genetic variation may confer an advantage to individuals, leading to natural selection. It describes how other variations may become common in, or lost from, populations even though no particular advantage, or disadvantage, is derived from them.

The potential for genetic variation in individuals of a population is virtually unlimited. When generations of individuals of a population undergo sexual reproduction, there is a constant shuffling of alleles. The shuffling of alleles and the interaction of resulting phenotypes with the environment have one of two consequences for the population. Either the relative frequencies of alleles change across generations or they do not. In the former case, evolution has occurred. In chapter 4, a change in the relative frequency of genes in a population across generations was defined as microevolution. In the next sections of this chapter, circumstances that favor microevolution are discussed in more detail.

Section 5.1 Thinking Beyond the Facts

Based on what you know about natural selection, why does evolution occur in groups of individuals of the same species over many generations (i.e., within populations)?

5.2 MUST EVOLUTION HAPPEN?

LEARNING OUTCOME

1. Justify the statement "most populations are evolving."

Evolution is central to biology, but is evolution always occurring in a particular population? Sometimes the rate of evolution is slow, and sometimes it is rapid. But are there times when evolution does not occur at all? The answer to this question lies in the theories of **population genetics,** the study of the genetic events in gene pools.

The Hardy-Weinberg Theorem

In 1908, English mathematician Godfrey H. Hardy and German physician Wilhelm Weinberg independently derived a mathematical model describing what happens to the relative frequency of alleles in a sexually reproducing population over time. Their combined ideas became known as the **Hardy-Weinberg theorem.** It states that the mixing of alleles at meiosis and their subsequent recombination do not alter the relative frequencies of the alleles in future generations, if certain assumptions are met. Stated another way, if certain assumptions are met, evolution will not occur because the relative allelic frequencies will not change from generation to generation, even though the specific mixes of alleles in individuals may vary.

The assumptions of the Hardy-Weinberg theorem are as follows:

1. The population size must be large. Large size ensures that gene frequency will not change by chance alone.
2. Individuals cannot migrate into, or out of, the population. Migration may introduce new alleles into the gene pool or add or delete copies of existing alleles.
3. Mutations must not occur. If they do, mutational equilibrium must exist. Mutational equilibrium exists when mutation from the wild-type allele to a mutant form is balanced by mutation from the mutant form back to the wild type. In either case, no new genes are introduced into the population from this source.
4. Sexual reproduction within the population must be random. Every individual must have an equal chance of mating with any other individual in the population. If this condition is not fulfilled, then some individuals are more likely to reproduce than others, and natural selection may occur.

These assumptions must be met if allelic frequencies are not changing–that is, if evolution is not occurring. Clearly, these assumptions are restrictive, and few, if any, real populations meet them. This means that most populations are evolving. The Hardy-Weinberg theorem, however, does provide a useful theoretical framework for examining changes in allelic frequencies in populations.

The next section explains how, when the assumptions are not met, microevolutionary change occurs.

Section 5.2 Thinking Beyond the Facts

Why is it accurate to say that evolution does not have to be occurring in all populations but most populations are evolving?

5.3 EVOLUTIONARY MECHANISMS

LEARNING OUTCOMES

1. Explain the four mechanisms of evolutionary change.
2. Compare the founder effect and the bottleneck effect, explaining why each is an example of neutral evolution and genetic drift.
3. Explain how selection pressure operates in each mode of natural selection.
4. Critique the statement "Natural selection always eliminates deleterious alleles from populations."

Evolution is neither a creative force working for progress nor a dark force sacrificing individuals for the sake of the group. It is neither moral nor immoral. It has neither a goal nor a mind to conceive a goal. Such goal-oriented thinking is said to be teleological. Evolution is simply a result of some individuals in a population surviving and being more effective at reproducing than others in the population, leading to changes in relative allelic frequencies. This section examines some of the situations when the Hardy-Weinberg assumptions are not met—situations in which gene frequencies change from one generation to the next and evolution occurs.

	Loss of genetic diversity through genetic drift
Parental generation	AA Aa Aa aa
Parental gametes	4/8 A A A A 4/8 a a a a
	Chance selection of 8 gametes ↓
F_1 generation	Aa Aa aa aa
F_1 gametes	2/8 A A 6/8 a a a a a a
	Chance selection of 8 gametes ↓
F_2 generation	Aa aa aa aa
F_2 gametes	1/8 A 7/8 a a a a a a a

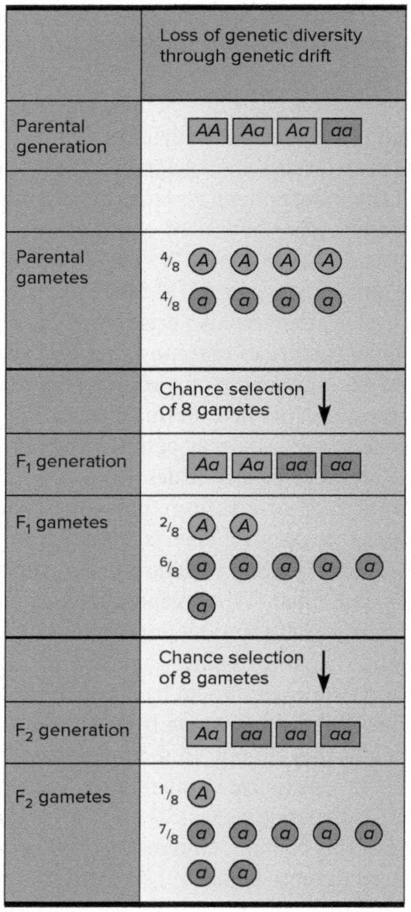

FIGURE 5.1

Genetic Drift. Genetic diversity may be lost as a result of genetic drift. Assume that alleles *a* and *A* are equally adaptive. Allele *a* might be incorporated into gametes more often than *A*, or it could be involved in more fertilizations. In either case, the frequency of *a* increases and the frequency of *A* decreases because of random events operating at the level of gametes.

Neutral Theory and Genetic Drift

Evolution does not always occur because one allele provides a selective advantage over a second allele. Many times two alleles provide neither advantage nor disadvantage to animals in a population. These alleles are said to be selectively neutral. **Neutral theory** describes mechanisms of evolutionary change when alleles are selectively neutral.

Chance often plays an important role in the perpetuation of genes in a population, and the smaller the population, the more significant the chance may be. Fortuitous circumstances, such as a chance encounter between reproductive individuals, may promote reproduction. Some traits of a population survive not because they convey increased fitness, but because they happen to be in gametes involved in fertilization. Chance events influencing the frequencies of genes in populations result in **genetic drift.**

The process of genetic drift is analogous to flipping a coin. The likelihood of getting a head or a tail is equal. The 50:50 ratio of heads and tails is most likely in a large number of tosses. In only 10 tosses, for example, the ratio may be a disproportionate 7 heads and 3 tails. Similarly, the chance of one or the other of two equally adaptive alleles being incorporated into a gamete, and eventually into an individual in a second generation, is equal. Gamete sampling in a small population may show unusual proportions of alleles in any one generation of gametes because meiotic events, like tossing a coin, are random. Assuming that both alleles have equal fitness, these unusual proportions are reflected in the genotypes of the next generation. These chance events may result in a particular allele increasing or decreasing in frequency (figure 5.1). In small populations, inbreeding is also common. Genetic drift and inbreeding are likely to reduce genetic variation within a small population.

In many cases, one member of a pair of alleles is lost in a population. When this happens the surviving allele is said to be fixed in the population. Fixation of an allele can occur within just a few generations of random mating. The likelihood of genetic drift occurring in small populations suggests that a Hardy-Weinberg equilibrium will not occur and evolution is occurring. Two special cases promote genetic drift and are described next.

The Founder Effect

When a few individuals from a parental population colonize new habitats, they seldom carry alleles in the same frequency as the alleles in the gene pool from which they came. The new colony that emerges from the founding individuals is likely to have a distinctive genetic makeup with far less variation than the larger population. This form of genetic drift is the **founder effect.**

The importance of the founder effect in influencing natural populations is controversial. Founder events are rarely observed and difficult to study. In 2004, a hurricane swept through the Bahamas and submerged several islands that were the homes of the brown anole lizard, *Anolis sagrei*. Lizards on these islands were all killed. In 2005, researchers from the University of California, Davis, placed one male and one female lizard on each of these islands from a nearby island that was not submerged. Over the next four years, offspring lizards on the formerly submerged islands were studied and compared to populations from the source island and other islands that were not submerged. These comparisons demonstrated that allelic variation within larger populations on the unsubmerged islands allowed natural selection to influence traits like leg length. The same traits in descendants of founding lizards on the formerly submerged islands were less responsive to natural selection because of limited genetic variation in these populations.

(a)

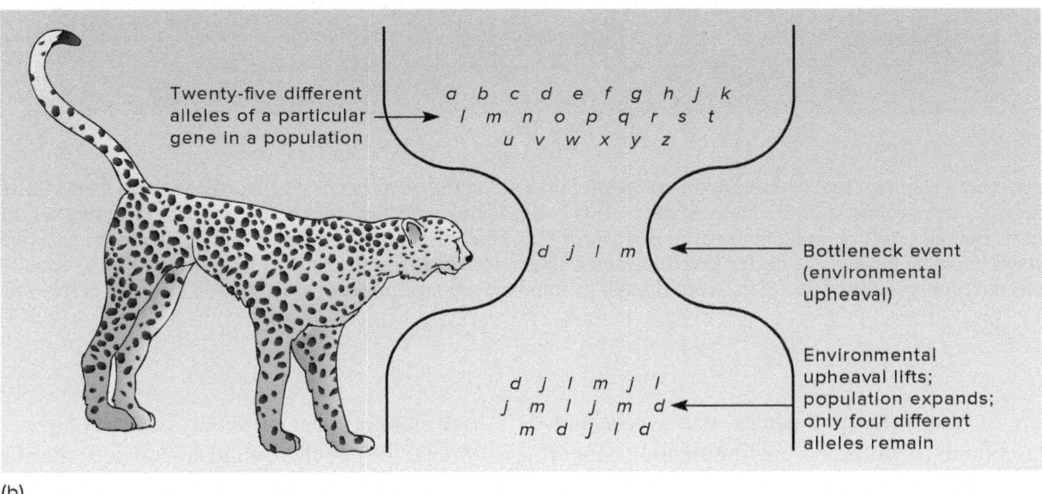

(b)

FIGURE 5.2

Bottleneck Effect. (*a*) Cheetahs (*Acinonyx jubatus*) of South and East Africa are endangered. (*b*) Severe reduction in the original population has caused a bottleneck effect. Even if the population size recovers, genetic diversity has been significantly reduced.
Source: Gary M. Stolz/U.S. Fish & Wildlife Service

The Bottleneck Effect

A second special case of genetic drift can occur when the number of individuals in a population is drastically reduced. For example, cheetah populations in South and East Africa are endangered. During the Pleistocene era, about one million years ago, there were four subspecies of cheetahs in North America, Europe, Asia, and Africa. About 12,000 years ago, they underwent a near extinction event brought about by climate change and human hunting that also affected many other large mammal species. Cheetahs are now found only in eastern and southern Africa, and their populations continue to decline (100,000 in 1900, 10,000 today) due to loss of habitat from farming and bush encroachment, population fragmentation, and declining prey density. Cheetah nucleotide diversity is very low (0.182%) compared to leopards (1.3%) and other large cat species. This genetic diversity is so low that even if population sizes are restored, cheetahs will have only a remnant of the original gene pool. This form of genetic drift is called the **bottleneck effect** (figure 5.2). A similar example concerns the northern elephant seal (*Mirounga angustirostris*), which was hunted to near extinction in the late 1800s for its blubber, which was used to make prized oil (figure 5.3). The population was reduced to about 100 individuals. Because males compete for reproductive rights, very few males actually passed their genes on to the next generation. Legislation to protect the elephant seal was enacted in 1922, and now the population is greater than 100,000 individuals. In spite of this relatively large number, the genetic variability in the population is very low. One study showed no genetic variation at 24 protein-coding loci.

The effects of bottlenecks are not well understood. The traditional interpretation is that decreases in genetic diversity make populations less likely to withstand environmental stress and more susceptible to extinction. That is, a population with high genetic

FIGURE 5.3

Bottleneck Effect. The northern elephant seal (*Mirounga angustirostris*) lives along the western coast of North America from Alaska to Baja, California. It gets its name from the very large proboscis of the male, which is used in producing very loud vocalizations during breeding off the coast of southern California and Baja. Males average 1,800 kg and females average 650 kg. This photograph of a large male and a group of females was taken on San Benito Island, Baja, Mexico. Males compete for females during breeding, and a single male may win the right to mate with up to 50 females. The northern elephant seal was severely overhunted in the late 1800s. Even though its numbers are now increasing, its genetic diversity is very low.

diversity is more likely to have some individuals with a combination of genes that allows them to withstand environmental changes (*see Wildlife Alert, chapter 3*).

Gene Flow

The Hardy-Weinberg theorem assumes that no individuals enter a population from the outside (immigrate) and that no individuals leave a population (emigrate). Immigration or emigration upsets the Hardy-Weinberg equilibrium, resulting in changes in relative allelic frequency (evolution). Changes in relative allelic frequency from the migration of individuals are called **gene flow.** Although some natural populations do not have significant gene flow, most populations do.

The effects of gene flow can differ, depending on the circumstances. The exchange of alleles between an island population and a neighboring continental population, for example, can change the genetic makeup of those populations. If gene flow continues and occurs in both directions, the two populations will become more similar, and this reduces the chances that speciation will occur. The absence of gene flow can lead to genetic isolation and, as discussed in section 5.4, the formation of a new species.

Natural barriers, such as mountain ranges and deserts, can limit gene flow. So can human influences such as the construction of barriers, like highways and The Great Wall of China. Highways can interrupt the connectivity of populations and have been linked to the loss of genetic diversity in desert bighorn sheep.* The melting of polar ice has changed historical patterns of gene flow within polar bear (*Ursus maritimus*) populations. In some populations, polar bears are migrating into more geographically restricted regions that maintain longer periods of continuous sea-ice cover (figure 5.4). In other subpopulations, gene flow is restricted by the absence of sea-ice connections between populations, which historically provided avenues for genetic exchange during the breeding season.

Mutation

Mutations are changes in the structure of genes and chromosomes (*see chapter 3*). The Hardy-Weinberg theorem assumes that no mutations occur or that mutational equilibrium exists. Mutations, however, are a fact of life. Most important, mutations are the origin of all new alleles and a source of variation that may be adaptive for an animal. Mutation counters the loss of genetic material from natural selection and genetic drift, and it increases the likelihood that variations will be present that allow some individuals to survive future environmental shocks. Mutations make extinction less likely.

* Epps CW, Palsbøll JD, Wehausen GK, Roderick GK, Ramey II RR, McCullough DR. 2005. Highways block gene flow and cause a rapid decline in genetic diversity of desert bighorn sheep. Ecology Letters. 8:1029–1038.

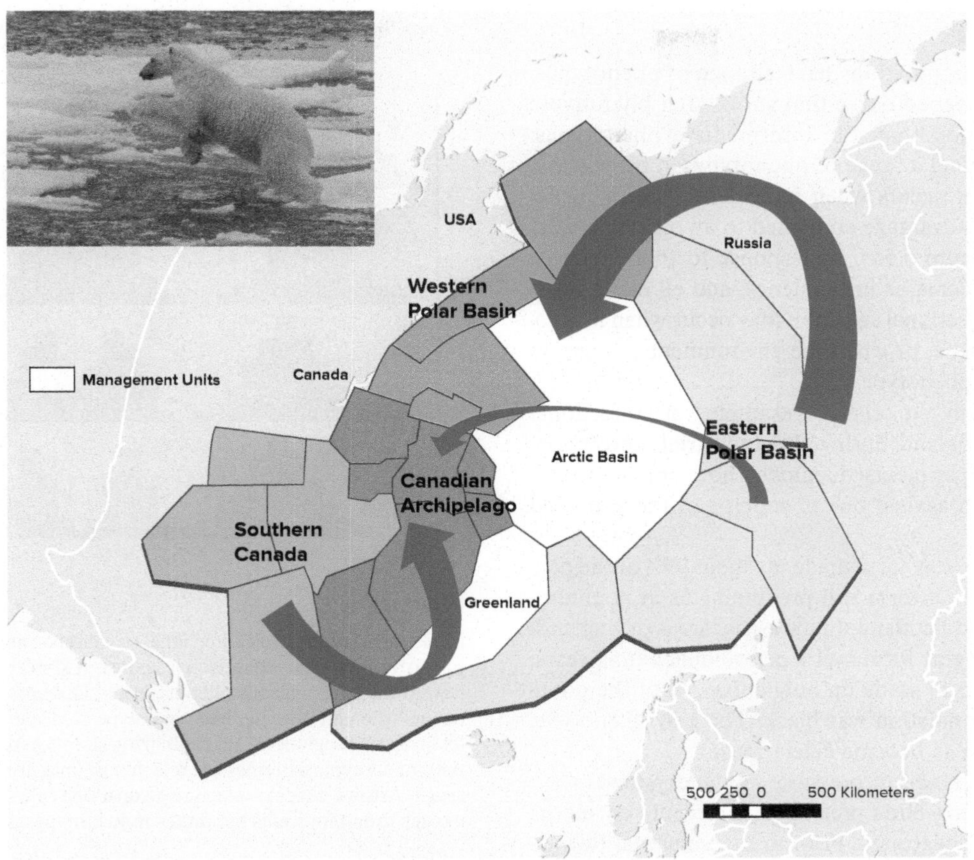

FIGURE 5.4

Gene Flow in Polar Bear (*Ursus maritimus*) Populations. Melting of sea ice has promoted the migration of polar bears from genetically distinct subpopulations into the Canadian Archipelago, which is somewhat insulated from sea-ice melting. This historically altered gene flow pattern has the potential to alter the genetic composition of the species. In small subpopulations within this larger range, the melting of sea ice breaks down ice corridors that historically promoted gene flow and higher genetic diversity.

Source: Peacock E, Sonsthagen SA, Obbard ME, Boltunov A, Regehr EV, et al. (2015) Implications of the Circumpolar Genetic Structure of Polar Bears for Their Conservation in a Rapidly Warming Arctic. PLOS ONE 10(1): e112021. doi:10.1371/journal.pone.0112021 http://journals.plos.org/plosone/article?id=10.1371/journal.pone.0112021

Mutations are random events, and the likelihood of a mutation is not affected by the mutation's usefulness. Organisms cannot filter harmful genetic changes from advantageous changes before they occur.

The effects of mutations vary enormously. Neutral mutations are neither harmful nor helpful to the organism. Neutral mutations may occur in regions of DNA that do not code for proteins. Other neutral mutations may change a protein's structure, but some proteins tolerate minor changes in structure without affecting the function of the protein. Genetic drift may permit the new allele to become established in the population, or the new allele may be lost because of genetic drift. Mutations that do affect protein function are more likely to be detrimental than beneficial. This is true because a random change in an established protein upsets millions of years of natural selection that occurred during the protein's evolution. Mutations in DNA that is incorporated into a gamete have the potential to affect the function of every cell in an individual in the next generation. These mutations are likely to influence the evolution of a group of organisms.

Mutational equilibrium exists when a mutation from the wild-type allele to a mutant form is balanced by a mutation from the mutant back to the wild type. This has the same effect on allelic frequency as if no mutation had occurred. Mutational equilibrium rarely exists, however. **Mutation pressure** is a measure of the tendency for gene frequencies to change through mutation.

Natural Selection Reexamined

The theory of natural selection remains preeminent in modern biology. Natural selection occurs whenever some phenotypes are more successful at leaving offspring than other phenotypes. The tendency for natural selection to occur—and upset Hardy-Weinberg equilibrium—is **selection pressure.** Although natural selection is simple in principle, it is diverse in operation.

Modes of Selection

For certain traits, many populations have a range of phenotypes, characterized by a bell-shaped curve that shows that phenotypic extremes are less common than the intermediate phenotypes. Natural selection may affect a range of phenotypes in three ways.

Directional selection occurs when individuals at one phenotypic extreme are at a disadvantage compared to all other individuals in the population (figure 5.5*a*). In response to this selection, the deleterious gene(s) decreases in frequency, and all other genes increase in frequency. Directional selection may occur when a mutation gives rise to a new gene, or when the environment changes to select against an existing phenotype.

Industrial melanism, a classic example of directional selection, occurred in England during the Industrial Revolution. Museum records and experiments document how environmental changes affected selection against one phenotype of the peppered moth, *Biston betularia*.

In the early 1800s, a gray form made up about 99% of the peppered moth population. That form still predominates in nonindustrial northern England and Scotland. In industrial areas of England, a black form replaced the gray form over a period of about 50 years. In these areas, the gray form made up only about 5% of the population, and 95% of the population was black. The gray phenotype, previously advantageous, had become deleterious.

The nature of the selection pressure was understood when investigators discovered that birds prey more effectively on moths resting on a contrasting background. Prior to the Industrial Revolution, gray moths were favored because they blended with the bark of trees on which they rested. The black moth contrasted with the lighter, lichen-covered bark and was easily spotted by birds (figure 5.6*a*). Early in the Industrial Revolution, however, factories used soft coal, and spewed soot and other pollutants into the air. Soot covered the tree trunks and killed the lichens where the moths rested. Bird predators now could easily pick out gray moths against the black background of the tree trunk, while the black form was effectively camouflaged (figure 5.6*b*).

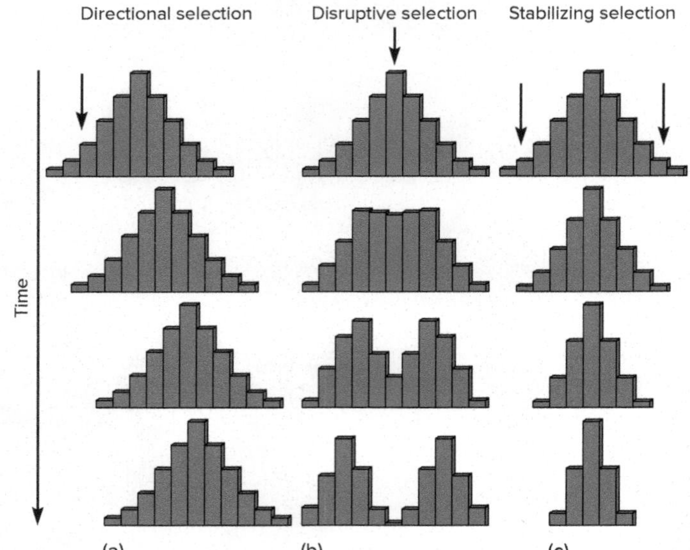

FIGURE 5.5

Modes of Selection. (*a*) Directional selection occurs when individuals at one phenotypic extreme are selected against. It shifts phenotypic distribution toward the advantageous phenotype. (*b*) Disruptive (diversifying) selection occurs when an intermediate phenotype is selected against. It produces distinct subpopulations. (*c*) Stabilizing selection occurs when individuals at both phenotypic extremes are selected against. It narrows at both ends of the range. Arrows indicate selection against one or more phenotypes. The X-axis of each graph indicates the range of phenotypes for the trait in question.

(a)

(b)

FIGURE 5.6

Directional Selection of the Peppered Moth, *Biston betularia*. Each photo shows two forms of the moth: black and gray. (*a*) Prior to the Industrial Revolution, bird predators easily spotted the black form of moth, and the gray form was camouflaged. (*b*) In industrial regions after the Industrial Revolution, selection was reversed because pollution killed lichens that covered the bark of trees where moths rested. Note how clearly the gray form is seen, whereas the black form is less visible.

In the 1950s, the British Parliament enacted air pollution standards that have reduced soot in the atmosphere. As expected, the gray form of the moth has experienced a small but significant increase in frequency.

Another form of natural selection involves circumstances selecting against individuals of an intermediate phenotype (figure 5.5*b*). **Disruptive** or **diversifying selection** produces distinct subpopulations (*see figure 1.2 and the accompanying discussion*). An interesting example of disruptive selection also illustrates another form of selection, sexual selection. **Sexual selection** occurs when individuals have varying success obtaining mates. It often results in the evolution of structures used in combat between males for mates, such as antlers and horns, or ornamentation that attracts individuals of the opposite sex, such as the brightly colored tail feathers of a peacock. Sexual selection is considered by some to be a form of natural selection, but others consider it separately from natural selection.

The plainfin midshipman (*Porichthys notatus*) inhabit depths of up to 400 m along the Pacific coast of North America. They are named for rows of bioluminescent photophores that reminded some of the rows of buttons on a naval uniform. Males move to shallower water for reproduction. The male establishes a nest in rock crevices and cares for young after attracting a female and spawning. Males of the species have one of two body forms (morphs) designated as type I and type II males. Type I males have an exaggerated morphology that probably arose as a result of sexual selection. Their very large heads and wide mouths (figure 5.7) are used in defending territories and nests under rocks. Their exaggerated sonic musculature is used to court females with loud humming vocalizations that can be heard by boaters and beachcombers. Aggressive vocalizations are used to ward off other males. The larger the males, the more effective they are in establishing large nests and providing long-term care for young. Type II males have a smaller, drab morphology and look similar to females. These males (*see figure 5.7*) are sometimes called "sneakers" because they dart into type I males' nests and fertilize eggs without having to invest their time and energy into nest construction, defense, and mate attraction. The type II morphology is apparently not recognized readily by type I males as a threat to their reproductive success. Type II males are tolerated near the nests of type I males, which improves the chances that type II males succeed with their "sneaker" strategy. Disruptive selection favors the retention of the reduced type II morphology because of the energy savings in not having to establish and defend a nest and in not supporting a larger body size. Disruptive selection also favors the retention of the exaggerated type I morphology because that is the body form that can establish territories and nests, defend them from other type I males, and attract females.

When both phenotypic extremes are deleterious, a third form of natural selection—**stabilizing selection**—narrows the phenotypic range (figure 5.5*c*). During long periods of environmental constancy, new variations that arise, or new combinations of genes that occur, are unlikely to result in more fit phenotypes than the genes that have allowed a population to survive for thousands of years, especially when the new variations are at the extremes of the phenotypic range.

A good example of stabilizing selection is the horseshoe crab (*Limulus*), which lives along the Atlantic coast of the United States (*see figure 14.8*). Comparison of the fossil record with living forms indicates that this body form has changed little over 200 million years. Apparently, the combination of characteristics present in this group of animals is adaptive for the horseshoe crab's environment.

FIGURE 5.7

Disruptive Selection. Males of the plainfin midshipman (*Porichthys notatus*) have two body forms. Type I males (center) have exaggerated heads and mouths, which are used in defending territories and nests from other type I males. Type II males (far left) lack the exaggerated heads and mouths and resemble females (far right). These "sneaker" males are tolerated near the nests of type I males and attempt to fertilize eggs deposited by females in type I male nests by darting into the nest during spawning. Disruptive selection has resulted in the maintenance of both of these body forms in plainfin midshipman populations.
©Margaret Marchaterre, Cornell University

Neutralist/Selectionist Controversy

Most biologists recognize that both natural selection and neutral evolution occur, but they may not be equally important in all circumstances. For example, during long periods when environments are relatively constant, and stabilizing selection is acting on phenotypes, genetic drift may operate at the molecular level. Certain genes could be randomly established in a population. Occasionally, however, the environment shifts, and directional or disruptive selection begins to operate, resulting in gene frequency changes (often fairly rapid).

The relative importance of neutral evolution and natural selection in natural populations is debated and is an example of the kinds of debates that occur among evolutionists. These debates concern the mechanics of evolution and are the foundations of science. They lead to experiments that will ultimately present a clearer understanding of evolution.

Balanced Polymorphism and Heterozygote Superiority

Polymorphism occurs in a population when two or more distinct forms exist without a range of phenotypes between them. **Balanced polymorphism** (Gr. *poly,* many + *morphe,* form) occurs when different phenotypes are maintained at relatively stable frequencies in the population and may resemble a population in which disruptive selection operates.

Sickle-cell anemia results from a change in the structure of the hemoglobin molecule. Some of the red blood cells of persons with the disease are misshapen, reducing their ability to carry oxygen. In the heterozygous state, the quantities of normal and sickled cells are roughly equal. Sickle-cell heterozygotes occur in some African populations with a frequency as high as 0.4. The maintenance of the sickle-cell heterozygotes and both homozygous genotypes at relatively unchanging frequencies makes this trait an example of a balanced polymorphism.

Why hasn't natural selection eliminated such a seemingly deleterious allele? The sickle-cell allele is most common in regions of Africa that are heavily infected with the malarial parasite *Plasmodium falciparum.* This parasite is transmitted by mosquitoes and has a life cycle that involves the invasion of red blood cells and liver cells (*see figure 6, appendix C*). Symptoms of the disease include recurring bouts of chills and fever, and it remains one of the greatest killers of humanity. Sickle-cell heterozygotes are less susceptible to malarial infections; if infected, they experience less severe symptoms than do homozygotes without sickled cells. Individuals homozygous for the normal allele are at a disadvantage because they experience more severe malarial infections, and individuals homozygous for the sickle-cell allele are at a disadvantage because they suffer from the severe anemia that the sickle cells cause. The heterozygotes, who usually experience no symptoms of anemia, are more likely to survive than either homozygote. This system is also an example of heterozygote superiority–when the heterozygote is more fit than either homozygote. Heterozygote superiority can lead to balanced polymorphism because perpetuation of the alleles in the heterozygous condition maintains both alleles at a higher frequency than would be expected if natural selection acted only on the homozygous phenotypes.

The processes that change relative allelic frequencies in populations can, over geological time periods (measured in thousands of years), result in the formation of new species. Species can become extinct very quickly when climatic or geological events cause abrupt environmental changes. The formation of new species and the extinction of species were defined in chapter 4 as macroevolution. Even though the formation of new species is difficult to observe directly because of the long time frames involved, the evidence presented in chapter 4 has convinced the vast majority of scientists that macroevolution occurs. The rest of this chapter continues the theme of macroevolutionary change.

SECTION 5.3 THINKING BEYOND THE FACTS

Conservation biologists attempt to preserve genetic diversity within populations. Which of the four evolutionary mechanisms described in this section has/have the potential to increase genetic diversity of populations? What does this say about the challenge faced by conservation biologists?

5.4 SPECIES AND SPECIATION

LEARNING OUTCOMES

1. Assess the usefulness of definitions of a species.
2. Compare the isolating mechanisms involved in each of the three forms of speciation.

Taxonomists classify organisms according to their similarities and differences (*see chapters 1 and 7*). The fundamental unit of classification is the species. Unfortunately, formulating a universally applicable definition of species is difficult. According to the **biological species concept,** a **species** is a group of populations in which genes are actually, or potentially, exchanged through interbreeding.

Although concise, this definition has problems associated with it. Taxonomists often work with morphological characteristics, and the reproductive criterion must be assumed based on morphological and ecological information. Also, some organisms do not reproduce sexually. Obviously, other criteria need to be applied in these cases. Another problem concerns fossil material. Paleontologists describe species of extinct organisms, but how can they test the reproductive criterion? Finally, populations of similar organisms may be so isolated from each other that gene exchange is geographically impossible. To test a reproductive criterion, biologists can transplant individuals to see if mating can occur, but mating of transplanted individuals does not necessarily prove what would happen if animals were together in a natural setting. In chapter 7, an increasingly popular phylogenetic species concept will be discussed. This concept depicts a species as a group of populations that have evolved independently of other groups of populations. Species groups are studied using phylogenetic analysis principles, which are discussed in chapter 7.

Rather than trying to establish a definition of a species that solves all these problems, it is probably better to simply understand the problems associated with the biological definition. In describing species, taxonomists use morphological, physiological, embryological, behavioral, molecular, and ecological criteria, realizing that all of these have a genetic basis.

Speciation is the formation of new species. A requirement of speciation is that subpopulations are prevented from interbreeding. For some reason, gene flow between populations or subpopulations does not occur. This is called **reproductive isolation.** When populations are reproductively isolated, natural selection and genetic drift can result in evolution taking a different course in each subpopulation. Reproductive isolation can occur in different ways.

Some forms of isolation may prevent mating from occurring. For example, barriers such as rivers or mountain ranges may separate subpopulations. Other forms of isolation may be behavioral. If the courtship behavior patterns of two animals are not mutually appropriate, or mating periods differ, mating does not occur. Other forms of isolation prevent successful fertilization and development even though mating may have occurred. Conditions within a female's reproductive tract may prevent fertilization by sperm of another species. The failure of hybrids to produce offspring of their own is a form of isolation that occurs even though mating and fertilization do occur. This hybrid nonviability is caused by structural differences between chromosomes–preventing the formation of viable gametes because chromosomes cannot synapse properly during meiosis.

Allopatric Speciation

Allopatric (Gr. *allos,* other + *patria,* fatherland) **speciation** occurs when subpopulations become geographically isolated from one another. For example, a mountain range or river may permanently separate members of a population. Adaptations to different environments or genetic drift in these separate populations may result in members not being able to mate successfully with each other, even if experimentally reunited. Many biologists believe that allopatric speciation is the most common kind of speciation.

The finches that Darwin saw on the Galápagos Islands are a classic example of allopatric speciation, as well as adaptive radiation (*see Evolutionary Insights, page 82, and chapter 4*). Adaptive radiation occurs when a number of new forms diverge from an ancestral form, usually in response to the opening of major new habitats.

Fourteen species of finches evolved from the original finches that colonized the Galápagos Islands. Ancestral finches, having emigrated from the mainland, probably were distributed among a few of the islands of the Galápagos. Populations became isolated on various islands over time, and though the original population probably displayed some genetic variation, even greater variation arose. The original finches were seed eaters, and after their arrival, they probably filled their preferred habitats rapidly. Variations within the original finch population may have allowed some birds to exploit new islands and habitats where no finches had been. Mutations changed the genetic composition of the isolated finch populations, introducing further variations. Natural selection favored the retention of the variations that promoted successful reproduction.

The combined forces of isolation, mutation, and natural selection allowed the finches to diverge into a number of species with specialized feeding habits (*see figure 4.4*). Of the 14 species of finches, 6 have beaks specialized for crushing seeds of different sizes. Others feed on flowers of the prickly pear cactus or in the forests on insects and fruit.

Parapatric Speciation

Another form of speciation, called **parapatric** (Gr. *para,* beside) **speciation,** occurs in small, local populations, called **demes.** For example, all of the frogs in a particular pond or all of the sea urchins in a particular tidepool make up a deme. Individuals of a deme are more likely to breed with one another than with other individuals in the larger population, and because they experience the same environment, they are subject to similar selection pressures. Demes are not completely isolated from each other because individuals, developmental stages, or gametes can move among demes of a population. On the other hand, the relative isolation of a deme may mean that its members experience different selection pressures than other members of the population. If so, speciation can occur. Although most evolutionists theoretically agree that parapatric speciation is possible, no certain cases are known. Parapatric speciation is therefore considered of less importance in the evolution of animal groups than allopatric speciation.

Sympatric Speciation

A third kind of speciation, called **sympatric** (Gr. *sym,* together) **speciation,** occurs within a single population (*see Evolutionary Insights, page 82*). Even though organisms are sympatric, they still may be reproductively isolated from one another. In order to demonstrate sympatric speciation, researchers must demonstrate that two species share a common ancestor and then that the two species arose without any form of geographic isolation. The latter is especially difficult to demonstrate. The driving forces for speciation are difficult to reconstruct because current ecological and selective factors may not reflect those present in the evolutionary past.

In spite of these difficulties, evidence is mounting that sympatric speciation plays a larger role in speciation than previously thought. Studies of indigobirds from Africa suggest sympatric speciation. Indigobirds lay their eggs in the nests of other bird species. They are called brood parasites. When eggs hatch, indigobird chicks learn the song of the host species that rears them. Mating is then more likely to occur between indigobirds reared by the same host species. Molecular evidence suggests genetic differences between species that are compatible with recent origins and sympatric speciation.

Sympatric speciation has also been important within the cat family (Felidae). Molecular and ecological evidence suggests that over 50% of speciation events in cat evolution have been sympatric in nature. Closely related cat species partition resources based on 24-h activity patterns, preferred habitats, and preferred foods. Such partitioning apparently led to reproductive isolation in spite of shared geographical locations.

Section 5.4 Thinking Beyond the Facts

How does reproductive isolation occur in each of the three forms of speciation described in this section?

5.5 RATES OF EVOLUTION

LEARNING OUTCOME

1. Compare phyletic gradualism and punctuated equilibrium models of evolution.

Charles Darwin perceived evolutionary change as occurring gradually over millions of years. This concept, called **phyletic gradualism,** has been the traditional interpretation of the tempo, or rate, of evolution.

EVOLUTIONARY INSIGHTS

Speciation of Darwin's Finches

When Charles Darwin visited the Galápagos Islands in 1835, he observed the dark-bodied finches whose adaptive radiation has become a classic example of speciation (box figure 5.1). Studies of these finches have provided insight into some of the ways in which speciation can occur. Peter R. and B. Rosemary Grant have been studying these finches for more than 40 years. They have directly observed microevolutionary change reflected in bill morphology in response to changes in rainfall and food availability. Other molecular studies have also contributed to our knowledge of the adaptive radiation of this group of birds.

Molecular studies of mitochondrial DNA have identified the most likely South American relatives of Darwin's finches, members of the grassquit genus *Tiaris*. Comparisons of the mitochondrial DNA of this group with Darwin's finches suggest that the latter colonized some of the Galápagos Islands not more than 3 million years ago (mya). A very rapid adaptive radiation occurred, with the number of finch species doubling approximately every 750,000 years. No other group of birds studied has undergone a more rapid evolutionary diversification (*see figure 4.4*). Darwin's finches have served as a model to answer questions of how and why species diverge.

The traditional explanation of speciation within Darwin's finches is based on the allopatric model discussed in this chapter. This explanation is based upon differences in food resources, and the observations of the Grants have provided support for this model. Geographic isolation of populations of finches occurred as finches competed for limited food resources. Finch survival depended on food availability, bill shape and size, and the presence or absence of competing species. Species diverged as bill morphology changed and as populations became ecologically isolated (*see figure 4.4*). Interbreeding became less likely. Molecular studies have complemented these observations. Genes have been identified that regulate bill size and shape and are being tracked through finch species.

The Grants have discovered, however, that the allopatric model is not the entire explanation for finch adaptive radiation. Three million years ago, the Galápagos Islands were much simpler than they are today. In fact, there were fewer islands when they were first colonized by finches. Apparently, the number of finch species increased as the number of islands increased as a result of volcanic activity. The increasing number of islands and oscillations in temperature and precipitation naturally affected vegetation. Habitats available for finches became more diverse and complex. The original warm, wet islands favored long, narrow bills that were used in gathering nectar and insects. The islands' moisture now fluctuates and the climate is more seasonal. The increasing diversity in habitats and food supply over 3 million years apparently promoted very rapid speciation among the finch populations.

The Grants have also discovered that sympatric forces probably have also promoted speciation. Different species of finches that live on the same island rarely hybridize because visual and acoustic cues are used in mate choice. In these finches, only the males sing, and both male and female offspring respond to and learn the song of their fathers. The young associate the song and the bill shape of their fathers. Females tend to mate with males that have the bill shape and song of their fathers. The fact that learned behaviors are influencing speciation introduces new sets of variables that may influence speciation. Errors in learning, variations in the vocal apparatus of individuals, and characteristics of sound transmission through the environment could all result in changes in song characteristics and could influence mate choice.

Observations of Darwin's finches have helped revolutionize biology. They played an important role in the development of Darwin's theory of evolution by natural selection, and they continue to provide important evidence of how evolution occurs at ecological and molecular levels.

(a)

(b)

BOX FIGURE 5.1 Speciation of Darwin's Finches. Speciation and adaptive radiation of Darwin's finches has been used as a classic example of allopatric speciation. Isolation of finches on different islands, and differences in food resources on those islands, selected for morphological differences in finch bills. For example, (*a*) the warbler finch (*Certhidea olivacea*) has a bill that is adapted for probing for insects and (*b*) the large ground finch (*Geospiza magnirostris*) has a bill that is adapted for crushing seeds. Studies show that increasing numbers of islands over the last 3 million years and changes in temperature and precipitation resulted in very rapid speciation. In addition, sympatric influences regarding the role of the males' song and bill shape probably also promoted speciation.

Some evolutionary changes, however, happen very rapidly. Studies of the fossil record show that many species do not change significantly over millions of years. These periods of stasis (Gr. *stasis,* standing still), or equilibrium, are interrupted when a group encounters an ecological crisis, such as a change in climate or a major geological event. Over the next 10,000 to 100,000 years, a variation that previously was selectively neutral or disadvantageous might now be advantageous. Alternatively, geological events might result in new habitats becoming available. (Events that occur in 10,000 to 100,000 years are almost instantaneous in an evolutionary time frame.) This geologically brief period of change "punctuates" the previous million or so years of equilibrium and eventually settles into the next period of stasis (figure 5.8). Long periods of stasis interrupted by brief periods of change characterize the **punctuated equilibrium model** of evolution.

Biologists have observed such rapid evolutionary changes in small populations. In a series of studies over a 20-year period, Peter R. Grant has shown that natural selection results in rapid morphological changes in the bills of Galápagos finches. A long, dry period from the middle of 1976 to early January 1978 resulted in birds with larger, deeper bills. Early in this dry period, birds quickly consumed smaller, easily cracked seeds. As they were forced to turn to larger seeds, birds with weaker bills were selected against, resulting in a measurable change in the makeup of the finch population of the island Daphne Major.

The evolution of multiple species of cichlid fish in Lake Victoria (*see chapter 1*) in the last 14,000 years is another example of rapid evolutionary change leading to speciation. One cichlid species (*Haplochromis pyrrhocephalus*) was nearly extirpated from Lake Victoria following the introduction of the Nile perch. Subsequent fishing pressure on the Nile perch allowed this cichlid population to recover somewhat; however, decreased water quality and oxygen levels placed additional selection pressures on its survival. Over a brief span of 20 years, researchers have documented a 64% increase in gill surface area and corresponding changes in head morphology that are allowing this cichlid population to recover once again.

Periods of stasis in the punctuated equilibrium model may be the result of stabilizing selection during times when environmental conditions are not changing. The ability of some organisms to escape changing environments through migration when environments are changing may also promote stasis.

One advantage of the punctuated equilibrium model is its explanation for the fossil record not always showing transitional stages between related organisms. The absence of transitional forms can often be attributed to fossilization being an unlikely event; thus, many transitional forms disappeared without leaving a fossil record. Because punctuated equilibrium involves rapid changes in small, isolated populations, preservation of intermediate forms in the fossil record is even less likely. The rapid pace (geologically speaking) of evolution resulted in apparent "jumps" from one form to another.

Phyletic gradualism and punctuated equilibrium are both valid models that explain evolutionary rates. Gradualism best describes the evolutionary history of some groups (e.g., mammals). Punctuated equilibrium best describes the evolutionary history of others (e.g., some marine invertebrates). For still other groups, evolution has been accented by periods when gradualism prevailed and other periods of rapid change and stasis.

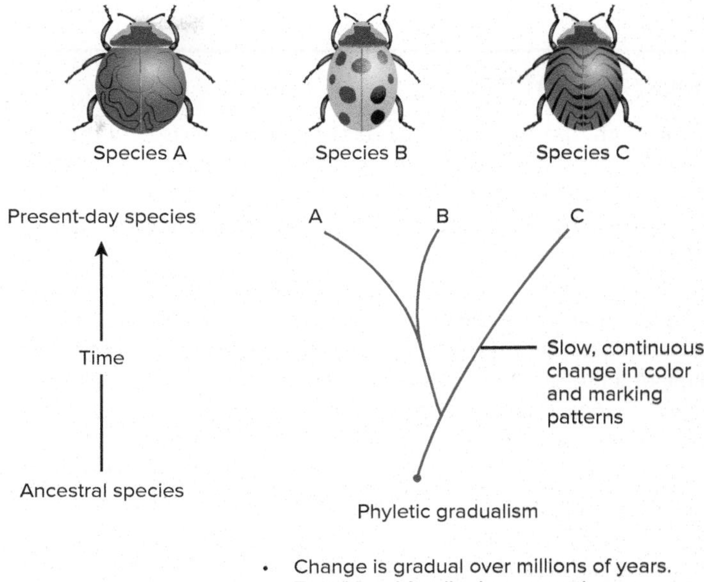

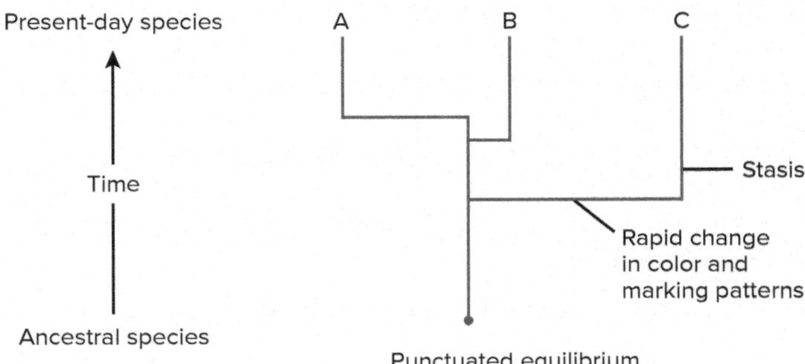

FIGURE 5.8

Rates of Evolution. A comparison of phyletic gradualism and punctuated equilibrium in three hypothetical beetle species. (*a*) In the phyletic gradualism model of evolution, changes are gradual over long time periods. Note that this tree implies a gradual change in color and marking patterns in the three beetle species. (*b*) In the punctuated equilibrium model of evolution, rapid periods of change interrupt long periods of stasis. This tree implies that the color and marking patterns in the beetles changed rapidly and did not change significantly during long periods of stabilizing selection (stasis).

Section 5.5 Thinking Beyond the Facts

How would you explain the presence of gaps in the fossil record of a group of organisms to someone challenging evolution? List at least three reasons why gaps appear.

5.6 MOLECULAR EVOLUTION

LEARNING OUTCOMES

1. Hypothesize the differences between a comparison of the nonconserved DNA sequences of a horse and a zebra, and the nonconserved DNA sequences of a frog and a fish.
2. Explain the role of gene duplication in the evolution of new genes.

Many evolutionists study changes in animal structure and function that are observable on a large scale–for example, changes in the shape of a bird's bill or in the length of an animal's neck. All evolutionary change, however, results from changes in the base sequences in DNA and amino acids in proteins. Molecular evolutionists investigate evolutionary relationships among organisms by studying DNA and proteins. For example, cytochrome *c* is a protein present in the cellular respiration pathways in all eukaryotic organisms (table 5.1). Cellular respiration is the set of metabolic pathways that convert energy in organic molecules, such as the simple sugar glucose, into energy tied up in the bonds of adenosine triphosphate (ATP). ATP is the form of chemical energy immediately useful in cells. Organisms that other research has shown to be closely related have similar cytochrome *c* molecules. That cytochrome *c* has changed so little during hundreds of millions of years does not suggest that mutations of the cytochrome *c* gene do not occur. Rather, it suggests that mutations of the cytochrome *c* gene are nearly always detrimental and are selected against. Because it has changed so little, cytochrome *c* is said to have been conserved evolutionarily and is very useful for establishing relationships among distantly related organisms.

TABLE 5.1
AMINO ACID DIFFERENCES IN CYTOCHROME C FROM DIFFERENT ORGANISMS

ORGANISMS	NUMBER OF VARIANT AMINO ACID RESIDUES
Cow and sheep	0
Cow and whale	2
Horse and cow	3
Rabbit and pig	4
Horse and rabbit	5
Whale and kangaroo	6
Rabbit and pigeon	7
Shark and tuna	19
Tuna and fruit fly	21
Tuna and moth	28
Yeast and mold	38
Wheat and yeast	40
Moth and yeast	44

Not all proteins are conserved as rigorously as cytochrome *c*. Some regions of DNA that do not code for proteins can change without detrimental effects and accumulate base changes over relatively short periods of time. Comparing these regions of DNA can provide information on the relationships among closely related organisms.

Gene Duplication

Recall that most mutations are selected against. Sometimes, however, an extra copy of a gene is present. One copy may be modified, but as long as the second copy furnishes the essential protein, the organism is likely to survive. Gene duplication, the accidental duplication of a gene on a chromosome, is one way that extra genetic material can arise.

Vertebrate hemoglobin and myoglobin are believed to have arisen from a common ancestral molecule (*see figure 4.14*). Hemoglobin carries oxygen in red blood cells, and myoglobin is an oxygen storage molecule in muscle. The ancestral molecule probably carried out both functions. However, about 800 mya, gene duplication followed by mutation of one gene resulted in the formation of two polypeptides: myoglobin and hemoglobin. Further gene duplications over the last 500 million years probably explain why most vertebrates, other than primitive fishes, have hemoglobin molecules consisting of four polypeptides.

SECTION 5.6 THINKING BEYOND THE FACTS

There are six subspecies of wild turkeys* (Meleagris gallopavo) *in North America. If one wanted to investigate evolutionary relationships among these subspecies, would it be better to use the cytochrome* c *gene or a non-protein-coding region of DNA? Explain.

5.7 MOSAIC EVOLUTION

LEARNING OUTCOME

1. Explain the concept of mosaic evolution.

As discussed earlier, rates of evolution can vary both in populations and in molecules and structures. A species is a mosaic of different molecules and structures that have evolved at different rates. Some molecules or structures are conserved in evolution; others change more rapidly. The basic design of a bird provides a simple example. All birds are easily recognizable as birds because of highly conserved structures, such as feathers, bills, and a certain body form. Particular parts of birds, however, are less conservative and have a higher rate of change. Wings have been modified for hovering, soaring, and swimming. Similarly, legs have been modified for wading, swimming, and perching. These are examples of **mosaic evolution.**

SECTION 5.7 THINKING BEYOND THE FACTS

What is another example of mosaic evolution?

Summary

5.1 **Populations and Gene Pools**

- Organic evolution is a change in the frequency of alleles in a population. Virtually unlimited genetic variation, in the form of new alleles and new combinations of alleles, increases the chances that a population will survive future environmental changes.

5.2 **Must Evolution Happen?**

- Population genetics is the study of events occurring in gene pools. The Hardy-Weinberg theorem states that if certain assumptions are met, gene frequencies of a population remain constant from generation to generation.

5.3 **Evolutionary Mechanisms**

- The assumptions of the Hardy-Weinberg theorem, when not met, define circumstances under which evolution will occur.
- Neutral theory describes mechanisms of evolutionary change when alleles are selectively neutral. Genetic drift occurs when chance events cause allelic frequencies to change. Two forms of genetic drift are the founder effect and the bottleneck effect.
- Gene flow occurs when allelic frequencies change as a result of migration into or out of a population.
- Mutations are changes in the structure of genes and chromosomes. They are the source of new alleles and genetic variation. Mutational equilibrium rarely exists, and thus, mutations usually result in changing allelic frequencies.
- The tendency for allelic frequencies to change through natural selection is called selection pressure. Selection may be directional, disruptive, or stabilizing.
- Balanced polymorphism occurs when two or more phenotypes are maintained in a population. Heterozygote superiority can lead to balanced polymorphism.

5.4 **Species and Speciation**

- According to a biological definition, a species is a group of populations within which there is potential for the exchange of genes. Significant problems are associated with the application of this definition.
- Speciation requires reproductive isolation.
- Allopatric speciation occurs when subpopulations become geographically isolated. It is probably the most common form of speciation.
- Parapatric speciation occurs in small local populations, called demes.
- Sympatric speciation occurs within a single population. Isolation occurs as a result of changing activity patterns, courtship behaviors, and partitioned habitats.

5.5 **Rates of Evolution**

- Phyletic gradualism is a model of evolution that depicts change as occurring gradually, over millions of years. Punctuated equilibrium is a model of evolution that depicts long periods of stasis interrupted by brief periods of relatively rapid change.

5.6 **Molecular Evolution**

- The study of rates of molecular evolution helps establish evolutionary interrelationships among organisms. A mutation may modify a duplicated gene, which then may serve a function other than its original role.

5.7 **Mosaic Evolution**

- A species is a mosaic of different molecules and structures that have evolved at differing rates.

Concept Review Questions

1. Groups of individuals of the same species occupying a given area at the same time and sharing a common set of genes are called
   a. clades.
   b. demes.
   c. populations.
   d. species units.
2. The Hardy-Weinberg theorem predicts that allele frequencies will remain constant in populations (evolution will not occur) when all of the following are true, except one. Select the exception.
   a. The population must be large so that genetic drift is not occurring.
   b. Migration into a population ensures new alleles are randomly distributed within a population.
   c. All individuals within the population have an equal opportunity for reproduction.
   d. No mutations are occurring or mutational equilibrium exists.
3. If genetic drift occurs in a population, then
   a. the population is probably large.
   b. the population will likely suffer a loss of alleles and become more genetically uniform.
   c. directional selection is occurring.
   d. gene flow will prevent loss of alleles.
4. A community of ground nesting birds and lizards experiences an environmental change that expands the area of arid habitat favored by the lizards and that is less usable by the birds for nesting. Which of the following scenarios would be most likely for this community?
   a. Directional selection could result in an increased prevalence of alleles that promote drought tolerance in the birds.
   b. Disruptive selection could result in the formation of two species of birds.
   c. Stabilizing selection could promote the formation of two species of lizards.
   d. Directional selection could promote the formation of two species of lizards.
5. Rapid periods of genetic change followed by extended periods of stabilizing selection and evolutionary stasis describe
   a. phyletic gradualism.
   b. punctuated equilibrium.
   c. parapatric speciation.
   d. sympatric speciation.

Analysis and Application Questions

1. Can natural selection act on variations that are not inherited? (Consider, for example, physical changes that arise from contracting a disease.) If so, what is the effect of that selection on subsequent generations?
2. In what way does overuse of antibiotics and pesticides increase the likelihood that these chemicals will eventually become ineffective? This is an example of which one of the three modes of natural selection?
3. What are the implications of the "bottleneck effect" for wildlife managers who try to help endangered species, such as the whooping crane, recover from near extinction?
4. What does it mean to think of evolutionary change as being goal-oriented? Explain why this way of thinking is wrong.
5. Imagine that two species of butterflies resemble one another closely. One of the species (the model) is distasteful to bird predators, and the other species (the mimic) is not. How could directional selection have resulted in the mimic species evolving a resemblance to the model species?

Animals like these giraffes *(Giraffa camelopardalis)* interact with their environment every moment they are alive. Interactions with their physical and biotic environments help define the limits and possibilities of their lives.

6 Ecology: Preserving the Animal Kingdom

Chapter Outline

6.1 Animals and Their Abiotic Environment
Energy
Temperature
Other Abiotic Factors
6.2 Biotic Factors: Populations
Population Growth
Population Regulation
Intraspecific Competition
6.3 Biotic Factors: Interspecific Interactions
Herbivory and Predation
Interspecific Competition
Coevolution
Symbiosis
Other Interspecific Adaptations
6.4 Communities
The Ecological Niche
Community Stability
6.5 Trophic Structure of Ecosystems
6.6 Cycling within Ecosystems
6.7 Ecological Problems
Human Population Growth
Earth's Resources and Global Inequality

All animals have certain requirements for life. In searching out these requirements, animals come into contact with other organisms and their physical environment. These encounters result in a multitude of interactions among organisms and alter even the physical environment. Understanding basic ecological principles helps us understand why animals live in certain places, why animals eat certain foods, and why animals interact with other animals in specific ways. It is also the key to understanding how human activities can harm animal populations and what we must do to preserve animal resources. The following discussion focuses on ecological principles that are central to understanding how animals live in their environment.

6.1 ANIMALS AND THEIR ABIOTIC ENVIRONMENT

LEARNING OUTCOMES

1. Differentiate biotic and abiotic ecological factors in an animal's habitat.
2. Describe how energy is used by a heterotroph.
3. Contrast the survival strategies of endotherms and ectotherms when environmental conditions become unfavorable and food resources become scarce.

Ecology (Gr. *oikos*, house + *logia*, the study of) is the study of the relationships between organisms and their environment and to other organisms. It is the study of the interactions of an animal with its **habitat**, which includes all living (biotic) and nonliving (abiotic) characteristics of the area in which the animal lives. Abiotic characteristics of a habitat include the availability of oxygen and inorganic ions, light, temperature, and current or wind velocity. Physiological ecologists who study abiotic influences have found that animals live within a certain range of values, called the **tolerance range,** for any environmental factor. At either limit of the tolerance range, one or more essential functions cease. A certain range of values within the tolerance range, called the **range of optimum,** defines the conditions under which an animal is most successful (figure 6.1).

Combinations of abiotic factors are necessary for an animal to survive and reproduce. When one of these is out of an animal's tolerance range, it becomes a **limiting factor.** For example, even though a stream insect may have the proper substrate for shelter, adequate current to bring in food and aid in dispersal, and the proper ions to ensure growth and development, inadequate supplies of oxygen make life impossible.

Often, an animal's response to an abiotic factor is to orient itself with respect to it; such orientation is called **taxis.** For example, a response to light is called phototaxis. If an animal favors well-lighted environments and moves toward a light source, it is displaying positive phototaxis. If it prefers low light intensities and moves away from a light source, it displays negative phototaxis.

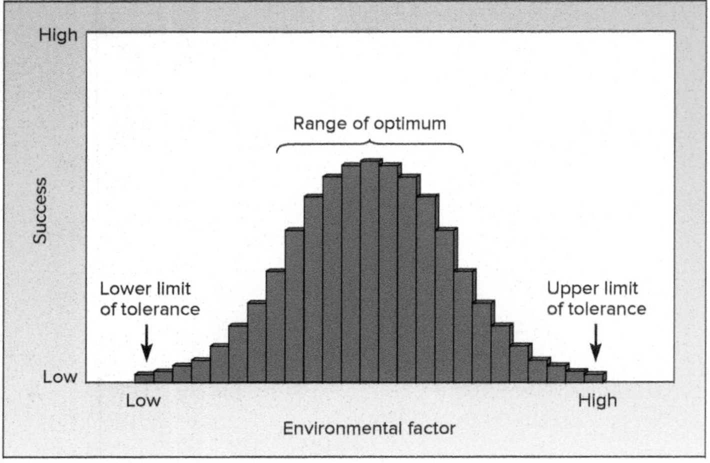

FIGURE 6.1

Tolerance Range of an Animal. Plotting changes in an environmental factor versus some index of success (perhaps egg production, longevity, or growth) shows an animal's tolerance range. The graphs that result are often, though not always, bell-shaped. The range of optimum is the range of values of the factor within which success is greatest. The range of tolerance and range of optimum may vary, depending on an animal's stage of life, health, and activity.

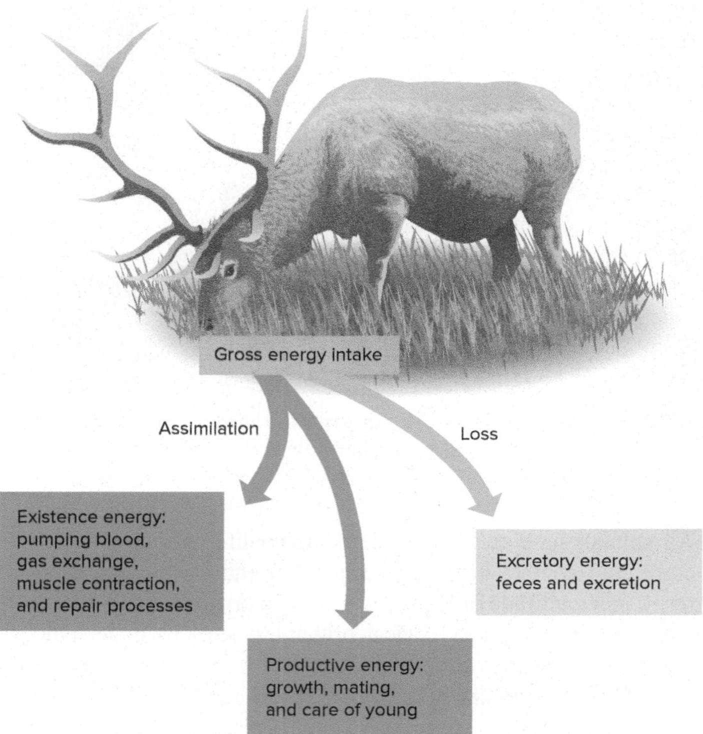

FIGURE 6.2

Energy Budgets of Animals. The gross energy intake of an animal is the sum of energy lost in excretory pathways plus energy assimilated for existence and productive functions. The relative sizes of the boxes in this diagram are not necessarily proportional to the amount of energy devoted to each function. An animal's gross energy intake, and thus the amount of energy devoted to productive functions, depends on various internal and external factors (e.g., time of year and reproductive status).

Energy

Energy is the ability to do work. For animals, work includes everything from foraging for food to moving molecules around within cells. To supply their energy needs, animals ingest other organisms; that is, animals are **heterotrophic** (Gr. *hetero,* other + *tropho,* feeder). **Autotrophic** (Gr. *autos,* self + *tropho,* feeder) organisms (e.g., plants, algae, and some protists) carry on photosynthesis or other carbon-fixing activities that supply their food source. An accounting of an animal's total energy intake and a description of how that energy is used and lost is an **energy budget** (figure 6.2).

The total energy contained in the food an animal eats is the gross energy intake. Some of this energy is lost in feces and through excretion (excretory energy); some of this energy supports minimal maintenance activities, such as pumping blood, exchanging gases, and supporting repair processes (existence energy); and any energy left after existence and excretory functions can be devoted to growth, mating, nesting, and caring for young (productive energy). Survival requires that individuals acquire enough energy to supply these productive functions. Favorable energy budgets are sometimes difficult to attain, especially in temperate regions where winter often makes food supplies scarce.

Temperature

An animal expends part of its existence energy in regulating body temperature (*see chapter 28*). Temperature influences the rates of chemical reactions in animal cells (metabolic rate) and affects the animal's overall activity. The body temperature of an animal seldom remains constant because of an inequality between heat loss and heat gain. Heat energy can be lost to objects in an animal's surroundings as infrared and heat radiation, to the air around the animal through convection, and as evaporative heat. On the other hand, heat is gained from solar radiation, infrared and heat radiation from objects in the environment, and relatively inefficient metabolic activities that generate heat as a by-product of cellular functions. Thermoregulatory needs influence many habitat requirements, such as the availability of food, water, and shelter.

In order to survive when food becomes scarce, when an animal is not feeding, or when environmental conditions become unfavorable, many animals conserve energy by allowing their body temperatures to fall and metabolic activities to decrease. Birds and mammals are **endothermic**. They control body temperatures using metabolically produced heat. In these animals, periods of unusually low body temperatures and metabolic rates are referred to as **controlled hypothermia** (Gr. *hypo*, under + *therme*, heat). Controlled hypothermia can take the form of daily torpor, hibernation, and aestivation. These processes are now considered different forms of the same set of physiological processes. They differ by the extent to which body temperature falls, the duration of the state, and the season in which it occurs.

Daily torpor is a time of decreased metabolism and lowered body temperature that may occur daily in bats, hummingbirds, and some other small birds and mammals who must feed almost constantly when they are active. Torpor allows these animals to survive brief periods when they do not feed.

Hibernation is a time of decreased metabolism and lowered body temperature that may last for weeks or months. Hibernation occurs in mammals, such as rodents, shrews, and bats. It also occurs in a few birds (e.g., the common poorwill, *Phalaenoptilus nuttallii*). In preparation for hibernation, an animal stores energy as body fat. During hibernation, an animal's metabolic rate drops, heart and respiratory rates fall, and the set point of its thermoregulatory center usually drops to near environmental temperature. Thermoregulation is not suspended. For example, a woodchuck's (*Marmota monax*) body temperature may be maintained at 7°C when air temperature averages 6°C. The winter sleep of bears is an exception. The body temperature of a hibernating bear only drops from 37°C to about 30°C, and bears can quickly wake and become active. Because the metabolic rate of a sleeping bear dramatically falls, as is the case for other hibernators, the winter sleep of bears is now considered another variation on the hibernation process.

Aestivation is a period of inactivity in some animals that must withstand extended periods of heat and drying. The animal usually enters a burrow as its environment begins to dry. It generally does not eat or drink and emerges again after moisture returns. Aestivation is common in many invertebrates, reptiles, and amphibians (*see figure 19.14b*). Lungfish (*see figure 18.11*) also enter aestivation when their aquatic habitats dry. Aestivation is less common in mammals, but is thought to occur in some lemurs and hedgehogs.

Animals like amphibians and nonavian reptiles are **ectothermic**. They derive most of their body heat from their environment. When environmental temperatures fall, they also become hypothermic. Unlike endotherms, they usually do not use metabolic processes to regulate body temperature. In temperate climates, their body temperatures and metabolic rates fall as environmental temperature falls. Amphibians and turtles often burrow into the mud of a pond or lake to overwinter. Snakes and lizards may congregate in a common site, where heat loss from a group of individuals is reduced as compared to single animals. Because body temperature is not regulated metabolically, animals can die from freezing during very harsh winters. This uncontrolled hypothermia in ectotherms is called **brumation**.

Other Abiotic Factors

Other important abiotic factors for animals include moisture, light, geology, and soils. All life's processes occur in the watery environment of the cell. Water that is lost must be replaced. The amount of light and the length of the light period in a 24-hour time span is an accurate index of seasonal change. Animals use light for timing many activities, such as reproduction and migration. Geology and soils often directly or indirectly affect organisms living in an area. Characteristics such as texture, amount of organic matter, fertility, and water-holding ability directly influence the number and kinds of animals living either in or on the soil. These characteristics also influence the plants upon which animals feed.

Section 6.1 Thinking Beyond the Facts

Those of us living in the United States comprise about 5% of the world's population. We are responsible for consuming 25% of the world's energy. Look again at the energy budget in figure 6.2. What part of our personal energy budgets contribute to our overconsumption?

6.2 BIOTIC FACTORS: POPULATIONS

Learning Outcomes

1. Compare survivorship attributes of primate and grasshopper populations.
2. Compare populations of animals during exponential growth phases to populations of animals during carrying capacity phases of logistic growth.
3. Differentiate between density-independent and density-dependent factors in population regulation.
4. Explain why intraspecific competition is often intense.

Biotic characteristics of a habitat include interactions that occur within an individual's own species as well as interactions with organisms of other species. Examples of biotic characteristics include how populations grow and how growth is regulated, food availability and competition for that food, and numerous other interactions between species that are the result of shared evolutionary histories.

Populations are groups of individuals of the same species that occupy a given area at the same time and have unique attributes. Two of the most important attributes involve the potential for population growth and the limits that the environment places on population growth.

Population Growth

Animal populations change over time as a result of birth, death, and dispersal. One way to characterize a population is with regard to how the chances of survival of an individual in the population change with age (figure 6.3). The Y-axis of a survivorship graph is a logarithmic plot of numbers of survivors, and the X-axis is a linear plot of age. There are three kinds of survivorship curves. Individuals in type I (convex) populations survive to an old age, and then die rapidly. Environmental factors are relatively unimportant in influencing mortality, and most individuals live their potential life span. Most primate populations approach type I survivorship. Individuals in type II (diagonal) populations have a constant probability of death throughout their lives. The environment has an important influence on death and is no harsher on the young than on the old. Populations of birds and rodents often have type II survivorship curves. Individuals in type III (concave) populations experience very high juvenile mortality. Those reaching adulthood, however, have a much lower mortality rate. Fishes and many invertebrates display type III survivorship curves.

An important difference in populations characterized by these three survivorship attributes is the level of parental care in each. In type I populations, relatively few offspring are produced, but parents invest a large portion of their resources to the care and protection of their offspring. At the opposite extreme, type III parents provide

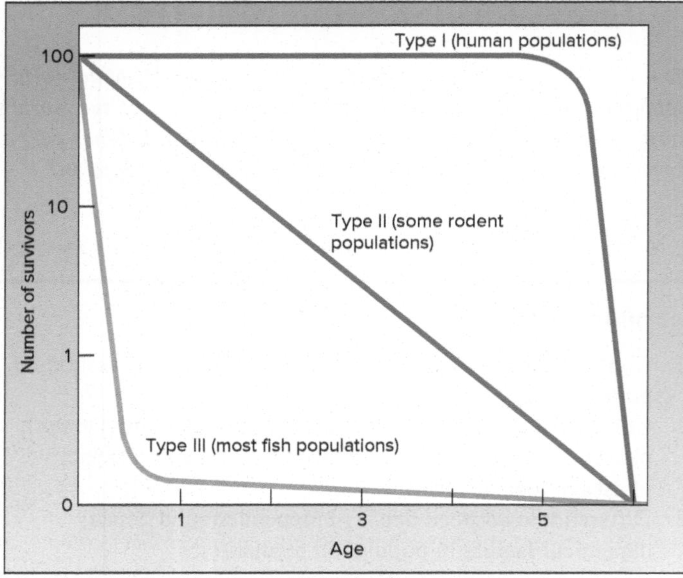

FIGURE 6.3

Survivorship. Survivorship curves are plots of the number of survivors (usually a logarithmic plot) versus relative age. Type I curves apply to populations in which individuals are likely to live out their potential life span. Type II curves apply to populations in which mortality rates are constant throughout age classes. Type III curves apply to populations in which mortality rates are the highest for the youngest cohorts.

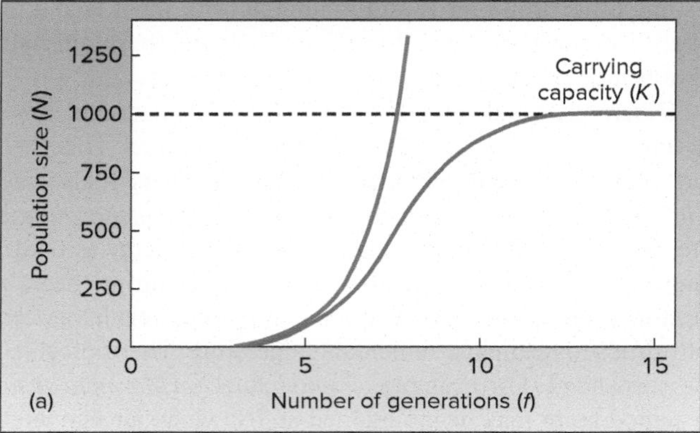

FIGURE 6.4

Exponential and Logistic Population Growth. (*a*) Exponential growth occurs when a population increases by the same ratio per unit time (red). A logistic growth curve reflects limited resources placing an upper limit on population size (blue). At carrying capacity (*K*), population growth levels off, creating an S-shaped curve. (*b*) Soay sheep (*Ovis aries*) trace their ancestry to domestic sheep that have become feral and unmanaged on the island of Soay off the coast of Scotland. Studies of these sheep have provided insights into population growth and regulation, including the demonstration of oscillations around carrying capacity (*K*).

little or no care or protection for large numbers of offspring and utilize resources for their own energy functions. In type II populations, parental care is usually intermediate between these extremes.

A second attribute of populations concerns population growth. The potential for a population to increase in numbers of individuals is remarkable. Rather than increasing by adding a constant number of individuals to the population in every generation, the population increases by the same ratio per unit time. In other words, populations experience **exponential growth** (figure 6.4*a*). Not all populations display the same capacity for growth. Such factors as the number of offspring produced, the likelihood of survival to reproductive age, the duration of the reproductive period, and the length of time it takes to reach maturity all influence reproductive potential.

Exponential growth cannot occur indefinitely because space, food, water, and other resources are limited. The constraints that climate, food, space, and other environmental factors place on a population are called **environmental resistance.** The population size that a particular environment can support is the environment's **carrying capacity** and is symbolized by *K*. In these situations, growth curves assume a sigmoid, or flattened S, shape, and the population growth is referred to as **logistic population growth** (*see figure 6.4a*).

The effects of environmental resistance are often not instantaneous. Growing populations may exceed carrying capacity. Eventually, increased death rate will cause these populations to decrease to, or go below, *K*. Growth curves may then fluctuate around *K*; or they may first exceed, and then fall back to, and stabilize at *K*. An unmanaged population of feral sheep on the island of Soay, off the coast of Scotland, displays fluctuations around *K* (figure 6.4*b*). When populations are small, ample resources allow birth rates to increase, and the number of sheep increases above *K*. The larger population then experiences increased mortality, especially in winter when resources are limited, and the population falls below *K* again. Other factors, such as weather, storm patterns, and sheep body weight, also contribute to this population oscillating around *K*.

Population Regulation

The conditions that an animal must meet to survive are unique for every species. What many species have in common, however, is that population density and competition affect populations in predictable ways.

Population Density

Density-independent factors influence the number of animals in a population without regard to the number of individuals per unit space (density). For example, weather conditions often limit populations. An extremely cold winter with little snow cover may devastate a population of lizards sequestered beneath the litter of the forest floor. Regardless of the size of the population, a certain percentage of individuals will freeze to death. Human activities, such as construction and deforestation, often affect animal populations in a similar fashion.

Density-dependent factors are more severe when population density is high than they are at other densities. Animals often use territorial behavior, song, and scent marking to tell others to look elsewhere for reproductive space. These actions become more pronounced as population density increases and are thus density dependent. Other density-dependent factors include competition for resources, disease, predation, and parasitism. Very low population density can also be detrimental. Low densities may result in an inability to deter predators or an inability to find mates. Animals experiencing near extinction events can face these difficulties.

Intraspecific Competition

Competition occurs when animals utilize similar resources and in some way interfere with each other's procurement of those resources. Competition among members of the same species, called **intraspecific competition,** is often intense because the resource requirements of individuals of a species are nearly identical. Intraspecific competition may occur without individuals coming into direct contact. (The "early bird that gets the worm" may not actually see later arrivals.) In other instances, the actions of one individual directly affect another. Territorial behavior is a common form of intraspecific competition. Male northern elephant seals (*Mirounga angustirostris, see figure 5.3*) establish territories and compete in violent clashes with other males for the right to mate with up to 50 females.

Section 6.2 Thinking Beyond the Facts

Some animals produce many offspring that require very little parental care. Other animals produce few offspring that require intensive parental care. What is the evolutionary trade-off involved with each strategy?

6.3 BIOTIC FACTORS: INTERSPECIFIC INTERACTIONS

LEARNING OUTCOMES

1. Discuss how herbivory, predation, and interspecific competition influence populations.
2. Explain coevolution.
3. Compare the different forms of symbiosis.
4. Assess the usefulness of visual appearance, odors, sounds, and behaviors either to hide one animal from another animal or to advertise properties of one animal to another animal.

Members of other species can affect all characteristics of a population. Interspecific interactions include herbivory, predation, competition, coevolution, and symbiosis. These artificial categories that zoologists create, however, rarely limit animals. Animals often do not interact with other animals in only one way. The nature of interspecific interactions may change as an animal matures, or as seasons or the environment changes.

Herbivory and Predation

Animals that feed on plants by cropping portions of the plant, but usually not killing the plant, are herbivores. This conversion provides food for predators that feed by killing and eating other organisms. Interactions between plants and herbivores, and predators and prey, are complex, and many characteristics of the environment affect them. Many of these interactions are described elsewhere in this text.

Interspecific Competition

When members of different species have similar resource requirements (e.g., food, shelter, nest sites), they compete for these resources. When the resource requirements are identical, one species may be forced to move or become extinct. This concept is referred to as the **competitive exclusion principle**. Competitive exclusion has been demonstrated in a few instances, but it is very unusual for two species to have nearly identical resource requirements, and some form of coexistence of species usually results from interspecific competition.

Coexistence can occur when species utilize resources in slightly different ways, for example, by using slightly different food resources or different parts of their environment for shelter. Robert MacArthur studied five species of warblers that all used the same caterpillar prey. Warblers partitioned their spruce tree habitats by dividing a tree into preferred regions for foraging and nesting. Although foraging regions overlapped, competition was limited, and the five species coexisted (figure 6.5).

Coevolution

The evolution of ecologically related species is sometimes coordinated such that each species exerts a strong selective influence on the other. This is **coevolution.**

Coevolution may occur when species are competing for the same resource or during predator–prey interactions. In the evolution of predator–prey relationships, for example, natural selection favors the development of protective characteristics in prey species. Similarly, selection favors characteristics in predators that allow them to become better at catching and immobilizing prey. Predator–prey relationships coevolve when a change toward greater predator efficiency is countered by increased elusiveness of prey.

Coevolution is obvious in the relationships between some flowering plants and their animal pollinators. Flowers attract pollinators with a variety of elaborate olfactory and visual adaptations. Insect-pollinated flowers are usually yellow or blue because insects see these wavelengths of light best. In addition, petal arrangements often provide perches for pollinating insects. Flowers pollinated by hummingbirds, on the other hand, are often tubular and red. Hummingbirds have a poor sense of smell but see red very well. The long beak of hummingbirds is an adaptation that allows them to reach far into tubular flowers. Their hovering ability means that they have no need for a perch.

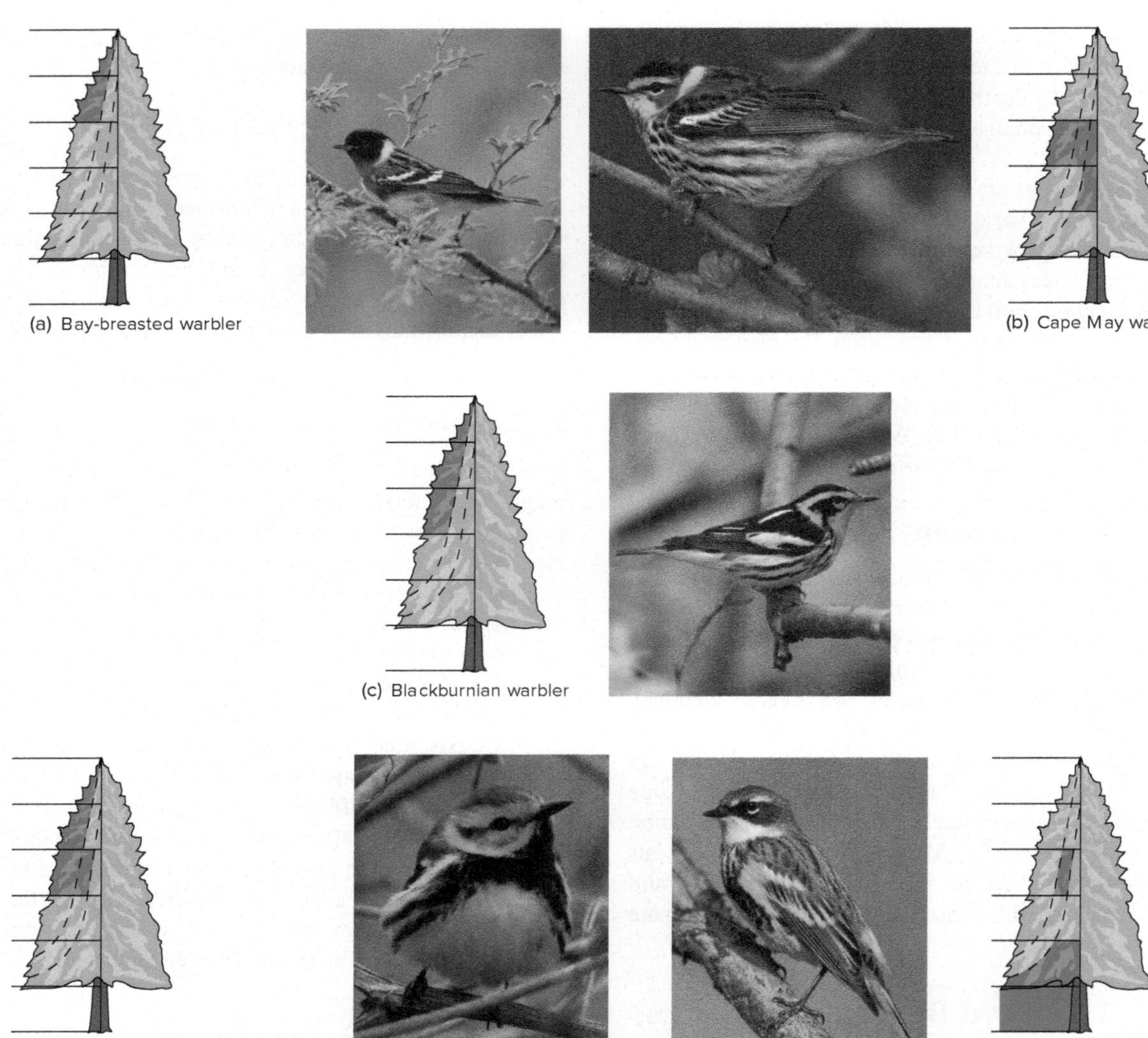

FIGURE 6.5

Coexistence of Competing Species. Robert MacArthur found that five species of warblers (*a–e*) coexisted by partitioning spruce trees into preferred foraging regions (*shown in dark green*). (*a*) Bay-breasted warbler (*Setophaga castanea*). (*b*) Cape May warbler (*S. tigrina*). (*c*) Blackburnian warbler (*S. fusca*). (*d*) Black-throated green warbler (*S. virens*). (*e*) Myrtle warbler (*S. coronata*).

Symbiosis

Some of the best examples of adaptations arising through coevolution come from two different species living in continuing, intimate associations, called **symbiosis** (Gr. *sym,* together + *bio,* life). Such interspecific interactions influence the species involved in dramatically different ways. You will encounter many examples of the following types of symbiosis in chapters 9–22.

Parasitism is a common form of symbiosis in which one organism lives in or on a second organism, called a host. The host usually survives at least long enough for the parasite to complete one or more life cycles. The relationships between a parasite and its host(s) are often complex. Some parasites have life histories involving multiple hosts. The definitive or final host is the host that harbors the sexual stages of the parasite. A fertile female in a definitive host may produce and release hundreds of thousands of eggs in her lifetime. Each egg gives rise to an immature stage that may be a parasite of a second host. This second host is called an intermediate host, and asexual reproduction may occur in this host. Some life cycles may have more than one intermediate host and more than one immature stage. For the life cycle to be completed, the final immature stage must have access to a definitive host (*see figure 10.13*).

Commensalism is a symbiotic relationship in which one member of the relationship benefits and the second is neither helped nor harmed. The distinction between parasitism and commensalism is somewhat difficult to apply in natural situations. Whether or not the host is harmed often depends on factors such as the host's nutritional state. Thus, symbiotic relationships may be commensalistic in some situations and parasitic in others (*see figure 15.10*).

Mutualism is a symbiotic relationship that benefits both members. Examples of mutualism abound in the animal kingdom, and many examples are described elsewhere in this text (*see figures 9.18b and 15.18*).

Other Interspecific Adaptations

Interspecific interactions have shaped many other characteristics of animals. **Crypsis** (L. *crypticus,* hidden), broadly defined, includes all instances of animals avoiding detection. We usually think of visual forms of crypsis as described later, but crypsis also includes chemical and auditory crypsis. Some lepidopteran caterpillars (larval butterflies and moths) emit chemicals that mimic odors of their host plants, making the caterpillars difficult to detect by ant predators. On the other hand, some African ants emit chemicals that mimic odors of their termite prey, making the ants difficult to detect by their prey. The pirate perch, *Aphredoderus sayanus,* secretes chemicals that make the fish chemically invisible to a variety of aquatic insect prey species. Auditory crypsis has been difficult to demonstrate conclusively, but it has been described in tiger moths, which emit sounds that may jam bat echolocation radar.

Visual crypsis takes on a variety of forms. Animals may take on color patterns that resemble their surroundings, such as the case of the peppered moth (*Biston betularia*) and industrial melanism discussed in chapter 5 (*see figure 5.6*). Disruptive coloration is used by some predators. Spots or stripes break up outlines or other features, like eyes, thus camouflaging the predator within its environment and helping the predator approach prey without detection (figure 6.6). Self-decoration is used by a variety of species of decorator crabs. Changeable skin or coat patterns and color is used by the Arctic fox (*Vulpes lagopus*). Its white coat in winter (*see figure 22.6*) and its brown coat in summer help it avoid detection. Similarly, the snowshoe hare (*Lepus americanus*) molts from white in winter (*see figure 4.6*) to a blue/gray coat in summer that blends with the rocks and vegetation of its habitat. Countershading is a kind of crypsis common in frog and toad eggs. These eggs are darkly pigmented on top and lightly pigmented on the bottom. When a bird or other predator views the eggs from above, the darkness of the top side hides the eggs from detection against the darkness below. When a fish views the eggs from below, the light undersurface of the eggs blends with the bright air-water interface.

FIGURE 6.6

Camouflage. The color pattern of this tiger (*Panthera tigris*) provides effective camouflage that helps when stalking prey.

FIGURE 6.7

Mimicry. The viceroy butterfly (*Limenitis archippus,* top) and the monarch butterfly (*Danaus plexippus,* bottom) are both distasteful to bird predators. When a bird tastes either species it avoids feeding on individuals of both species. This example is a form of mimicry called Müllerian mimicry, in which two species serve as co-mimics.

Resembling conspicuous animals may also be advantageous. **Mimicry** (L. *mimus,* to imitate) occurs when a species resembles one, or sometimes more than one, other species and gains protection by the resemblance. The resemblance may be based on visual appearance, behavior, sounds, or scent; and the mimicry can take a number of forms depending on the nature of the sharing of perceived characteristics (figure 6.7).

Some animals that protect themselves by being dangerous or distasteful to predators advertise their condition by conspicuous coloration. The sharply contrasting white stripe(s) of a skunk and bright colors of venomous snakes give similar messages. These color patterns are examples of warning or **aposematic coloration** (Gr. *apo,* away from + *sematic,* sign).

Section 6.3 Thinking Beyond the Facts

Parasitism results in harm being done to the host. Under what circumstances would the weakening or death of a host benefit a parasite?

6.4 COMMUNITIES

LEARNING OUTCOMES

1. Explain the concept of an ecological community.
2. Explain how the concept of an ecological niche is valuable in helping visualize the role of an animal in the environment.
3. Differentiate between seral and climax community stages in terms of community stability and biodiversity.

All populations living in an area make up a **community.** Communities are not just random mixtures of species; instead, they have a unique organization. Most communities have certain members that have overriding importance in determining community characteristics. They may be important because of their abundance or activity in the community. For example, a stream community may have a large population of rainbow trout that helps determine the makeup of certain invertebrate populations on which the trout feed. The trout reduce competition among prey species, allowing the prey to share similar resources. Removal of the trout could allow one of the invertebrate species to become dominant and crowd out other invertebrate species. The trout thus control the community characteristics and are called a **keystone species.**

Communities are also characterized by the variety of animals they contain. This variety is called **community (species) diversity** or richness. Factors that promote high diversity include a wide variety of resources, high productivity, climatic stability, moderate levels of predation, and moderate levels of disturbance from outside the community. Pollution often reduces the species diversity of ecosystems.

The Ecological Niche

The **ecological niche** is an important concept of community structure. The niche of any species includes all the attributes of an animal's lifestyle: where it looks for food, what it eats, where it nests, and what conditions of temperature and moisture it requires. Competition results when the niches of two species overlap. Interspecific competition often restricts the environments in which a species lives so that the actual, or realized niche, is smaller than the potential, or fundamental, niche of the species (figure 6.8). The acorn barnacle, *Chthamalus stellatus*, lives in shallow rocky intertidal regions of Great Britain and Ireland. In the absence of a competing barnacle, *Semibalanus balanoides, C. stellatus* occupies regions from the high-tide mark to below the low-tide mark. In the presence of *S. balanoides*, *C. stellatus* is forced out of its fundamental niche and is restricted to the region between low and high tides.

Although the niche concept is difficult to quantify, it is valuable for perceiving community structure. It illustrates that community members tend to complement each other in resource use. Partitioning resources allows competing species to survive in the same community. The niche concept is also helpful for visualizing the role of an animal in the environment.

Community Stability

As with individuals, communities are born and they die. Between those events is a time of continual change. Some changes are the result of climatic or geological events. Members of the community may be responsible for others. In one model of community change, the dominant members of the community change a community in predictable ways in a process called **succession** (L. *successio,* to follow) (figure 6.9). Communities may begin in areas nearly devoid of life. The first community to become established in an area is called the **pioneer community.** A consequence of life in a successional stage is that organisms use resources, die, and decay; and these changes make the environment less suitable for their own kind and more favorable for the next successional stage. Each successional stage is called a **seral stage,** or **sere**

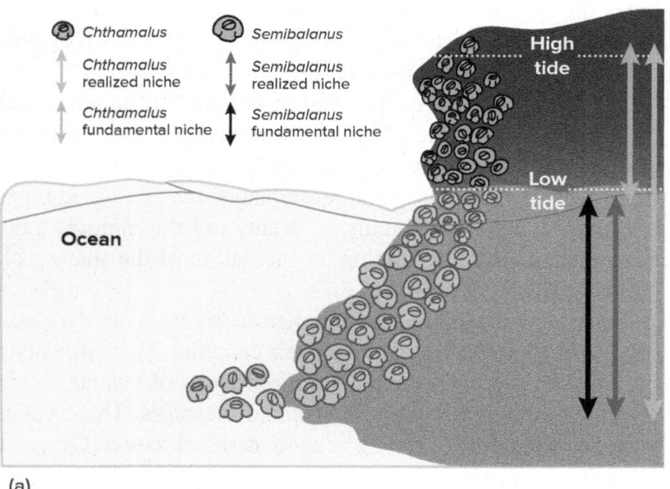

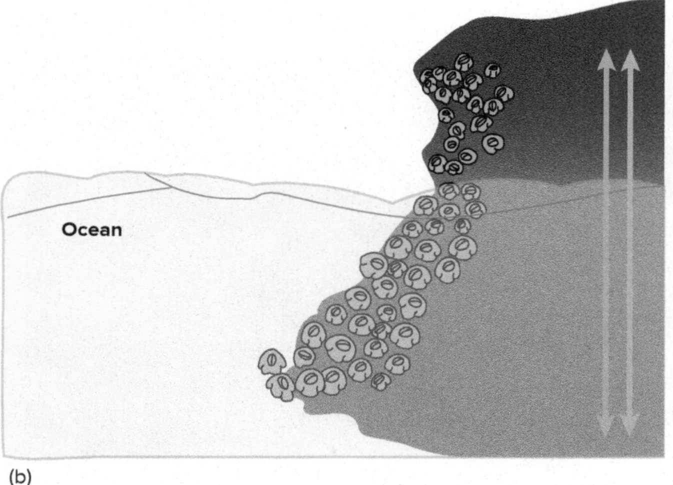

FIGURE 6.8

Interspecific Competition and Niche Partitioning. (*a*) The realized niche of *Chthamalus stellatus* does not include habitats below the low-tide mark in the presence of competition by *Semibalanus balanoides.* (*b*) In the absence of competition from *Semibalanus balanoides, Chthamalus stellatus* occupies habitats above and below the low-tide mark.

FIGURE 6.9
Succession. Primary succession on a sand dune. Beach grass is the first species to become established. It stabilizes the dune so that shrubs, and eventually trees, can grow.

(ME *seer,* to wither). Early seres often have higher growth rates and greater biomass (*see the next column*) than later seres, but they usually have lower species diversity. The final community is the **climax community.** It is different from the seral stages that preceded it because it can tolerate its own reactions. Accumulation of the products of life and death no longer make the area unfit for the individuals living there. Climax communities usually have complex structure and high species diversity.

Climax communities are not permanent. Natural disturbances like fires, storms, and climatic changes can alter the course of successional change or prevent the establishment of a climax community. For example, periodic fires and grazing by bison kept tallgrass prairies of North America in their seral prairie stage throughout presettlement times. After fires were controlled and bison were eliminated, these communities began to progress toward later seral stages characterized by shrubs and other woody vegetation. Consequently, certain animal populations associated with these tallgrass prairie communities became increasingly rare.

Natural landscapes are not characterized by single climax or seral communities. Biologically rich landscapes are characterized by a mosaic of different communities that harbor a rich, natural diversity of organisms. As one community juxtaposes from one stage to the next, an adjacent region may be renewed. This dynamic allows pioneer, subclimax, and climax community specialists (animals with relatively narrow fundamental niches) to persist in a given landscape. A wide variety of anthropogenic disturbances can interfere with natural community dynamics and drastically reduce species diversity.

Section 6.4 Thinking Beyond the Facts

Do you think it is possible to completely describe the niche for any species? Explain.

6.5 Trophic Structure of Ecosystems

Learning Outcomes

1. Compare ecosystem and community concepts.
2. Use the laws of thermodynamics to justify the observation that energy pathways in food webs are short.
3. Assess the vulnerability of animals at various trophic levels regarding their risk from heavy metal pollution.

Communities and their physical environment are called **ecosystems.** One important fact of ecosystems is that energy is constantly being used, and once it leaves the ecosystem, this energy is never reused. Energy supports the activities of all organisms in the ecosystem. It usually enters the ecosystem in the form of sunlight and is incorporated into the chemical bonds of organic compounds within living tissues. The total amount of energy converted into living tissues in a given area per unit time is called **primary production.** The primary production supports all organisms within an ecosystem. The total mass of all organisms in an ecosystem is the ecosystem's **biomass.** As energy moves through the ecosystem it is eventually lost as heat through the metabolic activities of producers and through various levels of consumer organisms.

The sequence of organisms through which energy moves in an ecosystem is called a **food chain.** One relatively simple food chain might look like the following:

grass → grasshopper → shrews → owls

Complexly interconnected food chains, called **food webs,** that involve many kinds of organisms are more realistic (figure 6.10). Because food webs can be complex, it is convenient to group organisms according to the form of energy used. These groupings are called **trophic levels.**

Producers (autotrophs) obtain nutrition (complex organic compounds) from inorganic materials (such as carbon, nitrogen, and phosphorus) and an energy source. They form the first trophic level of an ecosystem. The most familiar producers are green plants that carry on photosynthesis. Other trophic levels are made up of consumers (heterotrophs). Consumers eat other organisms to obtain energy. Herbivores (primary consumers) eat producers. Some carnivores (secondary consumers) eat herbivores, and other carnivores (tertiary consumers) eat the carnivores that ate the herbivores. Consumers also include scavengers that feed on large chunks of dead and decaying organic matter. Decomposers break down dead organisms and feces by digesting organic matter extracellularly and absorbing the products of digestion.

The efficiency with which the animals of a trophic level convert food into new biomass depends on the nature of the food (figure 6.11). Biomass conversion efficiency averages 10%, although efficiencies range from less than 1% for some herbivores to 35% for some carnivores.

In addition to helping us understand how energy moves through ecosystems, food webs help us understand what happens when substances that are not degraded by biological processes are

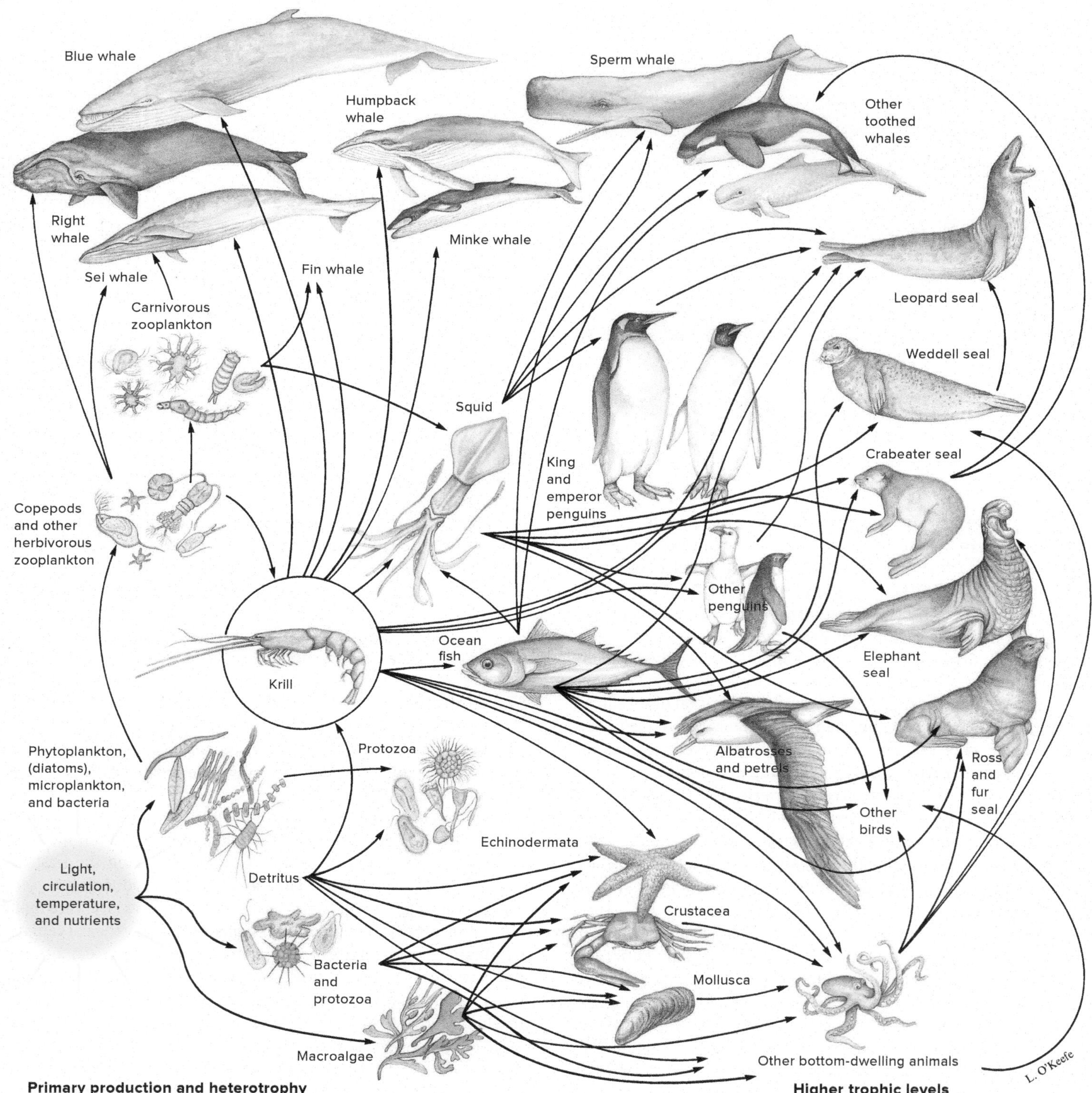

FIGURE 6.10

Food Webs. An Antarctic food web. Small crustaceans called krill support nearly all life in Antarctica. Six species of baleen whales, 20 species of squid, more than 100 species of fish, 35 species of birds, and 7 species of seals eat krill. Krill feed on algae, protozoa, other small crustaceans, and various larvae. To appreciate the interconnectedness of food webs, trace the multiple paths of energy from light (lower left), through krill, to the leopard seal.

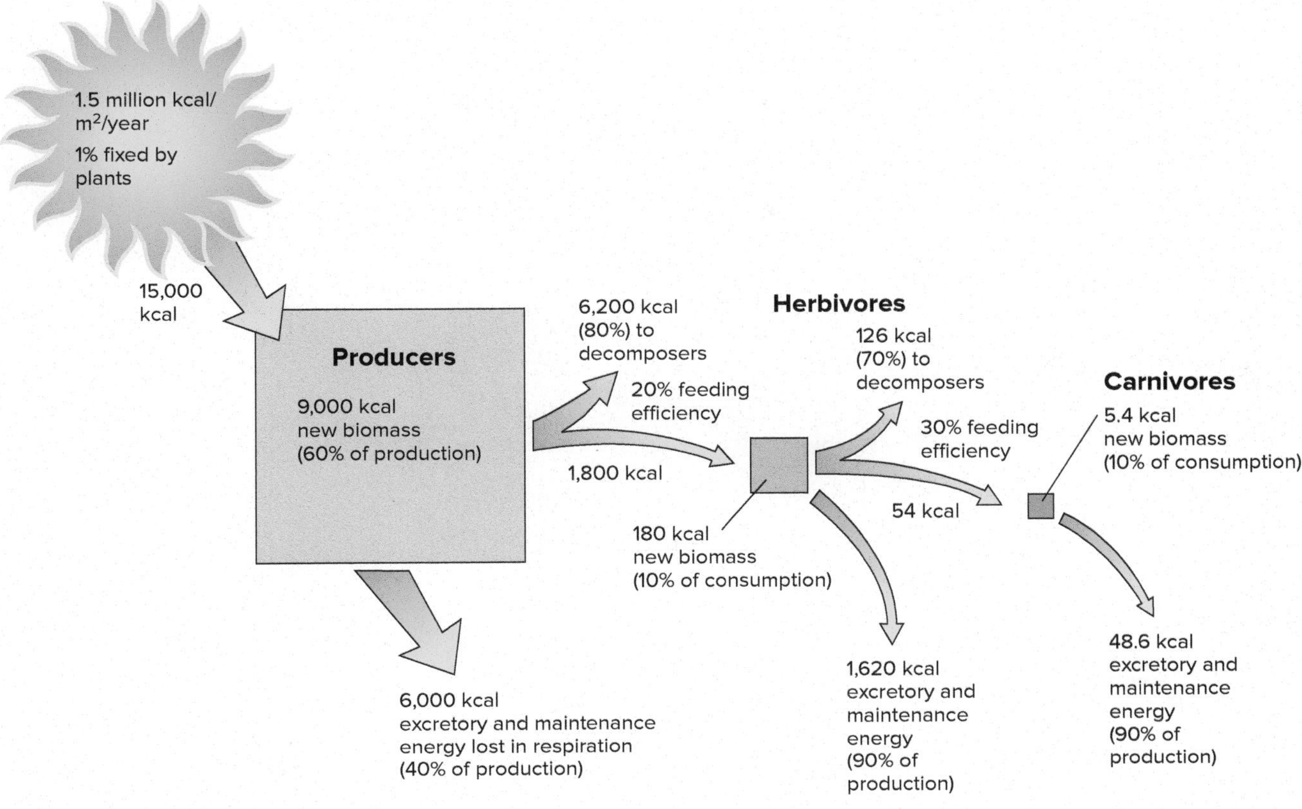

FIGURE 6.11

Energy Flow through Ecosystems. Approximately 1.5 million kcal of radiant energy strikes a square meter of the earth's surface each year. Plants convert less than 1% (15,000 kcal/m^2/year) into chemical energy. Of this, approximately 60% is converted into new biomass, and 40% is lost in respiration. The herbivore trophic level harvests approximately 20% of net primary production, and decomposers get the rest. Of the 1,800 kcal moving into the herbivore trophic level, 10% (180 kcal) is converted to new biomass, and 90% (1,620 kcal) is lost in respiration. Carnivores harvest about 30% of the herbivore biomass, and 10% of that is converted to carnivore biomass. At subsequent trophic levels, harvesting efficiencies of about 30% and new biomass production of about 10% can be assumed. All of these percentages are approximations. Absolute values depend on the nature of the primary production (e.g., forest versus grassland) and characteristics of the herbivores and carnivores (e.g., ectothermic versus endothermic).

introduced into ecosystems. **Biomagnification** is the increase in concentration of a substance in the tissues of organisms at higher levels of food webs. Heavy metals, like mercury, are elements and cannot be broken down. Novel organic compounds, called persistent organic pollutants (POPs), cannot be broken down biologically because evolution has not had time to build enzyme systems to process them. These substances persist in animal tissues when the intake of the substance exceeds the excretion of the substance by the animal. Persistence of substances in animal tissues is especially severe when the substances are lipid soluble, and they accumulate in an animal's fat deposits. Long-lived animals that feed at higher trophic levels accumulate these toxins throughout their lives, often poisoning them and organisms that feed on them. For example, mercury is a heavy metal that is a waste product of burning coal, gold mining, oil refining, and cement production. It accumulates in aquatic ecosystems and has resulted in wildlife poisonings (e.g., recent findings of toxic mercury levels in bald eagles) and warnings concerning human consumption of tilefish, swordfish, shark, mackerel, grouper, tuna, northern pike, and other species.

Section 6.5 Thinking Beyond the Facts

What lessons regarding energy and food webs are most important when we consider today's ecological problems?

6.6 CYCLING WITHIN ECOSYSTEMS

LEARNING OUTCOMES

1. Explain the differences between hydrological, gaseous, and sedimentary biogeochemical cycles.
2. Analyze the effect of extravagant burning of fossil fuels and deforestation on the carbon cycle.

Did you ever wonder where the calcium atoms in your bones were 100 or even 100 million years ago? Perhaps they were in the bones of an ancient reptile or in the sediments of prehistoric seas. Unlike energy, all matter is cycled from nonliving reservoirs to living systems and then back to nonliving reservoirs. This is the second important

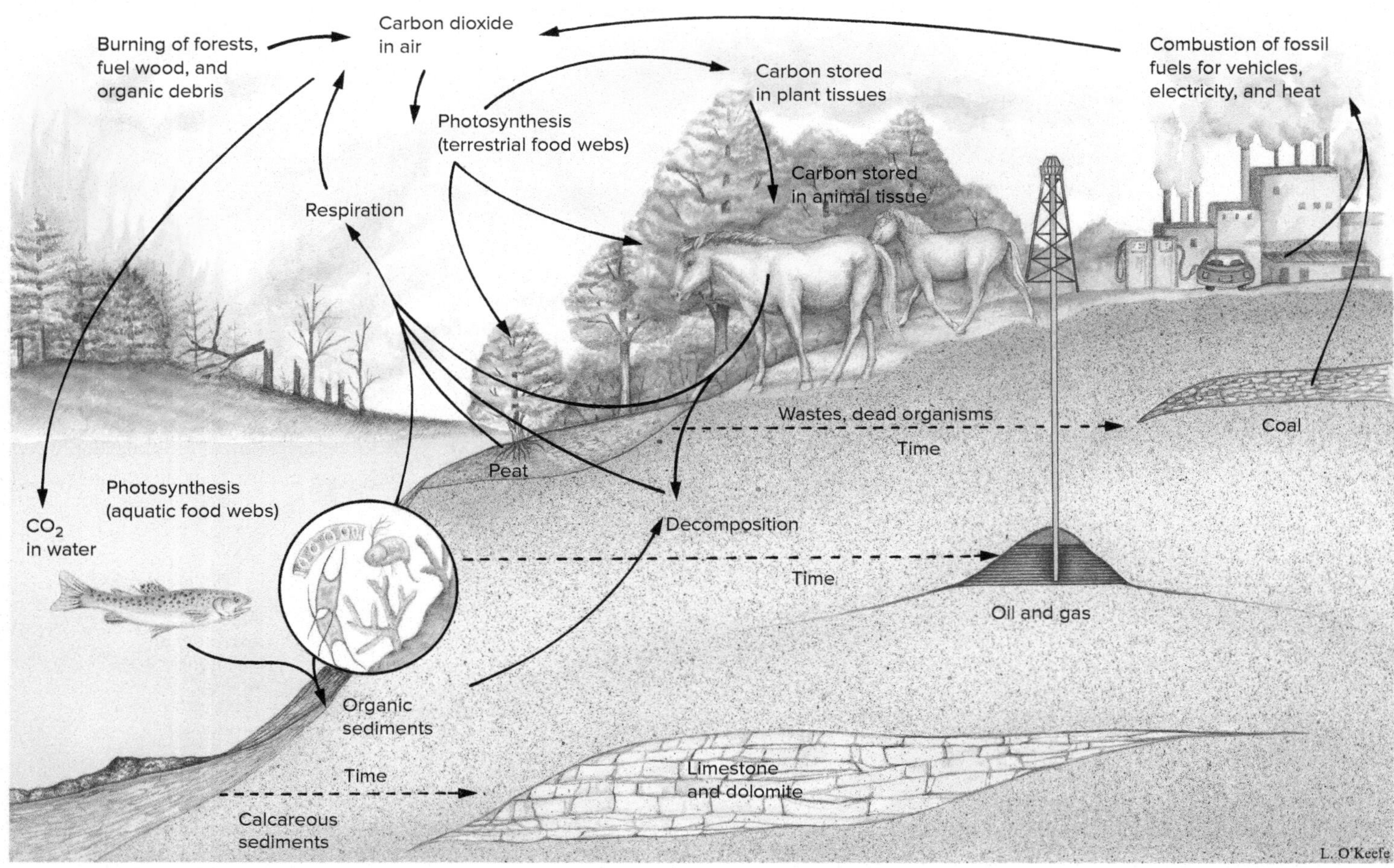

FIGURE 6.12

Carbon Cycle. Carbon cycles among its reservoir in the atmosphere, living organisms, fossil fuels, and limestone bedrock.

lesson learned from the study of ecosystems—matter is constantly recycled within ecosystems. Matter moves through ecosystems in **biogeochemical cycles.**

A nutrient is any element essential for life. Approximately 97% of living matter is made of oxygen, carbon, nitrogen, and hydrogen. Gaseous cycles involving these elements use the atmosphere or oceans as a reservoir. Elements such as sulfur, phosphorus, and calcium are less abundant in living tissues than are those with gaseous cycles, but they are no less important in sustaining life. The nonliving reservoir for these nutrients is the earth, and the cycles involving these elements are called sedimentary cycles. Water also cycles through ecosystems. Its cycle is called the hydrological cycle.

A good place to begin in considering any biogeochemical cycle is the point at which the nutrient enters living systems from the reservoir (atmosphere or earth). Nutrients with gaseous cycles require that the nutrient be captured as a gas and incorporated into living tissues. This is called fixation. In sedimentary cycles, the nutrient may enter living tissues by uptake with water, food, or other sources. Once the nutrient is incorporated into living tissues, it is cycled. Depending on the nutrient, it may be passed from plant tissue to herbivore, to carnivore, to decomposer and remain in the living portion of the biogeochemical cycle. The nutrient may cycle within living components of an ecosystem for thousands of years, or for a very short period. Eventually, the nutrient is returned to the reservoir. Ideally, the rate of return equals the rate of fixation within living systems. As discussed later, imbalance between return and fixation can result in severe ecological problems.

To help you understand the concept of a biogeochemical cycle, study the carbon cycle in figure 6.12. Carbon is plentiful on the earth and is rarely a limiting factor. A basic outline for the carbon cycle is fairly simple. The reservoir for carbon is carbon dioxide (CO_2) in the atmosphere or water. Carbon is fixed into organic matter by autotrophs, usually through photosynthesis, and enters aquatic and terrestrial food webs. Carbon returns to the reservoir when cellular respiration releases CO_2 into the atmosphere or water.

As is often the case with nutrient cycles, the pathway for carbon can be considerably more complex. In aquatic systems, some of the CO_2 combines with water to form carbonic acid ($CO_2 + H_2O \rightleftharpoons H_2CO_3$). Because this reaction is reversible, carbonic acid can supply CO_2 to aquatic plants for photosynthesis when CO_2 levels in the water decrease. Carbonic acid can also release CO_2 to the atmosphere. Some of the carbon in aquatic systems is tied up as calcium carbonate ($CaCO_3$) in the shells of molluscs and the skeletons of echinoderms. Accumulations of mollusc shells and echinoderm skeletons have resulted in limestone formations that are the bedrock of much of the United States. Geological uplift, volcanic activities, and weathering return much of this carbon to the earth's surface and the atmosphere. Other carbon is tied up in fossil fuels. Burning fossil fuels returns large quantities of this carbon to the atmosphere as CO_2 (*see figure 6.12 and the discussion of climate change on page 101*).

SECTION 6.6 THINKING BEYOND THE FACTS

Why does recycling aluminum cans, plastic, and other materials make sense in light of what you know about biogeochemical cycles?

6.7 ECOLOGICAL PROBLEMS

LEARNING OUTCOMES

1. Compare the age structure of a developed country and a developing country.
2. Explain the relationship between overpopulation and depletion of world resources.
3. Analyze the threats to Earth's biodiversity.

Barry Commoner (1917–2012) was a scientist, environmental and social activist, author, and presidential candidate (1980) who studied and taught at Washington University in St. Louis, Missouri. He is credited by many ecologists with being the father of the environmental movement, but his vision was wider than biology. He recognized that Earth's problems–ecological degradation, social injustice, and economic and national security–are all interconnected. He was one of the first to say that no permanent environmental solutions exist without social change. His four "laws of ecology" have become the cornerstones of the environmental movement. (1) Everything in the entire Earth ecosystem is interconnected. Damaging one part of an ecosystem has wide-ranging effects on the whole. (2) Everything must go somewhere. There is no place to put waste where it will not resurface in the future. (3) Nature knows best. Human-induced changes to ecosystems, what we think of as "improvements," are always detrimental to the ecosystems. (4) There is no "free lunch." Every environmental change has a consequence. The consequence can sometimes be delayed, but it cannot be avoided for long.

Now that you have studied some general ecological principles, it should be easier to understand why Commoner's rules are true. In this last section of chapter 6, we link basic ecological principles to conservation issues that we face.

Human Population Growth

An expanding human population is the root of virtually all environmental problems. Human populations, like those of other animals, tend to grow exponentially. The earth, like any ecosystem on it, has a carrying capacity and a limited supply of resources. When human populations achieve that carrying capacity, populations should stabilize. If they do not stabilize in a fashion that limits human misery, then war, famine, and/or disease are sure to take care of the problem.

What is the earth's carrying capacity? The answer is not simple. In part, it depends on the desired standard of living and on whether or not resources are distributed equally among all populations. The earth's population currently stands at 7.6 billion people. Virtually all environmentalists agree that the number is too high.

Efforts are being made to curb population growth in many countries, and these efforts have met with some success. Looking at the age characteristics of world populations helps explain why control measures are needed. The **age structure** of a population shows the proportion of a population in prereproductive, reproductive, and postreproductive classes. Age structure is often represented by an age pyramid. Figure 6.13 shows an age pyramid for a developed country and for a developing country. In developing countries like Kenya, the age pyramid has a broad base, indicating high birthrates. As in many natural populations, high infant mortality offsets these high birthrates. However, what happens when developing countries begin accumulating technologies that reduce prereproductive mortality and prolong the lives of the elderly? Unless reproductive practices change, a population explosion occurs and problems associated with housing, employment, education, food production, and health care are compounded.

The United Nations Population Division projects that the current world population of 7.6 billion will increase to 9.7 billion by 2050 and 10.9 billion by 2100. Average human fertility across the globe ranges between 4.5 children per woman (usually in the least developed countries) and 1.6 children per woman (usually in the more developed countries). Even in countries with the lowest fertility, populations are growing because of increased longevity and immigration from other parts of the world. The increase in older age classes is already straining the economics of elder care.

Earth's Resources and Global Inequality

Aldo Leopold (1887–1948), author of *A Sand County Almanac: And Sketches Here and There*, once said:

> A thing is right when it tends to preserve the integrity, stability and beauty of the biotic community. It is wrong when it tends otherwise.

It is right that we preserve Earth's resources for the sake of the earth. It is also right because we humans depend on those resources for our lives. Unfortunately there is a tremendous global inequality in the use of Earth's resources that has led to environmental degradation across the globe and human suffering in many regions of the world. It is estimated that the earth has approximately 1.9 hectares of productive resources (crop, grazing, forest, and fishing) per person. Supporting the lifestyles of people in the United States requires about 9.5 hectares of productive resources. (Western Europeans require about half of that area.) This amount is in contrast to the use of less than 0.5 hectares by the average Mozambican. To put it another way, 12% of the world's population lives in North American and western Europe, and these populations account for 60% of

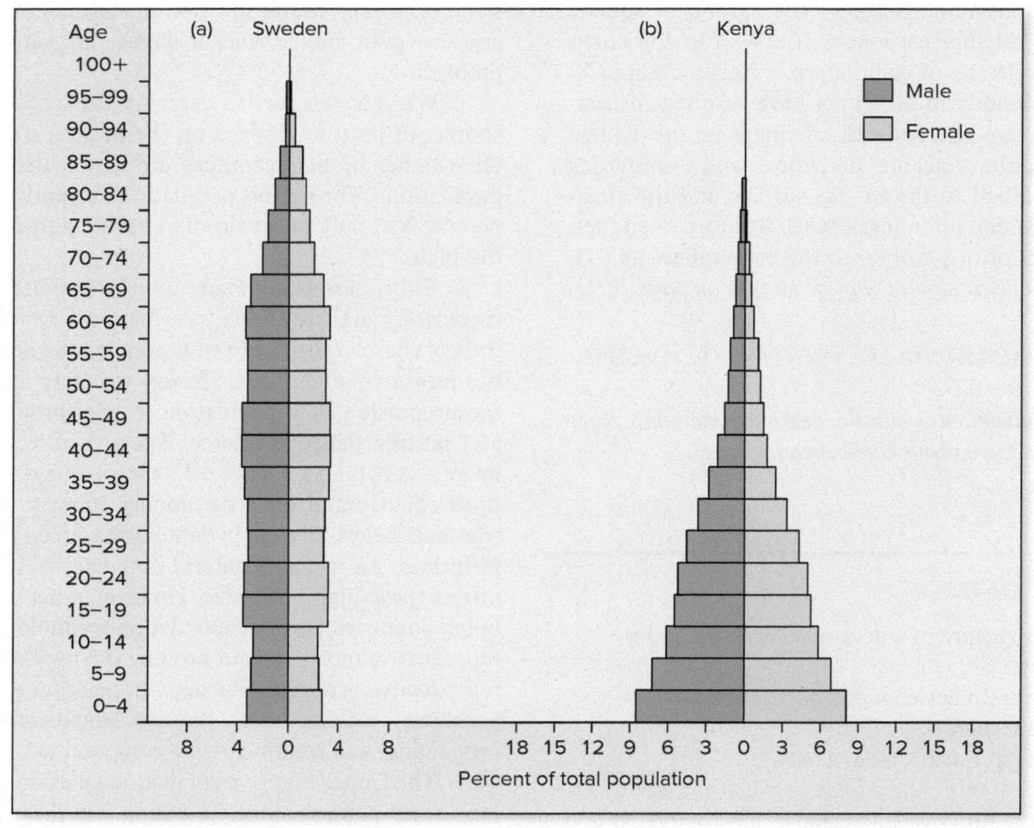

FIGURE 6.13

Human Age Pyramids (*a*) In developed countries, the age structure is parallel-sided because mortality in all age classes is relatively low. In this example, the slight widening of the pyramid in the 45 to 65 age range is because of the "baby boom" that occurred between 1945 and 1965. (*b*) In developing countries, a greater proportion of the population is in the prereproductive age classes. High mortality compensates for high birthrates, and the pyramid is triangular. As technologies reduce infant mortality and prolong the life span of the elderly, populations increase rapidly.

the world's consumption. In contrast, 33% of the world's population lives in southern Asia and sub-Saharan Africa and account for 3.2% of the world's consumption. The results of unbridled consumption and development have had devastating consequences for Earth's natural systems.

Biodiversity

The variety of living organisms in an ecosystem is called **biodiversity.** No one knows the number of species in the world. About 1.6 million species have been described, but taxonomists estimate that there may be up to 10 million total. A rich biodiversity is of inestimable value in natural systems. Every life-form, no matter how small, plays very important roles in Earth's ecosystems. Greater diversity of species creates natural sustainability for all our ecosystems. Healthy ecosystems are better able to withstand and recover from disturbances. Diverse ecosystems protect our water and soil resources; promote nutrient cycling and climate stability; and provide biological resources, including food, medicines, wood products, and novel genomes that benefit ecosystem and human populations. All of these functions require large, healthy populations. Large populations provide the genetic diversity required for surviving environmental changes. E.O. Wilson (ant biologist, naturalist, and father of the biodiversity movement) emphasized the importance of these ecosystem services when he said:

> Look closely at nature. Every species is a masterpiece, exquisitely adapted to the particular environment in which it has survived. Who are we to destroy or even diminish biodiversity?

The biodiversity of all natural areas of the world is threatened. In spite of greater awareness of environmental problems, the rate of biodiversity loss has not been reduced. The International Union for Conservation of Nature (IUCN) surveyed nearly 48,000 species and concluded that 46% resided in one of their extinction risk categories, ranging from extinct to a near threatened status. Biodiversity loss continues for five reasons discussed in the following sections.

Habitat Loss

Habitat loss and degradation displaces thousands of native plants and animals. Some of the most important threatened natural areas include tropical rain forests, coastal wetlands, and coral reefs (*see page 153*). Of these, tropical rain forests have probably received the most attention. Tropical rain forests cover only 7% of the earth's land surface, but they contain more than 50% of the world's species. They provide a wealth of economic and ecological services–providing novel medicines and foods, climate control through absorption of CO_2 from the atmosphere, and protection of delicate soil and water resources. Tropical rain forests are being destroyed rapidly, mostly for agricultural production. About 32 million hectares (an area the size of the state of Louisiana) are being cleared each year (*see figure 1.6*a,b). Loss of forest vegetation is accompanied by the loss of thousands of forest animals and forest soils. At current rates of destruction, most tropical rain forests will disappear this century.

Climate Change and Ocean Acidification

We have seen that cycling of material is a fundamental property of ecosystems. There are consequences to altering portions of a biogeochemical cycle, and understanding climate change requires understanding the carbon cycle (*see figure 6.12*). There is a clear and overwhelming consensus among climate scientists–climate change is real, and climate-warming trends of the past decades are due to human activities. Climate change is a by-product of our dependence on fossil fuels. China and the United States are responsible for over 40% of the greenhouse gas emissions (principally CO_2, *see table 1.4*). We have disrupted the carbon cycle by increasing the rate of CO_2 accumulation in the atmosphere from burning fossil fuels and by decreasing the rate of carbon fixation into organic matter through deforestation. Earth's temperature is determined by the relationship between the absorption of heat from the sun and radiation of heat from the earth back into space. Non-condensing greenhouse gases (principally CO_2) act as a thermostat to regulate the amount of water vapor and cloud cover in the atmosphere. Together, these gases create an insulating blanket that reduces the radiation of heat into space and cause global temperatures to increase. According to the Fourth Assessment Report by the United Nations Intergovernmental Panel on Climate Change (2007), global surface temperatures increased by 0.74°C during the last century, most of the change coming since 1950. A further increase in global temperature of 1.4–6.4°C is projected for the twenty-first century. Climate change promotes the melting of glaciers, loss of polar ice sheets, disruption of freshwater supplies, expansion of deserts, and alteration of regional weather systems. Climate change has many ecosystem effects. It disrupts food webs, shifts ranges of plants and animals (*see page 77 and figure 5.4.*), alters the timing of life-cycle events, and can promote the spread of pathogens (*see snowshoe hare, page 59*).

Ocean acidification is the partner to climate change. Increasing CO_2 in the atmosphere increases the amount of CO_2 that dissolves in water, thereby forming carbonic acid. Ocean waters have a natural pH of 8.2. Oceanic pH now averages 8.0 and is continuing to drop. A form of calcium carbonate, called aragonite, is used by many marine animals, including corals, echinoderms, molluscs, and crustaceans to build shells and skeletons. Increasing acidity either decreases availability of calcium carbonate for forming shells and skeletons or increases the rate of dissolution of calcium carbonate from shells and skeletons. Either way, calcifying species are adversely affected.

In 2015, 197 countries came together in Paris, France, to participate in the United Nations Framework Convention on Climate Change (UNFCCC). The Paris Agreement established the following goals:

- global average temperature will be kept well below 2°C above preindustrial levels (an increase of 1.5°C would significantly reduce impacts of climate change);
- greenhouse gas emissions will peak early in this century and reach net-zero emissions in the second half of this century;
- all parties are committed to contributing to climate change mitigation and adaptation;
- mitigation measures of individual countries will be expressed in nationally determined contributions (NDCs);
- NDCs will be reevaluated every five years; and
- countries will cooperate in achieving NDCs. (This goal recognizes the different starting points of participating countries and emphasizes the importance of developed countries leading and supporting the work of less developed countries.)

As of November, 2017, 170 of 197 participating countries have ratified the Paris Agreement.

How do we achieve these goals? The United States derives 88% of its energy from coal, oil, and natural gas. We can reduce CO_2 pollution through the use of alternative fuels and carbon capture and sequestering technologies. Worldwide renewable energy use includes ethanol and biodiesel (0.5% of total energy use), wind (2%), hydropower (2%), and photovoltaics (solar) (less than 0.05%). Power generation from nuclear reactors accounts for 13% of energy production. One of the biggest problems in the United States is the construction of east-west transmission lines required for long-distance transmission of electricity. Supergrids for energy transmission from regions that are reliably sunny or windy will take years to construct. Carbon capture technologies remove CO_2 from point sources, like power plants, and sequester that waste carbon in storage sites (geological formations) where it will not enter the atmosphere. These technologies are being tested around the world but are not without their drawbacks. They are expensive, require long-term storage of carbon waste, and require 10 to 40% of the energy processed by a plant. Clearly, carbon capture without conversion to renewable energy, energy conservation, and earth-friendly lifestyles (*see pages 102–103*) will not solve our climate-change problems.

Nutrient Load and Pollution

Excessive nutrient load and other forms of pollution make ecosystems uninhabitable. For example, consider the nutrients nitrogen and phosphorus. Nitrogen and phosphorus also undergo biogeochemical cycling. The reservoir for nitrogen is in the atmosphere (a gaseous cycle) and the reservoir for phosphorus is in the earth (a sedimentary cycle). Both nitrogen and phosphorus are fertilizers that promote algal growth in lakes, rivers, and oceans. Algal blooms and die-offs contribute to oxygen depletion in aquatic environments, which makes water uninhabitable for many animal species. Most nitrogen and phosphorus pollution comes from fertilizer use and agricultural run-off, including nitrogen and phosphorus pollution from concentrated animal feeding operations (CAFOs). The use

How Do We Know?

Local Steps to Alleviating Global Problems

Most of the readers of this textbook live in developed regions of the world, principally the United States. The environmental problems facing you in the years ahead no doubt seem formidable—more so with the realization that many of the problems are largely not of your making. On the other hand, your (our) life in the developed world has a far greater detrimental impact on global environmental health than you (we) probably realize. It is estimated that the average U.S. citizen uses 25 times more resources than a person living in a developing country. This statistic means that even though population growth rates are higher in developing countries, a family in a developing country would need to have 50 children to use the same resources as a family with two children in the United States. Those of us in developed countries bear much of the responsibility for responding to environmental problems. Solutions are both global and local. We can each do our parts by learning to live humbly and harmoniously with nature. Following are 10 relatively simple steps that we can all take right now to reduce the detrimental impact of our lives on the environment:

1. Contact your public officials, urging them to support environmental legislation. The Internet sites listed at the end of this box have up-to-date information on legislation currently being debated at national and state legislatures.
2. Learn more. The listed Internet sites have a wealth of information.
3. Share the knowledge. Write your local newspaper and other media about environmental issues. Monitor their coverage of the environment. The listed Internet sites provide strategies for sharing the information in schools and other organizations.
4. Reduce your home, apartment, or dormitory energy use. It is estimated that 21% of global warming pollution results from home energy use. Use compact fluorescent light bulbs; unplug battery chargers when they are not in use; use power strips to stop energy use by electronic devices when they are not in use; buy energy-efficient appliances; turn down the thermostat; and install programmable thermostats.
5. Drive a green-friendly car. Your choice of what car you drive is probably the single most important environmental decision you will make. Assuming you drive 12,000 miles per year and average 10 miles per gallon, you will add 13.6 tons of greenhouse gas to the atmosphere and spend $3,000 on gasoline (assuming $2.50 per gallon of gasoline). If you average 30 miles per gallon, you will add 4.5 tons of greenhouse gas to the atmosphere and spend $1,000. If you average 50 miles per gallon, you will add 2.7 tons of greenhouse gas to the atmosphere and spend $600.
6. Drive green-friendly. Lighten the load in your car. For every 100 pounds (45.5 kg) of extra weight in your car, you reduce gasoline mileage by 2%. Slow down and avoid rapid acceleration and braking. Driving 5 miles per hour over the speed limit reduces fuel economy by 6%. Rapid acceleration and braking wastes about 125 gallons of gasoline per year.
7. Maintain your car. Maintaining proper tire pressure and vehicle emissions devices saves fuel and reduces emissions.
8. Walk, ride a bicycle, or share a ride with a friend.
9. Plan ahead. Reduce the number of automobile trips you take to the store. Buying in bulk reduces the frequency of trips. Buying from local producers is usually more green-friendly.
10. Recycle. Take a few extra steps to use recycling containers. Recycling saves 70 to 90% of the energy and pollution that it would take to manufacture new materials from raw resources. Using 100 pounds of recycled copy paper saves 56 pounds of solid waste, 438 gallons of wastewater, 105 pounds of greenhouse gases, and a little less than 1 million BTUs of energy.

World Wide Web Sites

www.earthday.net Earthday Network website.

www.edf.org Environmental Defense Fund website; offers tools for calculating the environmental benefit of using recycled paper and for calculating your "carbon footprint."

www.eowilsonfoundation.org The E.O. Wilson Biodiversity Foundation is dedicated to fostering Earth stewardship through biodiversity research and education.

www.footprintnetwork.org Global Footprint Network provides

information on how lifestyles impact global resources. This site provides a calculator to assess your global footprint.
www.sierraclub.org Sierra Club website; look for links to events and environmental legislation in your state or region.
www.ucsusa.org Union of Concerned Scientists website.
www.worldpopulationbalance.org World Population Balance website—an organization devoted to population education.
www.350.org 350.org is a political action organization founded by Bill McKibben to work for solutions to climate change.

of corn-based ethanol-blend fuels, while reducing our use of fossil fuels, has increased the load of nitrogen flowing down the Mississippi River to the Gulf of Mexico by an estimated 30 to 40%. No-till farming, terracing, requiring treatment of effluent from CAFOs, eating less meat, and using alternative biofuels could reduce nitrogen and phosphorus pollution dramatically.

Overexploitation of Resources

Overexploitation of resources involves using resources and organisms at a greater rate than can be sustained by natural processes. Resources can be renewable or nonrenewable. **Renewable resources** are those that can replenish themselves when used within natural ecosystems or by humans. These include forest, fishery, and game resources and some energy resources like solar and wind power. **Nonrenewable resources** do not renew themselves over meaningful time frames. These include fossil fuels, soils, freshwater aquifers, and earth minerals and metals. When nonrenewable resources are used, they can never be replaced. Overexploitation includes unbridled use of both renewable and nonrenewable resources. Land and water are considered renewable in the sense that they can be used over-and-over again, and they can recover from some misuse. They are also considered nonrenewable in the sense that neither freshwater aquifers nor soils can be replaced after long-term exploitation by humans.

Land is an exploited resource. We use 35% of the earth's land surface for agricultural purposes, and expanding agriculture is one motivation for clearing more land. Urban sprawl, the spreading of a city and its suburbs into the surrounding countryside, puts pressure on farmland and natural areas. It has many negative consequences: promotes our dependence on automobiles; inflates costs for public transportation, per-person infrastructure, and per-person use of water and energy; and causes habitat destruction. Urban sprawl disrupts wildlife corridors that maintain connectivity between populations. Connectivity of populations promotes genetic exchange between populations. It permits animals to move (usually northward) in response to climate change. Zoning, more efficient agricultural practices, efficient food distribution, and reducing meat consumption in wealthy nations can help preserve our land.

Worldwide we draw 2,600 km^3 of freshwater annually from rivers, lakes, and groundwater. Irrigation accounts for 70% of this use, industry for 20%, and domestic use for 10%. Ground and surface water resources are drying. A major aquifer in the United States, the Ogallala Aquifer, has been depleted by about 9% since 1950. It is estimated that it would take about 6,000 years to recharge this aquifer through rainfall if it were to be completely depleted. Improvements in water-use efficiency include moving to more efficient drip and precision sprinkling irrigation systems and accurately monitoring soil moisture. Shifting away from the use of water in cooling power plants to dry cooling technology and replacing old appliances, toilets, and showerheads with water efficient ones are relatively easy improvements.

Wetlands are land areas that are permanently or seasonally saturated with water. They include swamps, marshes, and bogs. Wetlands provide many ecosystem services including flood control, groundwater replenishment, water purification, and shoreline stabilization. Wetlands are very important biodiversity reservoirs. They are home to thousands of animals. Approximately 200 new species of fish are described from U.S. wetlands each year. Unfortunately U.S. wetland losses are estimated to be in excess of 50% as compared to wetland areas present at the time of European settlement (1600s). Losses are the result of urban sprawl, floodplain development, agriculture, and road building.

One of many possible examples can illustrate the impacts of urban sprawl, freshwater use, and wetland destruction. The arroyo toad (*Anaxyrus californicus*) is a small greenish-gray toad with a spotty skin (figure 6.14). Arroyo toads were historically found in streams and river basins in southern California and Baja, Mexico.

FIGURE 6.14

The arroyo toad (*Anaxyrus californicus*). Arroyo toads are native to southern California and Baja, Mexico. They have been listed as endangered by the U.S. Fish and Wildlife Service. Habitat for these amphibians has been disrupted by developments that support human activities. This species' range has been reduced by 75%, including the loss of important migration corridors between interconnected habitat patches.

Arroyo toads live around shallow pools and sandy streams where they breed, and tadpoles mature into adults. Adults burrow into sandbars and stream banks for shelter during the day and during the dry season. Unfortunately for the toad, its home range is also home to about 20 million humans. The toad's habitat has become the location for highways, housing developments, water reservoirs, campgrounds, and off-road vehicle parks. Dam construction has been responsible for the destruction of 40% of this species' habitat. These developments chopped the toad's range into unconnected habitat patches, and the species range has been reduced by 75%. Its current population is estimated at 3,000 individuals, and it has been listed as an endangered species by the U.S. Fish and Wildlife Service. This listing has resulted in about 100,000 hectares of critical habitat being established for the arroyo toad. This habitat includes a mosaic of breeding, foraging, and shelter habitats–all interconnected by migration corridors. The response of this toad to these measures is uncertain.

Invasive Species

Invasive alien species are nonnative species that are introduced into an ecosystem. Invasive species can prey on native species and outcompete them for food and space (*see Nile perch, page 5*). These effects occur because ecosystems are naturally filled with organisms that have coevolved for millions of years. Native prey species may not have evolved defenses against foreign predators. On the other hand, prey species from one ecosystem that are introduced into a nonnative ecosystem may have no predators in their new environment. Uncontrolled by predators, these species often proliferate (*see page 330*). Invasive species may alter reproductive cycles of native species, introduce pathogens into ecosystems (*see page 351*), and alter ecosystem food webs. Invasive species have been accidentally released into ecosystems (*see pages 202-203 and 330*). They have also been introduced intentionally in attempts to control agricultural pests, in the form of ornamental plants, and as exotic pets that are no longer "loved."

Concluding Remarks from Two Ecological Giants

This section of chapter 6 began with quotes from Aldo Leopold and E.O. Wilson. We end with more of their wisdom. First, from *Round River: From the Journals of Aldo Leopold* (1972):

> The last word in ignorance is the man who says of an animal or plant, "What good is it?" If the land mechanism as a whole is good, then every part is good, whether we understand it or not. If the biota, in the course of aeons, has built something we like but do not understand, then who but a fool would discard seemingly useless parts? To keep every cog and wheel is the first precaution of intelligent tinkering.

Nature has inherent, inestimable value. Preserving every part of it should be a priority of every person who sees himself or herself as a part of nature–both for the survival of nature and for the survival of humanity. From E.O. Wilson in *The Diversity of Life* (1999):

> Humanity coevolved with the rest of life on this particular planet; other worlds are not in our genes. Because scientists have yet to put names on most kinds of organisms, and because they entertain only a vague idea of how ecosystems work, it is reckless to suppose that biodiversity can be diminished indefinitely without threatening humanity itself.

Section 6.7 Thinking Beyond the Facts

Building a four-lane highway through a natural area (e.g., a wetland) seems inconsequential to people who are unfamiliar with ecological principles. How can this common event be used to illustrate Barry Commoner's four laws of ecology?

Summary

6.1 **Animals and Their Abiotic Environment**

- Many abiotic factors influence where an animal may live. Animals have a tolerance range and a range of the optimum for environmental factors.
- Energy for animal life comes from consuming autotrophs or other heterotrophs. Energy is expended in excretory, existence, and productive functions.
- Temperature influences the rates of chemical reactions in animals and affects the animal's overall activity. When food resources become scarce, animals may enter daily torpor, hibernation, aestivation, or brumation.
- Water, light, geology, and soils are important abiotic environmental factors that influence animal lifestyles.

6.2 **Biotic Factors: Populations**

- Animal populations change in size over time. Changes can be characterized using survivorship curves.
- Animal populations grow exponentially until the carrying capacity of the environment is achieved, at which point constraints such as food, chemicals, climate, and space restrict population growth.
- Populations are regulated by density-independent factors and density-dependent factors.
- Intraspecific competition is often intense because resource requirements of individuals of the same species are nearly identical.

6.3 **Biotic Factors: Interspecific Interactions**

- Interspecific interactions affect all characteristics of a population. Herbivory occurs when animals feed on plants, and predation occurs when one animal kills and feeds on another animal.
- Interspecific competition results when different species compete for resources. Competing species either displace one another or they coexist by partitioning resources.
- Ecologically related species may exert strong selective influences on one another. This is coevolution.

- One form of coevolution occurs when two different species live in a continuing intimate association, called symbiosis. Symbiotic relationships may take the form of parasitism, commensalism, or mutualism.
- Animals avoid detection through crypsis. Mimicry occurs when one species gains protection by resembling another species. Some animals advertise their venomous or distasteful characteristics through aposmatic coloration.

6.4 **Communities**

- All populations living in an area make up a community. Species that help determine the makeup of a community are keystone species. Communities are characterized by community (species) diversity. Organisms have roles in their communities.
- The ecological niche concept helps ecologists visualize those roles.
- Communities often change in predictable ways. Successional changes may lead to stable climax communities, but this stability is often upset by anthropogenic and other environmental disturbances.

6.5 **Trophic Structure of Ecosystems**

- Energy in an ecosystem is not recyclable. Energy that is fixed by producers is eventually lost as heat. Biomagnification is the increase in concentration of a substance in the tissues of organisms at higher levels of food webs.

6.6 **Cycling within Ecosystems**

- Nutrients are cycled through ecosystems. Nutrients are elements important to the life of an organism (e.g., C, N, H, O, P, and S) and are constantly used, released, and reused throughout an ecosystem. Cycles involve movements of material from nonliving reservoirs in the atmosphere or earth to biological systems and back to the reservoirs again.

6.7 **Ecological Problems**

- Human population growth is the root of virtually all of our environmental problems. Trying to support too many people at the standard of living found in developed countries has resulted in air and water pollution and resource depletion.
- The variety of living organisms in an ecosystem is its biodiversity. Biodiversity of natural systems is threatened by habitat loss, climate change and ocean acidification, nutrient load and pollution, overexploitation of resources, and invasive species.

Concept Review Questions

1. In the energy budget of an animal, the gross energy intake is composed of all of the following, except one. Select the exception.
   a. Energy assimilated for functions such as pumping blood, gas exchange, and muscle contraction—this is existence energy.
   b. Energy assimilated in growth, mating, and care of young—this is productive energy.
   c. Energy left over after production and existence energy is used—this is storage energy.
   d. Energy lost in excretory pathways—this is excretory energy.
2. Which of the following is a period of inactivity during which an animal may withstand prolonged periods of heat and drying?
   a. Daily torpor
   b. Hibernation
   c. Brumation
   d. Aestivation
3. Most fish that spawn thousands of eggs in a single reproductive event would display a type ______ survivorship curve.
   a. I
   b. II
   c. III
   d. IV
4. A form of symbiosis in which one member of the relationship benefits and the second is neither helped nor harmed is called
   a. parasitism.
   b. mutualism.
   c. commensalism.
   d. mimicry.
5. Biomass conversion efficiency between ecosystem trophic levels averages
   a. 50%.
   b. 25%.
   c. 10%.
   d. 1%.

Analysis and Application Questions

1. Assuming a starting population of 10 individuals, a doubling time of one month, and no mortality, how long would it take a hypothetical population to achieve 10,000 individuals?
2. Section 6.4 briefly described how grassland ecosystems changed by the elimination of bison and fire. What ecosystem and community alterations (e.g., food webs, biogeochemical cycling, intraspecific and interspecific interactions) do you think would have accompanied the conversion of grassland to an ecosystem dominated by shrubs and woody vegetation?
3. Sea otters (*Enhydra lutris nereis*) are considered keystone species in kelp ecosystems (*see page 408*). Use the example of the sea otter to explain how habitat alterations that affect keystone species are especially damaging to ecosystems.
4. Overpopulation is a severe environmental problem, and most population growth is occurring in developing countries like India. Why is it wrong to focus primarily on population growth when considering overuse of natural resources by humans?

7

Animal Taxonomy, Phylogeny, and Organization

This Hawaiian spiny lobster (*Panulirus marginatus*) is a member of one of the 950,000 extant animal species that zoologists have named in a manner that creates order out of the tremendous diversity of animal forms.

Chapter Outline

7.1 Taxonomy and Phylogeny
- *A Taxonomic Hierarchy*
- *Nomenclature*
- *Taxonomic Methods*
- *Animal Systematics*

7.2 Patterns of Organization
- *Symmetry*
- *Other Patterns of Organization*

Biologists have identified approximately 1.6 million species of eukaryotic organisms, more than three-fourths of which are animals. Many zoologists spend their lives grouping animals according to shared characteristics. These groupings reflect the order found in living systems that is a natural consequence of shared evolutionary histories. Often, the work of these zoologists involves describing new species and placing them into their proper relationships with other species. Obviously, much work remains in discovering and classifying the approximately 8 million more undescribed species (*see chapter 6*).

Rarely do zoologists describe new taxa above the species level (*see figure 1.4*). In 1995, however, R. M. Kirstensen and P. Funch of the University of Copenhagen described a new animal species–*Symbion pandora*–on the mouthparts of Norway lobsters (*Nephrops norvegicus*). This species is so different that it was assigned to a new phylum–the broadest level of animal classification (figure 7.1). The description of this new phylum, Cycliophora, brought the total number of recognized extant animal phyla to 36. These same researchers later described a second new phylum of animals (Micrognathozoa) from springs in Greenland in 2000, and the total number of extant animal phyla increased to 37. Both of these groups of animals are discussed in chapter 10. Taxonomists have discovered that these two groups of animals are related to each other, and to other animals, in specific ways. This chapter describes the principles used by zoologists to investigate, and describe, relationships between groups of animals.

7.1 TAXONOMY AND PHYLOGENY

LEARNING OUTCOMES

1. Justify the statement that "taxonomy reflects phylogeny."
2. Explain how the taxonomic hierarchy and names of animals reflect evolutionary relationships.
3. Assess the kinds of data used in investigating animal phylogenies.
4. Compare the goals and methods of phylogenetic systematics and evolutionary systematics.

One of the characteristics of modern humans is our ability to communicate with a spoken language. Language not only allows us to communicate but also helps us encode and classify concepts, objects, and organisms that we encounter. To make sense out of life's diversity, we need more than just names for organisms. A potpourri of more than a million animal names is of little use to anyone. To be useful, a naming system must reflect the order and relationships that arise from evolutionary processes. The study of the kinds and diversity of organisms and of the evolutionary relationships among them is called **systematics** (Gr. *systema,*

FIGURE 7.1

A Recently Described Phylum, Cycliophora. Systematists group animals according to evolutionary relationships. Usually, the work of systematists results in newly described species being placed in higher taxonomic categories along with previously studied species. They rarely describe new higher taxonomic groups because finding an organism so different from any previously known organism is unlikely. *Symbion pandora* (shown here) was discovered in 1995 and was distinctive enough for the description of an entirely new phylum, Cycliophora. The individuals shown here are covering the mouthparts of a lobster and are about 0.3 mm long. Since 1995, additional *Symbion* species have been described from lobsters other than *Nephrops norvegicus*.
Courtesy of Professor Reinhardt Kristensen, University of Coppenhagen, Denmark

system + *ikos,* body of facts) or **taxonomy** (Gr. *taxis,* arrangement + L. *nominalis,* belonging to a name). These studies result in the description of new species and the organization of animals into groups (taxa) based on degree of evolutionary relatedness. The work of taxonomists involves describing inherited characteristics that animals share, ranking characteristics based on their order of appearance in the evolution of animals, and using this information to describe phylogenetic relationships among animals. A **phylogeny** (Gr. *phylon,* race + *geneia,* origin) is a description of the evolutionary history of a group of organisms and is usually depicted using tree diagrams that will be introduced in this chapter.

A Taxonomic Hierarchy

Modern taxonomy is rooted in the work of Karl von Linné (Carolus Linnaeus) (1707–1778). His binomial system (*see chapter 1*) is still used today. Von Linné also recognized that different species could be grouped into broader categories based on shared characteristics. Any grouping of animals that shares a particular set of characteristics forms an assemblage called a **taxon** (pl., taxa). For example, a housefly (*Musca domestica*), although obviously unique, shares certain characteristics with other flies (the most important of these being a single pair of wings). Based on these similarities, all true flies form a logical, more inclusive taxon. Further, all true flies share certain characteristics with bees, butterflies, and beetles. Thus, these animals form an even more inclusive taxon. They are all insects.

All animals are given names associated with eight taxonomic ranks arranged hierarchically (from broad to specific): **domain, kingdom, phylum, class, order, family, genus,** and **species** (table 7.1). As one moves down through the hierarchy from domain toward species, one is looking at groups derived from more recent ancestors and a smaller subset of more closely related animals (*see figure 1.4*). Taxonomists have the option of subdividing these ranks (e.g., subphylum, superclass, and infraclass) to express relationships between any two ranks.

Even though the work of von Linné predated modern evolutionary theory, many of his groupings reflect evolutionary relationships. Morphological similarities between two animals have a genetic basis and are the result of a common evolutionary history. Thus, in grouping animals according to shared characteristics, von Linné often grouped them according to their evolutionary relationships.

The Linnaean taxonomic hierarchy has limitations for modern biology. Above the species level, the definitions of what constitutes a particular taxon are not precise. For example, there is no definition of what constitutes a family. The cat family, Felidae, has 40 species and the ground beetle family, Carabidae, has more than 40,000 species. There are no criteria to establish that these two families represent the same level of divergence from a common ancestor or that the time frame for divergence in the two groups is related in any meaningful way.

As we will see later in this chapter, recently derived characteristics are more important than are ancestral characteristics in establishing evolutionary relationships. Traditional classification systems were established without distinguishing between recently derived and ancestral characteristics. The consequence of these limitations is that many older taxonomic hierarchies are not useful in making evolutionary predictions, and errors in older interpretations are being revealed.

TABLE 7.1
TAXONOMIC CATEGORIES OF A HUMAN AND A DOG

TAXONOMIC RANK*	HUMAN	DOMESTIC DOG
Domain	Eukarya	Eukarya
Kingdom	Animalia	Animalia
Phylum	Chordata	Chordata
Class	Mammalia	Mammalia
Order	Primates	Carnivora
Family	Hominidae	Canidae
Genus	*Homo*	*Canis*
Species	*Homo sapiens*	*Canis lupis*

*Taxonomists frequently subdivide these taxonomic ranks using prefixes like "super," "sub," and "infra." You will encounter taxonomic ranks like superphylum, subphylum, and infraclass in chapters that follow.

Nomenclature

Do you call certain freshwater crustaceans crawdads, crayfish, or crawfish? Do you call a common sparrow an English sparrow, a barn sparrow, or a house sparrow? The binomial system of nomenclature brings order to a chaotic world of common names. Common names have two problems. First, they vary from country to country, and from region to region within a country. Some species have literally hundreds of different common names. Biology transcends regional and nationalistic boundaries, and so must the names of what biologists study. Second, many common names refer to taxonomic categories higher than the species level. Most different kinds of pillbugs (class Crustacea, order Isopoda) or most different kinds of crayfish (class Crustacea, order Decapoda) cannot be distinguished from a superficial examination. A common name, even if you recognize it, often does not specify a particular species.

Nomenclature (L. *nominalis,* belonging to a name + *calator,* to call) is the assignment of a distinctive name to each species. The binomial system of nomenclature is universal and clearly indicates the level of classification involved in any description. No two kinds of animals have the same binomial name, and every animal has only one correct name, as required by the *International Code of Zoological Nomenclature,* thereby avoiding the confusion that common names cause. The genus of an animal begins with a capital letter, the species epithet begins with a lowercase letter, and the entire scientific name is italicized or underlined because it is derived from Latin or is latinized. Thus, the scientific name of humans is written *Homo sapiens.* When the genus is understood, the binomial name can be abbreviated *H. sapiens.*

Taxonomic Methods

Traditional systematic methods use phenotypic characteristics to arrange animals into phylogenetic trees. Many of the phylogenetic trees in chapters 7 to 22 depict relationships based on the use of observable traits. Traditional methods use many computational tools and taxonomic database resources to create and evaluate hypotheses of relatedness (phylogenetic trees) among animals based on similarities and differences in genetic traits present in a study group. Taxonomic database resources assemble taxonomic information including species descriptions, distribution information, ecology, taxonomic keys, and literature citations. For example, "FishBase" is a global species database for fishes that catalogues taxonomic information.

In recent years, molecular biological techniques have provided important information for taxonomic studies. The relatedness of animals is reflected in the gene products (proteins) animals produce and in the genes themselves (the sequence of nitrogenous bases in DNA). Because related animals have DNA derived from a common ancestor, genes and proteins of related animals are more similar than genes and proteins of distantly related animals. Gene sequencing technologies are used to determine base sequences in DNA. These technologies involve the use of the polymerase chain reaction (PCR) to derive thousands of copies of a DNA base sequence and automated gene sequencers to determine the sequence of the four nucleotide bases in the amplified gene product. Sequencing the nuclear DNA and the mitochondrial DNA of animals has become commonplace. Mitochondrial DNA is useful in taxonomic studies because mitochondria have their own genetic systems and are inherited cytoplasmically. That is, mitochondria are transmitted from parent to offspring through the egg cytoplasm and can be used to trace maternal lineages.

Studies of ribosomal RNA (rRNA) genes provide a wealth of data that have been used to study very old evolutionary relationships (10 million to billions of years [*see chapter 8*]). Ribosomal RNA is an ancient molecule, it is encoded within nuclear DNA, and it is present and retains its function in virtually all organisms. In addition, rRNA changes very slowly. Recall that ribosomal RNA makes up a portion of ribosomes—the organelle responsible for the translation of messenger RNA into protein. This slowness of change, called **evolutionary conservation,** indicates that the protein-producing machinery of a cell can tolerate little change and still retain its vital function. Evolutionary conservation of this molecule means that closely related organisms (recently diverged from a common ancestor) are likely to have similar ribosomal RNAs. Distantly related organisms are expected to have ribosomal RNAs that are less similar, but the differences are small enough that the relationships to some ancestral molecule are still apparent.

Animals also have DNA that does not code for functioning proteins. This extragenic DNA comprises a large portion of animal genomes. Because extragenic DNA does not code for functioning proteins, it can change without harming the animal—it is not conserved evolutionarily. Changes in extragenic DNA are not selected against by natural selection and can accumulate over relatively brief (e.g., 50,000 to a few million years) periods of time. A number of DNA analysis techniques (e.g., microsatellites and amplified fragment length polymorphisms [AFLP]) are used to examine extragenic DNA and investigate recent phylogenetic changes.

Analysis tools are used to study base sequences, identify the corresponding genes, and compare these genes to other base sequences in molecular databases. One commonly used database is maintained by the National Center for Biotechnology Information (NCBI). Similar tools are available for the study of microsatellite and AFLP data. Molecular systematists compare the base sequences or extragenic DNAs of different organisms. They enter these data into computer programs and examine all possible relationships among the different organisms. The systematists then decide which arrangement of the organisms best explains the data.

Although molecular techniques have proven to be extremely valuable to animal taxonomists, they will not replace traditional taxonomic methods. The most reliable phylogenetic hypotheses are created when traditional methods and molecular methods derive similar conclusions.

Animal Systematics

The goal of animal systematics is to arrange animals into groups that reflect evolutionary relationships. Ideally, these groups should include the most recent ancestral species and all of its descendants. Such a group is called a **monophyletic group** (figure 7.2). **Polyphyletic**

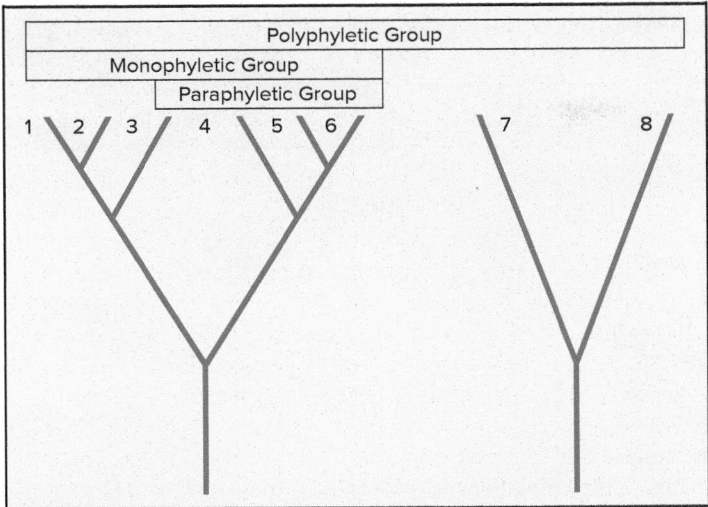

FIGURE 7.2

Evolutionary Groups. An assemblage of species 1–8 is a polyphyletic group because species 1–6 have a different ancestor than species 7 and 8. An assemblage of species 3–6 is a paraphyletic group because species 1 and 2 share the same ancestor as 3–6, but they have been left out of the group. An assemblage of species 1–6 is a monophyletic group because it includes all of the descendants of a single ancestor.

groups do not contain the most recent common ancestor of all members of the group. Members of a polyphyletic group have at least two phylogenetic origins. Since it is impossible for a group to have more than one most recent ancestor, a polyphyletic group reflects insufficient knowledge of the group. A **paraphyletic group** includes some, but not all, descendants of a most recent common ancestor. Paraphyletic groups may also result when knowledge of the group is insufficient and the relationships need clarification in genetic and evolutionary contexts (*see figure 7.2*).

In making decisions regarding how to group animals, taxonomists look for attributes called characters that indicate relatedness. A **character** is virtually anything that has a genetic basis and can be measured. Taxonomic characters may be morphological (e.g., structure of insect genitalia), embryological (e.g., early cleavage patterns of a fertilized egg), biogeographical (e.g., distribution of animals on opposite sides of mountain ranges), physiological (e.g., functions of similar hormones), ecological (e.g., habitat requirements), behavioral (e.g., courtship behaviors), or molecular (e.g., DNA base sequences). Two kinds of characters are recognized by taxonomists. Homologous characters (*see chapter 4*) are characters that are related through common descent. Vertebrate forelimbs in the form of arms, forelegs, and wings are homologous characters. Analogous characters are resemblances that result from animals adapting under similar evolutionary pressures but not as a result of common ancestry. The latter process is sometimes called convergent evolution. **Homoplasy** is a term applied to analogous resemblances. The similarity between the wings of birds and insects is a homoplasy. Homologies are useful in classifying animals, homoplasies are not. The presence of one or more homologous characters in two animals indicates some degree of relatedness between the animals. They had a common ancestor at some point in their evolutionary history.

As in any human endeavor, different approaches to solving problems are preferred by different groups of people. That is also the case with animal systematics. Two popular approaches to animal systematics include evolutionary systematics and phylogenetic systematics (cladistics).

Phylogenetic Systematics or Cladistics

Phylogenetic systematics (cladistics) is one approach to animal systematics. The goal of cladistics is the generation of hypotheses of genealogical relationships among monophyletic groups of organisms. Cladists believe that homologies of recent origin are most useful in phylogenetic studies. Attributes of species that are old and have been retained from a common ancestor are referred to as **ancestral character states** or **plesiomorphies** (Gr. *plesio,* near + *morphe,* form). In cladistic studies, these ancestral character states are common to all members of a group and indicate a shared ancestry. These common characters are called **symplesiomorphies** (Gr. *sym,* together + *plesio,* near + *morphe,* form). Because they are common to all members of a group, they cannot be used to describe relationships within the group. Characters that have arisen since common ancestry with the outgroup are called **derived character states** or **apomorphies** (Gr. *apo,* away + *morphe,* form). Derived characters shared by members of a group are called **synapomorphies** (Gr. *syn,* together + *apo,* away + *morphe,* form). Derived character states vary within study groups; therefore, they are useful in describing relationships within the group.

The work of cladists involves deciding what character(s) is ancestral for the group in question and distinguishing between derived and ancestral character states. In deciding what character is ancestral for a group of organisms, cladists look for a related group of organisms, called an **outgroup,** that is not included in the study group. The outgroup is used to determine whether a character is ancestral or has arisen within the study group. Examine figure 7.3. It depicts characters that can be used to determine relationships among five groups of vertebrates. The outgroup (Cephalochordata) shares two characters (pharyngeal slits and notochord) with the study group. These shared characters are, thus, symplesiomorphic and indicate common ancestry with the study group. The presence of vertebrae in all members of the study group, but their absence in the outgroup, tells us that this character is the shared ancestral character for the study group. Other characters shown in figure 7.3 are derived character states that can be used to determine relationships within these vertebrate groups and to construct a vertebrate phylogeny.

Figure 7.4 is a tree diagram called a **cladogram,** which depicts relationships within five groups of vertebrates. Cladograms depict a sequence in the origin of derived character states. Recall that the lines that depict evolutionary pathways are called branches of a phylogenetic tree, and the points of divergence between two or more branches are called nodes. Nodes represent points where two groups diverged from a common ancestor (*see chapter 4*). A cladogram is interpreted as a family tree depicting a hypothesis regarding a monophyletic lineage.

Character/ Animal	Pharyngeal slits and notochord	Vertebrae	Paired appendages used in swimming	Swim bladder or lung	Appendages with muscular lobes	Muscular limbs adapted for terrestrial locomotion	Sucker-like mouth and rasping tongue	Fins with dermal rays	Muscular limbs adapted for swimming
Cephalo-chordata (outgroup)	+	–	–	–	–	–	–	–	–
Lamprey	+	+	–	–	–	–	+	–	–
Sharks and related cartilaginous fish	+	+	+	–	–	–	–	–	–
Perch and related bony fishes	+	+	+	+	–	–	–	+	–
Lobe-finned fishes	+	+	+	+	+	–	–	–	+
Tetrapoda	+	+	+	+	+	+	–	–	–

FIGURE 7.3

Morphological Data for Five Groups of Vertebrates. A "+" indicates that a vertebrate group possesses the character listed at the top of the column. A "–" indicates that the character is absent and represents the ancestral character state. The derived character states shown are used to construct the phylogeny shown in the figure 7.4 cladogram.

Look again at the distribution of characters shown in figures 7.3 and 7.4. The presence of paired appendages and jaws are synapomorphies that are common to all vertebrate groups other than the lampreys. Figure 7.4 shows paired appendages used in swimming and jaws as being derived after the divergence of lampreys within vertebrate phylogeny. The derived character states for paired appendages involves the acquisition of muscular lobes (distinguishing lobe-finned fishes and tetrapods) or fins with dermal rays (distinguishing bony fishes). The character "appendages with muscular lobes" is a synapomorphy that creates a related subset of vertebrates, the lobe-finned fishes and the Tetrapoda. A related subset within a cladogram is called a **clade** (Gr. *klados,* branch). Notice that one can expand the tetrapod/lobe-finned fishes clade to include perch and related boy fishes by moving one's point of reference to the "swim bladder or lung" character.

Unlike the paired appendage characters, other characters in the cladogram seem to appear out of nowhere (e.g., sucker-like mouth and unique scales) and are not mentioned again. These are derived characters that originated within a lineage since divergence (or at the point of divergence) from a most recent common ancestor. In these cases, the absence of the character in one lineage represents the ancestral character state. Characters that originate within a lineage may define a particular taxon. For example, within the vertebrates, the sucker-like mouth is unique to the lampreys (*see figure 27.3*a). Characters that define a particular taxon are called **autapomorphies** (Gr. *aut*, self). Like synapomorphy, autapomorphy is a relative term that depends on the taxonomic rank being considered. A sucker-like mouth is an autapomorphy when considering the five vertebrate taxa in figure 7.4, but to a lamprey taxonomist this character would be shared and not useful in describing individual lamprey species.

Tetrapoda and lobe-finned fishes not only form a clade, but they are also sister groups. Two taxa are **sister groups** if they share a most recent common ancestor. Knowing that two taxa form a sister group ensures that one is considering a monophyletic clade. Perch and related fishes and lobe-finned fishes are not sister groups. A grouping that included only these two taxa would be paraphyletic because the most recent common ancestor of lobe-finned fishes is not shared with the perch and related fishes, but it is shared with the tetrapods.

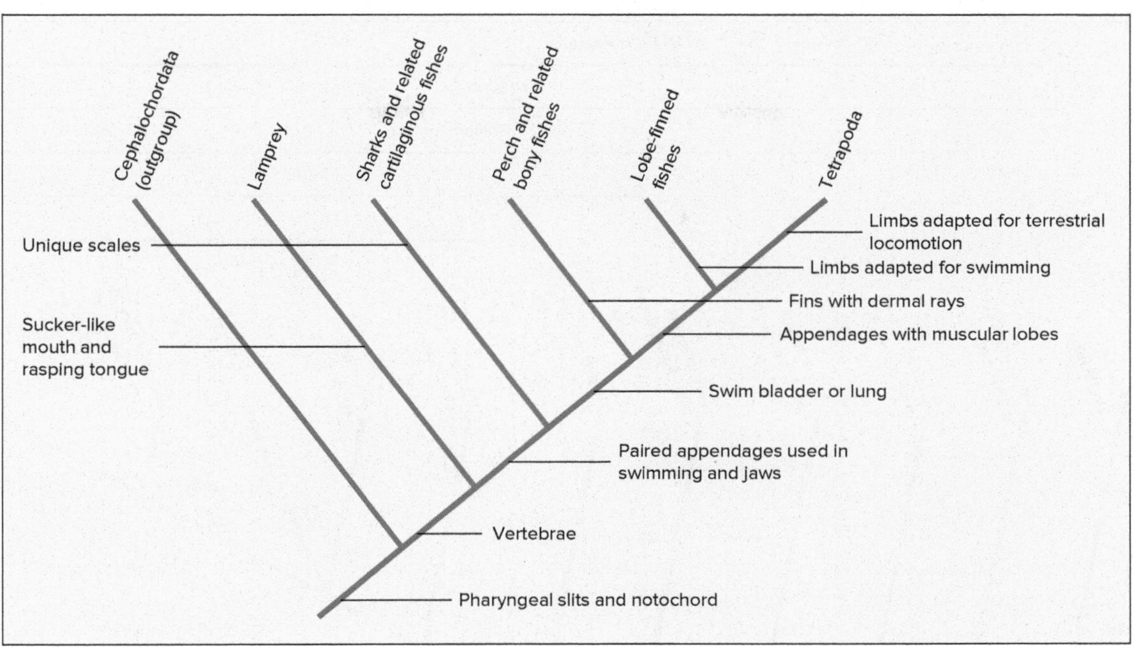

FIGURE 7.4

Interpreting Cladograms. This cladogram depicts an abbreviated vertebrate phylogeny. Five vertebrate taxa and the outgroup (Cephalochordata) are shown at the tips of the cladogram branches. Ancestral and derived character states are shown along the vertical axis. Branch points are called nodes and represent points of divergence between taxa. Pharyngeal slits and notochord are symplesiomorphic characters for the entire assemblage of animals. The presence of vertebrae is an ancestral character state for all vertebrates and distinguishes the vertebrates from the outgroup. Other characters, like swim bladder or lung, are shared derived (synapomorphic) characters that distinguish subsets (clades) within the vertebrate lineage. The swim bladder or lung character is common to the perch and related bony fishes, lobe-finned fishes, and Tetrapoda clade. The lobe-finned fishes and Tetrapoda form a sister group because they share a most recent common ancestor.

Figure 7.5 is a more detailed cladogram depicting the evolutionary relationships among the vertebrates. The cephalochordates are an outgroup for the entire vertebrate lineage. Notice that extraembryonic membranes is a synapomorphy used to define the clade containing the reptiles, birds, and mammals. These extraembryonic membranes are a shared character for these groups and are not present in any of the fish taxa or the amphibians. Distinguishing between reptiles, birds, and mammals requires looking at characters that are even more recently derived than extraembryonic membranes. A derived character, the shell, distinguishes turtles from all other members of the clade; skull characters distinguish the lizard/crocodile/bird lineage from the mammal lineage; and hair, mammary glands, and endothermy is a unique mammalian character combination. Note that a synapomorphy at one level of taxonomy may be a symplesiomorphy at a different level of taxonomy. Extraembryonic membranes is a synapomorphic character within the vertebrates that distinguishes the reptile/bird/mammal clade. It is symplesiomorphic for reptiles, birds, and mammals because it is ancestral for the clade and cannot be used to distinguish among members of these three groups.

As with the classification system as a whole, cladograms depict a hierarchy of relatedness. The grouping of organisms by derived characters results in a **hierarchical nesting,** which is shown in figure 7.5. Reptiles, birds, and mammals form a nested group defined by the presence of extraembryonic membranes. They are a part of a larger group of vertebrates including the amphibians and are characterized by the presence of four limbs adapted for terrestrial locomotion. These in turn are united with other vertebrates, the Gnathostomata, and defined by the presence of jaws formed from the mandibular arch. The less inclusive the nest is, the more closely related the organisms.

Evolutionary Systematics

A second approach to animal systematics is **evolutionary systematics.** It is an older, more traditional, approach to systematics, but evolutionary systematists have been relentless in integrating modern evolutionary and genetic theories into their approach to taxonomy. Two criteria used by evolutionary systematists in their work are recency of common descent and amount and nature of evolutionary change between branch points. Evolutionary systematists recognize and use plesiomorphic (ancestral) and apomorphic (derived) character states in a fashion similar to how phylogenetic systematists use character states. Derived character states are used to evaluate branching patterns within phylogenies. Unlike phylogenetic systematists, however, evolutionary systematists weigh some derived characters more heavily than other derived characters. In birds, for example, the set of characters that includes wings, feathers, and other flight

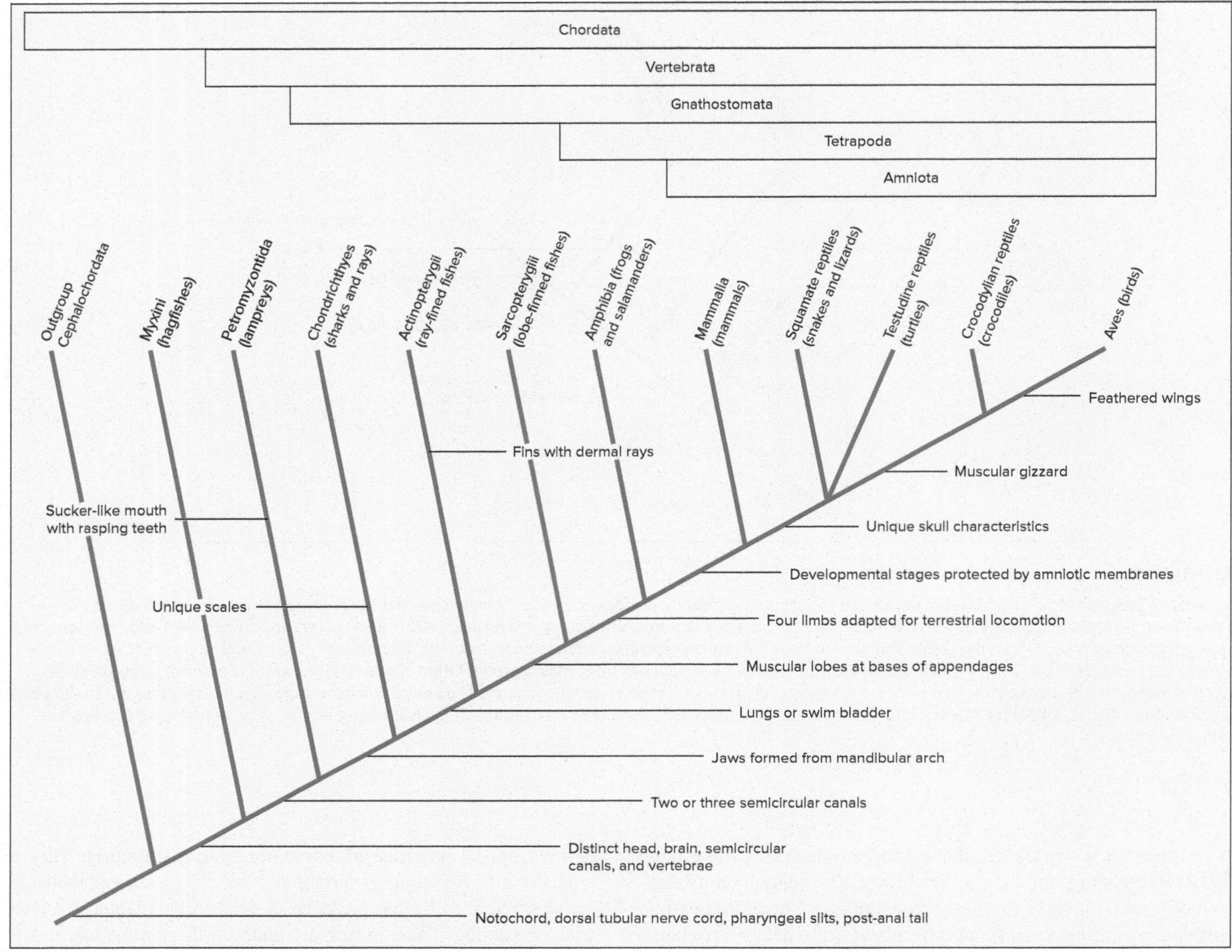

FIGURE 7.5

Cladogram Showing Vertebrate Phylogeny. A cladogram is constructed by identifying points at which two groups diverged. The following points are in reference to comparisons that will be made to the tree diagrams constructed by evolutionary systematists (*see figure 7.6*). Notice that timescales are not given or implied. The relative abundance of taxa is also not shown. Notice that this diagram shows the birds and crocodylians sharing a common branch, and that these two groups are more closely related to each other than either is to any other group of animals. Brackets at the top of the cladogram illustrate hierarchical nesting. Each higher bracket includes the brackets below it.

adaptations are particularly important in defining what it means to be a bird. These characters are weighted more heavily than other derived characters because they form an "adaptive zone," or a set of evolutionary changes that make the group unique. Thus, the unique characters of birds are considered more important to taxonomic decisions than other characters that imply close ties to dinosaurs and the crocodylians. The work of evolutionary systematists, like that of cladists, is represented by tree diagrams. Unlike cladograms, these diagrams are often integrated with information from the fossil record to depict time periods and relative abundance of taxa within a lineage (figure 7.6).

The Debate: Cladistics or Evolutionary Systematics

Cladists and evolutionary systematists debate the merits of their approaches to animal taxonomy. Zoologists widely accept cladistics. This acceptance has resulted in some nontraditional interpretations of animal phylogeny. A comparison of figures 7.5 and 7.6 shows one example of different interpretations derived through evolutionary systematics and cladistics. Recall that generations of taxonomists have assigned class-level status (Aves) to birds. Reptiles also have had class-level status (Reptilia). Cladistic analysis has shown,

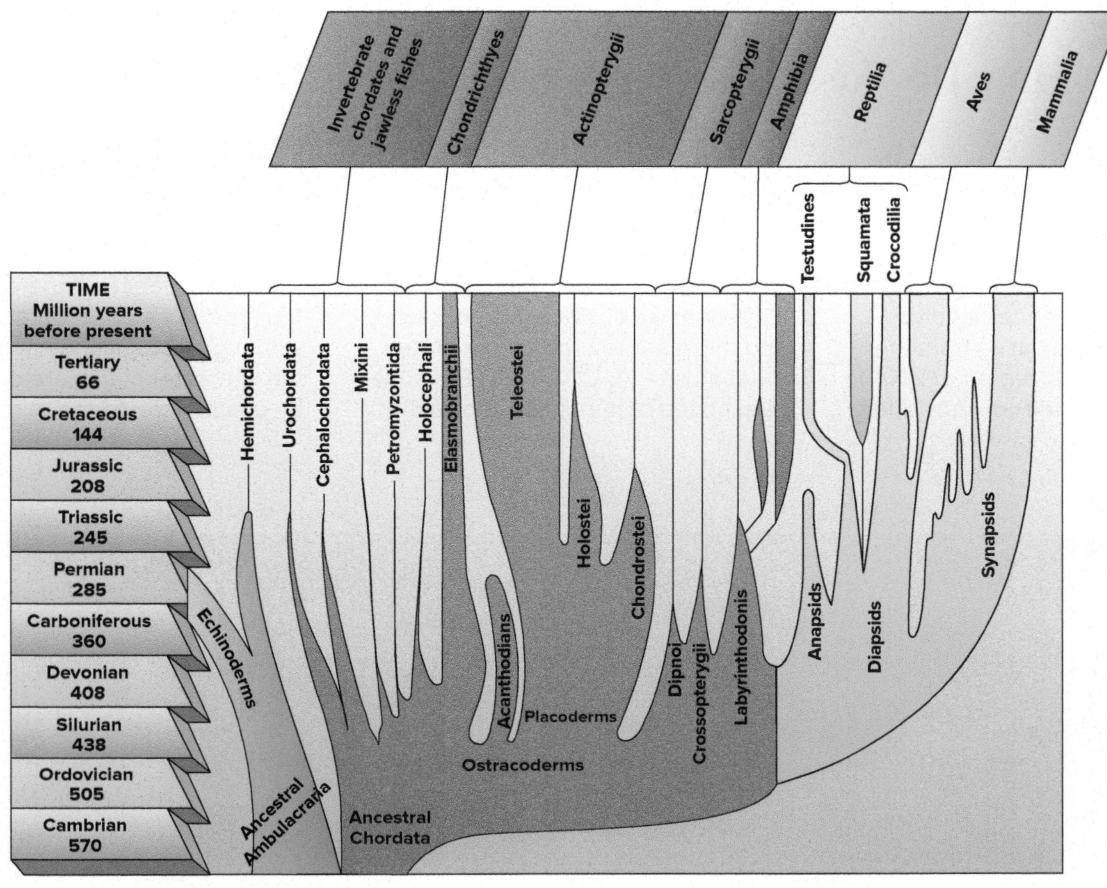

FIGURE 7.6

Phylogenetic Tree Showing Vertebrate Phylogeny. A phylogenetic tree derived from evolutionary systematics depicts the degree of divergence since branching from a common ancestor, which is indicated by the time periods on the vertical axis. The width of the branches indicates the number of recognized genera for a given time period. Note that this diagram shows the birds (Aves) as being closely related to the reptiles (Reptilia), and both groups as having class-level status.

How Do We Know Tree Diagrams Are Accurate?

The grouping of organisms by characters is not arbitrary. If one were to classify screws based on length, head type, and metal composition, one would begin at an arbitrary starting point, for example, by placing all brass screws in one pile and all steel screws in another pile. Then one might arbitrarily decide to subgroup within each composition grouping by length and then by head type. Someone else could reclassify the same screws using length as a starting point and end up with an entirely different set of nesting relationships. Neither classification would be incorrect because the characteristics used in this exercise do not reflect ancestral or derived states. Biological classification is unique. Not all starting points are correct—one must begin with an ancestral character and work upward through increasingly derived characters. Modern taxonomic methods involve testing and retesting data from different sources (e.g., morphological and various molecular sources). The relationships derived from all sources of evidence should be very similar. This congruence is evidence that a tree diagram accurately depicts evolutionary relationships.

How Do We Know—Phylogenies from Base Sequences

Phylogenetic studies rely on the analysis of nucleotide sequences from nuclear and mitochondrial DNA in addition to morphological data. The analysis of nucleotide sequences involves aligning (matching) bases from the same (homologous) region of DNA in two or more taxa or from individuals from different populations of the same species. Alignment software determines the best fit between sequences from the taxa being studied. Each nucleotide site in an alignment is considered a character, and there are four character states for each site (A, C, G, or T). Because sequences of hundreds of bases (characters) are analyzed, it is easy to see why these analyses are powerful phylogenetic tools.

Box figure 7.1*a* shows a very simple sequence alignment from four species and an outgroup. As with morphological data, the ancestral character states are determined by comparison to the outgroup.

Site	1	2	3	4	5	6	7	8	9	10
Outgroup	A	T	T	C	C	G	C	A	T	A
Species 1	G	T	C	C	C	G	T	A	T	A
Species 2	G	T	T	C	A	G	T	A	T	T
Species 3	A	T	T	C	C	G	T	T	T	A
Species 4	G	T	T	C	A	G	T	A	C	A

(a)

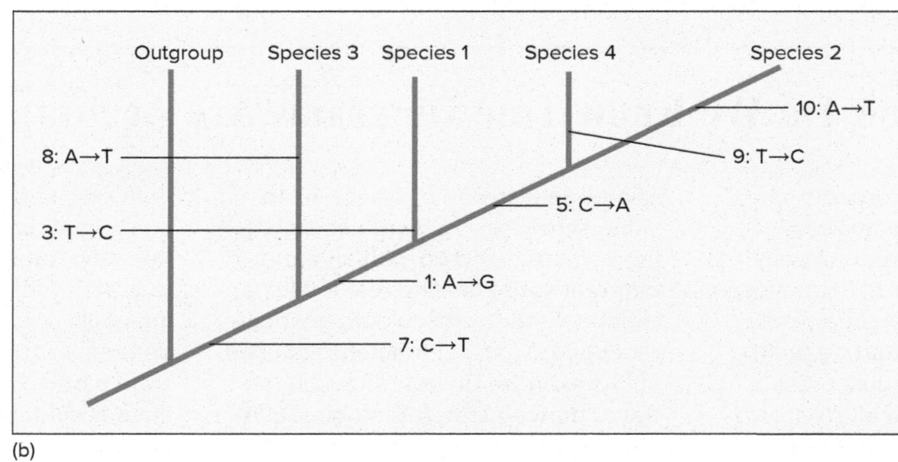

BOX FIGURE 7.1 A Phylogeny from a Base Sequence Alignment. (*a*) A base sequence alignment for four species and an outgroup. (*b*) A phylogeny derived from the base sequences shown in (*a*). The notation 7: C → T is interpreted as a change in the base cytosine (C) to the base thymine (T) at character site 7 in the base sequence.

HOW DO WE KNOW—PHYLOGENIES FROM BASE SEQUENCES *Continued*

Character sites 2, 4, and 6 in the alignment are in the ancestral state for all species. The study group (species 1–4) differs from the outgroup by a change at character site 7 from C to T. This is the character change that separates the study group from the group's most recent ancestor (the outgroup). Species 1, 2, and 4 share a synapomorphy at site 1 (A to G), and species 3 is defined by a change at character site 8 from A to T (it is autapomorphic). Further base changes in the sequence at character sites 5, 9, and 10 define the relationships of species 1, 2, and 4. Box figure 7.1*b* shows a cladogram derived from these base sequence data.

In real data applications, the sequences are much more complex. It is usually possible to derive more than one tree diagram (branching pattern or topology) from one set of data. Most hypotheses assume that the most accurate tree diagram is the one that requires the fewest number of base changes to derive it. (It is said to be parsimonious.) Computer programs can estimate tree branch lengths based on the number of base changes that occur along a particular branch to represent genetic distances. (More base changes correspond to greater genetic distance.) Other computational tools allow the phylogeneticist to calculate confidence parameters. (How good is the data?) Still other tools can be used to calculate time since divergence from a common ancestor using molecular clock models (*see page 61*).

however, that birds are more closely tied by common ancestry to the alligators and crocodiles than to any other group of living vertebrates. According to the cladistic interpretation, birds and crocodiles should be assigned to a group that reflects this close common ancestry. Birds would become a subgroup within a larger group that included both birds and reptiles. Crocodiles would be depicted as more closely related to the birds than they would be to snakes and lizards. Traditional evolutionary systematists maintain that the traditional interpretation is still correct because it takes into account the greater importance of the "adaptive zone" of birds (e.g., feathers and endothermy) that makes the group unique. Cladists support their position by pointing out that the designation of "adaptive zone" involves value judgments that cannot be tested.

A widely used biological species concept was described in chapter 5. The widespread acceptance of cladistics methods is partly responsible for increased popularity of a newer definition of a species. If we apply the **phylogenetic species concept,** we define a species on the basis of a common phylogenetic history. As described by Joel Cracraft (1983), a species is "the smallest diagnosable monophyletic group of populations within which there is a parental pattern of ancestry and descent." In other words, a species group shares a single ancestor and can be distinguished from other groups by one or more synapomorphies. The phylogenetic species concept, in practice, often makes use of molecular tools to identify monophyletic groups. This concept has the advantage of being applicable with sexual and asexual taxa as well as taxa that are known from the fossil record. Critics of the phylogenetic species concept, including many evolutionary systematists, point out that molecular tools are so powerful that they reveal many differences that are evolutionarily neutral. How does one distinguish between branching patterns that reflect variation within species as compared to branching patterns between species? Using the phylogenetic species concept causes the number of species within a group to proliferate and makes the naming of species with unique binomial names cumbersome.

As debates between cladists and evolutionary systematists continue, our knowledge of evolutionary relationships among animals will become more complete. Debates like these are the fuel that forces scientists to examine and reexamine old hypotheses. Animal systematics is certain to be a lively and exciting field in future years.

Chapters 9 through 22 are a survey of the animal kingdom. The organization of these chapters reflects the traditional taxonomy that makes most zoologists comfortable. Cladograms are usually included in "Further Phylogenetic Considerations" at the end of most chapters, and any different interpretations of animal phylogeny implicit in these cladograms are discussed.

SECTION 7.1 THINKING BEYOND THE FACTS

Why are derived characteristics more useful in establishing evolutionary relationships than are shared characteristics? Use two subgroups of mammals to illustrate your answer (**see table 22.1*****). For example, compare horses (order Perissodactyla) and camels (order Artiodactyla).***

7.2 PATTERNS OF ORGANIZATION

LEARNING OUTCOMES

1. Analyze the selective advantages that asymmetry, radial symmetry, and bilateral symmetry provide animals that possess these patterns of organization.
2. Evaluate the statement "Most animals possess either diploblastic or triploblastic tissue-level organization."
3. Differentiate three forms of triploblastic tissue organization.
4. Analyze the reasons for the prevalence of body cavities in animals.

One of the most strikingly ordered series of changes in evolution is reflected in body plans in the animal kingdom. On one hand, evolutionary changes in animal body plans may not reflect a series of progressive changes from simpler to more complex or from "primitive" to "advanced." A relatively "simple" organization can be the product of millions of years of evolution from more "complex" ancestors. This is the case for many parasites that have adapted to their hosts through the loss of certain organ systems that became irrelevant or detrimental to their survival as parasites (e.g., the loss of digestive systems). On the other hand, one sees in animal organization evolutionary patterns that characterize groups of animals. Examining these patterns of organization helps us understand why related groups of animals have similar body forms and how animal body forms may reflect adaptations to animals' environments.

Symmetry

The bodies of animals and protists are organized into almost infinitely diverse forms. Within this diversity, however, are certain patterns of organization. The concept of symmetry is fundamental to understanding animal organization. **Symmetry** describes how the parts of an animal are arranged around a point or an axis (table 7.2).

Asymmetry, which is the absence of a central point or axis around which body parts are equally distributed, characterizes many sponges (figure 7.7). Asymmetrical organisms do not develop complex communication, sensory, or locomotor functions. The absence of symmetry in sponges allows them to grow tightly attached to their substrate and mold themselves onto virtually any available substrate of any shape (*see figure 7.7*). Asymmetrical body forms have allowed sponges to flourish for hundreds of millions of years.

A sea anemone can move along a substrate, but only very slowly. How does it gather food? How does it detect and protect itself from predators? For this animal, a blind side would leave it vulnerable to attack and cause it to miss many meals. The sea anemone, as is the case for most sedentary animals, has sensory and feeding structures uniformly distributed around its body. Sea anemones do not have distinct head and tail ends. Instead, one point of reference is the end of the animal that possesses the mouth (the oral end), and a second point of reference is the end opposite the mouth (the aboral end). Animals such as the sea anemone are radially symmetrical. **Radial symmetry** is the arrangement of body parts such that any plane passing through the central oral-aboral axis divides the animal into mirror images (figure 7.8).

FIGURE 7.7

Asymmetry. Sponges display a cell-aggregate organization, and as this red encrusting sponge (*Monochora barbadensis*) shows, many are asymmetrical.

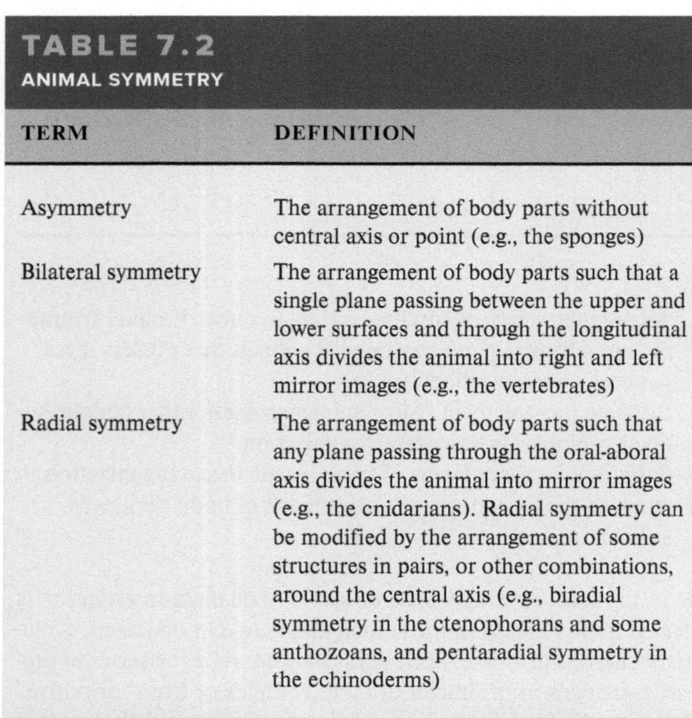

TABLE 7.2
ANIMAL SYMMETRY

TERM	DEFINITION
Asymmetry	The arrangement of body parts without central axis or point (e.g., the sponges)
Bilateral symmetry	The arrangement of body parts such that a single plane passing between the upper and lower surfaces and through the longitudinal axis divides the animal into right and left mirror images (e.g., the vertebrates)
Radial symmetry	The arrangement of body parts such that any plane passing through the oral-aboral axis divides the animal into mirror images (e.g., the cnidarians). Radial symmetry can be modified by the arrangement of some structures in pairs, or other combinations, around the central axis (e.g., biradial symmetry in the ctenophorans and some anthozoans, and pentaradial symmetry in the echinoderms)

FIGURE 7.8

Radial Symmetry. Planes that pass through the oral-aboral axis divide radially symmetrical animals, such as this tube coral polyp (*Tubastraea* sp.), into equal halves. Certain arrangements of internal structures modify the radial symmetry of sea anemones.

HOW DO WE KNOW—PHYLOGENIES FROM BASE SEQUENCES *Continued*

Character sites 2, 4, and 6 in the alignment are in the ancestral state for all species. The study group (species 1–4) differs from the outgroup by a change at character site 7 from C to T. This is the character change that separates the study group from the group's most recent ancestor (the outgroup). Species 1, 2, and 4 share a synapomorphy at site 1 (A to G), and species 3 is defined by a change at character site 8 from A to T (it is autapomorphic). Further base changes in the sequence at character sites 5, 9, and 10 define the relationships of species 1, 2, and 4. Box figure 7.1*b* shows a cladogram derived from these base sequence data.

In real data applications, the sequences are much more complex. It is usually possible to derive more than one tree diagram (branching pattern or topology) from one set of data. Most hypotheses assume that the most accurate tree diagram is the one that requires the fewest number of base changes to derive it. (It is said to be parsimonious.) Computer programs can estimate tree branch lengths based on the number of base changes that occur along a particular branch to represent genetic distances. (More base changes correspond to greater genetic distance.) Other computational tools allow the phylogeneticist to calculate confidence parameters. (How good is the data?) Still other tools can be used to calculate time since divergence from a common ancestor using molecular clock models (*see page 61*).

however, that birds are more closely tied by common ancestry to the alligators and crocodiles than to any other group of living vertebrates. According to the cladistic interpretation, birds and crocodiles should be assigned to a group that reflects this close common ancestry. Birds would become a subgroup within a larger group that included both birds and reptiles. Crocodiles would be depicted as more closely related to the birds than they would be to snakes and lizards. Traditional evolutionary systematists maintain that the traditional interpretation is still correct because it takes into account the greater importance of the "adaptive zone" of birds (e.g., feathers and endothermy) that makes the group unique. Cladists support their position by pointing out that the designation of "adaptive zone" involves value judgments that cannot be tested.

A widely used biological species concept was described in chapter 5. The widespread acceptance of cladistics methods is partly responsible for increased popularity of a newer definition of a species. If we apply the **phylogenetic species concept,** we define a species on the basis of a common phylogenetic history. As described by Joel Cracraft (1983), a species is "the smallest diagnosable monophyletic group of populations within which there is a parental pattern of ancestry and descent." In other words, a species group shares a single ancestor and can be distinguished from other groups by one or more synapomorphies. The phylogenetic species concept, in practice, often makes use of molecular tools to identify monophyletic groups. This concept has the advantage of being applicable with sexual and asexual taxa as well as taxa that are known from the fossil record. Critics of the phylogenetic species concept, including many evolutionary systematists, point out that molecular tools are so powerful that they reveal many differences that are evolutionarily neutral. How does one distinguish between branching patterns that reflect variation within species as compared to branching patterns between species? Using the phylogenetic species concept causes the number of species within a group to proliferate and makes the naming of species with unique binomial names cumbersome.

As debates between cladists and evolutionary systematists continue, our knowledge of evolutionary relationships among animals will become more complete. Debates like these are the fuel that forces scientists to examine and reexamine old hypotheses. Animal systematics is certain to be a lively and exciting field in future years.

Chapters 9 through 22 are a survey of the animal kingdom. The organization of these chapters reflects the traditional taxonomy that makes most zoologists comfortable. Cladograms are usually included in "Further Phylogenetic Considerations" at the end of most chapters, and any different interpretations of animal phylogeny implicit in these cladograms are discussed.

SECTION 7.1 THINKING BEYOND THE FACTS

Why are derived characteristics more useful in establishing evolutionary relationships than are shared characteristics? Use two subgroups of mammals to illustrate your answer (**see table 22.1*****). For example, compare horses (order Perissodactyla) and camels (order Artiodactyla).***

7.2 PATTERNS OF ORGANIZATION

LEARNING OUTCOMES

1. Analyze the selective advantages that asymmetry, radial symmetry, and bilateral symmetry provide animals that possess these patterns of organization.
2. Evaluate the statement "Most animals possess either diploblastic or triploblastic tissue-level organization."
3. Differentiate three forms of triploblastic tissue organization.
4. Analyze the reasons for the prevalence of body cavities in animals.

One of the most strikingly ordered series of changes in evolution is reflected in body plans in the animal kingdom. On one hand, evolutionary changes in animal body plans may not reflect a series of progressive changes from simpler to more complex or from "primitive" to "advanced." A relatively "simple" organization can be the product of millions of years of evolution from more "complex" ancestors. This is the case for many parasites that have adapted to their hosts through the loss of certain organ systems that became irrelevant or detrimental to their survival as parasites (e.g., the loss of digestive systems). On the other hand, one sees in animal organization evolutionary patterns that characterize groups of animals. Examining these patterns of organization helps us understand why related groups of animals have similar body forms and how animal body forms may reflect adaptations to animals' environments.

Symmetry

The bodies of animals and protists are organized into almost infinitely diverse forms. Within this diversity, however, are certain patterns of organization. The concept of symmetry is fundamental to understanding animal organization. **Symmetry** describes how the parts of an animal are arranged around a point or an axis (table 7.2).

Asymmetry, which is the absence of a central point or axis around which body parts are equally distributed, characterizes many sponges (figure 7.7). Asymmetrical organisms do not develop complex communication, sensory, or locomotor functions. The absence of symmetry in sponges allows them to grow tightly attached to their substrate and mold themselves onto virtually any available substrate of any shape (*see figure 7.7*). Asymmetrical body forms have allowed sponges to flourish for hundreds of millions of years.

A sea anemone can move along a substrate, but only very slowly. How does it gather food? How does it detect and protect itself from predators? For this animal, a blind side would leave it vulnerable to attack and cause it to miss many meals. The sea anemone, as is the case for most sedentary animals, has sensory and feeding structures uniformly distributed around its body. Sea anemones do not have distinct head and tail ends. Instead, one point of reference is the end of the animal that possesses the mouth (the oral end), and a second point of reference is the end opposite the mouth (the aboral end). Animals such as the sea anemone are radially symmetrical. **Radial symmetry** is the arrangement of body parts such that any plane passing through the central oral-aboral axis divides the animal into mirror images (figure 7.8).

TABLE 7.2
ANIMAL SYMMETRY

TERM	DEFINITION
Asymmetry	The arrangement of body parts without central axis or point (e.g., the sponges)
Bilateral symmetry	The arrangement of body parts such that a single plane passing between the upper and lower surfaces and through the longitudinal axis divides the animal into right and left mirror images (e.g., the vertebrates)
Radial symmetry	The arrangement of body parts such that any plane passing through the oral-aboral axis divides the animal into mirror images (e.g., the cnidarians). Radial symmetry can be modified by the arrangement of some structures in pairs, or other combinations, around the central axis (e.g., biradial symmetry in the ctenophorans and some anthozoans, and pentaradial symmetry in the echinoderms)

FIGURE 7.7

Asymmetry. Sponges display a cell-aggregate organization, and as this red encrusting sponge (*Monochora barbadensis*) shows, many are asymmetrical.

FIGURE 7.8

Radial Symmetry. Planes that pass through the oral-aboral axis divide radially symmetrical animals, such as this tube coral polyp (*Tubastraea* sp.), into equal halves. Certain arrangements of internal structures modify the radial symmetry of sea anemones.

Radial symmetry is often superficial. It may be modified internally or both internally and externally. Radial symmetry is frequently accompanied by the arrangement of some structures in pairs, or in other combinations, around the central oral-aboral axis. The paired arrangement of some structures in radially symmetrical animals is called biradial symmetry. The arrangement of structures in fives around a radial animal is called pentaradial symmetry.

Although the sensory, feeding, and locomotor structures in radially symmetrical animals could never be called "simple," they are not comparable to the complex sensory, locomotor, and feeding structures in many other animals. The evolution of such complex structures in radially symmetrical animals would require repeated distribution of specialized structures around the animal.

Bilateral symmetry is the arrangement of body parts such that a single plane, passing between the upper and lower surfaces and through the longitudinal axis of an animal, divides the animal into right and left mirror images (figure 7.9). Bilateral symmetry is characteristic of active, crawling, or swimming animals. Because bilateral animals move primarily in one direction, one end of the animal is continually encountering the environment. The end that meets the environment is usually where complex sensory, nervous, and feeding structures evolve and develop. These developments result in the formation of a distinct head and are called **cephalization** (Gr. *kephale,* head). Cephalization occurs at an animal's anterior end. Posterior is opposite anterior; it is the animal's tail end. Other important terms of direction and terms describing body planes and sections apply to bilateral animals. These terms are for locating body parts relative to a point of reference or an imaginary plane passing through the body (tables 7.2 and 7.3; figure 7.9).

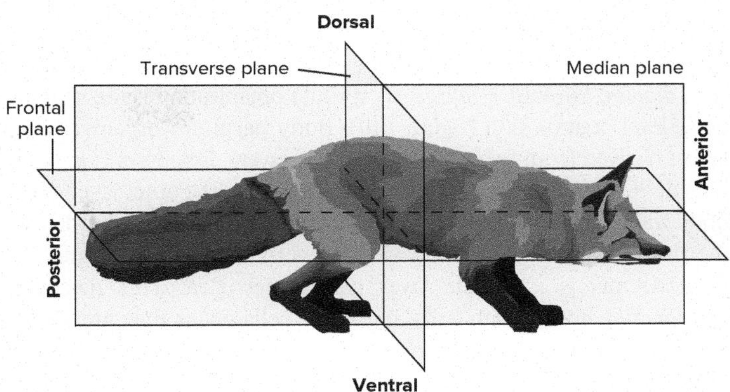

FIGURE 7.9

Bilateral Symmetry. Planes and terms of direction useful in locating parts of a bilateral animal. A bilaterally symmetrical animal, such as this fox, has only one plane of symmetry. An imaginary median plane is the only plane through which the animal could be cut to yield mirror-image halves.

TABLE 7.3
TERMS OF DIRECTION

TERM	DESCRIPTION
Aboral	The end opposite the mouth of a radially symmetrical animal
Oral	The end containing the mouth of a radially symmetrical animal
Anterior	The head end; usually the end of a bilateral animal that meets its environment
Posterior	The tail end
Caudal	Toward the tail
Cephalic	Toward the head
Distal	Away from the point of attachment of a structure on the body (e.g., the toes are distal to the knee)
Proximal	Toward the point of attachment of a structure on the body (e.g., the hip is proximal to the knee)
Dorsal	The back of an animal; usually the upper surface; synonymous with *posterior* for animals that walk upright
Ventral	The belly of an animal; usually the lower surface; synonymous with *anterior* for animals that walk upright
Inferior	Below a point of reference (e.g., the mouth is inferior to the nose in humans)
Superior	Above a point of reference (e.g., the neck is superior to the chest)
Lateral	Away from the plane that divides a bilateral animal into mirror images
Medial (median)	On or near the plane that divides a bilateral animal into mirror images

Other Patterns of Organization

In addition to body symmetry, other patterns of animal organization are recognizable. The patterns described in this section concern the organization of animal bodies based on tissue layers and body cavities.

The Unicellular (Cytoplasmic) Level of Organization

Organisms whose bodies consist of single cells or cellular aggregates display the unicellular level of organization. Unicellular body plans are characteristic of the protists. Some zoologists prefer to use the designation "cytoplasmic" to emphasize that all living functions are carried out within the confines of a single plasma membrane. Unicellular organization is not "simple." All unicellular organisms must provide for the functions of locomotion, food acquisition, digestion, water and ion regulation, sensory perception, and reproduction in a single cell.

Cellular aggregates (colonies) consist of loose associations of cells that exhibit little interdependence, cooperation, or coordination of function–therefore, cellular aggregates cannot be considered tissues (*see chapter 2*). In spite of the absence of interdependence, these organisms show some division of labor. Some cells may be specialized for reproductive, nutritive, or structural functions.

Diploblastic Organization

Cells are organized into tissues in most animal phyla. **Diploblastic** (Gr. *diplóos,* twofold + *blaste,* to sprout) organization is the simplest tissue-level organization (figure 7.10). Body parts are organized into layers derived from two embryonic tissue layers. **Ectoderm** (Gr. *ektos,* outside + *derm,* skin) gives rise to the epidermis, the outer layer of the body wall. **Endoderm** (Gr. *endo,* within) gives rise to the gastrodermis, the tissue that lines the gut cavity. Between the epidermis and the gastrodermis is a middle layer called mesoglea. This mesoglea may or may not contain cells, but when cells occur they are always derived from ectoderm or endoderm. When cells are present, this middle layer is sometimes referred to as mesenchyme. The term "mesoglea" is then reserved for the acellular condition. This text uses the term "mesoglea" for the middle layer and specifies the acellular or cellular condition.

The cells in each tissue layer are functionally interdependent. The gastrodermis consists of nutritive (digestive) and muscular cells, and the epidermis contains epithelial and muscular cells. The feeding movements of *Hydra* or the swimming movements of a jellyfish are only possible when groups of cells cooperate, showing tissue-level organization.

Triploblastic Organization

Animals described in chapters 10 to 22 are **triploblastic** (Gr. *treis,* three + *blaste,* to sprout); that is, their tissues are derived from three embryological layers. As with diploblastic animals, ectoderm forms the outer layer of the body wall, and endoderm lines the gut. A third embryological layer is sandwiched between the ectoderm and endoderm. This layer is **mesoderm** (Gr. *meso,* in the middle), which gives rise to supportive, contractile, and blood cells. Most triploblastic animals have an organ-system level of organization. Tissues are organized to form excretory, nervous, digestive, reproductive, circulatory, and other systems. Triploblastic animals are usually bilaterally symmetrical (or have evolved from bilateral ancestors) and are relatively active.

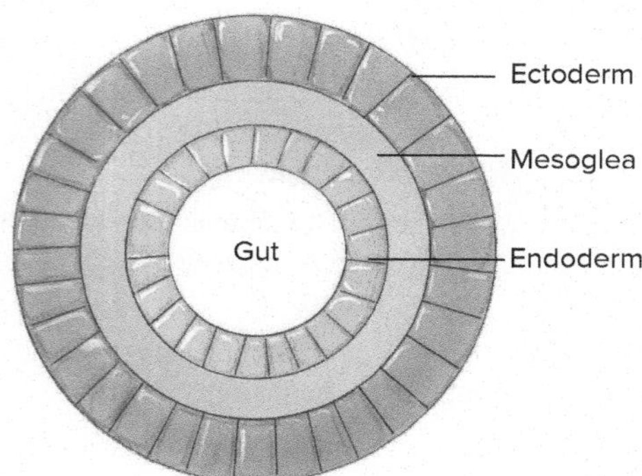

FIGURE 7.10

Diploblastic Body Plan. Diploblastic animals have tissues derived from ectoderm and endoderm. Between these two layers is a noncellular mesoglea. This image is a diagrammatic representation of a cross section through a hypothetical animal.

Triploblastic animals are organized into several subgroups based on the presence or absence of a body cavity and, for those that possess one, the kind of body cavity present. A body cavity is a fluid-filled space in which the internal organs can be suspended and separated from the body wall. Body cavities are advantageous because they

1. Provide more room for organ development.
2. Provide more surface area for diffusion of gases, nutrients, and wastes into and out of organs.
3. Provide an area for storage.
4. Often act as hydrostatic skeletons.
5. Provide a vehicle for eliminating wastes and reproductive products from the body.
6. Facilitate increased body size.

Of these, the hydrostatic skeleton deserves further comment. Body-cavity fluids give support, while allowing the body to remain flexible (*see figure 23.10*). Hydrostatic skeletons can be illustrated with a water-filled balloon, which is rigid yet flexible. Because the water in the balloon is incompressible, squeezing one end causes the balloon to lengthen. Compressing both ends causes the middle of the balloon to become fatter. In a similar fashion, body-wall muscles, acting on coelomic fluid, are responsible for movement and shape changes in many animals.

The Triploblastic Acoelomate Pattern Triploblastic animals whose mesodermally derived tissues form a relatively solid mass of cells between ectodermally and endodermally derived tissues are called **acoelomate** (Gr. *a,* without + *kilos,* hollow) (figure 7.11*a*). Some cells between the ectoderm and endoderm of acoelomate animals are loosely packed cells called parenchyma. Parenchymal cells are not specialized for a particular function, but they do form an internal supportive matrix against which muscles can act to change body shape and promote efficient locomotion. Triploblastic acoelomate animals are usually very small or very flat because the parenchyma limits the exchange of nutrients between the digestive tract and the outer cellular layers of the body and diffusion of gases and wastes between body cells and the environment. The evolution of body cavities helped to alleviate these exchange problems and allowed animals to become much larger.

The Triploblastic Pseudocoelomate Pattern A **pseudocoelom** (Gr. *pseudes,* false) is a body cavity not entirely lined by mesoderm (figure 7.11*b*). Muscle is associated with the inner body wall, but no muscular or connective tissues are associated with the gut tract, no mesodermal sheet covers the inner surface of the body wall, and no membranes suspend organs in the body cavity. Organs are free within the body cavity. Pseudocoeloms develop from a cavity, called the blastocoele, that develops in the early embryo. This cavity is retained as the pseudocoelom as the endoderm forms from invaginating ectodermal cells. Mesodermal cells form between ectoderm and endoderm and associate with the inner surface of the ectodermal layer. The pseudocoelomate pattern evolved convergently in multiple animal lineages (*see chapters 8, 10, and 13*).

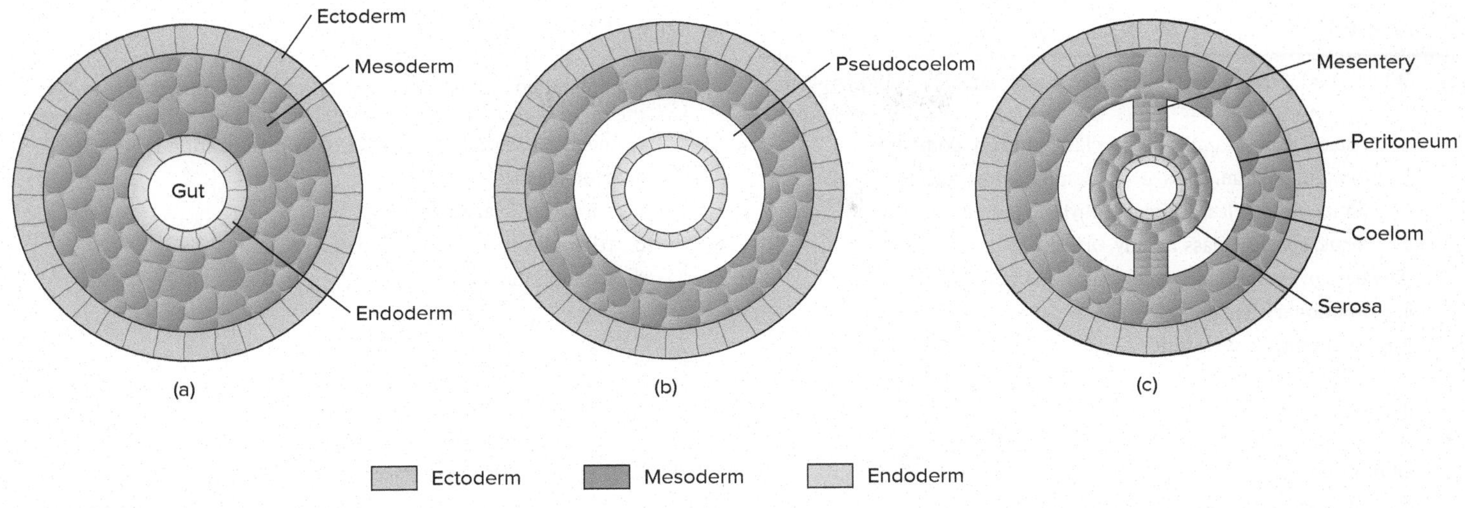

FIGURE 7.11

Triploblastic Body Plans. Triploblastic animals have tissues derived from ectoderm, mesoderm, and endoderm. (*a*) Triploblastic acoelomate pattern. (*b*) Triploblastic pseudocoelomate pattern. Note the absence of mesodermal lining on the gut tract. (*c*) Triploblastic coelomate pattern. Mesodermally derived tissues completely surround the coelom. These images are diagrammatic representations of cross sections through hypothetical animals.

The Triploblastic Coelomate Pattern A **coelom** is a body cavity completely surrounded by mesoderm (figure 7.11*c*). A thin mesodermal sheet, the peritoneum, lines the inner body wall and is continuous with the serosa, which lines the outside of visceral organs. The peritoneum and the serosa are continuous and suspend visceral structures in the body cavity. These suspending sheets are called mesenteries. Having mesodermally derived tissues, such as muscle and connective tissue, associated with internal organs enhances the function of virtually all internal body systems.

The coelomate pattern characterizes two major lineages in animal phylogeny. The coelom forms differently in each lineage. In one lineage (Protostomia), the coelom forms when mesodermal masses split, forming a cavity within the mesoderm (*see figure 8.11*a–e). In the second lineage (Deuterostomia), the coelom forms when mesodermal outpockets of the embryonic gut enlarge and pinch off from the gut tract. This coelom is a remnant of the embryonic gut cavity (*see figure 8.11*f–g).

Section 7.2 Thinking Beyond the Facts

Triploblastic animals live in all of Earth's environments. Diploblastic animals are virtually all freshwater or marine. Why are there no diploblastic animals in terrestrial environments?

Summary

7.1 **Taxonomy and Phylogeny**

- Systematics is the study of the evolutionary history and classification of organisms. Traditional classification systems reflect a taxonomic hierarchy in which organisms are grouped into ever broadening categories based on shared characteristics and evolutionary relationships.
- Nomenclature is the assignment of a distinctive name to each species.
- Systematists use shared genetic characteristics (characters) that include morphological, physiological, biogeographical, and molecular traits to evaluate evolutionary relationships.
- The two modern approaches to systematics are phylogenetic systematics (cladistics) and evolutionary systematics. The ultimate goal of systematics is to establish evolutionary relationships in monophyletic groups. Phylogenetic systematists (cladists) look for shared, derived characteristics that can be used to investigate evolutionary relationships. Cladists do not attempt to weigh the importance of different characteristics. Evolutionary systematists use homologies and rank the importance of different characteristics in establishing evolutionary relationships. These taxonomists take into consideration differing rates of evolution in taxonomic groups. Wide acceptance of cladistic methods has resulted in some nontraditional taxonomic groupings of animals.

7.2 **Patterns of Organization**

- The bodies of animals are organized into almost infinitely diverse forms. Within this diversity, however, are certain patterns of organization. Symmetry describes how the parts of an animal are arranged around a point or an axis. Animals may be asymmetrical, radially symmetrical, or bilaterally symmetrical.
- Other patterns of organization reflect how cells associate into tissues, and how tissues organize into organs and organ systems. Animals may be diploblastic, triploblastic acoelomate, triploblastic pseudocoelomate, or triploblastic coelomate.

Concept Review Questions

1. Which one of the following represents a hierarchical ordering from broader to more specific?
   a. Species, genus, family, order, class, phylum, domain
   b. Domain, phylum, class, order, family, genus, species
   c. Family, order, class, domain, phylum, species, genus
   d. Genus, species, class, family, order, phylum, domain
2. Slowness of evolutionary change in a characteristic is called
   a. evolutionary constancy.
   b. evolutionary conservation.
   c. monophyly.
   d. paraphyly.
3. A grouping of animals that includes a single common ancestor and all of its descendants is a _______________ group.
   a. conserved
   b. paraphyletic
   c. monophyletic
   d. polyphyletic
4. Attributes of groups that have been retained from a common ancestor are referred to as
   a. symplesiomorphies.
   b. derived characters.
   c. synapomorphies.
   d. nodal characters.
5. An animal possesses a body cavity, a layer of muscle that underlies the outer body wall, and a gut track without associated muscle or connective tissue. This animal's body organization is
   a. diploblastic.
   b. triploblastic acoelomate.
   c. triploblastic pseudocoelomate.
   d. triploblastic coelomate.

Analysis and Application Questions

1. In one sense, the animal taxonomy above the species level is artificial. In another sense, however, it is real. Explain this paradox.
2. Give proper scientific names to six hypothetical animal species. Assume that you have three different genera represented in your group of six. Be sure your format for writing scientific names is correct.
3. Describe hypothetical synapomorphies that would result in an assemblage of one order and two families (in addition to the three genera and six species from question 2).
4. Construct a cladogram, similar to that shown in figure 7.6, using your hypothetical animals from questions 2 and 3. Make drawings of your animals.

Earth was a violent, molten planet during the Hadean eon (4.5–4.0 billion years ago [bya]). The atmosphere was devoid of molecular oxygen and composed of carbon dioxide, hydrogen, and water vapor. Cooling through the Hadean eon, and into the Archean eon (4.0–2.5 bya), set the stage for the origin of life, the accumulation of molecular oxygen, and eventually the origin of animals (800 million years ago [mya]).

8

Animal Origins and Phylogenetic Highlights

Chapter Outline

8.1 Earth's Beginning and Evidence of Early Life
8.2 Life's Beginning and the First 3 Billion Years
Life's Origins and Early Evolution
Three Domains
Endosymbiosis and the Origin of Eukarya
8.3 Multicellularity and Animal Origins
Protist/Animal Crossroads
Animal Radiation and the Cambrian Explosion
8.4 Phylogenetic Highlights of Animalia

To understand animals requires that we understand when and where animals originated. That understanding requires that we look back in time about 800 million years. But animals were not the first organisms, and to understand the relationships of animals to other organisms we must look back to the very beginning of life on Earth. This chapter presents an overview of current ideas regarding how life arose and how early life-forms diversified. The chapter then describes what we know about the beginning of multicellularity and the animal kingdom. It ends with a discussion of phylogenetic highlights of the animal kingdom. Chapter 8 sets the stage for chapters 9 to 22, which provide more detailed and fascinating accounts of animal structure, function, and evolution.

8.1 EARTH'S BEGINNING AND EVIDENCE OF EARLY LIFE

LEARNING OUTCOMES

1. Describe the conditions on Earth before life appeared.
2. Evaluate the evidence that give clues to the presence of the earliest life on Earth.

Imagine our planet Earth 4.6 billion years ago (bya). Earth had just formed—hot, volcanic, and partially molten. It was a violent place. Solar system bodies frequently collided with this new Earth. One huge collision took place about 4.5 bya. A body the size of Mars hit the earth and sent debris into orbit around Earth. This debris eventually coalesced to form our moon. Earth gradually cooled as carbon dioxide (CO_2) from the atmosphere was absorbed into the molten Earth and sequestered into calcium carbonate. Cooling resulted in the formation of the oldest zircon crystals present in the rock of modern western Australia. Water vapor began to condense into liquid water, and rigid slabs of rock started to form a crust and began a history of tectonic shifting. Crustal plate shifting, a few centimeters each year, has influenced the geology of the earth and life on the earth ever since. This period of history, called the Hadean eon, lasted until about 4 bya (*see appendix B*).

The 4-billion-year threshold marks the beginning of the Archaean eon (4–2.5 bya). Over the next 1.5 billion years Earth's temperature cooled to near modern levels, and liquid water became prevalent. Deep ocean basins formed, and archaean rock formations were laid down in Greenland, Canada, northwestern United States, Australia, and southern Africa.

The Archaean eon began with an atmosphere that consisted primarily of CO_2, hydrogen, and water vapor, but the atmosphere slowly changed. Rock formations that are 3 billion years old show banded (oxidized) iron. The formation of banded iron requires atmospheric molecular oxygen (O_2), and the only source of molecular oxygen is photosynthesis–this means life! Microfossils of bacteria-like single cells have been found in rocks 3.45 billion years old. Stromatolites, sedimentary deposits held together by microbial mats, date into the Archaean eon. Evidence of organic carbon is found in rocks 2.7 billion years old. (Living systems leave carbon signatures through higher C^{12}/C^{14} isotope ratios than nonliving systems.) All of this evidence means that life arose about 3.8 bya. Since that time, the evolution of life-forms has shaped the history of the earth.

Section 8.1 Thinking beyond the Facts

Why are oxidized iron deposits in 3 billion-year-old rock formations a signature left by early life-forms?

8.2 Life's Beginning and the First 3 Billion Years

Learning Outcomes

1. Compose a scenario for the spontaneous origin of life given the atmospheric and oceanic conditions present on the archaean earth.
2. Describe the relationships between the three domains of living organisms, and assess the roles of horizontal gene transfer models and endosymbiosis in establishing these relationships.
3. Assess the roles of endosymbiosis, cellular compartmentalization, energy processing, and molecular oxygen in the evolution of the first eukaryotic cells.

Although we do not know how life originated, we do know that it originated on the earth in a reducing environment. That is, the atmosphere or the watery environment, was devoid of molecular oxygen (O_2) and contained carbon dioxide (CO_2), water or water vapor (H_2O), hydrogen gas (H_2), and probably compounds like ammonia (NH_3), formaldehyde (CH_2O), methane (CH_3), hydrogen cyanide (HCN), and hydrogen sulfide (H_2S). The first step in explaining the origin of life is to understand how organic molecules originated within this atmosphere and in Earth's oceans. To complete our understanding of the origin of life, scientists need to explain how the first organisms acquired the ability to replicate their genome, undergo translation to make proteins, support metabolic and transport systems, and accomplish a variety of regulatory tasks. Scientists are beginning to answer some of the questions wrapped up in these formidable goals.

Life's Origins and Early Evolution

Stanley Miller and Harold Urey were the first scientists to demonstrate experimentally that organic molecules could form spontaneously in a reducing atmosphere and in the presence of an energy source, such as lightning (figure 8.1). Similar experiments over the following decade, with slightly different starting assumptions, have confirmed the Miller/Urey conclusions that organic molecules could originate spontaneously from inorganic molecules present on the archaean earth. In 2009, John Sutherland and colleagues demonstrated the formation of nucleic acids from HCN, H_2S, and ultraviolet light using metal catalysts that would have been present in archaean substrates. These observations lend support to life originating in an "RNA world." RNA is capable of carrying genetic information and functioning as enzymes (ribozymes) that could have synthesized proteins from amino acid building blocks. If first life-forms were RNA based, the "chicken-or-egg" paradox of which came first–DNA or protein (enzymes)–disappears. (DNA codes for protein, but protein is required in the form of enzymes for DNA to function.) The transition from RNA-based life-forms to DNA-based life-forms would have occurred because the double-stranded DNA molecule (*see chapter 3*) is much more stable than the single-stranded RNA. Other geochemists suggest that organic molecules may have (also) had extraterrestrial origins. During the first billion years of Earth's existence, it was bombarded by thousands of meteors and comets. The analysis of meteorites and comets reveals a significant organic compound composition, and these compounds are similar to those found on Earth.

Living cells require membranes and must process energy to live. Membranes confine organic compounds within spaces where they can interact, and membranes serve as surfaces upon which reactions of life can occur. Membranes of early life-forms may have been comprised of fatty acids rather than phospholipids (*see chapter 2*). Membranes may have also set up proton gradients like those we described in chapter 2 when describing mitochondrial chemiosmosis. Proton gradients may have powered the origin of life, and alkaline hydrothermal vents that opened into acidic oceans may have been an ideal location for their establishment. (Hydrothermal vents are found where geothermally heated water spews from cracks in the earth's crust, for example, in deep ocean basins where tectonic plates are moving apart and in volcanically active regions. Oceans were probably acidic due to the absorption of CO_2 into the water forming carbonic acid [H_2CO_3].) At some point, membranes enclosed metabolic bubbles, and the simplest form of bacterial life began.

Three Domains

In chapter 7 we described how animal systematists use phenotypic characters of animals to group the animals by the characters that they share. These studies reveal evolutionary relationships that are often represented by cladograms and phylogenetic trees. When we want to understand the deepest roots of phylogeny, we must look at bacteria and other single-celled organisms. These organisms are too small to compare many phenotypic features of organization, thus one must apply the molecular approaches that were also described in chapter 7.

Studies of ribosomal RNA genes have led systematists to conclude that all life shares a common ancestor and that there are three major evolutionary lineages, which we refer to as domains

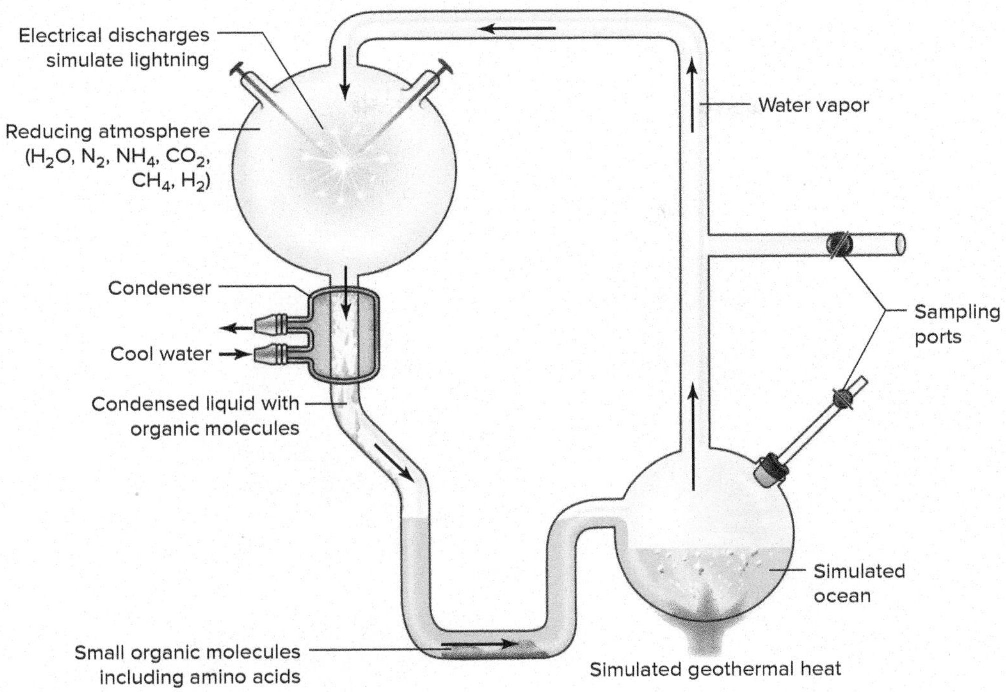

FIGURE 8.1

The Miller-Urey Experiment. The apparatus and experimental procedure used by Miller and Urey (1950s) simulated the assumed conditions in the primitive atmosphere of Earth. Geothermally heated ocean water produced water vapor. Simulated lightning strikes through a reducing atmosphere resulted in the formation of small organic molecules. These molecules were in detectable concentrations after a seven-day experimental period. Variations on this experiment were conducted through subsequent decades assuming slightly different atmospheric conditions with similar results. Some alternative experiments assume the conditions present in deep oceanic vents and also produce organic compounds including nucleic acids.

(figure 8.2). **Bacteria** is the domain containing the most abundant organisms. Recent studies have uncovered many new species and lineages, and the tree of life is a tree dominated by bacteria (*see figure 8.2*). That is not surprising since bacterial ancestors were the first life-forms. Over 90 bacterial phyla have been described, and microbial systematists are not finished. Seven of these lineages have species that are human pathogens. The second branch of the rRNA tree is shared by two domains: Archaea and Eukarya. **Archaea** are distinct from bacteria in genetic structure and function. The Archaea is more similar to the Eukarya in regard to the structure of chromatin and regulation of gene function. Also like the Eukarya, the Archaea have an actin-based cytoskeleton. The Archaea have a cell wall structure that is different from the Bacteria. These characteristics unite a diverse group of microbes. Some of the most notable for us are those that live in extreme environments. Some of these "extremophiles" are able to live in high-temperature environments (up to 80°C). Others live at very cold temperatures within glacial ice. Still others live in ocean depths at pressures 800 times atmospheric pressure. **Eukarya** is the third domain. It contains organisms (eukaryons) with compartmentalized cells (*see chapter 2*). Compartmentalization permits the evolution of specialization within cells. In the Eukarya, the nuclear envelope separates transcription and translation events. Mitochondrial and chloroplast membranes compartmentalize energy processing. Notice in figure 8.2 that Eukarya forms a very small branch within the tree of life and branches from a stem common with the Archaea. Within the eukaryotic branch are five lineages that include single-celled or colonial eukaryons called the **protists** (*see appendix C*), and three kingdoms of multicellular organisms, which are derived from two of these five lineages. These kingdoms are the **Plantae** (green plants–multicellular and photosynthetic), the **Fungi** (yeast, molds, and mushrooms–multicellular or unicellular, saprophytic, and spore producing), and the **Animalia** (animals–multicellular, motile, and heterotrophic). True multicellularity, tissues, organs, and organ systems evolved only in the Eukarya.

Animation Three Domains

Taxonomies are traditionally built assuming that genes are passed between generations in a species lineage through sexual or asexual reproduction, a process called **vertical gene transfer**. Recent rRNA studies have found evidence that genes have moved between species, a process called **horizontal (lateral) gene transfer (HGT or LGT)**. HGT results in species that are in different lineages sharing genes (figure 8.3). HGT was prevalent in the early history of life, and this exchange of genes between species was a principal mechanism of evolution among bacteria. As we will see shortly, it also profoundly influenced the evolution of the first animals. HGT persists today as a means by which bacteria pass antibiotic-resistant genes between species. As a result of HGT, evolutionary biologists

FIGURE 8.2

The Tree of Life. The three domains of life are Bacteria, Archaea, and Eukarya. Bacteria comprise the overwhelming majority of organisms on Earth. Eukaryons comprise the overwhelming biomass of living organisms because many are large organisms. Eukarya is divided into five supergroups. Animals are members of the supergroup Opisthokonta.

Source: Hug, LA et al. 2016. A new view of the tree of life. *Nat Micro*, 16048, doi:10:1038/nmicrobiol.216.48.

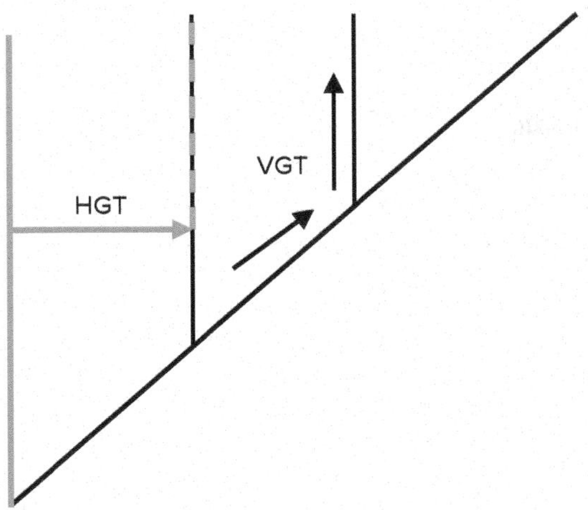

FIGURE 8.3

Horizontal Gene Transfer versus Vertical Gene Transfer. Biologists have traditionally thought of evolution as occurring as genes are passed between members of populations through sexual or asexual reproduction (*see chapter 5*). This type of gene transfer is called vertical gene transfer (VGT). Recent rRNA studies have shown that genes also pass between species in different lineages through horizontal (lateral) gene transfer (HGT). HGT was prevalent in the evolution of bacteria and archaeans, and it continues today. It also occurred in the early evolution of animals. HGT results in images of evolutionary pathways that are complex nets rather than simple branching trees.

view the base of the tree of life as a web or net rather than a set of two or three distinct lineages. The current view is that all life originated from a set of primitive cells that evolved together approximately 3.5 bya. These primitive cells had relatively few genes that were freely swapped through HGT. By 2.1 bya, the three domains of life had emerged from these earliest cells.

Endosymbiosis and the Origin of Eukarya

For about 1.5 billion years (and until about 2.1 bya), life on Earth was comprised of the Bacteria and Archaea. The evolution of cellular compartmentalization, mitochondria, and chloroplasts were hallmark events in the origin of the Eukarya. The endomembrane system of eukaryons (*see chapter 2*) compartmentalizes genetic transcription, protein synthesis and processing, waste processing, and many other cellular functions. All of these membranes arose by infolding of an archaean plasma membrane. Archaean cytoskeletal elements promoted cell movement (figure 8.4).

Virtually all eukaryotic cells contain mitochondria. The idea that mitochondria (and chloroplasts) originated through a symbiotic (*see chapter 6*) relationship between bacterial cells is quite old. In the 1960s, this idea was revived when it was discovered that these organelles had their own DNA, mRNA, tRNA, and ribosomes; and that their DNA was similar to the circular DNA found in bacteria. They are self-replicating organelles, and they are only partially dependent on nuclear DNA for their functions. The **endosymbiont theory** was proposed by Lynn Margulis (1938–2011), a biologist at the University of Massachusetts, Amherst. She proposed that eukaryotes formed when large, primitive cells engulfed smaller and simpler cells. (An endosymbiont is an organism that can live only inside another organism, forming a relationship that benefits both partners.)

Early bacteria survived anaerobically using metabolic pathways similar to glycolysis and variations on Krebs cycle pathways that did not require O_2 as a final electron acceptor (*see chapter 2*). They derived energy by metabolizing organic compounds produced spontaneously in the environment as previously described. As environments became depleted of these organic compounds, a few early bacteria developed photosynthetic pigments capable of converting CO_2 and water into simple sugars using solar energy. About 2.5 bya, O_2 became a small part of the atmosphere and Earth's oceans (0.1% modern oceanic O_2 levels) as photosynthetic bacteria (cyanobacteria) pumped O_2 into the oceans and primitive atmosphere as a by-product of photosynthesis.

Free oxygen in the oceans and atmosphere was not a good thing for many life-forms. It reacts with other molecules, producing harmful by-products (free radicals) that can disrupt normal biological functions. Some oceanic bacteria probably died from oxygen

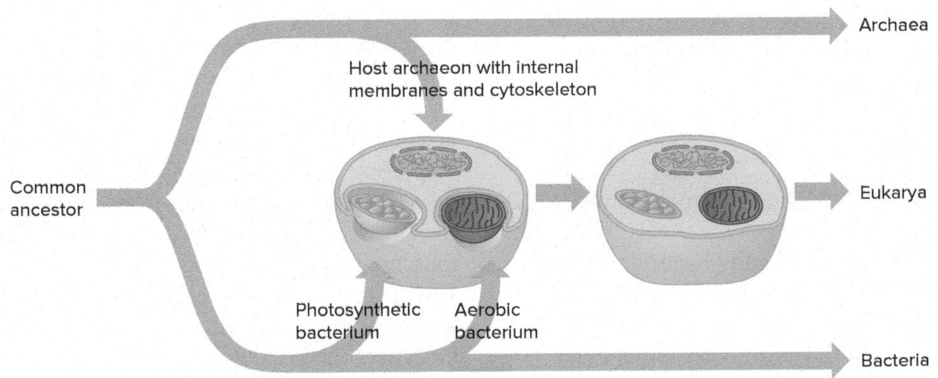

FIGURE 8.4

The Origin of Eukaryotic Cells. Eukaryotic cells arose about 2.1 billion years ago. The endomembrane system, including the nuclear envelope, originated by an infolding of the plasma membrane of a large archaeon. According to the endosymbiont theory, eukaryotic cells may have originated when aerobic bacterial cells were engulfed by archaean cells. The captured bacteria eventually became mitochondria. Similarly, captured cyanobacteria became chloroplasts of photosynthetic eukaryans. When some of the genetic material of the captured cells moved to the archaeon nucleus, the smaller cells became dependent on their host.

poisoning, but other bacteria (aerobic bacteria) that could tolerate free oxygen began to flourish. The aerobic metabolic pathways that use molecular oxygen as the final electron acceptor in the electron transport system (*see chapter 2*) evolved.

Next, endosymbiosis occurred. One way for a large cell, such as an archaeon, to survive in an oxygen-rich environment would be to engulf an aerobic bacterium in an inward-budding vesicle of its plasma membrane. (Recall that archaea possess actin-based cytoskeletal elements that would have promoted movement and engulfing functions.) This captured bacterium would then contribute biological reactions to detoxify the free oxygen and radicals. Eventually the membrane of the enveloped vesicle became the outer membrane of the mitochondrion. The cell membrane of the engulfed aerobic bacterium became the inner membrane system of the mitochondrion. The small bacterium thus found a new home in the larger cell, and the host cell could survive in an oxygenated environment and make use of Krebs cycle and electron transport system pathways for more efficient metabolism of organic molecules. In a similar fashion, archaean cells that picked up cyanobacteria obtained the forerunners of chloroplasts and became the ancestors of the green plants. Once these ancient cells acquired their endosymbiont organelles, genetic changes impaired the ability of the captured cells to live on their own outside the host cells. Over many millions of years, the larger cells and the captured cells became mutually interdependent and formed the first eukaryotic organisms (2.1 bya).

Animation
Endosymbiosis

Section 8.2 Thinking beyond the Facts

Bacterial life-forms appeared about 1 billion years after Earth formed. It then took another 2 billion years for the origin of eukaryotic organisms. What evolutionary events are thought to explain the origin of eukaryons, and why do you think it took so long for these events to occur?

8.3 MULTICELLULARITY AND ANIMAL ORIGINS

LEARNING OUTCOMES

1. Explain why multicellularity is selectively advantageous for large eukaryotic organisms.
2. Assess the evidence that links animal origins to the opisthokont clade of Eukarya.
3. Explain the contribution of Ediacaran and Cambrian period fossils to our knowledge of animal origins.

The earliest eukaryotic organisms (2.1 bya) were comprised of single cells, and many single-celled eukaryons (the protists) exist today (*see appendix C*). Relationships among the eukaryons are under active investigation, resulting in the classification of these organisms into five supergroups based on morphological, biochemical, and physiological data (*see figure 8.2*). The supergoup that concerns us is Opisthokonta, which includes animals, fungi, and their close protist relatives. It is characterized by a particular mitochondrial morphology, cells with a single posterior flagellum, and rRNA characters.

Multicellular eukaryons first originated about 1.2 bya when cells of a dividing protist remained associated following cell division. Multicellularity then originated over-and-over within at least four of the five supergroup lineages. Apparently there was more than one way to become multicellular. The fact that multicellularity arose multiple times suggests that there must be selective advantages in multicellular existence. The selective advantages probably include defense. A large organism was less vulnerable to predation by predatory protists. In addition, exchanges with the environment were more efficient in organisms made of more, smaller cells. Diffusion distances increase as cells grow larger, which limits the size of single cells (*see figure 2.2 and accompanying discussion*). Finally, multicellularity permits subdivision of labor in organisms. Cells of multicellular organisms can be specialized for specific functions like reproduction, feeding and digestion, sensory perception, and communication.

Protist/Animal Crossroads

The origins of multicellularity are difficult to investigate because they occurred so very deep in the evolutionary past. Scientists, using comparative genomics, are screening the genes of multicellular/unicellular cousin organisms to find genes that could account for the multicellular condition of one species versus the unicellular condition of the close relative. These investigations are beginning to pay off!

Even though eukaryotic multicellularity arose 1.2 bya, it did not happen that early in the opisthokont clade. Animal multicellularity originated between 800 and 635 million years ago (mya). Correlation is not proof, but low levels of O_2 (0.1% modern oceanic O_2 levels) through the Proterozoic eon (*see appendix B*) may have delayed the origin of animals. The emergence of animal multicellularity corresponds to the accumulation of relatively high levels of O_2 in the atmosphere and within the depths of the oceans (1 to 2% of modern oceanic O_2 levels) where animals first flourished. Molecular oxygen apparently needed to be in high enough concentrations to diffuse through cell layers of early animals, and O_2 is also required in the synthesis of a universal structural protein of animals–collagen. In addition to collagen, all animals share other important characters that were probably not related to O_2 accumulation: eggs and sperm, certain genes required for immune responses, genes regulating embryonic developmental pathways, monoflagellated cells, asters that function in cell division, and epithelial tissues (*see chapter 2*) with basement membranes and common intercellular junctions. These conserved features provide very strong support for the conclusion that all animals arose from a single ancestor. As you will see next, additional evidence indicates that this ancestor was a protist.

In 1874 Ernst Haeckel first suggested that a hollow colony of undifferentiated cells could arise when a single cell divided but remained linked to neighboring cells by intercellular bridges. Haeckel may have been on the right track. Flagellated protist opisthokonts, called choanoflagellates, form colonies (figure 8.5). Colonies are embedded in a jellylike extracellular matrix, and cells can

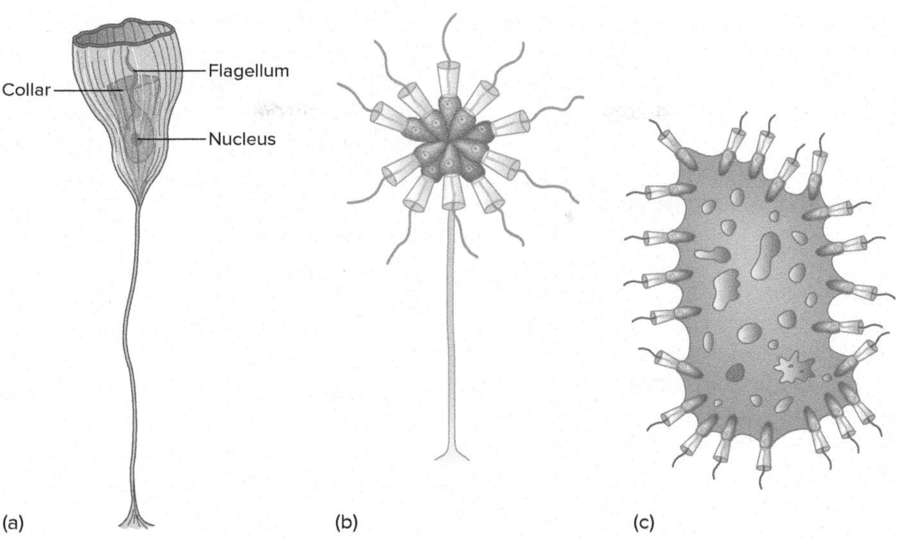

FIGURE 8.5

Choanoflagellate Diversity. (*a*) *Stephanoeca.* (*b*) *Codosiga*, a colonial species. (*c*) *Proterospongia*, another colonial species, with individuals embedded in a thick, gelatinous extracellular matrix.

remain connected by this matrix and intercellular bridges. Choanoflagellates hold other clues to animal origins. Choanoflagellates resemble the feeding cells of sponges, called choanocytes (*see figure 9.3*b). They possess a single flagellum that circulates water through a collar of actin-filled microvilli, which traps food particles–principally bacteria. The food is transported to the base of the collar and engulfed by phagocytosis. Sponges also feed on bacteria in this manner. The similarity between choanoflagellates and sponge choanocytes is confirmed by comparative genomics, which indicates that these groups share genes encoding cell adhesion proteins, extracellular matrix proteins, and a surface receptor protein (a tyrosine kinase receptor). (Collar-type cells have also been described in some cnidarians [sea anemones and their relatives, *see chapter 9*], echinoderms [sea stars and their relatives, *see chapter 16*], and hemichordates [acorn worms and pterobranchs, *see chapter 16*]. Homologies of these collar-type cells to choanoflagellates and sponge choanocytes have not been established.) The name **Apoikozoa** (Gr. *apoikia*, colony + *zoa*, animals) has been proposed for this monophyletic choanoflagellate/animal clade of opisthokonts (figure 8.6).

Interestingly, the bacterial prey of an ancestral choanoflagellate/animal crossroads organism may have had a significant influence on the advent of animals. Animals and choanoflagellates share membrane components, signaling pathways, and other genetically regulated functions that are probably bacterial in origin. (Cell signaling is the complex set of activities that cells use to govern cellular functions and coordinate cellular activities. It is involved in normal homeostasis as well as tissue repair and immune responses.) How did the choanoflagellate/animal ancestor acquire these bacterial genes? It is very likely that these genes were acquired from bacterial prey through horizontal gene transfer (HGT). Gene acquisition from ingested prey is common among heterotrophic protists. It is estimated that more than 4% of the choanoflagellate genome was acquired through HGT.

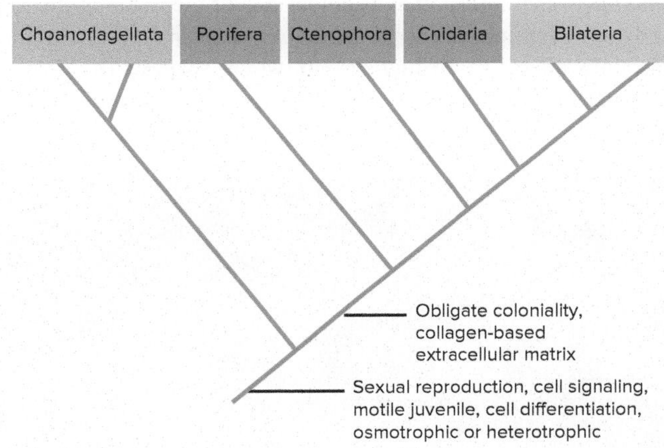

FIGURE 8.6

Clade Apoikozoa. The monophyletic clade Apoikozoa has been proposed by Budd and Jensen (2015) to describe the relationships within the choanoflagellate/animal branch of the supergroup Opisthokonta. Basal phyla other than Porifera, Cnidaria, and Ctenophora are omitted. After Budd GE, Jensen S. 2015. The origin of the animals and a 'Savannah' hypothesis for early bilaterian evolution. *Biol Rev*, 92 (2017): 446–473.

What then is our image of the last common ancestors (LCAs) of all animals? They were probably obligately colonial with a collagenous extracellular matrix. This character is what would have distinguished them from their choanocyte relatives whose colonial existence is usually not required. Some, almost certainly, fed in their microbial world through the use of collar cells. Others would have been osmotrophic (fed by absorbing dissolved nutrients from the water). They were probably large enough that they lived on the substrate of their ocean environment and would have been living on, or surrounded by, large microbial-mat colonies that covered the seafloor like plastic wrap. Because virtually all animals reproduce sexually, these LCAs were likely to reproduce using eggs and motile sperm that were produced by meiotic cell division. Motility at some level (perhaps in a larval stage) would have evolved.

Animal Radiation and the Cambrian Explosion

The last 95 million years of the Proterozoic eon (*see appendix B*) is called the Ediacaran period (635–541 mya). The Ediacaran period is critical to our understanding of the earliest stages of animal evolution. Although it is named after a site in Australia, the Ediacaran formation is worldwide in distribution. The earliest Ediacaran fossils date to about 579 mya, and these early fossils present a very poor record of early animal evolution. The scarcity of fossils from this period is almost certainly a result of unlikely fossilization of soft-bodied animals. The fact that some unlikely fossils exist suggests a fairly widespread presence of these organisms in ancient seas. Later Ediacaran fossils (about 550 mya) are less scarce but are interpreted differently by different investigators. It is sometimes difficult to determine whether these fossils are whole-animal fossils or represent animal parts. Many of these fossils are radially symmetrical and have been interpreted as cnidarians or members of the modern phylum Ctenophora (comb jellies, *see chapter 9*). Some of these fossils show an internal canal system, which suggests the presence of internal cavities, perhaps digestive cavities (figure 8.7). Upper and lower epithelial surfaces separated by an extracellular matrix or other inert material are present in some fossils, which suggests that diploblastic organization may have been present. One of the most famous later Ediacaran fossils is *Kimberella*—found in Australian and Russian deposits. *Kimberella* specimens range in size up to 15 cm in length. *Kimberella* is now accepted as an early bilaterally symmetrical animal (figure 8.8). Interestingly, there are no fossils that are convincingly spongelike. Many systematists are not convinced that any of our modern phyla can be traced back into the depths of the Ediacaran period.

The Proterozoic eon ended 541 mya and the Phanerozoic eon began (*see appendix B*). The Phanerozoic began with the Cambrian period. Early Cambrian fossils include a mix of sessile late-Ediacarans, early bilateral animals, and extensive microbial mats. Many of these animals still fed on bacteria like their LCAs did. Cambrian fossils reveal an upsurge in animals that burrowed through and under microbial mats. Some of these burrowers extracted nutrients from the sediment, some animals created vertical burrows to feed in the water column while anchored in microbial mats, and some burrowers were protected from predators and adverse environmental conditions. Like modern earthworms, these burrowers reworked oceanic substrates and diversified oceanic niches, making the ocean-floor resources more heterogeneous. These organisms were the seeds of an amazing evolutionary radiation—what has been called the Cambrian explosion.

A fossil formation, called the Burgess Shale, in Yoho National Park in the Canadian Rocky Mountains of British Columbia (figure 8.9) provides evidence of the Cambrian explosion. About 508 mya, mudslides rapidly buried thousands of marine animals

FIGURE 8.7

An Ediacaran Fossil. Fossils from the early Ediacaran period are rare. Ediacaran fossils become more abundant as one approaches the Cambrian boundary. *Dickinsonia* is a well-known late Ediacaran species (555 mya). Its taxonomic affinities are uncertain. It was probably motile, as evidenced by preserved tracks in the vicinity of many fossils. The branching diverticulae seen in this photograph may be internal canals that functioned in food distribution.

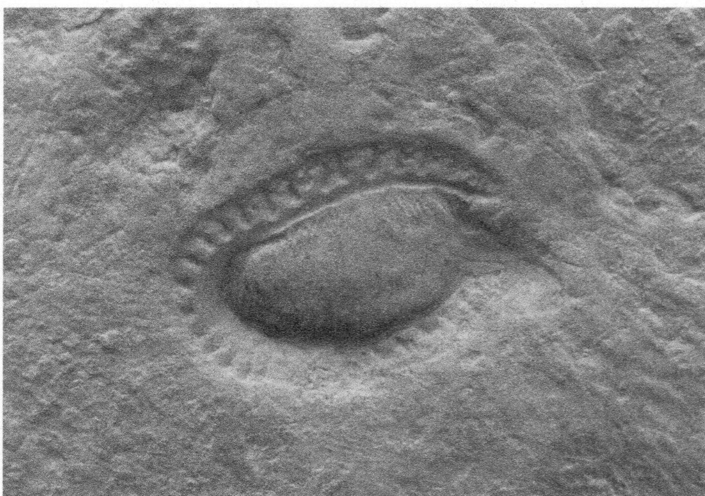

FIGURE 8.8

An Ediacaran Fossil. *Kimberella* is a well-known Ediacaran fossil (555 mya) that is widely accepted as a bilaterian. It ranges in length up to 15 cm. Fossils have been interpreted as showing the presence of muscle fibers, a nonmineralized shell or dorsal cover, and a head region. *Kimberella* is considered by most paleontologists to have been at least a stem bilaterian, and some investigators make a more controversial claim that it was a stem member of the phylum Mollusca.

FIGURE 8.9

An Artist's Reconstruction of the Burgess Shale. The Burgess Shale contained numerous unique forms of animal life as well as representatives of the animal phyla described in this textbook. *Opabinia* (unknown taxonomic affinity) is shown in the upper center. The soft-bodied *Wiwaxia* is shown at the lower left. It probably had mollusc affinities (*see chapter 11*). The creature with six or seven pairs of legs and conical spines is *Hallucigenia*, probably an early arthropod or onychophoran (*see chapters 14 and 15*). *Pikaia* (lower center) was an early chordate (*see Evolutionary Insights, chapter 17*). Trilobites (*see chapter 14*), cnidarians (*see chapter 9*), and poriferans (*see chapter 9*) are also shown.

under conditions that favored fossilization. These fossil beds provide evidence of virtually all the 37 extant animal phyla, plus about 20 other animal body forms that are so different from any modern animals that they cannot be assigned to any one of the modern phyla. These include small, soft-bodied (up to 5 cm) species in the genus *Wiwaxia*. These animals had dorsal spines that may have served as protection from predators and scraping mouthparts that may have been used to rasp food from its microbial mat home. Predators flourished during the early Cambrian period, and they probably contributed to the extinction of many Ediacaran species. These include an unassignable animal–the large (up to 1 m) swimming predator called *Anomalocaris*. *Opabinia* (up to 8 cm) was a soft-bodied predator that used a long, hollow proboscis for capturing prey or probing the substrate for burrowing animals. The Burgess Shale formation also has fossils of many extinct representatives of modern phyla. For example, a well-known Burgess Shale animal called *Sidneyia* (5–13 cm) is a representative of an extinct group of arthropods (e.g., insects, spiders, mites, and crabs). It was a predator that walked along the ocean floor feeding on other arthropods. Fossil formations like the Burgess Shale show that evolution cannot always be thought of as a slow progression. The Cambrian explosion involved rapid evolutionary diversification, followed by the extinction of many unique animals, including the extinction of some survivors from the Ediacaran period like *Kimberella*.

Why was this evolution so rapid? No one really knows. Within a few million years, oceanic oxygen levels increased to 10% of modern levels. Some geobiologists hypothesize that this was the most significant event in the evolution of life on Earth. Oxygen is used to drive energy-requiring muscles and nervous systems. In addition to being used in the synthesis of collagen, it is also used in the construction of shells and exoskeletons. Many zoologists hypothesize that animal evolution was rapid because so many ecological niches were available with virtually no competition from existing species. Others suggest that the presence of mineralized skeletons and predatory lifestyles that mark the beginning of the Cambrian period promoted rapid evolutionary change. Molecular studies may be providing clues as to how so many body forms emerged in a brief period of time. Variations in the development of body plans is controlled by a group of genes called homeobox (*Hox*) genes (*see page 68*). These genes specify the identity of body parts and the sequence in which body parts develop. Small changes in a few genes can produce dramatically different body forms. The rapid emergence of different body forms early in the Cambrian period may reflect changes that occurred in the evolution of the *Hox* gene complex.

While the exact ancestral pathways involved are uncertain, the organisms that emerged out of the oceanic Cambrian ooze formed the base of one eukaryotic lineage, the kingdom Animalia. What follows is a brief description of higher animal taxonomy and helps one to visualize relationships among the 37 extant animal phyla that emerged within this lineage.

Section 8.3 Thinking Beyond the Facts

How would you describe an ancient seafloor at the boundary between the Ediacaran and Cambrian periods?

8.4 PHYLOGENETIC HIGHLIGHTS OF ANIMALIA

LEARNING OUTCOMES

1. Explain what is implied in the term "basal phyla."
2. Assess the embryological and morphological evidence for the establishment of two major groups of bilateral animal phyla.

Taxonomic levels between kingdom and phylum are used to represent hypotheses of relatedness between animal phyla. The relationships between phyla have been very difficult to establish with certainty. Morphological and embryological evidence is being reinterpreted based on molecular evidence, principally from rRNA studies (figure 8.10 and *see appendix A*). Five **basal phyla** probably originated independently from other animal groups. These include the Ctenophora, Porifera, Placozoa, Acoelomorpha, and Cnidaria. The Cnidaria and Ctenophora are commonly grouped together (Radiata) based on the presence of radial symmetry and diploblastic organization in both groups. Recent studies suggest that the Ctenophora may, in fact, be triploblastic, and their apparent radial symmetry is secondary. This independent origin of the five phyla is reflected in figure 8.10.

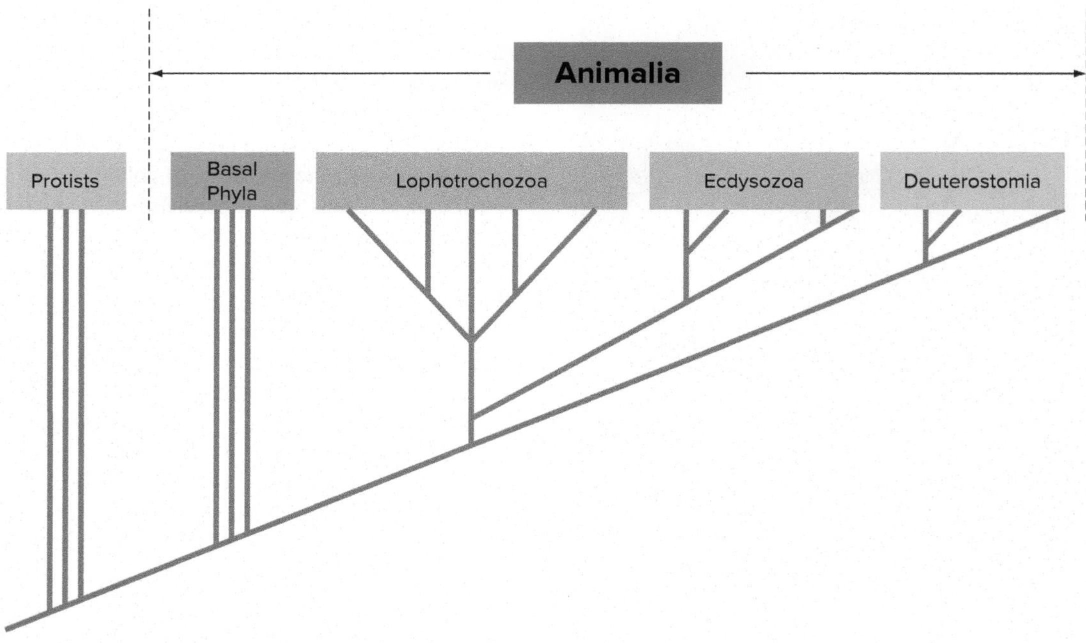

FIGURE 8.10

Animal Taxonomy. The use of molecular data has altered zoologists' interpretations of animal taxonomy. This tree depicts an independent origin of the phyla Ctenophora, Porifera, Placozoa, Acoelomorpha, and Cnidaria. Members of the phyla containing bilaterally symmetrical animals are divided into two lineages. Protostomia is further divided into two clades: Lophotrochozoa and Ecdysozoa. The Lophotrochozoa are protostomes that do not molt. They include the annelids, molluscs, and others. The Ecdysozoa are protostomes that molt. They include the nematodes, arthropods, and others. The Deuterostomia includes the echinoderms, hemichordates, and chordates.

Animals other than those just mentioned are bilaterally symmetrical (Bilateria) and share a common ancestry. Bilateral animals have been grouped into two branches based on embryological characteristics, including early cleavage patterns and the method of coelom formation. These groups are confirmed by molecular evidence.

Protostomia traditionally includes animals in the phyla Platyhelminthes, Nematoda, Mollusca, Annelida, Arthropoda, and others. Protostomia does not represent a formal taxonomic rank, but the protostomes do share important developmental characteristics and evolutionary ties (figure 8.11*a–e*). One characteristic is the pattern of early cleavages of the zygote. In spiral cleavage, the mitotic spindle is oriented obliquely to the axis of the zygote. This orientation produces an eight-celled embryo in which the upper tier of cells is twisted out of line with the lower cells. A second characteristic common in many protostomes is that early cleavage is determinate, meaning that the fate of the cells is established very early in development. If cells of the two- or four-celled embryo are separated, none develops into a complete organism. Other characteristics of protostome development include the manner in which the embryonic gut tract and the coelom form. Many protostomes have a top-shaped larva, called a **trochophore larva**. Molecular studies suggest that the protostomes include two major monophyletic lineages that are now considered superphyla–Lophotrochozoa (Spiralia) and Ecdysozoa (*see figure 8.10*). The **Lophotrochozoa** includes animals like the annelids (segmented worms) and molluscs (bivalves, snails, and their relatives). The name Lophotrochozoa is derived from the presence of a type of feeding structure (a lophophore) and trochophore larvae in some members of this group (*see chapters 10 to 12*). (The term "Spiralia" is preferred by some authors; however, the equivalency of the two terms is debated.) The **Ecdysozoa** includes animals like the arthropods (insects and their relatives, *see chapters 14 and 15*) and nematodes (roundworms, *see chapter 13*) that possess an outer covering called a cuticle that is shed or molted (a process called ecdysis) periodically during growth.

The other traditional group, **Deuterostomia**, includes animals in the phyla Echinodermata, Hemichordata, and Chordata (*see chapters 16 to 22*). Figure 8.11*f–i* shows the developmental characteristics that unite these phyla. Radial cleavage occurs when the mitotic spindle is oriented perpendicular to the axis of the zygote and results in embryonic cells directly over one another. Cleavage is indeterminate, meaning that the fate of the embryonic cells is determined late in development, and if embryonic cells are separated, they can develop into entire individuals. The manner of gut tract and coelom formation differs from that of protostomes. There is no characteristic larval stage within the group. Some small phyla formerly considered deuterostomes have been moved out of that group. These phyla include Chaetognatha, Brachiopoda, Ectoprocta, and Phoronida. This textbook considers the latter three phyla to be lophotrochozoans. The position of the Chaetognatha in higher animal taxonomy is under active investigation.

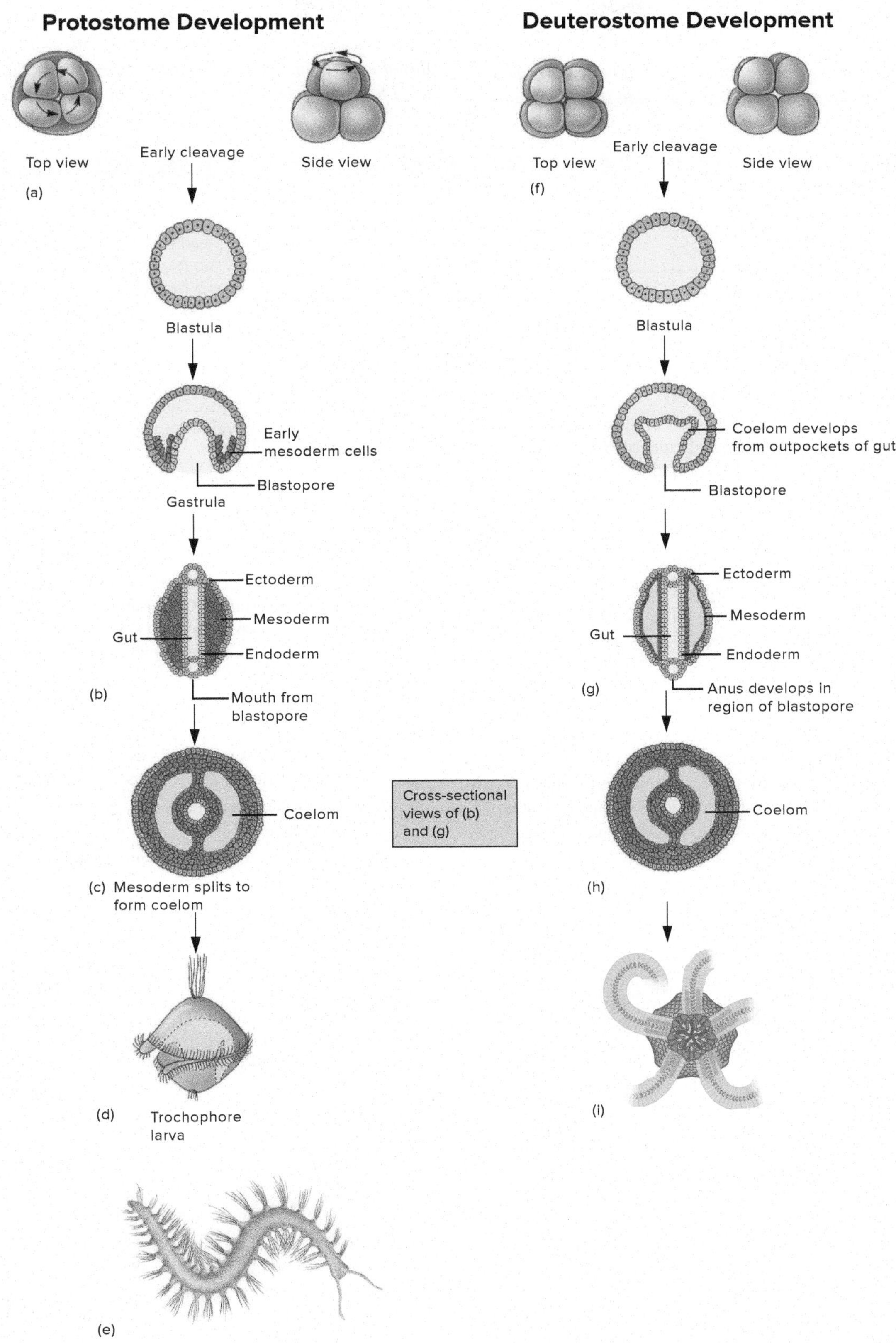

FIGURE 8.11

Developmental Characteristics. Protostome development is characterized by (*a*) spiral and determinate cleavage, (*b*) a mouth that forms from an embryonic blastopore, (*c*) schizocoelous coelom formation, and (*d*) a trochophore larva. A polychaete worm (phylum Annelida) is an example of an adult protostome (*e*). Deuterostome development is characterized by (*f*) radial and indeterminate cleavage, (*g*) an anus that forms in the region of the embryonic blastopore, and (*h*) enterocoelous coelom formation (vertebrates are an exception). Members of the phylum Echinodermata are deuterostomes (*i*).

Section 8.4 Thinking Beyond the Facts

The repetition of body parts, segmentation, was once believed to be a character that united the Arthropoda and the Annelida into a common lineage. Segmentation is also present in the Chordata. It is not present in Echinodermata, Nematoda, Rotifera, and other bilateral phyla. Based on this information and the tree diagram in appendix A, what must be true of the evolutionary origin of segmentation?

Summary

8.1 Earth's Beginning and Evidence of Early Life

- The Hadean eon (4.6–4 bya) was a time when molten Earth was cooling and Earth's crust was forming. Life originated in the Archaean eon (4–2.5 bya) about 3.8 billion years ago.

8.2 Life's Beginning and the First 3 Billion Years

- Life originated 3.8 billion years ago in a reducing environment. It was preceded by the origin of organic molecules such as amino acids and RNA. First life may have been RNA based, which was later replaced by a DNA-based genome. The evolution of life required the evolution of membranes to confine organic compounds and proton gradients to supply energy for living organisms. Hydrothermal vents may have provided environments favorable for the origin of life.
- There are three major lineages of organisms. Bacteria are the most abundant organisms. Archaea are distinct in structure and function from bacteria. Some archaea inhabit extreme environments. Eukarya is the third domain. It shares a branch of the rRNA tree with Archaea and contains organisms with compartmentalized cells. It includes the protists, fungi, plants, and animals. Horizontal gene transfer was important in the early history of life.
- Eukaryons arose from the archaean branch of the rRNA tree. An endomembrane system arose from an infolding of an archaean plasma membrane. The hypothesis that mitochondria, and later chloroplasts, originated through endosymbiosis is accepted by most biologists.

8.3 Multicellularity and Animal Origins

- Multicellularity arose multiple times within the eukaryons. Multicellularity leading to the first animals occurred in the eukaryotic supergroup, Opisthokonta. Animal multicellularity arose between 800 and 635 mya, and it is traced back to a common ancestor of choanoflagellate protists and animals that form a monophyletic clade called Apoikozoa. The early evolution of these lineages was influenced by horizontal gene transfer from bacterial prey.
- Fossils from the Ediacaran and Cambrian periods provide some evidence of the radiation of the first animals. Ediacaran fossils (560–541 mya) are often difficult to place within modern phyla. The beginning of the Cambrian period (541 mya) is represented in the fossil record with a mix of sessile and bilateral animals that fed on microbial mats that covered the ocean floor. These fossils reveal an upsurge in burrowing, which helped diversify oceanic niches. Fossils from the Burgess Shale (508 mya) reveal a Cambrian explosion that provides evidence of all modern animal phyla and numerous extinct phyla. The diversification of ecological niches, the presence of mineralized skeletons, predatory lifestyles, and an abundant supply of molecular oxygen may account for this period of very rapid animal diversification.

8.4 Phylogenetic Highlights of Animalia

- The Ctenophora, Porifera, Placozoa, Acoelomorpha, and Cnidaria probably arose independently of one another and are referred to as the basal phyla. Bilateral animals are monophyletic. They are placed into one of two large groups, Protostomia and Deuterostomia. Protostomes have common embryological characteristics (e.g., spiral, determinant cleavage) and unique rRNA characters. Protostomia is comprised of two lineages. Many lophotrochozoans have trochophore larval stage and a feeding structure called a lophophore. Ecdysozoans possess an outer cuticle that is shed or molted during growth. Deuterostomes are characterized by radial, indeterminant cleavage and unique rRNA characters.

Concept Review Questions

1. Which of the following statements is TRUE regarding the conditions on Earth 3.8 bya—the time frame for the origin of first life-forms?
   a. The atmosphere and oceanic environments were reducing in nature and contained CO_2, water or water vapor (H_2O), hydrogen gas (H_2), molecular oxygen (O_2), and probably compounds like ammonia (NH_3), formaldehyde (CH_2O), methane (CH_3), hydrogen cyanide (HCN), and hydrogen sulfide (H_2S).
   b. An extraterrestrial origin of some early organic compounds has been discounted.
   c. Deep oceanic hydrothermal vents may have provided an environment favorable for the origin of organic compounds and the first life-forms.
   d. An increase in atmospheric O_2 spurred the evolution of the first life-forms.
2. Which one of the following statements is FALSE regarding members of the three domains of life?
   a. Bacteria is the domain with the largest number of organisms. The tree of life is dominated by bacteria.
   b. Archaea are similar to eukaryons in aspects of chromatin structure and regulation of gene function. Some archaea are able to live in extreme environments.
   c. Eukarya comprises a very small branch of the tree of life. It includes the protists, Fungi, Plantae, and Animalia.
   d. Bacteria and Archaea are very closely related groups.
3. Which of the following accurately reflects modern ideas of the origin of the Eukarya?
   a. Eukaryotic cells originated from a large archaeon through an infolding of the plasma membrane, which compartmentalized the cell and gave rise to the endomembrane system.
   b. The first eukaryons were photosynthetic. Chloroplasts originated through endosymbiosis involving an archaeon and a cyanobacterium.
   c. Mitochondria could not have originated through endosymbiosis because they are dependent on nuclear DNA for all their functions and their replication.
   d. Mitochondria probably allowed early eukaryons to flourish when mitochondrial metabolic pathways established oxygen as an electron acceptor in the electron transport system.
   e. Both a and d are correct.

4. Which of the following is a monophyletic clade that includes only animals and choanoflagellates?
   a. Opisthokonta
   b. Apoikozoa
   c. Protostomia
   d. Lophotrochozoa
5. Which of the following designates animal phyla that probably originated independently from other animal phyla near the base of the animal phylogenetic tree?
   a. Protostomia
   b. Deuterostomia
   c. Ecdysozoa
   d. Basal phyla

Analysis and Application Questions

1. The Earth's atmosphere and oceanic environments began without molecular oxygen. What were the events that resulted in the accumulation of molecular oxygen in oceans, and what were the possible influences of that accumulation on animal evolution?
2. What would life on the seafloor have looked like 2.5 bya? How would it have changed over the next 2 billion years?
3. What evidence supports the endosymbiont theory?
4. Eukaryotic organisms dominate Earth in terms of biomass (total mass of living organisms). Why do microbiologists claim that the tree of life is essentially a bacterial tree?
5. Why are Ediacaran and Cambrian fossils so very important to our knowledge of animal origins?

9

The Basal Animal Phyla

Multicellularity arose one time in the lineage leading to the Animalia (Metazoa). The organization of groups of cells into tissues that provided for defense, reproduction, sensory perception, and communication followed quickly. This organization helps define what it means to be an animal. The sea anemone (*Urticina*) has all these functions occurring within a radially symmetrical body form.

Chapter Outline

9.1 Evolutionary Perspective
9.2 Phylum Porifera
The Body Wall, Cell Types, and Skeletons
Water Currents and Body Forms
Maintenance Functions (Porifera)
Reproduction
9.3 Phylum Cnidaria
The Body Wall and Nematocysts
Alternation of Generations
Maintenance Functions (Cnidaria)
Reproduction
Class Hydrozoa
Class Staurozoa
Class Scyphozoa
Class Cubozoa
Class Anthozoa
9.4 Phylum Ctenophora
9.5 Further Phylogenetic Considerations

9.1 EVOLUTIONARY PERSPECTIVE

LEARNING OUTCOMES

1. Explain the morphological and genomic features common to all animals.
2. Contrast the problems encountered in assigning a body-organization status to the basal animal phyla.

Many of us have had experiences walking along ocean shorelines and wading, swimming, or snorkeling in marine ecosystems. If you share these experiences, you have undoubtedly encountered some members of the phyla covered in this chapter. You may have felt the sting of a jellyfish's (phylum Cnidaria) stinging cells (nematocysts). You may have encountered the usually transparent comb jellies (phylum Ctenophora) that swim slowly in bays or are carried by currents and waves in near-shore habitats, and perhaps you have witnessed the bioluminescent glow that many comb jellies produce during late fall and winter nights. For many people, these animals are intriguing curiosities, but for zoologists these phyla–along with the sponges (phylum Porifera) and two others–have a much deeper significance. These phyla trace their roots deep into animal phylogeny. Their body forms and their genomes help us understand the earliest animals. These animals comprise the basal animal phyla.

The origin of animals within the opisthokont clade was described in chapter 8. Animal genomes encode unique cell adhesion proteins, extracellular matrix proteins, a surface receptor protein (a tyrosine kinase receptor), and other molecules. Animals share common structures including eggs and sperm, certain monoflagellated cells, asters that function in cell division, and epithelial tissues (*see chapter 2*) associated with a collagenous extracellular matrix. In one form or another, these features are present in five animal phyla that arose from the base of the animal phylogenetic tree. These basal phyla include the sponges (phylum Porifera), jellyfish and anemones (phylum Cnidaria), comb jellies (phylum Ctenophora), placozoans (phylum Placozoa), and acoelomorphs (phylum Acoelomorpha). (The latter two phyla are described briefly at the end of this chapter–*see table 9.4.*) The genomic and morphological differences in these phyla reflect their separate origins, which occurred more than 500 million years ago (*see appendixes A and B*)–a time frame that allows for substantial evolutionary change from stem ancestors (figure 9.1). The evolutionary relationships between these phyla will be discussed at the end of this chapter.

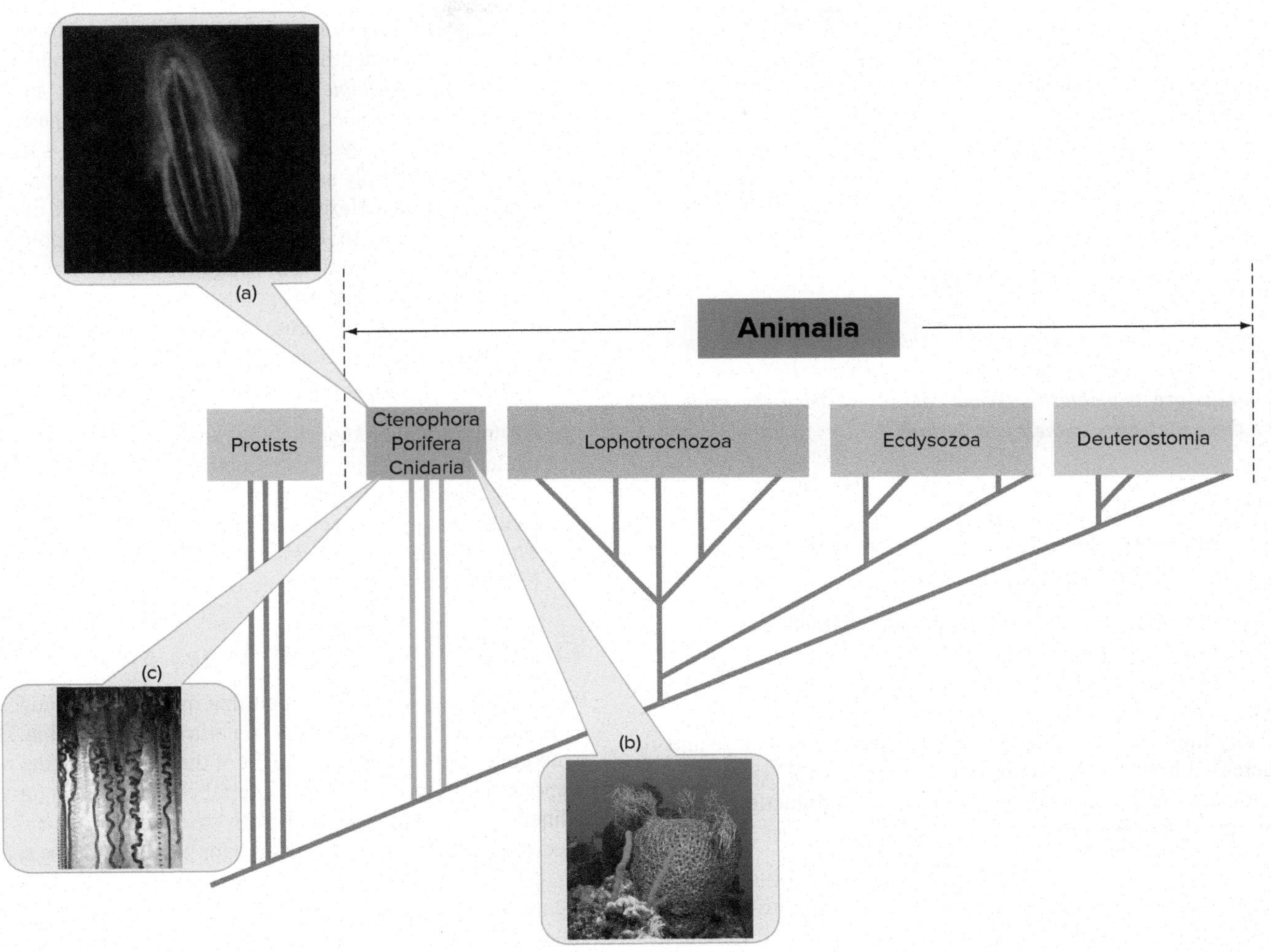

FIGURE 9.1

Evolutionary Relationships among the Basal Phyla. This figure shows one interpretation of the relationships of the Ctenophora, Porifera, and Cnidaria to other members of the animal kingdom. Evidence for these relationships is based on modern developmental and molecular biology. (*a*) Members of the phylum Ctenophora are the comb jellies. New information suggests that ancestral members of the phylum Ctenophora may be closest to the root of this animal phylogeny. Other interpretations place the sponges (phylum Porifera) at the base of the animal phylogenetic tree. *Mnemiopsis leidyi* is shown here. It is native to the western Atlantic Ocean. (*b*) Members of the phylum Porifera are probably derived from ancestral choanoflagellate stocks. (The red finger sponge, *Haliclona rubens,* and a bowl sponge, *Xestospongia,* are shown here. They are both found in Caribbean waters.) (*c*) Members of the phylum Cnidaria arose very early in animal evolution—probably from radially symmetrical ancestors. *Physalia physalis,* the Portuguese man-of-war, is shown here. The tentacles can be up to 9 m long and are laden with nematocysts that are lethal to small vertebrates and dangerous to humans. A bluish float at the surface of the water is about 12 cm long. It is not shown in this photograph. *Physalia physalis* occurs throughout the Caribbean and southern Atlantic Coast.

In addition to sharing the basic animal characteristics described in chapter 8 and in the previous paragraph, all of these basal phyla share another feature of organization—tissue-level organization (*see figures 7.10 and 7.11*). After hundreds of years of zoological study, one would think the nature of their body wall organization would be clearly understood. Unfortunately, that is not the case. Sponges are often represented as lacking tissues. As we will see, sponges have body walls comprised of cell layers separated by a collagenous, extracellular matrix called mesenchyme. This is essentially the definition of an epithelial tissue (*see chapter 2*), and zoologists specializing in poriferan biology have referred to these cell layers as epithelial tissues for many years. Some authorities prefer the term "incipient tissue" or "partially differentiated tissue" when referring to this organization because sponge tissues are different from other animal tissues. We will give a nod to poriferan specialists and refer to the sponges as having tissue-grade organization even though not all zoologists will agree with this choice. Similarly, the cnidarians and ctenophorans are usually considered diploblastic animals (*see figure 7.10*). Ctenophorans and some cnidarians have cells within their "noncellular" gelatinous mesoglea. Are the animals in these phyla triploblastic (*see figure 7.11*)? The origin of these mesogleal cells could help us decide. Mesoderm is derived in an embryo from endoderm. If these cells bud off endoderm, then it would probably be more accurate to represent these groups as being triploblastic. The case is not settled, but it seems to be more acute for the ctenophorans. Placozoan organization is similarly problematic. They possess two cell layers but lack a well-developed extracellular matrix in spite of having genes that encode extracellular matrix proteins. Acoelomorphs are the first animals we encounter that are clearly triploblastic. Observations like these are important because they influence our interpretations of the phylogeny of these groups.

Section 9.1 Thinking Beyond the Facts

Assume that zoologists make discoveries that lead them to the conclusion that poriferans, cnidarians, and placozoans are diploblastic and ctenophorans are triploblastic. What could have been the evidence that led these zoologists to their conclusions?

9.2 PHYLUM PORIFERA

LEARNING OUTCOMES

1. Describe the ecological distribution and characteristics of members of the phylum Porifera.
2. Analyze the functions carried out by components of the sponge body wall.
3. Justify the statement that "increased poriferan body size and increased body wall complexity go hand-in-hand."
4. Compare the forms of sexual and asexual reproduction present in members of the Porifera.

The Porifera (po-rif′er-ah) (L. *porus,* pore + *fera,* to bear), or sponges, are primarily marine animals that are very different in structure from any other group of animals. (figure 9.2; table 9.1). The approximately 9,000 species of sponges vary in size from less than a centimeter to a mass that would more than fill your arms.

Sponges live in all oceans and at all depths, although the greatest number occur at depths less than 200 m. One family in the class Demospongiae (*see table 9.1*) contains freshwater species. Most sponges are found in quiet, relatively clear water that permits a water-filtering existence. They are often attached to firm substrates, but they are also commonly associated with mangroves and sea grasses. Members of the class Hexactinellida (*see table 9.1*) are often found at depths exceeding 200 m, except in the South Atlantic near Antarctica where they are common in shallower water. Sponges are stable and long-lived organisms. As climate change, pollution, and disease threaten coral reef ecosystems, faster-growing sponge communities have become increasingly dominant (*see Wildlife Alert, pages 153–154*).

Characteristics of the phylum Porifera include the following:

1. Asymmetrical or superficially radially symmetrical
2. Skeleton composed of calcareous or siliceous spicules and/or the collagenous protein, spongin
3. Central cavity, or a series of branching chambers, through which water circulates during filter feeding
4. Epithelial tissues present; no organs

The Body Wall, Cell Types and Skeletons

Sponges are mostly sessile animals that move water through canal systems and filter food (principally bacteria) from the water. Their body structure is best understood in light of their water-filtering way of life. A sponge is comprised of an outer epithelial layer, a canal system that is lined by cells that move and filter water, and connected epithelial-lined spaces that form exit pathways for water out of the sponge.

The outer body wall of a sponge is called the **pinacoderm** (figure 9.3*a*). It is comprised of thin, flat, tightly connected cells called **pinacocytes.** The pinacoderm is underlain by a collagenous **mesohyl** (Gr. *meso,* middle + *hyl,* matter). The pinacoderm is the outer epidermis of a sponge, and its structure varies in different taxa. In some hexactinellids (*see table 9.1*), it is syncytial (lacks cell

(a)

(b)

FIGURE 9.2

Phylum Porifera. Many sponges are brightly colored with hues of red, orange, green, or yellow. (*a*) *Verongia* sp. (*b*) The elephant ear sponge (*Agelas clathrodes*).

(a) ©Nancy Sefton/Science Source (b) Source: NURC/UNCW and NOAA/FGBNMS

TABLE 9.1
CLASSIFICATION OF THE PORIFERA

Phylum Porifera (po-rif'er-ah)
The animal phylum whose members are sessile filter feeders and either asymmetrical or radially symmetrical; body organized around a system of water canals and chambers; skeletal elements may be (spicules) composed of calcium carbonate or silicon dioxide (silica); spongin present in some; tissue-grade organization. Approximately 9,000 species.

Class Calcarea (kal-kar'e-ah)
Spicules composed of calcium carbonate; spicules are needle shaped or have three or four rays; ascon, leucon, or sycon body forms; all marine. Calcareous sponges. *Grantia* (=*Scypha*), *Leucosolenia.*

Class Hexactinellida (hex-act"in-el'id-ah)
Spicules composed of silica and six rayed; spicules often fused into an intricate lattice; cup or vase shaped; syncytial epithelia; sycon or leucon body form; often found at 450 to 900 m depths in tropical West Indies and eastern Pacific. Glass sponges. *Euplectella* (Venus flower-basket).

Class Demospongiae (de-mo-spun'je-e)
Brilliantly colored sponges with needle-shaped or four-rayed siliceous spicules or spongin or both; leucon body form; up to 1 m in height and diameter. Includes one family of freshwater sponges, Spongillidae, and the bath sponges. *Cliona, Spongilla.*

Class Homoscleromorpha (ho-mo'skle-ro-morf-ah) Anatomically simple and encrusting in form. Siliceous spicules small and simple in shape or absent. Occur at depths ranging from shallow marine shelves to depths of 1,000 m. *Oscarella, Plakina.*

boundaries). In some sponges the mesohyl is very thin (3 μm), and in other sponges the mesohyl is much thicker.

The body wall of sponges is perforated by openings that function as inlets for water into the sponge body and are called **dermal pores** or **ostia** (L. sing. *ostium*, door). They are formed either by an embryonic cell that flattens and rolls leaving an opening through a single cell, or by several cells that surround an opening between them.

The inner epithelial layer of a sponge is called the **choanoderm.** This epithelium rests on the interior surface of the mesohyl and is comprised of a single layer of choanocytes or collar cells. **Choanocytes** (Gr. *choane,* funnel + *cyte,* cell) are flagellated cells that have a collarlike ring of microvilli surrounding a flagellum. The beating of the flagellum creates a low pressure area at the base of the flagellum that draws water through the microvilli near the cell body and pushes water along the flagellum away from the cell body (figure 9.3*b*). Microfilaments connect the microvilli, forming a filtering mesh within the collar. Choanocyte flagella usually beat continuously, collectively creating the water currents that circulate through a sponge. The exception to this is in some hexactinellid sponges (*see table 9.1*) whose syncytial epithelium is able to conduct action potentials (nerve-like electrical signals; *see figure 24.4*) to stop and restart flagellar beating in response to tissue damage or irritation. Other sponges slow incurrent water circulation by closing dermal pores in the presence of sediment or other unfavorable conditions. The cellular mechanisms responsible for these movements are being studied and are probably based on Ca^{2+} signaling within the pinacoderm. After being filtered, water moves into the canal system of a sponge and exits the sponge through one or more external openings called **oscula** (L. sing. osculum, little mouth; *see figure 9.3a*). Contractile pinacocytes lining canals and oscula often regulate filtering by closing or opening these excurrent water pathways.

Cells of sponges carry out many other functions. **Archeocytes** are ameboid stem cells found in the mesohyl that can differentiate into virtually any other cell type. In addition, they accept food vacuoles from choanocytes and distribute food throughout the sponge body. Other cells secrete collagenous fibers and skeletal elements. **Myocytes** are contractile cells that contribute to body movements.

Sponges are supported by a skeleton that may consist of microscopic needlelike spikes called **spicules.** Spicules are formed by amoeboid cells, are made of calcium carbonate or silica, and may take on a variety of shapes (figure 9.4). Alternatively, the skeleton may be made of **spongin** (a fibrous protein made of collagen). A commercial sponge is prepared by drying, beating, and washing a spongin-supported sponge until all cells are removed. The nature of the skeleton is an important characteristic in sponge taxonomy.

Water Currents and Body Forms

The life of a sponge depends on the water currents that choanocytes create. Water currents bring food and oxygen to a sponge and carry away metabolic and digestive wastes. Zoologists have described three sponge body forms. These body forms are not phylogenetically significant. The three body forms do not portray a sequence in the evolution of sponges, but they do help us visualize sponge body organization and how water circulates through sponges.

The simplest and least common sponge body form is the **ascon** (figure 9.5*a*). Ascon sponges are vaselike. Dermal pores lead directly to a chamber called the spongocoel. Choanocytes line the spongocoel, and their flagellar movements draw water into the spongocoel through the dermal pores. Water exits the sponge through the osculum, which is a single, large opening at the top of the sponge.

In the **sycon** body form, the sponge wall appears folded (figure 9.5*b*). Water enters a sycon sponge through dermal pores. Dermal pores of sycon sponges are the openings of invaginations of the body wall, called incurrent canals. Pores in the body wall connect incurrent canals to radial canals, and the radial canals lead to the spongocoel. Choanocytes line radial canals (rather than the spongocoel), and the beating of choanocyte flagella moves water from the dermal pores, through incurrent and radial canals, to the spongocoel, and out the osculum.

Leucon sponges have an extensively branched canal system (figure 9.5*c*). Water enters the sponge through dermal pores and moves through branched incurrent canals, which lead to choanocyte-lined chambers. Canals leading away from the chambers are called excurrent canals. Proliferation of chambers and canals has resulted in the absence of a spongocoel, and often, multiple exit points (oscula) for water leaving the sponge.

In complex sponges, an increased surface area for choanocytes results in large volumes of water being moved through the

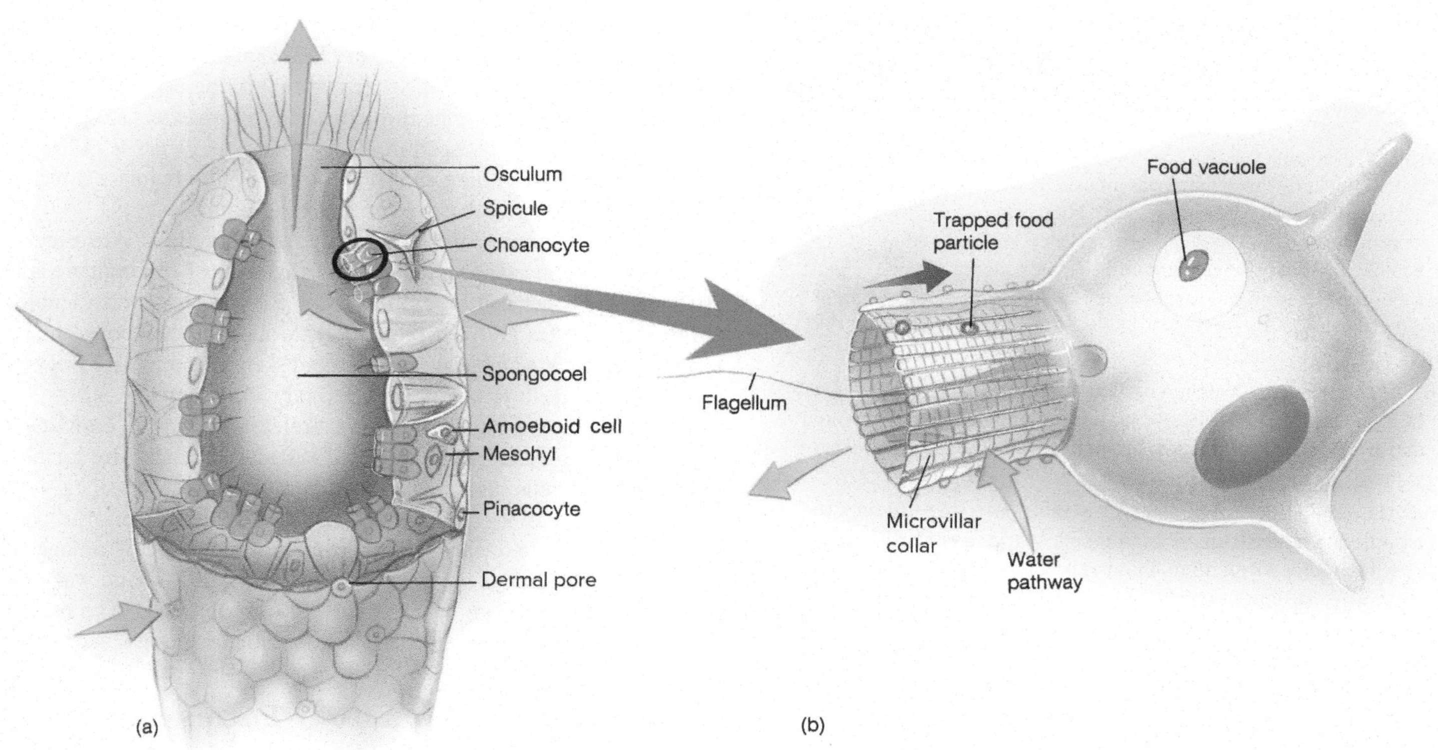

FIGURE 9.3

Morphology of a Simple Sponge. (*a*) In this example, pinacocytes form the outer body wall, and mesenchyme cells and spicules are in the mesohyl. Cells that extend through the body wall form dermal pores. (*b*) Choanocytes are cells with a flagellum surrounded by a collar of microvilli that traps food particles. Food moves toward the base of the cell, where it is incorporated into a food vacuole and passed to amoeboid mesenchyme cells, where digestion takes place. Blue arrows show water flow patterns. The brown arrow shows the direction of movement of trapped food particles.

sponge and greater filtering capabilities. Although the evolutionary pathways in the phylum are complex and incompletely described, most pathways have resulted in the leucon body form.

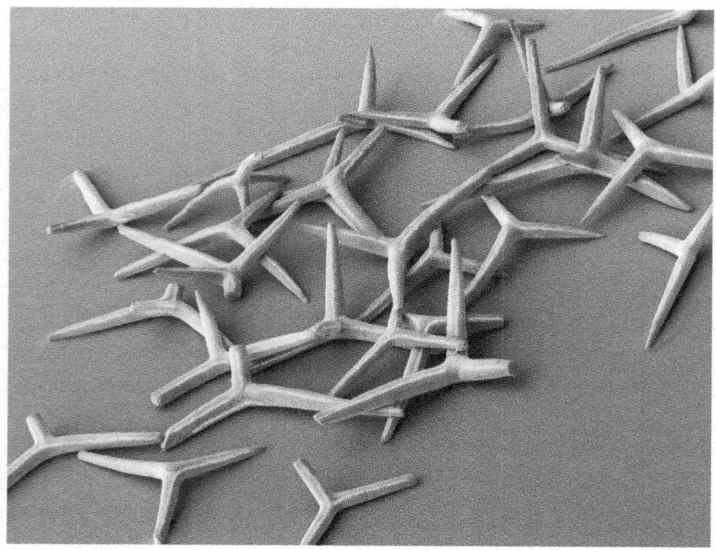

FIGURE 9.4

Scanning electron micrograph of triaxon spicules (Calcarea, SEM X20).

Maintenance Functions (Porifera)

Sponges feed on particles that range in size from 0.1 to 50 μm. Their food consists of bacteria, microscopic algae, protists, and other suspended organic matter. The prey are slowly drawn into the sponge and consumed. Large populations of sponges play an important role in reducing the turbidity of coastal waters. A single leucon sponge, 1 cm in diameter and 10 cm high, can filter in excess of 20 l of water every day.

Choanocytes filter small, suspended food particles. Water then moves into a sponge chamber at the open end of the collar. Suspended food is trapped on the collar and moved along microvilli to the base of the collar, where it is incorporated into a food vacuole (*see figures 9.3*b *and 2.20*). Digestion begins in the food vacuole by lysosomal enzymes and pH changes (*see figures 2.19 and 27.1*). Partially digested food is passed to amoeboid cells, which distribute it to other cells. A few sponges are carnivorous. These deep-water sponges (*Asbestopluma*) can capture small crustaceans using spicule-covered filaments.

Filtration is not the only way that sponges feed. Pinacocytes lining incurrent canals may phagocytize larger food particles (up to 50 μm). Sponges also may absorb by active transport nutrients dissolved in seawater.

Because of extensive canal systems and the circulation of large volumes of water through sponges, all sponge cells are in close

How Do We Know about Sponge Defenses?

Sponges are soft-bodied and sessile animals. They are a rich source of protein and calories for predators that can feed on sponge tissues. Some sponges possess spicules and tissue metabolites that reportedly serve as sponge defense mechanisms. Scientists have used feeding experiments and examined the gut contents of sponge predators (hawksbill [*Eretmochelys imbricata*] and other turtles, angelfish [family Pomacanthidae], and parrotfish [family Scaridae]) and have found that about 70% of Caribbean sponge species are chemically defended. Spicules, on the other hand, are not effective defensive structures. Fish predators are not deterred by spicules. They have strong jaws and pharyngeal teeth that can even crush limestone of reef habitats. Sea stars and crabs are also not deterred by spicules. Apparently, spicules have minimal defensive value against well-equipped predators. (Spicules still may be valuable in preventing encroachment by anemones and other organisms that may compete for sponge living space.)

The nature of the chemicals that provide sponge defense is often investigated using feeding experiments in which extracts of sponge tissues are analyzed chemically and infused into a predator's food, and then the food is offered to the predator. A variety of sponge metabolites have been found to have protective characteristics against turtles, fish, sea stars, and hermit crabs. Brominated alkaloids and terpenoid glycosides are examples of large organic molecules that serve as deterrent compounds. Metabolites like these also function as deterrents to fouling by bacterial and algal growth.

Predation strongly influences sponge distribution. Chemically defended species grow in open areas attached to reefs and rocky outcroppings. Apparently they heavily invest food resources in their defense mechanisms and grow slowly. Undefended species invest more of their resources in growth and repair. They grow rapidly and flourish in habitats protected from predators, like mangrove-covered shorelines, sea grass ecosystems, and in protected crevices in coral reef ecosystems. In these habitats, faster growth and repair rates of unprotected species allow them to thrive and outcompete chemically protected sponge species.

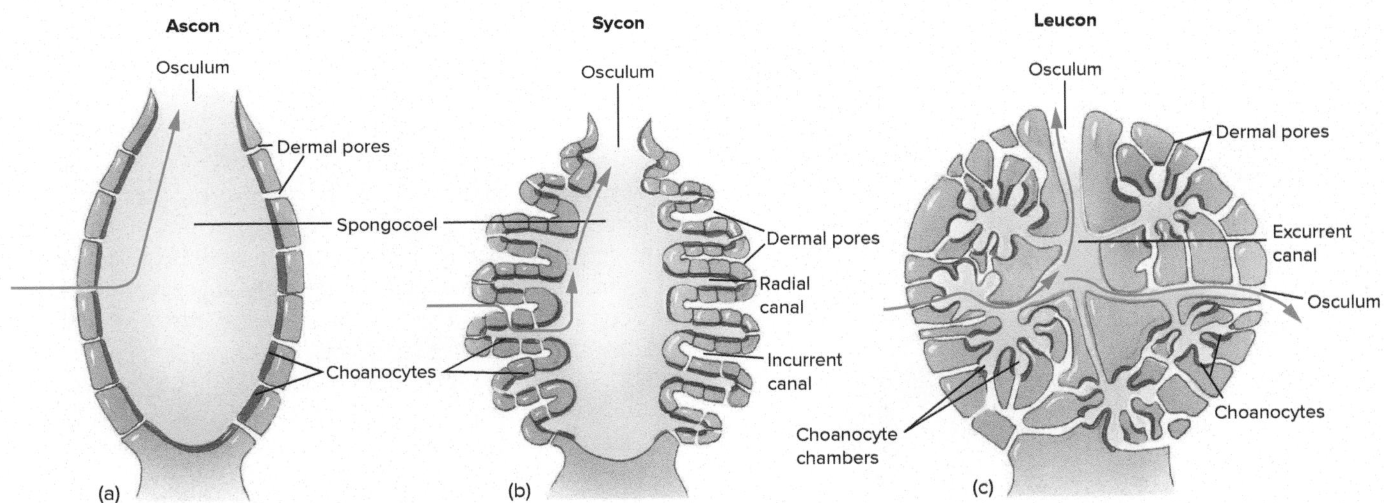

FIGURE 9.5

Sponge Body Forms. (*a*) An ascon sponge. Choanocytes line the spongocoel in ascon sponges. (*b*) A sycon sponge. The body wall of sycon sponges appears folded. Choanocytes line radial canals that open into the spongocoel. (*c*) A leucon sponge. The proliferation of canals and chambers results in the loss of the spongocoel as a distinct chamber. Multiple oscula are frequently present. Blue arrows show the direction of water flow.

contact with water. Thus, nitrogenous waste (principally ammonia) removal and gas exchange occur by diffusion.

Sponges do not have nerve cells to coordinate body functions. Most reactions result from individual cells responding to a stimulus. For example, water circulation through some sponges is at a minimum at sunrise and at a maximum just before sunset because light inhibits the constriction of cells surrounding dermal pores, keeping incurrent canals open. Cellular responses resulting in the constriction of ostia, canals, or oscula occur slowly and help to regulate filtration rates. Other reactions, however, suggest some communication among cells.

As described previously, electrical signals transmitted across the pinacoderm of hexactinellid sponges can cause choanocyte flagellar beating to cease very quickly.

Reproduction

Most sponges are **monoecious** (both sexes occur in the same individual) but do not usually self-fertilize because individual sponges produce eggs and sperm at different times. Certain choanocytes lose their collars and flagella and undergo meiosis to form flagellated sperm. Other choanocytes (and amoeboid cells in some sponges) probably undergo meiosis to form eggs. Sperm and eggs are released from sponge oscula. Fertilization occurs in the ocean water, and planktonic larvae develop.

In a few sponges, eggs are retained in the mesohyl of the parent. Sperm cells exit one sponge through the osculum and enter another sponge with the incurrent water. Sperm are trapped by choanocytes and incorporated into a vacuole. The choanocytes lose their collar and flagellum, become amoeboid, and transport sperm to the eggs.

In some sponges, early development occurs in the mesohyl. Cleavage of a zygote results in the formation of a flagellated larval stage. (A **larva** is an immature stage that may undergo a dramatic change in structure before attaining the adult body form.) The larva breaks free, and water currents carry the larva out of the parent sponge. After no more than two days of a free-swimming existence, the larva settles to the substrate and begins to develop into the adult body form (figure 9.6*a* and *b*).

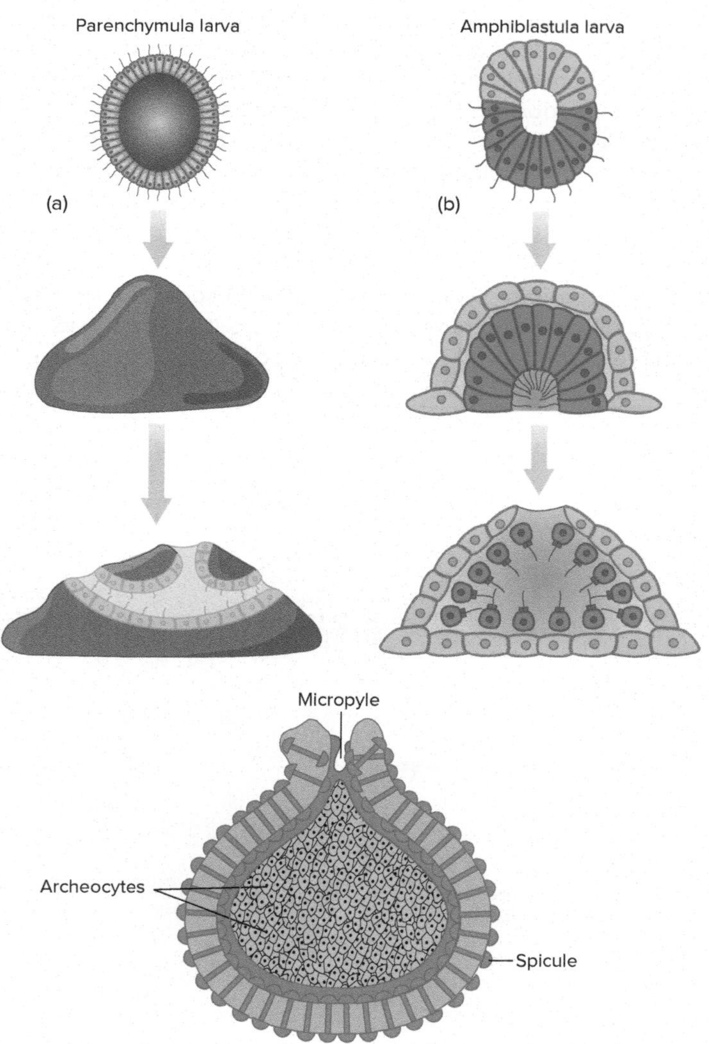

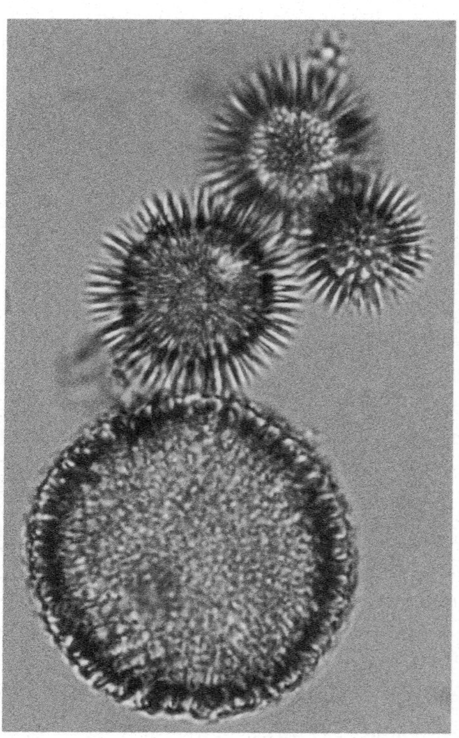

FIGURE 9.6

Development of Sponge Larval Stages. (*a*) Most sponges have a parenchymula larva (0.2 mm). Flagellated cells cover most of the larva's outer surface. After the larva settles and attaches, the outer cells lose their flagella and choanocytes develop internally from archeocytes. (*b*) Some sponges have an amphiblastula larva (0.2 mm), which is hollow and has half of the larva composed of flagellated cells. On settling, the flagellated cells invaginate into the interior of the embryo and form choanocytes. Nonflagellated cells overgrow the choanocytes and form the pinacocytes. (*c*) Gemmules (0.9 mm) are resistant capsules containing masses of archeocytes. Gemmules are released when a parent sponge dies (e.g., in the winter), and archeocytes form a new sponge when favorable conditions return. (*d*) Photomicrograph of a gemmule of the marine sponge (*Grantia* sp).

Asexual reproduction of freshwater and some marine sponges involves the formation of resistant capsules, called **gemmules,** containing masses of amoeboid archeocytes (figure 9.6*c* and *d*). When the parent freshwater sponge dies in the winter, it releases gemmules, which can survive both freezing and drying. When favorable conditions return in the spring, amoeboid cells stream out of a tiny opening, called the micropyle, and organize into a sponge.

Some sponges possess remarkable powers of regeneration. Portions of a sponge that are cut or broken from one individual regenerate new individuals.

Section 9.2 Thinking Beyond the Facts

Sponges are often represented as lacking tissue-level organization. How would you support the contention that sponges should be considered tissue-level animals?

9.3 PHYLUM CNIDARIA

LEARNING OUTCOMES

1. Describe characteristics of members of the phylum Cnidaria.
2. Explain how the diploblastic body wall of members of the phylum Cnidaria is used in support and locomotion.
3. Compare feeding and digestion strategies within the cnidarian classes.
4. Explain the function of cnidarian nervous and sensory structures.
5. Compare the life histories of members of the five cnidarian classes.

Members of the phylum Cnidaria (ni-dar′e-ah) (*Gr. knide,* nettle) possess radial or biradial symmetry. Biradial symmetry is a modification of radial symmetry in which a single plane, passing through a central axis, divides the animal into mirror images. It results from the presence of a single or paired structure in a basically radial animal and differs from bilateral symmetry in that dorsal and ventral surfaces are not differentiated. Radially symmetrical animals have no anterior or posterior ends. Thus, terms of direction are based on the position of the mouth opening. The end of the animal that contains the mouth is the oral end, and the opposite end is the aboral end. Radial symmetry is advantageous for sedentary animals because sensory receptors are evenly distributed around the body. These organisms can respond to stimuli from all directions.

The Cnidaria include over 10,000 species, are mostly marine, and are important in coral reef ecosystems (table 9.2).

Characteristics of the phylum Cnidaria include the following:

1. Radial symmetry or modified as biradial symmetry
2. Diploblastic, tissue-level organization
3. Gelatinous mesoglea between the epidermal and gastrodermal tissue layers
4. Gastrovascular cavity
5. Nerve cells organized into a nerve net
6. Specialized cells, called cnidocytes, used in defense, feeding, and attachment

TABLE 9.2
CLASSIFICATION OF THE CNIDARIA

Phylum Cnidaria (ni-dar′e-ah)
Radial or biradial symmetry, diploblastic organization, a gastrovascular cavity, and cnidocytes. More than 10,000 species.

Class Hydrozoa (hi″dro-zo′ah)
Cnidocytes present in the epidermis; gametes produced epidermally and always released to the outside of the body; mesoglea is largely acellular; medusae usually with a velum; many polyps colonial; mostly marine with some freshwater species. *Hydra, Obelia, Gonionemus, Physalia.*

Class Scyphozoa (si″fo-zo′ah)
Medusa prominent in the life history; polyp small; gametes gastrodermal in origin and released into the gastrovascular cavity; cnidocytes present in the gastrodermis as well as epidermis; medusa lacks a velum; mesoglea with wandering mesenchyme cells of epidermal origin, marine. *Aurelia.*

Class Staurozoa (sto-ro-zo′ah′)
Medusae absent; develop from benthic planula larvae; eight tentacles surrounding the mouth; attachment to substrate by adhesive disk; sexual reproduction only; marine. *Haliclystis.*

Class Cubozoa (ku″bo-zo′ah)
Medusa prominent in life history; polyp small; gametes gastrodermal in origin; medusa cuboidal in shape with tentacles that hang from each corner of the bell; marine. *Chironex.*

Class Anthozoa (an″tho-zo′ah)
Colonial or solitary polyps; medusae absent; cnidocytes present in the gastrodermis; cnidocils absent; gametes gastrodermal in origin; gastrovascular cavity divided by mesenteries that bear nematocysts; internal biradial or bilateral symmetry present; mesoglea with wandering mesenchyme cells; tentacles solid; marine. Anemones and corals. *Metridium.*

The Body Wall and Nematocysts

Cnidarians possess diploblastic, tissue-level organization (*see figure 7.10*). Cells organize into tissues that carry out specific functions, and all cells are derived from two embryological layers. The ectoderm of the embryo gives rise to an outer layer of the body wall, called the **epidermis,** and the inner layer of the body wall, called the **gastrodermis,** is derived from endoderm (figure 9.7). Cells of the epidermis and gastrodermis differentiate into a number of cell types for protection, food gathering, coordination, movement, digestion, and absorption. Between the epidermis and gastrodermis is a jelly-like layer called **mesoglea.** Cells are present in the middle layer of some cnidarians, but they have their origin in either the epidermis or the gastrodermis.

One kind of cell is characteristic of the phylum. Epidermal and/or gastrodermal cells called **cnidocytes** produce structures called cnidae, which are used for attachment, defense, and feeding. A **cnida** is a fluid-filled, intracellular capsule enclosing a coiled, hollow tube (figure 9.8). A lidlike operculum caps the capsule at one

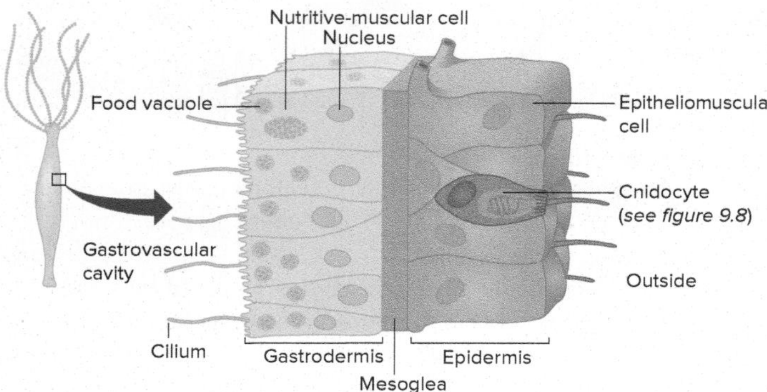

FIGURE 9.7

Body Wall of a Cnidarian (*Hydra*). Cnidarians are diploblastic (two tissue layers). The epidermis is derived embryologically from ectoderm, and the gastrodermis is derived embryologically from endoderm. Between these layers is mesoglea. Mesoglea is normally acellular in the Hydrozoa, but it contains wandering mesenchyme cells in members of the other classes. In the Hydrozoa, cnidocytes are present only in the epidermis. In members of other classes, they are present in both the epidermis and endodermis.

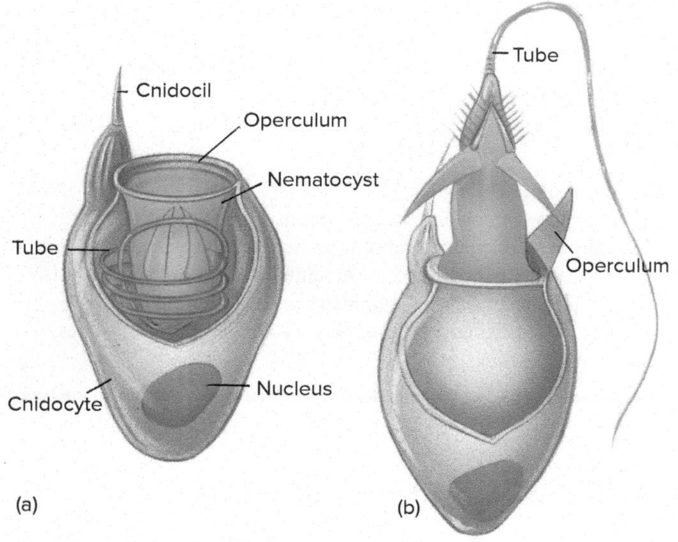

FIGURE 9.8

Cnidocyte Structure and Nematocyst Discharge. (*a*) A nematocyst is one type of cnida that develops in a capsule in the cnidocyte. The capsule is capped at its outer margin by an operculum (lid) that is displaced upon discharge of the nematocyst. The triggerlike cnidocil is responsible for nematocyst discharge. (*b*) A discharged nematocyst. When the cnidocil is stimulated, a rapid (osmotic) influx of water causes the nematocyst to evert, first near its base, and then progressively along the tube from base to tip. The tube revolves at enormous speeds as the nematocyst is discharged. In nematocysts armed with barbs, the advancing tip of the tube is aided in its penetration of the prey as barbs spring forward from the interior of the tube and then flick backward along the outside of the tube.

end. The cnidocyte usually has a modified cilium, called a cnidocil. Stimulation of the cnidocil forces open the operculum, discharging the coiled tube–as you would evert a sweater sleeve that had been turned inside out.

Zoologists have described nearly 30 kinds of cnidae. **Nematocysts** are a type of cnida used in food gathering and defense that may discharge a long tube armed with spines that penetrates the prey. The spines have hollow tips that deliver paralyzing toxins. Other cnidae contain unarmed tubes that wrap around prey or a substrate. Still other cnidae have sticky secretions that help the animal anchor itself. Six or more kinds of cnidae may be present in one individual.

Alternation of Generations

Many cnidarians possess two body forms in their life histories (figure 9.9). The **polyp** is usually asexual and sessile. It attaches to a substrate at the aboral end, and has a cylindrical body, called the column, and a mouth surrounded by food-gathering tentacles. The **medusa** (pl., *medusae*) is **dioecious** (sexes occur in different individuals) and free swimming. It is shaped like an inverted bowl, and tentacles dangle from its margins. The mouth opening is centrally located, facing downward, and the medusa swims by gentle pulsations of the body wall. The mesoglea is more abundant in a medusa than in a polyp, giving the former a jellylike consistency. When a cnidarian life cycle involves both polyp and medusa stages, the phrase "alternation of generations" is often applied.

Maintenance Functions (Cnidaria)

The gastrodermis of all cnidarians lines a blind-ending **gastrovascular cavity.** This cavity functions in digestion, the exchange of respiratory gases and metabolic wastes, and the discharge of gametes. Food, digestive wastes, and reproductive stages enter and leave the gastrovascular cavity through the mouth.

The food of most cnidarians consists of very small crustaceans, although some cnidarians feed on small fish. Nematocysts entangle and paralyze prey, and contractile cells in the tentacles cause the tentacles to shorten, which draws food toward the mouth. As food enters the gastrovascular cavity, gastrodermal gland cells secrete lubricating mucus and enzymes, which reduce food to a soupy broth. Certain gastrodermal cells, called nutritive-muscular cells, phagocytize partially digested food and incorporate it into food vacuoles, where digestion is completed. Nutritive-muscular cells also have circularly oriented contractile fibers that help move materials into or out of the gastrovascular cavity by peristaltic contractions. During peristalsis, ringlike contractions move along the body wall, pushing contents of the gastrovascular cavity ahead of them, expelling undigested material through the mouth.

Cnidarians derive most of their support from the buoyancy of water around them. In addition, a hydrostatic skeleton aids in support and movement. A **hydrostatic skeleton** is water or body fluids confined in a cavity of the body and against which contractile elements of the body wall act (*see figure 23.10*). In the Cnidaria, the water-filled gastrovascular cavity acts as a hydrostatic skeleton. Certain cells of the body wall, called epitheliomuscular cells, are contractile and aid in movement. When a polyp closes its mouth (to prevent water from escaping) and contracts longitudinal epitheliomuscular cells on one side of the body, the polyp bends toward that

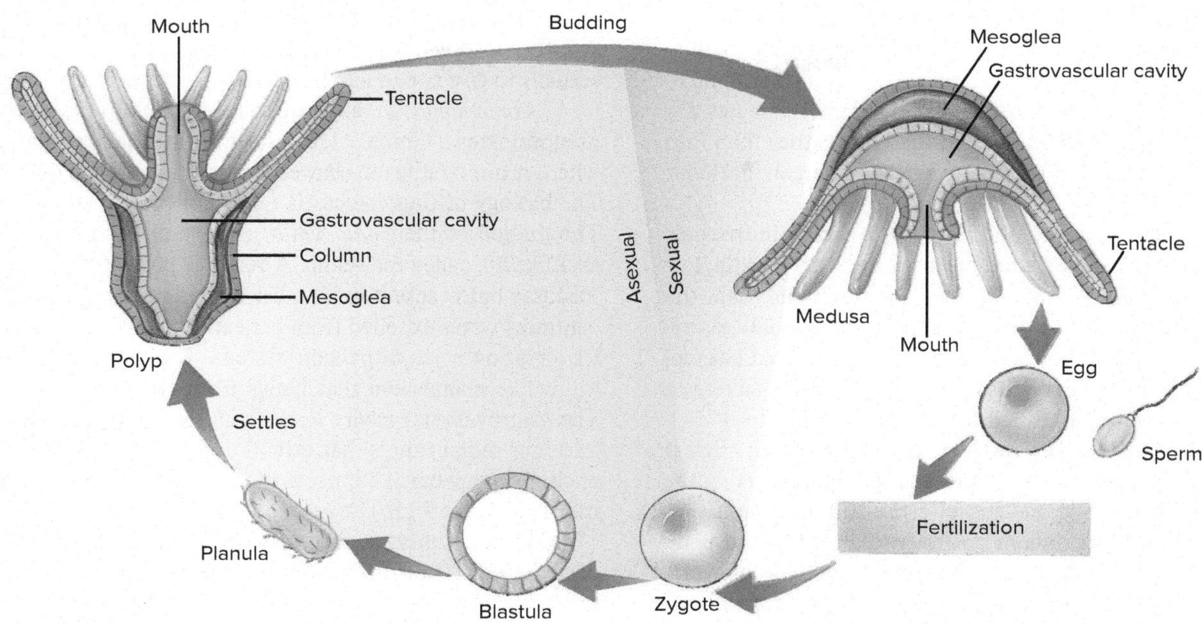

FIGURE 9.9

Generalized Cnidarian Life Cycle. This figure shows alternation between medusa and polyp body forms. Dioecious medusae produce gametes that may be shed into the water for fertilization. Early in development, a ciliated planula larva forms. After a brief free-swimming existence, the planula settles to the substrate and forms a polyp. Budding of the polyp produces additional polyps and medusa buds. Medusae break free of the polyp and swim away. The polyp or medusa stage of many species is either lost or reduced, and the sexual and asexual stages have been incorporated into one body form.

side. If these cells contract while the mouth is open, water escapes from the gastrovascular cavity, and the polyp collapses. Contraction of circular epitheliomuscular cells causes constriction of a part of the body and, if the mouth is closed, water in the gastrovascular cavity is compressed, and the polyp elongates.

Polyps use a variety of forms of locomotion. They may move by somersaulting from base to tentacles and from tentacles to base again, or move in an inchworm fashion, using their base and tentacles as points of attachment. Polyps may also glide very slowly along a substrate while attached at their base or walk on their tentacles.

Medusae move by swimming and floating. Water currents and wind are responsible for most horizontal movements. Vertical movements are the result of swimming. Contractions of circular and radial epitheliomuscular cells cause rhythmic pulsations of the bell and drive water from beneath the bell, propelling the medusa through the water.

Cnidarian nerve cells have been of interest to zoologists for many years because they may be the most primitive nervous elements in the animal kingdom (*see figure 24.6*a). By studying these cells, zoologists may gain insight into the evolution of animal nervous systems. Nerve cells are located below the epidermis, near the mesoglea, and interconnect to form a two-dimensional nerve net. This net conducts nerve impulses around the body in response to a localized stimulus. The extent to which a nerve impulse spreads over the body depends on stimulus strength. For example, a weak stimulus applied to a polyp's tentacle may cause the tentacle to be retracted. A strong stimulus at the same point may cause the entire polyp to withdraw.

Sensory structures of cnidarians are distributed throughout the body and include receptors for perceiving touch and certain chemicals. More specialized receptors are located at specific sites on a polyp or medusa.

Cnidarians have large surface-area-to-volume ratios. A consequence of this large surface area is that all cells are a short distance from the body surface, and oxygen, carbon dioxide, and nitrogenous wastes are exchanged with the environment by diffusion.

Reproduction

Most cnidarians are dioecious. Sperm and eggs may be released into the gastrovascular cavity or to the outside of the body. In some instances, eggs are retained in the parent until after fertilization.

A blastula forms early in development, and migration of surface cells to the interior fills the embryo with cells that will eventually form the gastrodermis. The embryo elongates to form a ciliated, free-swimming larva, called a **planula.** The planula attaches to a substrate, interior cells split to form the gastrovascular cavity, and a young polyp develops (*see figure 9.9*).

Medusae nearly always form by budding from the body wall of a polyp, and polyps may form other polyps by budding. Buds may detach from the polyp, or they may remain attached to the parent to contribute to a colony of individuals. Variations on this general pattern are discussed in the survey of cnidarian classes that follows.

Class Hydrozoa

Hydrozoans (hi″dro-zo′anz) are small, relatively common cnidarians. The vast majority are marine, but this is the one cnidarian class with freshwater representatives. Most hydrozoans have life cycles that display alternation of generations; however, in some, the medusa stage is lost, while in others, the polyp stage is very small.

Three features distinguish hydrozoans from other cnidarians: (1) nematocysts are only in the epidermis; (2) gametes are epidermal and released to the outside of the body rather than into the gastrovascular cavity; and (3) the mesoglea is largely acellular (*see table 9.2*).

Most hydrozoans have colonial polyps in which individuals may be specialized for feeding, producing medusae by budding, or defending the colony. In *Obelia,* a common marine cnidarian, the planula develops into a feeding polyp, called a **gastrozooid** (gas′tra-zo′oid) or hydranth (hi″dranth) (figure 9.10). The gastrozooid has tentacles, feeds on microscopic organisms in the water, and secretes a skeleton of protein and chitin, called the perisarc, around itself.

Growth of an *Obelia* colony results from budding of the original gastrozooid. Rootlike processes grow into and horizontally along the substrate. They anchor the colony and give rise to branch colonies. The entire colony has a continuous gastrovascular cavity and body wall, and is a few centimeters high. Gastrozooids are the most common type of polyp in the colony; however, as an *Obelia* colony grows, gonozooids are produced. A **gonozooid** (gon′o-zo′oid) or gonangium (go′nanj″e-um) is a reproductive polyp that produces medusae by budding. *Obelia's* small medusae form on a stalklike structure of the gonozooid. When medusae mature, they break free of the stalk and swim out an opening at the end of the gonozooid. Medusae reproduce sexually to give rise to more colonies of polyps.

Gonionemus is a hydrozoan in which the medusa stage predominates (figure 9.11*a*). It lives in shallow marine waters, where it often clings to seaweeds by adhesive pads on its tentacles. The biology of *Gonionemus* is typical of most hydrozoan medusae. The margin of the *Gonionemus* medusa projects inward to form a shelflike lip, called the velum. A velum is present on most hydrozoan medusae but is absent in all other cnidarian classes. The velum concentrates water expelled from beneath the medusa to a smaller outlet, creating a jet-propulsion system. The mouth is at the end of a tubelike **manubrium** that hangs from the medusa's oral surface. The gastrovascular cavity leads from the inside of the manubrium into four radial canals that extend to the margin of the medusa. An encircling ring canal connects the radial canals at the margin of the medusa (figure 9.11*b*).

In addition to a nerve net, *Gonionemus* has a concentration of nerve cells, called a nerve ring, that encircles the margin of the medusa. The nerve ring coordinates swimming movements. Embedded in the mesoglea around the margin of the medusa are sensory structures called statocysts. A **statocyst** consists of a small sac surrounding a calcium carbonate concretion called a statolith. When

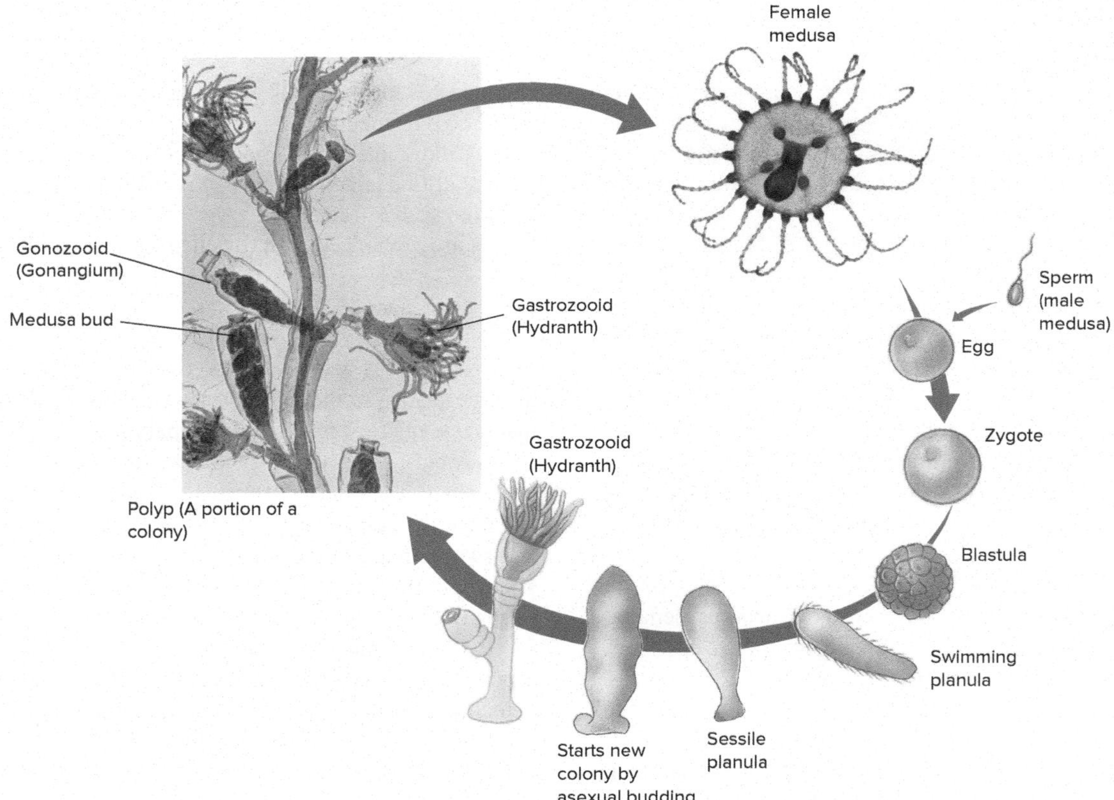

FIGURE 9.10

***Obelia* Structure and Life Cycle.** *Obelia* alternates between polyp and medusa stages. An entire polyp colony can stand 1–30 cm tall, depending on the species. A mature medusa is about 1 mm in diameter, and the planula is about 0.2 mm long. Unlike *Obelia,* the majority of colonial hydrozoans have medusae that remain attached to the parental colony, and they release gametes or larval stages through the gonozooid. The medusae often degenerate and may be little more than gonadal specializations in the gonozooid.

Gonionemus tilts, the statolith moves in response to the pull of gravity. This initiates nerve impulses that may change the animal's swimming behavior.

Gonads of *Gonionemus* medusae hang from the oral surface, below the radial canals. *Gonionemus* is dioecious and sheds gametes directly into seawater. A planula larva develops and attaches to the substrate, eventually forming a polyp (about 5 mm tall). The polyp reproduces by budding to make more polyps and medusae.

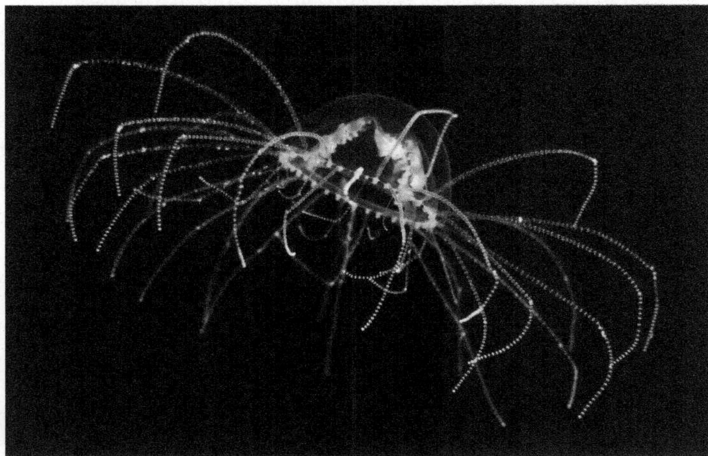

(a)

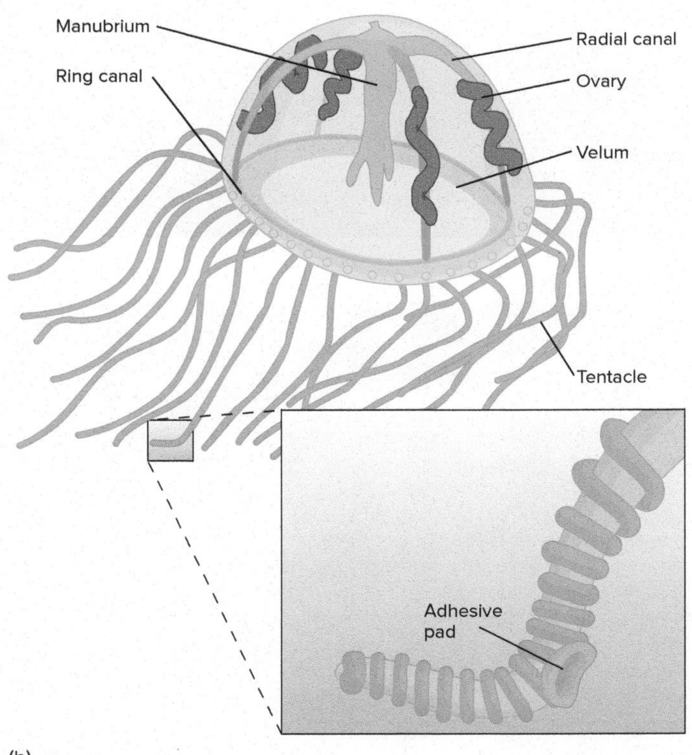

(b)

FIGURE 9.11

A Hydrozoan Medusa. (*a*) A *Gonionemus vertens* medusa. (*b*) Structure of *Gonionemus.*

Hydra is a common freshwater hydrozoan that hangs from the underside of floating plants in clean streams and ponds. *Hydra* lacks a medusa stage and reproduces both asexually by budding from the side of the polyp and sexually. Hydras are somewhat unusual hydrozoans because sexual reproduction occurs in the polyp stage. Testes are conical elevations of the body surface that form from the mitotic division of certain epidermal cells, called interstitial cells. Sperm form by meiosis in the testes. Mature sperm exit the testes through temporary openings. Ovaries also form from interstitial cells. One large egg forms per ovary. During egg formation, yolk is incorporated into the egg cell from gastrodermal cells. As ovarian cells disintegrate, a thin stalk of tissue attaches the egg to the body wall. After fertilization and early development, epithelial cells lay down a resistant chitinous shell. The embryo drops from the parent, overwinters, hatches in the spring, and develops into an adult.

Large oceanic hydrozoans belong to the order Siphonophora. These colonies are associations of numerous polypoid and medusoid individuals. Individual polyps and medusae are called zooids and originate through budding and remain attached to form the larger organism. Zooids have specific structures and functions. For example, some polyps, called dactylozooids, possess a single, long (up to 9 m) tentacle armed with cnidocytes for capturing prey. Other polyps are specialized for digesting prey. One type of medusoid individual, called a nectophore, cannot feed and is dependent on feeding polyps, but it performs swimming functions for the colony. Other medusoid individuals form sac floats, oil floats, leaflike defensive structures, and gonads. All of these zooids are arranged in specific patterns to make each type of siphonophoran unique and recognizable. Even though zoologists think of siphonophorans as colonies, each kind is described and named as a species.

Siphonophors are predators that wait for their prey (e.g., fish and crustaceans) to brush into their widely dispersed, nematocyst-laden tentacles. *Physalia physalis* is the Portuguese man-of-war (*see figure 9.1*c) and is commonly encountered in near-shore habitats. Deep-sea siphonophores live in darkness and in regions where prey are widely distributed. Many deep-sea species have bioluminescent lures that draw prey to their tentacles (figure 9.12).

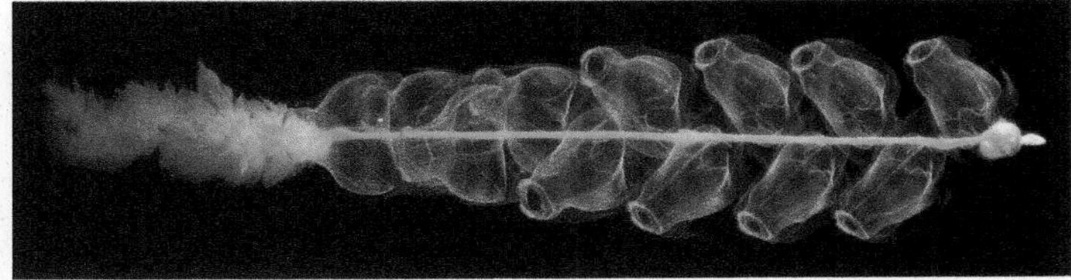

FIGURE 9.12

Order Siphonophora. Siphonophores are colonial, often bioluminescent, hydrozoans comprised of numerous polypoid and medusoid individuals. *Marrus orthocanna* occurs in the Arctic and northern Pacific Oceans. The orange pneumatophore on the right serves as a gas-filled float. Contractile swimming bells (nectophores) provide jet-like locomotion. The orange siphosome to the left is comprised of tentacles of feeding zooids.

Class Staurozoa

Members of the class Staurozoa (sto-ro-zo′ah′) are all marine. They were formerly classified into an order (Stauromedusae) within the class Scyphozoa. Even though staurozoans lack a medusa stage, the former order name is derived from the resemblance of the oral end of the polyp to a medusa. The body form is in the shape of a goblet with a series of eight tentacle clusters attached to the margin of the goblet (figure 9.13). The aboral end (the stem of the goblet) attaches to its substrate, usually rock or seaweed. Sexual reproduction results in the formation of a nonciliated, crawling planula larva, probably with very limited dispersal ability. The planula attaches to a substrate and matures into the adult. Even though the planula's ability to disperse may be limited, adults have been observed somersaulting by alternately attaching their base and tentacles. Rarely, they have been observed drifting freely in the water. There are about 100 described species of staurozoans. They are found in higher latitudes of the Atlantic Ocean and the northwestern Pacific coast of North America. Others have been found in Antarctic waters, and two species have been described from abyssal depths in the Pacific Ocean.

Class Scyphozoa

Members of the class Scyphozoa (si″fo-zo′ah) are all marine and are "true jellyfish" because the dominant stage in their life history is the medusa (figure 9.14). Unlike hydrozoan medusae, scyphozoan medusae lack a velum, the mesoglea contains amoeboid mesenchyme cells, cnidocytes occur in the gastrodermis as well as the epidermis, and gametes are gastrodermal in origin (*see table 9.2*).

Many scyphozoans are harmless to humans; others can deliver unpleasant and even dangerous stings. For example, *Mastigias quinquecirrha,* the so-called stinging nettle, is a common Atlantic scyphozoan whose populations increase in late summer and become hazardous to swimmers (figure 9.14*a*). A rule of thumb for swimmers is to avoid helmet-shaped jellyfish with long tentacles and fleshy lobes hanging from the oral surface.

Aurelia is a common scyphozoan in both Pacific and Atlantic coastal waters of North America (figure 9.14*b*). The margin of its medusa has a fringe of short tentacles and is divided by notches. The mouth of *Aurelia* leads to a stomach with four gastric pouches, which contain cnidocyte-laden gastric filaments. Radial canals lead from gastric pouches to the margin of the bell. In *Aurelia,* but not all scyphozoans, the canal system is extensively branched and

FIGURE 9.13

Class Staurozoa. *Lucernaria janetae* is a staurozoan from abyssal depths of the eastern Pacific. This species is larger than most staurozoans, about 10 cm across. *Image courtesy of J. Voight with support of the National Science Foundation.*

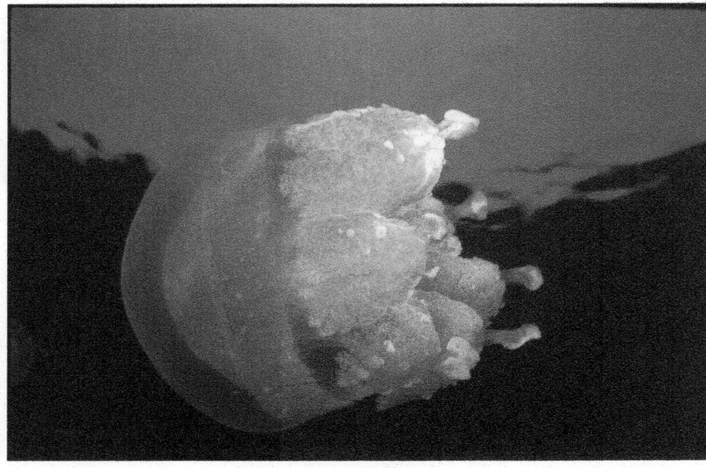

(a)

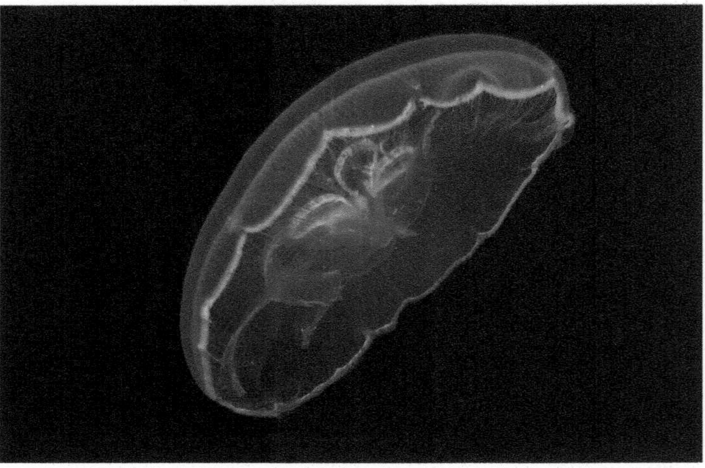

(b)

FIGURE 9.14

Representative Scyphozoans. (*a*) *Mastigias* (*b*) *Aurelia aurita.*

leads to a ring canal around the margin of the medusa. Gastrodermal cells of all scyphozoans possess cilia to continuously circulate seawater and partially digested food.

Aurelia is a plankton feeder. At rest, it sinks slowly in the water and traps microscopic animals in mucus on its epidermal surfaces. Cilia carry this food to the margin of the medusa. Four fleshy lobes, called oral lobes, hang from the manubrium and scrape food from the margin of the medusa (figure 9.15*a*). Cilia on the oral lobes carry food to the mouth.

In addition to sensory receptors on the epidermis, *Aurelia* has eight specialized structures, called rhopalia, in the notches at the margin of the medusa. Each **rhopalium** consists of sensory structures surrounded by rhopalial lappets. Two sensory pits (presumed to be olfactory) are associated with sensory lappets. A statocyst and photoreceptors, called ocelli, are associated with rhopalia (figure 9.15*b*). *Aurelia* displays a distinct negative phototaxis, coming to the surface at twilight and descending to greater depths during bright daylight.

Scyphozoans are dioecious. *Aurelia's* eight gonads are in gastric pouches, two per pouch. Gametes are released into the gastric pouches. Sperm swim through the mouth to the outside of the medusa. In some scyphozoans, eggs are fertilized in the female's gastric pouches, and early development occurs there. In *Aurelia,* eggs lodge in the oral lobes, where fertilization and development to the planula stage occur.

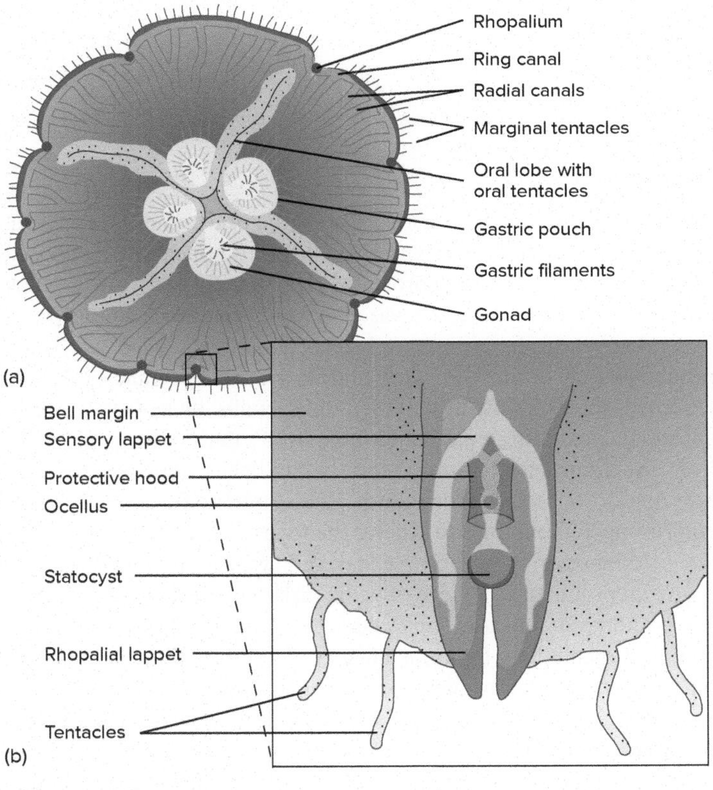

FIGURE 9.15

Structure of a Scyphozoan Medusa. (*a*) Internal structure of *Aurelia.* (*b*) A section through a rhopalium of *Aurelia.* Each rhopalium consists of two sensory (olfactory) lappets, a statocyst, and a photoreceptor called an ocellus.

(b) Source: Hyman, LH. 1940. *Biology of the Invertebrates*. McGraw-Hill Publishing Co.

The planula develops into a polyp called a **scyphistoma** (figure 9.16). The scyphistoma lives a year or more, during which time budding produces miniature medusae, called **ephyrae.** The budding scyphistoma is often called a strobila. Repeated budding of the scyphistoma results in ephyrae being stacked on the polyp—as you might pile saucers on top of one another. After ephyrae are released, they gradually attain the adult form.

Class Cubozoa

Members of the class Cubozoa (ku″bo-zo′ah) are called box jellyfish. They are distributed throughout tropical waters of the Atlantic, Pacific, and Mediterranean Oceans. Their class and common names are derived from the cuboidal shape of the medusa. Tentacles hang from the corners of the bell, and they possess a velarium (similar to the velum of hydrozoans) that reduces the aperture of the bell and creates an efficient jet-propulsion swimming locomotion. They have rhopalia located on the flat surfaces of the bell. Unlike those of other cnidarians, the rhopalia photoreceptors possess retinas, lenses, and corneas—which suggest image-forming capabilities. Other photoreceptors are simple ocelli that probably detect light and dark differences. Their nervous organization is as complex as any other cnidarians', including a nerve ring around the perimeter of the bell that coordinates swimming movements.

Box jellyfish are active predators. Unlike other cnidarians that rely on chance encounters with prey, box jellyfish actively pursue their fish prey. The venom of some Indo-Pacific species (figure 9.17) is extremely potent—dangerous, even fatal, to humans. (It acts by causing potassium ion leakage from cells, and may cause heart failure within minutes of being stung.) The venom of other species results in short-lived pain. Their unusually potent venoms are thought to function in defense against predation, although hawksbill turtles (*Eretmochelys imbricata*) seem to relish box jellyfish and are apparently unaffected by the venom.

Class Anthozoa

Members of the class Anthozoa (an′tho-zo′ah) are colonial or solitary, and lack medusae. Their cnidocytes lack cnidocils. They include anemones and stony and soft corals. Anthozoans are all marine and are found at all depths.

Anthozoan polyps differ from hydrozoan polyps in three respects: (1) the mouth of an anthozoan leads to a pharynx, which is an invagination of the body wall that leads into the gastrovascular cavity; (2) mesenteries (membranes) that bear cnidocytes and gonads on their free edges divide the gastrovascular cavity into sections; and (3) the mesoglea contains amoeboid mesenchyme cells (*see table 9.2*).

Externally, anthozoans appear to show perfect radial symmetry. Internally, the mesenteries and other structures convey biradial symmetry to members of this class.

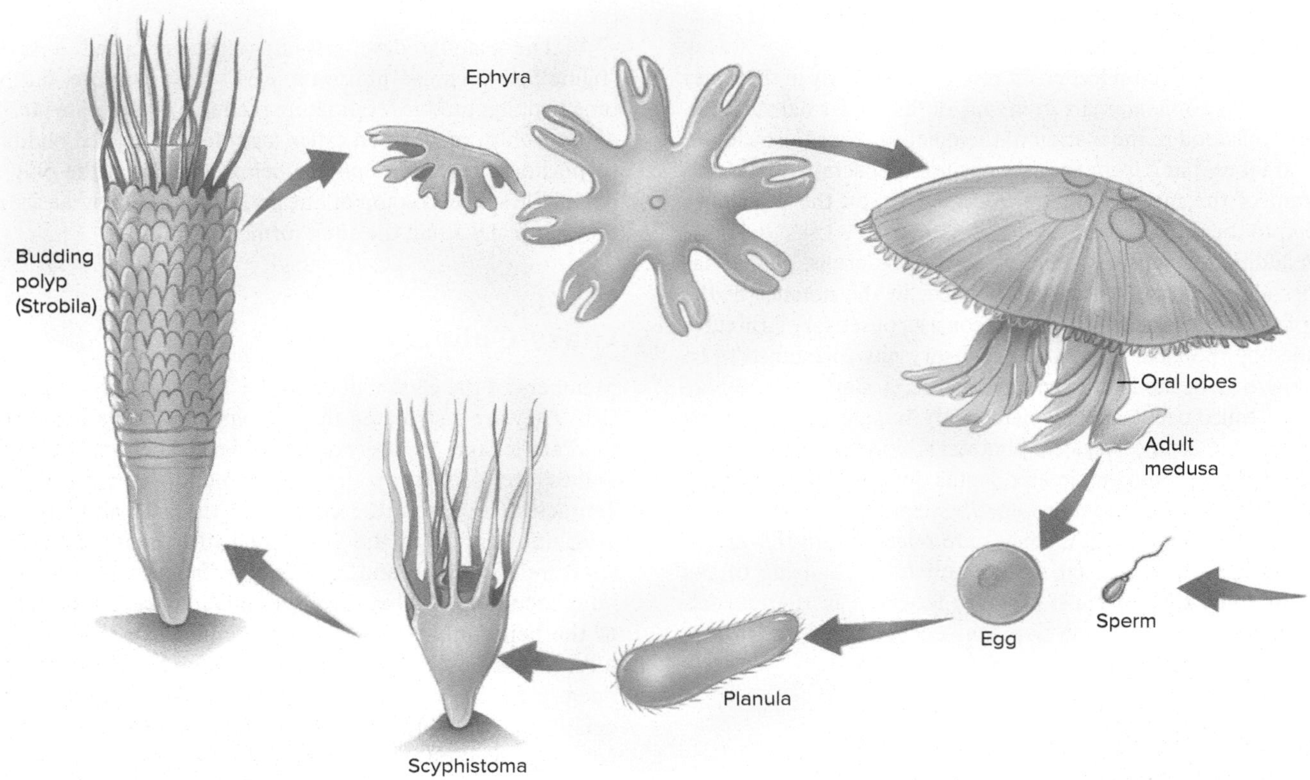

FIGURE 9.16

***Aurelia* Life History.** *Aurelia* is dioecious, and as with all scyphozoans, the medusa (10 cm) predominates in the organism's life history. The planula (0.3 mm) develops into a polyp called a scyphistoma (4 mm), which produces young medusae, or ephyrae, by budding.

Sea anemones are solitary, frequently large, and colorful (figure 9.18*a*). Some attach to solid substrates, some burrow in soft substrates, and some live in symbiotic relationships (figure 9.18*b*). The polyp attaches to its substrate by a pedal disk (figure 9.19). An oral disk contains the mouth and solid, oral tentacles. At one or both ends of the slitlike mouth is a siphonoglyph, which is a ciliated tract that moves water into the gastrovascular cavity to maintain the hydrostatic skeleton.

Mesenteries are arranged in pairs. Some attach at the body wall at their outer margin and to the pharynx along their inner margin. Other mesenteries attach to the body wall but are free along their entire inner margin. Openings in mesenteries near the oral disk permit water to circulate between compartments the mesenteries set off. The free lower edges of the mesenteries form a trilobed mesenterial filament. Mesenterial filaments bear cnidocytes, cilia that aid in water circulation, gland cells that secrete digestive enzymes, and cells that absorb products of digestion. Threadlike acontia at the ends of mesenterial filaments bear cnidocytes. Acontia subdue live prey in the gastrovascular cavity and can be extruded through small openings in the body wall or through the mouth when an anemone is threatened.

Muscle fibers are largely gastrodermal. Longitudinal muscle bands are restricted to the mesenteries. Circular muscles are in the gastrodermis of the column. When threatened, anemones contract their longitudinal fibers, allowing water to escape from the gastrovascular cavity. This action causes the oral end of the column to fold over the oral disk, and the anemone appears to collapse. Reestablishment of the hydrostatic skeleton depends on gradual uptake of water into the gastrovascular cavity via the siphonoglyphs.

Anemones have limited locomotion. They glide on their pedal disks, crawl on their sides, and walk on their tentacles. When disturbed, some "swim" by thrashing their bodies or tentacles. Some anemones float using a gas bubble held within folds of the pedal disk.

Anemones feed on invertebrates and fishes. Tentacles capture prey and draw it toward the mouth. Radial muscle fibers in the mesenteries open the mouth to receive the food.

Anemones show both sexual and asexual reproduction. In asexual reproduction, a piece of pedal disk may break away from the polyp and grow into a new individual in a process called pedal laceration. Longitudinal fission is quite common in anthozoans, and it often results in dense aggregations of genetically identical individuals (*see figure 7.8*). During longitudinal fission, the two halves of an individual appear to simply move in opposite directions, and the halves eventually pinch into two individuals. Longitudinal fission occurs over a one- to several-week period. Longitudinal fission is the primary form of reproduction for many anthozoan species. Transverse fission has also been reported in anthozoans, but it is much less common.

Sexual reproduction in anemones is quite variable, depending on the species involved. It usually occurs seasonally, but the season

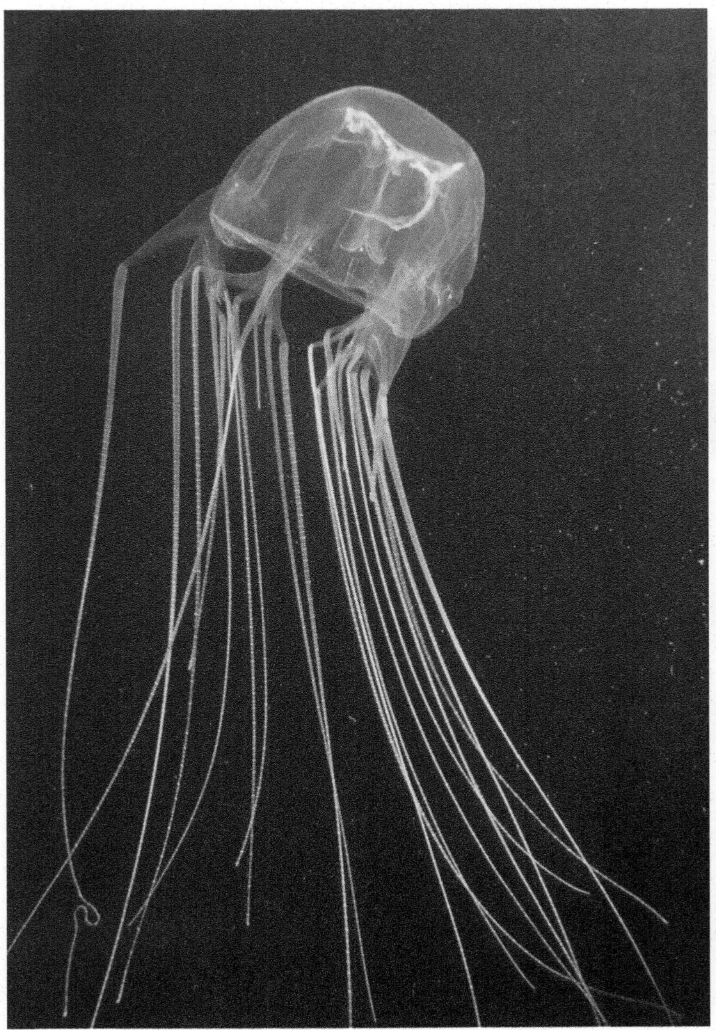

FIGURE 9.17

Class Cubozoa. The sea wasp, *Chironex fleckeri*. The medusa is cuboidal, and tentacles hang from the corners of the bell. *Chironex fleckeri* has caused more human suffering and death off Australian coasts than the Portuguese man-of-war has in any of its home waters. Death from heart failure and shock is not likely unless the victim is stung repeatedly.
©Karen Gowlett-Holmes/Getty Images

(a)

(b)

FIGURE 9.18

Representative Sea Anemones. (*a*) Sunburst or starburst anemone (*Anthopleura sola*). This anemone has specialized tentacles that it uses to aggressively defend its territory against other genetically dissimilar anemones. (*b*) This sea anemone (*Dardanus calidus*) lives in a mutualistic relationship with a hermit crab (*Eupagurus*). Hermit crabs lack a heavily armored exoskeleton over much of their bodies and seek refuge in empty snail shells. When this crab outgrows its present home, it will take its anemone with it to a new snail (whelk) shell. This anemone, riding on the shell of the hermit crab, has an unusual degree of mobility. In turn, the anemone's nematocysts protect the crab from predators.
(a) Source:Claire Fackler, CINMS, NOAA (b) ©Sami Sarkis/Getty Images

can range from early spring through fall. Unlike other cnidarians, anemones may be either monoecious or dioecious. In monoecious species, male gametes mature earlier than female gametes so that self-fertilization does not occur. This is called **protandry** (Gr. *protos,* first + *andros,* male). Gonads occur in longitudinal bands behind mesenterial filaments. Fertilization may be external or within the gastrovascular cavity. Cleavage results in the formation of a planula. The planula may leave the parent anemone, swim, feed, settle, and eventually form a new anemone. Alternatively, many anemones brood planula larvae and miniature anemones within the gastrovascular cavity of the parent until the young anemones emerge through the oral opening.

Other anthozoans are corals. Stony corals form coral reefs and, except for lacking siphonoglyphs, are similar to the anemones. Their common name derives from a cuplike calcium carbonate exoskeleton that epithelial cells secrete around the base and the lower portion of the column (figure 9.20). When threatened, polyps retract into their protective exoskeletons. Sexual reproduction is similar to that of anemones, and asexual budding produces other members of the colony.

Many cnidarians have developed close symbiotic relationships with unicellular algae. In marine cnidarians these algae usually reside in the epidermis or gastrodermis and are called zooxanthellae (*see figure 9.20*). Stony corals have large populations of these algae. Photosynthesis by dinoflagellate zooxanthellae often

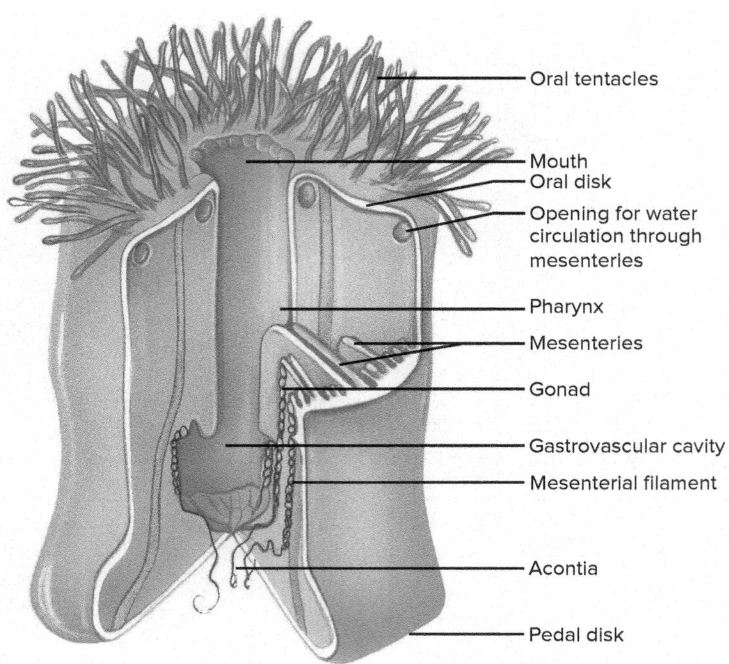

FIGURE 9.19

Class Anthozoa. Structure of the anemone, *Metridium* sp.

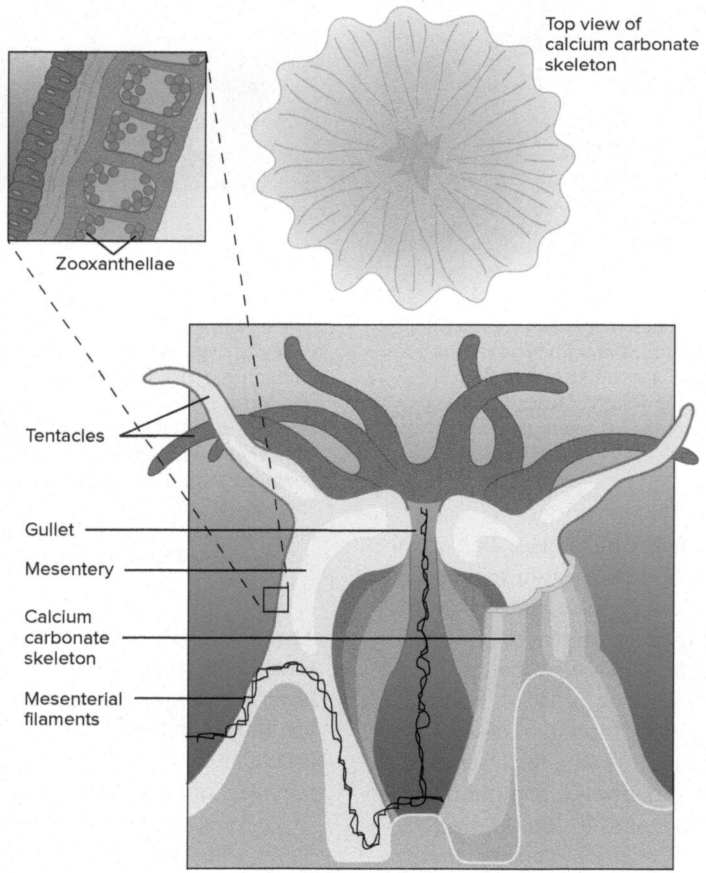

FIGURE 9.20

Class Anthozoa. A stony coral polyp in its calcium carbonate skeleton (longitudinal section).

(a)

(b)

FIGURE 9.21

Representative Octacorallian Corals. (*a*) Fleshy sea pen (*Ptilosaurus gurneyi*). (*b*) Purple sea fan (*Gorgonia ventalina*).
(a) ©Nature Sea Diver/Shutterstock (b) ©Diane Nelson

provides a significant amount of organic carbon for the coral polyps, and metabolism by the polyps provides algae with nitrogen and phosphorus by-products. Studies suggest that zooxanthellae aid in building coral reefs by promoting exceptionally high rates of calcium carbonate deposition. As zooxanthellae remove CO_2 from the environment of the polyp, associated pH changes induce the precipitation of dissolved calcium carbonate as aragonite (coral limestone). It is thought that the 90-m depth limit for reef building corresponds to the limits to which sufficient light penetrates to support zooxanthellae photosynthesis. Environmental disturbances, such as increased water temperature, can stress and kill zooxanthellae and result in coral bleaching (*see Wildlife Alert, pages 153–154*).

The colorful octacorallian corals are common in warm waters. They have eight pinnate (featherlike) tentacles, eight mesenteries, and one siphonoglyph. The body walls of members of a colony are connected, and mesenchyme cells secrete an internal skeleton of protein or calcium carbonate. Sea fans, sea pens, sea whips, red corals, and organ-pipe corals are members of this group (figure 9.21).

Section 9.3 Thinking Beyond the Facts

Cnidarians have two distinct tissue layers. They also have structures like gonads and a nerve net. Why is this considered tissue-level, not organ-level, organization?

9.4 PHYLUM CTENOPHORA

LEARNING OUTCOMES

1. Describe the ecological distribution and characteristics of members of the phylum Ctenophora.
2. Compare the body organization of a ctenophoran to a scyphozoan medusa.
3. Explain the feeding and reproductive functions in members of the phylum Ctenophora.

Animals in the phylum Ctenophora (ti-nof'er-ah) (Gr. *kteno,* comb + *phoros,* to bear) are called sea walnuts or comb jellies (table 9.3).

TABLE 9.3
CLASSIFICATION OF THE CTENOPHORA

Phylum Ctenophora (ti-nof′er-ah)
The animal phylum whose members are biradially symmetrical, diploblastic or possibly triploblastic, usually ellipsoid or spherical in shape, possess colloblasts, and have meridionally arranged comb rows. Class names are not listed here as higher-level taxonomy is currently under revision. Traditional class designations do not reflect monophyletic groups.

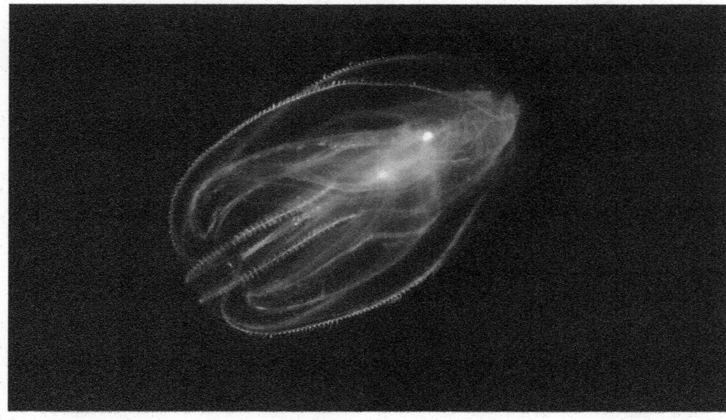

(a)

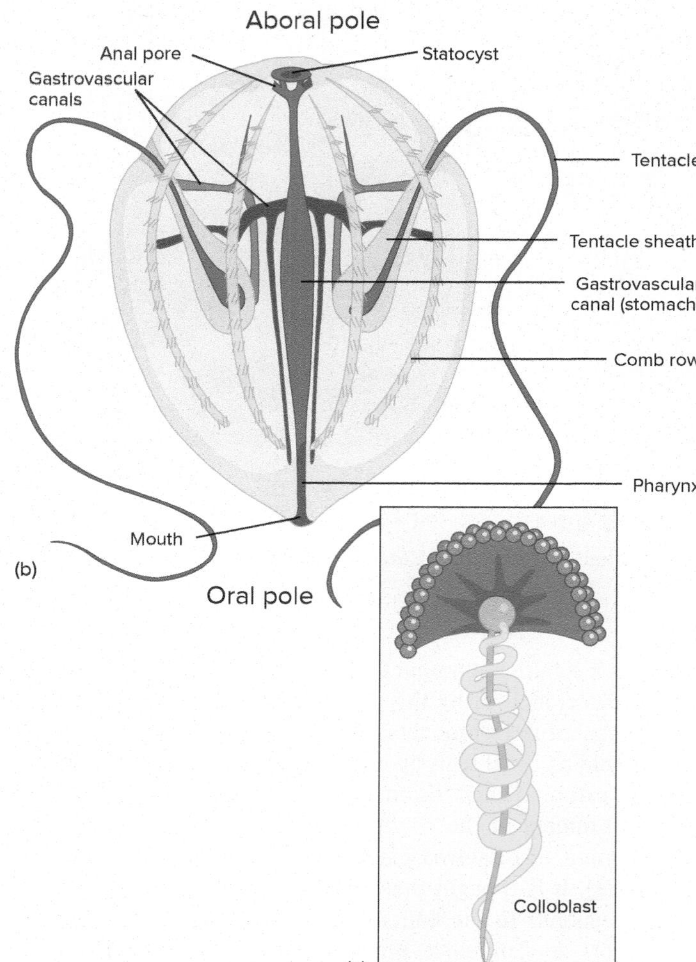

FIGURE 9.22
Phylum Ctenophora. (*a*) The ctenophore *Mnemiopsis leidyi*. Ctenophorans are well known for their bioluminescence. Light-producing cells are in the walls of their digestive canals, which are beneath comb rows. The beating of the comb-row cilia also produces iridescent color changes unrelated to bioluminescence. (*b*) The structure of *Pleurobranchia* sp. The animal usually swims with the oral end forward or upward. (*c*) Colloblasts consist of a hemispherical sticky head that connects to the core of the tentacle by a straight filament. A contractile spiral filament coils around the straight filament. Straight and spiral filaments prevent struggling prey from escaping.

The approximately 200 described species are all marine (figure 9.22). Most ctenophorans have a spherical form, although several groups are flattened and/or elongate. Their symmetry is described as biradial symmetry because of the presence of two tentacles in many of the spherical ctenophores.

There are eight orders of ctenophorans that have been divided into two classes, based on the presence or absence of tentacles. These orders are not monophyletic and have been rejected by virtually all zoologists. Higher-level taxonomy is under revision.

Ctenophorans occur in all oceans from coastal regions to depths up to 2,000 m, where they make up a significant component of the planktonic biomass. They are planktonic predators, feeding on small crustaceans (*see chapter 15*) and rotifers (*see chapter 10*). They are also prey for invertebrate predators, including other species of ctenophores, and vertebrates like fish, turtles, and whales. A few ctenophores live on the ocean floor.

Characteristics of the phylum Ctenophora include the following:

1. Diploblastic or possibly triploblastic, tissue-level organization
2. Biradial symmetry
3. Gelatinous, cellular mesoglea between the epidermal and gastrodermal tissue layers
4. True muscle cells develop within the mesoglea
5. Gastrovascular cavity
6. Nervous system in the form of a nerve net
7. Adhesive structures called colloblasts
8. Eight rows of ciliary bands, called comb rows, for locomotion

Ctenophorans have traditionally been classified as having diploblastic organization. Their "mesoglea," however, is always highly cellular and contains true muscle cells. This observation has led some zoologists to conclude that the ctenophoran middle layer should be considered a true tissue layer, which would make these animals triploblastic acoelomates. Their muscle cells function in maintaining body shape and assisting in feeding movements. Some muscle fibers are oriented in a spokelike radial pattern and run between the gastrovascular cavity and the outer epithelium. Other fibers encircle the body like latitude markers on a globe. Still other fibers loop through the mesoglea. Muscle cells are also associated with tentacles.

Pleurobranchia has a spherical or ovoid, transparent body about 2 cm in diameter. It occurs in the colder waters of the Atlantic and Pacific Oceans (figure 9.22*b*). *Pleurobranchia,* like most ctenophorans, has eight meridional bands of cilia, called **comb rows,**

WILDLIFE ALERT

Coral Reefs

Coral reefs are among the most threatened marine habitats. Along with tropical rain forests, they are among the most diverse ecosystems on the earth. Coral reefs comprise approximately 0.25% of marine habitats but, in spite of this small percentage, they are home to animals from at least 32 of the 37 extant animal phyla. About 25% of fish species are found on coral reefs, and approximately 100,000 species of reef invertebrates have been described (box figure 9.1). This diversity gives coral reefs tremendous intrinsic and economic value. Their highly productive waters yield 4 to 8 million tons of fishes for commercial fisheries. This is one-tenth of the world's total fish harvest, from an area that represents a small fraction of the ocean surface (box figure 9.2). Coral reefs attract billions of dollars' worth of tourist trade each year. It is estimated that coral reefs contribute $375 billion to the global economy each year. The ecological, aesthetic, and economic reasons for preserving coral reefs are overwhelming.

30°N

30°S

BOX FIGURE 9.2 Coral Reefs. Coral reefs are found throughout the tropical and subtropical regions of the world between 30°N and 30°S latitudes.

Disturbances of coral reefs can be devastating, because reefs grow very slowly. Normally a coral reef is alive with color. A disturbed reef turns white as a result of the death of anthozoan polyps, zooxanthellae (dinoflagellate protists that live in a mutualistic relationship with the anthozoans), and coralline algae (box figure 9.3). This bleaching reaction of a coral reef, if it results from a local disturbance, can be reversed rather quickly. Large-scale disturbances, however, can result in the death of large expanses of coral reef, which requires thousands of years to recover. In recent years, massive bleaching has been reported in tropical waters of the Atlantic, Caribbean, Pacific, and Indian Oceans.

Reefs require clean, constantly warm, shallow water to support the growth of zooxanthellae, which sustain coral anthozoans. Changing water levels, water temperature, and turbidity can adversely affect reef growth. Sedimentation from mining, dredging, and logging, or clearing mangrove swamps that trap sediment from coastal run-off, can block sunlight and result in the death of zooxanthellae. Some island communities mine coral reefs to extract limestone for concrete. Coastal development results in sewage and industrial pollution, which have damaged coral reefs. Oil spills are toxic to coral organisms. Ships that run aground damage large sections of coral reefs. Altered ecological relationships have resulted in the proliferation of the crown-of-thorns sea star (*Acanthaster planci*), which feeds on coral polyps and devastates reef communities of the South Pacific. The introduction of lionfish (*Pterois volitans*) has created a serious threat to Atlantic and Caribbean reef systems (*see figure 18.1 and page 330*). Snorklers and scuba divers who walk across reef surfaces, break off pieces of reef, or anchor their boats on

BOX FIGURE 9.1 A Coral Reef Ecosystem.
©McGraw-Hill Education/Barry Barker

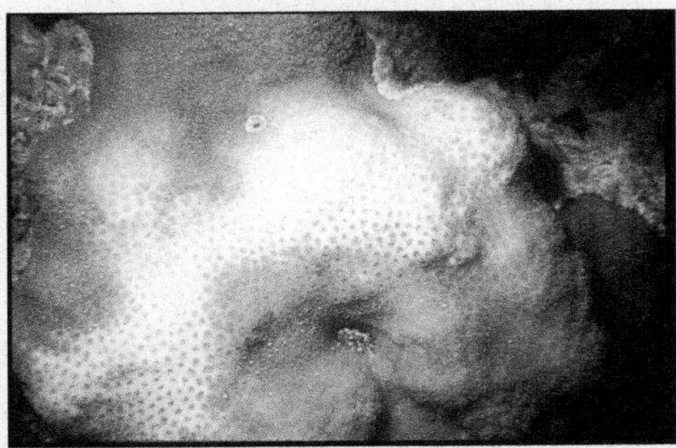

BOX FIGURE 9.3 Coral Bleaching. The bleached portion of the coral is shown in the lower portion of this photograph. The polyps in the upper portion of the photograph are still alive.
©Dr. Diane R. Nelson

WILDLIFE ALERT *Continued*

reefs similarly threaten reef life. Climate change (*see chapter 6*) has been implicated in reef bleaching. The results of global warming—changing water temperature, changing water levels, and increased frequency of tropical storms—have the potential to damage coral reefs by altering environmental conditions favorable for reef survival and growth.

Overfishing of large species such as sharks, turtles, and groupers has occurred in about 80% of Earth's reef ecosystems. Reductions in these populations allow smaller prey fish and invertebrate species to expand. This expansion can result in changes in algal communities important to reef ecosystems.

The threats to coral reefs seem almost overwhelming. Approximately 27% of our reef ecosystems are lost. At current rates of destruction we can expect another 60% to be gone within 30 years. Fortunately, biologists are finding that coral reefs are resilient ecosystems. If water quality is good, coral reefs can recover from local disturbances. National and international policies are needed that will prevent disturbances, including those from coastal sources; manage coral reefs as resources; and curtail greenhouse emissions that result in global warming. The Great Barrier Reef Marine Park off the Australian coast is the largest reef preserve in the world. It includes 2,900 reef formations and is managed by the Great Barrier Reef Marine Park Authority (GBRMPA). The goal of this management is to regulate the use of the reef. GBRMPA monitors water quality and coastal development with the purpose of conserving the reef biodiversity while sustaining commercial, educational, and recreational uses. Although other reef systems would require other approaches to management, the preservation steps taken by the GBRMPA should serve as a model for managing other reefs.

between the oral and aboral poles. Comb rows are locomotor structures that are coordinated through a statocyst at the aboral pole and a subepidermal nerve net that is superficially similar to that found in members of the Cnidaria. *Pleurobranchia* normally swims with its aboral pole oriented downward. The statocyst detects tilting, and the comb rows adjust the animal's orientation. Two long, branched tentacles arise from pouches near the aboral pole. Tentacles possess contractile fibers that retract the tentacles, and adhesive cells, called **colloblasts,** that capture prey (figure 9.22*c*).

Ingestion occurs as the tentacles wipe the prey across the mouth. The mouth leads to a branched gastrovascular canal system. Some canals are blind; however, two small anal canals open to the outside near the apical sense organ. Thus, unlike the cnidarians, ctenophores have an anal opening. Some undigested wastes are eliminated through these canals, and some are probably also eliminated through the mouth (*see figure 9.22*b).

Pleurobranchia is monoecious, as are all ctenophores. Two bandlike gonads are associated with the gastrodermis. One of these is an ovary, and the other is a testis. Gametes are shed through the mouth, fertilization is external, and a slightly flattened larva develops.

Section 9.4 Thinking Beyond the Facts

What anatomical features indicate that ctenophorans are triploblastic?

9.5 FURTHER PHYLOGENETIC CONSIDERATIONS

LEARNING OUTCOMES

1. Assess the statement that members of the phylum Ctenophora may comprise a sister group to all other animals.
2. Assess the evolutionary pressures that influenced the evolution of sponge body forms.
3. Describe evolutionary relationships within the Cnidaria.

The evolutionary relationships of the phyla covered in this chapter are subject to debate. As described in chapter 8, the Apoikozoa is a clade of opisthokonts that includes the choanoflagellates and animal phyla. The morphological and molecular similarities between choanoflagellates and the sponges are very convincing evidence of the basal position of ancestral poriferans within Animalia. A recent alternative to this conclusion is the presentation of evidence that ctenophorans may be the basal animal group. This conclusion is based on the fact that ctenophorans lack the *Hox* genes that control patterning along the body axis of virtually all other animals. In addition, the genes that control the formation of muscle (and perhaps mesoderm) in ctenophorans are different enough from similar genes in other animals that they may have had a separate evolutionary origin. Finally, the genes that control the formation and function of the ctenophoran nervous elements are unique. This fact suggests that the nervous elements of ctenophorans and those of cnidarians and bilaterians evolved in a parallel fashion rather than as a matter of descent. All of this suggests that the ctenophorans comprise a sister taxon to all other animals (*see appendix A*). This conclusion is debated. If it is true, complex animal characters like mesodermal tissues, muscles, and nervous elements would have evolved more than one time in animals (once in Ctenophora and once in all other animals), which seems unlikely to many zoologists. This debate is likely to continue for years to come.

Because of the presence of their skeletal elements, sponges are represented in the oldest fossil deposits—the Ediacaran formation (*see chapter 8*). Increased size would have selected for increased complexity of cell types and the unique system of water canals and chambers, which is the most important synapomorphy of sponges. The increased surface-to-volume ratio found in the syconoid and leuconoid body forms probably evolved in response to these selection pressures. Evolutionary relationships between poriferan classes remain very controversial, and no depiction of the relationships is attempted here.

Members of the phylum Cnidaria arose very early and are also represented in the Ediacaran formation. They are traditionally

TABLE 9.4
LESSER KNOWN BASAL PHYLA

PLACOZOA

Example	*Trichoplax adhaerens* *T. adhaerens* may comprise eight morphologically similar species.
Description	The Placozoa (means "flat animals") are very simple nonparasitic animals. *Trichoplax* is about 1 mm in diameter and, like an amoeba, it has no regular outline. The outer body consists of (dorsal) simple epithelium, which encloses a loose sheet of stellate cells resembling the mesenchyme of more complex animals. The ventral epithelium is composed of monocilated cells and nonciliated glandular cells. The animal uses the ciliated cells to creep along the seafloor. It feeds by secreting digestive enzymes into organic detritus and absorbing products of digestion. *Trichoplax* reproduces asexually, budding off smaller individuals, and the lower surface may also bud eggs into the mesenchyme.
Phylogenetic Relationships	There is no convincing fossil record of the Placozoa. The presence of extracellular matrix genes suggests that placozoans might be considered diploblastic. Placozoan organization may be a result of secondary loss. Molecular data suggest placozoans are closely related to sponges and the choanoflagellates.

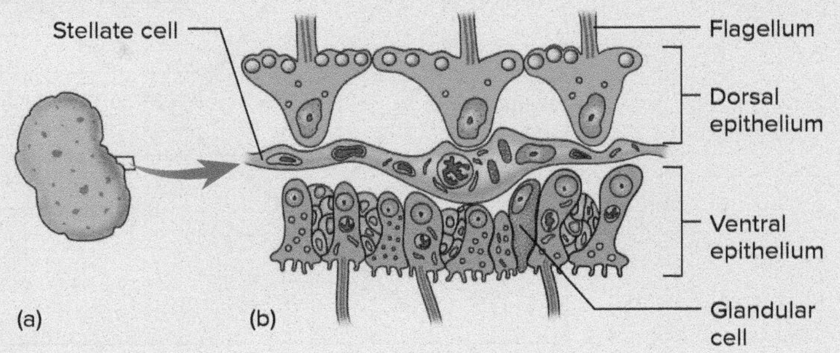

TABLE 9.4 FIGURE 1

Phylum Placozoa. (*a*) Drawing of *T. adhaerens.* (*b*) Drawing of a section through *T. adhaerens* illustrating the histological structure of this platelike animal.

ACOELOMORPHA

Example	*Waminoa* Approximately 350 species.
Description	Acoelomorphs are triploblastic and acoelomate. Acoelomorphs are less than 5 mm in length and are mostly free-living in marine sediments. They have a simple pharynx, incomplete gut, and mesodermal muscle cells. Sense organs include statocysts and ocelli. Reproduction is either by asexual fragmentation or by internal fertilization by monoecious pairs. No excretory or respiratory structures are present but these animals do have a radial arrangement of nerves in their elongated body.
Phylogenetic Relationships	The phylogenetic position of Acoelomorpha is controversial. Acoels have only four or five *Hox* genes, which suggests a basal position among bilaterally symmetrical triploblasts. Some zoologists consider the group to be a superphylum, which includes multiple clades. Other zoologists cite molecular studies and have concluded that Xenoturbellida (*see table 16.2*) and Acoelomorpha should be united into a new basally positioned phylum, Xenacoelomorpha.

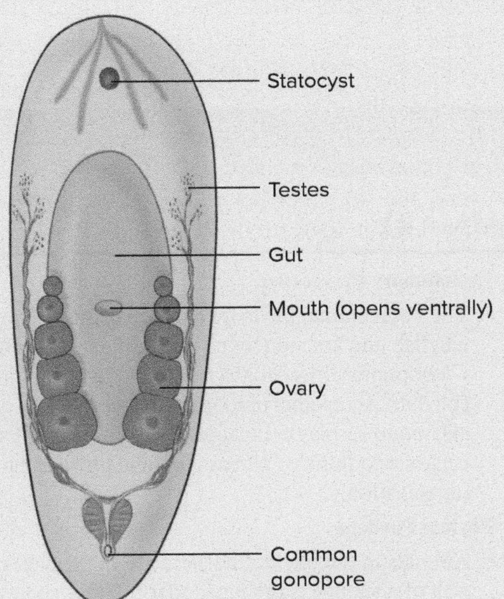

TABLE 9.4 FIGURE 2

Phylum Acoelmorpha. A drawing showing the generalized anatomy.

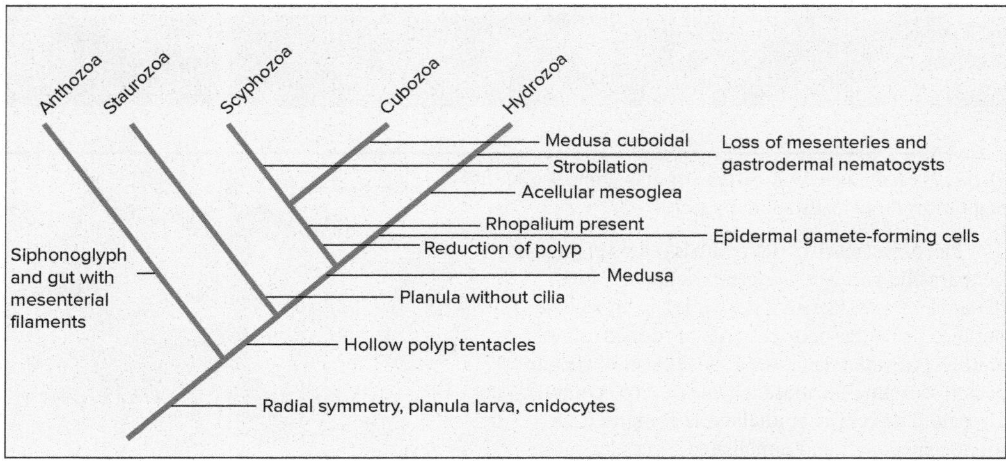

FIGURE 9.23

Cladogram Showing Cnidarian Taxonomy. Selected synapomorphic characters are shown. Most zoologists consider anthozoans to be ancestral to other cnidarians.

considered to have arisen from a radially symmetrical ancestor. Molecular data suggest that primitive anthozoans were closely related to the ancestral cnidarian stock. This interpretation is shown in figure 9.23. The Anthozoa are distinguished by the presence of the siphonoglyph and a gut with mesenterial filaments. All other cnidarians possess hollow polyp tentacles. The Staurozoa lack a medusa stage and have planula larvae without cilia. The Scyphozoa and Cubozoa are distinguished from the Anthozoa by the evolutionary reduction of the polyp stage. The Hydrozoa are distinguished by the loss of mesenteries, the loss of nematocysts in the gastrodermis, and the presence of gamete-forming cells in the epidermis.

Two additional phyla are positioned phylogenetically alongside Ctenophora, Porifera, and Cnidaria as basal animal phyla. The Placozoa and Acoelomorpha are described briefly in table 9.4.

SECTION 9.5 THINKING BEYOND THE FACTS

What probably explains the evolution of increased complexity in sponge body forms?

SUMMARY

9.1 **Evolutionary Perspective**

- The basal animal phyla include the sponges (phylum Porifera), jellyfish and anemones (phylum Cnidaria), ctenophores (phylum Ctenophora), placozoans (phylum Placozoa), and acoelomorphs (phylum Acoelomorpha). Sponges have traditionally been considered to display cellular-level organization. It is probably more accurate to think of all basal animal phyla as having tissue-level organization.

9.2 **Phylum Porifera**

- Animals in the phylum Porifera are the sponges. The outer body wall of a sponge is the pinacoderm and is separated from the inner choanoderm by a collagenous mesohyl. Cells of sponges are specialized to create water currents, filter food, produce gametes, form skeletal elements, and line the sponge body wall.
- Sponges circulate water through their bodies to bring in food and oxygen and to carry away wastes and reproductive products. Evolution has resulted in most sponges having complex canal systems and large water-circulating capabilities.
- Filtered food particles are incorporated into food vacuoles at the base of the choanocyte's collar and distributed to the rest of the organism. Sponges lack nerve cells but some communication among cells occurs.
- Sponges are monoecious. Eggs and sperm are usually released from a sponge and flagellated planktonic larvae develop and mature into the adult body form. Asexual reproduction can occur through the production of gemmules.

9.3 **Phylum Cnidaria**

- Members of the phylum Cnidaria are radially symmetrical and possess diploblastic, tissue-level organization. Cells are specialized for food gathering, defense, contraction, coordination, digestion, and absorption. Contractile cells in the epidermis and gastrodermis of the body wall act on water confined within the gastrovascular cavity to create a hydrostatic compartment used to accomplish support and movement.

- Cnidarian life histories include an attached polyp stage, which is often asexual, and a medusa stage, which is dioecious and free-swimming.
- Hydrozoans differ from other cnidarians in having ectodermal gametes, mesoglea without mesenchyme cells, and nematocysts only in their epidermis. Most hydrozoans have well-developed polyp and medusa stages.
- Staurozoans lack a medusa stage and their planula larvae lack cilia. They have a goblet-shaped body and attach to seaweed or rocks of marine habitats.
- The class Scyphozoa contains the jellyfish. The polyp stage of scyphozoans is usually very small.
- Members of the class Cubozoa live in warm, tropical waters. Some possess dangerous nematocysts.
- The Anthozoa lack the medusa stage. They include sea anemones and corals.

9.4 **Phylum Ctenophora**

- Members of the phylum Ctenophora are biradially symmetrical and diploblastic or possibly triploblastic. They possess comb rows that are used in locomotion and tentacles with adhesive colloblasts that are used in prey capture. They are monoecious, with external fertilization resulting in the development of a ciliated larval stage.

9.5 **Further Phylogenetic Considerations**

- Poriferans are traditionally thought of as the basal animal phylum, having close ties to the choanoflagellates within the opisthokonts. New molecular evidence suggests that ancestral ctenophorans diverged from the animal lineage separate from other animals and represent a sister group to other animals. The Porifera may have evolved from ancestral choanoflagellate protists. Within the Cnidaria, the ancient anthozoans are accepted as the stock from which modern anthozoans and other cnidarians evolved.

Concept Review Questions

1. Strong support for monophyly of the animal kingdom comes from all of the following, except one. Select the exception.
   a. The presence of flagellated cells in all animals
   b. The presence of asters used in cell division in all animals
   c. The presence of bilateral symmetry in all animals
   d. Molecular data
2. Members of the phylum Porifera are characterized by all of the following, except one. Select the exception.
   a. Asymmetry or superficial radial symmetry
   b. Pinacocytes
   c. Mesenchyme cells
   d. Cnidocytes
3. In the sycon sponge body form, choanocytes line the
   a. spongocoel.
   b. gastrovascular cavity.
   c. radial canals.
   d. incurrent canals.
4. A medusa stage is found in at least some members of the
   a. Anthozoa, Staurozoa, and Hydrozoa.
   b. Anthozoa only.
   c. Scyphozoa, Hydrozoa, and Cubozoa.
   d. Scyphozoa, Staurozoa, and Cubozoa.
5. Members of the cnidarian class ________________ are probably most closely related to the ancestor of the phylum.
   a. Hydrozoa
   b. Staurozoa
   c. Scyphozoa
   d. Cubozoa
   e. Anthozoa

Analysis and Application Questions

1. Colonies are defined in chapter 7 as loose associations of independent cells. Why are sponges considered to have surpassed that level of organization?
2. Most sponges and sea anemones are monoecious, yet sexual reproduction does not typically involve self-fertilization. Why is this advantageous for these animals? What ensures that sea anemones do not self-fertilize?
3. The Cnidaria and the Ctenophora are often collectively referred to as the "Radiata." What is the basis for this term? How would you assess the phylogenetic significance of its use?
4. Do you think that the polyp stage or the medusa stage predominated in the ancestral cnidarian? Support your answer. What implications does your answer have when interpreting the evolutionary relationships among the cnidarian classes?

10

The Smaller Lophotrochozoan Phyla

This tiger flatworm (*Pseudoceros crozerri*) is feeding on an assemblage of nearly transparent sea squirts (phylum Chordata). Flatworms are members of the phylum Platyhelminthes, which is one of the six phyla within the clade Platyzoa. The clade Platyzoa and other smaller lophotrochozoan phyla are discussed in this chapter.
©Diane Nelson

Chapter Outline

10.1 Evolutionary Perspective
10.2 Platyzoa: Phylum Platyhelminthes
Class Turbellaria
Class Trematoda
Class Monogenea
Class Cestoidea
10.3 Platyzoa: Smaller Phyla
Phylum Gastrotricha
Phylum Micrognathozoa
Phylum Gnathostomulida
Phylum Rotifera
Phylum Acanthocephala
10.4 Other Lophotrochozoans
Phylum Nemertea (Rhynchocoela)
Phylum Cycliophora
Phylum Ectoprocta (Bryozoa)
Phylum Brachiopoda
10.5 Further Phylogenetic Considerations

10.1 EVOLUTIONARY PERSPECTIVE

LEARNING OUTCOME

1. Assess the evidence that supports the establishment of Lophotrochozoa as a monophyletic lineage.

The animal phyla described in this chapter are the first members of the Bilateria (*see appendix A and chapter 8*; recall that the phylum Acoelomorpha was briefly described in table 9.4). As discussed in chapter 7 (*see figure 7.9*), bilateral symmetry is accompanied by cephalization and is characteristic of active, crawling or swimming animals. As you study this chapter, and those that follow, you will begin to see the many lifestyle changes that accompany this body form.

There is little disagreement among zoologists that the major animal linages are monophyletic. Disagreements that existed in the past regarding how animal phyla are related to one another are being settled by the use of morphological evidence and molecular evidence from analysis of ribosomal RNA, mitochondrial DNA, and nuclear DNA. As described in chapter 8, evidence supports the three major clades of bilaterally symmetrical animals: Lophotrochozoa (Spiralia), Ecdysozoa, and Deuterostomia (figure 10.1 and *see appendix A*). As zoologists continue to revise this tree, the overall message remains the same: the great diversity found in the animal kingdom arose through the process of evolution, and all animals exhibit features that give clues to their evolutionary history.

All of the bilateral tripoblastic animals discussed in this chapter are protostomes and lophotrochozoans. Recall from chapter 8 that protostome development is through spiral cleavage, and the mouth of the adult animal develops from the blastopore or from an opening near it (*see figure 8.11*). The lophotrochozoan phyla are covered in chapters 10 through 12. Lophotrochozoan relationships are indicated, not only by shared genetic makeup but also by the fact that many share either a horseshoe-shaped feeding structure called the **lophophore** (Gr. *lophos,* tuft, the "lopho" part of the name) or a larval form called the trochophore (Gr. *trochiscus,* a small wheel or disk, the "troch" part of the name) (figure 10.2). The ciliated lophophore will be discussed later in this chapter in connection with the bryozoans and brachiopods, and the trochophore will be noted in those phyla that possess this larval stage.

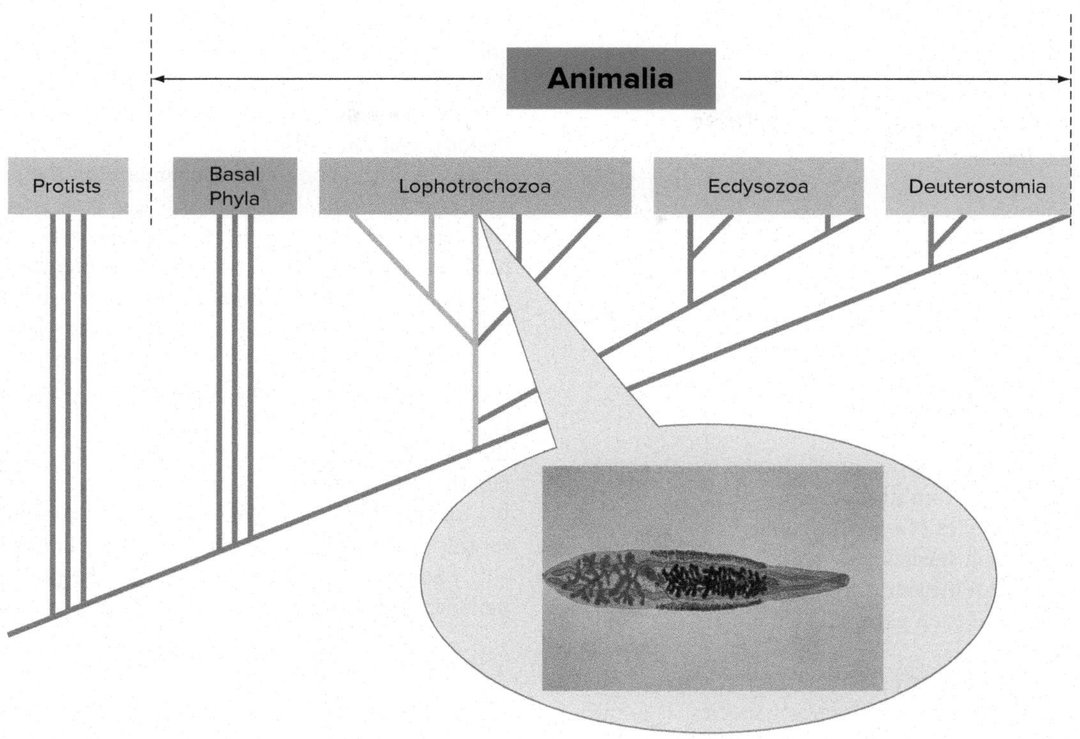

FIGURE 10.1

Lophotrochozoan Relationships. Ten lophotrochozoan phyla are covered in this chapter. Six of these phyla (including Platyhelminthes) comprise the clade Platyzoa. Two very large lophotrochozoan phyla, Mollusca and Annelida, will be discussed in chapters 11 and 12. The Chinese liver fluke (*Clonorchis sinensis*) is a member of the phylum Platyhelminthes.

©Biophoto Associates/Science Source

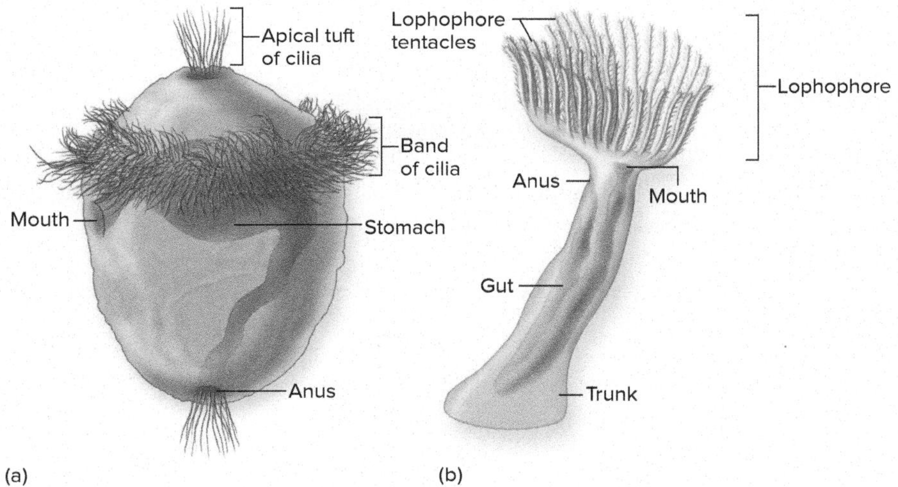

FIGURE 10.2

Lophotrochozoan Hallmarks. Many animals in the lophotrochozoan lineage have either a trochophore larval stage or a lophophore. (*a*) The trochophore larval form has a band of cilia around the middle of the larva that propels the larva through the water. (*b*) A lophophore is composed of a crown of cilated tentacles that generates water currents. Food particles are trapped by the tentacles and transported by cilia into the mouth.

This chapter covers some of the smaller lophotrochozoan phyla. Platyzoa is a traditional grouping of phyla: Platyhelminthes, Gastrotricha, Micrognathozoa, Gnathostomulida, Rotifera, and Acanthocephala, and these phyla will be discussed first. As will be discussed at the end of this chapter, some evidence suggests that Platyzoa may be paraphyletic. The discussion of the Platyzoa is followed by a description of four other lophotrochozoan phyla: Nemertea, Cycliophora, Ectoprocta, and Brachiopoda.

Section 10.1 Thinking Beyond the Facts

Some, but not all, lophotrochozoans possess a trochophore larval stage or a lophophore. Why do we include some animals that lack both of these features in this lineage?

10.2 Platyzoa: Phylum Platyhelminthes

Learning Outcomes

1. Describe characteristics of members of the phylum Platyhelminthes.
2. Assess the extent of the development of platyhelminth characteristics in each of the phylum's four classes.
3. Describe how the life cycles of symbiotic platyhelminths promote the survival of each species.
4. Propose possible control measures for platyhelminth parasites based on a knowledge of each parasite's life cycle.

The phylum Platyhelminthes (plat″e-hel-min′thez) (Gr. *platys,* flat + *helmins,* worm) contains over 34,000 animal species. It is difficult to characterize the phylum precisely because of the absence of synapomorphic characters (*see section 10.5*). Flatworms range in adult size from 1 mm or less to 25 m (*Taeniarhynchus saginatus; see figure 10.17*) in length. Their mesodermally derived tissues include a loose tissue called **parenchyma** (Gr. *parenck,* anything poured in beside) that fills spaces between other more specialized tissues, organs, and the body wall. Depending on the species, parenchyma may provide skeletal support, nutrient storage, motility, reserves of regenerative cells, transport of materials, structural interactions with other tissues, modifiable tissue for morphogenesis, oxygen storage, and perhaps other functions yet to be determined. This is the first phylum covered that has an organ-system level of organization—a significant evolutionary advancement over the tissue level of organization. The phylum is divided into four classes (table 10.1): (1) the Turbellaria consist of mostly free-living flatworms, whereas the (2) Monogenea, (3) Trematoda, and (4) Cestoidea contain species that engage in some form of symbiosis (*see chapter 6*). Turbellaria is a paraphyletic group. We will continue to use the class designation for this group until the taxonomy of these lineages is settled.

Some general characteristics of the phylum Platyhelminthes include the following:

1. Usually flattened dorsoventrally, triploblastic, acoelomate, bilaterally symmetrical
2. Unsegmented worms (members of the class Cestoidea are strobilated)

TABLE 10.1
CLASSIFICATION OF THE PLATYHELMINTHES*

Phylum Platyhelminthes (plat″e-hel-min′thez)
Flatworms; bilateral acoelomates. More than 34,000 species.

Class Turbellaria* (tur″bel-lar′e-ah)
Mostly free living and aquatic; external surface usually ciliated; predaceous; possess rhabdites, protrusible proboscis, frontal glands, and many mucous glands; mostly hermaphroditic. *Convoluta, Notoplana, Dugesia.* More than 3,000 species.

Class Monogenea (mon″oh-gen′e-ah)
Monogenetic flukes; mostly ectoparasites on vertebrates (usually on fishes; occasionally on turtles, frogs, copepods, squids); one life-cycle form in only one host; bear opisthaptor. *Disocotyle, Gyrodactylus, Polystoma*. About 1,100 species.

Class Trematoda (trem″ah-to′dah)
Trematodes; all are parasitic; several holdfast devices present; have complicated life cycles involving both sexual and asexual reproduction. More than 10,000 species.

Subclass Aspidogastrea (=Aspidobothrea)
Mostly endoparasites of molluscs; possess large opisthaptor; most lack an oral sucker. *Aspidogaster, Cotylaspis, Multicotyl.* About 32 species.

Subclass Digenea
Adults endoparasites in vertebrates; at least two different life-cycle forms in two or more hosts; have oral sucker and acetabulum. *Schistosoma, Fasciola, Clonorchis*. About 1,350 species.

Class Cestoidea (ses-toid′e-ah)
All parasitic with no digestive tract; have great reproductive potential; tapeworms. About 3,500 species.

Subclass Cestodaria
Body not subdivided into proglottids; larva in crustaceans, adult in fishes. *Amphilina, Gyrocotyle.* About 15 species.

Subclass Eucestoda
True tapeworms; body divided into scolex, neck, and strobila; strobila composed of many proglottids; both male and female reproductive systems in each proglottid; adults in digestive tract of vertebrates. *Proteocephalus, Taenia, Echinococcus, Taeniarhynchus; Diphyllobothrium*. About 1,000 species.

*Turbellaria is traditionally considered a class within Platyhelminthes. Recent molecular data indicate that it is a paraphyletic assemblage.

EVOLUTIONARY INSIGHTS

The Flatworms as Early Hunters

Since the first animals inhabited the surface of the Precambrian seabed, what would they have had available to them as potential food? The answer can only be unicellular or colonial protists, heterotrophic bacteria, and cyanobacteria.

The first animals were probably filter feeders, like modern sponges. Cnidarian and ctenophoran ancestors probably captured prey as these radial or biradial animals (or their prey) drifted with ocean currents. Extant free-living flatworms are mobile animals that attack, kill, and consume their prey—they are hunters or scavengers. Their earliest ancestors must have survived in a similar fashion and contributed to the burrows and tracks evidenced in Ediacaran fossil beds (see *chapter 8*).

The simplest surviving small flatworms, and presumably the ancestral forms, are and were consumers of protists and bacteria associated in some way with the bottom sediments and rocks. Protists and bacteria are abundant in marine sediments. Ediacaran seabeds were composed of dense microbial mats, but as animals diversified their activities reworked the microbial mats, and seabeds became a patchwork of different life-forms. This has had two fundamental repercussions on animal lifestyles. First, animals must be mobile in order to find new food supplies when local food has been exhausted. Second, a consumer of small, widely scattered microorganisms must itself be relatively small. The larger an animal is, the greater will be its metabolic requirements and the more food it must obtain per unit time. Thus, a large animal could not subsist on a diet of protists and bacteria because it could not find enough of them to eat.

Larger size is likely to offer selective advantages for three reasons: (1) larger animals tend to be able to produce more offspring than smaller animals; (2) a larger size can confer greater protection from consumption by other organisms; and (3) larger animals tend to be able to displace smaller animals from limited shared resources. It is not surprising that most extant flatworms are larger than the protists and bacteria-consuming species and that they have turned to hunting larger food items—other animals—rather than protists and bacteria.

3. Incomplete gut usually present (gut absent in Cestoidea)
4. Somewhat cephalized, with an anterior cerebral ganglion and often longitudinal nerve cords
5. Protonephridia serve in osmoregulation and excretion
6. Monoecious with internal fertilization; symbiotic species with complex life cycles
7. Nervous tissues form a pair of anterior ganglia and paired, interconnected longitudinal nerves

Class Turbellaria

Members of the class Turbellaria (tur″bel-lar′e-ah) (L. *turbellae,* a commotion + *aria,* like) are mostly free-living bottom dwellers in freshwater and marine environments, where they crawl on stones, sand, or vegetation. Turbellarians are named for the turbulence that their beating cilia create in the water. Over 3,000 species have been described. Turbellarians are predators and scavengers (*see Evolutionary Insights, above*). The few terrestrial turbellarians known live in the humid tropics and subtropics. Although most turbellarians are less than 1 cm long, the terrestrial, tropical ones may reach 60 cm in length. Coloration is mostly in shades of black, brown, and gray, although some groups display brightly colored patterns.

Body Wall

As in the Cnidaria, the ectodermal derivatives include an epidermis that is in direct contact with the environment (figure 10.3). Some epidermal cells are ciliated, and others contain microvilli. A basement membrane of connective tissue separates the epidermis from mesodermally derived tissues. An outer layer of circular muscle and an inner layer of longitudinal muscle lie beneath the basement membrane. Other muscles are located dorsoventrally and obliquely between the dorsal and ventral surfaces. Between the longitudinal muscles and the gastrodermis are the loosely organized parenchymal cells.

The innermost tissue layer is the endodermally derived gastrodermis. It consists of a single layer of cells that comprise the digestive cavity. The gastrodermis secretes enzymes that aid in digestion, and it absorbs the end products of digestion.

On the ventral surface of the body wall are several types of glandular cells of epidermal origin. **Rhabdites** are rodlike cells that swell and form a protective mucous sheath around the body, possibly in response to attempted predation or desiccation. **Adhesive glands** open to the epithelial surface and produce a chemical that attaches part of the turbellarian to a substrate. **Releaser glands** secrete a chemical that dissolves the attachment as needed.

Locomotion

Turbellarians were early bilaterally symmetrical animals. Bilateral symmetry is usually characteristic of animals with an active lifestyle—those that move from one locale to another. Turbellarians are primarily bottom dwellers that glide over the substrate. They move using both cilia and muscular undulations, with the muscular undulations being more important in their movement. Just beneath the epithelium are layers of muscle cells. The outer layer runs in a circular direction and the inner layer in a longitudinal direction. Muscles also run vertically and obliquely, making agile bending and twisting movements possible. (The dorsoventral muscles are essential for maintaining the flatness of flatworms. This flattened shape provides adequate surface area for diffusion of respiratory gases and metabolic wastes across body surfaces.) As they move, turbellarians lay down a sheet of mucus that aids in adhesion and helps

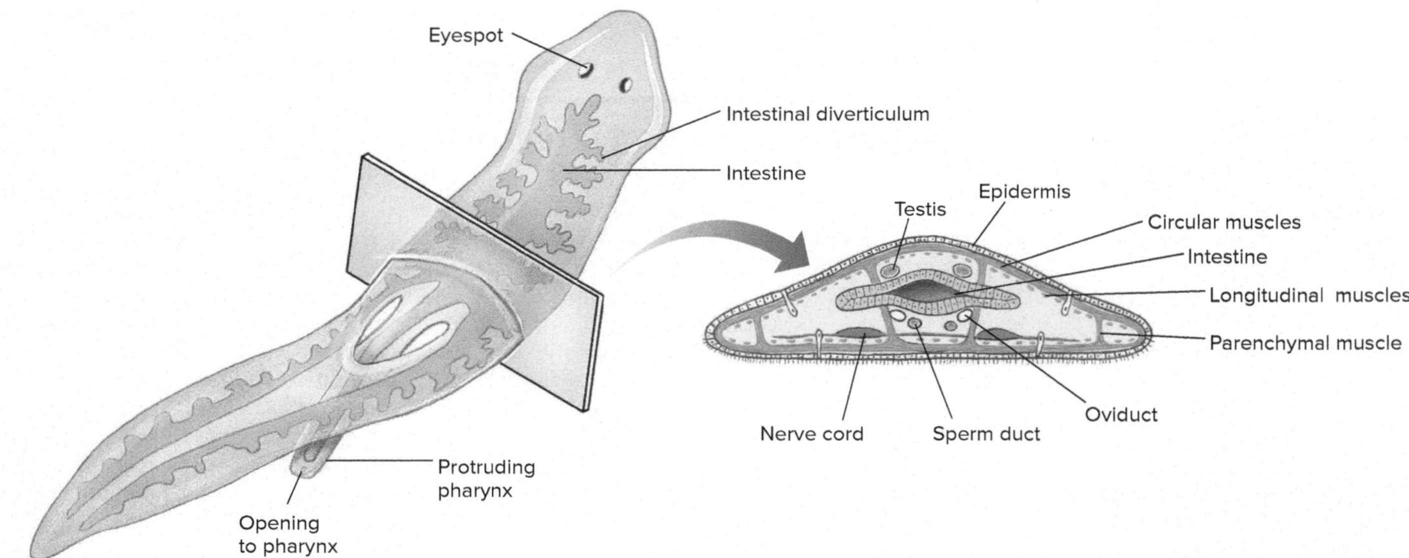

FIGURE 10.3

Phylum Platyhelminthes: Class Turbellaria. Cross section through the body wall of a sexually mature turbellarian (the planarian *Dugesia*), showing the relationships of the various body structures.

the cilia gain some traction. The densely ciliated ventral surface and the flattened bodies of turbellarians enhance the effectiveness of this locomotion. All of these movements thus result from two mechanisms: (1) The gliding is both muscular and ciliary. (2) The rapid movements pass from the head backward, propelling the animal forward, and are wholly muscular.

Digestion and Nutrition

The digestive tract of turbellarians is incomplete–it has a mouth opening but lacks an anus. This blind cavity varies from a simple, unbranched chamber (figure 10.4*a*) to a highly branched system of digestive tubes (figure 10.4*c* and *d*). Other turbellarians have digestive tracts that are lobed (figure 10.4*b*). Highly branched digestive systems result in more gastrodermis closer to the sites of digestion and absorption, reducing the distance nutrients must diffuse. This aspect of digestive tract structure is especially important in some of the larger turbellarians and partially compensates for the absence of a circulatory system.

The turbellarian pharynx functions as an ingestive organ. It varies in structure from a simple, ciliated tube to a complex organ developed from the folding of muscle layers. In the latter, the free end of the tube lies in a pharyngeal sheath and can project out of the mouth when feeding (figure 10.5).

Most turbellarians, such as the common planarian, are carnivores and feed on small, live invertebrates or scavenge on larger, dead animals; some are herbivores and feed on algae that they scrape from rocks. Sensory cells (chemoreceptors) on their heads help them detect food from a considerable distance.

Food digestion is partially extracellular. Pharyngeal glands secrete enzymes that help break down food into smaller units that can be taken into the pharynx. In the digestive cavity, phagocytic cells engulf small units of food, and digestion is completed in intracellular vesicles.

Exchanges with the Environment

The turbellarians do not have respiratory organs; thus, respiratory gases (CO_2 and O_2) are exchanged by diffusion through the body wall. Most metabolic wastes (e.g., ammonia) are also removed by diffusion through the body wall.

In marine environments, invertebrates are often in osmotic equilibrium with their environment. In freshwater, invertebrates are hypertonic to their aquatic environment and thus must regulate the osmotic concentration (water and ions) of their body tissues. The evolution of osmoregulatory structures in the form of protonephridia enabled turbellarians to invade freshwater.

Protonephridia (Gr. *protos,* first + *nephros,* kidney) (sing., protonephridium) are networks of fine tubules that run the length of the turbellarian, along each of its sides (figure 10.6*a*). Numerous, fine side branches of the tubules originate in the parenchyma as tiny enlargements called **flame cells** (figure 10.6*b*). Flame cells (so named because, in the living organism, they resemble a candle flame) have numerous cilia that project into the lumen of the tubule. Slitlike fenestrations (openings) perforate the tubule wall surrounding the flame cell. The beating of the cilia drives fluid down the tubule, creating a negative pressure in the tubule. As a result, fluid from the surrounding tissue is sucked through the fenestrations into the tubule. The tubules eventually merge and open to the outside of the body wall through a minute opening called a **nephridiopore.**

Nervous and Sensory Functions

Turbellarians have subepidermal nervous tissues. In some cases, nerves are netlike and fibers coalesce to form cerebral ganglia (figure 10.7*a*). The nervous tissues of most other turbellarians, such as the planarian *Dugesia,* consist of a subepidermal nerve net and several pairs of long nerve cords (figure 10.7*b*). Lateral

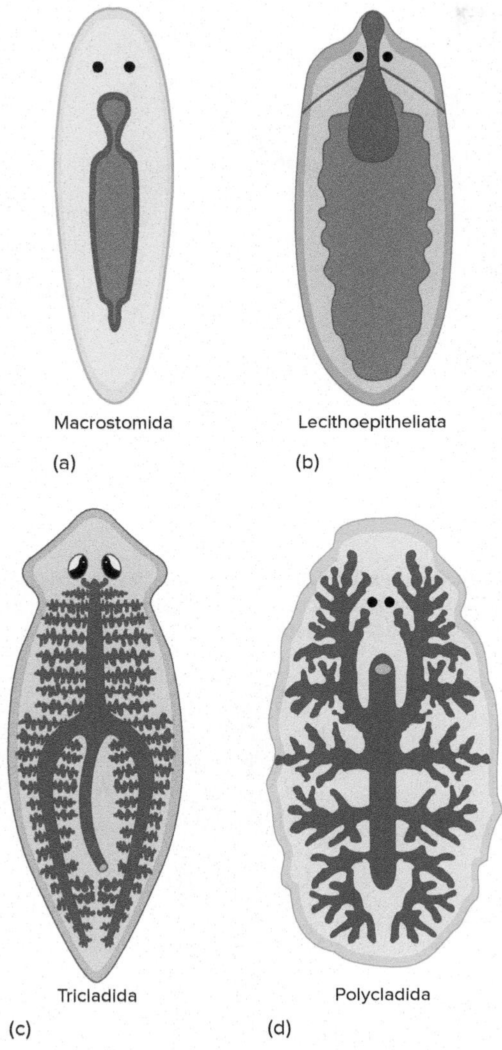

FIGURE 10.4

Digestive Systems in Some Turbellarians. (*a*) A simple pharynx and straight digestive cavity. (*b*) A simple pharynx and lobed digestive cavity. (*c*) A branched digestive cavity. (*d*) An extensively branched digestive cavity in which the branches reach almost all parts of the body.

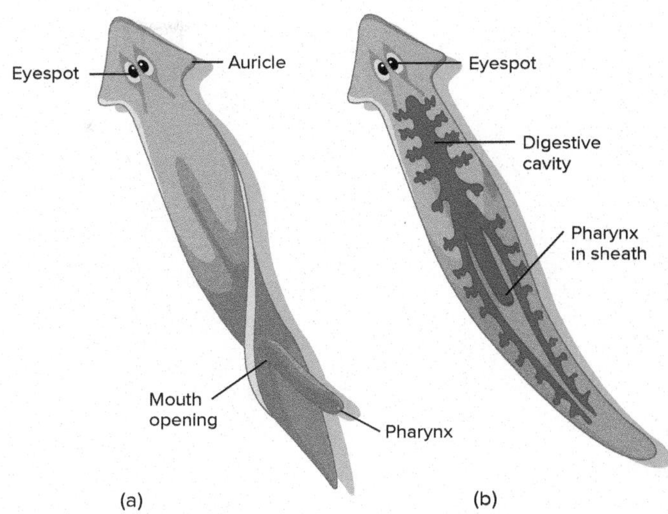

FIGURE 10.5

The Turbellarian Pharynx. A planarian turbellarian with its pharynx: (*a*) extended in the feeding position and (*b*) retracted within the pharyngeal sheath.

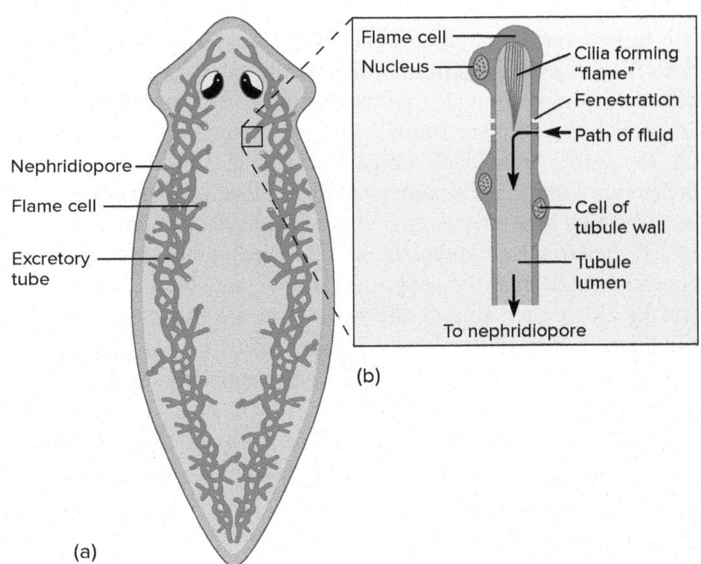

FIGURE 10.6

Protonephridial Excretion in a Turbellarian. (*a*) The protonephridial system lies in the parenchyma and consists of a network of fine tubules that run the length of the animal on each side and open to the surface by minute nephridiopores. (*b*) Numerous, fine side branches from the tubules originate in the parenchyma in enlargements called flame cells. The black arrows indicate the direction of fluid movement.

branches called commissures (points of union) connect the nerve cords. Nerve cords and their commissures give a ladder-like appearance to the turbellarian nervous organization. Neurons are organized into sensory (going to the primitive brain), motor (going away from the primitive brain), and association (connecting) types—an important evolutionary adaptation with respect to the nervous organization. Anteriorly, the nervous tissue concentrates into a pair of cerebral ganglia (sing., ganglion) called a primitive brain.

Turbellarians respond to a variety of stimuli in their external environment. Many tactile and sensory cells distributed over the body detect touch, water currents, and chemicals. **Auricles** (sensory lobes) may project from the side of the head (figure 10.7*b*). Chemoreceptors that aid in food location are especially dense in these auricles.

Most turbellarians have two simple eyespots called **ocelli** (sing., ocellus). These ocelli orient the animal to the direction of light. (Most turbellarians are negatively phototactic and move away from light.) Each ocellus consists of a cuplike depression lined with black pigment. Photoreceptor nerve endings in the cup are part of the neurons that leave the eye and connect with a cerebral ganglion.

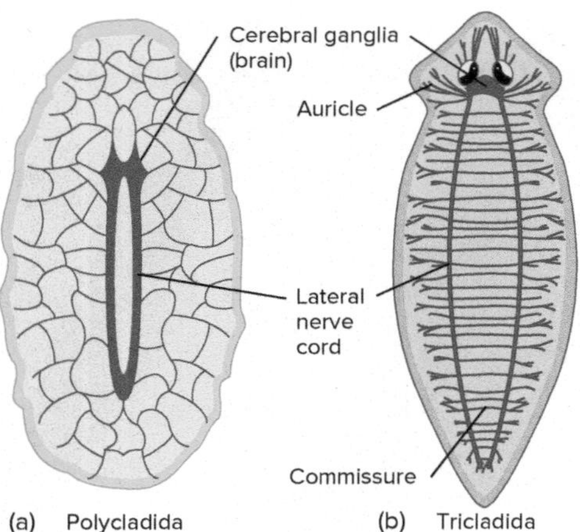

FIGURE 10.7

Nervous Organization in Two Orders of Turbellaria. (*a*) The nerve net in a turbellarian in the order Polycladida has cerebral ganglia and two lateral nerve cords. (*b*) The cerebral ganglia and nerve cords in the planarian *Dugesia.*

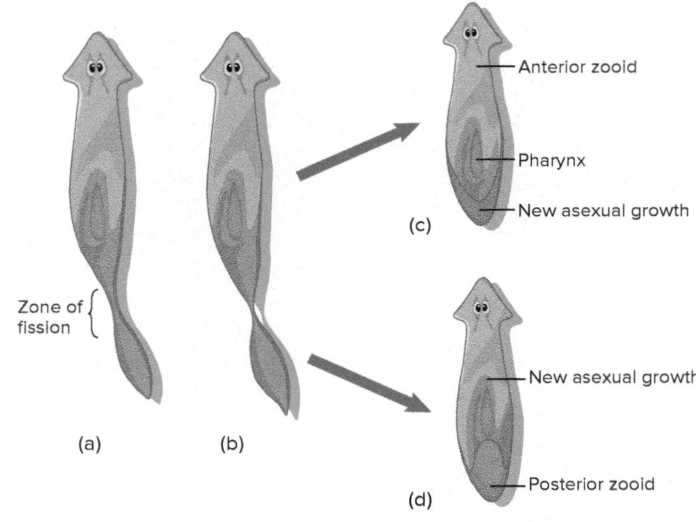

FIGURE 10.8

Asexual Reproduction in a Turbellarian. (*a*) Just before division and (*b*) just after. The posterior zooid soon develops a head, pharynx, and other structures. (*c, d*) Later development.

Reproduction and Development (Turbellaria)

Many turbellarians reproduce asexually by transverse fission. Fission usually begins as a constriction behind the pharynx (figure 10.8). The two (or more) animals that result from fission are called **zooids** (Gr., *zoon*, living being or animal), and they regenerate missing parts after separating from each other. Sometimes, the zooids remain attached until they have attained a fairly complete degree of development, at which time they detach as independent individuals.

Turbellarians are monoecious, and reproductive systems arise from the mesodermal tissues in the parenchyma. Numerous paired testes lie along each side of the worm and are the sites of sperm production. Sperm ducts (vas deferens) lead to a seminal vesicle (a sperm storage organ) and a protrusible penis (figure 10.9). The penis projects into a genital chamber.

The female system has one to many pairs of ovaries. Oviducts lead from the ovaries to the genital chamber, which opens to the outside through the genital pore.

Even though turbellarians are monoecious, reciprocal sperm exchange between two animals is usually the rule. This cross-fertilization ensures greater genetic diversity than does self-fertilization. During cross-fertilization, the penis of each individual is inserted into the copulatory sac of the partner. After copulation, sperm move from the copulatory sac to the genital chamber and then through the oviducts to the ovaries, where fertilization occurs. Yolk may either be directly incorporated into the egg during egg formation or yolk cells may be laid around the zygote as it passes down the female reproductive tract past the vitellaria (yolk glands).

Eggs are laid with or without a gel-like mass. A hard capsule called a **cocoon** (L., *coccum*, eggshell) encloses many turbellarian eggs. These cocoons attach to the substrate by a stalk and contain several embryos per capsule. Two kinds of capsules are laid. Summer capsules hatch in two to three weeks, and immature animals emerge. Autumn capsules have thick walls that can resist freezing and drying, and they hatch after overwintering.

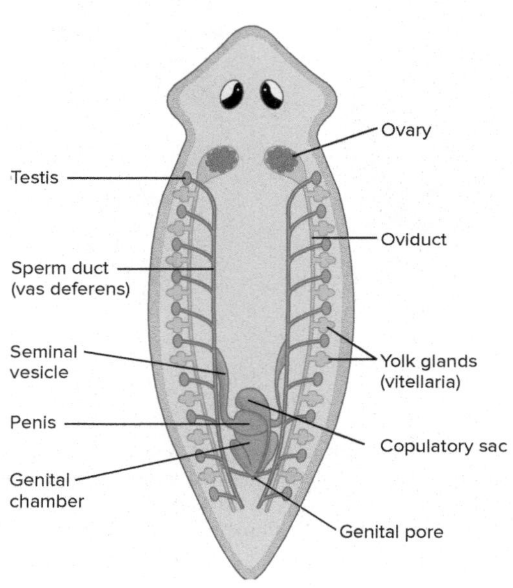

FIGURE 10.9

Triclad Turbellarian Reproductive System. Note that this single individual has both male and female reproductive organs.

Development of most turbellarians is direct–a gradual series of changes transforms embryos into adults. A few turbellarians have a free-swimming stage called a **Müller's larva.** It has ciliated extensions for feeding and locomotion. The larva eventually settles to the substrate and develops into a young turbellarian.

Class Trematoda

The approximately 10,000 species of parasitic flatworms in the class Trematoda (trem″ah-to′dah) (Gr. *trematodes,* perforated form) are collectively called **flukes,** which describes their wide, flat shape. Almost all adult flukes are parasites of vertebrates, whereas immature stages may be found in vertebrates or invertebrates, or encysted on plants. Many species are of great economic and medical importance.

Most flukes are flat and oval to elongate, and range from less than 1 mm to 6 cm in length (figure 10.10). They feed on host cells and cell fragments. The digestive tract includes a mouth and a muscular, pumping pharynx. Posterior to the pharynx, the digestive tract divides into two blind-ending, variously branched pouches called cecae (sing., cecum). Some flukes supplement their feeding by absorbing nutrients across their body walls.

Body-wall structure is similar for all flukes and represents an evolutionary adaptation to the parasitic way of life. The epidermis consists of an outer layer called the **tegument** (figure 10.11), which forms a syncytium (a continuous layer of fused cells). The outer zone of the tegument consists of an organic layer of proteins and carbohydrates called the glycocalyx. The glycocalyx aids in the transport of nutrients, wastes, and gases across the body wall, and protects the fluke against enzymes and the host's immune system. Also found in this zone are microvilli that facilitate nutrient exchange. Cytoplasmic bodies that contain the nuclei and most of the organelles lie below the basement membrane. Slender cell processes called cytoplasmic bridges connect the cytoplasmic bodies with the outer zone of the tegument.

There are two subclasses of trematodes. The subclass Aspidogastrea is a small group of flukes that are endoparasites of molluscs, and in some cases a second host may be a fish or turtle. The subclass Digenea contains the vast majority of flukes and will be covered in the following discussion.

Subclass Digenea

The flukes that comprise the subclass Digenea (Gr. *di,* two + *genea,* birth) include many medically important species. In this subclass, at least two different forms, an adult and one or more larval stages, develop—a characteristic from which the name of the subclass was derived. Because digenetic flukes require at least two different hosts to complete their life cycles, these animals possess the most complex life cycles in the entire animal kingdom. As adults, they are all endoparasites in the bloodstreams, digestive tracts, ducts of the digestive organs, or other visceral organs in a wide variety of vertebrates that serve as definitive, or final, hosts. The one or more intermediate hosts (the hosts that harbor immature stages) may harbor several different larval stages. The adhesive organs are two large suckers. The anterior sucker is the **oral sucker** and surrounds the mouth. The other sucker, the **acetabulum,** is located below the oral sucker on the middle portion of the body (*see figure 10.10*).

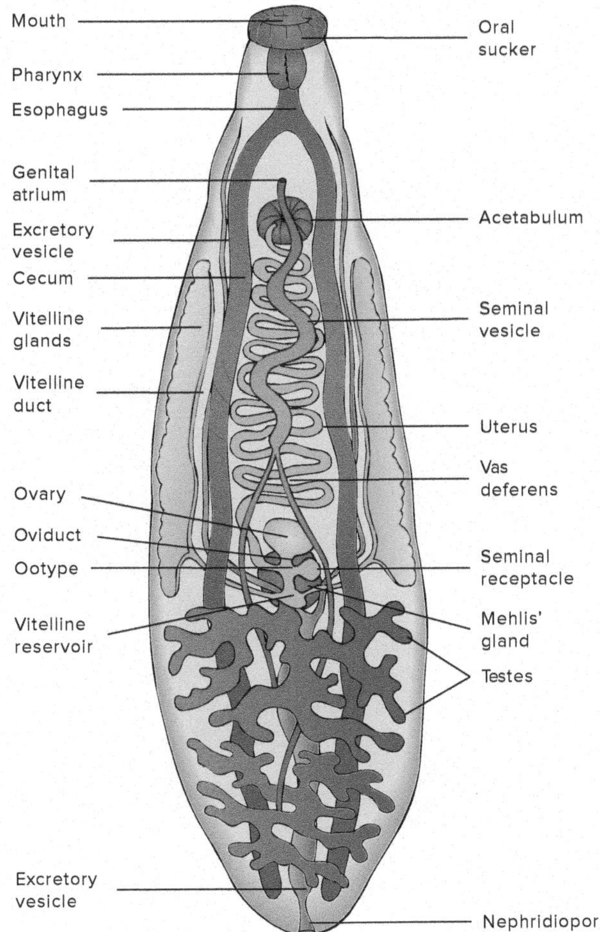

FIGURE 10.10

Generalized Fluke (Digenetic Trematode). Note the large percentage of the body devoted to reproduction. The Mehlis' gland is a conspicuous feature of the female reproductive tract; its function in trematodes is uncertain.

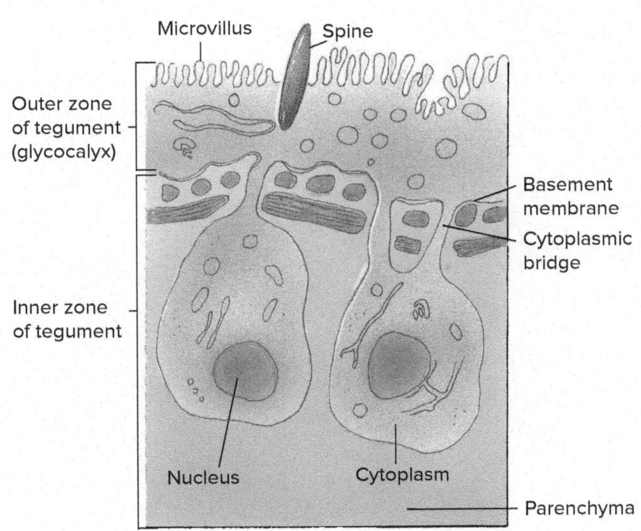

FIGURE 10.11

Trematode Tegument. The fine structure of the tegument of a fluke. The tegument is an evolutionary adaptation that is highly efficient at absorbing nutrients and effective for protection.

The eggs of digenetic trematodes are oval and usually have a lidlike hatch called an **operculum** (figure 10.12*a*). When an egg reaches freshwater, the operculum opens, and a ciliated larva called a **miracidium** (pl., miracidia) swims out (figure 10.12*b*). The miracidium swims until it finds a suitable first intermediate host (a snail) to which it is chemically attracted. The miracidium penetrates the snail, loses its cilia, and develops into a **sporocyst** (figure 10.12*c*). (Alternately, the miracidium may remain in the egg and hatch after a snail eats it.) Sporocysts are baglike and contain embryonic cells that develop into either **daughter sporocysts** or **rediae** (sing., redia) (figure 10.12*d*). At this point in the life cycle, asexual reproduction first occurs. From a single miracidium, hundreds of daughter sporocysts, and in turn, hundreds of rediae, can form by asexual reproduction. Embryonic cells in each daughter sporocyst or redia produce hundreds of the next larval stage, called **cercariae** (sing., cercaria) (figure 10.12*e*). (This phenomenon of producing many cercariae is called polyembryony. It greatly enhances the chances that one or two of these cercaria will further the life cycle.) A cercaria has a digestive tract, suckers, and a tail. Cercariae leave the snail and swim freely until they encounter a second intermediate or final host, which may be a vertebrate, invertebrate, or plant. The cercaria penetrates this host and encysts as a **metacercaria** (pl., metacercariae) (figure 10.12*f*). When the definitive host eats the second intermediate host, the metacercaria excysts and develops into an adult (figure 10.12*g*).

Some Important Trematode Parasites of Humans

The Chinese liver fluke *Clonorchis sinensis* is a common parasite of humans in Asia, where more than 30 million people are infected. The adult lives in the bile ducts of the liver, where it feeds on epithelial tissue and blood (figure 10.13*a*). The adults release embryonated eggs into the common bile duct. The eggs make their way to the intestine and are eliminated with feces (figure 10.13*b*). The miracidia are released when a snail ingests the eggs. Following the sporocyst and redial stages, cercariae emerge into the water. If a cercaria contacts a fish (the second intermediate host), it penetrates the epidermis of the fish, loses its tail, and encysts. The metacercaria develops into an adult in a human who eats raw or poorly cooked fish, a delicacy in Asian countries and gaining in popularity in the Western world (e.g., sushi, sashimi, ceviche).

Fasciola hepatica is called the sheep liver fluke (*see figure 10.12*a–g) because it is common in sheep-raising areas and uses

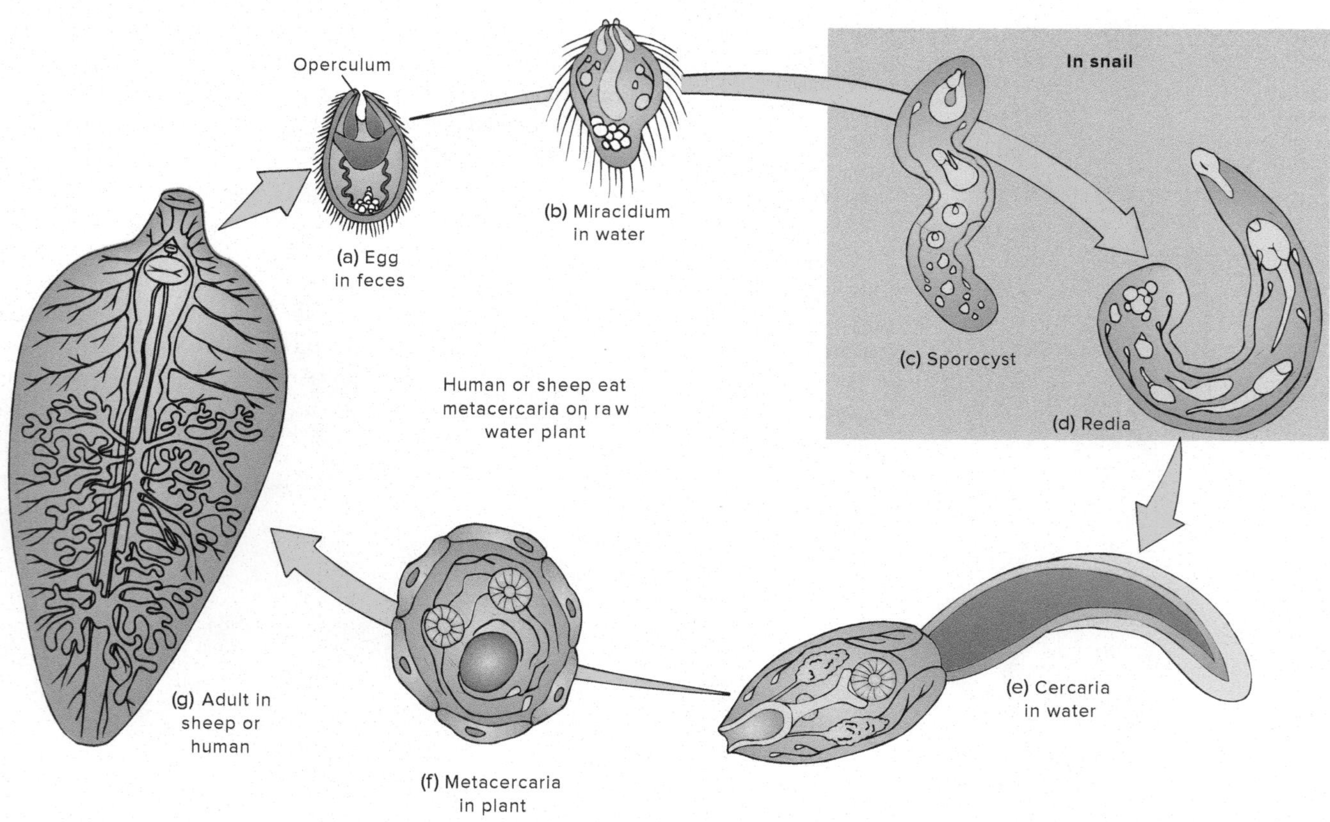

FIGURE 10.12

Class Trematoda: Subclass Digenea. The life cycle of the digenetic trematode *Fasciola hepatica* (the common liver fluke). The adult is about 30 mm long and 13 mm wide. The cercaria is about 0.5 mm long.

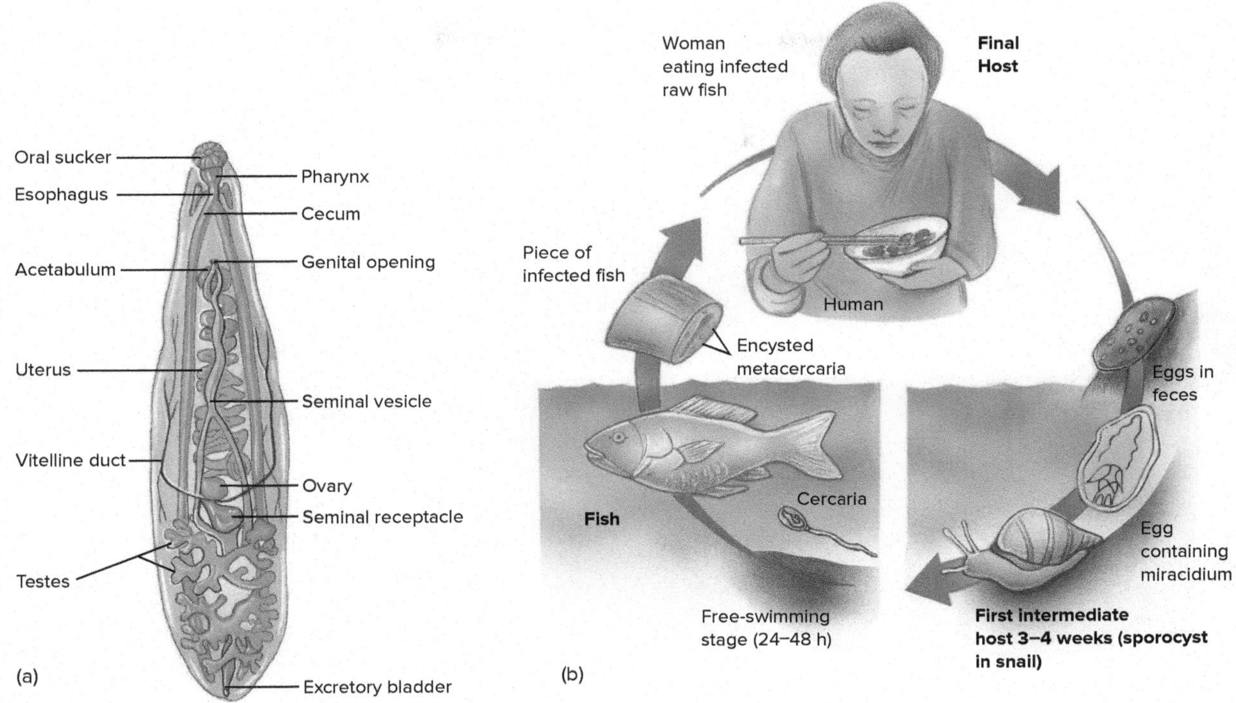

FIGURE 10.13
Chinese Liver Fluke *Clonorchis sinensis*. (*a*) Dorsal view. (*b*) Life cycle. The adult worm is 10 to 25 mm long and 1 to 5 mm wide (*see figure 10.1*).

sheep or humans as its definitive host. The adults live in the bile duct of the liver. Eggs pass via the common bile duct to the intestine, from which they are eliminated. Eggs deposited in freshwater hatch, and the miracidia must locate the proper species of snail. If a snail is found, miracidia penetrate the snail's soft tissue and develop into sporocysts that develop into rediae and give rise to cercariae. After the cercariae emerge from the snail, they encyst on aquatic vegetation. Sheep or other animals become infected when they graze on the aquatic vegetation. Humans may become infected with *Fasciola hepatica* by eating a freshwater plant called watercress that contains the encysted metacercaria.

Schistosomes are blood flukes with vast medical significance. The impact these flukes have had on history is second only to that of *Plasmodium (see appendix C)*. They infect more than 200 million people throughout the world. Infections are most common in Africa (*Schistosoma haematobium* and *S. mansoni*), South and Central America (*S. mansoni*), and Southeast Asia (*S. japonicum*). The adult dioecious worms live in the human bloodstream (figure 10.14*a*). The male fluke is shorter and thicker than the female, and the sides of the male body curve under to form a canal along the ventral surface (schistosoma means "split body"). The female fluke is long and slender and is carried in the canal of the male (figure 10.14*b*). Copulation is continuous, and the female produces thousands of eggs over her lifetime. Each egg contains a spine that mechanically aids it in moving through host tissue until it is eliminated in either the feces or urine (figure 10.14*c*). Unlike other flukes, schistosome eggs lack an operculum. The miracidium escapes through a slit that develops in the egg when the egg reaches freshwater (figure 10.14*d*). The miracidium seeks, via chemotaxis, a snail (figure 10.14*e*). The miracidium penetrates it, and develops into a sporocyst, then daughter sporocysts, and finally fork-tailed cercariae (figure 10.14*f*). There is no redial generation. The cercariae leave the snail and penetrate the skin of a human (figure 10.14*g*). Anterior glands that secrete digestive enzymes aid in penetration. Once in a human, the cercariae lose their tails and develop into adults in the intestinal veins, skipping the metacercaria stage.

Class Monogenea

Monogenetic flukes are so named because they have only one generation in their life cycle; that is, one adult develops from one egg. Monogeneans are mostly external parasites (ectoparasites) of freshwater and marine fishes, where they attach to the gill filaments and feed on epithelial cells, mucus, or blood. A large, posterior attachment organ called an **opisthaptor** facilitates attachment (figure 10.15). Adult monogeneans produce and release eggs that have one or more sticky threads that attach the eggs to the fish gill. Eventually, a ciliated larva called an **oncomiracidium** hatches from the egg and swims to another host fish, where it attaches by its opisthaptor and develops into an adult. Although monogeneans

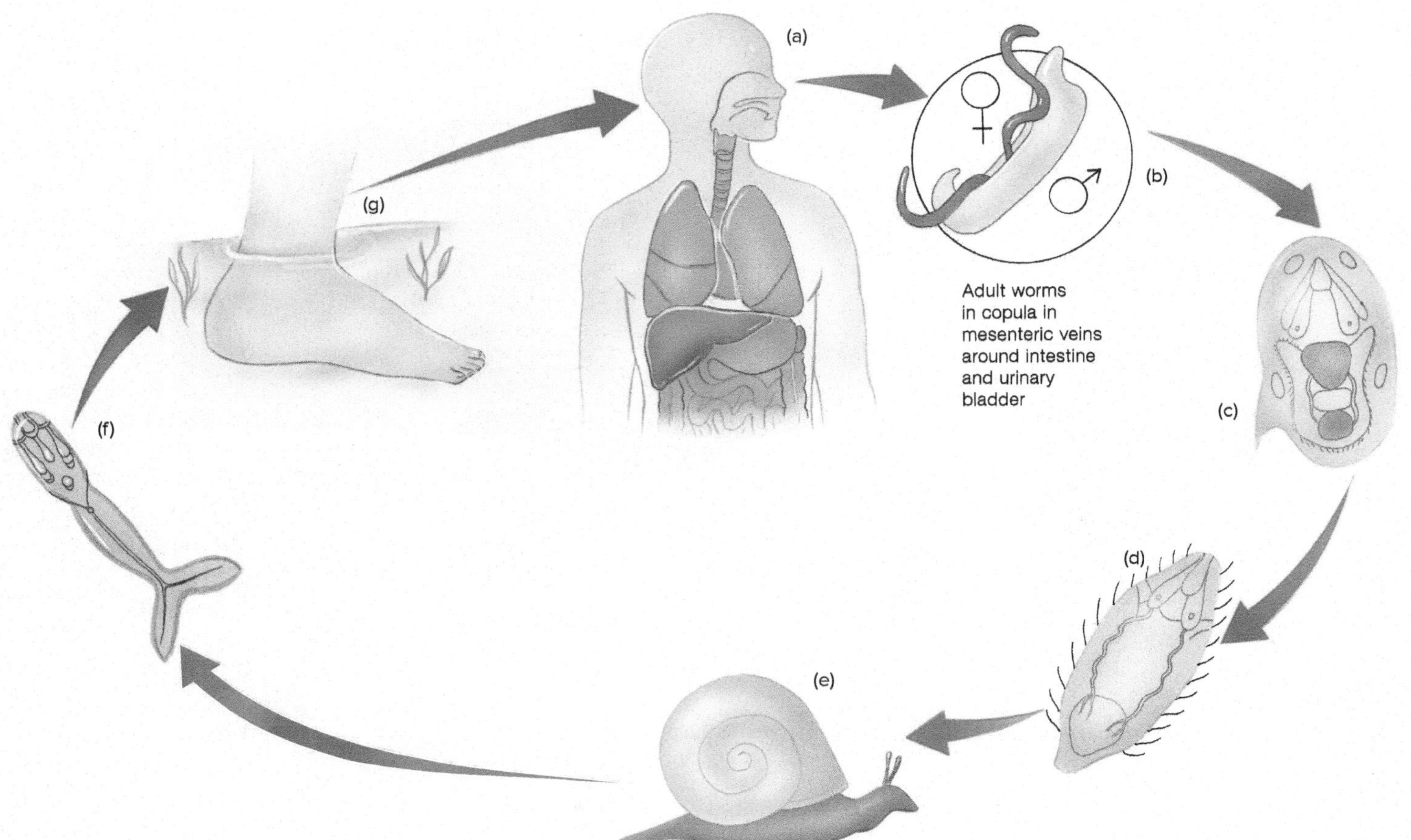

FIGURE 10.14

Life Cycle of a Schistosome Fluke. The cycle begins in a human (*a*) when the female fluke lays eggs (*b, c*) in the thin-walled, small vessels of the large or small intestine (*S. mansoni* and *S. japonicum*) or urinary bladder (*S. haematobium*). Secretions from the eggs weaken the walls, and the blood vessels rupture, releasing eggs into the intestinal lumen or urinary bladder. From there, the eggs leave the body. If they reach freshwater, the eggs hatch into ciliated, free-swimming larvae called miracidia (*d*). A miracidium burrows into the tissues of an aquatic snail (*e*), losing its cilia in the process, and develops into a sporocyst, then daughter sporocysts. Eventually, fork-tailed larvae (cercariae) are produced (*f*). After the cercariae leave the snail, they actively swim about. If they encounter human skin (*g*), they attach to it and release tissue-degrading enzymes. The larvae enter the body and migrate to the circulatory system, where they mature. They end up at the vessels of the intestines or urinary bladder, where sexual reproduction takes place, and the cycle begins anew. The adult worms are 10 to 20 mm long.

have been traditionally aligned with the trematodes, some structural and chemical evidence suggests that they are more closely related to tapeworms than to trematodes.

Class Cestoidea

The most highly specialized class of flatworms are members of the class Cestoidea (ses-toid'e-ah) (Gr. *kestos,* girdle + *eidos,* form), commonly called either tapeworms or cestodes. All of the approximately 3,500 species are endoparasites that usually reside in the vertebrate digestive system. Because they lack pigment as adults, their color is often white with shades of yellow or gray. Adult tapeworms range from 1 mm to 25 m in length.

Two unique adaptations to a parasitic lifestyle characterize tapeworms: (1) Tapeworms lack a mouth and digestive tract in all of their life-cycle stages; they absorb nutrients directly across their body wall. (2) Most adult tapeworms consist of a long series of repeating units called **proglottids.** Each proglottid contains one or two complete sets of reproductive structures.

As with most endoparasites, adult tapeworms live in a very stable environment. The vertebrate intestinal tract has very few environmental variations that would require the development of great anatomical or physiological complexity in any single tapeworm body system. The physiology of the tapeworm's host maintains the tapeworm's homeostasis (internal constancy). In adapting to such a specialized environment, tapeworms have lost some of the structures that were likely present in ancestral flatworms. Tapeworms are, therefore, a good example of evolution not always resulting in greater complexity.

There are two subclasses of tapeworms. The subclass Cestodaria contains about 15 species of fish parasites. The subclass Eucestoda contains medically important tapeworms.

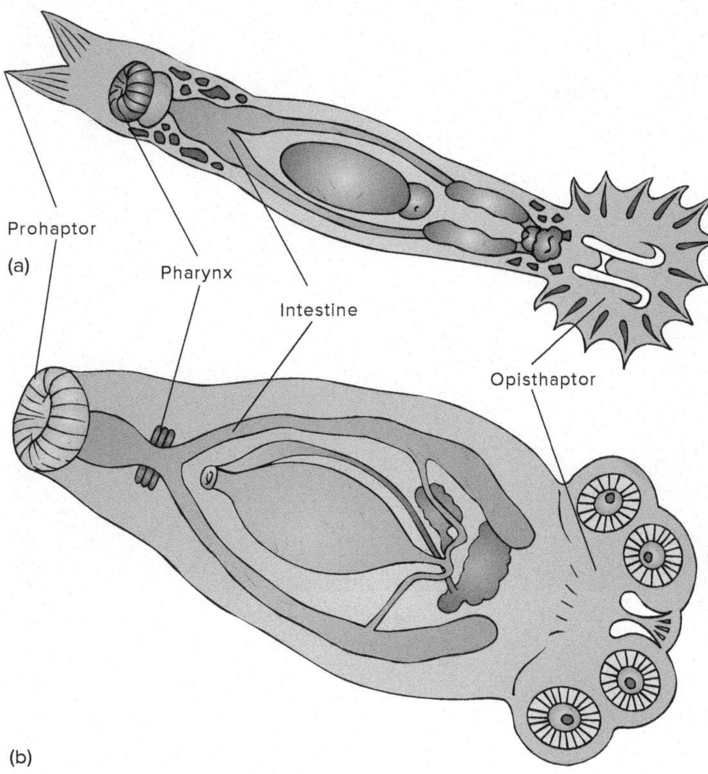

FIGURE 10.15
Class Monogenea. Two monogeneid trematodes. (*a*) *Gyrodactylus.* (*b*) *Sphyranura.* Note the opisthaptors by which these ectoparasites cling to the gills of their fish hosts. Both of these monogenea are about 1 cm long.

Subclass Eucestoda

Almost all of the cestodes belong to the subclass Eucestoda and are called true tapeworms. They represent the ultimate degree of specialization of any parasitic animal. The body is divided into three regions (figure 10.16*a*). At one end is a holdfast structure called the **scolex** that contains circular or leaflike suckers and sometimes a rostellum of hooks (figure 10.16*b*). With the scolex, the tapeworm firmly anchors itself to the intestinal wall of its definitive vertebrate host.

Although the scolex is not a "head," it narrows into what is often referred to as the "neck." Transverse constrictions in the neck give rise to the third body region, the **strobila** (Gr. *strobilus,* a linear series) (pl., strobilae). The strobila consists of a series of linearly arranged proglottids, which function primarily as reproductive units. As a tapeworm grows, new proglottids are added in the neck region, and older proglottids are gradually pushed away from the scolex, and proglottids mature and begin producing eggs. Thus, proglottids near the neck are said to be immature, those in the midregion of the strobila are mature, and those at the opposite end that have accumulated eggs are gravid (L., *gravida,* heavy, loaded, pregnant).

The outer body wall of tapeworms consists of a tegument similar in structure to that of trematodes (*see figure 10.11*). It plays a vital role in nutrient absorption because tapeworms have no digestive system. The tegument even absorbs some of the host's own enzymes to facilitate digestion.

With the exception of the reproductive systems, the body systems of tapeworms are reduced in structural complexity. The nervous system consists of only a pair of lateral nerve cords that arise from a nerve mass in the scolex and extend the length of the strobila. A protonephridial system also runs the length of the tapeworm (*see figure 10.6*).

Tapeworms are monoecious, and most of their physiology is devoted to producing large numbers of eggs. Each proglottid contains one or two complete sets of male and female reproductive organs (figure 10.16*a*). Numerous testes are scattered throughout the proglottid and deliver sperm via a duct system to a copulatory organ called a cirrus. The cirrus opens through a genital pore, which is an opening shared with the female system. The male system of a proglottid matures before the female system, so that copulation usually occurs with another mature proglottid of the same tapeworm or with another tapeworm in the same host. As previously mentioned, the avoidance of self-fertilization leads to hybrid vigor.

A single pair of ovaries in each proglottid produces eggs. Sperm stored in a seminal receptacle fertilize eggs as the eggs move through the oviduct. Vitelline cells from the vitelline gland are then dumped onto the eggs in the ootype. The ootype is an expanded region of the oviduct that shapes capsules around the eggs. The ootype is also surrounded by the Mehlis' gland, which aids in the formation of the egg capsule. Most tapeworms have a blind-ending uterus, where eggs accumulate (*see figure 10.16*a). As eggs accumulate, the reproductive organs degenerate; thus, gravid proglottids can be thought of as "bags of eggs." Eggs are released when gravid proglottids break free from the end of the tapeworm and pass from the host with the host's feces. In a few tapeworms, the uterus opens to the outside of the worm, and eggs are released into the host's intestine. Sometimes proglottids are not continuously lost, and some adult tapeworms usually become very long, such as the beef tapeworm (*Taeniarhynchus saginatus*).

Some Important Tapeworm Parasites of Humans

One medically important tapeworm of humans is the beef tapeworm *Taeniarhynchus saginatus* (figure 10.17). Adults live in the small intestine and may reach lengths of 25 m. About 80,000 eggs per proglottid are released as proglottids break free of the adult worm. As an egg develops, it forms a six-hooked (hexacanth) larva called the **oncosphere.** As cattle (the intermediate host) graze in pastures contaminated with human feces, they ingest oncospheres (or proglottids). Digestive enzymes of the cattle free the oncospheres, and the larvae use their hooks to bore through the intestinal wall into the bloodstream. The bloodstream carries the larvae to skeletal muscles, where they encyst and form a fluid-filled bladder called a **cysticercus** (pl., cysticerci) or **bladder worm.** When a human eats infected meat (termed "measly beef") that is raw or improperly cooked, the cysticercus is released from the meat, the scolex attaches to the human intestinal wall, and the tapeworm matures.

A closely related tapeworm, *Taenia solium* (the pork tapeworm), has a life cycle similar to that of *Taeniarhynchus saginatus,*

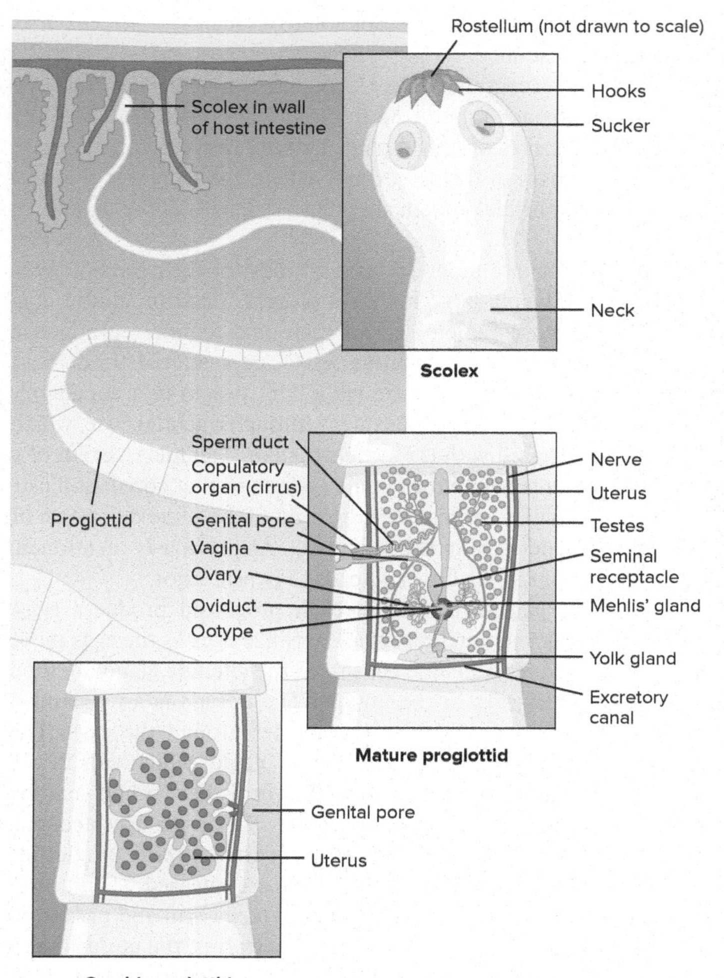

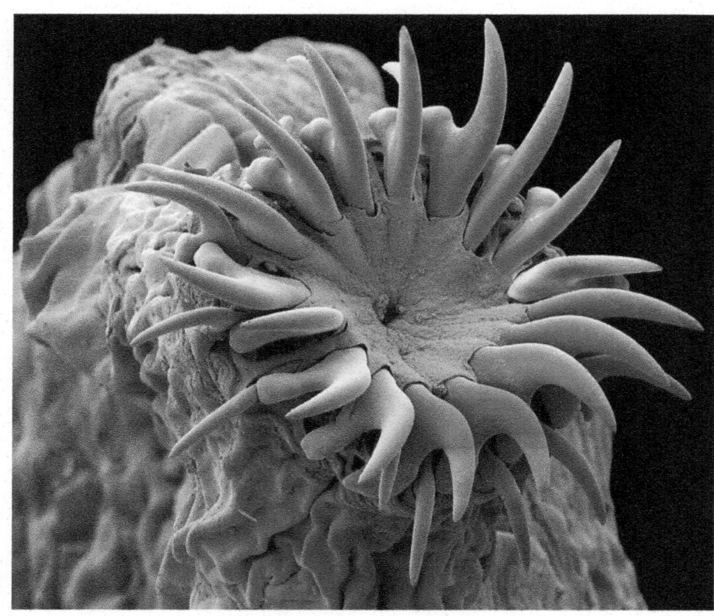

FIGURE 10.16

Class Cestoidea: A Tapeworm. (*a*) The scolex, neck, and proglottids of the pork tapeworm, *Taenia solium.* The adult worm attains a length of 2 to 7 m. Included is a detailed view of a mature proglottid with a complete set of male and female reproductive structures.
(*b*) The scolex of the cestode *Taenia solium* (SEM ×100). Notice the rostellum with two circles of 22 to 32 hooks.

except that the intermediate host is the pig. The strobila has been reported as being 10 m long, but 2 to 3 m is more common. The pathology is more serious in the human than in the pig. Gravid proglottids frequently release oncospheres before the proglottids have had a chance to leave the small intestine of the human host. When these larvae hatch, they move through the intestinal wall, enter the bloodstream, and are distributed throughout the body, where they eventually encyst in human tissue as cysticerci. The disease that results is called **cysticercosis** and can be fatal if the cysticerci encyst in the brain.

The broad fish tapeworm *Diphyllobothrium latum* is relatively common in the northern parts of North America, in the Great Lakes area of the United States, and throughout northern Europe. This tapeworm has a scolex with two longitudinal grooves (bothria; sing., bothrium) that act as holdfast structures (figure 10.18). The adult worm may attain a length of 10 m and shed up to a million eggs a day. Many proglottids release eggs through uterine pores. When eggs are deposited in freshwater, they hatch, and ciliated larvae called **coracidia** (sing., coracidium) emerge. These coracidia swim about until small crustaceans called copepods ingest them. The larvae shed their ciliated coats in the copepods and develop into **procercoid larvae.** When fish eat the copepods, the procercoids burrow into the muscle of the fish and become **plerocercoid larvae.** Larger fishes that eat smaller fishes become similarly infected with plerocercoids. When humans (or other carnivores) eat infected, raw, or poorly cooked fishes, the plerocercoids attach to the small intestine and grow into adult worms.

SECTION 10.2 THINKING BEYOND THE FACTS

Why is it important for an adult trematode or cestode to produce thousands of eggs?

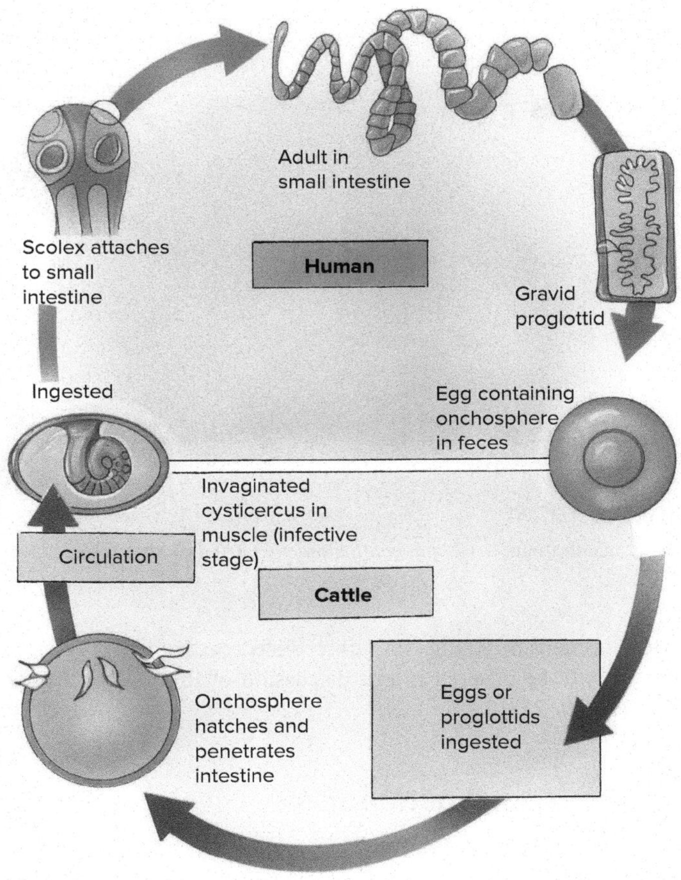

FIGURE 10.17

Life Cycle of the Beef Tapeworm *Taeniarhynchus saginatus*. Adult worms may attain a length of 25 m.
Source: [CDC] Centers for Disease Control (US). Atlanta, (GA).

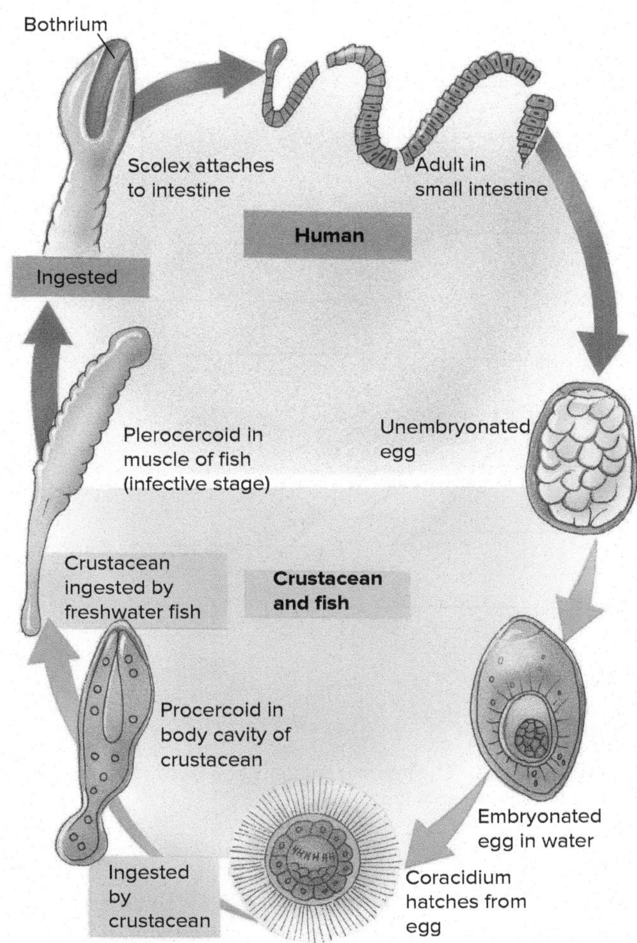

FIGURE 10.18

Life Cycle of the Broad Fish Tapeworm *Diphyllobothrium latum*. Adult worms may be 3 to 10 m long.
Source: [CDC] Centers for Disease Control (US). Atlanta, (GA)

10.3 PLATYZOA: SMALLER PHYLA

LEARNING OUTCOME

1. Describe one salient feature of each of the following smaller platyzoan phyla: Gastrotricha, Micrognathozoa, Gnathostomulida, Rotifera, and Acanthocephala.

In addition to the Platyhelminthes, the Platyzoa contains five other diverse phyla. None of the animals in this group of five is of great importance from the standpoint of human health or welfare, but each one is a fascinating testimonial to the diversity within the animal kingdom.

Phylum Gastrotricha

The gastrotrichs (gas-tro-tri′ks) (Gr. *gastros,* stomach + *trichos,* hair) are members of a small phylum of about 500 free-living marine and freshwater species that inhabit interstitial spaces (the space between ocean floor or lake bottom substrate particles). They range from 0.01 to 4 mm in length. Gastrotrichs use cilia on their ventral surface to move over the substrate. The phylum contains a single class divided into two orders.

The dorsal cuticle often contains scales, bristles, or spines, and a forked adhesive tube (tail) is often present (figure 10.19). A syncytial epidermis is beneath the cuticle. Sensory structures include tufts of long cilia and bristles on the rounded head. The nervous system includes a brain and a pair of lateral nerve trunks. The digestive system is a straight tube with a mouth, a muscular pharynx, a stomach-intestine, and an anus. The action of the pumping pharynx allows the ingestion of microorganisms and organic detritus from the bottom sediment and water. Digestion is mostly extracellular. Adhesive glands in the forked tail secrete materials that anchor the animal to solid objects. Paired protonephridia occur in freshwater species, rarely in marine ones. Gastrotrich protonephridia, however, are morphologically different from those found in other acoelomates. Each protonephridium possesses a single flagellum instead of the cilia found in flame cells.

Most of the marine species reproduce sexually and are monoecious. Most of the freshwater species reproduce asexually by

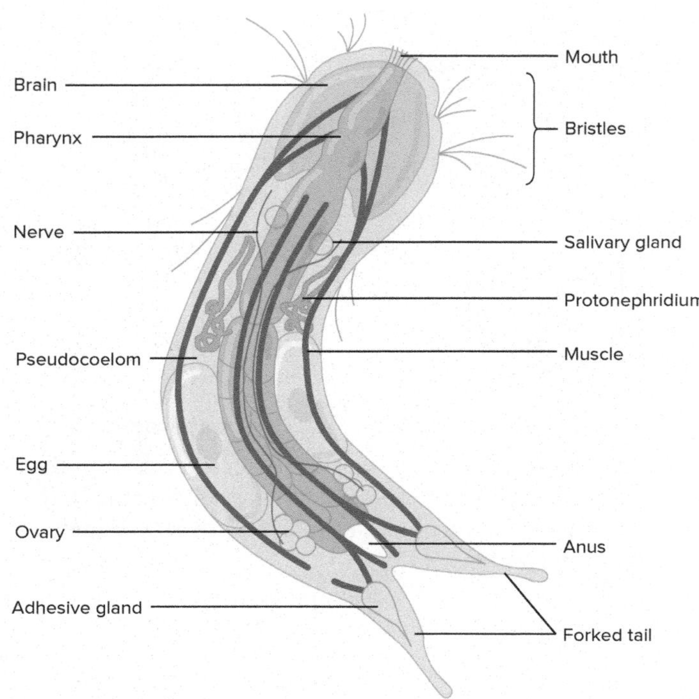

FIGURE 10.19

Phylum Gastrotricha. The internal anatomy of a freshwater gastrotrich. This animal is about 3 mm long.

parthenogenesis; the females can lay two kinds of unfertilized eggs. Thin-shelled eggs hatch into females during favorable environmental conditions, whereas thick-shelled resting eggs can withstand unfavorable conditions for long periods before hatching into females. There is no larval stage; development is direct, and the juveniles have the same form as the adults.

Phylum Micrognathozoa

Micrognathozoa (Gr. *micro,* small + *gnathos*, small jaws + *zoa,* animal) is a recently discovered (1994) animal phylum. *Limnognathia maerski* lives interstitially in cold, homeothermic (constant temperature) springs on Disko Island, Greenland. With an average length of one-tenth of a millimeter, they are one of the smallest animals.

The body of *L. maerski* is divided into three regions. The head is in two parts and has a complicated jaw system (figure 10.20). The jaws are composed of 15 separate elements, some of which are extended outside of its mouth while eating. It has a large ganglion, or "brain," and paired nerve cords extend along the lower side of the body. The head, thorax, and abdomen possess stiff sensory bristles that are made up of one to three cilia. Two rows of ciliated cells move the animal, and a ventral ciliated pad is adhesive in function.

Internally, there is a simple, complete gut. The anus opens to the outside only periodically. There are two pairs of protonephridia (*see figure 10.6*). It is likely that this species reproduces parthenogentically, as no males have been collected. **Parthenogenesis**

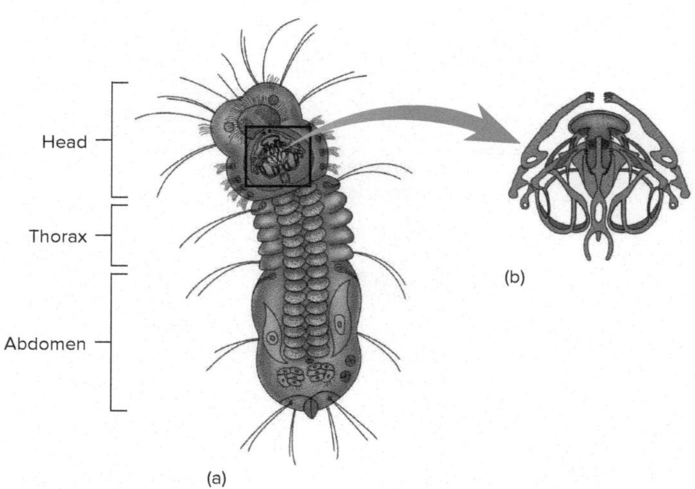

FIGURE 10.20

A Micrognathozoan. (*a*) *Limnognathia maerski.* (*b*) **Jaw structure.**

is development of young from unfertilized eggs, and a well-known example will be covered in our discussion of the phylum Rotifera (*see figure 10.23*).

Phylum Gnathostomulida

The phylum Gnathostomulida (Gr. *gnathos,* jaw + *stoma,* mouth + L. *ulus,* dim. suffix) is composed of over 100 species in 18 genera. Gnathostomulids are minute (less than 2 mm long), slender to threadlike animals (figure 10.21). They are commonly called "jaw worms" because of their unique jawed pharyngeal apparatus. Gnathostomulids are found interstitially in marine sands, often occurring in high densities in anoxic (low oxygen), sulfide-rich conditions. They occupy depths from the intertidal zone to more than 100 m. They have been found worldwide.

The elongated body of gnathostomulids is divided into head, trunk, and narrow tail regions. Gnathostomulids possess monociliated epithelial cells. The nervous system is composed of sensory cilia and ciliary pits located on the head. The gut is incomplete. Since gnathostomulids have no special excretory, circulatory, or gas-exchange structures, they probably depend largely on diffusion for circulation, excretion, and gas exchange.

Gnathostomulids are monoecious with cross-fertilization. In some cases, protandry occurs, in which male sex organs develop

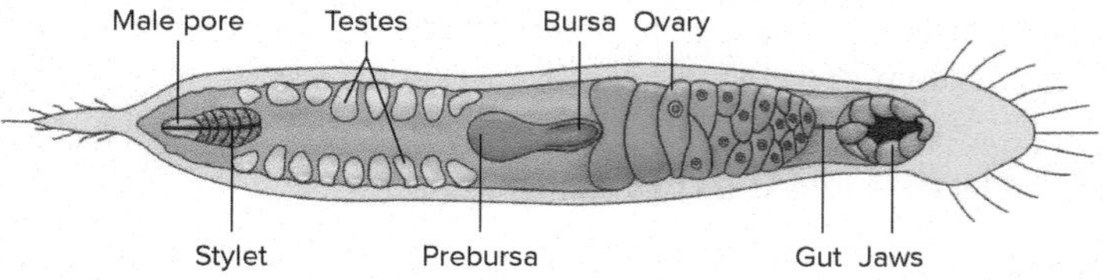

FIGURE 10.21

Phylum Gnathostomulida. Gnathostomulids are worldwide occupants of interstitial spaces of marine substrates. Their unique jaw apparatus is responsible for their common name "jaw worms."

and function before female sex organs appear. Each individual possesses a single ovary and one or two testes. Development is direct.

Phylum Rotifera

The Rotifera (ro'tif-er-ah) (L. *rota*, wheel + *fera*, to bear) are characterized by the presence of a ciliated organ called the **corona** (figure 10.22*a*). The ciliated lobes of the corona function in locomotion and food gathering. The cilia of the corona do not beat in synchrony; instead, each cilium is at a slightly earlier stage in the beat cycle than the next cilium in the sequence. A wave of beating cilia thus appears to pass around the periphery of the ciliated lobes and gives the impression of a pair of spinning wheels. (Interestingly, the rotifers were first called "wheel animalicules.")

Rotifers are small animals (0.1 to 3 mm in length) that are abundant in most freshwater habitats; a few (less than 10%) are marine. The approximately 2,000 species are traditionally divided into three classes (table 10.2). This taxonomy is under revision. The body has approximately 1,000 cells. Rotifers are usually solitary, free-swimming animals, although some colonial forms are known. Others occur between grains of sand.

Characteristics of the phylum Rotifera include the following:

1. Triploblastic, pseudocoelomate
2. Complete digestive system
3. Ciliated corona usually present
4. Toes with adhesive glands
5. Epidermally derived cuticle
6. Protonephridia with flame cells
7. Reproduction by parthenogenesis is common; males often small and seasonal

External Features

An epidermally secreted cuticle covers a rotifer's external surface. In many species, the cuticle thickens to form an encasement called a **lorica** (L. *corselet*, a loose-fitting case). The cuticle or lorica provides protection and is the main supportive element, although fluid in the pseudocoelom also provides hydrostatic support. The epidermis is syncytial; that is, no plasma membranes are between nuclei.

The head contains the corona, mouth, sensory organs, and brain (figure 10.22*b*). The corona surrounds a large, ciliated area called the buccal field. The trunk is the largest part of a rotifer, and is elongate

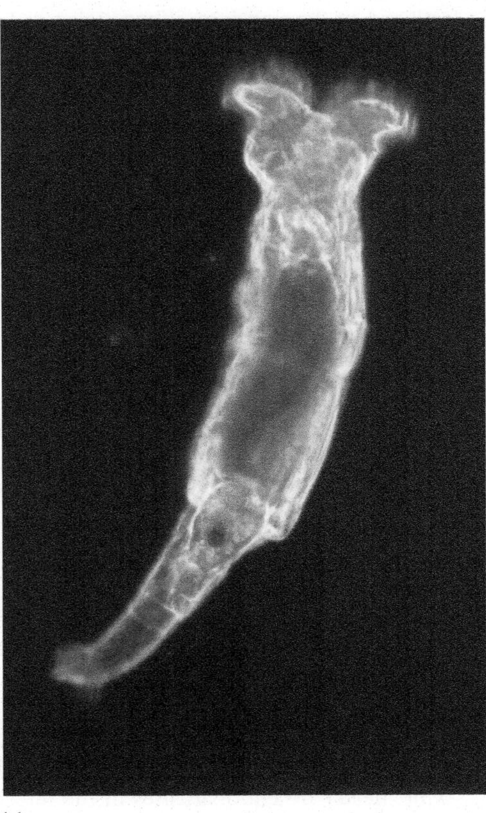

(a)

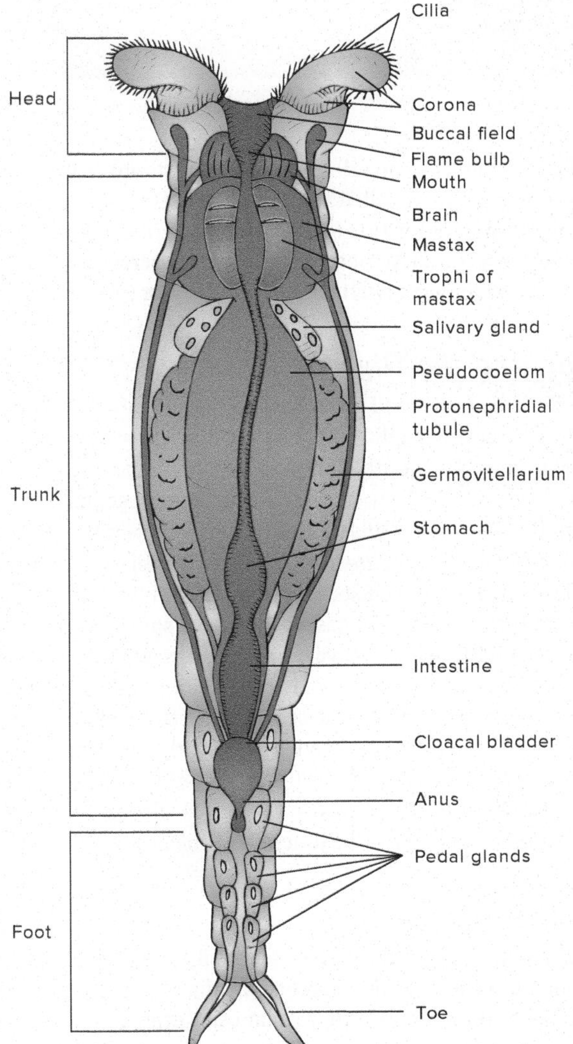

(b)

FIGURE 10.22

Phylum Rotifera. (*a*) A rotifer, *Brachionus* (LM 3150). (*b*) Internal anatomy of a typical rotifer, *Philodina*. This rotifer is about 2 mm long.

Table 10.2
CLASSIFICATION OF THE ROTIFERA

Phylum Rotifera (ro'tif-er-ah)
A ciliated corona surrounding a mouth; muscular pharynx (mastax) present with jawlike features; nonchitinous cuticle; parthenogenesis is common; both freshwater and marine species. About 2,000 species.

Class Seisonidea (sy"son-id'e-ah)*
A single genus of marine rotifers that are commensals of crustaceans; large and elongate body with reduced corona. *Seison.* Only two species.

Class Bdelloidea (del-oid'e-ah)
Anterior end retractile and bearing two trochal disks; mastax adapted for grinding; paired ovaries; cylindrical body; males absent. *Adineta, Philodina, Rotaria*. About 590 species.

Class Monogononta (mon"o-go-non'tah)
Rotifers with one ovary; mastax not adapted for grinding; produce mictic and amictic eggs. Males appear only sporadically. *Conochilus, Collotheca, Notommata*. About 1,400 species.

*Some authorities exclude Seisonidea from the Rotifera but include it within a clade encompassing Bdelloidea and Monogononta (true rotifers) and Acanthocephala (*see section 10.5*).

and saclike. The anus occurs dorsally on the posterior trunk. The posterior narrow portion is called the foot. The terminal portion of the foot usually bears one or two toes. At the base of the foot are many pedal glands whose ducts open on the toes. Secretions from these glands aid in the temporary attachment of the foot to a substratum.

Feeding and the Digestive System

Most rotifers feed on small microorganisms and suspended organic material. The coronal cilia create a current of water that brings food particles to the mouth. The pharynx contains a unique structure called the **mastax** (jaws). The mastax is a muscular organ that grinds food. The inner walls of the mastax contain several sets of jaws called trophi (*see figure 10.22*b). The trophi vary in morphological detail, and taxonomists use them to distinguish species.

From the mastax, food passes through a short, ciliated esophagus to the ciliated stomach. Salivary and digestive glands secrete digestive enzymes into the pharynx and stomach. Complete extracellular digestion and absorption of food occur in the stomach. In some species, a short, ciliated intestine extends posteriorly and becomes a cloacal bladder, which receives water from the protonephridia and eggs from the ovaries, as well as digestive waste. The cloacal bladder opens to the outside via an anus at the junction of the foot with the trunk.

Other Organ Systems

All visceral organs lie in a pseudocoelom filled with fluid and interconnecting amoeboid cells. Protonephridia that empty into the cloacal bladder function in osmoregulation. Rotifers, like other pseudocoelomates, exchange gases and dispose of nitrogenous wastes across body surfaces. The nervous system is composed of two lateral nerves and a bilobed, ganglionic brain on the dorsal surface of the mastax. Sensory structures include numerous ciliary clusters and sensory bristles concentrated on either one or more short antennae or the corona. One to five photosensitive eyespots may be on the head.

Reproduction and Development (Rotifera)

Some rotifers reproduce sexually, although several types of parthenogenesis occur in most species. Smaller males appear only sporadically in one class (Monogononta), and no males are known in another class (Bdelloidea). In the class Seisonidea, fully developed males and females are equally common in the population. Most rotifers have a single ovary and an attached syncytial vitellarium, which produces yolk that is incorporated into the eggs. The ovary and vitellarium often fuse to form a single germovitellarium (*see figure 10.22*b). After fertilization, each egg travels through a short oviduct to the cloacal bladder and out its opening.

In males, the mouth, cloacal bladder, and other digestive organs are either degenerate or absent. A single testis produces sperm that travel through a ciliated vas deferens to the gonopore. Male rotifers typically have an eversible penis that injects sperm, like a hypodermic needle, into the pseudocoelom of the female (hypodermic impregnation).

In one class (Seisonidea), the females produce haploid eggs that must be fertilized to develop into either males or females. In another class (Bdelloidea), all females are parthenogenetic and produce diploid eggs that hatch into diploid females. The third class (Monogononta) produces two different types of eggs (figure 10.23). **Amictic** (Gr. *a,* without + *miktos,* mixed or blended; thin-shelled summer eggs) **eggs** are produced by mitosis, are diploid, cannot be fertilized, and develop directly into amictic females. Thin-shelled, **mictic eggs** are haploid. If the mictic egg is not fertilized, it develops parthenogenetically into a male; if fertilized, mictic eggs secrete a thick, heavy shell and become dormant or resting winter eggs. Dormant eggs always hatch with melting snows and spring rains into amictic females, which begin a first amictic cycle, building up large populations quickly. By early summer, some females have begun to produce mictic eggs, males appear, and dormant eggs are produced. Another amictic cycle, as well as the production of more dormant eggs, occurs before the yearly cycle is over. Winds or birds often disperse dormant eggs, accounting for the unique distribution patterns of many rotifers. Most females lay either amictic or mictic eggs, but not both. Apparently, during oocyte development, the physiological condition of the female determines whether her eggs will be amictic or mictic.

Phylum Acanthocephala

Adult acanthocephalans (a-kan"tho-sef'a-lans) (Gr. *akantha,* spine or thorn + *kephale,* head) are endoparasites in the intestinal tract of vertebrates (especially fishes). Two hosts are required to complete the life cycle. The juveniles are parasites of crustaceans and insects. Acanthocephalans are generally small (less than 40 mm long), although one important species, *Macracanthorhynchus hirudinaceus,* which occurs in pigs, can be up to 80 cm long. The body of the adult is elongate and composed of a short anterior proboscis, a neck region, and a trunk (figure 10.24*a*). The proboscis is covered

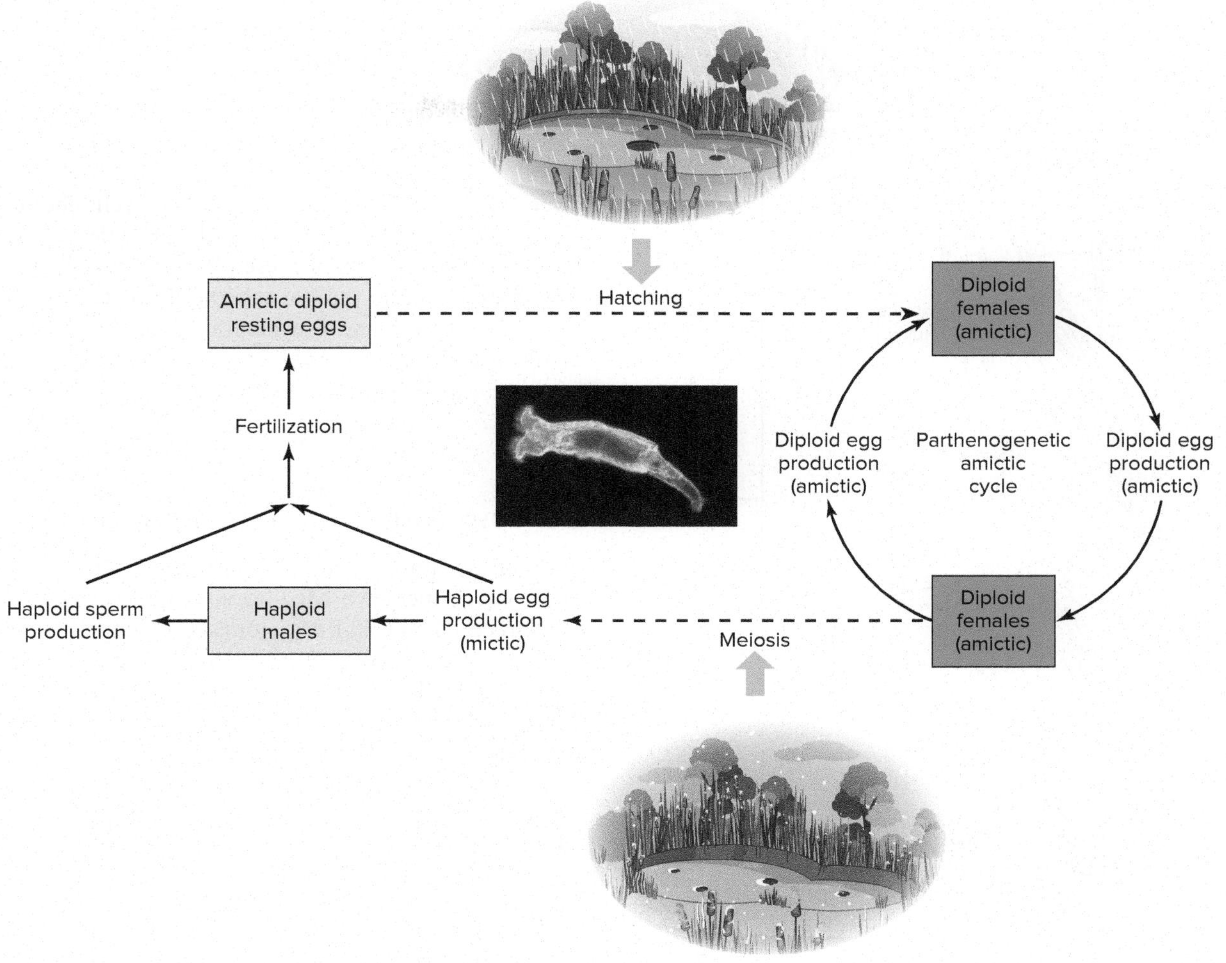

FIGURE 10.23

Life Cycle of a Monogonont Rotifer. Dormant, diploid, resting eggs hatch in response to environmental stimuli (e.g., melting snows and spring rains) to begin an amictic cycle that involves the parthenogenetic development of diploid eggs. This amictic cycle can be interrupted by environmental stimuli such as stagnating water, which initiates a summer mictic cycle that results in meiosis and the production of haploid eggs. If unfertilized, the mictic eggs develop into males that produce haploid sperm by mitosis. Fertilization of mictic eggs by sperm results in resistant diploid (amictic) resting eggs that withstand the unfavorable conditions. The return of favorable conditions results in the hatching of resting eggs and another amictic cycle. The production of resistant eggs also occurs in the fall with stimuli like freezing temperatures, and these eggs hatch in response to thawing and spring rains.

with recurved spines (figure 10.24*b*), hence, the name "spiny-headed worms." The retractable proboscis provides the means of attachment in the host's intestine. Females are always larger than males, and zoologists have identified about 1,000 species.

A living syncytial tegument that has been adapted to the parasitic way of life covers the body wall of acanthocephalans. A glycocalyx (*see figure 2.3*) consisting of mucopolysaccharides and glycoproteins covers the tegument and protects against host enzymes and immune defenses. No digestive system is present; acanthocephalans absorb food directly through the tegument from the host by specific membrane transport mechanisms and pinocytosis. Protonephridia may be present. The nervous system is composed of a ventral, anterior ganglionic mass from which anterior and posterior nerves arise. Sensory organs are poorly developed.

The sexes are separate, and the male has a protrusible penis. Fertilization is internal, and eggs develop in the pseudocoelom. The biotic potential of certain acanthocephalans is great; for example, a gravid female *Macroacanthorhynchus hirudinaceus* may contain up to 10 million embryonated eggs. The eggs pass out of the host with the feces and must be eaten by certain insects (e.g., cockroaches or grubs [beetle larvae]) or by aquatic crustaceans (e.g., amphipods, isopods, ostracods). Once in the invertebrate, the larva emerges from the egg and is now called an **acanthor.** It burrows through the gut wall and lodges in the hemocoel, where it develops into an **acanthella** and, eventually, into a **cystacanth.** When a mammal, fish, or bird eats the intermediate host, the cystacanth excysts and attaches to the intestinal wall with its spiny proboscis.

SECTION 10.3 THINKING BEYOND THE FACTS

What is the value of parthenogenesis in unstable habitats like in temperate ponds and seasonal pools?

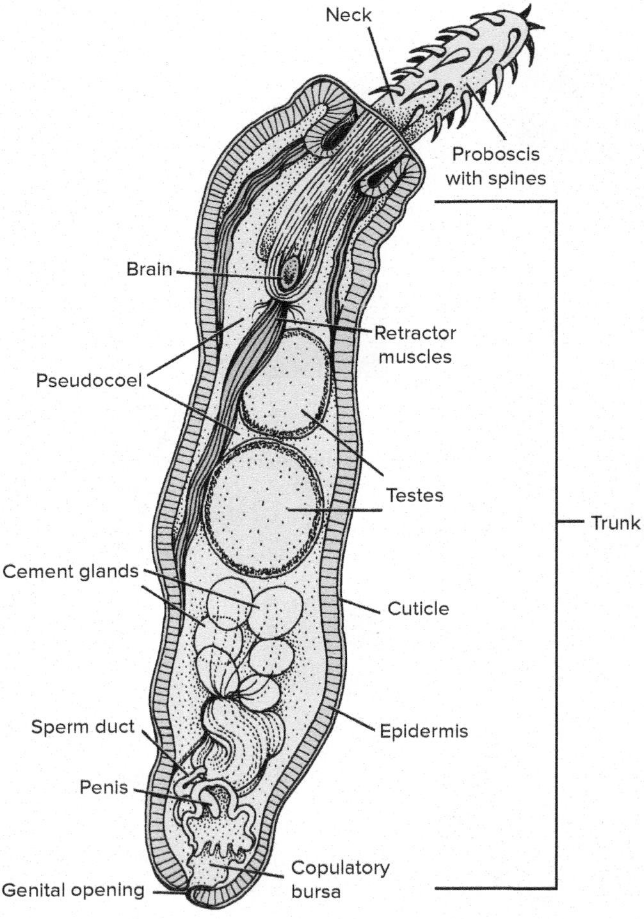

FIGURE 10.24

Phylum Acanthocephala. (*a*) Dorsal view of an adult male. (*b*) This acanthocephalan, *Oncicola canis*, is a parasite of wild carnivorous mammals and domestic cats and dogs. The intermdiate hosts for acanthor, acanthella, and cystacanth stages, as with most acanthocephalans, are arthropods.

10.4 OTHER LOPHOTROCHOZOANS

LEARNING OUTCOME

1. Describe one distinctive feature for members of the phyla Cycliophora, Nemertea, Ectoprocta, and Brachiopoda.

Four other lophotrochozoan phyla–Cycliophora, Nemertea, Ectoprocta, and Brachiopoda–will conclude this chapter. Phylogenetic relationships of these phyla to other lophotrochozoans will be described in the last section of this chapter. Two of these phyla (Ectoprocta and Brachiopoda) possess a lophophore–part of the reason for the clade designation Lophotrochozoa. Three additional lesser-known lophotrochozoan phyla are described in table 12.2.

Phylum Nemertea (Rhynchocoela)

Most of the approximately 900 species of nemerteans (nem-er′te-ans) (Gr. *Nemertes,* a Mediterranean sea nymph; the daughter of Nereus and Doris) are elongate, flattened worms found in marine mud and sand. Due to a long proboscis, nemerteans are commonly called proboscis worms. Adult worms range in size from a few millimeters to more than 30 m in length. Most nemerteans are pale yellow, orange, green, or red.

Characteristics of the phylum Nemertea include the following:

1. Long proboscis enclosed within a cavity called a rhynchocoel
2. Triploblastic, bilaterally symmetrical, unsegmented worms possessing a ciliated epidermis containing mucous glands
3. Complete digestive tract
4. Protonephridia
5. Cerebral ganglion, longitudinal nerve cords, and transverse commissures
6. Closed circulatory system
7. Body musculature organized into two or three layers

The most distinctive feature of nemerteans is a long proboscis held in a sheath called a **rhynchocoel** (figure 10.25). The proboscis may be tipped with a barb called a stylet. Carnivorous species use the proboscis to capture annelid (segmented worms) and crustacean prey.

Nemerteans have a complete one-way digestive tract. They have a mouth for ingesting food and an anus for eliminating digestive wastes. This enables mechanical breakdown of food, digestion, absorption, and feces formation to proceed sequentially in an anterior to posterior direction–a major evolutionary innovation found in most bilateral animals.

Another major adaptation found in most bilateral animals evolved in the nemerteans–a circulatory system consisting of two lateral blood vessels and, often, tributary vessels that branch from lateral vessels. However, no heart is present, and contractions of the walls of the large vessels propel blood. Blood does not circulate but simply moves forward and backward through the longitudinal vessels. Blood cells are present in some species. This combination of blood vessels with their capacity to serve local tissues and a one-way digestive system with its greater efficiency

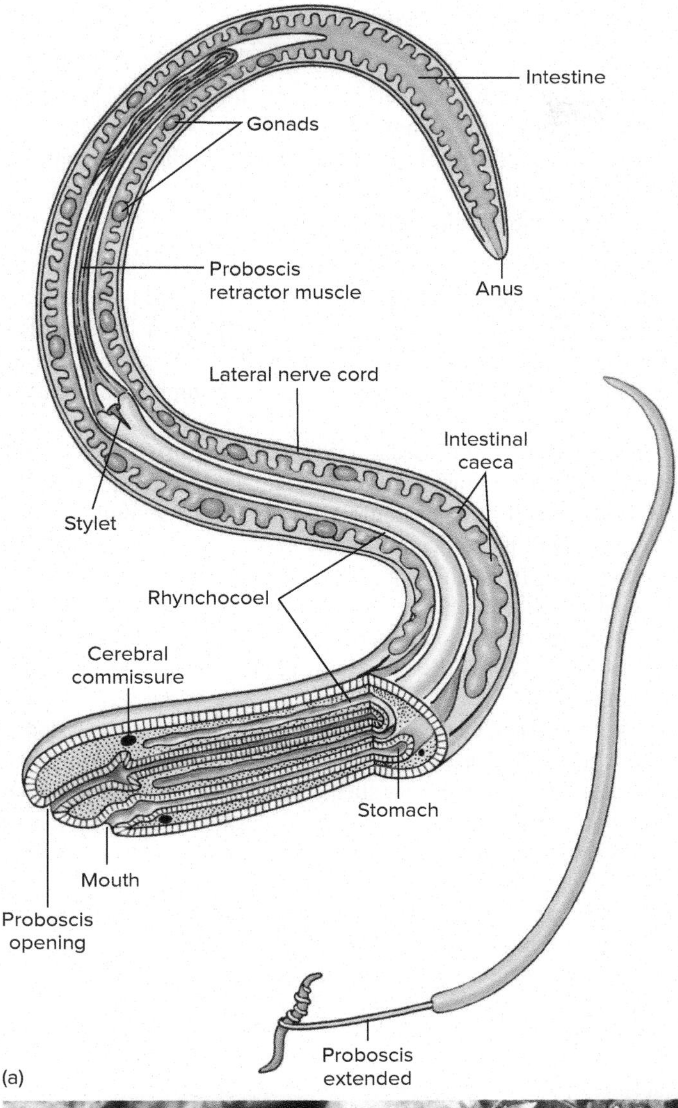

FIGURE 10.25

Phylum Nemertea. (*a*) Longitudinal section of a nemertean, showing the tubular gut and proboscis. (*b*) A ribbon worm, *Lineus* (blue and yellow), crawling on coral (orange). Nemerteans are the simplest animals that possess a complete digestive system, one that has two separate openings, a mouth and an anus.

Source: (a) Turbeville and Ruppert. 1983. Zoomorphology. 103:103. (b) ©Kjell B. Sandved/Science Source

at processing nutrients allows nemerteans to grow much larger than most flatworms.

Nemerteans are dioecious. Male and female reproductive structures develop from parenchymal cells along each side of the body. External fertilization results in the formation of a helmet-shaped, ciliated **pilidium larva.** After a brief free-swimming existence, the larva develops into a young worm that settles to the substrate and begins feeding.

When they move, adult nemerteans glide on a trail of mucus. Cilia and peristaltic contractions of body muscles provide the propulsive forces.

Nemerteans are notable for including the longest of known invertebrate animals. For example, *Lineus longissimus* regularly attains 30 m in length, and some individuals probably can achieve twice this length when fully extended.

Phylum Cycliophora

Along with Micrognathoza, Cycliophora is another recently described phylum. Cycliophorans live on the mouthparts (figure 10.26; *see figure 7.1*) of claw lobsters on both sides of the North Atlantic. These are very tiny animals, only 0.35 mm long and 0.10 mm wide. They attach to the mouthparts with an adhesive disc on the end of an acellular stalk. When the lobster to which they are attached starts to molt, the tiny symbiont begins a bizarre form of sexual reproduction. Dwarf males emerge, composed of nothing but nervous tissue and reproductive organs. Each dwarf male seeks out another female symbiont on the molting lobster and fertilizes its eggs, generating free-swimming individuals that can seek out another lobster and begin their life cycle anew.

Phylum Ectoprocta (Bryozoa)

Members of this phylum live in sessile colonies of individuals called zooids (Gr. *zoon,* living animal; an individual member of a colony

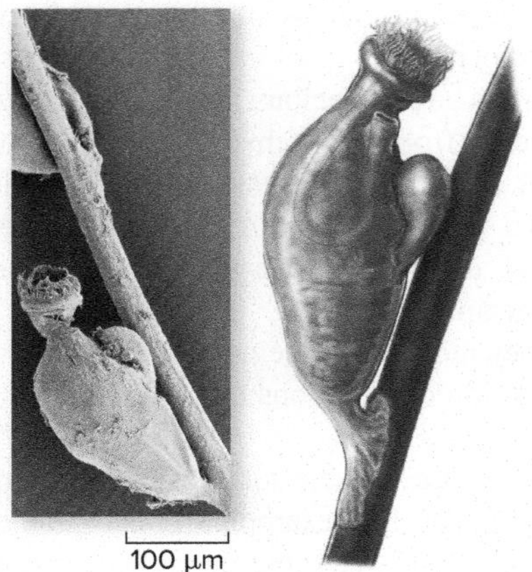

FIGURE 10.26

Phylum Cycliophora. About the size of the period at the end of this sentence, these lophotrochozoans live on the mouthparts of claw lobsters. One feeding stage (and part of a second one near the top of the photo) of the species *Symbion pandora* is shown attached to the mouthpart of a lobster.

of animals produced by incomplete budding or fission) living in marine and freshwater environments (figure 10.27). Marine ectoprocts are known from all depths and latitudes, mostly attaching to solid substrata (including boat hulls). Only a few species occur in fresh or brackish water. Colony members are usually less than 0.5 mm long, but colonies can be very large–in excess of 0.5 m. A colony's general plantlike appearance earned these animals the common name "moss animals." As a group, the Ectoprocta is a very successful lineage of suspension feeders. Approximately 5,000 living species are known.

Characteristics of the phylum Ectoprocta (Bryozoa) include the following:

1. Coelomate, with U-shaped or circular lophophore
2. Sessile, colonial, marine or freshwater
3. Gut U-shaped; anus opens outside base of lophophore tentacles
4. Asexual budding forms new colonies
5. Gas exchange by diffusion

The ectoproct colony is referred to as a **zoarium**. It is comprised of individual zooids. Each zooid secretes, and is surrounded by, an exoskeleton called the **zoecium**. The **polypide** is comprised of soft viscera (muscles, digestive tract, and nerve centers) and the lophophore. The opening through which the lophophore can be extended is called the **orifice.**

The nature of the zoecium differs among bryozoans–it may be gelatinous, chitinous, or calcified. Those species that produce a calcified or chitinous zoecium are important reef-builders. The different growth patterns among the bryozoans result in a variety of colony shapes, such as gelatinous masses, fans, bushes, and sheets. Mineralized skeletons of bryozoans first appeared in rocks from the Early Ordovician period (*see appendix B*) making the Ectoprocta (Bryozoa) the last major phylum to appear in the fossil record.

In most cases, each colony is the product of asexual reproduction from a single sexually produced individual. Most bryozoans are hermaphroditic. Some species shed eggs into the seawater, but most brood their eggs; cleavage is radial. Freshwater species produce a resistant statoblast that survives winter temperatures, which kill the parent colony.

These animals use a retractable lophophore in filter feeding. When the lophophore is extended, cilia on the tentacles create water currents that draw water into the center of the extended tentacles.

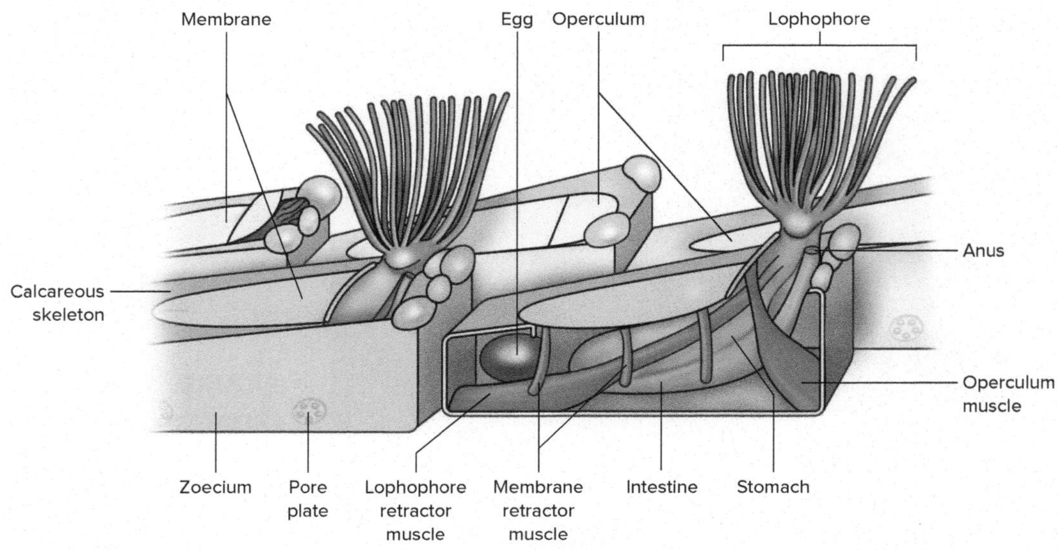

FIGURE 10.27

Ectoprocta. Two zooids are shown with their lophophores extended in the feeding position.

Suspended food particles are trapped by the cilia and transported to the mouth at the base of the tentacles. Digestion occurs in a U-shaped, complete digestive tract (*see figure 10.27*).

Phylum Brachiopoda

The phylum Brachiopoda (Gr. *brachion,* arm + *podos,* foot) is an ancient group of marine animals, which once flourished in the Paleozoic and Mesozoic seas. Brachiopods are called "lamp shells" based on pottery oil-lamps of ancient Greece and Rome. Modern species (335 species) have changed little from fossil species (26,000 species). All living brachiopods are relatively small (<10 cm shell length or width) and benthic in habitat.

Characteristics of the phylum Brachiopoda include the following:

1. Triploblastic, coelomate, bilaterally symmetrical
2. U-shaped gut with or without an anus
3. Body enclosed in dorsal and ventral valves
4. Nervous system with a ganglionated circumesophageal ring
5. Open circulatory system with one or more hearts
6. No gas exchange organs
7. Horseshoe-shaped lophophore in anterior mantle cavity

Brachiopods are completely enclosed, apart from the stalk, within a bivalve shell that is secreted by a mantle. Unlike the bivalve molluscs, whose valves are laterally positioned (*see chapter 11*), the two valves of brachiopods are dorsal and ventral. The ventral valve is slightly larger than the dorsal valve. Brachiopods may be unattached or cemented to the substratum by means of a fleshy stalk called a pedicel (figure 10.28*a* and *b*).

Two classes of brachiopods are recognized. Members of the class Inarticulata have shell valves composed of calcium phosphate and chitin, and valves are held together by muscles. Members of the class Articulata possess calcium carbonate valves, and valves articulate by means of teeth present on the ventral valves, which insert into sockets on the dorsal valve.

The horseshoe-shaped lophophore is located in the anterior portion of the mantle cavity. Ciliary currents carry food particles through slightly gaped valves. Tentacles trap suspended food particles and transport food to the mouth.

There are three coelomic cavities in brachiopods that are called the protocoel, the mesocoel, and the metacoel. The posterior metacoel contains the viscera. Brachiopods have an open circulatory system with one or more hearts. Gas exchange occurs across the mantle and lophophore surfaces by simple diffusion. Brachiopods have two pairs of metanephritic excretory structures.

Most species are dioecious and gametes are shed into the metacoel where they are usually brooded. Inarticulate brachiopods have a juvenile stage that resembles the adult. Articulate brachiopods have a larval stage that undergoes metamorphosis to the adult form.

Section 10.4 Thinking Beyond the Facts

How are the lophophores of ectoprocts and brachiopods used in feeding?

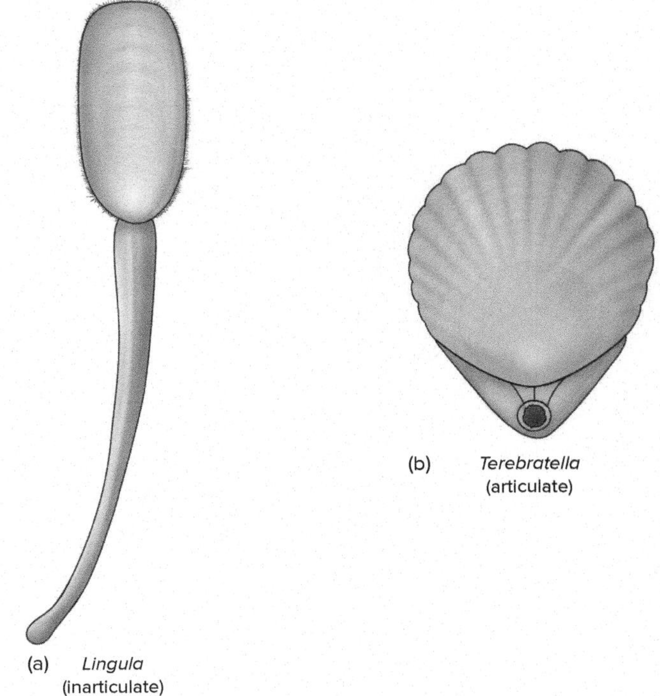

FIGURE 10.28

Brachiopods. (*a*) The inarticulate brachiopod *Ligula. Ligula* uses its pedicel to draw itself into its burrow. (*b*) Drawing of an articulate brachiopod, *Terebratella.* The valves have a tooth-and-socket articulation, and the ventral valve is slightly larger than the dorsal valve. A short pedicel anchors the brachiopod to its substrate.

10.5 Further Phylogenetic Considerations

Learning Outcomes

1. Describe the state of knowledge regarding evolutionary relationships among the Lophotrochozoa.
2. Assess the validity of the traditional platyhelminth classes.

Evolutionary relationships among the Lophotrochozoa are under active investigation. Lophotrochozoa is a very diverse clade of animals, thus differing interpretations of phylogenetic relationships should be expected. Depending upon the sources of data used (nuclear genes, mitochondrial genes, rRNA genes, and morphological characteristics), different phylogenetic interpretations do exist. All of the phyla discussed in this chapter, as well as the lophotrochozoans discussed in chapters 11 (Mollusca) and 12 (Annelida), are treated as monophyletic groups. As discussed later, there has been some debate as to whether or not this is true for Platyhelminthes.

Figure 10.29 is based upon a traditional interpretation of molecular and morphological data. Some recent studies do not support the validity of the clade Platyzoa. An increasing body of molecular data suggests that Gnathostomulida, along with Gastrotricha, have basal positions within the Lophotrochozoa. The Rotifera

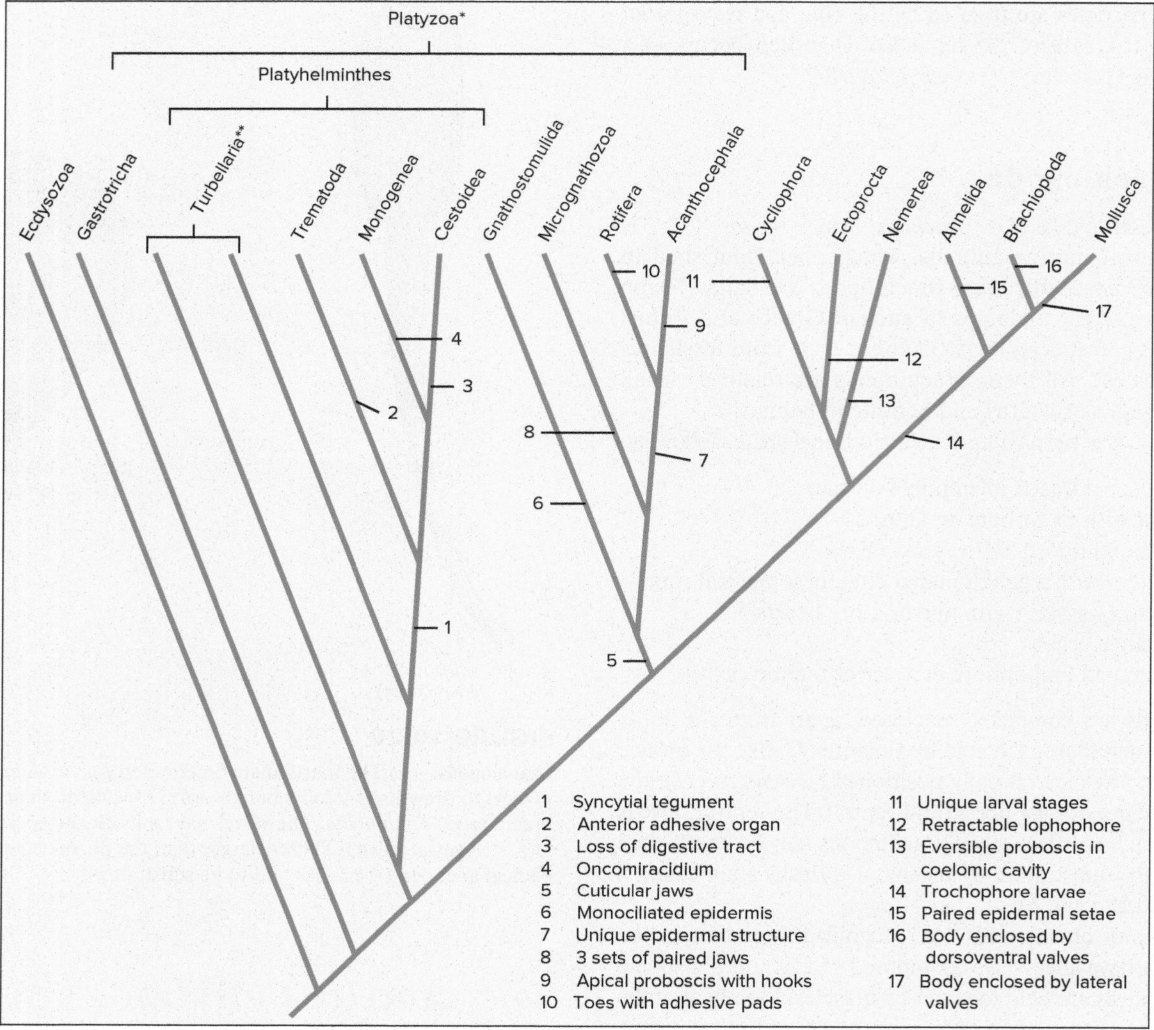

FIGURE 10.29

Phylogeny of the Lophotrochozoa. One interpretation of the phylogeny of the Lophotrochozoa based on molecular and morophological data. Synapomorphies are based on Brusca and Brusca (2003). *Invertebrates.* Sinauer. Molecular-based relationships are based on Paps, J. et al. (2009). Lophotrochozoa internal phylogeny: New insights from an up-to-date analysis of nuclear ribosomal genes. *Proc. R. Soc. B.* 276(1660): 1245–1254.
*Some molecular phylogenies show "Platyzoa" as a paraphyletic group. **Turbellaria is a paraphyletic group.

and Acanthocephala are closely related in all analyses, and some authorities suggest that these two groups should be united into a single phylum. Bdelloidea and Monogononta are considered true rotifers, and the Seisonidea and Acanthocephala join them in a clade called Syndermata. New data also suggest that Nemertea is more closely allied with Annelida, Mollusca, and Brachiopoda and that Cycliophora and Ectoprocta are allied with Platyhelminthes. Because of the very tentative nature of lophotrochzoan phylogeny, we can expect the interpretations described here to change as more work emerges that integrates molecular and morphological data.

As suggested in the previous text, there is considerable debate as to the status of the phylum Platyhelminthes. Taxonomists usually look for one unique synapomorphy when characterizing animal phyla. Because there is no such synapomorphy for this phylum, some researchers doubt whether or not it is monophyletic. Molecular studies seem to support monophyly for the vast majority of platyhelminth taxa. The inclusion of the traditional class Turbellaria within the phylum presents an additional problem. There is no synapomorphy that unites turbellarians with the remaining flatworms, and molecular studies have clearly demonstrated that this "class" is paraphyletic. The syncytial tegument of members of the Trematoda, Monogenea, and Cestoidea establish this group as a monophyletic clade. Molecular studies reinforce this conclusion. Molecular data also support a closer relationship between Monogenea and Cestoidea than previously assumed. More information will undoubtedly result in a revision of platyhelminth taxonomy.

Section 10.5 Thinking Beyond the Facts

Explain why many zoologists doubt that Platyhelminthes is a valid phylum.

Summary

10.1 **Evolutionary Perspective**

- Members of the clade Lophotrochoza are a diverse group of bilaterally symmetrical animals that are united by shared molecular characteristics. Many members of these phyla possess either a trochophore larval stage or a lophophore.

10.2 **Platyzoa: Phylum Platyhelminthes**

- In the Platyhelminthes, mesodermally derived parenchyma may provide support, nutrient storage, motility, reserves of regenerative cells, transport, oxygen storage, and other functions.
- Members of the class Turbellaria are free-living Plathyhelminthes. Most turbellarians move entirely by cilia and muscles and are predators and scavengers. Digestion is initially extracellular and then intracellular. Protonephridia are present in many flatworms and are involved in osmoregulation. A primitive brain and nerve cords are present. Turbellarians are monoecious with the reproductive systems adapted for internal fertilization.
- The class Trematoda is divided into two subclasses (Aspidogastrea and Digenea). Digenetic flukes are internal parasites of vertebrates and include medically important species. A gut is present, and most of these flukes are monoecious. Stages in the life cycle of a fluke include eggs, miracidia, sporocysts, cercariae, and adults. The monogenetic flukes (class Monogenea) are mostly ectoparasites of fishes.
- Cestodes, or tapeworms, are gut parasites of vertebrates. They possess a scolex with attachment organs, a neck region, and a strobila, which consists of a chain of segments (proglottids) budded off from the neck region. A gut is absent, and the reproductive system is repeated in each proglottid.

10.3 **Platyzoa: Smaller Phyla**

- Gastrotrichs are marine and freshwater. They inhabit interstitial spaces in bottom sediments. They have a complete digestive tract and are hermaphroditic.
- Micrognathozoans inhabit interstitial spaces of cold-water springs of Greenland. They have a complex jaw system and cilia function in sensory perception, locomotion, and attachment. They reproduce through parthenogenesis.
- Gnathostomulids occupy interstitial spaces in marine environments. They have an incomplete digestive tract and are monoecious.
- Rotifers live mostly in freshwater. They feed using a ciliated corona, have a complete digestive tract, and possess adhesive glands. They are dioecious, and many reproduce parthenogenetically.
- Acanthocephalans are endoparasites of vertebrates. They attach to their host intestinal wall with a spine-covered proboscis. Reproduction involves the development of larval stages within invertebrate intermediate hosts.

10.4 **Other Lophotrochozoans**

- Nemerteans possess a long proboscis held in a sheath. The proboscis is tipped with a barb called a stylet that is used to capture prey.
- Cycliophorans live attached to mouthparts of claw lobsters.
- Ectoprocts (bryozoans) are colonial and live on a variety of submerged marine substrata. Colonies are commonly many centimeters in width or height. Each individual (zooid) lives in a chamber (zoecium), which is a secreted exoskeleton of chitin, calcium, or a gelatinous material. They feed using a lophophore.
- Brachiopods possess a mantle, which secretes a shell comprised of dorsal and ventral valves. Brachiopods are usually attached to the substratum directly or by a pedicel. They feed using a lophophore.

10.5 **Further Phylogenetic Considerations**

- Interpretations of lophotrochozoan phylogeny lead to different conclusions. The validity of the phylum Platyhelminthes is being reevaluated. The traditional class Turbellaria is paraphyletic and probably will be abandoned as a valid taxonomic group.

Concept Review Questions

1. In the life cycle of the fluke responsible for schistosomiasis, the larva that leaves the intermediate host enters
   a. a human.
   b. a fish.
   c. a snail.
   d. another fluke.
   e. an unknown host.
2. Common intermediate hosts for most flukes are
   a. clams.
   b. flies.
   c. mice.
   d. snails.
   e. mosquitoes.
3. In the flatworm, flame cells are involved in what metabolic process?
   a. Reproduction
   b. Digestion
   c. Locomotion
   d. Osmoregulation
   e. All of these (a–d)
4. Parthenogenesis is present in the life cycles of
   a. rotifers.
   b. acanthocephalans.
   c. cycliophorans.
   d. micrognathozoans.
   e. both a and d are probably correct.
5. According to recent molecular information, _______________ is a paraphyletic assemblage of animals and probably should be abandoned as a taxonomic group.
   a. Cestoidea
   b. Trematoda
   c. Ectoprocta
   d. Brachiopoda

Analysis and Application Questions

1. What benefit would being a hermaphrodite confer on a parasitic species?
2. Does the absence of a digestive system indicate that tapeworms are primitive, ancestral forms of Platyhelminthes?
3. What kind of information must become available in order to have a clearer picture of lophotrochozoan phylogeny?
4. Why is the scolex of a tapeworm not considered a head like other animals have?
5. If the lophophore or the trochophore larval form is characteristic of the lophotrochozoa, why are some animals included in this group that have neither?

This marine nudibranch (*Chromodoris kuniei*) is a member of one of the most successful animal phyla. Members of the Mollusca have adapted to nearly every habitat on the earth. This chapter describes the remarkable diversity and adaptations of members of this phylum.

11

Molluscan Success

Chapter Outline

11.1 Evolutionary Perspective
Relationships to Other Animals
11.2 Molluscan Characteristics
11.3 Class Gastropoda
Torsion
Gastropod Shells
Locomotion (Gastropoda)
Feeding and Digestion (Gastropoda)
Other Maintenance Functions (Gastropoda)
Reproduction and Development (Gastropoda)
Gastropod Diversity
11.4 Class Bivalvia
Shell and Associated Structures
Gas Exchange, Filter Feeding, and Digestion
Other Maintenance Functions (Bivalvia)
Reproduction and Development (Bivalvia)
Bivalve Diversity
11.5 Class Cephalopoda
Cephalopod Shells
Locomotion (Cephalopoda)
Feeding Digestion (Cephalopoda)
Other Maintenance Functions (Cephalopoda)
Learning
Reproduction and Development (Cephalopoda)
11.6 Class Polyplacophora
11.7 Class Scaphopoda
11.8 Class Monoplacophora
11.9 Class Solenogastres
11.10 Class Caudofoveata
11.11 Further Phylogenetic Considerations

11.1 EVOLUTIONARY PERSPECTIVE

LEARNING OUTCOMES

1. Give examples of, and describe the prevalence of, members of the phylum Mollusca.
2. Discuss the relationship of the Mollusca to other animal phyla.

Octopuses, squids, and cuttlefish (the cephalopods) are some of the invertebrate world's most adept predators. Predatory lifestyles have resulted in the evolution of large brains (by invertebrate standards), complex sensory structures (by any standards), rapid locomotion, grasping tentacles, and tearing mouthparts. In spite of these adaptations, cephalopods rarely make up a major component of any community. Once numbering about 9,000 species, the class Cephalopoda now includes only about 700 species (figure 11.1).

Zoologists do not know why the cephalopods have declined so dramatically. Vertebrates may have outcompeted cephalopods because the vertebrates were also making their appearance in prehistoric seas, and some vertebrates (e.g., bony fish) acquired active, predatory lifestyles.

This evolutionary decline has not been the case for all molluscs. Overall, this group has been very successful. If success is measured by numbers of species, the molluscs are twice as successful as vertebrates! The vast majority of the nearly 100,000 living species of molluscs belongs to two classes: Gastropoda, the snails and slugs; and Bivalvia, the clams and their close relatives.

Molluscs are triploblastic, as are all the remaining animals covered in this text. In addition, they are the first animals described in this text that possess a coelom, although the coelom of molluscs is only a small cavity (the pericardial cavity) surrounding the heart, nephridia, and gonads. A coelom is a body cavity that arises in mesoderm and is lined by a sheet of mesoderm called the peritoneum (*see figure 7.11*c).

Relationships to Other Animals

Molluscs are members of the Lophotrochozoa. Their lophotrochozoan relatives include the annelids, along with other phyla discussed in chapter 10 (*see appendix A*).

SECTION 11.1 THINKING BEYOND THE FACTS

How do the molluscs illustrate the idea that complexity does not always ensure evolutionary success?

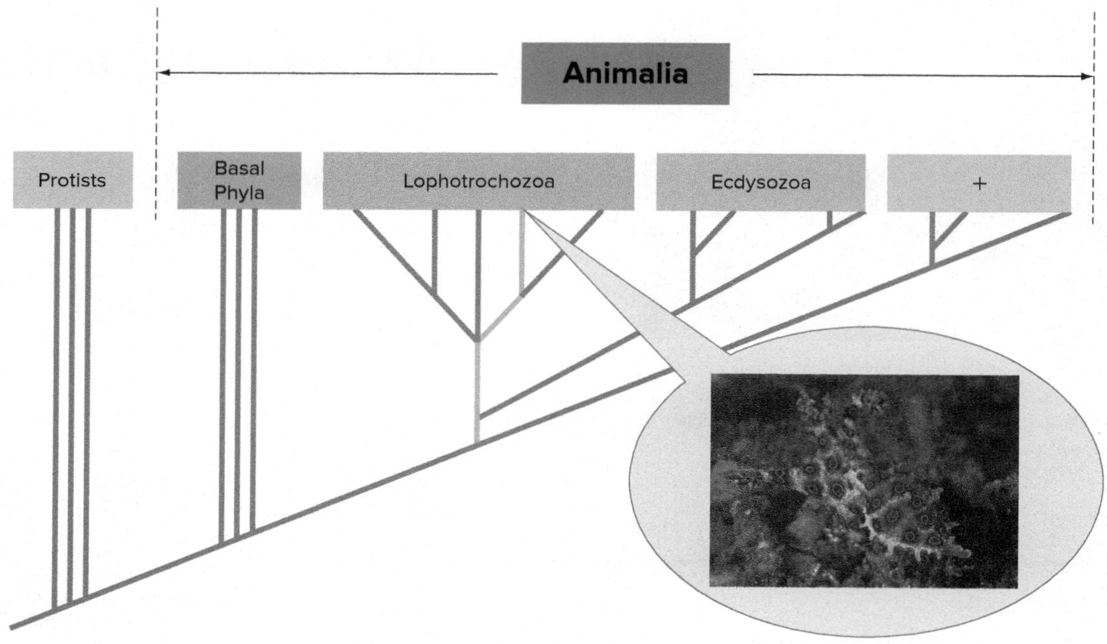

FIGURE 11.1

Evolutionary Relationships of Molluscs to Other Animals. The figure shows one interpretation of the relationship of the Mollusca to other members of the animal kingdom. The relationships depicted here are based on evidence from developmental and molecular biology. Molluscs are placed within the Lophotrochozoa along with the Annelida, Plathyhelminthes, Rotifera, and others (*see appendix A*). The phylum Mollusca includes nearly 100,000 living species, including members of the class Cephalopoda–some of the invertebrates' most adept predators. The blue-ringed octopus (*Hapalochlaena*) is shown here. Four species of *Hapalochlaena* are found in the Pacific Ocean from Japan to Australia. This very small octopus has a very powerful venom containing tetrodotoxin, the same found in cone snails and pufferfish. The venom causes respiratory paralysis and is powerful enough to kill a human.

11.2 MOLLUSCAN CHARACTERISTICS

LEARNING OUTCOME

1. Hypothesize about the structure of a hypothetical newly discovered class of molluscs.

Molluscs range in size and body form from the largest of all invertebrates, the giant squid (*Architeuthis*), measuring 18 m in length, to the smallest garden slug, less than 1 cm long. In spite of this diversity, the phylum Mollusca (mol-lus'kah) (L. *molluscus,* soft) is not difficult to characterize (table 11.1).

Characteristics of the phylum Mollusca include the following:

1. Body of two parts: head-foot and visceral mass
2. Mantle that secretes a calcareous shell and covers the visceral mass
3. Mantle cavity functions in excretion, gas exchange, elimination of digestive wastes, and release of reproductive products
4. Bilateral symmetry
5. Trochophore larvae, spiral cleavage, and schizocoelous coelom formation
6. Coelom reduced to cavities surrounding the heart, nephridia, and gonads
7. Open circulatory system in all but one class (Cephalopoda)
8. Radula usually present and used in scraping food

The body of a mollusc has two main regions–the head-foot and the visceral mass (figure 11.2). The **head-foot** is elongate with an anterior head, containing the mouth and certain nervous and sensory structures, and an elongate foot, used for attachment and locomotion. The **visceral mass** contains the organs of digestion, circulation, reproduction, and excretion and is positioned dorsal to the head-foot.

The **mantle** of a mollusc usually attaches to the visceral mass, enfolds most of the body, and may secrete a shell that overlies the mantle. The shell of a mollusc is secreted in three layers (figure 11.3). The outer layer of the shell is called the periostracum. Mantle cells at the mantle's outer margin secrete this protein layer. The middle layer of the shell, called the prismatic layer, is the thickest of the three layers and consists of calcium carbonate mixed with organic materials. Cells at the mantle's outer margin also secrete this layer. The inner layer of the shell, the nacreous layer, forms from thin sheets of calcium carbonate alternating with organic matter. Cells along the entire epithelial border of the mantle secrete the nacreous layer. Nacre secretion thickens the shell.

Between the mantle and the foot is a space called the **mantle cavity.** The mantle cavity opens to the outside and functions in gas

TABLE 11.1
CLASSIFICATION OF THE MOLLUSCA

Phylum Mollusca (mol-lus′kah)
The coelomate animal phylum whose members possess a head-foot, visceral mass, mantle, and mantle cavity. Most molluscs also possess a radula and a calcareous shell. Nearly 100,000 species.

Class Caudofoveata (kaw′do-fo″ve-a″ta)
Wormlike molluscs with a cylindrical, shell-less body and scale-like, calcareous spicules; lack eyes, tentacles, statocysts, crystalline style, foot, and nephridia. Deep-water, marine burrowers. *Chaetoderma*. Approximately 150 species.

Class Solenogastres (so-len″ah-gas′trez)
Shell, mantle, and foot lacking; worm-like; ventral (pedal) groove; head poorly developed; surface dwellers on coral and other substrates. Marine. *Neomenia*. Approximately 250 species.

Class Polyplacophora (pol″e-pla-kof′o-rah)
Elongate, dorsoventrally flattened; head reduced in size; shell consisting of eight dorsal plates. Marine, on rocky intertidal substrates. *Chiton*. Approximately 950 species.

Class Monoplacophora (mon″o-pla-kof′o-rah)
Molluscs with a single arched shell; foot broad and flat; certain structures serially repeated. Marine. *Neopilina*. Approximately 25 species.

Class Gastropoda (gas-trop′o-dah)
Shell, when present, usually coiled; body symmetry distorted by torsion; some monoecious species. Marine, freshwater, terrestrial. *Nerita, Orthaliculus, Helix*. More than 65,000 species.

Class Cephalopoda (sef″ah-lah″po′dah)
Foot modified into a circle of tentacles and a siphon; shell reduced or absent; head in line with the elongate visceral mass. Marine. *Octopus, Loligo, Sepia, Nautilus*. Approximately 700 species.

Class Bivalvia (bi″val′ve-ah)
Body enclosed in a shell consisting of two valves, hinged dorsally; no head or radula; wedge-shaped foot. Marine and freshwater. *Anodonta, Mytilus, Venus*. Approximately 15,000 species.

Class Scaphopoda (ska-fop′o-dah)
Body enclosed in a tubular shell that is open at both ends; tentacles used for deposit feeding; no head. Marine. *Dentalium*. More than 600 species.

This taxonomic listing reflects a phylogenetic sequence. The discussions that follow, however, begin with molluscs that are familiar to most students.

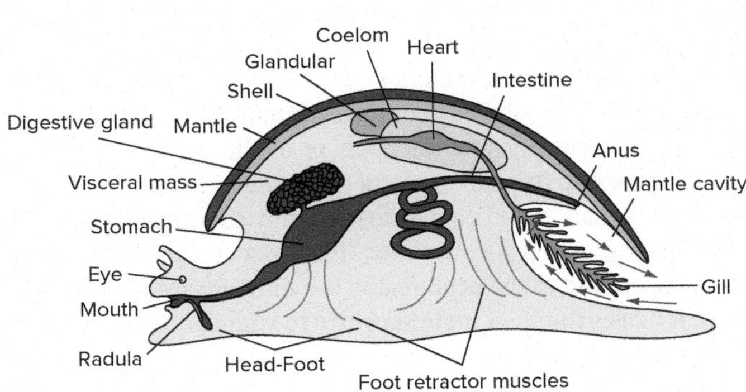

FIGURE 11.2

Molluscan Body Organization. This illustration of a hypothetical mollusc shows three features unique to the phylum. The head-foot is a muscular structure usually used for locomotion and sensory perception. The visceral mass contains organs of digestion, circulation, reproduction, and excretion. The mantle is a sheet of tissue that enfolds the rest of the body and secretes the shell. Arrows indicate the flow of water through the mantle cavity. As will be discussed later, the representation of the shell and muscular foot is not accurate for the Caudofoveata and Solenogastres (*see table 11.1*).

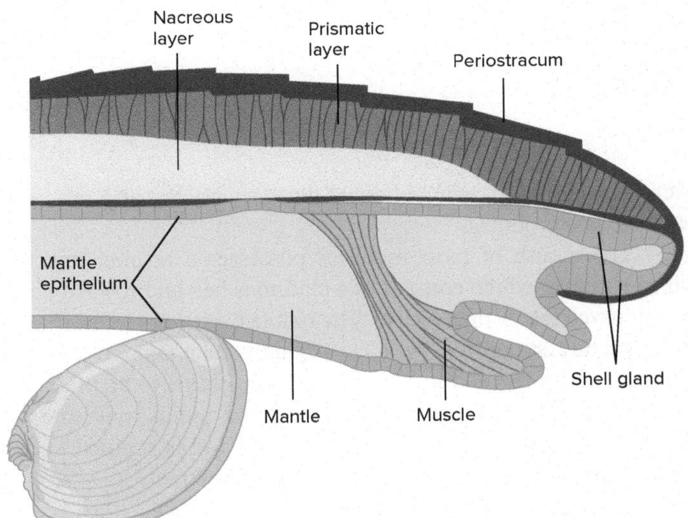

FIGURE 11.3

Molluscan Shell and Mantle. A transverse section of a bivalve shell and mantle shows the three layers of the shell and the portions of the mantle responsible for shell secretion.

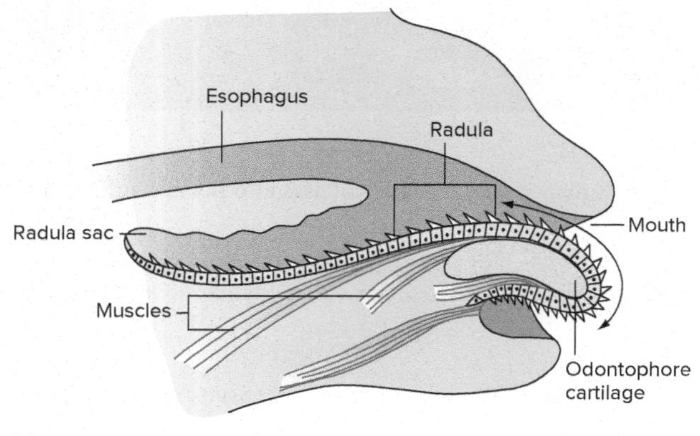

(a)

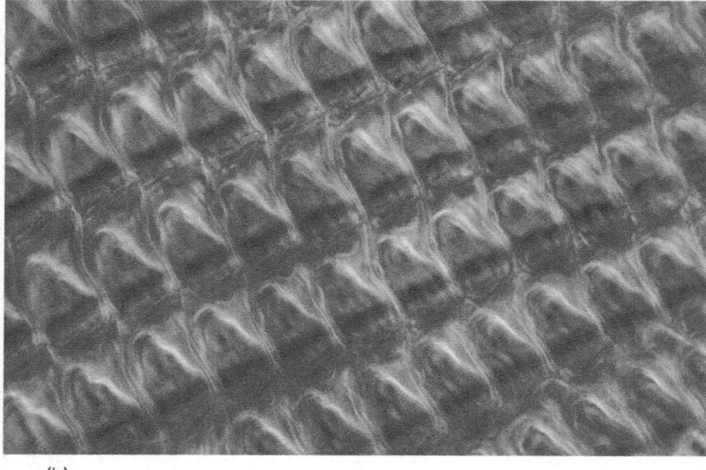

(b)

FIGURE 11.4

Radular Structure. (*a*) The radular apparatus lies over the cartilaginous odontophore. Muscles attached to the radula move the radula back and forth over the odontophore (*see arrows*). (*b*) Scanning electron micrograph of a gastropod radula. Radular structure is an important character used in gastropod taxonomy.

(a) Source: Russell-Hunter WD. 1979. *A Life of Invertebrates*. New York (NY): Macmillan.
(b) ©Comstock Images/Getty Images

exchange, excretion, elimination of digestive wastes, and release of reproductive products.

The mouth of most molluscs possesses a rasping structure called a **radula,** which consists of a chitinous belt and rows of posteriorly curved teeth (figure 11.4). The radula overlies a fleshy, tonguelike structure supported by a cartilaginous **odontophore.** Muscles associated with the odontophore permit the radula to be protruded from the mouth. Muscles associated with the radula move the radula back and forth over the odontophore. Food is scraped from a substrate and passed posteriorly to the digestive tract.

Section 11.2 Thinking Beyond the Facts

Which of the molluscan characteristics listed at the beginning of this section are unique to the molluscs?

11.3 CLASS GASTROPODA

LEARNING OUTCOMES

1. Compare members of the class Gastropoda to the generalized molluscan body form.
2. Explain the structure, function, and reproduction of members of the class Gastropoda.
3. Describe the distribution and ecology of prosobranch, opisthobranch, and pulmonate gastropods.

The class Gastropoda (gas-trop′o-dah) (Gr. *gaster,* gut + *podos,* foot) includes the snails, limpets, and slugs. With more than 65,000 living species (*see table 11.1*), Gastropoda is the largest and most varied molluscan class. Its members occupy a wide variety of marine, freshwater, and terrestrial habitats. Most people give gastropods little thought unless they encounter *Helix pomatia* (*escargot*) in a French restaurant or are pestered by garden slugs and snails. One important impact of gastropods on humans is that gastropods are intermediate hosts for some medically important trematode parasites of humans (*see chapter 10*).

Torsion

One of the most significant modifications of the molluscan body form in the gastropods occurs early in development. **Torsion** is a 180°, counterclockwise twisting of the visceral mass, mantle, and mantle cavity. Torsion positions the gills, anus, and openings from the excretory and reproductive systems just behind the head and nerve cords, and twists the digestive tract into a U shape (figure 11.5).

The adaptive significance of torsion is speculative; however, three advantages are plausible. First, without torsion, withdrawal into the shell would proceed with the foot entering first and the more vulnerable head entering last. With torsion, the head enters the shell first, exposing the head less to potential predators. In some snails, a proteinaceous, and in some calcareous, covering, called an **operculum,** on the dorsal, posterior margin of the foot enhances protection. When the gastropod draws the foot into the mantle cavity, the operculum closes the opening of the shell, thus preventing desiccation when the snail is in drying habitats. A second advantage of torsion concerns an anterior opening of the mantle cavity that allows clean water from in front of the snail to enter the mantle cavity, rather than water contaminated with silt stirred up by the snail's crawling across its substrate. The twist in the mantle's sensory organs around to the head region is a third advantage of torsion because it makes the snail more sensitive to stimuli coming from the direction in which it moves.

Note in figure 11.5*d* that, after torsion, the anus and nephridia empty dorsal to the head and create potential fouling problems. However, a number of evolutionary adaptations seem to circumvent this problem. Various modifications allow water and the wastes it carries to exit the mantle cavity through notches or openings in the mantle and shell posterior to the head. Some gastropods undergo detorsion, in which the embryo undergoes a full 180° torsion and then untwists approximately 90°. The mantle cavity thus opens on the right side of the body, behind the head.

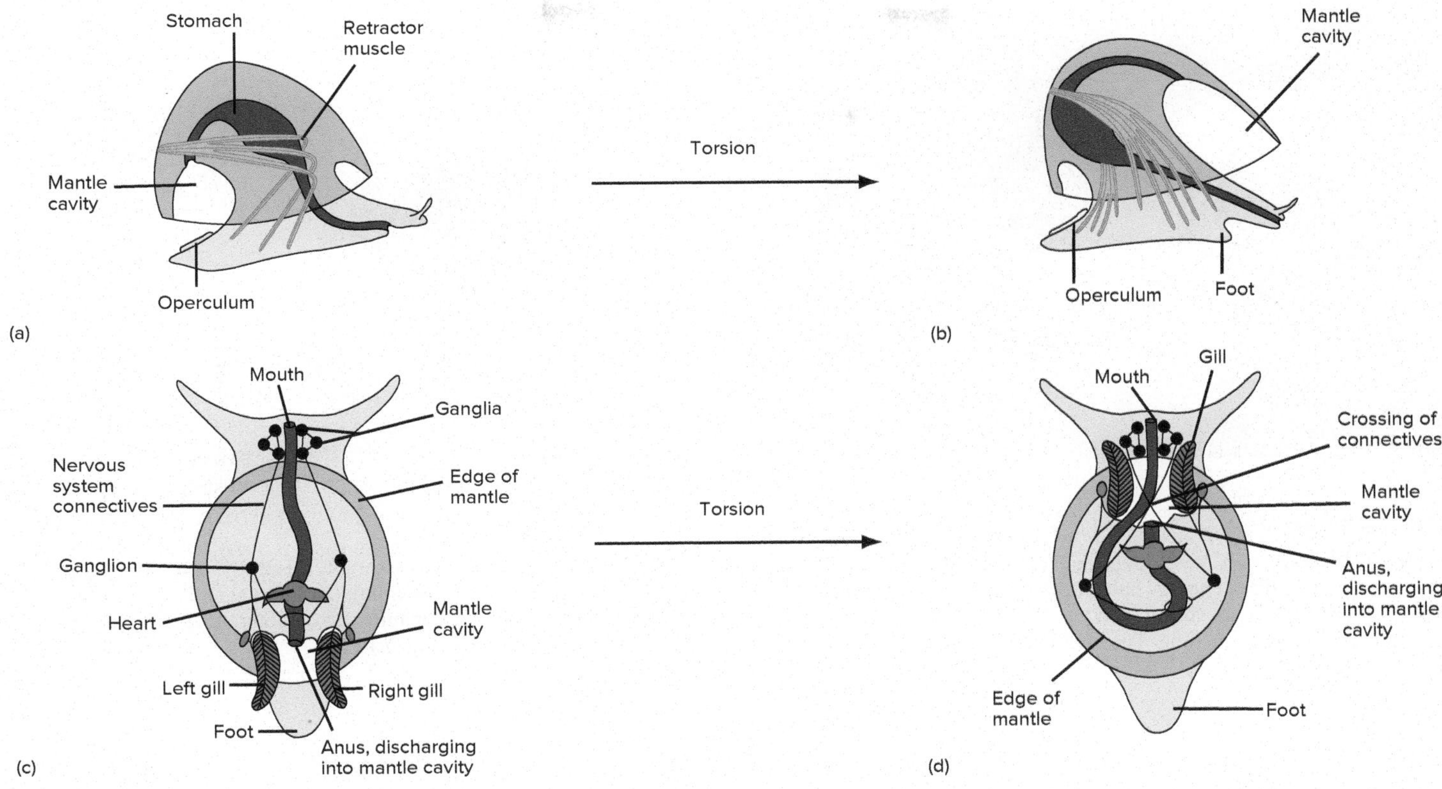

FIGURE 11.5

Torsion in Gastropods. (*a*) A pretorsion gastropod larva. Note the posterior opening of the mantle cavity and the untwisted digestive tract. (*b*) After torsion, the digestive tract is looped, and the mantle cavity opens near the head. The foot is drawn into the shell last, and the operculum closes the shell opening. (*c*) A hypothetical adult ancestor, showing the arrangement of internal organs prior to torsion. (*d*) Modern adult gastropods have an anterior opening of the mantle cavity and the looped digestive tract.

Source:] Hyman, L. 1967. *The Invertebrates*. Volume VI. McGraw-Hill, Inc.

Gastropod Shells

Most gastropods have a shell, and the names "snail," "limpet," and "slug" are descriptive of the type of shell present. These common names have no taxonomic significance–they do not reflect monophyletic groups. The term "snail" usually refers to a gastropod that can withdraw entirely into its shell. A "limpet" has a shallowly conical or dome-like shell and a broad muscular foot that it uses to cling to its substrate. (One family of marine gastropods, Patillidae, is a group referred to as "true limpets.") A "slug" is a gastropod with no apparent shell, although slugs usually retain an internalized shell that is a reservoir of calcium salts located near digestive glands (*see figure 11.6*).

The earliest fossil gastropods had a shell that was coiled in one plane. This arrangement is not common in later fossils, probably because growth resulted in an increasingly cumbersome shell. (Some modern snails, however, have secondarily returned to this shell form.)

Most modern snail shells are asymmetrically coiled into a more compact form, with successive coils or whorls slightly larger than, and ventral to, the preceding whorl (figure 11.6*a*).

This pattern leaves less room on one side of the visceral mass for certain organs, which means that organs that are now single were probably paired ancestrally. This asymmetrical arrangement of internal organs is described further in the descriptions of particular body systems.

Shells have many unique sculpturing and whorl patterns that are taxonomically significant and often reflect adaptations for deterring specific predators. The cone-shaped shell of cone snails (*Conus*) prevents crab predators from getting a grip on this potential prey. (Interestingly, cone snails themselves use a powerful venom when preying on fish, other molluscs, or marine worms.) One species of terrestrial snail (*Karaftohelix gainesi*) apparently swings its shell like a club to deter its ground beetle predators. The dome-shaped shell of marine limpets (Patellidae) provides these gastropods protection from violent surf action as they cling to rocks in their intertidal habitats.

Locomotion (Gastropoda)

Nearly all gastropods have a flattened foot that is often ciliated, covered with gland cells, and used to creep across the substrate (figure 11.6*b*). The smallest gastropods use cilia to propel themselves over a mucous trail. Larger gastropods use waves of muscular contraction that move over the foot. The foot of some gastropods is modified for clinging, as in abalones and limpets, or for swimming, as in sea butterflies and sea hares.

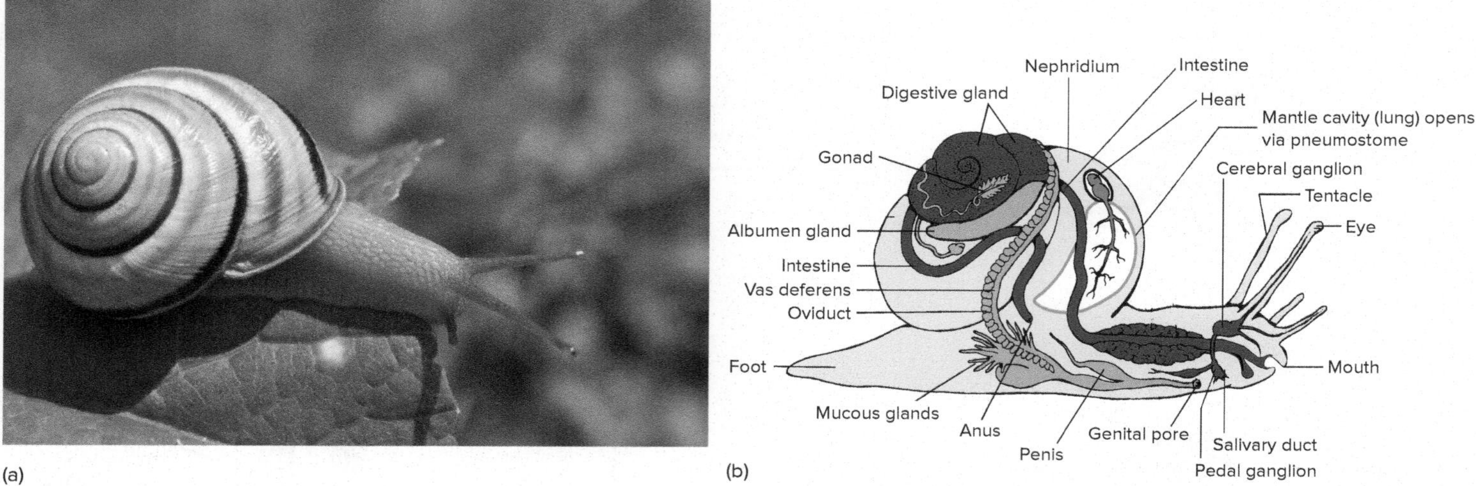

FIGURE 11.6

Gastropod Structure. (*a*) The Roman snail (*Helix pomatia*) is a pulmonate gastropod prized for preparation as escargot. It is distributed throughout Europe, where it is threatened by habitat destruction. Efforts to raise it commercially have been largely unsuccessful. (*b*) Internal structure of a pulmonate gastropod. The shell is not shown, and the visceral mass would be covered by the mantle. The portion of the mantle surrounding the mantle cavity is vascular and functions as a lung. The mantle cavity opens to the outside through a pneumostome. (*see figure 11.7*c)

Feeding and Digestion (Gastropoda)

Most gastropods feed by scraping algae or other small, attached organisms from their substrate using their radula. Others are herbivores that feed on larger plants, scavengers, or predators.

The anterior portion of the digestive tract may be modified into an extensible proboscis, which contains the radula. This structure is important for some predatory snails that must extract animal flesh from hard-to-reach areas. The digestive tract of gastropods, like that of most molluscs, is ciliated. Food is trapped in mucous strings and incorporated into a mucoid mass called the **protostyle,** which extends to the stomach and is rotated by cilia. A digestive gland in the visceral mass releases enzymes and acid into the stomach, and food trapped on the protostyle is freed and digested. Wastes form fecal pellets in the intestine.

Other Maintenance Functions (Gastropoda)

Gas exchange always involves the mantle cavity. Primitive gastropods had two gills; modern gastropods have lost one gill because of coiling. Some gastropods have a rolled extension of the mantle, called a **siphon,** that serves as an inhalant tube. Burrowing species extend the siphon to the surface of the substrate to bring in water. Gills are lost or reduced in land snails (pulmonates), but these snails have a richly vascular mantle for gas exchange between blood and air. Mantle contractions help circulate air and water through the mantle cavity (*see figure 26.13*).

Gastropods, like most molluscs, have an **open circulatory system.** During part of its circuit around the body, blood leaves the vessels and directly bathes cells in tissue spaces called sinuses. Molluscs typically have a heart consisting of a single, muscular ventricle and two auricles. Most gastropods have lost one member of the pair of auricles because of coiling and thus have a single auricle and a single ventricle (*see figure 11.6*b).

In addition to transporting nutrients, wastes, and gases, the blood of molluscs acts as a hydraulic skeleton. A **hydraulic skeleton** consists of fluid under pressure that may be confined to tissue spaces to extend body structures and to support the body. Molluscs contract muscles to force fluid, in this case blood, into a distant structure to push it forward. For example, snails have sensory tentacles on their heads, and if a tentacle is touched, retractor muscles can rapidly withdraw it. However, no antagonistic muscles exist to extend the tentacle. The snail slowly extends the tentacle by contracting distant muscles to squeeze blood into the tentacle from adjacent blood sinuses.

The nervous system of primitive gastropods is characterized by six ganglia located in the head-foot and visceral mass. In primitive gastropods, torsion twists the nerves that link these ganglia. The evolution of the gastropod nervous system has resulted in the untwisting of nerves and the concentration of nervous tissues into fewer, larger ganglia, especially in the head (*see figure 11.6*b).

Gastropods have well-developed sensory structures. Eyes may be at the base or at the end of tentacles. They may be simple pits of photoreceptor cells or they may consist of a lens and cornea. Statocysts are in the foot. **Osphradia** (sing. osphradium) are chemoreceptors in the anterior wall of the mantle cavity that detect sediment and chemicals in inhalant water or air. The osphradia of predatory gastropods help detect prey.

Primitive gastropods possessed two nephridia. In modern species, the right nephridium has disappeared, probably because of shell coiling. The nephridium consists of a sac with highly folded walls and connects to the reduced coelom, the pericardial cavity. Excretory wastes are derived largely from fluids filtered and secreted into the coelom from the blood. The nephridium modifies this waste by selectively reabsorbing certain ions and organic molecules. The nephridium opens to the mantle cavity or, in land snails, on the right side of the body adjacent to the mantle cavity and anal opening. Aquatic gastropod species excrete ammonia because they have access to water in which the toxic ammonia is diluted. Terrestrial snails must convert ammonia to a less-toxic form–uric acid. Because uric acid is relatively insoluble in water and less toxic, it can be excreted in a semisolid form, which helps conserve water.

Reproduction and Development (Gastropoda)

Many marine snails are dioecious. Gonads lie in spirals of the visceral mass (*see figure 11.6*b). Ducts discharge gametes into the sea for external fertilization.

Many other snails are monoecious, and internal, cross- fertilization is the rule. Copulation may result in mutual sperm transfer, or one snail may act as the male and the other as the female. A penis has evolved from a fold of the body wall, and portions of the female reproductive tract have become glandular and secrete mucus, a protective jelly, or a capsule around the fertilized egg. Some monoecious snails are protandric in that testes develop first, and after they degenerate, ovaries mature.

Eggs are shed singly or in masses for external fertilization. Internally fertilized eggs are deposited in gelatinous strings or masses. The large, yolky eggs of terrestrial snails are deposited in moist environments, such as forest-floor leaf litter, and a calcareous shell may encapsulate them. In most marine gastropods, spiral cleavage results in a free-swimming **trochophore larva** (*see figures 10.2*a *and 11.13*a) that develops into another free-swimming larva with foot, eyes, tentacles, and shell, called a **veliger larva.** Sometimes, the trochophore is suppressed, and the veliger is the primary larva. Torsion occurs during the veliger stage, followed by settling and metamorphosis to the adult.

Gastropod Diversity

Gastropods are traditionally divided into three subclasses. As we have encountered with other traditional group names (e.g., Trematoda, ctenophoran classes, and the lophotrochozan clade Syndermata), molecular analyses have prompted a reevaluation of these subclass taxa. Morphological characters used in evaluating gastropod relationships have included shell structure, internal anatomical features, and radular structure. Molecular analyses suggest that there has been considerable convergence (*see homoplasy; Chapter 7*) in these characters at the higher taxonomic levels, and this convergence means that the subclasses are polyphyletic groupings. The characters previously mentioned do seem to be useful at levels of classification below family. Phylogeneticists are working to group gastropods into unranked clades (without assigning taxonomic rankings) between class and family levels. In spite of being phylogenetically obsolete, the former subclass names (Prosobranchia, Opisthobranchia, and Pulmonata) are still used in reference to gastropod groups. The

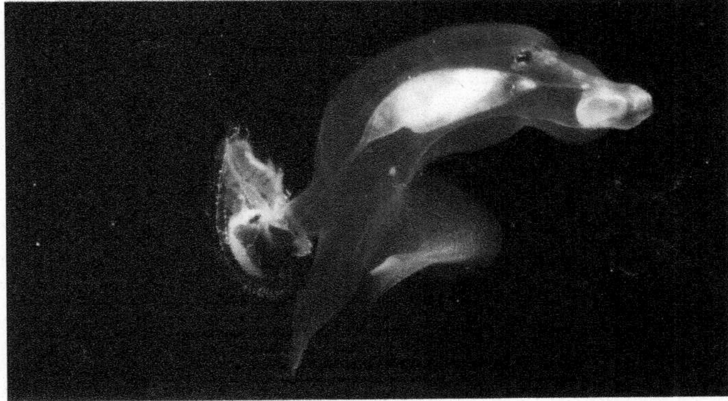

(a)

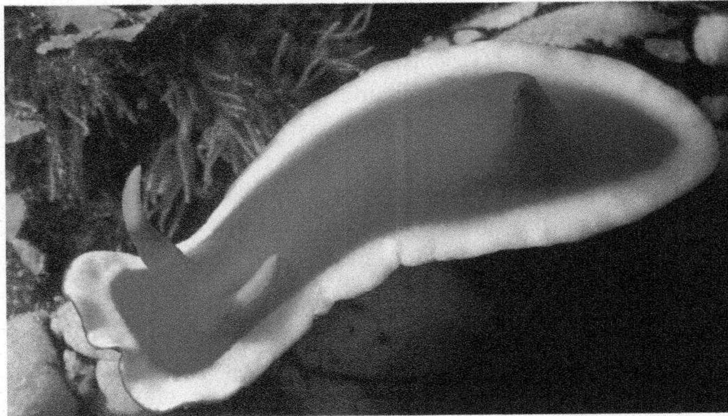

(b)

(c)

FIGURE 11.7

Variations in the Gastropod Body Form. (*a*) Subclass Prosobranchia. This heteropod (*Carinaria*) is a predator that swims upside down in the open ocean. Its body is nearly transparent. The head is at the left and the reduced shell is at the right. Heteropoda is a superfamily of prosobranchs composed of open-ocean, swimming snails with a finlike foot and reduced shell. (*b*) Subclass Opisthobranchia (*Hypselodoris*). Colorful nudibranchs have no shell or mantle cavity. The projections on the dorsal surface are used in gas exchange. In some nudibranchs, the dorsal projections are armed with nematocysts for protection. Nudibranchs prey on sessile animals, such as soft corals and sponges. (*c*) Subclass Pulmonata. Terrestrial slugs like this one (*Ariolimax columbianus*) lack an external shell. Note the opening to the lung (pneumostome). This photograph was taken in the temperate rainforest ecosystem of Olympic National Park, Washington.

terms are descriptive and familiar to generations of zoologists, and they are pervasive in gastropod literature. With about 65,000 species and over 600 families, the gastropods are second only to the insects (*see chapter 15*) in animal species diversity.

Prosobranch (Gr. *pro,* before + *branchia,* gills) gastropods are so named because their gills are positioned in the mantle cavity in front of their heart. Most prosobranchs are herbivores or deposit feeders; however, some are carnivorous. Cone snails are carnivorous species that inject venom into their fish, mollusc, or annelid prey with a radula modified into a hollow, harpoonlike structure. Prosobranch gastropods include most of the familiar marine snails and the abalone. This subclass also includes the heteropods. Heteropods are voracious predators, with very small shells or no shells. Their foot is modified into an undulating "fin" that propels the animal through the water (figure 11.7*a*).

Opisthobranchs (Gr. *opistho,* behind) are so named because their gills are positioned to the side of their heart. They include sea hares, sea slugs, and their relatives (figure 11.7*b*). They are mostly marine, and their shell, mantle cavity, and gills are often reduced or lost in these animals, but they are not defenseless. Many acquire undischarged nematocysts (*see figure 9.8*) from their cnidarian prey, which they use to ward off predators. The pteropods have a foot modified into thin lobes for swimming.

The pulmonates (L. *pulmo,* lung) are predominantly freshwater or terrestrial species (*see figure 11.6*). These snails are mostly herbivores and have a long radula for scraping plant material. The mantle cavity of pulmonate gastropods is highly vascular and serves as a lung. Air or water moves in or out of the opening of the mantle cavity, the **pneumostome.** In addition to typical freshwater or terrestrial snails, the pulmonates include terrestrial slugs (figure 11.7*c*).

SECTION 11.3 THINKING BEYOND THE FACTS

Most gastropods have shells for protection from desiccation and predators. Slugs, however, do not have shells. What is a slug's protective mechanism?

11.4 CLASS BIVALVIA

LEARNING OUTCOMES

1. Compare members of the class Bivalvia to the generalized molluscan body form.
2. Explain the structure, function, and reproduction of members of the class Bivalvia.
3. Describe bivalve diversity and ecology.
4. Analyze the effect of sediment build-up resulting from erosion on river ecosystems containing bivalve populations.

With close to 15,000 species, the class Bivalvia (bi'val"ve-ah) (L. *bis,* twice + *valva,* leaf) is the second largest molluscan class. This class includes the clams, oysters, mussels, and scallops (*see table 11.1*). A sheetlike mantle and a shell consisting of two valves (hence, the class name) cover these laterally compressed animals. Many bivalves are edible, and some form pearls. Because most bivalves are filter feeders, they are valuable in removing bacteria and other microorganisms from polluted water.

Shell and Associated Structures

The two convex halves of the shell are called **valves.** Along the dorsal margin of the shell is a proteinaceous hinge and a series of tongue-and-groove modifications of the shell, called teeth, that prevent the valves from twisting (figure 11.8). The oldest part of the shell is the **umbo,** a swollen area near the shell's anterior margin. Although bivalves appear to have two shells, embryologically, the shell forms as a single structure. The shell is continuous along its dorsal margin, but the mantle, in the region of the hinge, secretes relatively greater quantities of protein and relatively little calcium carbonate. The result is an elastic hinge ligament. The elasticity of the hinge ligament opens the valves when certain muscles relax.

Adductor muscles at either end of the dorsal half of the shell close the shell. Anyone who has tried to force apart the valves of a bivalve mollusc knows the effectiveness of these muscles. This is important for bivalves because their primary defense against predatory sea stars is to tenaciously refuse to open their shells. Chapter 16 explains how sea stars have adapted to meet this defense strategy.

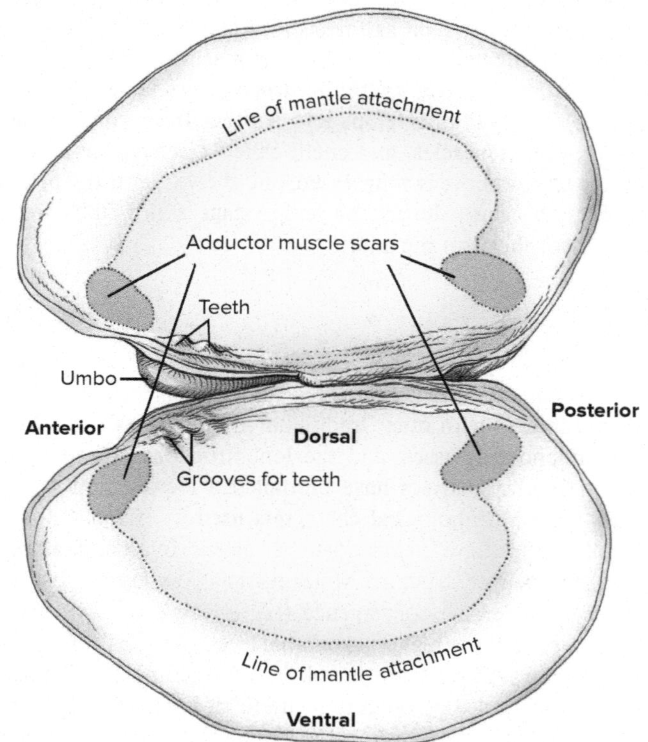

FIGURE 11.8

Inside View of a Bivalve Shell. The umbo is the oldest part of the bivalve shell. As the bivalve grows, the mantle lays down more shell in concentric lines of growth.

The bivalve mantle attaches to the shell around the adductor muscles and near the shell margin. If a sand grain or a parasite lodges between the shell and the mantle, the mantle secretes nacre around the irritant, gradually forming a pearl. The Pacific oysters, *Pinctada margaritifera* and *Pinctada mertensi,* form the highest-quality pearls.

Gas Exchange, Filter Feeding, and Digestion

Bivalve adaptations to sedentary, filter-feeding lifestyles include the loss of the head and radula and, except for a few bivalves, the expansion of cilia-covered gills. Gills form folded sheets (lamellae), with one end attached to the foot and the other end attached to the mantle. The mantle cavity ventral to the gills is the inhalant region, and the cavity dorsal to the gills is the exhalant region (figure 11.9*a*). Cilia move water into the mantle cavity through an incurrent opening of the mantle. Sometimes, this opening is at the end of a siphon, which is an extension of the mantle. A bivalve buried in the substrate can extend its siphon to the surface and still feed and exchange gases. Water moves from the mantle cavity into small pores in the surface of the gills, and from there, into vertical channels in the gills, called water tubes. In moving through water tubes, blood and water are in close proximity, and gases exchange by diffusion (figure 11.9*b*). Water exits the bivalve through a part of the mantle cavity at the dorsal aspect of the gills, called the suprabranchial chamber, and through an excurrent opening in the mantle (figure 11.9*a*).

The gills trap food particles brought into the mantle cavity. Zoologists originally thought that ciliary action was responsible for the trapping. However, the results of a recent study indicate that cilia and food particles have little contact. The food-trapping mechanism is unclear, but once food particles are trapped, cilia move them to ciliated tracts called **food grooves** along the dorsal and ventral margins of the gills. These ciliated tracts move food toward the mouth (figure 11.10). Cilia covering leaflike **labial palps** on either

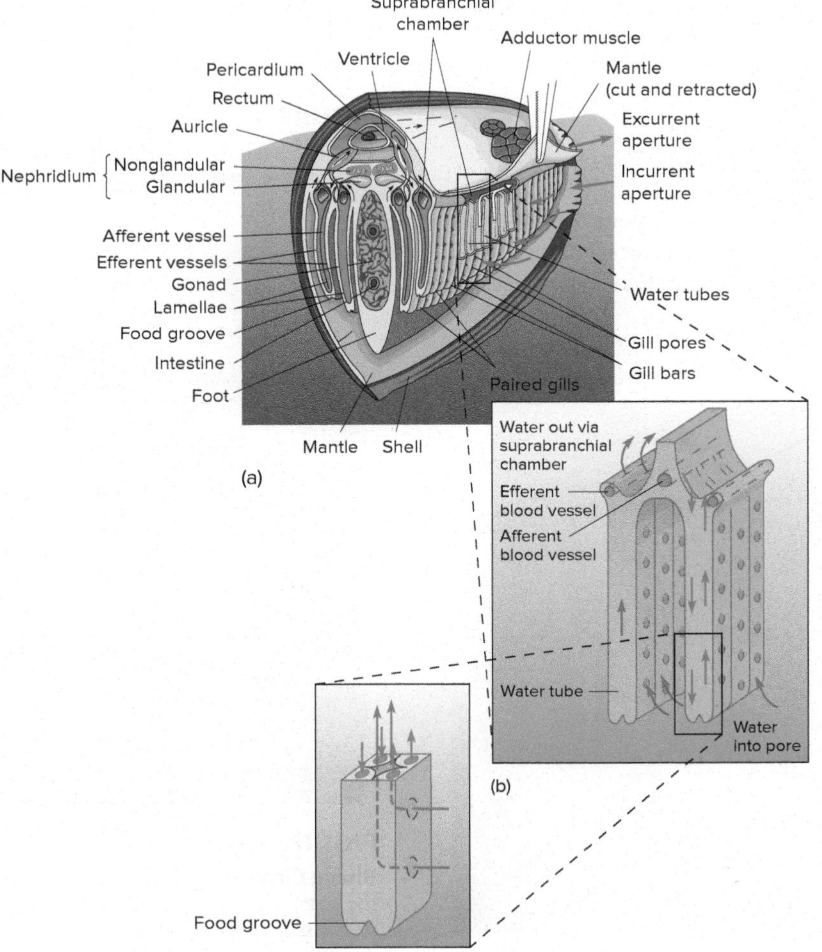

FIGURE 11.9

Lamellibranch Gill of a Bivalve. (*a*) Blue arrows indicate incurrent and excurrent water currents. Food is filtered as water enters water tubes through pores in the gills. (*b*) Cross section through a portion of a gill. Water passing through a water tube is in close proximity to blood. Water and blood exchange gases in the water tubes. Blue arrows show the path of water. Red arrows show the path of blood.

side of the mouth also sort filtered food particles. Cilia carry small particles into the mouth and move larger particles to the edges of the palps and gills. This rejected material, called pseudofeces, falls, or is thrown, onto the mantle, and a ciliary tract on the mantle transports the pseudofeces posteriorly. Water rushing out when valves are forcefully closed washes pseudofeces from the mantle cavity.

The digestive tract of bivalves is similar to that of other molluscs (figure 11.11*a*). Food entering the esophagus entangles in a mucoid food string, which extends to the stomach and is rotated by cilia lining the digestive tract. A consolidated mucoid mass, the **crystalline style,** projects into the stomach from a diverticulum, called the style sac (figure 11.11*b*). Enzymes for carbohydrate and fat digestion are incorporated into the crystalline style. Cilia of the style sac rotate the style against a chitinized **gastric shield.** This abrasion and acidic conditions in the stomach dislodge enzymes. The mucoid food string winds around the crystalline style as it rotates, which pulls the food string farther into the stomach from the esophagus. This action and the acidic pH in the stomach dislodge food particles in the food string. Further sorting separates fine particles from undigestible coarse materials. The latter are sent on to the intestine. Partially digested food from the stomach enters a digestive gland for intracellular digestion. Cilia carry undigested wastes in the digestive gland back to the stomach and then to the intestine. The intestine empties through the anus near the excurrent opening, and excurrent water carries feces away.

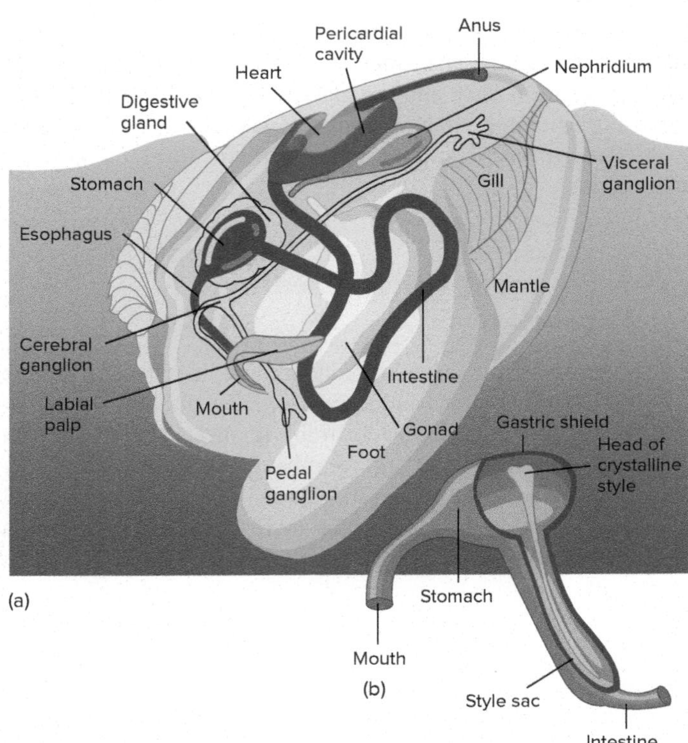

FIGURE 11.11

Bivalve Structure. (*a*) Internal structure of a bivalve. (*b*) Bivalve stomach, showing the crystalline style and associated structures.

Other Maintenance Functions (Bivalvia)

Bivalves have an open circulatory system. Blood flows from the heart to tissue sinuses, nephridia, gills, and back to the heart (figure 11.12). The mantle is an additional site for oxygenation. In some bivalves, a

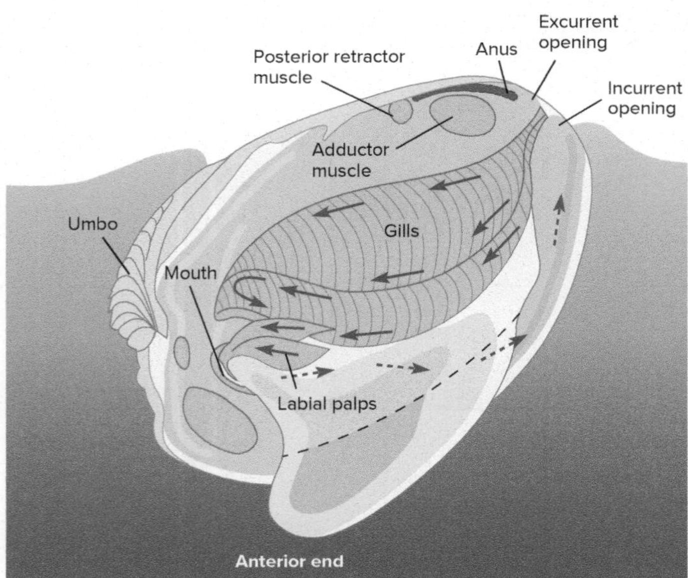

FIGURE 11.10

Bivalve Feeding. Solid purple arrows show the path of food particles after the gills filter them. Dashed purple arrows show the path of particles that the gills and the labial palps reject.

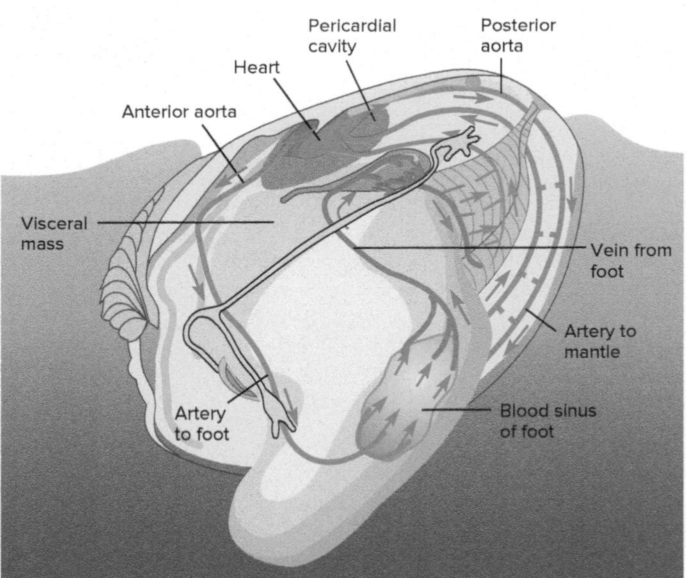

FIGURE 11.12

Bivalve Circulation. Blood flows (red arrows) from the single ventricle of the heart to tissue sinuses through anterior and posterior aortae. Blood from tissue sinuses flows to the nephridia, to the gills, and then to the auricles of the heart. In all bivalves, the mantle is an additional site for oxygenation. In some bivalves, a separate aorta delivers blood to the mantle. This blood returns directly to the heart. The ventricle of bivalves is always folded around the intestine. Thus, the pericardial cavity (the coelom) encloses the heart and a portion of the digestive tract.

separate aorta delivers blood directly to the mantle. Two nephridia are below the pericardial cavity (the coelom). Their duct system connects to the coelom at one end and opens at nephridiopores in the anterior region of the suprabranchial chamber (*see figure 11.11*).

The circulatory system of a bivalve is also used in locomotion and burrowing. Blood pumped into the foot causes the foot to extend into the substrate. Muscles in the foot contract to cause the foot to swell into an anchor. Finally, retractor muscles (*see figure 11.10*) attached from the shell to the visceral mass and foot then pull the body and shell into the substrate.

The nervous system of bivalves consists of three pairs of interconnected ganglia associated with the esophagus, the foot, and the posterior adductor muscle. The margin of the mantle is the principal sense organ. It always has sensory cells, and it may have sensory tentacles and photoreceptors. In some species (e.g., scallops), photoreceptors are in the form of complex eyes with a lens and a cornea. Other receptors include statocysts near the pedal ganglion and an osphradium in the mantle, beneath the posterior adductor muscle.

Reproduction and Development (Bivalvia)

Most bivalves are dioecious. A few are monoecious, and some of these species are protandric. Gonads are in the visceral mass, where they surround the looped intestine. Ducts of these gonads open directly to the mantle cavity or by the nephridiopore to the mantle cavity.

Most bivalves exhibit external fertilization. Gametes exit through the suprabranchial chamber of the mantle cavity and the exhalant opening. Development proceeds through trochophore and veliger stages (figure 11.13*a,b*). When the veliger settles to the substrate, it assumes the adult form.

Most freshwater bivalves brood their young. Fertilization occurs in the mantle cavity by sperm brought in with inhalant water. Some brood their young in maternal gills through reduced trochophore and veliger stages. Young clams are shed from the gills. Freshwater bivalves in the family Unionidae brood their young to a modified veliger stage called a **glochidium,** which is parasitic on fishes (figure 11.13*c*). These larvae possess two tiny valves,

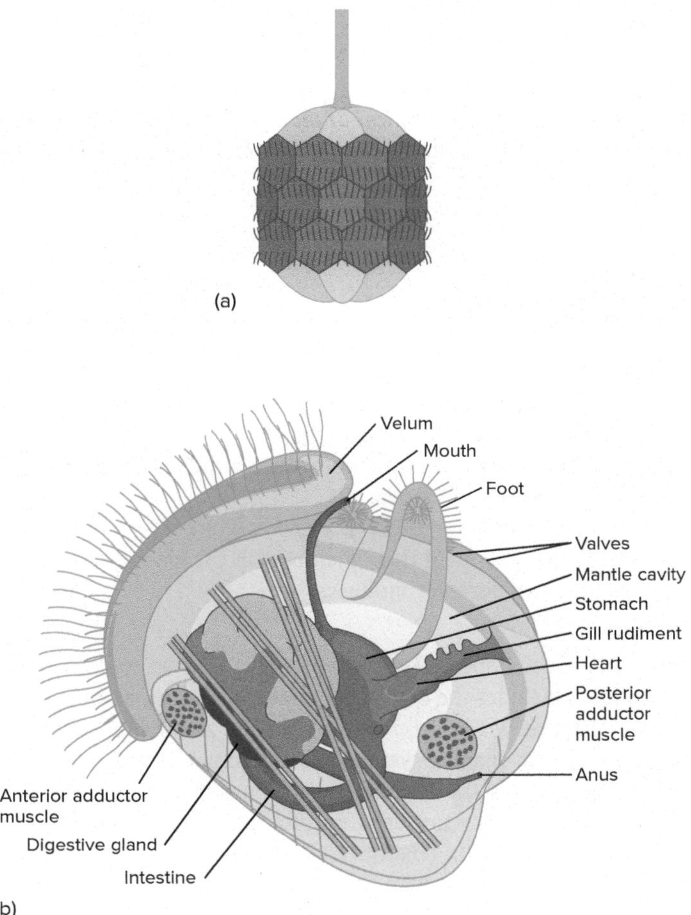

FIGURE 11.13

Larval Stages of Bivalves. (*a*) Trochophore larva (0.4 mm) of the marine clam *Yoldia limatula*. (*b*) Veliger (0.5 mm) of an oyster. (*c*) Glochidia (1.0 mm) of a freshwater clam. Note the two toothlike hooks on upper glochidium used to attach to fish gills.

and some species have toothlike hooks. Larvae exit through the exhalant aperture and sink to the substrate. Most die. If a fish contacts a glochidium, however, the larva attaches to the gills, fins, or another body part and may begin to feed on host tissue. The fish may form a cyst around the larva. The mantles of some freshwater bivalves have elaborate modifications that present a fishlike lure to entice predatory fish. When a fish attempts to feed on the lure, the bivalve ejects glochidia onto the fish (figure 11.14). After several weeks of larval development, during which a glochidium begins acquiring its adult structures, the miniature clam falls from its host and takes up its filter-feeding lifestyle. The glochidium is a dispersal stage for an otherwise sedentary animal and probably has little effect on the fish.

Bivalve Diversity

The class Bivalvia, like Gastropoda, has undergone recent taxonomic revisions. Molecular and morphological studies of the past 10 years have resulted in the description of four subclasses and 110 extant families of Bivalvia. Bivalves are worldwide in distribution and live in nearly all freshwater and marine habitats (figure 11.15). They may completely or partially bury themselves in sand or mud; attach to solid substrates; or bore into submerged wood, coral, or limestone. They occur at all depths, from intertidal zones to the deepest regions of the world's oceans.

The mantle margins of burrowing bivalves are frequently fused to form distinct openings in the mantle cavity (siphons). This fusion helps direct the water washed from the mantle cavity during burrowing and helps keep sediment from accumulating in the mantle cavity.

The term "mussel" is usually reserved for bivalves that secrete byssal threads from a byssal gland associated with their foot. A byssal thread is comprised of keratin and other proteins, secreted as a liquid onto a hard substrate, and hardened through polymerization and other chemical modifications. Byssal threads attach mussels to substrates in wave-swept intertidal regions. *Mytilus* is a common edible muscle that occurs in large masses in the upper intertidal regions. Many freshwater bivalves are also referred to as mussels. Some of these, like the invasive zebra mussel (*Dreissena polymorpha*; *see Wildlife Alert, pages 202–203*), use byssal threads throughout their lives. All native American freshwater mussels use byssal threads only during their juvenile stages. Oyster (family Ostreidae) veligers attach to a substrate using drops of adhesive from a byssal gland. Adult oysters become permanently attached by cementation of one valve to a hard surface.

Boring bivalves live beneath the surface of limestone, clay, coral, wood, and other substrates. Boring begins when the larvae settle to the substrate, and the anterior margin of their valves mechanically abrades the substrate. Acidic secretions from the mantle margin that dissolve limestone sometimes accompany physical abrasion. As the bivalve grows, it is often imprisoned in its rocky burrow because the most recently bored portions of the burrow are larger in diameter than portions bored earlier.

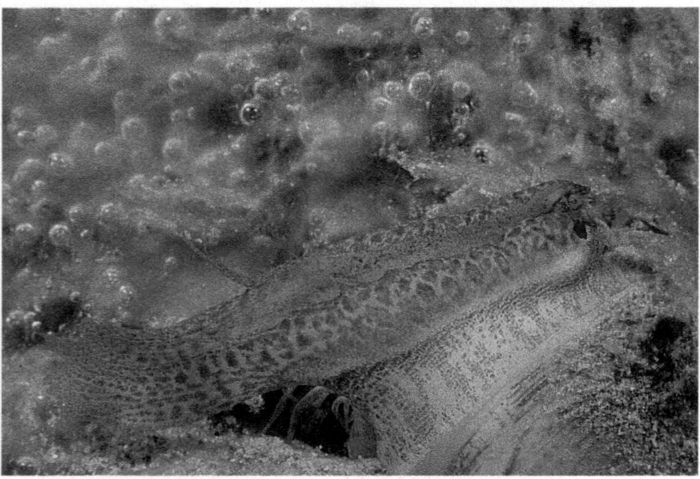

FIGURE 11.14

Class Bivalvia. This photograph shows a modification of the mantle of a freshwater bivalve (*Lampsilis reeviana*) into a lure. The edge of the bivalve shell is shown in the lower right corner of the photograph. When a fish approaches and bites at the lure, glochidia are released onto the fish.
Courtesy of Dr. William Roston

Section 11.4 Thinking Beyond the Facts

What anatomical feature typical of most molluscs and bivalve ancestors is absent in modern bivalves, and what accounts for its loss in members of this class?

11.5 CLASS CEPHALOPODA

LEARNING OUTCOMES

1. Compare members of the class Cephalopoda to the generalized molluscan body form.
2. Explain the structure, function, and reproduction of members of the class Cephalopoda.
3. Describe cephalopod diversity and ecology.
4. Justify the statement that "members of the class Cephalopoda are the most complex molluscs."

The class Cephalopoda (sef′ah-lop″o-dah) (Gr. *kephale,* head + *podos,* foot) includes the octopuses, squid, cuttlefish, and nautiluses (figure 11.16; *see table 11.1; figure 11.1*). They are the most complex molluscs and, in many ways, the most complex invertebrates. The anterior portion of the cephalopod foot has been modified into a circle of tentacles or arms used for prey capture, attachment, locomotion, and copulation (figure 11.17). The foot is also incorporated into a funnel associated with the mantle cavity and is used for jetlike locomotion. The molluscan body plan is further modified in that the cephalopod head is in line with the visceral mass. Cephalopods have a highly muscular mantle that encloses all of the body except the head and tentacles. The mantle acts as a pump to bring large quantities of water into the mantle cavity.

FIGURE 11.15

Bivalve Diversity. (*a*) This giant clam (*Tridacna*) is one of two genera and nine species of giant clams. Giant clams occur in association with coral reefs throughout the tropical Indo-Pacific region. Giant clams are unusual in that they derive a substantial portion of their nutrition from a relationship with photosynthetic, symbiotic algae (zooxanthellae) that live in their large, fleshy mantle. Their mantles are typically brightly colored as a result of this association. (*b*) The giant rock scallop (*Hinnites*) occurs on the Pacific coast from British Columbia to central Baja California. As a mature adult, it is large (up to 25 cm) and is attached to a hard substrate by mantle secretions. Before attachment, however, a young rock scallop can swim in a jet-propulsion fashion by opening its valves and clapping them closed. The brightly colored mantle is highly sensory. (*c*) The geoduck (pronounced "gooey duck") (*Panopea generosa*) is the largest burrowing bivalve. Most mature individuals have a mass of about 1 kg, but unusually large individuals have attained masses of over 6 kg and have siphons of 2 m in length. It is also the longest lived bivalve, with some individuals living more than 100 years. It burrows in soft mud and extends its siphon to the surface for filter feeding and dispersing gametes.

(a)

(b)

FIGURE 11.16

Class Cephalopoda. (*a*) Chambered nautilus (*Nautilus*). (*b*) A cuttlefish (*Sepia*).

(a) ©Michael Aw/Photodisc/Getty Images (b) ©Purestock/SuperStock

Cephalopod Shells

Ancestral cephalopods probably had a conical shell. The only living cephalopod that possesses an external shell is the nautilus (six species, family Nautilidae; *see figure 11.16*a). Septa subdivide its coiled shell. As the nautilus grows, it moves forward, secreting new shell around itself and leaving an empty septum behind. Only the last chamber is occupied. When formed, these chambers are fluid filled. A cord of tissue called a siphuncle perforates the septa, absorbing fluids by osmosis and replacing them with metabolic gases. The amount of gas in the chambers is regulated to alter the buoyancy of the animal.

In all other cephalopods, the shell is reduced or absent. In cuttlefish (order Sepiida), the shell is internal and laid down in thin layers, leaving small, gas-filled spaces that increase buoyancy. Cuttlefish shell, called cuttlebone, is used to make powder for polishing and is fed to pet birds to supplement their diet with calcium. The shell of a squid (order Teuthida) is reduced to an internal, chitinous structure called the pen or gladius. (The latter name is based on its resemblance to the Roman short sword of the same name). In addition, squid also have cartilaginous plates in the mantle wall, neck, and head that support the mantle and protect the brain. The shell is absent or reduced to a pair of stylets in most octopuses, although one small suborder—the Cirrina—possesses cartilaginous fin supports that are thought of as remnants of a shell.

Locomotion (Cephalopoda)

As predators, cephalopods depend on their ability to move quickly using a jet-propulsion system. The mantle of cephalopods contains radial and circular muscles. When circular muscles contract, they decrease the volume of the mantle cavity and close collarlike valves to prevent water from moving out of the mantle cavity between the head and the mantle wall. Water is thus forced out of a narrow siphon. Muscles attached to the siphon control the direction of the animal's movement. Radial mantle muscles bring water into the mantle cavity by increasing the cavity's volume. Posterior fins act as stabilizers in squid and also aid in propulsion and steering in cuttlefish. "Flying squid" (family Onycoteuthidae) have been clocked at speeds of 30 km/hr. Octopuses are more sedentary animals. They may use jet propulsion in an escape response, but normally, they crawl over the substrate using their tentacles. In most cephalopods, the use of the mantle in locomotion coincides with the loss of an external shell, because a rigid external shell would preclude the jet-propulsion method of locomotion described. The agility that accompanied the loss of a shell, and visual trickery that will be described shortly, may help explain how cephalopod populations have survived as successful hunters even though they are prey for many bony fishes, sharks, and marine mammals.

Feeding and Digestion (Cephalopoda)

Cephalopods are predators, and they usually locate their prey by sight and capture prey with tentacles that have adhesive cups. In squid, the margins of these cups are reinforced with tough protein and sometimes possess small hooks (figure 11.18). Cuttlefish and nautiluses feed on small invertebrates on the ocean floor. Octopuses are nocturnal hunters and feed on snails, fishes, and crustaceans. Squid feed on fishes and shrimp, which they kill by biting across the back of the head.

All cephalopods have jaws and a radula. The jaws are powerful, beaklike structures for tearing food, and the radula rasps food, forcing it into the mouth cavity. Virtually all cephalopods studied are venomous. They capture prey with their tentacles, pierce shells and other body parts with their jaws, and eject venom from their salivary glands onto (or into) their prey. In at least one species, the extremely venomous blue-ringed octopus (*Hapalochlaena*), the venom is produced by endosymbiotic bacteria living within the octopus's salivary glands (*see figure 11.1*).

The digestive tract of cephalopods is muscular, and peristalsis (coordinated muscular waves) replaces ciliary action in moving food. Most digestion occurs in a stomach and a large cecum. Digestion is primarily extracellular, with large digestive glands supplying enzymes. An intestine ends at the anus, near the funnel, and exhalant water carries wastes out of the mantle cavity.

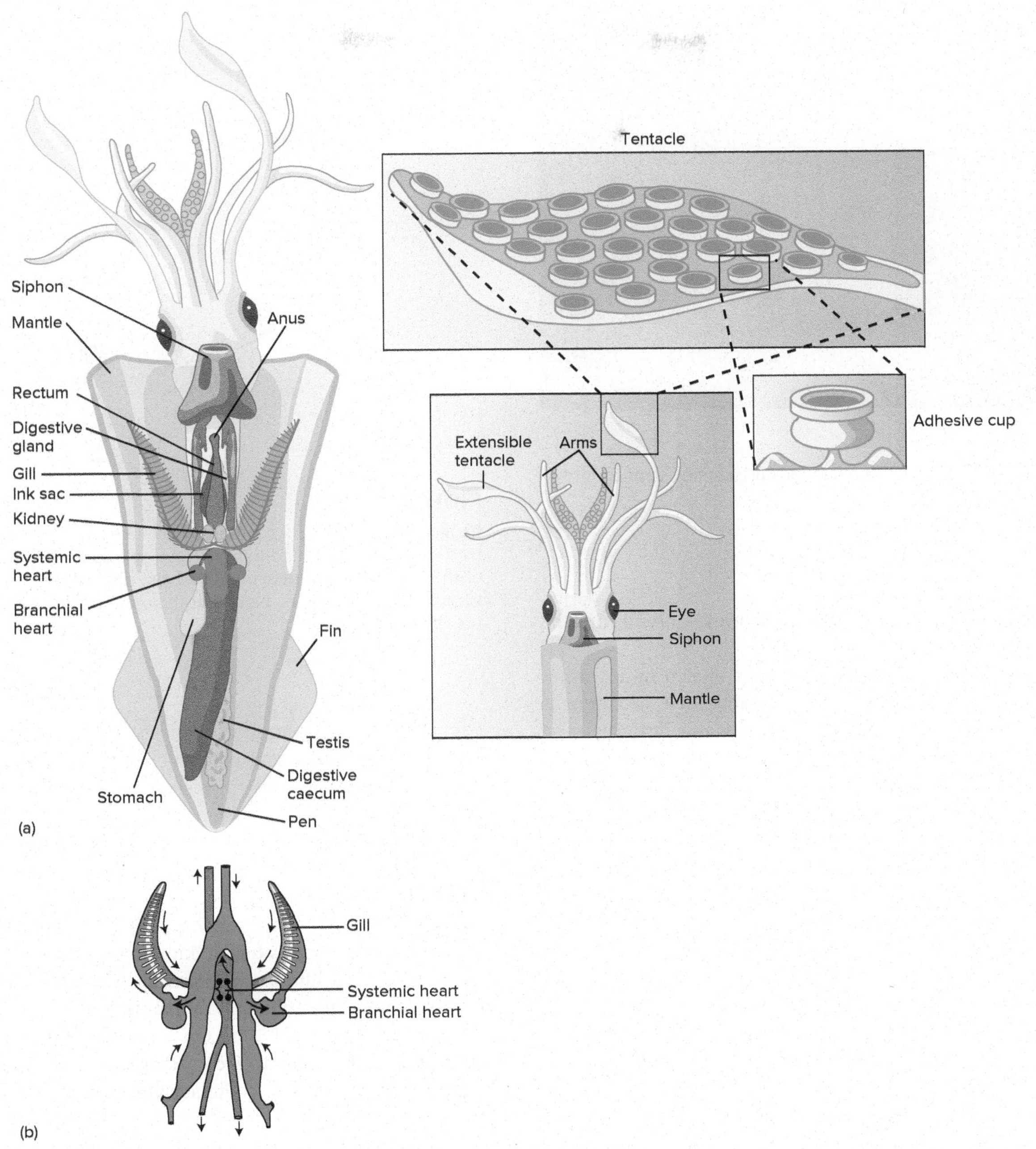

FIGURE 11.17

Internal Structure of the Squid, *Loligo*. The shell of most cephalopods is reduced or absent, and the foot is modified into a funnel-shaped siphon and a circle of tentacles and/or arms that encircle the head. (*a*) The dissected anatomy of a squid. The mantle is shown cut open revealing the visceral mass and gills within the mantle cavity. The inset to the right shows the undissected anatomy of the squid including the structure of a tentacle and adhesive cups. (*b*) The systemic and branchial hearts of a squid. Black arrows show the path of blood flow to and from the gills.

Other Maintenance Functions (Cephalopoda)

Cephalopods, unlike other molluscs, have a **closed circulatory system.** Blood is confined to vessels throughout its circuit around the body. Capillary beds connect arteries and veins, and exchanges of gases, nutrients, and metabolic wastes occur across capillary walls. In addition to having a systemic heart consisting of two auricles and one ventricle, cephalopods have contractile arteries and structures called branchial hearts. The latter are at the base of each gill and help move blood through the gill. These modifications increase blood pressure and the rate of blood flow—necessary for active animals with relatively high metabolic rates. Large quantities of water circulate over the gills at all times. Cephalopods exhibit greater excretory efficiency because of the closed circulatory system. A close association of blood vessels with nephridia

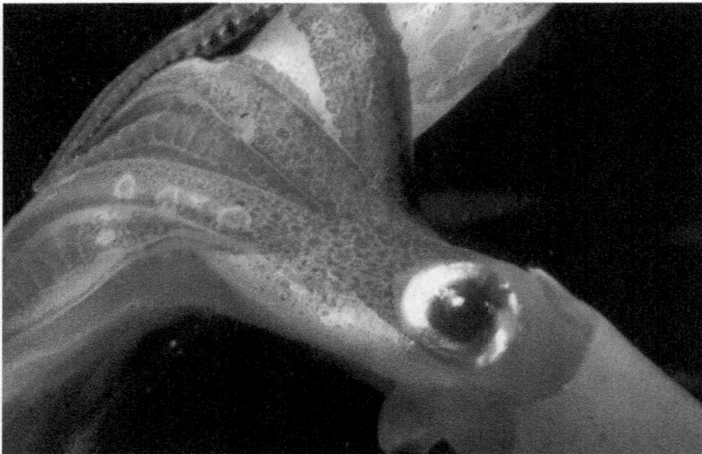

FIGURE 11.18

Cephalopod Arms and Tentacles. Cephalopods use suction cups for prey capture and as holdfast structures.

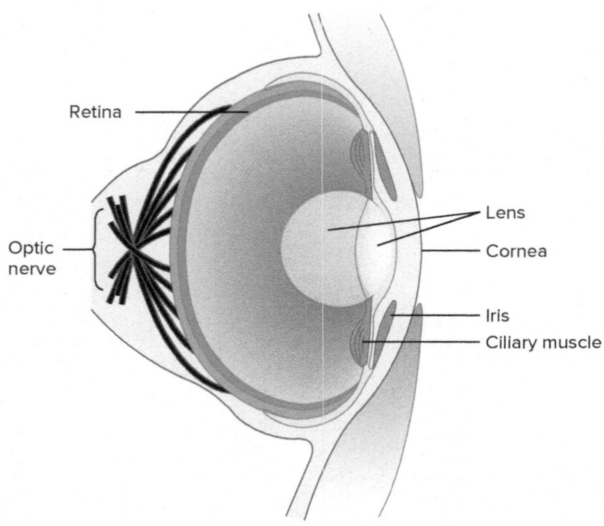

FIGURE 11.19

Cephalopod Eye. The eye is immovable in a supportive and protective socket of cartilages. It contains a rigid, spherical lens. An iris in front of the lens forms a slitlike pupil that can open and close in response to varying light conditions. Note that the optic nerve comes off the back of the retina.

allows wastes to filter and secrete directly from the blood into the excretory system.

The cephalopod nervous system is unparalleled in any other invertebrate. Cephalopod brains are large, and their evolution is directly related to cephalopod predatory habits and dexterity. The brain forms by a fusion of ganglia. Large areas are devoted to controlling muscle contraction (e.g., swimming movements and sucker closing), sensory perception, and functions such as memory and decision making. Research on cephalopod brains has provided insight into human brain functions.

The eyes of octopuses, cuttlefish, and squid are similar in structure to vertebrate eyes, and vision is a primary sense that is used in finding prey and in intraspecific interactions (figure 11.19; *see figures 11.18 and 24.16*c). (This similarity is an excellent example of convergent evolution.) In contrast to the vertebrate eye, nerve cells leave the eye from the outside of the eyeball, so that no blind spot exists. (The blind spot of the vertebrate eye is a region of the retina where no photoreceptors exist because of the convergence of nerve cells into the optic nerve. When light falls on the blind spot, no image is perceived.) Like many aquatic vertebrates, cephalopods focus by moving the lens back and forth. Cephalopods can form images, distinguish shapes, and discriminate brightness and patterns. One species of squid can discriminate some colors. The nautiloid eye is less complex. It lacks a lens, and the interior is open to seawater: thus, it acts as a pinhole camera. Nautiloids apparently rely on olfaction to a greater extent than do other cephalopods.

Cephalopod statocysts respond to gravity and acceleration, and are embedded in cartilages next to the brain. Osphradia are present only in *Nautilus.* Tactile receptors and additional chemoreceptors are widely distributed over the body.

In spite of most species being color-blind, cephalopods use color and pattern changes in remarkable ways. Cephalopods have pigment cells called **chromatophores,** which are located in their mantle and body wall. When tiny muscles attached to these pigment cells contract, the chromatophores quickly expand and change the color of the animal. Color and pattern changes, in combination with ink discharge, function in alarm responses. In defensive displays, color changes may spread in waves over the body to form large, flickering patterns. Pattern changes may also help cephalopods to blend with their background. The cuttlefish, *Sepia,* can even make a remarkably good impression of a checkerboard background. Color changes are also involved with courtship displays. Some species combine chromatophore displays with bioluminescence. Light emission is the result of a symbiotic relationship with bioluminescent bacteria located within the cephalopod's mantle cavity.

The firefly squid, *Watasenia scintillans*, is found in the Northwest Pacific. It is a small (8 cm), deep-water (200 to 400 m) species whose members come to the surface by the millions from March through May for mating and feeding. They produce a blue bioluminescence using chromatophores located nearly everywhere on their bodies. Their incredible light show is thought to function in attracting mates, confusing predators, and attracting prey. Their eyes possess three types of photoreceptor cells (cones), which suggests that they probably have color vision. (All other cephalopods investigated have a single type of photoreceptor cell.)

All cephalopods except nautiloids and members of the octopus suborder Cirrina possess an ink gland that opens just behind the anus. Ink is a brown or black fluid containing melanin and other chemicals that is discharged into exhalent water from the siphon. Discharged ink confuses a predator, allowing the cephalopod to escape. For example, *Sepiola* reacts to a predator by darkening itself with chromatophore expansion prior to releasing ink. After ink discharge, *Sepiola* changes to a lighter color again to assist its escape. The predator is left with a mouthful of ink.

Learning

The complex nervous system of cephalopods contrasts sharply with that of other molluscs (*see figure 24.6*e). Octopuses and cuttlefish have larger brains relative to body weight than any other invertebrate, fish, or amphibian. The octopus brain is modified into complex lobes that serve as visual and tactile centers. Early scientific work on cephalopod learning began in the late 1940s at Stazione Zoologica in Naples, Italy. Experiments with *Octopus vulgaris* demonstrated that this octopus could be trained to attack, kill, and feed on a crab when presented with certain visual stimuli. Surgical removal of parts of the brain demonstrated that regions of the brain called vertical and superior frontal lobes were the learning and memory centers for visual stimuli. Since these early experiments, cephalopods have been trained to negotiate mazes; distinguish shapes, sizes, and patterns in objects; and remember what they have learned. (Octopuses can remember learned information for up to four months.) Cephalopods use both chemical and auditory stimuli in their behaviors. There have been reports of a higher form of learning in octopuses. Observational learning involves an animal learning by observing other animals performing a task. Initial reports of observational learning have not been reproduced, and these investigations continue. Interpreting all of this information has been very difficult.

The questions of how and why intelligence evolved in the cephalopods are intriguing. The evolution of intelligence, for example in primate mammals, is usually associated with long lives and social interactions. Cephalopods have neither long lives nor complex social structure. Most cephalopods live about one year (some octopuses live up to four years). Octopuses are solitary, and squid and cuttlefish school with little social structure. Cephalopod brains and intelligence may have evolved in response to avoiding predators while living as active predators themselves. Variable food resources present in changing habitats could select for increased intelligence. Cephalopods avoid predators by posturing–dangling their arms and tentacles among seaweed to mimic their surroundings. Their color changes are used in camouflage and interactions with other cephalopods. The male Caribbean reef squid, *Sepioteuthis sepioldea,* uses one grey color display to attract females and a striping display to ward off competitor males. If a male is positioned between a female and another male, the side of his body facing the female displays the courtship pattern and the side facing the male displays the striping pattern. While color changes are not unusual in animals, they are usually hormonally controlled. Chromatophore changes in cephalopods are controlled by the nervous system and can occur in less than one second. This combination of nervous system functions is truly unique among the invertebrates and would not be possible without their large, complex brains.

Reproduction and Development (Cephalopoda)

Cephalopods are dioecious with gonads in the dorsal portion of the visceral mass. The male reproductive tract consists of testes and structures for encasing sperm in packets called **spermatophores.** The female reproductive tract produces large, yolky eggs and is modified with glands that secrete gel-like cases around eggs. These cases frequently harden on exposure to seawater.

One tentacle of male cephalopods, called the **hectocotylus,** is modified for spermatophore transfer. In *Loligo* and *Sepia,* the hectocotylus has several rows of smaller suckers capable of picking up spermatophores. During copulation, male and female tentacles intertwine, and the male removes spermatophores from his mantle cavity. The male inserts his hectocotylus into the mantle cavity of the female and deposits a spermatophore near the opening to the oviduct. Spermatophores have an ejaculatory mechanism that frees sperm from the baseball-bat-shaped capsule. Eggs are fertilized as they leave the oviduct and are deposited singly or in stringlike masses. They usually attach to some substrate, such as the ceiling of an octopus's den. Octopuses clean developing eggs of debris with their arms and squirts of water.

Cephalopods develop in the confines of the egg membranes, and the hatchlings are miniature adults. Young are never cared for after hatching.

Section 11.5 Thinking Beyond the Facts

What anatomical feature(s) typical of most molluscs and cephalopod ancestors are absent or highly modified in modern cephalopods, and what accounts for differences in members of this class?

11.6 CLASS POLYPLACOPHORA

LEARNING OUTCOME

1. Compare members of the class Polyplacophora to the generalized molluscan body form.

The class Polyplacophora (pol″e-pla-kof′o-rah) (Gr. *polys,* many + *plak,* plate + *phoros,* to bear) contains about 950 species of chitons. Chitons are common inhabitants of hard substrates in shallow marine water. Early Native Americans ate chitons. Chitons have a fishy flavor but are tough to chew and difficult to collect.

Chitons have a reduced head, a flattened foot, and a shell that divides into eight articulating dorsal valves (figure 11.20*a*). A muscular mantle that extends beyond the margins of the shell and foot covers the broad foot (figure 11.20*b*). The mantle cavity is restricted to the space between the margin of the mantle and the foot. Chitons crawl over their substrate in a manner similar to gastropods. The muscular foot allows chitons to securely attach to a substrate and withstand strong waves and tidal currents. When chitons are disturbed, the edges of the mantle tightly grip the substrate, and foot muscles contract to raise the middle of the foot, creating a powerful vacuum that holds the chiton in place. Articulations in the shell allow chitons to roll into a ball when dislodged from the substrate.

A linear series of gills is in the mantle cavity on each side of the foot. Cilia on the gills create water currents that enter below the anterior mantle margins and exit posteriorly. The digestive, excretory, and reproductive tracts open near the exhalant area of the mantle cavity, and exhalant water carries products of these systems away.

Most chitons feed on attached algae. A chemoreceptor, the subradular organ, extends from the mouth to detect food, which the radula rasps from the substrate. Mucus traps food, which then enters

(a)

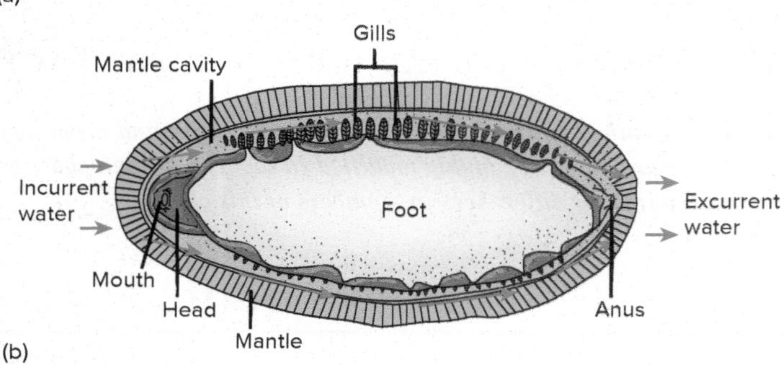

(b)

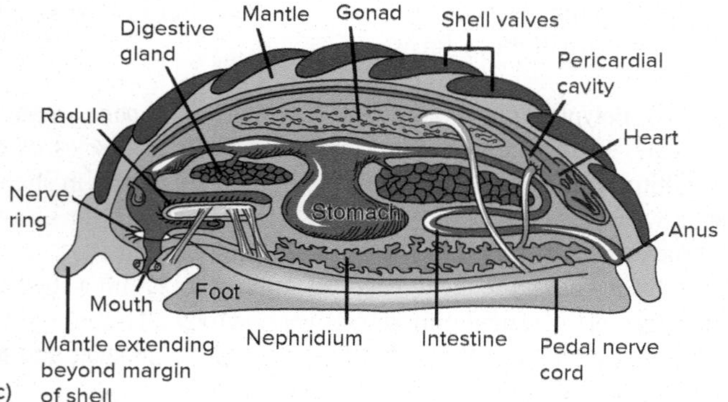

(c)

FIGURE 11.20

Class Polyplacophora. (*a*) Dorsal view of a chiton (*Tonicella lineata*). Note the shell consisting of eight valves and the mantle extending beyond the margins of the shell. (*b*) Ventral view of a chiton. The mantle cavity is the region between the mantle and the foot. Blue arrows show the path of water moving across gills in the mantle cavity. (*c*) Internal structure of a chiton.
(a) ©Randimal/Shutterstock

the esophagus by ciliary action. Extracellular digestion and absorption occur in the stomach, and wastes move on to the intestine (figure 11.20*c*).

The nervous system is ladderlike, with four anteroposterior nerve cords and numerous transverse nerves. A nerve ring encircles the esophagus. Sensory structures include osphradia, tactile receptors on the mantle margin, chemoreceptors near the mouth, and statocysts in the foot. In some chitons, photoreceptors dot the surface of the shell.

Sexes are separate in chitons. External fertilization and development result in a swimming trochophore that settles and metamorphoses into an adult without passing through a veliger stage.

Section 11.6 Thinking Beyond the Facts

What features of members of the class Polyplacophora (if any) would be considered atypical of molluscs in general?

11.7 CLASS SCAPHOPODA

LEARNING OUTCOME

1. Compare members of the class Scaphopoda to the generalized molluscan body form.

Members of the class Scaphopoda (ska-fop′o-dah) (Gr. *skaphe,* boat + *podos,* foot) are called tooth shells or tusk shells. The over 600 species are all burrowing marine animals that inhabit moderate depths. Their most distinctive characteristic is a conical shell that is open at both ends. The head and foot project from the wider end of the shell, and the rest of the body, including the mantle, is greatly elongate and extends the length of the shell (figure 11.21). Scaphopods live mostly buried in the substrate with head and foot oriented down and with the apex of the shell projecting into the water above. Incurrent and excurrent water enters and leaves the mantle cavity through the opening at the apex of the shell. Functional gills are absent, and gas exchange occurs across mantle folds. Scaphopods have a radula and tentacles, which they use in feeding on foraminiferans. Sexes are separate, and trochophore and veliger larvae are produced.

Section 11.7 Thinking Beyond the Facts

What features of members of the class Scaphopoda (if any) would be considered atypical of molluscs in general?

11.8 CLASS MONOPLACOPHORA

LEARNING OUTCOME

1. Compare members of the class Monoplacophora to the generalized molluscan body form.

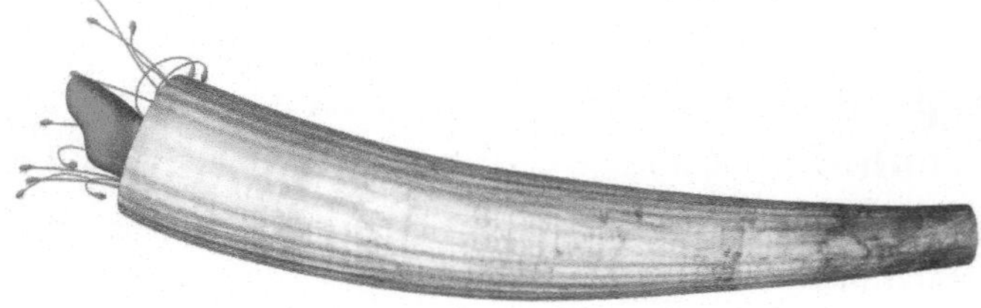

FIGURE 11.21

Class Scaphopoda, *Dentalium*. This conical shell is open at both ends. In its natural environment, the animal is mostly buried, with the apex of the shell (far right) projecting into the water.
©SciencePics/Shutterstock

Members of the class Monoplacophora (mon″o-pla-kof′o-rah) (Gr. *monos,* one + *plak,* plate + *phoros,* to bear) have an undivided, arched shell; a broad, flat foot; a radula; and serially repeated pairs of gills and foot-retractor muscles. They are dioecious; however, nothing is known of their embryology. This group of molluscs was known only from fossils until 1952, when a limpet-like monoplacophoran, named *Neopilina,* was dredged up from a depth of 3,520 m off the Pacific coast of Costa Rica (figure 11.22). Approximately 25 species have been described since the discovery of *Neopilina.*

Section 11.8 Thinking Beyond the Facts

Shortly after the discovery of* Neopilina, *many biologists suggested that the molluscs and annelids might share a common ancestry. What would have led these biologists to this hypothesis?

11.9 CLASS SOLENOGASTRES

LEARNING OUTCOME

1. Compare members of the class Solenogastres to the generalized molluscan body form.

There are approximately 250 species in the class Solenogastres (sole′no-gas′trez) (Gk., *solen,* channel + *gaster,* gut). These cylindrical molluscs lack a shell and crawl on their ventral foot, which is modified into a pedal groove (figure 11.23). Solenogasters lack a shell. This condition is thought to represent the ancestral state for the phylum. Instead of a shell, their bodies are covered by minute embedded calcareous spicules. Some solenogasters have secondarily lost the radula. True gills are absent; however, gill-like structures are usually present. Solenogasters are surface dwellers on corals and other marine substrates, and are carnivores, frequently feeding on cnidarian polyps. Solenogasters are monoecious.

Section 11.9 Thinking Beyond the Facts

What evidence is there that the shell and a large muscular foot are derived molluscan characteristics?

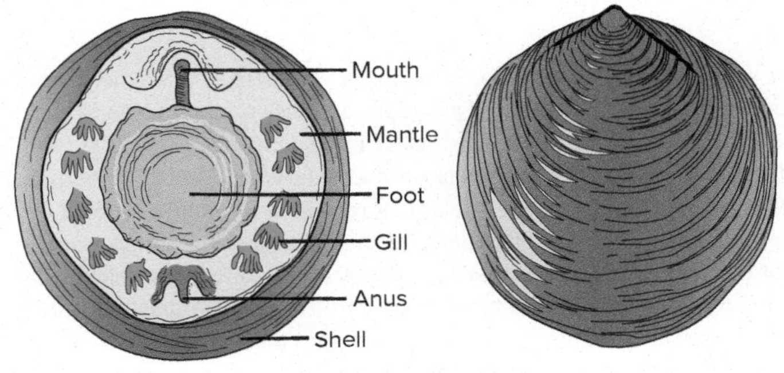

FIGURE 11.22

Class Monoplacophora. (*a*) Ventral and (*b*) dorsal views of *Neopilina.*

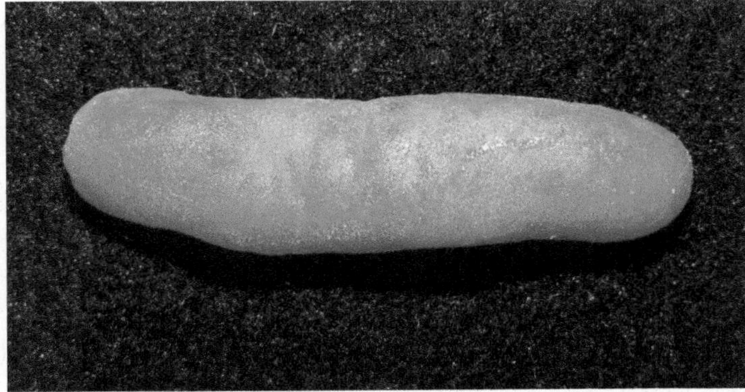

(a)

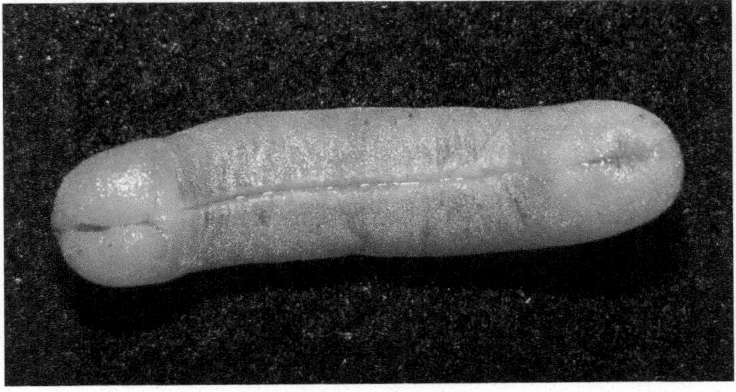

(b)

FIGURE 11.23

Class Solenogastres. (*a*) Photomicrograph of a dorsal view of *Neomenia yamamotoi,* a very large solenogaster collected from a depth of 250 m in the Gulf of Alaska. This species ranges in length from 7 to 11 cm. Most solenogasters are less than 50 mm in length. Minute calcareous spicules are embedded in the mantle, and a shell is absent. (*b*) This ventral view of *N. yamamotoi* shows the pedal groove that may be formed from a rolling of the mantle margins over the edges of the foot.

11.10 CLASS CAUDOFOVEATA

LEARNING OUTCOME

1. Compare members of the class Caudofoveata to the generalized molluscan body form.

Members of the class Caudofoveata (kaw′do-fo′ve-a″ta) (L. cauda, *tail* + fovea, *depression*) are wormlike molluscs that range in size from 2 mm to 14 cm and live in vertical burrows on the deep-sea floor (figure 11.24). They feed on foraminiferans and are dioecious. They have scalelike spicules on the body wall and feed using a radula. The absence of a shell, a muscular foot, and nephridia suggests that this group may resemble the ancestral mollusc. Zoologists have described approximately 150 species, but little is known of their ecology.

Section 11.10 Thinking Beyond the Facts

Members of the class Caudofoveata lack certain molluscan characteristics. Is this absence of characteristics an ancestral or derived state? Why is this distinction important?

11.11 FURTHER PHYLOGENETIC CONSIDERATIONS

LEARNING OUTCOME

1. Analyze the relationships among molluscan classes.

Fossil records of molluscan classes indicate that the phylum is more than 500 million years old. Molluscs are represented in fossils from later stages of the Ediacaran period and early Cambrian period (*see chapter 8 and appendix B*). Molecular data and shared protostome characteristics are interpreted as placing the Mollusca within the Lophotrochozoa along with the Annelida and other phyla (*see chapters 10 and 12, figure 11.1, and appendix A*). The segmental appearance of gills and other structures found in monoplacophorans, such as *Neopilina*, is now considered a very different form of segmentation from that found in Annelida or any other group of animals. The evolutionary ties between the Mollusca and Annelida are probably very distant.

The shell and muscular foot that characterize most modern molluscs were probably not present in the first molluscs. The mantle of solenogasters and caudofoveates is associated with a cuticle containing embedded calcium carbonate spicules, and this may be similar to the ancestral condition. The "girdle" surrounding the shell and covering the edge of the mantle of polyplacophorans is considered by some to be a remnant of this cuticle (*see figure 11.20*). The large muscular foot of most modern molluscs is first seen in the polyplacophora. The polyplacophoran shell, consisting of eight dorsal plates, is probably intermediate between the calcareous spicules of caudofoveates and solenogasters and the single shell of more derived molluscs.

The diversity of body forms and lifestyles in the phylum Mollusca is an excellent example of adaptive radiation. Molluscs began as slow-moving, marine bottom dwellers, but the evolution of unique molluscan features allowed them to diversify relatively quickly. By the end of the Cambrian period, some were filter feeders, some were burrowers, and others were swimming predators. Later, some molluscs became terrestrial and invaded many habitats, from tropical rain forests to arid deserts.

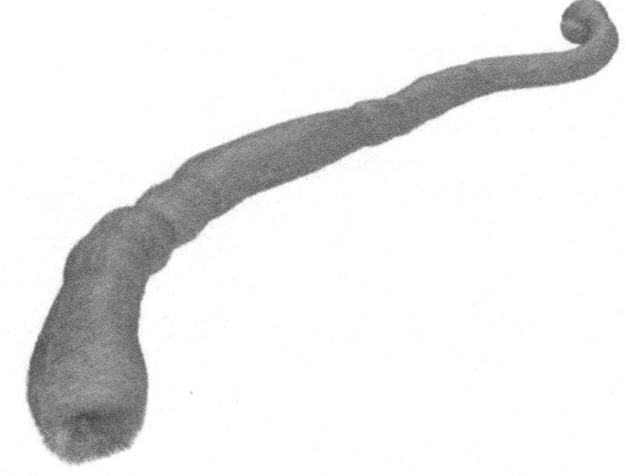

FIGURE 11.24

Class Caudofoveata. *Chaetoderma* is a genus of Caudofoveata widely distributed in deep seas of the world.

Figure 11.25 shows one interpretation of molluscan phylogeny. It reflects the idea that the muscular foot and shell are not ancestral, and shows Caudofoveata and Solenogastres as most closely resembling the molluscan ancestor. All other molluscs have a shell or are derived from shelled ancestors. The multipart shell distinguishes the Polyplacophora from other classes. Other selected synapomorphies, discussed earlier in this chapter, are noted in the cladogram. There are, of course, other interpretations of molluscan phylogeny. The extensive adaptive radiation of this phylum has made higher taxonomic relationships difficult to discern.

SECTION 11.11 THINKING BEYOND THE FACTS

What features of members of the class Polyplacophora suggest that they were derived after solenogasters and caudofoveates, but before all other molluscs?

WILDLIFE ALERT
Freshwater Bivalves

Classification: Phylum Mollusca, class Bivalvia, order Lamellibranchia
Range: North America, especially the Midwest and the Mississippi River Basin
Habitat: Freshwater lakes, rivers, and streams
Number Remaining: Unknown
Status: Threatened and Endangered

NATURAL HISTORY AND ECOLOGICAL STATUS

North America has approximately 300 species of freshwater bivalves, the highest bivalve diversity in the world. The U.S. Midwest and the Mississippi River Basin is home for much of this diversity. Most freshwater bivalves are within the order Unionida. Virtually all members of this order reproduce through the production of glochidia, and they often use mantle lures to attract their fish hosts (box figures 11.1 and 11.2, *see figures 11.13*c *and 11.14*). Freshwater bivalves are relatively sedentary and live on the bottom of lakes, rivers, and streams. They often remain in a small home area for their entire lives–up to 100 years–and their filter-feeding lifestyle means that their soft tissues are exposed to whatever is dissolved or suspended in the water. As a result, bivalves are very sensitive to environmental disturbances, and The Nature Conservancy estimates that about 70% of North American freshwater bivalves are extinct or imperiled.

The problems that freshwater bivalves face stem from economic exploitation, habitat destruction, pollution, and invasion of foreign

WILDLIFE ALERT *Continued*

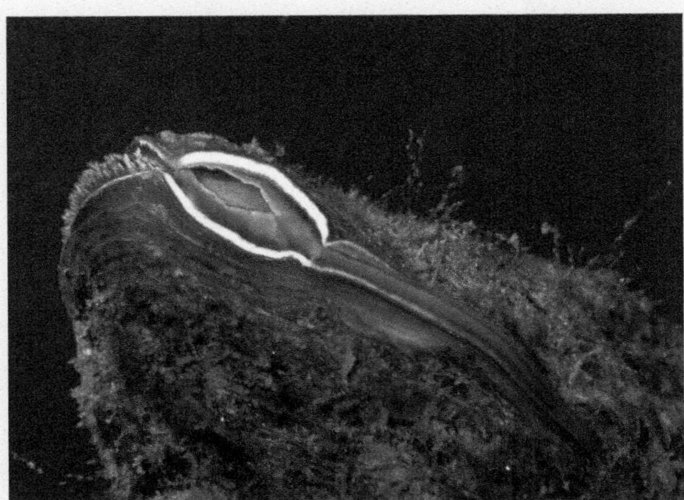

BOX FIGURE 11.1 **The Threatened Rabbitsfoot Mussel (*Quadrula cylindrica*).** The orange and white mantle that surrounds the excurrent aperture of the rabbitsfoot apparently attracts the attention of the shiners (e.g., *Cyprinella lutrensis*) that are hosts for rabbitsfoot glochidia. The rabbitsfoot is listed as threatened throughout its Mississippi River Basin range by the U.S. Fish and Wildlife Service.

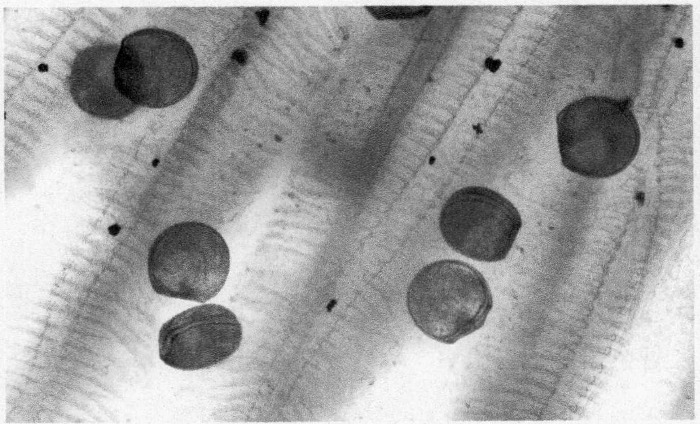

BOX FIGURE 11.2 **Glochidia of Rabbitsfoot (*Quadrula cylindrica*) Attached to Shiner Gills.** Rabbitsfoot glochidia are brooded in a female bivalve's marsupial gills in aggregates of eggs called conglutinates. Conglutinates are released from the gills to the mantle cavity when they mature. Conglutinates are released through the excurrent aperture when a host shiner (*Cyprinella*) investigates the colored mantle lure. The conglutinates break apart spontaneously or when bitten by the fish. Glochidia that are released attach to the host's gills, where they develop into juvenile bivalves and eventually drop from the fish to the riverbed.

bivalves. The pearl-button industry began harvesting freshwater bivalves in the late 1800s. In the early 1900s, 196 pearl-button factories were operating throughout the Mississippi River Basin. Thick-shelled species like the rabbitsfoot (*Quadrula cylindrica, see box figure 11.1*) were harvested in large numbers. After harvesting and cleaning, bivalve shells were drilled to make buttons (box figure 11.3). In 1912, this industry produced more than $6 million in buttons. Since that year, however, the harvest has declined. In the 1940s, plastic buttons replaced pearl buttons, but bivalve harvesters soon discovered a new market for freshwater bivalve shells. Small pieces of bivalve shell placed into pearl oysters are a nucleus for the formation of a cultured pearl. A renewed impetus for harvesting freshwater bivalves in the 1950s resulted in overharvesting. In 1966, harvesters took 3,500 tons of freshwater bivalves.

Freshwater mussels require shallow, stable, sandy, or gravelly substrates, although a few species prefer mud. Channelization for barge traffic and flood control purposes has destroyed many bivalve habitats. Siltation from erosion has replaced stable substrates with soft, mucky river bottoms. A variety of pollutants, especially agricultural run-off containing pesticides, threaten bivalves directly. Bivalves are especially sensitive to spikes in un-ionized ammonia and heavy metal concentrations. These alterations also threaten host fish upon which the bivalves depend.

Two invasive bivalves threaten native species. The Asian clam (*Corbicula fluminea*)–native to Asian, Mediterranean, and African freshwaters–was introduced into Washington State in the 1930s and has spread throughout North America. It is a serious competitor for native bivalve habitats. The zebra mussel (*Dreissena polymorpha*) was introduced into the Great Lakes of North America, probably in ship ballast water, from Eastern Europe and Russia in the 1980s. Zebra mussels have the unfortunate habit of using native bivalve shells as a substrate for attachment by byssal threads. They can cover a native bivalve so densely that the native bivalve cannot feed.

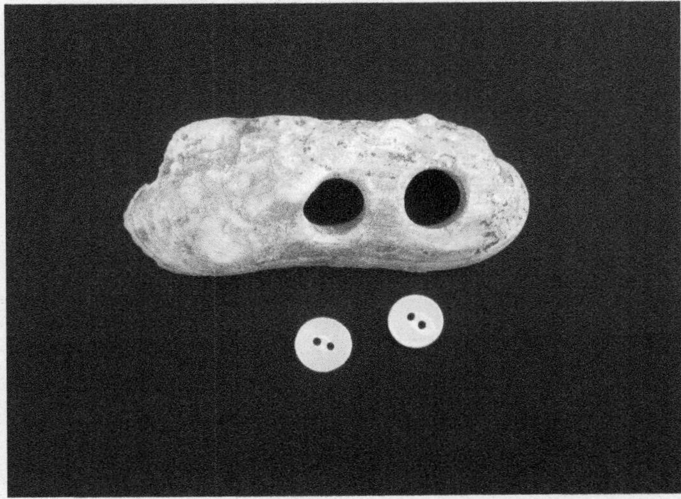

BOX FIGURE 11.3 **The Shell of a Rabbitsfoot Drilled to Make Pearl Buttons.** This mussel was harvested from the Neosho River in Kansas approximately 100 years ago.

Source: Edwin J. Miller, Kansas Dept. of Wildlife, Parks & Tourism

Steps are being taken to help freshwater bivalves survive these threats. Commercial bivalve harvesting has been eliminated by some states and is closely regulated by others. Habitat restoration efforts include reducing erosion and pesticide run-off from farmlands and urban areas, removing dams and impoundments, and managing invasive species. Laboratories, like those at Missouri State University, are actively involved with research aimed at culturing at-risk species and reintroducing them into reclaimed habitats (http://courses.missouristate.edu/chrisbarnhart/home/).

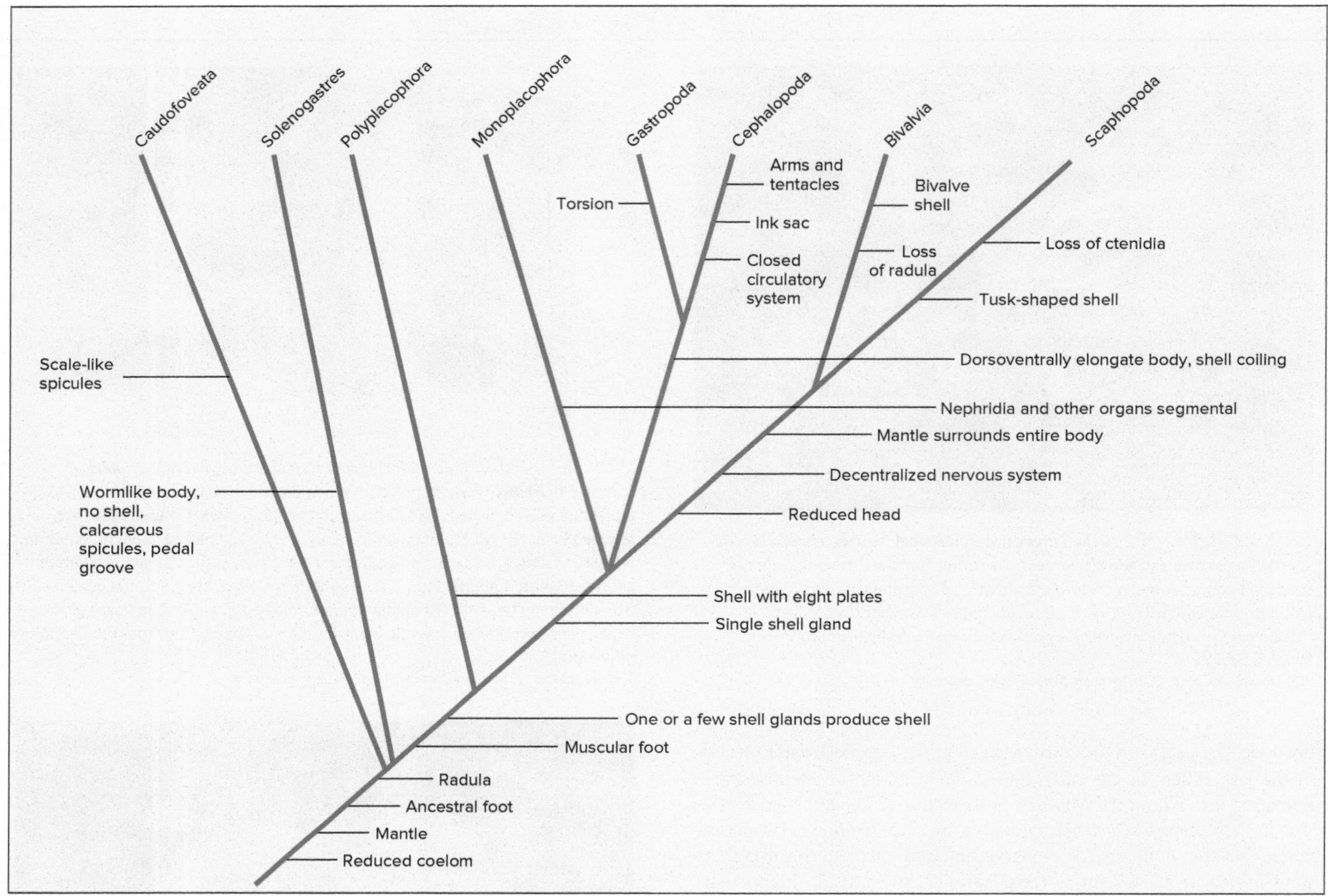

FIGURE 11.25

Molluscan Phylogeny. Cladogram showing possible evolutionary relationships among the molluscs.

Summary

11.1 **Evolutionary Perspective**

- Members of the phylum Mollusca include octopuses and their relatives, snails, bivalves, and others. They have evolutionary ties to the Annelida and other lophotrochozoans.

11.2 **Molluscan Characteristics**

- The body of a mollusc is divided into two regions. The head-foot contains the mouth, nervous structures, and a muscular foot used in attachment and locomotion. The visceral mass contains organs of digestion, reproduction, and excretion. Distinctive molluscan structures also include the mantle, mantle cavity, and the radula.

11.3 **Class Gastropoda**

- The class Gastropoda is the largest of the molluscan classes. Snails and slugs live in a wide variety of habitats and serve as intermediate hosts for important human parasites.
- Torsion is a gastropod developmental process that changes the orientation of the visceral mass and head-foot.
- Shells, when present, are usually coiled.
- Most gastropods feed by scraping algae and other attached organisms from their substrate. Their digestive system is ciliated and food is incorporated into a protostyle.
- Gas exchange occurs across gills or a vascular mantle. Gastropods have an open circulatory system, a hydraulic skeleton, well-developed circulatory and nervous systems, and an excretory system consisting of a single nephridium.
- Gastropods may be either monoecious or dioecious with internal or external fertilization. Fertilization either results in large, yolky terrestrial eggs or trochophore and veliger larvae.
- Gastropod taxonomy is being reevaluated. The three traditional subclasses are not monophyletic.

11.4 **Class Bivalvia**

- The class Bivalvia includes the clams, oysters, mussels, and scallops. Bivalves lack a head and are covered by a sheetlike mantle and a shell consisting of two valves.
- Most bivalves use expanded gills for filter feeding. Labial palps direct food to the mouth. The digestive tract is ciliated and food digestion is aided by the crystalline style.
- Bivalves have an open circulatory system and a nervous system consisting of three pairs of interconnected ganglia.
- Most bivalves are dioecious with external fertilization and trochophore and veliger larvae. Most freshwater bivalves brood their larvae and release a glochidium, which is parasitic on fishes.
- Bivalves live in all-aquatic habitats and may be surface-dwelling or bury themselves in sand or mud.

11.5 **Class Cephalopoda**

- Members of the class Cephalopoda are the octopuses, squids, cuttlefish, and nautiluses. The anterior portion of their foot has been modified into a circle of tentacles.
- Except for nautiluses, cephalopods have a reduced shell.
- Cephalopods can use a jet-propulsion mechanism in swimming and escape responses.
- Cephalopods locate prey by sight; capture prey using tentacles; and use jaws, venom, and a radula to kill and tear their prey. Their digestive system is muscular, and food is processed through peristalsis.
- Cephalopods have a closed circulatory system including systemic and branchial hearts. Their nervous and sensory systems are highly developed. They use color and pattern changes in camouflage, mating behaviors, and attracting prey.
- Cephalopods are capable of learned behaviors directed at finding prey and avoiding predators.
- Cephalopods are dioecious. Sperm are transferred to females using spermatophores. Eggs are fertilized within a female's mantle cavity and deposited singly or in stringlike masses.

11.6 **Class Polyplacophora**

- Members of the class Polyplacophora are the chitons. They have a muscular foot for attachment to solid substrates and a shell consisting of eight dorsal plates.

11.7 **Class Scaphopoda**

- Members of the class Scaphopoda are the tooth shells or tusk shells. Their shell is open at both ends and the apex of the shell extends to the surface of the substrate as these animals move through soft marine sediments.

11.8 **Class Monoplacophora**

- Members of the class Monoplacophora have an undivided arched shell and a broad, flat foot. They live at great depths in the Pacific Ocean and were known only as fossils until 1952.

11.9 **Class Solenogastres**

- Solenogasters live associated with coral, where they feed on coral polyps. They lack a shell and muscular foot. Their foot is modified into a ventral pedal groove.

11.10 **Class Caudofoveata**

- Members of the class Caudofoveata are burrowers in deep-water marine substrates. They lack a shell and muscular foot.

11.11 **Further Phylogenetic Considerations**

- Molluscs are lophotrochozoans with distant ties to the Annelida. The caudofoveates and solenogasters are probably most similar to the mollusc ancestor. Adaptive radiation in the molluscs has resulted in their presence in most ecosystems of the earth.

Concept Review Questions

1. All of the following may be functions of the mantle cavity of molluscs, except one. Select the exception.
   a. Gas exchange
   b. Serving as a hydrostatic skeleton
   c. Excretion
   d. Elimination of digestive wastes
   e. Release of reproductive products
2. Members of all of the following classes possess an open circulatory system, except one. Select the exception.
   a. Bivalvia
   b. Gastropoda
   c. Cephalopoda
   d. Polyplacophora
3. In the gastropods, food is trapped in a mucoid mass called the ___________________, which extends to the stomach and is rotated by cilia.
   a. style
   b. protostyle
   c. crystalline style
   d. radula
   e. odontophore
4. Which of the following is the correct sequence of structures encountered by food entering the mantle cavity of a bivalve?
   a. Incurrent opening, gill filaments, food groove, labial palps, mouth
   b. Incurrent opening, labial palps, food groove, gill filaments, mouth
   c. Incurrent opening, food groove, gill filaments, labial palps, mouth
   d. Incurrent opening, gill filaments, labial palps, food groove, mouth
5. All of the following statements regarding the evolution of the Mollusca are true except one. Select the exception.
   a. Gastropods and cephalopods are more closely related to each other than to other mollusc groups.
   b. Solenogastres and Caudofoveata are the classes whose members most closely resemble the mollusc ancestor.
   c. The muscular foot and shell were characteristics of the mollusc ancestor.
   d. Molluscs are lophotrochozans and distantly related to the Annelida.

Analysis and Application Questions

1. Compare and contrast the hydraulic skeletons of molluscs with the hydrostatic skeletons of cnidarians and pseudocoelomates.
2. Review the functions of body cavities presented in chapter 7. Which of those functions, if any, apply to the coelom of molluscs?
3. Students often confuse torsion and shell coiling. Describe each and its effect on gastropod structure and function.
4. Bivalves are often used as indicators of environmental quality. Based on your knowledge of bivalves, describe why bivalves are useful environmental indicator organisms.

12

Annelida: The Metameric Body Form

The Samoan palolo worm (*Eunice viridis*) is one of approximately 15,000 species in the phylum Annelida. The palolo's reproductive habits are unusual, but effective. As you will see in this chapter, "unusual" is not the exception for this taxonomically challenging phylum.

Chapter Outline

12.1 Evolutionary Perspective
- *Relationships to Other Animals*
- *Metamerism and Tagmatization*

12.2 Annelid Structure and Function
- *External Structure and Locomotion*
- *Feeding and the Digestive System*
- *Gas Exchange and Circulation*
- *Nervous and Sensory Functions*
- *Excretion*
- *Regeneration, Reproduction, and Development*

12.3 Clade (Class) Errantia
- *Nereis (Neanthes, Alitta)*
- *Glycera*
- *Fireworms*

12.4 Clade (Class) Sedentaria
- *Tubeworms*
- *Siboglinidae*
- *Echiura*
- *Clitellata*

12.5 Basal Annelid Groups
- *Chaetopteridae*
- *Sipuncula*

12.6 Further Phylogenetic Considerations

12.1 EVOLUTIONARY PERSPECTIVE

LEARNING OUTCOMES

1. Describe the relationships of members of the Annelida to other animal phyla.
2. Explain the revisions of annelid taxonomy brought about through molecular analyses.
3. Explain the benefits of metamerism for an annelid.

At the time of the November full moon on islands near Samoa in the South Pacific, people rush about preparing for one of their biggest yearly feasts. In just one week, the sea will yield a harvest that can be scooped up in nets and buckets (figure 12.1). Worms by the millions transform the ocean into what one writer called "vermicelli soup!" Celebrants gorge themselves on worms that have been cooked or wrapped in breadfruit leaves. The Samoan palolo worm (*Eunice viridis, alternatively Palola viridis*) spends its entire adult life in coral burrows at the sea bottom. Each November, one week after the full moon, this worm emerges from its burrow, and specialized body segments devoted to sexual reproduction break free and float to the surface, while the rest of the worm is safe on the ocean floor. The surface water is discolored as gonads release their countless eggs and sperm. The natives' feast is short lived, however; these reproductive swarms last only two days and do not recur for another year.

The Samoan palolo worm is a member of the phylum Annelida (ah-nel'i-dah) (*L. annellus,* ring). Other members of this phylum include countless marine worms, the soil-building earthworms, and predatory leeches (table 12.1).

Characteristics of the phylum Annelida include:

1. Body metameric, bilaterally symmetrical, and worm-like
2. Spiral cleavage, trochophore larvae (when larvae are present), and schizocoelous coelom formation
3. Paired, epidermal setae (chaetae)
4. Closed circulatory system
5. Dorsal suprapharyngeal ganglia and ventral nerve cord(s) with ganglia
6. Metanephridia (usually) or protonephridia

Relationships to Other Animals

It has been clear for many years that Annelida is a monophyletic assemblage of marine, freshwater, and terrestrial worms. They are lophotrochozoans and, thus, share common ancestry with Mollusca, Brachiopoda, Bryozoa, Nemertea, and others (figure 12.2 and

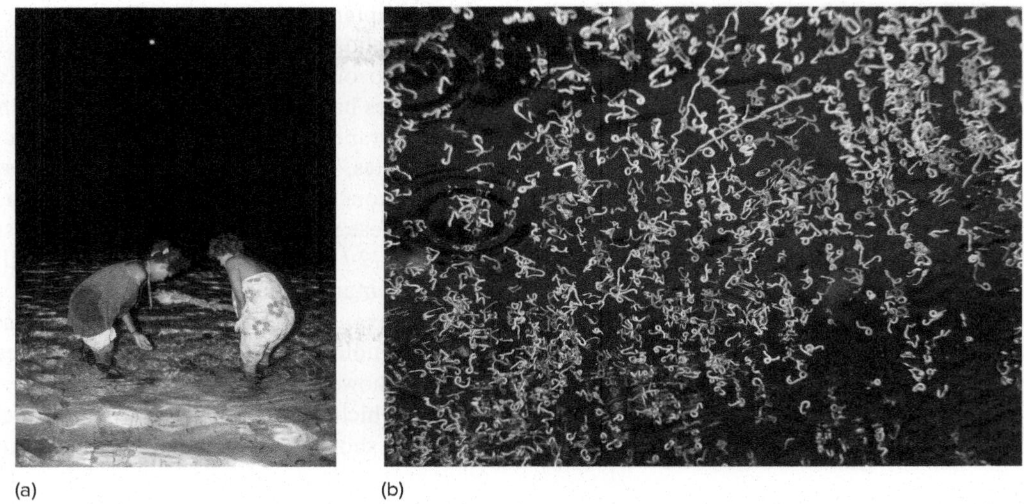

FIGURE 12.1

Palolo Feasts. (*a*) Women from the Nggela Group (Florida Islands) of the Solomon Islands use light from a torch to attract reproductive segments of *Eunice viridis* (family Eunicidae) during collecting. Photoreceptors on the reproductive segments elicit the response of the worms to light. (*b*) A swarm of worms as viewed through the lens of a diver's camera. The photograph on page 206 shows palolo worms (referred to by islanders as "odu") ready for feasting.

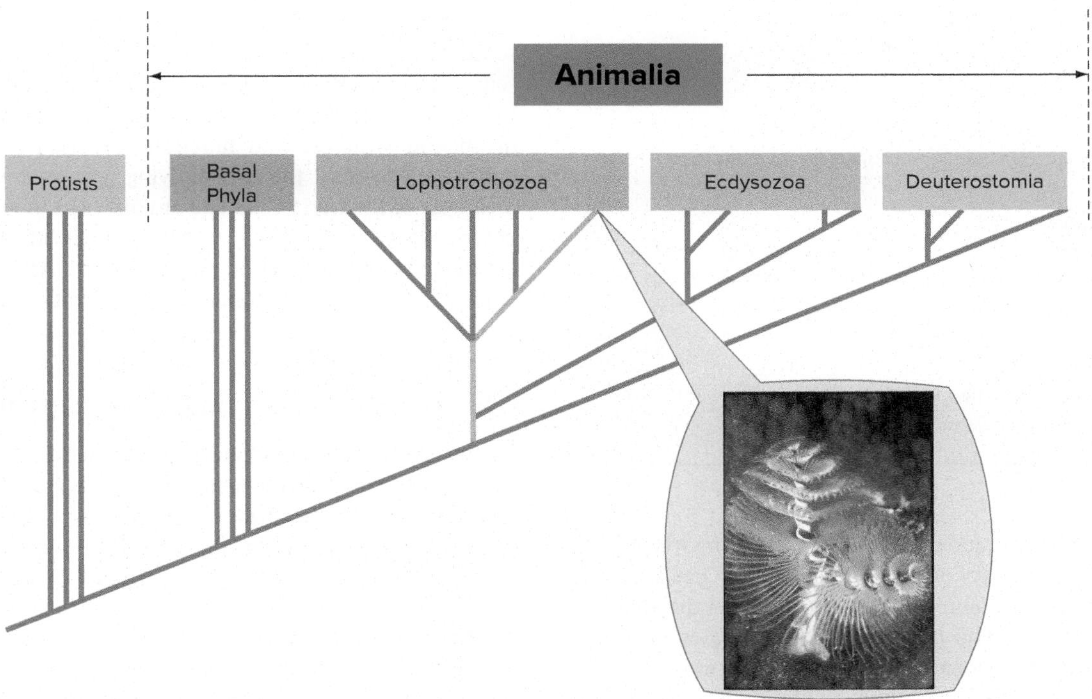

FIGURE 12.2

Evolutionary Relationships of Annelids to Other Animals. This figure shows an interpretation of the relationship of the Annelida to other members of the animal kingdom. The relationships depicted here are based on evidence from developmental and molecular biology. Annelids are placed within the Lophotrochozoa along with the Mollusca, Plathyhelminthes, Rotifera, and others (*see appendix A*). The phylum includes approximately 15,000 species of segmented worms. Most of these are marine and traditionally classified into the class Polychaeta. As discussed in the text, this placement is being reevaluated based on evidence from molecular biology. The Christmas-tree worm (*Spirobranchus giganteus*) (family Serpulidae) shown here is a member of the clade Sendentaria. The spiral fans of this tube-dwelling annelid are derived from prostomial palps that surround the mouth and are specialized for feeding and gas exchange.

TABLE 12.1
Classification of the Phylum Annelida

Phylum Annelida (ah-nel'i-dah)
The phylum of triploblastic, coelomate animals whose members are metameric (segmented), elongate, and cylindrical or oval in cross section. Annelids have a complete digestive tract; paired, epidermal setae; and a ventral nerve cord. Approximately 15,000 species of annelids have been described.

Clade (Class) Errantia (er-ran'tiah)
Marine annelids; parapodia with prominent lobes supported by internalized chaetae and ventral cirri; palps well developed. This clade comprises the majority of marine annelids. Members of three clades nested within Errantia are described in this chapter:

Nereidae, the clam worm (*Nereis*)
Glyceridae, bloodworms (*Glycera*)
Eunicida, the palolo (*Eunice*) and fireworms (*Eurythoe*)

Clade (Class) Sedentaria (sed-en-ter'iah)
Marine, freshwater, and terrestrial annelids; parapodia with reduced lobes or parapodia lacking; setae associated with the stiff body wall to facilitate anchoring in tubes and burrows; palps reduced. Members of five clades nested within Sedentaria are described in this chapter:

Serpulidae, Christmas-tree worms (*Spirobranchus*)
Sabelllidae, featherduster worms (*Bispira*)
Siboglinidae, beardworms (*Riftia*)
Echirua, spoon worms (*Bonellia*)
Clitellata, earthworms (*Lumbricus*) and members of the subset clade Hirudinea (leeches, *Hirudo* and others)

Other Annelid Taxa
Sipuncula (si-pun'ku-lah)
Marine; unsegmented body; retractable anterior trunk called an introvert. Formerly phylum Sipuncula. Inclusion of this group within the Annelida remains somewhat controversial. *Themiste.*

Chaetopteridae (ke-top-ter'ida)
Marine; three-part body; lives in a U-shaped tube, appendages for feeding and creating water currents. *Chaetopterus.*

see chapters 10 and 11). Our understanding of phylogeny within the phylum, however, has a history of contentious debate. The application of modern phylogenetic analysis is helping unravel the annelid taxonomic-tangle. Anyone who has followed the debates involved with elucidating annelid phylogeny will have a better appreciation of the dynamics within the field of animal taxonomy. These debates underscore the importance of taxonomy in helping zoologists understand the evolution of bilateral morphology (*see the discussion of metamerism and tagmatization in the next section*).

Annelida has traditionally been divided into two classes: Polychaeta (marine worms) and Clitellata (leeches, earthworms, and others). Recent molecular phylogenetic work has shown that the entire annelid assemblage is encompassed by what was being called "Polychaeta." As will be discussed in more detail at the end of this chapter, "Polychaeta" has been revealed to be synonymous with Annelida, and the former should be discarded as a class name. Two older terms used to differentiate groups within the polychaetes have been resurrected to represent two major annelid clades: Errantia and Sedentaria. Sedentaria now includes some "polychaetes," leeches, earthworms, and even worms that were formerly considered separate phyla (Echiura and Pogonophora). Both of these major clades are comprised of smaller nested clades (*see table 12.1*). For example, Clitellata is still a valid clade within the Sendentaria composed of earthworms, leeches, and a few smaller taxa. The leeches form an even smaller monophyletic clade (Hirudinea) within the Clitellata. The earthworms and their relatives, however, are not monophyletic. Thus the name "Oligochaeta," which was formerly considered a subclass name within the Clitellata, should be abandoned as a taxonomic designation. Two other groups lie outside of Sedentaria and Errantia. One of these, Chaetopteridae, was formerly considered to be a family within the "Polychaeta." The other, Sipuncula, was another phylum outside of Annelida. The hierarchical taxonomic-categories associated with these clades have not been established. We treat Errantia and Sedentaria as clades with a parenthetical "class" designation, which seems a logical outcome of ongoing and future research.

Chapter 12 is now organized to reflect this new phylogenetic work. In the next section, we describe metamerism and tagmatization. This discussion is followed by coverage of annelid structure and function. The term "polychaete" is used in a nontaxonomic sense to refer to a host of marine annelids (both errantians and sedentarians) when adaptations related to their common marine habitat result in similar structures and functions. After the basics of annelid structure and function are discussed, the two major annelid clades and the two outlying groups are described. Since members of the clade Clitellata are different in many respects from other sedentarians, unique aspects of their structure and function are described in the section on Sedentaria. As always, this chapter ends with "Further Phylogenetic Considerations."

Metamerism and Tagmatization

Earthworm bodies are organized into a series of ringlike segments. What is not externally obvious, however, is that the body is divided internally as well. Segmental arrangement of body parts in an animal is called **metamerism** (Gr. *meta,* after + *mere,* part).

Metamerism profoundly influences virtually every aspect of annelid structure and function, such as the anatomical arrangement of organs that are coincidentally associated with metamerism. For example, the compartmentalization of the body has resulted in each segment having its own excretory, nervous, and circulatory structures. In most modern annelids two related functions are probably the primary adaptive features of metamerism: flexible support and efficient locomotion. These functions depend on the metameric arrangement of the coelom and can be understood by examining the development of the coelom and the arrangement of body-wall muscles.

During embryonic development, the body cavity of annelids arises by a segmental splitting of a solid mass of mesoderm that occupies the region between ectoderm and endoderm on either side of the embryonic gut tract. Enlargement of each cavity forms a double-membraned septum on the anterior and posterior margins of each coelomic space and dorsal and ventral mesenteries associated with the digestive tract (figure 12.3).

Muscles also develop from the mesodermal layers associated with each segment. A layer of circular muscles lies below the epidermis, and a layer of longitudinal muscles, just below the circular muscles, runs between the septa that separate each segment. In addition, some marine annelids have oblique muscles, and the leeches have dorsoventral muscles.

One advantage of the segmental arrangement of coelomic spaces and muscles is the creation of hydrostatic compartments, which allow a variety of advantageous locomotor and supportive functions not possible in nonmetameric animals that use a hydrostatic skeleton. Each segment can be controlled independently of distant segments, and muscles can act as antagonistic pairs within a segment. The constant volume of coelomic fluid provides a hydrostatic skeleton against which muscles operate. Resultant localized changes in the shape of groups of segments provide the basis for swimming, crawling, and burrowing.

A second advantage of metamerism is that it lessens the impact of injury. If one or a few segments are injured, adjacent segments, set off from injured segments by septa, may be able to maintain nearly normal functions, which increases the likelihood that the worm, or at least a part of it, will survive the trauma.

A third advantage of metamerism is that it permits the modification of certain regions of the body for specialized functions, such as feeding, locomotion, and reproduction. The specialization of body regions in a metameric animal is called **tagmatization** (Gr. *tagma,* arrangement). The specialization of posterior segments of the palolo worm for reproductive functions (*see figure 12.1*) is an example of annelid tagmatization. Metamerism is not unique to the Annelida. It is also present in the Arthropoda (insects, arachnids, and their relatives) and Chordata (vertebrates, including humans). Interestingly metamerism is absent in the annelid taxa Echiura and Sipuncula. It was probably lost in these lineages. This evolutionary convergence of the metameric body form in these three very successful phyla means that one or more of the previously described advantages have been a major influence in the evolution of animal body forms.

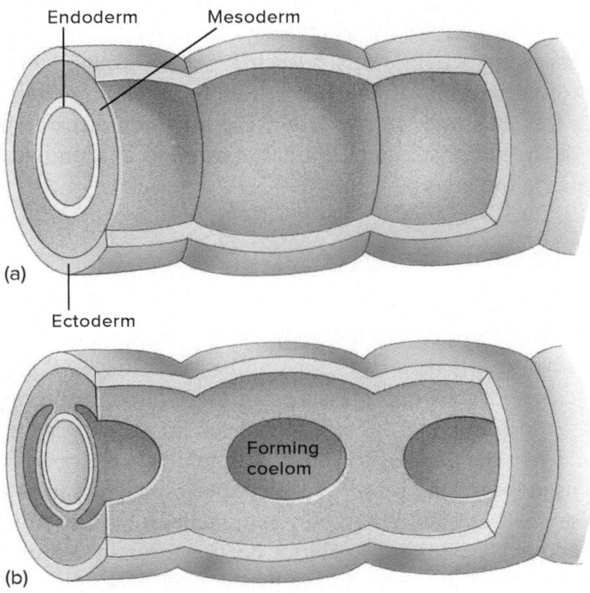

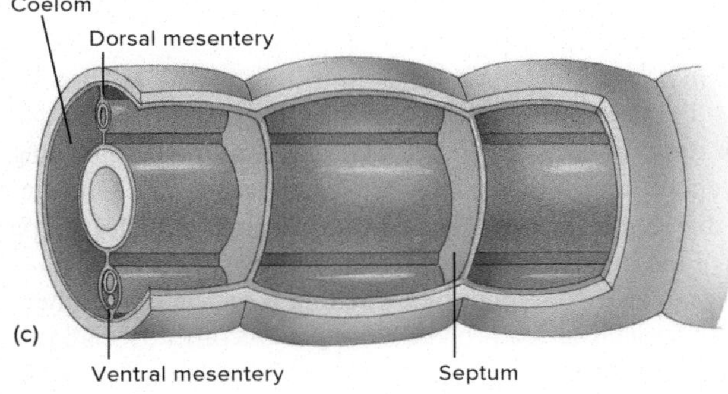

FIGURE 12.3

Development of Metameric, Coelomic Spaces in Annelids. (*a*) A solid mesodermal mass separates ectoderm and endoderm in early embryological stages. (*b*) Two cavities in each segment form from the mesoderm splitting on each side of the endoderm (schizocoelous coelom formation, *see figure 8.11*). (*c*) These cavities spread in all directions. Enlargement of the coelomic sacs leaves a thin layer of mesoderm applied against the outer body wall (the parietal peritoneum) and the gut tract (the visceral peritoneum), and dorsal and ventral mesenteries form. Anterior and posterior expansion of the coelom in adjacent segments forms the double-membraned septum that separates annelid metameres.

SECTION 12.1 THINKING BEYOND THE FACTS

How has evidence from molecular biology forced the reevaluation of classification of the Annelida?

12.2 ANNELID STRUCTURE AND FUNCTION

LEARNING OUTCOMES

1. Describe how metamerism influences annelid structure and locomotion.
2. Characterize digestive, nervous, and excretory systems of annelids.
3. Compare the closed circulatory system of an annelid worm to the open circulatory system of a bivalve mollusc.
4. Describe reproductive strategies of members of the Annelida.

Polychaetes are mostly marine, and are usually between 5 and 10 cm long. Polychaetes have adapted to a variety of habitats. Errantian polychaetes live on the ocean floor, under rocks and shells, and within the crevices of coral reefs. Many members of both major annelid clades are burrowers and move through their substrate by peristaltic contractions of the body wall. A bucket of intertidal sand normally yields vast numbers and an amazing variety of these burrowing annelids. Many sedentarians construct tubes of cemented sand grains or secreted organic materials. Mucus-lined tubes serve as protective retreats and feeding stations. Other annelids, like oligochaetes and leeches, have adapted to freshwater and terrestrial environments.

Sedentarians and errantians share many features of structure and function because they share a common ancestry. On the other hand, adaptations to diverse environments within these groups mean that there are many variations on the annelid theme. These common features, and many variations, will be described as we progress through this section.

External Structure and Locomotion

In addition to metamerism, the most distinctive feature of many annelids is the presence of lateral extensions called **parapodia** (Gr. *para,* beside + *podion,* little foot) (figure 12.4). In the Errantia, chitinous rods support the parapodia, and numerous setae project from the parapodia. Parapodia are reduced or absent (clade Clitellata) in the Sedentaria. Setae are present in most sedentarians but they are in closer proximity to the body wall. (Setae are absent in most leeches.) **Setae** (L. *saeta,* bristle) (also called **chaetae**) are bristles secreted from invaginations of the distal ends of parapodia. They aid locomotion by digging into the substrate (Errantia) and also hold a worm in its burrow or tube (Sedentaria).

The **prostomium** (Gr. *pro,* before + *stoma,* mouth) of a polychaete is a lobe that projects dorsally and anteriorly to the mouth and contains numerous sensory structures, including eyes, antennae, palps, and ciliated pits or grooves, called nuchal organs. The palps of some sedentarians are highly modified into filtering fan-like structures (*see figures 12.2 and 12.12*). The first body segment, the **peristomium** (Gr. *peri,* around), surrounds the mouth and bears sensory tentacles or cirri.

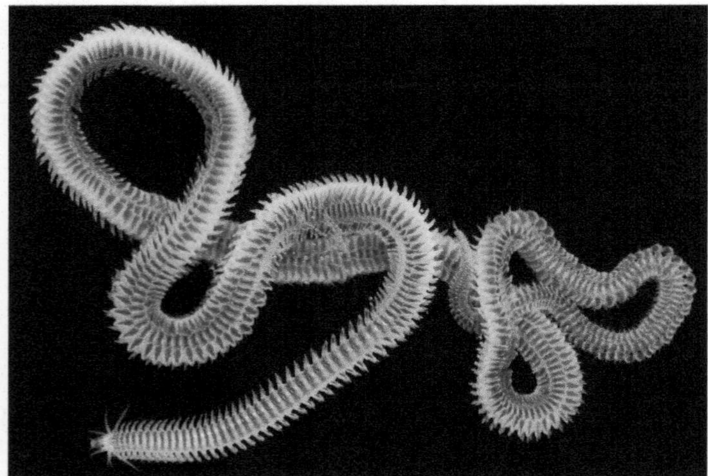

FIGURE 12.4

Clade Errantia. External structure of *Nereis*. Note the numerous parapodia.

The epidermis of annelids consists of a single layer of columnar cells that secrete a protective, nonliving **cuticle.** Some annelids have epidermal glands that secrete luminescent compounds.

Various species of errantian annelids are capable of walking, fast crawling, or swimming. To enable them to do so, the longitudinal muscles on one side of the body act antagonistically to the longitudinal muscles on the other side of the body so that undulatory waves move along the length of the body from the posterior end toward the head. The propulsive force is the result of parapodia and setae acting against the substrate or water. Parapodia on opposite sides of the body are out of phase with one another. When longitudinal muscles on one side of a segment contract, the parapodial muscles on that side also contract, stiffening the parapodium and protruding the setae for the power stroke (figure 12.5*a*). As a polychaete changes from a slow crawl to swimming, the period and amplitude of undulatory waves increase (figure 12.5*b*).

Burrowing sedentarians push their way through sand and mud by contractions of the body wall or by eating their way through the substrate. In the latter, the annelids digest organic matter in the substrate and eliminate absorbed and undigestible materials via the anus.

Feeding and the Digestive System

The digestive tract of most annelids is a straight tube that mesenteries and septa suspend in the body cavity. The anterior region of the digestive tract is modified into a proboscis that special protractor muscles and coelomic pressure can evert through the mouth. Retractor muscles bring the proboscis back into the peristomium. In some, when the proboscis is everted, paired jaws are opened and may be used for seizing prey. Predatory species may not leave their burrow or coral crevice. When prey approaches a burrow entrance, the worm quickly extends its anterior portion, everts the proboscis, captures prey with its jaws, and pulls the prey back into the burrow. Some annelids have venom glands at the base of the jaw. Other annelids are herbivores and scavengers and use jaws for tearing food. Deposit-feeding polychaetes (e.g., *Arenicola,* the sedentarian lugworm) extract organic matter from the marine sediments they ingest. The digestive tract consists of a pharynx that, when everted, forms the proboscis; a storage sac, called a crop; a grinding gizzard; and a long, straight intestine (*see figure 12.12*). Organic matter is digested extracellularly, and the inorganic particles are passed through the intestine and released as "castings."

Many sedentary and tube-dwelling polychaetes are filter feeders. They usually lack a proboscis but possess other specialized feeding structures. Some tube dwellers, called fanworms, possess radioles that form a spiral-shaped or funnel-shaped fan (*see figures 12.2 and 12.12*). Cilia on the radioles circulate water through the fan, trapping food particles. Trapped particles are carried along a food groove at the axis of the radiole. During transport, a sorting mechanism rejects the largest particles and transports the finest particles to the mouth. Another filter feeder, *Chaetopterus,* lives in a U-shaped tube and secretes a mucous bag that collects food particles, which may be as small as 1 μm. The parapodia of segments 14 through 16 are modified into fans that create filtration currents. When full, the entire mucous bag is ingested (*see figure 12.22*).

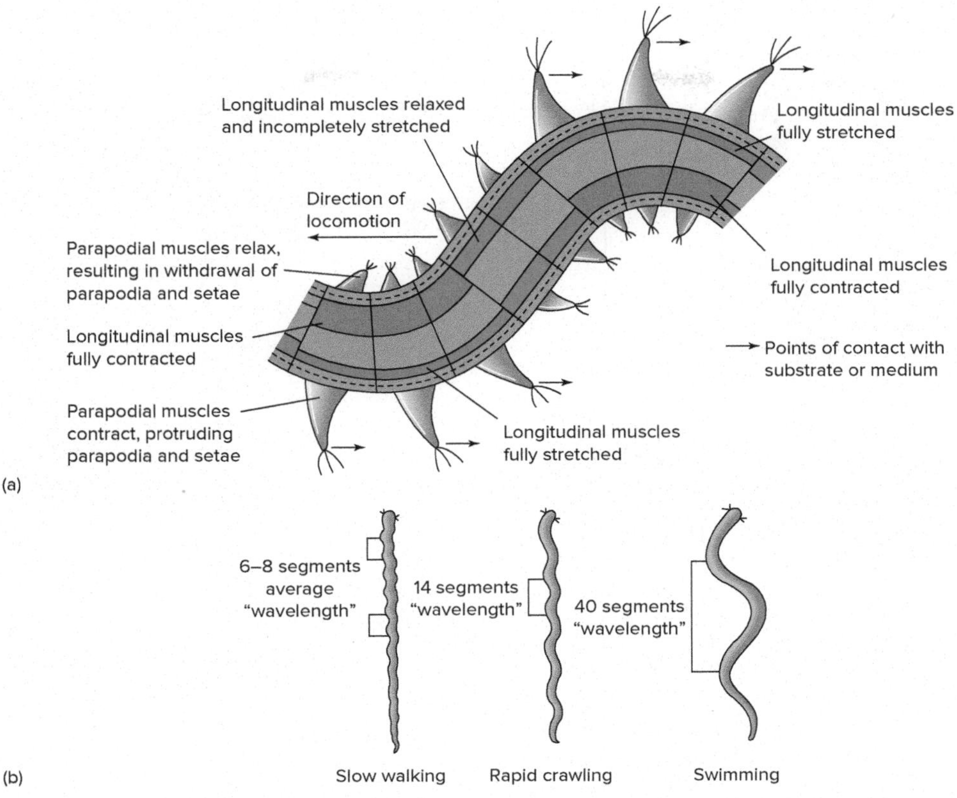

FIGURE 12.5

Annelid Locomotion. (*a*) Dorsal view of a primitive errantian annelid, showing the antagonism of longitudinal muscles on opposite sides of the body and the resultant protrusion and movement of parapodia. (*b*) Both the period and amplitude of locomotor waves increase as the annelid changes from a "slow walk" to a swimming mode.

Source: Russell-Hunter, WD. 1979. *A life of invertebrates*. New York (NY): Macmillan.

Elimination of digestive waste products can be a problem for tube-dwelling polychaetes. Those that live in tubes that open at both ends simply have wastes carried away by water circulating through the tube. Those that live in tubes that open at one end must turn around in the tube to defecate, or they may use ciliary tracts along the body wall to carry feces to the tube opening.

Polychaetes that inhabit substrates rich in dissolved organic matter can absorb as much as 20 to 40% of their energy requirements across their body wall as sugars and other organic compounds. This method of feeding occurs in other animal phyla, too, but rarely accounts for more than 1% of their energy needs.

Gas Exchange and Circulation

The respiratory gases of most annelids simply diffuse across the body wall, and parapodia increase the surface area for these exchanges. In many annelids, parapodial gills further increase the surface area for gas exchange.

Annelids have a closed circulatory system. Oxygen is usually carried in combination with molecules called respiratory pigments, which are usually dissolved in the plasma rather than contained in blood cells. Blood may be colorless, green, or red, depending on the presence and/or type of respiratory pigment.

Contractile elements of annelid circulatory systems consist of a dorsal aorta that lies just above the digestive tract and propels blood from rear to front, and a ventral aorta that lies ventral to the digestive tract and propels blood from front to rear. Running between these two vessels are two or three sets of segmental vessels that receive blood from the ventral aorta and break into capillary beds in the gut and body wall. Capillaries coalesce again into segmental vessels that deliver blood to the dorsal aorta (figure 12.6; *see figures 12.17 and 12.18*).

Nervous and Sensory Functions

Nervous systems are similar in virtually all annelids. The annelid nervous system includes a pair of suprapharyngeal ganglia, which connect to a pair of subpharyngeal ganglia by circumpharyngeal connectives that run dorsoventrally along either side of the pharynx. A double ventral nerve cord runs the length of the worm along the ventral margin of each coelomic space, and a paired segmental ganglion is in each segment. The double ventral nerve cord and paired segmental ganglia may fuse to varying extents in different taxonomic groups. Lateral nerves emerge from each segmental ganglion, supplying the body-wall musculature and other structures of that segment (figure 12.7*a*).

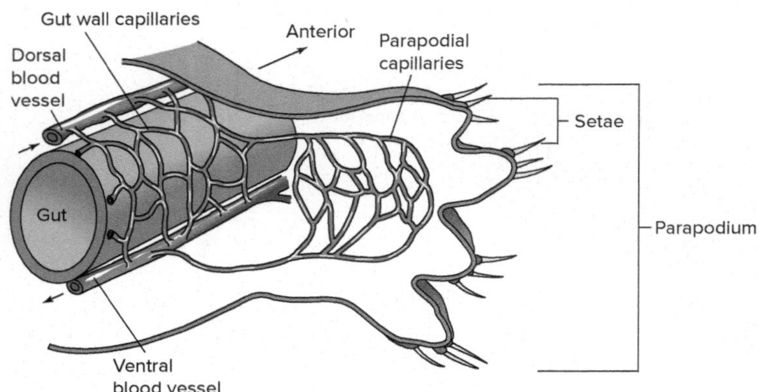

FIGURE 12.6

Circulatory System of a Polychaete. Cross section through the body and a parapodium. In the closed circulatory system shown here, blood passes posterior to anterior in the dorsal vessel and anterior to posterior in the ventral vessel. The direction of blood flow is indicated by black arrows. Capillary beds interconnect dorsal and ventral vessels.

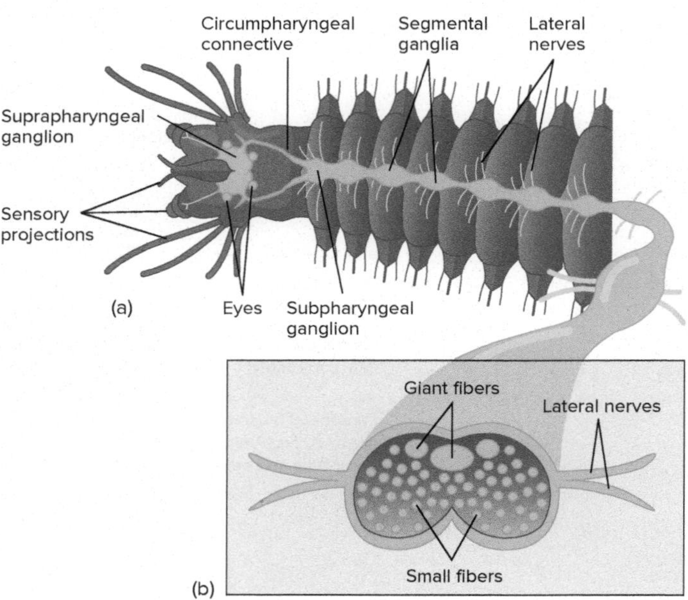

FIGURE 12.7

Annelid Nervous System. (*a*) Connectives link suprapharyngeal and subpharyngeal ganglia. Segmental ganglia and lateral nerves occur along the length of the worm. (*b*) Cross section of the ventral nerve cord, showing giant fibers.

Segmental ganglia coordinate swimming and crawling movements in isolated segments. (Anyone who has used portions of worms as live fish bait can confirm that the head end—with the pharyngeal ganglia—is not necessary for coordinated movements.) Each segment acts separately from, but is closely coordinated with, neighboring segments. The subpharyngeal ganglia help mediate locomotor functions requiring coordination of distant segments. The suprapharyngeal ganglia probably control motor and sensory functions involved with feeding, and sensory functions associated with forward locomotion.

In addition to small-diameter fibers that help coordinate locomotion, the ventral nerve cord also contains giant fibers involved with escape reactions (figure 12.7*b*). For example, a harsh stimulus, such as a fishhook, at one end of a worm causes rapid withdrawal from the stimulus. Giant fibers are approximately 50 μm in diameter and conduct nerve impulses at 30 m/s (as opposed to 0.5 m/s in the smaller, 4-μm-diameter annelid fibers).

Annelids have various sensory structures. Two to four pairs of eyes are on the surface of the prostomium. They vary in complexity from dermal photoreceptor cells; to simple cups of receptor cells; to structures made up of a cornea, lens, and vitreous body. Most polychaetes react negatively to increased light intensities. Fanworms, however, react negatively to decreasing light intensities. If shadows cross them, fanworms retreat into their tubes. This response is thought to help protect fanworms from passing predators. Earthworms lack well-developed eyes, which is not surprising, given their subterranean lifestyle. They do possess a dermal light sense that arises from photoreceptor cells scattered over dorsal and lateral surfaces of the body. Scattered photoreceptors mediate a negative phototaxis in strong light (evidenced by movement away from the light source) and positive phototaxis in weak light (evidenced by movement toward the light source).

Other sense organs mediate responses to chemicals, gravity, and touch. Nuchal organs are pairs of ciliated sensory pits or slits in the head region. Nerves from the suprapharyngeal ganglia innervate nuchal organs, which are thought to be chemoreceptors for food detection. Statocysts are in the head region of polychaetes, and ciliated tubercles, ridges, and bands, all of which contain receptors for tactile senses, cover the body wall.

Excretion

Annelids excrete ammonia, and because ammonia diffuses readily into the water, most nitrogen excretion probably occurs across the body wall. Excretory organs of annelids are more active in regulating water and ion balances, although these abilities are limited. Most marine annelids, if presented with extremely diluted seawater, cannot survive the osmotic influx of water and the resulting loss of ions. The evolution of efficient osmoregulatory abilities has allowed only a few polychaetes to invade freshwater. Freshwater annelids excrete copious amounts of very dilute urine, although they retain vital ions, which is important for organisms living in environments where water is plentiful but essential ions are limited. In addition to ammonia, earthworms excrete urea, a less toxic nitrogenous waste.

The excretory organs of annelids, like those of many invertebrates, are called nephridia. Annelids have two types of nephridia. A protonephridium consists of a tubule with a closed bulb at one end and a connection to the outside of the body at the other end. Protonephridia have a tuft of flagella in their bulbular end that drives fluids through the tubule (figure 12.8*a*; *see also figure 10.6*). Some primitive annelids possess paired, segmentally arranged protonephridia that have their bulbular end projecting through the anterior septum into an adjacent segment and the opposite end opening through the body wall at a nephridiopore.

Most annelids possess a second kind of nephridium, called a metanephridium. A **metanephridium** consists of an open, ciliated funnel, called a nephrostome, that projects through an anterior septum into the coelom of an adjacent segment. At the opposite end, a tubule opens through the body wall at a nephridiopore or, occasionally, through the intestine (figure 12.8*b* and *c*). There is usually one pair of metanephridia per segment, and tubules may be extensively coiled, with one portion dilated into a bladder. A capillary bed is usually associated with the tubule of a metanephridium for active transport of ions between the blood and the nephridium (figure 12.8*d*; *see also figures 12.17 and 12.18*).

Most annelids have chloragogen tissue associated with the digestive tract. Chloragogen tissue surrounds the dorsal blood vessel and lies over the dorsal surface of the intestine (*see figure 12.18*). Chlorogogen tissue acts similarly to the vertebrate liver in that it deaminates amino acids and, in earthworms, converts ammonia into urea. It also converts excess carbohydrates into energy-storage molecules of glycogen and fat.

Regeneration, Reproduction, and Development

Many annelids have remarkable powers of regeneration. They can replace lost parts, and some species have break points that allow worms to sever themselves when a predator grabs them. Lost segments are later regenerated.

Some polychaetes reproduce asexually by budding or by transverse fission; however, sexual reproduction is much more common. Most polychaetes are dioecious. Gonads develop as masses of gametes and project from the coelomic peritoneum. Primitively, gonads occur in every body segment, but most polychaetes have gonads limited to specific segments. Gametes are shed into the coelom, where they mature. Mature female worms are often packed with eggs. Gametes may exit worms by entering nephrostomes of metanephridia and exiting through the nephridiopore, or they may be released, in some polychaetes, after the worm ruptures. In these cases, the adult soon dies. Only a few polychaetes have separate gonoducts, a condition believed to be primitive (*see figure 12.8*a–c).

Fertilization is external in most polychaetes, although a few species copulate. A unique copulatory habit has been reported in *Platynereis megalops* from Woods Hole, Massachusetts. Toward the end of their lives, male and female worms cease feeding, and their intestinal tracts begin to degenerate. At this time, gametes have accumulated in the body cavity. During sperm transfer, male and female worms coil together, and the male inserts his anus into the mouth of the female. Because the digestive tracts of the worms have degenerated, sperm transfer directly from the male's coelom to the egg-filled coelom of the female. This method ensures fertilization of most eggs, after which the female sheds eggs from her anus. Both worms die soon after copulation.

Epitoky is the formation of a reproductive individual (an epitoke) that differs from the nonreproductive form of the species (an atoke). Frequently, an epitoke has a body that is modified into two body regions. Anterior segments carry on normal maintenance functions, and posterior segments are enlarged and filled with gametes. The epitoke may have modified parapodia for more efficient swimming.

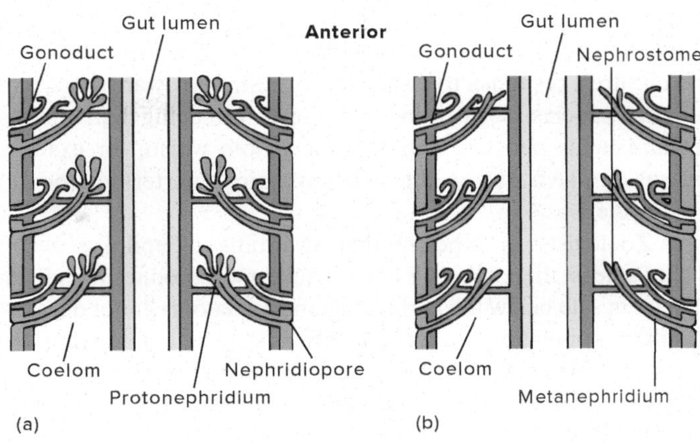

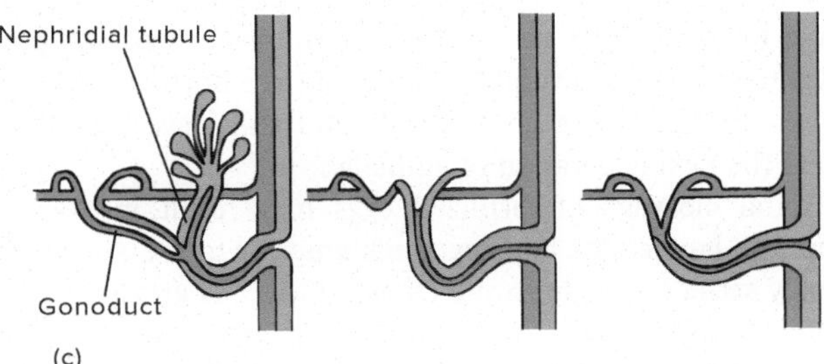

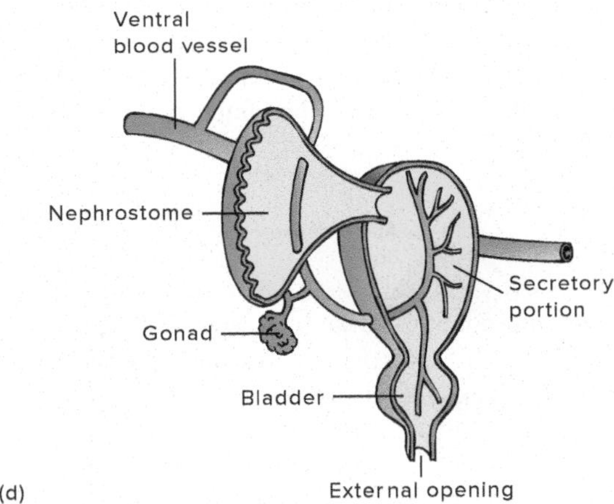

FIGURE 12.8

Annelid Nephridia. (*a*) Protonephridium. The bulbular ends of this nephridium contain a tuft of flagella that drives wastes to the outside of the body. In primitive annelids, a gonoduct (coelomoduct) carries reproductive products to the outside of the body. (*b*) Metanephridium. An open ciliated funnel (the nephrostome) drives wastes to the outside of the body. (*c*) In modern annelids, the gonoduct and the nephridial tubules undergo varying degrees of fusion. (*d*) Nephridia of modern annelids are closely associated with capillary beds for secretion, and nephridial tubules may have an enlarged bladder.

Source: Russell-Hunter, WD. 1979. *A life of invertebrates*. New York (NY): Macmillan.

This chapter begins with an account of the reproductive swarming habits of *Eunice viridis* (the Samoan palolo worm) and one culture's response to those swarms. Similar swarming occurs in other species, usually in response to changing light intensities and lunar periods. The Atlantic palolo worm, for example, swarms at dawn during the first and third quarters of the July lunar cycle.

Zoologists hypothesize that swarming of epitokes accomplishes at least three things. First, because nonreproductive individuals remain safe below the surface waters, predators cannot devastate an entire population. Second, external fertilization requires that individuals become reproductively active at the same time and in close proximity to one another. Swarming ensures that large numbers of individuals are in the right place at the proper time. Third, swarming of vast numbers of individuals for brief periods provides a banquet for predators. However, because vast numbers of prey are available for only short periods during the year, predator populations cannot increase beyond the limits of their normal diets. Therefore, predators can dine gluttonously and still leave epitokes that will yield the next generation of animals.

Spiral cleavage of fertilized eggs may result in planktonic trochophore larvae that bud segments anterior to the anus. Larvae eventually settle to the substrate (figure 12.9). As growth proceeds, newer segments continue to be added posteriorly. Thus, the anterior end of a polychaete is the oldest end. Many other polychaetes lack a trochophore and display direct development or metamorphosis from another larval stage.

Sexual reproduction within the clade Clitellata is markedly different from that described above. These worms are monoecious and have direct development (no larval stages) within a cocoon. These differences will be described later.

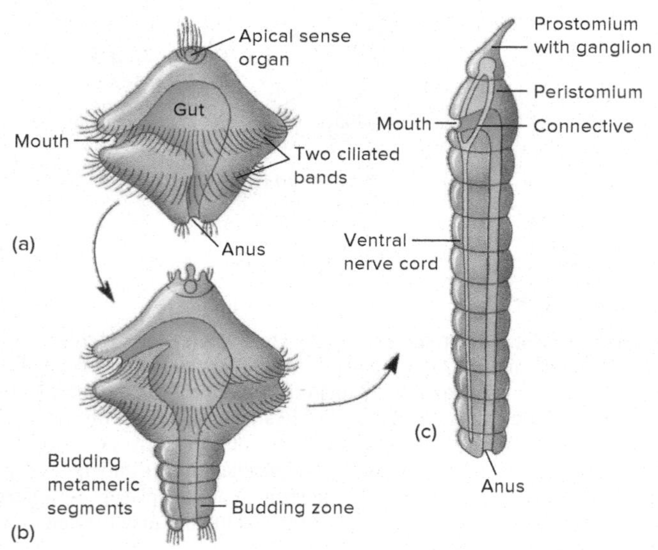

FIGURE 12.9

Polychaete Development. (*a*) Trochophore. (*b*) A later planktonic larva, showing the development of body segments. As more segments develop, the larva settles to the substrate. (*c*) Juvenile worm.
Source: Russell-Hunter, WD. 1979. *A life of invertebrates*. New York (NY): Macmillan.

SECTION 12.2 THINKING BEYOND THE FACTS

Why does the association of a capillary bed with a metanephridium provide more efficient waste processing as compared to the function of a protonephridium?

12.3 CLADE (CLASS) ERRANTIA

LEARNING OUTCOMES

1. Characterize members of the clade Errantia.
2. Describe the life histories of *Nereis* and *Glycera*.
3. Hypothesize on the importance of the timing and occurrence of reproductive swarming in many errantians.

Members of the clade Errantia (L. *errant*, to wander) are mostly marine annelids. The clade designation is currently unranked, although some zoologists assign it class-level status. They have parapodia with prominent lobes, relatively long setae, and well-developed palps. As their name suggests, these features are adaptations for active, crawling and swimming lives. Errantians comprise the majority of the Annelida. The Samoan palolo worm (*Eunice viridis*) (*see page 206 and figure 12.1*) is a member of the clade Errantia.

Nereis (*Neanthes, Alitta*)

Sandworms, ragworms, and clam worms are all common names for an annelid genus studied in general zoology laboratories as representative annelids (*see figure 12.4*). Some authorities recognize the name *Neanthes,* and others the name *Alitta,* as the valid senior synonym for this genus. There are numerous species in this genus that are similar in appearance and biology. Most species burrow in sand and mud of temperate marine habitats and have been collected from depths up to 90 m. *Nereis* (*see figure 12.4*) is common in mudflats where it reaches lengths greater than 40 cm. Clam worms spend most of the daylight hours in mucus-lined burrows and emerge at night to feed. They use their eversible proboscis and large jaws to feed on marine vegetation and to capture nematodes and small arthropods. Clam worms, like most polychaetes, reproduce only once during their life, and reproduction is correlated with the lunar cycle. Males and females metamorphose into a reproductive (epitoke) stage. Using enlarged swimming parapodia, the worms swim to the surface in large numbers and spawn gametes through breaks in their body walls. Adults die after releasing gametes. Fertilization results in the development of trochophore larvae and then juvenile worms (*see figure 12.9*). Although these worms are cosmopolitan, at least one species (*Nereis virens*) is threatened from overharvesting by bait collectors. Attempts are currently underway to raise clam worms commercially for the bait industry.

Glycera

There are a variety of species within the genus *Glycera* (figure 12.10). Like *Nereis, Glycera* is studied in general zoology laboratories because it is readily available from commercial bait suppliers based

FIGURE 12.10
Clade Errantia. This bloodworm (*Glycera*) is a common burrowing errantian in shallow marine water. They are harvested extensively along the northeastern coast of North America as fish bait.
Source: NOAA Central Library

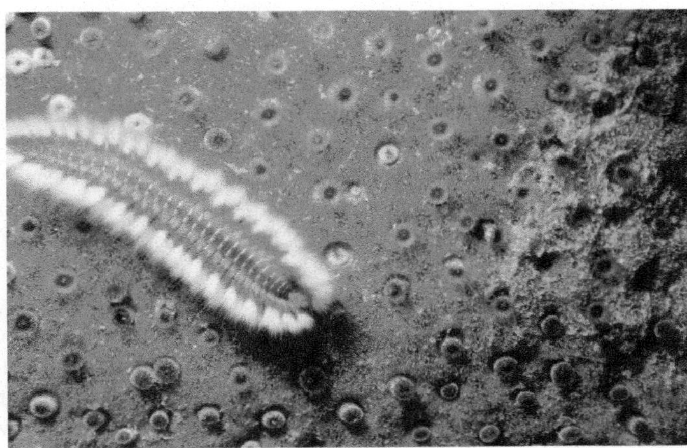

FIGURE 12.11
Clade Errantia. The orange fireworm (*Eurythoe complanta*) is found in the Caribbean. It is shown here feeding on coral polyps. The white setae along the margins of the worm are hollow and venomous.
Source: NOAA Central Library

along the Atlantic coast of Canada and Maine. *Glycera* is called the bloodworm because of the presence of hemoglobin-containing coelomocytes distributing oxygen throughout the body cavity. Septa separating metameric compartments are incomplete; thus, the coelom is continuous through the length of the worm, and body-wall movements are responsible for distributing coelomocytes and metabolic products through the body. Protonephridia are used in excretion since there is no pumping vessel or other vasculature required for the function of metanephridia.

Glycerids burrow in soft sediments and feed on small marine invertebrates. They have a large eversible proboscis that is tipped with four jaws. The proboscis is used in burrowing and, in combination with the jaws, in capturing prey.

Reproduction follows a pattern similar to that described for *Nereis*. *Glycera* reproduces once in its lifetime, and reproductive swarms are correlated to the lunar cycle. Gametes are released into the body cavity. When worms are gravid the body walls of both sexes rupture, and gametes are released for external fertilization and the development of trochophore larvae. Obviously, the adults do not survive their brief reproductive flurry.

Fireworms

Fireworms comprise a variety of genera within the Errantia (family Amphinomidae). They feed on soft and hard coral polyps and small crustaceans. *Eurythoe* is a common genus that occurs in the Caribbean and other oceans of the world (figure 12.11). The white setae fringing this annelid are hollow and venom-filled. The setae easily penetrate the skin of a careless swimmer or predator, and a powerful neurotoxin causes pain, redness, and swelling. The bright colors of fireworms are an example of aposematic (warning) coloration (*see chapter 6*). Some species are bioluminescent and use light in mating. Females secrete a bioluminescent protein that creates a glowing green mucus coating over the worm. This secretion occurs during a precisely timed swarming that is triggered by the lunar cycle and apparently attracts males for spawning. Juvenile worms produce bioluminescent flashes that are thought to distract predators.

Section 12.3 Thinking Beyond the Facts

How is the method of gamete release during spawning of both* Nereis *and* Glycera *different from that of some other polychaete annelids?

12.4 CLADE (CLASS) SEDENTARIA

LEARNING OUTCOMES

1. Contrast the clades Errantia and Sedentaria.
2. Compare the oligochaete body form and the leech body form.
3. Analyze the earthworm body form in terms of adaptations to a largely subterranean lifestyle.
4. Compare and contrast the methods of fertilization and development of *Nereis* with members of the clade Clitellata.

Members of the clade Sedentaria include a variety of marine tubeworms, the siboglinids, the echiurans, and members of the clade Clitellata. The latter includes the leeches (Hirudinea), which is a monophyletic taxon. It also includes a variety of terrestrial and freshwater annelids formerly referred to as a subclass "Oligochaeta." The oligochaetes are not monophyletic, thus the name is used here to designate a variety of taxa within Clitellata that display some common structural and functional features. Sedentarians have parapodia with reduced lobes, or parapodia are completely lacking. Setae are closely associated with the stiff body wall to facilitate anchoring in tubes and burrows, and palps are reduced. The name Sedentaria was originally descriptive of the relatively sedentary "polychaete" tubeworms. That description no longer adequately represents the lifestyle of members of this group. Some zoologists think of the clade name as a class-level designation.

Tubeworms

The common name, tubeworm, is applied to a variety of sedentarian taxa that construct tubes, which are parchment-like, calcareous, or composed of cemented sand grains. Others construct burrows in sand or mud. Some tubeworms feed on particulate matter in seawater using cilia and mucus to trap and transport suspended organic matter. Featherduster or fanworms (Sabellidae) construct parchment tubes and protrude a crown of arm-like radioles from the open end of the tube (figure 12.12). Ciliary currents move water upward through the radioles and organic particles are trapped in mucus. Cilia then move the food particles toward the base of the radioles to the mouth. Christmas-tree worms (Serpulidae) use ciliary feeding in a similar fashion (*see figure 12.2*). Other tubeworms that live in burrows extract organic matter from marine sediments by sweeping tentacles through the sediment or creating water currents that bring organic matter into their burrow.

FIGURE 12.12

Clade Sedentaria. The social featherduster (*Bispira brunnea*) is a member of the family Sabellidae and is native to the Caribbean Sea. Cilia on the tentacle-like radioles protruding from the tube opening produce water currents that trap suspended food particles. The cilia then transport the food particles to the mouth of the worm located at the bases of the radioles.
©Diane R. Nelson

Siboglinidae

Siboglinids (beardworms) consist of about 120 species of tube-dwelling marine worms (figure 12.13). Formerly classified into the phylum Pognophora, members of this group are now formally annelids within the polychaete family Siboglinidae. Their tubes are embedded in soft marine sediments in cold, deep (over 100 m), and nutrient-poor waters. Siboglinids have no mouth or digestive tract. Nutrient uptake is via the outer cuticle and from endosymbiotic bacteria that siboglinids harbor in the posterior part (trophosome) of the body. These bacteria fix carbon dioxide into organic compounds that both the host and the symbiont can use.

Echiura

Echiurans (spoon worms) consist of about 130 species of animals that burrow in the mud or sand of shallow marine waters throughout the world. Some live protected in rock crevices. Their bodies are covered only by a thin cuticle. As a result, the animals keep to the safety of their burrows or crevices even when feeding (figure 12.14). An echiuran feeds by sweeping organic material into its spatula-shaped proboscis (thus the name "spoon worm"). Individual echiurans are from 15 to 50 cm in length, but the extensible proboscis may increase their length up to 2 m. Unlike most annelids, echiurans are not segmented.

Clitellata

Members of the clade Clitellata (kli-te'la-tah) (L. *clitellae,* saddle) include the earthworms, other "oligochaetes," and the leeches. Cladistic studies have established that the presence of a clitellum used in cocoon formation, monoecious direct development, and few or no setae are symplesiomorphic characters of this clade. Molecular data have provided very strong support for the monophyly of the Clitellata.

Lumbricus—*A Representative "Oligochaete"*

Oligochaetes are found throughout the world in freshwater and terrestrial habitats (*see table 12.1*). A few oligochaetes are estuarine,

FIGURE 12.13

Clade Sedenatria. Siboglinids were formerly considered to comprise a phylum, Pogonophora. The giant red siboglinid, *Riftia,* is shown here inside their tubes.
Source: NOAA Okeanos Explorer Program, Galapagos Rift Expedition 2011

FIGURE 12.14

Clade Sedentaria. Echiurans were originally classified as annelids and later reclassified within their own phylum. They are now likely returning to the Annelida. This photograph shows the proboscis of an echiuran extending from a burrow.

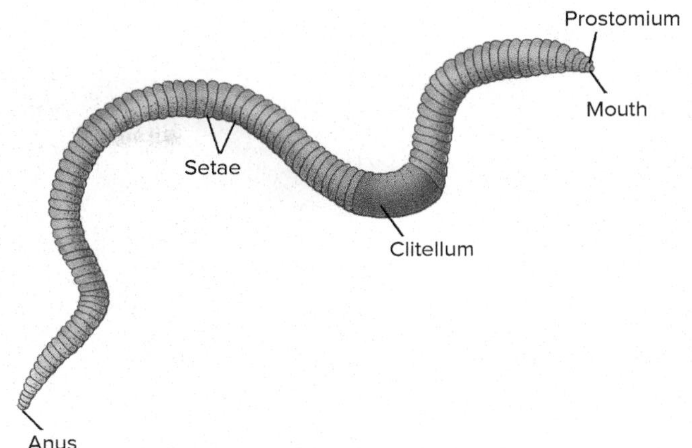

FIGURE 12.15

Subclass Oligochaeta. External structures of the earthworm, *Lumbricus terrestris*.

and some are marine. Aquatic species live in shallow water, where they burrow in mud and debris. Terrestrial species live in soils with high organic content and rarely leave their burrows. In hot, dry weather, they may retreat to depths of 3 m below the surface. The soil-conditioning habits of earthworms are well known. *Lumbricus terrestris* is commonly used in zoology laboratories because of its large size. It was introduced to the United States from northern Europe and has flourished. Common native species like *Eisenia foetida* and various species of *Allolobophora* are smaller.

External Structure and Locomotion (Clitellata)

Oligochaetes (Gr. *oligos,* few + *chaite,* hair) have setae, but fewer than are found in polychaetes (thus, the derivation of the class name). Earthworms lack parapodia because parapodia and long setae would interfere with their burrowing lifestyles. *Lumbricus* does have short setae associated with its integument. The prostomium consists of a small lobe or cone in front of the mouth and lacks sensory appendages. A series of segments in the anterior half of an oligochaete is usually swollen into a girdlelike structure called the **clitellum** that secretes mucus during copulation and forms a cocoon (figure 12.15).

Oligochaete locomotion involves the antagonism of circular and longitudinal muscles in groups of segments. Neurally controlled waves of contraction move from rear to front.

Segments bulge and setae protrude when longitudinal muscles contract, providing points of contact with the burrow wall. In front of each region of longitudinal muscle contraction, circular muscles contract, causing the setae to retract, and the segments to elongate and push forward. Contraction of longitudinal muscles in segments behind a bulging region pulls those segments forward. Thus, segments move forward relative to the burrow as waves of muscle contraction move anteriorly on the worm (figure 12.16).

Burrowing is the result of coelomic hydrostatic pressure being transmitted toward the prostomium. As an earthworm pushes its way through the soil, it uses expanded posterior segments and protracted setae to anchor itself to its burrow wall. Any person pursuing fishing worms experiences the effectiveness of this anchor when trying to extract a worm from its burrow. Contraction of circular muscles transforms the prostomium into a conical wedge, 1 mm in diameter at its tip. Contraction of body-wall muscles generates coelomic pressure that forces the prostomium through the soil. During burrowing, earthworms swallow considerable quantities of soil.

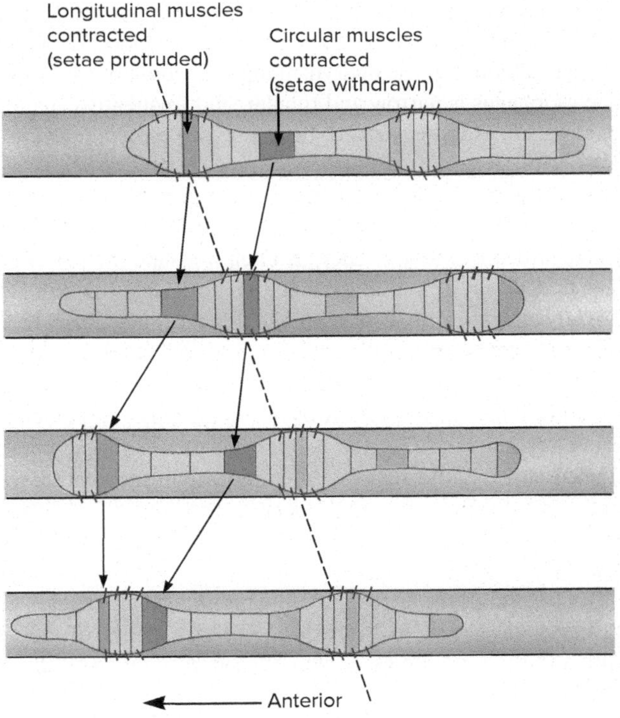

FIGURE 12.16

Earthworm Locomotion. Arrows designate activity in specific segments of the body, and broken lines indicate regions of contact with the substrate.

Source: Russell-Hunter, WD. 1979. *A life of invertebrates*. New York (NY): Macmillan.

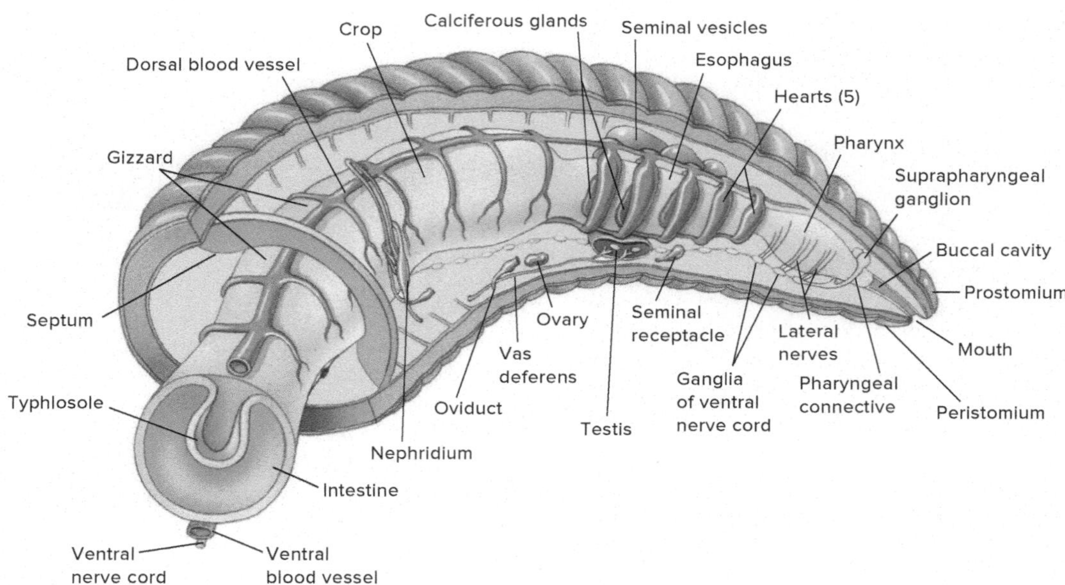

FIGURE 12.17

Earthworm Structure. Lateral view of the internal structures in the anterior segments of an earthworm. A single complete septum is shown.

Maintenance Functions (Clitellata) Oligochaetes are scavengers and feed primarily on fallen and decaying vegetation, which they drag into their burrows at night. The digestive tract is similar to that described in section 12.2 (figure 12.17). The mouth leads to a muscular pharynx. In the earthworm, pharyngeal muscles attach to the body wall. The pharynx acts as a pump for ingesting food. The mouth pushes against food, and the pharynx pumps the food into the esophagus. The esophagus is narrow and tubular, and frequently expands to form a stomach, crop, or gizzard; the latter two are common in terrestrial species. A crop is a thin-walled storage structure, and a gizzard is a muscular, cuticle-lined grinding structure. Calciferous glands are evaginations of the esophageal wall that rid the body of excess calcium absorbed from food. They also help regulate the pH of body fluids. A dorsal fold of the lumenal epithelium called the typhlosole substantially increases the surface area of the intestine (figure 12.18).

Earthworm respiratory and circulatory functions are as described for polychaetes. Some segmental vessels expand and may be contractile. In the earthworm, for example, expanded segmental vessels surrounding the esophagus propel blood between dorsal and ventral blood vessels and anteriorly in the ventral vessel toward the mouth. Even though these are sometimes called "hearts," the main propulsive structures are the dorsal and ventral vessels (*see figure 12.17*).

The ventral nerve cords and all ganglia of oligochaetes have undergone a high degree of fusion. Other aspects of nervous structure and function are essentially the same as those of annelids described in section 12.2.

Reproduction and Development (Clitellata) All oligochaetes are monoecious and exchange sperm during copulation. One or two pairs of testes and one pair of ovaries are located on the anterior septum of certain anterior segments. Both the sperm ducts and the oviducts have ciliated funnels at their proximal ends to draw gametes into their respective tubes.

Testes are closely associated with three pairs of **seminal vesicles,** which are sites for maturation and storage of sperm prior

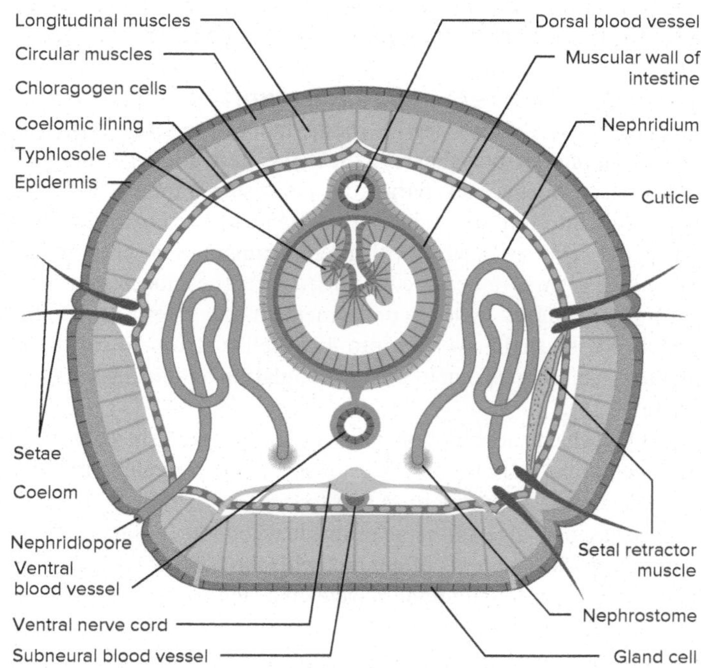

FIGURE 12.18

Earthworm Cross Section. The nephrostomes shown here would actually be associated with the next anterior segment. The ventral pair of setae on one side has been omitted to show the nephridiopore opening of a nephridium.

to their release. **Seminal receptacles** receive sperm during copulation. A pair of very small ovisacs, associated with oviducts, are sites for the maturation and storage of eggs prior to egg release (figure 12.19).

During copulation of *Lumbricus,* two worms line up facing in opposite directions, with the ventral surfaces of their anterior ends in contact with each other. This orientation lines up the clitellum of one worm with the genital segments of the other worm. A mucous sheath that the clitellum secretes envelops the anterior halves of both worms and holds the worms in place. Some species also have penile structures and genital setae that help maintain contact between worms. In *Lumbricus,* the sperm duct releases sperm, which travel along the external, ventral body wall in sperm grooves formed when special muscles contract. Muscular contractions along this groove help propel sperm toward the openings of the seminal receptacles. In other oligochaetes, copulation results in the alignment of sperm duct and seminal receptacle openings, and sperm transfer is direct. Copulation lasts 2 to 3 h, during which both worms give and receive sperm (*see figure 29.3*).

Following copulation, the clitellum forms a cocoon for the deposition of eggs and sperm. The cocoon consists of mucoid and chitinous materials that encircle the clitellum. The clitellum secretes a food reserve, albumen, into the cocoon, and the worm begins to back out of the cocoon. Eggs are deposited in the cocoon as the cocoon passes the openings to the oviducts, and sperm are released as the cocoon passes the openings to the seminal receptacles. Fertilization occurs in the cocoon, and as the worm continues backing out, the ends of the cocoon are sealed, and the cocoon is deposited in moist soil.

Spiral cleavage is modified, and there are no larval forms. Hatching occurs in one to a few weeks, depending on the species, when young worms emerge from one end of the cocoon.

Freshwater oligochaetes also reproduce asexually. Asexual reproduction involves transverse division of a worm, followed by the regeneration of missing segments.

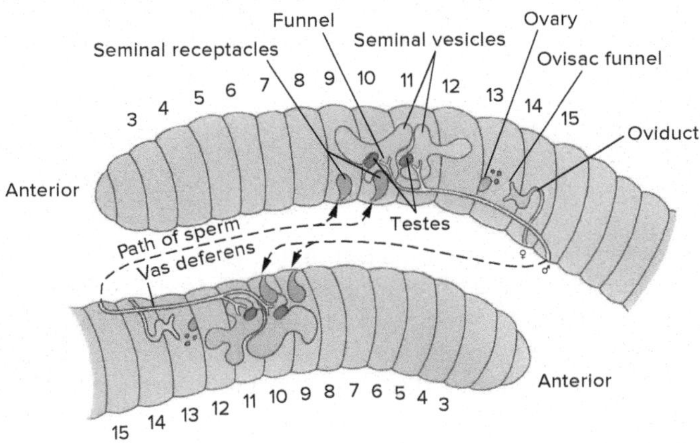

FIGURE 12.19

Earthworm Reproduction. Mating earthworms, showing arrangements of reproductive structures and the path sperm take during sperm exchange (shown by arrows).

Hirudinea

The clade Hirudinea (hi″ru-din′e-ah) (L. *hirudin,* leech) contains approximately 500 species of leeches (*see table 12.1*). Most leeches are freshwater; others are marine or completely terrestrial. Leeches prey on small invertebrates or feed on the body fluids of vertebrates.

Maintenance Functions (Hirudinea) Leeches lack parapodia and head appendages. Setae are absent in most leeches. In a few species, setae occur only on anterior segments. Leeches are dorsoventrally flattened and taper anteriorly. They have 34 segments, but the segments are difficult to distinguish externally because they have become secondarily divided. Several secondary divisions, called **annuli,** are in each true segment. Anterior and posterior segments are usually modified into suckers (figure 12.20).

Modifications of body-wall musculature and the coelom influence patterns of leech locomotion. The musculature of leeches is more complex than that of other annelids. A layer of oblique muscles is between the circular and longitudinal muscle layers. In addition, dorsoventral muscles are responsible for the typical leech flattening. The leech coelom has lost its metameric partitioning. Septa are lost, and connective tissue has invaded the coelom, resulting in a series of interconnecting sinuses.

These modifications have resulted in altered patterns of locomotion. Rather than being able to utilize independent coelomic compartments, the leech has a single hydrostatic cavity and uses it in a looping type of locomotion. Figure 12.21 describes the mechanics of this locomotion. Leeches also swim using undulations of the body.

Many leeches feed on body fluids or the entire bodies of other invertebrates. Some feed on the blood of vertebrates, including human blood. Leeches are sometimes called ectoparasites; however, the association between a leech and its host is relatively

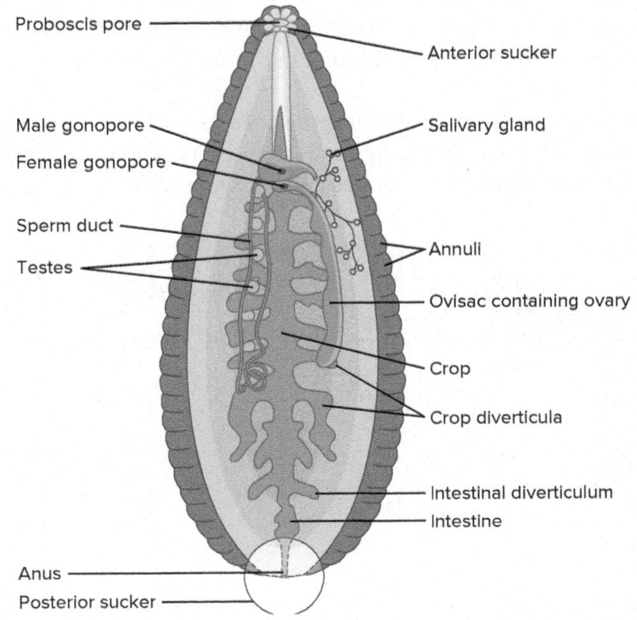

FIGURE 12.20

Internal Structure of a Leech. Annuli subdivide each true segment. Septa do not subdivide the coelom.

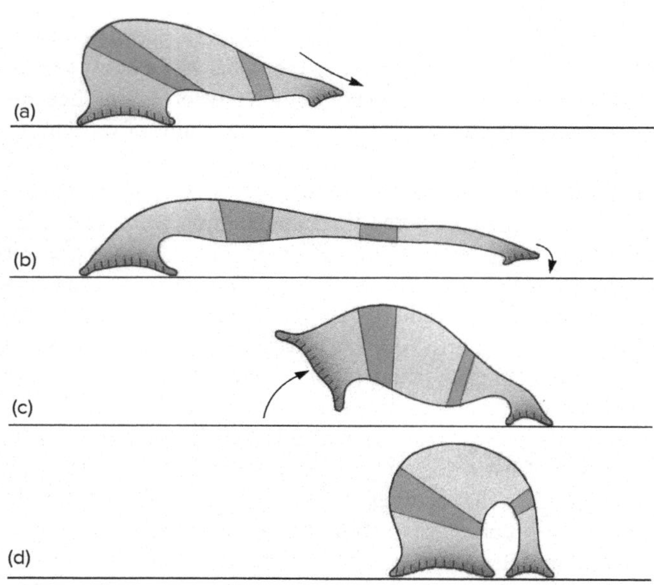

FIGURE 12.21

Leech Locomotion. (*a* and *b*) Attachment of the posterior sucker causes reflexive release of the anterior sucker, contraction of circular muscles, and relaxation of longitudinal muscles. This muscular activity compresses fluids in the single hydrostatic compartment, and the leech extends. (*c* and *d*) Attachment of the anterior sucker causes reflexive release of the posterior sucker, the relaxation of circular muscles, and the contraction of longitudinal muscles, causing body fluids to expand the diameter of the leech. The leech shortens, and the posterior sucker again attaches.
Source: Russell-Hunter, WD. 1979. *A life of invertebrates*. New York (NY): Macmillan.

brief. Therefore, describing leeches as predatory is probably more accurate. Leeches are also not species specific, as are most parasites. (Leeches are, however, class specific. That is, a leech that preys on a turtle may also prey on an alligator, but probably would not prey on a fish or a frog.)

The mouth (or proboscis pore) of a leech opens in the middle of the anterior sucker. In some leeches, the anterior digestive tract is modified into a protrusible proboscis, lined inside and outside by a cuticle. In others, the mouth is armed with three chitinous jaws. While feeding, a leech attaches to its prey by the anterior sucker and either extends its proboscis into the prey or uses its jaws to slice through host tissues. Salivary glands secrete an anticoagulant called hirudin that prevents blood from clotting.

Behind the mouth is a muscular pharynx that pumps body fluids of the prey into the leech. The esophagus follows the pharynx and leads to a large stomach with lateral cecae. Most leeches ingest large quantities of blood or other body fluids and gorge their stomachs and lateral cecae, increasing their body mass 2 to 10 times. After engorgement, a leech can tolerate periods of fasting that may last for months. The digestive tract ends in a short intestine and anus (*see figure 12.20*).

Leeches exchange gases across the body wall. Some leeches retain the basic annelid circulatory pattern, but in most leeches, it is highly modified, and coelomic sinuses replace vessels. Coelomic fluid has taken over the function of blood and, except in two orders, respiratory pigments are lacking.

The leech nervous system is similar to that of other annelids. The suprapharyngeal and subpharyngeal ganglia and the pharyngeal connectives all fuse into a nerve ring that surrounds the pharynx. Ganglia at the posterior end of the animal fuse in a similar way.

Most leeches have photoreceptor cells in pigment cups (2 to 10) along the dorsal surface of the anterior segments. Normally, leeches are negatively phototactic, but when they are searching for food, the behavior of some leeches changes, and they become positively phototactic, which increases the likelihood of contacting prey that happen to pass by.

Hirudo medicinalis, the medicinal leech, has a well developed temperature sense, which helps it to detect the higher body temperature of its mammalian prey. Other leeches are attracted to extracts of prey tissues.

All leeches have sensory cells with terminal bristles in a row along the middle annulus of each segment. These sensory cells, called sensory papillae, are of uncertain function but are taxonomically important.

Leeches have 10 to 17 pairs of metanephridia, one per segment in the middle segments of the body. Their metanephridia are highly modified and possess, in addition to the nephrostome and tubule, a capsule thought to be involved with the production

How Do We Know about Feeding of the Medicinal Leech *Hirudo medicinalis?*

The medicinal leech uses both temperature and chemical senses to detect prey and initiate a feeding response. A feeding response involves a series of stereotyped behaviors that includes probing, attaching, biting, and ingestion. Elliot (1986) found that a permeable bag filled with water warmed to 38°C elicited the probing response but none of the other behaviors. A warmed artificial blood mixture containing small molecular components of blood elicited all of the feeding behaviors. Eliminating either NaCl or arginine (an amino acid) from the mixture prevented the last three responses. A warmed bag containing NaCl and arginine initiated all feeding responses. Surgical removal of the lip area just dorsal to the proboscis pore removed the feeding responses. In control animals, surgery to other regions of the leech body did not interfere with feeding.

Source: Elliott, EJ. 1986. Chemosensory stimuli in feeding behavior of the leech *Hirudo medicinalis, Journ of Comp Phys* 159(3): 391–401.

of coelomic fluid. Chloragogen tissue proliferates through the body cavity of most leeches.

Reproduction and Development (Hirudinea) All leeches reproduce sexually and are monoecious. None are capable of asexual reproduction or regeneration. They have a single pair of ovaries and from four to many testes. Leeches have a clitellum that includes three body segments. The clitellum is present only in the spring, when most leeches breed.

Sperm transfer and egg deposition usually occur in the same manner as described for the earthworm. A penis assists sperm transfer between individuals. A few leeches transfer sperm by expelling a spermatophore from one leech into the integument of another, a form of hypodermic impregnation. Special tissues within the integument connect to the ovaries by short ducts. Cocoons are deposited in the soil or are attached to underwater objects. There are no larval stages, and the offspring are mature by the following spring.

SECTION 12.4 THINKING BEYOND THE FACTS

What is a parasite* (see chapter 6)*? Why is it more accurate to describe leeches as predators rather than ectoparasites?

12.5 BASAL ANNELID GROUPS

LEARNING OUTCOME

1. Describe the phylogenetic status and lifestyles of the Chaetopteridae and the sipunculans.

The last two annelid groups covered in this chapter have been phylogenetic enigmas. In one case, the Chaetopteridae, we have a group that generations of zoologists had considered to be a family within the "Polychaeta." In the other case, Sipuncula, we have a group of worms that had been given phylum status. As we will see in the next section, recent molecular evidence is causing reconsideration of both of these taxonomic assignments.

Chaetopteridae

Chaetopterus, the parchment worm, is found worldwide in relatively shallow temperate and tropical marine habitats. It is a permanent resident of its tough, parchment U-shaped tube that lies buried in sandy substrates or is attached to hard substrates (figure 12.22). Tubes can get as long as 80 cm, and worms are typically 15–20 cm. Both ends of the tube are open and protrude from the substrate. The body of the worm is divided into three regions. The anterior region of the worm has a shovel-like mouth and a series of segments with highly modified bristle-like parapodia. Modified parapodia in the middle segments of the worm create water currents, which flow anterior to posterior through the tube. Parapodia on the 12th segment of the worm, called wings, form a mucous net used for trapping food particles suspended in the water flowing through the tube. The mucus, along with trapped food particles, is rolled into a 3 mm ball and passed anteriorly toward the mouth by a dorsal ciliated groove. Reproduction involves external fertilization, development of planktonic trochophore larvae, and settling of juvenile worms to the substrate. Chaetopterids are also bioluminescent. The chemistry behind this bioluminescence is poorly understood. Bioluminescent chemicals are emitted with secreted mucus and produce fleeting green flashes and a long-lasting blue glow. The function of this bioluminescence is poorly understood. Some researchers think it may help deter predators.

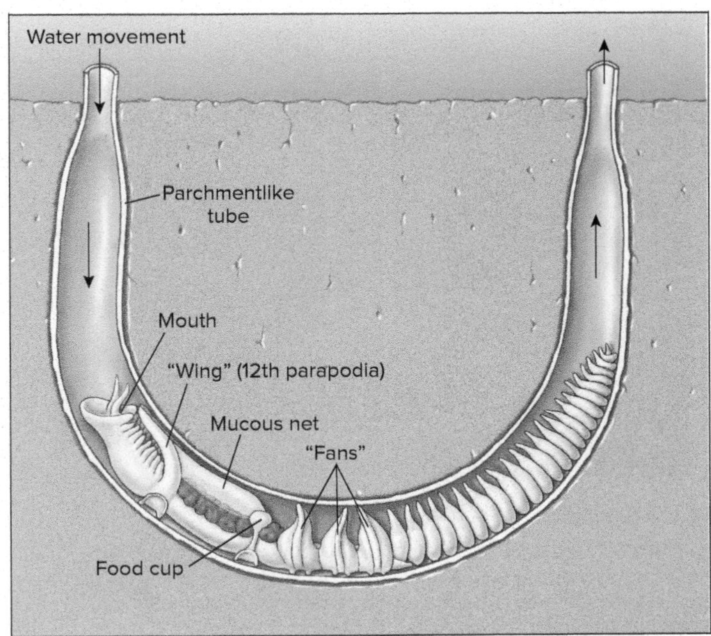

FIGURE 12.22

Chaetopterus. *Chaetopterus* lives in its U-shaped, parchment tube. Water circulates through the tube by the action of modified parapodia, called "fans." The wing-like parapodia on the 12th segment secrete a mucous net that filters food from the circulating water. The food cup rolls mucus and filtered food into a ball that is passed to the mouth.

Sipuncula

Sipunculans (peanut worms) (si-pun'ku-lah) (L. *siphunculus*, small tube) consist of about 350 species of burrowing worms found in oceans throughout the world. These worms live in mud, sand, or any protected retreat. Their name is derived from their habit of retracting into a peanut-shape when disturbed. The anterior portion of the body, the introvert, can be extended and a group of tentacles surrounding the mouth is used in feeding. Sipunculans range in size from 2 mm to 75 cm (figure 12.23). Like the echiruans, sipunculans lack segmentation and parapodia. Gametes are released through metanephridial tubules, and external fertilization and development results in the formation of trochophore larvae.

SECTION 12.5 THINKING BEYOND THE FACTS

In what ways are the Chaetopteridae and sipunculans "unusual annelids?" What features of their biology are annelid-like?

FIGURE 12.23

Sipunculans. Sipunculans are found throughout the world's oceans, where they live in mud, sand, or rock crevices. *Themiste pyroides* is shown here. It is found in the substrate of temperate oceans at depths of 0–36 m.

12.6 FURTHER PHYLOGENETIC CONSIDERATIONS

LEARNING OUTCOME

1. Formulate a conversation between a modern taxonomist and a taxonomist who worked 100 years ago as they compare their perceptions of annelid phylogeny.

Taxonomic relationships within the Annelida are the subject of intense research. The original classification of the Annelida can be traced back to Jean Baptiste Lamarck (1809), who established the taxon. He recognized the similarities between the oligochaetes and the polychaeters, but the leeches were left out of the group until the mid-1800s. Since that time, many taxonomic studies have attempted to sort out relationships within the phylum and to other phyla. These studies have resulted in radically different, and sometimes conflicting, interpretations of annelid phylogenetics.

As described earlier, recent molecular studies affirm the monophyly of the Annelida. They place the Chaetopteridae and Sipuncula near the base of the annelid phylogeny and outside of the two major annelid clades, Errantia and Sedentaria (figure 12.24). The inclusion of sipunculans as annelids is still controversial, but we have done so in this chapter because of the strength of this molecular evidence.

One of the objections to including sipunculans in Annelida is the absence of segmentation in the group. Neither the echiuran nor the sipunculans are segmented, and the segmentation of the siboglinids and Chaetopteridae is highly modified. If one assumes that segmentation is an ancestral characteristic of the Annelida, then it must have been independently lost or modified very early in annelid evolution (Sipuncula and Chaetopteridae) and later in more derived groups (Echiura and Siboglinidae).

The Clitellata are believed to have evolved from a group of polychaetes that invaded freshwater. A few species of freshwater polychaetes remain today. This freshwater invasion required the ability to regulate the salt and water content of body fluids. In addition, direct development inside of a cocoon rather than as free-swimming larval stages promoted the invasion of terrestrial environments.

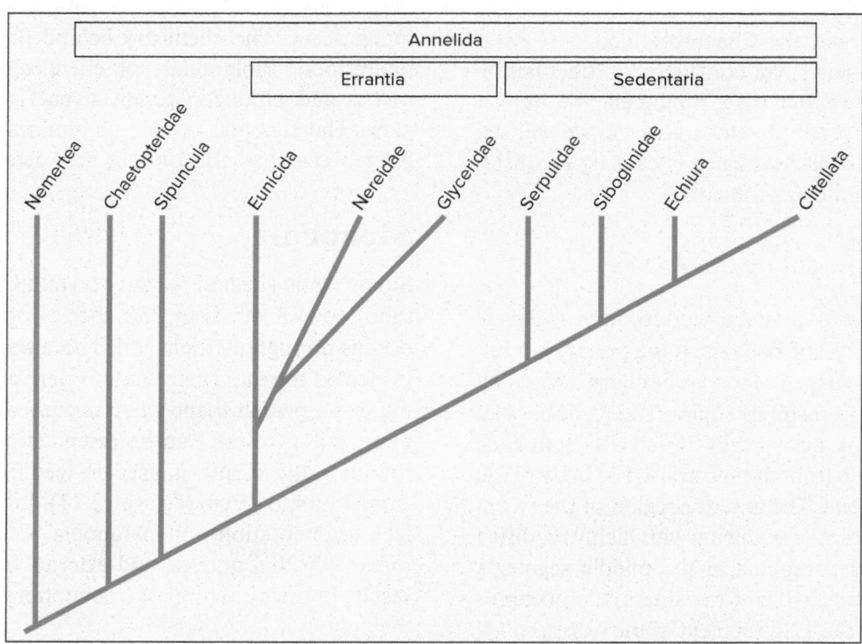

FIGURE 12.24

Annelid Phylogeny. This molecular phylogeny shows one interpretation of the relationships among the annelids. It depicts the Annelida as being comprised of two major clades: Errantia and Sedentaria. Each of these has nested minor clades. (The nested clades shown here are those with representatives discussed in this chapter.) Chaetopteridae and Sipuncula are shown as basal annelids. The inclusion of the sipunculans within the Annelida is controversial, but supported by a growing body of evidence.
Source: Struck, TH et al. 2007. Annelid Phylogeny and the Status of Sipuncula and Echiura. *BMC Evol Bio* 7:57. Available from: http://www.biomedcentral.com/1471-2148-7-57.

During the Cretaceous period, approximately 100 mya, some sedentarian annelids invaded moist, terrestrial environments. This period saw the climax of the giant land reptiles, but more important, it was a time of proliferation of flowering plants. The reliance of modern oligochaetes on deciduous vegetation can be traced back to their ancestors' exploitation of this food resource. Some of the early freshwater sedentarians gave rise to leeches. Ancestral leeches then colonized marine and terrestrial habitats from freshwater. Chapters 10 through 12 presented information on most of the better-known lophotrochozoan phyla. Brief descriptions of three additional lesser-known lophotrochozoan phyla are presented in table 12.2.

SECTION 12.6 THINKING BEYOND THE FACTS

Discuss the statement that segmentation is a definitive, yet variable, characteristic of the Annelida.

TABLE 12.2
LESSER KNOWN LOPHOTROCHOZOANS

ENTOPROCTA	
Examples	*Urnatella gracilis, Pedicellina cernua.* Approximately 200 species.
Description	With the exception of one genus (*Urnatella*), entoprocts are marine. They live attached to firm substrates from the intertidal zone to depths of 500 m. They appear radially symmetrical; however, they are actually bilateral animals (0.1–7 mm, figure *a*). They are comprised of a cuplike calyx and a stalk. The calyx and stalk attach to each other at the dorsal aspect of the animal, and the ventral surface is oriented upward and ringed by a crown of ciliated tentacles. Cilia create water currents that move dorsal to ventral through the tentacles (figure *b*). Plankton is trapped in mucous sheets and is moved by cilia to the mouth. The gut is U-shaped, and the anus opens ventrally inside the crown of tentacles (Entoprocta, *Gr. entos*, inside + *proktos*, anus). A few entoprocts are solitary, but most are colonial. Colony growth occurs by budding. Sexual reproduction occurs when sperm released by a male are taken into a female with ciliary currents. (Many species are protandrous. *See chapter 29.*) Fertilization and protostome-type development occur within the female's brood chamber, and a planktonic, ciliated larval stage (resembling a trochophore) eventually settles and undergoes metamorphosis to the adult.
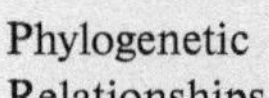Phylogenetic Relationships	The fossil record for entoprocts is scanty, but they probably arose during the Precambrian with most other animal phyla. Some molecular phylogenies place the entoprocts as a sister group to the cycliophorans (*see chapter 10*).

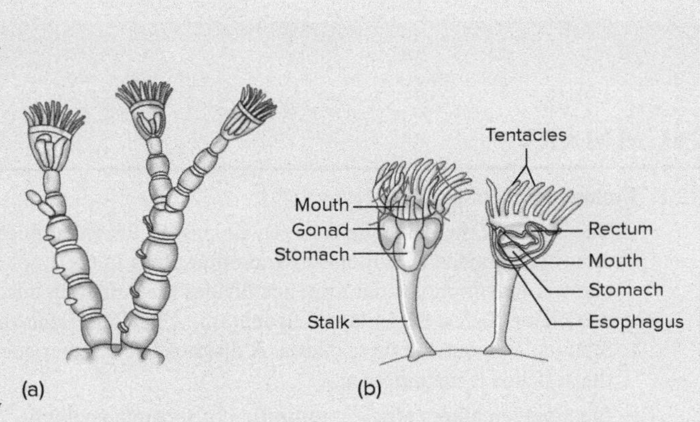

PHORONIDA	
Examples	*Phoronis architecha, Phoronopsis.* Approximately 20 species.
Description	Phoronids are vermiform, marine, and benthic tube dwellers. They use a lophophore in filter feeding. Their bodies (about 2 cm long) are divided into a flaplike epistome, a lophophore-bearing mesostome, and an elongate trunk (metastome). The gut is U-shaped and the anus is close to the mouth. All phoronids reproduce sexually.
Phylogenetic Relationships	There are scanty fossil records for Phoronida. Molecular phylogenies usually depict phoronids and brachiopods as sister groups within Lophotrochozoa (*see chapter 10*).

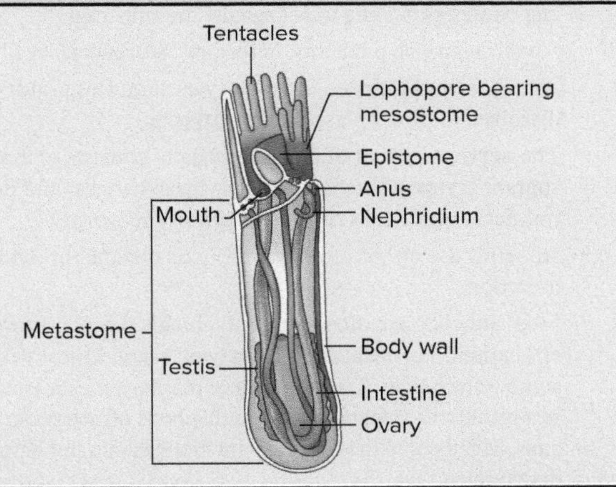

(*Continued*)

TABLE 12.2 Continued

MESOZOA	
Examples	*Dicyema, Pseudicyema, and Dicyemennea*. Approximately 50 species.
Description	Mesozoans (1-5 mm) are symbionts of marine invertebrates. They are comprised of two layers of cells that are not differentiated into tissues. Two mesozoan clades include Rhombozoa (parasites of cephalopod molluscs) and Orthonectida (symbionts of echinoderms, molluscs, and annelids). Their life cycles can be complex, involve sexual and asexual stages, and are not completely understood.
Phylogenetic Relationships	It is not surprising that fossils of these tiny, delicate animals are unknown. Molecular data, and very different morphologies and life cycles, suggest that the two mesozoan clades mentioned above had separate origins. If so, they should be considered separate phyla. "Mesozoa" would then be an informal designation.

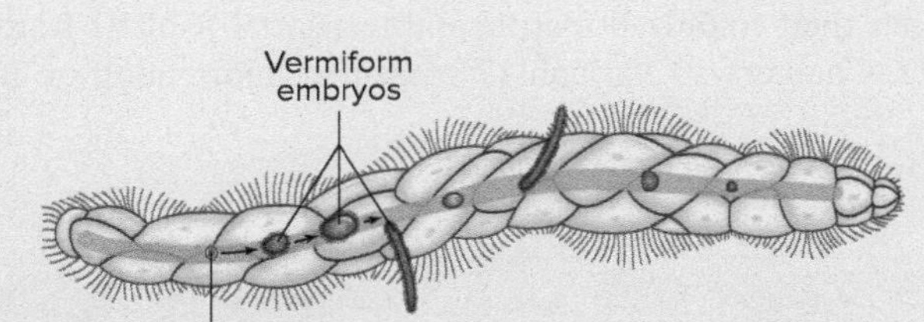

SUMMARY

12.1 **Evolutionary Perspective**

- The origin of the Annelida is largely unknown. Recent studies unite the annelids with molluscs and other phyla in the Lophotrochozoa. Recent molecular evidence divides the Annelida into two major clades, Errantia and Sedentaria. Chaetopteridae and Sipuncula lie outside these clades. A diagnostic characteristic of the annelids is metamerism.
- Metamerism allows efficient utilization of separate coelomic compartments as a hydrostatic skeleton for support and movement. Metamerism also lessens the impact of injury and makes tagmatization possible.

12.2 **Annelid Structure and Function**

- Most annelids possess parapodia with numerous setae. Locomotion involves the antagonism of longitudinal muscles on opposite sides of the body, which creates undulatory waves along the body wall and causes parapodia to act against the substrate.
- Annelids may be predators, herbivores, scavengers, or filter feeders.
- Annelids have a closed circulatory system. Respiratory pigments dissolved in blood plasma carry oxygen.
- The nervous system of annelids usually consists of a pair of suprapharyngeal ganglia, subpharyngeal ganglia, and double ventral nerve cords that run the length of the worm.
- Annelids use either protonephridia or metanephridia in excretion.
- Most annelids are dioecious, and gonads develop from coelomic epithelium. Fertilization is usually external. Epitoky occurs in some polychaetes. Development of marine annelids usually results in a planktonic trochophore larva that buds off segments near the anus. Members of the Clitellata are monoecious and have direct development.

12.3 **Clade (Class) Errantia**

- Members of the clade (class) Errantia are mostly marine. They possess relatively long setae and well developed palps.
- *Nereis*, the clam worm, burrows in mud and sand substrates. It uses its proboscis and large jaws to feed on marine vegetation and invertebrates. Reproduction occurs through swarming and external fertilization. Development includes a trochophore larval stage.
- *Glycera*, the bloodworm, has hemoglobin-containing coelomocytes, burrows in mud and sand, and feeds on marine invertebrates. Reproduction occurs in a fashion similar to reproduction by *Nereis*.
- Fireworms feed on coral polyps and small crustaceans. Hollow setae are venomous, and their bright colors are an example of aposmatic coloration.

12.4 **Clade (Class) Sedentaria**

- Sedentarians include a variety of marine tubeworms, the siboglinids, the echiurans, and the Clitellata.
- Tubeworms construct tubes or live in burrows. They feed on organic matter in seawater, which they filter using cilia and mucus. Featherduster worms have a crown of arm-like radioles that create water currents and trap food particles.
- Siboglinids live at great depths. They harbor symbiotic bacteria that fix carbon dioxide into organic compounds.
- Echirurans are spoon worms. They live in marine burrows or rock crevices. They feed by sweeping organic material into their spatula-shaped proboscis.
- The Clitellata includes earthworms, other oligochaetes, and leeches (Hirudinea). They possess a clitellum used in cocoon formation. The earthworms and other "oligochaetes" are primarily freshwater and terrestrial annelids. Oligochaetes possess few setae, and they lack a head and parapodia. Earthworms are scavengers that feed on dead and decaying vegetation. Their digestive tract is tubular and straight, and it frequently has modifications for

storing and grinding food and for increasing the surface area for secretion and absorption. Earthworms are monoecious and exchange sperm during copulation.

- Members of the Hirudinea are the leeches. Complex arrangements of body-wall muscles and the loss of septa influence patterns of locomotion. Leeches are predatory and feed on body fluids, the entire bodies of other invertebrates, and the blood of vertebrates. Leeches are monoecious, and reproduction and development occur as in earthworms.

12.5 **Basal Annelid Groups**

- Chaetopteridae are the parchment worms. They live in U-shaped parchment tubes and use highly modified parapodia in mucous-based filter feeding. Reproduction involves external fertilization and the development of trochophore larvae.
- Sipunculans (the peanut worms) are marine burrowers that use an introvert in feeding. They are unsegmented.
- Three small phyla of lesser known lophotrochozoans include Entoprocta, Phoronida, and Mesozoa.

12.6 **Further Phylogenetic Considerations**

- Taxonomic studies affirm monophyly of the Annelida. Within the Annelida, Chaetopteridae and Sipuncula are basal groups, and Errantia and Sedentaria are the two major derived clades. Within the Sedentaria, members of the clade Clitellata invaded freshwater and terrestrial habitats.

Concept Review Questions

1. Current evidence indicates that
   a. Annelida is a monophyletic group composed of two major clades, Errantia and Sedentaria.
   b. Annelida is a paraphyletic group.
   c. Annelida is a monophyletic group, and its two clades (Polychaeta and Clitellata) are each clearly single-lineage groups.
   d. Annelida is a polyphyletic group, and the name should not be used as a phylum designation.
2. Which of the following statements is true regarding metamerism?
   a. It arose only once in animal evolution.
   b. It is found only in the Annelida and Chordata.
   c. It permits a variety of locomotor and supportive functions not possible in nonmetameric animals.
   d. Its main disadvantage is that it increases the likelihood that injury will result in death of an animal.
3. Which of the following statements about annelids is true?
   a. They have an open circulatory system.
   b. Their dorsal nerve cord begins anteriorly at suprapharyngeal ganglia.
   c. Most gas exchange occurs as a result of diffusion of gases across the body wall and parapodia. Some annelids have parapodial gills.
   d. Most adult annelids use protonephridia in excretion.
4. Like many molluscs, annelid development usually involves a _________ larval stage.
   a. veliger
   b. trochophore
   c. glocidium
   d. planula
5. After being released from the _________ of one earthworm, sperm is temporarily stored in the _________ of a second earthworm.
   a. chloragogen tissue; nephridium
   b. seminal vesicles; seminal receptacles
   c. testes; ovary
   d. seminal vesicles; coelom

Analysis and Application Questions

1. Distinguish between a protonephridium and a metanephridium. Name a group of annelids whose members may have protonephridia. What other phylum have we studied whose members also had protonephridia? Do you think that metanephridia would be more useful for an animal with an open or a closed circulatory system? Explain.
2. In what annelid groups are septa between coelomic compartments lost? What advantages does this loss give each group?
3. What differences in nephridial function might you expect in freshwater and marine annelids?
4. Relatively few annelids have invaded freshwater. Can you think of a reasonable explanation for this?

13

The Smaller Ecdysozoan Phyla

These nematode worms were passed from the intestine of a young boy in Kenya, Africa. *Ascaris lumbricoides* is one of a number of important human parasites in the phylum Nematoda, but the parasitic species represent a small portion of this large and important phylum. Chapters 13 through 15 discuss Nematoda and related animals in the clade Ecdysozoa.
Source: James Gathany/CDC

Chapter Outline

13.1 Evolutionary Perspective
13.2 Phylum Nematoda (Roundworms)
Structure and Function
Reproduction and Development
Some Important Nematode Parasites of Humans
13.3 Other Ecdysozoan Phyla
Phylum Nematomorpha
Phylum Kinorhyncha
Phylum Priapulida
Phylum Loricifera
13.4 Further Phylogenetic Considerations

13.1 EVOLUTIONARY PERSPECTIVE

LEARNING OUTCOMES

1. Describe the unifying features that define the clade Ecdysozoa.
2. Describe one common function of the cuticle in ecdysozoan phyla covered in this chapter.

"Beauty is in the eye of the beholder"–the Western philosopher, Plato, said it first around 300 B.C. He was probably not thinking about nematodes when he made that statement. For some zoologists, however, it applies. What phylum of animals has representatives that live at the freezing point in Arctic ice (*Cryonema*)? What phylum has members that live in sulfide-rich marine sediments (Stilbonematinae), in millipede guts (*Zalophora*), on beer coasters in Germany (*Panagrellus*), or in the placentas of sperm whales (*Placentonema*)? In what phylum can one count over 200 species in a few cubic centimeters of marine mud or 90,000 individuals in a single rotting apple? What phylum is composed of decomposers and predators in aquatic sediment and soil? What phylum has parasitic members that have killed and maimed across all other animal phyla? The answer to all of these questions is the phylum Nematoda. The nematode structure that you will study in this chapter is deceptively simple–this is an introductory textbook. A lifetime of detailed study, however, could never uncover the endless variation found within this highly diverse phylum.

Nematoda is just one of the eight phyla covered in chapters 13 through 15 that are members of the clade Ecdysozoa. Members of these phyla are united by common molecular features and a single morphological character–they all possess a nonliving, secreted **cuticle** (L. *cutis*, skin) that is molted periodically as they grow (figure 13.1). Molting of the cuticle is called **ecdysis** (Gr. *ekdysis*, getting out). Chapter 13 describes members of the phylum Nematoda and four other smaller ecdysozoan phyla: Nematomorpha, Kinorhyncha, Priapulida, and Loricifera. Chapters 14 and 15 describe the largest animal phylum, Arthropoda, and two smaller phyla, Tardigrada and Onychophora. These three phyla comprise the ecdysozoan clade Panarthropoda. Phylogenetic relationships within the ecdysozoan phyla are discussed at the end of chapters 13 and 15.

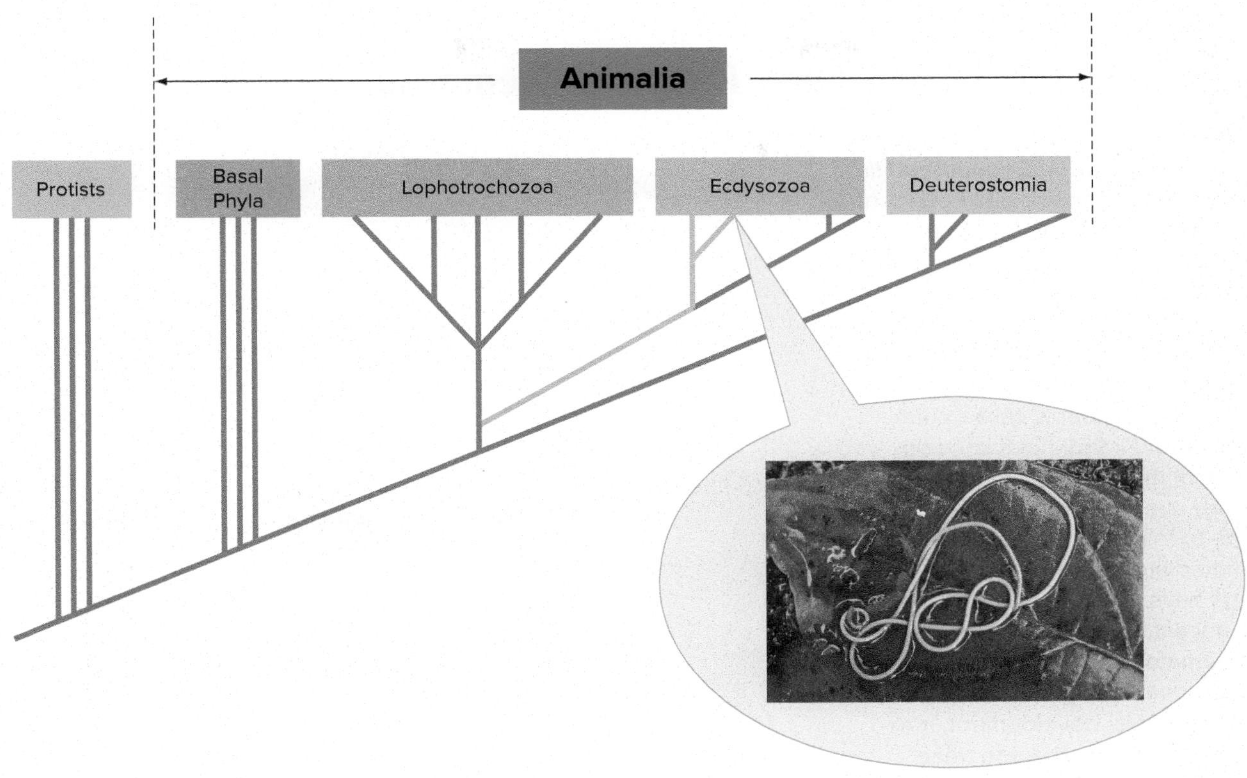

FIGURE 13.1

Ecdysozoan Phylogeny. Ecdysozoans are united by molecular characteristics and the presence of a secreted, nonliving cuticle. The cuticle is molted to accommodate growth—a process called ecdysis. Chapters 13 through 15 cover the ecdysozoan phyla. The Gordian worm, *Gordius robustus* (*see inserted photograph*), is a member of the ecdysozoan phylum Nematomorpha.

The function of the cuticle is somewhat different in the phyla covered in chapters 13 through 15. The arthropods have a reduced coelom (chapters 14 and 15) and the cuticle forms an armor-like exoskeleton. This exoskeleton provides support and protection and is jointed for mobility. It is one of the major reasons for the success of the phylum. The cuticles of the phyla described in chapter 13 are composed of either collagen or chitin—an important taxonomic difference that will be discussed at the end of this chapter. The phyla in chapter 13 are also pseudocoelomate (*see figure 7.11*). One of the major functions of the pseudocoelom interacting with the cuticle (*see chapter 7*) is to act as a hydrostatic skeleton. Body movements occur as a result of contractions of body wall musculature acting on the fluid or gel within the pseudocoelom. The tough cuticle prevents tearing of the body wall with changes in pseudocoelomic pressure. The pseudocoelom also serves as a vehicle for the distribution of nutrients, respiratory gases, and metabolic wastes. The pseudocoelom is of little phylogenetic significance. It is present in phyla scattered through Lophotrochozoa and Ecdysozoa.

Section 13.1 Thinking Beyond the Facts

Ecdysis, as defined in this section, involves shedding a nonliving, secreted, cuticle in order to accommodate growth. Ecdysis occurs in all ecdysozoans* (see chapters 13–15). *A molting process, sometimes also called ecdysis, occurs in reptiles and other vertebrates (deuterostomes) to accommodate growth. Examine figures 13.1 and 13.17 to evaluate the relationship, if any, between these molting processes.

13.2 PHYLUM NEMATODA (ROUNDWORMS)

LEARNING OUTCOMES

1. Explain how the body wall and pseudocoelom of a nematode influence locomotion of a nematode.
2. Describe the body systems involved in maintenance functions of nematodes.
3. Describe the reproductive system and development of nematodes.
4. Contrast the life cycles of common nematode parasites with the life cycles of digenetic flukes.

Nematodes (nem′a-todes) (Gr. *nematos,* thread) or roundworms are some of the most abundant animals on the earth—some 5 billion may be in every acre (4,046 m^2) of fertile garden soil. Zoologists estimate that the number of roundworm species may be

How Do We Know: Eutely in Nematoda

The nematode worm *Caenorhabditis elegans* is a 1-mm-long soil resident. The worms are very easy to maintain in the laboratory and grow from fertilized egg to an adult in just three days. The somatic cell number in all adults is constant for the entire animal and for each given organ in all individuals of the species. This constancy of cell number is called eutely (Gr. *euteia*, thrift). All 959 adult somatic cells arise from the zygote in virtually the same way for every *C. elegans.* Using a microscope to follow all the cell divisions starting immediately after a zygote forms, researchers have been able to constuct the entire ancestry of every cell in the adult body, the nematode's cell lineage. This is possible because *C. elegans* is transparent at all stages of its development, making it possible for researchers to trace the lineage of every cell, from the zygote to the adult worm. A half day after the first cell division, the nematode is a larva of 558 cells. Further rounds of cell division and the death of precisely 113 cells sculpt the 959-celled adult worm. A cell lineage diagram like the one in the figure is a type of fate map, a representation of the fate of various parts of a developing embryo. The much simplified fate map shown to the right indicates the development of the major tissues of the nematode's body.

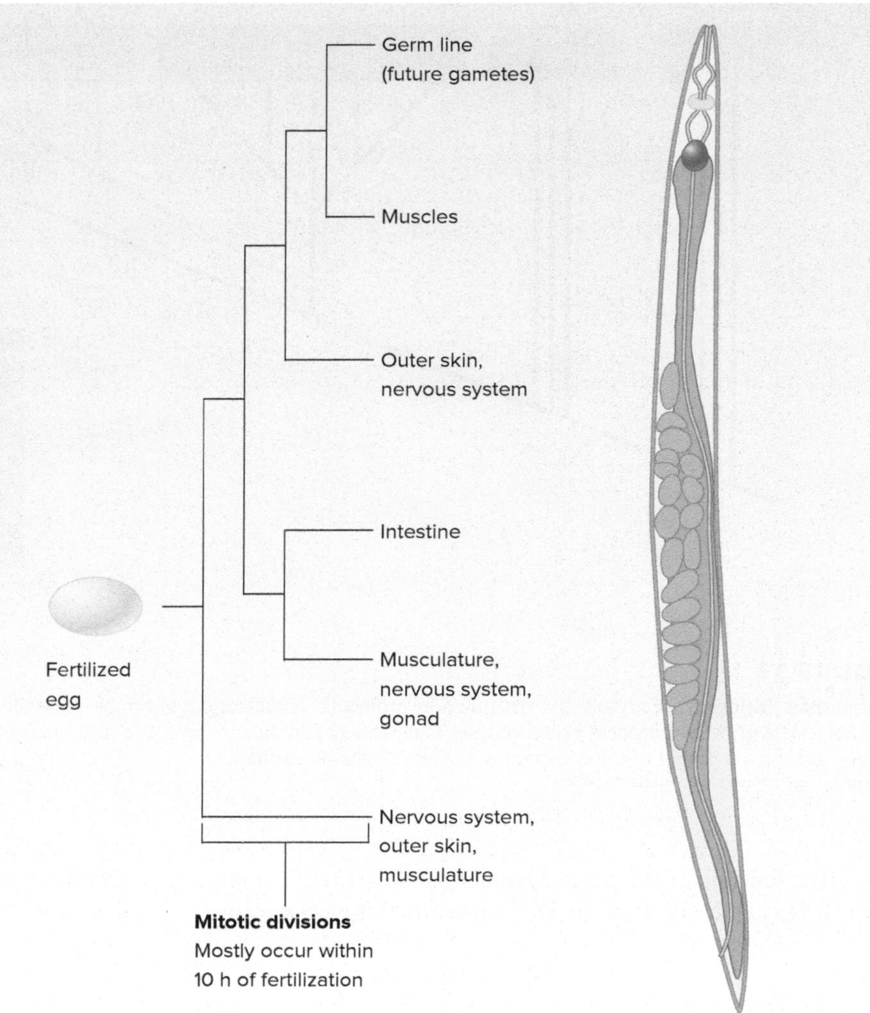

as high as 500,000. Roundworms feed on every conceivable source of organic matter–from rotting substances to the living tissues of other invertebrates, vertebrates, and plants. They range in size from microscopic to several meters long. Many nematodes are parasites of plants or animals; most others are free living in marine, freshwater, or soil habitats. Some nematodes play an important role in recycling nutrients in soils and bottom sediments.

Except in their sensory structures, nematodes lack cilia, a characteristic they share with arthropods. Also in common with some arthropods, the sperm of nematodes are amoeboid.

Taxonomy within the phylum Nematoda is a subject of intensive study. Two classes have been recognized based upon morphological studies: Secernentea and Adenophorea. Molecular studies suggest that these class designations do not accurately reflect nematode phylogeny. Additional information on nematode phylogeny is presented at the end of this chapter.

Characteristics of the phylum Nematoda include the following:

1. Triploblastic, bilateral, vermiform (resembling a worm in shape; long and slender), unsegmented, pseudocoelomate
2. Body wall round in cross section with longitudinal muscles only
3. Ecdysis of the collagenous cuticle accompanies growth of juvenile stages
4. Complete digestive tract; mouth usually surrounded by lips bearing sense organs
5. Excretory system usually composed of collecting tubules or renette cells

Structure and Function

A nematode body is slender, elongate, cylindrical, and tapered at both ends (figure 13.2*a* and *b*). Much of the success of nematodes is due to their outer, noncellular, collagenous cuticle (figure 13.2*c*) that is continuous with the foregut, hindgut, sense organs, and parts of the female reproductive system. The collagenous cuticle may be smooth, or it may contain spines, bristles, papillae (small, nipple-like projections), warts, or ridges, all of which are of taxonomic significance. Three primary layers make up the cuticle: cortex, matrix layer, and basal layer. The cuticle maintains internal hydrostatic pressure, provides mechanical protection, and, in parasitic species of nematodes, resists digestion by the host. The cuticle is usually molted four times during maturation.

Beneath the cuticle is the epidermis, or hypodermis, which surrounds the pseudocoelom (figure 13.2*d*). The epidermis may be syncytial, and its nuclei are usually in the four epidermal cords (one dorsal, one ventral, and two lateral) that project inward. The longitudinal muscles are the principal means of locomotion in nematodes. Contraction of these muscles results in undulatory waves that pass from the anterior to the posterior end of the animal, creating characteristic thrashing movements. Nematodes lack circular muscles and therefore cannot crawl as do worms with more complex musculature.

Some nematodes have lips surrounding the mouth, and some species bear spines or teeth on or near the lips (figure 13.3). In others, the lips have disappeared. Some roundworms have head shields that afford protection. Sensory organs include amphids, phasmids, or ocelli. **Amphids** are anterior depressions in the cuticle that contain modified cilia and function in chemoreception. **Phasmids** are near the anus and also function in chemoreception. The presence or absence of these organs are taxonomically important. Paired ocelli (eyes) are present in aquatic nematodes.

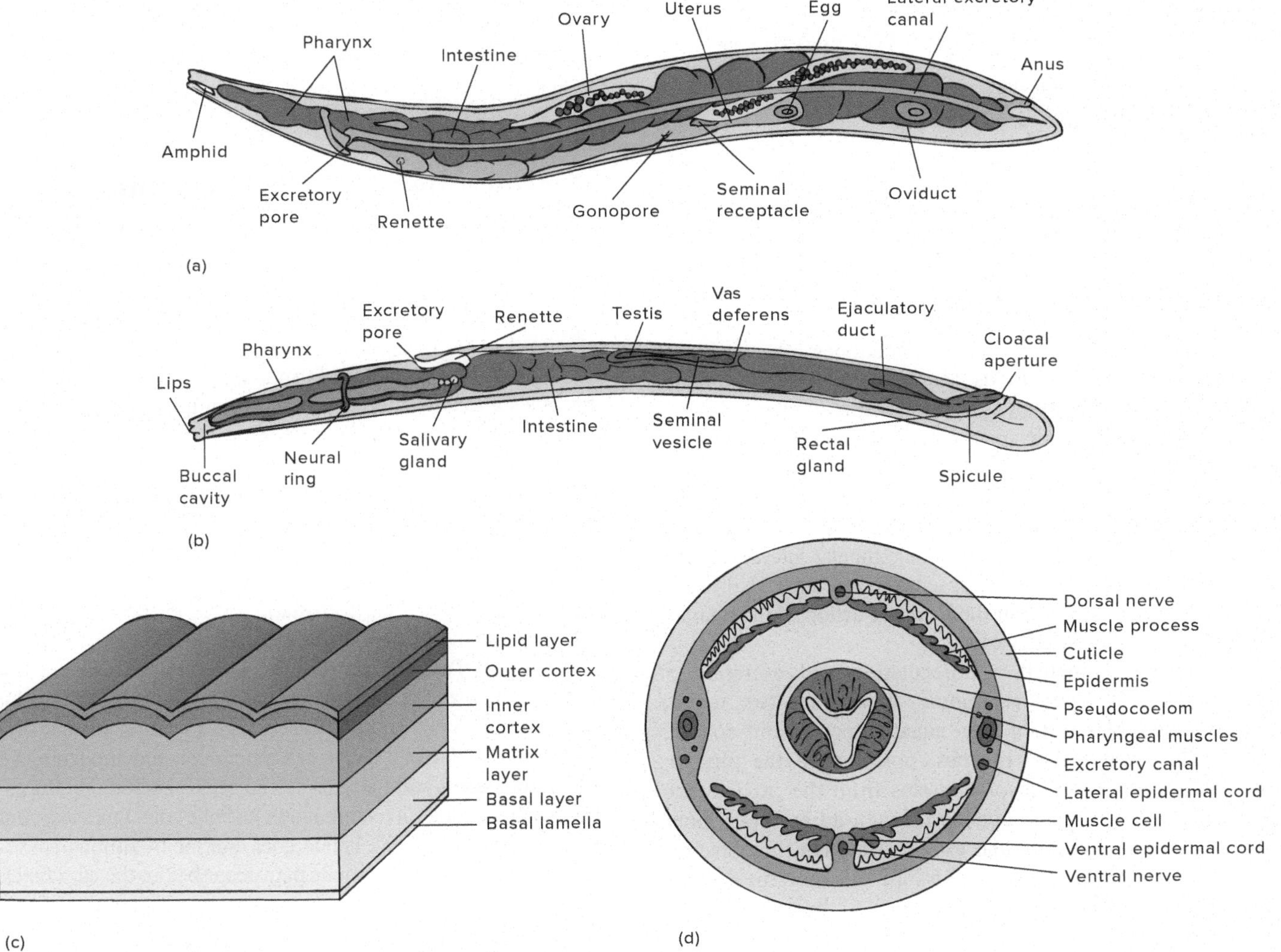

FIGURE 13.2

Phylum Nematoda. Internal anatomical features of an (*a*) female and (*b*) male *Rhabditis*. (*c*) Section through a nematode cuticle, showing the various layers. (*d*) Cross section through the region of the muscular pharynx of a nematode. The hydrostatic pressure in the pseudocoelom maintains the rounded body shape of a nematode and also collapses the intestine, which helps move food and waste material from the mouth to the anus.

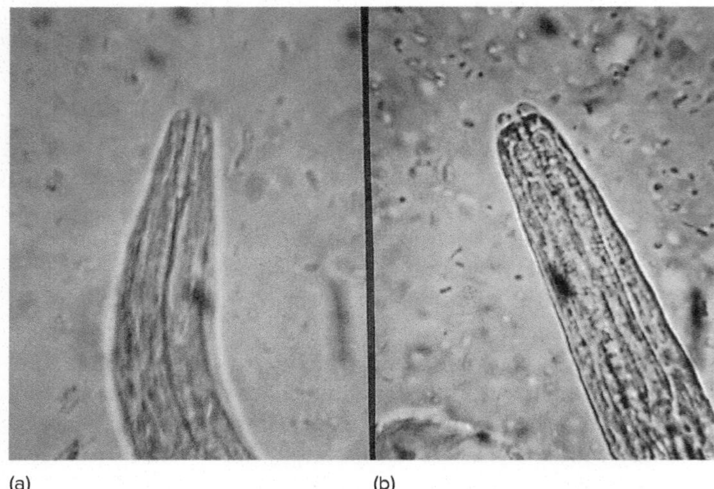

FIGURE 13.3

Nematode Mouth Structure. Mouthparts of nematodes are adapted for various purposes. (*a*) Head of the rhabditiform larval stage of the hookworm (*Necator*). Adult hookworms have tooth-like plates that are used to attach to the intestinal wall of their host (*see figure 13.9*). (*b*) Head of the threadworm *Strongyloides*. This nematode tunnels through the intestinal mucosa of its host (humans or other mammals, depending on the species of threadworm).

Source: Courtesy of the CDC

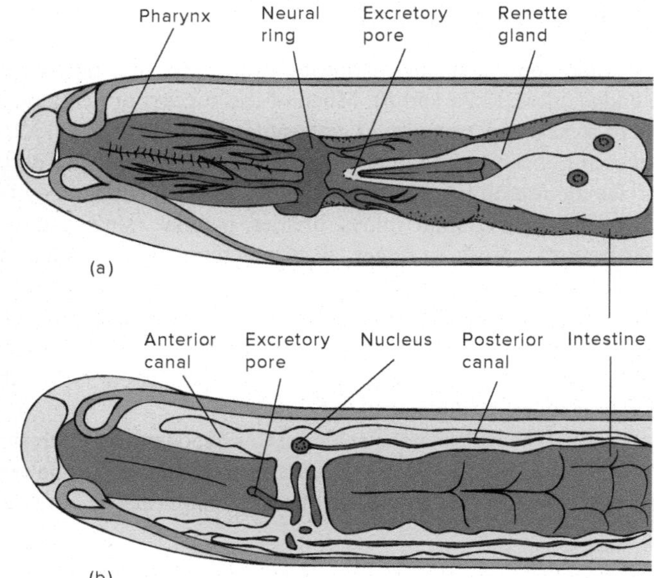

FIGURE 13.4

Nematode Excretory Systems. (*a*) Glandular, as in *Rhabditis.* (*b*) Tubular, as in *Ascaris.*

The nematode pseudocoelom is a fluid-filled cavity that contains the visceral organs and forms a hydrostatic skeleton. All nematodes are round because the body muscles contracting against the pseudocoelomic fluid generate an equal outward force in all directions (*see figure 13.2*d).

Depending on the environment, nematodes are capable of feeding on a wide variety of foods; they may be carnivores, herbivores, omnivores, or saprobes (saprotrophs) that consume decomposing organisms, or parasitic species that feed on blood and tissue fluids of their hosts.

Nematodes have a complete digestive system consisting of a mouth, which may have teeth, jaws, or stylets (sharp, pointed structures); buccal cavity; muscular pharynx; long, tubular intestine where digestion and absorption occur; short rectum; and anus. Hydrostatic pressure in the pseudocoelom and the pumping action of the pharynx push food through the alimentary canal.

Nematodes accomplish osmoregulation and excretion of nitrogenous waste products (ammonia, urea) with two unique systems. The glandular system is in aquatic species and consists of ventral gland cells, called **renettes,** posterior to the pharynx (figure 13.4*a*). Each gland absorbs wastes from the pseudocoelom and empties them to the outside through an excretory pore. Parasitic nematodes have a tubular excretory system that develops from the renette system (figure 13.4*b*). In this system, the renettes unite to form a large canal, which opens to the outside via an excretory pore.

The nervous system consists of an anterior neural ring (*see figures 13.2*b *and 13.4*a). Nerves extend anteriorly and posteriorly; many connect to each other via commissures. Certain neuroendocrine secretions are involved in growth, molting, cuticle formation, and metamorphosis.

Reproduction and Development

Most nematodes are dioecious and dimorphic, with the males being smaller than the females. The long, coiled gonads lie free in the pseudocoelom.

The female system consists of a pair of convoluted ovaries (figure 13.5*a*). Each ovary is continuous with an oviduct whose proximal end is swollen to form a semnal receptacle. Each oviduct becomes a tubular uterus; the two uteri unite to form a vagina that opens to the outside through a genital pore.

The male system consists of a single testis, which is continuous with a vas deferens that eventually expands into a seminal vesicle and ends at the cloaca. Males usually possess spicules that are extruded from the cloaca and inserted into the female genital pore to aid in copulation (figure 13.5b, *see figure 13.7*).

After copulation, hydrostatic forces in the pseudocoelom (*see figure 13.2*d) move each fertilized egg to the gonopore (genital pore). The number of eggs produced varies with the species; some nematodes produce only several hundred, whereas others may produce hundreds of thousands daily. Some nematodes give birth to larvae (ovoviviparity). External factors, such as temperature and moisture, influence the development and hatching of the eggs. Hatching produces a larva (some parasitologists refer to it as a juvenile) that has most adult structures. The larva (juvenile) undergoes four molts, although in some species, the first one or two molts may occur before the eggs hatch.

Some Important Nematode Parasites of Humans

Parasitic nematodes show a number of evolutionary adaptations to their way of life. These include a high reproductive potential, life cycles that increase the likelihood of transmission from one host to

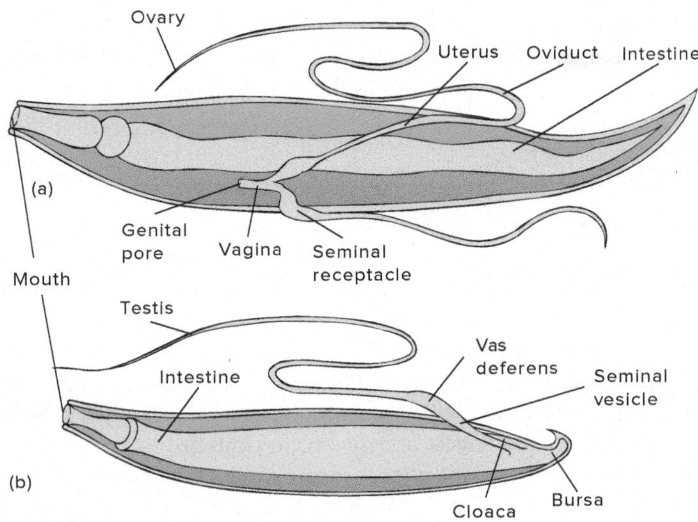

FIGURE 13.5

Nematode Reproductive Systems. The reproductive systems of (*a*) female and (*b*) male nematodes, such as *Ascaris.* Some male nematodes possess a flap of tissue called the bursa, which aids in copulation. The sizes of the reproductive systems are exaggerated to show details.

another, an enzyme-resistant cuticle, resistant eggs, and encysted larvae. Nematode life cycles are not as complicated as those of cestodes or trematodes because only one host is usually involved. Discussions of the life cycles of five important human parasites follow.

Ascaris lumbricoides: *The Giant Intestinal Roundworm of Humans*

As many as 800 million people throughout the world may be infected with *Ascaris lumbricoides.* Adult *Ascaris* (Gr. *askaris,* intestinal worm) live in the small intestine of humans. They produce large numbers of eggs that exit with the feces (figure 13.6). A first-stage larva develops rapidly in the egg, molts, and matures into a second-stage larva, the infective stage. When a human ingests embryonated eggs, they hatch in the intestine. The larvae penetrate the intestinal wall and are carried via the circulation to the lungs. They molt twice in the lungs, migrate up the trachea, and are swallowed. The worms attain sexual maturity in the intestine, mate, and begin egg production (figure 13.7).

Enterobius vermicularis: *The Human Pinworm*

Pinworms (*Enterobius;* Gr. *enteron,* intestine + *bios,* life) are the most common roundworm parasites in the United States. Adult *Enterobius vermicularis* become established in the lower region of the large intestine. At night, gravid females migrate out of the rectum to the perianal area, where they deposit eggs containing a first-stage larva (figure 13.8) and then die. The females and eggs produce an itching sensation. When a person scratches the itch, the hands and bedclothes are contaminated with the eggs. When the hands touch the mouth and the eggs are swallowed, the eggs hatch. The larvae molt four times in the small intestine and migrate to the large intestine. Adults mate, and females soon begin egg production.

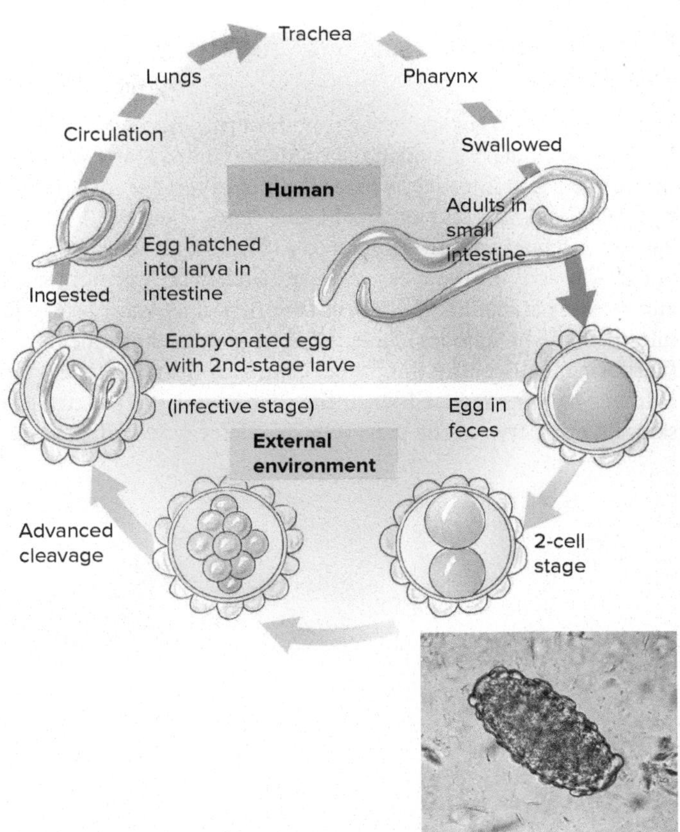

FIGURE 13.6

Life Cycle of *Ascaris lumbricoides*. Adults live in the small intestine of humans. Fertilized and embryonated eggs are deposited in soil or water with human feces. When another human ingests embryonated eggs, larvae are released, carried to the lungs, and eventually swallowed to the intestine where adults mature and reproduce. The inset photograph shows an unfertilized *Ascaris* egg. Note the thick resistant egg capsule. This capsule allows embryonated eggs to remain viable in the soil for extended periods of time. *Source: Centers for Disease Control, Atlanta, GA.*

Source: [CDC] Centers for Disease Control (US). Atlanta, (GA).

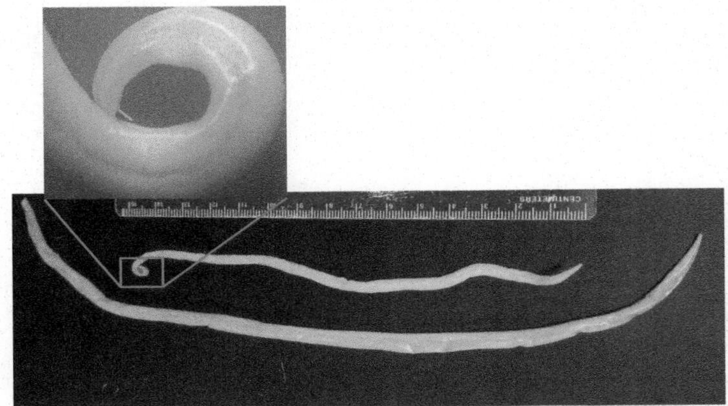

FIGURE 13.7

Ascaris. Ascarids inhabit the intestines of virtually all vertebrates. They all display sexual dimorphism. The male is smaller and has a hooked posterior end. The inset photograph shows an extruded copulatory spicule. The male human ascarid, *Ascaris lumbricoides*, measures 15 to 31 cm, and the female measures 20 to 35 cm.

(a) ©Todd Tupper, Phd., Northern Virginia Community College (b) ©Todd Tupper, Phd., Northern Virginia Community College

Necator americanus: The New World Hookworm

The New World or American hookworm, *Necator americanus* (L. *necator,* killer), is found in the southern United States. The adults live in the small intestine, where they hold onto the intestinal wall with teeth and feed on blood and tissue fluids (figure 13.9). Individual females may produce as many as 10,000 eggs daily, which pass out of the body in the feces. An egg hatches on warm, moist soil and releases a small rhabditiform (the first- and second-stage juveniles of some nematodes) larva. It molts and becomes the infective filariform (the infective third-stage larva of some nematodes) larva. Humans become infected when the filariform larva penetrates the skin, usually between the toes. (Outside defecation and subsequent walking barefoot through the immediate area maintains the life cycle in humans.) The larva burrows through the skin to reach the circulatory system. The rest of its life cycle is similar to that of *Ascaris* (*see figure 13.6*).

Trichinella spiralis: The Porkworm

Adult *Trichinella* (Gr. *trichinos,* hair) *spiralis* live in the mucosa of the small intestine of humans and other carnivores and omnivores (e.g., the pig). In the intestine, adult females give birth to young larvae that then enter the circulatory system and are carried to skeletal (striated) muscles of the same host (figure 13.10). The young larvae encyst in the skeletal muscles and remain infective for many years. The disease this nematode causes is called **trichinosis.** Another host must ingest infective meat (muscle) to continue the life cycle. Humans most often become infected by eating improperly cooked pork products. Once ingested, the larvae excyst in the stomach and make their way to the small intestine, where they molt four times and develop into adults.

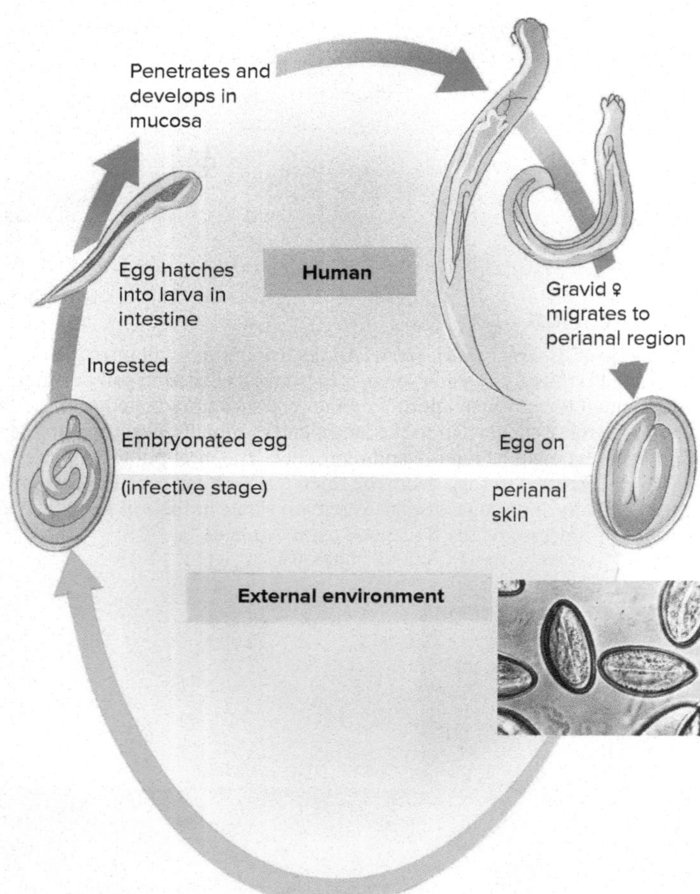

FIGURE 13.8

Life Cycle of *Enterobius vermicularis*. The direct transmission of embryonated eggs between hosts has allowed this species to become a very common human parasite in North America and other developed regions of the world. The inset photograph shows eggs of *Enterobius*. Eggs are thin-walled and D-shaped. Direct transmission between hosts does not require a thick capsule wall. The distinctive shape of the egg allows diagnosis of an infection by applying transparent tape to the perianal area of the host and then applying the tape to a microscope slide for examination. *Source: Centers for Disease Control, Atlanta, GA.*
Source: [CDC] Centers for Disease Control (US). Atlanta, (GA).

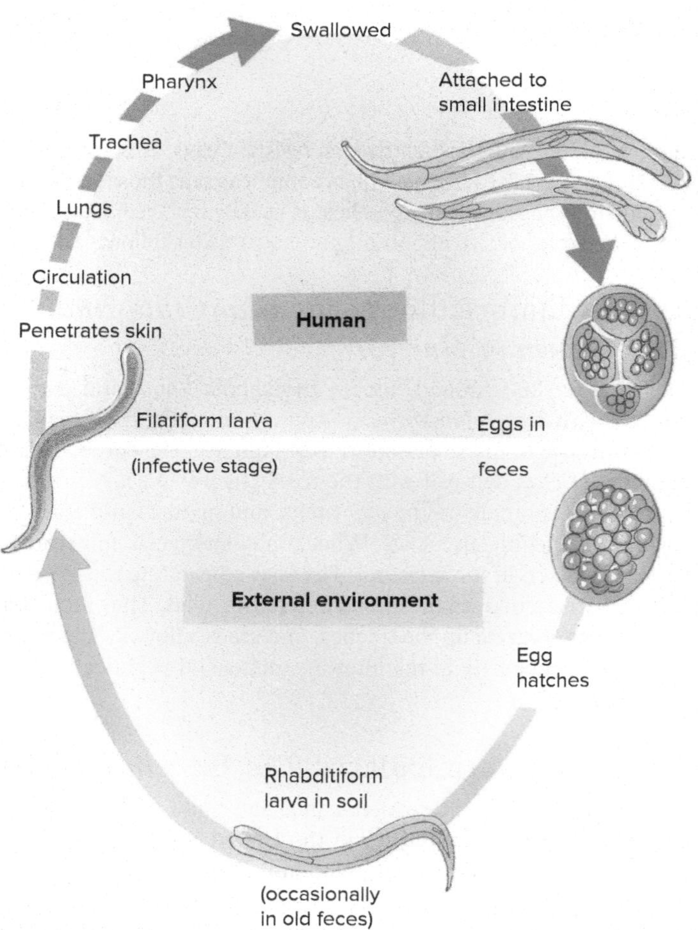

FIGURE 13.9

Life Cycle of *Necator americanus*. *Necator* adults live in the small intestines of their hosts and feed on blood and tissue fluids. Eggs are released in feces, deposited in soil, and hatch into larvae that penetrate the skin of their next host. *Source: Centers for Disease Control, Atlanta, GA.*
Source: [CDC] Centers for Disease Control (US). Atlanta, (GA).

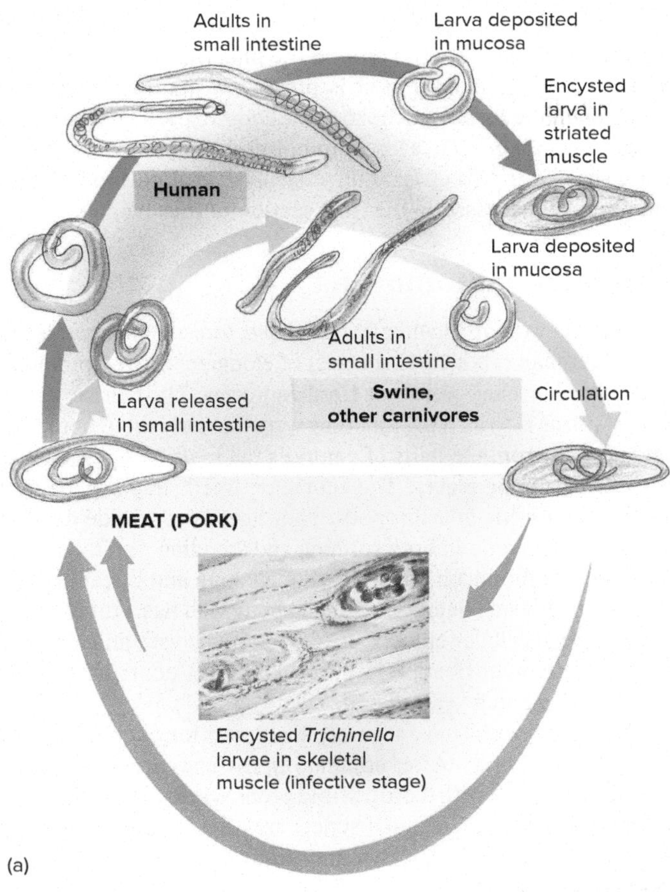

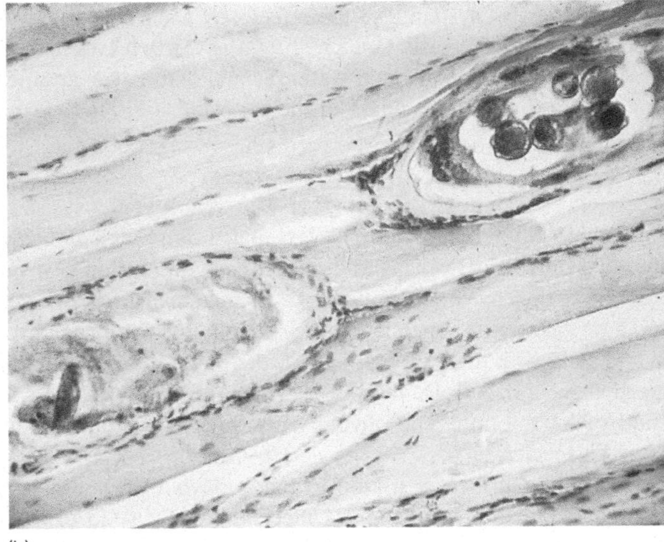

FIGURE 13.10

Life Cycle of *Trichinella spiralis*. (*a*) *Trichinella* adults live in the small intestines of mammalian carnivores and omnivores. Eggs hatch within the female worm's reproductive tract. Juveniles are released and travel through the blood stream of the host to muscle, where the juveniles encyst. The infection is passed to another host when muscle is eaten raw or poorly cooked. (*b*) An enlargement of the insert in (*a*), showing two encysted larvae in skeletal muscle (LM 3450).

(*a*) *Source: Centers for Disease Control, Atlanta, GA.*

The Filarial Worms

In tropical countries, over 250 million humans are infected with filarial (L. *filium,* thread) worms. Two examples of human filarial worms are *Wuchereria bancrofti* and *W. malayi.* These elongate, threadlike nematodes live in the lymphatic system, where they block the vessels. Because lymphatic vessels return tissue fluids to the circulatory system, when the filiarial nematodes block these vessels, fluids and connective tissue tend to accumulate in peripheral tissues. This fluid and connective tissue accumulation causes the enlargement of various appendages, a condition called **elephantiasis** (figure 13.11).

In the lymphatic vessels, filarial nematode adults copulate and produce larvae called **microfilariae** (figure 13.12). The microfilariae are released into the bloodstream of the human host and migrate to the peripheral circulation at night. When a mosquito feeds on a human, it ingests the microfilariae. The microfilariae migrate to the mosquito's thoracic muscles, where they molt twice and become infective. When the mosquito takes another meal of blood, the mosquito's proboscis injects the infective third-stage larvae into the blood of the human host. The final two molts take place as the larvae enter the lymphatic vessels.

A filarial worm prevalent in the United States is *Dirofilaria immitis,* a parasite of dogs. Because the adult worms live in the heart and large arteries of the lungs, the infection is called **heartworm**

FIGURE 13.11

Elephantiasis Caused by the Filarial Worm *Wuchereria bancrofti*. It takes years for elephantiasis to advance to the degree shown in this photograph.

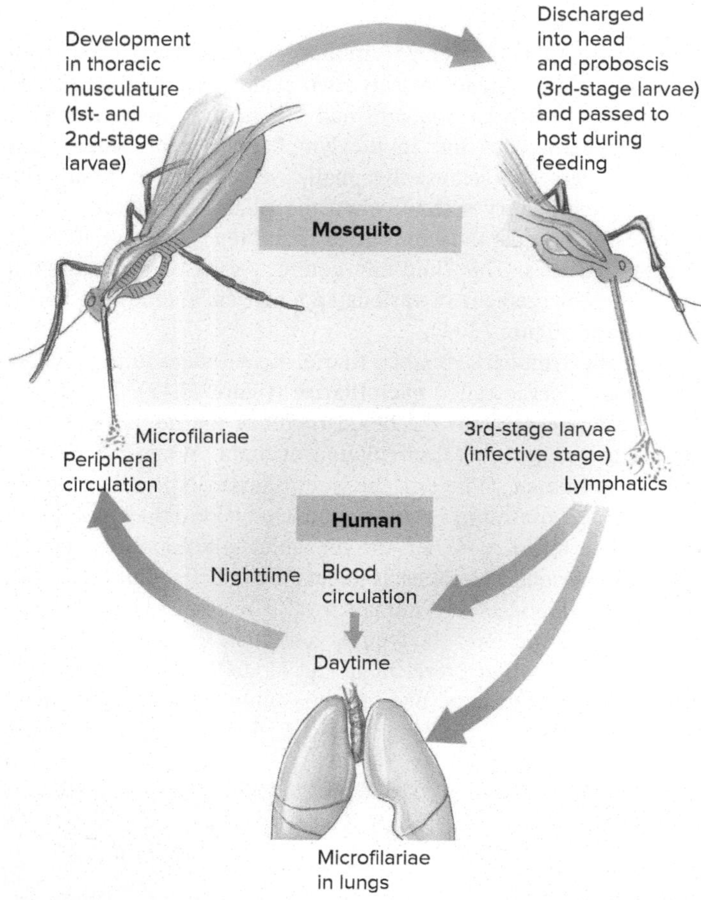

FIGURE 13.12

Life Cycle of *Wuchereria*. *Wuchereria* adults live in the lymphatic vessels of their human hosts. Microfilarial larvae are picked up by mosquitoes. After development in the mosquitoes, infective larvae are injected into the blood stream of another human host. *Source: Centers for Disease Control, Atlanta, GA.* Source: [CDC] Centers for Disease Control (US). Atlanta, (GA).

disease. Once established, these filarial worms are difficult to eliminate, and the condition can be fatal. Prevention with heartworm medicine is thus advocated for all dogs.

SECTION 13.2 THINKING BEYOND THE FACTS

Many parasitic nematodes are intestinal parasites whose eggs are released with feces of their host. How does this observation regarding nematode life cycles inform control measures directed at nematode parasites?

13.3 OTHER ECDYSOZOAN PHYLA

LEARNING OUTCOME

1. Characterize members of the phyla Nematomorpha, Kinorhyncha, Priapulida, and Loricifera.

Members of the remaining four phyla are a diverse group of ecdysozoans. The primary morphological feature they share with all ecdysozoans is the cuticle. Two other characteristics common to all phyla covered in this chapter are the pseudocoelom and a collar-shaped neural ring ("brain") around the pharynx. The former is of no taxonomic significance as it is present in multiple phyla scattered through the Lophotrochozoa and Ecdysozoa. The neural ring (*see figures 13.2*b *and 13.4*a) is an important synapomorphy for the phyla discussed in this chapter. Phylogenetic relationships of these groups will be discussed in the final section of this chapter.

Phylum Nematomorpha

Nematomorphs (nem″a-to-mor′fs) (Gr. *nema,* thread + *morphe,* form) are a small group (about 300 species) of elongate worms commonly called either **horsehair worms** or **Gordian worms.** The hairlike nature of these worms is so striking that they were formerly thought to arise spontaneously from the hairs of a horse's tail in drinking troughs or other stock-watering places. The adults are free living, but the juveniles are all parasitic in arthropods. They have a worldwide distribution and can be found in both running and standing water.

The nematomorph body is extremely long and threadlike and has no distinct head (figure 13.13). The body wall has a thick collagenous cuticle, a cellular epidermis, longitudinal cords, and a muscle layer of longitudinal fibers. The nervous system contains an anterior neural ring and a ventral cord.

Nematomorphs have separate sexes; two long gonads extend the length of the body. After copulation, the eggs are deposited in water. A small larva with a protrusible proboscis armed with spines hatches from the egg. Terminal stylets are also present on the proboscis. The larva must quickly enter an arthropod (e.g., a beetle, cockroach) host, either by penetrating the host or by being eaten. Lacking a digestive system, the larva feeds by absorbing material directly across its body wall. Once mature, the worm leaves its host only when the arthropod is near water. Sexual maturity is attained during the free-living adult phase of the life cycle.

Phylum Kinorhyncha

Kinorhynchs (kin′o-rinks) are small (less than 1 mm long), elongate, bilaterally symmetrical worms found exclusively in marine environments, where they live in mud and sand. Because they have no external cilia or locomotor appendages, they simply burrow through the mud and sand with their snouts. In fact, the phylum takes its name (Kinorhyncha, Gr. *kinein,* motion + *rhynchos,* snout) from this method of locomotion. The phylum Kinorhyncha contains about 150 known species.

The body surface of a kinorhynch is devoid of cilia and is composed of 13 or 14 definite units called **zonites** (figure 13.14). The head of a kinorhynch is zonite 1 and is called the **introvert.** It bears the mouth and oral cone, and it is retractable. When it is retracted, the anterior end is covered by chitinous spines on the neck (zonite 2). These spines are called **scalids** and some are modified into plate-like **placids.** Scalids grip the substrate during burrowing. The trunk consists of the remaining 11 or 12 zonites and terminates with the anus. Each trunk zonite bears a pair of lateral spines and one dorsal spine.

The body wall consists of a chitinous cuticle, epidermis, and two pairs of muscles: dorsolateral and ventrolateral. The pseudocoelom is large and contains amoeboid cells.

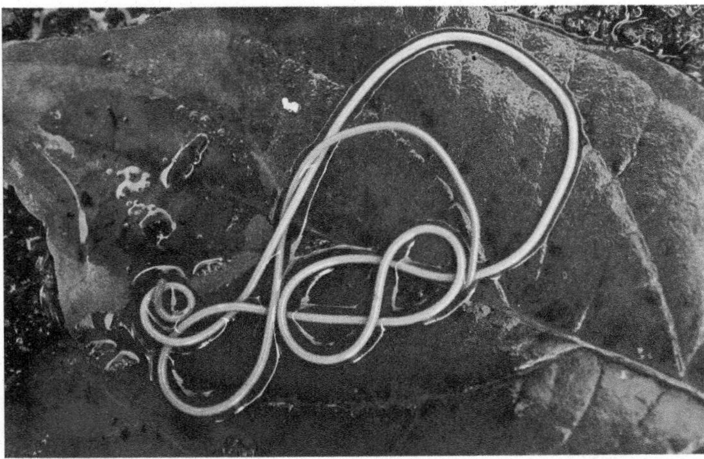

FIGURE 13.13

Phylum Nematomorpha. This adult nematomorph (*Gordius robustus*) is about 25 cm long and emerged from its cricket host after a rain. These worms tend to twist and turn upon themselves, giving the appearance of complicated knots—thus, the name "Gordian worms." (Legend has it that King Gordius of Phrygia tied a formidable knot—the Gordian knot—and declared that whoever might undo it would be the ruler of all Asia. No one could accomplish this until Alexander the Great cut through it with his broadsword.)

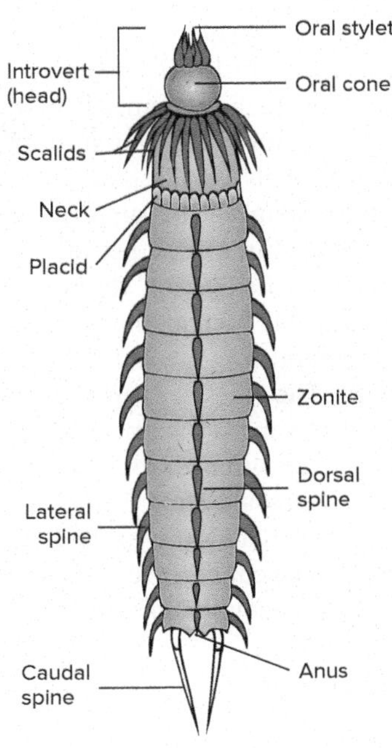

FIGURE 13.14

Phylum Kinorhyncha. External anatomy of an adult kinorhynch (dorsal view).

A complete digestive system is present, consisting of a mouth, buccal cavity, muscular pharynx, esophagus, stomach, intestine (where digestion and absorption take place), and anus. Most kinorhynchs feed on diatoms, algae, and organic matter.

A pair of protonephridia is in zonite 11. The nervous system consists of a neural ring that encircles the pharynx and single ventral nerve cord with a ganglion (a mass of nerve cells) in each zonite. Some species have eyespots and sensory bristles. Kinorhynchs are dioecious with paired gonads. Several spines that may be used in copulation surround the male gonopore. The young hatch into larvae that do not have all of the zonites. As the larvae grow and molt, the adult morphology appears. Once adulthood is attained, molting no longer occurs.

Phylum Priapulida

The priapulids (pri′a-pyu-lids) (Gr. *priapos,* phallus + *ida,* pleural suffix) are a small group (only 16 species) of marine worms found in cold waters. They live buried in the mud and sand of the seafloor, where they feed on small annelids and other invertebrates.

EVOLUTIONARY INSIGHTS

What Are Worms?

We use the term "worm" in a very generic sense—for example, free-living flatworms, parasitic flatworms, soft-bodied worms, tapeworms, and segmented worms. This generic use of the word "worm" also occurs in this chapter: roundworms, porkworm, filarial worms, and horsehair or Gordian worms. In other chapters you will encounter the terms acorn-worms, arrow-worms, and others. Therefore, what are worms? Worms covered in this zoology book can be defined as animals that do not possess legs, are not covered by a protective shell, do not bear a lophophore, are soft-bodied, have bodies from two or three to more than 15,000 times longer than wide, and are either flattened or rounded in section. This means that "worm" is an indefinite, though suggestive, term popularly applied to any animal that is not obviously something else. Worms do not form a natural group. This term is applied for convenience based on the absence of features that are used to describe other animal body forms. Obviously, other positive features are used to group vermiform animals into the taxonomically valid taxa covered in chapters 10, 12, and 13.

Gold Mine Treasure

Single-cell organisms are now known to live deep in the earth, more than 9,000 feet below the surface in extreme conditions. Until recently, it was thought that these conditions were too extreme for animals. Geoscientists have recently reported their discovery of a small nematode in the shaft of a gold mine in Africa. The nematode, *Halicephalobus mephisto*, is only 0.5 mm long. This nematode feeds on bacteria and can tolerate temperatures above 38°C.

The priapulid body is cylindrical and ranges in length from 2 mm to about 8 cm (figure 13.15). The anterior part of the body is an introvert (proboscis), which priapulids can draw into the longer, posterior trunk. The introvert functions in burrowing, and spines surround it. A thin chitinous cuticle that bears spines covers the muscular body, and the trunk bears superficial annuli. A straight digestive tract is suspended in a large pseudocoelom that acts as a hydrostatic skeleton. In some species, the pseudocoelom contains amoeboid cells that probably function in gas transport. The nervous system consists of a neural ring around the pharynx and a single midventral nerve cord. The sexes are separate but not superficially distinguishable. A pair of gonads is suspended in the pseudocoelom and shares a common duct with the protonephridia. The duct opens near the anus, and gametes are shed into the sea. Fertilization is external, and the eggs eventually sink to the bottom, where the larvae develop into adults. The cuticle is repeatedly molted throughout life. The most commonly encountered species is *Priapulus caudatus.*

Phylum Loricifera

Members of the phylum Loricifera (lor′a-sif-er-ah) (L. *lorica*, clothed in armor + *fero*, to bear) live in spaces between marine gravel. A characteristic species is *Nanaloricus mysticus*. It is a small, bilaterally symmetrical worm with a spiny head called an **introvert,** a thorax, and an abdomen surrounded by a cuticular lorica (figure 13.16). Loriciferans can retract both the introvert and thorax into the anterior end of the lorica. The introvert bears eight oral stylets that surround the mouth. The lorical cuticle is periodically molted. A pseudocoelom is present and contains a short digestive system, neural ring that encircles the pharynx, and several ganglia. Loriciferans are dioecious with paired gonads. Zoologists have described more than 30 species, and many prospective species have been collected and are awaiting study.

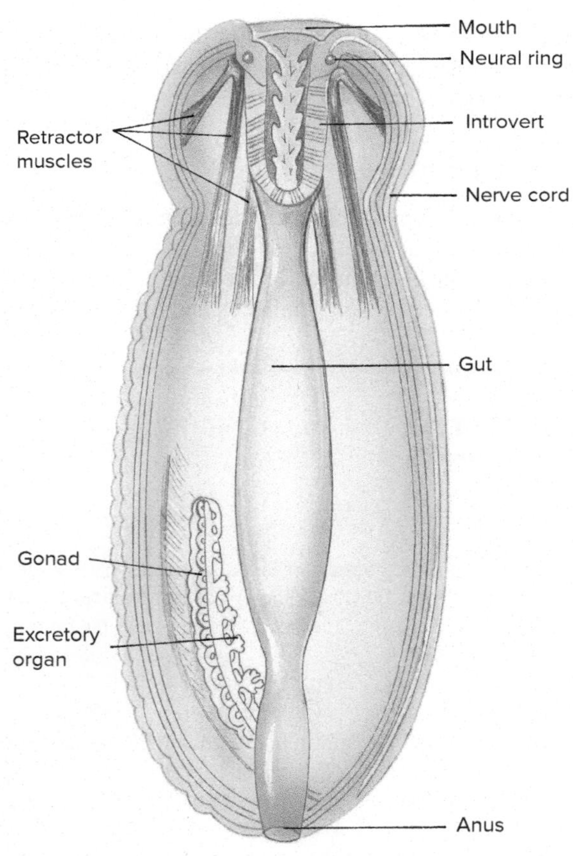

FIGURE 13.15

Phylum Priapulida. Internal anatomy of the priapulid *Priapulus caudatus,* with the introvert withdrawn into the body.

Section 13.3 Thinking Beyond the Facts

Adult nematomorphs are aquatic, but sometimes they are observed on sidewalks or lawns following a rain. How can you explain these observations?

13.4 FURTHER PHYLOGENETIC CONSIDERATIONS

LEARNING OUTCOMES

1. Describe the relationships of the ecdysozoan phyla and controversies associated with this phylogeny.
2. Describe phylogeny within the Nematoda.

The advent of molecular techniques revolutionized our interpretations of the phylogenetic relationships of animals described in chapters 13 through 15. As described earlier, these phyla are

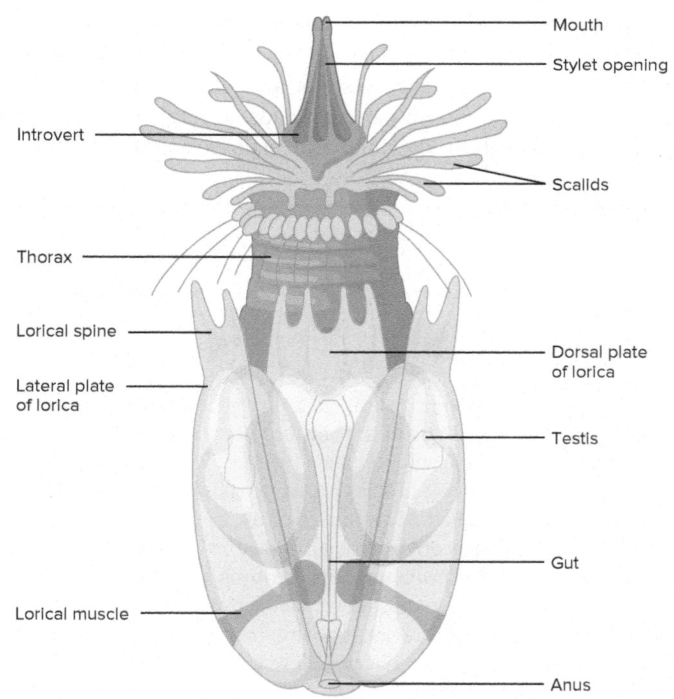

FIGURE 13.16

Phylum Loricifera. Dorsal view of the anatomy of an adult male *Nanaloricus mysticus.*

characterized by a secreted cuticle that is shed through ecdysis. The importance of this unifying character has been confirmed by molecular analyses many times. The relationships within the Ecdysozoa are more contentious. Ideally, a combination of morphological and molecular data will eventually sort out the uncertainties that exist. One interpretation of ecdysozoan phylogeny is shown in figure 13.17 and described below.

Some molecular studies have resulted in taxonomists concluding that there are two clades within the Ecdysozoa. The Arthropoda, Tardigrada, and Onychophora are covered in chapters 14 and 15 and comprise the clade Panarthropoda. The phylogentic relationships within the Panarthropoda will be discussed at the end of chapter 15.

The five phyla described in this chapter are often designated as the clade Cycloneuralia. They all possess a collar-shaped neural ring ("brain") around the pharynx. Within this clade the Nematoda and Nematomorpha are sister groups. Kinorhyncha and Priapulida also share common ancestry. These relationships are supported by morphological evidence. The Nematoda and Nematomorpha possess a collagenous cuticular structure and the Kinorhyncha and Priapulida share a chitinous cuticular structure and an introvert with scalids. The relationship of the Loricifera to these other four phyla is less certain. Because loriciferans also possess a chitinous cuticle and an introvert with scalids, they are considered by some authorities to be a sister group to the kinorhynch/priapulid clade. The molecular data that could support this conclusion are scanty, and some of these data that do exist suggest that loriciferans are more closely related to the nematode/nematomorph clade. More studies are needed to resolve this uncertainty.

The largest and most important phylum covered in this chapter is Nematoda. Phylogenetic relationships within the phylum are even more contentious than the relationships within the Ecdysozoa as a whole. The traditional classes (Secernentea and Adenophorea) clearly do not reflect nematode phylogeny. Molecular studies reveal three major clades. Nematodes are usually assumed to have arisen in marine habitats, and members of some groups within the basal clade, Chromadorea, are predominantly marine. This clade diverged into multiple orders including many terrestrial and freshwater groups. The important nematode parasites described earlier are all members of a single chromadorean group, Spirurina. Two other major nematode clades are Enoplia and Dorylaimia. Dorylaimians are absent from marine habitats. Both clades are composed of a variety of predators, decomposers, and plant and animal parasites.

Section 13.4 Thinking Beyond the Facts

What uncertainties exist in our understanding of phylogentic relationships within the Ecdysozoa, and what type of evidence is required in order to resolve these uncertainties?

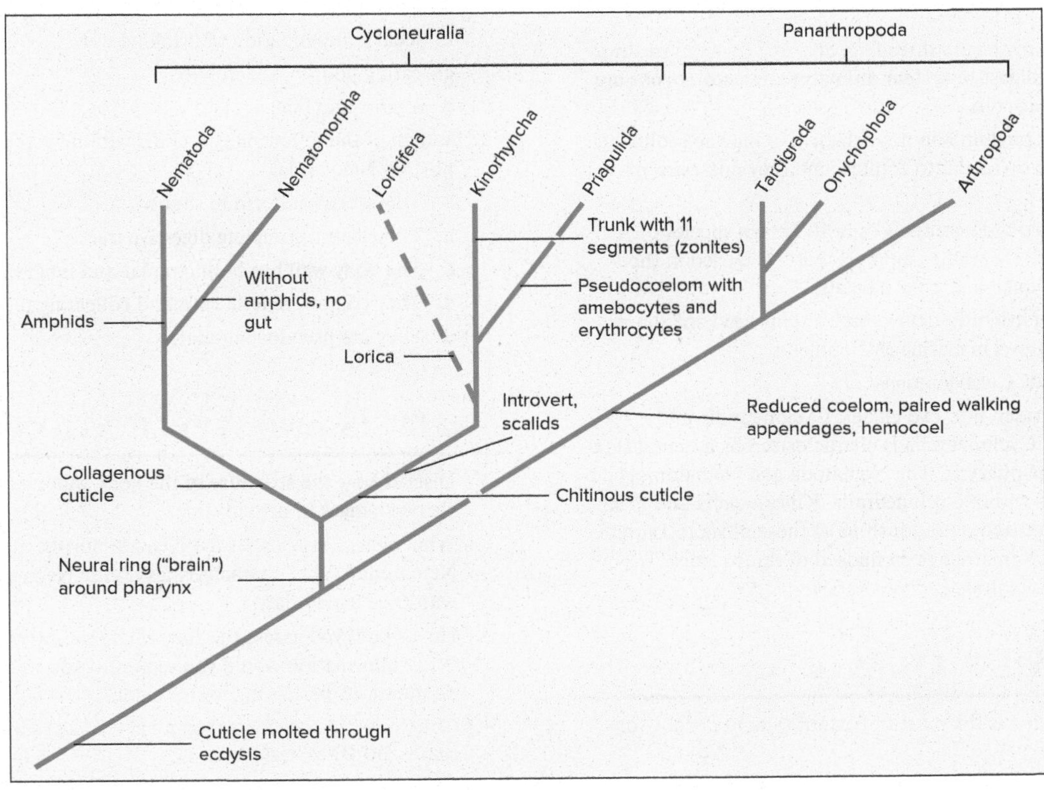

FIGURE 13.17

One Interpretation of Ecdysozoan Phylogeny. Some molecular data suggest that Ecdysozoa is comprised of two clades, Cycloneuralia and Panarthropoda. The position of the Loricifera is uncertain, although loriciferans are usually thought to be more closely related to the kinorhynchs and priapulids than they are to the nematodes and nematomorphs. Synapomorphic characters for Panarthropoda are presented in chapter 15.

Source: Dunn et al. 2008. Broad phylogenomic sampling improves resolution of the animal tree of life. Nature 452(7188): 745–749.

Summary

13.1 **Evolutionary Perspective**

- Ecdysozoans are united by molecular characteristics and in the presence of a cuticle that is molted during growth (ecdysis). Ecdysozoa is composed of Nematoda, Nematomorpha, Kinorhyncha, Priapulida, Loricifera, and the Panarthropoda (Arthropoda, Tardigrada, and Onychophora).
- The cuticle of the ecdysozoans discussed in chapter 13 functions with the pseudocoelom in hydrostatic skeletal functions.

13.2 **Phylum Nematoda (Roundworms)**

- Members of the phylum Nematoda are the roundworms. They are elongate, slender, and circular in cross section.
- Nematodes have a complete digestive tract, longitudinal body-wall muscles, and renette cells that function in excretion. The nervous system consists of an anterior neural ring and nerves that extend anteriorly and posteriorly from the ring.
- Nematodes are dioecious and dimorphic. Copulation results in the fertilization of eggs, which may be released to the environment, or they may hatch within the female reproductive tract.
- Nematodes include many important human parasites. These include *Ascaris lumbricoides*, *Enterobius vermicularis*, *Necator americanus*, and the filarial worms.

13.3 **Other Ecdysozoan Phyla**

- Adult Nematomorpha are threadlike and free living in freshwater. They lack a digestive system. Immature nematomorphs are parasites of arthropods.
- Members of the phylum Kinorhyncha live in marine sediments. Their bodies are divided into zonites, and they possess a retractable introvert.
- The phylum Priapulida contains only 16 known species of cucumber-shaped, worm-like animals that live buried in the bottom sand and mud in marine habitats.
- Members of the phylum Loricifera have a spiny head and thorax, and they live in gravel in marine environments.

13.4 **Further Phylogenetic Considerations**

- Ecdysozoa is divided into two clades: Cycloneuralia and Panarthropoda. Cycloneuralia is characterized by a neural ring that encircles the pharynx. The Nematoda and Nematomorpha are sister groups within Cycloneuralia. Kinorhyncha and Priapulida are closely related. Relationships of these phyla to Loricifera are less certain. Nematoda is composed of three clades. Taxonomy within Nematoda is controversial.

Concept Review Questions

1. Which of the following is the most important synapomorphy for Ecdysozoa?
   a. They possess a fluid-filled body cavity.
   b. The following germ layers are present: ectoderm, endoderm, and mesoderm.
   c. They all have a complete digestive system.
   d. All members molt in order to grow.
   e. All members are microscopic in size.
2. All of the phyla discussed in this chapter are united by the presence of
   a. an introvert.
   b. a collagenous cuticle.
   c. a chitinous cuticle.
   d. a neural ring that encircles the pharynx.
   e. scalids.
3. Members of this phylum are free living in freshwater as adults, but they are parasites of arthropods in their larval stage.
   a. Nematoda
   b. Nematomorpha
   c. Loricifera
   d. Kinorhyncha
   e. Priapulida
4. The most likely route for an infection by *Trichinella spiralis* is
   a. wading in infested streams.
   b. fecal contamination of drinking water.
   c. eating poorly cooked pork.
   d. a mosquito bite.
5. Which of the following is FALSE with respect to members of the phylum Nematoda?
   a. They are vermiform in shape.
   b. They have a complete digestive tract.
   c. The body wall has both circular and longitudinal muscles.
   d. They contain renette cells and collecting tubules.
   e. They are pseudocoelomate.

Analysis and Application Questions

1. Discuss how the structure of the body wall places limitations on shape changes in nematodes.
2. What characteristics set the Nematomorpha apart from the Nematoda? What characteristics do the Nematomorpha share with the Nematoda?
3. How would you assess the state of taxonomy within the Ecdysozoa? What information would you request if you were to resolve phylogenetic questions pertaining to this clade?
4. Compare effective control strategies for infestations of *Necator americanus* and *Wuchereria*.

Arthropods are among the most misunderstood of all the animals. Studying chapters 14 and 15, free of blemished preconceptions, will help you understand how studying members of this phylum has enthralled professional and amateur zoologists for hundreds of years.

14

The Arthropods: Blueprint for Success

Chapter Outline

14.1 Evolutionary Perspective
Classification and Relationships to Other Animals
14.2 Metamerism and Tagmatization
14.3 The Exoskeleton
14.4 The Hemocoel
14.5 Metamorphosis
14.6 Subphylum Trilobitomorpha
14.7 Subphylum Chelicerata
Class Merostomata
Class Arachnida
Class Pycnogonida
14.8 Subphylum Myriapoda
Class Diplopoda
Class Chilopoda
Classes Pauropoda and Symphyla
14.9 Further Phylogenetic Considerations

14.1 EVOLUTIONARY PERSPECTIVE

LEARNING OUTCOMES

1. Describe characteristics of members of the phylum Arthropoda.
2. Explain the phylogenetic relationships of arthropods to other phyla.

To be misrepresented and misunderstood–what an awful way to live one's life. The image of a spider conjures up fears that we have learned, not from encounters with spiders, but from relatives and friends who learned their fears in a similar fashion. Virtually no one became fearful of spiders from being bitten–spider bites are rare and seldom serious. While spiders do have fangs, they are adapted for preying on insects and other arthropods, and most spider fangs are unable to pierce human skin. The rare spider bite usually occurs by unknowingly placing a hand into some spider abode. Then, if the spider can, it may bite to defend itself or its egg sac. Bites, sores, and rashes may be diagnosed as "spider bites," but often mistakenly so. Other arthropods–like fleas, ticks, and mosquitoes–can bite and leave sores and rashes that are misdiagnosed. Bacterial infections and reactions to plant toxins may also be diagnosed as "spider bites."

To misunderstand and misrepresent–that's also an unfortunate way to live one's life. There are a few spiders around the world that are venomous to humans, and two of these will be discussed later in this chapter. There are nearly 50,000 species of spiders that provide the service of insect control. They are among the most numerous of all predators of insects, although their value in this regard has been very difficult to quantify. They are a valuable food resource of other predators in terrestrial food webs. Very importantly, they are teaching us novel ways to use, and even to produce, the silk that makes up their very impressive webs. Finally, once we set aside our preconceived and falsely placed fears, we can begin to appreciate the intricate structure and beauty of these remarkable creatures (figure 14.1).

Spiders are one of the many groups of animals belonging to the phylum Arthropoda (ar″thrah-po′dah) (Gr. *arthro,* joint + *podos,* foot). Crayfish, lobsters, mites, scorpions, and insects are also arthropods. Zoologists have described about 1 million species of arthropods and estimate that millions more are undescribed. In this chapter and chapter 15, you will discover the many ways in which some arthropods are considered among the most successful of all animals.

Characteristics of the phylum Arthropoda include the following:

1. Metamerism modified by the specialization of body regions for specific functions (tagmatization)
2. Chitinous exoskeleton that provides support and protection and is modified to form sensory structures

FIGURE 14.1

Class Arachnida, Order Araneae. Members of the family Araneidae, the orb weavers, produce some of the most beautiful and intricate spider webs. Many species are relatively large, like this garden spider–*Argiope*. A web is not a permanent construction. When webs become wet with rain or dew, or when they age, they lose their stickiness. The entire web, or at least the spiraled portion, is then eaten and replaced.

3. Paired, jointed appendages
4. Growth accompanied by ecdysis or molting
5. Ventral nervous system
6. Coelom reduced to cavities surrounding gonads and sometimes excretory organs
7. Open circulatory system in which blood is released into tissue spaces (hemocoel) derived from the blastocoel
8. Complete digestive tract
9. Metamorphosis often present; reduces competition between immature and adult stages

Classification and Relationships to Other Animals

Arthropoda is a monophyletic taxon that is a part of the protosome clade Ecdysozoa. Arthropods are thus related to the Nematoda, Nematomorpha, Kinorhyncha, and others (figure 14.2 and *see chapter 13*). Synapomorphies for this clade include a cuticle, loss of epidermal cilia, and shedding the cuticle in a process called ecdysis. Two smaller phyla within the Ecdysozoa, (Onychophora and Tardigrada) are sister groups with the Arthropoda. Arthropoda, Onycophora, and Tardigrada form the clade Panarthropoda (*see figure 13.17*). The onychophorans and tardigrades, and their position within the Panarthropoda, will be discussed at the end of chapter 15.

There has been an explosion of new information regarding the evolutionary relationships within the phylum Arthropoda that is causing zoologists to reexamine current and older hypotheses regarding arthropod phylogeny. Living arthropods are traditionally divided into four subphyla: Chelicerata, Myriapoda, Hexapoda, and Crustacea. All members of a fifth subphylum, Trilobitomorpha, are extinct (table 14.1). Ideas regarding the evolutionary relationships among these subphyla, including the paraphyletic status of the subphylum Crustacea, are discussed at the end of chapter 15. This chapter examines Trilobitomorpha,

TABLE 14.1

CLASSIFICATION OF THE PHYLUM ARTHROPODA

Phylum Arthropoda (ar″thrah-po′dah)
Animals that show metamerism with tagmatization, a jointed exoskeleton, and a ventral nervous system.

Subphylum Trilobitomorpha (tri″lo-bit″o-mor′fah)
Marine, all extinct; lived from Cambrian to Carboniferous periods; bodies divided into three longitudinal lobes; head, thorax, and abdomen present; one pair of antennae and biramous appendages.

Subphylum Chelicerata (ke-lis″er-ah′tah)
Body usually divided into prosoma and opisthosoma; first pair of appendages piercing or pincerlike (chelicerae) and used for feeding.

Class Merostomata (mer″o-sto′mah-tah)
Marine, with book gills on opisthosoma. Two subclasses: Eurypterida, a group of extinct arthropods called giant water scorpions, and Xiphosura–the horseshoe crabs. *Limulus.*

Class Arachnida (ah-rak′nĭ ′-dah)
Mostly terrestrial, with book lungs, tracheae, or both; usually four pairs of walking legs in adults. Spiders, scorpions, ticks, mites, harvestmen, and others.

Class Pycnogonida (pik″no-gon′ĭ ′-dah)
Reduced abdomen; no special respiratory or excretory structures; four to six pairs of walking legs; common in all oceans. Sea spiders.

Subphylum Myriapoda (mir″e-a-pod′ah) (Gr. *myriad,* ten thousand + *podus,* foot) Body divided into head and trunk; four pairs of head appendages; uniramous appendages. Millipedes and centipedes.

Class Diplopoda (dip″lah-pod′ah)
Two pairs of legs per apparent segment; body usually round in cross section. Millipedes.

Class Chilopoda (ki″lah-pod′ah)
One pair of legs per segment; body oval in cross section; poison claws. Centipedes.

Class Pauropoda (por″o-pod′ah)
Small (0.5 to 2 mm), soft-bodied animals; 11 segments; 9 pairs of legs; live in leaf mold. Pauropods.

Class Symphyla (sim-fi′lah)
Small (2 to 10 mm); long antennae; centipede-like; 10 to 12 pairs of legs; live in soil and leaf mold. Symphylans.

***Subphylum Crustacea** (krus-tā s′e-ah)
Most aquatic, head with two pairs of antennae, one pair of mandibles, and two pairs of maxillae; biramous appendages.

***Subphylum Hexapoda** (hex″sah-pod′ah)
Body divided into head, thorax, and abdomen; five pairs of head appendages; three pairs of uniramous appendages on the thorax. Insects and their relatives.

*"Crustacea" is a paraphyletic assemblage that encompasses the monophyletic clade "Hexapoda." Together, they comprise a clade called "Pancrustacea," which is the subject of chapter 15.

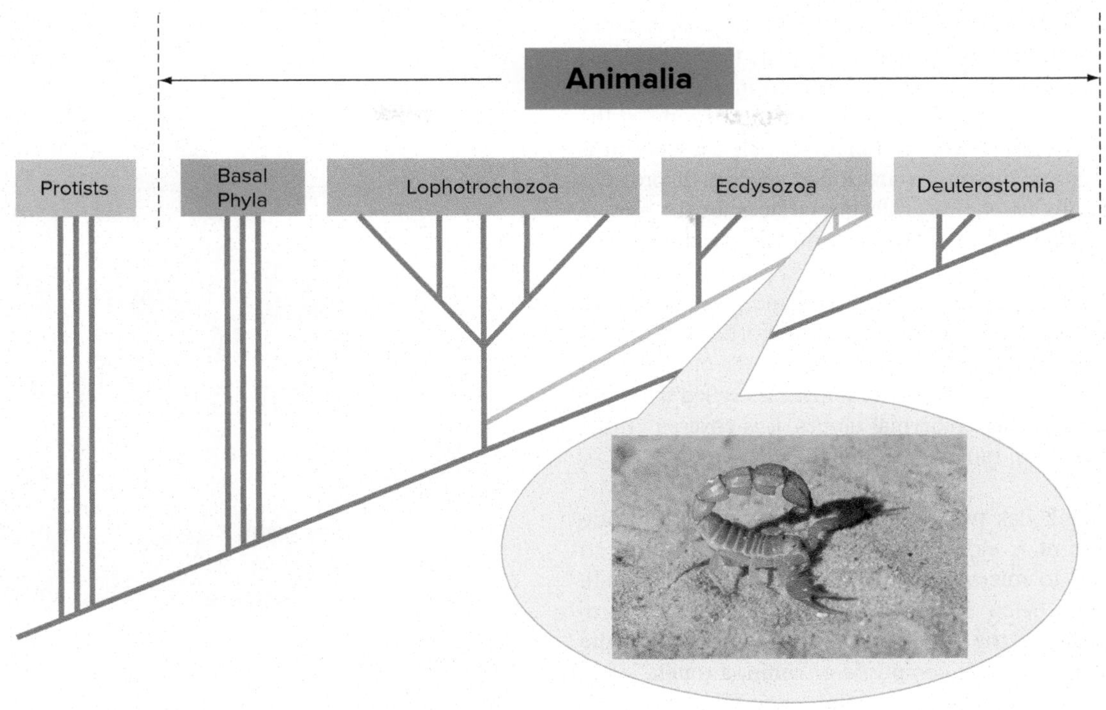

FIGURE 14.2

Evolutionary Relationships of the Arthropods to Other Animals. This figure shows one interpretation of the relationships of the Arthropoda to other members of the animal kingdom. The relationships depicted here are based on evidence from developmental and molecular biology. Arthropods are placed within the Ecdysozoa along with the Nematoda, Nematomorpha, Onychophora, and others (*see appendix A*). The phylum includes over 1 million species. This scorpion (*Buthus*) belongs to the subphylum Chelicerata–one of the four arthropod subphyla containing living species. This photo was taken in the Sahara desert.

Chelicerata, and Myriapoda, and chapter 15 covers Crustacea and Hexapoda.

Section 14.1 Thinking Beyond the Facts

Why is metamerism, which is common to the Annelida and Arthropoda, no longer considered to be evidence of close evolutionary ties between the two phyla?

14.2 Metamerism and Tagmatization

Learning Outcome

1. Compare arthropod metamerism with annelid metamerism.

Four aspects of arthropod biology have contributed to their success. One of these is metamerism. Metamerism of arthropods is most evident externally because the arthropod body is often composed of a series of similar segments, each bearing a pair of appendages. Internally, however, septa do not divide the body cavity of an arthropod, and most organ systems are not metamerically arranged. The reason for the loss of internal metamerism is speculative; however, the presence of metamerically arranged hydrostatic compartments would be of little value in the support or locomotion of animals enclosed by an external skeleton (discussed under "The Exoskeleton").

As discussed in chapter 12, metamerism permits the specialization of regions of the body for specific functions. This regional specialization is called tagmatization. In arthropods, body regions, called tagmata (sing., tagma), are specialized for feeding and sensory perception, locomotion, and visceral functions.

Section 14.2 Thinking Beyond the Facts

How is the metamerism of arthropods different from that observed in the Annelida* (see chapter 12)*?

14.3 The Exoskeleton

Learning Outcomes

1. Compare the structure and function of the arthropod exoskeleton or cuticle to that present in other ecdysozoans.
2. Compare two methods of hardening of the arthropod procuticle.
3. Describe the processes involved with arthropod ecdysis.
4. Assess the statement: Before, during, and after ecdysis, an arthropod is never without its exoskeleton.

An external, jointed skeleton, called an **exoskeleton** or **cuticle,** encloses arthropods. In ecdysozoans other than arthropods (*see chapter 13*) a major function of the collagenous cuticle is to interact with the pseudocoelom to create a hydrostatic skeleton. In arthropods, the chitinous cuticle has a larger array of functions. The exoskeleton is often cited as the major reason for arthropod success. It provides structural support, protection, impermeable surfaces for the prevention of water loss, and a system of levers for muscle attachment and movement.

The exoskeleton covers all body surfaces and invaginations of the body wall, such as the anterior and posterior portions of the gut tract. It is nonliving and is secreted by a single layer of epidermal cells (figure 14.3). The epidermal layer is sometimes called the hypodermis because, unlike other epidermal tissues, it is covered on the outside by an exoskeleton, rather than being directly exposed to air or water.

The exoskeleton has two layers. The epicuticle is the outermost layer. Made of a waxy lipoprotein, it is impermeable to water and a barrier to microorganisms and pesticides. The bulk of the exoskeleton is below the epicuticle and is called the procuticle. (In crustaceans, the procuticle is sometimes called the endocuticle.) The procuticle is composed of **chitin,** a tough, leathery polysaccharide, and several kinds of proteins. The procuticle hardens through a process called sclerotization and sometimes by impregnation with calcium carbonate. Sclerotization is a tanning process in which layers of protein are chemically cross-linked with one another–hardening and darkening the exoskeleton. In insects and most other arthropods, this bonding occurs in the outer portion of the procuticle. The exoskeleton of crustaceans hardens by sclerotization and by the deposition of calcium carbonate in the middle regions of the procuticle. Some proteins give the exoskeleton resiliency. Distortion of the exoskeleton stores energy for such activities as flapping wings and jumping. The inner portion of the procuticle does not harden.

Hardening in the procuticle provides armorlike protection for arthropods, but it also necessitates a variety of adaptations to allow arthropods to live and grow within their confines. Invaginations of the exoskeleton form firm ridges and bars for muscle attachment. Another modification of the exoskeleton is the formation of joints. A flexible membrane, called an articular membrane, is present in regions where the procuticle is thinner and less hardened (figure 14.4). Other modifications of the exoskeleton include sensory receptors, called sensilla, in the form of pegs, bristles, and lenses, and modifications of the exoskeleton that permit gas exchange.

The growth of an arthropod would be virtually impossible unless the exoskeleton were periodically shed, such as in the molting process called **ecdysis** (Gr. *ekdysis,* getting out). Ecdysis is divided into four stages: (1) Enzymes, secreted from hypodermal glands, begin digesting the old procuticle to separate the hypodermis and the exoskeleton (figure 14.5*a,b*); (2) new procuticle and epicuticle are secreted (figure 14.5*c,d*); (3) the old exoskeleton splits open along predetermined ecdysal lines when the animal stretches by air or water intake; pores in the procuticle secrete additional epicuticle (figure 14.5*e*); and (4) finally, calcium carbonate deposits and/or sclerotization harden the new exoskeleton (figure 14.5*f*).

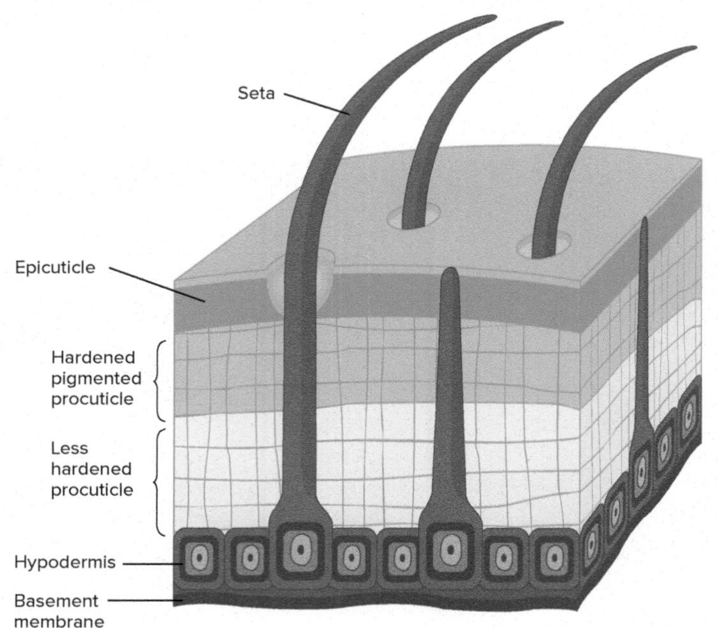

FIGURE 14.3

Arthropod Exoskeleton. The epicuticle is made of a waxy lipoprotein and is impermeable to water. Calcium carbonate deposition and/or sclerotization harden the outer layer of the procuticle. Chitin, a tough, leathery polysaccharide, and several kinds of proteins make up the bulk of the procuticle. The hypodermis secretes the entire exoskeleton.

During the few hours or days of the hardening process, the arthropod is vulnerable to predators and remains hidden. The nervous and endocrine systems control all of these changes; the controls are discussed in more detail later in this chapter.

SECTION 14.3 THINKING BEYOND THE FACTS

What properties of the exoskeleton provide for each of the following functions: protection from mechanical injury, protection against desiccation, storing energy for jumping, attachment of muscles, and flexibility?

14.4 THE HEMOCOEL

LEARNING OUTCOME

1. Compare the functions of the arthropod hemocoel to the functions of the annelid coelom.

A third arthropod characteristic is the presence of a **hemocoel.** The hemocoel is derived from an embryonic cavity called the blastocoel that forms in the blastula (*see figure 8.11*). The hemocoel provides an internal cavity for the open circulatory system of arthropods. Internal organs are bathed by body fluids in the hemocoel to provide for the exchange of nutrients, wastes, and sometimes gases. The coelom, which forms by splitting blocks of mesoderm later in the development of most protostomes, was

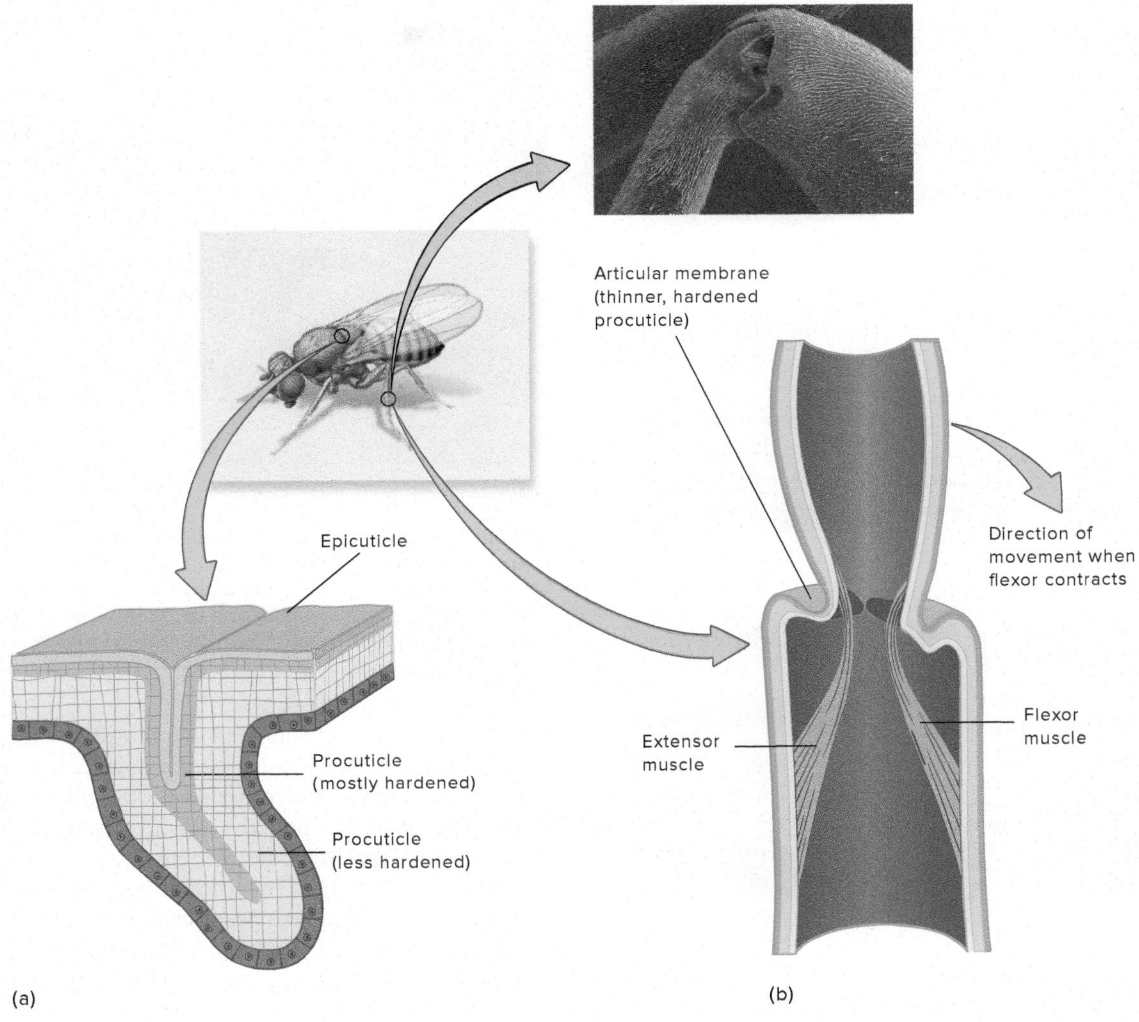

FIGURE 14.4

Modifications of the Exoskeleton. (*a*) Invaginations of the exoskeleton result in firm ridges and bars when the outer procuticle in the region of the invagination remains thick and hard. These are muscle attachment sites. (*b*) Regions where the outer procuticle is thinned are flexible and form membranes and joints.

Source: Russell-Hunter, WD. 1979. *A life of invertebrates*. New York (NY): Macmillan.

reduced in ancestral arthropods. The presence of the rigid exoskeleton and body wall means that the coelom is no longer used as a hydrostatic compartment. In modern arthropods, the coelom forms small cavities around the gonads and sometimes the excretory structures.

Section 14.4 Thinking Beyond the Facts

How is a hemocoel different from a coelom* (see chapter 7)*?

14.5 METAMORPHOSIS

LEARNING OUTCOME

1. Explain how metamorphosis contributed to arthropod success.

A fourth characteristic that has contributed to arthropod success is a reduction of competition between adults and immature stages because of metamorphosis. Metamorphosis is a radical change in body form and physiology as an immature stage, usually called a larva, becomes an adult. The evolution of arthropods has resulted in an increasing divergence of body forms, behaviors, and habitats between immature and adult stages. Adult crabs, for example, usually prowl the sandy bottoms of their marine habitats for live prey or decaying organic matter, whereas larval crabs live and feed in the plankton. Similarly, the caterpillar that feeds on leafy vegetation eventually develops into a nectar-feeding adult butterfly or moth. Having different adult and immature stages means that the stages do not compete with each other for food or living space. In some arthropod and other animal groups, larvae also serve as the dispersal stage. (*See chapters 15 and 25* for more details on arthropod metamorphosis.)

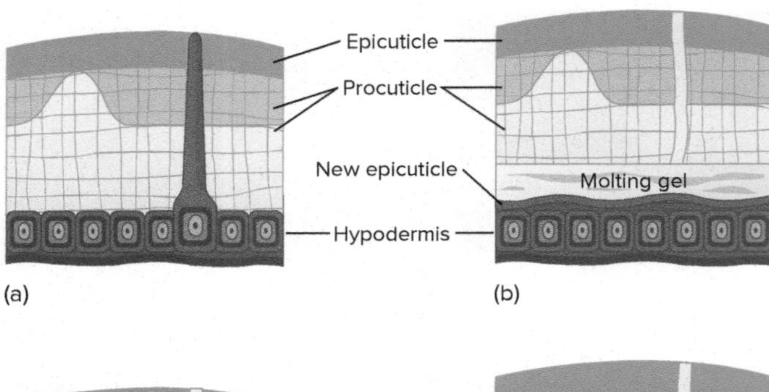

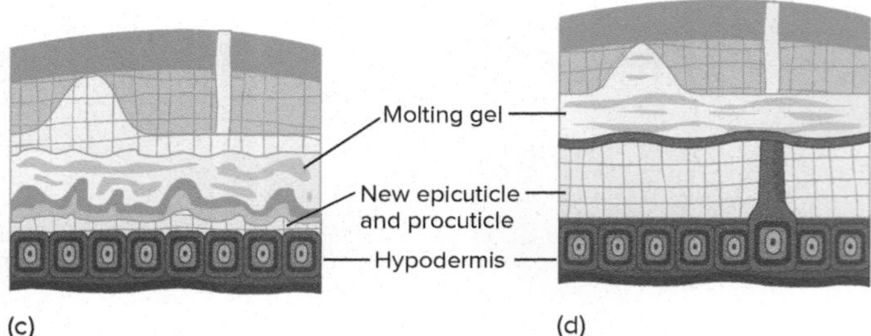

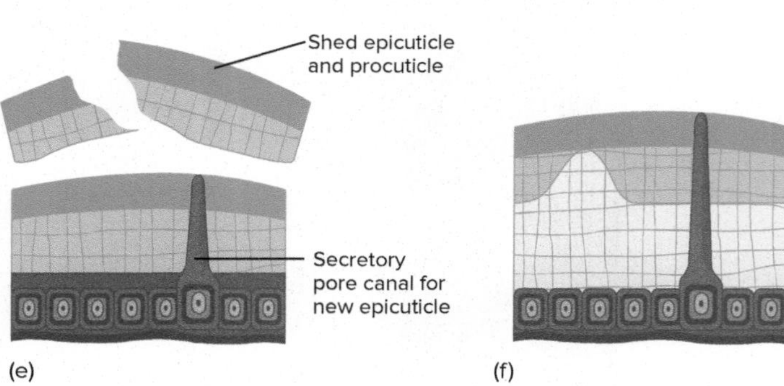

FIGURE 14.5

Events of Ecdysis. (*a* and *b*) During preecdysis, the hypodermis detaches from the exoskeleton, and the space between the old exoskeleton and the hypodermis fills with a fluid called molting gel. (*c* and *d*) The hypodermis begins secreting a new epicuticle, and a new procuticle forms as the old procuticle is digested. The products of digestion are incorporated into the new procuticle. Note that the new epicuticle and procuticle are wrinkled beneath the old exoskeleton to allow for increased body size after ecdysis. (*e*) Ecdysis occurs when the animal swallows air or water, and the exoskeleton splits along predetermined ecdysal lines. The animal pulls out of the old exoskeleton. (*f*) After ecdysis, the new exoskeleton hardens by calcium carbonate deposition and/or sclerotization, and pigments are deposited in the outer layers of the procuticle. Additional material is added to the epicuticle.

Section 14.5 Thinking Beyond the Facts

What animals, other than arthropods, do you know that undergo metamorphosis during development?

14.6 SUBPHYLUM TRILOBITOMORPHA

LEARNING OUTCOME

1. Compose a response to someone showing you a trilobite fossil and asking "What is this?"

FIGURE 14.6

Trilobite Structure. The trilobite body had three longitudinal sections (thus, the subphylum name). It was also divided into three tagmata. A head, or cephalon, bore a pair of antennae and eyes. The trunk, or thorax, bore appendages for swimming or walking. A series of posterior segments formed the pygidium, or tail. *Cheirusus ingricus* is shown here.

Members of the subphylum Trilobitomorpha (tri″lo-bit″o-mor′fah) (Gr. *tri,* three + *lobos,* lobes) were a dominant form of life in the oceans from the Cambrian period (600 million years ago [mya]) to the Carboniferous period (345 mya) (*see appendix B*). They crawled along the substrate feeding on annelids, molluscs, and decaying organic matter. The trilobite body was oval, flattened, and divided into three tagmata: head (cephalon), thorax, and pygidium (figure 14.6). All body segments articulated so that the trilobite could roll into a ball to protect its soft ventral surface. Most fossilized trilobites are found in this position. Trilobite appendages consisted of two lobes or rami, and are called **biramous** (L. *bi,* twice + *ramus,* branch) **appendages.** The inner lobe was a walking leg, and the outer lobe bore spikes or teeth that may have been used in digging or swimming or as gills in gas exchange.

Section 14.6 Thinking Beyond the Facts

Trilobites appeared very early in the fossil record, but their fossils are preceded by even more ancient, Ediacaran arthropods. Trilobites are considered a sister group to members of the remaining subphyla. What trilobite features are so unique, in comparison to other arthropods, that this group is set apart from other subphyla?

14.7 SUBPHYLUM CHELICERATA

LEARNING OUTCOMES

1. Describe the body forms of members of the subphylum Chelicerata.
2. Contrast the function of the following pairs of arachnid structures: coxal glands vs. Malpighian tubules and book lungs vs. tracheae.

3. Describe the structure and function of arachnid circulatory, nervous, and sensory systems.
4. Describe adaptations of arachnids for terrestrial habitats.
5. Characterize members of the arachnid orders Scorpionida, Araneae, Opiliones, and Acarina.

One arthropod lineage, the subphylum Chelicerata (ke-lis″er-ah′tah) (Gr. *chele,* claw + *ata,* plural suffix), includes familiar animals, such as spiders, mites, and ticks, and less familiar animals, such as horseshoe crabs and sea spiders. These animals have two tagmata. The **prosoma** or **cephalothorax** is a sensory, feeding, and locomotor tagma. It usually bears eyes, but unlike in other arthropods, never has antennae. Paired appendages attach to the prosoma. The first pair, called **chelicerae,** is often pincerlike or chelate, and is most often used in feeding. They may also be specialized as hollow fangs or for a variety of other functions. The second pair, called **pedipalps,** is usually sensory but may also be used in feeding, locomotion, or reproduction. Paired walking legs follow pedipalps. Posterior to the prosoma is the **opisthosoma,** which contains digestive, reproductive, excretory, and respiratory organs.

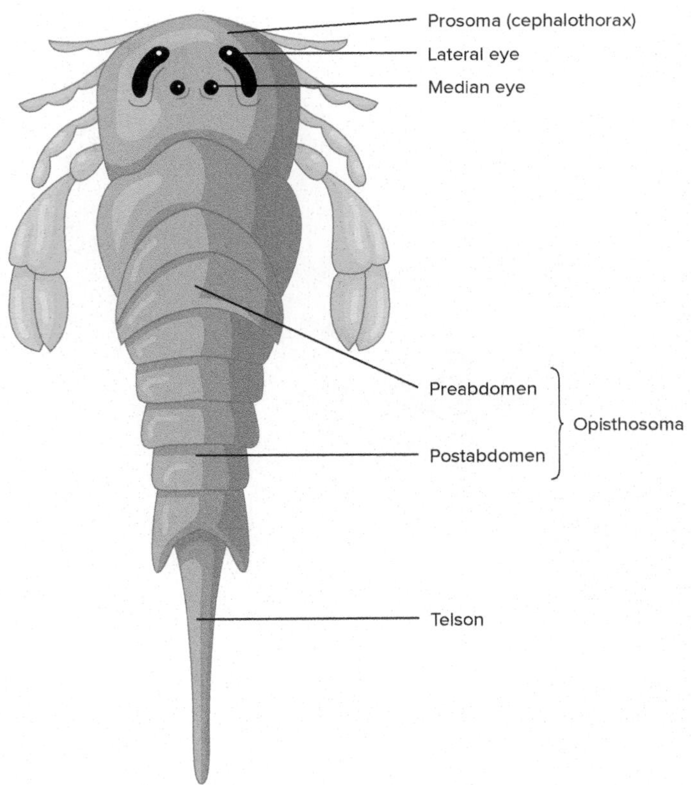

FIGURE 14.7

Class Merostomata. A eurypterid, *Euripterus remipes.*

Class Merostomata

Members of the class Merostomata (mer″o-sto′mah-tah) are divided into two subclasses. The Xiphosura are the horseshoe crabs, and the Eurypterida are the giant water scorpions (figure 14.7). The latter are extinct, having lived from the Cambrian period (600 mya) to the Permian period (280 mya, *see appendix B*).

Only four species of horseshoe crabs are living today. One species, *Limulus polyphemus,* is widely distributed in the Atlantic Ocean and the Gulf of Mexico (figure 14.8*a*). Horseshoe crabs scavenge sandy and muddy substrates for annelids, small molluscs, and other invertebrates. Their body form has remained virtually unchanged for more than 200 million years, and they were cited in chapter 5 as an example of stabilizing selection.

A hard, horseshoe-shaped carapace covers the prosoma of horseshoe crabs. The chelicerae, pedipalps, and first three pairs of walking legs are chelate and are used for walking and food handling. The last pair of appendages has leaflike plates at its tips and is used for locomotion and digging (figure 14.8*b*).

The opisthosoma of a horseshoe crab includes a long, unsegmented telson. If wave action flips a horseshoe crab over, the crab arches its opisthosoma dorsally, which helps it to roll to its side and flip right side up again. The first pair of opisthosomal appendages covers genital pores and is called the genital operculum. The remaining five pairs of appendages are **book gills.** The name is derived from the resemblance of these platelike gills to the pages of a closed book. Gases are exchanged between the blood and water as blood circulates through the book gills. Horseshoe crabs have an open circulatory system, as do all arthropods. Blood circulation in horseshoe crabs is similar to that described later in this chapter for arachnids and crustaceans.

Horseshoe crabs are dioecious. During reproductive periods, males and females congregate in intertidal areas. The male mounts the female and grasps her with his pedipalps. The female excavates shallow depressions in the sand, and as she sheds eggs into the depressions, the male fertilizes them. Fertilized eggs are covered with sand and develop unattended.

Class Arachnida

Members of the class Arachnida (ah-rak′nĭ-dah) (Gr. *arachne,* spider) are some of the most misrepresented members of the animal kingdom. Their reputation as fearsome and grotesque creatures is vastly exaggerated. The majority of spiders, mites, ticks, scorpions, and related forms are either harmless or very beneficial to humans.

Arachnids probably arose from the eurypterids and were early terrestrial inhabitants. The earliest fossils of aquatic scorpions date from the Silurian period (405 to 425 mya), fossils of terrestrial scorpions date from the Devonian period (350 to 400 mya), and fossils of all other arachnid groups are present by the Carboniferous period (280 to 345 mya, *see appendix B*).

Water conservation is a major concern for any terrestrial organism, and their relatively impermeable exoskeleton preadapted ancestral arachnids for terrestrialism. **Preadaptation** occurs when a structure present in members of a species proves useful in promoting reproductive success when an individual encounters new environmental situations. Later adaptations included the evolution of efficient excretory structures, internal surfaces for gas exchange,

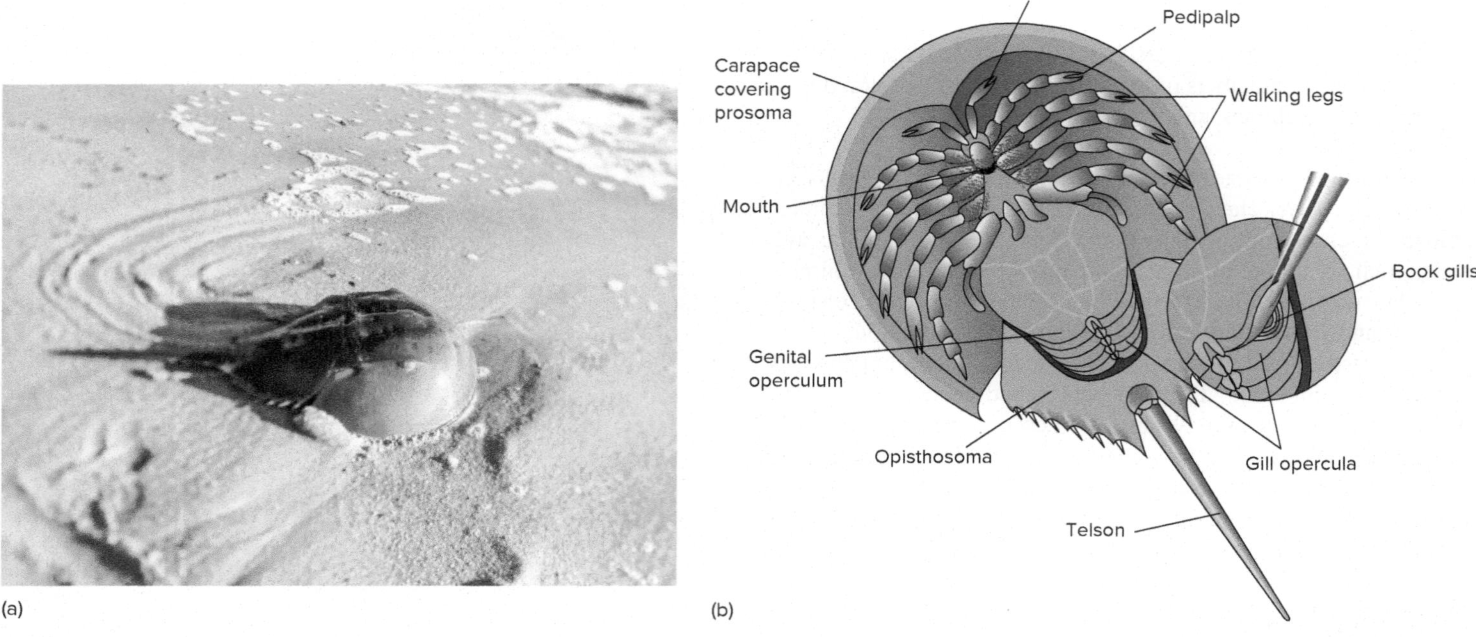

FIGURE 14.8

Class Merostomata. (*a*) Dorsal view of the horseshoe crab *Limulus polyphemus.* (*b*) Ventral view.

appendages modified for locomotion on land, and greater deposition of wax in the epicuticle.

Form and Function

Most arachnids are carnivores. They hold small arthropods with their chelicerae while enzymes from the gut tract pour over the prey. Partially digested food is then taken into the mouth. Others inject enzymes into prey through hollow chelicerae (e.g., spiders) and suck partially digested animal tissue. The gut tract of arachnids is divided into three regions. The anterior portion is the foregut, and the posterior portion is the hindgut. Both develop as infoldings of the body wall and are lined with cuticle. A portion of the foregut is frequently modified into a pumping stomach, and the hindgut is frequently a site of water reabsorption. The midgut between the foregut and hindgut is noncuticular and lined with secretory and absorptive cells. Lateral diverticula increase the area available for absorption and storage.

Arachnids use coxal glands and/or Malpighian tubules for excreting nitrogenous wastes. **Coxal glands** are paired, thin-walled, spherical sacs bathed in the blood of body sinuses. Coxal glands are probably homologous to nephridia. Nitrogenous wastes are absorbed across the wall of the sacs, transported in a long, convoluted tubule, and excreted through excretory pores at the base of the posterior appendages. Arachnids that are adapted to dry environments possess blind-ending diverticula of the gut tract that arise at the juncture of the midgut and hindgut. These tubules, called **Malpighian tubules,** absorb waste materials from the blood and empty them into the gut tract. Excretory wastes are then eliminated with digestive wastes. The major excretory product of arachnids is uric acid. As discussed in chapter 28, uric acid excretion is advantageous for terrestrial animals because uric acid is excreted as a semisolid with little water loss.

Gas exchange also occurs with minimal water loss because arachnids have few exposed respiratory surfaces. Some arachnids possess structures, called **book lungs,** that are assumed to be modifications of the book gills in the Merostomata. Book lungs are paired invaginations of the ventral body wall that fold into a series of leaflike lamellae (figure 14.9). Air enters the book lung through a slitlike opening and circulates between lamellae. Respiratory gases diffuse between the blood moving among the lamellae and the air in the lung chamber. Other arachnids possess a series of branched, chitin-lined tubules that deliver air directly to body tissues. These tubule systems, called **tracheae** (sing., trachea), open to the outside through openings called **spiracles** along the ventral or lateral aspects of the abdomen. (Tracheae are also present in insects but had a separate evolutionary origin. Aspects of their physiology are described in chapters 15 and 26.)

The circulatory system of arachnids, like that of most arthropods, is an open system in which a dorsal contractile vessel (usually called the dorsal aorta or "heart") pumps blood into tissue spaces of the hemocoel. Blood bathes the tissues and then returns to the dorsal aorta through openings in the aorta called ostia. Arachnid blood contains the dissolved respiratory pigment hemocyanin and has amoeboid cells that aid in clotting and body defenses.

The nervous system of all arthropods is ventral and, in ancestral arthropods, must have been laid out in a pattern similar

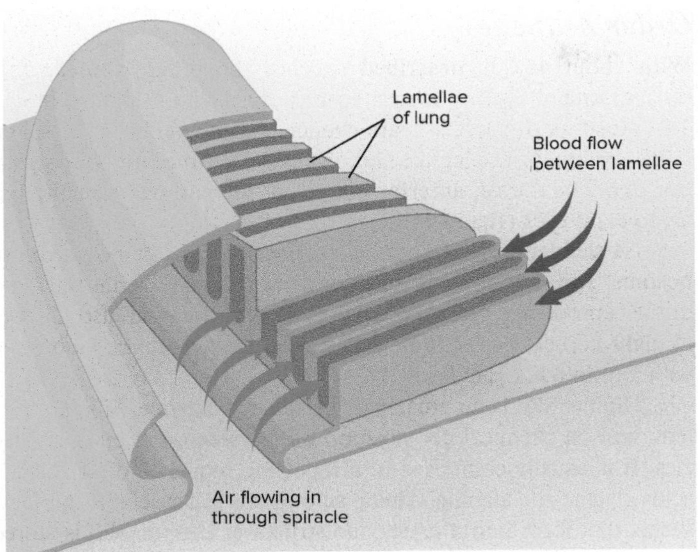

FIGURE 14.9

Arachnid Book Lung. Air and blood moving on opposite sides of a lamella of the lung exchange respiratory gases by diffusion. Figure 14.12 shows the location of book lungs in spiders.

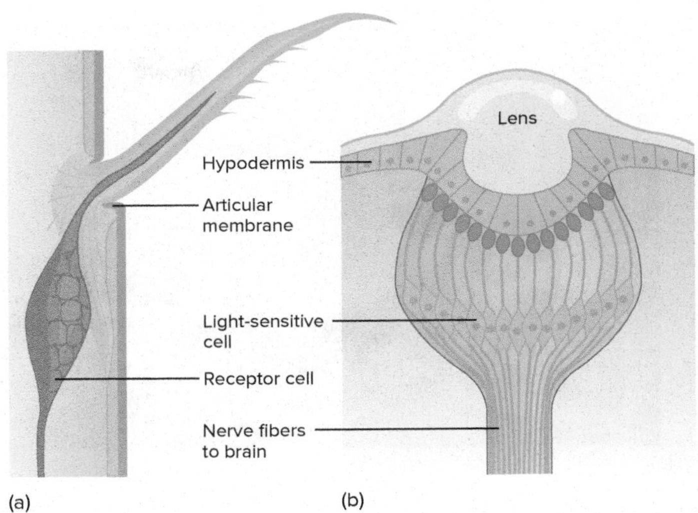

FIGURE 14.10

Arthropod Seta and Eye (Ocellus). (*a*) A seta is a hairlike modification of the cuticle set in a membranous socket. Displacement of the seta initiates a nerve impulse in a receptor cell (sensillum) associated with the base of the seta. (*b*) The lens of this spider eye is a thickened, transparent modification of the cuticle. Below the lens and hypodermis are light-sensitive sensillae with pigments that convert light energy into nerve impulses.

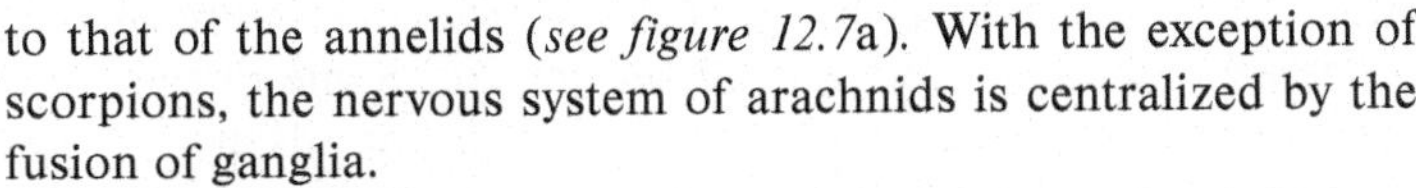
to that of the annelids (*see figure 12.7*a). With the exception of scorpions, the nervous system of arachnids is centralized by the fusion of ganglia.

The body of an arachnid has a variety of sensory structures. Most mechanoreceptors and chemoreceptors are modifications of the exoskeleton, such as projections, pores, and slits, together with sensory and accessory cells. Collectively, these receptors are called **sensilla.** For example, setae are hairlike, cuticular modifications that may be set into membranous sockets. Displacement of a seta initiates a nerve impulse in associated receptor and nerve cells (figure 14.10*a*). Vibration receptors are very important to some arachnids. Spiders that use webs to capture prey, for example, determine both the size of the insect and its position on the web by the vibrations the insect makes while struggling to free itself. The chemical sense of arachnids is comparable to taste and smell in vertebrates. Small pores in the exoskeleton are frequently associated with peglike, or other, modifications of the exoskeleton, and they allow chemicals to stimulate nerve cells. Arachnids possess one or more pairs of eyes, which they use primarily for detecting movement and changes in light intensity (figure 14.10*b*). The eyes of some hunting spiders probably form images.

Arachnids are dioecious. Paired genital openings are on the ventral side of the second abdominal segment. Sperm transfer is usually indirect. The male often packages sperm in a spermatophore, which is then transferred to the female. Courtship rituals confirm that individuals are of the same species, attract a female to the spermatophore, and position the female to receive the spermatophore. In some taxa (e.g., spiders), copulation occurs, and sperm is transferred via a modified pedipalp of the male. Development is direct, and the young hatch from eggs as miniature adults. Many arachnids tend their developing eggs and young during and after development.

Order Scorpionida

Members of the order Scorpionida (skor″pe-ah-ni′dah) are the scorpions (figure 14.11*a*). There are about 1,800 species of scorpions that are common from tropical to warm temperate climates. Scorpions are secretive and nocturnal, hiding during most daylight hours under logs and stones.

Scorpions have a prosoma that is fused into a shield-like carapace, and small chelicerae project anteriorly from the front of the carapace (figure 14.11*b*). A pair of enlarged, chelate pedipalps is posterior to the chelicerae. The opisthosoma is divided. An anterior preabdomen contains the slitlike openings to book lungs, comblike tactile and chemical receptors called pectines, and genital openings. The postabdomen (commonly called the tail) is narrower than the preabdomen and is curved dorsally and anteriorly over the body when aroused. At the tip of the postabdomen is a sting. The sting has a bulbular base that contains venom-producing glands and a hollow, sharp, barbed point. Smooth muscles eject venom during stinging. Only a few scorpions have venom that is highly toxic to humans. Species in the genera *Androctonus* (northern Africa) and *Centuroides* (Mexico, Arizona, and New Mexico) have been responsible for human deaths. Other scorpions from the southern and southwestern areas of North America give stings comparable to wasp stings.

Prior to reproduction, male and female scorpions have a period of courtship that lasts from 5 min to several hours. Male and female scorpions face each other and extend their abdomens high into the air. The male seizes the female with his pedipalps, and they

(a)

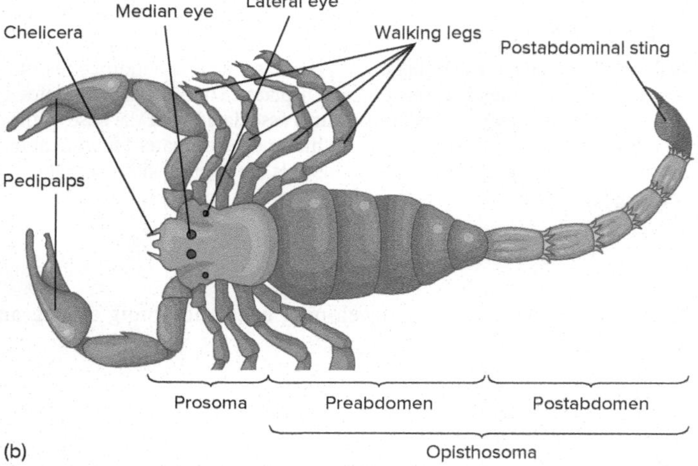

(b)

FIGURE 14.11

Order Scorpionida. (*a*) The desert hairy scorpion (*Hardrurus arizonensis*) is the largest scorpion in North America (up to 14 cm). (*b*) External anatomy of a scorpion.

repeatedly walk backward and then forward. The lower portion of the male reproductive tract forms a spermatophore that is deposited on the ground. During courtship, the male positions the female so that the genital opening on her abdomen is positioned over the spermatophore. Downward pressure of the female's abdomen on a triggerlike structure of the spermatophore releases sperm into the female's genital chamber.

Most arthropods are **oviparous;** females lay eggs that develop outside the body. Many scorpions and some other arthropods are **ovoviviparous;** development is internal, although large, yolky eggs provide all the nourishment for development. Some scorpions, however, are **viviparous,** meaning that the mother provides nutrients to nourish the embryos. Eggs develop in diverticula of the ovary that are closely associated with diverticula of the digestive tract. Nutrients pass from the digestive tract diverticula to the developing embryos. Development requires up to 1.5 years, and 20 to 40 young are brooded. After birth, the young crawl onto the mother's back, where they remain for up to a month.

Order Araneae

With about 46,000 described species, the order Araneae (ah-ran′a-e) one of the two largest groups of arachnids (figure 14.12). The prosoma of spiders bears chelicerae with venom glands and fangs. Pedipalps are leglike and, in males, are modified for sperm transfer. The dorsal, anterior margin of the carapace usually has six to eight eyes (figure 14.13).

A slender, waistlike pedicel attaches the prosoma to the opisthosoma. The abdomen is swollen or elongate and contains openings to the reproductive tract, book lungs, and tracheae. It also has two to eight conical projections, called spinnerets, that are associated with multiple silk glands.

Spider silk is an amazingly versatile substance. Silk is a protein, and its chemical composition differs somewhat between species. It is usually composed of a repeating sequence of the amino acids glycine and alanine. Amino acid chains self-assemble into beta sheets that form into a crystalline structure. Unspun silk is stored as a gel in a storage chamber. Solid fibers are formed as water is removed, lipids and thiol (sulfur containing) groups are added, and the gel is forced through fine channels of the spinnerettes. Other compounds modify the properties of the silk. Pyrrolidine compounds are hygroscopic and keep the silk moist. Potassium hydrogen phosphate releases hydrogen ions and creates an acidic pH to protect the silk from fungal and bacterial degradation.

Spiders produce several kinds of silk, each with its own material property and use. A web like the orb weaver web (*see figure 14.1*) consists of a number of types of silk threads. Some threads form the frame that borders the web. Other radial threads are laid in a sunburst pattern from the middle of the web to the frame. Still other threads are called the catching spiral. These sticky threads spiral outward from the central hub of the web. In addition to forming webs for capturing prey, silk is used to line retreats, to lay a safety line that fastens to the substrate to interrupt a fall, and to wrap eggs into a case for development. Air currents catch silk lines that young spiders produce and disperse them. Silk lines have carried spiders at great altitudes for hundreds of kilometers. This is called ballooning. Dragline silk exhibits greater tensile strength than Kevlar, the strongest synthetic polymer known (which in turn is stronger than steel). In addition, spider silk is elastic. Elasticity and tensile strength enable the web to stop the motion of a flying insect without damage to the web. Webs also resist damage from the wind, movements of the anchoring points of the web, and struggling movements of trapped prey.

Most spiders feed on insects and other arthropods that they hunt or capture in webs. A few (e.g., tarantulas or "bird spiders") feed on small vertebrates. Spiders bite their prey to paralyze them and then sometimes wrap prey in silk. They puncture the prey's body wall and inject enzymes. The spider's pumping stomach then sucks predigested prey products into the spider's digestive tract. The venom of most spiders is harmless to humans. Widow spiders (*Lactrodectus*) and brown recluse spiders (*Loxosceles*) are exceptions, since their venom is toxic to humans (figure 14.14).

Mating of spiders involves complex behaviors that include chemical, tactile, and/or visual signals. Females deposit chemicals called pheromones on their webs or bodies to attract males.

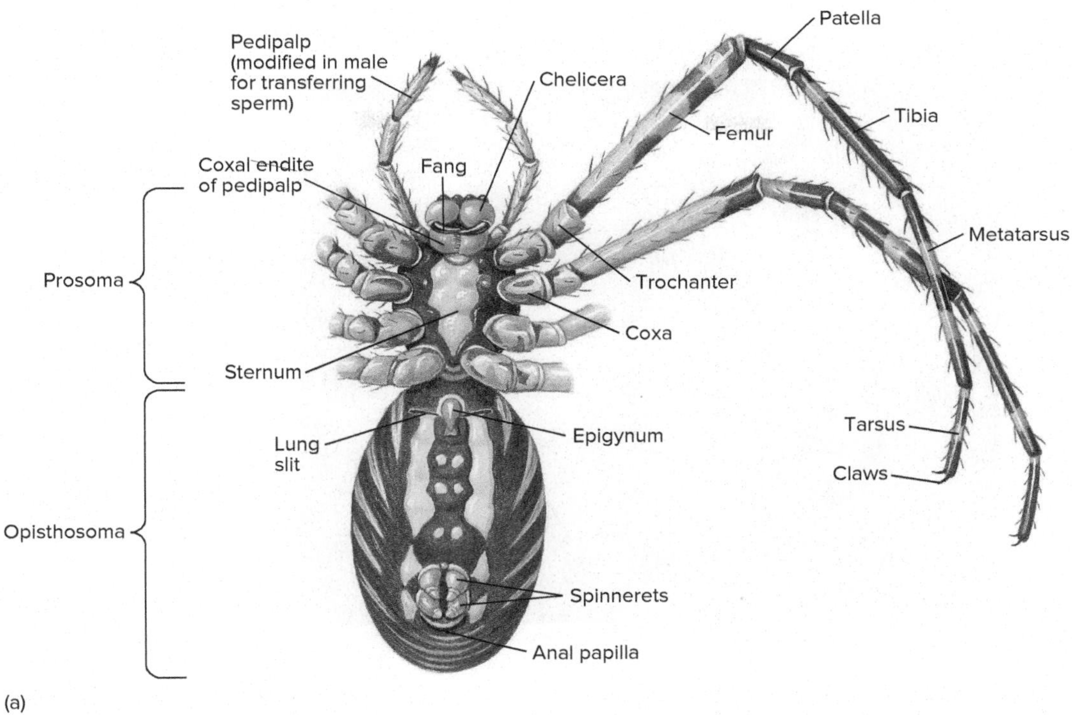

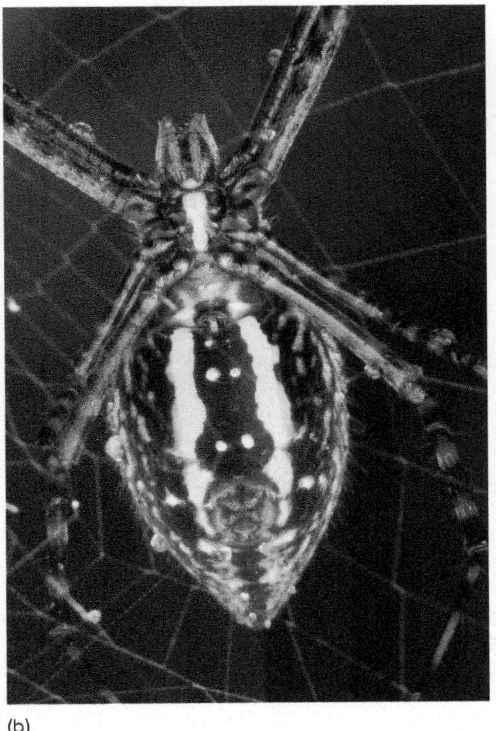

FIGURE 14.12

External Structure of *Argiope*. (*a*) Ventral view. (*b*) How many of the structures in (*a*) can you identify in this photograph?

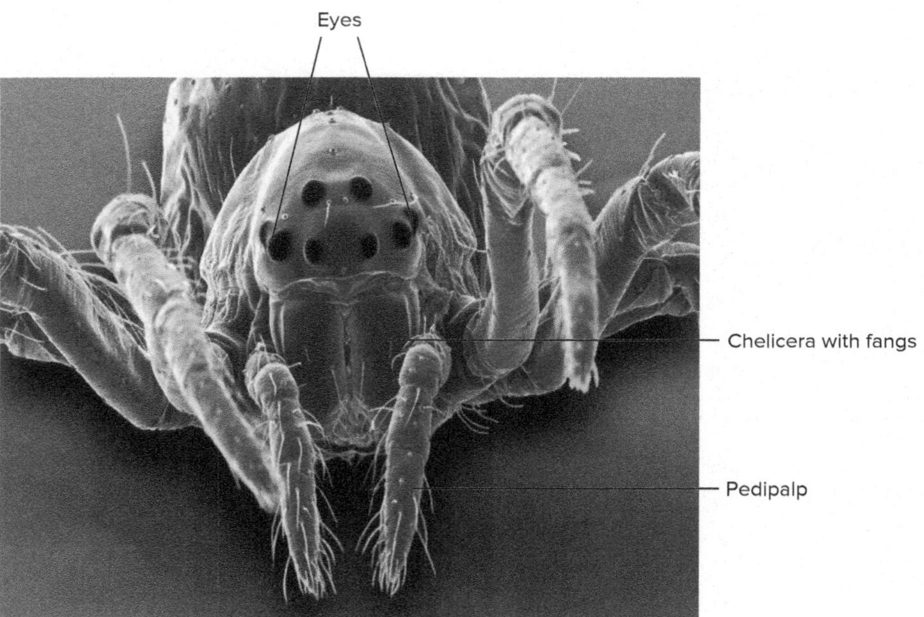

FIGURE 14.13

Prosoma of a Spiderling. This scanning electron micrograph of the prosoma of a spiderling clearly shows eight eyes, pedipalps, and fanged chelicerae (×20). The number and arrangement of eyes are important taxonomic characters.

(a)

(b)

FIGURE 14.14

Two Venomous Spiders. (*a*) A female black widow spider (*Lactrodectus mactans*) is recognized by its shiny black body with a red hourglass pattern on the ventral surface of its opisthosoma. Males have the same body shape but are somewhat smaller, less shiny, and lack the red hourglass pattern. There are more than 30 species of widow spiders distributed worldwide (except in Arctic and Antarctic environments). Their venom is a neurotoxin that causes pain and nausea. (*b*) A brown recluse spider (*Loxosceles reclusa*) is recognized by the dark brown, violin-shaped mark on the dorsal aspect of its prosoma (black arrow). Note that the neck of the "violin" points toward the opisthosoma. The legs are uniformly light colored and lack large setae. Brown recluse spiders are found in the central and southern portions of North America (north/south: Iowa into Mexico and east/west: Kansas through Tennessee, Kentucky, and Georgia). Their venom is histolytic and causes tissue necrosis.

(Pheromones are chemicals that one individual releases into the environment to create a behavioral change in another member of the same species.) A male may attract a female by plucking the strands of a female's web. The pattern of plucking is species specific and helps identify and locate a potential mate and prevents the male spider from becoming the female's next meal. The tips of a male's pedipalps possess a bulblike reservoir with an ejaculatory duct and a penislike structure called an embolus. Prior to mating, the male

How Do We Know about Spider Silk?

Tensile strength is the amount of force a material can withstand without breaking. The tensile strength of spider silk is measured by miniature strain gauges that pull on a sample of silk from both ends and measure the force required to cause the silk to break. Current research on spider silk is attempting to describe the properties of silk and, using molecular techniques, to produce artificial spider silk. Researchers have sequenced the gene coding for spider silk and have placed these genes into host organisms, including bacteria, plants, and goats. Silk proteins have been produced by these host organisms. The next step is to learn how to spin these proteins into silk threads. Imagine the usefulness of a thread that is stronger than Kevlar and steel, yet elastic and biodegradable. Potential uses include body armor, parachutes, fishing nets, extremely fine sutures for microsurgery of eyes and nerves, and artificial ligaments and tendons. What scientists are learning from more than 300 million years of spider silk evolution has the potential for developing a new life-saving and life-changing technology.

fills the reservoir of his pedipalps by depositing sperm on a small web and then collecting sperm with his pedipalps. During mating, the pedipalp is engorged with blood, the embolus is inserted into the female's reproductive opening, and sperm are discharged. The female deposits up to 3,000 eggs in a silken egg case, which she then seals and attaches to webbing, places in a retreat, or carries with her.

Order Opiliones

Members of the order Opiliones (o′pi-le″on-es) are the harvestmen or daddy longlegs (figure 14.15). This order is comprised of about 7,000 species. The prosoma broadly joins to the opisthosoma, and thus the body appears ovoid. Legs are very long and slender. Many harvestmen are omnivores (they feed on a variety of plant and animal material), and others are strictly predators. They seize prey with their pedipalps and ingest prey as described for other arachnids. Digestion is both external and internal. Sperm transfer is direct, as males have a penislike structure. Females have a tubular ovipositor that projects from a sheath at the time of egg laying. Females deposit hundreds of eggs in damp locations on the ground.

FIGURE 14.15

Order Opiliones. Harvestmen or daddy longlegs are abundant in vegetation in moist, humid environments. They do not produce silk or venom but feed on a variety of plant and animal materials. The widespread belief that harvestmen produce venom that is toxic to humans is not true. In temperate regions, the harvestmen appear in large numbers in autumn, thus the common name. *Phalagium opilo* is shown here on a white umbellifer flower.

Order Acarina

Members of the order Acarina (ak′ar-i″nah) are the mites and ticks. The number of described species in this order rivals that of Araneae (approximately 50,000). The number of described species in this order rivals that of Araneae (approximately 50,000). Many are ectoparasites (parasites on the outside of the body) on humans and domestic animals. Others are free living in both terrestrial and aquatic habitats. Of all arachnids, acarines have had the greatest impact on human health and welfare.

Mites are 1 mm or less in length. The prosoma and opisthosoma are fused and covered by a single carapace. An anterior projection called the capitulum carries mouthparts. Chelicerae and pedipalps are variously modified for piercing, biting, anchoring, and sucking, and adults have four pairs of walking legs.

Free-living mites may be herbivores or scavengers. Herbivorous mites, such as spider mites, damage ornamental and agricultural plants. Scavenging mites are among the most common animals in soil and in leaf litter. These mites include some pest species that feed on flour, dried fruit, hay, cheese, and animal fur (figure 14.16).

Parasitic mites usually do not permanently attach to their hosts, but feed for a few hours or days and then drop to the ground. One mite, the notorious chigger or red bug (*Trombicula*), is a parasite during one of its larval stages on all groups of terrestrial vertebrates. A larva enzymatically breaks down and sucks host skin, causing local inflammation and intense itching at the site of the bite. The chigger larva drops from the host and then molts to the next immature stage, called a nymph. Nymphs eventually molt to adults, and both nymphs and adults feed on insect eggs.

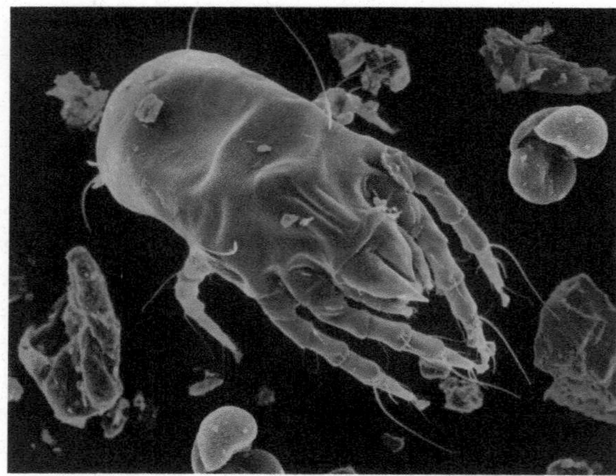

FIGURE 14.16

Order Acarina. *Dermatophagoides pteronyssinus* (0.2–0.3 mm) is a mite that is common in homes and grain storage areas. It is thought to be a major cause of dust allergies. Colorized SEM.
©Science Photo Library/Alamy Stock Photo

A few mites are permanent ectoparasites. The follicle mite, *Demodex folliculorum,* is common (but harmless) in the hair follicles of most of the readers of this text. Itch mites cause scabies in humans and other animals. *Sarcoptes scabei* is the human itch mite. It tunnels in the epidermis of human skin, where females lay about 20 eggs each day. Secretions of the mites irritate the skin, and infections are acquired by contact with an infected individual.

Ticks are ectoparasites during their entire life history. They may be up to 3 cm in length but are otherwise similar to mites. Hooked mouthparts are used to attach to their hosts and to feed on blood. The female ticks, whose bodies are less sclerotized than those of males, expand when engorged with blood. Copulation occurs on the host, and after feeding, females drop to the ground to lay eggs. Eggs hatch into six-legged immatures called seed ticks. Immatures feed on host blood and drop to the ground for each molt. Some ticks transmit diseases to humans and domestic animals. For example, *Dermacentor andersoni* transmits the bacteria that cause Rocky Mountain spotted fever and tularemia, and *Ixodes scapularis* transmits the bacteria that cause Lyme disease (figure 14.17).

Other orders of arachnids include whip scorpions, whip spiders, pseudoscorpions, and others.

Class Pycnogonida

Members of the class Pycnogonida (pik″no-gon′i-dah) are the sea spiders. All of the 1,300 species are marine and worldwide, but are most common in cold waters (figure 14.18). Pycnogonids live on the ocean floor and frequently feed on cnidarian polyps and ectoprocts. Some sea spiders feed by sucking prey tissues through a proboscis. Others tear at prey with their first pair of appendages, called chelifores.

Pycnogonids are dioecious. Gonads are U-shaped, and branches of the gonads extend into each leg. Gonopores are on one of the pairs of legs. As the female releases eggs, the male fertilizes

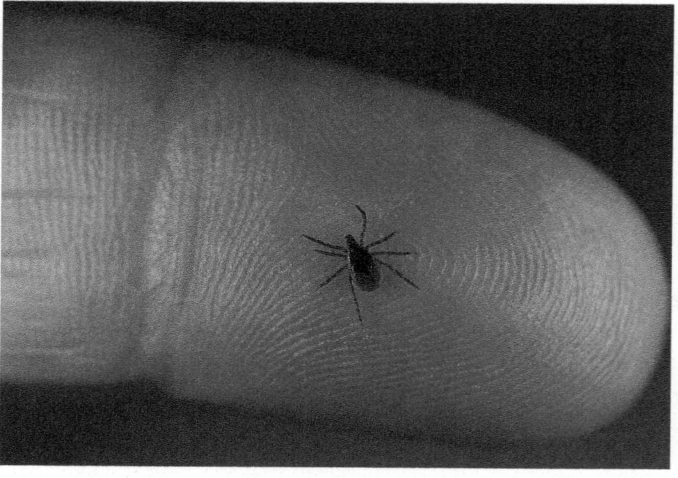

(a) (b)

FIGURE 14.17

Order Acarina. (*a*) *Ixodes scapularis,* the tick that transmits the bacteria that cause Lyme disease. (*b*) The adult (shown here) is about the size of a sesame seed, and the nymph is the size of a poppy seed. People walking in tick-infested regions should examine themselves regularly and remove any ticks found on their skin because ticks can transmit diseases, such as Rocky Mountain spotted fever, tularemia, and Lyme disease.
(a) Source: Centers for Disease Control and Prevention; ©Scott Camazine/Science Source

FIGURE 14.18

Class Pycnogonida. Sea spiders are often found in intertidal regions feeding on cnidarian polyps.
©Marevision/Getty Images

them, and the fertilized eggs are cemented into spherical masses and attached to a pair of elongate appendages of the male, called ovigers, where they are brooded until hatching.

Section 14.7 Thinking Beyond the Facts

Describe arachnid adaptations studied in this section that promote life apart from standing water.

14.8 SUBPHYLUM MYRIAPODA

LEARNING OUTCOMES

1. Describe the characteristics of members of the class Diplopoda.
2. Describe the characteristics of members of the class Chilopoda.
3. Characterize the body forms and habitats of members of the classes Symphyla and Pauropoda.

The subphylum Myriapoda (mir″e-a-pod′ah) (Gr. *myriad,* ten thousand + *podus,* foot) is divided into four classes: Diplopoda (millipedes), Chilopoda (centipedes), Symphyla (symphylans), and Pauropoda (pauropodans) (*see table 14.1*). They are characterized by a body consisting of two tagmata (head and trunk) and uniramous appendages. All modern myriapods are terrestrial.

Class Diplopoda

The class Diplopoda (dip′lah-pod′ah) (Gr. *diploos,* twofold + *podus,* foot) contains the millipedes. With 16 orders and over 12,000 species, Diplopoda is the most diverse of the four myriapod classes. Ancestors of this group appeared on land during the Devonian period (400 mya, see *appendix B*) and were among the first terrestrial animals. Millipedes have 11 to 100 trunk segments derived from an embryological and evolutionary fusion of primitive metameres. Each apparent segment represents a fusion of two embryological segments and is referred to as a diplosegment. An obvious result of this fusion is the occurrence of two pairs of appendages on each diplosegment. Fusion is also reflected internally by two ganglia, two pairs of ostia, and two pairs of tracheal trunks per apparent segment. Most millipedes are round in cross section, although some are more flattened (figure 14.19*a*).

Millipedes are worldwide in distribution and are nearly always found in or under leaf litter, humus, or decaying logs. Their epicuticle does not contain much wax; therefore, their choice of habitat is important to prevent desiccation. Their many legs, simultaneously pushing against the substrate, help millipedes bulldoze through the habitat. Millipedes feed on decaying plant matter using their mandibles in a chewing or scraping fashion. A few millipedes have mouthparts modified for sucking plant juices.

(a)

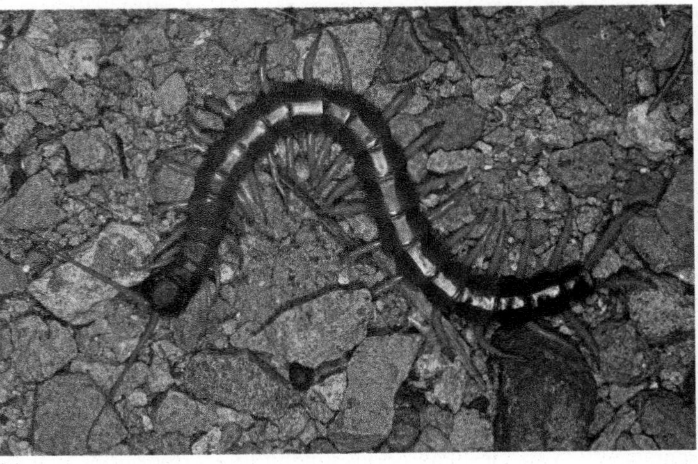

(b)

FIGURE 14.19

Myriapods. (*a*) A woodland millipede (*Ophyiulus pilosus*). (*b*) A giant Sonoran centipede (*Scolopendra heros*).
(a) ©Aminart/Getty Images; (b) ©C. Allan Morgan/Getty Images

Millipedes roll into a ball when faced with desiccation or when disturbed. Many also possess repugnatorial glands that produce hydrogen cyanide, which repels other animals. Hydrogen cyanide is not synthesized and stored as hydrogen cyanide because it is caustic and would destroy millipede tissues. Instead, a precursor compound and an enzyme mix as they are released from separate glandular compartments. Repellants increase the likelihood that the millipede will be dropped unharmed and decrease the chances that the same predator will try to feed on another millipede.

Male millipedes transfer sperm to female millipedes with modified trunk appendages, called gonopods, or in spermatophores. Eggs are fertilized as they are laid and hatch in several weeks. Immatures acquire more legs and segments with each molt until they reach adulthood.

Class Chilopoda

Members of the class Chilopoda (ki″lah-pod′ah) (Gr. *cheilos*, lip + *podus*, foot) are the centipedes. Chilopoda is divided into five orders with about 5,000 species. Most centipedes are nocturnal and scurry about the surfaces of logs, rocks, or other forest-floor debris. Like millipedes, most centipedes lack a waxy epicuticle and therefore require moist habitats. Their bodies are flattened in cross section, and they have a single pair of long legs on each of their 15 or more trunk segments. The last pair of legs is usually modified into long sensory appendages.

Centipedes are fast-moving predators–the only predatory myriapods (figure 14.19*b*). Food usually consists of small arthropods, earthworms, and snails; however, some centipedes feed on frogs and rodents. Venom claws (modified first-trunk appendages called maxillipeds) kill or immobilize prey. Maxillipeds, along with mouth appendages, hold the prey as mandibles chew and ingest the food. Most centipede venom is essentially harmless to humans, although many centipedes have bites that are comparable to wasp stings; a few human deaths have been reported from large, tropical species.

Centipede reproduction may involve courtship displays in which the male lays down a silk web using glands at the posterior tip of the body. He places a spermatophore in the web, which the female picks up and introduces into her genital opening. Eggs are fertilized as they are laid. A female may brood and guard eggs by wrapping her body around the eggs, or they may be deposited in the soil. Young are similar to adults except that they have fewer legs and segments. Legs and segments are added with each molt.

Classes Pauropoda and Symphyla

Members of the class Pauropoda (por″o-pod′ah) (Gr. *pauros*, small + *podus*, foot) are soft-bodied animals with 11 segments and branching antennae (figure 14.20*a*). The class is comprised of two orders and 835 species. Pauropods inhabit forest floor litter, where they consume fungi, humus, and other decaying organic matter. Their very small size (2 mm) and thin, moist exoskeleton allow gas exchange across the body surface and diffusion of nutrients and wastes in the body cavity.

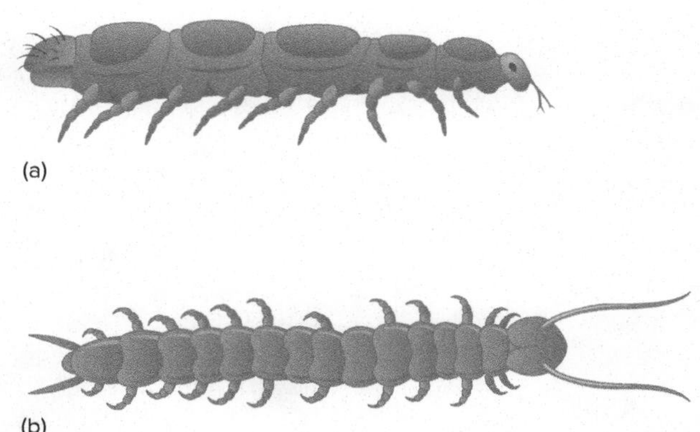

FIGURE 14.20

Subphylum Myriapoda. (*a*) A member of the class Pauropoda (*Pauropus*). (*b*) A member of the class Symphyla (*Scutigerella*).

Members of the class Symphyla (sim-fi′lah) (Gr. *sym*, same + *phyllos*, leaf) are small arthropods (2 to 10 mm in length) that occupy soil and leaf mold, superficially resemble centipedes, and are often called garden centipedes (figure 14.20*b*). There are 195 species of Symphyla that are grouped into a single order. They lack eyes and have 12 leg-bearing trunk segments and long filamentous antennae. The posterior segment may have one pair of spinnerets or long, sensory bristles. Symphylans normally feed on decaying vegetation; however, some species are pests of vegetables and flowers.

Section 14.8 Thinking Beyond the Facts

Why is it important to use the phrase "legs per apparent segment" when describing the millipedes?

14.9 FURTHER PHYLOGENETIC CONSIDERATIONS

LEARNING OUTCOME

1. Explain why ancestral chelicerates are so very important in arthropod evolution.

As this chapter indicates, the arthropods have been very successful. This is evidenced by the diverse body forms and lifestyles of the arachnids and myriapods studied in this chapter.

The subphylum Chelicerata is a very important group of animals from an evolutionary standpoint, even though they are less numerous in terms of numbers of species and individuals than the subphyla covered in chapter 15. Their arthropod exoskeleton and the evolution of excretory and respiratory systems that minimize water loss resulted in ancestral members of this subphylum becoming some of the first terrestrial animals. The ancient myriapods quickly joined the chelicerates on land during the Silurian period (*see appendix B*). Chelicerates and myriapods, however, are not the only terrestrial arthropods. In terms of numbers of species and numbers of individuals, chelicerates are dwarfed in terrestrial environments by the insects (subphylum Hexapoda). Members of the fourth arthropod subphylum (Crustacea) have, for the most part, never ventured out onto land. Instead, they have become the predominant arthropods in marine and freshwater environments. These two subphyla are the subjects of chapter 15. The evolutionary relationships within the entire phylum are also covered at the end of the next chapter.

Section 14.9 Thinking Beyond the Facts

Why are the chelicerates an important group from an evolutionary standpoint?

Summary

14.1 **Evolutionary Perspective**

- Arthropods include crayfish, lobsters, spiders, insects, and others. They are characterized by metamerism with tagmatization, a chitinous exoskeleton, and paired jointed appendages.
- Arthropods are members of the Ecdysozoa and are closely related to the Nematoda, Nematomorpha, Kinorhyncha, and others. Living arthropods are divided into four subphyla: Chelicerata, Crustacea, Hexapoda, and Myriapoda. All members of a fifth subphylum, Trilobitomorpha, are extinct.

14.2 **Metamerism and Tagmatization**

- Metamerism of arthropods is most evident externally where body segmentation is most obvious. Internal metamerism is often reduced. Tagmatization has resulted in specialization of the arthropod body for specific functions.

14.3 **The Exoskeleton**

- The exoskeleton covers the entire body surface of an arthropod. It provides structural support and protection. It also helps prevent water loss and provides surfaces for muscle attachment. The exoskeleton consists of two layers, the epicuticle and the partially hardened procuticle. The growth of an arthropod is accompanied by periodic molting of the exoskeleton, a process called ecdysis. The exoskeleton must be periodically shed to facilitate growth.

14.4 **The Hemocoel**

- The hemocoel of arthropods develops from the embryonic blastocoel and provides a cavity for bathing of the internal organs by the open circulatory system. The coelom of arthropods forms small cavities around gonads and sometimes excretory structures.

14.5 **Metamorphosis**

- Metamorphosis is a change in form and physiology that occurs when an animal transitions between larval and adult stages. Metamorphosis has contributed to arthropod success by reducing competition between immature and adult arthropods.

14.6 **Subphylum Trilobitomorpha**

- Members of the extinct subphylum Trilobitomorpha had oval, flattened bodies that consisted of three tagmata and three longitudinal lobes. Appendages were biramous. Trilobites were a dominant form of life 600 to 345 mya.

14.7 **Subphylum Chelicerata**

- The subphylum Chelicerata has members whose bodies are divided into a prosoma and an opisthosoma. They also possess chelicerae and pedipalps.
- The horseshoe crabs and the giant water scorpions belong to the class Merostomata.
- The class Arachnida includes spiders, mites, ticks, scorpions, and others. Their exoskeleton partially preadapted the arachnids for their terrestrial habitats. Coxal glands, Malpighian tubules and book lungs, along with efficient circulatory and nervous systems also contribute to arachnid success in terrestrial habitats. Most arachnids are predators on other invertebrates.
- The sea spiders are the only members of the class Pycnogonida.

14.8 **Subphylum Myriapoda**

- The subphylum Myriapoda includes four classes of arthropods.
- Members of the class Diplopoda (the millipedes) are characterized by apparent segments bearing two pairs of legs.
- Members of the class Chilopoda (the centipedes) are characterized by a single pair of legs on each of their 15 segments and a body that is flattened in cross section.
- The class Pauropoda contains soft-bodied animals that feed on fungi and decaying organic matter in forest-floor litter. Members of the class Symphyla are centipede-like arthropods that live in soil and leaf mold, where they feed on decaying vegetation.

14.9 **Further Phylogenetic Considerations**

- The exoskeleton, and efficient respiratory and excretory systems, preadapted ancient members of the subphylum Chelicerata for life on land. Ancient myriapods invaded terrestrial environments shortly after the chelicerates became terrestrial.

Concept Review Questions

1. All of the following are grouped with the Arthropods in the Ecdysozoa, except one. Select the exception.
   a. Nematoda
   b. Mollusca
   c. Nematomorpha
   d. Kinorhyncha

2. The inner layer of the exoskeleton of an arthropod is called the _____________. It contains chitin, and its outer region is hardened. The outermost layer of the exoskeleton is waxy and impermeable to water. Which one of the following terms correctly fills the blank in the first sentence?
   a. epicuticle
   b. procuticle
   c. hypodermis
   d. basement membrane
3. Members of the class _____________ include the spiders, scorpions, ticks, and mites.
   a. Merostomata
   b. Arachnida
   c. Pycnogonida
   d. Branchiopoda
   e. Malacostraca
4. Centipedes are members of the subphylum ___________ and the class ___________.
   a. Myriapoda; Diplopoda
   b. Hexapoda; Chilopoda
   c. Myriapoda; Chilopoda
   d. Chelicerata; Diplopoda
5. Members of this class appear to have two pairs of legs on each body segment. This condition is misleading because each apparent segment results from a fusion of two segments.
   a. Chilopoda
   b. Arachnida
   c. Diplopoda
   d. Pycnogonida

Analysis and Application Questions

1. What is tagmatization, and why is it advantageous for metameric animals?
2. Explain how, in spite of being an armorlike covering, the exoskeleton permits movement and growth.
3. Why is the arthropod exoskeleton often cited as the major reason for arthropod success?
4. Explain why excretory and respiratory systems of ancestral arachnids probably preadapted these organisms for terrestrial habitats.

Insects, like this preying mantid (class Insecta, order Mantodea), have become the most successful class of animals in terrestrial habitats. Crustaceans have been nearly as successful in marine and freshwater habitats. This chapter will help you understand the reasons for the success of these two groups, which comprise the clade Pancrustacea.

15

The Pancrustacea: Crustacea and Hexapoda

Chapter Outline

15.1 Evolutionary Perspective
15.2 Subphylum Crustacea
Class Malacostraca
Class Branchiopoda
Class Maxillopoda
15.3 Subphylum Hexapoda
Class Insecta: Structure and Function
Insect Reproduction and Development
Insect Behavior
Insects and Humans
Insect Orders: The Big Four
15.4 Further Phylogenetic Considerations
Clade Panarthropoda
Phylum Onychophora
Phylum Tardigrada
Arthropod Phylogeny

15.1 EVOLUTIONARY PERSPECTIVE

LEARNING OUTCOME

1. Describe the factors that promoted the evolutionary dominance of crustaceans in freshwater and marine environments and the dominance of insects in terrestrial habitats.

By almost any criterion, the insects and crustaceans have been enormously successful. Zoologists have described approximately 900,000 species of insects and about 70,000 species of crustaceans. Insects comprise 80% of all living eukaryotic species! In terms of number of species, crustaceans are behind the second-place arthropod subphylum, Chelicerata (approximately 100,000 species). In terms of numbers of individuals and diversity of body forms, however, the crustaceans are unparalleled.

Crustaceans are the arthropod masters of the seas and freshwaters. They range in size from the Japanese spider crab (*Macrocheira kaempferi*) (4 m measured from tip-to-tip of its outstretched legs) to the microscopic zooplankton species that fill our lakes and oceans. Copepods (class Maxillopoda) and krill (order Euphausiacea) are among the most abundant of all animals. Copepods (figure 15.1) are food for herring, sardines, and mackerel. Krill are food for the baleen whales (gray whales, right whales, blue whales, and humpbacks), a variety of Antarctic fishes, and sea birds. Both copepods and krill occupy critical positions in marine food webs (*see figure 6.10*). Marine predators are not the only consumers of crustacean flesh. About 5 million tons of crustaceans are harvested annually from marine fisheries or farms for human consumption (lobsters, shrimp, crayfish, etc.).

Insects are the arthropod masters of terrestrial environments. Arachnids preceded insects into terrestrial environments but did not dominate land environments for long. By the early Devonian period (about 400 million years ago [mya], *see appendix B*), when herbaceous plants and the first forests were beginning to flourish and enough ozone had accumulated to filter ultraviolet radiation from the sun, insects appeared on land. Since that time, insects have become the dominant land arthropod—many would argue the land dominant animal.

What is responsible for the dominance of these two groups in their respective environments? In both cases, the exoskeleton is probably a large factor in their success. For both groups, this armor-like covering provides an unparalleled level of support and protection against predators and other external threats. Metamorphosis allows the

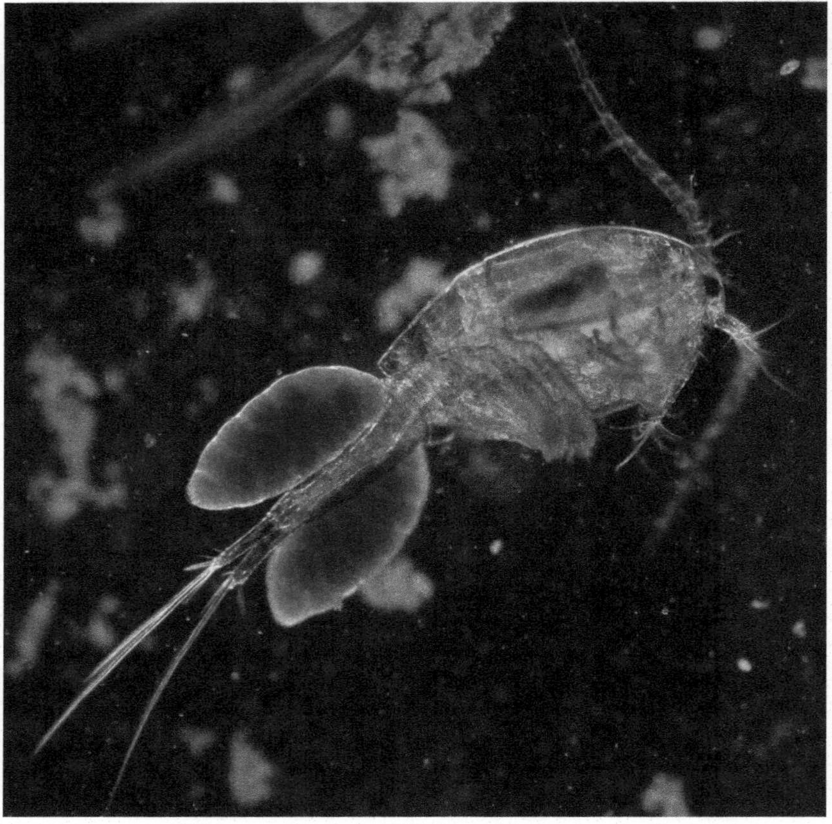

FIGURE 15.1

The Most Abundant Animal? Copepod crustaceans (class Maxillopoda) are small (1-2 mm) but abundant in oceans and freshwaters of the world and form important links in aquatic food webs. Note the purple egg sacs attached to the abdomen of this female. The long antennae are used in an oarlike fashion, along with abdominal appendages, to propel her through the water. A member of the order Cyclopoida is shown here.

immature members of both groups to exploit different resources than do adults. For the insects, the exoskeleton's waxy epicuticle enhanced the exoskeleton's water-conserving properties. Most importantly, the exoskeleton and associated nerves and muscles form flight mechanisms that allow insects to use widely scattered food resources, to invade new habitats, and to escape unfavorable environments. These factors–along with desiccation-resistant eggs, metamorphosis, high reproductive potential, and diversification of mouthparts and feeding habits–permitted insects to become the dominant class of organisms on the earth.

This chapter covers two arthropod subphyla: Crustacea and Hexapoda (table 15.1). As discussed at the conclusion of this chapter, they form a clade Pancrustacea. As you move through this chapter you should begin to appreciate the adaptations responsible for the tremendous success of this group of animals.

Section 15.1 Thinking Beyond the Facts

Some zoologists argue that "crustaceans" are the dominant arthropod on land and in the sea. What taxonomic revision must be accepted for this to be correct?

15.2 SUBPHYLUM CRUSTACEA

LEARNING OUTCOMES

1. Describe the characteristics of members of the subphylum Crustacea.
2. Describe adaptations for aquatic habitats seen in the decapod malacostracans.
3. Compare members of the orders Euphausiacea, Isopoda, Amphipoda, and Cladocera.
4. Compare members of the classes Branchiopoda and Maxillopoda.

Some members of the subphylum Crustacea (krus-tā s′e-ah) (L. *crustaceus,* hard shelled), such as crayfish, shrimp, lobsters, and crabs, are familiar to nearly everyone. Many others are lesser-known but very common taxa. These include copepods, cladocerans, fairy shrimp, isopods, amphipods, and barnacles. Except for some isopods and crabs, crustaceans are all aquatic.

Crustaceans differ from other living arthropods in two ways. They have two pairs of antennae, whereas all other arthropods have one pair or none. In addition, crustaceans possess biramous

TABLE 15.1

CLASSIFICATION OF THE PANCRUSTACEA (CRUSTACEA AND HEXAPODA)*

Phylum Arthropoda (ar″thra-po′dah)
Animals with metamerism and tagmatization, a jointed exoskeleton, and a ventral nervous system.

Subphylum Crustacea (krus-tās′e-ah)
Most aquatic, head with two pairs of antennae, one pair of mandibles, and two pairs of maxillae; biramous appendages.

Class Remipedia (re-mi-pe′de-ah)
Cave-dwelling crustaceans from the Caribbean basin, Indian Ocean, Canary Islands, and Australia; body with approximately 30 segments that bear uniform, biramous appendages.

Class Cephalocarida (sef″ah-lo-kar′ĭ-dah)
Small (3 mm) marine crustaceans with uniform, leaflike, and triramous appendages.

Class Branchiopoda (brang″ke-o-pod′ah)
Flattened, leaflike appendages used in respiration, filter feeding, and locomotion, found mostly in freshwater. Fairy shrimp, brine shrimp, clam shrimp, and water fleas.

Class Malacostraca (mal-ah-kos′trah-kah)
Appendages possibly modified for crawling, feeding, swimming. Lobsters, crayfish, crabs, shrimp, krill, and isopods (some terrestrial).

Class Maxillopoda (maks″il-ah-pod′ah)
Five head, six thoracic, and four abdominal somites plus a telson; thoracic segments variously fused with the head; abdominal segments lack typical appendages; abdomen often reduced. Barnacles and copepods.

Subphylum Hexapoda (hex″sah-pod′ah) (Gr. *hexa,* six + *podus,* foot)
Body divided into head, thorax, and abdomen; five pairs of head appendages; three pairs of uniramous appendages on the thorax. Insects and their relatives.

Class Entognatha (en″to-na′tha) (Gr. *entos,* within + *gnathos,* jaw)
Mouth appendages hidden within the head; mandibles with single articulation; legs with one undivided tarsus.

Order Collembola (col-lem′bo-lah)
Antennae with four to six segments; compound eyes absent; abdomen with six segments, most with springing appendage on fourth segment; inhabit soil and leaf litter. Springtails.

Order Protura (pro-tu′rah)
Minute, with cone-shaped head; antennae, compound eyes, and ocelli absent; abdominal appendages on first three segments; inhabit soil and leaf litter. Proturans.

Order Diplura (dip-lu′rah)
Head with many segmented antennae; compound eyes and ocelli absent; cerci multisegmented or forcepslike; inhabit soil and leaf litter. Diplurans.

Class Insecta (in-sekt′ah) (L. *insectum,* to cut up)
Mouth appendages exposed and projecting from head; mandibles usually with two points of articulation; well-developed Malpighian tubules.

Subclass Archaeognatha (ar″ke-ona′tha)

Order Archaeognatha
Small, wingless, cylindrical and scaly body; mandibles with single articulation; abdomen 11 segmented with 3 to 8 pairs of styli and 3 caudal filaments; ametabolous metamorphosis. Jumping bristletails.

Subclass Zygentoma (xi-gen′to-mah)

Order Thysanura (thi-sa-nu′rah)
Tapering abdomen; flattened; scales on body; terminal cerci; long antennae; ametabolous metamorphosis. Silverfish.

Subclass Pterogota (ter-i-go′tah)
Wings on second and third thoracic segments; wings may be modified or lost; no pregenital abdominal appendages; direct sperm transfer.

Infraclass Palaeoptera (pa″le-op′ter-ah)
Wings incapable of being folded at rest, held vertically above the body or horizontally out from the body; wings with many veins and cross-veins; antennae reduced or vestigial in adults.

Order Ephemeroptera (e-fem-er-op′ter-ah)
Elongate abdomen with two or three tail filaments; two pairs of membranous wings with many veins; forewings triangular; short, bristlelike antennae; hemimetabolous metamorphosis. Mayflies.

Order Odonata (o-do-nat′ah)
Elongate, membranous wings with netlike venation; abdomen long and slender; compound eyes occupy most of head; hemimetabolous metamorphosis. Dragonflies and damselflies.

(*Continued*)

TABLE 15.1 Continued

Infraclass Neoptera (ne-op′ter-ah)
Wings folded at rest; venation reduced.

Order Plecoptera (ple-kop′ter-ah)
Adults with reduced mouthparts; elongate antennae; long cerci; nymphs aquatic with gills; hemimetabolous metamorphosis. Stoneflies.

Order Mantodea (man-to′deah)
Prothorax long; prothoracic legs long and armed with strong spines for grasping prey; predators; hemimetabolous metamorphosis. Mantids.

Order Blattaria (blat-tar′eah)
Body oval and flattened; head concealed from above by a shieldlike extension of the prothorax; hemimetabolous metamorphosis. Cockroaches.

Order Isoptera (i-sop′ter-ah)
Workers white and wingless; front and hindwings of reproductives of equal size; reproductives and some soldiers may be sclerotized; abdomen broadly joins thorax; social; hemimetabolous metamorphosis. Termites.

Order Dermaptera (der-map′ter-ah)
Elongate; chewing mouthparts; threadlike antennae; abdomen with unsegmented forcepslike cerci; hemimetabolous metamorphosis. Earwigs.

Order Orthoptera (or-thop′ter-ah)
Forewing long, narrow, and leathery; hindwing broad and membranous; chewing mouthparts; hemimetabolous metamorphosis. Grasshoppers, crickets, and katydids.

Order Phasmida (fas′mi-dah)
Body elongate and sticklike; wings reduced or absent; some tropical forms are flattened and leaflike; hemimetabolous metamorphosis. Walking sticks and leaf insects.

Order Phthiraptera (fthi-rap′ter-ah)
Small, wingless ectoparasites of birds and mammals; body dorsoventrally flattened; white; hemimetabolous metamorphosis. Sucking and chewing lice.

Order Hemiptera (hem-ip′ter-ah)
Piercing-sucking mouthparts; mandibles and first maxillae styletlike and lying in grooved labium; wings membranous; hemimetabolous metamorphosis. Bugs, cicadas, leafhoppers, and aphids.

Order Thysanoptera (thi-sa-nop′ter-ah)
Small bodied; sucking mouthparts; wings narrow and fringed with long setae; plant pests; hemimetabolous metamorphosis. Thrips.

Order Neuroptera (neu-rop′ter-ah)
Wings membranous, hindwings held rooflike over body at rest; holometabolous metamorphosis. Lacewings, snakeflies, antlions, and dobsonflies.

Order Coleoptera (ko-le-op′ter-ah)
Forewings sclerotized, forming covers (elytra) over the abdomen; hindwings membranous; chewing mouthparts; the largest insect order; holometabolous metamorphosis. Beetles.

Order Trichoptera (tri-kop′ter-ah)
Mothlike with setae-covered antennae; chewing mouthparts; wings covered with setae and held rooflike over abdomen at rest; larvae aquatic and often dwell in cases that they construct; holometabolous metamorphosis. Caddisflies

Order Lepidoptera (lep-i-dop′ter-ah)
Wings broad and covered with scales; mouthparts formed into a sucking tube; holometabolous metamorphosis. Moths, butterflies.

Order Diptera (dip′ter-ah)
Mesothoracic wings well developed; metathoracic wings reduced to knoblike halteres; variously modified but never chewing mouthparts; holometabolous metamorphosis. Flies.

Order Siphonaptera (si-fon-ap′ter-ah)
Laterally flattened, sucking mouthparts; jumping legs; parasites of birds and mammals; holometabolous metamorphosis. Fleas.

Order Hymenoptera (hi-men-op′ter-ah)
Wings membranous with few veins; well-developed ovipositor, sometimes modified into a stinger; mouthparts modified for biting and lapping; social and solitary species; holometabolous metamorphosis. Ants, bees, and wasps.

*Selected orders of inserts are described. Crustacea is a paraphyletic assemblage (*see section 15.4*).

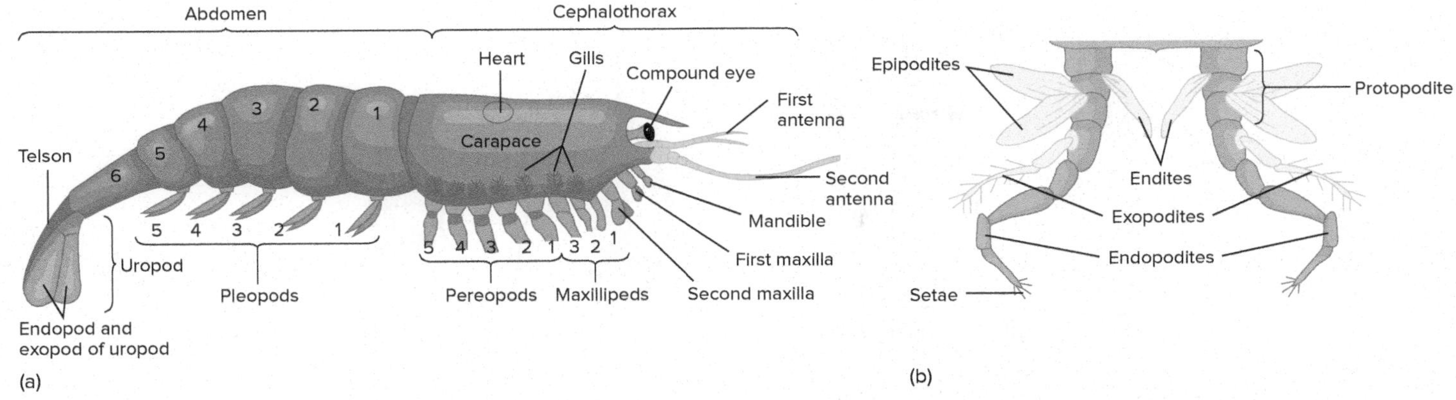

FIGURE 15.2

Crustacean Body Form. (*a*) External anatomy of a generalized crustacean. Gills are formed as outgrowths of the body wall and protected under extensions of the exoskeleton called the carapace. (*b*) One set of paired appendages, showing the generalized biramous structure. A protopodite attaches to the body wall. An exopodite (a lateral ramus) and an endopodite (a medial ramus) attach at the end of the protopodite. In modern crustaceans, both the distribution of appendages along the length of the body and the structure of appendages are modified for specialized functions.

appendages, each of which consists of a basal segment, called the **protopodite,** with two rami (distal processes that give the appendage a Y shape) attached. The medial ramus is the **endopodite,** and the lateral ramus is the **exopodite** (figure 15.2). Trilobites had similar structures, and the phylogenetic significance of arthropod appendage structure is discussed at the end of this chapter. There are five classes of crustaceans (*see table 15.1*) and numerous orders. The three most common classes, and selected orders within those classes, are covered next.

Class Malacostraca

Malacostraca (mal-ah-kos′trah-kah) (Gr. *malakos,* soft + *ostreion,* shell) is the largest class of crustaceans. It includes crabs, lobsters, crayfish, shrimp, mysids, shrimplike krill, isopods, and amphipods.

The order Decapoda (dek-i-pod′ah) is the largest order of crustaceans and includes shrimp, crayfish, lobsters, and crabs. Shrimp have a laterally compressed, muscular abdomen and pleopods for swimming. Lobsters, crabs, and crayfish are adapted to crawling on the surface of the substrate (figure 15.3). The abdomen of crabs is greatly reduced and is held flexed beneath the cephalothorax.

Crayfish illustrate general crustacean structure and function. They are convenient to study because of their relative abundance and large size (figure 15.4). The body of a crayfish is divided into two regions. A cephalothorax is derived from the developmental fusion of a sensory and feeding tagma (the head) with a locomotor tagma (the thorax). The exoskeleton of the cephalothorax extends laterally and ventrally to form a shieldlike carapace. The abdomen is posterior to the cephalothorax, has locomotor and visceral functions, and, in crayfish, takes the form of a muscular "tail."

Paired appendages are present in both body regions (figure 15.5). The first two pairs of cephalothoracic appendages are the first and second antennae. The third through fifth pairs of appendages are associated with the mouth. During crustacean evolution, the third pair of appendages became modified into chewing or grinding structures called **mandibles.** The fourth and fifth pairs of appendages, called **maxillae,** are for food handling. The second maxilla bears a gill and a thin, bladelike structure, called a scaphognathite (gill bailer), for circulating water over the gills. The sixth through the eighth cephalothoracic appendages are called maxillipeds and are derived from the thoracic tagma. They are accessory sensory and food-handling appendages. The last two pairs of maxillipeds also bear gills. Appendages 9 to 13 are thoracic appendages called pereopods (walking legs). The first pereopod, known as the cheliped, is enlarged and chelate (pincherlike) and used in defense and capturing food. All but the last pair of appendages of the abdomen are called pleopods (swimmerets) and are used for swimming. In females, developing eggs attach to pleopods, and the embryos are brooded until after hatching. In males, the

FIGURE 15.3

Order Decapoda. The lobsters, shrimp, crayfish, and crabs comprise the largest crustacean order. The lobster *Homarus americanus* is shown here.

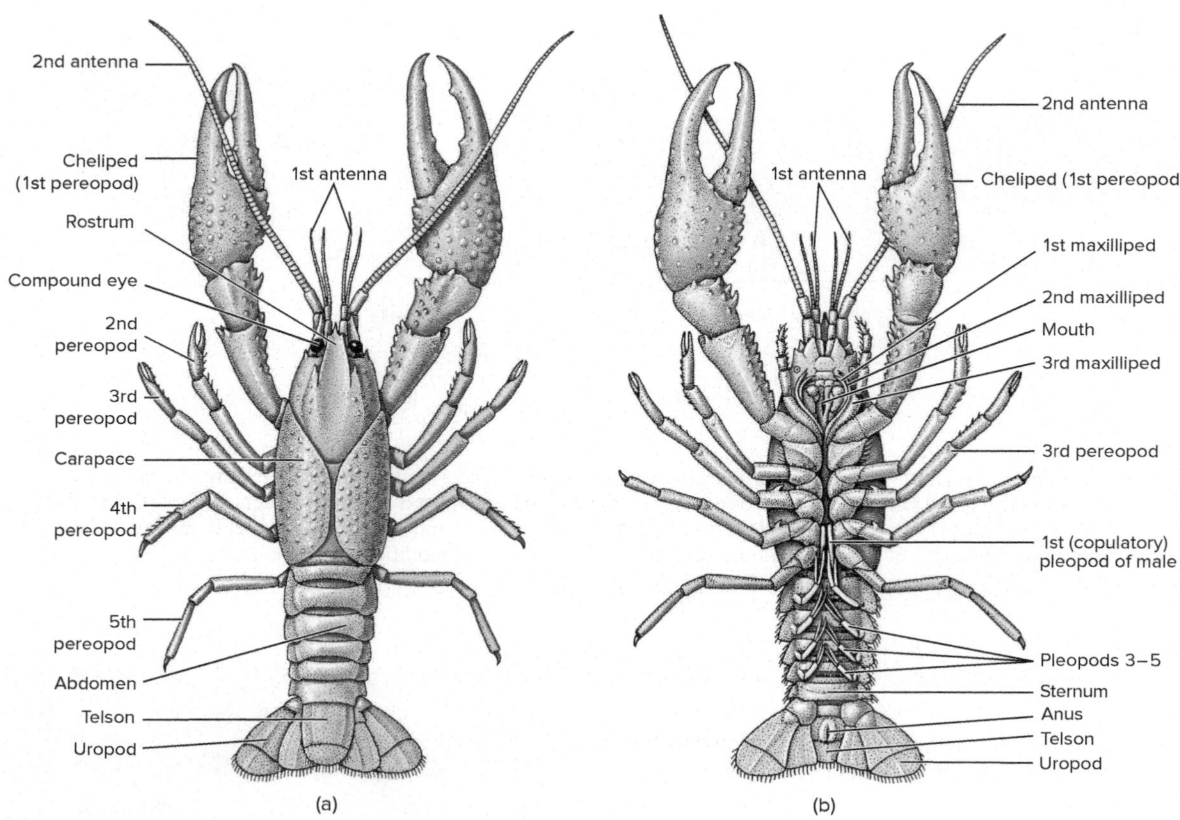

FIGURE 15.4

External Structure of a Male Crayfish. (*a*) Dorsal view. (*b*) Ventral view.

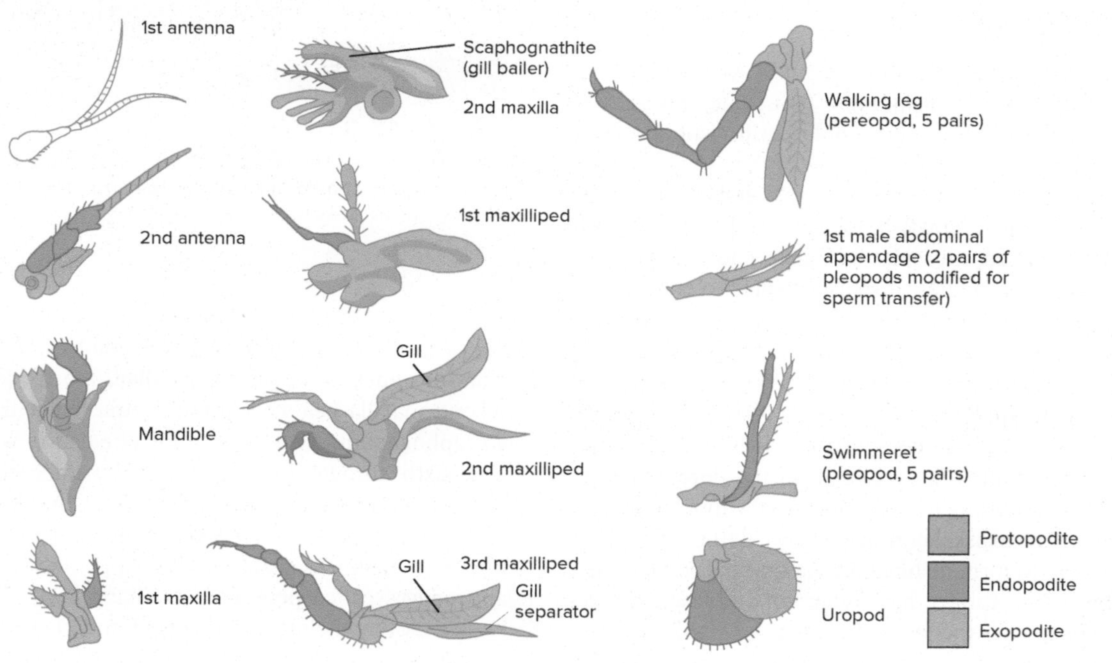

FIGURE 15.5

Crayfish Appendages. Individual appendages are arranged in sequence. Homologies regarding the structure of appendages are color coded. The origin and homology of the first antennae are uncertain.

first two pairs of pleopods are modified into gonopods (claspers) used for sperm transfer during copulation. The abdomen ends in a median extension called the telson. The telson bears the anus and is flanked on either side by flattened, biramous appendages of the last segment, called uropods. The telson and uropods make an effective flipperlike structure used in swimming and in escape responses.

All crustacean appendages, except the first antennae, have presumably evolved from an ancestral biramous form, as evidenced by their embryological development, in which they arise as simple two-branched structures. (First antennae develop as uniramous appendages and later acquire the branched form. The crayfish and their close relatives are unique in having branched first antennae.) Structures, such as the biramous appendages of a crayfish, whose form is based on a common ancestral pattern and have similar development in the segments of an animal, are said to be **serially homologous.**

Crayfish prey upon other invertebrates, eat plant matter, and scavenge dead and dying animals. The foregut includes an enlarged stomach, part of which is specialized for grinding. A digestive gland secretes digestive enzymes and absorbs products of digestion. The midgut extends from the stomach and is often called the intestine. A short hindgut ends in an anus and is important in water and salt regulation (figure 15.6*a*).

As previously described, the gills of a crayfish attach to the bases of some cephalothoracic appendages. Gills are in a branchial (gill) chamber, the space between the carapace and the lateral body wall (figure 15.6*b*). The beating of the scaphognathite of the second maxilla drives water anteriorly through the branchial chamber. Oxygen and carbon dioxide are exchanged between blood and water across the gill surfaces, and a respiratory pigment, hemocyanin, carries oxygen in blood plasma.

Circulation in crayfish is similar to that of most arthropods. Dorsal, anterior, and posterior arteries lead away from a muscular heart. Branches of these vessels empty into sinuses of the hemocoel. Blood returning to the heart collects in a ventral sinus and enters the gills before returning to the pericardial sinus, which surrounds the heart (figure 15.6*b*).

Crustacean nervous systems show trends similar to those in annelids and arachnids. Primitively, the ventral nervous system is ladderlike. Higher crustaceans show a tendency toward centralization and cephalization. Crayfish have supraesophageal and subesophageal ganglia that receive sensory input from receptors in the head and control the head appendages. The ventral nerves and segmental ganglia fuse, and giant neurons in the ventral nerve cord function in escape responses (*see figure 15.6*a). When nerve impulses are conducted posteriorly along giant nerve fibers of a crayfish, powerful abdominal flexor muscles of the abdomen contract alternately with weaker extensor muscles, causing the abdomen to flex (the propulsive stroke) and then extend (the recovery stroke). The telson and uropods form a paddlelike "tail" that propels the crayfish posteriorly.

In addition to antennae, the sensory structures of crayfish include compound eyes, statocysts, chemoreceptors, proprioceptors, and tactile setae. Chemical receptors are widely distributed over the appendages and the head. Many of the setae covering the mouthparts and antennae are chemoreceptors used in sampling food and

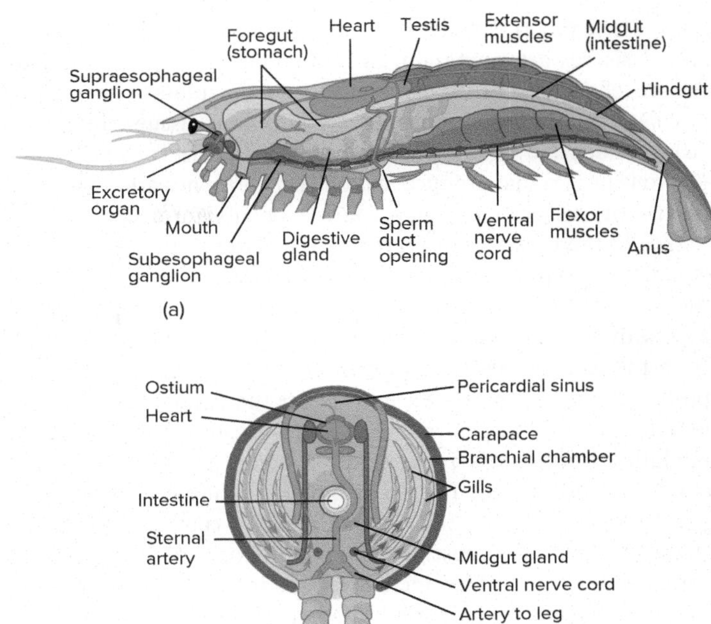

FIGURE 15.6

Internal Structure of a Crayfish. (*a*) Lateral view of a male. In the female, the ovary is in the same place as the testis of the male, but the gonoducts open at the base of the third pereopod. (*b*) Cross section of the thorax in the region of the heart. In this diagram, gills are shown attached higher on the body wall than they actually occur to show the path of blood flow (arrows) through them.

detecting pheromones. A statocyst is at the base of each of the paired first antennae (*see figure 24.14*). A statocyst is a pitlike invagination of the exoskeleton that contains setae and a group of cemented sand grains called a statolith. Crayfish movements move the statolith and displace setae. Statocysts provide information regarding movement, orientation with respect to the pull of gravity, and vibrations of the substrate. Because the statocyst is cuticular, it is replaced with each molt. Sand is incorporated into the statocyst when the crustacean is buried in sand. Other receptors involved with equilibrium, balance, and position senses are tactile receptors on the appendages and at joints. When a crustacean is crawling or resting, stretch receptors at the joints are stimulated. Crustaceans detect tilting from changing patterns of stimulation. These widely distributed receptors are important to crustaceans that lack statocysts.

Crayfish have compound eyes mounted on movable eyestalks. The lens system consists of 25 to 14,000 individual receptors called ommatidia. Compound eyes also occur in insects, and their structure and function are discussed later in this chapter. Larval crustaceans have a single, median photoreceptor consisting of a few sensilla. These simple eyes, called ocelli, allow larval crustaceans to orient toward or away from the light, but do not form images. Many larvae are planktonic and use their ocelli to orient toward surface waters.

The endocrine system of a crayfish controls functions such as ecdysis, sex determination, and color change. Endocrine glands release chemicals called hormones into the blood, where they circulate and cause responses at certain target tissues. In crustaceans, endocrine functions are closely tied to nervous functions. Nervous tissues that produce and release hormones are called neurosecretory tissues (*see figure 25.6*). X-organs are neurosecretory tissues in the eyestalks of crayfish. Associated with each X-organ is a sinus gland that accumulates and releases the secretions of the X-organ. Other glands, called Y-organs, are not directly associated with nervous tissues. They are near the bases of the maxillae. Both the X-organ and the Y-organ control ecdysis. The X-organ produces molt-inhibiting hormone, and the sinus gland releases it. The target of this hormone is the Y-organ. As long as molt-inhibiting hormone is present, the Y-organ is inactive. Certain conditions prevent the release of molt-inhibiting hormone; when these conditions exist, the Y-organ releases ecdysone hormone, leading to molting. (These "certain conditions" are often complex and species specific. They include factors such as nutritional state, temperature, and photoperiod.) Other hormones that facilitate molting have also been described. These include, among others, a molt-accelerating factor.

Androgenic glands in the cephalothorax of males mediate another endocrine function. (Females possess rudiments of these glands during development, but the glands never mature.) Normally, androgenic hormone(s) promotes the development of testes and male characteristics, such as gonopods. The removal of androgenic glands from males results in the development of female sex characteristics, and if androgenic glands are experimentally implanted into a female, she develops testes and gonopods.

Hormones probably regulate many other crustacean functions. Some that have been investigated include the development of female brooding structures in response to ovarian hormones, the seasonal regulation of ovarian functions, and the regulation of heart rate and body color changes by eyestalk hormones.

The excretory organs of crayfish are called antennal glands (green glands) because they are at the bases of the second antennae and are green in living crayfish. In other crustaceans, they are called maxillary glands because they are at the bases of the second maxillae. They are structurally similar to the coxal glands of arachnids and presumably had a common evolutionary origin. Excretory products form by the filtration of blood. Ions, sugars, and amino acids are reabsorbed in the tubule before the diluted urine is excreted. As with most aquatic animals, ammonia is the primary excretory product. However, crayfish do not rely solely on the antennal glands to excrete ammonia. Ammonia also diffuses across thin parts of the exoskeleton. Even though it is toxic, ammonia is water soluble, and water rapidly dilutes it. All freshwater crustaceans face a continual influx of freshwater and loss of ions. Thus, the elimination of excess water and the reabsorption of ions become extremely important functions. Gill surfaces are also important in ammonia excretion and water and ion regulation (osmoregulation).

Crayfish, and all other crustaceans except the barnacles, are dioecious. Gonads are in the dorsal portion of the thorax, and gonoducts open at the base of the third (females) or fifth (males) pereopods. Mating occurs just after a female has molted. The male turns the female onto her back and deposits nonflagellated sperm near the openings of the female's gonoducts. Fertilization occurs after copulation, as the eggs are shed. The eggs are sticky and securely fasten to the female's pleopods. Fanning movements of the pleopods over the eggs keep the eggs aerated. The development of crayfish embryos is direct, with young hatching as miniature adults. Many other crustaceans have a planktonic, free-swimming larva called a nauplius (figure 15.7*a*). In some, the nauplius develops into a miniature adult. Crabs and their relatives have a second larval stage called a zoea (figure 15.7*b*). When all adult features are present, except sexual maturity, the immature crab is called the postlarva.

The order Euphausiacea (yah-fah-see-a′see-ay) includes the krill. Euphausiids are important members of the zooplankton in all oceans of the world. They undertake daily vertical migrations from ocean depths during daylight hours to surface waters at night. Swarming is common and most are bioluminescent. Their bioluminescence is probably acquired from the bioluminescent dinoflagellates that they eat. Feeding on phytoplankton at the base of the food web, krill serve as food for many other organisms. Antarctic krill are the food source for 6 species of baleen whales, more than 100 species of fish, 35 species of birds, 7 species of seals, and 20 species of squid (*see figure 6.10*). It is estimated that the biomass of Antarctic krill is approximately 380 million metric tons, and more than one-half of this biomass is eaten by whales and other marine organisms annually. Krill reproduction must replace this harvest by wildlife every year. Worldwide, commercial fishing also harvests in excess of 200,000 metric tons of krill that are used for aquaculture (e.g., salmon farming), aquarium food, and human consumption. Omega-3 rich krill oil is marketed as a dietary supplement in spite of the fact that omega-3 oils are easily available through sustainable dietary habits. In Japan, krill are eaten as *okiami,* and krill are processed for sale in the health-food industry worldwide. Norway and the Republic of Korea are responsible for about 75% of this harvest. The 1990s saw a drastic decline in krill populations around Antarctica and Japan. The Convention on the Conservation of Antarctic Marine Living Resources (CCAMLR) is a consortium of 24 member countries that has set catch quotas for krill to ensure a long-term sustainable krill fishery.

Two other orders of malacostracans have members familiar to most humans. Members of the order Isopoda (i″so-pod′ah) include "pillbugs." Isopods are dorsoventrally flattened, may be either aquatic or terrestrial, and scavenge decaying plant and animal material. Some have become modified for clinging to and feeding on other animals. For example, the tongue-eating louse (*Cymothoa exigua*) is a marine isopod that enters a fish's mouth through the gill chamber of the fish. The female isopod attaches to the fish's tongue, and the male attaches to the fish's gill arches. The isopods then feed on blood. In the process they sever blood vessels leading to the tongue. The lack of blood causes the tongue to atrophy and fall away. Apparently, little other harm comes to the fish as the fish then uses the attached isopod in swallowing functions—as a substitute tongue. Terrestrial isopods live under rocks and logs and in leaf litter (figure 15.8*a*). Members of the order Amphipoda (am″fi-pod′ah) have a laterally compressed body that gives them a shrimplike appearance. Amphipods move by crawling or swimming on their

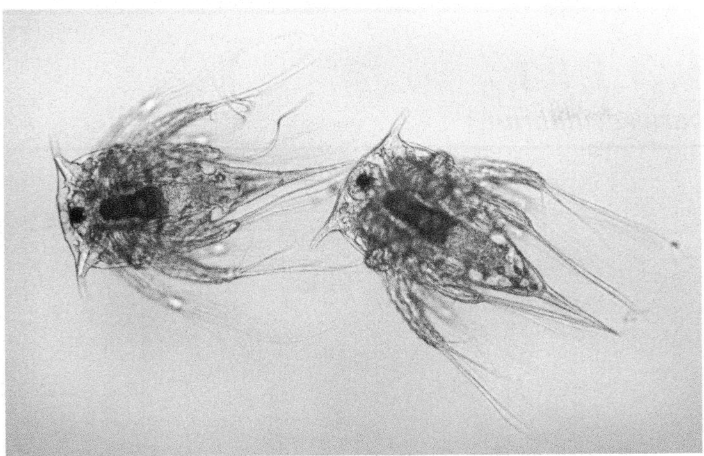

(a)

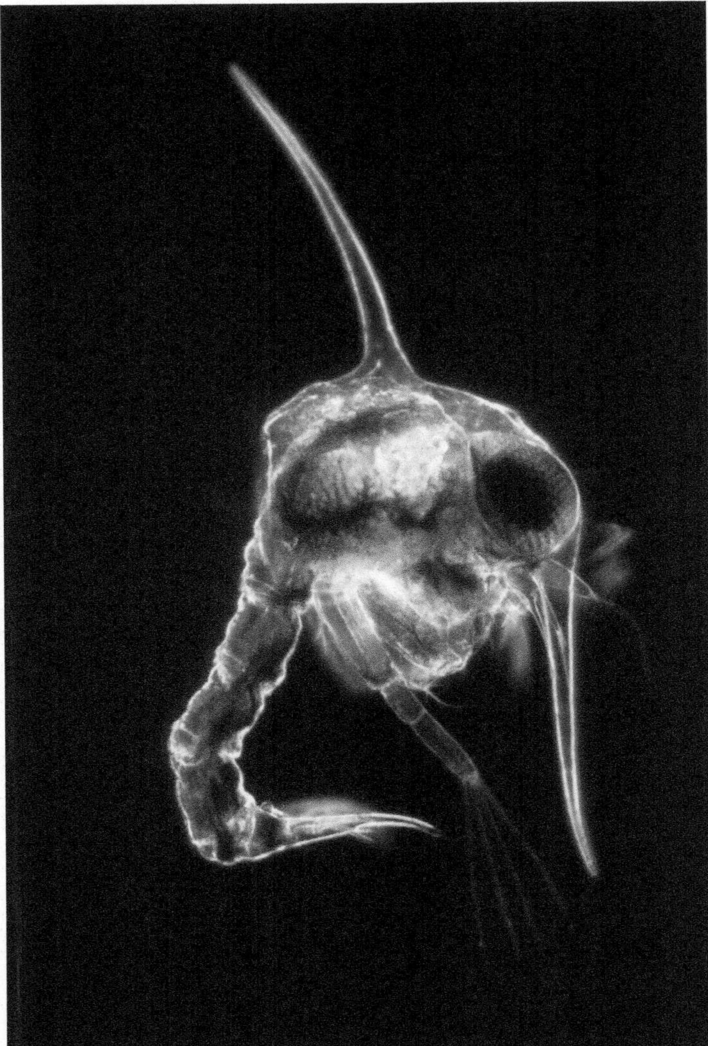

(b)

FIGURE 15.7

Crustacean Larvae. (*a*) Nauplius larva of a barnacle (0.5 mm). (*b*) Zoea larvae (1 mm) of a crab (*Carcinus maenas*).

(a) ©Nnehring/iStock/Getty Images Plus (b) ©FLPA/Alamy

sides along the substrate. Some species are modified for burrowing, climbing, or jumping (figure 15.8*b*). Amphipods are scavengers, and a few species are parasites.

Class Branchiopoda

Members of the class Branchiopoda (brang″ke-o-pod′ah) (Gr. *branchio,* gill + *podos,* foot) primarily live in freshwater. All branchiopods possess flattened, leaflike appendages used in respiration, filter feeding, and locomotion.

Fairy shrimp and brine shrimp comprise the order Anostraca (an-ost′ra-kah). Fairy shrimp usually live in temporary ponds that spring thaws and rains form. Eggs are brooded, and when the female dies, and the temporary pond begins to dry, the embryos

(a)

(b)

FIGURE 15.8

Orders Isopoda and Amphipoda. (*a*) Some isopods roll into a ball when disturbed or threatened with drying—thus the name "pillbug." Terrestrial isopods (suborder Oniscidea) are commonly called woodlice. (*b*) This amphipod (*Orchestoidea californiana*) spends some time out of the water hopping along beach sands—thus the name "beachhopper."

(a) ©George Grall/Getty Images (b) ©Peter Bryant

WILDLIFE ALERT
A Cave Crayfish (*Cambarus aculabrum*)

VITAL STATISTICS

Classification: Phylum Arthropoda, class Malacostraca, order Decapoda
Habitat: Two limestone caves of northwest Arkansas
Number Remaining: 1,000 to 2,500
Status: Endangered

NATURAL HISTORY AND ECOLOGICAL STATUS

So very little is known of this crayfish species that it does not even have a common name that is distinct from other cave crayfish. It lives only in caves, and like many other obligate cave dwellers, it lacks pigmentation and functional eyes, but other sensory organs are greatly enhanced (box figure 15.1). The origins of such morphological adaptations to caves, called "troglomorphy," are the subject of debate by evolutionary biologists. Some cave biologists (biospeleologists) believe that the lack of food resources selects for traits that conserve energy. Others believe that caves lack predators and lack selective pressures that maintain functional organs (such as vision to help detect predators), and thus unneeded structures tend to degenerate due to the accumulation of mutations. This species has been found in four disconnected caves in Benton County, Arkansas (box figure 15.2). In one cave, there is a 2-km stream and an underground lake. Up to 40 crayfish have been observed in one visit to this cave by researchers. The other caves are smaller and fewer individuals have been observed. Researchers estimate that the total population size for this species may be between 1,000 and 2,500 individuals with fewer than 200 of these being reproductively mature.

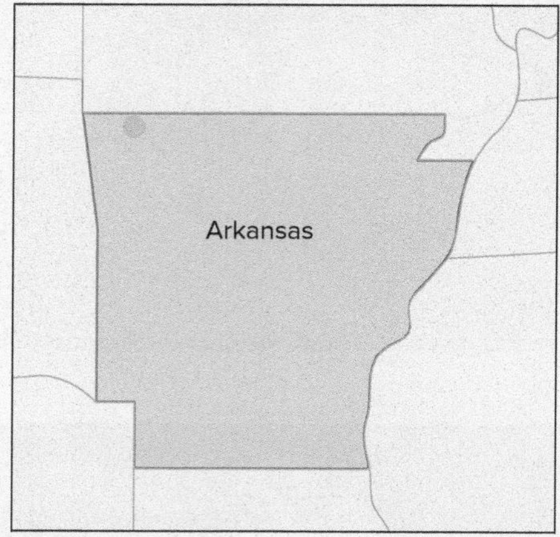

BOX FIGURE 15.2 The distribution of *Cambarus aculabrum*.

The lifespan of *Cambarus aculabrum* approaches 75 years. This longevity means that it takes many years for an individual to reach reproductive maturity, and late maturity makes population recovery more tenuous. Females carrying eggs or young have never been observed. Reproductive males are present between October and January.

The major threat to this cave crayfish is groundwater pollution. The primary source of this pollution is from agricultural operations and septic systems near caves. Soils in the area are extremely shallow, and pollutants can pass directly into groundwater reservoirs within fractured limestone bedrock. Pesticide use and accidental spills of toxic chemicals are also potential threats. One cave is within the drainage system of more than 100 confined agricultural feeding operations (swine and poultry). Another cave is surrounded by a retirement community development consisting of more than 36,000 lots. Septic tank pollution and alteration of the hydrology by extensive pavement are important concerns for crayfish in this cave. The endangered status of this crayfish is warranted because present and future environmental contamination could quickly eliminate such small, local populations. Most work being done on this crayfish is currently being carried out by the Subterranean Biodiversity Project, Department of Biological Sciences, University of Arkansas.

BOX FIGURE 15.1 **A Cave Crayfish (*Cambarus aculabrum*).** Note the absence of pigmentation and reduced eyes that are often characteristic of cave-dwelling (troglobitic) invertebrates.

become dormant in a resistant capsule. Embryos lie on the forest floor until the pond fills again the following spring, at which time they hatch into nauplius larvae. Animals, wind, or water currents may carry the embryos to other locations. Their short and uncertain life cycle is an adaptation to living in ponds that dry up. The vulnerability of these slowly swimming and defenseless crustaceans probably explains why they live primarily in temporary ponds, a habitat that contains few larger predators. Brine shrimp also form resistant embryos. They live in salt lakes and ponds (e.g., the Great Salt Lake in Utah).

Members of the order Cladocera (kla-dos′er-ah) are called water fleas (figure 15.9). A large carapace covers their bodies, and they swim by repeatedly thrusting their second antennae downward to create a jerky, upward locomotion. Females reproduce

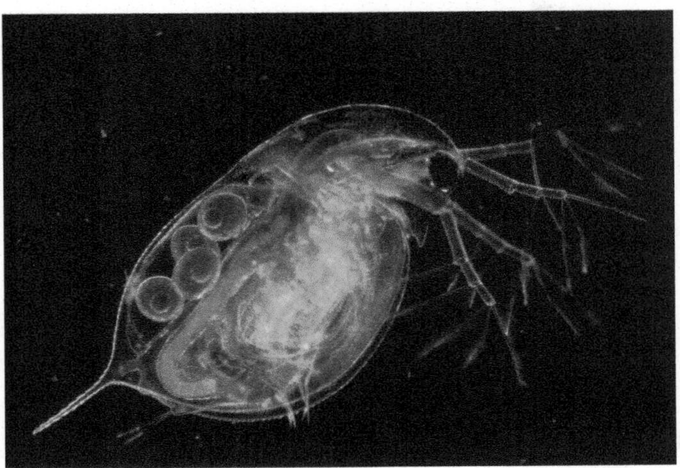

FIGURE 15.9

Class Branchiopoda. The cladoceran water flea (*Daphnia*). Lateral view (2 mm). Note the large second antennae that are used in swimming and the eggs being carried beneath the carapace.

parthenogenetically (without fertilization) in spring and summer, and can rapidly populate a pond or lake. Eggs are brooded in an egg case beneath the carapace. At the next molt, the egg case is released and either floats or sinks to the bottom of the pond or lake. In response to decreasing temperature, changing photoperiod, or decreasing food supply, females produce eggs that develop parthenogenetically into males. Sexual reproduction produces resistant "winter eggs" that overwinter and hatch in the spring.

Class Maxillopoda

Members of the class Maxillopoda include a variety of small (with the exception of the barnacles) and sometimes bizarre crustaceans that are recognized by their short bodies and the unique combination of five head, six thoracic, and four abdominal segments, plus a telson. Interestingly, one subclass, Pentastomida, is made up of parasites of the respiratory passages of reptiles, birds, and mammals. Until recently, these animals were not recognized as crustaceans and were placed in their own phylum. Their inclusion in the Maxillopoda is the result of molecular studies of DNA coding for ribosomal RNA. Two groups of common maxillopods, the copepods and the barnacles, are described next.

Subclass Copepoda

Members of the subclass Copepoda (ko″pe-pod′ah) (Gr. *kope,* oar + *podos,* foot) include some of the most abundant crustaceans (*see figure 15.1*). There are both marine and freshwater species. Copepods have a cylindrical body and a median ocellus that develops in the nauplius stage and persists into the adult stage. The first antennae (and the thoracic appendages in some) are modified for swimming, and the abdomen is free of appendages. Most copepods are planktonic and use their second maxillae for filter feeding. Their importance in marine food webs was noted in the "Evolutionary Perspective" that opens this chapter. A few copepods live on the substrate, a few are predatory, and others are commensals or parasites of marine invertebrates, fishes, or marine mammals.

Subclass Thecostraca

The barnacles are members of the infraclass Cirripedia (sir″ĭ-ped′e-ah). They are sessile and highly modified as adults (figure 15.10*a*). They are exclusively marine and include about 1,000 species. Most barnacles are monoecious. The planktonic nauplius of barnacles is followed by a planktonic larval stage, called a cypris larva, which has a bivalved carapace. Cypris larvae attach to the substrate by their first antennae and metamorphose to adults. In the process of metamorphosis, the abdomen is reduced, and the gut tract becomes U-shaped. Thoracic appendages are modified for filtering and moving food into the mouth. Calcareous plates cover the larval carapace in the adult stage.

Barnacles attach to a variety of substrates, including rock outcroppings, ship bottoms, whales (figure 15.10*b*), and other animals. Some barnacles attach to their substrate by a stalk. Others are

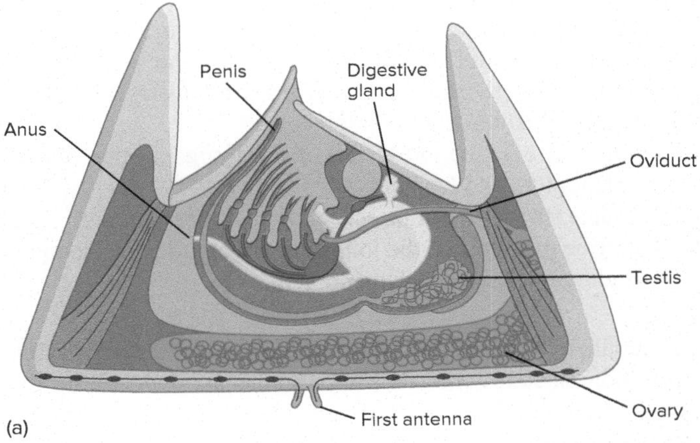

FIGURE 15.10

Class Maxillopoda, Infraclass Cirripedia. (*a*) Internal structure of a stalkless (acorn) barnacle. (*b*) Whale/barnacle commensalism. Barnacles are sedentary as adults and attach to a variety of substrates, including whales like this humpback whale (*Megaptera novaeangliae*). These acorn barnacles (*Coronula diadema*) gain mobility and protection from its whale host. In most cases, the whale is neither benefited nor harmed in this relationship. Most copepods are between 1 and 5 mm in length. A member of the order Cyclopoida is shown here.

nonstalked and are called acorn barnacles. Barnacles that colonize ship bottoms reduce both ship speed and fuel efficiency. Much time, effort, and money have been devoted to research on keeping ships free of barnacles.

Some barnacles have become highly modified parasites. The evolution of parasitism in barnacles is probably a logical consequence of living attached to other animals.

Section 15.2 Thinking Beyond the Facts

How is the concept of serial homology related to the concept of homology, which was introduced in chapter 4?

15.3 SUBPHYLUM HEXAPODA

LEARNING OUTCOMES

1. Explain how you would distinguish an insect from any other arthropod.
2. Compare the function of body regions of insects to the function of body regions of crustaceans.
3. Describe the maintenance functions of insects including the following: nutrition and digestion, gas exchange, circulation and temperature regulation, nervous and sensory functions, and reproduction and development.
4. Hypothesize on the prevalence of social organization in the Hymenoptera and Isoptera.
5. Evaluate the impact of insects on human populations.
6. Describe members of the following insect orders: Coleoptera, Lepidoptera, Diptera, Hymenoptera.

The subphylum Hexapoda (hex″sah-pod′ah) (Gr. *hexa,* six + *podus,* foot) includes animals whose bodies are divided into three tagmata, have five pairs of head appendages, and have three pairs of legs on the thorax. The subphylum is divided into two classes. Entognatha (en″to-na′tha) (Gr. *entos,* within + *gnathos,* jaw) includes the collembolans, proturans, and diplurans (*see table 15.1*). Members of this class have mouthparts that are hidden inside the head capsule, thus the class name. It is probably not a monophyletic grouping because the entognathous mouthparts of the diplurans are apparently not homologous to the mouthparts of the other two orders. Insecta (L. *insectum,* to cut up) includes the 30 orders of insects. Table 15.1 provides a partial listing of these orders. They are all characterized by mouthparts that project from the head capsule.

Class Insecta: Structure and Function

Members of the class Insecta are, in terms of numbers of species and individuals, the most successful land animals. In spite of obvious diversity, common features make insects easy to recognize. Many insects have wings and one pair of antennae, and virtually all adults have three pairs of legs.

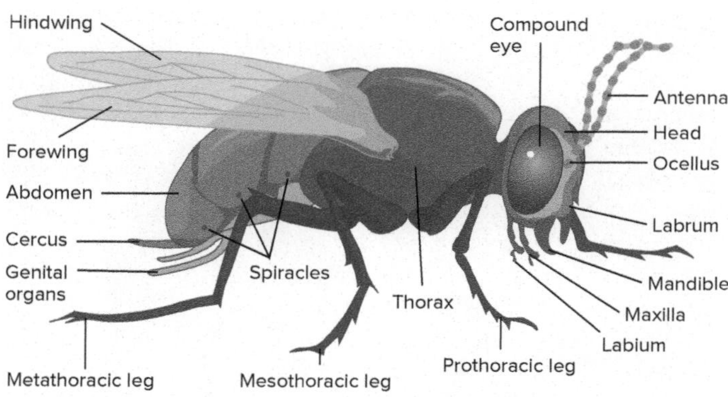

FIGURE 15.11

External Structure of a Generalized Insect. Insects are characterized by a body divided into head, thorax, and abdomen; three pairs of legs; and two pairs of wings.

External Structure and Locomotion

The body of an insect is divided into three tagmata: head, thorax, and abdomen (figure 15.11). The head bears a single pair of antennae, mouthparts, compound eyes, and zero, two, or three ocelli. The thorax consists of three segments. These segments are, from anterior to posterior, the **prothorax,** the **mesothorax,** and the **metathorax.** One pair of legs attaches along the ventral margin of each thoracic segment, and a pair of wings, when present, attaches at the dorsolateral margin of the mesothorax and metathorax. Wings have thickened, hollow veins. These veins originate in immature insects as blood vessels that supply blood to the developing wings. In adults, the wings dry and become mostly cuticular. The wing veins remain for increased wing strength. The thorax also contains two pairs of spiracles, which are openings to the tracheal system. Most insects have 10 or 11 abdominal segments, each of which has a lateral fold in the exoskeleton that allows the abdomen to expand when the insect has gorged itself or when it is full of mature eggs. Each abdominal segment has a pair of spiracles. Also present are genital structures used during copulation and egg deposition, and sensory structures called cerci. Gills are present on abdominal segments of certain immature aquatic insects.

Insect Flight Insects move in diverse ways. From an evolutionary perspective, however, flight is the most important form of insect locomotion. Insects were the first animals to fly. One of the most popular hypotheses on the origin of flight states that wings may have evolved from rigid, gill-like lateral outgrowths of the thorax. Later, these fixed lobes could have been used in gliding from the tops of tall plants to the forest floor. The ability of the wing to flap, tilt, and fold back over the body probably came later.

Another requirement for flight was the evolution of limited thermoregulatory abilities. Thermoregulation is the ability to maintain body temperatures at a level different from environmental temperatures. Relatively high body temperatures, perhaps 25°C or greater, are needed for flight muscles to contract rapidly enough for flight.

Flight mechanisms are named based on whether flight muscles act directly at the bases of the wings or indirectly to change the shape of the thoracic exoskeleton. Alternative names reflect whether or not there is a one-to-one correspondence between nerve impulses and wing beats. Some insects use a **direct** or **synchronous flight** mechanism, in which muscles acting directly on the bases of the wings contract to produce a downward thrust, and muscles attaching indirectly on the dorsal and ventral aspect of the exoskeleton contract to produce an upward thrust (figure 15.12*a*). The synchrony of direct flight mechanisms depends on the nerve impulse to the flight muscles that must precede each wingbeat. Butterflies (Lepidoptera), dragonflies (Odonata), and grasshoppers (Orthoptera) are examples of insects with a synchronous flight mechanism.

Other insects use an **indirect** or **asynchronous flight** mechanism. Indirectly attaching muscles act to change the shape of the exoskeleton for both upward and downward wing strokes. Dorsoventral muscles pulling the dorsal exoskeleton (tergum) downward produce the upward wing thrust. The downward thrust occurs when longitudinal muscles contract and cause the exoskeleton to arch upward (figure 15.12*b*). The resilient properties of the exoskeleton enhance the power and velocity of these strokes. During a wingbeat, the thorax is deformed, storing energy in the exoskeleton. At a critical point midway into the downstroke, stored energy reaches a maximum, and at the same time, resistance to wing movement suddenly decreases. The wing then "clicks" through the rest of the cycle, using energy stored in the exoskeleton. Asynchrony of this flight mechanism arises from the lack of one-to-one correspondence between nerve impulses and wingbeats. A single nerve impulse can result in approximately 50 cycles of the wing, and frequencies of 1,000 cycles per second (cps) have been recorded in some midges (order Diptera)! The asynchrony between wingbeat and nerve impulses is dependent on flight muscles being stretched during the "click" of the thorax. The stretching of longitudinal flight muscles during the upward beat of the wing initiates the subsequent contraction of these muscles. Similarly, stretching during the downward beat of the wing initiates subsequent contraction of dorsoventral flight muscles. Indirect flight muscles are frequently called **fibrillar flight muscles.** Flies (order Diptera) and wasps (order Hymenoptera) are examples of insects with an asynchronous flight mechanism.

Simple flapping of wings is not enough for flight. The tilt of the wing must be controlled to provide lift and horizontal propulsion. In most insects, muscles that control wing tilting attach to sclerotized plates at the base of the wing.

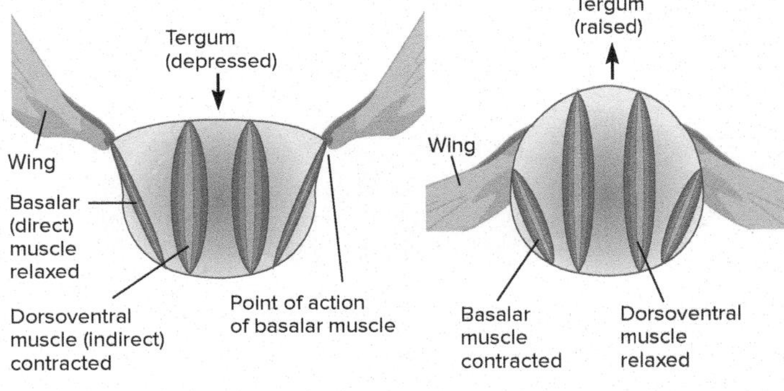

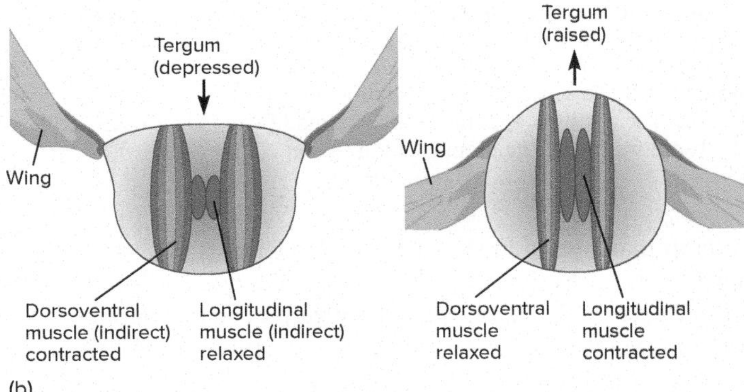

FIGURE 15.12

Insect Flight. (*a*) Muscle arrangements for the direct or synchronous flight mechanism. Note that muscles responsible for the downstroke attach at the base of the wings. (*b*) Muscle arrangements for the indirect or asynchronous flight mechanism. Muscles changing the shape of the thorax cause wings to move up and down.

Other Forms of Locomotion Insects walk, run, jump, or swim across the ground or other substrates. When they walk, insects have three or more legs on the ground at all times, creating a very stable stance. When they run, fewer than three legs may be in contact with the ground. A fleeing cockroach (order Blattaria) reaches speeds of about 5 km/h, although it seems much faster when trying to catch one. The apparent speed is the result of their small size and ability to quickly change directions. Jumping insects, such as grasshoppers (order Orthoptera), usually have long, metathoracic legs in which leg musculature is enlarged to generate large, propulsive forces. Energy for a flea's (order Siphonaptera) jump is stored as elastic energy of the exoskeleton. Muscles that flex the legs distort the exoskeleton. A catch mechanism holds the legs in this "cocked" position until special muscles release the catches and allow the stored energy to quickly extend the legs. This action hurls the flea for distances that exceed 100 times its body length (*see figure 23.11*). A comparable distance for a human long jumper would be the length of two football fields!

Nutrition and the Digestive System

The diversity of insect feeding habits parallels the diversity of insects themselves. There are many variations on mouthparts of insects, but the mouthparts are based on a common arrangement of structures shown in figure 15.13, which shows the biting-chewing mouthparts of a grasshopper. An upper liplike structure is called the labrum. It is sensory, and unlike the remaining mouthparts, is not derived from segmental, paired appendages. Mandibles are sclerotized chewing mouthparts. They usually bear teeth for grinding and cutting and have a side-to-side movement. The maxillae have cutting surfaces and palps that are sensory and food-holding structures. The **labium**

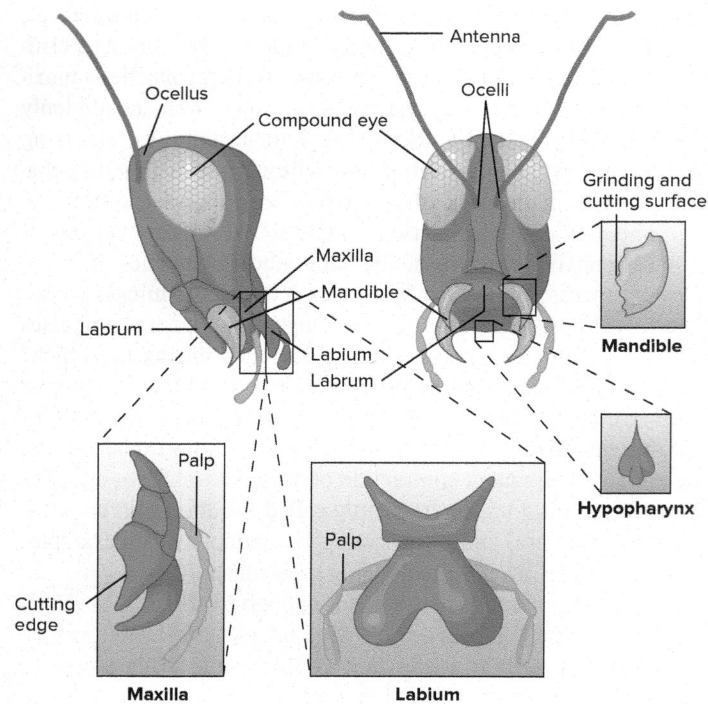

FIGURE 15.13

Head and Mouthparts of a Grasshopper. All mouthparts except the labrum are derived from segmental appendages. The labrum is a sensory upper lip. The mandibles are heavily sclerotized and used for tearing and chewing. The maxillae have cutting edges and a sensory palp. The labium forms a sensory lower lip. The hypopharynx is a sensory, tonguelike structure.

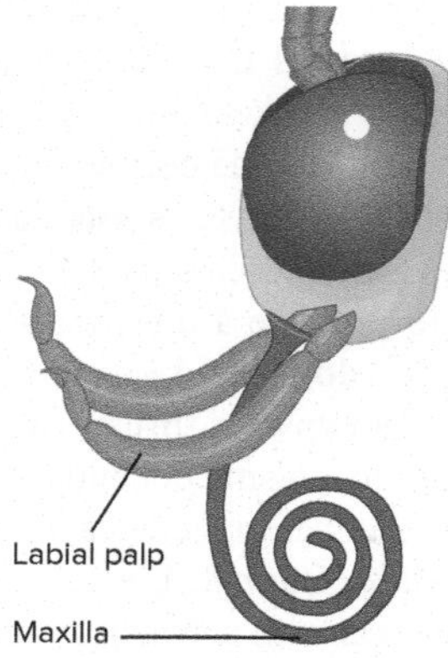

FIGURE 15.14

Specialization of Insect Mouthparts. The mouthparts of insects are often highly specialized for specific feeding habits. For example, the sucking mouthparts of a butterfly consist of modified maxillae that coil when not in use. Mandibles, labia, and the labrum are reduced in size. A portion of the anterior digestive tract is modified as a muscular pump for drawing liquids through the mouthparts.

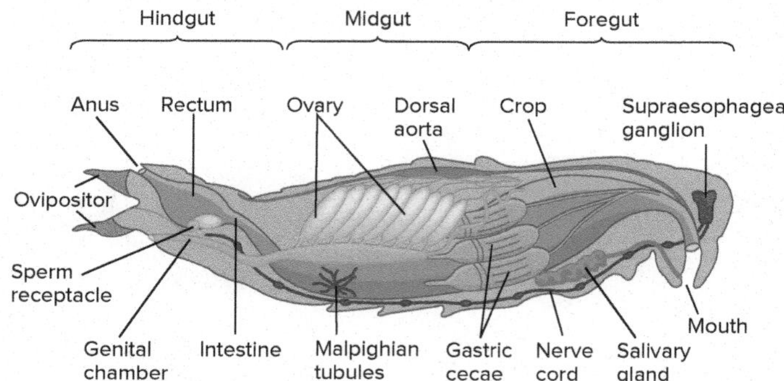

FIGURE 15.15

Internal Structure of a Generalized Insect. Salivary glands produce enzymes but may be modified for the secretion of silk, anticoagulants, or pheromones. The crop is an enlargement of the foregut and stores food. The proventriculus is a grinding and/or straining structure at the junction of the foregut and midgut. The location of the proventriculus is covered by gastric cecae in this illustration. Gastric cecae secrete digestive enzymes. The intestine and the rectum are modifications of the hindgut that absorb water and the products of digestion.

is a sensory lower lip and its palps are also used in holding food. As its structure suggests, the labium forms from an embryological and evolutionary fusion of paired head appendages. A hypopharynx is a tonguelike sensory structure. The efficiency of these biting-chewing mouthparts is obvious in watching a caterpillar feeding on a leaf or considering a termite feeding on wooden structures.

The structure of mouthparts is modified in insects that suck liquid food, although the basic arrangement of mouthparts is retained. There are many variations in sucking mouthparts. In mosquitoes (order Diptera), six stylets are formed from the labrum, hypopharynx, mandibles, and maxillae and are used by females to pierce flesh and suck blood. The labium does not actually penetrate flesh but rests on the skin to form a guide for the other six mouthparts. In the butterflies and moths, the maxillae form a long, coiled tube that is used to suck nectar from flowers (figure 15.14).

The housefly has sponging mouthparts. The labium is expanded into a labellum. Saliva is secreted from the mouth, and minute channels on the labellum provide pathways for liquefied food to move into the mouth through capillary action.

The digestive tract, as in all arthropods, is long and straight and consists of three regions: a foregut, a midgut, and a hindgut (figure 15.15). The foregut is often modified into a muscular pharynx. In sucking insects, the pharynx (or in some, the oral cavity) is used for sucking fluids into the digestive tract. Behind the pharynx is a crop that is used in storage. A proventriculus or gizzard regulates movement to the midgut and may function in grinding food. The midgut provides the surfaces for digestion and absorption, and gastric cecae increase the surface area for these functions. The hindgut or intestine is primarily involved with the reabsorption of water.

Gas Exchange

Gas exchange with air requires a large surface area for the diffusion of gases. In terrestrial environments, these surfaces are also avenues for water loss. Respiratory water loss in insects, as in some arachnids, is reduced through the invagination of respiratory surfaces to form highly branched systems of chitin-lined tubes, called tracheae.

Tracheae open to the outside of the body through spiracles, which usually have a chamber called the atrium, a dust filtering mechanism, and some kind of closure device to prevent excessive water loss. Spiracles lead to tracheal trunks that branch, eventually giving rise to smaller branches, the tracheoles. Taenidia are internal rings or spiral thickenings of tracheal trunks that keep tracheae from collapsing and allow lengthwise expansion with body movements. Tracheoles end intracellularly and are especially abundant in metabolically active tissues, such as flight muscles. No cells are more than 2 or 3 μm from a tracheole (figure 15.16).

Most insects have ventilating mechanisms that move air into and out of the tracheal system. For example, contracting flight muscles alternately compress and expand the larger tracheal trunks and thereby ventilate the tracheae. In some insects, carbon dioxide that metabolically active cells produce is sequestered in the hemocoel as bicarbonate ions (HCO_3^-). As oxygen diffuses from the tracheae to the body tissues, and is not replaced by carbon dioxide, a vacuum is created that draws more air into the spiracles. This process is called passive suction. Periodically, the sequestered bicarbonate ions are converted back into carbon dioxide, which escapes through the tracheal system. Other insects contract abdominal muscles in a pumplike fashion to move air into and out of their tracheal systems.

In many aquatic insects, spiracles are nonfunctional and gases diffuse across the body wall. In others (some aquatic coleopterans and hemipterans), a bubble of air covers the spiracles and is carried underwater with the insect, which may periodically surface to refresh the air. Alternatively, gases may diffuse into and out of the bubble directly from the water. Some aquatic insects (immature mayflies and some immature stoneflies) have tracheal gills. Gases diffuse across the gill surface into branches of the tracheal system that extend into the gills.

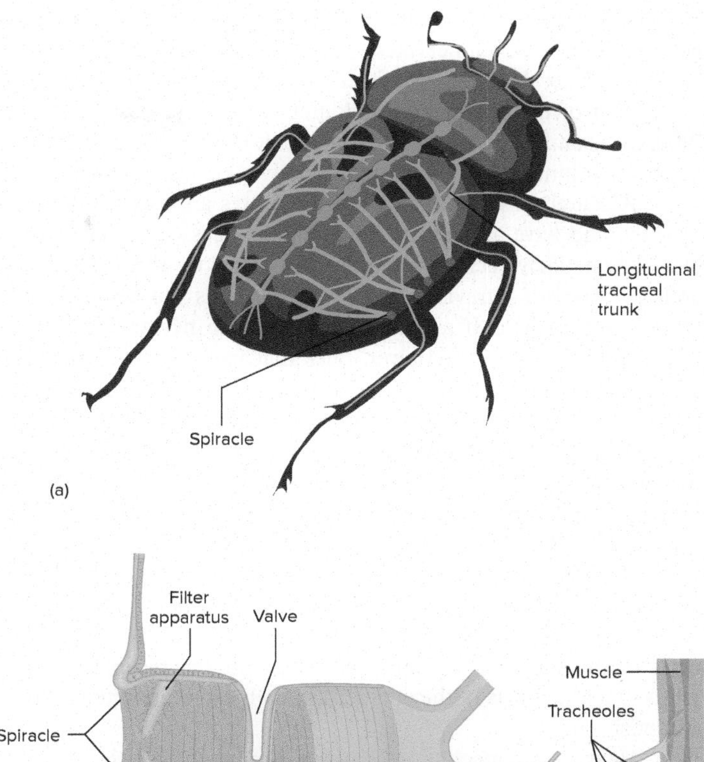

FIGURE 15.16

Tracheal System of an Insect. (*a*) Major tracheal trunks. (*b*) Tracheoles end in cells, and the terminal portions of tracheoles are fluid filled.

Circulation and Temperature Regulation

The circulatory system of insects, like that of all other arthropods, is an open system in which the dorsal contractile vessel (the "heart") pumps blood into tissue spaces of the hemocoel. Blood bathes tissues and then returns to the dorsal aorta through ostia. Blood distributes nutrients, hormones, and wastes, and amoeboid blood cells participate in body defense and repair mechanisms. Blood is not important in gas transport.

As described earlier, thermoregulation is a requirement for flying insects. Virtually all insects warm themselves by basking in the sun or resting on warm surfaces. Because they use external heat sources in temperature regulation, insects are generally considered ectotherms. Other insects (e.g., some moths, alpine bumblebees, and beetles) can generate heat by rapid contraction of flight muscles, a process called shivering thermogenesis. Metabolic heat generated in this way can raise the temperature of thoracic muscles from near 0 to 35°C. Because some insects rely to a limited extent on metabolic heat sources, they have a variable body temperature and are sometimes called heterotherms. Insects can also cool themselves by seeking cool, moist habitats.

Honeybees (*Apis*, order Hymenoptera) regulate their own body temperature and the temperature of their hive. Honeybees need an internal body temperature of 35°C for flight. Contraction of flight muscles generates heat, and regurgitation of fluid through the mouth for evaporative cooling regulates body temperature to the ideal 35°C. Honeybees require that 35°C also be maintained within the hive to sustain larval development and form wax. Hive temperature is increased by the formation of tight clusters of bees within the hive. Bees in the clusters generate heat through muscle contraction, and heat can be dissipated from the hive as bees beat their wings at the entrance of the hive, thus circulating cooler outside air through the hive. In the winter, heat is generated within the hive by "winter clusters." Shivering thermogenesis within these clusters maintains hive temperatures to no lower than about 20°C in spite of subfreezing outside air temperatures.

Nervous and Sensory Functions

The nervous system of insects is similar to the pattern described for annelids and other arthropods (*see figure 15.15*). The supraesophageal ganglion is associated with sensory structures of the head.

Connectives join the supraesophageal ganglion to the subesophageal ganglion, which innervates the mouthparts and salivary glands and has a general excitatory influence on other body parts. Segmental ganglia of the thorax and abdomen fuse to various degrees in different taxa. Insects also possess a well-developed visceral nervous system that innervates the gut, reproductive organs, and heart.

Research has demonstrated that insects are capable of some learning and have a memory. For example, bees (order Hymenoptera) instinctively recognize flowerlike objects by their shape and ability to absorb ultraviolet light, which makes the center of the flower appear dark. If a bee is rewarded with nectar and pollen, it learns the odor of the flower. Bees that feed once at artificially scented feeders choose that odor in 90% of subsequent feeding trials. Odor is a very reliable cue for bees because it is more constant than color and shape. Wind, rain, and herbivores may damage the latter.

Sense organs of insects are similar to those found in other arthropods, although they are usually specialized for functioning on land. Mechanoreceptors perceive physical displacement of the body or of body parts. Setae are distributed over the mouthparts, antennae, and legs (*see figure 14.10*a). Touch, air movements, and vibrations of the substrate can displace setae. Stretch receptors at the joints, on other parts of the cuticle, and on muscles monitor posture and position.

Hearing is a mechanoreceptive sense in which airborne pressure waves displace certain receptors. All insects can respond to pressure waves with generally distributed setae; others have specialized receptors. For example, **Johnston's organs** are in the base of the antennae of most insects, including mosquitoes and midges (order Diptera). Long setae that vibrate when certain frequencies of sound strike them cover the antennae of these insects. Vibrating setae move the antenna in its socket, stimulating sensory cells. Sound waves in the frequency range of 500 to 550 cycles per second (cps) attract and elicit mating behavior in the male mosquito *Aedes aegypti.* These waves are in the range of sounds that the wings of females produce. **Tympanal (tympanic) organs** are in the legs of crickets and katydids (order Orthoptera), in the abdomen of grasshoppers (order Orthoptera) and some moths (order Lepidoptera), and in the thorax of other moths. Tympanal organs consist of a thin, cuticular membrane covering a large air sac. The air sac acts as a resonating chamber. Just under the membrane are sensory cells that detect pressure waves. Grasshopper tympanal organs can detect sounds in the range of 1,000 to 50,000 cps. (The human ear can detect sounds between 20 and 20,000 cps.) Bilateral placement of tympanal organs allows insects to discriminate the direction and origin of a sound.

The tympanal organs in moths of the family Noctuidae are sensitive to sounds in the 3,000- to 150,000-cps frequency range. This range encompasses the ultrasonic frequencies (sound frequencies too high to be heard by humans) emitted by bats using echolocation to find their prey. During echolocation, bats emit ultrasonic sounds that reflect from flying insects back to the bats' unusually large external ears. Using this information, bats can determine the exact location of an insect and can even distinguish the kind of insect. The tympanal organs of a noctuid moth can determine both distance from the bat and direction of the sounds as the sounds bounce off the moth's body. The bilateral placement of tympanal organs means that sound arriving from the moth's right side strikes the right tympanal organ more strongly because the moth's body shades the left tympanal organ from the sound. Thus, the moth can determine the approximate location of the predator and take evasive action.

Insects use chemoreception in feeding, selection of egg-laying sites, mate location, and, sometimes, social organization. Chemoreceptors are usually abundant on the mouthparts, antennae, legs, and ovipositors, and take the form of hairs, pegs, pits, and plates that have one or more pores leading to internal nerve endings. Chemicals diffuse through these pores and bind to and excite nerve endings.

All insects are capable of detecting light and may use light in orientation, navigation, feeding, or other functions. **Compound eyes** are well developed in most adult insects. They are similar in structure and function to those of other arthropods, and recent evidence points to their homology (common ancestry) with those of crustaceans. Compound eyes consist of a few to 28,000 receptors, called **ommatidia,** that fuse into a multifaceted eye. The outer surface of each ommatidium is a lens and is one facet of the eye. Below the lens is a crystalline cone. The lens and the crystalline cone are light-gathering structures. Certain cells of an ommatidium, called retinula cells, have a special light-collecting area, called the rhabdom. The rhabdom converts light energy into nerve impulses. Pigment cells surround the crystalline cone, and sometimes the rhabdom, and prevent the light that strikes one rhabdom from reflecting into an adjacent ommatidium (figure 15.17).

Although many insects form an image of sorts, the concept of an image has no real significance for most species. The compound eye is better suited for detecting movement. Movement of a point of light less than 0.1° can be detected as light successively strikes adjacent ommatidia. For this reason, bees are attracted to flowers blowing in the wind, and predatory insects select moving prey.

Compound eyes detect wavelengths of light that the human eye cannot detect, especially in the ultraviolet end of the spectrum. In the honeybee compound eye, two rhabdoms in each ommatidium are specialized for detecting ultraviolet (UV) radiation. Many flowers observed under UV illumination take on a bull's-eye pattern, which is a common mutualistic (*see chapter 6*) adaptation that leads pollinating insects to the reproductive organs of the flower (figure 15.18). One rhabdom of each ommatidium of the honeybee eye can detect the direction of polarization of UV radiation. Bees navigate to and from food sources based on the direction of the food source relative to the angle of incoming sunlight. Having the ability to detect the direction of polarization of UV radiation allows honeybees to navigate even when the sun is obscured by clouds.

Ocelli consist of 500 to 1,000 receptor cells beneath a single cuticular lens (*see figure 14.10*b). Ocelli are sensitive to changes in light intensity and may be important in the regulation of daily rhythms.

The complexity of some insect behavior is deceptive. It may seem as if insects make conscious decisions in their actions; however, this is seldom the case. As with a noctuid moth's evasive responses to a bat's cries and a honeybee's ability to navigate, most insect behavior patterns are reflexes programmed by specific interconnections of nerve cells.

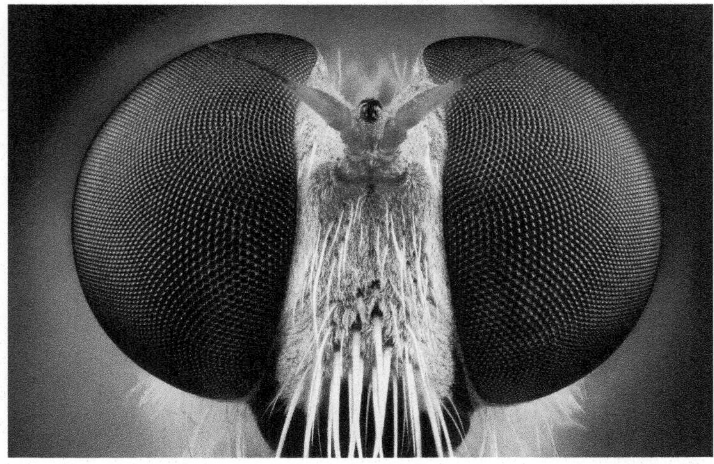

(a)

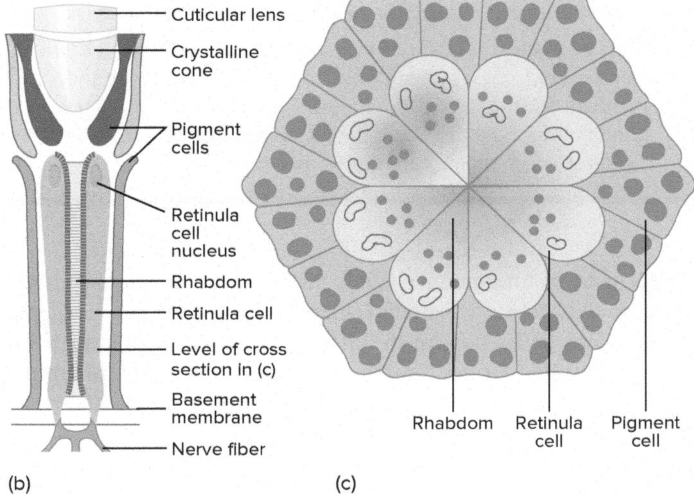

FIGURE 15.17

Compound Eye of an Insect. (*a*) Compound eye of a robber fly (Asilidae). (*b*) Structure of an ommatidium. The lens and the crystalline cone are light-gathering structures. Retinula cells have light-gathering areas, called rhabdoms. Pigment cells prevent light in one ommatidium from reflecting into adjacent ommatidia. In insects that are active at night, the pigment cells are often migratory, and pigment can be concentrated around the crystalline cone. In these insects, low levels of light from widely scattered points can excite an ommatidium. (*c*) Cross section through the rhabdom region of an ommatidium.

Excretion

The primary insect excretory structures are the Malpighian tubules and the rectum. Malpighian tubules end blindly in the hemocoel and open to the gut tract at the junction of the midgut and the hindgut. Microvilli cover the inner surfaces of their cells. Various ions are actively transported into the tubules, and water passively follows. Uric acid is secreted into the tubules and then into the gut, as are amino acids and ions (figure 15.19). In the rectum, water, certain ions, and other materials are reabsorbed, and the uric acid is eliminated. Malpighian tubules of hexapods are not homologous to those found in some arachnids. In the latter, they attach at the hindgut.

FIGURE 15.18

Ultraviolet Vision in Insects. Many flowers observed under UV illumination take on a bull's-eye pattern, which is a common mutualistic adaptation that leads pollinating insects to the reproductive organs of the flower. The upper photo is of a marsh marigold (*Caltha palustris*) under normal illumination. The lower photo is of the same flower using an ultraviolet filter to simulate what an insect might see.

As described in chapter 14, the excretion of uric acid is advantageous for terrestrial animals because it minimizes water loss. There is, however, an evolutionary trade-off. The conversion of primary nitrogenous wastes (ammonia) to uric acid is energetically costly. Nearly half of the food energy a terrestrial insect consumes may be used to process metabolic wastes! In aquatic insects, ammonia simply diffuses out of the body into the surrounding water.

Chemical Regulation

The endocrine system controls many physiological functions of insects, such as cuticular sclerotization (*see chapter 14*), osmoregulation, egg maturation, cellular metabolism, gut peristalsis, and heart rate. As in all arthropods, ecdysis is under neuroendocrine control. In insects, the dorsolateral portions of the supraesophageal ganglion and two endocrine glands—the corpora allata and the prothoracic glands—control these activities (*see figure 25.7*).

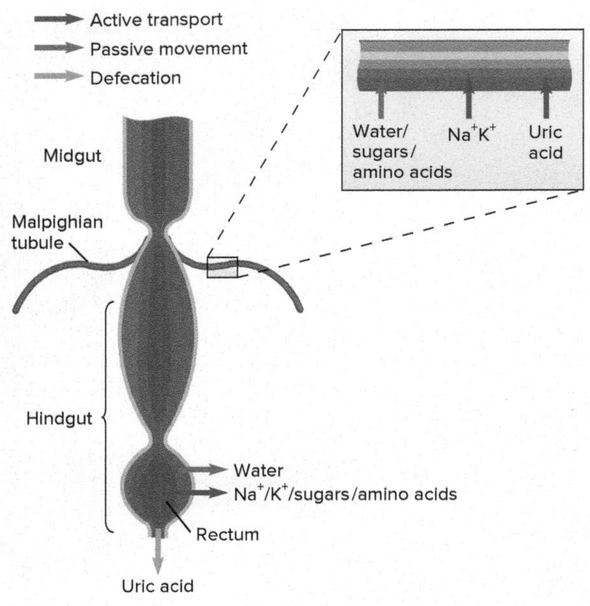

FIGURE 15.19

Insect Excretion. Malpighian tubules remove nitrogenous wastes from the hemocoel. Various ions are actively transported across the outer membranes of the tubules. Water follows these ions into the tubules and carries amino acids, sugars, and some nitrogenous wastes along passively. Some water, ions, and organic compounds are reabsorbed in the basal portion of the Malpighian tubules and the hindgut; the rest are reabsorbed in the rectum. Uric acid moves into the hindgut and is excreted.

TABLE 15.2

FUNCTIONS OF INSECT PHEROMONES

Sex pheromones–Excite or attract members of the opposite sex; accelerate or retard sexual maturation. Example: Female moths produce and release pheromones that attract males.

Caste-regulating pheromones–Used by social insects to control the development of individuals in a colony. Example: The amount of "royal jelly" fed to a female bee larva determines whether the larva will become a worker or a queen.

Aggregation pheromones–Produced to attract individuals to feeding or mating sites. Example: Certain bark beetles aggregate on pine trees during an attack on a tree.

Alarm pheromones–Released to warn other individuals of danger; may cause orientation toward the pheromone source and elicit a subsequent attack or flight from the source. Example: A sting from one bee alarms other bees in the area, who are likely to attack.

Trailing Pheromones–Laid down by foraging insects to help other members of a colony identify the location and quantity of food found by one member of the colony. Example: Ants often trail on a pheromone path to and from a food source. The pheromone trail is reinforced each time an ant travels over it.

Neurosecretory cells of the supraesophageal ganglion manufacture prothoracicotropic hormone (PTTH). This hormone travels in neurosecretory cells through a pair of structures called the corpora cardiaca to the corpora allata. The corpora allata then release PTTH to the hemolymph, which stimulates the prothoracic gland to secrete ecdysone. Ecdysone initiates the reabsorption of the inner portions of the procuticle and the formation of the new exoskeleton. Chapter 25 discusses these events further. Other hormones are also involved in ecdysis. The recycling of materials absorbed from the procuticle, changes in metabolic rates, and pigment deposition are a few of probably many functions that hormones control.

In immature stages, the corpora allata produces and releases juvenile hormone. The amount of juvenile hormone circulating in the hemocoel determines the nature of the next molt. Large concentrations of juvenile hormone result in a molt to a second immature stage, intermediate concentrations result in a molt to a third immature stage, and low concentrations result in a molt to the adult stage. Decreases in the level of circulating juvenile hormone also lead to the degeneration of the prothoracic gland so that, in most insects, molts cease once adulthood is reached. Interestingly, after the final molt, the level of juvenile hormone increases again, but now it promotes the development of accessory sexual organs, yolk synthesis, and the egg maturation.

Pheromones are chemicals an animal releases that cause behavioral or physiological changes in another member of the same species (*see figure 25.1*e). Zoologists have described many different insect uses of pheromones (table 15.2). Pheromones are often so specific that the stereoisomer (chemical mirror image) of the pheromone may be ineffective in initiating a response. Wind or water may carry pheromones several kilometers, and a few pheromone molecules falling on a chemoreceptor of another individual may be enough to elicit a response.

Insect Reproduction and Development

One of the reasons for insects' success is their high reproductive potential. Reproduction in terrestrial environments, however, has its risks. Temperature, moisture, and food supplies vary with the season. Internal fertilization requires highly evolved copulatory structures because gametes dry quickly on exposure to air. In addition, mechanisms are required to bring males and females together at appropriate times.

Complex interactions between internal and external environmental factors regulate sexual maturity. Internal regulation includes interactions between endocrine glands (primarily the corpora allata) and reproductive organs. External regulating factors may include the quantity and quality of food. For example, the eggs of mosquitoes (order Diptera) do not mature until after the female takes a meal of blood, and the number of eggs produced is proportional to the quantity of blood ingested. Many insects use the photoperiod (the relative length of daylight and darkness in a 24-h period) for timing reproductive activities because it indicates seasonal changes. Population density, temperature, and humidity also influence reproductive activities.

A few insects, including silverfish (order Thysanura) and springtails (order Collembola), have indirect fertilization. The male deposits a spermatophore that the female picks up later. Most insects have complex mating behaviors for locating and recognizing a potential mate, for positioning a mate for copulation, or for pacifying an aggressive mate. Mating behavior may involve pheromones (moths, order Lepidoptera), visual signals (fireflies, order Coleoptera), and auditory signals (cicadas, order Hemiptera; and grasshoppers, crickets, and katydids, order Orthoptera). Once other stimuli have brought the male and female together, tactile stimuli from the antennae and other appendages help position the insects for mating.

Abdominal copulatory appendages of the male usually transfer the sperm to an outpocketing of the female reproductive tract, the sperm receptacle (*see figure 15.15*). Eggs are fertilized as they leave the female and are usually laid near the larval food supply. Females may use an **ovipositor** to deposit eggs in or on some substrate. Eggs often include a tough capsule and may be laid in groups within tough proteinaceous coverings called oothecae, which permit embryonic development in desiccating environments.

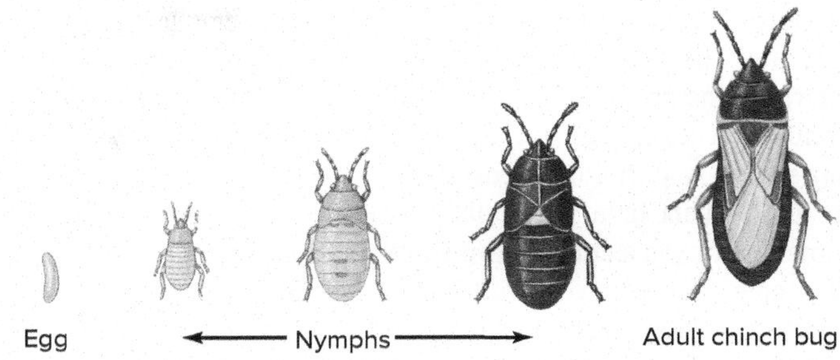

FIGURE 15.20

Hemimetabolous Development of the Chinch Bug, *Blissus leucopterus* (Order Hemiptera). Eggs hatch into nymphs. Note the gradual increase in nymph size and the development of external wing pads. In the adult stage, the wings are fully developed, and the insect is sexually mature.

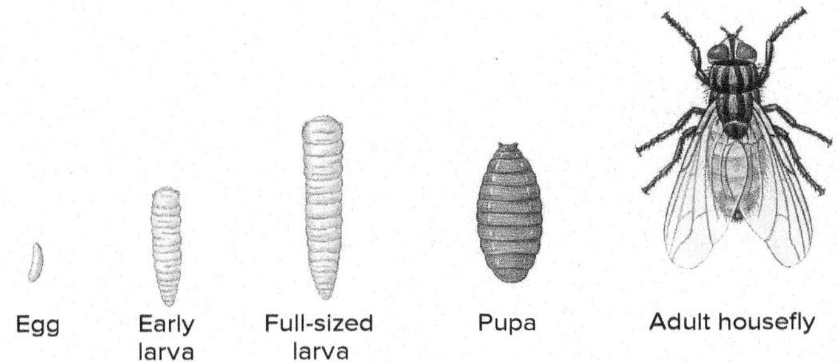

FIGURE 15.21

Holometabolous Development of the Housefly, *Musca domestica* (Order Diptera). The egg hatches into a larva that is different in form and habitat from the adult. After a certain number of larval instars, the insect pupates. During the pupal stage, all adult characteristics form.

Insect Development and Metamorphosis

Insect evolution has resulted in the divergence of immature and adult body forms and habits. Immature stages are a time of growth and accumulation of energy reserves for the transition to adulthood. Immature stages are accompanied by molting from one stage or **instar** to the next. The adult stage is associated with reproduction and dispersal. The evolutionary trend is for insects to spend a greater part of their lives in immature stages. The developmental patterns of insects reflect degrees of divergence between immatures and adults and are classified into three (or sometimes four) categories. Complex interactions between juvenile hormone and ecdysone regulate the transition between immature and adult stages and the differences in the following developmental patterns (*see figure 25.7*).

In insects that display **ametabolous** (Gr. *a,* without + *metabolos,* change) **metamorphosis,** the primary differences between adults and larvae are body size and sexual maturity. Both adults and larvae are wingless. The number of molts in the ametabolous development of a species varies, and unlike most other insects, molting continues after sexual maturity. Silverfish (order Thysanura) have ametabolous metamorphosis. The earliest insects, which resembled silverfish, probably were ametabolous.

After the evolution of flight, early winged insects developed hemimetabolous metamorphosis. **Hemimetabolous** (Gr. *hemi,* half) **metamorphosis** involves a species-specific number of molts between egg and adult stages, during which immatures gradually take on the adult form. The external wings develop (except in those insects, such as lice, that have secondarily lost wings), adult body size and proportions are attained, and the genitalia develop during this time. Immatures are called **nymphs.** Grasshoppers (order Orthoptera) and chinch bugs (order Hemiptera) show hemimetabolous metamorphosis (figure 15.20). When immature stages are aquatic, they often have gills (e.g., mayflies, order Ephemeroptera; dragonflies, order Odonata). These immatures are called **naiads** (L. *naiad,* water nymph).

The most successful (in terms of numbers of species) of all insects display holometabolous metamorphosis. In **holometabolous** (Gr. *holos,* whole) **metamorphosis,** immatures are called larvae because they are very different from the adult in body form, behavior, and habitat (figure 15.21). Larval stages are usually times of voracious feeding, rapid growth, and molting between instars. The number of larval instars is species specific, and the last larval molt forms the **pupa.** The pupa is a time of apparent inactivity but is actually a time of radical cellular change, during which all characteristics of the adult insect develop. A protective case may enclose the pupal stage. The last larval instar (e.g., moths, order Lepidoptera) constructs a **cocoon** partially or entirely from silk. The **chrysalis** (e.g., butterflies, order Lepidoptera) and **puparium** (e.g., flies, order Diptera) are the last larval exoskeletons and are retained through the pupal stage. Other insects (e.g., mosquitoes, order Diptera) have pupae that are unenclosed by a larval exoskeleton, and the pupa may be active. The final molt to the adult stage usually occurs within the cocoon, chrysalis, or puparium, and the adult then exits, frequently using its mandibles to open the cocoon or other enclosure. This final process is called emergence or eclosion. Adult stages are primarily devoted to sexual reproduction.

Seventy-five percent or more of all described insect species have holometabolous metamorphosis (*see table 15.1*). Two evolutionary advantages of holometabolous metamorphosis may help explain the success of these groups. First, holometabolous metamorphosis often reduces competition between larval instars and adults. The radically different body form and feeding habits of a caterpillar as compared to the adult butterfly mean that the larval and adult stages do not compete with each other for food or living space. A particular ecosystem can therefore support many more individuals of a particular species. Second, one stage of a holometabolous insect may have an adaptation for a particular role, for example, dispersal, overwintering, or survival during very hot or very dry conditions. For instance, the pupae of some moths (order Lepidoptera, family Saturniidae) enter an arrested stage of development when the period of daylight in a 24-h period drops below 14 h. This physiological state of dormancy triggered by environmental conditions is called **diapause.** A translucent cuticle in the pupa allows it to detect when the period of daylight again exceeds 14 h and development resumes. Warming after a period of cold also signals the end of diapause. Thus, the pupal stage in these moths serves as the overwintering stage.

Insect Behavior

Insects have many complex behavior patterns. Most of these are innate (genetically programmed). For example, a newly emerged queen in a honeybee hive will search out and try to destroy other queen larvae and pupae in the hive. This behavior is innate because no experiences taught the potential queen that her survival in the hive required the death or dispersal of all other potential queens. Similarly, no experience taught her how queen-rearing cells differ from the cells containing worker larvae and pupae. Some insects are capable of learning and remembering, and these abilities play important roles in insect behavior.

Social Insects

Social behavior has evolved in many insects and is particularly evident in those insects that live in colonies. Usually, different members of the colony are specialized, often structurally as well as behaviorally, for performing different tasks. Social behavior is most highly evolved in the bees, wasps, and ants (order Hymenoptera) and in termites (order Isoptera). Each kind of individual in an insect colony is called a **caste.** Often, three or four castes are present in a colony. Reproductive females are referred to as queens. Workers may be sterile males and females (termites) or sterile females (order Hymenoptera), and they support and protect the colony. Their reproductive organs are often degenerate. Reproductive males inseminate the queen(s) and are called kings or drones. Soldiers are usually sterile and may possess large mandibles to defend the colony.

Honeybees (*Apis*, order Hymenoptera) have three of these castes in their colonies (figure 15.22). A single queen lays all the eggs. Workers are female, and they construct the comb out of wax that they produce. They also gather nectar and pollen, feed the queen and drones, care for the larvae, and guard and clean the hive. These tasks are divided among workers according to age. Younger workers take care of jobs around the hive, and older workers forage for nectar and pollen. Except for those that overwinter, workers live for about one month. Drones develop from unfertilized eggs, do not work, and are fed by workers until they leave the hive to attempt mating with a queen.

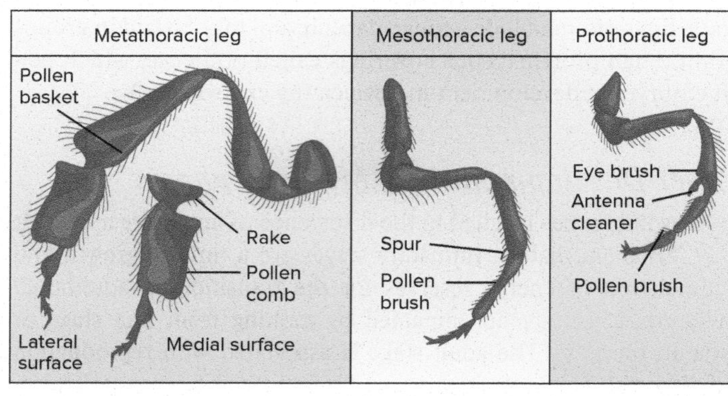

FIGURE 15.22

Honeybees (*Apis*, Order Hymenoptera). Honeybees have a social organization consisting of three castes. Eye size and overall body size distinguish the castes. (*a*) A worker bee. (*b*) A drone bee. (*c*) A queen bee marked with blue to identify her. (*d*) The inner surface of metathoracic legs have setae, called the pollen comb, that remove pollen from the mesothoracic legs and the abdomen. Pollen is then compressed into a solid mass by being squeezed in a pollen press and moved to a pollen basket on the outer surface of the leg, where the pollen is carried. The mesothoracic legs gather pollen from body parts. The prothoracic legs of a worker bee clean pollen from the antennae and body.

A pheromone that the queen releases controls the honeybee caste system. Workers lick and groom the queen and other workers. In so doing, they pick up and pass to other workers a caste-regulating pheromone. This pheromone inhibits the workers from rearing new queens. As the queen ages, or if she dies, the amount of caste-regulating pheromone in the hive decreases. As the pheromone decreases, workers begin to produce and feed the food for queens ("royal jelly") to several female larvae developing in the hive. This food contains chemicals that promote the development of queen characteristics. The larvae that receive royal jelly develop into queens, and as they emerge, the new queens begin to eliminate each other until only one remains. The queen that remains goes on a mating flight and returns to the colony, where she lives for several years.

The evolution of social behavior involving many individuals leaving no offspring, and sacrificing individuals for the perpetuation of the colony may seem contrary to natural selection, where the driving force in life is getting one's own alleles into the next generation. In fact, a worker that tends the eggs of the queen is passing more of her genes on to the next generation than if she reared her own young in a typical diploid mating system.

Many Hymenoptera (bees, wasps, and ants) have a haplodiploid sex-determination system. This system has males that are haploid, having developed from an unfertilized egg. Males have a mother, but no father. Females develop from fertilized eggs, with half of their genes coming from their mother and the other half coming from their father. All daughters of that father receive an identical set of genes from the father's sperm cells because he is haploid and his sperm cells are produced by mitosis. The other half of a daughter's genes are from her mother (the queen). The combination of genes in an egg is produced by meiosis and displays more variability (*see chapter 3*). On average, two daughters produced in a haplodiploid sex-determination system will share 75% of their genes because all daughters share the exact same set of paternal genes. Two daughters produced in a normal diploid sex-determination system will share 50% of their genes because gametes in both parents are produced through meiosis. These percentages mean that a worker bee whose tasks help ensure the survival of her mother's (the queen) offspring contributes a greater proportion of its common genes to the next generation than if the worker bee could mate in a normal diploid sex-determination system. The evolutionary theory that helps us understand these processes, which select for the survival of close relatives at a cost to an individual's own survival, is called **kin selection.**

Insects and Humans

Only about 0.5% of insect species adversely affect human health and welfare. Many others have provided valuable services and commercially valuable products, such as wax, honey, and silk, for thousands of years. Insects are responsible for the pollination of approximately 65% of all plant species. Insects and flowering plants have coevolutionary relationships that directly benefit humans. The annual value of insect-pollinated crops is estimated at $29 billion per year in the United States.

In the fall of 2006, beekeepers began reporting a mysterious disappearance of bees from their hives. Given that honeybees pollinate about one-third of the crop plants that make up the diet of the population of the United States, the problem should concern us all. Researchers have not identified a single cause of Colony Collapse Disorder (CCD). Instead, there seems to have been a "perfect storm" of problems. The combined effects of viral and bacterial pathogens, parasitic mites, management stress, and environmental stress contribute synergistically to CCD. Management and environmental stress factors include inadequate supplemental nutrition and hive overcrowding during winter months and during hive transport, low diversity of pollen and nectar sources, and pesticide exposure. The CCD problem is continuing today, but it is somewhat lessened by an effort to increase the number of bee colonies to offset losses. In 2017, the number of bee colonies in North America, Europe, and Australia increased about 3%; and worldwide, the number of managed colonies (80 million) has remained relatively stable over the past 20 years.

Insects are also agents of biological control. The classic example of one insect regulating another is the vedalia (lady bird) beetles' control of cottony-cushion scale. The scale insect *Icerya purchasi* was introduced into California in the 1860s. Within 20 years, the citrus industry in California was virtually destroyed. The vedalia beetle (*Rodolia cardinalis*) was brought to the United States in 1888 and 1889 and cultured on trees infested with scale. In just a few years, the scale was under control, and the citrus industry began to recover.

Many other insects are also beneficial. Soil-dwelling insects play important roles in aeration, drainage, and turnover of soil, and they promote decay processes. Other insects serve important roles in food webs. Insects are used in teaching and research, and have contributed to advances in genetics, population ecology, and physiology. Insects have also given endless hours of pleasure to those who collect them and enjoy their beauty.

Some insects, however, are parasites and vectors of disease. Parasitic insects include head, body, and pubic lice (order Anoplura); bedbugs (order Hemiptera); and fleas (order Siphonaptera). Other insects transmit disease-causing microorganisms, nematodes, and flatworms. Insect-transmitted diseases, such as malaria, yellow fever, bubonic plague, encephalitis, leishmaniasis, and typhus, have changed the course of history.

Other insects are pests of domestic animals and plants. Some reduce the health of domestic animals and the quality of animal products. Insects feed on crops and transmit plant diseases, such as Dutch elm disease, potato virus, and asters yellow. Annual lost revenue from insect damage to crops or insect-transmitted diseases in North America is approximately $7.5 billion.

Insect Orders: The Big Four

Describing the details of structure, function, diversity, and ecological relationships of 30 insect orders is impossible in a general zoology textbook (*see table 15.1*). There are, however, four orders of insects that, by virtue of the number of species they contain, are of overriding importance. These orders are Coleoptera (the beetles), Lepidoptera (the moths and butterflies), Diptera (the flies), and Hymenoptera (the ants, bees, and wasps).

Order Coleoptera: The Beetles

Members of the order Coleoptera (ko-le-op′ter-ah) (Gr. *koleos*, sheath + *ptera*, wing) are the beetles. They are holometabolous insects and comprise 40% of all insect species and 25% of all species of organisms. There are approximately 350,000 described species, but entomologists estimate there are many more undescribed species (5 million or more), especially in tropical rain forest ecosystems. They live in all natural habitats except polar ecosystems. One family, the rove beetles (Staphylinidae), has species that have become adapted to intertidal marine environments.

Adults are characterized by a very hard exoskeleton and hardened, protective forewings (the elytra) that usually cover the dorsal

FIGURE 15.23

Order Coleoptera. Bombardier beetles (*Brachinus*) expel chemicals from two glands in their abdomen. Mixing of these chemicals, along with catalysts, creates a violent exothermic reaction that can be fatal to predatory insects and cause burns to human skin.

aspect of the thorax when the insect is not in flight. While flying, the forewings are raised to allow the hindwings to expand and power flight. Depending on the species, both adults and larvae consume a variety of living and dead plant and animal material. June beetle (Scarabaeidae, *Phyllophaga*) larvae feed on roots of grasses and other plants and adults feed on foliage of trees and shrubs. Most rove beetles (Staphylinidae) are predators of other insects and invertebrates as both larvae and adults. Larvae are voracious feeders that eventually pupate, usually within the larval habitat. Adults and larvae employ a variety of defensive strategies, including camouflage, mimicry, toxicity, and more active defense. Bombardier beetles (figure 15.23) expel chemicals from two glands in their abdomen. One gland produces hydroquinone and the other hydrogen peroxide. Mixing of these chemicals, along with catalysts, creates a violent exothermic reaction that expels boiling, foul-smelling products with a loud popping sound. These reactions can be fatal to predatory insects and cause burns to human skin.

Some beetles are considered pests of agricultural crops. The Colorado potato beetle (Chrysomelidae, *Leptinotarsa decemlineata*) can completely defoliate potato plants, has caused millions of dollars in damage to the potato crop in the United States and Europe annually, and has acquired significant resistance to chemicals used in its control. The boll weevil (Curculionidae, *Anthonomus grandis*) has caused billions of dollars of losses to cotton growers. Beetles as a whole, however, are very beneficial in most ecosystems. They enhance decomposition processes and are predators of other insects and invertebrates. Ladybird beetles (Coccinellidae) are very beneficial predators of aphids in gardens and orchards.

Order Lepidoptera: Moths and Butterflies

Members of the order Lepidoptera (lep-i-dop′ter-ah) (Gr. *lepido*, scale + *ptera*, wing) are the moths and butterflies. They are holometabolous insects and include approximately 180,000 described species. Their larvae are caterpillars, which have chewing mouthparts, three pairs of true thoracic legs, and up to five pairs of abdominal prologs. Caterpillars consume plant tissues and feed almost constantly. Adults usually have scale-covered wings and mouthparts adapted for sucking nectar. The distinction between "moth" and "butterfly" is based on the characteristics that have limited taxonomic usefulness, as some moths are more closely related to butterflies than to other moths. Lepidopterans employ a number of defense strategies. Chemicals ingested and sequestered in tissues during larval feeding often make both larvae and adults unpalatable to predators. This defense strategy is used by the monarch butterfly (Danaidae, *Danaus plexippus*), as chemicals ingested by larvae feeding on milkweeds are retained and protect the adult from predation by birds. Mimicry, in which different species of unpalatable butterflies resemble one another in order to deter predators, is common (*see figure 6.7*). Cryptic coloration (*see chapter 6*), in which caterpillars resemble the leaves or twigs upon which they feed or crawl, is also common.

Lepidopterans have had important influences on human culture and economics. From the young child who gives "butterfly kisses," to serious photographers and painters, to hobbyist collectors, to research scientists, moths and butterflies have inspired millions of people. Moths and butterflies are also important pollinators, contributing millions of dollars to world economies. Some lepidopterans are also serious pests. The codling moth (Tortricidae, *Cydia pomonella*) was introduced from Europe. Its larvae are serious pests of apples, pears, walnuts, and other fruits. The gypsy moth (Lymantriidae, *Lymantria dispar*) was also introduced from Europe in the 1860s by a French scientist interested in silk production by the moth's larvae. It was accidentally released and has become a notorious pest of hardwood (especially oak) forests.

Order Diptera: The Flies

Members of the order Diptera (dip′ter-ah) (Gr. *di*, two + *ptera*, wings) are the flies. They are holometabolous insects and include approximately 125,000 described species (figure 15.24). Flies inhabit almost every imaginable habitat, including extreme habitats like hot springs, saline lakes, tundra ponds, and marine tide pools.

Flies are noted for their visual capability and agility in flight. True flies, as opposed to other insects with "fly" in their name (e.g., mayfly, caddisfly, and dragonfly), are characterized by having a single pair of wings on their mesothorax. The metathorax bears a pair of knoblike structures called halteres that are derived from ancestral metathoracic wings and vibrate rapidly during flight. They act as gyroscopes to provide stability (e.g., prevent tumbling) during flight. In addition, halteres modify wing movements in response to information from the eyes to promote stability during direction change. Adult mouthparts are modified for fluid diets. Dipterans feed by sponging, sucking, or lapping plant or animal fluids. Depending on the group of flies, larvae may either have complete or partially complete head capsules and fleshy bodies or, in the case of maggots, the larval head is completely reduced to a pair of mouth hooks.

Flies occupy all trophic levels. They include herbivores, detritivores, coprophages (dung feeders), predators, and parasites. Many of the flies are beneficial, serving as predators and parasites of other

FIGURE 15.24

Order Diptera. Many flies are important horticultural pests. This melon fly (*Bactrocera cucurbitae*) is an important pest of beans, melons, cucumbers, pumpkins, squashes, tomatoes, and other hosts. The melon fly was introduced to Hawaii from Japan in 1895. Damage is caused by adult females depositing eggs in the vegetative parts of the host and larvae feeding within the plant tissue. Larval feeding often results in secondary bacterial and fungal infections that cause the fruit to rot. Quarantine laws regulate the movement of fruit and vegetables from areas where the fly is established to other regions.
Source: USDA, ARS, Scott Bauer photographer

insects and pollinators of plants. Chocolate lovers can thank the cacao tree midges (various species of *Forcipomyia* and *Euprojoannisia*, Ceratopogonidae) for helping to provide their favorite indugence. These tiny flies are apparently the only pollinators small enough to work their way into the intricate cacao tree flowers. Others are vectors of some of the most devastating of all diseases of humans, such as malaria (*see appendix C*), yellow fever, encephalitis, trypanosomiasis, Zika virus–related disorders (microcephaly and other nervous-system birth defects), filariasis (*see figures 13.11 and 13.12*), and many others. Mosquitoes and the diseases they carry have caused more human misery than any other group of organisms. Mosquito-borne malaria led to the defeat of barbarian invaders of Rome. Later, malaria halted the Roman invasion of Scotland. Yellow fever, also a mosquito-borne disease, halted work on the Panama Canal by the French and allowed the United States to take over the task. The development of antimalarial medicines by the Allies in World War II gave the Allies advantage over the Japanese in the South Pacific. These, and many other, examples suggest that the flies have influenced human history like no other single order of nonhuman animals.

Order Hymenoptera: Ants, Bees, and Wasps

Members of the order Hymenoptera (hi-men-op′ter-ah) (Gr. *hymen*, membrane + *ptera*, wings) are the ants, bees, and wasps. They are holometabolous insects and include approximately 150,000 described species. Hymenopterans inhabit every continent except Antarctica. They are characterized by two pairs of membranous wings coupled by hooklike hamuli. Mouthparts may be modified for feeding on nectar, although mandibles usually remain functional chewing structures. The metathoracic segment is reduced and fused with the first abdominal segment. In some wasps (sawflies and their relatives), the metathorax is broadly joined to the abdomen, but most hymenopterans have a slender "waistlike" second abdominal segment. In males, the posterior end of the abdomen bears complex genitalia. In females, the posterior abdomen bears an ovipositor that may be modified for piercing or sawing plant tissue for egg deposition or for stinging. Larvae of sawflies and their relatives are caterpillarlike, but most hymenopteran larvae are grublike.

The more primitive members of this order are the sawflies, horntails, and their relatives. Sawflies use their sawlike ovipositor for cutting into plant tissue for depositing eggs. Sawfly outbreaks have seriously damaged forests and cultivated plants, for example, wheat. Other members of the order Hymenoptera include bees (Apidae), whose structure, function, colony organization, and economic importance were described earlier (*see figure 15.22*). The order also includes the wasps, hornets, and yellow jackets (Vespidae). Parasitic wasps (Braconidae and Ichneumonidae) lay eggs on the larvae of other insects (usually larvae of coleopterans, lepidopterans, or dipterans (figure 15.25). Ants (Formicidae) occupy every landmass on earth. Like most hymenopterans, ants have highly evolved social systems. The ant social systems, and the accompanying caste systems, are every bit as complex as that described previously for the honeybee. Most of the approximately 12,000 described species of ants are tropical. Noted naturalist, evolutionary biologist, and ant biologist E. O. Wilson estimates that the biomass of ants in the Amazon rain forest exceeds the biomass of all Amazonian vertebrates by a factor of four.

Section 15.3 Thinking Beyond the Facts

How would your life be different one year from now if all insect life were to disappear from the earth today?

15.4 FURTHER PHYLOGENETIC CONSIDERATIONS

LEARNING OUTCOMES

1. Describe the clade Panarthropoda.
2. Explain the evolutionary relationships among the arthropod subphyla.
3. Describe what is known about the origin and early evolution of the insects.

Monophyly of Arthropoda has never been questioned by most zoologists and has been confirmed by recent molecular studies. Relationships within the phylum, and relationships of arthropods to other phyla, are more tentative. Recent molecular data, and reinterpretations of morphological data, are providing new insights into—but not the final word on—arthropod phylogeny.

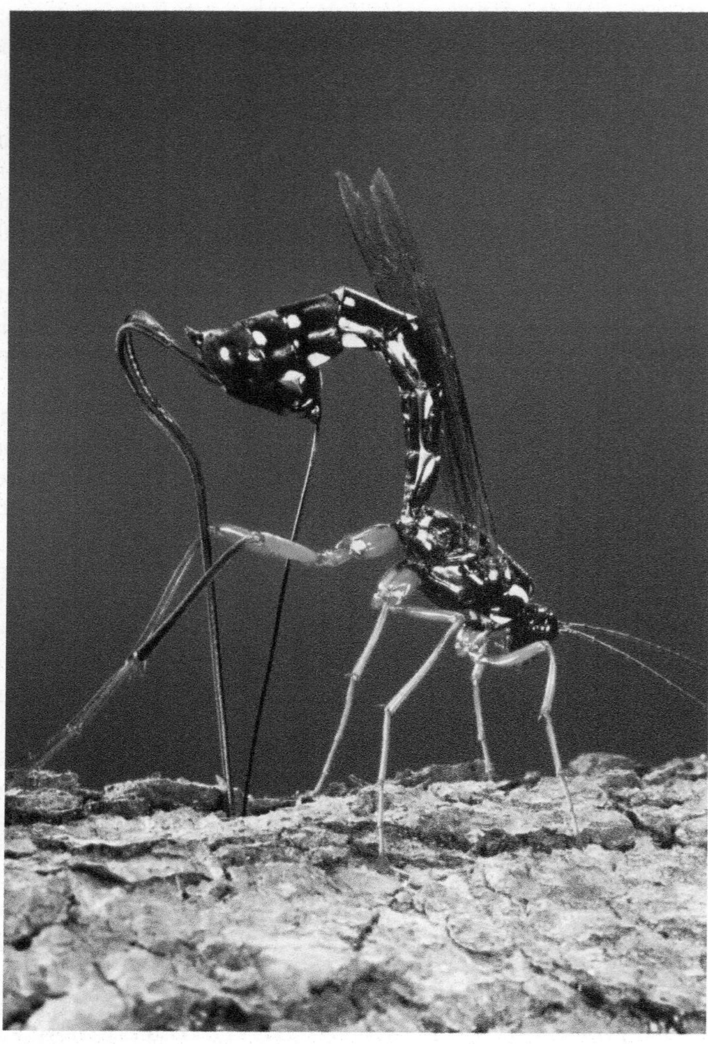

FIGURE 15.25

Order Hymenoptera. Wasps in the family Ichneumonidae are parasites of other insects. This giant ichneumon wasp (*Rhyssa persuasoria*) uses its antennae to locate the larva of its prey, in this case a wood wasp, within a fir tree trunk. She uses her ovipositor to drill into the log and lay eggs on or near a larva. When an egg hatches, the wasp larva enters the host larva, where the wasp larva feeds and eventually pupates. Because the wasp larva virtually always kills its host, it is referred to as a parasitoid.

Clade Panarthropoda

The clade Panarthropoda is composed of Arthropoda and two smaller ecdysozoan phyla: Onychophora and Tardigrada. Molecular evidence suggests that these phyla are sister taxa with the Onychophora being the closest relatives of the arthropods (figure 15.26). In addition, the Onychophora and Tardigrada share a set of morphological characteristics with the arthropods: a similar form of metamerism, metamerically arranged paired appendages, a cuticle shed by ecdysis, and the presence of a hemocoel.

Phylum Onychophora

Members of the phylum Onychophora (on-e-kof′o-rah) are known as velvet or walking worms. They include about 200 free-living species (2–10 cm in length). Extant forms are terrestrial and found in humid, tropical, and subtropical regions. Several well-known fossils have been found in marine fossil formations, including the 508-million-year-old Burgess Shale (*see chapter 8*). They are nocturnal and use their unjointed legs to crawl (figure 15.27). Most species are predaceous and feed on small invertebrates. Onychophorans secrete two streams of adhesive slime up to 30 cm from oral papillae. The slime entangles prey. Salivary secretions are then released into the prey with mandibles, partially predigesting the prey's tissues. The partially predigested meal is then sucked into the mouth. Sexes are separate, fertilization is internal, and development is either oviparous or viviparous.

Phylum Tardigrada

Members of the phylum Tardigrada (tar-di-grad′dah) are known as water bears because of their "teddy bear" appearance and the way they lumber over aquatic vegetation (figure 15.28). They live in marine interstitial areas, in freshwater detritus, and in the water film on terrestrial lichens, liverworts, and mosses. There are about 1,200 species of tardigrades. They usually feed on plant or animal fluids by piercing cells with a pair of oral stylets. Tardigrades can enter a period of suspended animation termed cryptobiosis. This ability offers great survival benefit to these animals, which live in habitats where conditions can suddenly become adverse. If a tardigrade begins to dry out (desiccate), it contracts into a shape that produces an ordered packing of organs and tissues to minimize mechanical damage that the desiccation causes. Rehydration reverses these events. Repeated periods of cryptobiosis can extend a life span of approximately 1 year to 60 to 70 years.

Arthropod Phylogeny

Arthropods were some of the earliest animals to evolve. They are represented in the fossils of the Ediacaran fauna (*see chapter 8*), and therefore date to Precambrian times about 600 mya. Earliest fossil arthropods were not trilobites, but the trilobites appeared in the early Cambrian (*see appendix B*) and comprised the first major lineage to diverge from the ancestral arthropods. Trilobites are considered the sister group to all other arthropods (*see figure 15.26*).

The remaining arthropod lineages also diverged during late Precambrian times. The chelicerates form one monophyletic lineage. They were common in the Ordovician period about 500 mya (*see appendix B*), and by the Silurian period (about 450 mya) chelicerates had become the first land animals.

The pancrustaceans include the crustaceans and the hexapods (*see figure 15.26*). The Pancrustacea is the subject of chapter 15, and, after studying this chapter, you should see obvious differences between the insects and the crustaceans: uniramous versus biramous appendages, terrestrial versus aquatic adaptations, one pair versus two pairs of antennae, wings versus no wings. No wonder

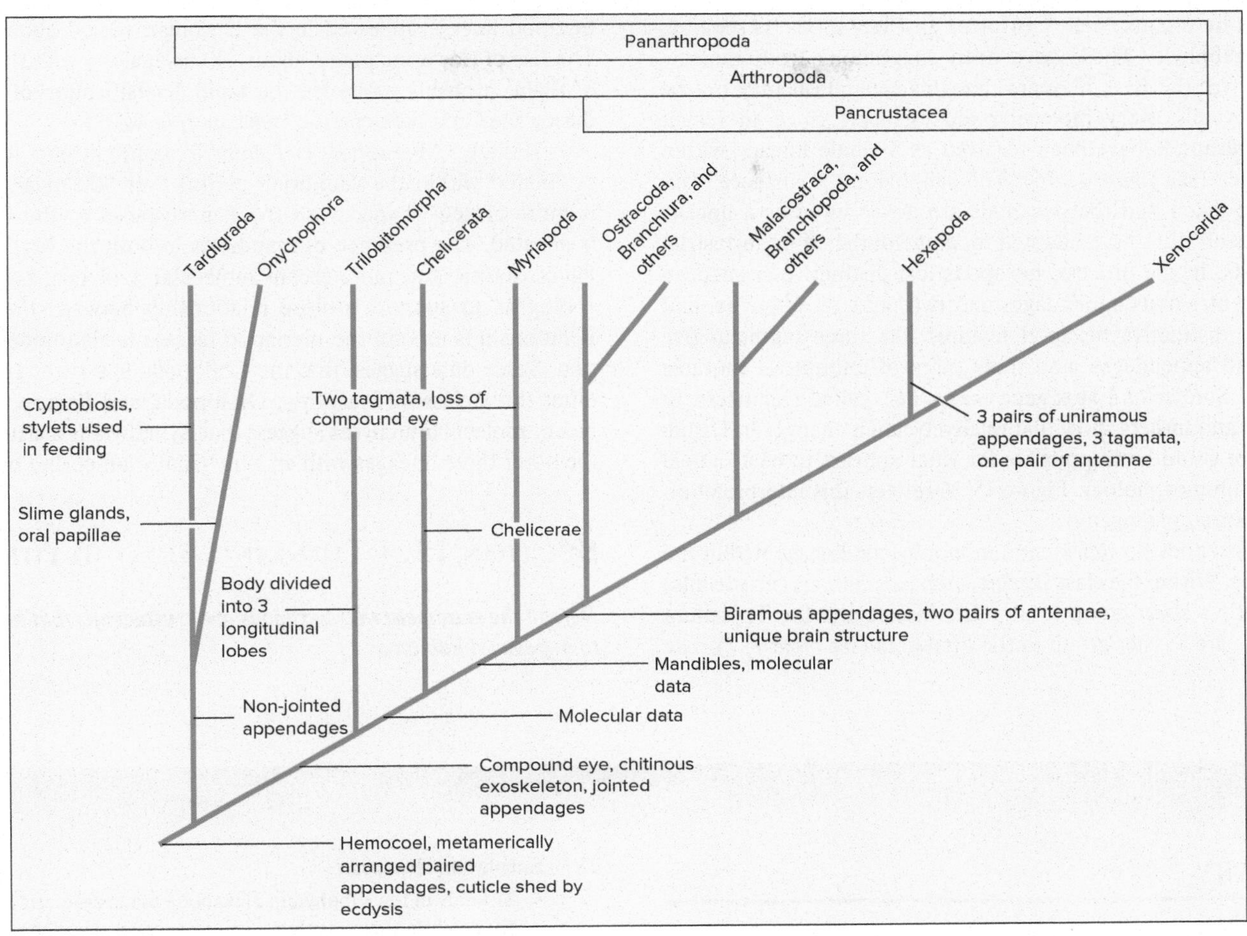

FIGURE 15.26

Phylogeny of the Panarthropoda. This phylogeny is strongly supported by molecular data. A few synapomophic characters discussed in the text are included. Some crustacean lineages have been omitted to simplify the representation. Panarthropoda is comprised of the Tardigrada, Onychophora, and Arthropoda. The Trilobitomorpha is the first arthropod lineage to diverge from ancestral arthropods. The Chelicerata comprises a monophyletic lineage. The Myriapods are probably a sister group to the Pancrustacea (Crustacea and Hexapoda). This phylogeny, if it remains supported by future research, means that the subphylum Crustacea is a paraphyletic group that includes the monophyletic hexapod lineage.

FIGURE 15.27

Phylum Onychophora. *Peripatus* (5 cm).

FIGURE 15.28

Phylum Tardigrada. Rendering from an SEM (0.5 mm).

taxonomists have considered Crustacea and Hexapoda as separate arthropod subphyla. There have been suggestions from molecular studies over the past 10 years that this separation may not be valid. In virtually every molecular analysis performed in recent years, the hexapods have been resolved as a single lineage within the Crustacea (see *figure 15.26*). The subphylum "Crustacea" is a paraphyletic taxon and the hexapods are now viewed as a lineage of "crustaceans" that have adapted to, and flourished in, terrestrial environments. In the process, hexapods lost distinctive crustacean features like biramous appendages and two pairs of antennae, and they gained distinctive hexapod features like three tagmata, five pairs of head appendages, and three pairs of uniramous thoracic appendages. Studies of a Hox gene (*see p. 68*), called Distal-less, in crustaceans and insects show that relatively small changes in Distal-less structure could be responsible for what appears to be a radical change in limb morphology. Figure 15.26 reflects this interpretation of pancrustacean phylogeny.

The hexapods do represent a monophyletic lineage within the Pancrustacea. Within the class Insecta, archaeognathans (bristletails) are apparently a sister group to all other insects and the Thysanura (silverfishes) are a sister group to the Pterogota (*see table 15.1*). The hexapod lineage appeared in the Devonian period about 400 mya. The rise of flowering plants about 130 mya, along with the evolution of flight, probably promoted the rapid diversification of the insects during the Cretaceous period (*see appendix B*).

Finally, Myriapoda is clearly a monophyletic lineage that originated within the Cambrian period over 500 mya. Whether it is more closely aligned with the Pancrustacea or the Chelicerata is debated. The presence of mandibles in both the Myriapoda and Pancrustacea, and more recent molecular evidence, has led some zoologists to support a close relationship between these groups. Relationships among the myriapod classes is also under investigation. Some data suggest that the Chilopoda is a sister group to the other three classes (Symphyla, Diplopoda, and Pauropoda). Other recent molecular analyses suggest that Symphyla is a sister group to the other three lineages with an early Cambrian period origin.

Section 15.4 Thinking Beyond the Facts

Defend the statement that hexapods are crustaceans that have adapted to terrestrial habitats.

Summary

15.1 **Evolutionary Perspective**

- Crustaceans are the dominant arthropods in marine and freshwater environments. Hexapods are the dominant arthropods on land. The exoskeleton and metamorphosis are keys to their success. The evolution of flight was an important adaptation for many hexapods.

15.2 **Subphylum Crustacea**

- The subphylum Crustacea contains animals characterized by two pairs of antennae and biramous appendages. All crustaceans, except for some isopods, are primarily aquatic.
- The class Malacostraca includes the crabs, lobsters, crayfish, shrimp, isopods, and amphipods. This is the largest crustacean class in terms of numbers of species and contains the largest crustaceans. Malacostracans, like most crustaceans, have appendages that are biramous and based on a common ancestral (serially homologous) form. They have an open circulatory system and a ventral nervous system. Their endocrine system controls ecdysis, sex determination, color change, and other functions. Excretory organs are called green (antennal) glands. Malacostracans are dioecious with external fertilization and development.
- Members of the class Branchiopoda have flattened, leaflike appendages. Examples are fairy shrimp, brine shrimp, and water fleas.
- Members of the class Maxillopoda include the copepods and the barnacles.

15.3 **Subphylum Hexapoda**

- Animals in the subphylum Hexapoda are characterized by bodies divided into three tagmata, five pairs of head appendages, and three pairs of legs. Hexapoda includes two classes, Entognatha and Insecta. Insect flight involves either a direct (synchronous) flight mechanism or an indirect (asynchronous) flight mechanism. Mouthparts of insects are adapted for chewing, piercing, and/or sucking, and the gut tract may be modified for pumping, storage, digestion, and water conservation.
- In insects, gas exchange occurs through a tracheal system. The insect nervous system is similar to that of other arthropods. Sensory structures include tympanal organs, compound eyes, and ocelli. Malpighian tubules transport uric acid to the digestive tract. Conversion of nitrogenous wastes to uric acid conserves water but is energetically expensive. Hormones regulate many insect functions, including ecdysis and metamorphosis. Pheromones are chemicals emitted by one individual that alter the behavior of another member of the same species.
- Insect adaptations for reproduction on land include resistant eggs, external genitalia, and behavioral mechanisms that bring males and females together at appropriate times. Metamorphosis of an insect may be ametabolous, hemimetabolous, or holometabolous. Neuroendocrine and endocrine secretions control metamorphosis.
- Insects show both innate and learned behavior.
- Many insects are beneficial to humans, and a few are parasites and/or transmit diseases to humans or agricultural products. Others attack cultivated plants and stored products.

- The four largest insect orders are Coleoptera, Lepidoptera, Diptera, and Hymenoptera.

15.4 **Further Phylogenetic Considerations**

- The clade Panarthropoda is comprised of Arthropoda, Onychophora, and Tardigrada. Onychophora and Tardigrada are sister taxa to Arthropoda.
- Members of the phylum Onychophora are the velvet worms. They are terrestrial and live in humid tropical and subtropical climates.
- Members of the phylum Tardigrada are the water bears. They live in marine interstitial areas, in freshwater detritus, and in the water film on terrestrial lichens, liverworts, and mosses.
- Within the Arthropoda the Trilobitomorpha was the first lineage to diverge from ancestral arthropods. The Chelicerata forms a second monophyletic lineage. Crustacea and Hexapoda comprise a pancrustacean lineage. "Crustacea" is probably paraphyletic. Myriapoda is a monophyletic lineage that may be closely related to either the Pancrustacea or Chilopoda.

Concept Review Questions

1. Members of the class ____________ include the lobsters, shrimp, and krill.
   a. Merostomata
   b. Arachnida
   c. Pycnogonida
   d. Branchiopoda
   e. Malacostraca
2. Removal of the X-organ of a crayfish might
   a. prevent it from undergoing ecdysis.
   b. prevent it from becoming an adult.
   c. promote premature ecdysis.
   d. convert a male into a female.
3. The wings of insects are never found on the
   a. prothorax.
   b. mesothorax.
   c. prothorax and mesothorax.
   d. metathorax.
4. Flight muscles act to change the shape of the thorax of an insect and accomplish both the upward and downward wing strokes in the ____________ flight mechanism.
   a. direct
   b. synchronous
   c. indirect
   d. labial
5. Insect development that involves a species-specific number of molts between egg and adult stages, the external development of wings (when wings are present), and immature stages (nymphs) that resemble adults is called
   a. ametabolous metamorphosis.
   b. hemimetabolous metamorphosis.
   c. holometabolous metamorphosis.
   d. complete metamorphosis.

Analysis and Application Questions

1. What problems are associated with living and reproducing in terrestrial environments? Explain how insects overcome these problems.
2. List as many examples as you can of how insects communicate with each other. In each case, what is the form and purpose of the communication?
3. In what way does holometabolous metamorphosis reduce competition between immature and adult stages? Give specific examples.
4. What role does each stage play in the life history of holometabolous insects?

16

Ambulacraria: Echinoderms and Hemichordates

This sponge brittle star (*Ophiothrix suensonii*) is a member of the phylum Echinodermata. It is found in the Caribbean Sea, the Bahamian waters, and around the island of Bermuda. It represents one of the five classes of extant echinoderms, a remnant of this phylum's evolutionary past. This brittle star is associated with a rope sponge (*Amphimedon compressa*).

Chapter Outline

16.1 Evolutionary Perspective
Relationships to Other Animals
16.2 Phylum Echinodermata
Echinoderm Characteristics
Class Asteroidea
Class Ophiuroidea
Class Echinoidea
Class Holothuroidea
Class Crinoidea
16.3 Phylum Hemichordata
Class Enteropneusta
Class Pterobranchia
16.4 Further Phylogenetic Considerations

16.1 EVOLUTIONARY PERSPECTIVE

LEARNING OUTCOMES

1. Compare the evolutionary relationships between the sea stars and chordates, such as fishes and mammals, versus the evolutionary relationships between the sea stars and crustaceans.
2. Explain the relationships between the Echinodermata and the Hemichordata.

If you could visit 400-million-year-old Paleozoic seas, you would see representatives of nearly every phylum studied in the previous seven chapters of this text. In addition, you would observe many representatives of the phylum Echinodermata (i-ki″na-dur′ma-tah) (Gr. *echinos,* spiny + *derma,* skin + *ata,* to bear). Many ancient echinoderms attached to their substrates and probably lived as suspension feeders–a feature found in only one class of modern echinoderms. Today, the relatively common sea stars, sea urchins, sand dollars, and sea cucumbers represent this phylum. In terms of numbers of species, echinoderms may seem to be a declining phylum. Fossil records indicate that about 12 of 18 classes of echinoderms have become extinct. That does not mean, however, that living echinoderms are of minor importance. Members of three classes of echinoderms have flourished and often make up a major component of the biota of marine ecosystems. Echinoderms are found in all of Earth's oceans at virtually all depths. They can account for the majority of species present in some ecosystems, and they are often keystone species (*see chapter 6*). As described in later sections, they often help maintain fragile coral and kelp forest ecosystems and recycle seafloor nutrients.

Another phylum, Hemichordata, is much less familiar, but you may have seen evidence of these animals during a walk along a seashore at low tide. Coiled castings (sand, mud, and excrement) at the opening of U-shaped burrows are evidence of worm-like animals that are members of the phylum Hemichordata. Other members of this phylum include equally unfamiliar filter feeders called pterobranchs (table 16.1).

Relationships to Other Animals

Members of the phyla Echinodermata and Hemichordata are the first members of the clade Deuterostomia covered in this textbook (figure 16.1). The other major phylum of deuterostomes is Chordata, the subject of chapters 17 to 22 (*see chapter 8 and appendix A*). All deuterostomes probably evolved from a filter-feeding ancestor. This life-form was lost in

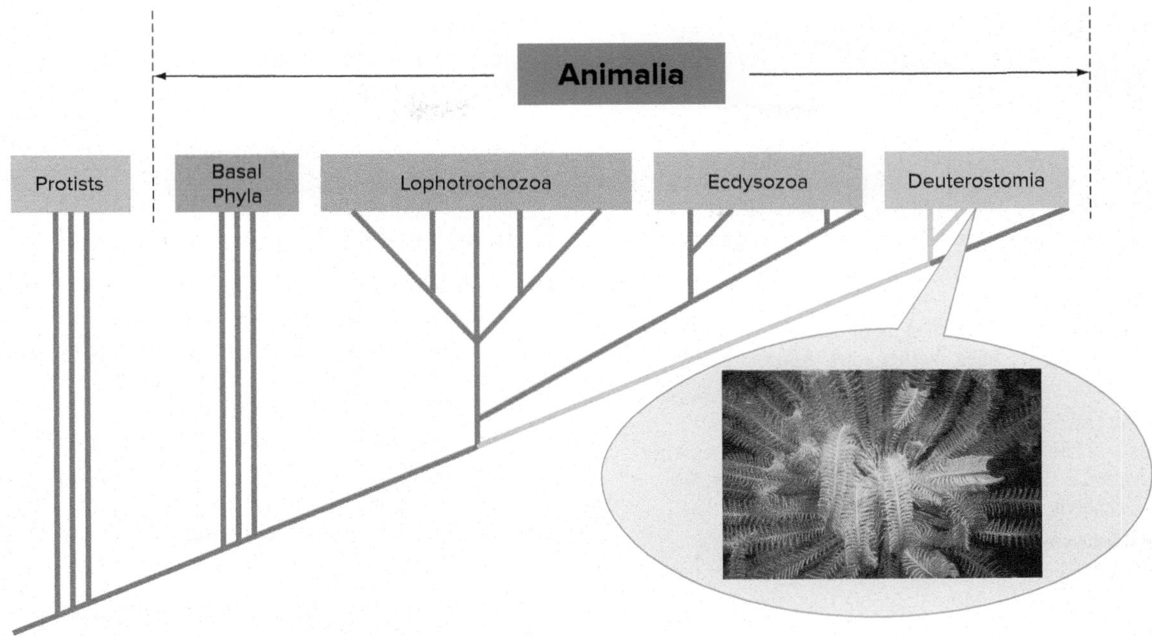

FIGURE 16.1

Evolutionary Relationships of Echinoderms to Other Animals. This figure shows one interpretation of the relationships of the Echinoderms to other members of the animal kingdom (*see appendix A*). The relationships depicted here are based on evidence from developmental and molecular biology. Echinoderms are placed within the Deuterostomia along with the Chordata, Hemichordata, and possibly others (*see table 16.2*). The feather star shown here (*Comanthina schlegeli*) uses its highly branched arms in suspension feeding. Although this probably reflects the original use of echinoderm appendages, most modern echinoderms use arms for locomotion, capturing prey, and scavenging the substrate for food. Feather stars can detach from the substrate and also use their arms in swimming and crawling.

the vertebrate lineage. Deuterostomes are all coelomate, and they typically share embryological features such as radial, indeterminant cleavage; an anus that forms in the region of the blastopore; and enterocoelous coelom formation (*see chapter 8 and figure 8.11*). There are, of course, exceptions to these generalizations. While invertebrate chordates (*see chapter 17*) undergo enterocoelous coelom formation, the body cavities of vertebrate chordates form in a schizocoelous manner. The wealth of evidence supporting deuterostome affinities of chordates suggests that this is a derived characteristic within the vertebrate lineage. The echinoderms and hemichordates are covered in one chapter because they comprise the deuterostome clade Ambulacraria. Studies of *Hox* genes, rRNA genes, and mitochondrial DNA have led researchers to agree on the common ancestry of members of these two phyla. In addition, Echinodermata and Hemichordata are united by two important synapomorphies: a shared larval morphology and tripartite (three-part) coelomic cavities.

Although echinoderm adults are radially symmetrical, they almost certainly evolved from bilaterally symmetrical ancestors. Evidence for this relationship includes bilaterally symmetrical echinoderm larval stages and extinct forms that were not radially symmetrical (*see Further Phylogenetic Considerations*). Recent behavioral studies suggest that adult echinoderms retain behaviors that are manifested during responses to escape danger and reflect a bilateral ancestry. One arm of sea stars tends to be the leading arm as a sea star moves away from threats. Similarly, sea star centers of gravity and responses that accomplish righting themselves after being flipped to their aboral surfaces suggest the presence of an anterior/posterior body axis. *Hox* genes (*see page 68*) are known to control body axis and symmetry development in deuterostomes, and these same genes may be involved in controlling the development of pentaradial symmetry from an ancestral bilaterian structure.

SECTION 16.1 THINKING BEYOND THE FACTS

What evidence unites the Hemichordata and the Echinodermata into the clade Ambulacraria?

16.2 PHYLUM ECHINODERMATA

LEARNING OUTCOMES

1. Characterize members of the phylum Echinodermata.
2. Contrast the body forms present in the echinoderm classes.
3. Compare the water-vascular systems of members of the classes Echinoidea, Asteroidea, and Crinoidea.
4. Contrast the following maintenance functions in members of the echinoderm classes: nutrition and digestion, gas exchange and internal transport, and nervous functions.
5. Describe echinoderm reproductive functions, including larva forms and development.

Extant members of the phylum Echinodermata are divided into five classes. Echinoderms include the very familiar sea stars and sea

TABLE 16.1

CLASSIFICATION OF THE AMBULACRARIA

Phylum Echinodermata (i-ki″na-dur′ma-tah)
The phylum of triploblastic, coelomate animals whose members are pentaradially symmetrical as adults and possess a water-vascular system and an endoskeleton covered by epithelium. Pedicellaria often present.

Class Crinoidea (krin-oi′de-ah)
Free living or attached by an aboral stalk of ossicles; flourished in the Paleozoic era. Sea lilies; feather stars. Approximately 630 living species.

Class Asteroidea (as″te-roi′de-ah)
Rays not sharply set off from central disk; ambulacral grooves with tube feet; suction disks on tube feet; pedicellariae present. Sea stars. Approximately 1,800 species.

Class Ophiuroidea (o-fe-u-roi′de-ah)
Arms sharply marked off from the central disk; tube feet without suction disks. Brittle stars. More than 2,000 species.

Class Echinoidea (ek″i-noi′de-ah)
Globular or disk shaped; no rays; movable spines; skeleton (test) of closely fitting plates. Sea urchins, sand dollars. Approximately 1,000 species.

Class Holothuroidea (hol″o-thu-roi′de-ah)
No rays; elongate along the oral-aboral axis; microscopic ossicles embedded in a muscular body wall; circumoral tentacles. Sea cucumbers. Approximately 1,700 species.

Phylum Hemichordata (hem″i-kor-da′tah)
Widely distributed in shallow, marine, tropical waters and deep, cold waters; softbodied and worm-like; diffuse epidermal nervous system; most with pharyngeal slits.

Class Enteropneusta (ent″er-op-nus′tah)
Shallow-water, worm-like animals; inhabit burrows on sandy shore lines; body divided into three regions: proboscis, collar, and trunk. Acorn worms (*Balanoglossus,* and *Saccoglossus*). Approximately 100 species.

Class Pterobranchia (ter″o-brang′ke-ah)
With or without pharyngeal slits; two or more arms; often colonial, living in an externally secreted encasement. *Rhabdopleura.* Approximately 30 species.

This listing reflects a phylogenetic sequence; however, the discussion that follows begins with the echinoderms that are familiar to most students.

urchins. Other echinoderms are less well known, like the sea daisies and feather stars. This section describes the distinctive characteristics of the phylum and features unique to each class.

Characteristics of members of the phylum Echinodermata include the following:

1. Calcareous endoskeleton in the form of ossicles that arise from mesodermal tissue
2. Adults with pentaradial symmetry and larvae with bilateral symmetry
3. Water-vascular system composed of water-filled canals used in locomotion, attachment, and/or feeding
4. Complete digestive tract that may be secondarily reduced
5. Hemal system derived from coelomic cavities
6. Nervous system consisting of a nerve net, nerve ring, and radial nerves

Echinoderm Characteristics

The approximately 7,000 species of living echinoderms are exclusively marine and occur at all depths in all oceans. Modern adult echinoderms have a form of radial symmetry, called **pentaradial symmetry,** in which body parts are arranged in fives, or a multiple of five, around an oral-aboral axis (figure 16.2*a*). Radial symmetry is adaptive for sedentary or slowly moving animals because it allows a uniform distribution of sensory, feeding, and other structures around the animal. Some modern mobile echinoderms, however, have secondarily returned to a basically bilateral form.

(a)

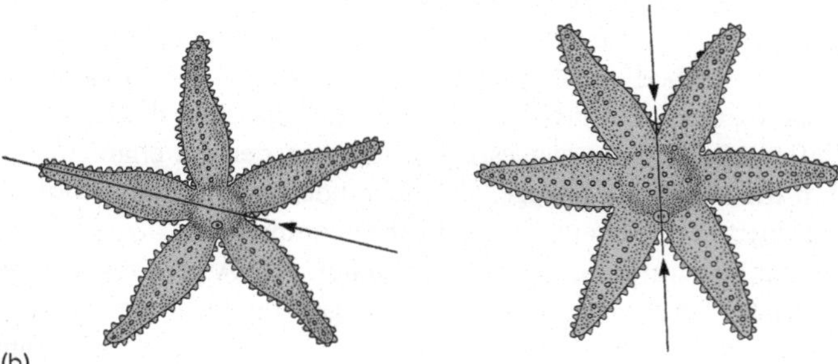

(b)

FIGURE 16.2

Pentaradial Symmetry. (*a*) Echinoderms exhibit pentaradial symmetry, in which body parts are arranged in fives around an oral-aboral axis. Note the madreporite between the bases of two arms. *Pisaster ochraceus* is found along the Pacific coast of North America. It is a keystone species (*see chapter 6*) preying on bivalves (*Bivalvia*), chitons (*Polyplacophora*), barnacles (*Maxillopoda*), and other intertidal organisms. Its few natural predators include sea otters (*Enhydra lutris*). (*b*) Comparison of hypothetical penta- and hexaradial echinoderms. The five-part organization may be advantageous because joints between skeletal ossicles are never directly opposite one another, as they would be with an even number of parts. Having joints on opposite sides of the body in line with each other (arrows) could make the skeleton weaker.

The echinoderm skeleton consists of a series of calcium carbonate plates called ossicles. These plates are derived from mesoderm, held in place by connective tissues, and covered by an epidermal layer. If the epidermal layer is abraded away, the skeleton may be exposed in some body regions. The skeleton is frequently modified into fixed or articulated spines that project from the body surface.

The evolution of the skeleton may be responsible for the pentaradial body form of echinoderms. The joints between two skeletal plates represent a weak point in the skeleton (figure 16.2*b*). By not having weak joints directly opposite one another, the skeleton is made stronger than if the joints were arranged opposite each other.

The **water-vascular system** of echinoderms is a series of water-filled canals, and their extensions are called tube feet. It originates embryologically as a modification of the coelom and is ciliated internally. The water-vascular system includes a ring canal that surrounds the mouth (figure 16.3). The ring canal usually opens to the outside or to the body cavity through a stone canal and an opening called the madreporite. In sea stars, the madreporite is a sievelike plate. In others it is a simple opening. The madreporite may serve as an inlet to replace water lost from the water-vascular system and may help equalize pressure differences between the water-vascular system and the outside. Tiedemann bodies are swellings often associated with the ring canal. They are believed to be sites for the production of phagocytic cells, called coelomocytes, whose functions are described later in this chapter. Polian vesicles are sacs that are also associated with the ring canal and function in fluid storage for the water-vascular system.

Five (or a multiple of five) radial canals branch from the ring canal. Radial canals are associated with arms of star-shaped echinoderms. In other echinoderms, they may be associated with the body wall and arch toward the aboral pole. Many lateral canals branch off each radial canal and end at the tube feet.

Tube feet are extensions of the canal system and usually emerge through openings in skeletal ossicles (*see figure 16.2*a). Internally, tube feet usually terminate in a bulblike, muscular ampulla. When an ampulla contracts, it forces water into the tube foot, which then extends. Valves prevent the backflow of water from the tube foot into the lateral canal. A tube foot often has a suction cup at its distal end. When the foot extends and contacts solid substrate, muscles of the suction cup contract and create a vacuum. In some taxa, tube feet have a pointed or blunt distal end. These echinoderms may extend their tube feet into a soft substrate to secure contact during locomotion or to sift sediment during feeding.

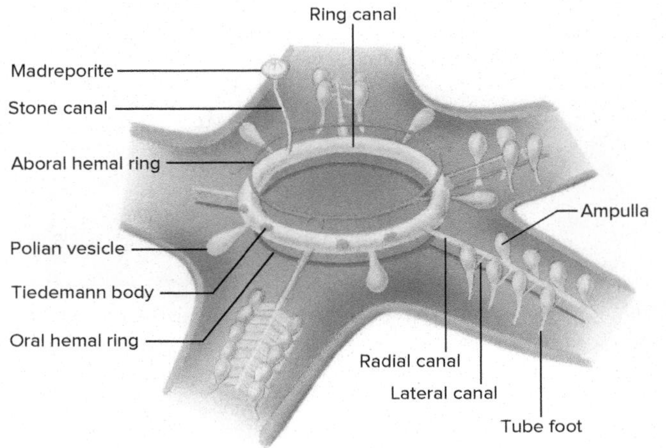

FIGURE 16.3

Water-Vascular System of a Sea Star. The ring canal gives rise to radial canals that lead into each arm. It opens to the outside or to the body cavity through a stone canal that ends at a madreporite on the aboral surface. Polian vesicles and Tiedemann bodies are often associated with the ring canal.

The water-vascular system has other functions in addition to locomotion. As is discussed at the end of this chapter, the original function of water-vascular systems was probably feeding, not locomotion. In addition, the soft membranes of the tube feet permit the exchange of respiratory gases and nitrogenous wastes with the environment. Tube feet also have sensory functions.

A **hemal system** consists of strands of tissue that encircle an echinoderm near the ring canal of the water-vascular system and run into each arm near the radial canals (*see figure 16.3*). The hemal system is derived from the coelom and circulates fluid using cilia that line its channels. The function of the hemal system is largely unknown, but it probably helps distribute nutrients absorbed from the digestive tract. It may aid in the transport of large molecules, hormones, or coelomocytes, which are cells that engulf and transport waste particles within the body.

Class Asteroidea

The sea stars make up the class Asteroidea (as″te-roi′de-ah) (Gr. *aster,* star + *oeides,* in the form of) and include about 1,800 species. They often live on hard substrates in marine environments, although some species also live in sandy or muddy substrates. Sea stars often serve as keystone species (*see chapter 6*) in their ecosystems. For example, predation by the sea star *Pisaster ochraceus* is a major factor in regulating species diversity in intertidal habitats of the northwestern Pacific coast of the United States. The impact of sea stars on marine ecosystems can be huge and very negative when ecological relationships involving sea stars are changed, usually due to human ecosystem alterations (*see chapter 9, Wildlife Alert: Coral Reefs*). Sea stars may be brightly colored with red, orange, blue, or gray. *Asterias* is an orange sea star common along the Atlantic coast of North America and is frequently studied in introductory zoology laboratories.

Sea stars usually have five arms that radiate from a central disk. The oral opening, or mouth, is in the middle of one side of the central disk. It is normally oriented downward, and movable oral spines surround it. Movable and fixed spines project from the skeleton and roughen the aboral surface. Thin folds of the body wall, called **dermal branchiae** (sing. dermal branchia) or **papulae** (sing. papula), extend between ossicles and function in gas exchange (figure 16.4). In some sea stars, the aboral surface has numerous pincherlike structures called **pedicellariae** (sing. pedicellaria), which clean the body surface of debris and have protective functions. Pedicellariae may be attached on a movable spine, or they may be immovably fused to skeletal ossicles.

A series of ossicles in the arm form an **ambulacral groove** that runs the length of the oral surface of each arm. The ambulacral groove houses the radial canal, and paired rows of tube feet protrude through the body wall on either side of the ambulacral groove. Tube

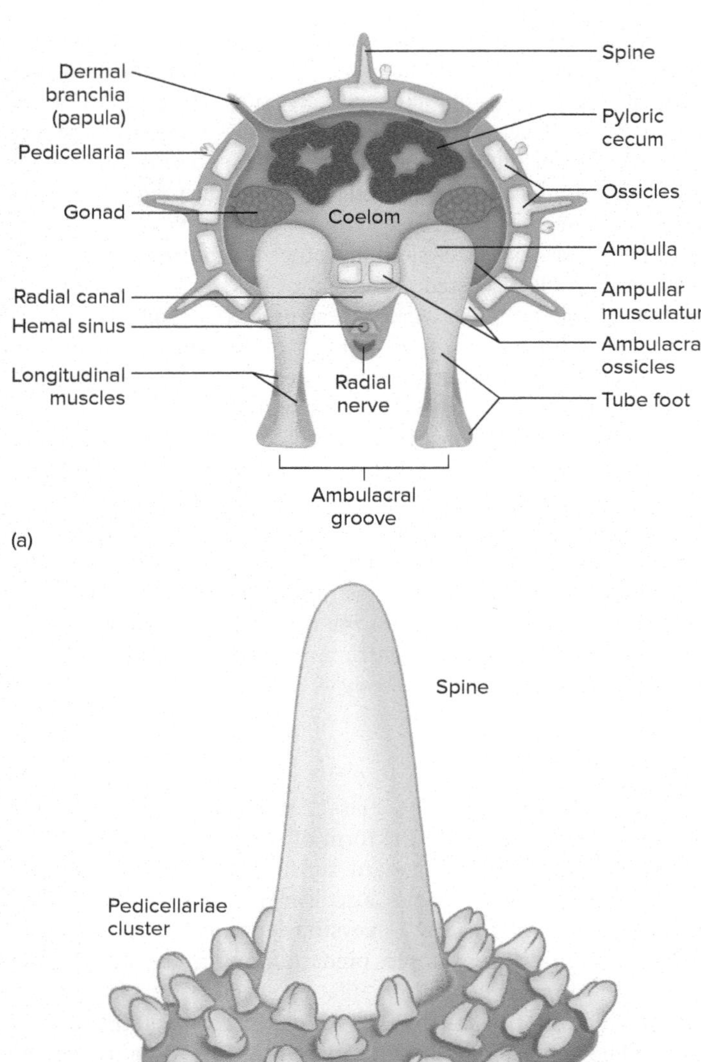

FIGURE 16.4

Body Wall and Internal Anatomy of a Sea Star. (*a*) A cross section through one arm of a sea star shows the structures of the water-vascular system and the tube feet extending through the ambulacral groove. (*b*) A spine with a cluster of pedicellariae.

feet of sea stars move in a stepping motion. Alternate extension, attachment, and contraction of tube feet move sea stars across their substrate. The nervous system coordinates the tube feet so that all feet move the sea star in the same direction; however, the tube feet do not move in unison. The suction disks of tube feet are effective attachment structures, allowing sea stars to maintain their position, or move from place to place, in spite of strong wave action.

Maintenance Functions

Sea stars feed on snails, bivalves, crustaceans, polychaetes, corals, detritus, and a variety of other food items. The mouth opens to a short esophagus and then to a large stomach that fills most of the coelom of the central disk. The stomach is divided into two regions. The larger, oral stomach, sometimes called the cardiac stomach, receives ingested food (figure 16.5). It joins the smaller, aboral stomach, sometimes called the pyloric stomach. The aboral (pyloric) stomach gives rise to ducts that connect to secretory and absorptive structures called pyloric cecae. Two pyloric cecae extend into each arm. A short intestine leads to rectal cecae (uncertain functions) and to a nearly nonfunctional anus, which opens on the aboral surface of the central disk.

Some sea stars ingest whole prey, which are digested extracellularly within the stomach. Undigested material is expelled through the mouth. Many sea stars feed on bivalves by forcing the valves apart. (Anyone who has tried to pull apart the valves of a bivalve shell can appreciate that this is a remarkable accomplishment.) When a sea star feeds on a bivalve, it wraps itself around the bivalve's ventral margin. Tube feet attach to the outside of the shell, and the body-wall

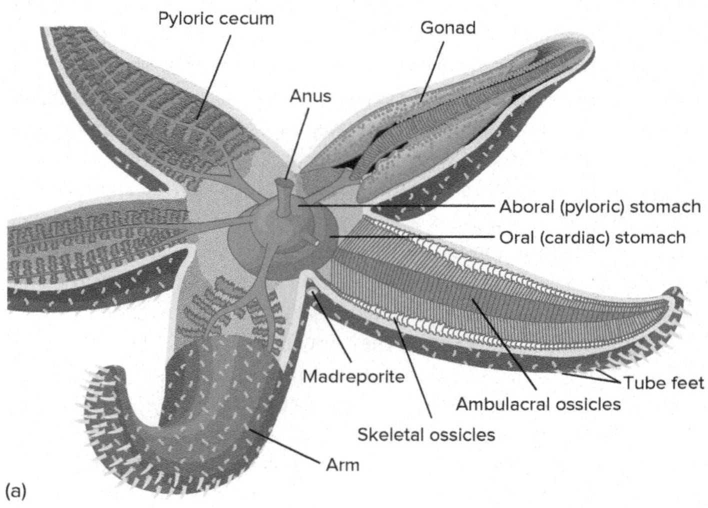

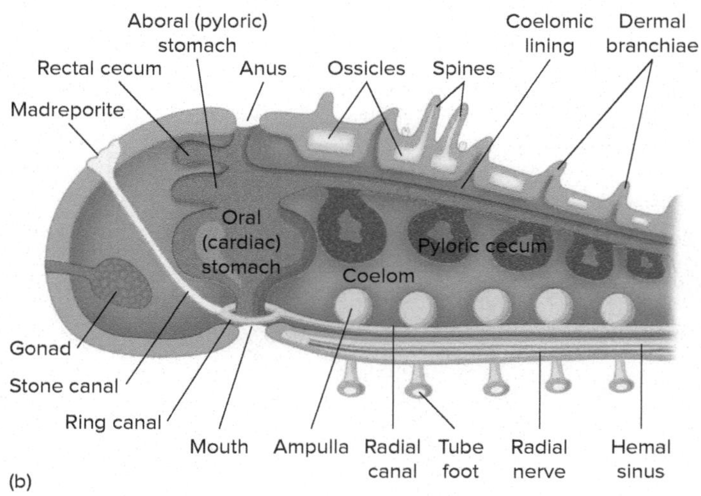

FIGURE 16.5

Internal Structure of a Sea Star. (*a*) Aboral view. Pyloric cecae have been removed in one arm to show gonad structure, and pyloric cecae and gonads have been removed in another arm to show ambulacral ossicles. (*b*) Lateral view through the central disk and one arm.

musculature forces the valves apart. (This is possible because the sea star changes tube feet when the muscles of engaged tube feet begin to tire.) When the valves are opened about 0.1 mm, increased coelomic pressure everts the oral (cardiac) portion of the sea star's stomach into the bivalve shell. Digestive enzymes are released, and partial digestion occurs in the bivalve shell. This digestion further weakens the bivalve's adductor muscles, and the shell eventually opens completely. Partially digested tissues are taken into the aboral (pyloric) portion of the stomach, and into the pyloric cecae for further digestion and absorption. After feeding and initial digestion, the sea star retracts the stomach, using stomach retractor muscles.

Gases, nutrients, and metabolic wastes are transported in the coelom by diffusion and by the action of ciliated cells lining the body cavity. Gas exchange and excretion of metabolic wastes (principally ammonia) occur by diffusion across dermal branchiae, tube feet, and other membranous structures. A sea star's hemal system consists of strands of tissue that encircle the mouth near the ring canal, extend aborally near the stone canal, and run into the arms near radial canals (*see figure 16.3*).

The nervous system of sea stars consists of a nerve ring that encircles the mouth and radial nerves that extend into each arm. Radial nerves lie within the ambulacral groove, just oral to the radial canal of the water-vascular system and the radial strands of the hemal system (*see figure 16.4*). Radial nerves coordinate the functions of tube feet. Other nervous elements are in the form of a nerve net associated with the body wall. This nerve net is probably an adaptation to the pentaradial body form and not homologous to the nerve net studied in the basal phyla (*see chapter 9*).

Most sensory receptors are distributed over the surface of the body and tube feet. Sea stars respond to light, chemicals, and various mechanical stimuli. They often have specialized photoreceptors at the tips of their arms. These are actually tube feet that lack suction cups but have a pigment spot surrounding a group of photoreceptors called ocelli.

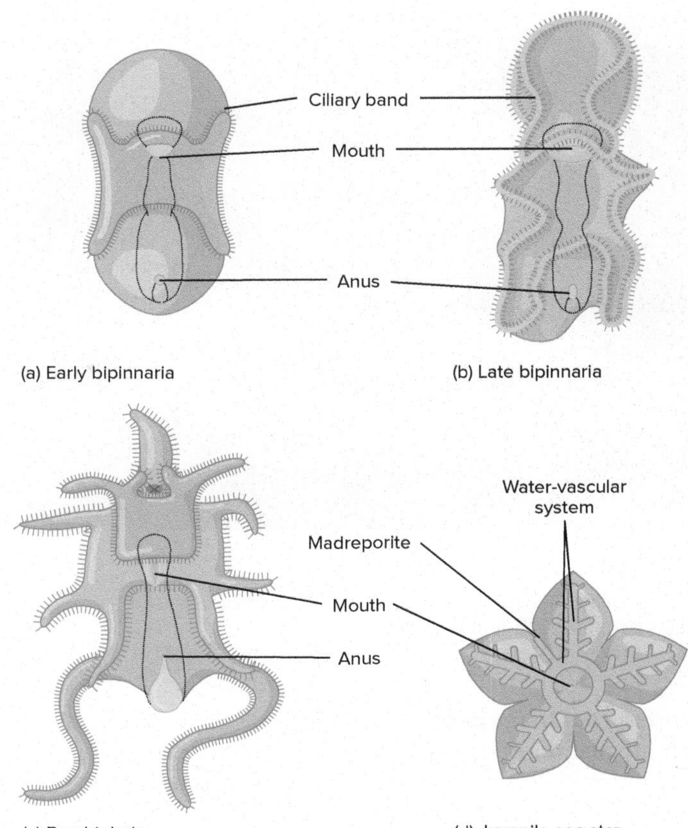

FIGURE 16.6

Development of a Sea Star. Later embryonic stages are bilaterally symmetrical, ciliated, and swim and feed in the plankton. In a few species, embryos develop from yolk stored in the egg during gamete formation. Following blastula and gastrula stages, larvae develop. (*a*) Early bipinnaria larva (0.5 mm). (*b*) Late bipinnaria larva (1 mm). (*c*) Brachiolaria larva (1 mm). (*d*) Juvenile sea star (1 to 2 mm).

Regeneration, Reproduction, and Development

Sea stars are well known for their powers of regeneration. They can regenerate any part of a broken arm. In a few species, an entire sea star can be regenerated from a broken arm if the arm contains a portion of the central disk. Regeneration is a slow process, taking up to a year for complete regeneration. Asexual reproduction occurs in some asteroids and involves division of the central disk, followed by regeneration of each half.

Most sea stars are dioecious, but sexes are indistinguishable externally. Two gonads are present in each arm, and these enlarge to nearly fill an arm during the reproductive periods. Gonopores open between the bases of each arm.

The embryology of echinoderms has been studied extensively because of the relative ease of inducing spawning and maintaining embryos in the laboratory. External fertilization is the rule. Because gametes cannot survive long in the ocean, maturation of gametes and spawning must be coordinated if fertilization is to take place. The photoperiod (the relative length of light and dark in a 24-hr period) and temperature are environmental factors used to coordinate sexual activity. In addition, gamete release by one individual is accompanied by the release of spawning pheromones, which induce other sea stars in the area to spawn, increasing the likelihood of fertilization.

Embryos are planktonic, and cilia are used in swimming (figure 16.6). After gastrulation, bands of cilia differentiate, and a bilaterally symmetrical larva, called a bipinnaria larva, forms. The larva usually feeds on planktonic protists. Because the larval stages are planktonic, they can be dispersed long distances by ocean currents. The development of larval arms results in a brachiolaria larva, which settles to the substrate, attaches, and metamorphoses into a juvenile sea star.

Sea Daisies

One group of very unusual echinoderms has previously been assigned to its own class, Concentricycloidea. The current consensus is that these echinoderms are highly modified members of the class Asteroidea. Two species of sea daisies have been described (figure 16.7). They lack arms and are less than 1 cm in diameter. The most distinctive features of this group are the two circular water-vascular rings that encircle the disklike body. The inner of the two rings probably corresponds to the ring canal of other asteroids. The outer ring contains tube feet and ampullae and probably

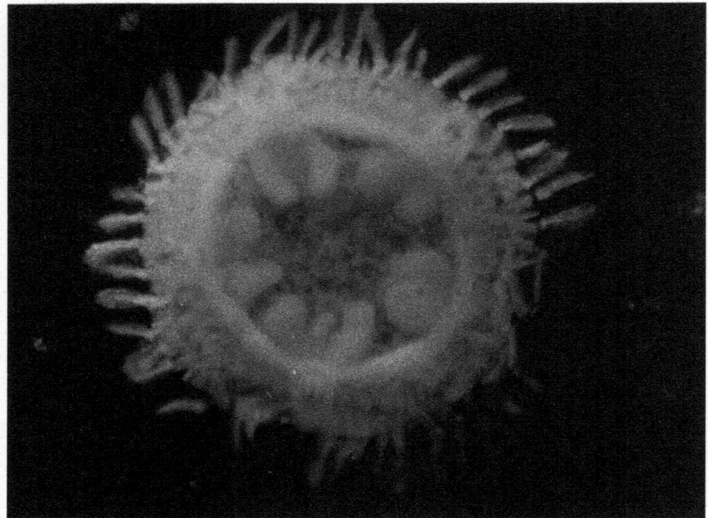

FIGURE 16.7

A Sea Daisy. A preserved sea daisy (*Xyloplax medusiformis*). This specimen is 3 mm in diameter.
©Dr Alan Baker, New Zealand

corresponds to the radial canals of other asteroids. Sea daisies lack an internal digestive system. Instead, a thin membrane, called a velum, covers the surface of the animal that is applied to the substrate (e.g., decomposing organic matter) and digests and absorbs nutrients. Internally, five pairs of brood pouches hold embryos during development. No free-swimming larval stages are known.

Class Ophiuroidea

The class Ophiuroidea (o-fe-u-roi′de-ah) (Gr. *ophis,* snake + *oura,* tail + *oeides,* in the form of) includes the basket stars and the brittle stars or serpent stars. With over 2,000 species, this is the most diverse group of echinoderms. Ophiuroids, however, are often overlooked because of their small size and their tendency to occupy crevices in rocks and coral or to cling to algae.

The arms of ophiuroids are long and, unlike those of asteroids, are sharply set off from the central disk, giving the central disk a pentagonal shape. Brittle stars have unbranched arms, and most have a central disk that ranges in size from 1 to 3 cm (figure 16.8*a*). Basket stars have arms that branch repeatedly (figure 16.8*b*). Neither dermal branchiae nor pedicellariae are present in ophiuroids. The tube feet of ophiuroids lack suction disks and ampullae, and the contraction of muscles associated with the base of a tube foot extends the tube foot. Unlike the sea stars, the madreporite of ophiuroids is on the oral surface.

The water-vascular system of ophiuroids is not used for locomotion. Instead, the skeleton is modified to permit a unique form of grasping and movement. Superficial ossicles, which originate on the aboral surface, cover the lateral and oral surfaces of each arm. The ambulacral groove—containing the radial nerve, hemal strand, and radial canal—is thus said to be "closed." Ambulacral ossicles are in the arm, forming a central supportive axis. Successive ambulacral ossicles articulate with one another and are acted upon by relatively large muscles to produce snakelike movements (hence the derivation of the class name) that allow the arms to curl around a stalk of algae or to hook into a coral crevice. During locomotion, the central disk is held above the substrate, and two arms pull the animal along, while other arms extend forward and/or trail behind the animal.

Maintenance Functions

Ophiuroids may prey on small crustaceans or marine worms, but most are scavengers. They use their arms and tube feet in sweeping motions to collect particulate matter, which are then transferred to the mouth. Basket stars are suspension feeders that wave their arms and trap plankton on mucus-covered tube feet. Trapped plankton is

How Do We Know about Echinoderm Regeneration?

Simple observations have revealed that all echinoderms have the ability to regenerate lost parts, but regeneration is best developed in the Asteroidea, Ophiuroidea, and Crinoidea. In the case of some asteroids, the entire organism may be regenerated from a body part. Regeneration allows the replacement of a body part lost by predation or, in the asteroids and ophiuroids, regeneration following fission is a form of asexual reproduction. In some cases a pool of undifferentiated stem cells is retained in adults, and these stem cells have the ability to give rise to any adult structure that is lost. In other cases, adult cells actually revert to a stemcell-like form and acquire the ability to develop into any lost adult structure. Studies of the genetic control for regeneration have identified a series of genes related to those that function in wound repair in vertebrates. Some of the chemicals that mediate regeneration include common neurotransmitters (*see chapter 24*) found in the nervous systems of most animals. Because both echinoderms and humans are deuterostomes, we share many genes that control development. Understanding echinoderm regeneration may provide clues to help us understand why some tissues, such as human heart muscle and certain nervous tissues, cannot repair themselves. This understanding could lead to treatments that restore functions to damaged tissues.

(a)

(b)

FIGURE 16.8

Class Ophiuroidea. (*a*) This ruby brittle star (*Ophioderma rubicundum*) is found on coral reefs throughout the Caribbean. It uses its long, snakelike arms for crawling along its substrate and curling around objects in its environment. (*b*) Basket stars have five highly branched arms. They wave the arms in the water and with the mucus-covered tube feet capture planktonic organisms. *Gorgonocephalus eucnemis* (shown here) is commonly found in northern Atlantic and Pacific Oceans where it commonly occurs at depths of 15 to 2,000 m.

(a) ©Amar and Isabelle Guillen - Guillen Photo LLC/Alamy; (b) Source: NOAA/CBNMS

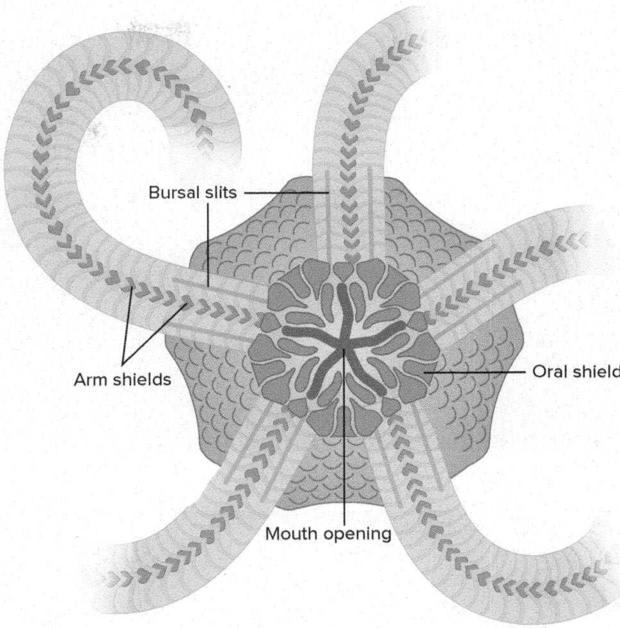

FIGURE 16.9

Class Ophiuroidea. Oral view of the disk of the brittle star *Ophiomusium.*

Source: Hyman, L. 1967. *The Invertebrates*. Volume VI. McGraw-Hill, Inc.

passed from tube foot to tube foot along the length of an arm until it reaches the mouth.

The mouth of ophiuroids is in the center of the central disk, and five triangular jaws form a chewing apparatus. The mouth leads to a saclike stomach. There is no intestine, and no part of the digestive tract extends into the arms.

The coelom of ophiuroids is reduced and is mainly confined to the central disk, but it still serves as the primary means for the distribution of nutrients, wastes, and gases. Coelomocytes aid in the distribution of nutrients and the expulsion of particulate wastes. Ammonia is the primary nitrogenous waste product, and it is lost by diffusion across tube feet and membranous sacs, called **bursae,** that invaginate from the oral surface of the central disk. Slits in the oral disk, near the base of each arm, allow cilia to move water into and out of the bursae (figure 16.9).

Regeneration, Reproduction, and Development

Like sea stars, ophiuroids can regenerate lost arms. If a brittle star is grasped by an arm, the contraction of certain muscles may sever and cast off the arm—hence the common name brittle star. This process, called autotomy (Gr. *autos,* self + *tomos,* to cut), is used in escape reactions. The ophiuroid later regenerates the arm. Some species also have a fission line across their central disk. When an ophiuroid splits into halves along this line, two ophiuroids regenerate.

Ophiuroids are dioecious. Males are usually smaller than females, who often carry the males. The gonads are associated with each bursa, and gametes are released into the bursa. Eggs may be shed to the outside or retained in the female's bursa. In the latter case, sperm released by the male swim into the female's bursal slits, and eggs are fertilized and held through early development. Embryos are protected in the bursa and are sometimes nourished by the parent. A larval stage, called an ophiopluteus, is planktonic. Its long arms bear ciliary bands used to feed on plankton, and it undergoes metamorphosis before sinking to the substrate.

Class Echinoidea

The sea urchins, sand dollars, and heart urchins make up the class Echinoidea (ek″i-noi′de-ah) (Gr. *echinos,* spiny + *oeides,* in the form of). The approximately 1,000 species are widely distributed in nearly all marine environments. Sea urchins are specialized for living on hard substrates, often wedging themselves into crevices and holes in rock or coral (figure 16.10*a*). Sand dollars and heart urchins

(a)

(b)

FIGURE 16.10

Class Echinoidea. (*a*) Sea urchins (*Strongylocentrotus franciscanus*). (*b*) Sand dollars are specialized for living in soft substrates, where they are often partially buried. The purple sand dollar (*Echinodiscus auritus*) is shown here among a group of solitary corals (Cnidaria, *Heterocyathus aequicostatus*).
(a) ©Index Stock/Alamy; (b) ©Matthew Banks/Alamy

usually live in sand or mud, and they burrow just below the surface (figure 16.10*b*). They use tube feet to catch organic matter settling on them or passing over them. Sand dollars often live in dense beds, which favors efficient reproduction and feeding.

Echinoids play important roles in marine ecosystems. They are important food items in the diets of sea otters (*Enhydra lutris, see page 408*), California sheephead (*Semicossyphus pulcher*), and wolf eels (*Anarrhichthys ocellatus*). Human interference in marine ecosystems that reduces populations of these predators leaves sea urchin populations unchecked. Populations of sea urchins can flourish, and the resulting overgrazing on algae by sea urchins can leave "urchin barrens" nearly devoid of algae. This type of overgrazing has threatened the highly productive kelp forests of northern latitudes that contribute to the mitigation of global warming and ocean acidification by conversion of CO_2 into organic matter. Conservation of sea urchin predators, like the sea otter, is essential to maintaining healthy marine ecosystems.

Sea urchins are rounded, and their oral end is oriented toward the substrate. Their skeleton, called a test, consists of 10 sets of closely fitting plates that arch between oral and aboral ends. Five rows of ambulacral plates have openings for tube feet, and alternate with five interambulacral plates, which have tubercles for the articulation of spines. The base of each spine is a concave socket, and muscles at its base move the spine. Spines are often sharp and sometimes hollow, and they may contain venom. The pedicellariae of sea urchins have either two or three jaws and connect to the body wall by a relatively long stalk (figure 16.11*a*). They clean the body of debris. Pedicellariae of some sea urchins contain venom sacs and are grooved or hollow to inject venom into a predator, such as a sea star.

The water-vascular system is similar to that of other echinoderms. Radial canals run along the inner body wall between the oral and the aboral poles. Tube feet possess ampullae and suction cups, and the water-vascular system opens to the outside through many pores in one aboral ossicle that serves as a madreporite.

Echinoids move by using spines for pushing against the substrate and tube feet for pulling. Sand dollars and heart urchins use spines to help them to burrow in soft substrates. Some sea urchins

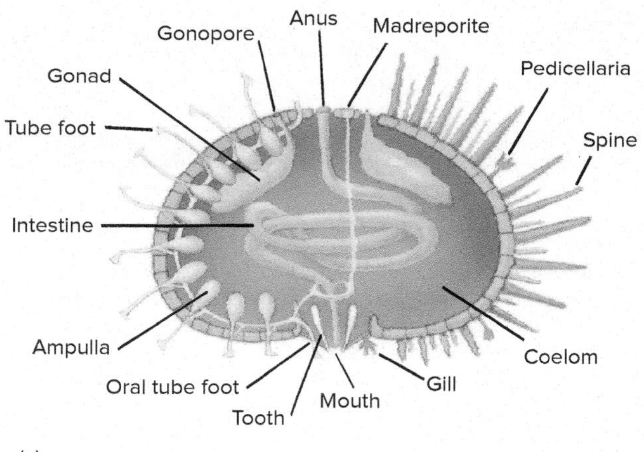

(a)

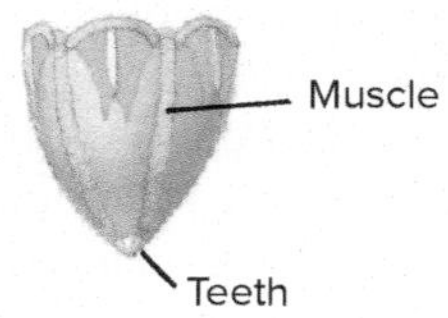

(b)

FIGURE 16.11

Internal Anatomy of a Sea Urchin. (*a*) Sectional view. (*b*) Aristotle's lantern is a chewing structure consisting of about 35 ossicles and associated muscles.

burrow into rock and coral to escape the action of waves and strong currents. They form cup-shaped depressions and deeper burrows, using the action of their chewing Aristotle's lantern, which is described next.

Maintenance Functions

The diet of echinoids is varied. They feed on algae, marine carrion, and sessile animals such as coral polyps and ectoprocts. Oral tube feet surrounding the mouth manipulate food. A chewing apparatus, called **Aristotle's lantern,** can be projected from the mouth (figure 16.11*b*). It consists of about 35 ossicles and attached muscles and cuts food into small pieces for ingestion. The mouth cavity leads to a pharynx, an esophagus, and a long, coiled intestine that ends aborally at the anus.

Echinoids have a large coelom, and coelomic fluids are the primary circulatory medium. Small gills, found in a thin membrane surrounding the mouth, are outpockets of the body wall and are lined by ciliated epithelium. Gas exchange occurs by diffusion across this epithelium and across the tube feet. Ciliary currents, changes in coelomic pressure, and the contraction of muscles associated with Aristotle's lantern move coelomic fluids into and out of gills. Excretory and nervous functions are similar to those described for asteroids.

Reproduction and Development

Echinoids are dioecious. Gonads are on the internal body wall of the interambulacral plates. During breeding season, they nearly fill the spacious coelom. One gonopore is in each of the five ossicles, called genital plates, located at the aboral end of the echinoid. The sand dollars are an exception, usually having only four gonads and gonopores. Gametes are shed into the water, and fertilization is external. Development eventually results in a pluteus larva that spends several months in the plankton and eventually undergoes metamorphosis to the adult.

Class Holothuroidea

The class Holothuroidea (hol″o-thu-roi′de-ah) (Gr. *holothourion,* sea cucumber + *oeides,* in the form of) has approximately 1,700 species, whose members are commonly called sea cucumbers. Sea cucumbers are found at all depths in all oceans, where they crawl over hard substrates or burrow through soft substrates (figure 16.12).

Sea cucumbers have no arms, and they are elongate along the oral-aboral axis. They lie on one side, which is usually flattened as a permanent ventral side, giving them a secondary bilateral symmetry. Tube feet surrounding the mouth are enlarged, highly modified, and referred to as tentacles. Most adults range in length between 10 and 30 cm. Their body wall is thick and muscular, and it lacks protruding spines or pedicellariae. Beneath the epidermis is the dermis, a thick layer of connective tissue with embedded ossicles. Sea cucumber ossicles are microscopic and do not function in determining body shape. Larger ossicles form a calcareous ring that encircles the oral end of the digestive tract, serving as a point of attachment for body-wall muscles (figure 16.13). Beneath the dermis is a layer of circular muscles overlying longitudinal muscles. The body wall of sea cucumbers, when boiled and dried, is known as trepang in Asian countries. It may be eaten as a main-course item or added to soups as flavoring and a source of protein.

FIGURE 16.12

Class Holothuroidea. *Pseudocolochirus axiologus* (Australian sea apple) is native to waters of northern Australia. It reaches lengths up to 30 cm and uses its tentacles to feed on plankton. This species is fished extensively for the aquarium trade (*see Wildlife Alert, pages 296 to 297*).

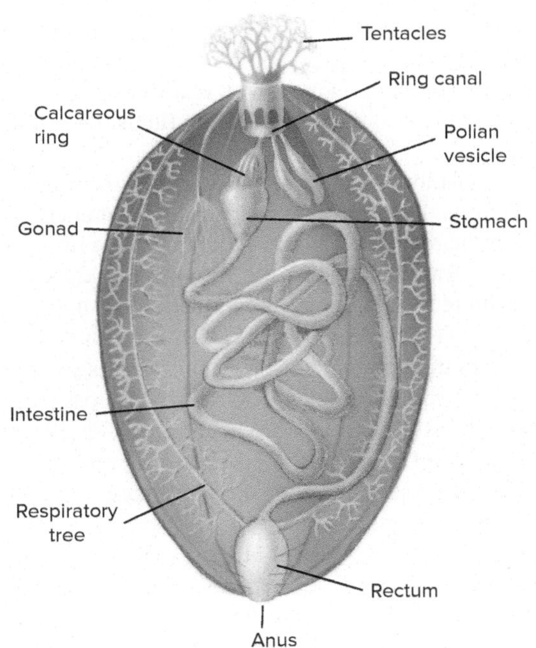

FIGURE 16.13

Internal Structure of a Sea Cucumber, *Thyone*. The mouth leads to a stomach supported by a calcareous ring. The calcareous ring is also the attachment site for longitudinal retractor muscles of the body. Contractions of these muscles pull the tentacles into the anterior end of the body. The stomach leads to a looped intestine. The intestine continues to the rectum and anus. (The anterior portion of the digestive tract is displaced aborally in this illustration.)

The madreporite of sea cucumbers is internal, and the water-vascular system is filled with coelomic fluid. The ring canal encircles the oral end of the digestive tract and gives rise to 1 to 10 Polian vesicles. Five radial canals and the canals to the tentacles branch from the ring canal. Radial canals and tube feet, with suction cups and ampullae, run between the oral and aboral poles. The side of a sea cucumber resting on the substrate contains three of the five rows of tube feet, which are primarily used for attachment. The two rows of tube feet on the upper surface may be reduced in size or may be absent.

Sea cucumbers are mostly sluggish burrowers and creepers, although some swim by undulating their bodies from side to side. Locomotion using tube feet is inefficient, because the tube feet are not anchored by body-wall ossicles. Locomotion more commonly results from contractions of body-wall muscles that produce wormlike, locomotor waves that pass along the length of the body.

Maintenance Functions

Most sea cucumbers ingest particulate organic matter using their tentacles. Mucus covering the tentacles traps food as the tentacles sweep across the substrate or are held out in seawater. The digestive tract consists of a stomach; a long, looped intestine; a rectum; and an anus (*see figure 16.13*). Sea cucumbers thrust tentacles into their mouths to wipe off trapped food. During digestion, coelomocytes move across the intestinal wall, secrete enzymes to aid in digestion, and engulf and distribute the products of digestion.

The coelom of sea cucumbers is large, and the cilia of the coelomic lining circulate fluids throughout the body cavity, distributing respiratory gases, wastes, and nutrients. The hemal system of sea cucumbers is well developed, with relatively large sinuses and a network of channels containing coelomic fluids. Its primary role is food distribution.

A pair of tubes called **respiratory trees** attach at the rectum and branch throughout the body cavity of sea cucumbers. The pumping action of the rectum circulates water into these tubes. When the rectum dilates, water moves through the anus into the rectum. Contraction of the rectum, along with contraction of an anal sphincter, forces water into the respiratory tree. Water exits the respiratory tree when tubules of the tree contract. Respiratory gases and nitrogenous wastes move between the coelom and seawater across these tubules.

The nervous system of sea cucumbers is similar to that of other echinoderms but has additional nerves supplying the tentacles and pharynx. Some sea cucumbers have statocysts, and others have relatively complex photoreceptors.

Casual examination suggests that sea cucumbers are defenseless against predators. Many sea cucumbers, however, produce toxins in their body walls that discourage predators. Other sea cucumbers can evert tubules of the respiratory tree, called Cuverian tubules, through the anus. These tubules contain sticky secretions and toxins capable of entangling and immobilizing predators. In addition, contractions of the body wall may result in the expulsion of one or both respiratory trees, the digestive tract, and the gonads through the anus. This process, called evisceration, occurs in response to chemical and physical stress and may be a defensive adaptation that discourages predators. Regeneration of lost parts follows.

Reproduction and Development

Most sea cucumbers are dioecious. They possess a single gonad, located anteriorly in the coelom, and a single gonopore near the base of the tentacles. Spawning is often correlated with lunar cycles and is often seasonal. External fertilization results in the development of planktonic larvae. Larval periods last up to one month. Metamorphosis precedes settling to the substrate. In some species, a female's tentacles trap eggs as the eggs are released. After fertilization, eggs are transferred to the body surface, where they are brooded. Although rare, coelomic brooding also occurs. Eggs are released into the body cavity, where fertilization (by an unknown mechanism) and early development occur. The young leave through a rupture in the body wall. Sea cucumbers frequently live 5 to 10 years and often require years to grow to sexual maturity. Slow growth and maturation rates contribute to the problems that sea cucumbers experience in recovering from anthropogenic disturbances (*see Wildlife Alert, pages 296 to 297*). Sea cucumbers can also reproduce by transverse fission, followed by regeneration of lost parts.

Class Crinoidea

Members of the class Crinoidea (krin-oi′de-ah) (Gr. *krinon,* lily + *oeides,* in the form of) include the sea lilies and the feather stars. They are the most primitive of all living echinoderms and are very different from any covered thus far. Approximately 630 species are living today; however, an extensive fossil record indicates that many more were present during the Paleozoic era, 200 to 600 million years ago (mya).

Sea lilies attach permanently to their substrate by a stalk (figure 16.14). The attached end of the stalk bears a flattened disk or rootlike extensions that are fixed to the substrate. Disklike ossicles of the stalk appear to be stacked on top of one another and are held together by connective tissues, giving a jointed appearance. The stalk

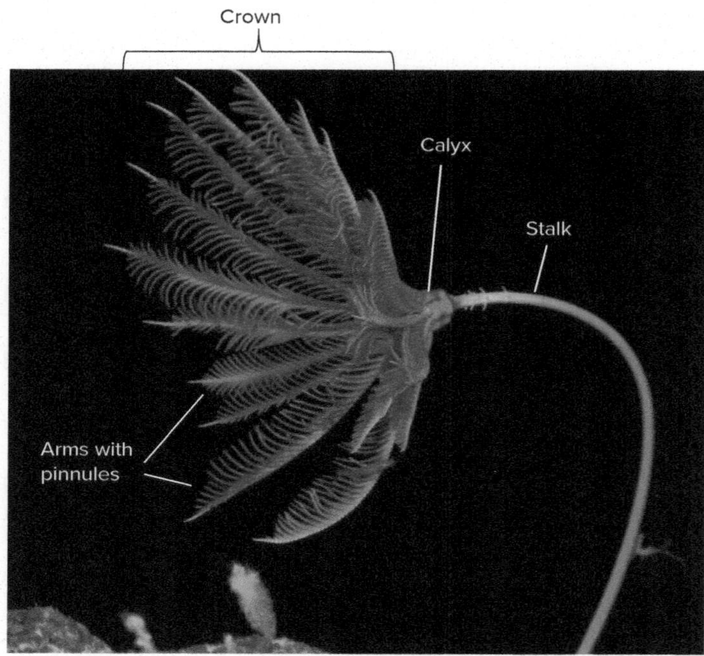

FIGURE 16.14

Class Crinoidea. *Proisocrinus ruberrimus* (red stalked sea lily or 'Moulin Rouge') is found in the central Pacific Ocean at depths approaching 2,000 meters. The stalk of crinoids is thought to lift their filtering crowns above the substrate where water currents are more pronounced, thus increasing filtering efficiency.

Source: NOAA Okeanos Explorer Program, INDEX-SATAL 2010

usually bears projections, or cirri, arranged in whorls. The unattached end of a sea lily is called the crown. The aboral end of the crown attaches to the stalk and is supported by a set of ossicles, called the **calyx.** Five arms also attach at the calyx. They are branched, supported by ossicles, and bear smaller branches (pinnules)–giving them a featherlike appearance. Tube feet are in a double row along each arm. Ambulacral grooves on the arms lead toward the mouth. The mouth and anus open onto the upper (oral) surface.

Feather stars are similar to sea lilies, except they lack a stalk and are swimming and crawling animals (figure 16.15). The aboral end of the crown bears a ring of rootlike cirri, which cling when the animal is resting on a substrate. Feather stars swim by raising and lowering the arms, and they crawl over substrate by pulling with the tips of the arms.

Maintenance Functions

Circulation, gas exchange, and excretion in crinoids are similar to these functions in other echinoderms. In feeding, however, crinoids use outstretched arms for suspension feeding. A planktonic organism that contacts a tube foot is trapped, and cilia in ambulacral grooves carry it to the mouth. Although this method of feeding is different from how other modern echinoderms feed, it probably reflects the original function of the water-vascular system.

Crinoids lack the nerve ring found in most echinoderms. Instead, a cup-shaped nerve mass below the calyx gives rise to radial nerves that extend through each arm and control the tube feet and arm musculature.

Reproduction and Development

Many crinoids, like other echinoderms, are dioecious. Others are monoecious, with male gametes developing before female gametes. This sequence of gamete development is called protandry and ensures that cross-fertilization will occur. Gametes form from germinal epithelium in the coelom and are released through ruptures in the walls of the arms. Some species spawn in seawater, where fertilization and development occur. Other species brood embryos on the outer surface of the arms. Metamorphosis occurs after larvae attach to the substrate. Like other echinoderms, crinoids can regenerate lost parts.

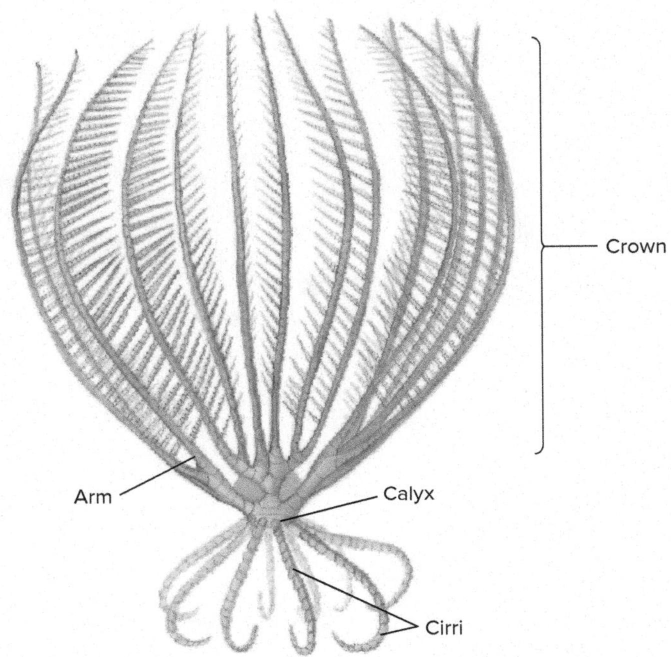

FIGURE 16.15
Class Crinoidea. A feather star (*Neometra*).

SECTION 16.2 THINKING BEYOND THE FACTS

How is pentaradial symmetry manifested in members of each of the five echinoderm classes?

16.3 PHYLUM HEMICHORDATA

LEARNING OUTCOMES

1. Describe the characteristics of the members of the phylum Hemichordata.
2. Compare the body forms and maintenance functions of members of the classes Enteropneusta and Pterobranchia.
3. Contrast reproduction and development of members of the classes Enteropneusta and Pterobranchia.

The phylum Hemichordata (hem′i-kor-da″tah) (Gr. *hemi,* half + L. *chorda,* cord) includes the acorn worms (class Enteropneusta) and the pterobranchs (class Pterobranchia) (*see table 16.1*). Members of both classes live in or on marine sediments.

Characteristics of the phylum Hemichordata include the following:

1. Marine, deuterostomate animals with a body divided into three regions: proboscis, collar, and trunk; coelom divided into three cavities (tripartite coelom)
2. Ciliated pharyngeal slits
3. Open circulatory system
4. Complete digestive tract
5. Dorsal, sometimes tubular, nerve cord

Class Enteropneusta

Members of the class Enteropneusta (ent′er-op-nus″tah) (Gr. *entero,* intestine + *pneustikos,* for breathing) are marine worms that usually range in size between 10 and 40 cm, although some can be as long as 2 m. Zoologists have described over 100 species, and most occupy U-shaped burrows in sandy and muddy substrates between the limits of high and low tides. The list of deep-sea enteropneust species is growing longer. The common name of the enteropneusts–acorn worms–is derived from the appearance of the proboscis, which is a short, conical projection at the worm's anterior end. A ringlike collar is posterior to the proboscis, and an elongate trunk is the third division of the body (figure 16.16). Along with the three body regions, the enteropneusts have a coelom divided into three cavities (*see figure 16.16*). This tripartite coelom is a feature the hemichordates share with the echinoderms. A ciliated epidermis and gland cells cover acorn worms. The mouth is located ventrally between the proboscis and the collar. Varying numbers of pharyngeal slits, from a few to several hundred, are positioned laterally on the trunk. Pharyngeal

WILDLIFE ALERT

Imperiled Sea Cucumbers

In chapter 6 and previous Wildlife Alerts, you have seen that there are common drivers that threaten animals with extinction. Animals that occupy the highest trophic levels, those that have slow growth and slow reproductive rates, those that occupy very small geographical ranges, and those whose habitats are disrupted by human activities are vulnerable to extinction. These factors often direct our attention toward large, charismatic animals like lowland gorillas (*Gorilla gorilla*), black rhinos (*Diceros bicornis*), Pacific bluefin tunas (*Thunnus orientalis*), and California condors (*Gymnogyps californianus*). These drivers of extinction, although very important, do not necessarily apply to all threatened and endangered animals. Invertebrates comprise 95% of all marine species, and many of these are facing extinction too. Less charismatic than many vertebrates, they garner less attention; and too often less attention means that we know much less about their taxonomic status, ecology, population statistics, and why extinction may be imminent.

Sea cucumbers are in trouble throughout Earth's oceans, and they certainly comprise one of Earth's less charismatic animal groups. Sea cucumbers provide valuable ecosystem services that one cannot accurately represent with dollar signs. They are often called "earthworms of the sea" because they feed on detritus and turn over the seafloor, much like earthworms do on land. They recycle marine nutrients by hastening decomposition processes, and recent evidence suggests that their alkaline defecations may help buffer coral reef ecosystems from damage caused by ocean acidification. Some of the same "extinction drivers" described in the previous text apply to sea cucumbers (habitat destruction and global warming), but the primary driver that threatens sea cucumber populations is economic in nature.

Of the 1,700 species of sea cucumbers, nearly 100 are harvested to supply Asian markets. Sea cucumber muscles are served in sushi bars, and dried sea cucumbers (bêche-de-mer or trepang) are added to soup and sautéed with vegetables and meats. The intestine is used to prepare a gourmet Japanese dish called konowata. Alleged medicinal uses for sea cucumbers include treatments for ulcers, cuts, arthritis, and sexual impotency. The annual sea cumber harvest from the world's oceans is estimated between 50,000 and 75,000 tonnes (1 metric ton [t] = 1,000 kg or 2.2 tons).

Most of the sea cucumber harvest is derived from the most common species that reside in easily accessible coastal habitats (box figure 16.1), and this harvest is destined for markets in China and Hong Kong. Market prices for bêche-de-mer vary with the species being sold, but they begin at about $140 per kilogram. The most highly valued species (*Holothuria scabra* and *Apostichopus japonicus*, box figure 16.2*a*, *b*) have brought prices between $1,500 and $3,000 per kilogram. These kinds of prices, the accessibility of shallow-water (less than 20 m) species, and low technology harvesting techniques (e.g., breath-holding diving) have been a boon for local coastal fisheries. Fisheries in western Central Pacific islands have shifted the bulk of their efforts to harvesting and processing sea cucumbers. That shift is also occurring, but to a lesser extent, in temperate northern latitudes around the world (*see box figure 16.1*).

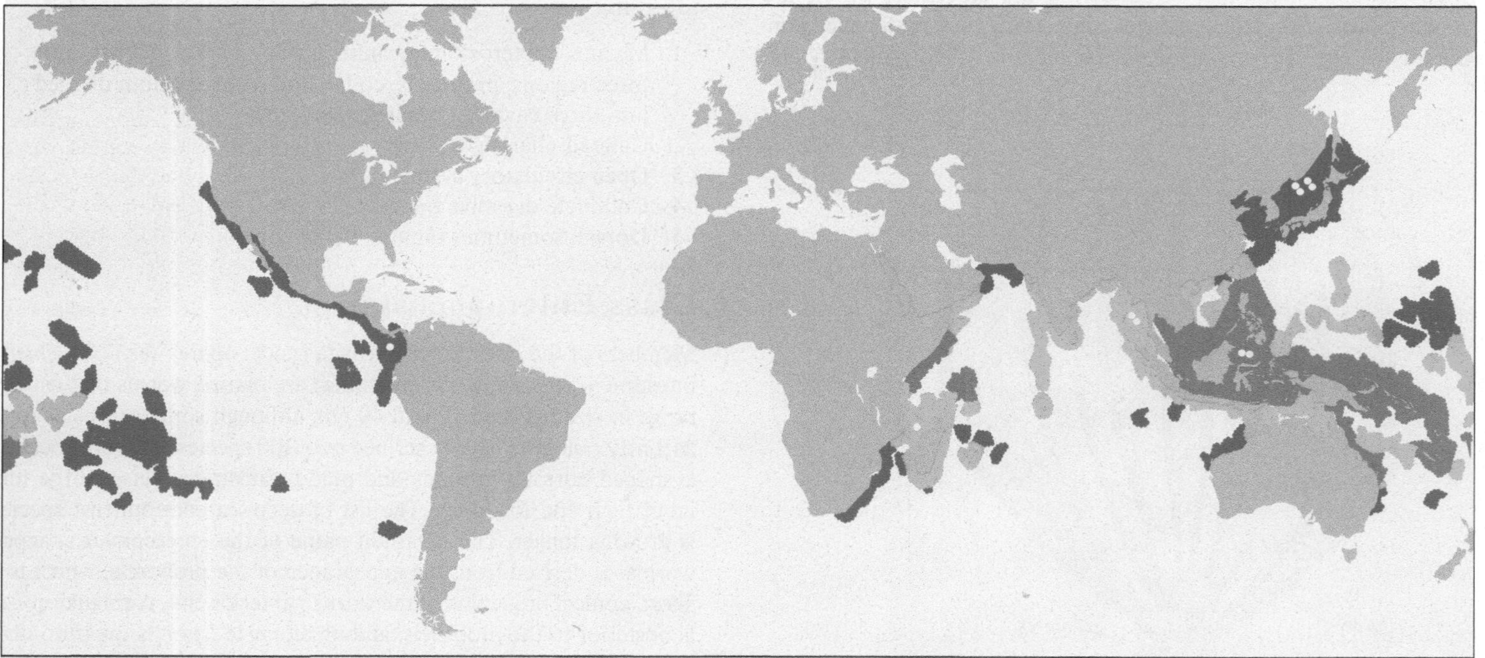

BOX FIGURE 16.1 World-Wide Incidence of Threatened Sea Cucumbers (Order Aspidochirotida). Aspidochirotida is one of five orders within Holothuroidea. It contains 377 species, of which 60 are harvested. Note that most threatened species occur in coastal regions of tropical seas.

Source: Purcell SW, Polidoro BA, Hamel J, Gamboa RU, Mercier A. 2014. The cost of being valuable: predictors of extinction risk in marine invertebrates exploited as luxury seafood. *Proc of the Royal Soc B* [Internet] [cited 2017 Aug 242]; 281: 20133296. Available from: rspb.royalsocietypublishing.org/content/281/1781/20133296.

WILDLIFE ALERT *Continued*

(a)

(b)

BOX FIGURE 16.2 **Threatened Sea Cucumbers.** These high-demand species are members of the order Aspidochirotida. (*a*) *Holothuria scabra.* (b) *Apostichopus japonicus.*
(a) ©Juan Antonio Orihuela Sanchez/Alamy; (b) ©Andrey Nekrasov/Alamy

Evaluating effects of overfishing on sea cucumber populations is difficult because taxonomic status, ecology, and population statistics are unknown for many species. It is estimated by the IUCN (International Union for Conservation of Nature) and CITES (Convention on International Trade in Endangered Species of Wild Fauna and Flora) that populations of the most valuable species have undergone a 60% reduction. Locally, these population reductions are even greater. The IUCN has red listed some sea cucumber species as endangered (e.g., populations of *Isostichopus fuscus* around the Galápagos Islands have undergone an 80% reduction).

There are efforts to regulate sea cucumber harvesting. In the 1990s Ecuador's government imposed a ban on sea cucumber harvesting in Galápagos waters. The ban was modified, but restrictive harvesting was very difficult to monitor because of the lack of enforcement resources. Similar moratoria were enacted in the Solomon Islands, Fiji, and Tonga. Harvesting regulations focus on quotas, seasons, equipment, and sea cucumber size limits. Regulations are in place in many Pacific nations, including the Polynesian islands and Australia, Mexico, and Canada. In the United States, regulations are state-sponsored with Washington State and Alaska leading the way in regulatory efforts. Studies conducted during the past two decades suggest that these regulations have had limited success in restoring already depleted sea cucumber populations. Limited success is due to sea cucumber longevity (10 years or more), slow growth rates, low reproductive rates, and illegal fishing. The U.S. Fish and Wildlife Service frequently encounters sea cucumbers, illegally fished in Mexico, being brought into the United States in small quantities by individual carriers. The smuggled sea cucumbers are then assembled into large containers for shipment to China. One Arizona-based operation was indicted in 2017 for shipping $17.5 million worth of illegal Mexican sea cucumbers to China from San Diego, California. Chinese aquaculture efforts to reduce fishing demands are supplying about 10,000 t of sea cucumbers per year.

Clearly, the exploitation of sea cucumbers is a worldwide problem that involves demands for animal products in distant countries, the economic and survival interests of native populations, and scientific conservation efforts. Balancing all of these interests requires multinational and multidisciplinary approaches to conservation. It also reinforces the need for more research to understand the biology of even the least charismatic species that comprise much of Earth's biodiversity.

slits are openings between the anterior region of the digestive tract, called the pharynx, and the outside of the body.

Maintenance Functions

Cilia and mucus assist acorn worms in feeding. Detritus and other particles adhere to the mucus-covered proboscis. Tracts of cilia transport food and mucus posteriorly and ventrally. Ciliary tracts converge near the mouth and form a mucoid string that enters the mouth. Acorn worms may reject some substances trapped in the mucoid string by pulling the proboscis against the collar. Ciliary tracts of the collar and trunk transport rejected material and discard it posteriorly.

The digestive tract of enteropneusts is a simple tube. Food is digested as diverticula of the gut, called hepatic sacs, release enzymes. The worm extends its posterior end out of the burrow during defecation. At low tide, coils of fecal material, called castings, lie on the substrate at burrow openings.

The nervous system of enteropneusts is ectodermal in origin and lies at the base of the ciliated epidermis. It consists of dorsal and ventral nerve tracts and a network of epidermal nerve cells, called a nerve plexus. In some species, the dorsal nerve is tubular and usually contains giant nerve fibers that rapidly transmit impulses. There are no major ganglia. Sensory receptors are unspecialized and widely distributed over the body.

Because acorn worms are small, respiratory gases and metabolic waste products (principally ammonia) probably are exchanged by diffusion across the body wall. In addition, respiratory gases are exchanged at the pharyngeal slits. Cilia associated with pharyngeal slits circulate water into the mouth and out of the body through the pharyngeal slits. As water passes through the pharyngeal slits, gases are exchanged by diffusion between water and blood sinuses surrounding the pharynx.

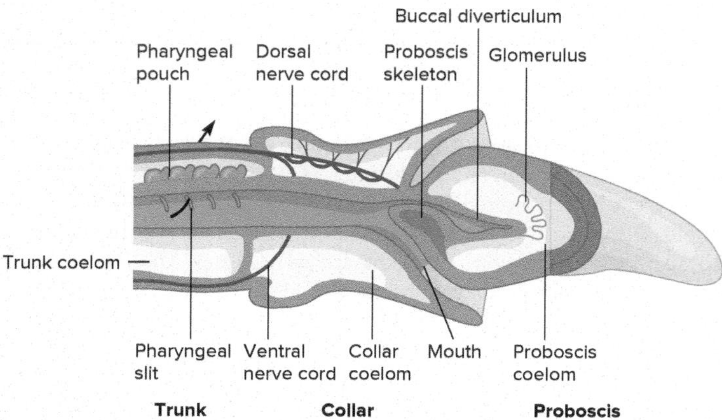

FIGURE 16.16

Class Enteropneusta. Longitudinal section showing the proboscis, collar, pharyngeal region, and internal structures. The black arrow shows the path of water through a pharyngeal slit.

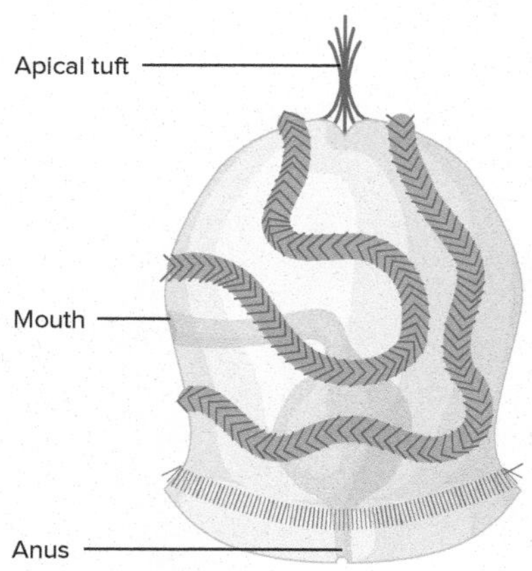

FIGURE 16.17

Tornaria Larva of an Enteropneust (*Balanoglossus*). When larval development is complete, a tornaria locates a suitable substrate, settles, and begins to burrow and elongate (1 mm). *See figure 16.6* for a comparison to echinoderm larval stages.

The circulatory system of acorn worms consists of one dorsal and one ventral contractile vessel. Blood moves anteriorly in the dorsal vessel and posteriorly in the ventral vessel. Branches from these vessels lead to open sinuses. All blood flowing anteriorly passes into a series of blood sinuses, called the glomerulus, at the base of the proboscis. A small, somewhat rigid diverticulum of the gut tract called the buccal diverticulum extends into the proboscis and appears to antagonize contractile elements that pump the blood. The buccal diverticulum is a synapomorphy that unites the Enteropneusta and Pterobranchia within the Hemichordata (*see figure 16.19*). Excretory wastes may be filtered through the glomerulus, into the coelom of the proboscis, and released to the outside through one or two pores in the wall of the proboscis. The blood of acorn worms is colorless, lacks cellular elements, and distributes nutrients and wastes.

Reproduction and Development

Enteropneusts are dioecious. Two rows of gonads lie in the body wall in the anterior region of the trunk, and each gonad opens separately to the outside. Fertilization is external. Spawning by one worm induces others in the area to spawn—behavior that suggests the presence of spawning pheromones. Ciliated larvae, called **tornaria,** swim in the plankton for several days to a few weeks (figure 16.17). The larvae settle to the substrate and gradually transform into the adult form. In addition to shared deuterostome characteristics, the tripartite coelom, a diffuse epidermal nervous system, and a wealth of molecular information, the similarity of the tornaria larva to echinoderm larvae provides additional evidence of the close evolutionary ties between the hemichordates and echinoderms (*see figure 16.6*).

Class Pterobranchia

Pterobranchia (ter′o-brang″ke-ah) (Gk. *pteron,* wing or feather + *branchia,* gills) is a small class of hemichordates found mostly in deep, oceanic waters of the Southern Hemisphere. A few live in European coastal waters and in shallow waters near Bermuda. Zoologists have described approximately 30 species of pterobranchs.

Pterobranchs are small, ranging in size from 0.1 to 5 mm. Most live in secreted tubes in asexually produced colonies. As in enteropneusts, the pterobranch body is divided into three regions. The proboscis is expanded and shieldlike (figure 16.18). It secretes the tube and aids in movement in the tube. The collar possesses two to nine arms with numerous ciliated tentacles. The trunk is U-shaped.

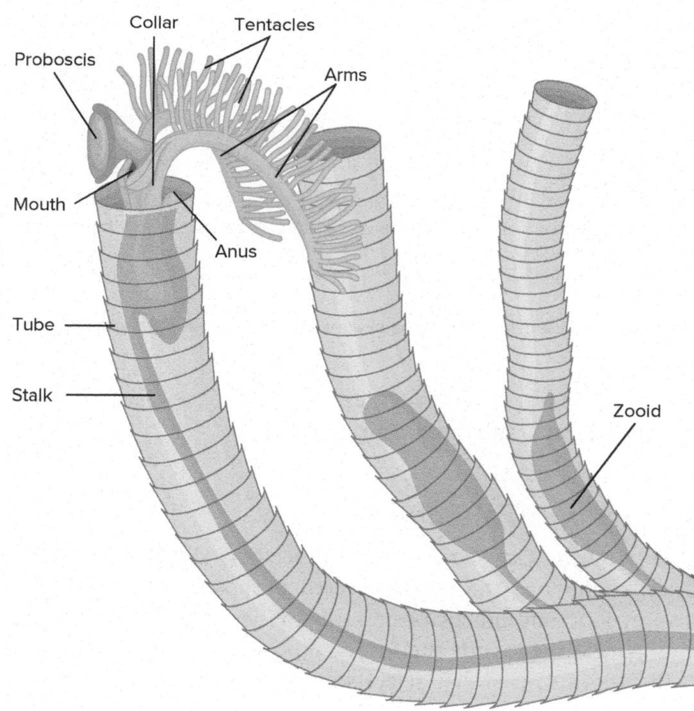

FIGURE 16.18

External Structure of the Pterobranch *Rhabdopleura*. Ciliated tracts on tentacles and arms direct food particles toward the mouth (5 mm).

Maintenance Functions

Pterobranchs use water currents that cilia on their arms and tentacles generate to filter feed. Cilia trap and transport food particles toward the mouth. Although one genus has a single pair of pharyngeal slits, respiratory and excretory structures are unnecessary in animals as small as pterobranchs because gases and wastes exchange by diffusion.

Reproduction and Development

Asexual budding is common in pterobranchs and is responsible for colony formation. Pterobranchs also possess one or two gonads in the anterior trunk. Most species are dioecious, and external fertilization results in the development of a planula-like larva that lives for a time in the tube of the female. This nonfeeding larva eventually leaves the female's tube, settles to the substrate, forms a cocoon, and metamorphoses into an adult.

Section 16.3 Thinking Beyond the Facts

How are the larval stages of enteropneusts similar to those found in the echinoderms? Why might these similarities be important?

16.4 FURTHER PHYLOGENETIC CONSIDERATIONS

LEARNING OUTCOMES

1. Explain the evidence that supports the relationships between the Hemichordata, Echinodermata, and Chordata.
2. Explain why the mouth-down orientation seen in most echinoderms is believed to be a derived character state.
3. Describe the evolutionary relationships among echinoderm classes.

As described at the beginning of this chapter, the echinoderms and the hemichordates comprise the deuterostome clade Ambulacraria (figure 16.19). The similarity of the larval stages that move and feed by ciliated bands (*see figures 16.6 and 16.17*), first noted over 100 years ago, has been used as a homology that reflects the phylogenetic ties between these two phyla. This homology has been confirmed through molecular techniques.

The ties between the amulacrarian clade and the phylum Chordata (*see chapters 17 through 22*) are no less certain. In addition to molecular evidence and deuterostome characteristics like radial cleavage and enterocoelous coelom formation, there is growing evidence that the pharyngeal slits described for the Hemichordata evolved very

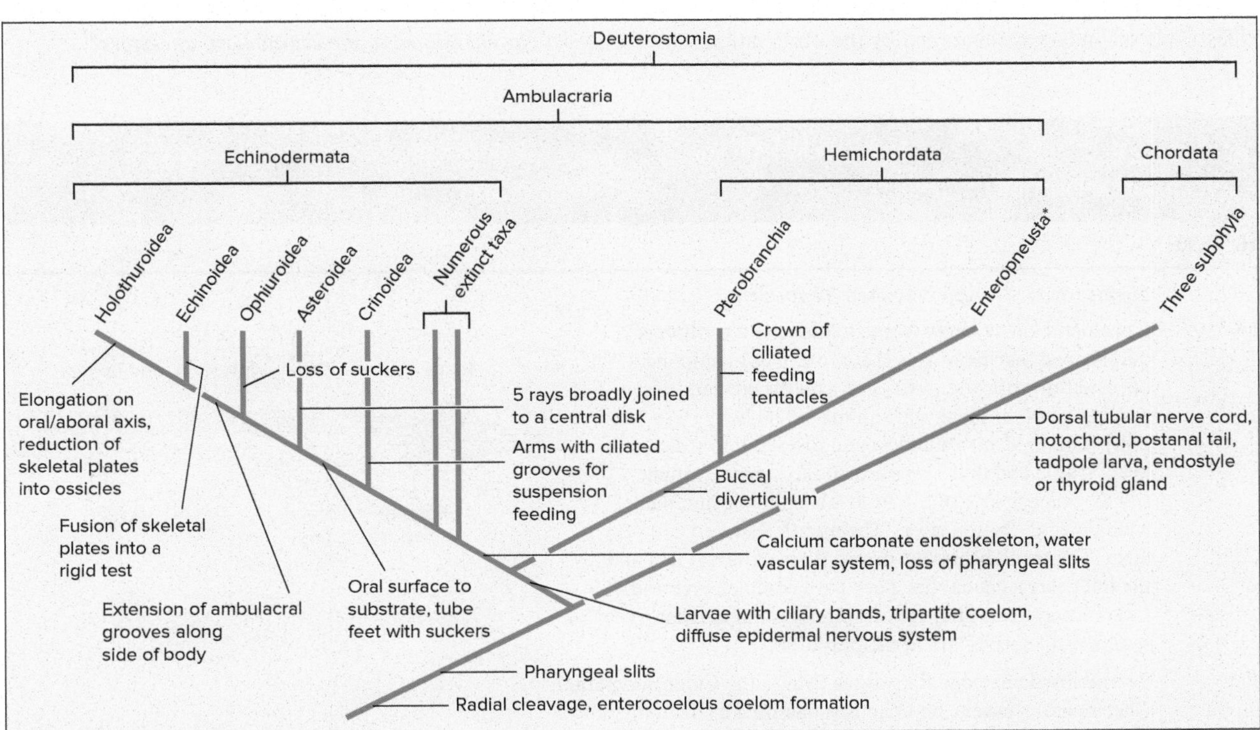

FIGURE 16.19

Deuterostome Phylogeny. The clade Ambulacraria and the phylum Chordata comprise the Deuterostome lineage. The ancestral deuterostome was probably bilaterally symmetrical and used pharyngeal slits in filter feeding. Pharyngeal slits were lost early in the echinoderm lineage. Most zoologists agree that the Crinoidea are closely related to ancestral echinoderms. This interpretation shows a relatively distant relationship between the Asteroidea and Ophiuroidea. Some taxonomists interpret the five-rayed body form as a synapomorphy that links these two groups to a single ancestral lineage. Phylogeny within the Enteropneusta (*) is under investigation. This lineage is probably not monophyletic.

early in the deuterostome lineage. Pharyngeal slits were probably used in filter feeding by deuterostome ancestors. Gene expression studies support the homology of pharyngeal slits in the Hemichordata and Chordata. These structures were evidently lost early in the echinoderm lineage. A group of fossils, called carpoid fossils, are from a diverse assemblage of mid-Cambrian animals (*see appendix B*). Some of these fossils show evidence of a calcium carbonate skeleton (an echinoderm characteristic) and pharyngeal slits. It is tempting to speculate that some carpoids may be stem echinoderms, thus further supporting the deuterostome antiquity of pharyngeal slits.

Of all living echinoderms, the crinoids most closely resemble the oldest fossils. Because crinoids use their water-vascular system primarily for suspension feeding rather than locomotion, the former was probably the original function of the water-vascular system. As do crinoids, early echinoderms probably assumed a mouth-up position and attached aborally. They may have used arms and tube feet to capture food and move it to the mouth. The calcium carbonate endoskeleton may have evolved to support extended filtering arms and to protect these sessile animals.

Many modern echinoderms are more mobile. This mobile lifestyle is probably secondarily derived, as is the mouth-down orientation of most echinoderms. The mouth-down position is advantageous for predatory and scavenging lifestyles. Similarly, changes in the water-vascular system, such as the evolution of ampullae, suction disks, and feeding tentacles, can be interpreted as adaptations for locomotion and feeding in a more mobile lifestyle. The idea that the mobile lifestyle is secondary is reinforced by the observation that some echinoderms, such as the irregular echinoids and the holothuroids, have bilateral symmetry imposed upon a pentaradial body form.

The evolutionary relationships among the echinoderms are not clear. Numerous fossils date into the Cambrian period, but no interpretation of the evolutionary relationships among living and extinct echinoderms is definitive. Inclusion of extinct taxa would result in 15 or more classes. Figure 16.19 shows one interpretation of the evolutionary relationships among extant (with living members) echinoderm classes. The resemblance of crinoids to oldest echinoderm fossils and the analysis of ribosomal RNA and mitochondrial DNA sequences provide strong support for the basal position of the crinoids among extant echinoderms. Most taxonomists agree that the echinoids and holothuroids are closely related. Whether the ophiuroids are more closely related to the echinoid/holothuroid lineage or the asteroid lineage is debated.

Chapters 17 through 22 describe the Chordata—the last phylum covered in the animal-survey portion of this book. Before moving on, notice that there are two additional phyla of lesser-known invertebrates: Chaetognatha and Xenoturbellida (table 16.2). Placing our treatment of these phyla at the end of this chapter is a matter of convenience as their phylogenetic relationships are poorly understood.

SECTION 16.4 THINKING BEYOND THE FACTS

Why are the crinoids considered the group of extant echinoderms most closely related to the echinoderm ancestors?

TABLE 16.2
PHYLA OF UNCERTAIN AFFINITIES

CHAETOGNATHA	
Examples	*Sagitta, Spadella.* Approximately 12 species.
Description	Chaetognathans (arrow worms) comprise a phylum of dart-shaped marine worms that are a major component of plankton worldwide. All chaetognaths are carnivorous, preying on other planktonic animals. The body (2 to 120 mm) is covered with a cuticle and divided into a distinct head, trunk, and tail. There are hooked, prey-grasping spines, on each side of the head—thus the phylum name, which means "bristle jaws." The mouth is armed with tiny teeth. The trunk bears one or two pairs of lateral fins and the tail bears a caudal fin. They have complex eyes and a nervous system. They lack respiratory and circulatory systems. All species are hermaphroditic. 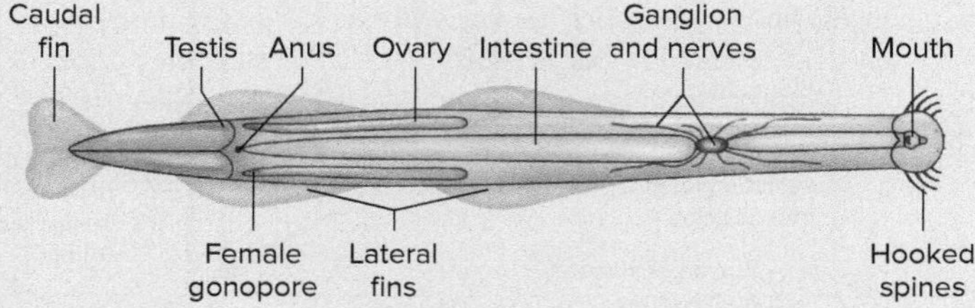
Phylogenetic Relationships	Several fossil chaetognath species date to the Cambrian period. Chaetognaths have been variously classified as deuterostomes, ecdysozoan protostomes, and a group basal to both lophotrochozoan and ecdysozoan lineages. Possible deuterostome affinities are based on deuterostome-like development of chaetognath embryos. Ecdysozan affinities are suggested by the presence of a cuticle and other similarities to nematode muscular and nervous organization. Molecular studies place Chaetognatha close to the base of the protostome tree or within the Ecdysozoa.

XENOTURBELLIDA	
Examples	Two species: *Xenoturbella bocki* and *Xenoturbella westbladi*
Description	*Xenoturbella* is a genus of worm-like (up to 4 cm) bilaterian animals that live in North Sea mud. These animals have no brain, an incomplete gut, no excretory system, and no organized gonads. Sexual reproduction does occur, as gametes are produced from unstructured gonadal tissue. Xenoturbellids possess cilia and a diffuse nervous system. Their larval stage develops as an internal parasite of certain molluscs. The adults feed on bivalves and bivalve eggs.
Phylogenetic Relationships	There are two radically different interpretations of phylogeny for Xenoturbellida. Morphological and molecular studies suggest a close relationship with Acoelomorpha (*see table 9.4*) near the base of the bilaterian tree, thus they may be united with the acoels in a phylum Xenacoelomorpha. Other molecular studies suggest that *Xenoturbella* is a primitive basal deuterostome or in a sister-group relationship with the echinoderms and hemichordates. The latter interpretation is reflected in appendix A; however this representation is extremely tentative.

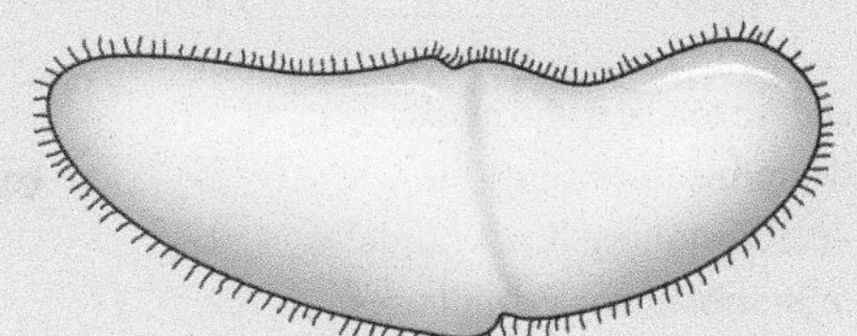

SUMMARY

16.1 **Evolutionary Perspective**

- Echinoderms, hemichordates, and chordates comprise the deuterstome lineage. Ambulacraria is a clade containing Echinodermata and Hemichordata.

16.2 **Phylum Echinodermata**

- Members of the phylum Echinodermata are characterized by the presence of pentaradial symmetry, an internal calcium carbonate skeleton, a water-vascular system, and a hemal system.
- Asteroidea includes the sea stars. They have arms that are broadly attached to a central disk. Sea stars are predators that ingest whole prey, which are digested extracellularly. Gases, nutrients, and metabolic wastes are transported through the coelom by diffusion and ciliated cells lining the body cavity. The nervous system is pentaradially organized. Sea stars are dioecious, and external fertilization results in the formation of planktonic bipinnaria and brachiolaria larvae.
- Ophiuroidea includes basket stars and brittle stars. They have arms sharply set off from a central disk. Ophiuroids are predators and scavengers that use arms and feet to sweep food to their mouths. The coelom of ophiuroids is reduced. It is the primary means of distribution of nutrients, wastes, and gases. Ophiuroids are dioecious. Externally fertilized eggs may either develop in the plankton, or they may be brooded.
- Echinoidea includes the sea urchins and sand dollars. They have round, flattened, or globular tests that are covered with spines, which are used in locomotion. They have a specialized chewing structure, called Aristotle's lantern. External fertilization results in a planktonic pluteus larva.
- Holothuroidea includes sea cucumbers. They have a muscular body with microscopic ossicles. They lack arms and have secondarily derived bilateral symmetry. Sea cucumbers ingest particulate organic matter using their tentacles. Their coelom is large, and cilia distribute respiratory gases, wastes, and nutrients. Respiratory gases and wastes are exchanged across the tubules of respiratory trees. Sea cucumbers are dioecious, and fertilization and development are external.
- Crinoidea includes sea lilies and feather stars. They have a crown of arms connected to a calyx. Arms are used in suspension feeding. A stalk (sea lilies) or cirri (feather stars) is used for attachment to the substrate. Crinoids are dioecious, and fertilization and development are external.

16.3 **Phylum Hemichordata**

- Members of the phylum Hemichordata include the acorn worms and the pterobranchs.
- Acorn worms are burrowing marine worms. They live in U-shaped burrows in intertidal sandy and muddy substrates. They feed on detritus, which clings to their conical proboscis. Reproduction involves external fertilization and the development of a planktonic tornaria larval stage.
- Pterobranchs live in secreted tubes in shallow marine waters. They filter feed using cilia-covered arms and tentacles. Asexual budding and external fertilization with the development of a planula-like larval stage are common forms of reproduction.

16.4 **Further Phylogenetic Considerations**

- The Echinodermata and the Hemichordata comprise the deuterostome clade Ambulacraria. Their relationship is supported by the presence of homologous larval stages and recent molecular evidence. Their closest relatives are members of the phylum Chordata. Relationships within the Echinodermata are controversial. The water-vascular system probably evolved for suspension feeding, similar to its use for that purpose by crinoids. The evolution of a more mobile lifestyle has resulted in the use of the water-vascular system for locomotion and in the assumption of a mouth-down position.

Concept Review Questions

1. Echinoderms are ______________. Their closest relatives include the ______________.
   a. ecdysozoans; arthropods
   b. deuterostomes; annelids
   c. deuterostomes; hemichordates
   d. lophotrochozoans; molluscs
2. All of the following are characteristics of members of the phylum Echinodermata, except one. Select the exception.
   a. Calcareous endoskeleton
   b. Pentaradial symmetry
   c. Water-vascular system
   d. Ventral nervous system with an anterior brain
3. Thin folds of the body wall that function in gas exchange in echinoderms are called
   a. dermal branchiae.
   b. hemal folds.
   c. tube feet.
   d. ossicles.
4. Sea stars comprise the class
   a. Asteroidea.
   b. Holothuroidea.
   c. Crinoidea.
   d. Echinoidea.
5. Members of this group of hemichordates mostly live in deep waters of the Southern Hemisphere and inhabit secreted tubes in asexually produced colonies.
   a. Enteropneusta
   b. Pterobranchia
   c. Ascidiacea
   d. Thaliacea

Analysis and Application Questions

1. What is pentaradial symmetry, and why is it adaptive for echinoderms?
2. Compare and contrast the structure and function of the water-vascular systems of asteroids, ophiuroids, echinoids, holothuroids, and crinoids.
3. In which of the groups in question 2 is the water-vascular system probably most similar in form and function to an ancestral condition? Explain your answer.
4. What physical process is responsible for gas exchange and excretion in all echinoderms? What structures facilitate these exchanges in each echinoderm class?

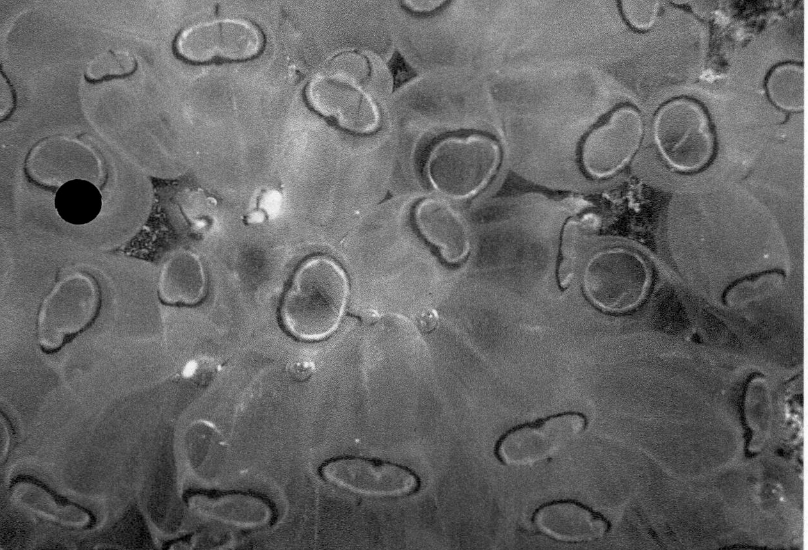

These tunicates (*Clavelina picta, subphylum Urochordata*) look nothing like members of the phylum Chordata, but they are! Careful examination of their larval stage reveals all the features that are hallmarks of the phylum. The pair of openings seen on each tunicate are siphons, which are the openings for incurrent and excurrent water that circulates through this filter-feeding animal.

17

Chordata: Urochordata and Cephalochordata

Chapter Outline

17.1 Evolutionary Perspective
Phylogenetic Relationships
17.2 Phylum Chordata
Subphylum Urochordata
Subphylum Cephalochordata
17.3 Further Phylogenetic Considerations

17.1 EVOLUTIONARY PERSPECTIVE

LEARNING OUTCOME

1. Describe the evolutionary relationships of the Chordata to other animal phyla.

Some members of the phylum discussed in this chapter are more familiar to you than members of any other group of animals. This familiarity is not without good reason, for you yourself are a member of this phylum–Chordata. Other members of the Chordata, however, are much less familiar. During a walk along a seashore you could see animals clinging to rocks exposed by low tide. At first glance, you might describe them as jellylike masses with two openings at their unattached end. Some live as solitary individuals; others live in colonies. If you handle these animals, you may be rewarded with a stream of water squirted from their openings. Casual observations provide little evidence that these small filter feeders, called sea squirts or tunicates, are chordates. However, detailed studies have made that conclusion certain. Tunicates and a small group of fishlike animals called lancelets are often called the invertebrate chordates because they lack a vertebral column (figure 17.1).

Phylogenetic Relationships

Animals in the phylum Chordata share the classic deuterostome characteristics shown in figures 8.11 and 16.19 with echinoderms and hemichordates as well as pharyngeal slits and a host of molecular characters. Most zoologists, therefore, accept the conclusion that ancestral representatives of these phyla were derived from a common deuterostome ancestor (*see figure 17.1*). As discussed at the end of chapter 16, and in the sections that follow, pharyngeal slits play a very important role in our understanding deuterostome relationships and in the lives of members of the Chordata.

SECTION 17.1 THINKING BEYOND THE FACTS

What important character is shared by deuterostomes (but was lost early in the echinoderm lineage)?

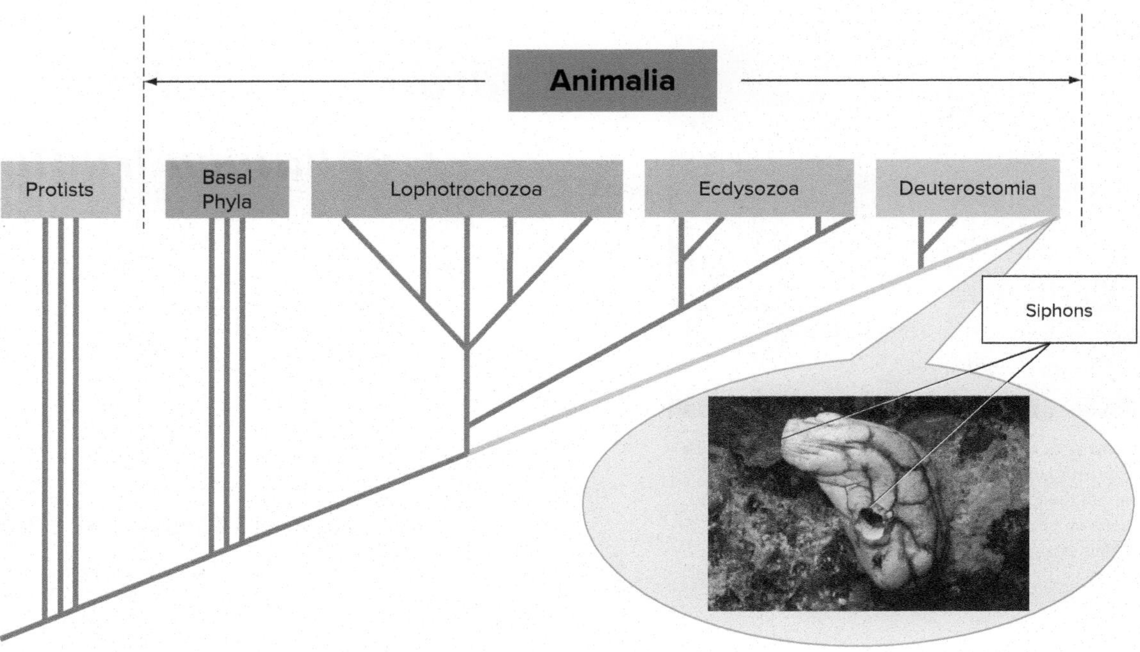

FIGURE 17.1

Evolutionary Relationships of Chordata to Other Animals. This figure shows one interpretation of the relationships of the Chordata to other members of the animal kingdom (*see appendix A*). The relationships depicted here are based on evidence from developmental and molecular biology. Chordates are placed within the Deuterostomia along with the Echinodermata and Hemichordata. This tunicate, or sea squirt (*Polycarpa aurata*), is an invertebrate chordate that attaches to substrates in marine environments. Note the two siphons for circulating water through a filter-feeding apparatus.

Source: Dr. Dwayne Meadows, NOAA/NMFS/OPR

17.2 PHYLUM CHORDATA

LEARNING OUTCOMES

1. Describe the characteristics of members of the phylum Chordata.
2. Compare adult tunicates to the generalized chordate body form.
3. Compare adult cephalochordates to the generalized chordate body form.

Although the phylum Chordata (kor-dat′ah) (L. *chorda,* cord) does not have an inordinately large number of species (about 65,000), its members have been very successful at adapting to aquatic and terrestrial environments throughout the world. Sea squirts, members of the subphylum Urochordata, are briefly described in the "Evolutionary Perspective" that opens this chapter. Other chordates include lancelets (subphylum Cephalochordata) and the vertebrates (subphylum Craniata) (table 17.1). The major characteristics of the phylum Chordata include the following:

1. Bilaterally symmetrical, deuterostomate animals
2. A unique combination of five characteristics present at some stage in development: notochord, pharyngeal slits or pouches, dorsal tubular nerve cord, postanal tail, and an endostyle or thyroid gland
3. Complete digestive tract
4. Ventral, contractile blood vessel (heart)

The combination of five characteristics listed in number 2 is distinctive of chordates, and these characteristics are discussed further in the paragraphs that follow (figure 17.2).

The phylum is named after the **notochord** (Gr. *noton,* the back + L. *chorda,* cord), a supportive rod that extends most of the length of the animal dorsal to the body cavity and into the tail. It consists of a connective-tissue sheath that encloses cells, each of which contains a large, fluid-filled vacuole. This arrangement gives the notochord some turgidity, which prevents compression along the anteroposterior axis. At the same time, the notochord is flexible enough to allow lateral bending, as in the lateral undulations of a fish during swimming. In most adult vertebrates, cartilage or bone partly or entirely replaces the notochord.

Pharyngeal slits are a series of openings in the pharyngeal region between the digestive tract and the outside of the body. As described previously, pharyngeal slits are also found in the Hemichordata. These slits are not unique to the chordates, but they are adapted for use in important and distinctive ways in this phylum. In some chordates, diverticula from the gut in the pharyngeal region never break through to form an open passageway to the outside. These diverticula are then called pharyngeal pouches. The earliest chordates used the slits for filter feeding; some living chordates still use them for feeding. Other chordates have developed gills in the pharyngeal pouches for gas exchange. The pharyngeal slits of terrestrial vertebrates are mainly embryonic features and may be incomplete.

TABLE 17.1
CLASSIFICATION OF THE CHORDATA

Phylum Chordata (kor-dat′ah)
Occupy a wide variety of marine, freshwater, and terrestrial habitats. A notochord, pharyngeal slits, a dorsal tubular nerve cord, a postanal tail, and an endostyle or thyroid gland are all present at some time in chordate life histories. About 65,000 species.

Subphylum Urochordata (u″ro-kor-da′tah)
Notochord, nerve cord, and postanal tail present only in free-swimming larvae; adults sessile, or occasionally planktonic, and enclosed in a tunic that contains some cellulose; marine. Sea squirts or tunicates.

Class Ascidiacea (as-id″e-as′e-ah)
All sessile as adults; solitary or colonial; colony members interconnected by stolons. Sea squirts (*Ascidia, Ciona*). (Ascidiacea is comprised of four clades: Stolidobranchia, Molgulidae, Phlebobranchia, and Aplousobranchiata.)

Class Appendicularia (a-pen″di-ku-lar′e-ah) **(Larvacea)** (lar-vas′e-ah)
Planktonic; adults retain tail and notochord; lack a cellulose tunic; epithelium secretes a gelatinous covering of the body. Appendicularians (*Fritillaria*).

Class Thaliacea (tal″e-as′e-ah)
Planktonic; adults are tailless and barrel shaped; oral and atrial openings are at opposite ends of the tunicate; muscular contractions of the body wall produce water currents. Salps (*Salpa, Thetys*).

Subphylum Cephalochordata (sef″a-lo-kor-dat′ah)
Body laterally compressed and transparent; fishlike; all five chordate characteristics persist throughout life. Amphioxus (*Branchiostoma*).

Subphylum Craniata (kra″ne-ah′tah)*
Skull surrounds the brain, olfactory organs, eyes, and inner ear. Unique embryonic tissue, neural crest, contributes to a variety of adult structures including sensory nerve cells, and some skeletal and other connective tissue structures.

Infraphylum Hyperotreti (hi″per-ot′re′te)
Fishlike; skull consisting of cartilaginous bars; jawless; no paired appendages; mouth with four pairs of tentacles; olfactory sacs open to mouth cavity; 5 to 15 pairs of pharyngeal slits; ventrolateral slime glands. Hagfishes.

Infraphylum Vertebrata (ver″te-bra′tah)
Notochord, nerve cord, postanal tail, and pharyngeal slits present at least in embryonic stages; vertebrae surround nerve cord and serve as primary axial support.

Class Petromyzontida (pet′ro-mi-zon″tid-ah)
Fishlike; jawless; no paired appendages; cartilaginous skeleton; sucking mouth with teeth and rasping tongue. Lampreys.

Class Chondrichthyes (kon-drik′thi-es)
Fishlike; jawed; paired appendages and cartilaginous skeleton; no swim bladder. Skates, rays, sharks.

Class Actinopterygii (ak″tin-op′te-rig-e-i)
Bony fishes having paired fins supported by dermal rays; basal portions of paired fins not especially muscular, tail fin with approximately equal upper and lower lobes (homocercal tail); blind olfactory saca; pneumatic sacs function as swim bladder. Ray-finned fishes.

Class Sarcopterygii (sar-kop-te-rig′e-i)
Bony fishes having paired fins with muscular lobes; pneumatic sacs function as lungs; atria and ventricles at least partially divided. Lungfishes, coelacanths, and tetrapods.**

(Class) Amphibia (am-fib′e-ah)
Skin with mucoid secretions; possess lungs and/or gills; moist skin serves as respiratory organ; aquatic developmental stages usually followed by metamorphosis to an adult. Frogs, toads, salamanders.

(Class) Reptilia (rep-til′e-ah)
Dry skin with epidermal scales; amniotic eggs; terrestrial embryonic development. Snakes, lizards, alligators.

(Class) Aves (a′vez)
Feathers used for flight; efficiently regulate body temperature (endothermic); amniotic eggs. Birds.

(Class) Mammalia (mah-ma′le-ah)
Bodies at least partially covered by hair; endothermic; young nursed from mammary glands; amniotic eggs. Mammals.

*Members of the Craniata are discussed in chapters 18 through 22.
**Cladistic analysis has shown that the following four traditional class designations for tetrapods represent a paraphyletic assemblage. Aves is a lineage within Reptilia, and all tetrapods comprise a clade within Sarcopterygii (*see chapters 18 and 19*).

The **tubular nerve cord** and its associated structures are largely responsible for chordate success. The nerve cord runs along the longitudinal axis of the body, just dorsal to the notochord, and usually expands anteriorly as a brain. This central nervous system is associated with the development of complex systems for sensory perception, integration, and motor responses.

A fourth chordate characteristic is a **postanal tail.** (A postanal tail extends posteriorly beyond the anal opening.) Either the notochord or vertebral column supports the tail.

The fifth characteristic unique to chordates is the presence of an **endostyle** or **thyroid gland.** An endostyle is present on the ventral aspect of the pharynx in tunicates, cephalochordates, and larval

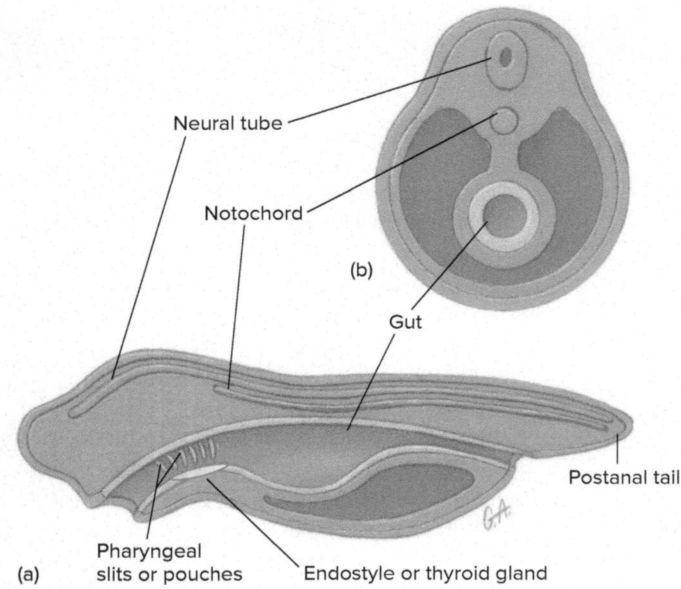

FIGURE 17.2

Chordate Body Plan. The development of all chordates involves the formation of a neural tube, the notochord, pharyngeal slits or pouches, a postanal tail, and an endostyle or thyroid gland. Derivatives of all three primary germ layers are present. (*a*) Lateral view. (*b*) Cross section.

lampreys. It secretes mucus that helps trap food particles during filter feeding. In adult lampreys and other chordates, the endostyle is transformed into an endocrine structure, the thyroid gland. (*See "How Do We Know about the Evolution of the Thyroid Gland from the Endostyle?" page 308.*)

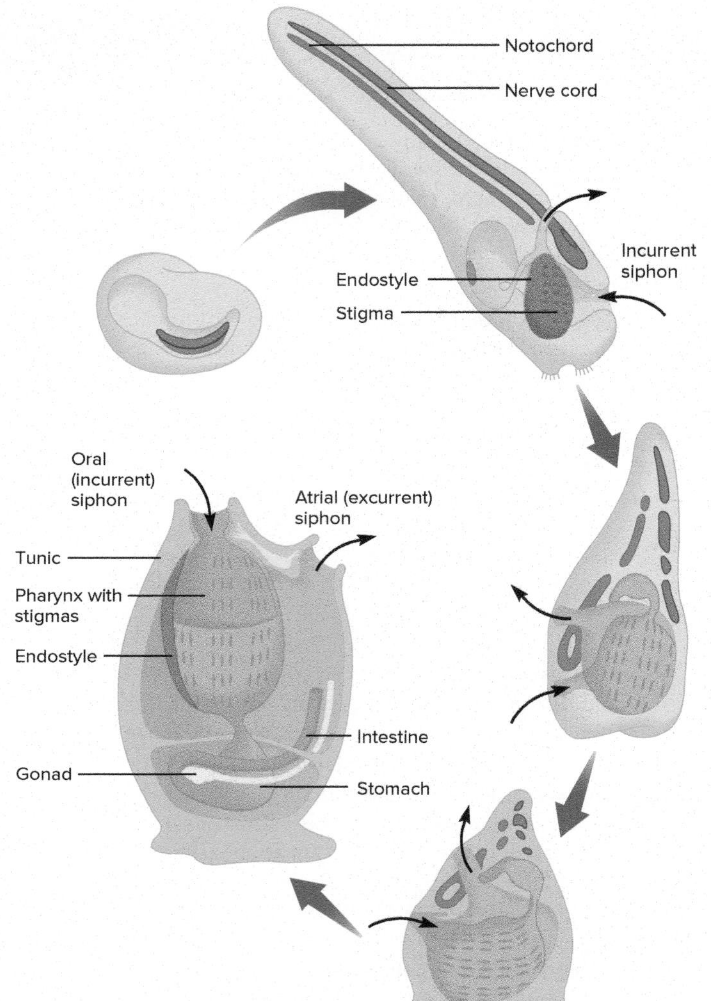

FIGURE 17.3

Tunicate Metamorphosis. Small black arrows show the path of water through the body.

Subphylum Urochordata

Members of the subphylum Urochordata (u″ro-kor-dah′tah) (Gr. *uro,* tail + L. *chorda,* cord) are the tunicates or sea squirts. Phylogenetic analysis has demonstrated the presence of six urochordate clades that are currently grouped into three classes. Ascidiacea (four of six clades) is the largest class of tunicates (2,300 species, *see table 17.1 and chapter opener and figure 17.1 photographs*). They are sessile as adults and are either solitary or colonial. In some localities, tunicates occur in large enough numbers to be considered a dominant life-form. Ascidians attach their saclike bodies to rocks, pilings, ship hulls, and other solid substrates. The unattached end of urochordates contains two siphons that permit seawater to circulate through the body. One siphon is the oral siphon, which is the inlet for water circulating through the body and is usually directly opposite the attached end of the ascidian (figure 17.3). It also serves as the mouth opening. The second siphon, the atrial siphon, is the opening for excurrent water.

The body wall of most tunicates (*L. tunicatus,* to wear a tunic or gown) is a connective-tissue-like covering, called the tunic, that appears gel-like but is often quite tough. Secreted by the epidermis, it is composed of proteins, various salts, and cellulose. Some mesodermally derived tissues, including blood vessels and blood cells, are incorporated into the tunic. Rootlike extensions of the tunic, called stolons, help anchor a tunicate to the substrate and may connect individuals of a colony.

A second class of urochordates is Appendicularia (approximately 70 species, figure 17.4*a*). They are planktonic in the pelagic zone (near the surface of the open ocean to just above the seafloor). Adults retain the tadpole body form typical of larval urochordates, and they are usually less than 1cm in length (excluding the tail). Appendicularians secrete a test of protein and cellulose that surrounds the animal (like a bubble) and includes a delicate filter-feeding mesh. The mesh is frequently clogged, and the test is discarded and rebuilt several times each day. These tests contribute to "marine snow," suspended organic particles that are continually raining down through oceanic water columns. Appendicularians are common and abundant, but they are seldom noticed because they are so very small and delicate.

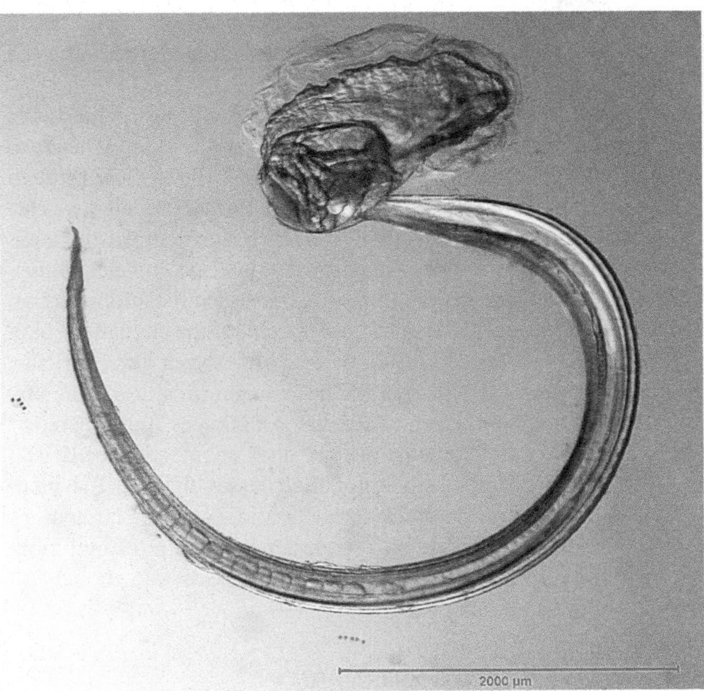

(a)

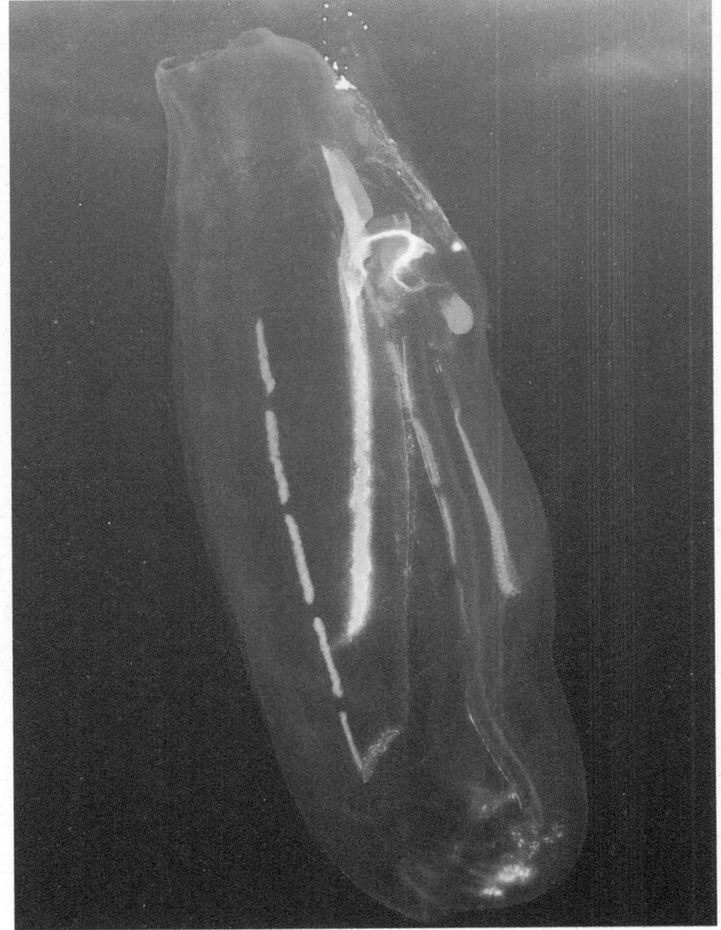

(b)

FIGURE 17.4

Subphylum Urochordata. (*a*) Members of the class Appendicularia are planktonic and have a tail and notochord that persist into the adult stage. *Oiko labradorensis* is shown here without its enclosing test. (*b*) The thaliaceans, or salps, are barrel-shaped, planktonic urochordates. Oral and atrial siphons are at opposite ends of the body, and muscles of the body wall contract to create a form of weak jet propulsion. *Cyclosalpa* individuals form ringlike colonies or colonies comprised of chains of rings. A single individual is shown here. Note the blue-green bioluminescent organs on the sides of this individual. The central yellow stripe is the intestinal loop.

The third class of urochordates is Thaliacea (also approximately 70 species, figure 17.4*b*). Thaliaceans are free swimming, pelagic filter feeders. Some species (e.g., pyrosomes) form bioluminescent colonies of hundreds or thousands of individuals that can be several meters in length. Other species are comprised of solitary individuals. Thaliaceans are also common and comprise an important component of marine food webs.

Maintenance Functions

Longitudinal and circular muscles below the body-wall epithelium help change the shape of the adult tunicate. They act against the elasticity of the tunic and the hydrostatic skeleton that seawater confined to internal chambers creates.

The nervous system of tunicates is largely confined to the body wall. It forms a nerve plexus with a single ganglion located on the wall of the pharynx between the oral and atrial openings (figure 17.5*a*). This ganglion is not vital for coordinating bodily functions. Tunicates are sensitive to many kinds of mechanical and chemical stimuli, and receptors for these senses are distributed over the body wall, especially around the siphons. There are no complex sensory organs.

The most obvious internal structures of the urochordates are a very large pharynx and a cavity, called the atrium, that surrounds the pharynx laterally and dorsally (figure 17.5*b*). The pharynx of tunicates originates at the oral siphon and is continuous with the remainder of the digestive tract. The oral margin of the pharynx has tentacles that prevent large objects from entering the pharynx. Numerous pharyngeal slits called stigmas perforate the pharynx. Cilia associated with the stigmas cause water to circulate into the pharynx, through the stigmas, and into the surrounding atrium. Water leaves the tunicate through the atrial siphon.

The digestive tract (gut) of adult tunicates continues from the pharynx and ends at the anus near the atrial siphon. The endostyle is a ventral ciliated groove that forms a mucous sheet (figure 17.5*b*). Cilia move the mucous sheet dorsally across the pharynx. Food particles, brought into the oral siphon with incurrent water, are trapped in the mucous sheet and passed dorsally. Food is incorporated into a string of mucus that by ciliary action moves into the next region of the gut tract. Digestive enzymes are secreted in the stomach, and most absorption occurs across the walls of the intestine. Excurrent water carries digestive wastes from the anus out of the atrial siphon.

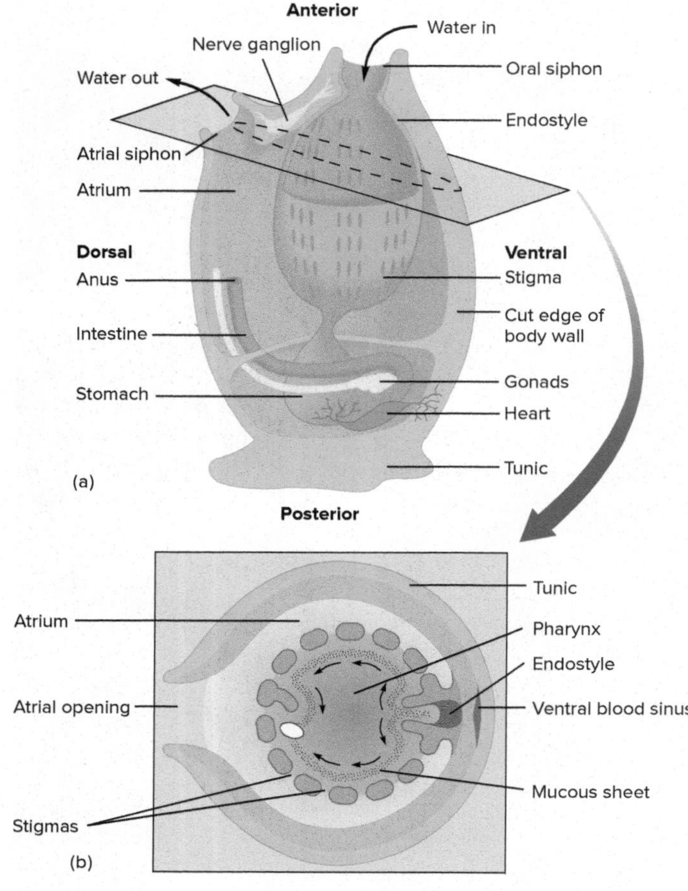

FIGURE 17.5

Internal Structure of a Tunicate. (*a*) Longitudinal section. Black arrows show the path of water. (*b*) Cross section at the level of the atrial siphon. Small black arrows show movement of food trapped in mucus that the endostyle produces.

In addition to its role in feeding, the pharynx also functions in gas exchange. Gases are exchanged as water circulates through the tunicate.

The tunicate heart lies at the base of the pharynx. One vessel from the heart runs anteriorly under the endostyle, and another runs posteriorly to the digestive organs and gonads. Blood flow through the heart is not unidirectional. Peristaltic contractions of the heart may propel blood in one direction for a few beats; then the direction is reversed. The significance of this reversal is not completely understood. One intriguing suggestion is that one-way circulation through the heart would mean that tissues near the end of the circulatory path would only contact blood already depleted of oxygen and nutrients. Periodic reversal of flow through the heart might provide a more uniform distribution of oxygen and nutrients to all tissues. Tunicate blood plasma is colorless and contains various kinds of amoeboid cells.

Ammonia diffuses into water that passes through the pharynx and is excreted. In addition, amoeboid cells of the circulatory system accumulate uric acid and sequester it in the intestinal loop. Pyloric glands on the outside of the intestine are also thought to have excretory functions.

Reproduction and Development

Urochordates are monoecious. Gonads are located near the loop of the intestine, and genital ducts open near the atrial siphon. Gametes may be shed through the atrial siphon for external fertilization, or eggs may be retained in the atrium for fertilization and early development. Although self-fertilization occurs in some species, cross-fertilization is the rule. Development results in the formation of a tadpolelike larva with all five chordate characteristics. Metamorphosis begins after a brief free-swimming larval existence, during which the larva does not feed. The larva settles to a firm substrate and attaches by adhesive papillae located below the mouth. During metamorphosis, the outer epidermis shrinks and pulls the notochord and other tail structures internally for reorganization into adult tissues. The internal structures rotate 180°, positioning the oral siphon opposite the adhesive papillae and bending the digestive tract into a U-shape (*see figure 17.3*).

How Do We Know about the Evolution of the Thyroid Gland from the Endostyle?

The study of the development of one group of vertebrates, the lampreys, has provided insight into the endostyle's evolutionary history. In addition to producing mucus for filter feeding in invertebrate chordates, the endostyle can bind the amino acid tyrosine. This is true in larval lampreys. When a larval lamprey metamorphoses to an adult and becomes a predator, the mucus-secreting functions of the endostyle become secondary, and the endostyle is transformed into the thyroid gland. The iodine-containing secretions of the thyroid gland regulate metamorphosis and metabolic rate. It is widely accepted that the development of the lamprey's thyroid gland demonstrates the evolutionary events that led to the vertebrate thyroid gland. The endostyle of vertebrate ancestors probably had both mucus-secreting and endocrine functions. With the evolution of jaws and a more active, predatory lifestyle, endocrine functions would have been favored.

Subphylum Cephalochordata

Members of the subphylum Cephalochordata (sef″a-lo-kor-dah′tah) (Gr. *kephalo,* head + L. *chorda,* cord) are called lancelets. Lancelets clearly demonstrate the five chordate characteristics, and for that reason they are often studied in introductory zoology courses.

The cephalochordates consist of three extant genera, *Epigonichthys*, *Branchiostoma* (amphioxus), and *Asymmetron,* and about 32 species. They are distributed throughout the world's oceans in shallow waters that have clean sand substrates.

Cephalochordates are small (up to 5 cm long), fishlike animals. They are elongate, laterally flattened, and nearly transparent. In spite of their streamlined shape, cephalochordates are relatively weak swimmers and spend most of their time in a filter feeding position–partly to mostly buried with their anterior end sticking out of the sand (figure 17.6*a*).

The notochord of cephalochordates extends from the tail to the head, giving them their name. Unlike the notochord of other chordates, most of the cells are muscle cells, making the notochord somewhat contractile. Both of these characteristics are probably adaptations to burrowing. Contraction of the muscle cells increases the rigidity of the notochord by compressing the fluids within, giving additional support when pushing into sandy substrates. Relaxation of these muscle cells increases flexibility for swimming.

Segmentally arranged muscle cells on either side of the notochord cause undulations that propel the cephalochordate through the water. Longitudinal, ventrolateral folds of the body wall help stabilize cephalochordates during swimming, and a median dorsal fin and a caudal fin also aid in swimming.

An oral hood projects from the anterior end of cephalochordates. Ciliated, fingerlike projections, called cirri, hang from the ventral aspect of the oral hood and are used in feeding. The posterior wall of the oral hood bears the mouth opening that leads to a large pharynx. Numerous pairs of pharyngeal slits perforate the pharynx and are supported by cartilaginous gill bars. Large folds of the body wall extend ventrally around the pharynx and fuse at the ventral midline of the body, creating the atrium, a chamber that surrounds the pharyngeal region of the body. It may protect the delicate, filtering surfaces of the pharynx from bottom sediments. The opening from the atrium to the outside is called the atriopore (*see figure 17.6*a).

Maintenance Functions

Cephalochordates are filter feeders. During feeding, they are partially or mostly buried in sandy substrates with their mouths pointed upward. Ciliary bands positioned on folded epithelium on the lateral surfaces of gill bars sweep water into the mouth. These ciliary bands beat in waves creating the impression of a rotating wheel, thus the name "wheel organ" (figure 17.6*b*). Water passes from the pharynx, through pharyngeal slits to the atrium, and out of the body through the atriopore. Food is initially sorted at the cirri. Larger materials catch on cilia of the cirri. As these larger particles accumulate, contractions of the cirri throw them off. Smaller, edible particles are pulled into the mouth with water and are collected by cilia on the gill bars and in mucus secreted by the endostyle. As in tunicates, the endostyle is a ciliated groove that extends longitudinally along the midventral aspect of the pharynx. Cilia move food and mucus dorsally, forming a food cord to the gut. A ring of cilia rotates the food cord, dislodging food. Digestion is both extracellular and intracellular. A diverticulum off the gut, called the midgut cecum, extends anteriorly. It ends blindly along the right side of the pharynx and secretes digestive enzymes. An anus is on the left side of the ventral fin.

Cephalochordates do not possess a true heart. Contractile waves in the walls of major vessels propel blood. Blood contains amoeboid cells and bathes tissues in open spaces.

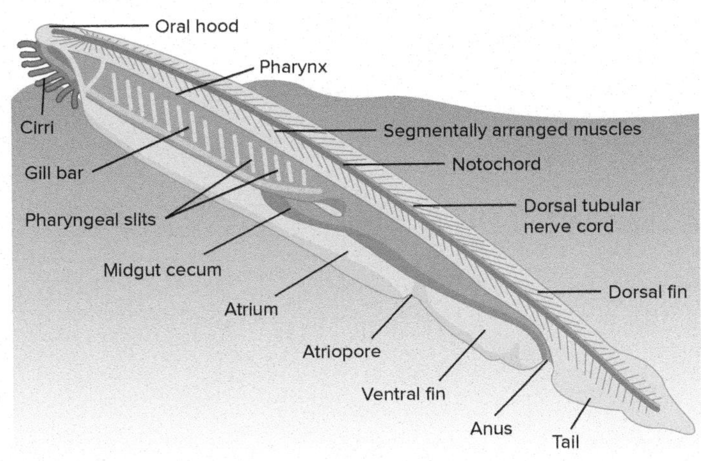

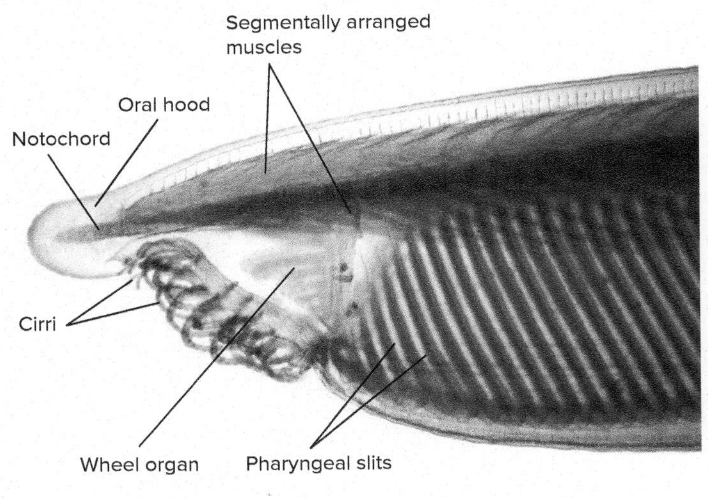

FIGURE 17.6

Subphylum Cephalochordata. (*a*) Internal structure of *Branchiostoma* (amphioxus) shown in its partially buried feeding position. (*b*) Oral hood and anterior pharyngeal region of *Branchiostoma*.

Excretory tubules are modified coelomic cells closely associated with blood vessels. This arrangement suggests active transport of materials between the blood and the excretory tubules.

The coelom of cephalochordates is reduced, compared to that of most other chordates. It is restricted to canals near the gill bars, the endostyle, and the gonads.

Reproduction and Development

Cephalochordates are dioecious. Gonads bulge into the atrium from the lateral body wall. Gametes are shed into the atrium and leave the body through the atriopore. External fertilization leads to a bilaterally symmetrical larva. Larvae are free swimming, but they eventually settle to the substrate before metamorphosing into adults.

SECTION 17.2 THINKING BEYOND THE FACTS

How are the five distinctive chordate characteristics represented in tunicates and cephalochordates?

17.3 FURTHER PHYLOGENETIC CONSIDERATIONS

LEARNING OUTCOMES

1. Describe the relationships between members of the three chordate subphyla.
2. Characterize members of the largest craniate infraphylum, Vertebrata.
3. Hypothesize on the body forms of the basal deuterostome and the common deuterostome ancestor of the three chordate clades.

Even though the evolutionary pathways leading from the common deuterostome ancestor to the Echinodermata, Hemichordata, and Chordata are speculative; evidence linking these phyla is well established (figure 17.7, *see Evolutionary Insights, page 311*). Fossil evidence has now clearly demonstrated that the use of pharyngeal slits in filter feeding is a very old deuterostome characteristic that was subsequently lost in the echinoderm lineage.

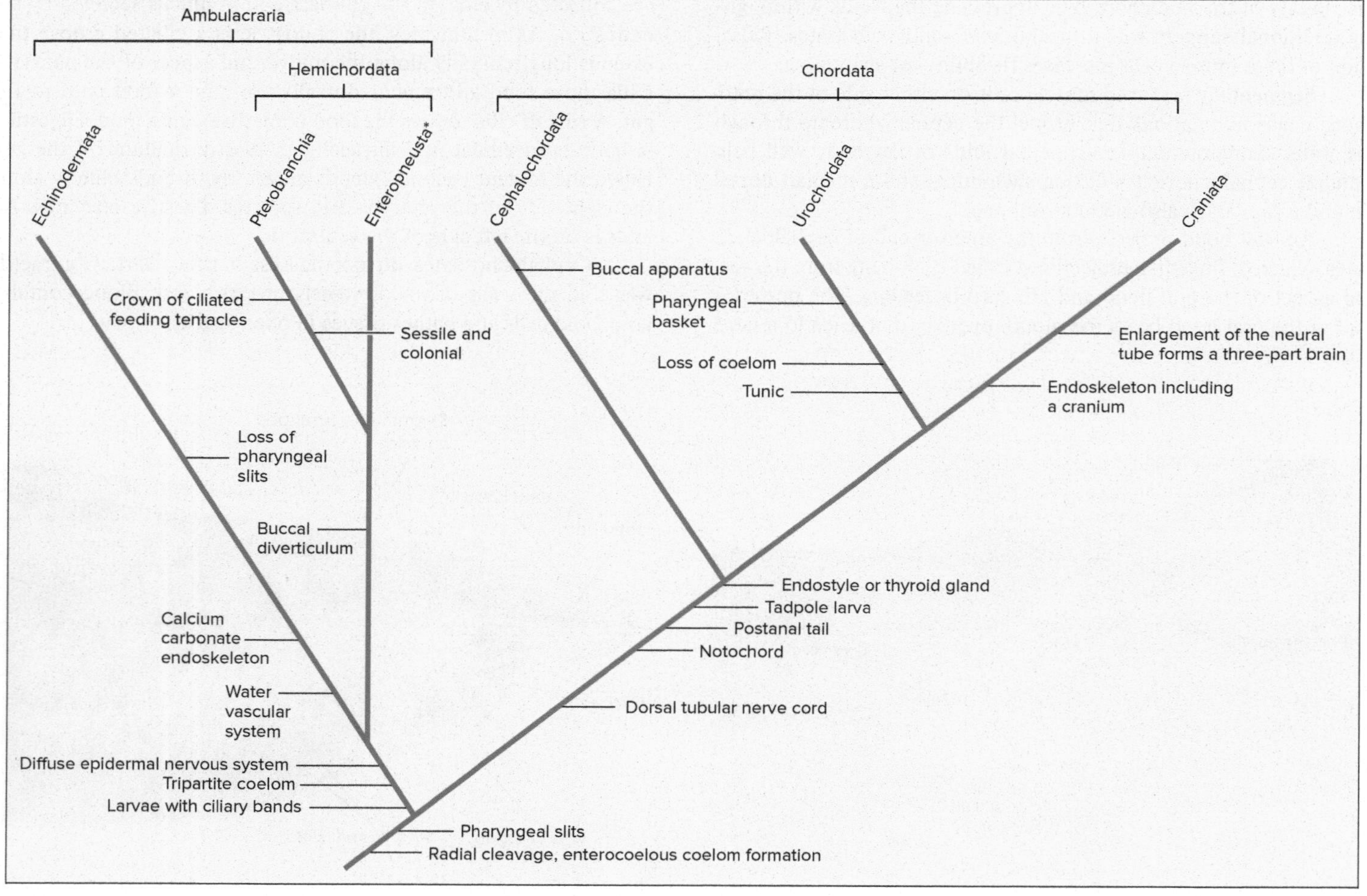

FIGURE 17.7

One Interpretation of Deuterostomate Phylogeny. Developmental and molecular evidence link the echinoderms and hemichordates into the clade Ambulacraria. The Enteropneusta (*) is probably not monophyletic. The dorsal tubular nerve cord, notochord, postanal tail, and endostyle (thyroid gland in adult vertebrates) are important characteristics distinctive of the Chordata. Some of the synapomorphies that distinguish the chordate subphyla are shown.

EVOLUTIONARY INSIGHTS

Early Deuterostome Evolution

Early Deuterostomes

Understanding the origin and diversification of the deuterostomes involves compiling and analyzing evidence from sophisticated methodologies used in the fields of paleontology, molecular biology, and developmental biology. Hypotheses derived from these data are tentative, and they will likely be revised as new evidence is discovered and analyzed.

Presently, the deuterostome fossil record extends to about 520 million years, and this record reveals that all five deuterostome groups (Echinodermata, Hemichordata, Urochordata, Cephalochordata, and Craniata) were present in the mid-Cambrian seas (*see appendix B*). Most molecular clock data (*see chapter 4, How Do We Know Evolutionary Timescales?*) suggest basal deuterostomes arose in the early Cambrian period (about 550 mya). To look into this additional 30 million years, we must rely on molecular and developmental evidence.

Ample developmental evidence indicates that echinoderms and hemichordates share a common ancestry (clade Ambulacraria). All deuterostomes share a set of 14 *HOX* genes (*see chapter 4, How Do We Know about Evolution—"Evo-Devo"?*) that regulate anterior/posterior patterning during embryonic development. In ambulacrarians there is a shared duplication in three of these genes. Different genes control the development of the dorsoventral axis of deuterostomes. In the ambulacrarians, this axis is oriented more similarly to that of the protostomes. In chordates, the axis is inverted. This inversion results in the pharyngeal bars being positioned ventrally in chordates as compared to their dorsal position in hemichordates and in fossil echinoderms that show pharyngeal slits. Additionally, ambulacrarian development involves dipleurula-like larvae, which many scientists consider a homologous character among deuterostomes. The duplicated *HOX* genes, the differing orientation of the gill slits, and the homologous dipleurula-like larvae all point to a very early divergence of the ambulacrarian and chordate clades (*see figure 17.7*).

Two hypotheses may help us visualize the basal deuterostome ancestor, and both hypotheses involve a hemichordate-like animal. A pterobranch-like model for the basal deuterostome ancestor is favored by some zoologists because of the similarity of pterobranch tentacles and lophophores of some lophotrochozoans (*see appendix A and figure 10.2*b). This model links protostome and deuterostome lineages, but is only weakly supported by molecular data. A second hypothesis involves an enteropneust-like ancestor with filter-feeding pharyngeal slits, well-developed muscles, and a simple nerve-net structure. How one interprets the basal deuterostome ancestor would then influence one's interpretation of the ancestral ambulacrarian that was derived from that basal ancestor.

Soft-tissue fossils of the mid-Cambrian provide somewhat better resolution of later ambulacrarian and chordate (*see the following text*) lineages. There is some ambiguity in interpretation of soft-tissue fossils because preservation is incomplete; however, they do document the Cambrian diversification of pterobranchs and echinoderms.

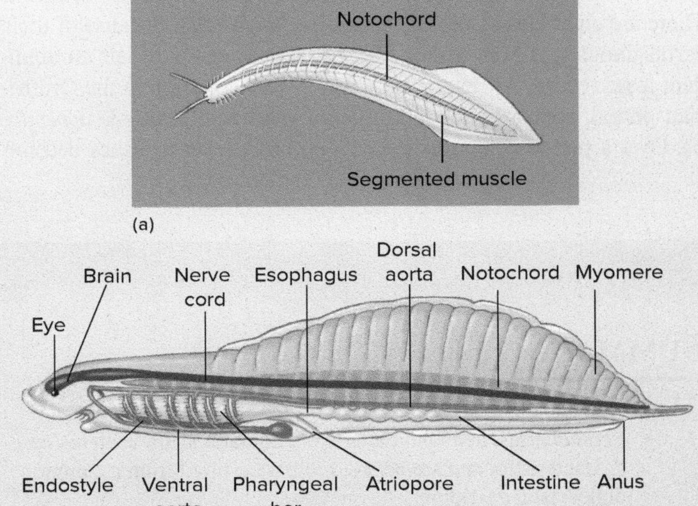

BOX FIGURE 17.1 **Early Chordates.** (*a*) This drawing is based on a fossil of a 520-million-year-old chordate, *Pikaia gracilens,* from the Burgess Shale of British Columbia. (*b*) *Haikouella* is the oldest chordate fossil (530 million years old) found in fossil beds in China.

Early Chordates

Because molecular data favor a sister relationship between tunicates and craniates, many zoologists expect earliest chordate fossils to show shared cephalochordate/craniate characters: cephalization (a brain and eyes), a notochord, segmentally arranged muscle bundles (myotomes), and pharyngeal slits or bars (*see figure 17.7*). Various soft-tissue fossils show one or more of these features, but two fossils have generated considerable interest regarding chordate origins. *Pikaia* was unearthed from the Burgess fossil beds (520 mya, *see chapter 8*), and *Haikouella* has been described from older (530 mya) fossil beds in China (box figure 17.1). The former shows evidence of myomeres and a notochord but not pharyngeal slits. The latter shows evidence of the chordate characteristics listed previously. The chordate affinity of these animals is generally accepted, but their placement in chordate phylogeny is uncertain.

Early deuterostome phylogeny remains unresolved, but progress is being made. The future surely promises new inspired zoologists, new fossil discoveries, new techniques for the analysis of soft-tissue fossils, and new molecular and developmental techniques that will work together to reveal more about our oldest ancestors.

All molecular, morphological, and developmental evidence point to the chordates as a monophyletic clade. The third chordate subphylum is Craniata (*see table 17.1 and figure 17.7*). The origin of the chordate clade is tentative as is the relationship between the tunicates, cephalochordates, and craniates. Studies of rRNA genes, and some misinterpreted Paleozoic fossils now classified as stem echinoderms, support a sister-group relationship between the cephalochordates and the craniates. Recent molecular evidence strongly supports a tunicate/craniate sister relationship, which is depicted in figure 17.7.

The largest and most successful craniates belong to the infraphylum Vertebrata. Bony or cartilaginous vertebrae that completely or partially replace the notochord characterize the vertebrates. The development of the anterior end of the nerve cord into a three-part brain (forebrain, midbrain, and hindbrain) and the development of specialized sense organs on the head are evidence of a high degree of cephalization. The skeleton is modified anteriorly into a skull or cranium. There are eight classes of vertebrates (*see table 17.1*). Because of their cartilaginous and bony endoskeletons, vertebrates have left an abundant fossil record. Ancient jawless fishes were common in the Ordovician period, approximately 500 million years ago (mya, *see appendix B*). Over a period of approximately 100 million years, fishes became the dominant vertebrates. Near the end of the Devonian period, approximately 400 mya, terrestrial vertebrates made their appearance. Since that time, vertebrates have radiated into most of the earth's habitats. Chapters 18 through 22 give an account of these events.

Section 17.3 Thinking Beyond the Facts

What evidence supports the clade Ambulacraria? What evidence is used to establish relationships among the three chordate clades? What new evidence could clarify our understanding of the basal deuterostome ancestor?

Summary

17.1 **Evolutionary Perspective**

- Echinoderms, hemichordates, and chordates share deuterostome characteristics and are believed to have evolved from a common diploblastic or triploblastic ancestor.

17.2 **Phylum Chordata**

- Chordates have five distinctive characteristics. A notochord is a supportive rod that extends most of the length of the animal. Pharyngeal slits are a series of openings between the digestive tract and the outside of the body. The tubular nerve cord lies just above the notochord and expands anteriorly into a brain. A postanal tail extends posteriorly to the anus and is supported by the notochord or the vertebral column. An endostyle functions in filter feeding or a thyroid gland functions as an endocrine organ.
- Members of the subphylum Urochordata are the tunicates or sea squirts. Ascidians are sessile filter feeders that attach their saclike bodies to solid substrates. Two siphons permit seawater to circulate through their bodies. Their bodies are covered by a connective-tissue-like tunic. Appendicularians are pelagic filter feeders that are tadpolelike as adults. Thaliaceans are colonial or solitary urochordates that are also pelagic filter feeders. Some are colonial and bioluminescent. The endostyle of urochordates secretes mucus that is used to trap filtered food within their large pharynx. Urochordates are monoecious with external fertilization and the development of a tadpolelike larva.
- The subphylum Cephalochordata includes small, tadpolelike filter feeders that live in shallow marine waters with clean, sandy substrates. Their notochord extends from the tail into the head and is somewhat contractile. Cephalochordates are filter feeders. Cilia on the lateral surfaces of gill bars sweep food into the mouth, and food is filtered as water passes though pharyngeal slits and into the atrium. Cephalochordates are dioecious, and fertilization and larval development are external.

17.3 **Further Phylogenetic Considerations**

- Echinoderms, hemichordates, and chordates comprise Deuterostomia. Chordata is a monophyletic assemblage consisting of three subphyla: Urochordata, Cephalochordata, and Craniata. Urochordata is probably a sister-group to Craniata. The vertebrates are the largest and most successful craniates.
- The origin of the deuterostome lineage is unresolved, but recent evidence is shedding light on lingering questions. Molecular data suggest that basal deuterostomes originated about 550 mya. The fossil record shows that by 502 mya all five extant deuterostome phyla were in existence. The common deuterostome ancestor may have been pterobranch-like or enteropneust-like. Developmental studies provide convincing evidence for the ambulacrarian clade. Fossils showing shared cephalochordate/chordate characters are dated to about 530 mya.

Concept Review Questions

1. Which one of the following would represent a valid sister-group for the phylum Chordata?
   a. Ambulacraria
   b. Hemichordata
   c. Echinodermata
   d. Enteropneusta
2. Which one of the following characters is distinctive of the Chordata but not necessarily unique to members of the phylum?
   a. Dorsal tubular nerve cord
   b. Postanal tail
   c. Pharyngeal slits
   d. Endostyle or thyroid gland
   e. Notochord
3. Members of the phylum Chordata possess all of the following characteristics, except one. Select the exception.
   a. Endostyle or thyroid gland
   b. Dorsal tubular nerve cord
   c. Postanal tail
   d. Pharyngeal slits
   e. Open circulatory system

4. Members of which of the following groups are planktonic as adults and possess a tail and notochord that persist into the adult stage?
   a. Ascidiacea
   b. Thaliacea
   c. Appendicularia
   d. Enteropneusta
5. The endostyle of urochordates was the evolutionary forerunner of the __________ of vertebrates.
   a. thymus
   b. pituitary gland
   c. pancreas
   d. thyroid gland

Analysis and Application Questions

1. What evidence links echinoderms, hemichordates, and chordates to the same evolutionary lineage?
2. What evidence of chordate affinities is present in adult tunicates? In larval tunicates? If you only could examine an adult, would you be able to identify it as being a chordate?
3. Discuss the role of filter feeding in deuterostome evolution. At what point in chordate evolution does feeding become foraging or predatory in nature?
4. Compare and contrast metamorphosis in the Echinodermata and the Urochordata. Do any similarities you describe indicate common ancestry of the two groups?

18

The Fishes: Vertebrate Success in Water

When you hear the word "fish," you probably conjure an image of an animal like the rainbow parrotfish (*Scarus guacamaia*) shown here. This chapter covers a variety of taxa, in multiple craniate lineages that are tagged by the name "fish." We hope that this coverage will impress you with the remarkable diversity of forms and adaptations found in these groups.

Chapter Outline

18.1 Evolutionary Perspective
Phylogenetic Relationships
18.2 Survey of Fishes
Infraphylum Hyperotreti—Class Myxini
Infraphylum Vertebrata—Lampreys and Gnathostome Fishes
18.3 Evolutionary Pressures
Locomotion
Nutrition and the Digestive System
Circulation and Gas Exchange
Nervous and Sensory Functions
Excretion and Osmoregulation
Reproduction and Development
18.4 Further Phylogenetic Considerations

18.1 EVOLUTIONARY PERSPECTIVE

LEARNING OUTCOMES

1. Explain the phylogenetic relationships among classes of extant fishlike craniates.
2. Justify the conclusion that the subphylum Craniata is at least 500 million years old.
3. Hypothesize on the evolutionary events and modifications that led to diversification of the fishes.

Water, a buoyant medium that resists rapid fluctuations in temperature, covers 73% of the earth's surface. Because life began in water, and living tissues are made mostly of water, it might seem that nowhere else would life be easier to sustain. This chapter describes why that is not entirely true.

You do not need to wear scuba gear to appreciate that fishes are adapted to aquatic environments in a fashion that no other group of animals can surpass. If you spend recreational hours with hook and line, visit a marine theme park, or simply glance into a pet store when walking through a shopping mall, you can attest to the variety and beauty of fishes. This variety is evidence of adaptive radiation that began more than 500 million years ago (mya) and shows no sign of ceasing. Fishes dominate many watery environments and are also the ancestors of all other members of the infraphylum Vertebrata.

Phylogenetic Relationships

Fishes are included, along with the chordates in chapters 19 through 22, in the subphylum Craniata (kra′ne-ah′tah) (*see table 17.1*; figure 18.1). This name is descriptive of a skull that surrounds the brain, olfactory organs, eyes, and inner ear. These animals are divided into two infraphyla. The infraphylum Hyperotreti (hi″per-ot′re-te) includes the hagfishes, and the infraphylum Vertebrata (ver″te-bra′tah) is divided into two groups based on the presence or absence of hinged jaws (table 18.1). The fishes span both of these infraphyla, and therefore represent a very diverse assemblage of animals.

In recent years, the classification of fishes has been undergoing substantial revision, and changes are likely to continue for some time. For example, traditional taxonomic groupings have combined the hagfishes, lampreys, and an extinct assemblage called ostracoderms into a single group called "agnatha" based on the absence of jaws in these groups. Modern cladistic analysis has revealed, however, that the lampreys share more characteristics with cartilaginous and bony fishes than with the hagfishes. As a result, many taxonomists consider

FIGURE 18.1

Fishes. Five hundred million years of evolution have resulted in unsurpassed diversity in fishes. The spines of this beautiful marine lionfish (*Pterois volitans*) are extremely venomous. In spite of its beauty, this amazing fish is causing huge ecosystem problems in the Caribbean Sea and the middle and southern Atlantic coasts of the United States (*see Wildlife Alert, p. 330*).

the lampreys to be vertebrates. The hagfishes would thus comprise their own infraphylum, Hyperotreti. In addition, jawed fishes with bony skeletons have been classified together in a single class, "Osteichthyes" (os″te-ik-the-es). This grouping is paraphyletic because the common ancestor of this group is also the ancestor of the tetrapods (amphibians, reptiles, birds, and mammals), sharing characteristics such as a bony skeleton, swim bladders or lungs, and other more technical features. Many recent classification systems have dropped "Osteichthyes" as a class name and elevated former subclasses, Sarcopterygii (sar-kop-te-rij-e-i) and Actinopterygii (ak″tin-op′te-rig-e-i), to class-level status. Strictly speaking, because the tetrapods are descendants of the Sarcopterygii, they should also be included within this group. This topic will be discussed further in chapter 19. One of the most well-recognized interpretations of fish classification reflects these changes and is presented in table 18.1 and figure 18.2.

Although zoologists do not definitely know what animals were the first craniates, Chinese researchers have recently identified a suite of very ancient (530–520 mya) fossils of small, lancelet-shaped chordate animals that possessed non-bony protective braincases (i.e., *Haikouichthys, Myllokunmingia, Zhongjianichthys*). These animals had large eyes and fishlike muscle blocks on the

TABLE 18.1

CLASSIFICATION OF LIVING FISHES*

Subphylum Craniata (kra″ne-ah′tah)
Skull surrounds the three-part brain, olfactory organs, eyes, and inner ear. Unique embryonic tissue, neural crest, contributes to a variety of adult structures, including sensory nerve cells and some skeletal and other connective tissue structures.

Infraphylum Hyperotreti (hi″per-ot′re-te)
Fishlike; skull consisting of cartilaginous bars; jawless; no paired appendages; mouth with four pairs of tentacles; olfactory sacs open to mouth cavity; 5 to 15 pairs of pharyngeal slits; ventrolateral slime glands. Hagfishes.

Infraphylum Vertebrata (ver″te-bra′tah)
Vertebrae surround nerve cord and serve as primary axial support.

Superclass Petromyzontomorphi (pet′ro-mi-zon″to-morf′e)
A large, sucker-like mouth, reinforced by cartilage. Gill arches with spine-shaped processes.

Class Petromyzontida (pet′ro-mi-zon″tid-ah)
Sucking mouth with teeth and rasping tongue; seven pairs of pharyngeal slits; blind olfactory sacs. Lampreys.

Superclass Gnathostomata (na′tho-sto′ma-tah)
Hinged jaws and paired appendages; vertebral column may have replaced notochord; three semicircular canals.

Class Chondrichthyes (kon-drik′thi-es)
Tail fin with large upper lobe (heterocercal tail); cartilaginous skeleton; lack opercula and a swim bladder or lungs. Sharks, skates, rays, ratfishes.

Subclass Elasmobranchii (e-laz″mo-bran′ke-i)
Cartilaginous skeleton may be partially ossified; placoid scales or no scales. Sharks, skates, rays.

Subclass Holocephali (hol″o-sef′a-li)
Operculum covers pharyngeal slits; lack scales; teeth modified into crushing plates; lateral-line receptors in an open groove. Ratfishes.

Class Actinopterygii (ak″tin-op′te-rig-e-i)
Paired fins supported by dermal rays; basal portions of paired fins not especially muscular; tail fin with approximately equal upper and lower lobes (homocercal tail); blind olfactory sacs; pneumatic sacs function as swim bladders. Ray-finned fishes.

Class Sarcopterygii (sar-kop-te-rij′e-i)
Paired fins with muscular lobes; pneumatic sacs function as lungs; atria and ventricles at least partly divided. Lungfishes, coelacanths, and tetrapodomorphs, including all tetrapods.

*For a recent alternative interpretation of fish classification see Nelson JS, Grande TC, Wilson MVH. 2016. Fishes of the world. Fifth Edition. Hoboken (NJ): John Wiley & Sons, Inc. 752 p.

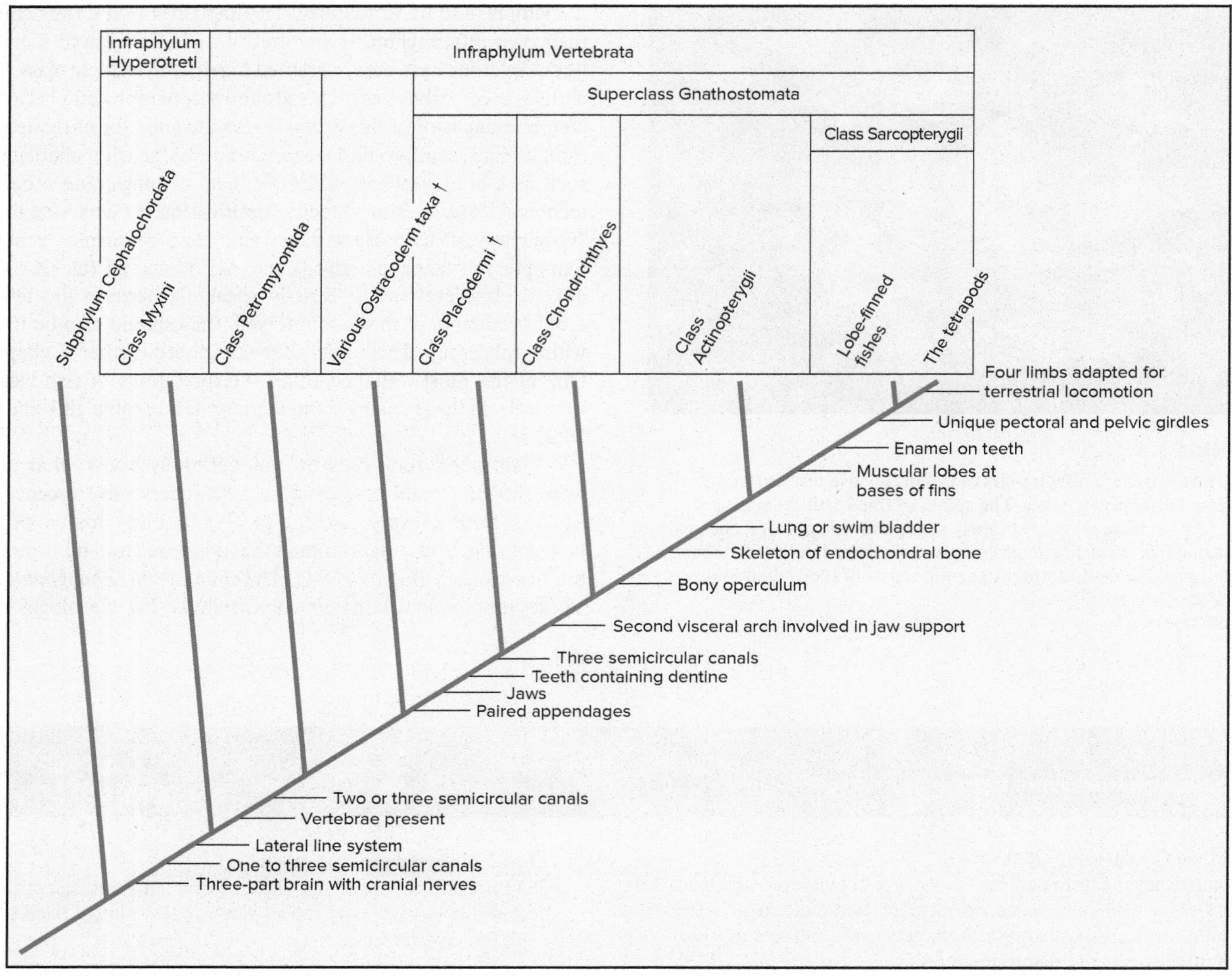

FIGURE 18.2

One Interpretation of the Phylogeny of the Craniata with Emphasis on the Fishes. The evolutionary relationships among fishes are unsettled. This cladogram shows a few selected ancestral and derived characters. Each lower taxon has numerous synapomorphies that are not shown. Most zoologists consider the ostracoderms a paraphyletic group (having multiple lineages). Their representation as a monophyletic group is an attempt to simplify this presentation. Daggers (†) indicate groups whose members are extinct. The inclusion of the tetrapods within the class Sarcopterygii reflects the origin of the tetrapods, but it is difficult to reconcile with traditional classification schemes.

side of the body. Thus, they were likely active swimmers and visual predators that pursued prey. *Haikouichthys, Myllokunmingia* and *Zhongjianichthys* may be among the first craniates to exist.

The origin of bone in craniates is equally intriguing. A group of ancient eel-like craniates, the conodonts, is known from fossils that date back about 510 million years (figure 18.3). They had two large eyes and a mouth filled with toothlike structures made of dentine—a component found in the craniate skeleton. These structures may represent the earliest occurrence of bone in vertebrates. Thus, some zoologists consider conodonts to be early vertebrates. Other hypotheses on the origin of bone suggest that it may have arisen as denticles in the skin (similar to those of sharks), in association with certain sensory receptors, or as structures for mineral (especially calcium phosphate) storage.

Although hypotheses on its origins are still somewhat tentative, bone was well developed by 500 mya in a group of fishes called ostracoderms (figure 18.4). These bottom-dwelling fishes were sluggish and heavily armored with bony plates used for defense. Ostracoderms were jawless and did not possess paired fins. Placoderms, another group of prehistoric fishes, were also heavily armored. However, this group contained representatives that possessed both jaws and paired appendages. Prior to 2012, it was largely accepted that paired pelvic appendages evolved within the gnathostome lineage prior to the appearance of movable jaws. However, recent data

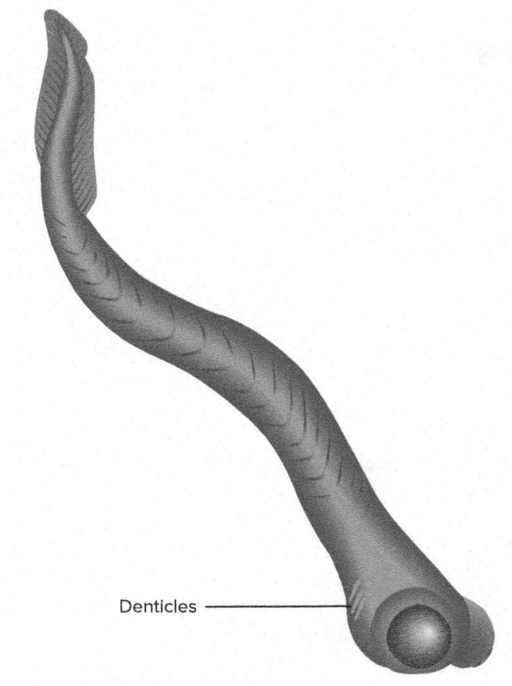

(a)

(b)

FIGURE 18.3

A Conodont (*Clydagnathus*) Reconstruction. (*a*) A wealth of conodont fossils has been found that date to 510 mya. Recent information has led many zoologists to accept them as some of the very early vertebrates. They were 1 cm in length, and their two large eyes, eel-like body, and toothlike denticles suggest that they lived as predators in prehistoric seas. (*b*) A photomicrograph showing fossilized denticles of a conodont. These denticles were probably used for crushing food.

derived from the fossil placoderm, *Parayunnanolepis xitunensis*, has influenced hypotheses on the origins of paired pelvic fins. It is now thought that pelvic fins may have evolved before the appearance of movable jaws. Regardless of the order of appearance, the evolution of jaws and paired appendages was necessary for the subsequent adaptive radiation and diversification of the fishes (*see discussion in the following text*). Extant fishes are a very diverse assemblage of animals: They comprise nearly half of all extant chordate species.

Did ancestral fishes live in freshwater or in the sea? The answer to this question is not simple. The first vertebrates were probably marine, because ancient stocks of other deuterostome phyla were all marine. Vertebrates, however, adapted to freshwater very early, and much of the evolution of fishes occurred there. Apparently, early vertebrate evolution involved the movement of fishes back and forth between marine and freshwater environments. The majority of the evolutionary history of some fishes took place in ancient seas, and most of the evolutionary history of others occurred in freshwater. The importance of freshwater in the evolution of fishes is evidenced by the fact that more than 41% of all fish species are found in freshwater, even though freshwater habitats represent only a small percentage (0.0093% by volume) of the earth's water resources.

SECTION 18.1 THINKING BEYOND THE FACTS

What are the key evolutionary changes that helped to facilitate the diversification and adaptive radiation of the fishes?

18.2 SURVEY OF FISHES

LEARNING OUTCOMES

1. Explain how you would determine whether or not an eel-like fish presented to you was a member of the class Myxini, Petromyzontida, or Actinopterygii.
2. Describe the characteristics of members of the class Chondrichthyes.
3. Distinguish between members of the class Sarcopterygii and members of the class Actinopterygii.

The taxonomy of fishes has been the subject of debate for many years. Modern cladistic analysis has resulted in complex revisions in the taxonomy of this group of vertebrates. Figure 18.2 includes the tetrapods as members of the superclass Gnathostomata. Inclusion of the tetrapods in this cladogram emphasizes the relationship of the fishes to land vertebrates and makes the lineage monophyletic. The tetrapods are discussed in chapters 19 through 22.

Infraphylum Hyperotreti—Class Myxini

Hagfishes are members of the class Myxini (mik'si-ne) (Gr. *myxa*, slime). According to some analyses, there are over 70 species divided into six genera. Their heads are supported by cartilaginous bars and their brains are enclosed in a fibrous sheath. They lack vertebrae and retain the notochord as the axial supportive structure. They have rudimentary eyespots and four pairs of sensory tentacles surrounding their mouths. They also possess ventrolateral slime glands that store specialized protein molecules that explode upon mixing

FIGURE 18.4

Artist's Rendering of an Ancient Silurian Seafloor. Two ostracoderms, *Pteraspis* and *Anglaspis,* are in the background. Ostracoderms are extinct agnathans from several classes. Most were filter feeders that filtered plankton and marine sediments. Others used bony plates surrounding the mouth to crack open exoskeletons and shells of molluscs.

Animation
Early Vertebrates

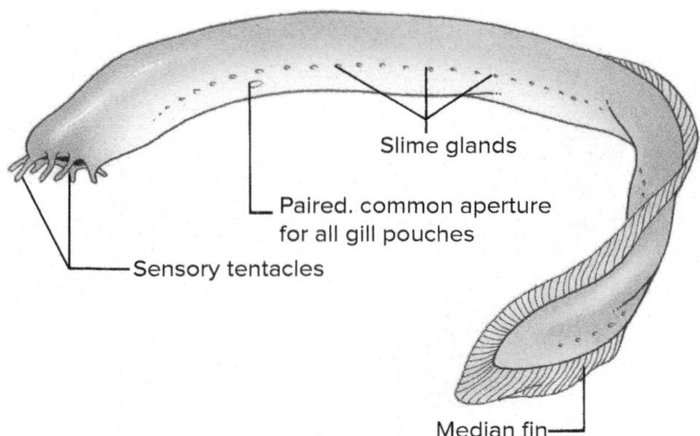

FIGURE 18.5

Class Myxini. Hagfish external structure.

with seawater, forming copious amounts of slime (figure 18.5). The slime is an effective defensive mechanism. It can rapidly suffocate a predatory fish by covering the predator's gill filaments. Hagfishes are found in cold-water marine habitats of both the Northern and Southern Hemispheres. Most zoologists now consider the hagfishes to be the most primitive group of extant craniates.

Hagfishes live buried in the sand and mud of marine environments, where they feed on soft-bodied invertebrates and scavenge dead and dying vertebrates. Hagfishes may enter a carcass suitable for consumption by using a rasping tongue and specialized body contortions called knots. They press their body knots against the carcass while simultaneously pulling hunks of flesh away with their rasping tongue. This action eventually creates a pathway into the carcass. Once inside, hagfishes will eat the internal contents of the carcass, leaving only a sac of skin and bones. Hagfishes may also enter carcasses through existing orifices. Hagfish slime is also used in this process to prevent abrasion while moving across sharp bits of broken bone that may be present inside the carcass.

Hagfishes are scavengers that clean the ocean floor of dead and decaying animals. Thus, they are important components of deep-sea food webs and nutrient cycles. Some populations of hagfishes (e.g., *Eptatretus stoutii*) are in decline due to overfishing for their unique skin (sold as "eel skin"). Hagfish declines likely have broad impacts due to their important role in deep-sea systems.

Little is known of reproduction in hagfishes. Some species are hermaphroditic, all are oviparous, and no larval stage is present.

Infraphylum Vertebrata—Lampreys and Gnathostome Fishes

The vertebrates are characterized by vertebrae that surround a nerve cord and serve as primary axial support. Today, most vertebrates are members of the superclass Gnathostomata. They include the jawed fishes and the tetrapods. About 400 mya, that was not the case—the jawless ostracoderms were a very early and successful group of vertebrates. A third group of vertebrates, the lampreys, is also jawless and lives in both marine and freshwater environments.

Class Petromyzontida

Lampreys are agnathans in the class Petromyzontida (pet′ro-mi-zon″tid-ah) (Gr. *petra,* rock + *myzo,* suckle + *odontos,* teeth). They are common inhabitants of marine and freshwater environments in temperate regions. Most adult lampreys prey on other fishes, and the larvae are filter feeders. The mouth of an adult is suckerlike and surrounded by lips that have sensory and attachment functions. Numerous epidermal teeth line the mouth and cover a movable tonguelike structure (figure 18.6). Adults attach to prey with their lips and teeth and use their tongues to rasp away scales. Lampreys have salivary glands with anticoagulant secretions and feed mainly on the blood of their prey. Some lampreys, however, are not predatory. For example, some members of the genus *Lampetra* are called brook lampreys. The larval stages of brook lampreys last for about three years, and the adults neither feed nor leave their stream. They reproduce soon after metamorphosis and then die.

Adult sea lampreys (*Petromyzon marinus*) live in the ocean or the Great Lakes. Near the end of their lives, they migrate—sometimes hundreds of miles—to a spawning bed in a freshwater stream. Once lampreys reach their spawning site, usually in relatively shallow water with swift currents, they begin building a nest

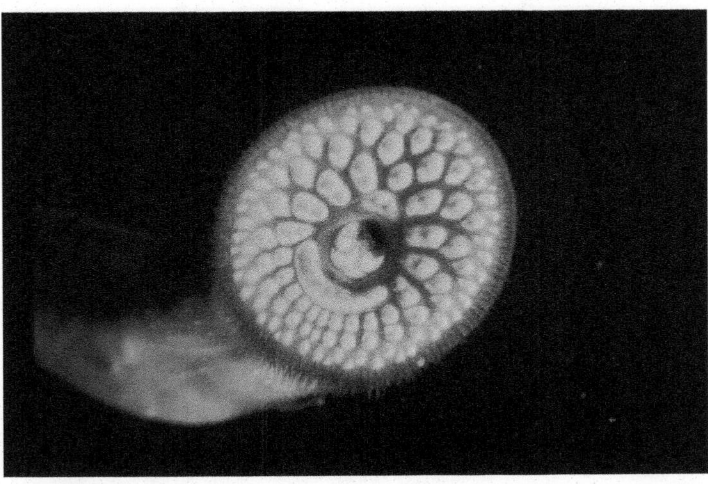

FIGURE 18.6

Class Petromyzontida. A lamprey (*Petromyzon marinus*). Note the sucking mouth and teeth used to feed on other fish.
Source: U.S. Fish & Wildlife Service

by making small depressions in the substrate. When the nest is prepared, a female usually attaches to a stone with her mouth. A male uses his mouth to attach to the female's head and wraps his body around the female (figure 18.7). Eggs are shed in small batches over a period of several hours, and fertilization is external. The relatively sticky eggs are then covered with sand.

Eggs hatch in approximately three weeks into ammocoete larvae. The larvae drift downstream to softer substrates, where they bury themselves in sand and mud and filter feed in a fashion similar to amphioxus (*see figure 17.6*).

Ammocoete larvae grow from 7 mm to about 17 cm over 3–7 years. During later developmental stages, the larvae metamorphose to the adult over a period of several months. The mouth becomes suckerlike, and the teeth, tongue, and feeding musculature develop. Lampreys eventually leave the mud permanently and begin a journey to the sea to begin life as predators. Adults return only once to the headwaters of their stream to spawn and die.

Superclass Gnathostomata—Jawed Vertebrates

Jaws of vertebrates evolved from the most anterior pair of pharyngeal arches (the skeletal supports for the pharyngeal slits). This extremely important event in vertebrate evolution permitted more efficient gill ventilation and the capture and ingestion of a variety of food sources. Similarly, paired appendages were extremely important evolutionary developments. Their origin has been debated for many years, and the debate remains unresolved. In the absence of paired appendages, fishes must have been relatively sluggish bottom dwellers. Increased activity without paired appendages would have led to instability. Paired appendages can be used to counter the tendency to roll during locomotion. They can be used to control the tilt or pitch of the swimming fish and can be used in lateral steering. Pectoral fins of fishes are appendages usually just behind the head, and pelvic fins are usually located ventrally and more posteriorly.

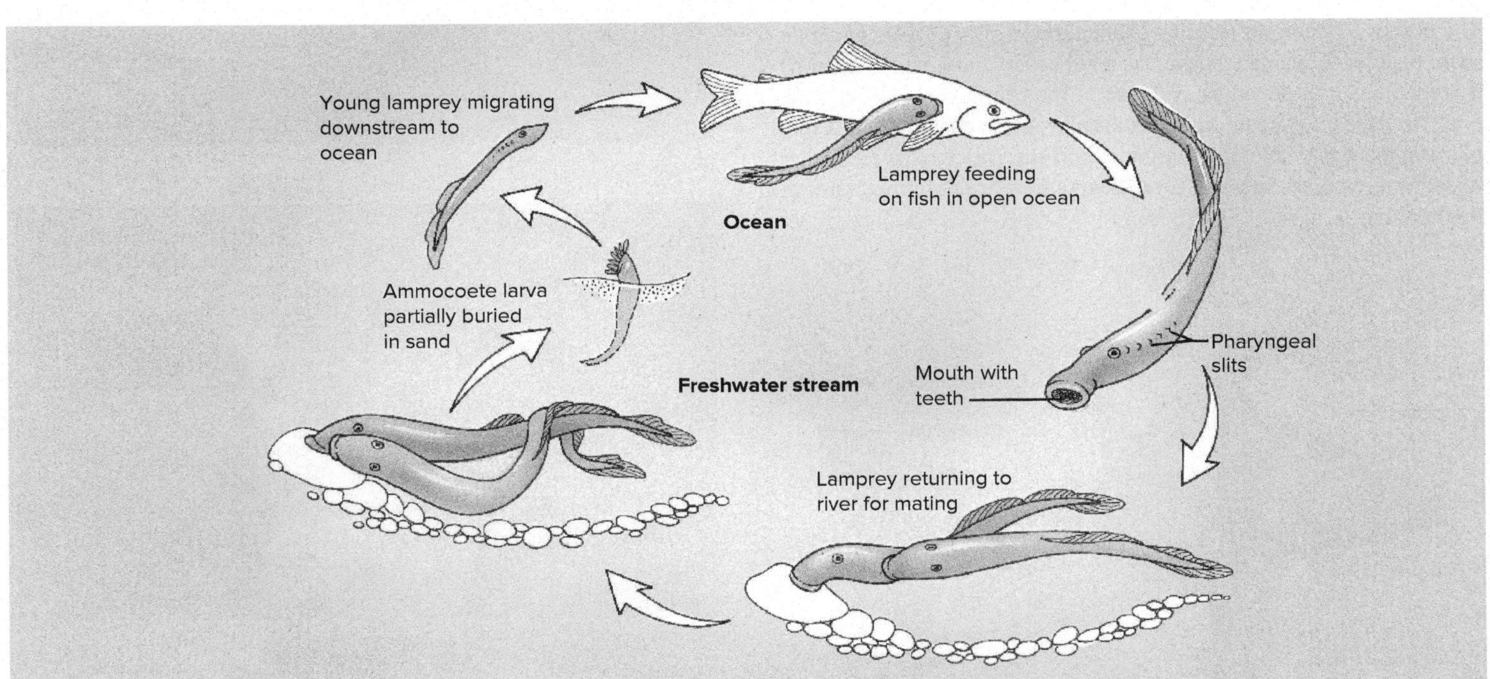

FIGURE 18.7

Life History of a Sea Lamprey. Sea lampreys feed in the open ocean and, near the end of their lives, migrate into freshwater streams, where they mate. They deposit eggs in nests on the stream bottom, and young ammocoete larvae hatch in about three weeks. Ammocoete larvae live as filter feeders until they attain sexual maturity.

In modern bony fishes, the pelvic fins are usually positioned just behind the pectoral fins (figure 18.8). The evolution of jaws and paired appendages must have revolutionized the life of early fishes. Both contributed to the evolution of the predatory lifestyles of many fishes. The ability to feed efficiently allowed fishes to produce more offspring and exploit new habitats. These new habitats, in turn, fostered the adaptive radiation of fishes. The results of this adaptive radiation are described in this chapter.

Three classes of gnathostomes still have living members: the cartilaginous fishes (class Chondrichthyes) and two groups of bony fishes (classes Actinopterygii and Sarcopterygii). As mentioned, placoderms contained the earliest jawed fishes (*see figure 18.2*). They are now extinct and apparently left no descendants. A fourth group of ancient, extinct fishes, the acanthodians, are known from late Ordovician deposits. They are among the oldest fossils with features present in more advanced jawed fishes (both cartilaginous and bony). They are considered to be the last common ancestor of modern gnathostomes.

Class Chondrichthyes Members of the class Chondrichthyes (kon-drik′thi-es) (Gr. *chondros,* cartilage + *ichthyos,* fish) include the sharks, skates, rays, and ratfishes (*see table 18.1*). Most chondrichthians are carnivores or scavengers, and most are marine. In addition to their biting mouthparts and paired appendages, chondrichthians possess placoid scales and a cartilaginous endoskeleton.

The subclass Elasmobranchii (e-laz′mo-bran′ke-i) (Gr. *elasmos,* plate metal + *branchia,* gills), which includes the sharks, skates, and rays, has about 1,100 species (figure 18.9). Sharks arose from early jawed fishes midway through the Devonian period, about 375 mya. The absence of certain features characteristic of bony fishes (e.g., a swim bladder to regulate buoyancy, a gill cover, and a bony skeleton) is sometimes interpreted as evidence of the primitiveness of elasmobranchs. This interpretation is mistaken, as these characteristics simply resulted from different adaptations in the two groups to similar selection pressures. Some of these adaptations are described later in this chapter.

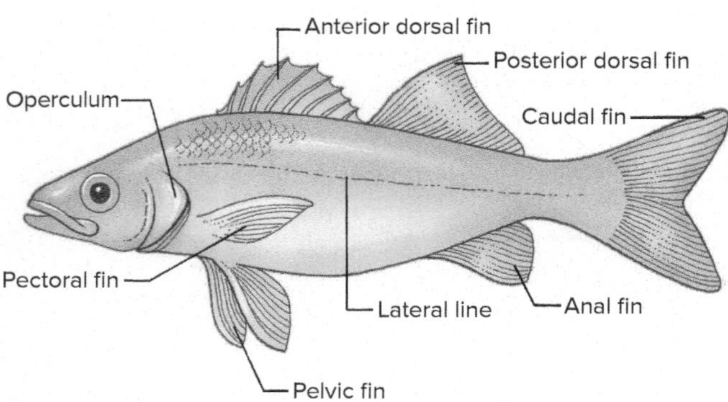

FIGURE 18.8

Paired Pectoral and Pelvic Appendages. Appendages of a member of the gnathostomes. These appendages are secondarily reduced in some species.

(a)

(b)

(c)

FIGURE 18.9

Class Chondrichthyes. (*a* and *b*) Subclass Elasmobranchii. (*a*) A Caribbean reef shark (*Carcharhinus perezi*) and (*b*) a bluespotted ribbontail ray (*Taeniura lymma*). This stingray is found in the Indian and western Pacific Oceans. (*c*) Subclass Holocephali. The ratfish (*Hydrolagus colliei*).

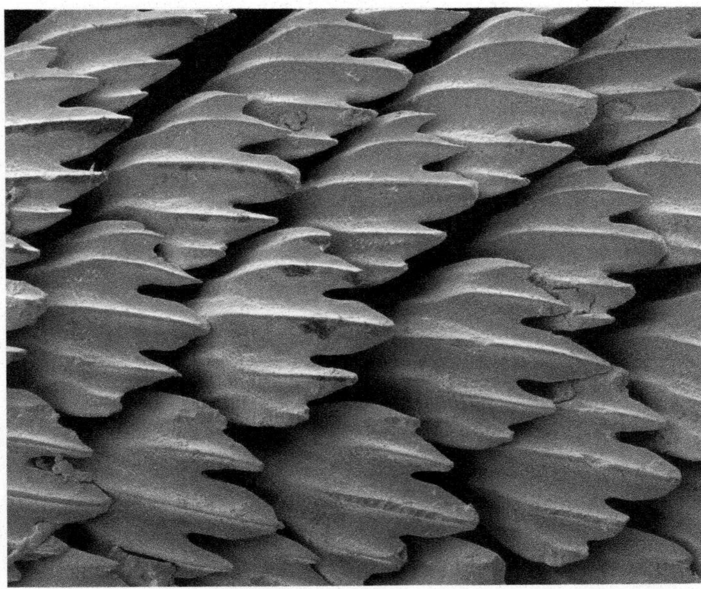
(a)

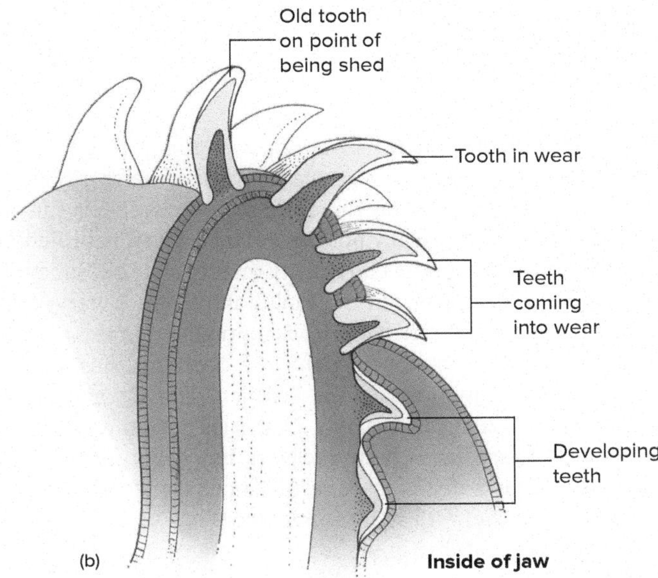

(b)

FIGURE 18.10

Scales and Teeth of Sharks. (*a*) Section of hammerhead shark (*Sphyrnidae*) skin magnified to show posteriorly pointing placoid scales. (*b*) The teeth of sharks develop as modified placoid scales. Newer teeth that move from the inside to the outside of the jaw continuously replace older teeth. (b) Source: Steel R. 1985. *Sharks of the world*. London (UK): Cassell PLC.

Tough skin with dermal, placoid scales covers sharks (figure 18.10*a*). These scales project posteriorly and give the skin a tough, sandpaper texture. (In fact, dried shark skin has been used for sandpaper.) Posteriorly pointed scales also reduce friction with the water as a shark swims.

Shark teeth are actually modified placoid scales. The row of teeth on the outer edge of the jaw is backed up by rows of teeth attached to a ligamentous band that covers the jaw cartilage inside the mouth. As the outer teeth wear and become useless, newer teeth move into position from inside the jaw and replace them. In young sharks, this replacement is rapid, with a new row of teeth developing every seven or eight days (figure 18.10*b*). Crowns of teeth in different species may be adapted for shearing prey or for crushing the shells of molluscs.

Sharks range in size from less than 1 m (e.g., *Squalus acanthias,* the spiny dogfish shark) to greater than 10 m (e.g., basking [*Cetorhinus maximus*] and whale sharks [*Rhincodon typus*]). The largest living sharks are not predatory but are filter feeders. They have pharyngeal-arch modifications that strain plankton. Among the most well known predatory sharks are the great white (*Carcharodon carcharias*) and mako (*Isurus*). The great white is among the largest of the predatory sharks, while the makos are remarkably fast predatory fishes. Extinct specimens may have reached lengths of 15 m or more.

Skates and rays are specialized for life on the ocean floor. They usually inhabit shallow water, where they use their blunt teeth to feed on invertebrates. Their most obvious modification for life on the ocean floor is a lateral expansion of the pectoral fins into winglike appendages. Locomotion results from dorsoventral muscular waves that pass posteriorly along the fins. Frequently, elaborate color patterns on the dorsal surface of these animals provide effective camouflage. The sting ray (*Dasyatis*) has a tail modified into a defensive lash; a group of modified placoid scales persists as a venomous spine (*see figure 18.9*b). Also included in this group are the electric rays (*Narcine* and *Torpedo*) and manta rays (*Manta*).

A second major group of chondrichthians, in the subclass Holocephali (hol″o-sef′a-li) (Gr. *holos,* whole + *kephalidos,* head), contains just over 30 species. Unlike the elasmobranchs, the upper jaw of holocephalan fishes is affixed to the braincase (this jaw structure is the basis for the subclass name). A frequently studied example, *Chimaera,* has a large head with a small mouth surrounded by large lips. A narrow, tapering tail has resulted in the common name "ratfish" (*see figure 18.9*c). Holocephalans diverged from other chondrichthians nearly 350 mya. Since that time, specializations not found in other elasmobranchs have evolved, including a gill cover, called an **operculum,** and teeth modified into continually growing, large plates for crushing mollusc shells. Holocephalans lack scales.

Bony Fishes

Bony fishes are characterized by having at least some bone in their skeleton and/or scales, bony operculum covering the gill openings, and lungs or a swim bladder. Any group that has over 28,000 species and is a major life-form in most of the earth's vast aquatic habitats must be judged very successful from an evolutionary perspective.

The first fossils of bony fishes are from late Silurian deposits (approximately 405 million years old). By the Devonian period (350 mya), the two classes (Sarcopterygii and Actinopterygii) were in the midst of their adaptive radiations (*see table 18.1; see also figure 18.2*).

Class Sarcopterygii Members of the class Sarcopterygii (sar-kop-te-rij′e-i) (Gr. *sark,* flesh + *pteryx,* fin) have muscular lobes associated with their fins and usually use lungs in gas exchange. One group of sarcopterygians are the lungfishes. Only three genera survive today, and all live in regions where seasonal droughts are common. When freshwater lakes and rivers begin to stagnate and dry, these fishes use lungs to breathe air (figure 18.11). Some (*Neoceratodus*) inhabit the freshwaters of Queensland, Australia. They survive stagnation by breathing air, but they normally use gills and cannot withstand total drying. Others are found in freshwater rivers and lakes in tropical Africa (*Protopterus*) and tropical South America (*Lepidosiren*). They can survive when rivers or lakes are dry by burrowing into the mud. They keep a narrow air pathway open by bubbling air to the surface. After the substrate dries, the only evidence of a lungfish burrow is a small opening in the earth. Lungfishes may remain in aestivation (*see chapter 28*) for six months or more. When rain again fills the lake or riverbed, lungfishes emerge from their burrows to feed and reproduce.

A second group of sarcopterygians contains the coelacanths. The most recent coelacanth fossils are more than 70 million years old. In 1938, however, people fishing in deep water off the coast of South Africa brought up one fish that was identified as a coelacanth (figure 18.12). Since then, numerous other specimens have been caught in deep water around the Comoro Islands off Madagascar. The discovery of this fish, *Latimeria chalumnae,* was a milestone event because coelacanths had previously been known only from the fossil record. It is large—up to 80 kg—and has heavy scales. A second species of coelacanth, *Latimeria menadoensis,* was discovered in 1997 off the coast of Indonesia. Ancient coelacanths lived in freshwater lakes and rivers; thus, the ancestors of *Latimeria* must have moved from freshwater habitats to the deep sea.

A third group of sarcopterygians, the Tetrapodomorpha, became extinct before the close of the Paleozoic period. This group includes the ancestors of ancient amphibians and all tetrapods. They are discussed further at the end of this chapter.

FIGURE 18.11

Class Sarcopterygii. The Queensland lungfish (*Neoceratodus forsteri*) is an ancient surviving member of the lungfish lineage. Lungfish have lungs that allow them to withstand stagnation of its habitat.

FIGURE 18.12

A Sarcopterygian, the Coelacanth. Two species of *Latimeria* are the only known surviving coelocanths.

Class Actinopterygii The class Actinopterygii (ak″tin-″op″te-rig-e-i) (Gr. *aktis,* ray + *pteryx,* fin) contains fishes that are sometimes called the ray-finned fishes because their fins lack muscular lobes. They usually possess **swim bladders,** gas-filled sacs along the dorsal wall of the body cavity that regulate buoyancy. The many points of divergence in the evolution of Actinopterygii have resulted in the great diversity of fishes represented in this class.

One group of actinopterygians, the chondrosteans, contains many species that lived during the Permian, Triassic, and Jurassic periods (215 to 120 mya), but only 27 species remain today. Ancestral chondrosteans had a bony skeleton, but living members, the sturgeons and paddlefishes, have cartilaginous skeletons. Chondrosteans also have a tail with a large upper lobe.

Most sturgeons live in the sea and migrate into rivers to breed (figure 18.13*a*). (Some sturgeons live in freshwater but maintain the migratory habits of their marine relatives.) They are large (up to 1,000 kg), and bony plates cover the anterior portion of the body. Heavy scales occur along the lateral surface. The sturgeon mouth is small, and jaws are weak. Sturgeons feed on invertebrates that they stir up from the sea or riverbed using their snouts. Because sturgeons are valued for their eggs (caviar), they have been severely overfished.

Paddlefishes are large, freshwater chondrosteans. They have a large, paddlelike rostrum that is innervated with sensory organs believed to detect weak electrical fields (figure 18.13*b*). They swim through the water with their large mouths open, filtering crustaceans and small fishes. They are found mainly in lakes and large rivers of the Mississippi River basin and are also found in China.

The largest group of actinopterygians (Neopterygii) flourished in the Jurassic period and succeeded most chondrosteans. Two very primitive genera occur in temperate to warm freshwaters

(a)

(b)

FIGURE 18.13

Class Actinopterygii, the Chondrosteans. (*a*) Shovelnose sturgeons (*Scaphirhynchus platorynchus*). Sturgeons are covered anteriorly by heavy bony plates and posteriorly by scales. (*b*) The distinctive rostrum of a paddlefish (*Polydon spathula*) is densely innervated with sensory structures that are probably used to detect minute electrical fields. Note the mouth in its open, filter-feeding position.

(a) ©Ken Lucas/ardea.com (b) ©Saran Jantraurai/Shutterstock

of North America. *Lepisosteus,* the gars, have thick scales and long jaws that they use to catch fish prey. *Amia calva* is commonly referred to as the dogfish or bowfin. Most living fishes are members of this group and are referred to as teleosts or modern bony fishes. After their divergence from ancient marine actinopterygians in the late Triassic period, teleosts experienced a remarkable evolutionary diversification and adapted to nearly every available aquatic habitat (figure 18.14). The number of teleost species exceeds 26,000.

The next section in this chapter explores some of the keys to the evolutionary success of this largest vertebrate group. A part of the answer to why the modern bony fishes have become so diverse and plentiful lies in the fact that 73% of the earth's surface is covered by water and an abundance of aquatic habitats are available. That, however, cannot be the whole answer because many other aquatic animals have been much less successful. The keys to teleost success lie in their ability to adapt to a demanding environment. Highly

(a)

(b)

(c)

FIGURE 18.14

Class Actinopterygii, the Teleosts. (*a*) Flat fish, such as this winter flounder (*Pseudopleuronectes americanus*), have both eyes on the right side of the head, and they often rest on their side fully or partially buried in the substrate. (*b*) The yellowtail snapper (*Ocyurus chrysurus*) is a popular sport fish and food item found in offshore tropical water. It reaches a length of 75 cm and a mass of 2.5 kg. (*c*) The sarcastic fringehead (*Neoclinus blanchardi*) retreats to holes on the mud bottom of the ocean. Males engage in impressive territorial battles. They determine dominance through wrestling, where combatants vigorously press their gaping mouths against one another.

(a) Source: Jerry Prezioso, NEFSC/NOAA (b) ©Humberto Ramirez/Moment/Getty Images (c) ©Greg Amptman/Shutterstock

efficient respiratory systems allow fishes to extract oxygen from an environment that holds little oxygen per unit volume; efficient locomotor structures allow fishes to move through a buoyant, but viscous medium; highly efficient sensory systems include typical vertebrate systems, and also a lateral-line system that detects low-pressure waves and electroreception; and efficient reproductive mechanisms have the potential to produce overwhelming numbers of offspring.

SECTION 18.2 THINKING BEYOND THE FACTS

What major characteristics distinguish each of the classes of fishes?

18.3 EVOLUTIONARY PRESSURES

LEARNING OUTCOMES

1. Discuss adaptations to life in water that are present fish locomotor, digestive, nervous, and sensory systems.
2. Explain how adaptations to life in water influenced the structure and physiology of exchange systems (circulation, gas exchange, ion regulation, and excretion) of fishes.
3. Describe and reproductive strategies observed in selected marine and freshwater fishes.

Why is a fish fishlike? This apparently redundant question is unanswerable in some respects because some traits of animals are selectively neutral and thus neither improve nor detract from overall fitness. On the other hand, aquatic environments have physical characteristics that are important selective forces for aquatic animals. Although animals have adapted to aquatic environments in different ways, you can understand many aspects of the structure and function of a fish by studying the fish's habitat. This section will help you appreciate the many ways that a fish is adapted for life in water.

Locomotion

For many animals, the density of water makes movement through it difficult and costly. For a fish, however, swimming is less energetically costly than running is for a terrestrial organism. The streamlined shape of a fish and the mucoid secretions that lubricate its body surface reduce friction between the fish and the water. Water's buoyant properties also contribute to the efficiency of a fish's movement through the water. A fish expends little energy in support against the pull of gravity.

Fishes move through the water using their fins and body wall to push against the incompressible surrounding water. Specialized muscle bundles of most fishes are arranged in a ⋛ pattern (*see figure 23.19*). Because these muscles extend posteriorly and anteriorly in a zigzag fashion, contraction of each muscle bundle can affect a relatively large portion of the body wall. Very efficient, fast-swimming fishes, such as tuna and mackerel, supplement body movements with a vertical caudal (tail) fin that is tall and forked. The forked shape of the caudal fin reduces surface area that could cause turbulence and interfere with forward movement.

Nutrition and the Digestive System

The earliest fishes were probably filter feeders and scavengers that sifted through the mud of ancient seafloors for decaying organic matter, annelids, molluscs, or other bottom-dwelling invertebrates. Fish nutrition dramatically changed when the evolution of jaws transformed early fishes into efficient predators.

Most fishes have teeth that are simple cone-shaped structures. They are uniform along the length of the jaw (**homodont** condition) and seated into a very shallow depression at the summit of the jawbone (**acrodont** condition) by a cement-like material.

Most modern fishes are predators and spend much of their lives searching for food. Their prey vary tremendously. Some fishes feed on invertebrate animals floating or swimming in the plankton or living in or on the substrate. Many feed on other vertebrates. Similarly, the kinds of food that one fish eats at different times in its life varies. For example, as a larva, a fish may feed on plankton; as an adult, it may switch to larger prey, such as annelids or smaller fish. Fishes usually swallow prey whole. Teeth capture and hold prey, and some fishes have teeth that are modified for crushing the shells of molluscs or the exoskeletons of arthropods. To capture prey, fishes often use the suction that closing the opercula and rapidly opening the mouth creates, which develops a negative pressure that sweeps water and prey inside the mouth.

Other feeding strategies have also evolved in fishes. Herring (Clupeidae), paddlefishes (Polyodontidae), and whale sharks (*Rhincodon*) are filter feeders. Long gill processes, called **gill rakers,** trap plankton while the fish is swimming through the water with its mouth open (*see figure 18.13*b). Other fishes, such as carp (Cyprinidae), feed on a variety of plants and small animals. A few, such as the lamprey, are ectoparasites for at least a portion of their lives. A few are primarily herbivores, feeding on plants.

The fish digestive tract is similar to that of other vertebrates. An enlargement, called the stomach, stores large, often infrequent, meals. The small intestine, however, is the primary site for enzyme secretion and food digestion. Sharks and other elasmobranchs have a spiral valve in their intestine, and bony fishes possess outpockets of the intestine, called pyloric ceca, that increase absorptive and secretory surfaces.

Circulation and Gas Exchange

All vertebrates have a closed circulatory system in which a heart pumps blood, with red blood cells containing hemoglobin, through a series of arteries, capillaries, and veins. The evolution of lungs in fishes was paralleled by changes in vertebrate circulatory systems. These changes are associated with the loss of gills, delivery of blood to the lungs, and separation of oxygenated and unoxygenated blood in the heart.

The vertebrate heart develops from four embryological enlargements of a ventral aorta. In fishes, blood flows from the venous system through the thin-walled sinus venosus into the thin-walled, muscular atrium. From the atrium, blood flows into a larger, more muscular ventricle. The ventricle is the primary pumping structure. Anterior to the ventricle is the conus arteriosus, which connects to the ventral aorta. In teleosts, the conus arteriosus is replaced by an expansion of the ventral aorta called the bulbus arteriosus (figure 18.15*a*). Blood is carried by the ventral aorta to afferent vessels leading to the gills. These vessels break into capillaries and blood is oxygenated. Blood is then collected by efferent vessels, delivered to the dorsal aorta, and distributed to the body, where it enters a second set of capillaries. Blood then returns to the heart through the venous system.

In most fishes, blood passes through the heart once with every circuit around the body. A few fishes (e.g. the lungfishes) have lungs, and the pattern of circulation is altered. Understanding the pattern of circulation in these fishes is important because it was an important preadaptation for terrestrial life. In the lungfish, circulation to gills continues, but a vessel to the lungs has developed as a branch off aortic arch VI (figure 18.15*b*). This vessel is now called the pulmonary artery. Blood from the lungs returns to the heart through pulmonary veins and enters the left side of the heart. The atrium and ventricle of the lungfish heart are partially divided. These partial divisions help keep less oxygenated blood from the body separate from the oxygenated blood from the lungs. A spiral valve in the conus arteriosus helps direct blood from the right side of the heart to the pulmonary artery and blood from the left side of the heart to the remaining aortic arches. Thus, the lungfishes show a distinction between a pulmonary circuit and a systemic circuit.

Gas Exchange

Fishes live in an environment that contains less than 2.5% of the oxygen present in air. To maintain adequate levels of oxygen in their bloodstream, fishes must pass large quantities of water across gill surfaces and extract the small amount of oxygen present in the water.

Most fishes have a muscular pumping mechanism to move the water into the mouth and pharynx, over the gills, and out of the fish through gill openings. Muscles surrounding the pharynx and the opercular cavity, which is between the gills and the operculum, power this pump.

Some elasmobranchs and open-ocean bony fishes, such as the tuna (Scombridae), maintain water flow by holding their mouths open while swimming. This method is called **ram ventilation.** Elasmobranchs do not have opercula to help pump water, and therefore some sharks must keep moving to survive. Others move water over their gills with a pumping mechanism similar to that just described. Rather than using an operculum in the pumping process, however, these fishes have gill bars with external flaps that close and form a cavity functionally similar to the opercular cavity of other fishes. Spiracles are modified first pharyngeal slits that open just behind the eyes of elasmobranchs and are used as an alternate route for water entering the pharynx.

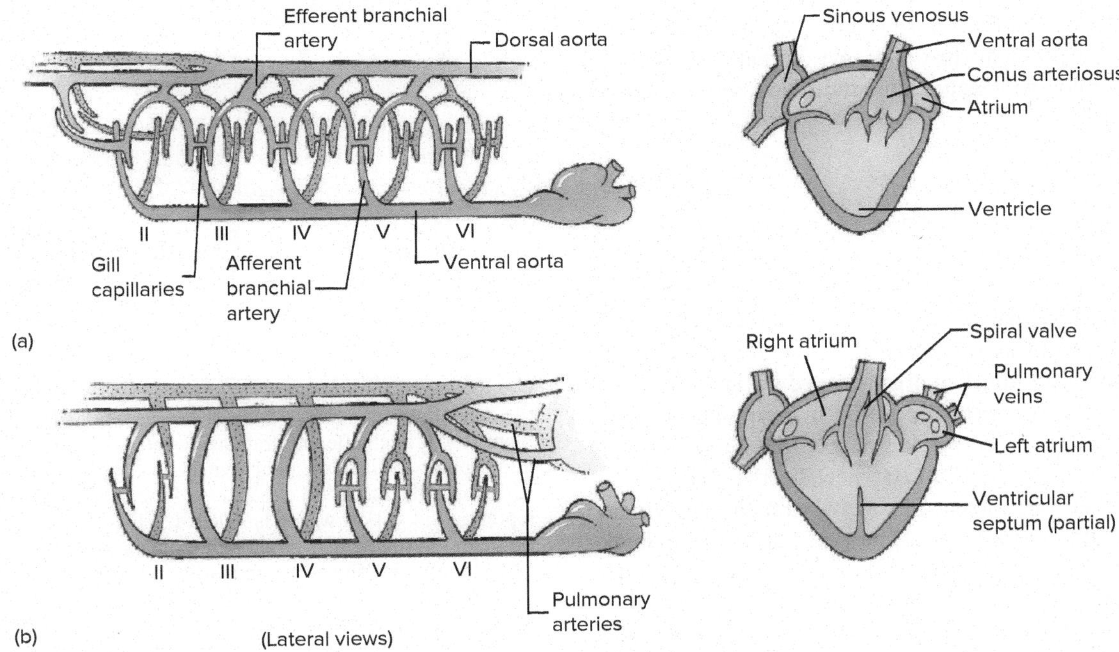

FIGURE 18.15

Circulatory System of Fishes. Diagrammatic representation of the circulatory systems of (*a*) bony fishes and (*b*) lungfishes. Hearts are drawn from a ventral view. The fish's posterior end would be to your right, and anterior end to your left. Major branches of arteries carrying blood to and from the gills are called branchial arteries (or embryologically, aortic arches) and are numbered with Roman numerals. They begin with II because aortic arch I is lost during embryonic development. In the bony fish (*a*) gill capillaries are present in branchial arches II–VI. The lungfish (*b*), however, lacks gill capillaries in branchial arches III and IV because they are responsible for taking blood that was oxygenated in lungs from the heart to the aorta. Valves that are present in branchial arches II, V, and VI close when using lungs to oxygenate blood.

Gas exchange across gill surfaces is very efficient. **Gill (visceral) arches** support gills. **Gill filaments** extend from each gill arch and include vascular folds of epithelium, called **pharyngeal lamellae** (figure 18.16*a* and *b*). Branchial arteries carry blood to the gills and into gill filaments. The arteries break into capillary beds in pharyngeal lamellae. Gas exchange occurs as blood and water move in opposite directions on either side of the lamellar epithelium. This **countercurrent exchange mechanism** provides very efficient gas exchange by maintaining a concentration gradient between the blood and the water over the entire length of the capillary bed (figure 18.16*c* and *d*).

Swim Bladders and Lungs

The Indian climbing perch (*Anabas testudineus*) spends its life almost entirely on land. These fish, like most bony fishes, have gas chambers called **pneumatic sacs.** In nonteleost fishes and some teleosts, a pneumatic duct connects the pneumatic sacs to the esophagus or another part of the digestive tract. Swallowed air enters these sacs, and gas exchange occurs across vascular surfaces. Thus, in the Indian climbing perch, lungfishes, and ancient rhipidistians (extinct sarcopterygians), pneumatic sacs function(ed) as lungs. In other bony fishes, pneumatic sacs act as swim bladders.

Lungs are likely more primitive than swim bladders. Much of the early evolution of bony fishes occurred in warm, freshwater lakes and streams during the Devonian period. These bodies of water frequently became stagnant and periodically dried. Having lungs in these habitats could have meant the difference between life and death. On the other hand, the later evolution of modern bony fishes occurred in marine and freshwater environments, where stagnation was not a problem. In these environments, the use of pneumatic sacs in buoyancy regulation would have been adaptive (figure 18.17).

Buoyancy Regulation

Did you ever consider why you can float in water? Water is a supportive medium, but that is not sufficient to prevent you from sinking. Even though you are made mostly of water, other constituents of tissues are more dense than water. Bone, for example, has a specific gravity twice that of water. Why, then, can you float? You can float because of two large, air-filled organs called lungs.

Fishes maintain their vertical position in a column of water in one or more of four ways. One way is to incorporate low-density compounds into their tissues. Fishes (especially their livers) are saturated with buoyant oils. A second way fishes maintain vertical position is to use fins to provide lift. The pectoral fins of a shark are planing devices that help create lift as the shark moves through the water. Also, the large upper lobe of a shark's caudal fin provides upward thrust for the posterior end of the body. A third adaptation is the reduction of heavy tissues in fishes. The bones of fishes are generally less dense than those of terrestrial vertebrates. One of the adaptive features of the elasmobranch cartilaginous skeleton probably results from cartilage being only slightly heavier than water. The fourth adaptation is the swim bladder. A fish regulates buoyancy by precisely controlling the volume of gas in its swim bladder. (You can mimic this adaptation while floating in water. How well do you float after forcefully exhaling as much air as possible?)

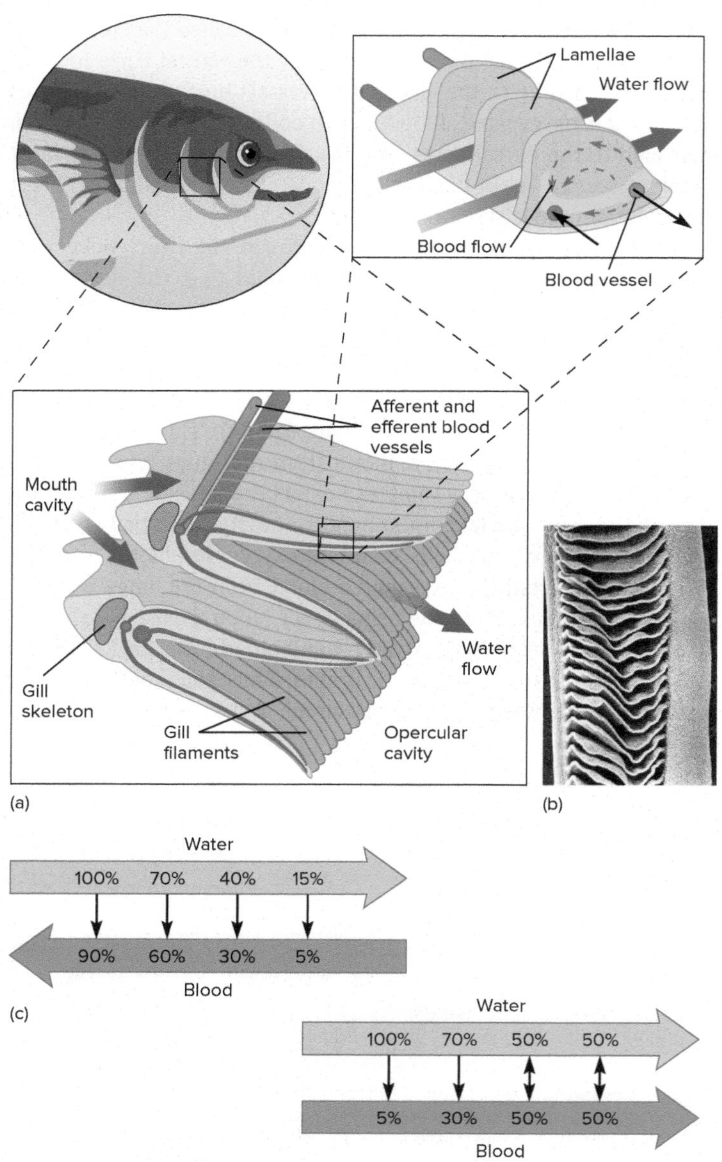

FIGURE 18.16

Gas Exchange at the Pharyngeal Lamellae. (*a*) The gill arches under the operculum support two rows of gill filaments. Blood flows into gill filaments through afferent branchial arteries, and these arteries break into capillary beds in the pharyngeal lamellae. Water and blood flow in opposite directions on either side of the lamellae. Dashed red lines represent blood flow within lamellae. (*b*) Electron micrograph of a flounder gill filament showing numerous lamellae (SEM ×300). This electron micrograph corresponds with the inset in (a). (*c* and *d*) A comparison of countercurrent and parallel exchanges. Water entering the spaces between pharyngeal lamellae is saturated with oxygen in both cases. In countercurrent exchange (*c*), this water encounters blood that is almost completely oxygenated, but a diffusion gradient still favors the movement of more oxygen from the water to the blood. As water continues to move between lamellae, it loses oxygen to the blood because it is continually encountering blood with a lower oxygen concentration. Thus, a diffusion gradient is maintained along the length of the lamellae. If blood and water moved in parallel fashion (*d*), oxygen would diffuse from water to blood only until the oxygen concentration in blood equaled the oxygen concentration in water, and the exchange would be much less efficient.

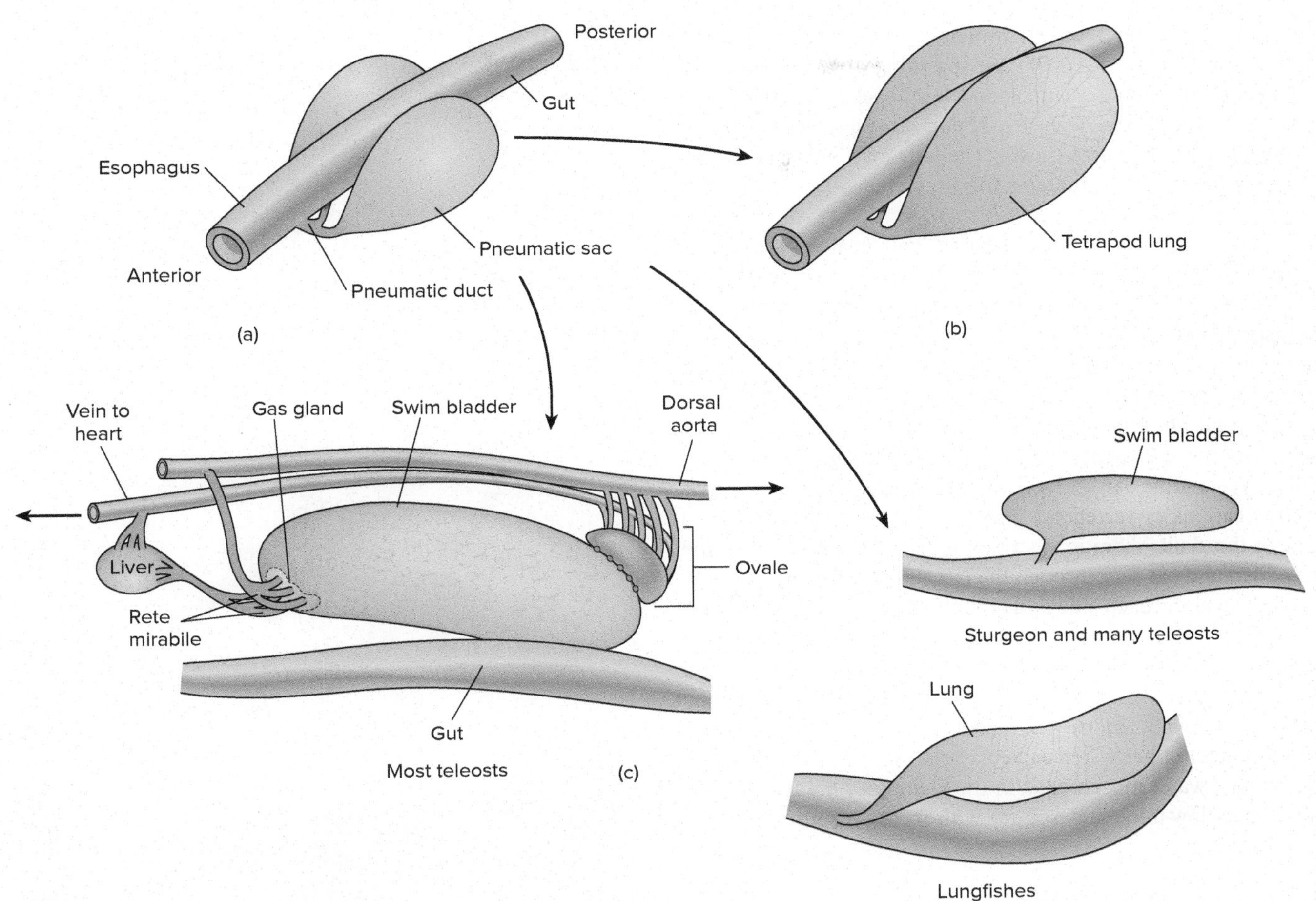

FIGURE 18.17
Possible Sequence in the Evolution of Pneumatic Sacs. (*a*) Pneumatic sacs may have originally developed from ventral outgrowths of the esophagus. Many ancient fishes probably used pneumatic sacs as lungs. (*b*) Primitive lungs developed further during the evolution of vertebrates. Internal compartmentalization increases surface area for gas exchange in land vertebrates. (*c*) In most bony fishes, pneumatic sacs are called swim bladders, and they are modified for buoyancy regulation. Swim bladders are dorsal in position to prevent a tendency for the fish to "belly up" in the water. Pneumatic duct connections to the esophagus are frequently lost, and gases transfer from the blood to the swim bladder through a countercurrent exchange mechanism called a rete mirabile and the gas gland. The ovale, at the posterior end of the swim bladder, returns gases to the bloodstream.

The pneumatic duct connects the swim bladders of garpike, sturgeons, and other primitive bony fishes to the esophagus or another part of the digestive tract. These fishes gulp air at the surface to force air into their swim bladders. Fishes with this type of swim bladder are termed "physostomous" (Gr. *phusa*, bladder + *stoma*, mouth).

Most teleosts have swim bladders that have lost a functional connection to the digestive tract. The blood secretes gases (various mixtures of nitrogen and oxygen) into the swim bladder using a countercurrent exchange mechanism in a vascular network called the rete mirabile ("miraculous net"). Gases are secreted from the rete mirabile into the swim bladder through a gas gland. Gases may be reabsorbed into the blood at the posterior end of the bladder, the ovale (*see figure 18.17*c). Fishes with this type of bladder are termed, "physoclistous" (Gr. *phusa*, bladder + *kleistos*, closed).

Nervous and Sensory Functions

The central nervous system of fishes, as in other vertebrates, consists of a brain and a spinal cord. Sensory receptors are widely distributed over the body. In addition to generally distributed receptors for touch and temperature, fishes possess specialized receptors for olfaction, vision, hearing, equilibrium and balance, and for detecting water movements.

Openings, called external nares, in the snouts of fishes lead to olfactory receptors. In most fishes, receptors are in blind-ending olfactory sacs. In a few fishes, the external nares open to nasal passages that lead to the mouth cavity. Research has revealed that some fishes rely heavily on their sense of smell. For example, salmon (Salmonidae) and lampreys return to spawn in the streams in which they hatched years earlier. Their migrations to these streams often involve distances of hundreds of kilometers, and the fishes' perception of the characteristic odors of their spawning stream guide them.

The eyes of fishes are similar in most aspects of structure to those in other vertebrates. They are lidless, however, and the lenses are round. Focusing requires moving the lens forward or backward in the eye. (Most other vertebrates focus by changing the shape of the lens.)

Receptors for equilibrium, balance, and hearing are in the inner ears of fishes, and their functions are similar to those of other vertebrates. Semicircular canals detect rotational movements, and other sensory patches help with equilibrium and balance by detecting the direction of the gravitational pull. Fishes lack the outer and/or middle ear, which conducts sound waves to the inner ear in other vertebrates. Anyone who enjoys fishing knows, however, that most fishes can hear. Vibrations may pass from the water through the bones of the skull to the middle ear, and a few fishes have chains of bony ossicles (modifications of vertebrae) that connect the swim bladder to the back of the skull. Vibrations strike the fish, are amplified by the swim bladder, and are sent through the ossicles to the skull.

Running along each side and branching over the head of most fishes is a lateral-line system. The **lateral-line system** consists of sensory pits in the epidermis of the skin that connect to canals that run just below the epidermis. In these pits are receptors that are stimulated by water moving against them (*see figure 24.19*). Lateral lines are used to detect either water currents or a predator or a prey that may be causing water movements, in the vicinity of the fish. Fishes may also detect low-frequency sounds with these receptors.

Electroreception and Electric Fishes

All organisms produce weak electrical fields from the activities of nerves and muscles. **Electroreception** is the detection of electrical fields that the fish or another organism in the environment generates. Electroreception and/or electrogeneration has been demonstrated in over 500 species of fishes in the classes Chondrichthyes and Actinopterygii. These fishes use their electroreceptive sense for detecting prey and for orienting toward or away from objects in the environment.

Prey detection with this sense is highly developed in the rays and sharks. Spiny dogfish sharks (*Squalus acanthias*) locate prey by electroreception. A shark can find and eat a flounder that is buried in sand, and it will try to find and eat electrodes that are creating electrical signals similar to those that the flounder emits. However, a shark cannot find a dead flounder buried in the sand or a live flounder covered by an insulating polyvinyl sheet (material that restricts movement of electrical currents). Electroreceptors are located on the heads of sharks and are called ampullary organs (*see figure 24.18*).

Some fishes are not only capable of electroreception but can also generate electrical currents. An electric fish (*Gymnarchus niloticus*) lives in freshwater systems of Africa. Muscles near its caudal fin are modified into organs that produce a continuous electrical discharge. This current spreads between the tail and the head. Pore-like perforations near the head contain electroreceptors. The electrical waves circulating between the tail and the head are distorted by objects in their field. This distortion is detected in changing patterns of receptor stimulation (figure 18.18). The electrical sense of

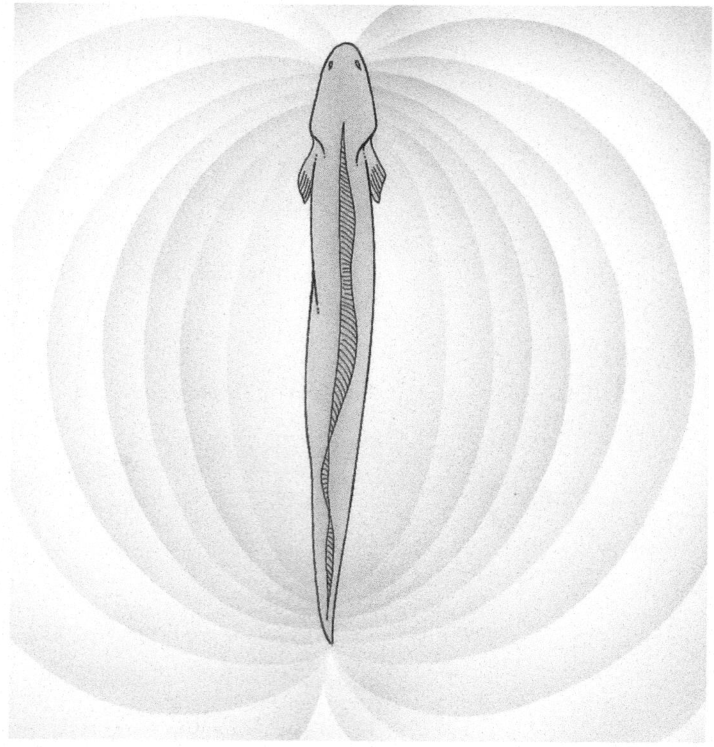

(a)

(b)

FIGURE 18.18

Electric Fishes. (*a*) The electrical field of a fish detects the presence of prey and other objects in the fish's murky environment. Currents circulate from electrical organs in the fish's tail to electroreceptors near its head. An object in this electrical field changes the pattern of stimulation of electroreceptors. (*b*) The electric fish (*Gymnarchus niloticus*).
©Tom McHugh/Science Source

Gymnarchus is an adaptation to living in murky freshwater habitats where eyes are of limited value.

The fishes best known for producing strong electrical currents are the electric eel (a bony fish) and the electric ray (an elasmobranch). The electric eel (*Electrophorus electricus*) occurs in rivers of the Amazon basin in South America. The organs for producing

FIGURE 18.19

Electric Fishes. A lesser electric ray (*Narcine brasiliensis*).
©Jonathan Bird/Getty Images

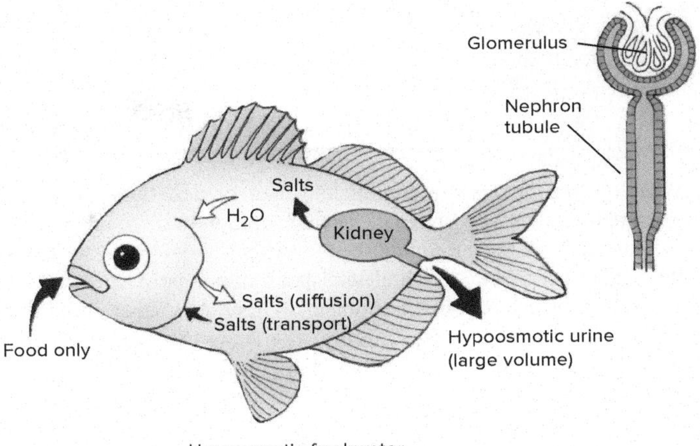

(a) Freshwater teleosts (hypertonic blood)

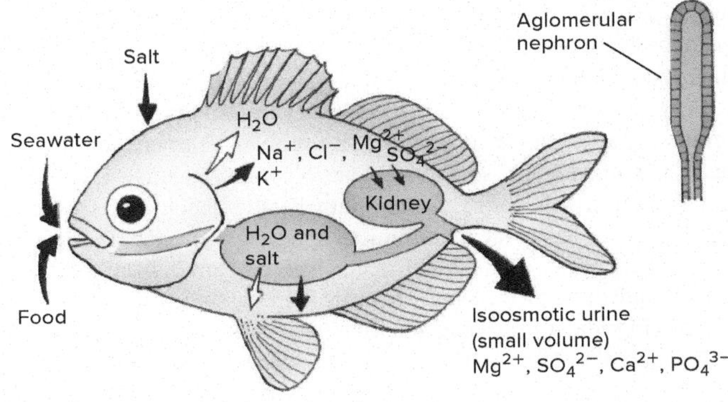

(b) Marine teleosts (hypotonic blood)

FIGURE 18.20

Osmoregulation by (*a*) Freshwater and (*b*) Marine Fishes. Large arrows indicate passive uptake or loss of water or electrolytes (ions) through ingestion and excretion. Small, solid arrows indicate active transport processes at gill membranes and kidney tubules. Small, open arrows indicate passive uptake or loss by diffusion through permeable surfaces. Insets of kidney nephrons depict adaptations within the kidney. Water, ions, and small organic molecules are filtered from the blood at the glomerulus of the nephron. Essential components of the filtrate can be reabsorbed within the tubule system of the nephron. Marine fishes conserve water by reducing the size of the glomerulus of the nephron, and thus reducing the quantity of water and ions filtered from the blood. Other ions can be secreted from the blood into the kidney tubules. Marine fishes can produce urine that is isoosmotic with the blood. Freshwater fishes have enlarged glomeruli and short tubule systems. They filter large quantities of water from the blood, and tubules reabsorb some ions from the filtrate. Freshwater fishes produce a hypoosmotic urine.

electrical currents are in the trunk of the electric eel and can deliver shocks in excess of 500 V. The electric ray (*Narcine*) has electric organs in its fins that are capable of producing pulses of 50 A at about 50 V (figure 18.19). Shocks that these fishes produce are sufficiently strong to stun or kill prey and discourage predators.

Excretion and Osmoregulation

Fishes, like all animals, must maintain a proper balance of electrolytes (ions) and water in their tissues. This osmoregulation is a major function of the kidneys and gills of fishes. Kidneys are located near the midline of the body, just dorsal to the peritoneal membrane that lines the body cavity. As with all vertebrates, the excretory structures in the kidneys are called **nephrons.** Nephrons filter bloodborne nitrogenous wastes, ions, water, and small organic compounds across a network of capillaries called a **glomerulus.** The filtrate then passes into a tubule system, where essential components may be reabsorbed into the blood. The filtrate remaining in the tubule system is then excreted.

Freshwater fishes live in an environment containing few dissolved substances. Osmotic uptake of water across gill, oral, and intestinal surfaces and the loss of essential ions by excretion and defecation are constant. To control excess water build-up and ion loss, freshwater fishes never drink and only take in water when feeding. Also, the numerous nephrons of freshwater fishes often have large glomeruli and relatively short tubule systems. Reabsorption of some ions and organic compounds follows filtration. Because the tubule system is relatively short, little water is reabsorbed. Thus, freshwater fishes produce large quantities of very dilute urine. Ions are still lost, however, through the urine and by diffusion across gill and oral surfaces. Active transport of ions into the blood at the gills compensates for this ion loss. Freshwater fishes also get some salts in their food (figure 18.20*a*).

Marine fishes face the opposite problems. Their environment contains 3.5% ions, and their tissues contain approximately 0.65% ions. Marine fishes, therefore, must combat water loss and accumulation of excess ions. They drink water and eliminate excess ions by excretion, defecation, and active transport across gill surfaces. The nephrons of marine fishes often possess small glomeruli and long tubule systems. Much less blood is filtered than in freshwater fishes, and water is efficiently, although not entirely, reabsorbed from the nephron (figure 18.20*b*).

WILDLIFE ALERT

Invasive Species–A Growing Problem in a Shrinking World

What do you do with unused fishing bait or aquarium fishes that become too much of a bother? The easy solution to disposing of these "unwanteds" has been to dump them over the side of the boat or flush them down the toilet. This careless disposal of nonnative species is one cause of growing ecosystem problems. Other nonnative species are introduced intentionally to "improve" the landscape or control other pest species. Still other nonnative species are introduced accidentally, for example, the zebra mussel (*Dreissena polymorpha, see Wildlife Alert, pages 202–203*) was accidentally introduced from Russia in ship ballast, and the sea lamprey (*Petromyzon marinus, see figures 18.6 and 18.7*) was permitted into the Great Lakes by the construction of the Welland Canal between Lakes Ontario and Erie. A species that is introduced into nonnative habitats may be removed from natural predators or other agents that keep its numbers in check. Under these circumstances, invasive species multiply rapidly, and tremendous harm can come to the alien species' new home.

The red lionfish (*Pterois volitans, see figure 18.1*) is an example of an invasive species that is creating massive problems throughout the Caribbean and middle and southern Atlantic coastal regions of the United States. This fish is native to the South Pacific and Indian Oceans. Its unique beauty has made it an attractive aquarium species. That is where the problem began. In 1992, Hurricane Andrew blew in from the Atlantic Ocean and through the Gulf of Mexico. In passing Biscayne Bay near Miami, Florida, a beachside aquarium was broken, and it released lionfishes into the Atlantic Ocean. Additional releases by hobbyists have occurred as well. Females of this species spawn between 12,000 and 15,000 eggs at a time, and in warm water, they can spawn every four days during their reproductive period, which is early in the year in Florida's coastal waters. The result has been the most rapid marine invasion ever recorded.

Lionfishes are voracious predators, consuming a diverse array of live fishes and crustaceans. In their native habitats, coevolution with Pacific and Indian Ocean predators keep lionfishes in check. In their invaded habitats, their venomous dorsal fin and anal fin spines make them prey to very few natural predators. When threatened, lionfishes erect their spines and swim forward or backward toward the threat to inflict stings. The venom is a mixture of acetylcholine and a neurotoxin that modifies neuromuscular transmission (*see figure 23.21*). A sting can result in pain, swelling, nausea, disorientation, and even paralysis. The lionfish's diverse diet, combined with virtually no pressure from predators, threatens coastal ecosystems. Their prey includes parrotfishes (*see this chapter's opening photo*), which feed on algae that can potentially overgrow coral reefs. They compete with other predators like snappers and groupers–the latter are already threatened by overfishing. These coastal ecosystems are severely threatened by this amazing fish.

Measures are being taken to control the invasive lionfish. These measures primarily include education regarding the release of exotic species and efforts to remove these fishes from coastal waters. Removal efforts include netting, spearing, and trapping. Hook-and-line fishing is not effective as lionfishes feed only on living prey, not dead bait at the end of a hook. The public is also being educated with the knowledge that lionfishes are good to eat. Lionfish venom is confined to their spines. Once spines are removed, and the fish is filleted, lionfishes are fit for human consumption.

Elasmobranchs have a unique osmoregulatory mechanism. They convert some of their nitrogenous wastes into urea in the liver. This in itself is somewhat unusual, because most fishes excrete ammonia rather than urea. Even more unusual, however, is that urea is sequestered in tissues all over the body. Enough urea is stored to make body tissues slightly hyperosmotic to seawater. (That is, the concentration of solutes in a shark's tissues is essentially the same as the concentration of ions in seawater.) Therefore, the problem most marine fishes have of losing water to their environment is much less severe for elasmobranchs. Energy that does not have to be devoted to water conservation can now be used in other ways. This adaptation required the development of tolerance to high levels of urea, because urea disrupts important enzyme systems in the tissues of most other animals.

In spite of this unique adaptation, elasmobranchs must still regulate the ion concentrations in their tissues. In addition to having ion-absorbing and secreting tissues in their gills and kidneys, elasmobranchs possess a rectal gland that removes excess sodium chloride from the blood and excretes it into the cloaca. (A **cloaca** is a common opening for excretory, digestive, and reproductive products.)

Diadromous fishes migrate between freshwater and marine environments. Salmon (*e.g., Oncorhynchus*) and marine lampreys (*Petromyzon*) migrate from the sea to freshwater to spawn, and the freshwater eel (*Anguilla*) migrates from freshwater to marine environments to spawn. Diadromous migrations require gills capable of coping with both uptake and secretion of ions. Osmoregulatory powers needed for migration between marine and freshwater environments may not be developed in all life-history stages. Young salmon, for example, cannot enter the sea until certain cells on the gills develop ion-secreting powers.

Fishes have few problems getting rid of the nitrogenous by-products of protein metabolism. Up to 90% of nitrogenous wastes are eliminated as ammonia by diffusion across gill surfaces. Even though ammonia is toxic, aquatic organisms can have it as an excretory product because ammonia diffuses into the surrounding water. The remaining 10% of nitrogenous wastes is excreted as urea, creatine, or creatinine. These wastes are produced in the liver and are excreted via the kidneys.

Reproduction and Development

Imagine, 45 kg of caviar from a single, 450-kg sturgeon! Admittedly, a 450-kg sturgeon (Acipenseridae) is a very large fish (even for a sturgeon), but a fish producing millions of eggs in a single season is not unusual. These numbers simply reflect the hazards of developing in aquatic habitats unattended by a parent. The vast majority of these millions of potential adults will never survive to reproduce. Many eggs will never be fertilized, many fertilized eggs may wash ashore and dry, currents and tides will smash many eggs and embryos, and others will be eaten. In spite of all of these hazards, if only four of the millions of embryos of each breeding pair survive and reproduce, the population will double.

Producing overwhelming numbers of eggs, however, is not the only way that fishes increase the chances that a few of their offspring will survive. Some fishes show mating behavior that helps ensure fertilization, or nesting behavior that protects eggs from predation, sedimentation, and fouling.

Mating may occur in large schools, and one individual releasing eggs or sperm often releases spawning pheromones that induce many other adults to spawn. Huge masses of eggs and sperm released into the open ocean ensure the fertilization of many eggs.

Most fishes are oviparous, meaning that eggs develop outside the female from stored yolk (*see chapter 29, Vertebrate Reproductive Strategies, page 546*). Some elasmobranchs are ovoviviparous, and their embryos develop in a modified oviduct of the female. Nutrients are supplied from yolk stored in the egg. Other elasmobranchs, including gray reef sharks (*Carcharhinus amblyrhynchos*) and hammerheads (Sphyrnidae), are viviparous. A placenta-like outgrowth of a modified oviduct diverts nutrients from the female to the yolk sacs of developing embryos. Internal development of viviparous bony fishes usually occurs in ovarian follicles, rather than in the oviduct. In guppies (*Lebistes*), eggs are retained in the ovary, and fertilization and early development occur there. Embryos are then released into a cavity within the ovary and development continues, with nourishment coming partly from yolk and partly from ovarian secretions.

Some fishes have specialized structures that aid in sperm transfer. Male elasmobranchs, for example, have modified pelvic fins called claspers. During copulation, the male inserts a clasper into the cloaca of a female. Sperm travel along grooves of the clasper. Fertilization occurs in the female's reproductive tract and usually results in a higher proportion of eggs being fertilized than in external fertilization. Thus, fishes with internal fertilization usually produce fewer eggs.

In many fishes, care of the embryos is limited or nonexistent. Some fishes, however, construct and tend nests (figure 18.21 and *see figure 5.7*), and some carry embryos during development. Clusters of embryos may be brooded in special pouches attached to some part of the body, or they may be brooded in the mouth. Some of the best-known brooders include the seahorses (*Hippocampus*) and pipefishes (e.g., *Syngnathus*). Males of these closely related fishes carry embryos throughout development in ventral pouches. The male Brazilian catfish (*Loricaria typhys*) broods embryos in an enlarged lower lip.

Sunfishes (Centrarchidae) and sticklebacks (Gasterosteidae) provide short-term care of posthatching young. Male sticklebacks assemble fresh plant material into a mass in which the young take refuge. If one offspring wanders too far from the nest, the male snaps it up in its mouth and spits it back into the nest. Sunfish males do the same for young that wander from schools of recently hatched fishes. The Cichlidae engage in longer term care (*see figures 1.1 and 1.2*). In some species, young are mouth brooded, and other species tend young in a nest. After hatching, the young venture from the parent's mouth or nest but return quickly when the parent signals danger with a flicking of the pelvic fins.

FIGURE 18.21

A Garibaldi (*Hypsypops rubicundus*). The male cultivates a nest of filamentous red algae and then entices a female to lay eggs in the nest. Males also defend the nest against potential predators.

Section 18.3 Thinking Beyond the Facts

How are each of the following processes or structures helpful in explaining why a fish is fishlike: countercurrent exchange mechanism, left atrium, pneumatic sacs, and lateral-line system?

18.4 FURTHER PHYLOGENETIC CONSIDERATIONS

LEARNING OUTCOME

1. Assess the importance of the ancient Tetrapodomorpha in our view of vertebrate evolution.

Two important series of evolutionary events occurred during the evolution of the bony fishes. One of these was an evolutionary explosion that began about 150 mya and resulted in the vast diversity of teleosts living today. The second series of events involves the evolution of terrestrialism. The presence of functional lungs in modern lungfishes has led to the suggestion that the lungfish lineage may have been ancestral to modern terrestrial vertebrates. Most cladistic and anatomical evidence indicates that the lungfish lineage gave rise to no other vertebrate taxa.

The explanation for the origin of terrestrial vertebrates involves the sarcopterygian clade, Tetrapodomorpha. This clade includes an order of amphibian-like fishes in the order Osteolepiformes. Osteolepiforms possessed several unique characteristics in common with early amphibians. These included structures of the jaw, teeth, vertebrae, and limbs, among others. These sarcopterygians probably represent early stages in the transition between fishes and tetrapods. Their basic limb structure shows homologies to that found in terrestrial vertebrates (Evolutionary Insights, pp. 332–333).

Another group of Tetrapodomorpha may be still closer to the tetrapod ancestor. In 2004, a new 375-million-year-old fossil, *Tiktaalik,* was discovered in the Canadian arctic (figure 18.22). This fish had fins, gills, and scales, and it also possessed many more tetrapod characters than other sarcopterygian fossils, including a dorsoventrally compressed and widened skull and striking tetrapod forelimb skeletal homologies. It lacked the opercular supports and dorsal and anal fins present in other sarcopterygians. *Tiktaalik* is the first sarcopterygian fossil that shows evidence of a pectoral girdle and a freely moveable neck. (The pectoral girdle of tetrapods attaches the forelegs to the vertebral column, and other fishes lack a neck.) The association of these fossils with a river channel suggests that these events occurred in freshwater environments. Skeletal features suggest that *Tiktaalik* swam and pushed its way through shallow water, where it probably preyed on small fishes and invertebrates. The front fins had a limited range of motion, so *Tiktaalik* is unlikely to have walked across land but could prop itself up on forelimbs at the water's edge. *Tiktaalik* may not be the direct ancestor of amphibians, but this "fishapod" certainly demonstrates the evolutionary preadaptations that allowed vertebrates to become

FIGURE 18.22

The Fishapod *Tiktaalik*. *Tiktaalik* was discovered on Ellismere Island, Canada, in 2004. This 375-million-year-old fossil helps us understand the transition between sarcopterygian fishes and tetrapods. Note the flattened, amphibian-like head. This reconstruction of *Tiktaalik* shows the use of foreappendages, supported by a pectoral girdle, for propping itself out of the water. A neck permitted head mobility not found in other sarcopterygian fishes. *Tiktaalik* probably used these adaptions for foraging the water's edge for small fishes and invertebrate prey. (*See figure 19.2* for comparison with an early amphibian.)

EVOLUTIONARY INSIGHTS

The Early Evolution of the Vertebrate Limb

Nowhere are evolutionary transitions more clearly documented than in the vertebrate lineage. It is important to remember that documenting evolutionary transitions does not necessarily mean that we trace the exact animal species involved in a series of evolutionary changes. Instead, paleontologists and biologists look for transitional stages in the development of structures represented in fossil and developmental records. One example is the documentation of changes in limb structure in the fish-to-amphibian transition.

The basic arrangement of bones in the limbs of terrestrial vertebrates was presented in chapter 4 (*see figure 4.11*) as an example of the concept of homology. In the vertebrate limb, a single proximal element, the humerus (femur in the hindlimb), articulates with two distal elements, the radius (fibula) and ulna (tibia). These are followed more distally by the wrist bones, the carpals (ankle, tarsals), and then the bones of the hand, the metacarpals (foot, metatarsals) and phalanges. With the exception of the bones of the hand, this basic pattern can be observed in the forelimb of an osteolepiform fish named *Eusthenopteron* (box figure 18.1*a*). The bones of the hand were not present in *Eusthenopteron.* Instead, small dermal elements called lepidotrichia supported the distal portion of the fin. The same pattern was present in early tetrapods such as *Acanthostega* and *Ichthyostega* (box figure 18.1*c* and *d*), except that the bones of the hand were present and the dermal elements were absent. *Acanthostega* had eight digits in the forelimb. *Ichthyostega* had seven digits in the hindlimb. (The number of bones in the forelimb is unknown.) The unusual number of digits was probably adaptive in forming a surface for propulsion through water. In later tetrapods, the usual number of digits was reduced to five, which is typical of most modern forms.

The origin of the bones of the hand/foot of tetrapods has puzzled researchers. One possible answer is emerging from the study of molecular biology. A group of genes, called homeotic genes (homeobox-containing genes or *Hox* genes), plays important roles in determining the identity and location of body structures, including limbs. Biologists now realize that small changes in *Hox* genes can have profound effects on body structures. Mutations that alter the function of one of these genes produce mouse embryos lacking digits. Could a similar change have resulted in the evolution of digits in tetrapod limbs? Most zoologists think so.

Recently, paleontologists have discovered fossils of a sarcopterygian fish, *Sauripterus.* This fish had the arrangement of proximal bones in its forelimbs as described for *Eusthenopteron.* Like *Eusthenopteron, Sauripterus* had lepidotrichia that supported the fin. Interestingly, there were also cartilage-derived bones similar to those that supported the hand and foot of *Acanthostega* and *Ichthyostega* (box figure 18.1*b*)! There are eight of these digitlike bones. They undoubtedly stiffened the distal portion of the fin. A stout shoulder girdle and stiffened fins suggest that these appendages could have been used in propulsion against either water or

(*Continued*)

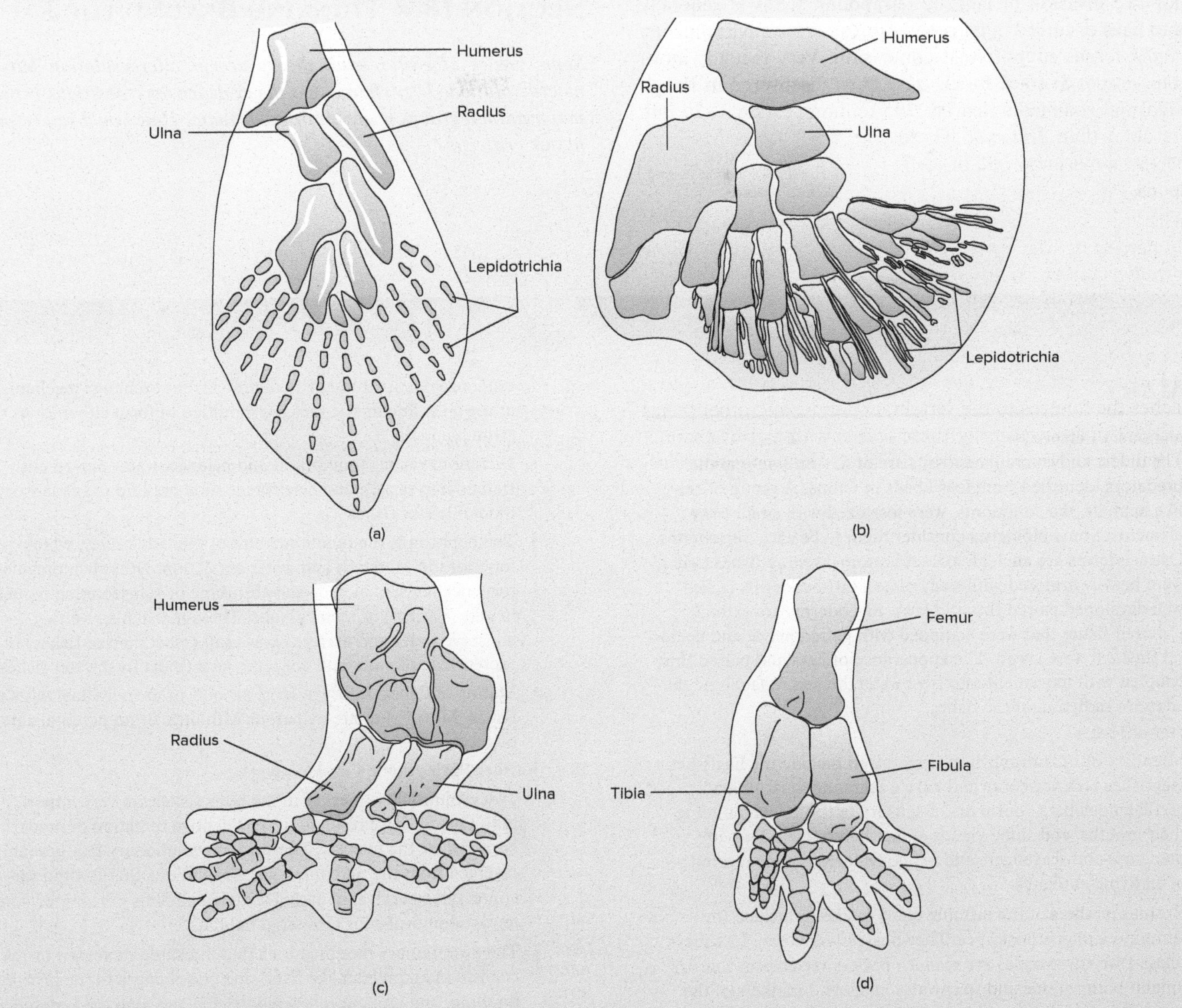

BOX FIGURE 18.1 The Evolution of Tetrapod Limbs. (*a*) The forelimb of the sarcopterygian fish *Eusthenopteron.* Lepidotrichia are dermal elements not found in tetrapods. Bones of the hand are absent. (*b*) The forelimb of the sarcopterygian fish *Sauripterus.* Note the presence of both lepidotrichia and eight digitlike bony elements in the distal portion of the fin. (*c*) The forelimb of the tetrapod *Acanthostega* had eight digits. (*d*) The hindlimb of the tetrapod *Ichthyostega* had seven digits. The forelimb structure is unknown. *Source: Shubin, Neil. University of Chicago. Chicago (IL).*

substrate. This discovery, along with the discovery of *Tiktaalik,* helps us understand the transition between fishes and amphibians.

Paleontologists have also noticed that even though fossils of *Sauripterus* and tetrapods such as *Acanthostega* and *Ichthyostega* have more than five digits (or digitlike elements), some of the digits are structurally identical (*see box figure 18.1*). In fact, there seem to be five basic digit types. These five digit types correspond to five genes in one *Hox* gene cluster. Experimental manipulation of this *Hox* gene cluster has transformed one digit type into another in developing chick embryos and altered the number of digits in mouse embryos. Could the presence of extra digits in the ancient tetrapods be explained by similar changes in the activity of a *Hox* gene cluster? Most zoologists think so.

Molecular biologists have observed that the same *Hox* gene clusters are expressed along the body wall of embryos in the locations of both fore and hind appendages in identical ways. Could this explain the evolution of the similar structure of fore- and hindlimbs of vertebrates? Most zoologists think so.

The transition between fishes and amphibians is well documented. Not only is paleontology documenting what changes happened but also developmental biology is providing answers to how the changes probably occurred. It is one example of the many exciting developments that are emerging as a result of combining traditional methods with molecular techniques.

terrestrial. The invasion of land by tetrapodomorphs is generally assumed to have occurred in freshwater or estuarine environments, and *Tiktaalik* fossils support that conclusion. Very recently, however, tetrapodomorph track fossils have been discovered in Polish marine tidal flat sediments that are 395 million years old—20 million years older than *Tiktaalik*. These fossils suggest a marine origin of earliest tetrapods.

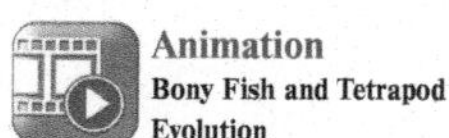

Section 18.4 Thinking Beyond the Facts

Some critics of evolutionary theory accept microevolution but not macroevolution. They charge that the evidence for transitions between major animal groups is absent or inadequate. How would you respond to this criticism?

Summary

18.1 **Evolutionary Perspective**

- The oldest known craniate fossils are of 530-million-year-old predators identified from fossil beds in China. A group of eel-like animals, the conodonts, were fossilized with small bony elements. Some biologists consider them to be early vertebrates. Ostracoderms are ancient, extinct bottom-dwelling fishes that were heavily armored with bony plates. Ostracoderms lacked well-developed paired fins and jaws. Placoderms are extinct, armored fishes that were equipped with paired pelvic and pectoral fins. All were jawed. The appearance of jaws and paired fins, coupled with movement into freshwater, helped to facilitate the adaptive radiation of the fishes.

18.2 Survey of Fishes

- Members of the infraphylum Hyperotreti include the hagfishes. Hagfishes lack vertebrae and have a cranium consisting of cartilaginous bars, four pairs of sensory tentacles surrounding their mouths, and slime glands along their bodies. They are predators and scavengers and are considered the most primitive of all living craniates.
- Extant members of the infraphylum Vertebrata include the lampreys and gnathostomes. They possess vertebrae. Lampreys (class Petromyzontida) are modern jawless vertebrates that inhabit both marine and freshwater habitats. Members of the superclass Gnathostomata are the jawed vertebrates. In addition to the tetrapods, this superclass includes three classes of fishes. Members of the class Chondrichthyes include the cartilaginous fishes, and members of the classes Sarcopterygii and Actinopterygii include the bony fishes. The class Sarcopterygii includes the lungfishes, the coelacanths, and the rhipidistians; and the class Actinopterygii includes the ray-finned fishes. In the Actinopterygii, the teleosts are the modern bony fishes. Members of this very large group have adapted to virtually every available aquatic habitat.

18.3 **Evolutionary Pressures**

- Fishes show numerous adaptations to living in aquatic environments. These adaptations include an arrangement of body-wall muscles that creates locomotor waves in the body wall.
- Most modern fishes have acrodont teeth. Most fish are predators, but some species are filter feeders or herbivores.
- Fishes have a heart that develops from four embryological chambers. Blood circulates from the heart through gills to body tissues. Blood returns to the heart in the venous system. Gas exchange occurs through a countercurrent exchange mechanism at the gills. Pneumatic sacs are modified to form either lungs or swim bladders.
- Olfaction, vision, equilibrium and balance, water movement (lateral-line sense), and electroreception are important sensory modalities for fishes.
- The nephron is the functional unit of the fish kidney, which functions in excretion and water regulation. Other osmoregulatory mechanisms include salt-absorbing or salt-secreting tissues in gills, kidneys, or rectal glands. These mechanisms either conserve water and excrete excess salt (most marine fishes) or excrete excess water and conserve ions (most freshwater fishes).
- Mating of fishes occurs in large schools or as individual pairs of fishes. Most fishes are oviparous with little or no parental care of young.

18.4 **Further Phylogenetic Considerations**

- Two evolutionary lineages in the bony fishes are very important. One of these resulted in the adaptive radiation of modern bony fishes, the teleosts. The second evolutionary line probably diverged from the Sarcopterygii. Adaptations that favored sarcopterygian survival in early Devonian streams preadapted some tetrapodomorphs for terrestrial habitats.
- The evolutionary modifications that indicated increased terrestrialism in amphibian-like fishes include changes to the proximal forelimb, and reduction of lepidotrichia and increased presence of digitlike bony elements in the distal portion of fins.

Concept Review Questions

1. According to recent changes in fish classification, the infraphylum Vertebrata includes all of the following, except one. Select the exception.
   a. Hagfishes
   b. Lampreys
   c. Cartilaginous fishes
   d. Bony fishes
2. One would expect to find an ammocoete larva of a lamprey
   a. clinging to a rock in a freshwater stream.
   b. partially buried in the substrate of a freshwater stream.
   c. attached to, and feeding on, a freshwater fish.
   d. attached to, and feeding on, a marine fish.

3. All of the following groups have members with an operculum, except one. Select the exception.
   a. Holocephali
   b. Elasmobranchii
   c. Sarcopterygii
   d. Actinopterygii
4. Fins with muscular lobes and lungs used in gas exchange are characteristic of members of the class
   a. Chondrichthyes.
   b. Myxini.
   c. Sarcopterygii.
   d. Actinopterygii.
5. Which of the following statements regarding pneumatic sacs is FALSE?
   a. Pneumatic sacs originated as ventral outgrowths of the esophagus.
   b. Pneumatic sacs function as lungs in many sarcopterygians.
   c. Pneumatic sacs lie dorsal to the digestive tract in modern fish.
   d. The more primitive function of pneumatic sacs is to serve as swim bladders for buoyancy regulation.

Analysis and Application Questions

1. What characteristic of water makes it difficult to move through, but also makes support against gravity a minor consideration? How is a fish adapted for moving through water?
2. Is the cartilaginous skeleton of chondrichthians a primitive characteristic? In what ways is this characteristic adaptive for these fish?
3. Would swim bladders with functional pneumatic ducts work well for a fish that lives at great depths? Why or why not?
4. Compare and contrast osmoregulatory problems of fish living in freshwater with those of fish living in the ocean. What are the solutions to these problems in each case?

19

Amphibians: The First Terrestrial Vertebrates

Amphibians live a "double life." They often move back-and-forth between water and land or live one stage of their lives on land and another stage in water. Male blue poison dart frogs (*Dendrobates tinctorius*) vocalize on the banks of tropical streams to attract females. Eggs and tadpole larvae develop within the stream.
©Tom Brakefield/Stockbyte/Getty Image

Chapter Outline

19.1 Evolutionary Perspective
Phylogenetic Relationships
19.2 Survey of Amphibians
Order Gymnophiona
Order Caudata
Order Anura
19.3 Evolutionary Pressures
External Structure and Locomotion
Nutrition and the Digestive System
Circulation, Gas Exchange, and Temperature Regulation
Nervous and Sensory Functions
Excretion and Osmoregulation
Reproduction, Development, and Metamorphosis
19.4 Amphibians in Peril
19.5 Further Phylogenetic Considerations

19.1 EVOLUTIONARY PERSPECTIVE

LEARNING OUTCOMES

1. Justify the statement that "any gaps in evidence documenting the fish-to-amphibian transition are essentially gone."
2. Hypothesize why *Ichthyostega* likely resembled stem tetrapods.
3. Describe the key features that link Temnospondyli to Lissamphibia.
4. Describe the extant vertebrate groups that comprise the tetrapod lineage.

Who, while walking along the edge of a pond or stream, has not been startled by the "plop" of an equally startled frog jumping to the safety of its watery retreat? Or who has not marveled at the sounds of a chorus of frogs breaking through an otherwise silent spring evening? These experiences and others like them have led some to spend their lives studying members of the class Amphibia (am-fib′e-ah) (L. *amphibia,* living a double life): frogs, toads, salamanders, and caecilians (figure 19.1). The class name implies that amphibians either move back-and-forth between water and land or live one stage of their lives in water and another on land. One or both of these descriptions is accurate for most amphibians.

Amphibians are **tetrapods** (Gr. *tetra,* four + *podos,* foot). The name is derived from the presence of four muscular limbs and feet with toes and fingers (digits). Some zoologists use the term "Tetrapoda" to formally refer to all sarcopterygian descendants that possess well-formed forelimbs and hindlimbs. Other zoologists reserve the term tetrapod for "crown-group animals." The tetrapod crown group includes the extant (living) tetrapods plus their most recent common ancestor. Tetrapod, in this sense, refers to living amphibians (often called **Lissamphibia** [lis′am-fib′e-ah]), the reptiles (including birds), mammals, and the common ancestor of these groups. A host of extinct "stem tetrapods" is excluded from this use of the term "Tetrapoda."

Phylogenetic Relationships

Chapter 18 described ideas regarding the origin of tetrapods from ancient sarcopterygians. Figure 18.2 and table 18.1 correctly describe all tetrapods as being included within the Tetrapodomorpha, a group of sarcopterygians that also includes lobe-finned fishes. This

FIGURE 19.1

Class Amphibia. Amphibians, like this red-eyed treefrog (*Agalychnis callidryas*), are common vertebrates in most terrestrial and freshwater habitats. Their ancestors were the first terrestrial vertebrates.

grouping is phylogenetically correct because the class Sarcopterygii is thus a monophyletic lineage. Unfortunately, this grouping presents problems for the traditional classification system that groups amphibians and other tetrapods into their own (paraphyletic) classes. We will continue to use the traditional tetrapod names with the "class" designation, realizing that taxonomists still have work to do.

The fossil record provides evidence of many extinct tetrapod taxa, and no one knows what animal was the first tetrapod. The best-known and some of the earliest fossils were first discovered in Greenland in 1932. They are of a 365-million-year-old group called Ichthyostegalia. *Ichthyostega* (figure 19.2) was not the ancestor of all tetrapods, but it has been influential in formulating ideas regarding what the ancestral animals must have been like. Important characteristics evident in these fossils include the loss of some cranial bones and the appearance of a mobile neck, the loss of opercular bones, a reduction of the notochord, the formation of a more rigid vertebral column, four muscular limbs with discrete digits, loss of fin rays, and the presence of a sacral vertebra that fuses the vertebral column and the pelvis (*see figure 18.22*).

Although the specific lineages are still debated, amphibian origins are fairly well understood. The anatomical changes that resulted from selective pressures are well documented in the paleontological record. This record, coupled with molecular data, has led to the development of three hypotheses describing the origins of Lissamphibia. Currently, the most widely accepted hypothesis indicates that Lissamphibia is derived from an early amphibian group called Temnospondyli. Temnospondyls existed in the early Carboniferous to the early Cretaceous (*see appendix B*) and ranged in size from a few centimeters to six meters. Many species had elongated and dorsoventrally flattened skulls, and one genus (*Mastodonsaurus*) possessed enlarged mandibular fangs. Temnospondyli had similarities to Lissamphibia, including a modified palate that allows retraction of the eyes into the skull, pedicillate teeth, two occipital condyles, and shortened ribs.

In addition to the amphibians, the tetrapod lineage includes the amniotes (reptiles [including birds] and mammals). The name "amniote" is derived from the presence of an amniotic egg that resists drying and allows development to occur in a terrestrial environment. The relationships between the amphibians and the amniotes are discussed at the end of this chapter, and the amniotes are covered in chapters 20 through 22.

Figure 19.3 shows one interpretation of the evolutionary relationships in the tetrapod lineage. It is important to reemphasize that these relationships are highly controversial. A host of extinct lineages are omitted, and the representation of two tetrapod lineages, "Reptiliomorpha" and "Amphibia," oversimplify the controversies surrounding these lineages. The importance of the reptiliomorph lineage is discussed at the end of this chapter.

Section 19.1 Thinking Beyond the Facts

Why is the use of the "class" designation for members of the Amphibia phylogenetically incorrect?

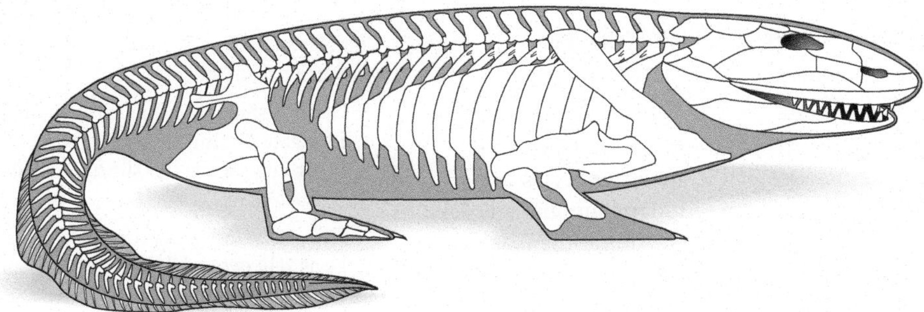

FIGURE 19.2

***Ichthyostega:* An Early Amphibian.** Fossils of this early amphibian were discovered in eastern Greenland in late Devonian deposits. The total length of the restored specimen is about 65 cm. Terrestrial adaptations are heavy pectoral and pelvic girdles and sturdy limbs that probably helped push the body across the ground. Strong jaws suggest that it was a predator in shallow water, perhaps venturing onto shore. Other features include a skull that is similar in structure to that of ancient sarcopterygian fishes and a finlike tail. Note that bony rays dorsal to the spines of the vertebrae support the tail fin. This pattern is similar to the structure of the dorsal fins of fishes and is unknown in any other tetrapod. The arrangement of bony elements in the distal portion of the foreleg is unknown.

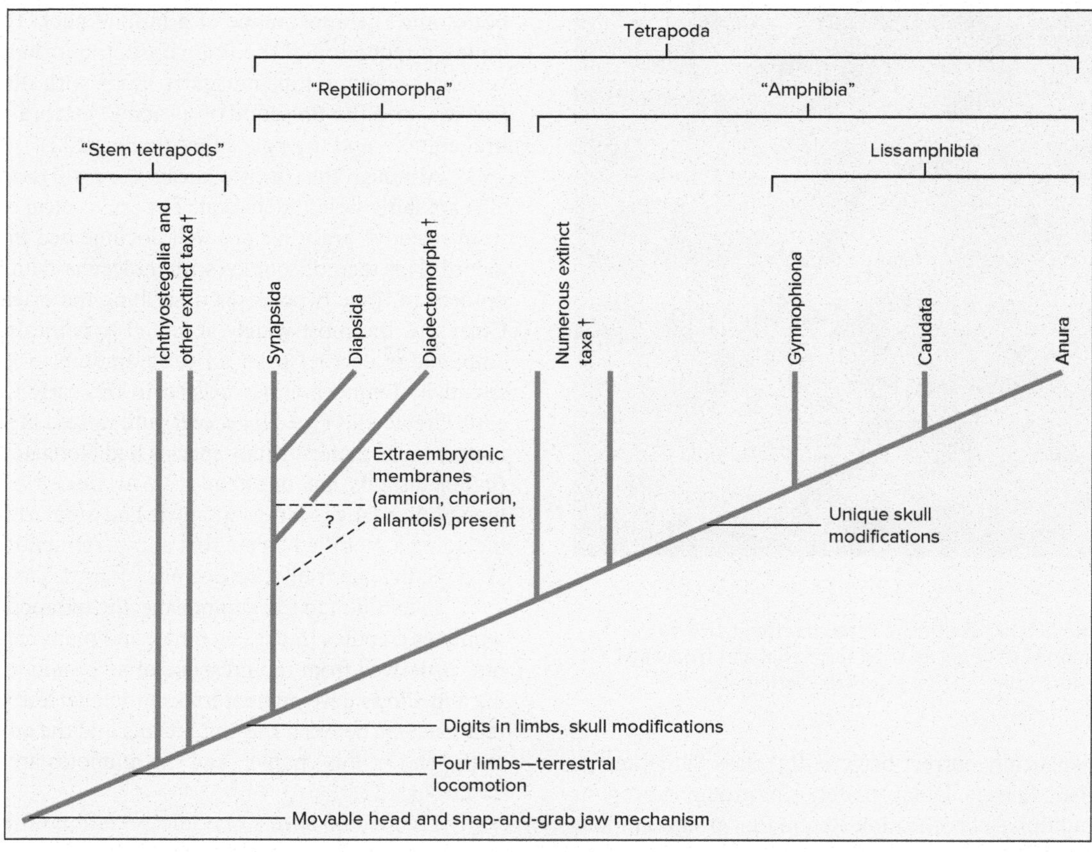

FIGURE 19.3

One Hypothesis of Evolutionary Relationships among the Tetrapods. The earliest amphibians arose during the Devonian period. There are three classes of modern amphibians and numerous extinct taxa (two of these are depicted here). It is widely accepted that Lissamphibia is derived from an earlier amphibian group, Temnospondyli. Temnospondyli would likely be positioned with the extinct taxa nearest the indication of "unique skull modifications." The reptiliomorph lineage of early tetrapods gave rise to the amniotes (avian and non-avian reptiles, mammals) and other extinct taxa. Synapomorphic characters for lower taxonomic groups are not indicated. Daggers (†) indicate extinct taxa. The Tetrapoda, as represented here, indicates the crown-group members of this lineage.

19.2 SURVEY OF AMPHIBIANS

LEARNING OUTCOMES

1. Describe characteristics of members of the order Gymnophiona.
2. Describe characteristics of members of the order Caudata.
3. Describe characteristics of members of the order Anura.

Amphibians occur on all continents except Antarctica, but they are absent from many oceanic islands. The 7,303 modern species are a mere remnant of this once-diverse group. Modern amphibians belong to three orders: Caudata, the salamanders; Anura, the frogs and toads; and Gymnophiona, the caecilians (table 19.1).

TABLE 19.1
CLASSIFICATION OF LIVING AMPHIBIANS

Class Amphibia (am-fib′e-ah)
Skin with mucoid secretions and lacking epidermal scales, feathers, or hair; larvae usually aquatic and undergo metamorphosis to the adult; two atrial chambers in the heart; one cervical and one sacral vertebra.

Order Gymnophiona (jim″no-fi′o-nah)
Elongate, limbless; segmented by annular grooves; specialized for burrowing; tail short and pointed; rudimentary left lung. Caecilians.

Order Caudata (kaw′dat-ah)
Long tail, two pairs of limbs; lack middle ear. Salamanders, newts, sirens, and amphiumas.

Order Anura (ah-noor′ah)
Tailless; elongate hindlimbs modified for jumping and swimming; five to nine presacral vertebrae with transverse processes (except the first); postsacral vertebrae fused into rodlike urostyle; tympanum and larynx well developed. Frogs, toads.

Order Gymnophiona

Members of the order Gymnophiona (jim″no-fi′ o-nah) (Gr. *gymnos*, naked + *ophineos*, snakelike) are the caecilians (figure 19.4). There are approximately 200 species that are mostly pantropical (across the tropics) in distribution except for central Africa, Madagascar, and

FIGURE 19.4

Order Gymnophiona. There are currently10 recognized caecilian families. A Bombay caecilian (*Ichthyophis bombayensis*) is shown here. Ichthyophids are distributed across parts of India and southeast Asia. Skin folds that overlie separations of muscle bundles are visible in this photograph.

FIGURE 19.5

Order Caudata. The red salamander, *Pseudotriton ruber*, is native to moist temperate forests, ponds, and streams in the eastern United States.

lands of the Papuan-Australian region. Although some caecilians are aquatic (i.e., Typhlonectidae), most are wormlike burrowers that feed on soil invertebrates. Caecilians are generally less than 1 m in length, but one species (*Caecilia thompsoni*) can grow to 1.5 m in total length. Caecilians appear segmented because of folds that overlie separations between muscle bundles. A retractile tentacle positioned between their eyes and nostrils contains olfactory receptors. Tentacles in some caecilians may also contain photoreceptors. Skin covers their eyes; thus caecilian eyes probably do not form images but may function in light/dark perception. No extant caecilians have limbs or girdles, and their skulls are modified for burrowing. Unlike all other extant amphibians, caecilians may be scaled. Scales are mineralized and contain collagen.

Fertilization is internal in the caecilians and is accomplished via a protrusible male copulatory organ called a phalodeum. Some species are viviparous. Larval stages in viviparous caecilians are passed in the female's reproductive tract, and they emerge as miniature adults. Larvae of viviparous caecilians possess feeding modifications called fetal teeth that allow them to feed on the lipid-rich inner lining of the mother's oviducts prior to metamorphosis. This larval feeding habit is termed dermophagy (Gr. *dermo*, skin + *phag*, eat) and does not occur in other tetrapods. Developing young in some viviparous caecilians may obtain nutrients by consuming eggs or embryos in the mother's oviducts. This feeding habit is termed oophagy (Gr. *oo*, egg + *phag*, eat). Other caecilians are oviparous. Like all other oviparous amphibians, they lay aquatic eggs that develop into aquatic larvae, or they lay terrestrial eggs that contain embryos from which young caecilians hatch. Eggs are laid in strings, and clutch size varies between 5 and 100 eggs. Females guard and maintain eggs until hatching. Because of their secretive nature, much is to be learned about some caecilian species. The International Union for Conservation of Nature lists seven caecilian species as either endangered or threatened.

A recently discovered fossil caecilian, *Chinlestegophis jenkinsi*, may alter hypotheses on the timing of amphibian evolution, pushing amphibian origins back some 15 million years. A temnospondyl group called Stereospondyli was formerly considered an evolutionary dead-end. However, due to the discovery of *C. jenkinsi*, sterospondyls are now thought to have given rise to caecilians.

Order Caudata

Members of the order Caudata (kaw'dat-ah) (L. *cauda*, tail + Gr. *ata*, to bear) are the salamanders. Most of the well over 650 species of salamanders are found in the Northern Hemisphere, with a secondary radiation into tropical and subtropical regions. They possess a tail throughout life, and both pairs of legs, when present, are unspecialized (figure 19.5). There are three major clades of salamanders, Cryptobranchoidea (Asian salamanders, hellbenders and giant salamanders), the Sirens (eel-like salamanders with reduced forelimbs and no hindlimbs), and Salamandroidea (the vast majority of salamander species; *see figures 19.5–19.7*).

Most terrestrial salamanders live in moist forest-floor litter and have aquatic larvae. Numerous families live in caves, where constant temperature and moisture conditions create a nearly ideal environment. The family Plethodontidae contains the most fully terrestrial salamanders. Most plethodontids lay their eggs on land, rather than in water. Young hatch as miniatures of the adults. The family Salamandridae, contains salamanders that we call newts. They spend most of their lives in water and frequently retain caudal fins. The red-spotted newt (*Notophthalmus viridescens*) of the eastern United States and southeastern Canada has a tripartite life history that involves an aquatic larval stage, a terrestrial "red-eft" stage, and a sexually mature aquatic stage. Salamanders range in length from only a few centimeters, to as large as 1.5 m (the Japanese giant salamander, *Andrias japonicas*). The largest North American salamander is the hellbender (*Cryptobranchus alleganiensis*). It reaches a total length of about 65 cm.

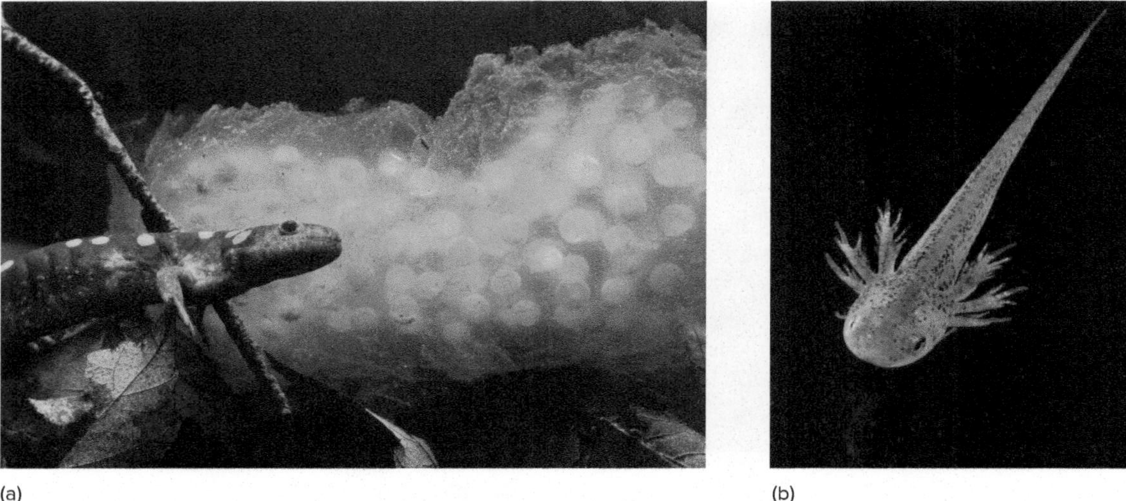

(a) (b)

(c)

FIGURE 19.6

Order Caudata. (*a*) A female spotted salamander, *Ambystoma maculatum*, after depositing eggs in a temporary, vernal pool. (*b*) A nine-day old spotted salamander larva. Larval spotted salamanders are omnivores. (*c*) Terrestrial adult spotted salamanders feed on worms and small arthropods.
(a) ©Rolf Nussbaumer Photography/Alamy; (b) ©Custom Life Science Images/Alamy; (c) ©Wirepec/Getty Images

Most salamanders have internal fertilization without copulation. Males produce a pyramidal, gelatinous spermatophore that is capped with sperm and deposited on the substrate. Females pick up the sperm cap with the cloaca and store the sperm in a special pouch, the spermatheca. Eggs are fertilized as they pass through the cloaca and are usually deposited singly, in clumps, or in strings (figure 19.6*a*). Larvae are similar to adults but smaller. They often possess external gills, a tail fin, larval dentition, and a rudimentary tongue (figure 19.6*b*). The aquatic larval stage usually metamorphoses into a terrestrial adult (figure 19.6*c*). Many other salamanders undergo incomplete metamorphosis and are paedomorphic (e.g., *Necturus*); that is, they become sexually mature while retaining ancestral larval characteristics. Obligate paedomorphosis occurs in mudpuppies (e.g., *Necturus,* figure 19.7). These species retain larval characteristics and have never been observed to undergo metamorphosis. Other salamanders undergo facultative paedomorphosis, for example, salamanders in the genus *Ambystoma*. Some of these salamanders retain larval characteristics as long as their pond habitats retain water. When the pond begins to dry, however, the salamander metamorphoses to its terrestrial form, losing its gills and respiring with lungs. The terrestrial form then can search for water, which is required for reproduction. Metamorphosis in salamanders, as in all amphibians, is controlled by the anterior pituitary and thyroid glands (*see figure 25.10*).

Order Anura

The order Anura (ah-noor′ ah) (Gr. *a*, without + *oura*, tail) is a diverse amphibian group that contains approximately 6,400 species of frogs and toads. Anurans can be found in most moist

FIGURE 19.7

The Red River Mudpuppy *Necturus maculosus.*
Mudpuppies are divided into two genera in the family Proteidae. They are found throughout North America and Europe (one species) and spend their entire lives underwater. They exchange respiratory gases through their gills, which they retain as adults from the larval stage, and across their skin. They prefer slow-moving streams and shallow lakes, where they feed on crayfish, snails, and insect larvae.
©Michael Graziano

environments, except in high latitudes and on some oceanic islands. There are exceptions of course. For example, the range of the wood frog *Lithobates sylvaticus* (*see figure 19.15*) extends into northern Alaska, and certain spadefoot toads (Scaphiopodidae) occur in desert habitats. Adult anurans lack tails. Although there is an anuran family named "tailed frogs" (Ascaphidae), the tail is a specialized reproductive organ and not a true tail. In anurans, caudal (tail) vertebrae fuse into a rodlike structure called the urostyle (*see figure 23.16*). Hindlimbs are long and muscular, and foot structure varies. Feet can have toes that bear adhesive pads modified for climbing, or feet may have keratinized spades used for digging. Hindfeet are always webbed to some degree.

Anurans have diverse life histories. There are aquatic, arboreal (tree dwelling), fossorial (burrowing/subterranean), and terrestrial (ground dwelling) forms. Although fertilization is almost always external, and eggs are typically aquatic, some species are far more aquatic than others. Larval stages, called tadpoles, have well developed and laterally compressed tails, one or two openings (spiracles) in the body that permit water flow over gills, and a proteinaceous, beaklike structure used to scrape up the plants and other organic matter on which they feed. Unlike adults, tadpoles are mostly herbivorous; however, they can also be opportunistic omnivores and detritivores. Anuran larvae undergo a drastic and sometimes rapid metamorphosis from the larval to adult body form. Larvae remain limbless until near metamorphosis.

The distinction between "frog" and "toad" is more vernacular than scientific. "Toad" usually refers to anurans with relatively dry and warty skin that are more terrestrial than other members of the order. However, numerous taxa referred to as "frogs" also have these characteristics. Anuran warts can be natural thickenings of the dermis or aggregations of cutaneous glands (*see figure 23.6*). They are not produced by the papillomaviruses that can lead to benign wartlike tumors in humans and other animals. True toads belong to the family Bufonidae (*see figure 19.16 and box figure 19.1*) and include such species as the American toad (*Anaxyrus americanus*), the European common toad (*Bufo bufo*), and the massive cane toad (*Rhinella marina*). Members of the family Ranidae are often called "true frogs," although the term "frog" is also correctly used in reference to other anurans that have relatively smooth skin and prefer more aquatic habitats. Members of the Ranidae include such species as the green (*Lithobates clamitans*), and common (*Rana temporaria*) frogs of North America and Europe, respectively. They have long and powerful hindlimbs with extensive webbing between the toes (*see figures 19.15 and 19.18*). As with toads, numerous families share these characteristics.

Section 19.2 Thinking Beyond the Facts

In what ways are the derivations of the names of the amphibian orders descriptive of each group of animals?

19.3 EVOLUTIONARY PRESSURES

LEARNING OUTCOMES

1. Justify the statement that "the double life of amphibians results from adaptations of virtually every body system of lissamphibians."
2. Justify the statement that "the single undivided ventricle of the amphibian heart might seem like an evolutionary step backward, but in fact it is an adaptation to the amphibian way of life."
3. Compare reproductive strategies in members of the class Caudata to reproductive strategies of members of the class Anura.

Most amphibians divide their lives between freshwater and land. This divided life is reflected in body systems that show adaptations to both environments. In the water, amphibians are supported by water's buoyant properties, they exchange gases with the water, and they face the same osmoregulatory problems as freshwater fishes. On land, amphibians support themselves against gravity, exchange gases with the air, and tend to lose water to the air.

External Structure and Locomotion

Vertebrate skin protects against infective microorganisms, ultraviolet light, desiccation, and mechanical injury. As discussed later in this chapter, the skin of amphibians also functions in defense, gas exchange, temperature regulation, and absorption and storage of water.

Amphibian skin lacks a covering of scales (with the exception of dermal scales in most caecilians), feathers, or hair. It is, however, highly glandular, and its secretions aid in protection. These glands keep the skin moist to prevent drying. They also produce sticky secretions that help a male cling to a female during

FIGURE 19.8

The Blue Poison Arrow Frog (*Dendrobates tinctorius*). This frog was formerly known as *D. azureus*. Its specific epithet was based on its brilliant blue azure color. It is now considered to be a morphological variant (or subspecies) of *D. tinctorius*. It is found in South American forests of Brazil and Suriname. During the rainy season, males establish territories, females fight over males, and eggs are deposited in water during amplexus. Males carry the tadpoles that develop to bromeliads and other water-trapping plant species where they eventually metamorphose to adults. Glandular secretions of the skin of members of the family Dendrobatidae are neurotoxins that protect these frogs from predators. The bright color pattern seen here is an example of aposematic coloration, which warns potential predators of this frog's toxicity.

mating and produce toxic chemicals that discourage potential predators. The skin of many amphibians is smooth, although epidermal thickenings may produce warts, claws, or sandpapery textures, which are usually the result of keratin deposits or the formation of hard, bony areas.

All amphibians possess glandular secretions that are noxious or toxic to varying degrees. The glands that produce these secretions, called granular glands, are distributed throughout the skin. They secrete a complex chemical mixture of biologically active compounds including alkaloids, peptides, biogenic amines, and steroids. Chemicals are secreted when the amphibian experiences stress, to discourage potential predators, and to protect from bacterial and fungal infections. Granular gland secretions can have neurotoxic, myotoxic, antibacterial, and antifungal effects (*see figure 23.6*). Neurotoxins in the skin of frogs in the family Dendrobatidae have been used by South American natives to tip their poison arrows (figure 19.8).

Chromatophores are specialized cells in the epidermis and dermis that are responsible for skin color and color changes. Color changes in amphibians are mostly endocrine driven and typically reflect behavioral state and body temperature. Cryptic coloration, aposematic coloration, and mimicry are all common in amphibians.

Support and Movement

Water buoys and supports aquatic animals. The skeletons of fishes function primarily in protecting internal organs, providing points of attachment for muscles, and keeping the body from collapsing during movement. In terrestrial vertebrates, however, the skeleton is modified to provide support against gravity, and it must be strong enough to support the relatively powerful muscles that propel terrestrial vertebrates across land. The amphibian skull is flattened, is relatively smaller, and has fewer bony elements than the skulls of fishes. These changes lighten the skull so it can be supported out of the water. Changes in jaw structure and musculature allow terrestrial vertebrates to crush prey held in the mouth.

The vertebral column of amphibians is modified to provide support and flexibility on land (figure 19.9). It acts somewhat like the arch of a suspension bridge by supporting the weight of the body between anterior and posterior paired appendages. Supportive processes called zygapophyses on each vertebra prevent twisting. Unlike fishes, amphibians have a neck. The first vertebra is a cervical vertebra, which moves against the back of the skull and allows the head to nod vertically. The last trunk vertebra is a sacral vertebra. This vertebra anchors the pelvic girdle to the vertebral column to provide increased support. A ventral plate of bone, called the sternum, is present in the anterior ventral trunk region and supports the forelimbs and protects internal organs. It is reduced or absent in the Anura.

The origin of the bones of vertebrate appendages is not precisely known; however, similarities in the structures of the bones of the amphibian appendages and the bones of the fins of ancient sarcopterygian fishes suggest possible homologies (*see chapter 18, Evolutionary Insights*). Joints at the shoulder, hip, elbow, knee, wrist, and ankle allow freedom of movement and better contact with the substrate. The pelvic girdle of amphibians consists of three bones (the ilium, ischium, and pubis) that firmly attach pelvic appendages to the vertebral column. These bones, which are present in all tetrapods, but not fishes, are important for support on land.

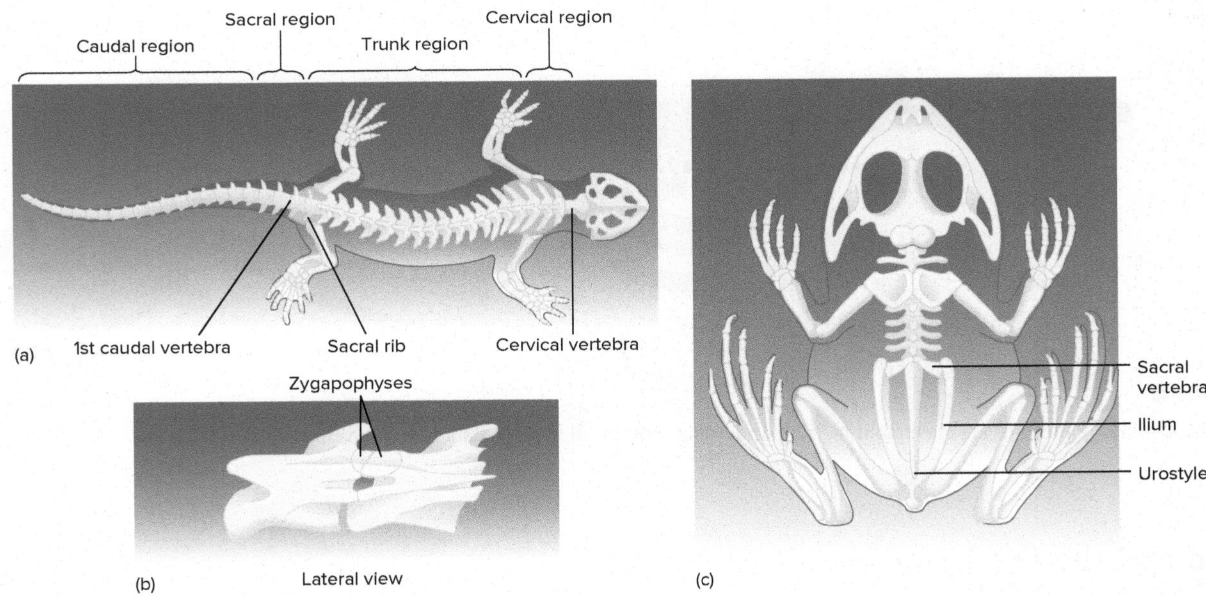

FIGURE 19.9

Skeletons of Amphibians. (*a*) The salamander skeleton is divided into four regions: cervical, trunk, sacral, and caudal. (*b*) Interlocking processes, called zygapophyses, prevent twisting between vertebrae. (*c*) A frog skeleton shows adaptations for jumping. Note the long back legs and the firm attachment of the back legs to the vertebral column through the ilium and urostyle.

Tetrapods depend more on appendages than on the body wall for locomotion. Thus, body-wall musculature is reduced, and appendicular musculature predominates. (Contrast, for example, what you eat in a fish dinner as compared to a plate of frog legs.)

Salamanders employ a relatively unspecialized form of locomotion that is reminiscent of the undulatory waves that pass along the body of a fish. Terrestrial salamanders also move by a pattern of limb and body movements in which the alternate movement of appendages results from muscle contractions that throw the body into a curve to advance the stride of a limb (figure 19.10). Caecilians have an accordion-like movement in which adjacent body parts push or pull forward at the same time. The long hindlimbs and the pelvic girdle of anurans are modified for jumping. The dorsal bone of the pelvis (the ilium) extends anteriorly and securely attaches to the vertebral column, and the urostyle extends posteriorly and attaches to the pelvis (*see figure 19.9*). These skeletal modifications stiffen the posterior half of the anuran. Long hindlimbs and powerful muscles form an efficient lever system for jumping. Elastic connective tissues and muscles attach the pectoral girdle to the skull and vertebral column and function as shock absorbers for landing on the forelimbs.

How Do We Know about Amphibian Skin Toxins?

Scientists and the pharmaceutical industry are very interested in the secretions of amphibian granular glands. Some of the secretions have shown antibacterial, antifungal, and even anticancer effects. Others are similar in structure to peptides that occur naturally in the vertebrate brain. The study of these secretions may help us to understand the functions of vertebrate brain chemicals and may aid in the development of novel drugs that, unlike antibiotics, could selectively kill microbes without inducing microbial resistance.

Early research required that the amphibians be killed in order to extract chemicals from their skin. Later it was learned that electrical stimulation of the skin would induce the secretion of chemicals that could be collected without harm to the animal. Unfortunately, electrical stimulation causes the release of a complex mixture of chemicals that is difficult to manipulate. Molecular work has demonstrated the feasibility of isolating messenger RNA (mRNA) from granular cells without harming the animal. The mRNA can be used to produce copy DNA that codes for the genes responsible for peptide toxins. Cloning this DNA could supply unlimited amounts of a purified chemical for study. This is great news. It could allow scientists to investigate a wealth of naturally occurring chemicals without harming animals, many of whom are disappearing at an alarming rate.

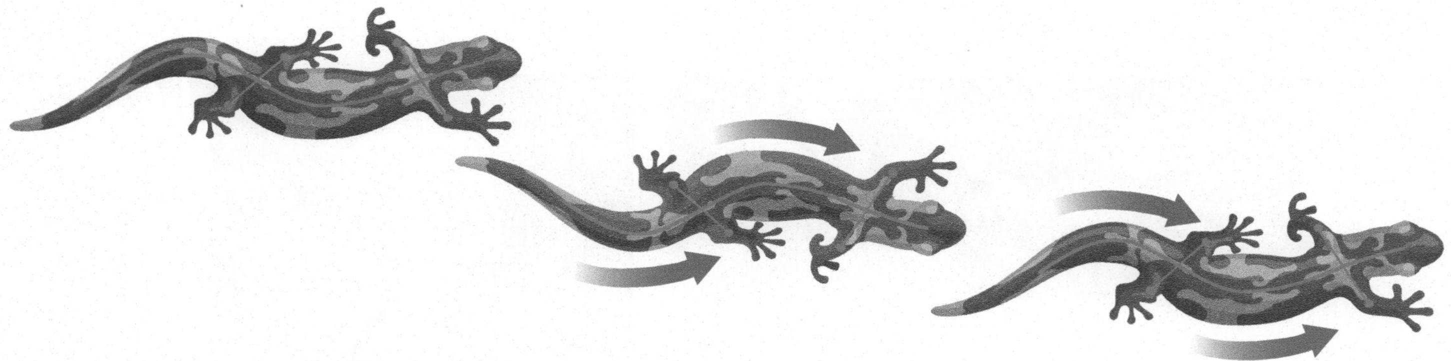

FIGURE 19.10

Salamander Locomotion. Pattern of leg movement in salamander locomotion. Blue arrows show leg movements.

Nutrition and the Digestive System

Adult amphibians are carnivores that feed on a wide variety of invertebrates. The diets of some anurans, however, are more diverse. For example, a bullfrog will prey on small mammals, birds, and other anurans. The main factors that determine what amphibians will eat are prey size and availability. Most larvae are herbivorous and feed on algae and other plant matter (*see description on page 341*). Most amphibians locate their prey by sight and simply wait for prey to pass by. Olfaction plays an important role in prey detection by aquatic salamanders and caecilians.

Many salamanders are relatively unspecialized in their feeding methods, using only their jaws to capture prey. Anurans and plethodontid salamanders, however, use their tongues and jaws to capture prey. The prey capture mechanism differs somewhat in the two groups. A true tongue is first seen in amphibians. (The "tongue" of fishes is simply a fleshy fold on the floor of the mouth [*see figure 27.3*b]. Fish food is swallowed whole and not manipulated by the "tongue.") The anuran tongue attaches at the anterior margin of the jaw and folds back over the floor of the mouth. Mucous and buccal glands on the tip of the tongue exude sticky secretions. When prey comes within range, an anuran lunges forward and flicks out its tongue (figure 19.11). The tongue turns over, and the lower jaw is depressed. The head tilts on its single cervical vertebra, which helps aim the strike. The tip of the tongue entraps the prey, and the tongue and prey are flicked back inside the mouth. All of this may happen in 0.05 to 0.15 s! The anuran holds the prey by pressing it against teeth on the roof of the mouth, and the tongue and other muscles of the mouth push food toward the esophagus. The eyes sink downward during swallowing and help force food toward the esophagus. Plethodontid salamanders' tongues are protruded using muscles associated with the hyoid bone that lies in the floor of the mouth.

FIGURE 19.11

Flip-and-Grab Feeding in a Toad. The tongue attaches at the anterior margin of the toad's jaw and flips out to capture a prey item on its sticky secretions.

Tooth structure of amphibians is unique. The jaws of ancient tetrapods like *Ichthyostega* (*see figure 19.2*) were lined by distinctive shallow-socketed, sharp conical teeth that had internal interconnected folds of enamel. Modern amphibians have pedicellate teeth in which the dentine of the upper tooth crown and the tooth base is calcified and hardened. The dentine between the crown and the base is uncalcified, which makes the tooth somewhat flexible—it can bend inward but not outward. As prey struggle, the inward bending of the teeth tends to draw prey further into the mouth cavity. Teeth are used for holding prey, not chewing.

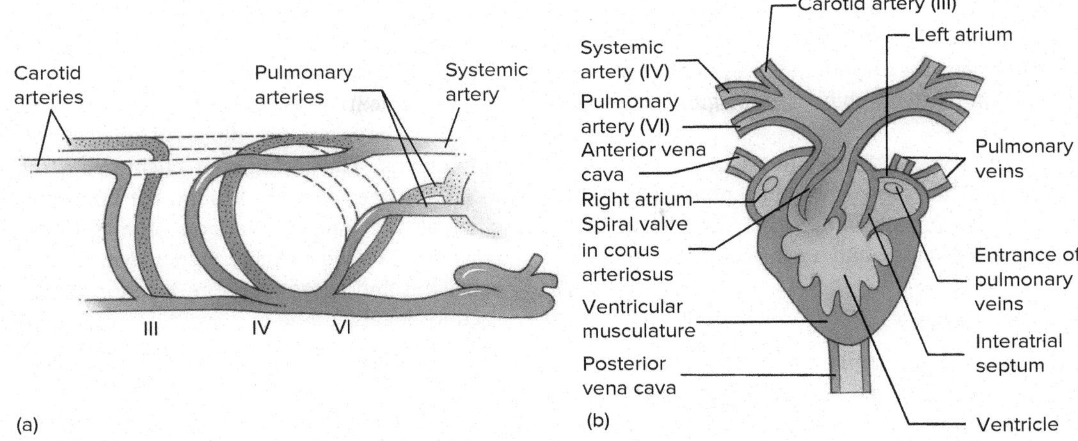

FIGURE 19.12

Diagrammatic Representation of an Anuran Circulatory System. (*a*) The Roman numerals indicate the various aortic arches. Vessels shown in dashed outline are lost during embryological development. (*b*) A ventral view of the heart.

Circulation, Gas Exchange, and Temperature Regulation

The circulatory system of amphibians shows remarkable adaptations for a life divided between aquatic and terrestrial habitats. The separation of pulmonary and systemic circuits in amphibians is similar to what was described for lungfishes (figure 19.12; *see also figure 18.15*b). The atria are partially divided in salamanders, but they are completely divided in the air-breathing anurans. After leaving the atria, blood enters the ventricle and then the conus arteriosus. The ventricle of amphibians is undivided, but it is not a wide-open chamber. It is laced with ribbons and cords of cardiac muscle. A spiral valve in the conus arteriosus or ventral aorta helps direct blood into pulmonary and systemic circuits. This anatomy is fairly efficient in keeping systemic venous blood returning to the right atrium separate from oxygenated blood returning to the heart from the lungs. Even though they cannot fully explain how this separation is accomplished, physiologists have calculated an 84% efficiency in keeping venous systemic blood separated from pulmonary return (in American bullfrogs, *Lithobates catesbeianus*).

As discussed later, gas exchange occurs across the skin of amphibians, as well as in the lungs. Therefore, blood entering the right side of the heart is often as well oxygenated as blood entering the heart from the lungs when present. (Most adult salamanders lack lungs.) When an amphibian is completely submerged, all gas exchange occurs across the skin and other moist surfaces; therefore, blood coming into the right atrium has a higher oxygen concentration than blood returning to the left atrium from the lungs. Under these circumstances, blood vessels leading to the lungs constrict, reducing blood flow to the lungs and conserving energy. This adaptation is especially valuable for those frogs and salamanders that overwinter in the mud at the bottom of a pond (*see figure 26.5*b).

Adult amphibians have fewer aortic arches than fishes. After leaving the conus arteriosus, blood may enter the carotid artery (aortic arch III), which takes blood to the head; the systemic artery (aortic arch IV), which takes blood to the body; or the pulmonary artery (aortic arch VI).

In addition to a vascular system that circulates blood, amphibians have a well-developed lymphatic system of blind-ending vessels that returns fluids, proteins, and ions filtered from capillary beds in tissue spaces to the circulatory system. The lymphatic system also transports water absorbed across the skin. Unlike other vertebrates, amphibians have contractile vessels, called lymphatic hearts, that pump fluid through the lymphatic system. Lymphatic spaces between body-wall muscles and the skin transport and store water absorbed across the skin.

Gas Exchange

Terrestrial animals expend much less energy moving air across gas-exchange surfaces than do aquatic organisms, because air contains 20 times more oxygen per unit volume than does water. On the other hand, exchanges of oxygen and carbon dioxide require moist surfaces, and the exposure of respiratory surfaces to air may result in rapid water loss.

Amphibian skin is typically moist and is richly supplied with capillary beds. These two factors permit the skin to function as a respiratory organ. Gas exchange across the skin is called **cutaneous respiration** and can occur either in water or on land. This ability allows a frog to spend the winter in the mud at the bottom of a pond. In salamanders, 30–90% of gas exchange occurs across the skin. In some amphibians (e.g., larval amphibians and plethodontid salamanders), gas exchange occurs across the moist surfaces of the mouth and pharynx. This **buccopharyngeal respiration** accounts for a small percentage of total gas exchange.

Most amphibians, except for plethodontid salamanders, possess lungs. The lungs of salamanders are relatively simple sacs. The lungs of anurans are subdivided, increasing the surface area for gas exchange. Pulmonary (lung) ventilation occurs by a **buccal pump** mechanism. Muscles of the mouth and pharynx create a positive pressure to force air into the lungs (*see figure 26.17*).

The contribution of cutaneous respiration to total gas exchange is relatively constant and cannot be increased when metabolic rates increase. Lungs can compensate for this limitation by increasing gas-exchange rates in proportion to an amphibian's changing metabolic demands.

Amphibian larvae and some adults respire using external gills. Cartilaginous rods that form between embryonic pharyngeal slits support three pairs of gills. During metamorphosis, the gills are usually reabsorbed, pharyngeal slits close, and lungs become functional.

Temperature Regulation

Amphibians are ectothermic. (They depend on external sources to maintain body temperature [*see chapter 28*].) Any poorly insulated aquatic animal, regardless of how much metabolic heat it produces, loses heat as quickly as it is produced because of powerful heat-absorbing properties of the water. Therefore, when amphibians are in water, they take on the temperature of their environment. On land, however, their body temperatures can differ from that of the environment.

Temperature regulation is mainly behavioral. Some cooling results from evaporative heat loss. In addition, many amphibians are nocturnal and remain in cooler burrows or under moist leaf litter during the hottest part of the day. Amphibians may warm themselves by basking in the sun or lying on warm surfaces. Body temperatures may rise 10°C above the air temperature. Basking after a meal is common because increased body temperature increases the rate of all metabolic reactions–including digestive functions, growth, and the fat deposition necessary to survive periods of dormancy.

Amphibians' daily and seasonal environmental temperatures often fluctuate widely, and therefore amphibians have correspondingly wide temperature tolerances. Critical temperature extremes for some salamanders lie between -2 and 27°C and for some anurans between 3 and 41°C. One anuran that can survive extremely low temperatures is the wood frog (*Lithobates sylvaticus*). The presence of cryoprotective agents in the blood, and the ability to redistribute water, allows this species to survive temperatures as low as -6°C.

Nervous and Sensory Functions

The nervous system of amphibians is similar to that of other vertebrates. The brain of adult vertebrates develops from three embryological subdivisions. In amphibians, the forebrain contains olfactory centers and regions that regulate color changes by releasing melanophore-stimulating hormone (*see figure 25.9*). It also regulates visceral functions. The midbrain contains a region called the optic tectum that assimilates sensory information and initiates motor responses. The midbrain also processes visual sensory information. The hindbrain functions in motor coordination and in regulating heart rate and the mechanics of respiration.

Many amphibian sensory receptors are widely distributed over the skin. Some of these are simply bare nerve endings that respond to heat, cold, and pain. The lateral-line system is similar in structure to that found in fishes, and it is present in all aquatic larvae, aquatic adult salamanders, and some highly aquatic adult anurans. Lateral-line organs are distributed singly or in small groups along the lateral and dorsolateral surfaces of the body, especially the head. These receptors respond to low-frequency vibrations in the water and movements of the water relative to the animal (*see figure 24.19*).

Chemoreception (olfaction) is an important sense for many amphibians. The location of chemoreceptors varies; they can be located in the nasal epithelium, the lining of the mouth, under the chin, on the tongue, and over other regions of the skin. Chemoreception is used in mate recognition in most amphibians and in courtship and spermatophore location by salamanders. It is also used for locating food and avoiding noxious chemicals.

Vision is one of the most important senses in amphibians because they are primarily sight feeders, often responding to the movements of their prey. (Caecilians are an obvious exception.) Numerous adaptations allow the eyes of amphibians to function in terrestrial environments (figure 19.13). The eyes of some amphibians (i.e., anurans and some salamanders) are on the front of the head, providing the binocular vision and well-developed depth perception necessary for capturing prey. Other amphibians with smaller lateral eyes (some salamanders) lack binocular vision. The lower eyelid is movable, and it cleans and protects the eye. Much of it is transparent and is called the **nictitating membrane.** When the eyeball retracts into the orbit of the skull, the nictitating membrane is drawn up over the cornea. In addition, orbital glands lubricate and wash the eye. Together, eyelids and glands keep the eye free of dust and other debris. The lens is large and nearly round. It is set back from the cornea, and a fold of epithelium called the iris surrounds it. The iris can dilate or constrict to control the size of the pupil.

Focusing, or accommodation, involves bending (refracting) light rays to a focal point on the retina. Light waves moving from air across the cornea are refracted because of the change in density between the two media. The lens provides further refraction. Like the eyes of most tetrapods, the amphibian eye focuses on distant objects when the eye is at rest. To focus on near objects, the protractor lentis muscle must move the lens forward (*see figure 19.13*). Receptors called rods and cones are in the retina. Because cones are associated with color vision in some other vertebrates, their occurrence suggests that amphibians can distinguish between some

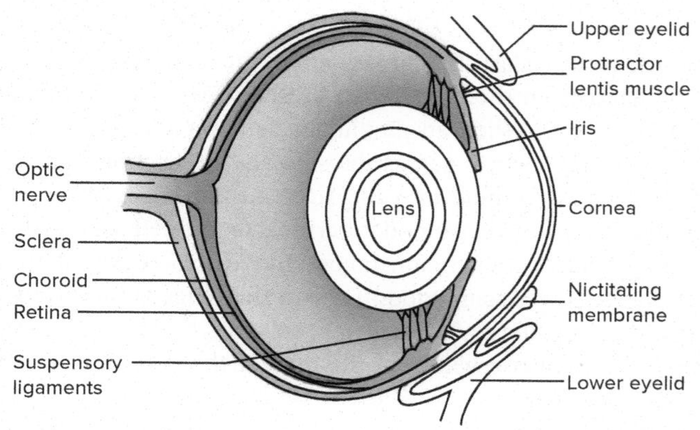

FIGURE 19.13

Amphibian Eye. Longitudinal section of an anuran eye.

wavelengths of light. The extent to which color vision is developed is unknown. The neuronal interconnections in the retina are complex and allow an amphibian to distinguish between terrestrial insect prey and newly metamorphosed young. These connections also permit an amphibian to distinguish shadows that may warn of an approaching predator from other movements created by blades of grass blowing in the wind or flying insect prey.

The auditory system of amphibians is clearly an evolutionary adaptation to life on land. It transmits both substrate-borne vibrations and, in anurans, airborne vibrations. The ears of anurans consist of a tympanic membrane, a middle ear, and an inner ear. The tympanic membrane is a piece of integument stretched over a cartilaginous ring that receives airborne vibrations and transmits them to the middle ear, which is a chamber beneath the tympanic membrane. Abutting the tympanic membrane is a middle-ear ossicle (bone) called the stapes (columella), which transmits vibrations of the tympanic membrane into the inner ear (*see figure 24.20*). High-frequency (1,000 to 5,000 Hz) airborne vibrations are transmitted to the inner ear through the tympanic membrane. Low-frequency (100 to 1,000 Hz) substrate-borne vibrations are transmitted through the front appendages and the pectoral girdle to the inner ear through a second ossicle, called the operculum.

Muscles attached to the operculum and stapes can lock either or both of these ossicles, allowing an anuran to screen out either high- or low-frequency sounds. This mechanism is adaptive because anurans use low- and high-frequency sounds in different situations. Mating calls are high-frequency sounds that are of primary importance for only a part of the year (the breeding season). At other times, low-frequency sounds may warn of approaching predators.

Salamanders lack a tympanic membrane and middle ear. They live in streams, ponds, caves, and beneath leaf litter. They have no mating calls, and the only sounds they hear are probably low-frequency vibrations transmitted through the substrate and skull to the stapes and inner ear.

The sense of equilibrium and balance is similar to that described for fishes in chapter 18. The inner ear of amphibians has semicircular canals that help detect rotational movements and other sensory patches that respond to gravity and detect linear acceleration and deceleration.

Excretion and Osmoregulation

The kidneys of amphibians lie on either side of the dorsal aorta on the dorsal wall of the body cavity. A duct leads to the cloaca, and a storage structure, the urinary bladder, is a ventral outgrowth of the cloaca.

The nitrogenous waste product that amphibians excrete is either ammonia or urea. Amphibians that live in freshwater excrete ammonia. It is the immediate end product of protein metabolism; therefore, no energy is expended converting it into other products. The toxic effects of ammonia are avoided because ammonia rapidly diffuses into the surrounding water. Amphibians that spend more time on land excrete urea that is produced from ammonia in the liver. Although urea is less toxic than ammonia, it still requires relatively large quantities of water for its excretion. Unlike ammonia, urea can be stored in the urinary bladder. Some amphibians excrete ammonia when in water and urea when on land.

One of the biggest problems that amphibians face is osmoregulation. In water, amphibians have the same osmoregulatory problems as freshwater fishes. They must rid the body of excess water and conserve essential ions. Amphibian kidneys produce large quantities of hypotonic urine, and the skin and walls of the urinary bladder transport Na^+, Cl^-, and other ions into the blood.

On land, amphibians must conserve water. Adult amphibians do not replace water by intentional drinking, nor do they have the impermeable skin characteristic of other tetrapods or kidneys capable of producing a hypertonic urine. Instead, amphibians limit water loss by behaviors that reduce exposure to desiccating conditions. Many terrestrial amphibians are nocturnal. During daylight hours, they retreat to areas of high humidity, such as under stones, or in logs, leaf mulch, or burrows. Water lost on nighttime foraging trips must be replaced by water uptake across the skin while in the retreat. Diurnal amphibians usually live in areas of high humidity and rehydrate themselves by entering the water. Many amphibians reduce evaporative water loss by reducing the amount of body surface exposed to air. They may curl their bodies and tails into tight coils and tuck their limbs close to their bodies (figure 19.14*a*). Individuals may form closely packed aggregations to reduce overall surface area.

Some amphibians have protective coverings that reduce water loss. Hardened regions of skin are resistant to water loss and may be used to plug entrances to burrows or other retreat openings to maintain high humidity in the retreat. Other amphibians prevent water loss by forming cocoons that encase the body during long periods of dormancy. Cocoons are made from outer layers of the skin that detach and become parchmentlike. These cocoons open only at the nares or the mouth and, in experimental situations, reduce water loss 20–50% over noncocooned individuals (figure 19.14*b*).

Paradoxically, the skin—the most important source of water loss and gas exchange—is also the most important structure for rehydration. When an amphibian flattens its body on moist surfaces, the skin, especially in the ventral pelvic region, absorbs water. The skin's permeability, vascularization, and epidermal sculpturing all promote water reabsorption. Minute channels increase surface area and spread water over surfaces not necessarily in direct contact with water.

Amphibians can also temporarily store water. Water accumulated in the urinary bladder and lymph sacs can be selectively reabsorbed to replace evaporative water loss. Amphibians living in very dry environments can store volumes of water equivalent to 35% of their total body weight.

Reproduction, Development, and Metamorphosis

Amphibians are dioecious, and ovaries and testes are located near the dorsal body wall. Fertilization is usually external (caecilians and most salamanders are exceptions), and because the developing eggs lack any resistant coverings, development is tied to moist habitats, usually water. A few anurans have terrestrial nests that are kept moist by being enveloped in foam or by being located near the water and subjected to flooding. In a few species, larval stages are passed in

(a)

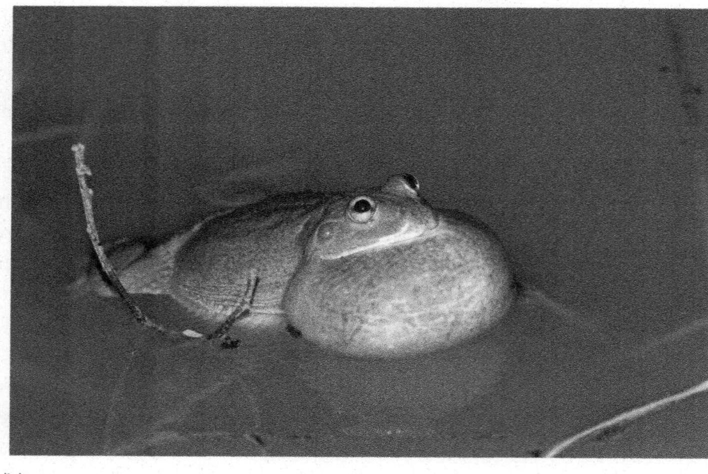

(b)

FIGURE 19.14

Water Conservation by Anurans. (*a*) Daytime sleeping posture of the green tree frog, *Hyla cinerea*. The closely tucked appendages reduce exposed surface area. (*b*) The water-holding frog (*Cyclorana platycephala*) burrows in sandy soil throughout most of Australia. During dry periods, this frog survives in burrows in a water-tight cocoon made from the outer layers of its skin. After a rain *C. platycephala* emerges from its cocoon to reproduce. The frog shown here has emerged from its burrow and has inflated its vocal sac during a mating call. Before returning to its burrow to survive the next dry period, it will engorge its urinary bladder and the lymphatic spaces under the skin with water.

the egg membranes, and the immatures hatch into an adultlike body. Only about 10% of all salamanders have external fertilization. All others produce spermatophores, and fertilization is internal. Eggs may be deposited in soil or water or retained in the oviduct during development. All caecilians have internal fertilization, and 75% have internal development. Amphibian development, which zoologists have studied extensively, usually includes larval stages called tadpoles. Amphibian tadpoles often differ from the adults in mode of respiration, form of locomotion, and diet. These differences reduce competition between adults and larvae.

Interactions between intrinsic neuroendocrine controls and extrinsic factors determine the timing of courtship and reproduction in both temperate and tropical amphibians. In temperate regions, temperature, photoperiod, and water availability are important extrinsic factors that induce breeding activities. In some temperate species, threshold temperatures required for breeding activities are not fixed; they may increase as the season progresses. This increase is presumably due to homeostatic adjustments to typical seasonal warming. Breeding periods are seasonal. Although there are a few exceptions, breeding activities in temperate species occur mostly in late winter, spring, and summer. In tropical regions, most amphibian breeding occurs in the rainy seasons.

Courtship behavior helps individuals locate breeding sites and identify potential mates. It also prepares individuals for reproduction and ensures that eggs are fertilized and deposited in locations that promote successful development.

Salamanders rely primarily on olfactory and visual cues in courtship and mating, whereas male vocalizations and tactile cues are important for anurans. Many species congregate in one location during times of intense breeding activity. Male vocalizations are species specific and function in the initial attraction and contact between mates. After that, tactile cues become more important.

FIGURE 19.15

Amplexus. In frogs, such as the wood frog (*Lithobates sylvaticus*), eggs are released and fertilized when the male (smaller frog) mounts and grasps the female under the forelimbs. This positioning is called axillary amplexus.

The male clings to the female dorsally by grasping her with his forelimbs—either just posterior to her forelimbs (figure 19.15) or just anterior to her hindlimbs). This position is termed **amplexus**; it puts the male and female in close contact and orients them in the same direction so that external fertilization is efficient. Amplexus usually last from 1 to 24 h, but it may last for days in some species. During the breeding season, males develop enlarged and roughened toe pads (called nuptial pads) to help cling to females during amplexus.

Little is known of caecilian breeding behavior. Males have an intromittent organ that is a modification of the cloacal wall, and fertilization is internal.

Vocalization

Sound production is primarily a reproductive function of male anurans. Advertisement calls attract females to breeding areas and announce to other males that a given territory is occupied. Advertisement calls are species specific, and the repertoire of calls for any one species is limited. The calls may also help induce behavioral and physiological readiness to breed. Release calls inform a partner that a frog is incapable of reproducing. Unresponsive females give release calls if a male attempts amplexus, as do males that have been mistakenly identified as female by another male. Distress calls are not associated with reproduction; either sex produces these calls in response to pain or being seized by a predator. The calls may be loud enough to cause a predator to release the frog. The distress call of the South American jungle frog *Leptodactylus pentadactylus* is a loud scream similar to the call of a cat in distress.

The sound-production apparatus of frogs consists of the larynx and its vocal cords. This laryngeal apparatus is well developed in males, who also possess a vocal sac. In the majority of frogs, vocal sacs develop as a diverticulum from the lining of the buccal cavity (figure 19.16). Air from the lungs is forced over the vocal cords and cartilages of the larynx, causing them to vibrate. Muscles control the tension of the vocal cords and regulate the frequency of the sound. Vocal sacs act as resonating structures and increase the volume of the sound.

The use of sound to attract mates is especially useful in organisms that occupy widely dispersed habitats and must come together for breeding. Finding a mate of the species can be difficult, because many species of anurans often converge at the same pond for breeding at the same time. Vocalizations help ensure intraspecific recognition.

Parental Care

Parental care increases the chances of any one egg developing, but it requires large energy expenditures on the part of the parent. The most common form of parental care in amphibians is attendance of the egg clutch by either parent. Maternal care occurs in species with internal fertilization (predominantly salamanders and caecilians), and paternal care may occur in species with external fertilization (predominantly anurans). It can involve aeration of aquatic eggs, cleaning and/or moistening of terrestrial eggs, protection of eggs from predators, or removal of dead and infected eggs.

Eggs are sometimes transported by a parent. Females of the genus *Pipa* carry eggs on their backs. Two species of *Rheobatrachus* were discovered in Australia within the last 30 years. They have not been observed in the wild since the 1980s and are presumed extinct. *Rheobatrachus* females brooded tadpoles in their stomachs, and the young emerged from the females' mouths (figure 19.17)! Unfortunately, we will never know whether the female swallowed fertilized eggs and all development occurred in her stomach or whether she swallowed tadpoles. During brooding, the female's stomach expanded to fill most of her body cavity, and the stomach stopped producing digestive secretions. Viviparity and ovoviviparity occur primarily in salamanders and caecilians.

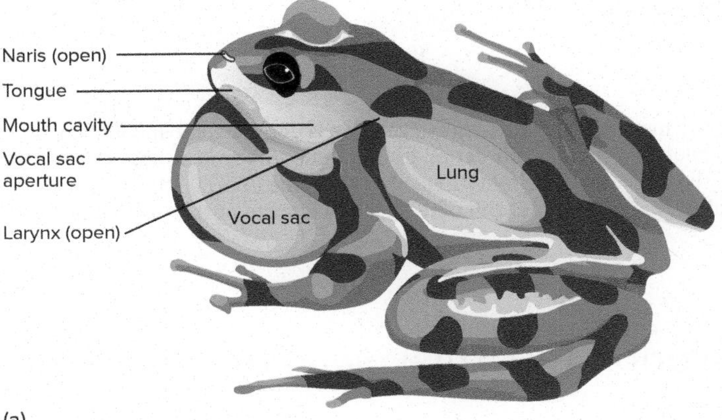

FIGURE 19.16

Anuran Vocalization. (*a*) Generalized vocal apparatus of an anuran. (*b*) Inflated vocal sac of the Great Plains toad (*Anaxyrus cognatus*).
©Roberta Olenick/Getty Images

FIGURE 19.17

Parental Care of Young. Female *Rheobatrachus* with young emerging from her mouth. This Australian frog species has not been observed in the wild since the 1980s and is now presumed extinct.
©ANT Photo Library/Science Source

FIGURE 19.18

Events of Metamorphosis in the California Red-Legged Frog *Rana draytonii*. (*a*) Before metamorphosis. In this stage there is development and growth of larval structures such as keratinized mouthparts, gills, spiracles, and caudal musculature and fins. This period lasts 5–7 months in *R. draytonii*. (*b–d*) Metamorphosis. The median eminence of the hypothalamus develops and initiates the secretion of thyroid-stimulating hormone (TSH). TSH causes the release of large quantities of T_4 and T_3, which promote the growth of limbs, reabsorption of the tail, and other changes of metamorphosis, resulting eventually in a young, adult frog.

(a-c) ©Dan Suzio/Science Source; (d) Source: Mark Jennings, U.S. Fish and Wildlife Service

Metamorphosis

Metamorphosis is a series of abrupt structural, physiological, and behavioral changes that transform a larva into an adult. Various environmental conditions, including crowding and food availability, influence the time required for metamorphosis. Most directly, however, metamorphosis is under the control of neurosecretions of the hypothalamus, hormones of the anterior lobe of the pituitary gland (the adenohypophysis), and the thyroid gland (*see figure 25.10*).

Morphological changes associated with the metamorphosis of caecilians and salamanders are relatively minor. Reproductive structures develop, gills are lost, and a caudal fin (when present) is lost. In the Anura, however, changes from the tadpole into the small frog are more dramatic (figure 19.18). Limbs and lungs develop, the tail is reabsorbed, the skin thickens, and marked changes in the head and digestive tract (associated with a new mode of nutrition) occur.

The mechanics of metamorphosis explain paedomorphosis in amphibians. Some salamanders are paedomorphic because cells fail to respond to thyroid hormones, whereas others are paedomorphic because they fail to produce the hormones associated with metamorphosis. In some salamander families, paedomorphosis is the rule. In other families, the occurrence of paedomorphosis is variable and influenced by environmental conditions.

SECTION 19.3 THINKING BEYOND THE FACTS

What evidence is there in the circulatory systems of amphibians and lungfish* (see figure 18.15) *of the sarcopterygian ancestry of amphibians?

19.4 AMPHIBIANS IN PERIL

LEARNING OUTCOMES

1. Explain why amphibians are especially vulnerable to environmental disturbances.
2. Assess possible conservation measures that can help preserve amphibian populations.
3. Hypothesize the ecological effects of large-scale amphibian extinctions.

Amphibians are disappearing at an alarming rate—with minimum extinction rate estimates some 25,000 times greater than background (natural) extinction rates. Recent data indicate that over one-third of the nearly 7,700 extant amphibian species are classified as critically endangered, endangered, or vulnerable by the

WILDLIFE ALERT
Golden Toad (*Incilius periglenes*)

VITAL STATISTICS

Classification: Phylum Chordata, class Amphibia, order Caudata, family Bufonidae

Range: Monteverde Cloud Forest of Costa Rica

Habitat: Within fallen leaves and moss in the cloud forest, more than 2,000 m above sea level

Number remaining: Probably none

Status: Extinct

NATURAL HISTORY AND ECOLOGICAL STATUS

Most Wildlife Alerts in this textbook focus on species that have either a "threatened" or "endangered" status. This Wildlife Alert is focused on a species that has not been observed in the wild since 1991. The disappearance of this species has been important in drawing attention to the plight of amphibians worldwide.

Golden toads were discovered in 1964. Males have a brilliant gold color (box figure 19.1), and females are dark brown with red patches fringed in yellow. They ranged in size from 40 to 55 mm in length. Golden toads occupied a few square kilometers in the Monteverde Cloud Forest of northern Costa Rica above 2,000 m (box figure 19.2). These toads probably spent most of the year under fallen leaves and in moss. They were actually observed only during the months of April through June (the rainy season), when they gathered in temporary ponds in an explosive breeding frenzy. The males, which outnumbered the females, apparently could not distinguish the sex or species of potential mates. Often 4–10 individuals would attempt amplexus, resulting in what has been called "toad balls." Females of the species responded to amplexus with a distinctive shiver or vibration that identified their sex and species. Successful mating resulted in the release of about 300 eggs. Tadpoles depended on the maintenance of temporary ponds for their five-week development. Little is known of other aspects of their biology.

BOX FIGURE 19.2 **The Monteverde Cloud Forest Preserve.** This mountain forest has a nearly constant cloud cover that envelops the canopy of trees. Most of the precipitation is in the form of fog drip, where water from fog condenses on leaves and drips to the forest floor. Much of the understory of the forest is characterized by ferns and mosses, ideal habitat for the golden toad and other amphibians.
©Kevin Schafer/Getty Images

BOX FIGURE 19.1 **Female and male golden toads (*Incilius periglenes*) in amplexus.** Females are dark brown with red patches fringed in yellow.
©Premaphotos/Alamy

Climate change is suspected in the demise of the golden toad. In 1986 and 1987, there was a drought and record high temperatures that dried temporary ponds before the tadpoles matured. These changes probably also dried the leaf and moss habitats where the adults lived. It is also possible that climate change increased the susceptibility of the toads to chytridomycosis and other diseases. Decreased habitat due to drought may have increased crowding as toads competed for living space. Crowding would have made disease transmission between toads easier. In recent years, members of other families of amphibians that shared habitat with the golden toad are also in decline—for example, the harlequin frog (*Atelopus varius*). The Monteverde Cloud Forest is being protected in hopes that there may be a few remaining golden toads and to preserve other rare and sensitive cloud-forest fauna.

International Union for Conservation of Nature. Data on some 25% of amphibian species are so scant that their populations cannot be assessed. More than 165 species have suffered extinction in the last 20 years. One of the reasons that amphibians are so sensitive to environmental changes has to do with their double life—on land and in the water—and their thin, permeable skin. Water and airborne pollutants quickly penetrate amphibian skin. Developmental stages that depend on moist environments are quickly killed by desiccation.

Local events can decimate amphibian populations. In addition to causing direct mortality, clear-cutting forests drastically alters amphibian habitats in a multitude of ways. One effect of clear-cutting is that an unnatural amount of sunlight reaches forest floors and dries the moist habitats that amphibians require. Mining, drilling, industrial and agricultural operations, and urban sprawl also directly destroy amphibians and their habitats. If not carefully regulated, toxic runoff and excess acid deposition from these activities can also permeate adjacent landscapes and watersheds, and cause amphibian population declines. Suburbanization, although probably not as harmful as the issues previously described, also contributes to population declines of certain amphibian species. Roadways in suburbanized landscapes negatively affect seasonal migrations; it takes only one vehicle to kill half of the migrating amphibians present on roads. Additionally, sod and concrete prevent burrowing, and lawn treatments, pesticides, and herbicides can contribute to amphibian mortality.

Although anthropogenic habitat alterations are a major reason for amphibian declines, amphibians are also disappearing from vast areas of the earth where local damage has not occurred. These declines are said to be "enigmatic" and are partly attributed to a pathogenic chytrid fungus. The chytrid fungus (*Batrachochytrium dendrobatidis* [*Bd*]) can lead to chytridiomycosis, a disease that causes abnormal thickening and ulceration of the skin. These skin abnormalities can lead to mortality either directly or indirectly, as skin is critical to both respiration and osmoregulation in amphibians. The fungus may also produce toxins that can be absorbed through the skin. Chytridiomycosis has been implicated in mass die-offs in parts of the Americas, Australia, New Zealand, and Spain. Researchers at the Centers for Disease Control hypothesize that *Bd* originated in South Africa and was introduced elsewhere via an asymptomatic carrier, the African clawed frog (*Xenopus laevis*). This animal has been transported around the world for use in research laboratories and as pets. Climate change is also implicated in enigmatic declines, but the precise mechanisms involved are not yet fully understood. Some studies show changes in amphibian physiology related to mild winters, and others indicate that changing temperatures and rainfall patterns are interfering with reproduction and development. Climate change is also likely responsible for the spread of *Bd* into formerly unoccupied areas and has been implicated in the extinction of the golden toad (*Incilius periglenes*) of Costa Rica's Monteverde Cloud Forest (*see Wildlife Alert, page 351*).

Urgent conservation action is needed to save amphibians from extinction. Diverse conservation efforts are needed to help protect amphibians. Population monitoring programs are needed to identify amphibian populations and monitor their health. Laws are needed to protect amphibians from unlicensed collecting and transport. Wetland conservation protects the delicate habitats that amphibians require. In the long term, laws to reverse climate change are urgently needed. In the short term, conservation groups are raising money for captive breeding programs, and zoos around the world are helping preserve the hundreds of amphibian species that are threatened with extinction. The plight of amphibians has implications that go beyond this one group of vertebrates. Due to their sensitivity to environmental changes, amphibians are ecosystem indicators–they are warning us of problems that threaten all species.

SECTION 19.4 THINKING BEYOND THE FACTS

In what ways are amphibians serving as an "environmental warning system"?

19.5 FURTHER PHYLOGENETIC CONSIDERATIONS

LEARNING OUTCOMES

1. Describe the three sets of evolutionary changes in the sarcopterygian lineage that allowed movement onto land.
2. Explain how the anatomical changes present in Anthracosauria and Diadectomorpha enhanced survival on land.
3. Describe the relationship between ancestral amphibians and early amniotes.

In the past, there has been some debate as to whether or not the three modern orders of amphibians (Lissamphibia) represent a monophyletic grouping. Common characteristics such as the stapes/operculum complex, the importance of the skin in gas exchange, aspects of the structure of the skull and teeth, and molecular evidence have convinced most zoologists of the close relationships within this group. The exact nature of these relationships, however, remains controversial. The relationships depicted in figure 19.3 represent one of a number of hypotheses.

Three sets of evolutionary changes in sarcopterygian lineages allowed movement onto land. Two of these occurred early enough that they are found in all amphibians. One was the set of changes in the skeleton and muscles that allowed greater mobility on land. A second change involved a jaw mechanism and movable head that permitted effective exploitation of insect resources on land. A jaw-muscle arrangement that permitted tetrapodomorph fishes to snap, grab, and hold prey was adaptive when early tetrapods began feeding on insects in terrestrial environments.

The third set of changes occurred in the amniote lineage–the development of an egg that was resistant to drying. Although the **amniotic egg** is not completely independent of water, the extraembryonic membranes that form during development protect the embryo from desiccation, store wastes, and promote gas exchange. In addition, this egg has a leathery or calcified shell that is protective, yet porous enough to allow gas exchange with the environment. Many amphibians, and even some fish, also lay eggs that resist desiccation. The particular structure of the amniotic egg, however, is an important feature that unites reptiles (including birds) and mammals and was one of the keys to the success of this lineage (*see figure 19.3*).

Tentative hypotheses regarding the origin of the amniotes involve a lineage, often called Reptiliomorpha (*see figure 19.3*). This lineage includes the amniotes as well as extinct anamniote members that share characteristics present in both contemporaneous amphibians and reptiles. These anamniotes are not considered reptiles as defined by modern taxonomy. Although the evolution from basal tetrapod to amniote was rapid and left an incomplete record (because extraembryonic membranes do not fossilize well), the axial and appendicular skeletal modifications of certain reptiliomorph groups provide strong evidence that they are on, or near, the basal amniote lineage.

Anthracosaurs are anamniote tetrapods considered to be ancestral to amniotes. They possess numerous features that occur in amniotes but not in early (Paleozoic) or later amphibian groups. These features include atlas/axis modifications that allow for a more mobile head, a suite of reptile-like vertebral characteristics, and (unlike tetrapodamorph ancestors–*see box figure 18.1*) five-toed foreappendages. Anthracosaurs had water-tight skin, but unlike known amniotes, they were bound to water for reproduction and had a primitive sprawling posture, a greater reliance on body wall musculature for locomotion, and an amphibian-like postcranial skeleton. Additionally, anthracosaurs possessed incomplete rib cages, indicating that they were probably reliant on a buccal pump (*see figure 26.17*) to ventilate lungs.

Diadectomorphs are considered to be basal amniotes (figure 19.19). They share a number of derived characteristics with early reptiles that are not seen in their fossilized ancestors. Vertebral and cranial adaptations strengthened the skeleton for life in the absence of water's buoyant support. The diadectomorph appendicular skeleton was less sprawling and more suitably jointed for terrestrial locomotion than their predecessors. Reproductive strategies of diadectomorphs are debated. They may have had nonshelled aquatic amniotic eggs with direct development or fully terrestrial eggs.

While the exact pathways are uncertain, we do know that two lineages of amniotes diverged from basal amniotes. One of these, the diapsid lineage, is represented by the reptiles, including birds (chapters 20 and 21). The other lineage, the synapsid lineage, is represented by the mammals (chapter 22) and their early predecessors (*see figure 22.2*).

FIGURE 19.19

Fossil of *Seymouria baylorensis*. This species was a reptiliomorph whose fossils date to about 280 million years ago (mya) and were first unearthed in Texas. Other fossils of *Seymouria* have been found in other parts of North America and in Europe.

Section 19.5 Thinking Beyond the Facts

Why is knowledge of the sarcopterygian lineage so important in understanding animal evolution?

Summary

19.1 **Evolutionary Perspective**

- Terrestrial vertebrates are called tetrapods and arose from sarcopterygians. The exact relationship among living and extinct amphibian groups is tentative. However, the most widely accepted hypothesis is that the ancient amphibian group Temnospondyli gave rise to Lissamphibia.

19.2 **Survey of Amphibians**

- The order Gymnophiona contains the caecilians. Caecilians are tropical, wormlike burrowers that inhabit aquatic or terrestrial habitats. They have internal fertilization and many are viviparous.
- Members of the order Caudata are the salamanders. Salamanders are widely distributed, usually have internal fertilization, and may have aquatic larvae or direct development.
- Frogs and toads comprise the order Anura. Anurans lack tails and possess adaptations for jumping and swimming. External fertilization results in tadpole larvae, which metamorphose to adults.

19.3 **Evolutionary Pressures**

- The skin of amphibians is moist and functions in gas exchange, water regulation, and protection. The skeletal and muscular systems of amphibians are adapted for movement on land.
- Adult amphibians are carnivores that capture prey in their jaws or seize them with their tongues.
- The circulatory system of amphibians is modified to accommodate the presence of lungs, gas exchange at the skin, and loss of gills in most adults. Gas exchange is cutaneous, buccopharyngeal, and pulmonary. A buccal pump accomplishes pulmonary ventilation. A few amphibians retain gills as adults.
- Sensory receptors of amphibians, especially the eye and ear, are adapted for functioning on land.
- Amphibians excrete ammonia or urea. Ridding the body of excess water when in water and conserving water when on land are functions of the kidneys, bladder, lymphatic vessels, and amphibian behavior.
- The reproductive habits of amphibians are diverse. Many have external fertilization and development. Others have internal fertilization and development. Courtship, vocalizations, and parental care are common in some amphibians. The nervous and endocrine systems control metamorphosis.

19.4 **Amphibians in Peril**

- Local habitat destruction, disease, climate change, and other causes are resulting in an alarming reduction in amphibian populations around the world.

19.5 **Further Phylogenetic Considerations**

- Reptilomorpha is an ancient reptile-like amphibian group that contains species with characteristics found in both modern amphibians and reptiles. Diadectomorpha are considered to be basal amniotes. It is unknown whether or not their amniotic eggs were shelled. The shelled amniotic egg liberated subsequent reptile groups from water and is found in modern reptiles (including birds) and mammals.

Concept Review Questions

1. The group of animals that contains only living amphibians is
   a. Stegocephalia.
   b. Tetrapoda.
   c. Lissamphibia.
   d. Amniota.
2. The order ________________ is composed of the salamanders.
   a. Caudata
   b. Anura
   c. Gymnophiona
3. The order ________________ is composed of the frogs and toads.
   a. Caudata
   b. Anura
   c. Gymnophiona
4. Most members of this order have internal fertilization without copulation and some are paedomorphic.
   a. Caudata
   b. Anura
   c. Gymnophiona
5. Adult amphibians are carnivores, feeding on a variety of invertebrates and, in some cases, small vertebrates. Larval amphibians virtually all feed on aquatic invertebrates.
   a. True
   b. False

Analysis and Application Questions

1. How are the skeletal and muscular systems of amphibians adapted for life on land?
2. Why is the separation of oxygenated and nonoxygenated blood in the heart not as important for amphibians as it is for other terrestrial vertebrates?
3. Explain how the skin of amphibians is used in temperature regulation, protection, gas exchange, and water regulation. Under what circumstances might cooling interfere with water regulation?
4. In what ways could anuran vocalizations have influenced the evolution of that order?
5. What steps should be taken to save imperiled amphibians? Name some things that you can do.

Megachirella wachtleri is a middle Triassic lizard-like diapsid that is part of a nonavian reptile assemblage that contains the Lepidosauria (tuataras, lizards, worm lizards, and snakes).

20

Nonavian Reptiles: Diapsid Amniotes

Chapter Outline

20.1 Evolutionary Perspective
Cladistic Interpretation of the Amniote Lineage
Early Amniote Evolution and Skull Structure
20.2 Survey of the Nonavian Reptiles
Testudines
Archosauria
Lepidosauria
20.3 Evolutionary Pressures
External Structure and Locomotion
Nutrition and the Digestive System
Circulation, Gas Exchange, and Temperature Regulation
Nervous and Sensory Functions
Excretion and Osmoregulation
Reproduction and Development
20.4 Further Phylogenetic Considerations

20.1 EVOLUTIONARY PERSPECTIVE

LEARNING OUTCOMES

1. Justify the statement that "the amniotic egg provided solutions that made development apart from external watery environments possible."
2. Compare amniote taxonomy before and after the application of cladistic methods.
3. Describe the hypotheses of amniote evolution.

The amphibians, although they venture onto land, are usually associated with moist environments. Their moist skin that functions in gas exchange and water regulation and their developmental stages keep them tied to moist habitats. In the Carboniferous period about 350 million years ago (mya) (*see appendix B*), ties to watery habitats were broken with the reptiliomorph ancestors of the amniote lineage. The **Amniota** (L. *amnion,* membrane around a fetus) is a monophyletic lineage that includes the animals in classes traditionally designated as Reptilia (the nonavian reptiles chapter 20, figure 20.1), Aves (the birds, chapter 21), and Mammalia (the mammals, chapter 22). This lineage is characterized by the presence of **amniotic eggs.** Amniotic eggs have extraembryonic membranes that protect the embryo from desiccation, cushion the embryo, promote gas transfer, and store waste materials (figure 20.2). The amniotic eggs of nonavian and birds also have leathery or hard shells that protect the developing embryo, albumen that cushions and provides moisture and nutrients for the embryo, and yolk that supplies food to the embryo. All of these features are adaptations for development on land. (The amniotic egg is not, however, the only kind of land egg: some arthropods, amphibians, and even a few fishes have eggs that develop on land.) The amniotic egg is the major synapomorphy that distinguishes the reptiles (both avian and nonavian) and mammals from other vertebrates. Even though the amniotic egg has played an important role in the nonavian reptile invasion of terrestrial habitats, it is just one of many adaptations that have allowed members of this group to flourish on land. Other adaptations for terrestrialism that will be described in this chapter include an impervious skin, horny nails for digging and locomotion, water-conserving kidneys, and enlarged lungs. Nonavian reptiles have also lost the lateral-line system of fishes and amphibians.

Cladistic Interpretation of the Amniote Lineage

Figure 20.3 shows one interpretation of amniote phylogeny. The mammals are represented as being most closely related to ancestral amniotes. The remaining taxa form a monophyletic group within the Amniota. The rules of cladistic analysis state that all descendants of a most

FIGURE 20.1

Class Reptilia. Members of the class Reptilia possess amniotic eggs, which develop free from standing or flowing water. Numerous other adaptations have allowed members of this class to flourish on land. The eastern snapping turtle (*Chelydra serpentina*) is shown here. Its range extends from southeastern Canada, west to the Rocky Mountains, and south to Florida. The species epithet refers to its long snake-like neck. Individuals may live up to 100 years of age.

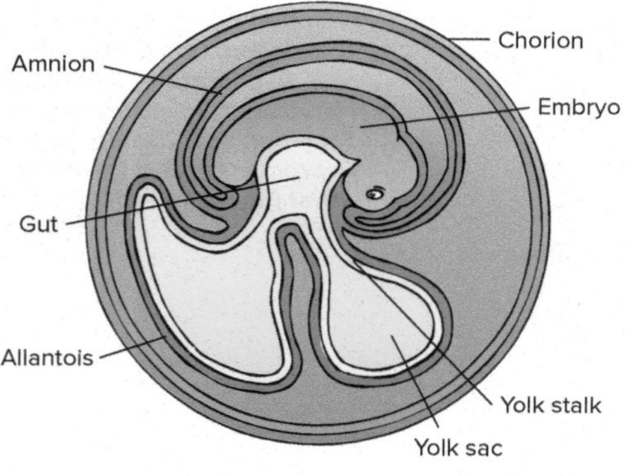

FIGURE 20.2

The Amniotic Egg. The amniotic egg provides a series of extraembryonic membranes that protect the embryo from desiccation. In reptiles, including birds, and one group of mammals, the egg is enclosed within a shell (not shown). The embryo develops at the surface of a mass of yolk. The amnion encloses the embryo in a fluid-filled sac and protects against shock and desiccation. The chorion is nearer the shell and becomes highly vascular and aids in gas exchange. The allantois is a ventral outgrowth of the gut and stores nitrogenous wastes (e.g., uric acid).

recent common ancestor must be included in a particular taxon (*see chapter* 7). Applying this rule requires that the birds (Aves) be included, along with their closest relatives–the dinosaurs, in this reptilian clade. There is little doubt in the minds of most zoologists that shared characteristics such as their single occipital condyle on the skull (the point of attachment between the skull and the first cervical vertebra), a single ear ossicle, lower jaw structure, and dozens of other morphological characteristics as well as compelling molecular evidence warrant the inclusion of birds as a part of the reptilian lineage. In this textbook, these relationships are recognized by referring to birds as "avian reptiles" and all other reptiles as "nonavian reptiles," terminology that is increasingly popular among zoologists. It is impossible, however, to ignore the traditional grouping of nonavian reptiles and avian reptiles into separate classes (Reptilia and Aves, respectively), even though this system creates a paraphyletic reptilian group. Partly out of a respect for this tradition, and partly to make the coverage of the reptiles more manageable, nonavian reptiles are covered in chapter 20 and avian reptiles are covered in chapter 21. The traditional classification of living nonavian reptiles is shown in table 20.1.

TABLE 20.1

CLASSIFICATION OF LIVING NONAVIAN REPTILES

Class Reptilia* (rep-til′e-ah)
Dry skin with epidermal scales; skull with one point of articulation with the vertebral column (occipital condyle); respiration via lungs; metanephric kidneys; internal fertilization; amniotic eggs.

Order Testudines (tes-tu′din-ez) or **Chelonia** (ki-lo′ne-ah)
Teeth absent in adults and replaced by a horny beak; short, broad body; shell consisting of a dorsal carapace and ventral plastron. Turtles.

Order Crocodylia (krok″o-dil′e-ah)
Elongate, muscular, and laterally compressed; tongue not protrusible; complete ventricular septum. Crocodiles, alligators, caimans, gavials.

Order Sphenodontia (sfen′o-dont″i-ah) or **Rhynchocephalia** (rin″ko-se-fa′le-ah)
Contains very primitive, lizardlike reptiles; well-developed parietal eye. Two species survive in New Zealand. Tuataras.

Order Squamata (skwa-ma′tah)
Recognized by specific characteristics of the skull and jaws (temporal arch reduced or absent and quadrate movable or secondarily fixed); the most successful and diverse group of living reptiles. Snakes, lizards, worm lizards.

*The class Reptilia as shown here is a paraphyletic grouping. A monophyletic representation would include the entire reptilian lineage, which would also include the birds (Aves).

Early Amniote Evolution and Skull Structure

Chapter 19 ended with a discussion of the reptilomorph ancestors of amniotes. It is not known exactly when the first animals that laid terrestrial, shelled amniotic eggs appeared because amniotic eggs do not fossilize well. However, pelycosaurs and romeriids were early terrestrial amniotes that are thought to have laid these types of eggs. Pelycosaurs were mostly large and lizardlike. One well-known genus, *Dimetrodon* could grow over 3 m in total length. This animal was equipped with elongated neural spines that formed a dorsal sail that

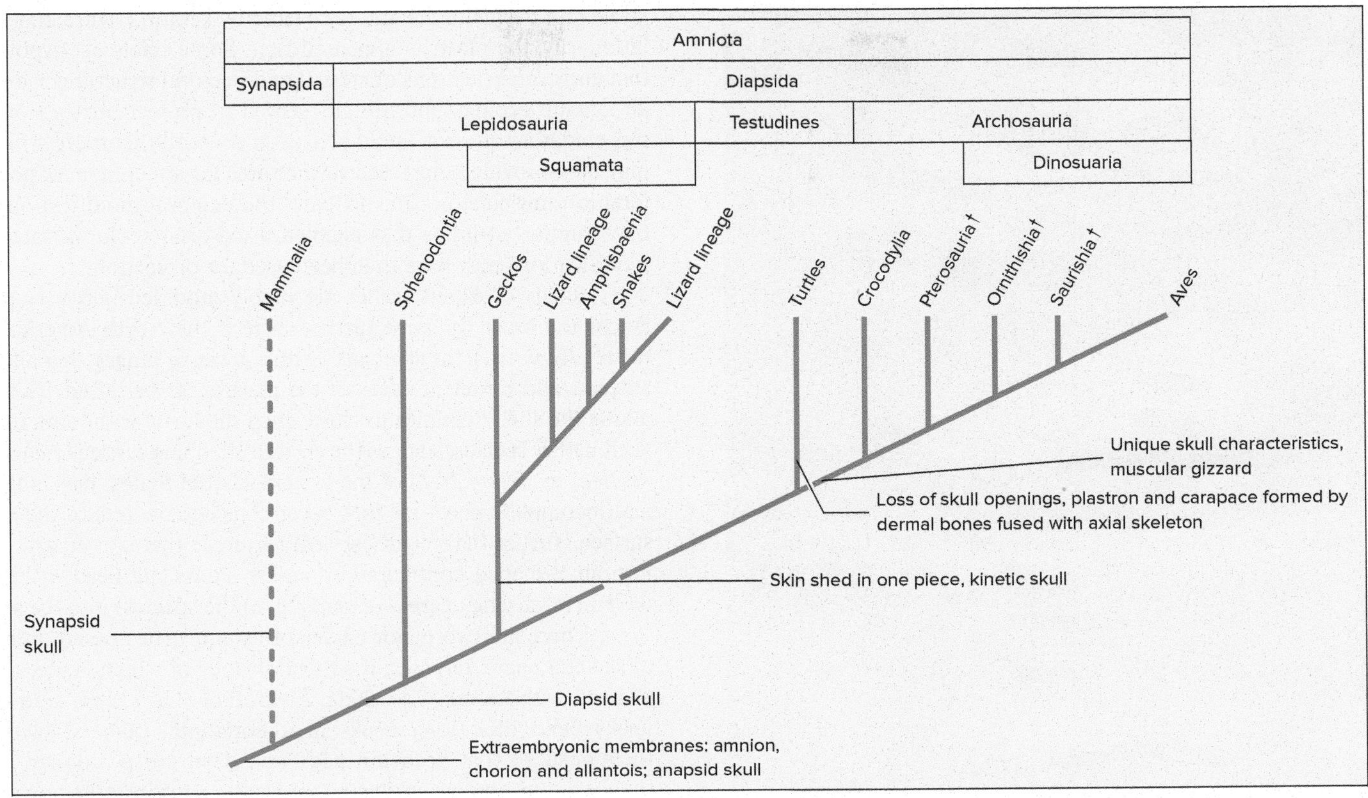

FIGURE 20.3

Amniote Phylogeny. This cladogram shows one interpretation of amniote phylogeny. Phylogenetic relationships within the amniotes are controversial. Strong molecular evidence suggests that the mammals (Synapsida) are closely related to ancestral amniotes. This lineage is shown using a dashed line. All other amniotes (Diapsida, including Aves) are a part of the reptillian linage, shown in solid lines. The traditional classification that excludes the birds from Reptilia is not valid because it results in paraphyletic groupings. In the interpretation shown here, the turtles are grouped with Diapsida. The absence of temporal fenestrae in their skulls must be a derived characteristic. Testudines is shown as a sister group to Archosauria, which is a subject of ongoing debate. Synapomorphies used to distinguish lower taxa are highly technical skeletal (usually skull) characteristics and are not shown. Daggers (†) indicate some extinct taxa. Other numerous extinct taxa are not shown.

was likely used for thermoregulation. Most romeriids were smaller than pelycosaurs and were agile and also lizardlike.

The adaptive radiation of amniotes corresponded with the adaptive radiation of their terrestrial insect prey. It began in the late Carboniferous and early Permian periods (*see appendix B*). The oldest amniote fossils show a divergence that resulted in two lineages. One lineage, Synapsida, led to mammals; and the second lineage, Diapsida, led to all reptiles, including the birds. The term "synapsid" (Gr. *syn*, with + *hapsis*, arch) refers to an amniote skull condition in which there is a single opening (fenestra) in the temporal (posterolateral) region of the skull (figure 20.4*a*). The term "diapsid" (Gr. *di*, two) refers to the amniote skull condition in which there are upper and lower temporal fenestrae (figure 20.4*b*). The common ancestor of these groups had an anapsid skull. The term "anapsid" (Gr. *an*, without) refers to a skull that lacks temporal fenestrae (figure 20.4*c*). Although turtles (Testudines) possess an anapsid skull, they are currently grouped within Diapsida. The cranial similarities between extinct anapsids and turtles are considered convergent, and the anapsid skull in Testudines likely resulted from an evolutionary loss of temporal fenestrae (*see figure 20.3*).

Section 20.1 Thinking Beyond the Facts

What are some examples and characteristics of ancestral amniotes, and what lineages resulted from their divergence?

20.2 SURVEY OF THE NONAVIAN REPTILES

LEARNING OUTCOMES

1. Describe the characteristics of the nonavian reptiles.
2. Compare the characteristics of the members of the orders Testudines, Crocodylia, and Squamata.
3. Justify the inclusion of superficially different snakes and lizards in a single order, Squamata.

Nonavian reptiles are characterized by a skull with one surface (condyle) for articulation with the first neck vertebra, respiration by lungs, metanephric kidneys, internal fertilization, and amniotic eggs.

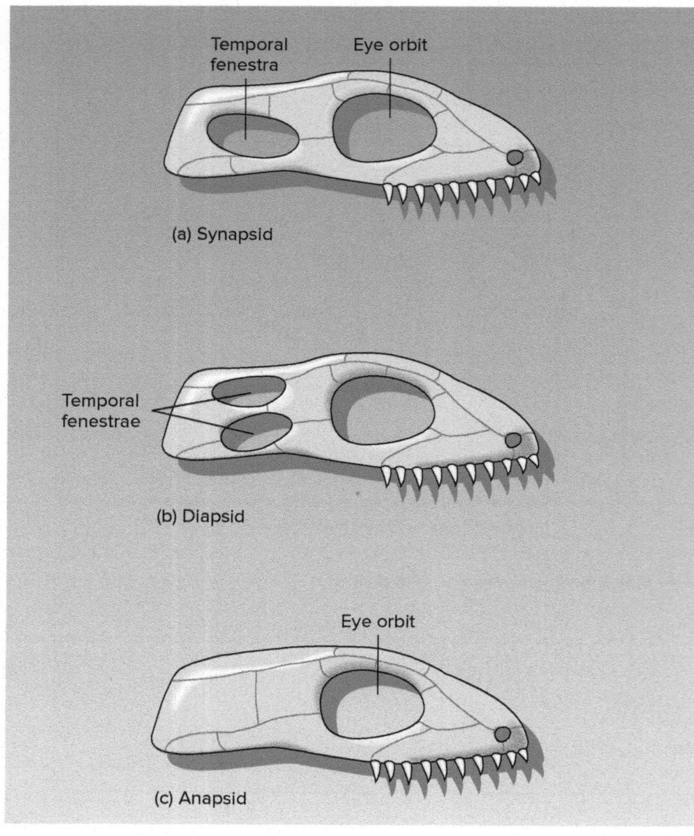

FIGURE 20.4

Amniote Skull Characteristics. Amniotes are classified according to skull characteristics and jaw muscle attachment. (*a*) Synapsid skull. The opening under each zygomatic process of your temporal bone and temporal process of your zygomatic bone (your cheekbone) are examples of the single (unpaired) temporal fenestrae of synapids. All extant synapsids are mammals. (*b*) Diapsid skulls. This kind of skull is characteristic of lizards, snakes, worm lizards, tuataras, and birds. (*c*) Anapsid skull. This kind of skull is characteristic of turtles and numerous extinct lineages.

Nonavian reptiles also have dry skin with keratinized epidermal scales. **Keratin** is a resistant protein found in epidermally derived structures of amniotes. It is protective, and when chemically bonded to phospholipids, prevents water loss across body surfaces. Members of three of the four orders described here live on all continents except Antarctica. However, nonavian reptiles dominate only tropical and subtropical ecosystems.

Testudines

Testudines (tes-tu'din-ez) (L. *testudo,* tortoise) is one of the traditional orders of extant nonavian reptiles. They are the turtles. The over 340 described species of turtles are characterized by a bony shell, limbs articulating internally to the ribs, and a keratinized beak rather than teeth. The dorsal portion of the shell is the **carapace**, which forms from a fusion of vertebrae, expanded ribs, and bones in the dermis of the skin. Keratin covers the bone of the carapace. The ventral portion of the shell is the **plastron**. It forms from bones of the pectoral girdle and dermal bone and is covered by keratin (figure 20.5). Evolutionarily, the plastron appeared first. Some scientists hypothesize that early modifications of sternal and pectoral structures facilitated an aquatic ecology and initially served to protect turtles from ventral predatory attacks. Other scientists contend that these structures may have provided increased surface area for attachment of powerful forelimb musculature. In this case, the pectoral modifications may have supported limbs that enabled a fossorial ecology. Later shell modifications may have then been used for protection.

Shells of extant turtles are highly modified and vary greatly across the order. In some turtles, such as the North American box turtle (*Terrapene*), the shell has flexible areas, or hinges, that allow the anterior and posterior edges of the plastron to be raised. The hinge allows the shell openings to close when the turtle withdraws into the shell. Other species, such as the eastern snapping turtle (*Chelydra serpentina*; *see figure 20.1*) of the eastern United States, have unhinged and incomplete plastrons that cover considerably less of the ventral surface. Turtles have eight cervical vertebrae that can be articulated into an S-shaped configuration, which allows the head to be withdrawn (to varying degrees depending on the species) into the shell.

There are two major clades of extant turtles based on a suite of skeletal characteristics; the most obvious of which is the method of head withdrawal. One clade, Pleurodira (Gr. *pleuro*, side + *dir*, neck) bend their long necks in a horizontal plane to withdraw their head so that both the neck and head are positioned across the forelimbs and are protected under the anteriormost portions of the carapace and plastron. One distinguished example is the South American matamata (*Chelus fimbriatus*). This species is a large, long-necked turtle with a broad, flat head and shell. Its neck and head are equipped with skin flaps and protuberances that make it highly cryptic, leaflike, and sensitive to tactile stimulation. The second clade, Cryptodira (Gr. *crypto*, hidden) contains species that can fully withdraw the head and neck in a vertical plane between the forelimbs. Cryptodira is diverse, and includes freshwater turtles, terrapins, tortoises, and sea turtles.

Turtles have long life spans. Most reach sexual maturity after seven or eight years and live 14 or more years. Large tortoises, like those of the Galápagos Islands, may live in excess of 100 years (*see figure 4.3*). (Tortoises are entirely terrestrial and lack webbing in their feet.) All turtles are oviparous (*see chapter 29, Vertebrate Reproductive Strategies, page 000*). Females use their hindlimbs to excavate nests in the soil. There they lay and cover with soil clutches of 5 to 100 eggs. Development takes from four weeks to one year, and the parent does not attend to the eggs during this time. The young are independent of the parent at hatching.

In recent years, turtle conservation programs have been enacted. Slow rates of growth and long juvenile periods make turtles vulnerable to extinction in the face of high mortality rates. Turtle hunting and predation on young turtles and turtle nests by dogs and other animals have severely threatened many species. Sea turtles, freshwater turtles, and tortoises are equally at risk. Conservation of tortoises and freshwater turtles is complicated by cultural practices that demand turtles for food and medicine. Conservation of sea turtles is complicated by their having ranges of thousands of square kilometers of ocean, so that protective areas must include waters under the jurisdiction of many different nations (figure 20.6).

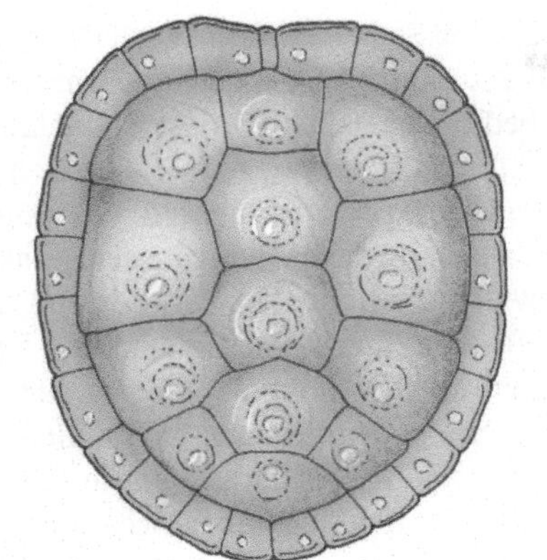

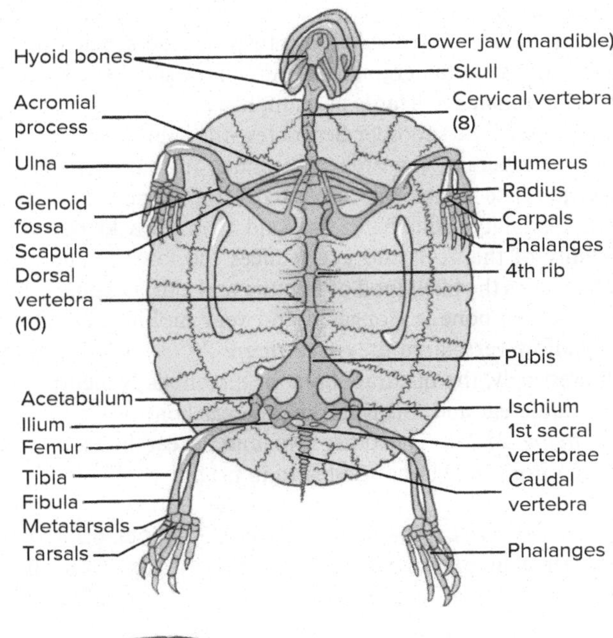

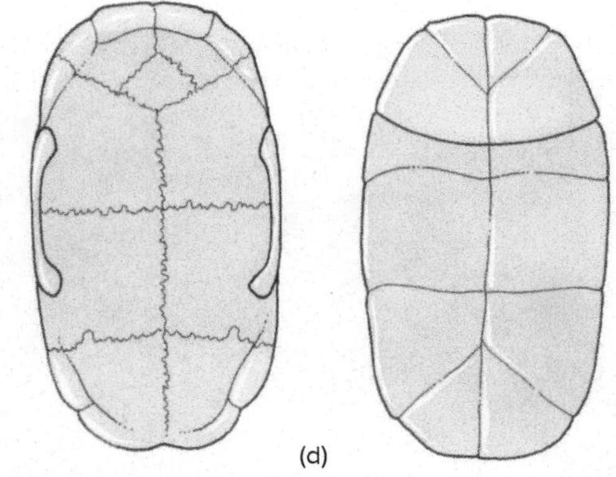

FIGURE 20.5

Skeleton of a Turtle. (*a*) Dorsal view of the carapace. (*b*) Ventral view of the carapace and appendicular skeleton. Keratin covers the carapace, which is composed of fused vertebrae, expanded ribs, and dermal bone. (*c*) Dorsal view of the plastron. (*d*) Ventral view of the plastron. The plastron forms from dermal bone and bone of the pectoral girdle. It is also covered by keratin.

FIGURE 20.6

Order Testudines. Green sea turtles (*Chelonia mydas*) are found in tropical and subtropical waters around the globe, with distinctive Atlantic and Pacific populations. They are considered endangered throughout their range. The common name is derived from green fat found beneath their carapace. Green sea turtles nest every two to four years and migrate many miles from feeding areas to the nesting beaches, often the beaches from which they hatched. Mating occurs in the sea, and females move onto the beach to dig nests with their hind flippers. A single female will deposit 100 to 200 eggs in a nest before returning to the sea. After about 120 days eggs hatch, and juveniles make the perilous trip across the beach to the sea. A high percentage of juveniles are preyed upon by gulls and crabs, and little is known of the life of juveniles that do reach the ocean.

Archosauria

Archosaurs are characterized by teeth set into sockets in the jaws (thecodont condition), muscular gizzards, and skull openings in front of the eyes. The first archosaurs probably appeared during the Permian period about 250 mya (*see appendix B*). They became a dominant terrestrial vertebrate during the Triassic period about 240 mya.

There were two lineages of Archosauria. Most members of these lineages became extinct during a mass extinction at the Triassic-Jurassic boundary approximately 200 mya. A few members of each of these two lineages survived. Descendants of one lineage, the alligators and crocodiles, survived a second mass extinction 65 mya (Cretaceous-Tertiary boundary). Surviving members of this lineage are described in the section that follows.

The descendants of the second archosaur lineage flourished as dinosaurs and pterosaurs during the Mesozoic era until the second mass extinction (Cretaceous-Tertiary boundary) erased all but one small group of dinosaurs called the theropods. This group rebounded and, as described in chapter 21, these surviving reptiles flourished. Today, we see them in virtually every terrestrial niche on earth—they are the birds.

Order Crocodylia

The order Crocodylia (krok″o-dil′e-ah) (Gr. *krokodeilos*, lizard) contains approximately 25 species of crocodiles, alligators, caiman, and gavials. In addition to the archosaur characteristics mentioned earlier, crocodylians can be distinguished from other reptiles by their triangular rather than circular eye orbits, laterally compressed teeth, four-chambered hearts, nonprotrusible tongues, and muscular, laterally compressed tails. The largest extant crocodylian is the saltwater crocodile (*Crocodylus porosus*). It can reach more than 7 m in total length. Gavials are medium-sized riverine crocodylians endemic to the Indo-Malay biogeographic realm. They have elongated and narrow snouts adapted for seizing fish prey. The smallest crocodylian is Cuvier's dwarf caiman (*Paleosuchus palpebrosus*). This South American species is typically less than 1.5 m in length.

Crocodylians have not changed much over their 215-million-year history. The snout is elongate and often used to capture food by a sideways sweep of the head. The nostrils are at the tip of the snout, so the animal can breathe while mostly submerged. Air passageways of the head lead to the rear of the mouth and throat, and a flap of tissue near the back of the tongue forms a watertight seal that allows breathing without inhaling water in the mouth. A plate of bone, called the secondary palate, evolved in the archosaurs and separates the nasal and mouth passageways. The muscular, elongate, and laterally compressed tail is used for swimming, offensive and defensive maneuvers, and attacking prey. Teeth are used only for seizing prey. Food is swallowed whole, but if a prey item is too large, crocodylians tear it apart by holding onto a limb and rotating their bodies wildly until the prey is dismembered. Crocodylians swallow rocks and other objects as abrasives for breaking apart ingested food. Crocodylians are oviparous and display parental care of hatchlings that parallels that of birds. Nesting behavior and parental care may be traced back to the common ancestor of both groups.

Lepidosauria

Lepidosauria is a monophyletic diapsid lineage that first appeared in the early Triassic (*see figure 20.3 and appendix B*). They were widespread 200 to 100 mya, and available fossil evidence suggests that the early Lepidosauria were dominated by Sphenodontida. Sphenodontida is now represented by a single genus that inhabits New Zealand. A second lepidosaur lineage is Squamata. These nonavian reptiles appeared in the middle Jurassic period and, like dinosaurs, underwent major extinctions 65 mya. Squamates that survived are the ancestors of today's lizards and snakes. One of the oldest known lepidosaur-like fossils is from *Megachirella wachtleri*, a middle Triassic nonavian reptile that is thought to be closely related to ancestral Lepidosauria (*see page 355*). Lepidosaurs are characterized by the presence of overlapping keratinized epidermal scales, regular shedding cycles of the outer epidermal layer, paired male reproductive organs in the tail base, and a transverse cloacal slit. Tails that can undergo autotomy (breaking along fracture planes followed by regeneration) and detailed skeletal characteristics are also common features of this group.

Order Sphenodontia

The two surviving species of the order Sphenodontia (sfen′o-dont″i-ia) (Gr. *sphen,* wedge + *odontos,* tooth) are the tuataras (*Sphenodon punctatus* and *S. guntheri*) (figure 20.7). These superficially lizard-like nonavian reptiles are what remains of a diverse lineage of Mesozoic lepidosaurs. Their skull structure distinguishes these nonavian reptiles. Unlike the Squamata, which will be described next, tuataras have an akinetic skull. An **akinetic skull** has an upper jaw that is firmly attached to the skull. This firm attachment provides a very powerful bite. In addition, two rows of acrodont teeth on the upper jaw and a single row of acrodont teeth (*see page 366*) in the lower jaw produce a shearing bite that can decapitate a small bird. Formerly more widely distributed in New Zealand, the tuataras fell prey to human influences and domestic animals. They are now present only on remote offshore islands and are protected by New Zealand law. They are oviparous and share underground burrows with ground-nesting seabirds. Tuataras venture out of their burrows at dusk and dawn to feed on insects or, occasionally, small vertebrates.

Order Squamata

The order Squamata (skwa-ma′tah) (L. *squama,* scale + *ata,* to bear) is traditionally divided into two suborders. The two suborder designations are undergoing taxonomic revision. The suborder Sauria, which includes lizards and amphisbaenias, is paraphyletic as it encompasses the suborder Serpentes–the snakes.

Members of the order are unique in that they possess movable quadrate bones and other skull modifications that increase skull flexibility. Squamates are said to have a **kinetic skull.** The mobility of the skull and jaw reduces the force of the bite, but it also reduces the likelihood of fracture and aids in feeding. In snakes, the quadrate bone is elongated and very mobile and allows them to swallow very large prey (*see figure 20.13 and box figure 27.1*). (Interestingly, the quadrate bone of mammals is incorporated into the middle ear and forms the middle ear bone known as the incus [*see figure 24.21*]. The other two middle ear bones are similarly derived from skull bones first seen in primitive fish.)

Suborder Sauria—The Lizards About 6,100 species of lizards are in suborder Sauria (sawr′e-ah) (Gr. *sauro,* lizard). In contrast to snakes, lizards usually have two pairs of limbs, eyelids,

FIGURE 20.7

Order Sphenodontia. The tuatara (*Sphenodon punctatus*).

external ear openings, their upper and lower jaws unite anteriorly, and their skulls are adapted to crush prey. The lizards that are legless retain remnants of a pelvic and/or pectoral girdle and sternum. Lizards vary in length from only a few centimeters to as large as 3 m. Many lizards live on surface substrates and retreat under logs or rocks when necessary. Others are fossorial, or semifossorial or arboreal. Most lizards are oviparous; some are ovoviviparous or viviparous. They usually oviposit under rocks or debris or in burrows. One phylogenetic interpretation of Sauria yields six major clades. Selected clades are discussed in the following text.

Gekkota is a large clade comprised of over 1,500 species collectively known as the geckos. They are often found on the walls of human dwellings in semitropical areas and are short and stout. Many geckos are nocturnal and, unlike most other lizards, many can vocalize. Their large eyes, with pupils that contract to narrow slits during the day and dilate widely at night, are adapted for night vision. Claws and adhesive pads on their digits (digital lamellae comprised of microvillus surfaces) aid in clinging to trees, rocks, and walls. Some geckos can lose (and later regenerate) large patches of skin when grasped. The tail of most geckos is autotomic (*see page 365*).

The clade Iguania also consists of over 1,500 species. Iguanians generally have robust bodies, short necks, and distinct heads. Representatives include the marine iguanas of the Galápagos Islands, the flying dragons of Southeast Asia, and the true chameleons of Africa and India. Marine iguanas (*Amblyrhynchus cristatus*) are algivorous (algae-eaters) throughout life, live in marine environments, and form large nocturnal sleeping aggregations to keep warm and help avoid predation. The flying dragons, (*Draco*) have lateral folds of rib-supported skin that expand to form gliding surfaces. When *Draco* launches itself from a tree, it can glide 30 m or more! The true chameleons are adapted for arboreal lifestyles and use a long, sticky, and rapidly protrusible tongue to capture insect prey. Unlike other lizards, chameleons have independently moving eyes and specialized feet that enable them to firmly grip branches while navigating foliage (*see figure 20.12*). True chameleons can change color in response to illumination, temperature, and behavioral state. *Anolis*, or the "pet store chameleon," can also change color. However, *Anolis* is a New World iguanian, not a true chameleon.

The monitor lizard, Mexican beaded lizard, and Gila monster are part of a lizard clade called Anguimorpha (L. *angui*, snake + Gr. *morph*, form). The largest of all lizards is a type of venomous monitor lizard native to the Indonesian archipelago. It's called the Komodo dragon (*Varanus komodoensis*) and reaches lengths of up to 9 m. The Mexican beaded lizard (*Heloderma horridum*) and Gila monster (*H. suspectum*) are stout lizards native to southwestern North America (figure 20.8). They produce small amounts of venom in glands of the lower jaw. Venom is released into grooves on the surface of teeth and is introduced into prey as the lizard chews. Their bites are seldom fatal to humans. Some species in this clade are entirely legless or have reduced limbs. They are known as slowworms, glass lizards, and legless lizards.

The group Amphisbaenia (am″fisbe′ne-ah) (Gr. *amphi*, double + *baen*, to walk) consists of nearly 190 species of specialized burrowers called worm lizards. All are fossorial. Although they have reduced limbs or are entirely legless, they differ from the aforementioned legless lizards in that their skulls are wedge or shovel shaped and they have a single median tooth in the upper jaw that sits between two lower teeth, forming a nipper that is used in burrowing. They also have ringlike folds in the skin called annuli that loosely attach to the body wall. Muscles of the skin cause it to telescope and bulge outward forming an anchor against a burrow wall. Amphisbaenians move easily forward and backward—thus their group name. They feed on worms and small insects, are oviparous, and live in the soils of Africa, South America, the Caribbean, and the Middle East (figure 20.9).

FIGURE 20.8

Order Squamata. The Gila monster (*Heloderma suspectum*) is a venomous lizard of southwestern North America.

Suborder Serpentes—The Snakes About 3,400 snake species are in the suborder Serpentes (ser-pen′tez) (L. *serpere*, to crawl). There are two major clades of snakes, one clade is small and contains snake species collectively referred to as "blindsnakes." The other clade contains all other snake species. Although most are not dangerous to humans, about 20% of snake species are venomous. Worldwide, it is estimated that there are between 20,000 and 30,000 human fatalities annually due to snakebite. Most of these deaths occur in sub-Saharan Africa and Southeast Asia. In the United States, only five or six people die each year from snakebite. Deaths are usually due to the lack of emergency medical health care. Some instances of venomous snakebite are the result of intentionally molesting the snake. Many populations are in decline due to intentional killing, disease, the pet trade, and habitat loss (*see Wildlife Alert, page 362*). The International Union for Conservation of Nature lists 269 species as either endangered, critically endangered, near threatened, or vulnerable.

Snakes are elongate and lack limbs, although vestigial pelvic appendages and pelvic girdles may be present. The skeleton may contain more than 200 vertebrae and pairs of ribs. Specialized joints between vertebrae make the body very flexible. Snakes possess elongate skulls with adaptations that facilitate swallowing large prey, rather than chewing it (*see Nutrition and the Digestive System, page 366*). Other differences between snakes and lizards include the

WILDLIFE ALERT

The Eastern Diamondback Rattlesnake (*Crotalus adamanteus*)*

Vital Statistics

Classification: Phylum Chordata, class Reptilia, order Squamata.

Range: The southeastern United States from southeastern North Carolina southward along the Atlantic coastal plain to Florida and the Florida Keys. Populations exist westward to the Gulf Coast including parts of southern Alabama and Mississippi.

Habitat: Associated with dry lowland palmetto and wiregrass flatwoods, pine or pine-oak forests, and coastal dune and maritime forest systems of the North American southeast.

Status: Petitioned for listing as federally threatened on 22 August 2011.

Natural History and Ecological Status

The eastern diamondback rattlesnakes are the largest venomous snakes in the United States (box figure 20.1). Although they are typically less than 1.8 m and under 4.5 kg, historical records have documented them reaching over 2.5 m and as much as 7 kg. These impressive squamatids are thickly built and bear prominent dark, dorsal diamond-shaped blotches for which they are named. Their wide, chunky heads are distinctly patterned with two lightly colored diagonally running facial stripes that extend from either side of the eye downward toward the neck, terminating near the hind jaw. Because of their impressively patterned skins, venomous bites, and large size, these snakes are one of the most heavily persecuted vertebrates in North America. Fear, ignorance, and a desire for their hides and rattles drive people to shoot and bludgeon staggering numbers of these animals (box figure 20.2).

BOX FIGURE 20.1 **Eastern Diamondback Rattlesnakes (*Crotalus adamanteus*).** These snakes live and hunt in coastal habitats in southeastern United States. Their cryptic coloration aids in the "sit-and-wait" predatory behavior seen in this photograph. Unfortunately, the species has experienced noted declines as a result of habitat destruction and degradation and intentional removal by people. In an effort to develop safe management practices for both people and snakes, the Georgia Sea Turtle Center Research Department and the Applied Wildlife Conservation Lab are conducting research to understand the habits of this important predator using radio telemetry. For more information on this, and other projects conducted by the Applied Wildlife Conservation Lab visit http://wildlifelab.wixsite.com/awcl. Photo ©Lance Padden.
©Lance Paden

BOX FIGURE 20.2 **Rattlesnake Bounty.** Bounties for rattlesnakes are not a thing of the past. This bounty poster was displayed in a store window in a rural southeastern U.S. town where these animals remain unprotected. The photograph was taken in March, 2014.
©Todd Tupper, Phd., Northern Virginia Community College

Habitat loss and degradation (fragmentation, urban sprawl, and pine farming), removal from backyard habitats, and road mortality (sometimes intentional) are the primary contributors resulting in a reduction in range and population size of eastern diamondbacks. The elimination of longleaf pine (*Pinus palustris*) savannas and coastal habitats is contributing most to the decline of the eastern diamondback. These snakes require open-canopy habitats that allow for the growth of herbaceous ground cover, a feature of longleaf pine forests and coastal environments. The presettlement secondary-growth longleaf pine forests also contained rot-resistant old-growth pine stumps, which provide the snakes with refugia. In the remaining patches of longleaf pine, many of the rot-resistant pine stumps have been harvested to produce pine oil, rosin, and turpentine. This stump harvesting leaves the eastern diamondbacks without shelter, even in preferred pinelands.

Longleaf pine forests originally covered 37 million hectares. Because of wildfire suppression and development pressures associated with human population growth, remaining longleaf pine patches have been replaced by later successional, closed-canopy loblolly (*Pinus taeda*) and slash (*Pinus ellottii*) pine forests—and remain that way because of current management practices. In 2001, it was estimated that less than 3% of the longleaf pine forest remains, making it one of the most endangered habitats in the world.

Additionally, rattlesnakes are victimized during annual "rattlesnake roundups." Roundups consist of evicting snakes from their burrows (oftentimes created and used by the federally threatened gopher tortoises, *Gopherus polyphemus*) by flooding them out with gasoline or ammonia. The snakes that do not succumb to the poison are captured, inhumanely transported, dumped into a pen, and eventually decapitated

*Guest Contributor: Kimberly M. Andrews, Jekyll Island Authority and University of Georgia, Jekyll Island, GA.

WILDLIFE ALERT *Continued*

or beaten to death in front of crowds of onlookers. Further, climate change is affecting their activity patterns which in turn influence their energetics and susceptibility to infectious diseases, such as snake fungal disease. The cumulative and pervasive threats have resulted in the range-wide declines that prompted specialists to submit a petition for federal listing. It is estimated that the number of eastern diamondbacks has been reduced from an approximated presettlement population of 3.08 million to fewer than 100,000.

A key to preserving this species is effective management of remnant habitats within its range and drastically reducing the number of individuals that are killed annually. There are a number of organizations seeking to stop intentional killing through education and others dedicated to protecting longleaf pine and coastal habitats. Scientists are working to understand this species' ecology and to evaluate the genetic health of remaining populations. For more information on the conservation and ecology of the eastern diamondback, see the following citations.

Adkins Giese CL, Greenwald DN, Means DB, Matturro B, Reis J. 2011. Petition to list the eastern diamondback rattlesnake (*Crotalus adamanteus*) as threatened under the Endangered Species Act.

Martin WH, Means DB. 2000. Distribution and habitat relationships of the eastern diamondback rattlesnake (*Crotalus adamanteus*). *Herpetol Nat His.* 7:9-34.

Means DB. 2009. The effects of rattlesnake roundups on eastern diamondback rattlesnake (*Crotalus adamanteus*). *Herpetol Conserv Bio.* 4: 132-141.

Ware S, Frost C, Doerr PO. 1993. Southern mixed hardwood forest: the former longleaf pine forest. In Martin WH, Boyce SG, Echternacht AC, eds. Biodiversity of the southeastern United States: lowland terrestrial communities. New York: J. Wiley and Sons; p 447-93.

FIGURE 20.9

Order Squamata. The Mexican mole lizard (*Bipes biporus*) belongs to the amphisbaenian family Bipedidae. This family is the only amphisbaenian group to retain stubby forelegs that it uses in burrowing. This species is endemic to the Sonoran desert of southern Baja, California.

mechanism for focusing the eyes and the morphology of the retina. Elongation and narrowing of the body has resulted in the reduction or loss of the left lung and displacement of the gallbladder, right kidney, and, often, gonads. Most snakes are oviparous, although many, such as the New World vipers, many boas, and many cobras, are either ovoviviparous or viviparous.

Although seemingly less diverse than other nonavian reptiles, snake anatomy and physiology can be highly specialized, and their habits diverse. They can be arboreal, fossorial, aquatic, or marine. They can even be found gliding up to 100 m between tall trees and understory vegetation (e.g., the flying snakes, *Chrysopelea sp.*). Some snakes are very small, at 10 cm or less (e.g., the Barbados threadsnake, *Leptotyphlops carlae*), while others are massive, reaching almost 230 kg (e.g., the green anaconda, *Eunectes murinus*), and well over 7 m (e.g., the reticulated python, *Python reticulatus*).

Hypotheses on snake origins are very tentative. This is, in part, because snake skulls do not fossilize well. The earliest known snake fossils date back some 167 million years. Five genera from 95 mya are important to the current understanding of snake evolution because they may demonstrate an evolutionary loss of limbs. All of these genera have hindlimbs only. However, only one genus (*Najash*) has hindlimbs that are attached to the axial skeleton (as can be seen in other limbed vertebrates). The remaining four genera have no direct connection between the appendicular and axial skeleton. It is not currently known if limblessness is ancestral to the appearance of the characteristic elongate skull, nor are the extrinsic evolutionary drivers of limblessness. Some zoologists hypothesize that the earliest snakes were fossorial and have ancestors that are similar to modern blindsnakes (Typhlopidae). Molecular data support this hypothesis, and the loss of appendages and corresponding changes in eye structure could parallel evolutionary changes seen in caecilians (*see figure 19.4*). Other zoologists contend that ancestral snakes were aquatic, or inhabited densely vegetated habitats. Fossils of a primitive (90-million-year-old) snake from Australia (*Alamitophis tingamarra*) without burrowing adaptations support this hypothesis.

SECTION 20.2 THINKING BEYOND THE FACTS

How would you explain the fact that the snakes, lizards, and worm lizards, although superficially very different in body form and ecology, are members of the same order?

20.3 EVOLUTIONARY PRESSURES

LEARNING OUTCOMES

1. Discuss the structural and physiological adaptations that make life apart from an abundant water supply possible in nonavian reptiles.
2. Compare the feeding mechanism of snakes to the feeding mechanisms of other nonavian reptiles.
3. Compare the reproductive biology of crocodiles to reproduction by other nonavian reptiles.

The lifestyles of most nonavian reptiles reveal striking adaptations for terrestrialism. For example, a lizard common to deserts of the southwestern United States—the chuckwalla (*Sauromalus obesus*)—survives during late summer when temperatures exceed 40°C (104°F) and when arid conditions wither plants and blossoms upon which chuckwallas browse (figure 20.10). To withstand these hot and dry conditions, chuckwallas disappear below ground and aestivate. Temperatures moderate during the winter, but little rain falls, so life on the desert surface is still not possible for the chuckwalla. The summer's sleep, therefore, merges into a winter's sleep. The chuckwalla does not emerge until March, when rain falls, and the desert explodes with greenery and flowers. The chuckwalla browses and drinks, storing water in large reservoirs under its skin. Chuckwallas are not easy prey. If threatened, a chuckwalla takes refuge in the nearest rock crevice. There, it inflates its lungs with air, increasing its girth and wedging itself against the rock walls of its refuge. Friction of its body scales against the rocks makes the chuckwalla nearly impossible to dislodge.

The adaptations that chuckwallas display are not exceptional for nonavian reptiles. This section discusses some of these adaptations that make life apart from an abundant water supply possible.

FIGURE 20.10

Chuckwalla (*Sauromalus*). Many nonavian reptiles, like this chuckwalla, possess adaptations that make life apart from standing or running water possible.

Source: NPS Photo by Robb Hannawacker.

External Structure and Locomotion

Unlike that of amphibians, the skin of nonavian reptiles has no respiratory functions. Nonavian reptile skin is thick, dry, and keratinized (*see figure 23.7*). Scales and scutes may be strengthened by bony plates and may be modified for various functions. For example, the large belly scales of snakes provide contact with the substrate during locomotion. Although nonavian reptile skin is much less glandular than that of amphibians, secretions include pheromones that function in sex recognition and defense.

Although other vertebrates shed epidermal cells, some nonavian reptiles (e.g., snakes) are well known for periodically shedding their entire skin at once to accommodate growth. (Shedding in reptiles is called ecdysis, and this term is also used to describe a similar, though unrelated, process in certain ecdysozoans; *see figure 14.5*). In vertebrates, the blood vessels of the skin do not innervate the epidermis. Thus, the cells of the outer epidermis die because they are not nourished by the underlying blood supply. The loss of dead epidermal cells in humans occurs as flaking of the body skin or scalp (i.e., dandruff). In lepidosaur reptiles, lymph and other enzymes uniformly separate the dead outer epidermal layer from the newly, and synchronously formed, underlying epidermis. Therefore, their dead epidermal layer is either lost in large fragments, or as one piece. Crocodylians and turtles shed their outer epidermis in smaller patches, or as single scales. The frequency of ecdysis varies from species to species but is typically greater in juveniles due to their higher growth rates.

The chromatophores of nonavian reptiles are primarily dermal in origin and function much like those of amphibians. Cryptic coloration, mimicry, and aposematic coloration occur in nonavian reptiles. Color and color change also function in sex recognition and thermoregulation.

Support and Movement

The skeletons of snakes, amphisbaenians, and turtles show modifications; however, in its general form, the nonavian reptile skeleton is based on one inherited from ancient amphibians. The skeleton is highly ossified to provide greater support. The skull is longer than that of amphibians, and a plate of bone, the **secondary palate,** partially separates the nasal passages from the mouth cavity (figure 20.11). As described earlier, the secondary palate evolved in archosaurs, where it was most likely an adaptation for breathing when the mouth was full of water or food. It is also present in other nonavian reptiles, although developed to a lesser extent. Longer snouts also permit greater development of olfactory epithelium and increased reliance on the sense of smell.

Nonavian reptiles have more cervical vertebrae than do amphibians. The first two cervical vertebrae (atlas and axis) provide greater freedom of movement for the head. An atlas articulates with a single condyle on the skull and facilitates nodding. An axis is modified for rotational movements. Variable numbers of other cervical vertebrae provide additional neck flexibility.

The ribs of nonavian reptiles may be highly modified. Those of turtles and the flying dragon were described previously. The ribs of snakes have muscular connections to large belly scales to aid locomotion. The cervical vertebrae of cobras have ribs that may be flared in aggressive displays.

Two or more sacral vertebrae attach the pelvic girdle to the vertebral column. The caudal vertebrae of many lizards possess a vertical fracture plane. When these lizards are grasped by the tail, caudal vertebrae can be broken, and a portion of the tail is lost. Tail loss, or **autotomy,** is an adaptation that allows a lizard to escape from a predator's grasp, or the disconnected, wiggling piece of tail may distract a predator from the lizard. The lizard later regenerates the lost portion of the tail.

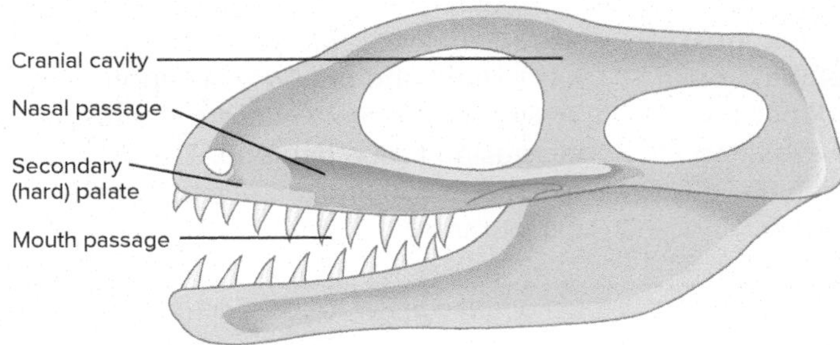

FIGURE 20.11

Secondary Palate. Sagittal section of the skull of a synapsid, showing the secondary palate that separates the nasal and mouth cavities. Extension of the bones of the anterior skull forms the anterior portion of the secondary palate (the hard palate), and skin and soft connective tissues form the posterior portion of the secondary palate (the soft palate).

Locomotion in primitive nonavian reptiles is similar to that of salamanders. The body is slung low between paired, stocky appendages, which extend laterally and move in the horizontal plane. The limbs of other nonavian reptiles are more elongate and slender and are held closer to the body. The knee and elbow joints rotate posteriorly; thus, the body is higher off the ground, and weight is supported vertically. Many prehistoric nonavian reptiles were bipedal, meaning that they walked on hindlimbs. They had a narrow pelvis and a heavy, outstretched tail for balance. Bipedal locomotion freed the front appendages, which became adapted for prey capture or flight in some animals.

Nutrition and the Digestive System

Most nonavian reptiles are carnivores, although turtles may be herbivores, carnivores, or omnivores depending on the species. The tongues of turtles and crocodylians are nonprotrusible and aid in swallowing. Like some anurans, some lizards and the tuatara have sticky tongues for capturing prey. As previously mentioned, the tongue of chameleons is long and exceeds their body length (figure 20.12).

Teeth of archosaurs (except birds) are **thecodont.** Thecodont teeth are deeply anchored in jaw sockets. Crocodylians have

FIGURE 20.12

Order Squamata. A chameleon (*Chamaeleo calyptratus*) using its tongue to capture prey. Note the prehensile tail and "zygodactyl feet" adapted for grasping branches.

How Do We Know about Snake Venom?

Dr. Bryan Fry of the University of Melbourne in Australia has analyzed the genes coding for more than 24 kinds of venoms. His conclusion is that different kinds of venoms evolved by gene duplication (*see chapter 5*) from a single ancestral gene. Venom apparently evolved before fangs between 60 and 80 mya. The proteins that comprise venoms trace their evolutionary history to ancestors of molecules that play other roles in animal systems. Some venoms contain a protein related to acetylcholinesterase, an enzyme that helps control muscle contraction (*see chapter 23*). In snake venom, a related molecule induces flaccid paralysis and difficulty breathing. Another protein, kallikrein, is released by certain white blood cells and causes vascular and blood pressure changes. The venomous version induces a dangerous drop in blood pressure. This work not only provides insight into the evolution of venomous snakes in general, but also illustrates again how conservative evolution is. It is apparently much easier for evolutionary events to act on, and modify, existing genes than to start from scratch.

What's the best emergency treatment for a venomous snake bite? It's a cell phone and rapid transport to an emergency room. Venom extraction kits don't work. These kits extract about 0.04% of venom. They actually make matters worse by increasing tissue damage. The use of ice and constriction bands is also discouraged.

approximately 80 cone-shaped teeth that are replaced up to about 50 times through their lives. This multiple replacement is the **polyphyodont** condition. Crocodylians are the only nonmammalian vertebrates with thecodont teeth. Squamate teeth are not set into sockets. Teeth are **acrodont** (Amphisbaenia) or **pleurodont** (most lizards, and snakes). In the former, teeth are attached along the ridge of the jaw, and in latter the teeth are attached along the medial edge of the jaw. Squamate teeth are not all simple cones. Some herbivorous lizards have teeth with multiple cusps. Probably, the most remarkable adaptations of snakes involve modifications of the skull for feeding. The bones of the skull and jaws loosely join and may spread apart to ingest prey much larger than a snake's normal head size (figure 20.13*a*). The bones of the upper jaw are movable on the skull, and ligaments loosely join the halves of both of the upper and lower jaws anteriorly. Therefore, each half of the upper and lower jaws can move independently of one another. After a prey is captured, opposite sides of the upper and lower jaws are alternately thrust forward and retracted. Posteriorly pointing teeth prevent prey escape and help force the food into the esophagus. The glottis, the respiratory opening, is far forward so that the snake can breathe while slowly swallowing its prey.

Vipers (family Viperidae) possess hollow fangs on the maxillary bone at the anterior margin of the upper jaw (figure 20.13*b*). These fangs connect to venom glands that inject venom when the viper bites. The maxillary bone (upper jaw bone) of vipers is hinged so that when the snake's mouth is closed, the fangs fold back and lie along the upper jaw. When the mouth opens, the maxillary bone rotates and causes the fangs to swing down (figure 20.13*c*). Because the fangs project outward from the mouth, vipers may strike at objects of any size. Rear-fanged snakes (in the family Colubridae) possess grooved rear teeth. In those that are venomous, venom is channeled along these grooves and worked into prey to quiet them during swallowing. These snakes usually do not strike; however, the African boomslang (*Dispholidus typus*) has caused human fatalities. Coral snakes, sea snakes, and cobras (Elapidae) have fangs that rigidly attach to the upper jaw in an erect position. When the mouth is closed, the fangs fit into a pocket in the outer gum of the lower jaw. Fangs are grooved or hollow, and contraction of muscles associated with venom glands injects venom into the fangs. Some cobras can "spit" venom at their prey; if not washed from the eyes, the venom may cause blindness.

Venom glands are modified salivary glands. Most snake venoms are mixtures of neurotoxins and hemotoxins. Contrary to popular belief, venom is usually not a snake's primary means of defense. Snake venoms are complex mixtures of zootoxic enzymes that evolved to rapidly immobilize prey. They have wide-ranging effects depending on the snake species and the snake's prey. Venoms can (among additional actions) destroy blood vessels, red blood cells, nerve cells, muscle cells, lipid membranes, and nucleic acids. They can also impair blood clotting and prevent neurotransmission in a number of ways. The venoms of elapids are highly neurotoxic due to their capacity to degrade nerve centers and cause respiratory paralysis of prey. Viper venoms are highly hemotoxic because they rapidly degrade blood cells and blood vessel linings.

Circulation, Gas Exchange, and Temperature Regulation

The circulatory system of nonavian reptiles is based on that of amphibians. Because nonavian reptiles are, on average, larger than amphibians, their blood must travel under higher pressures to reach distant body parts. To take an extreme example, the blood of the dinosaur *Brachiosaurus* had to be pumped a distance of about 6 m from the heart to the head—mostly uphill! (The blood pressure of a giraffe is about double that of a human to move blood the 2 m from the heart to the head.)

Like amphibians, nonavian reptiles possess two atria that are completely separated in the adult and have veins from the body and lungs emptying into them. Except for turtles, the sinus venosus is no longer a chamber but has become a patch of cells that acts as a pacemaker. The ventricle of most nonavian reptiles is incompletely divided (figure 20.14). (Only in crocodylians is the ventricular septum complete.) The ventral aorta and the conus arteriosus divide during development and become three major arteries that leave the heart. A pulmonary artery leaves the ventral side of the ventricle and takes blood to the lungs. Two systemic arteries, one from the ventral side of the heart and the other from the dorsal side of the heart, take blood to the lower body and the head.

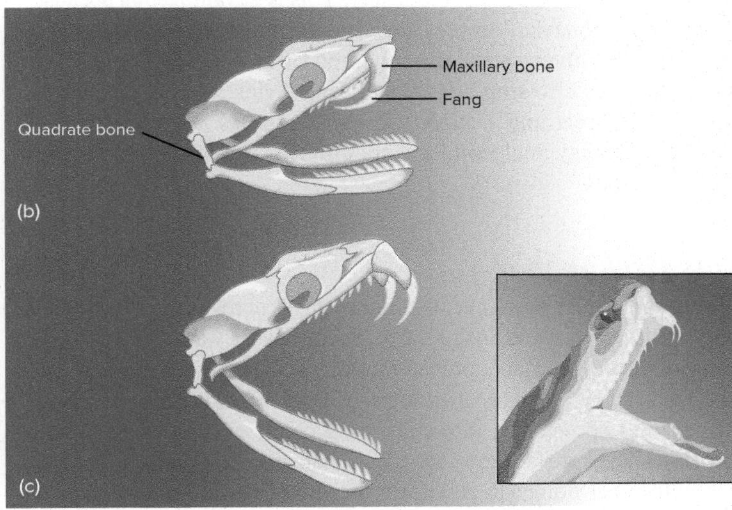

FIGURE 20.13

Feeding Adaptations of Snakes. (*a*) A copperhead (*Agkistrodon contortrix*) ingesting a prey. Flexible joints allow the bones of the kinetic skull to separate during feeding. Note the pit organ just anterior to the eye. (*b*) The skull of a viper. The hinge mechanism of the jaw allows upper and lower bones on one side of the jaw to slide forward and backward alternately with bones of the other side. Posteriorly curved teeth hold prey as it is worked toward the esophagus. (*c*) Note that the maxillary bone, into which the fang is embedded, swings forward when the mouth opens. The mobility of the quadrate bone is a unique characteristic of all squamates.

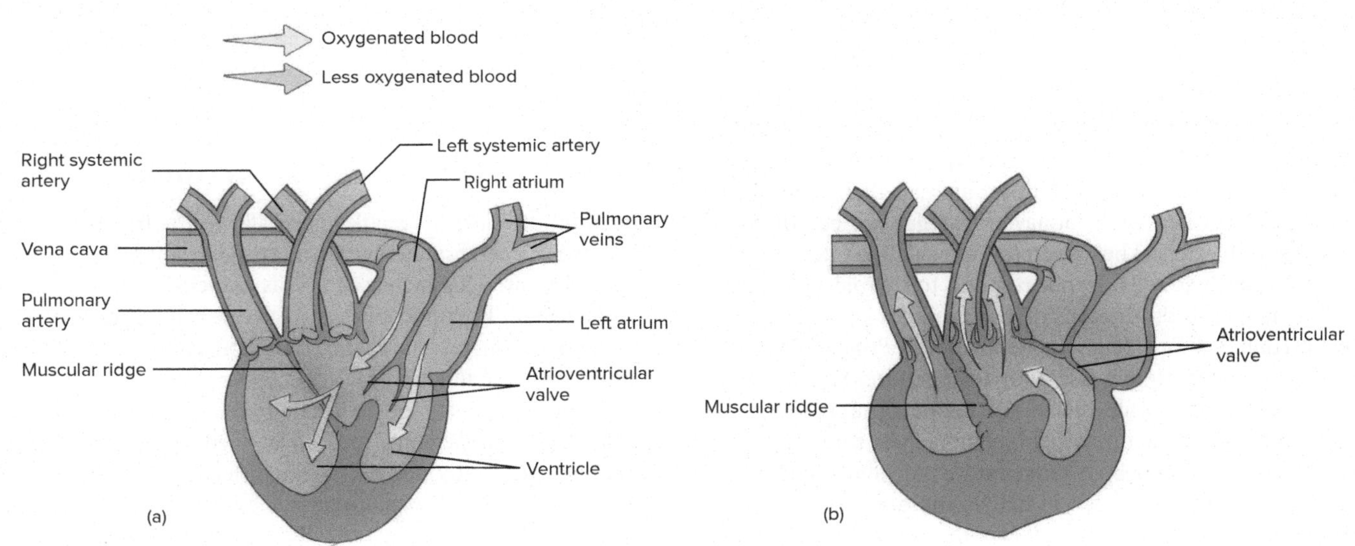

FIGURE 20.14

Heart and Major Arteries of a Lizard. (*a*) When the atria contract, blood enters the ventricle. An atrioventricular valve prevents the mixing of oxygenated and less oxygenated blood across the incompletely separated ventricle. (*b*) When the ventricle contracts, a muscular ridge closes to direct oxygenated blood to the systemic arteries and less oxygenated blood to the pulmonary artery.

Source: Heisler, N. 1983. The Company of Biologists Limited. Journ of Exp Bio. 105: 105.

Blood low in oxygen enters the ventricle from the right atrium and leaves the heart through the pulmonary artery and moves to the lungs. Blood high in oxygen enters the ventricle from the lungs via pulmonary veins and the left atrium, and leaves the heart through left and right systemic arteries. The incomplete separation of the ventricle permits shunting of some blood away from the pulmonary circuit to the systemic circuit by the constriction of muscles associated with the pulmonary artery. This is advantageous because virtually all nonavian reptiles breathe intermittently. When turtles withdraw into their shells, their method of lung ventilation cannot function. They also stop breathing during diving. During periods of apnea ("no breathing"), blood flow to the lungs is limited, which conserves energy and permits more efficient use of the pulmonary oxygen supply.

Gas Exchange

Nonavian reptiles exchange respiratory gases across internal respiratory surfaces to avoid losing large quantities of water. A larynx is present; however, vocal cords are usually absent. Cartilages support the respiratory passages of nonavian reptiles, and lungs are partitioned into spongelike, interconnected chambers. Lung chambers provide a large surface area for gas exchange.

In most nonavian reptiles, a negative-pressure mechanism is responsible for lung ventilation. A posterior movement of the ribs and the body wall expands the body cavity, decreasing pressure in the lungs and drawing air into the lungs. Air is expelled by elastic recoil of the lungs and forward movements of the ribs and body wall, which compress the lungs. The ribs of turtles are a part of their shell; thus, movements of the body wall to which the ribs attach are impossible. Turtles exhale by contracting muscles that force the viscera upward, compressing the lungs. They inhale by contracting muscles that increase the volume of the visceral cavity, creating negative pressure to draw air into the lungs.

Temperature Regulation

Unlike aquatic animals, terrestrial animals may face temperature extremes (−65 to 70°C) that are incompatible with life. Temperature regulation, therefore, is important for animals that spend their entire lives out of water. Most nonavian reptiles are ectotherms. Ectotherms rely on external heat sources for regulating internal temperatures because their cells do not respond to thyroid hormones in the same manner as do the cells of avian reptiles and mammals. However, certain pythons (e.g. Indian pythons, *Python molurus*) can use metabolic heat to increase body temperature. Females coil around their eggs and elevate their body temperature as much as 7.3°C above the air temperature using metabolic heat sources.

The ability to regulate body temperature independent of external heat sources was also present in many dinosaurs. Recent studies suggest that at least some dinosaurs were mesothermic. These dinosaurs could raise body temperature metabolically, but they did not maintain higher internal temperatures for long periods of time. Strict endothermy would be energetically costly for a large reptile like *Tyrannosaurus*. In all likelihood, an endothermic *Tyrannosaurus* would not have been able to find enough food to prevent starvation.

Some nonavian reptiles can survive wide temperature fluctuations (e.g., −2 to 41°C for some turtles). To sustain activity, however, body temperatures are regulated within a narrow range, between 25 and 37°C. If that is not possible, the reptile usually seeks a retreat where body temperatures are likely to remain within the range compatible with life.

Many thermoregulatory activities of nonavian reptiles are behavioral, and they are best known in the lizards. To warm itself, a lizard orients itself at right angles to the sun's rays, often on a surface inclined toward the sun, and presses its body tightly to a warm surface to absorb heat by conduction (*see figures 20.10 and 28.1*). To cool itself, a lizard orients its body parallel to the sun's rays, seeks shade or burrows, or assumes an erect posture (legs extended and tail arched) to reduce conduction from warm surfaces. In hot climates, many nonavian reptiles are nocturnal.

Various physiological mechanisms also regulate body temperature. As temperatures rise, some nonavian reptiles begin panting, which releases heat through evaporative cooling. (Little evaporative cooling occurs across the dry skin of nonavian reptiles.) Marine iguanas divert blood to the skin while basking in the sun and warm up quickly. On diving into the cool ocean, however, marine iguanas reduce heart rate and blood flow to the skin, which slows heat loss. Chromatophores also aid in temperature regulation. Dispersed chromatophores (thus, a darker body) increase the rate of heat absorption.

In temperate regions, many nonavian reptiles withstand cold winter temperatures by becoming inactive when body temperatures and metabolic rates decrease. Individuals that are usually solitary may migrate to a common site, called a hibernaculum, to spend the winter. Heat loss from individuals in hibernacula is reduced because the total surface area of many individuals clumped together is reduced compared to widely separated animals. Unlike true hibernators, the body temperatures of nonavian reptiles in torpor are not regulated, and if the winter is too cold or the retreat is too exposed, the animals can freeze and die. Death from freezing is an important cause of mortality for temperate reptiles.

Nervous and Sensory Functions

The brain of nonavian reptiles is similar to the brains of other vertebrates. The cerebral hemispheres are somewhat larger than those of amphibians. This increased size is associated with an improved sense of smell. The optic lobes and the cerebellum are also enlarged, which reflects increased reliance on vision and more refined coordination of muscle functions.

The complexity of nonavian reptile sensory systems is evidenced by a chameleon's method of feeding. Its protruding eyes swivel independently, and each has a different field of view. Initially, the brain keeps both images separate, but when an insect is spotted, both eyes converge on the prey. Binocular vision then provides the depth perception for determining whether or not the insect is within the range of the chameleon's tongue (*see figure 20.12*).

Vision is the dominant sense in most nonavian reptiles, and their eyes are similar to those of amphibians (*see figure 19.13*). Snakes focus on nearby objects by moving the lens forward. Contraction of the iris places pressure on the gel-like vitreous body in the posterior region of the eye, and displacement of this gel pushes the lens forward. All other nonavian reptiles focus on nearby objects when the normally elliptical lens is made more spherical, as a result of ciliary muscles pressing the ciliary body against the lens. Nonavian reptiles have a greater number of cones than do amphibians and probably have well-developed color vision.

Upper and lower eyelids, a nictitating membrane, and a blood sinus protect and cleanse the surface of the eye. In snakes and some lizards, the upper and lower eyelids fuse in the embryo to form a protective window of clear skin, called the spectacle. (During ecdysis, the outer layers of the spectacle become clouded and impair the vision of snakes.) The blood sinus, which is at the base of the nictitating membrane, swells with blood to help force debris to the corner of the eye, where it may be rubbed out. Horned lizards (*Phrynosoma*) squirt blood from their eyes by rupturing this sinus in a defensive maneuver to startle predators.

Some nonavian reptiles possess a **median (parietal) eye** that develops from outgrowths of the roof of the forebrain (*see figure 24.29*). The median eye is involved with regulating circadian rhythms and hormone production. In the tuatara, it is an eye with a lens, a nerve, and a retina. In other nonavian reptiles, the parietal eye is less developed. Parietal eyes are covered by skin and probably cannot form images. They can, however, differentiate light and dark periods and are used in orientation to the sun.

The structure of nonavian reptile ears varies. The ears of snakes detect substrate vibrations. They lack a middle ear cavity, an auditory tube, and a tympanic membrane. A bone of the jaw articulates with the stapes and receives substrate vibrations. Snakes can also detect airborne vibrations. In other nonavian reptiles, a tympanic membrane may be on the surface or in a small depression in the head. The inner ear of nonavian reptiles is similar to that of amphibians.

Olfactory senses are better developed in nonavian reptiles than in amphibians. In addition to the partial secondary palate providing more surface for olfactory epithelium, many nonavian reptiles possess blind-ending pouches that open through the secondary palate into the mouth cavity. These pouches, called **Jacobson's (vomeronasal) organs,** are in diapsid nonavian reptiles; however, they are best developed in the squamates. Jacobson's organs develop in embryonic crocodylians but are not present in adults of this group. Anapsids (turtles) lack these olfactory organs. The protrusible, forked tongues of snakes and lizards are accessory olfactory organs for sampling airborne chemicals. A snake's tongue flicks out and then moves to the Jacobson's organs, which perceive odor molecules. Tuataras use Jacobson's organs to taste objects held in the mouth.

Pit vipers have heat-sensitive **pit organs** on each side of the face between the eye and the nostril (*see figures 20.13*a *and 24.24*). These depressions are lined with sensory epithelium and are used to detect objects with temperatures different from the snake's surroundings. Pit vipers are usually nocturnal, and their pits help them to locate small, warm-blooded prey. Other snakes possess similar structures. They may be more numerous and are positioned nearer the mouth.

Sea turtles (e.g. Cheloniidae) can detect the earth's magnetic field and use it in navigation. Sea turtles such as the green sea turtle (*see figure 20.6*) hatch on nesting beaches, make their way across the beach to the sea, and then spend many years swimming the world's oceans. In the case of one population of green sea turtles, adults feed off the coast of Brazil, but nest 3,000 km away on tiny Ascension Island. Ten to fifteen years after hatching, green sea turtles return to the nesting beach from which they hatched, where females lay eggs. Not only do they orient using the earth's magnetic field, but also sea turtles can apparently distinguish between magnetic fields of different geographic locations. They use these abilities to find their way to the water as hatchlings and navigate toward their home beach as adults. Magnetic magnetite particles have been found in the brains of both turtles and birds, although a direct link between magnetite and navigation has not been established.

Excretion and Osmoregulation

The kidneys of embryonic nonavian reptiles are similar to those of fishes and amphibians. Life on land, increased body size, and higher metabolic rates, however, require kidneys capable of processing wastes with little water loss. A kidney with many more blood-filtering units, called nephrons, replaces the nonavian reptile embryonic kidney during development. The functional kidneys of adult nonavian reptiles are called metanephric kidneys. Their function depends on a circulatory system that delivers more blood at greater pressures to filter large quantities of blood.

Most nonavian reptiles excrete uric acid. It is nontoxic, and being relatively insoluble in water, it precipitates in the excretory system. The urinary bladder or the cloacal walls reabsorb water, and the uric acid can be stored in a pastelike form. Use of uric acid as an excretory product also made possible the development of embryos in terrestrial environments, because nontoxic uric acid can be concentrated in egg membranes.

In addition to the excretory system's reabsorption of water, internal respiratory surfaces and relatively impermeable exposed surfaces reduce evaporative water loss. The behaviors that help regulate temperature also help conserve water. Nocturnal habits and avoiding hot surface temperatures during the day by burrowing reduce water loss. When water is available, many nonavian reptiles (e.g., chuckwallas) store large quantities of water in lymphatic spaces under the skin or in the urinary bladder. Many lizards possess salt glands below the eyes for ridding the body of excess salt.

Reproduction and Development

Vertebrates could never be truly terrestrial until their reproduction and embryonic development became separate from standing or running water. For vertebrates, internal fertilization and the amniotic egg (*see figure 20.2*) made complete movement to land possible. The amniotic egg, however, is not completely independent of water. Pores in the eggshell permit gas exchange but also allow water to evaporate. Amniotic eggs require significant energy expenditures by parents. Parental care occurs in some nonavian reptiles and may involve maintaining relatively high humidity around the eggs. These eggs are often supplied with large quantities of yolk for long developmental periods, and parental energy and time are sometimes invested in the posthatching care of dependent young.

Accompanying the development of amniotic eggs is the necessity for internal fertilization. Fertilization must occur in the reproductive tract of the female before protective egg membranes are laid down around an egg. All male nonavian reptiles, except tuataras, possess an intromittent organ for introducing sperm into the female reproductive tract. Lizards and snakes possess paired hemipenes at the base of the tail that are erected by being turned inside out, like the finger of a glove.

Gonads lie in the abdominal cavity. In males, a pair of ducts delivers sperm to the cloaca. After copulation, sperm may be stored in a seminal receptacle in the female reproductive tract. Secretions of the seminal receptacle nourish the sperm and arrest their activity. Sperm may be stored for up to four years in some turtles, and up to six years in some snakes! In temperate latitudes, sperm can be stored over winter. Copulation may take place in the fall, when individuals congregate in hibernacula, and fertilization and development may occur in the spring, when temperatures favor successful development. Fertilization occurs in the upper regions of the oviduct, which leads from the ovary to the cloaca. Glandular regions of the oviduct secrete albumen and the eggshell. The shell is usually tough yet flexible. In some crocodylians, the eggshell is calcareous and rigid, like the eggshells of birds.

Parthenogenesis has been described in six families of lizards and three species of snakes. In these species, no males have been found. Populations of parthenogenetic females have higher reproductive potential than bisexual populations. A population that suffers high mortality over a cold winter can repopulate its habitat rapidly because all surviving individuals can produce offspring. This apparently offsets the disadvantages of genetic uniformity resulting from parthenogenesis.

Nonavian reptiles often have complex reproductive behaviors that may involve males actively seeking out females. As in other animals, courtship functions in sexual recognition and behavioral and physiological preparation for reproduction. Head-bobbing displays by some male lizards reveal bright patches of color on the throat and enlarged folds of skin. Courtship in snakes is based primarily on tactile stimulation. Tail-waving displays are followed by the male running his chin along the female, entwining his body around her, and creating wavelike contractions that pass posteriorly to anteriorly along his body. Lizards and snakes also use sex pheromones to assess the reproductive condition of a potential mate. Vocalizations are important only in crocodylians. During the breeding season, males are hostile and may bark or cough as territorial warnings to other males. Roaring vocalizations also attract females, and mating occurs in the water.

After they are laid, nonavian reptile eggs are usually abandoned (figure 20.15). Virtually all turtles bury their eggs in the ground or in plant debris. Other nonavian reptiles lay their eggs under rocks, in debris, or in burrows. About 100 species of nonavian reptiles have some degree of parental care of eggs. One example is the American alligator *(Alligator mississippiensis)* (figure 20.16). The female builds a mound of mud and vegetation about 1 m high and 2 m in diameter. She hollows out the center of the mound, partially fills it with mud and debris, deposits her eggs in the cavity, and then covers the eggs. Temperature within the nest influences the sex of the hatchlings. Temperatures at or below 31.5°C result in female offspring. Temperatures between 32.5 and 33°C result in male offspring. Temperatures around 32°C result in both male and female offspring. (Similar temperature effects on sex determination are known in some lizards and many turtles.) The female remains in the vicinity of the nest throughout development to protect the eggs from predation. She frees hatchlings from the nest in response to their high-pitched calls and picks them up in her mouth to transport them to water. She may scoop shallow pools for the young and remain with them for up to two years. Young feed on scraps of food the female drops when she feeds and on small vertebrates and invertebrates that they catch on their own.

FIGURE 20.15

***Nonavian Reptile* Eggs and Young.** This American alligator (*Alligator mississippiensis*) is hatching from its egg.

©Heiko Kiera/Shutterstock

FIGURE 20.16

Parental Care in *Nonavian Reptiles*. A female American alligator (*Alligator mississippiensis*) tending to her nest. Decomposition of the material that makes up the nest generates some of the heat necessary for incubation.

Source: NPSPhoto, Lori Oberhofer, 2005

Section 20.3 Thinking Beyond the Facts

Ecdysis, or molting, characterizes the ecdysozoans (chapters 13 to 15), and it is also a process that accommodates growth in many vertebrates. Compare and contrast the purposes and processes of ecdysis in these groups and evaluate the possibility that the processes might be homologous.

20.4 Further Phylogenetic Considerations

Learning Outcomes

1. Describe the evolutionary fate of the archosaur branch of the reptilian lineage.
2. Describe the evolutionary fate of the synapsid branch of the amniote lineage.

Divergence within the archosaur branch of the diapsid reptilian lineage occurred about 200 mya (*see figure 20.3*). These evolutionary events gave rise to the crocodylians, dinosaurs, and two groups of fliers. The pterosaurs (Gr. *pteros,* wing + *sauros,* lizard) ranged from sparrow size to animals with wingspans of 13 m. An elongation of the fourth finger supported their membranous wings, their sternum was adapted for the attachment of flight muscles, and their bones were hollow to lighten the skeleton for flight. As presented in chapter 21, these adaptations are paralleled by, though not identical to, adaptations in the birds—the descendants of the second lineage of flying archosaurs.

The synapsid branch of the amniote lineage diverged about 320 mya and gave rise to numerous extinct forms, and to the mammals. The legs of synapsids were relatively long and held their bodies off the ground. Teeth and jaws were adapted for effective chewing and tearing. Additional bones were incorporated into the middle ear. These and other mammalian characteristics developed between the Carboniferous and Triassic periods. The "Evolutionary Perspective" section of chapter 22 describes more about the nature of this transition.

Section 20.4 Thinking Beyond the Facts

What might explain why parental care of young is common in crocodylians and birds but not in other reptiles?

Summary

20.1 Evolutionary Perspective

- Amniota is a monophyletic lineage that includes both avian and nonavian reptiles, and the mammals. Amniotes are characterized by the presence of an amniotic egg that, among other functions, protects the embryo from desiccation. The amniotic egg facilitates reproduction on land.
- Adaptive radiation of primitive amniotes resulted in the mammalian (synapsid) and reptilian (diapsid) lineages. The diapsid lineage includes the avian reptiles (birds) and the nonavian reptiles (snakes, lizards, crocodiles, and tuataras). The traditional division of this reptilian lineage into the classes Reptilia and Aves incorrectly creates a paraphyletic group. Pelycosaurs and romeriids are examples of primitive amniotes. Although the anapsid skull of testudinates was thought to have descended from primitive anapsids, it is now thought to be a convergent characteristic that resulted from the loss of two pairs of temporal fenestrae. Thus, Testudines are classified within Diapsida.

20.2 Survey of Nonavian Reptiles

- The order Testudines contains the turtles. Turtles have a bony shell and lack teeth. All have anapsid skulls and are oviparous. Turtles may have evolved from diapsids that were either aquatic or fossorial.
- Archosauria includes crocodylians, birds, and dinosaurs. The order Crocodylia contains alligators, crocodiles, caimans, and gavials. These groups have a well-developed secondary palate and display nesting behaviors and parental care.
- Lepidosauria includes tuataras, lizards, amphisbaenias, and snakes. The order Sphenodontia contains two species of tuataras. They are found only on remote islands of New Zealand. This group was widespread 200–100 mya. The order Squamata contains the lizards, snakes, and worm lizards. Lizards usually have two pairs of legs, and most are oviparous. Snakes lack developed limbs and have skull adaptations for swallowing large prey. They may have evolved from either fossorial or aquatic ancestors. Worm lizards are specialized burrowers. They have a single median tooth in the upper jaw, and most are oviparous.

20.3 Evolutionary Pressures

- The skin of nonavian reptiles is dry and keratinized, and it provides a barrier to water loss. It also has epidermal scales and chromatophores. The nonavian reptile skeleton is modified for support and movement on land. Loss of appendages in snakes is accompanied by greater use of the body wall in locomotion.
- Nonavian reptiles have a tongue that may be used in feeding. Bones of the skulls of snakes are loosely joined and spread apart during feeding.
- The circulatory system of nonavian reptiles is divided into pulmonary and systemic circuits and functions under relatively high blood pressures. Blood may be shunted away from the pulmonary circuit during periods of apnea. Gas exchange occurs across convoluted lung surfaces. Ventilation of lungs occurs by a negative-pressure mechanism. Nonavian reptiles are ectotherms and mainly use behavioral mechanisms to thermoregulate.

- Vision is the dominant sense in most nonavian reptiles. Median (parietal) eyes, ears, Jacobson's organs, and pit organs are important receptors in some nonavian reptiles.
- Reptiles posses water-conserving metanephric kidneys. Because uric acid is nontoxic and relatively insoluble in water, nonavian reptiles can store and excrete it as a semisolid. Internal respiratory surfaces and dry skin also promote water conservation.
- The amniotic egg and internal fertilization permit development on land. They require significant parental energy expenditure. Some nonavian reptiles use visual, olfactory, and auditory cues for reproduction. Parental care is important in crocodylians.

20.4 **Further Phylogenetic Considerations**

- The archosaur branch of the reptilian lineage gave rise to the dinosaurs, crocodylians, pterosaurs, and the birds. The synapsid branch of the amniote linage gave rise to numerous extinct forms, and the mammals.

Concept Review Questions

1. All of the following are a part of the diapsid reptilian lineage, except one. Select the exception.
   a. Mammals
   b. Snakes
   c. Lizards
   d. Dinosaurs
   e. Birds
2. A researcher found a fossilized amniote skull with two temporal fenestrae. Which of the following statements regarding this skull would be TRUE?
   a. It would have been a member of the synapsid lineage and was probably an early mammal.
   b. It would have been a member of the diapsid lineage and could have been an ancestor of a modern mammal.
   c. It would have been a member of the diapsid lineage and could have been an ancestor of a modern lizard.
   d. It would have been anapsid—probably an ancient turtle.
3. Members of the order Squamata include the
   a. turtles.
   b. snakes.
   c. tuataras.
   d. crocodiles.
4. Nonavian reptiles periodically shed the outer, epidermal layers of the skin in a process called ecdysis. This process is not homologous to ecdysis in Ecdysozoa.
   a. True
   b. False
5. The partial septum in the ventricle of most nonavian reptiles is
   a. an evidence that reptiles are less highly evolved than birds and mammals, whose ventricles are completely divided.
   b. an adaptation that allows shunting of blood from the pulmonary to the systemic circuit during periods of intermittent breathing.
   c. a shunting mechanism that is especially well developed in turtles.
   d. Both b and c are correct.

Analysis and Application Questions

1. Explain the recent changes in the higher taxonomy of the amniotes. Do you think that the Reptilia should be retained as a formal class designation? If so, what groups of animals should it contain?
2. What characteristics of the life history of turtles make them vulnerable to extinction? What steps do you think should be taken to protect endangered turtle species?
3. The incompletely divided ventricle of nonavian reptiles is sometimes portrayed as an evolutionary transition between the heart of primitive amphibians and the completely divided ventricles of birds and mammals. Do you agree with this portrayal? Why or why not?
4. What effect could significant global warming have on the sex ratios of crocodylians? Speculate on what long-term effects might be seen in populations of crocodylians as a result of global warming.

These waved albatrosses (*Phoebastria irrorata*) are in courtship. These behaviors are common in modern birds. Behavioral links between birds and their ancestors are difficult to assess; however, many structural characteristics, once thought to be avian, are now providing clues to the reptilian ancestry of birds.

21

Birds: The Avian Reptiles

Chapter Outline

21.1 Evolutionary Perspective
Phylogenetic Relationships
Nonavian Theropods
The Ancient Birds: Archaeopteryx, Sinornis, and Eoalulavis
The Initial Uses of Feathers and the Origins of Flight
Diversity of Modern Birds
21.2 Evolutionary Pressures
External Structure and Locomotion
Nutrition and the Digestive System
Circulation, Gas Exchange, and Temperature Regulation
Nervous and Sensory Systems
Excretion and Osmoregulation
Reproduction and Development
Migration and Navigation

21.1 EVOLUTIONARY PERSPECTIVE

LEARNING OUTCOMES

1. Assess the anatomical similarities and differences between the nonavian and avian reptiles.
2. Explain how evolutionary changes in ancient theropods led to powered flight and the conserved body form of modern birds.

Drawings of birds on the walls of caves in Southern France and Spain, bird images of ancient Egyptian and American cultures, and the bird images in biblical writings are evidence that humans have marveled at birds and bird flight for thousands of years. From Leonardo da Vinci's early drawings of flying machines (1490) to Orville Wright's first successful powered flight on 17 December, 1903, humans have tried to take to the sky and experience soaring like a bird.

Birds' ability to navigate long distances between breeding and wintering grounds is just as impressive as flight. For example, Arctic terns have a migratory route that takes them from the Arctic to the Antarctic and back again each year, a distance of approximately 35,000 km (figure 21.1). Their rather circuitous route takes them across the northern Atlantic Ocean, to the coasts of Europe and Africa, and then across vast stretches of the southern Atlantic Ocean before they reach their wintering grounds.

Phylogenetic Relationships

Avian reptiles are traditionally classified as members of the class Aves (a′ves) (L. *avis,* bird). The major characteristics of this group are adaptations for flight, including appendages modified as wings, feathers, endothermy, a high metabolic rate, a vertebral column modified for flight, and bones lightened by numerous air spaces. In addition, modern birds possess a horny bill and lack teeth.

The similarities between birds and nonavian reptiles are so striking that, in the 1860s, T. H. Huxley described birds as "glorified reptiles" and included them in a single class Sauropsida. As zoologists and paleontologists learn more about the relationships between birds and other reptiles, many scientists advocate Huxley's original idea. Anatomical similarities include features such as a single occipital condyle on the skull (the point of articulation between the skull and the first cervical vertebra), a single ear ossicle, lower jaw structure, and dozens of other technical skeletal characteristics. Physiological characteristics, such as the presence of nucleated red blood cells and aspects of liver and kidney

(a)

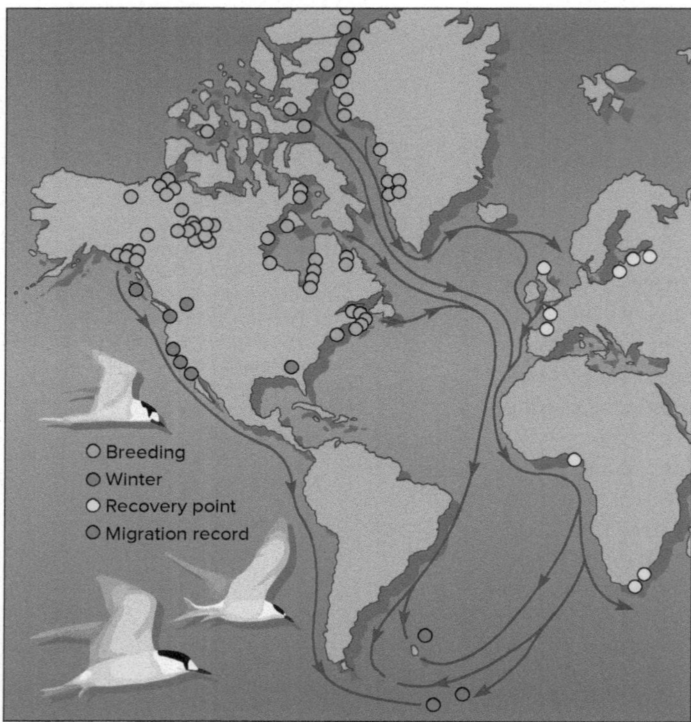

(b)

FIGURE 21.1

Class Aves. (*a*) The birds were derived from the archosaur lineage of ancient nonavian reptiles. Adaptations for flight include appendages modified as wings, feathers, endothermy, a high metabolic rate, a vertebral column modified for flight, and bones lightened by numerous air spaces. Flight has given birds, like this Arctic tern (*Sterna paradisaea*), the ability to exploit resources unavailable to other vertebrates. (*b*) Migration route of the Arctic tern. Arctic terns breed in northern North America, Greenland, and the Arctic. Migrating birds cross the Atlantic Ocean on their trip to Antarctica during the Northern Hemisphere's winter season. In the process, they fly about 35,000 km (22,000 mi) each year.

function, are shared by nonavian reptiles and birds. Some birds and other reptiles share behavioral characteristics, for example, those related to nesting and care of young. Even characteristics that were once thought to be characteristic of birds but not nonavian reptiles, such as endothermy (mesothermy), air spaces in bones, and the presence of feathers, have been demonstrated in some dinosaurs.

Nonavian Theropods

Modern birds likely descended from extinct feathered reptiles collectively referred to as theropods. These birdlike reptiles were part of the reptilian order Saurischia, which is an extinct group of dinosaurs nested within the reptilian clade Archosauromorpha. The only living archosauromorphs are the crocodylians and birds. Thus, birds are more closely related to crododylians than they are to any other extant animal group. Although prehistoric in appearance, modern crocodylians are not living dinosaurs. The only extant members of the clade Dinosauria are the birds (*see figure 20.3*).

Spectacular discoveries from 160-million-year-old fossil beds in northern China support the hypothesis of theropod ancestry. These fossils may not represent animals directly ancestral to birds, but more importantly they repeatedly show that ancestral features of birds were present in diverse species within one important lineage—the nonavian theropods. Fossils of at least a dozen theropod dinosaurs bearing feathers have been discovered (figure 21.2). One of the oldest fossil theropods, *Yi qi* (Chinese, wing + strange), lived

FIGURE 21.2

An Artist's Representation of Feathered Theropods and Ancient Birds. In the foreground center is *Caudipteryx*. It had symmetrical feathers and was about the size of a turkey. *Microraptor* is shown in flight in the background. Note the representation of feathers on all four appendages. No one is certain that *Microraptor* could fly. It may have climbed trees and glided, but it did possess asymmetrical feathers characteristic of bird flight feathers. *Jixiangornis* (tree branch) fossils from northeastern China show skeletal structures that suggest the ability to fly. The colors shown here, and the interactions between species that are implied, are the artist's interpretation.

during the middle or late Jurassic. It is unique among other known theropods because it had a feather-covered body and membranous wings similar in appearance to bat wings. It was probably a gliding dinosaur. *Yi qi* lacked wing flexibility and therefore could not flap. Cretaceous fossils from three other genera (*Sinosauropteryx, Caudipteryx, Microraptor*) were also important in formulating ideas about avian lineage. Their wings were more birdlike than batlike. *Sinosauropteryx* was a chicken-sized dinosaur that had small tubular structures, similar to feathers, in their early stages of development in modern birds. *Caudipteryx* was a turkey-sized theropod with symmetrical feathers on fore appendages and tail. It is assumed that neither of these theropods was a flier, because asymmetrical feathers are required for powered flight. Conversely, *Microraptor* had asymmetrical feathers on its fore and hind appendages and possessed a feathered tail. Other skeletal features of *Microraptor* suggest that it climbed into trees. Thus, *Microraptor* may have used its wings for gliding flight. However, some studies suggest that it may have been capable of generating lift. Additional dinosaur fossils showing a furcula, or wishbone, further support the theropod ancestry hypothesis. (The furcula is derived from a fusion of the clavicles and, as described later, is an adaptation for flight in birds.) Together these fossils demonstrate that feathers predate flight (specific purposes are addressed in the following text). Thus, the evolution of flight is, in part, due to an evolutionary repurposing of feathers and, given the unique anatomy of *Yi qi*, may have been achieved differently in the various groups of feathered, winged theropods.

The Ancient Birds: *Archaeopteryx, Sinornis,* and *Eoalulavis*

In 1861, an important vertebrate fossil was found in a slate quarry in Bavaria, Germany (figure 21.3). The fossil (named *Archaeopteryx* [Gr. *archaios*, ancient + *pteron*, wing]) was of a pigeon-sized animal that lived during the Jurassic period, about 150 million years ago (mya). It had a long, reptilian tail and clawed fingers. A complete head was not preserved, but imprints of feathers on the tail and on short, rounded wings were the main evidence that led to the interpretation that this was a fossil of an ancient bird. Later, more complete fossils were discovered, revealing teeth in beaklike jaws. These discoveries further reinforced the idea of theropod ancestry for birds. The clavicles (wishbone) of *Archaeopteryx* were well developed and probably provided points of attachment for wing muscles. However, unlike in modern birds, the sternum was flat, the tail was long and bony, and wing bones and other skeletal sites for attachment of muscles that would be used in flight were less developed. These observations indicate that *Archaeopteryx* may have primarily been a glider rather than a competent flier. Thus, although *Archaeopteryx* is sometimes considered to be an early avian theropod, it more likely represents a transitional form between nonavian theropods and early avians.

Fossils of a second ancient bird, *Sinornis*, were again discovered in China. *Sinornis* is 15 million years younger than *Archaeopteryx*, and although it possessed some primitive dinosaur-like characteristics, it

(a)

(b)

FIGURE 21.3

***Archaeopteryx*, an Ancient Bird.** (*a*) *Archaeopteryx* fossil. (*b*) Artist's representation. Some zoologists think that *Archaeopteryx* was a ground dweller rather than the tree dweller depicted here.

had features similar to those of modern birds. These include a short-ended body and tail and a sternum that could accommodate flight muscles. The claws were reduced, and the forelimbs were modified to permit the folding of wings at rest. These characteristics indicate that powered flight was well developed in birds 135 mya.

A third fossil bird from early Cretaceous deposits in Spain provides additional important information on the origin of flight. This bird, *Eoalulavis*, was found in 125-million-year-old deposits and had a wingspan of 17 cm (roughly the same as a goldfinch). This fossil showed that *Eoalulavis* had a wing structure called an alula. As described later in this chapter, the alula is present in many modern birds that engage in slow, hovering flight. Its presence in this fossil indicates that complex flight mechanisms associated with slow, hovering, and highly maneuverable flight evolved at least 125 mya.

The Initial Uses of Feathers and the Origins of Flight

Due to the complex and sometimes brilliant pigmentation of feathers, some zoologists think that the earliest feathers may have been used for intraspecific communication such as courtship display, thermoregulation, water repellency, camouflage, or as balancing devices while running along the ground. Regardless of the specific function in these early forms, feathers were later repurposed to support flight. The discovery of *Archaeopteryx* helped zoologists formulate hypotheses on the origin of avian flight. Some zoologists think that the clawed digits on the wings (of *Archaeopteryx* and, later, other species) may have been used to climb trees and cling to branches. A sequence in the evolution of flight may have involved jumping from branch to branch, or branch to ground. At some later point, gliding evolved. Still later, weak flapping supplemented gliding, and, finally, wing-powered flight evolved.

Other zoologists note that the hindlimb structure of the earliest birds suggests that they may have been bipedal, running and hopping along the ground. Their wings may have functioned in batting flying insects out of the air or in pouncing on and trapping insects and other prey against the ground. The teeth and claws, which resemble the talons of modern predatory birds, may have been used to grasp prey. Wings would have been useful in providing stability during horizontal jumps when pursuing prey, and they would also have allowed flight over short distances. The benefits of such flight may have led eventually to wing-powered flight.

Diversity of Modern Birds

The fossil record shows that theropods, including birds like *Eoalulavis,* were flying around mid-Jurassic ecosystems as early as 130 mya. A great diversity of theropods was present 115 mya. Some ancient theropods had wings with claws. Others had narrow-long wings. Many had teeth. Some were large and flightless; others were adapted for swimming. These ancient theropods filled a variety of ecological niches. Most of the lineages that these fossils represent became extinct, along with (other) dinosaurs, at the end of the Mesozoic era. An asteroid impact, possibly in combination with volcanism in the Deccan Traps region of India, created cataclysmic atmospheric and climatic changes 66 mya (*see appendix B*). These climatic changes resulted in dinosaur extinctions, including the extinction of bird lineages.

The toothless birds that survived into the Tertiary period were ancestors of modern, toothless birds (Neornithes). These bird ancestors underwent a very rapid radiation. Modern birds are divided into two groups. Paleognathae is a superorder composed of large, flightless birds (ostriches, rheas, and others). Neognathae is the lineage that includes our modern flying bird species. The phylogeny of modern birds has been controversial. Classical taxonomy relied on characteristic behaviors, songs, anatomical differences, and ecological niches to distinguish the bird orders. Recent whole-genome molecular analyses of birds representing 32 of the 35 avian orders, including all 30 neognath orders, are helping to resolve the controversies (table 21.1).

Section 21.1 Thinking Beyond the Facts

Why do zoologists hypothesize that ancestors of modern birds were theropod dinosaurs?

21.2 EVOLUTIONARY PRESSURES

LEARNING OUTCOMES

1. Describe how avian reptile body systems are adapted to support the energy requirements and mechanics of flight.
2. Describe the physiology of flight from takeoff to soaring and steering.
3. Hypothesize on how the production of precocial and altricial young has contributed to the evolutionary success of the avian reptiles.

Virtually every body system of a bird shows some adaptation for flight. Endothermy, feathers, acute senses, long, flexible necks, and lightweight bones are a few of the many adaptations described in this section.

External Structure and Locomotion

The covering of feathers on a bird is called plumage. Feathers have two primary functions essential for flight. They form flight surfaces that provide lift and aid steering, and they prevent excessive heat loss, permitting endothermic maintenance of high metabolic rates. Feathers also have roles in courtship, incubation, and waterproofing.

Developmentally, there are two types of feathers. Both of these can be subdivided based on how each developmental type is modified for specific functions. Only a few of these modifications will be described.

The bodies of birds are covered by flattened, tightly closed feathers that create aerodynamic surfaces, such as those of wings and tails. These are called **pennaceous feathers** (figure 21.4*a*). These feathers have a prominent shaft or rachis from which barbs branch. Barbules branch from the barbs and overlap one another. Tiny hooks, called hamuli (sing., hamulus), interlock with grooves in adjacent barbules to keep the feather firm and smooth. Pennaceous

TABLE 21.1

TAXONOMY OF THE AVIAN REPTILES

There are 35 orders of birds. Selected orders shown below are grouped by major clades. Based on Jarvis ED. 2014. Whole-genome analyses resolve early branches in the tree of life of modern birds. Science 346 (6215):1321–31. Proposed reclassifications to this scheme will be formalized by 2018.

Class Aves (a′ves) (L. *avis*, bird)
Adaptations for flight include: fore appendages modified as feathered wings, endothermic, high metabolic rate, flexible neck, fused posterior vertebrae, and bones lightened by numerous air spaces. The skull is lightened by a reduction in bone and the presence of a horny bill that lacks teeth. The birds.

Superorder Palaeognathae (pal′e-og″nath-e)

Order Struthioniformes (stroo″the-on-i-for′mez)
Large, flightless birds; wings with numerous fluffy plumes. Ostriches and rheas.

Superorder Neognathae (ne-og′nath-e)

Galloanseres (gal″lo-an′ser-ez)

Order Anseriformes (an″ser-i-for′mez)
Waterfowl; South American screamers, ducks, geese, and swans; the latter three groups possess a wide, flat bill and an undercoat of dense down; webbed feet.

Order Galliformes (gal″li-for′mez)
Landfowl; Short beak; short, concave wings; strong feet and claws. Curassows, grouse, quail, pheasants, turkeys.

Neoaves (ne″o-a′ves)

Order Podicipediformes (pod″i-si-ped″i-for′mez)
Short wings; soft and dense plumage; feet webbed with flattened nails. Grebes.

Phoenicopteriformes (fen″i-kop-ter-i-for′mez)
Oval-shaped bodies with pink or crimson-red feathers; black flight feathers; exceptionally long legs and necks; large bills curve downward in the middle, upper bill is smaller than the lower bill. Flamingos.

Order Columbiformes (co-lum″bi-for′mez)
Dense feathers loosely set in skin; well-developed crop. Pigeons, doves, sandgrouse.

Order Cuculiformes (ku-koo″li-for′mez)
Reversible fourth toe; soft, tender skin. Plantaineaters, roadrunners, cuckoos.

Order Caprimulgiformes (kap″ri-mul″ji-for′mez)
Owl-like head and plumage, but weak bill and feet; beak with wide gape; insectivorous. Whip-poor-wills, other goatsuckers. Swifts and hummingbirds were formerly grouped in a separate order, Apodiformes. Molecular studies have resulted in their being grouped with other Caprimulgiformes.

Order Gruiformes (gru″i-for′mez)
Order characteristics variable and not diagnostic. Marsh birds, including cranes, limpkins, rails, coots.

Order Charadriiformes (ka-rad″re-i-for′mez)
Order characteristics variable. Shorebirds, gulls, terns, auks.

Order Gaviiformes (ga″ve-i-for′mez)
Strong, straight bill; diving adaptations include legs far back on body, bladelike tarsus, webbed feet, and heavy bones. Loons.

Order Pelecaniformes (pel″e-can-i-for′mez)
Four toes joined in common web; nostrils rudimentary or absent; large gular sac. Pelicans, boobies, cormorants, anhingas, frigate-birds. Herons and egrets were formly grouped in a separate order, Ciconiiformes. Molecular studies have resulted in their being grouped with other Pelecaniformes.

Order Procellariiformes (pro-sel-lar-e-i-for′mez)
Tubular nostrils, large nasal glands; long and narrow wings. Albatrosses, shearwaters, petrels.

Order Sphenisciformes (sfe-nis″i-for′mez)
Heavy bodied; flightless, flipperlike wings for swimming; well insulated with fat. Penguins.

Order Accipitriformes (ak-cipi″tri-for′mez)*
Diurnal birds of prey. Strong, hooked beak; large wings; raptorial feet. Distinguished from Falconiformes by molecular characteristics. Hawks, eagles, vultures.

Order Strigiformes (strij″i-for′mez)*
Large head with fixed eyes directed forward; raptorial foot. Owls.

(Two related orders and the remaining avian orders are continued on the following page. Spaces between orders represent divisions between major clades described by Jarvis [2014]).*

TABLE 21.1 Continued

Order Piciformes (pis″i-for′mez)
Usually long, strong beak; strong legs and feet with fourth toe permanently reversed in woodpeckers. Woodpeckers, toucans, honeyguides, barbets.
Order Coraciiformes (kor″ah-si″ah-for′mez)
Large head; large beak; metallic plumage. Kingfishers, todies, bee eaters, rollers.

Order Falconiformes (fal″ko-ni-for′mez)
Strong, hooked beak; large wings; raptorial feet. Falcons. Distinguished from Accipitriformes by molecular characteristics.
Order Psittaciformes (sit″ta-si-for′mez)
Maxilla hinged to skull; thick tongue; reversible fourth toe; usually brightly colored. Parrots, lories, macaws.
Order Passeriformes (pas″er-i-for′mez)
Largest avian order; 69 families of perching birds; perching foot; variable external features. Swallows, larks, crows, titmice, nuthatches, and many others.

feathers are modified for a number of functions. For example, **flight feathers** line the tip and trailing edge of the wing. They are asymmetrical with longer barbs on one side of the shaft. **Contour feathers** are usually symmetrical and line the body and cover the base of the flight feathers. They provide waterproofing, insulation, and streamlining.

Plumulaceous feathers have a rudimentary shaft to which a wispy tuft of barbs and barbules is attached. Plumulaceous feathers include insulating **down feathers,** which lie below contour feathers (figure 21.4*b*). Feather formation begins as epidermal cells proliferate, forming a feather papilla (figure 21.5*a*). Epidermal cells at the base of the feather papilla proliferate downward, forming a tubular follicle that grows into the dermis. Dermal pulp within the feather sheath supports capillaries that provide nourishment for feather development (figure 21.5*b*). Epidermal cells grow upward from a region called the epidermal collar. These cells become keratinized and form a protective outer feather sheath and inner feather barbs. As barbs develop from the collar they spiral around each other. Some barbs fuse to form the shaft of the feather (figure 21.5*c*). Other barbs develop pigmentation and contribute to the feather vane. The sheath distal to the follicle disintegrates, and what remains of the sheath within the follicle contributes to the calamus (figure 21.5*d*). In a plumulaceous feather, the barbs do not spiral and the shaft forms only at the base of the feather.

Birds maintain a clean plumage to rid the feathers and skin of parasites. Preening, which is done by rubbing the bill over the feathers, keeps feathers smooth, clean, and in place. Hamuli that become dislodged can be rehooked by running a feather through the bill. Secretions from an oil gland (uropygial gland) at the base of the tail of many birds are spread over the feathers during preening to keep the plumage water resistant and supple. The secretions also lubricate the bill and legs to prevent chafing. Anting is a maintenance behavior common to many songbirds and involves picking up ants in the bill and rubbing them over the feathers. Formic acid secreted by ants is toxic to feather mites.

Feather pigments deposited during feather formation produce most colors in a bird's plumage. Other colors, termed structural colors, arise from irregularities on the surface of the feather that diffract white light. For example, blue feathers are never blue because of the presence of blue pigment. A porous, nonpigmented outer layer on a barb reflects blue wavelengths of light. The other wavelengths pass into the barb and are absorbed by the dark pigment melanin. Iridescence results from the interference of light waves caused by a flattening and twisting of barbules. An example of iridescence is the perception of interchanging colors on the neck and back of hummingbirds (*Archilochus*) and grackles (*Quiscalus*). Color patterns are involved in cryptic coloration, species and sex recognition, and sexual attraction.

Mature feathers receive constant wear; thus, all birds periodically shed and replace their feathers in a process called **molting.** The timing of molt periods varies in different taxa. After hatching, a typical Northern Hemisphere songbird chick is covered with down. Juvenile feathers replace the down at the juvenile molt. A postjuvenile molt in the fall results in plumage similar to that of the adult. Once sexual maturity is attained, a prenuptial molt occurs in late winter or early spring, prior to the breeding season. A postnuptial molt usually occurs between July and October. Flight feathers are frequently lost in a particular sequence so that birds are not wholly deprived of flight during molt periods. However, many ducks, coots, and rails cannot fly during molt periods and hide in thick marsh grasses.

The Skeleton

The bones of most birds are lightweight yet strong. Some bones, such as the humerus (forearm bone), have large air spaces and internal strutting (reinforcing bony bars), which increase strength without adding weight (figure 21.6*c*). Birds also have a reduced number of skull bones, and a lighter, keratinized sheath called a bill replaces the teeth. The demand for lightweight bones for flight is countered in some birds with other requirements. For example, some aquatic birds (e.g., loons [*Gavia*]) have dense bones, which help reduce buoyancy during diving.

The appendages involved in flight cannot manipulate nesting materials or feed young. The bill and very flexible neck and feet

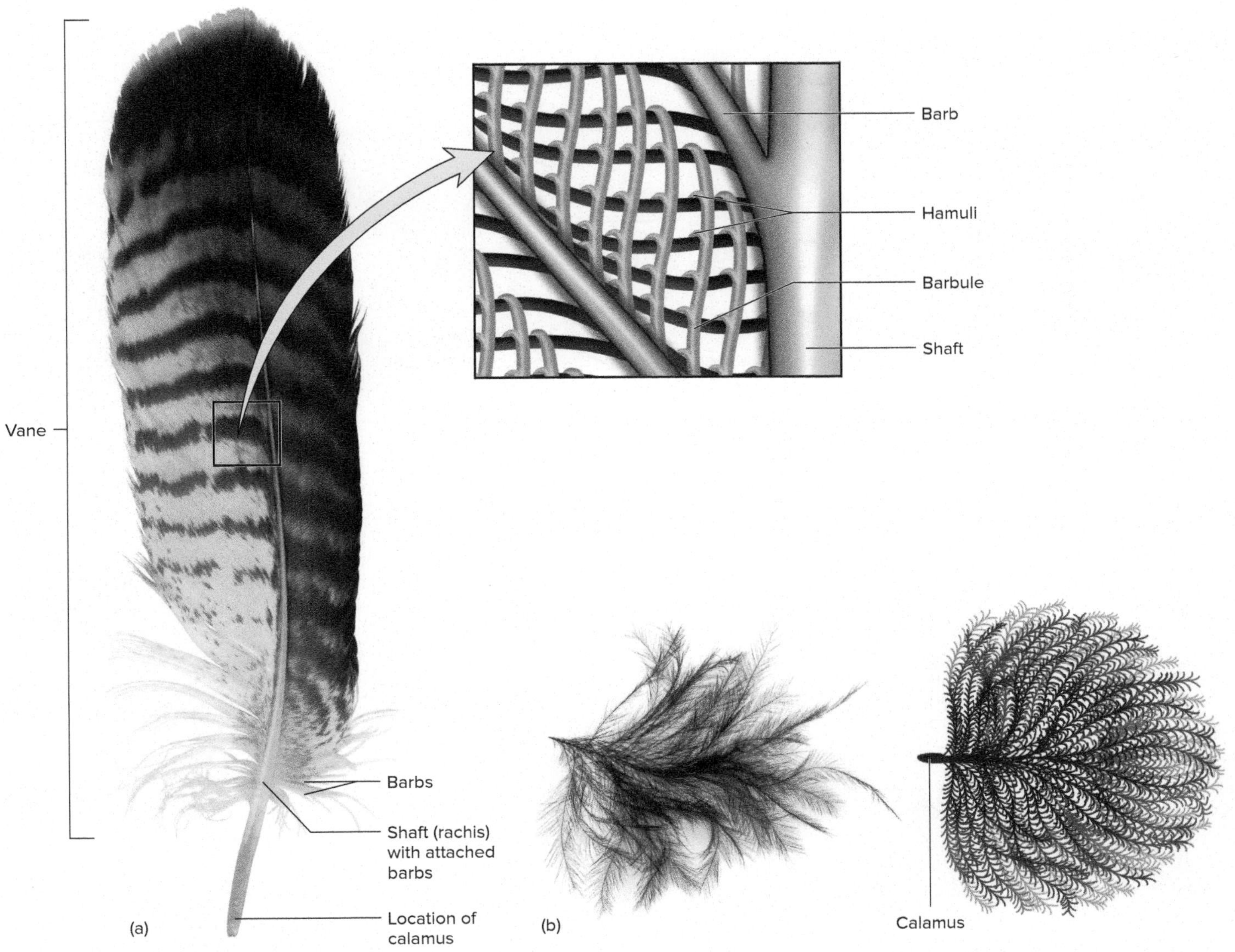

FIGURE 21.4

Developmental Feather Types and Anatomy. (*a*) Pennaceous feathers have a central shaft or rachis to which barbs attach. Barbs give rise to barbules that overlap. Small, hook-like hamuli are associated with barbules, and hamuli interlock with grooves on adjacent barbules. The result is a tightly closed vane that helps form the aerodynamic surfaces of birds. (*b*) Plumulaceous feathers have a rudimentary shaft. Barbs and barbules do not interlock, and they give the feather a wispy appearance. These include downy feathers that provide insulation for birds.

Source: Wynne, PJ. 2003. Which came first, the feather or the bird? Sci American, March 2003: 87–89.

make these activities possible. The cervical vertebrae have saddle-shaped articular surfaces that permit great freedom of movement. In addition, the first cervical vertebra (the atlas) has a single point of articulation with the skull (the occipital condyle), which permits a high degree of rotational movement between the skull and the neck. (The single occipital condyle is another characteristic that birds share with nonavian reptiles.) This flexibility allows the bill and neck to function as a fifth appendage.

The pelvic girdle, vertebral column, and ribs are strengthened for flight. The thoracic region of the vertebral column contains ribs, which attach to thoracic vertebrae. Most ribs have posteriorly directed uncinate processes that overlap the next rib to strengthen the rib cage (figure 21.6*a*). (These are also present on the ribs of most other reptiles further indicating common ancestry.) Posterior to the thoracic region is the lumbar region. The **synsacrum** forms by the fusion of the posterior thoracic vertebrae, all the lumbar and sacral vertebrae, and the anterior caudal vertebrae. Fusion of these bones helps maintain proper flight posture and supports the hind appendages during landing, hopping, and walking. The posterior caudal vertebrae are fused into a **pygostyle,** which helps support the tail feathers that are important in steering.

The sternum of most birds bears a large, median keel for attachment of flight muscles. (Exceptions to this include some flightless birds, such as ostriches.) Ribs keep the keel firmly

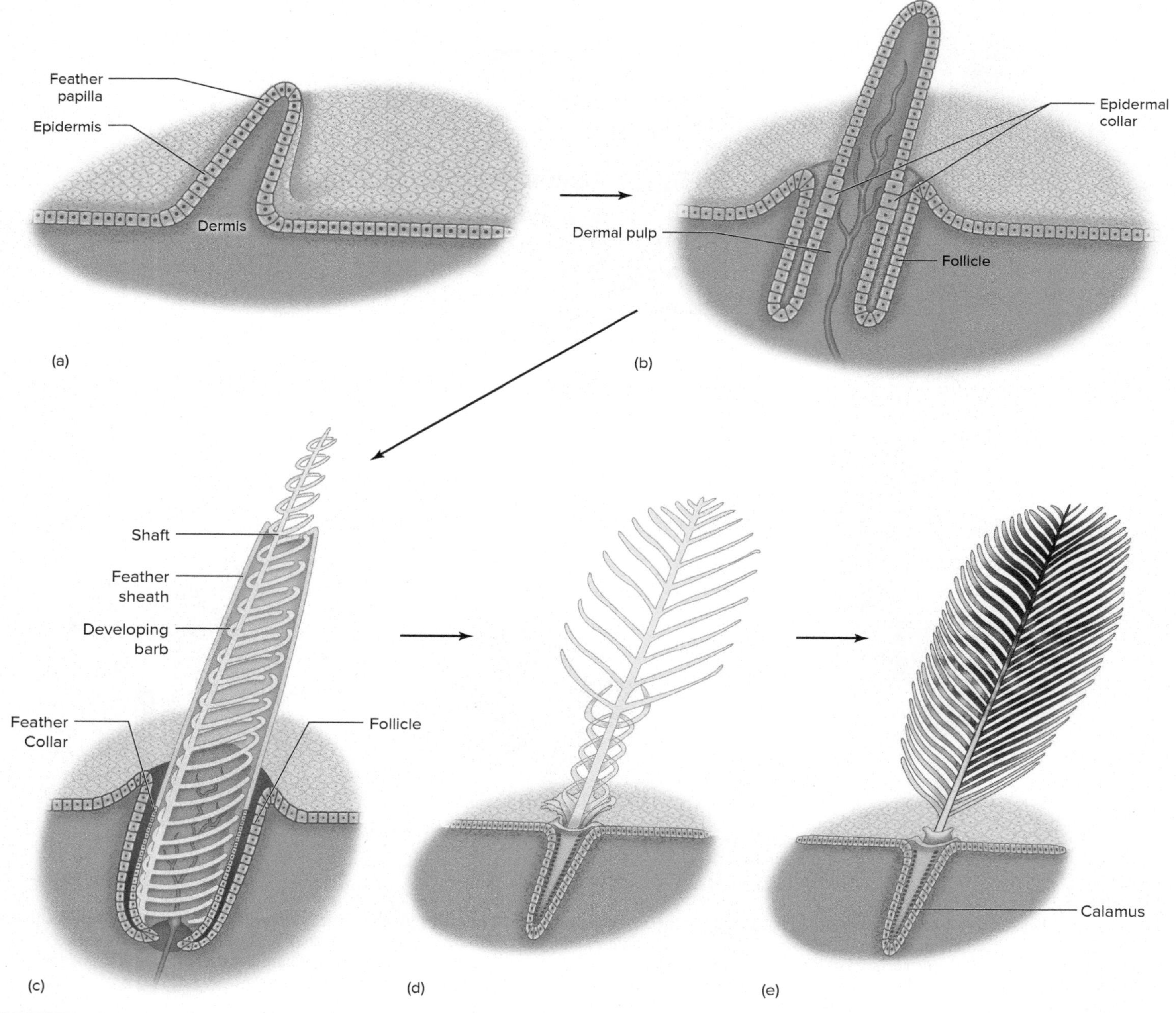

FIGURE 21.5

Formation of a Pennaceous Feather. (*a*) Epidermal cells proliferate, forming a feather papilla (bud). (*b*) Epidermal cells at the base of the feather papilla proliferate downward, forming a tubular feather follicle. Blood supply within the dermal pulp provides nutrients for feather development. Feather growth occurs from a ringlike group of follicular cells called the epidermal collar. (*c*) Epidermal collar cells grow upward, forming a keratinized feather sheath and feather barbs. Barbs from the collar spiral around each other. Some barbs fuse to form the shaft (rachis). Other barbs contribute to the feather vane. (*d, e*) The sheath distal to the follicle disintegrates. The sheath within the follicle contributes to the calamus. Pigmentation of the feather develops within maturing epidermal cells. The formation of a plumulaceous feather is similar to that shown here, except barbs do not spiral and the shaft forms only at the base of the feather.

Source: Wynne, PJ. 2003. Which came first, the feather or the bird? Sci American, March 2003: 87–89.

attached to the rest of the axial skeleton. Paired clavicles fuse medially and ventrally into a **furcula** (L. *furcae*, little fork; commonly called the wishbone). The furcula braces the pectoral girdle against the sternum and serves as an additional site for the attachment of flight muscles.

The appendages of birds have also been modified. Some bones of the front appendages have been lost or fused, and they are points of attachment for flight feathers. The rear appendages are used for hopping, walking, running, and perching. Perching tendons run from the toes across the back of the ankle joint to muscles of the lower

How Do We Know about Feather Evolution?

The sequence of events in feather development parallels feather evolution in theropod dinosaurs. The earliest feathers, such as those found in *Sinosauropteryx,* were mostly small and tubular. Plumulaceous feathers probably evolved before pennaceous feathers because in development, the shaft forms from the fusion of barbs. Some small tufted feathers are found on *Sinosauropteryx* and other theropods. Symmetrical pennaceous feathers are seen in fossils of *Caudipteryx* and other theropods. Finally, asymmetrical feathers similar to bird flight feathers are seen in *Microraptor*. Molecular data also support how evolution may have produced the sequence implied here. Feather development is controlled by *Hox* genes in a fashion similar to limb and digit development (*see chapter 18, Evolutionary Insights*). Feather development involves a sequence of genetic events in which one event depends on the previous event. These genetic events direct a sequence that proceeds from tubular feathers, to plumulaceous feathers, and finally to pennaceous feathers. Modest changes in pattern-forming *Hox* genes may be responsible for an evolutionary sequence of feather types. Interestingly, these same genes have been found in alligators (also in the archosaurian lineage). This discovery suggests that feathers, or feather prototypes, may be much older than has been projected up to this point.

leg. When the ankle joint is flexed, as in landing on a perch, tension on the perching tendons increases, and the foot grips the perch (figure 21.6*b*). This automatic grasp helps a bird perch even while sleeping. The muscles of the lower leg can increase the tension on these tendons, for example, when an eagle grasps a fish in its talons.

Muscles

The largest, strongest muscles of most birds are the flight muscles. They attach to the sternum and clavicles and run to the humerus. The muscles of most birds are adapted physiologically for flight. Flight muscles must contract quickly and fatigue very slowly. These muscles have many mitochondria and produce large quantities of ATP to provide the energy required for flight, especially long-distance migrations. Domestic fowl have been selectively bred for massive amounts of muscle ("white meat") that humans like as food. However, it is poorly adapted for flight because it lacks blood flow and contains few mitochondria.

Flight

The wings of birds are adapted for different kinds of flight. However, regardless of whether a bird soars, glides, or has a rapid flapping flight, the mechanics of staying aloft are similar. Bird wings form an **airfoil.** The anterior margin of the wing is thicker than the posterior margin. The upper surface of the wing is slightly convex, and the lower surface is flat or slightly concave. Air passing over the wing travels farther and faster than air passing under the wing, decreasing air pressure on the upper surface of the wing and creating lift (figure 21.7*a*). The lift wings create must overcome the bird's weight, and the forces that propel the bird forward must overcome drag that the bird moving through the air creates. Increasing the angle that the leading edge of the wing makes with oncoming air (the angle of attack)

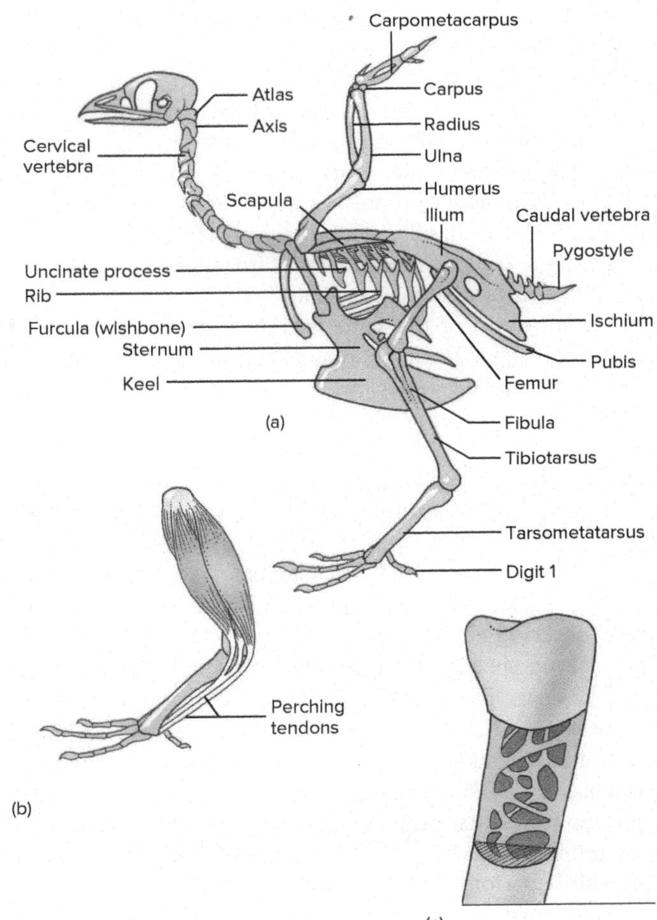

FIGURE 21.6

Bird Skeleton. (*a*) Skeleton of a pigeon. (*b*) Perching tendons run from the toes across the back of the ankle joint and cause the foot to grip a perch. (*c*) Internal structure of the humerus. Note the air spaces in this pneumatic bone.

increases lift. As the angle of attack increases, however, the flow of air over the upper surface becomes turbulent, reducing lift (figure 21.7*b*). Turbulence is reduced if air can flow rapidly through slots at the leading edge of the wing. Slotting the feathers at the wing tips and the presence of an alula on the anterior margin of the wing reduce turbulence. The **alula** is a group of small feathers supported by bones of the medial digit. During takeoff, landing, and hovering flight, the angle of attack increases, and the alula is elevated (figure 21.7*c and e*). During soaring and fast flight, the angle of attack decreases, and slotting is reduced.

The distal part of the wing generates most of the propulsive force of flight. Because it is farther from the shoulder joint, the distal part of the wing moves farther and faster than the proximal part of the wing. During the downstroke (the powerstroke), the leading edge of the distal part of the wing is oriented slightly downward and creates a thrust somewhat analogous to the thrust that an airplane propeller creates (figure 21.7*d*). During the upstroke (the recovery stroke), the distal part of the wing is oriented upward to decrease resistance. Feathers on a wing overlap so that, on the downstroke, air presses the feathers at the wing margins together, allowing little air to pass between them, enhancing both lift and propulsive forces. On the upstroke, wings fold inward slightly. Also, feathers part slightly, allowing air to pass between them, which reduces resistance during this recovery stroke.

The tail of a bird serves a variety of balancing, steering, and braking functions during flight. It also enhances lift that wings produce during low-speed flight. During horizontal flight, spreading the tail feathers increases lift at the rear of the bird and causes the head to dip for descent. Closing the tail feathers has the opposite effect. Tilting the tail sideways turns the bird. When a bird lands, its tail deflects downward, serving as an air brake.

Different birds, or the same bird at different times, use different kinds of flight. During gliding flight, the wing is stationary, and a bird loses altitude. Waterfowl coming in for a landing use gliding flight. Flapping flight generates the power for flight and is the most common type of flying. Many variations in wing shape and flapping patterns result in species-specific speed and maneuverability. Soaring flight allows some birds to remain airborne with little energy expenditure. During soaring, wings are essentially stationary, and the bird uses updrafts and air currents to gain altitude. Hawks (e.g., *Buteo*), vultures (e.g., *Cathartes*), and other soaring birds are frequently observed circling along mountain valleys, soaring downwind to pick up speed and then turning upwind to gain altitude. As the bird slows and begins to lose altitude, it turns downwind again. The wings of many soarers are wide and slotted to provide maximum maneuverability at relatively low speeds. Oceanic soarers, such as albatrosses (Diomedeidae) and frigate birds (*Fregata*), have long, narrow wings that provide maximum lift at high speeds, but they compromise maneuverability and ease of takeoff and landing. Hummingbirds perform hovering flight. They hover in still air by fanning their wings back and forth (50 to 80 beats per second) to remain suspended in front of a flower or feeding station. The wings move in a figure-eight pattern. As they move, the wings are flipped from right-side-up to upside-down, and lift is generated from both the top and bottom sides of the wings.

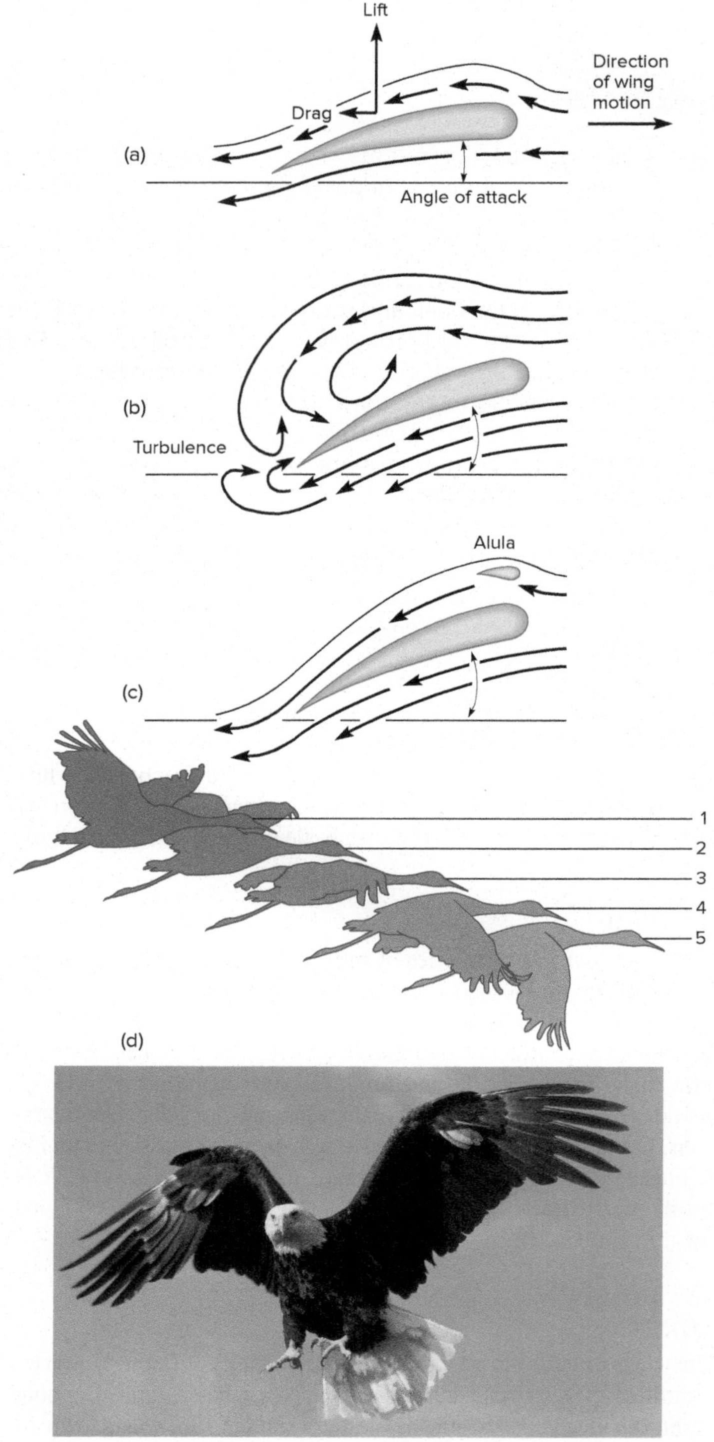

FIGURE 21.7

Mechanics of Bird Flight. (*a*) A bird's wing acts as an airfoil. Air passing over the top of the wing travels farther and faster than air passing under the wing, creating lift. (*b*) Increasing the angle of attack increases lift but also increases turbulence. (*c*) The alula reduces turbulence. (*d*) Wing orientation during a downstroke. (*e*) Note the alula on the right wing of this bald eagle (*Haliaeetus leucocephalus*) in hovering flight.

Nutrition and the Digestive System

Most birds have ravenous appetites! This appetite supports a high metabolic rate that makes endothermy and flight possible. For example, hummingbirds feed almost constantly during the day. In spite of high rates of food consumption, they often cannot sustain their rapid metabolism overnight, and they may become torpid, with reduced body temperature and respiratory rate, until they can feed again in the morning.

FIGURE 21.8

Bird Flight and Feeding Adaptations. This male ruby-throated hummming bird (*Archilochus colubris*) hovers while feeding on flower nectar. The ruby throat of the male glimmers in bright light but appears dark in indirect light. Hummingbird bills often match the length and curvature of the flower from which the birds extract nectar. The ruby-throated hummingbird, however, is a generalist that feeds at about 30 species of flowers. Its distribution ranges throughout eastern North America, including southern Canada, into Central America.
©Daniel Dempster Photography/Alamy

Bird bills and tongues are modified for a variety of feeding habits and food sources (figures 21.8 and 21.9). For example, a woodpecker's tongue is barbed for extracting grubs from the bark of trees (*see figure 27.3*d). Sapsuckers excavate holes in trees and use a brushlike tongue for licking the sap that accumulates in these holes. The tongues of hummingbirds and other nectar feeders roll into a tube for extracting nectar from flowers. Modifications of the bill reflect dietary habits of the various species. For example, the bill of an eagle is modified for tearing prey (hooked), the bill of a cardinal is specialized for cracking seeds, and the bill of a flamingo is used to strain food from the water.

In many birds, a diverticulum of the esophagus, called the crop, is a storage structure that allows birds to quickly ingest large quantities of locally abundant food. They can then seek safety while digesting their meal. The crop of pigeons (Columbidae) produces "pigeon's milk," a cheesy secretion formed by the proliferation and sloughing of cells lining the crop. Young pigeons (squabs) feed on pigeon's milk until they are able to eat grain. Cedar waxwings (*Bombycilla cedrorum*), vultures, and birds of prey use their esophagus for similar storage functions. Crops are less well developed in insect-eating birds because insectivorous birds feed throughout the day on sparsely distributed food.

The stomach of birds is modified into two regions. The proventriculus secretes gastric juices that initiate digestion (figure 21.10). The ventriculus (gizzard) has muscular walls to abrade and crush seeds or other hard materials. Birds may swallow sand and other abrasives to aid digestion. The bulk of enzymatic digestion and absorption occurs in the small intestine, aided by secretions from

(a)

(b)

(c)

FIGURE 21.9

Some Specializations of Bird Bills. (*a*) The bill of a bald eagle (*Haliaeetus leucocephalus*) is specialized for tearing prey. (*b*) The thick, powerful bill of this cardinal (*Cardinalis cardinalis*) cracks tough seeds. (*c*) The bill of a flamingo (*Phoenicopterus ruber*) strains food from the water in a head-down feeding posture. Large bristles fringe the upper and lower mandibles. As water is sucked into the bill, larger particles are filtered and left outside. Inside the bill, tiny inner bristles filter smaller algae and animals. The tongue removes food from the bristles.
(a) ©momnjax/Getty Images (b) Source: U.S. Fish and Wildlife Service/NCTC Image Library (c) ©Ed Reschke/Getty Images

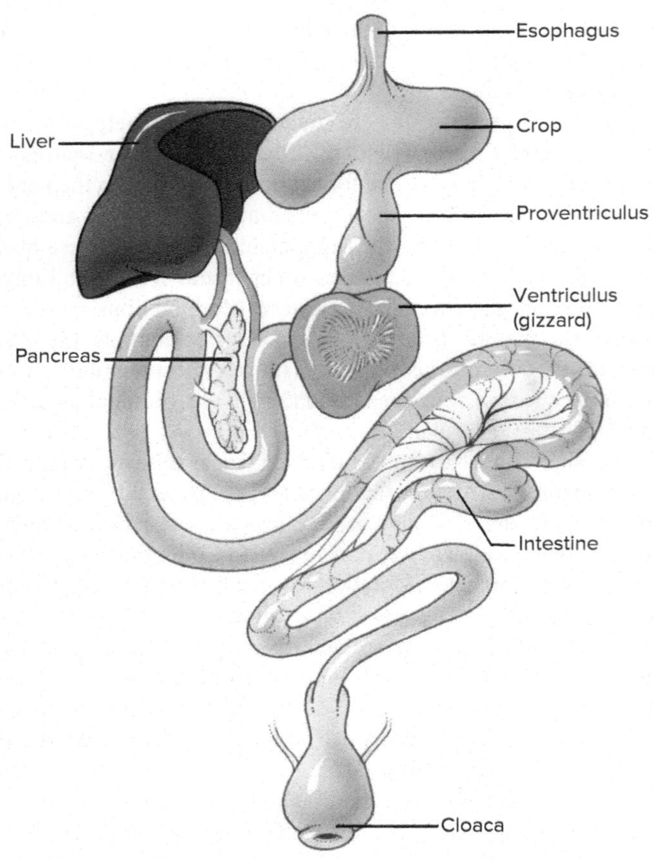

FIGURE 21.10

Digestive System of a Pigeon. Birds have high metabolic rates that require a nearly constant nutrient supply.

the pancreas and liver. Paired ceca may be located at the union of the large and small intestine. These blind-ending sacs contain bacteria that aid in cellulose digestion. Birds usually eliminate undigested food through the cloaca; however, owls form pellets of bone, fur, and feathers that are ejected from the ventriculus through the mouth. Owl pellets accumulate in and around owl nests and are useful in studying their food habits.

Birds are often grouped by their feeding habits. These groupings are somewhat artificial, however, because birds may eat different kinds of food at different stages in their life history, or they may change diets simply because of changes in food availability. American robins (*Turdus migratorius*), for example, feed largely on worms and other invertebrates when these foods are available. In the winter, however, robins may feed on berries.

In some of their feeding habits, birds directly conflict with human interests. Bird damage to orchard and grain crops is tallied in the millions of dollars each year. Flocking and roosting habits of some birds, such as European starlings (*Sturnus vulgaris*) and red-winged blackbirds (*Agelaius phoeniceus*), concentrate millions of birds in local habitats, where they devastate fields of grain. Recent monocultural practices tend to aggravate problems with grain-feeding birds by encouraging the formation of very large flocks.

Birds of prey have minimal impact on populations of poultry and game birds, and on commercial fisheries. Unfortunately, the mistaken impression that they are responsible for significant losses has led humans to poison and shoot them.

Circulation, Gas Exchange, and Temperature Regulation

Like the crocodylian heart, the avian heart has completely separated atria and ventricles, resulting in separate pulmonary and systemic circuits. This separation prevents any mixing of highly oxygenated blood with less oxygenated blood. In vertebrate evolution, the sinus venosus has gradually decreased in size. It is a separate chamber in fishes, amphibians, and turtles and receives blood from the venous system. In other reptiles, it is a group of cells in the right atrium that serves as the pacemaker for the heart. In birds, the sinus venosus also persists only as a patch of pacemaker tissue in the right atrium. The bird heart is relatively large (up to 2.4% of total body weight), and it beats rapidly. Rates in excess of 1,000 beats per minute have been recorded for hummingbirds under stress. Larger birds have relatively smaller hearts and slower heart rates. The heart rate of an ostrich (*Struthio*), for example, varies between 38 and 176 beats per minute. A large heart, rapid heart rate, and complete separation of highly oxygenated blood from less oxygenated blood are important adaptations for delivering the large quantities of blood required for endothermy and flight.

Gas Exchange

The respiratory system of birds is extremely complex and efficient. It consists of external nares, which lead to nasal passageways and the pharynx. Bone and cartilage support the trachea. A special voice box, called the **syrinx,** is located where the trachea divides into bronchi. The muscles of the syrinx and bronchi, as well as the characteristics of the trachea, produce bird vocalizations. The bronchi lead to a complex system of air sacs that occupy much of the body and extend into some of the bones of the skeletal system (figure 21.11*a*). The air sacs and bronchi connect to the lungs. The lungs of birds are made of small (400 μm) air tubes called **parabronchi.** Air capillaries about 10 μm in diameter branch from the parabronchi and are associated with capillary beds for gas exchange (figure 21.11c).

Inspiration and expiration result from increasing and decreasing the volume of the thorax and from alternate expansion and compression of air sacs during flight and other activities. During breathing, the movement of the sternum and the posterior ribs compresses the thoracic air sacs. X-ray movies of European starlings in a wind tunnel show that the contraction of flight muscles distorts the furcula. Alternate distortion and recoiling helps compress and expand air sacs between the bone's two shafts.

It takes two ventilatory cycles to move a particular volume of air through the respiratory system of a bird. During the first inspiration, air moves into the abdominal air sacs. At the same time, air already in the lungs moves through parabronchi into the thoracic air sacs (figure 21.11*b*, cycle 1). During expiration, the air in the thoracic air sacs moves out of the respiratory system, and the air in the abdominal air sacs moves into parabronchi. At the second

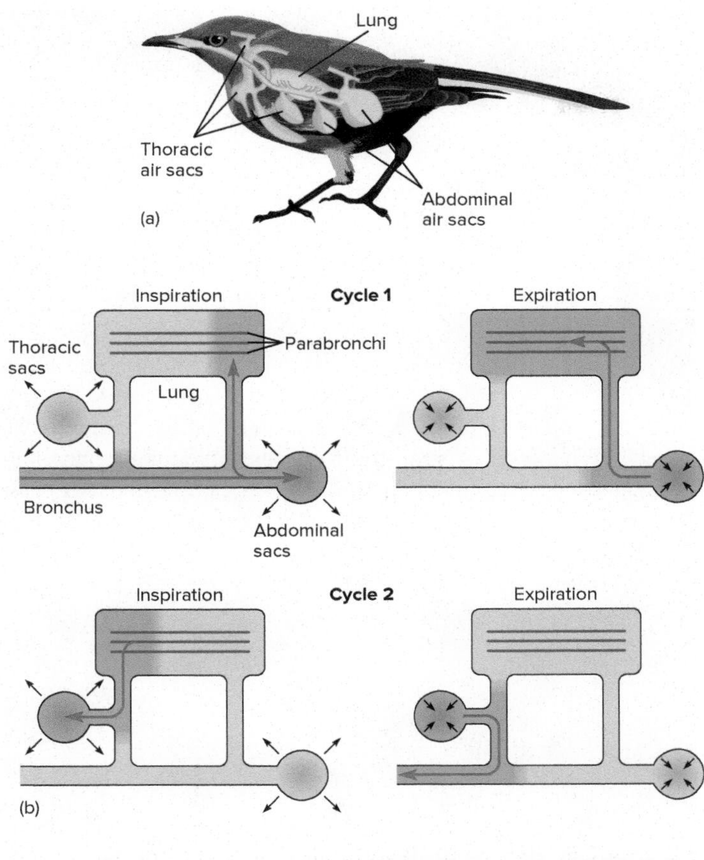

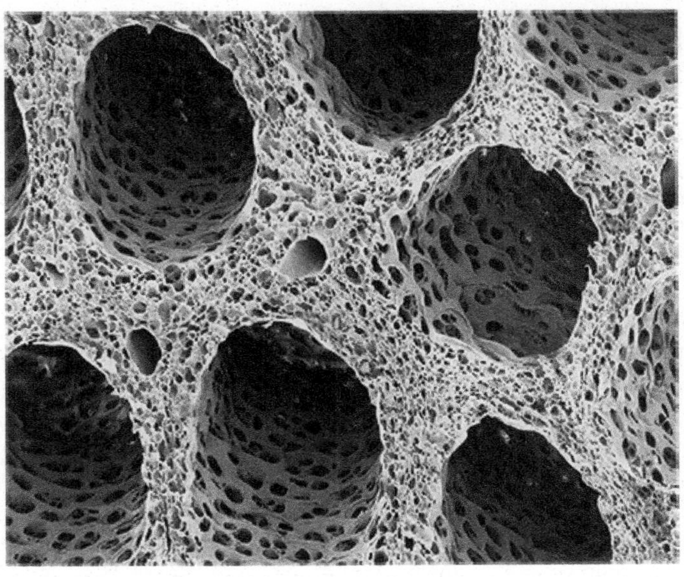

FIGURE 21.11

Respiratory System of a Bird. (*a*) Air sacs branch from the respiratory tree. (*b*) Air flow during inspiration and expiration. Air flows through the parabronchi during both inspiration and expiration. The shading represents the movement through the lungs of one inspiration. (*c*) Scanning electron micrograph of parabronchi. Parabronchi are approximately 400 μm in diameter. Tiny air capillaries branch from parabronchi. Blood capillaries are shown in cross section between parabronchi.

inspiration (figure 21.11*b*, cycle 2), the air moves into the thoracic air sacs, and it is expelled during the next expiration.

Because of high metabolic rates associated with flight, birds have a greater rate of oxygen consumption than any other vertebrate. When other tetrapods inspire and expire, air passes into and out of respiratory passageways in a simple back-and-forth cycle. Ventilation is interrupted during expiration, and much "dead air" (air not forced out during expiration) remains in the lungs. Because of the unique system of air sacs and parabronchi, bird lungs have a nearly continuous movement of oxygen-rich air over respiratory surfaces during both inspiration and expiration. The quantity of "dead air" in the lungs is sharply reduced, compared with other vertebrates. Interestingly, studies have found evidence of a similar unidirectional flow of air through crocodylian lung spaces. Although the systems in birds and crocodylians differ in details (crocodylians lack air sacs and pneumatic bones), this finding suggests that unidirectional lung ventilation may have evolved early in the archosaur lineage. Some scientists hypothesize that efficient gas exchange may have contributed to the survival of the archosaurs through the mass extinction that killed other dinosaurs.

This avian system of gas exchange is more efficient than that of any other tetrapod. In addition to supporting high metabolic rates, this efficient gas exchange system probably also explains how birds can live and fly at high altitudes, where oxygen tensions are low. During their migrations, bar-headed geese (*Anser indicus*) fly over the peaks of the Himalayas at altitudes of 9,200 m. A human can begin to feel symptoms of altitude sickness at 2,500 m. Experienced mountain climbers can function at altitudes up to about 7,500 m without auxiliary oxygen supplies.

Thermoregulation

Birds maintain body temperatures between 38 and 45°C. Lethal extremes are lower than 32 and higher than 47°C. In cold environments heat must be conserved and generated through metabolic processes. On a cold day, a resting bird fluffs its feathers to increase their insulating properties by increasing the dead air spaces within them. It also tucks its bill into its feathers to reduce heat loss from the respiratory tract. The most exposed parts of a bird are the feet and tarsi, which have neither fleshy muscles nor a rich blood supply. Temperatures in these extremities are allowed to drop near freezing to prevent heat loss. Countercurrent heat exchange between the warm blood flowing to the legs and feet, and the cooler blood flowing to the body core from the legs and feet, prevents excessive heat loss at the feet. Heat is returned to the body core before it goes to the extremities and is lost to the environment (*see figure 28.5*b). Shivering also generates heat in extreme cold. Increases in metabolism during winter months require additional food.

Birds may also need to dissipate excess heat. Flight produces large quantities of heat. Unlike mammals, birds do not have sweat glands, but they still rely on evaporation of water for cooling. Birds pant to dissipate heat through the respiratory tract. Some birds enhance evaporative cooling through "gular flutter," which is the rapid vibration of the upper throat and the floor of the mouth.

Some birds conserve energy on cool nights by allowing their body temperatures to drop. For example, whip-poor-wills

(*Antrostomus*) allow their body temperatures to drop from about 40 to near 16°C, and respiratory rates become very slow.

Nervous and Sensory Systems

A mouse, enveloped in the darkness of night, skitters across the floor of a barn. An owl in the loft overhead turns in the direction of the faint sounds the tiny feet make. As the sounds made by hurrying feet change to a scratchy gnawing of teeth on a sack of feed, the barn owl dives for its prey (figure 21.12). Fluted tips of flight feathers make the owl's approach imperceptible to the mouse, and the owl's ears, not its eyes, guide it to its prey. Barn owls successfully locate and capture prey in more than 75% of attempts! This ability is just one example of the many sensory adaptations of birds.

The forebrain of birds is much larger than that of nonavian reptiles due to the enlargement of the cerebral hemispheres, including a region of gray matter, the corpus striatum. The corpus striatum functions in visual learning, feeding, courtship, and nesting. A pineal body on the roof of the forebrain appears to stimulate ovarian development and to regulate other functions influenced by light and dark periods. The optic tectum (the roof of the midbrain), along with the corpus striatum, plays an important role in integrating sensory functions. The midbrain also receives sensory input from the eyes. As in other reptiles, the hindbrain of birds includes the cerebellum and the medulla oblongata, which coordinate motor activities and regulate heart and respiratory rates, respectively.

Vision is an important sense for most birds. The structures of bird eyes are similar to those of other vertebrates, but bird eyes are much larger relative to body size than those of other vertebrates (*see figure 19.13*). The eyes are usually somewhat flattened in an anteroposterior direction; however, the eyes of birds of prey protrude anteriorly because of a bulging cornea. Birds have a unique double-focusing mechanism. Padlike structures (similar to those of nonavian reptiles) control the curvature of the lens, and ciliary muscles change the curvature of the cornea. Double, nearly instantaneous focusing allows an osprey or other bird of prey to remain focused on a fish throughout a brief, but breathtakingly fast, descent.

The retina of a bird's eye is thick and contains both rods and cones. Rods are active under low light intensities and cones are active under high light intensities. Cones are especially concentrated ($1,000,000/mm^2$) at a focal point called the fovea. Unlike other vertebrates, some birds have two foveae per eye. The one at the center of the retina is sometimes called the "search fovea" because it gives the bird a wide angle of monocular vision. The other fovea is at the posterior margin of the retina. It functions with the posterior fovea of the other eye to allow binocular vision. The posterior fovea is called the "pursuit fovea" because binocular vision produces depth perception, which is necessary to capture prey. The words "search" and "pursuit" are not meant to imply that only predatory birds have these two foveae. Other birds use the "search fovea" to observe the landscape below them during flight and the "pursuit" fovea when depth perception is needed, as in landing on a branch of a tree.

The position of the eyes on the head also influences the degree of binocular vision (figure 21.13). Pigeons have eyes located well back on the sides of their heads, giving them a nearly 360° monocular field, but a narrow binocular field. They do not have to pursue their food (grain), and a wide monocular field of view helps them stay alert to predators while feeding on the ground. Hawks and owls have eyes farther forward on the head. This increases their binocular field of view and correspondingly decreases their monocular field of view.

FIGURE 21.12

Barn Owl (*Tyto alba*). A keen sense of hearing and large eyes that provide excellent night vision allow barn owls to find prey in spite of the darkness of night.
©Fuse/Getty Images

Like other reptiles, birds have a nictitating membrane that is drawn over the surface of the eye to cleanse and protect the eye.

Olfaction apparently plays a minor role in the lives of most birds. External nares open near the base of the beak, but the olfactory epithelium is poorly developed. Exceptions include vultures (in the genus *Cathartes*), which locate their dead and dying prey largely by smell.

In contrast, most birds have well-developed hearing. Loose, delicate feathers called auriculars cover the external ear opening. Middle- and inner-ear structures are similar to those of nonavian reptiles. The sensitivity of the avian ear (100 to 15,000 Hz) is similar to that of the human ear (16 to 20,000 Hz).

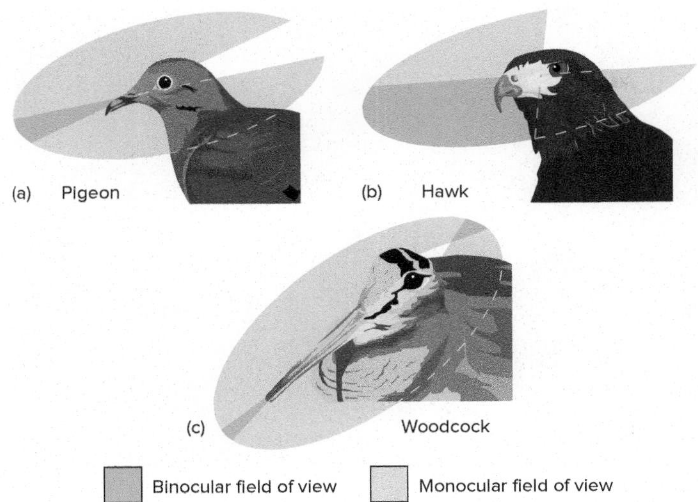

FIGURE 21.13

Avian Vision. The fields of view of (*a*) a pigeon, (*b*) a hawk, and (*c*) a woodcock. Woodcocks have eyes located far posteriorly and have a narrow field of binocular vision in front and behind. They can focus on predators circling above them while probing mud with their long beaks.

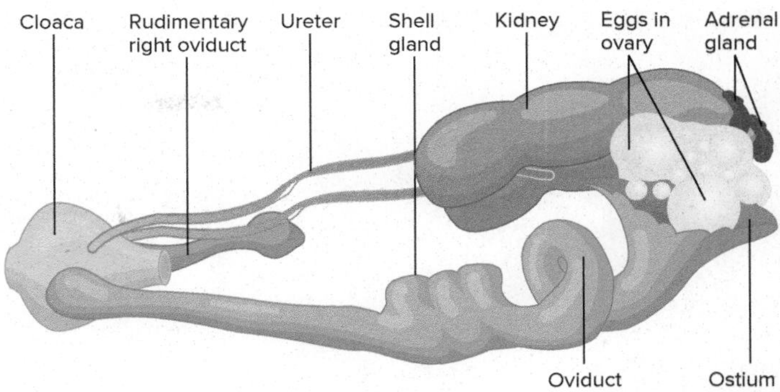

FIGURE 21.14

Urogenital System of a Female Pigeon. The right ovary and oviduct are rudimentary in most female birds.

Excretion and Osmoregulation

Birds and nonavian reptiles face essentially identical excretory and osmoregulatory demands. Like other reptiles, birds excrete uric acid, which is temporarily stored in the cloaca. Water is also reabsorbed in the cloaca. As with nonavian reptiles, the excretion of uric acid conserves water and promotes embryo development in terrestrial environments. In addition, some birds have supraorbital salt glands that drain excess sodium chloride through the nasal openings to the outside of the body (*see figure 28.17*). These are especially important in marine birds that drink seawater and feed on invertebrates containing large quantities of salt in their tissues. Salt glands can secrete salt in a solution that is two to three times more concentrated than other body fluids. Salt glands, therefore, compensate for the kidney's inability to concentrate salts in the urine. Similar glands are also present in certain nonavian reptiles that inhabit marine waters.

Reproduction and Development

The sexual activities of birds have been observed more closely than those of any other group of animals. These activities include establishing territories, finding mates, constructing nests, incubating eggs, and feeding young.

All birds are oviparous. Gonads are in the dorsal abdominal region, next to the kidneys. Testes are paired, and coiled tubules (vasa deferentia) conduct sperm to the cloaca. An enlargement of the vasa deferentia, the seminal vesicle, is a site for the temporary storage and maturation of sperm prior to mating. Testes enlarge during the breeding season. Except for certain waterfowl and ostriches, birds have no intromittent organ, and sperm are transferred by cloacal contact when the male briefly mounts the female.

In females, two ovaries form during development, but usually only the left ovary fully develops (figure 21.14). A large, funnel-shaped opening (the ostium) of the oviduct envelops the ovary and receives eggs after ovulation. The egg is fertilized in the upper portions of the oviduct, and albumen that glandular regions of the oviduct wall secrete gradually surrounds the zygote as it completes its passage. A shell gland in the lower region of the oviduct adds a shell and possibly pigmentation. The oviduct opens into the cloaca.

Many birds establish territories prior to mating. Although size and function vary greatly among species, territories generally allow birds to mate without interference. They provide nest locations and sometimes food resources for adults and offspring. Breeding birds defend their territories and expel intruders of the same sex and species. Threats are common, but actual fighting is minimal.

Mating may follow the attraction of a mate to a territory. For example, male woodpeckers (Picidae) drum on trees to attract females. Male ruffed grouse (*Bonasa umbellus*) fan their wings on logs and create sounds that can be heard for many miles. Cranes have a courtship dance that includes stepping, bowing, stretching, and jumping displays. Mating occurs when a mate's call or posture signals readiness. It happens quickly but repeatedly to assure fertilization of all the eggs that will be laid (*see chapter 21 opening photograph*).

Most birds are **monogamous.** A single male pairs with a single female during the breeding season. Some birds (swans, geese, and eagles) pair for life. Frequent mating apparently strengthens the pair bonds that develop. Monogamy is common when resources are widely and evenly distributed, and one bird cannot control the access to resources. Monogamy is also advantageous because both parents usually participate in nest building and care of the young. One parent incubates and protects the eggs or chicks while the other searches for food.

Some birds are **polygynous.** Males mate with more than one female, and the females care for the eggs and chicks. Polygyny tends to occur in species whose young are less dependent at hatching and in situations where patchy resource distribution may attract many females to a relatively small breeding area. Prairie chickens are polygynous, and males display in groups called leks. In prairie chicken leks, the males in the center positions are preferred and attract the majority of females (figure 21.15).

FIGURE 21.15

Courtship Displays. A male greater prairie chicken (*Tympanuchus cupido*) displaying in a lek.
Source: George Lavendowski/USFWS

A few bird species are **polyandrous,** and the females mate with more than one male. For example, female spotted sandpipers (*Actitis macularius*) are larger than males, and they establish and defend their territories from other females. They lay eggs for each male that is attracted to and builds a nest in their territory. If a male loses his eggs to a predator, the female replaces them. Polyandry results in the production of more eggs than in monogamous matings. It is thought to be advantageous when food is plentiful but, because of predation or other threats, the chances of successfully rearing young are low.

Nest construction usually begins after pair formation. The female usually initiates this instinctive behavior. A few birds do not make nests. Emperor penguins (*Aptenodytes forsteri*), for example, breed on the snow and ice of Antarctica, where no nest materials are available. Their single egg is incubated on the web of the foot (mostly the male's foot), tucked within a fold of abdominal skin.

(a)

(b)

FIGURE 21.16

Altricial and Precocial Chicks. (*a*) An American robin (*Turdus migratorius*) feeding nestlings. Robins have altricial chicks that are helpless at hatching. (*b*) Killdeer (*Charadrius vociferus*) have precocial chicks that are down covered and can move about.
(a) ©ivkuzmin/Getty Images (b) ©Flirt/Alamy

Nesting Activities

The nesting behavior of birds is often species specific. Some birds choose nest sites away from other members of their species, and other birds nest in large flocks. Unfortunately, predictable nesting behavior has led to the extinction of some species of birds.

The group of eggs laid and chicks produced by a female is called a **clutch.** Clutch size usually varies. Most birds incubate their eggs, and some birds have a featherless, vascularized incubation or brood patch (*see figure 25.13*) that helps keep the eggs at temperatures between 33 and 37°C. Birds turn the eggs to prevent egg membranes from adhering in the egg and deforming the embryo. Adults of some species sprinkle the eggs with water to cool and humidify them. The Egyptian plover (*Pluvianus aegyptius*) carries water from distant sites in the breast feathers. The incubation period lasts between 10 and 80 days and correlates with egg size and degree of development at hatching. One or two days before hatching, the young bird penetrates an air sac at the blunt end of its egg, inflates its lungs, and begins breathing. Hatching occurs as the young bird pecks the shell with a keratinized egg tooth on the tip of the upper jaw and struggles to free itself.

Some birds are helpless at hatching; others are more independent. Those that are entirely dependent on their parents are said to be **altricial** (L. *altricialis,* to nourish), and they are often naked at hatching (figure 21.16*a*). Altricial young must be brooded constantly at first because endothermy is not developed. They grow rapidly, and when they leave the nest, they are nearly as large as their parents. For example, American robins weigh 4 to 6 g at hatching and leave the nest 13 days later weighing 57 g. **Precocial** (L. *pracoci,* early ripe) young are alert and lively at hatching (figure 21.6*b*). They are usually covered with down and can walk, run, swim, and feed themselves–although one parent is usually present to lead the young to food and shelter.

Young altricial birds have huge appetites and keep one or both parents continually searching for food. They may consume a mass of food that equals their own weight each day. Adults bring food to the nest or regurgitate food stored in the crop or esophagus. A fixed action pattern (a predictable behavioral response by a parent to a behavioral stimulus initiated by the young) initiates feeding in certain avian species. Vocal signals or color patterns on the bills or throats of adults initiate feeding responses in the young. Parents instinctively feed gaping mouths, and many hatchlings have brightly colored mouth linings or spots that attract a parent's attention. The first-hatched young is fed first—most often because it is usually the largest and can stretch its neck higher than can its nestmates.

Life is usually brief for birds. Approximately 50% of eggs laid yield birds that leave the nest. Most birds, if kept in captivity, have a potential life span of 10 to 20 years. Natural longevity is much shorter. The average American robin lives 1.3 years, and the average black-capped chickadee (*Poecile atricapillus*) lives less than one year. Mortality is high in the first year from predators and inclement weather.

Migration and Navigation

More than 20 centuries ago, Aristotle described birds migrating to escape the winter cold and summer heat. He had the mistaken impression that some birds disappear during winter because they hibernate and that others transmutate to another species. It is now known that some birds migrate long distances. Modern zoologists study the timing of migration, stimuli for migration, and physiological changes during migration, as well as migration routes and how birds navigate over huge expanses of land or water.

Migration (as used here) refers to periodic round trips between breeding and nonbreeding areas. Most migrations are annual, with nesting areas in northern regions and wintering grounds in the south. (Migration is more pronounced for species found in the Northern Hemisphere because about 70% of the earth's land is in the Northern Hemisphere.) Migrations occasionally involve east/west movements or altitude changes. Migration allows birds to avoid climatic extremes and to secure adequate food, shelter, and space throughout the year.

Birds that migrate spend part of the year in regions where the abundance of resources in their breeding area vary from season to season but where the pattern of food availability is predictable. These birds may migrate to tropical regions in the winter, but return to a northern breeding area to take advantage of plentiful spring and summer resources. Annual migrant birds include flycatchers (Tyrannidae), thrushes (Turdidae), and hummingbirds (Trochilidae). Migration is a less desirable life-history characteristic when resources in the breeding area are predictably available all year. Birds such as cardinals, titmice, and woodpeckers find adequate food in the same region all year and are called resident bird species.

Birds migrate in response to species-specific physiological conditions. Innate (genetic) clocks and environmental factors influence their preparation for migration. The photoperiod is an important migratory cue for many birds, particularly for birds in temperate zones. The changing photoperiod initiates seasonal changes in gonadal development that often serve as migratory stimuli. Increasing day length in the spring promotes gonadal development, and decreasing day length in the fall initiates gonadal regression. In many birds, the changing photoperiod also appears to promote fat deposition, which acts as an energy reserve for migration. The anterior lobe of the pituitary gland and the pineal body have been implicated in mediating photoperiod responses.

The mechanics of migration are species specific. Some long-distance migrants may store fat equal to 50% of their body weight and make nonstop journeys. Other species that take a more leisurely approach to migration begin their journeys early and stop frequently to feed and rest. In clear weather, many birds fly at altitudes greater than 1,000 m, which reduces the likelihood of hitting tall obstacles. Purple martins (*Progne subis*) can cover more than 770 km per day. Many birds have very specific migration routes (*see figure 21.1*).

Navigation

Homing pigeons (*Columba livia*) have served for many years as a pigeon postal service. In ancient Egyptian times and as recently as World War II, pigeons returned messages from the battlefield.

Birds use two forms of navigation. Route-based navigation involves keeping track of landmarks (visual or auditory) on an outward journey so that those landmarks can be used in a reverse sequence on the return trip. Location-based navigation is based on establishing the direction of the destination from information available at the journey's site of origin. It involves the use of sun compasses, other celestial cues, and/or the earth's magnetic field.

Birds' lenses are transparent to ultraviolet light, and their photoreceptors respond to it, allowing them to orient using the sun, even on cloudy days. This orientation cue is called a sun compass. Because the sun moves through the sky between sunrise and sunset, birds use internal clocks to perceive that the sun rises in the east, is approximately overhead at noon, and sets in the west. The biological clocks of migratory birds can be altered. For example, birds ready for northward migration can be held in a laboratory in which the "laboratory sunrise" occurs later than the natural sunrise. When released to natural light conditions, they fly in a direction they perceive to be north, but which is really north-west. Night migrators can also orient using the sun by flying in the proper direction from the sunset.

Celestial cues other than the sun can be used to navigate. Humans recognize that in the Northern Hemisphere, the North Star lines up with the axis of rotation of the earth. The angle between the North Star and the horizon decreases as you move toward the equator. Birds may use similar information to determine latitude.

As mentioned, migratory birds can traverse massive distances. In addition to landmarks and celestial cues, birds also use magnetic cues for migration. Although the physiology is not yet completely understood, it is apparent that the earth's magnetic field serves as both a compass and map in certain long-distance migratory species.

The redundancy of bird navigational mechanisms suggests that under different circumstances, migratory birds probably use different sources of information to successfully migrate between breeding and wintering grounds.

Section 21.2 Thinking Beyond the Facts

Why is flight much more complicated than just flapping wings?

WILDLIFE ALERT

Red-Cockaded Woodpecker (*Picoides borealis*)

Vital Statistics

Classification: Phylum Chordata, class Aves, order Piciformes, family Picidae

Range: Fragmented, isolated populations where southern pines exist in the United States (Alabama, Arkansas, Florida, Georgia, Louisiana, Mississippi, North Carolina, Oklahoma, South Carolina, Tennessee Texas, and Virginia)

Habitat: Open stands of pines with a minimum age of 80 to 120 years

Number remaining: Approximately 15,000 birds

Status: Endangered throughout its range

Natural History and Ecological Status

Red-cockaded woodpeckers are 18 to 20 cm long with a wingspan of 35 to 38 cm. They have black-and-white horizontal stripes on the back, and their cheeks and underparts are white (box figure 21.1). Males have a small, red spot on each side of their black cap. After the first postfledgling molt, fledgling males have a red crown patch. The diet of these woodpeckers consists mostly of insects and wild fruit.

Eggs are laid from April through June, with females using their mate's roosting cavity as a nest. The average clutch size is three to five eggs. Most often, the parent birds and one or more male offspring from previous nests form a family unit called a group. A group may include one breeding pair and as many as seven other birds. Rearing the young becomes a shared responsibility of the group.

BOX FIGURE 21.1 **Red-cockaded woodpecker (*Picoides borealis*).**
Source: U.S Fish & Wildlife Service/Mark Ramirez

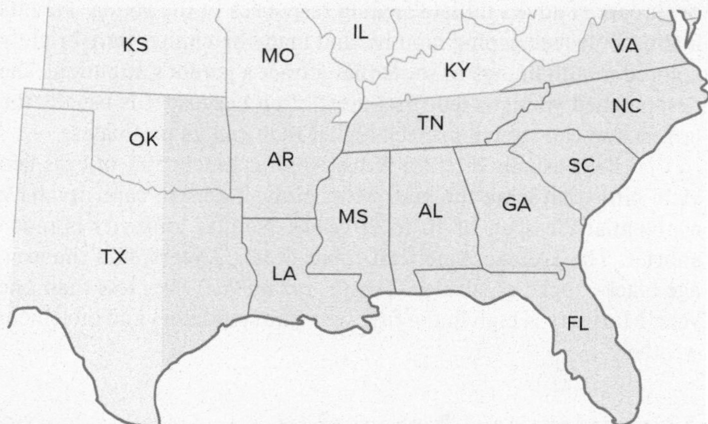

BOX FIGURE 21.2 **Distribution of the red-cockaded woodpecker (*Picoides borealis*).**

The range of red-cockaded woodpeckers is closely tied to the distribution of southern pines, with open stands of trees in the 80- to 120-year-old group being the favored nesting habitat (box figure 21.2). Dense stands or hardwoods are usually avoided. The woodpeckers excavate roosting cavities in living pines, usually those infected with a fungus that produces what is known as red-heart disease. The aggregate of cavity trees is called a cluster and may include 1 to 20 or more cavity trees on 3 to 60 acres. Completed cavities in active use have numerous small resin wells, which exude sap. The birds keep the sap flowing, apparently as a cavity defense mechanism against rat snakes (*Pantherophis alleghaniensis*) and other predators. The territory for a group averages about 200 acres.

Red-cockaded woodpecker population declines are due primarily to logging of their preferred pine forest habitat. This habitat contains trees that are 80 years or more old. When pines are cut down, hardwood understories proliferate. These hardwoods are unsuitable for red-cockaded woodpecker nesting. Recommendations for management and protection include: survey, monitor, and assess the status of individual populations and the species; protect and manage nesting and foraging habitats on federal lands; encourage protection and management on private lands; and inform and involve the public. The success of recovery efforts is reflected in population figures. Bird populations have increased from 4,700 birds in 1993 to approximately 15,000 in 2016.

SUMMARY

21.1 Evolutionary Perspective

- Birds are members of the archosaur lineage. A growing fossil record of ancestral theropods is documenting the origin of ancient birds. These fossils also give clues to the origin of flight and the origin of modern bird lineages.
- *Yi qi* is unique among theropods because it had a feather-covered body and membranous wings. It was incapable of powered flight. *Sinosauripteryx* and *Caudipteryx* were feathered gliders. *Microraptor* had four wings and possessed asymmetrical feathers. It may have been capable of weak-powered flight.
- *Archaeopteryx* was important in formulating hypotheses on the evolution of flight. It was likely a transitional form between avian and nonavian theropods. *Sinornis* and *Eoalulavis* were early birds.
- Feathers evolved before powered flight, and the initial uses for feathered wings may have been to enhance prey capture, or to glide from tree to tree.
- Most ancient avian lineages became extinct at the end of the Mesozoic era. The toothless birds that survived into the Tertiary period were ancestors of modern, toothless birds (Neornithes). Thirty-five avian orders are recognized today.

21.2 Evolutionary Pressures

- Feathers function in flight, insulation, sex recognition, and waterproofing. Feathers are maintained and periodically molted. The bird skeleton is light and made more rigid by the fusion of bones. Birds use the neck and bill as a fifth appendage. Bird wings form airfoils that provide lift. Tilting the wing during flapping generates propulsive force. Gliding, flapping, soaring, and hovering flight are used by different birds or by the same bird at different times.
- Birds feed on a variety of foods, as reflected in the structure of the bill and other parts of the digestive tract.
- The heart of birds consists of two atria and two ventricles. Rapid heart rate and blood flow support the high metabolic rate of birds. The respiratory system of birds provides one-way, nearly constant air movement across respiratory surfaces. Birds are able to maintain high body temperatures endothermically because of insulating fat deposits and feathers.
- The development of the corpus striatum enlarged the cerebral hemispheres of birds. Vision is the most important avian sense.
- Birds excrete uric acid as the primary metabolic waste. Supraorbital salt glands may aid in solute regulation.
- Birds are oviparous. Reproductive activities include the establishment and defense of territories, courtship, and nest building. Either or both bird parents incubate the eggs, and one or both parents feed the young. Altricial chicks are helpless at hatching, and precocial chicks are alert and lively shortly after hatching.
- Migration allows some birds to avoid climatic extremes and to secure adequate food, shelter, and space throughout the year. The photoperiod is the most important migratory cue for birds.
- Birds use both route-based navigation and location-based navigation. The earth's magnetic field is also important in migration.

CONCEPT REVIEW QUESTIONS

1. Endothermy, feathers, and flight are all characteristics unique to the avian reptiles.
   a. True
   b. False
2. These are plumulaceous feathers that are used for insulating a bird.
   a. Flight feathers
   b. Contour feathers
   c. Down feathers
   d. Pennaceous feathers
3. This tiny hook interlocks adjacent barbules of pennaceous feathers, keeping the feathers firm and smooth.
   a. Shaft
   b. Calamus
   c. Hamulus
   d. Vane
4. Which of the following is a modification of the bird skeleton that helps the bird maintain its proper flight posture?
   a. Pygostyle
   b. Furcula
   c. Alula
   d. Synsacrum
5. The proventriculus of a bird
   a. stores food prior to digestion.
   b. secretes gastric juices.
   c. grinds food (gizzard).
   d. is an organ for absorbing digestive products.

ANALYSIS AND APPLICATION QUESTIONS

1. Birds are sometimes called "glorified reptiles." Discuss why this description is appropriate.
2. What adaptations of birds promote endothermy? Why is endothermy important for birds?
3. Birds are, without exception, oviparous. Why do you think that is true?
4. What are the advantages that offset the great energy expenditure that migration requires?
5. In what ways are the advantages and disadvantages of monogamy, polygyny, and polyandry related to the abundance and use of food and other resources?

22

Mammals: Synapsid Amniotes

Hair, mammary glands, and specialized teeth—these are some of the hallmark traits that evolved in synapsid amniotes during the Mesozoic era. Spotted hyenas (*Crocuta crocuta*) are shown here consuming a scavenged carcass. Hyenas are opportunistic feeders, acting as predators and scavengers.

Chapter Outline

22.1 Evolutionary Perspective
22.2 Diversity of Mammals
22.3 Evolutionary Pressures
External Structure and Locomotion
Nutrition and the Digestive System
Circulation, Gas Exchange, and Temperature Regulation
Nervous and Sensory Functions
Excretion and Osmoregulation
Behavior
Reproduction and Development
22.4 Human Evolution
Who Are the Primates?
Evolution of Hominins
Cultural Evolution—A Distinctly Human Process of Change

22.1 EVOLUTIONARY PERSPECTIVE

LEARNING OUTCOMES

1. Describe some of the key evolutionary changes in synapsid anatomy that led to the eventual evolution of modern mammals.
2. Assess the importance of two mass-extinction events in the evolution of modern mammals.

The fossil record that documents the origin of the mammals from ancient reptilian ancestors is very complete and relatively uncontroversial. It is being used to test, and has confirmed, many macroevolutionary hypotheses (*see chapter 4*). The beginning of the Tertiary period, about 70 million years ago (mya), was the start of the "age of mammals." It coincided with the extinction of many reptilian lineages, which led to the adaptive radiation of the mammals. Tracing the roots of the mammals, however, requires returning to the Carboniferous period 320 mya, when the synapsid branch of the amniote lineage diverged from the reptilian branch of this lineage. The fossil record is conclusive—the synapsid lineage was the first amniote lineage to diversify, beginning about 320 mya (*see figure 20.3*). Synapsids quickly became very diverse and widespread. They were the dominant, large-bodied animals on the earth for more than 100 million years, through the remaining Carboniferous and Permian periods.

Mammalian characteristics evolved gradually over a period of 200 million years (figure 22.1). Most of what we know about early synapsids is based on skeletal characteristics. Other mammalian features like hair, mammary glands, and endothermy do not preserve well in the fossil record. Early synapsids had a sprawling gait and were probably ectothermic. The large sails on some, like *Dimetrodon* (figure 22.2*a*), probably helped these synapsids raise body temperature after a cool night. These sails are also an evidence that early synapsids lacked hair. Early synapsids were probably also egg-layers (they were oviparous). Some were herbivores; others showed skeletal adaptations reflecting increased effectiveness as predators.

The anterior teeth of the upper jaw were large and were separated from the posterior teeth by a gap that accommodated the enlarged anterior teeth of the lower jaw when the jaw closed. The palate was arched, which strengthened the upper jaw and allowed air to pass over prey held in the mouth.

By the middle of the Permian period, other successful synapsids had arisen. They were a diverse group of animals in the extinct order Therapsida. Some were predators, and others were herbivores. In the predatory therapsids, teeth were concentrated at the front of the mouth and enlarged for holding and tearing prey. The posterior teeth were reduced in size and number. The jaws of some therapsids were elongate and generated a large biting force when snapped closed. The teeth of the herbivorous therapsids were also mammal-like. Some had a large space, called the diastema, separating the anterior and the posterior teeth. The posterior teeth had ridges (cusps) and cutting edges that were probably used to shred plant material. Unlike other synapsids, therapsids held hindlimbs directly beneath the body and moved them

FIGURE 22.1

Class Mammalia. The decline of the ruling reptiles about 70 mya permitted mammals to radiate into diurnal habitats previously occupied by dinosaurs and other reptiles. Hair, endothermy, and mammary glands characterize mammals. The lowland gorilla (*Gorilla gorilla,* order Primates) is shown here.

parallel to the long axis of the body. Changes in the size and shape of the ribs suggest the separation of the trunk into thoracic and abdominal regions and a breathing mechanism similar to that of mammals.

About 240 mya, most of the very successful therapsids were wiped out during a major extinction event at the Permian–Triassic boundary—possibly as a result of huge Siberian volcanic events. Only a few therapsids in the suborder Cynodonta, including *Cynognathus* (figure 22.2*b*), survived this extinction event. By this time, however, the reptilian (diapsid) amniote lineage had also emerged (*see figure 20.3*). The archosaurs (dinosaurs, crocodiles, and eventually the birds) also survived this extinction event, and these reptiles became the dominant large animals on terrestrial landscapes through the Mesozoic era, which ended about 65 mya. Cynodonts became increasingly smaller (ranging in size from a mouse to a domestic cat), probably nocturnal, and more mammal-like. (The fact that most mammals lack color vision is also an indication that their ancestors were nocturnal). The smaller size and development of hair and endothermy were probably selected for as these mammal precursors exploited niches not occupied by much larger dinosaurs and smaller diurnal (day-active [L. *diurnalis*, daily]) reptiles living at the same time. Other mammalian characteristics evolved during the Jurassic period. There were changes in the structure of the middle ear and increased development in regions of the brain devoted to hearing and olfaction. A well-known representative cynodont possessing these features is *Hadrocodium*. *Hadrocodium* existed 195 mya and was a small, mouse-like creature estimated to weigh about 2 g (figure 22.2*c*). In other early mammals (mammaliforms), teeth became highly specialized to facilitate rapid food processing and to allow for the exploitation of varied food resources. The oldest preserved hair is found in cynodont fossils (e.g., *Megaconus, Castoracuada*) from about 160 mya. However, fossil evidence indicates that hair or hairlike structures may have existed some 100 million years prior to that in therapsids from the upper Permian. During the Mesozoic era, mammal populations were relatively diverse, with

FIGURE 22.2

Premammalian Synapsids. (*a*) *Dimetrodon* was a 3-m-long synapsid. It probably fed on other reptiles and amphibians. The large sail may have been a recognition signal and a thermoregulatory device. (*b*) *Cynognathus* probably foraged for small animals, much like a badger does today. The badger-sized animal was a cynodont within the order Therapsida, the stock from which mammals arose during the mid-Triassic period. (*c*) *Hadrocodium* was just over 3 cm long. This mammaliform was likely a nocturnal endotherm. It possessed a mammal-like middle ear.

over 300 genera present, but they were not particularly abundant. The first therapsids possessing most of the mammalian characteristics listed in table 22.1 were present in the Jurassic period, and ecologically diverse representatives of the extant mammalian orders were present in the late Mesozoic era.

About 65 mya another mass extinction occurred—probably associated with asteroid impact in what is now Central America. Dinosaurs, many ancient birds, and many other taxa became extinct, but at least some early mammals survived this second mass extinction event of synapsid history. This Cretaceous-Tertiary extinction allowed surviving mammals to continue the diversification that began in the Mesozoic era and expand into niches formerly occupied by the dinosaurs. The Tertiary period became the "age of mammals." This period is when the modern orders of mammals expanded and a group of animals called hominins arose. Hominin evolution would eventually result in a species that would drastically alter the biosphere with unprecedented breadth and speed—*Homo sapiens* (*see section 22.4*).

Section 22.1 Thinking Beyond the Facts

In what ways did the demise of the dinosaurs present evolutionary opportunities for mammals? What evidence of their early evolution is seen in modern mammals?

22.2 DIVERSITY OF MAMMALS

LEARNING OUTCOMES

1. Explain the role of continental movements in influencing mammalian evolution.
2. Assess the costs and benefits of the prototherian, metatherian, and eutherian reproductive strategies.

Hair, mammary glands, specialized teeth, three middle-ear ossicles, endothermy, and other characteristics listed in table 22.1 characterize the members of the class Mammalia (mah-ma'le-ah) (L. *mamma*, breast). There are over 5,400 species of mammals that range in size from the bumblebee bat (*Craseonycteris thonglongya*; 3 to 4 cm in length, 2 g) to the blue whale (*Balaenoptera musculus*; more than 30 m in length, 180 metric tons). They are the dominant large terrestrial animals on all continents of the earth, and some have extended their habitats into the oceans and the air.

There are two lineages of living mammals (figure 22.3). The subclass Prototheria (Gr. *protos*, first + *therion*, wild beast) contains the surviving infraclass Ornithodelphia (Gr. *ornis*, bird + *delphia*, birthplace), commonly called the monotremes (Gr. *monos*, one + *trema*, opening). These names refer to the fact that monotremes,

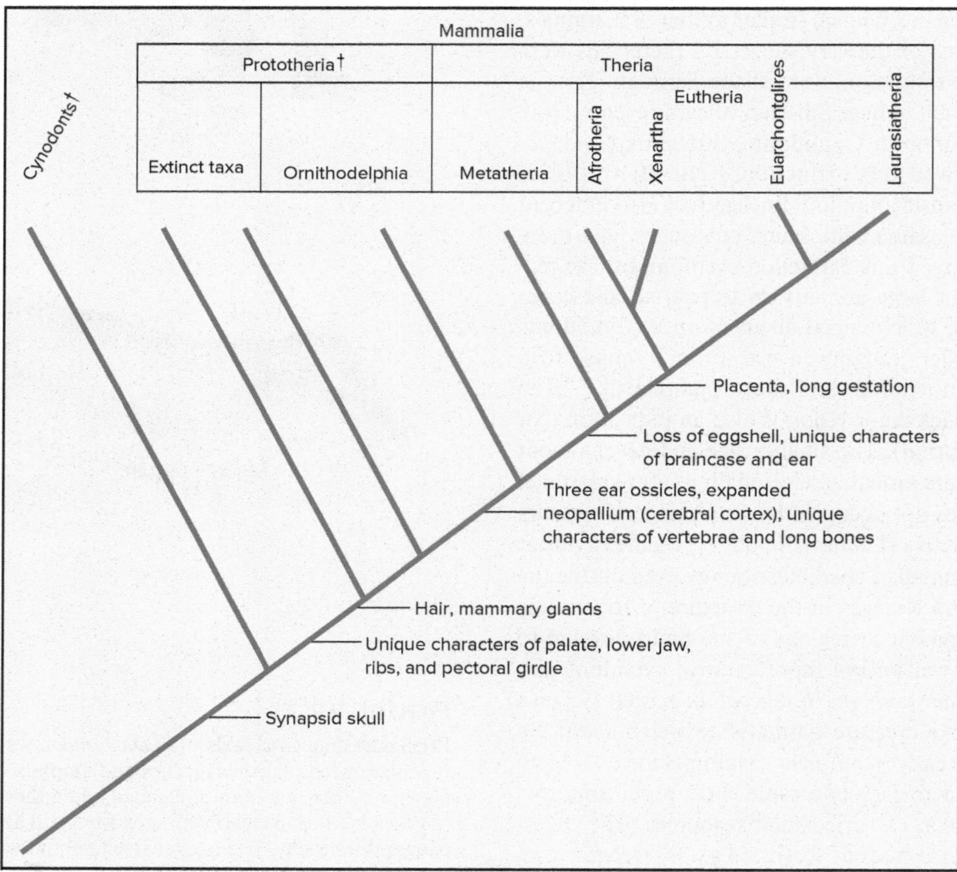

FIGURE 22.3

Mammalian Phylogeny. A cladogram showing one interpretation of the evolutionary relationships among mammals. Selected characters are shown. Daggers (†) indicate some extinct taxa. Numerous extinct groups have been omitted from the cladogram. One recent interpretation of the relationships between the eutherian superorders is shown. Evolutionary relationships between eutherian orders is tentative and not represented.

TABLE 22.1

CLASSIFICATION OF LIVING MAMMALS

Class Mammalia (mah-ma′le-ah)
Mammary glands; hair; diaphragm; three middle-ear ossicles; heterodont dentition; sweat, sebaceous, and scent glands; four-chambered heart; large cerebral cortex.

Subclass Prototheria (pro″to-ther′e-ah)
Oviparous; cloaca present.

Infraclass Ornithodelphia (or″ne-tho-del′fe-ah)
Technical characteristics of the skull distinguish members of this infraclass. Monotremes.

Subclass Theria (ther′e-ah)
Technical characteristics of the skull distinguish members of this subclass.

Infraclass Metatheria (met″ah-ther′e-ah)
Viviparous; primitive placenta; young are born early and often are carried in a marsupial pouch on the female's belly. Marsupials.

Infraclass Eutheria (u-ther′e-ah)*
Complex placenta; young develop to advanced stage prior to birth. Placentals.

Superorder Afrotheria (af″ro-ther′e-ah)

Order Proboscidea (pro″bah-sid′e-ah)
Long, muscular proboscis (trunk) with one or two finger-like processes at the tip; short skull with the second incisor on each side of the upper jaw modified into tusks; six cheek teeth are present in each half of each jaw; teeth erupt (grow into place) in sequence from front to rear, so that one tooth in each jaw is functional. African and Indian elephants.

Order Sirenia (si-re′ne-ah)
Large, aquatic herbivores that weigh in excess of 600 kg; nearly hairless, with thick, wrinkled skin; heavy skeleton; forelimb is flipperlike, and hindlimb is vestigial; horizontal tail fluke is present; horizontally oriented diaphragm; teeth lack enamel. Manatees (coastal rivers of the Americas and Africa), dugongs (western Pacific and Indian Oceans).

Superorder Xenarthra (ze′nar-thra)

Order Pilosa (pi-lo′suh)
Xenarthrous vertebrae; double vena cava; divided uterus in females similar to that of metatherians; combined urinary and genital duct; males have internal testes and no glans penis; reduced metabolic rate; toothless and feeding on termites and ants or arboreal herbivores that possess check teeth only, teeth lack enamel; strong claws for digging or climbing; Central and South America. Anteaters and tree sloths.

Order Cingulataa (sin′gyu′lat-uh)
Xenarthrous vertebrae; dorsolateral body surface covered with protective bony plates arranged into bands that are separated by softer skin; short, powerful legs, feet equipped with strong claws on toes; most species have little hair; opportunistic omnivores, cheek teeth only, teeth lack enamel; reduced metabolic rate; primarily inhabit South and Central America, one species exists in the United States; most primitive group of living eutherians. Armadillos.

Superorder Laurasiatheria (lo-rat″sha-ther′e-ah)

Order Eulipotyphla (u′li-po-tif″lah)
Small mammals with long, narrow mobile snouts. Feed on insects and earthworms. Formerly, these animals were included in an order, Insectivora, that included a variety of additional taxa, including tenrecs and golden moles. Insectivora was found to be polyphyletic. Tenrecs and golden moles are now separated into the order Afrosoricida, which is within the superorder Afrotheria. Hedgehogs, true moles, shrews.

Order Chiroptera (ki-rop′ter-ah)
Cosmopolitan, but especially abundant in the tropics; bones of the arm and hand are elongate and slender; flight membranes extend from the body, between digits of forelimbs, to the hindlimbs; most are insectivorous, but some are fruit eaters, fish eaters, and blood feeders; second-largest mammalian order. Bats.

Order Carnivora (kar-niv′o-rah)
Predatory mammals; usually have a highly developed sense of smell and a large braincase; premolars and molars modified into carnassial apparatus; three pairs of upper and lower incisors usually present, and canines are well developed. Dogs, cats, bears, raccoons, minks, sea lions, seals, walruses, otters.

Order Perissodactyla (pe-ris″so-dak′ti-lah)
Hoofed; axis of support passes through the third digit. Skull usually elongate, large molars and premolars; primarily grazers. (The Artiodactyla also have hoofs. Artiodactyls and perissodactyls are, therefore, called ungulates.) Odd-toed ungulates (L. *ungula*, hoof): horses, rhinoceroses, zebras, tapirs.

Order Artiodactyla (ar″te-o-dak′ti-lah)
Hoofed; axis of support passes between third and fourth digits; digits one, two, and five reduced or lost; primarily grazing and browsing animals (pigs are an obvious exception). Even-toed ungulates: pigs, hippopotamuses, camels, antelope, deer, sheep, giraffes, cattle.

Order Cetacea (se-ta′she-ah)
Streamlined, nearly hairless, and insulated by thick layers of fat (blubber); no sebaceous glands; forelimbs modified into paddlelike flippers for swimming; hindlimbs reduced and not visible externally; tail fins (flukes) flattened horizontally; external naris (blowhole) on top of skull. Toothed whales (beaked whales, narwhals, sperm whales, dolphins, porpoises, killer whales); toothless, filter-feeding baleen whales (right whales, gray whales, blue whales, and humpback whales).

Superorder Euarchontoglires (u-ark-on″to-gler′ez)

Order Lagomorpha (lag″o-mor′fah)
Two pairs of upper incisors; one pair of lower incisors; incisors are ever-growing and slowly worn down by feeding on vegetation. Rabbits, pikas.

Order Rodentia (ro-den′che-ah)
Largest mammalian order; upper and lower jaws bear a single pair of ever-growing incisors. Squirrels, chipmunks, rats, mice, beavers, porcupines, woodchucks, lemmings.

Order Primates (pri-ma′tez)
Adaptations of primates reflect adaptations for increased agility in arboreal (tree-dwelling) habitats; omnivorous diets; unspecialized teeth; grasping digits; freely movable limbs; nails on digits; reduced nasal cavity; enlarged stereoscopic eyes and cerebral hemispheres. Lemurs (Madagascar and the Comoro Islands), tarsiers (jungles of Sumatra and the East Indies), monkeys, gibbons, great apes (apes and humans).

*Selected eutherian orders are described.

(a)

(b)

(c)

FIGURE 22.4

Representatives of the Mammalian Infraclasses Ornithodelphia and Metatheria. The infraclass Ornithodelphia: (*a*) A duck-billed platypus (*Ornithorhychus anatinus*). (*b*) The short-beaked echidna (*Tachyglossus aculeatus*) is native to Australia. The infraclass Metatheria: (*c*) The koala (*Phascolarctos cinereus*) feeds on eucalyptus leaves in Australia.

(a) ©John Carnemolla/Shutterstock (b) ©H Lansdown/Alamy Stock Photo (c) ©Andras Deak/Getty Images

unlike other mammals, possess a cloaca and are oviparous. Recall that a cloaca is a common opening for excretory, reproductive, and digestive products and was found in all other vertebrates, including the reptilian amniote lineage.

Monotremes were formerly far more widespread and were distributed across what is now considered Africa, South America, and Antarctica. The five extant species of monotremes are found in Australia and New Guinea (figure 22.4).

The subclass Theria diverged into two infraclasses by the late Cretaceous period. The infraclass Metatheria (Gr. *meta,* after) contains the marsupial mammals. They are viviparous but have very short gestation periods. A protective pouch, called the marsupium, covers the female mammary glands. The young crawl into the marsupium after birth, where they feed and complete development. The oldest marsupial fossils are found in 125-million-year-old deposits in China.

Competition with eutherians drove many metatherian populations to extinction outside of Australia, where they remained isolated from eutherians because of shifting landmasses. Although metatherians were outcompeted in North America by the eutherians, a few metatherians from South America eventually reestablished populations in North America (via the Isthmus of Panama). About 330 species of marsupials currently live in the Australian region and the Americas (figure 22.4*c*; *see also figure 22.17*).

The other therian infraclass, Eutheria (Gr. *eu,* true), contains the placental mammals. They are usually born at an advanced stage of development, having been nourished within the uterus. Exchanges between maternal and fetal circulatory systems occur by diffusion across an organ called the **placenta,** which is composed of both maternal and fetal tissue. About 4,400 species of eutherians are classified into 21 orders (figures 22.5 and 22.6; *see also figures 22.11* and *22.15* through *22.17*). Selected eutherian orders are listed in table 22.1.

One interpretation of recent molecular data, along with traditional morphological studies, has resulted in the description of four eutherian clades, shown in table 22.1 as superorders. The evolution of these four clades was also strongly influenced by geological events. Between 250 and 100 mya, the earth's landmasses were combined into a single landmass called Pangaea. About 100 mya, a southern supercontinent that consisted of Africa, South America, and Antarctica (Gondwanaland) broke away from a northern supercontinent (Laurasia) that consisted of North America, Europe, and Asia. These continental movements isolated the ancestors of the southern placental mammals, listed in table 22.1 as afrotherians and xenarthans, from other ancestral groups on Laurasia. Later continental movements, such as the separation of South America from Africa, the rejoining of Africa with Europe and Asia, and the joining of North and South America, further isolated, or united, groups of mammals. One current hypothesis on the relationships between these superorders is shown in figure 22.3. Evolutionary relationships among the orders and major clades are unresolved.

(a)

(b)

FIGURE 22.5

Representatives from Orders Pilosa and Cingulata. (*a*) Order Pilosa: A giant anteater (*Myrmecophaga tridactyla*). Anteaters lack teeth. They use powerful forelimbs to tear into an insect nest and a long tongue covered with sticky saliva to capture prey. (*b*) Order Cingulata: An armadillo (*Dasypus novemcinctus*).
(a) ©Peter Schoen/Getty Images (b) ©Elva Dorn/Alamy

SECTION 22.2 THINKING BEYOND THE FACTS

How are biogeographic events important influences on mammalian evolution?

22.3 EVOLUTIONARY PRESSURES

LEARNING OUTCOMES

1. Justify the statement that "many of the characteristic (often unique) mammalian functions are tied to the structure and functions of mammalian integument and musculoskeletal systems."
2. Explain how mammalian nutritive, exchange, and nervous system functions have contributed to the evolutionarily success the mammals.
3. Justify the statement that "the uniqueness of mammals is tied to their behavioral and reproductive functions."

Mammals are naturally distributed on all continents except Antarctica, and they live in all oceans. This section discusses the many adaptations that have accompanied their adaptive radiation.

External Structure and Locomotion

The skin of a mammal, like that of other vertebrates, consists of epidermal and dermal layers. It protects from mechanical injury, invasion by microorganisms, and the sun's ultraviolet light. Skin is also important in temperature regulation, sensory perception, excretion, and water regulation (*see figure 23.9*).

Hair is a keratinized derivative of the epidermis of the skin and is uniquely mammalian. It is seated in an invagination of the epidermis called a hair follicle. A coat of hair, called pelage, usually consists of two kinds of hair. Long guard hairs protect a dense coat of shorter, insulating underhairs.

Because hair is composed largely of dead cells, it must be periodically molted. In some mammals (e.g., humans), molting occurs gradually and may not be noticed. In others, hair loss occurs rapidly and may result in altered pelage characteristics. In the fall, many mammals acquire a thick coat of insulating underhair, and the pelage color may change. For example, the Arctic fox takes on a white or cream color with its autumn molt, which helps conceal the fox in a snowy environment. With its spring molt, the Arctic fox acquires a gray and yellow pelage (*see figure 22.6*).

Hair is also important for the sense of touch. Mechanical displacement of a hair stimulates nerve cells associated with the hair root. Guard hairs may sometimes be modified into thick-shafted hairs called vibrissae. Vibrissae occur around the legs, nose, mouth, and eyes of many mammals. Their roots are richly innervated and very sensitive to displacement.

Air spaces in the hair shaft and air trapped between hair and the skin provide an effective insulating layer. A band of smooth muscle, called the arrector pili muscle, runs between the hair follicle and the lower epidermis. When the muscle contracts, the hair stands upright, increasing the amount of air trapped in the pelage and improving its insulating properties. Arrector pili muscles are under the control of the autonomic nervous system, which also controls a mammal's "fight-or-flight" response. In threatening situations, the hair (especially on the neck and tail) stands on end and may give the perception of increased size and strength.

Hair color depends on the amount of pigment (melanin) deposited in it and the quantity of air in the hair shaft. The pelage of most mammals is dark above and lighter underneath. This pattern makes them less conspicuous under most conditions. Some mammals advertise their defenses using aposematic (warning) coloration. The contrasting markings of a skunk are a familiar example.

FIGURE 22.6

Order Carnivora. An Arctic fox (*Vulpes lagopus*) with its winter pelage. With its spring molt, the Arctic fox acquires a gray-and-yellow-colored coat.

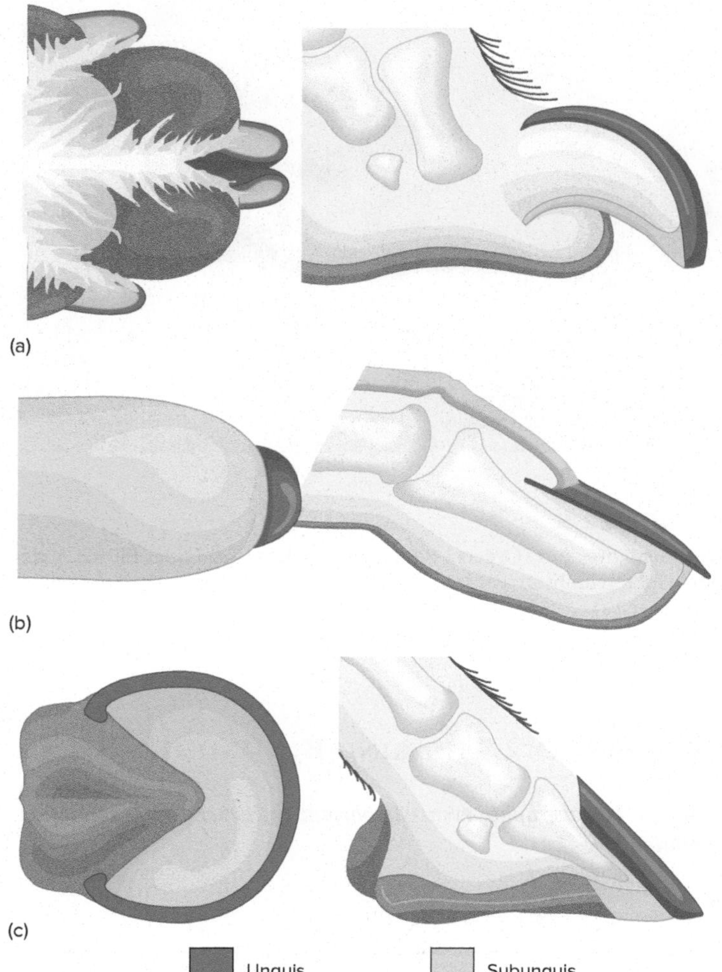

FIGURE 22.7

Structure of Claws, Nails, and Hooves. (*a*) Claws. (*b*) Nails are flat, broad claws found on the hands and feet of primates and are an adaptation for arboreal habits, where grasping is essential. (*c*) Hooves are characteristic of ungulate mammals. The number of toes is reduced, and the animals walk or run on the tips of the remaining digits. The unguis is a hard, keratinized dorsal plate, and the subunguis is a softer ventral plate.

Pelage is reduced in large mammals from hot climates (e.g., elephants and hippopotamuses) and in some aquatic mammals (e.g., whales) that often have fatty insulation. A few mammals (e.g., naked mole rats) have almost no pelage (*see box figure 28.1*).

Claws are present in all amniote classes. They are used for locomotion and offensive and defensive behavior. Claws form from accumulations of keratin that cover the terminal phalanx (bone) of the digits. In some mammals, they are specialized to form nails or hooves (figure 22.7).

Glands develop from the epidermis of the skin. **Sebaceous (oil) glands** are associated with hair follicles, and their oily secretion lubricates and waterproofs the skin and hair. Most mammals also possess **sudoriferous (sweat) glands**. Small sudoriferous glands (eccrine glands) release watery secretions used in evaporative cooling. Larger sudoriferous glands (apocrine glands) secrete a mixture of salt, urea, and water, which microorganisms on the skin convert to odorous products.

Scent or **musk glands** are around the face, feet, or anus of many mammals. These glands secrete pheromones, which may be involved with defense, species and sex recognition, and territorial behavior.

Mammary glands are functional in female mammals and are present, but nonfunctional, in males. The milk that mammary glands secrete contains water, carbohydrates (especially the sugar lactose), fat, protein, minerals, and antibodies. Mammary glands are probably derived evolutionarily from apocrine glands and usually contain substantial fatty deposits.

Monotremes have mammary glands that lack nipples. The glands discharge milk into depressions on the belly, where the young lap it up. In other mammals, mammary glands open via nipples or teats, and the young suckle for their nourishment (figure 22.8).

The Skull and Teeth

The skulls of mammals show important modifications of the reptilian pattern. One feature that zoologists use to distinguish reptilian from mammalian skulls is the method of jaw articulation. In reptiles, the jaw articulates at two small bones at the rear of the jaw. In mammals, these bones have moved into the middle ear, and along with the stapes, form the middle-ear ossicles. A single bone of the lower jaw articulates the mammalian jaw.

A secondary palate evolved twice in vertebrates—in the archosaur lineage (*see figure 20.3*) and in the synapsid lineage. In some therapsids, small, shelflike extensions of bone (the hard palate) partially separated the nasal and oral passageways (*see figure 20.11*). In mammals, the secondary palate extends posteriorly by a fold of skin, called the soft palate, which almost completely separates the nasal passages from the mouth cavity. Unlike other vertebrates that swallow food whole or in small pieces, some mammals chew their food. The more extensive secondary palate allows mammals to breathe while

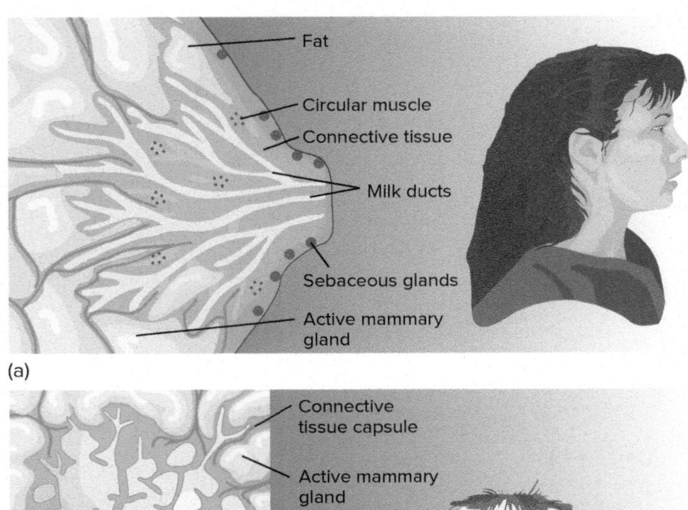

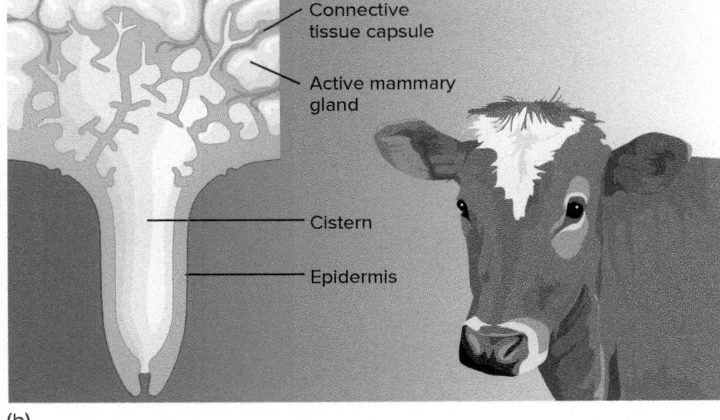

FIGURE 22.8

Mammary Glands. Mammary glands are specialized to secrete milk following the birth of young. (*a*) In humans and other primates, many ducts lead from the glands to a nipple. Parts of the duct system are enlarged to store milk. Suckling by an infant initiates a hormonal response that causes the mammary glands to release milk. (*b*) Some mammals (e.g., cattle) have teats that form by the extension of a collar of skin around the opening of mammary ducts. Milk collects in a large cistern prior to its release. The number of nipples or teats varies with the number of young produced.

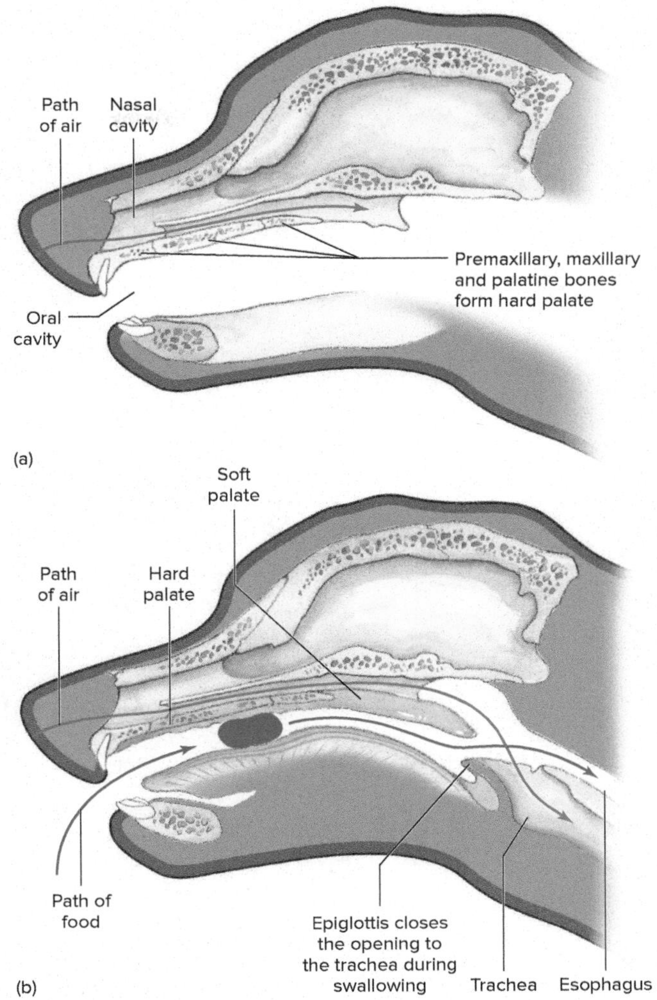

FIGURE 22.9

Secondary Palate. (*a*) The secondary palate (consisting of the hard and soft palates) of a mammal almost completely separates the nasal and oral cavities. (*b*) Breathing stops only momentarily during swallowing.

chewing. Breathing needs to stop only briefly during swallowing (figure 22.9). However, in some mammals the epiglottis is permanently tucked up above the posterior end of the soft palate, forming an airtight seal between oral and nasal cavities. Equines, some rodents, and lagomorphs have this condition and are thus referred to as obligate nasal breathers.

The structure and arrangement of teeth are important indicators of mammalian lifestyles. In reptiles, the teeth are uniformly conical, a condition referred to as homodont. In mammals, the teeth are often specialized for different functions, a condition called **heterodont.** Recall that in reptiles other than archosarus, teeth were attached along the top or inside of the jaw. In mammals (as well as crocodylians) teeth are thecodont (*see chapter 20*). That is, teeth are set into sockets of the jaw. Most mammals have two sets of teeth during their lives. The first teeth emerge before or shortly after birth and are called deciduous or milk teeth. These teeth are lost, and permanent teeth replace them. This single replacement of teeth is called the **diphyodont** condition. (Recall that most reptiles have polyphyodont teeth.)

Adult mammals have up to four kinds of teeth. Incisors are the most anterior teeth in the jaw. They are usually chisel-like and used for gnawing or nipping. Canines are often long, stout, and conical. They are usually used for catching, killing, and tearing prey. Canines and incisors have single roots. Premolars are positioned next to canines, and have one or two roots and truncated surfaces for chewing. Molars have broad chewing surfaces and two (upper molars) or three (lower molars) roots.

Mammalian species have characteristic numbers of each kind of adult tooth. Zoologists use a **dental formula** to characterize taxa. It is an expression of the number of teeth of each kind in one-half of the upper and lower jaws. The teeth of the upper jaw are listed above those of the lower jaw and in the following order: incisors, canine, premolars, and molars. For example:

Human	Beaver
2 • 1 • 2 • 3	1 • 0 • 1 • 3
2 • 1 • 2 • 3	1 • 0 • 1 • 3

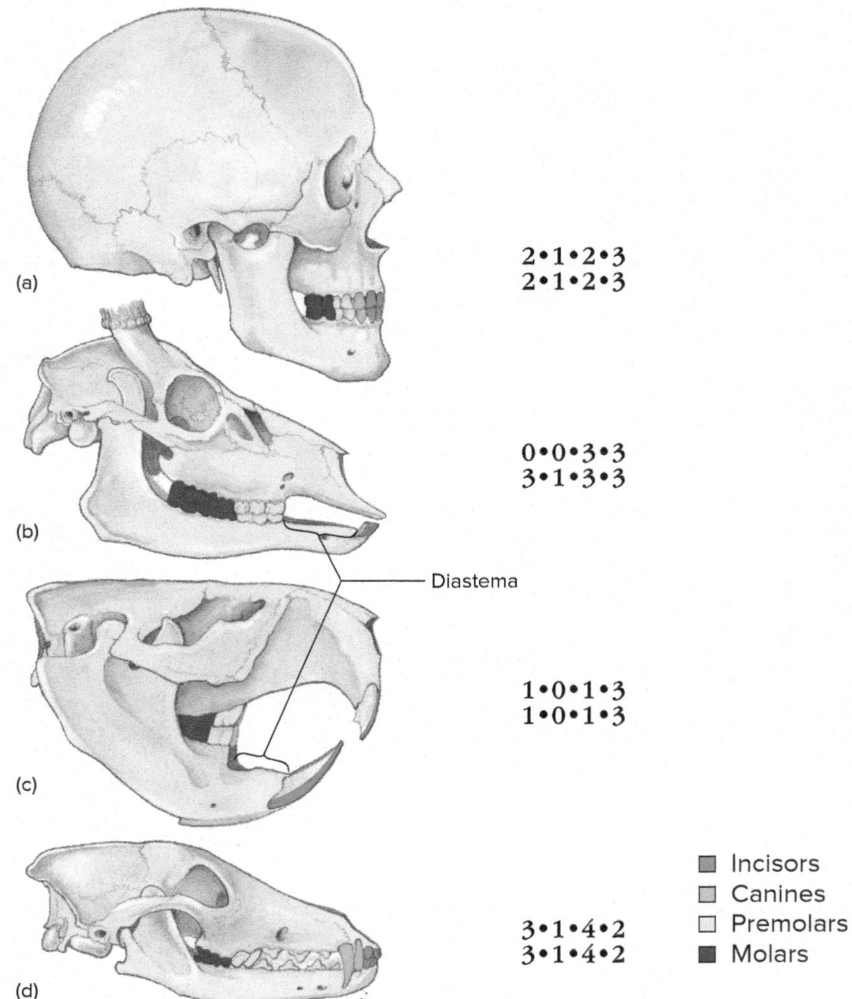

FIGURE 22.10

Specializations of Teeth. (*a*) An omnivore (*Homo sapiens*). (*b*) An herbivore, the male fallow deer (*Dama dama*). (*c*) A rodent, the North American beaver (*Castor canadensis*). (*d*) A carnivore, the coyote (*Canis latrans*).

Mammalian teeth (dentition) may be specialized for particular diets. In some mammals, the dentition is reduced, sometimes to the point of having no teeth. For example, armadillos and the giant anteater (order Edentata) feed on termites and ants, and their teeth are reduced.

Some mammals (e.g., humans, order Primates; and pigs, order Artiodactyla) are omnivorous; they feed on a variety of plant and animal materials. They have anterior teeth with sharp ripping and piercing surfaces, and posterior teeth with flattened grinding surfaces for rupturing plant cell walls (figure 22.10*a*).

Mammals that eat plant material often have flat, grinding posterior teeth and incisors, and sometimes have canines modified for nipping plant matter (e.g., horses, order Perissodactyla; deer, order Artiodactyla) or gnawing (e.g., rabbits, order Lagomorpha; beavers, order Rodentia) (figure 22.10*b* and *c*). In rodents, the incisors grow throughout life. Although most mammals have enamel covering the entire tooth, rodents have enamel only on the front surfaces of their incisors. The teeth are kept sharp by slower wear in front than in back. A gap called the diastema separates the anterior food-procuring teeth from the posterior grinding teeth. The diastema results from an elongation of the snout that allows the anterior teeth to reach close to the ground or into narrow openings to procure food. The posterior teeth have a high, exposed surface (crown) and continuous growth, which allows these teeth to withstand years of grinding tough vegetation.

Predatory mammals use canines and incisors for catching, killing, and tearing prey. In members of the order Carnivora (e.g., coyotes, dogs, and cats), the fourth upper premolars and first lower molars form a scissorlike shearing surface, called the carnassial apparatus, that is used for cutting flesh from prey (figure 22.10*d*).

How Do We Know about the Evolution of Mammary Glands?

Why do we think mammary glands are related to apocrine skin glands? Much of the evidence comes from similar developmental patterns and secretions of the two kinds of glands. Like mammary glands, some apocrine glands secrete lipids and other complex organic molecules.

Lactation may have evolved well before mammals were present. Even though there is no fossil record that traces the evolution of mammary glands, information from the study of changes in bones, teeth, and skin suggests that mammary glands may have been functioning in nonmammalian synapsids. Rather than having glands with a nipple, early synapsids probably had a mammary area similar to that found in monotremes. Monotreme young suck milk from mammary hairs following hatching. A similar function may have occurred in early synapsids, but other observations suggest another possible function. Monotreme eggs are covered with a moist, sticky secretion of unknown origin. It is possible that this secretion is produced by mammary glands and protects the eggs from desiccation. Hairs that develop in association with apocrine glands, distributed over a wide area, would help apply this secretion to a batch of eggs. (Hair is shed from the nipple area of mammals other than monotremes.) These hypotheses need testing but, if they are correct, the two hallmark mammalian characters—hair and mammary glands—first arose in synapsid reptiles for functions other than their primary use in modern mammals. More often than not, evolution does not reinvent—it recycles.

The Vertebral Column and Appendicular Skeleton

The vertebral column of mammals is divided into five regions. As with reptiles and birds, the first two cervical vertebrae are the atlas and axis. Five other cervical vertebrae usually follow. Even the giraffe and the whale have seven neck vertebrae, which are greatly elongated or compressed, respectively. In contrast, tree sloths have either six or nine cervical vertebrae, and the manatee has six cervical vertebrae.

The trunk is divided into thoracic and lumbar regions, as is the case for birds. In mammals, the division is correlated with their method of breathing. The thoracic region contains the ribs. Most ribs connect to the thoracic vertebrae and to the sternum via costal cartilages. Other ribs attach only to the thoracic vertebrae or to other ribs through cartilages. All ribs protect the heart and lungs. The articulation between the thoracic vertebrae provides the flexibility needed in turning, climbing, and lying on the side to suckle young. The interlocking processes of lumbar vertebrae are modified for added support and do not allow the same degree of mobility of the thoracic and cervical spine.

The appendicular skeleton of quadrupedal mammals rotates under the body so that the appendages are usually directly beneath the body. Joints usually limit the movement of appendages to a single anteroposterior plane, causing the tips of the appendages to move in long arcs. The bones of the pelvic girdle are fused in the adult, a condition that is advantageous for locomotion but presents problems during the birth of offspring. In a pregnant female, the ventral joint between the halves of the pelvis–the pubic symphysis–loosens before birth, allowing the pelvis to spread during birth.

Muscles

Because the appendages are directly beneath the bodies of most mammals, the skeleton bears the weight of the body. Muscle mass is concentrated in the upper appendages and girdles. Many running mammals (e.g., deer, order Artiodactyla) have little muscle in their lower leg that would slow leg movement. Instead, tendons run from muscles high in the leg to cause movement at the lower joints.

Nutrition and the Digestive System

The digestive tract of mammals is similar to that of other vertebrates but has many specializations for different feeding habits. Some specializations of teeth have already been described.

The feeding habits of mammals are difficult to generalize. Feeding habits reflect the ecological specializations that have evolved. For example, most members of the order Carnivora feed on animal flesh and are therefore carnivores. Other members of the order, such as bears, feed on a variety of plant and animal products and are omnivores. Some carnivorous mammals are specialized for feeding on arthropods or softbodied invertebrates, and are often referred to (rather loosely) as insectivores. These include animals in the orders Eulipotyphla (e.g., shrews), Chiroptera (bats), and Edentata (anteaters) (*see figure 22.5*a). Herbivores such as deer (order Artiodactyla) and zebras (order Perissodactyla) (figure 22.11) feed mostly on vegetation, but their diet also includes invertebrates inadvertently ingested while feeding.

Specializations in the digestive tracts of most herbivores reflect the difficulty of digesting food rich in cellulose. Horses, rabbits, and many rodents have an enlarged **cecum** at the junction of the

FIGURE 22.11

Order Perissodactyla. This plains zebra (*Equus quagga*) is native to the savannahs of eastern Africa.
©Harvey Lloyd/Getty Images

large and small intestines. A cecum is a fermentation pouch where microorganisms aid in cellulose digestion (*see figure 27.7*). Sheep, cattle, and deer are called ruminants (L. *ruminare,* to chew the cud). Their stomachs are modified into four chambers (*see figure 27.6*). The first three chambers are storage and fermentation chambers and contain microorganisms that synthesize a cellulose-digesting enzyme (cellulase). Gases that fermentation produces are periodically belched, and some plant matter (cud) is regurgitated and rechewed. Other microorganisms convert nitrogenous compounds in the food into new proteins.

Circulation, Gas Exchange, and Temperature Regulation

The hearts of birds and mammals are superficially similar. Both are four-chambered pumps that keep blood in the systemic and pulmonary circuits separate, and both evolved from the hearts of ancient tetrapodomorphs. Their similarities, however, are a result of adaptations to active lifestyles. Recall that the evolution of similar structures in different lineages is called convergent evolution. The mammalian heart evolved from the synapsid lineage, whereas the avian heart evolved within the diapsid archosaur lineage (figure 22.12).

One of the most important adaptations in the circulatory system of eutherian mammals concerns the distribution of respiratory gases and nutrients in the fetus (figure 22.13*a*). Exchanges between maternal and fetal blood occur across the placenta. Although maternal and fetal blood vessels are intimately associated, no blood actually mixes. Nutrients, gases, and wastes simply diffuse between fetal and maternal blood supplies.

Blood entering the right atrium of the fetus is returning from the placenta and is highly oxygenated. Because fetal lungs are not inflated, resistance to blood flow through the pulmonary arteries is high. Therefore, most of the blood entering the right atrium bypasses the right ventricle and passes instead into the left atrium through a valved opening between the atria (the foramen ovale). Some blood from the right atrium, however, does enter the right ventricle and the pulmonary artery. Because of the resistance at the uninflated lungs, most of this blood is shunted to the aorta through a vessel connecting the aorta and the left pulmonary artery (the ductus arteriosus). At birth, the placenta is lost, and the lungs are inflated. Resistance to blood flow through the lungs is reduced, and blood flow to them increases. Flow through the ductus arteriosus decreases, and the vessel is gradually reduced to a ligament. Blood flow back to the left atrium from the lungs correspondingly increases, and the valve of the foramen ovale closes and gradually fuses with the tissue separating the right and left atria (figure 22.13*b*).

Gas Exchange

High metabolic rates require adaptations for efficient gas exchange. Most mammals have separate nasal and oral cavities and longer snouts, which provide an increased surface area for warming and moistening inspired air. Respiratory passageways are highly branched, and large surface areas exist for gas exchange. Mammalian lungs resemble a highly vascular sponge, rather than the saclike structures of amphibians and a few reptiles.

Mammalian lungs, like those of reptiles, inflate using a negative-pressure mechanism. Unlike reptiles and birds, however, mammals possess a muscular **diaphragm** that separates the thoracic and abdominal cavities. Inspiration results from contraction of the diaphragm and expansion of the rib cage, both of which decrease the intrathoracic pressure and allow air to enter the lungs. Expiration is normally by elastic recoil of the lungs and relaxation of inspiratory muscles, which decreases the volume of the thoracic cavity. The contraction of other thoracic and abdominal muscles can produce forceful exhalation.

Temperature Regulation

Mammals are widely distributed over the earth, and some face harsh environmental temperatures. Nearly all face temperatures that require them to dissipate excess heat at some times and to conserve and generate heat at other times.

The heat-producing mechanisms of mammals are divided into two categories. Shivering thermogenesis is muscular activity that generates large amounts of heat but little movement. Nonshivering thermogenesis involves heat production by general cellular metabolism and the metabolism of special fat deposits called brown fat. Chapter 28 discusses these heat-generating processes in more detail.

Heat production is effective in thermoregulation because mammals are insulated by their pelage and/or fat deposits. Fat deposits are also sources of energy to sustain high metabolic rates.

Mammals without a pelage can conserve heat by allowing the temperature of surface tissues to drop. A walrus in cold, arctic waters has a surface temperature near 0°C; however, a few centimeters below the skin surface, body temperatures are about 35°C. Upon emerging from the icy water, the walrus quickly warms its skin by increasing peripheral blood flow. Most tissues cannot tolerate such rapid and extreme temperature fluctuations. Further investigations are likely to reveal some unique biochemical characteristics of these skin tissues.

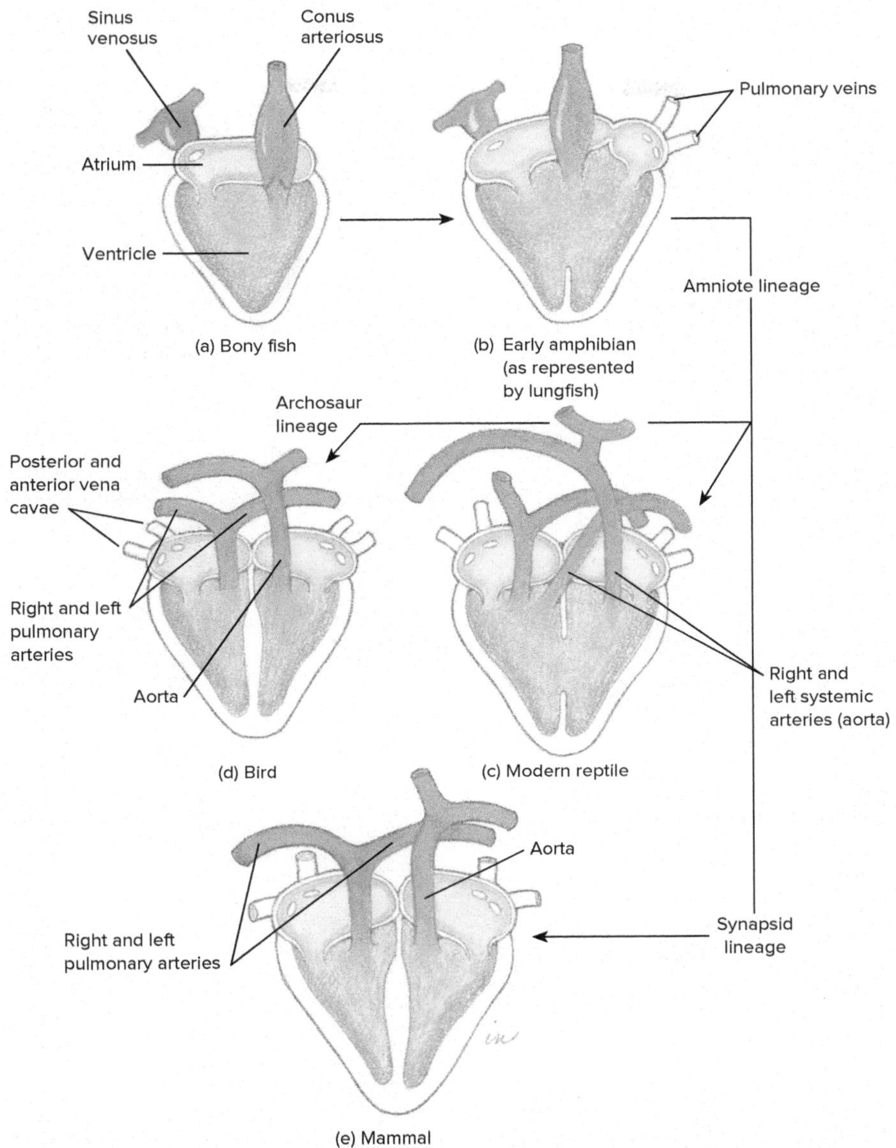

FIGURE 22.12

Possible Sequence in the Evolution of the Vertebrate Heart. (*a*) Diagrammatic representation of a bony fish heart. (*b*) In lungfish, partially divided atria and ventricles separate pulmonary and systemic circuits. This heart was probably similar to that in primitive amphibians and early amniotes. (*c*) The hearts of modern reptiles were derived from the pattern in (*b*). (*d*) The archosaur and (*e*) synapsid lineages resulted in completely separated, four-chambered hearts.

Even though most of the body of an arctic mammal is unusually well insulated, appendages often have thin coverings of fur as an adaptation to changing thermoregulatory needs. Even in winter, an active mammal sometimes produces more heat than is required to maintain body temperature. Patches of poorly insulated skin allow excess heat to be dissipated. During periods of inactivity or extreme cold, however, arctic mammals must reduce heat loss from these exposed areas, often by assuming heat-conserving postures. Mammals sleeping in cold environments conserve heat by tucking poorly insulated appendages and their faces under well-insulated body parts.

Countercurrent heat-exchange systems may help regulate heat loss from exposed areas (figure 22.14). Arteries passing peripherally through the core of an appendage are surrounded by veins that carry blood back toward the body. When blood returns to the body through these veins, heat transfers from arterial blood to venous blood and returns to the body rather than being lost to the environment. When excess heat is produced, blood is shunted away from the countercurrent veins toward peripheral vessels, and excess heat is radiated to the environment.

Mammals have few problems getting rid of excess heat in cool, moist environments. Heat can be radiated into the air from

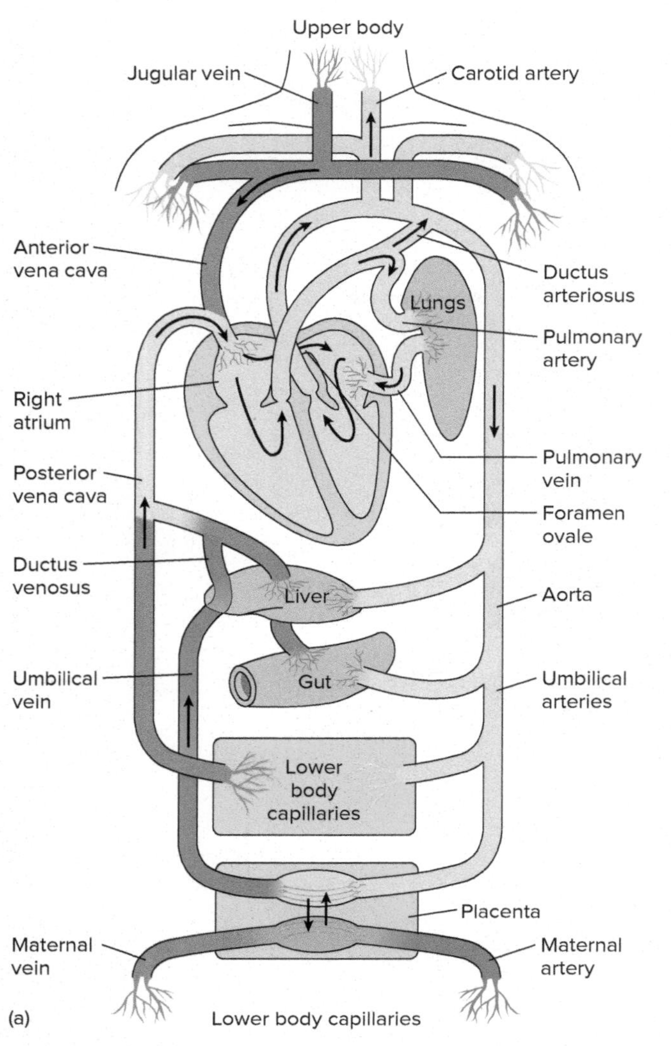

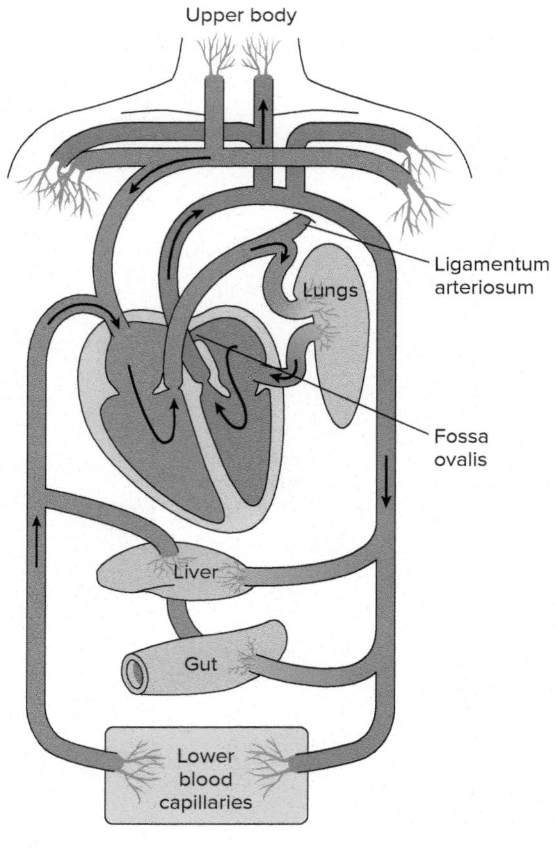

FIGURE 22.13

Mammalian Circulatory Systems. The circulatory patterns of (*a*) fetal and (*b*) adult mammals. Highly oxygenated blood is shown in red, and less oxygenated blood is shown in blue. In fetal circulation, highly oxygenated blood from the placenta mixes with less oxygenated blood prior to entering the right atrium. Thus, most arterial blood of the fetus is moderately oxygenated. The pale lavender color in (*a*) symbolizes this state of oxygenation.

vessels near the surface of the skin or lost by evaporative cooling from either sweat glands or respiratory surfaces during panting.

Hot, dry environments present far greater problems, because evaporative cooling may upset water balances. Jackrabbits (*Lepus*) and elephants (*Loxodonta*) use their long ears to radiate heat. Small mammals often avoid the heat by remaining in burrows during the day and foraging for food at night. Other mammals seek shade or watering holes for cooling.

Winter Sleep and Hibernation Mammals react in various ways to environmental extremes. Caribou (*Rangifer tarandus*) migrate to avoid extremes of temperature, and wildebeest migrate to avoid seasonal droughts. Other mammals retreat to burrows under the snow, where they become less active but are still relatively alert and easily aroused–a condition called **winter sleep.** For example, American black bears (*Ursus americanus*) and raccoons (*Procyon lotor*) retreat to dens in winter. Their body temperatures and metabolic rates decrease somewhat, but they do not necessarily remain inactive all winter.

Hibernation is a period of winter inactivity in which the hypothalamus of the brain slows the metabolic, heart, and respiratory rates. True hibernators include all Monotremata (echidna and duck-billed platypus) and many members of the Insectivora (e.g., moles and shrews), Rodentia (e.g., chipmunks and woodchucks), and Chiroptera (bats). In preparation for hibernation, mammals usually accumulate large quantities of body fat. After a hibernating mammal retreats to a burrow or a nest, the hypothalamus sets the body's thermostat to about 2°C. The respiratory rate of a hibernating ground squirrel (Sciuridae) falls from 100 to 200 breaths per minute to about four breaths per minute. The heart rate falls from 200 to 300 beats per minute to about 20 beats per minute. During hibernation, a mammal may lose a third to half of its body

(a)

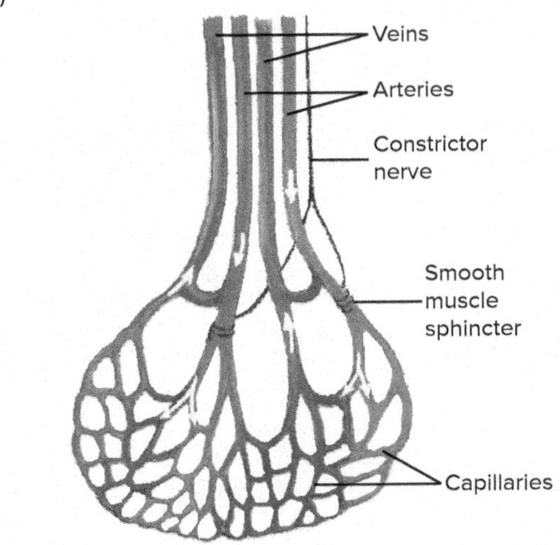

(b)

FIGURE 22.14

Countercurrent Heat Exchange. Countercurrent heat exchangers conserve body heat in mammals adapted to cold environments. (*a*) Systems similar to the one depicted here are found in the legs of reindeer (*Rangifer tarandus*) and in the flippers of dolphins (e.g., *Tursiops*). (*b*) Heat transferred from blood moving peripherally in arteries warms venous blood returning from an extremity. During winter, the lower part of a reindeer's leg may be at 10°C, while body temperature is about 40°C. White arrows indicate direction of blood flow.

weight. Arousal from hibernation occurs by metabolic heating, frequently using brown fat deposits (*see figure 28.7*), and it takes several hours to raise body temperature to near 37°C. As described in chapter 6, winter sleep and hibernation are forms of controlled hypothermia (lowered body temperature) and are different forms of the same set of physiological processes. They differ by the extent to which body temperature falls and the duration of hypothermic condition.

Nervous and Sensory Functions

The basic structure of the vertebrate nervous system is retained in mammals. The development of complex nervous and sensory functions goes hand-in-hand with active lifestyles and is most evident in the enlargement of the cerebral hemispheres and the cerebellum of mammals. Most integrative functions shift to the enlarged cerebral cortex (neocortex; *see figure 24.11*).

In mammals, the sense of touch is well developed. Receptors are associated with the bases of hair follicles and are stimulated when a hair is displaced.

Olfaction was apparently an important sense in early mammals, because fossil skull fragments show elongate snouts, which would have contained an epithelium with many olfactory receptor cells. Cranial casts of fossil skulls show enlarged olfactory regions. Olfaction is still an important sense for many mammals. Mammals can perceive olfactory stimuli over long distances during either the day or night to locate food, recognize members of the same species, and avoid predators.

Auditory senses were similarly important to early mammals. More recent adaptations include an ear flap (the pinna) and the external ear canal leading to the tympanum that directs sound to the middle ear. The middle ear contains three ear ossicles that conduct vibrations to the inner ear. The sensory patch of the inner ear that contains the sound receptors is long and coiled and is called the cochlea. This structure provides more surface area for receptor cells and gives mammals greater sensitivity to pitch and volume than is present in reptiles. Cranial casts of early mammals show well-developed auditory regions.

Vision is an important sense in many mammals, and eye structure is similar to that described for other vertebrates. Accommodation occurs by changing the shape of the lens (*see figure 24.28*). Color vision is less well developed in mammals than in reptiles and birds. Rods dominate the retinas of most mammals, which supports the hypothesis that early mammals were nocturnal. Certain primates have well-developed color vision. Squirrels (Sciuridae), rats (Muridae), rabbits (Leporidae), and a few other mammals are capable of seeing colors to a lesser extent.

Excretion and Osmoregulation

Mammals, like all amniotes, have a metanephric kidney. Unlike reptiles and birds, which excrete mainly uric acid, mammals excrete urea. Urea is less toxic than ammonia and does not require large quantities of water in its excretion. Unlike uric acid, however, urea is highly water soluble and cannot be excreted in a semisolid form; thus, some water is lost. Excretion in mammals is always a major route for water loss.

In the nephron of the kidney, fluids and small solutes are filtered from the blood through the walls of a group of capillary-like vessels, called the glomerulus. The remainder of the nephron consists of tubules that reabsorb water and essential solutes and secrete particular ions into the filtrate.

The primary adaptation of the mammalian nephron is a portion of the tubule system called the loop of the nephron. The transport processes in this loop and the remainder of the tubule system allow mammals to produce urine that is more concentrated than blood. For example, North American beavers produce urine that is twice as concentrated as blood, while Australian hopping mice (*Notomys*) produce urine that is 22 times more concentrated than blood. This accomplishes the same function that nasal and orbital salt glands do in reptiles and birds.

Water loss varies greatly, depending on activity, physiological state, and environmental temperature. Water is lost in urine and feces, in evaporation from sweat glands and respiratory surfaces, and during nursing. Mammals in very dry environments have many behavioral and physiological mechanisms to reduce water loss. The kangaroo rat, named for its habit of hopping on large hind legs, is capable of extreme water conservation (figure 22.15). It is native to the southwestern deserts of the United States and Mexico, and it survives without drinking water. Its feces are almost dry, and its nocturnal habits reduce evaporative water loss. Condensation as warm air in the respiratory passages encounters the cooler nasal passages minimizes respiratory water loss. A low-protein diet, which reduces urea production, minimizes excretory water loss. The nearly dry seeds that the kangaroo rat eats are rich sources of carbohydrates and fats. Metabolic oxidation of carbohydrates produces water as a by-product.

(a)

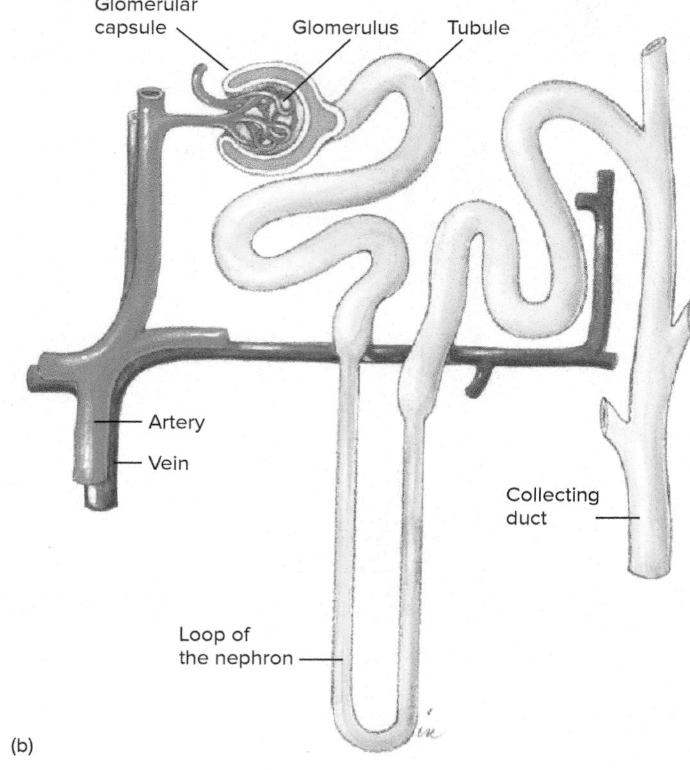

(b)

FIGURE 22.15

Order Rodentia. (*a*) The kangaroo rat (*Dipodomys ordii*). (*b*) The long loop of the nephron of this desert animal conserves water, preventing dehydration.

Behavior

Mammals have complex behaviors that enhance survival. Visual cues are often used in communication. The bristled fur, arched back, and open mouth of a cat communicate a clear message to curious dogs or other potential threats. A tail-wagging display of a dog has a similarly clear message. A wolf (*Canus lupus*) defeated in a fight with other wolves lies on its back and exposes its vulnerable throat and belly. Similar displays may allow a male already recognized as being subordinate to another male to avoid conflict within a social group.

Pheromones are used to recognize members of the same species, members of the opposite sex, and the reproductive state of a member of the opposite sex. Pheromones may also induce sexual behavior, help establish and recognize territories, and ward off predators. The young of many mammalian species recognize their parents, and parents recognize their young, by smell. Bull elk (*Cervus canadensis*) smell the rumps of females during the breeding season to recognize those in their brief receptive period. They also urinate on their own bellies and underhair to advertise their reproductive status to females and other males. Male mammals urinate on objects in the environment to establish territories and to allow females to become accustomed to their odors. Rabbits and rodents spray urine on a member of the opposite sex to inform the second individual of the first's readiness to mate. Skunks (*Mephitis*) use chemicals to ward off predators.

Auditory communication is also important in the lives of mammals. Herd animals stay together and remain calm as long as familiar sounds (e.g., bellowing, hooves walking over dry grasses and twigs, and rumblings from ruminating stomachs) are uninterrupted. Unfamiliar sounds may trigger alarm and flight.

Vocalizations and tactile communication are important in primate social interactions. Tactile communication ranges from precopulatory "nosing" that occurs in many mammals to grooming.

Grooming helps maintain a healthy skin and pelage, but also reinforces important social relationships within primate groups.

Territoriality

Many mammals mark and defend certain areas from intrusion by other members of the same species. When cats rub their faces and necks on humans or on furniture, the behavior is often interpreted as affection. Cats, however, are really staking claim to their territory, using odors from facial scent glands. Some territorial behavior attracts females to, and excludes other males from, favorable sites for mating and rearing young.

Male California sea lions (*Zalophus californianus*) establish territories on shorelines where females come to give birth to young. For about two weeks, males engage in vocalizations, displays, and sometimes serious fighting to stake claim to favorable territories (figure 22.16). Older, dominant bulls are usually most successful in establishing territories, and young bulls generally swim and feed just offshore. When they arrive at the beaches, females select a site for giving birth. Selection of the birth site also selects the bull that will father next year's offspring. Mating occurs approximately two weeks after the birth of the previous year's offspring. Development is arrested for the three months during which the recently born young do most of their nursing. This mechanism is called embryonic diapause. Thus, even though actual development takes about nine months, the female carries the embryo and fetus for a period of one year.

FIGURE 22.16

Order Carnivora. California sea lions (*Zalophus californianus*) on a rookery at Monterey, California. The adult male (center) is posturing and preparing to vocalize.
©Miroslav Halama/Shutterstock

Reproduction and Development

In no other group of animals has viviparity developed to the extent it has in mammals. Mammalian viviparity requires a large expenditure of energy on the part of the female during development and on the part of one or both parents caring for young after they are born. Viviparity is advantageous because females are not necessarily tied to a single nest site but can roam or migrate to find food or a proper climate. Viviparity is accompanied by the evolution of a portion of the reproductive tract where the young are nourished and develop. In viviparous mammals, the oviducts are modified into one or two uteri (sing., uterus).

Reproductive Cycles

Most mammals have a definite time or times during the year in which ova (eggs) mature and are capable of being fertilized. Reproduction usually occurs when climatic conditions and resource characteristics favor successful development. Mammals living in environments with few seasonal changes and those that exert considerable control over immediate environmental conditions (e.g., humans) may reproduce at any time of the year. However, they are still tied to physiological cycles of the female that determine when ova can be fertilized.

Most female mammals undergo an **estrus** (Gr. *oistros,* a vehement desire) **cycle**, which includes a time during which the female is behaviorally and physiologically receptive to the male. During the estrus cycle, hormonal changes stimulate the maturation of ova in the ovary and induce ovulation (release of one or more mature ova from an ovarian follicle). A few mammals such as rabbits (Leporidae) and ferrets and mink (Mustelidae) are induced ovulators; coitus (copulation) induces ovulation.

Hormones also mediate changes in the uterus and vagina. As the ova are maturing, the inner lining of the uterus proliferates and becomes more vascular in preparation for receiving developing embryos. External swelling in the vaginal area and increased glandular discharge accompany the proliferation of vaginal mucosa. During this time, males show heightened interest in females, and females are receptive to males. If fertilization does not occur, the changes in the uterus and vagina are reversed until the next cycle begins. No bleeding or sloughing of uterine lining usually occurs.

Many mammals are monestrus and have only a single yearly estrus cycle that is sharply seasonal. Wild dogs (*Canis*), bears (Ursidae), and sea lions are monestrus; domestic dogs are diestrus. Other mammals are polyestrus. Rats and mice have estrus cycles that repeat every four to six days.

The menstrual cycle of female humans, apes, and monkeys is similar to the estrus cycle in that it results in a periodic proliferation of the inner lining of the uterus and correlates with the maturation of an ovum. If fertilization does not occur before the end of the cycle, menses–the sloughing of the uterine lining–occurs. Chapter 29 describes human menstrual and ovarian cycles.

Fertilization usually occurs in the upper third of the oviduct within hours of copulation. In a few mammals, fertilization may be delayed. In some bats, for example, coitus occurs in autumn, but fertilization is delayed until spring. Females store sperm in the uterus for periods in excess of two months. This **delayed fertilization** is apparently an adaptation to winter dormancy. Fertilization can occur immediately after females emerge from dormancy rather than waiting until males attain their breeding state.

WILDLIFE ALERT

The Southern (California) Sea Otter (*Enhydra lutris nereis*)

VITAL STATISTICS

Classification: Phylum Chordata, class Mammalia, order Carnivora

Range: Southern California coast

Habitat: Kelp beds in near-shore waters

Number remaining: 2,000

Status: Threatened

NATURAL HISTORY AND ECOLOGICAL STATUS

Sea otters (*Enhydra lutris*) are divided into three subspecies based upon the morphological and molecular characteristics. Their historic range includes most of the northern Pacific rim from Hokkaido, Japan, to Baja California (box figure 22.1). Prior to the 1700s, the sea otter population probably numbered between 150,000 and 300,000 individuals. Of the three subspecies, the southern (California) sea otter (*E. lutris nereis*) has been in the greatest danger of extinction.

Sea otters are the smallest marine mammals (box figure 22.2). Mature males average 29 kg and mature females average 20 kg. They feed on molluscs, sea urchins, and crabs. They use shells and rocks to pry their prey from the substrate and to crack shells and tests of their food items. Unlike other marine mammals, they have no blubber for insulation from cold water. Their very thick fur, with about 150,000 hairs per cm^2, is their insulation. (The human head has about 42,000 hairs per cm^2.) Sea otters are considered a keystone predator. By preying on a variety of kelp herbivores, they enhance the productivity of kelp beds and increase the diversity of the kelp ecosystem. (The kelp ecosystem is one of the most diverse ecosystems in temperate regions of the earth.)

BOX FIGURE 22.2 The Southern (California) Sea Otter. *Enhydra lutris nereis.*
©Tory Kallman/Getty Images

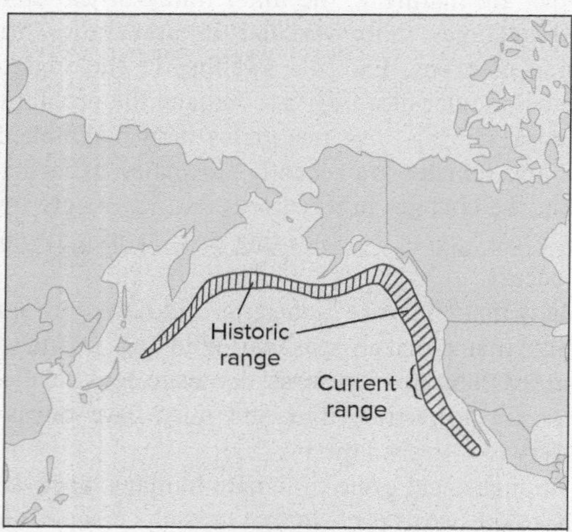

BOX FIGURE 22.1 Range of Sea Otters. The historic range of sea otters (shaded) probably consisted of a cline (a gradual transition between) of the three subspecies. Southern sea otters (*Enhydra lutris nereis*) now occupy a portion of the California coast between Half Moon Bay and Gaviota.

Southern sea otters have faced, and continue to face, pressures that threaten their survival. In the 1700s, they were hunted extensively for their thick fur. They are sensitive to contaminants in the ecosystem. Poisons such as pesticides, PCBs, and tributylin (a component of antifouling agents used on boat hulls) accumulate in their tissues and weaken the animals. When oil from tanker spills becomes trapped in an otter's thick fur, it destroys its insulating qualities and quickly kills the otter. Today the leading cause of sea otter mortality is disease. All of these pressures devastated southern sea otter populations. Historically, there were about 150,000 to 300,000 southern sea otters along their range, which extended along what is now the California coast. In the early 1900s, they were thought to be extinct until a small group of otters was observed on California's Big Sur coast.

Southern sea otters are now protected by the International Convention for the Preservation and Protection of Fur Seals, the Marine Mammal Protection Act, and the Endangered Species Act. This protection and other recovery efforts have protected the otters and sheltered other species in the kelp ecosystem. Since 1995, the population has fluctuated between 2,000 and 3,000 animals.

In many other mammals, fertilization occurs right after coitus, but development is arrested after the first week or two. This **embryonic diapause**, which was described previously for sea lions, also occurs in some bats, bears, martens (Mustelidae), and marsupials. The adaptive significance of embryonic diapause varies with species. In the sea lion, embryonic diapause allows the mother to give birth and mate within a short interval, but not have her resources drained by both nursing and pregnancy. It also allows

(a)

(b)

FIGURE 22.17

Order Marsupialia. (*a*) The Virginia opossum (*Didelphis virginiana*) is found throughout Central America and is the only opossum found in North America. (*b*) Opossum young nursing in a marsupial pouch.

young to be born at a time when resources favor their survival. In some bats, fertilization occurs in the fall before hibernation, but birth is delayed until resources become abundant in the spring.

Modes of Development

Monotremes are oviparous. The ovaries release ova with large quantities of yolk. After fertilization, shell glands in the oviduct deposit a shell around the ovum, forming an egg. Female echidnas incubate eggs in a ventral pouch. Platypus eggs are laid in their burrows.

In marsupials, embryos are initally enclosed by extraembryonic membranes and float in uterine fluid. After emerging from extraembryonic membranes, most nourishment for the fetus comes from "uterine milk" that uterine cells secrete. Some nutrients diffuse from maternal blood into a highly vascular yolk sac that makes contact with the uterus. This connection in marsupials is called a choriovitelline (yolk sac) placenta. This period of development is very brief. The marsupial **gestation period** (the length of time young develop within the female reproductive tract) varies between 8 and 40 days in different species. The gestation period is short because of marsupials' inability to sustain the production of hormones that maintain the uterine lining. After birth, tiny young crawl into the marsupium and attach to a nipple, where they suckle for an additional 60 to 270 days (figure 22.17).

In eutherian mammals, the embryo implants deeply into the uterine wall. Embryonic and uterine tissues grow rapidly and become highly folded and vascular, forming a chorioallantoic placenta. Although maternal and fetal blood do not mix, nutrients, gases, and wastes diffuse between the two bloodstreams. Gestation periods of eutherian mammals vary from 20 days (some rodents) to 19 months (the African elephant). Following birth, the placenta and other tissues that surrounded the fetus in the uterus are expelled as "afterbirth." The newborns of many species (e.g., humans) are helpless at birth (altricial); others (e.g., deer and horses) can walk and run shortly after birth (precocial).

Section 22.3 Thinking Beyond the Facts

What modifications of the general circulatory pattern of adult mammals are found in fetal mammals? What do these modifications accomplish?

22.4 HUMAN EVOLUTION

LEARNING OUTCOMES

1. Explain the global conditions that influenced the evolution of bipedal locomotion in early apes in Africa.
2. Describe a sequence of hominins and time frames that are important in understanding events of human evolution.
3. Describe the role of cultural evolution in the development of human societies.

Before modern theories of evolution appeared, questions about our origins absorbed human thought and fueled the fires of debate. Today paleontologists hunt fossil remains, paleoecologists study environmental constraints placed on early humans, molecular biologists study the genetic sequences of primates, cytologists study the chromosome composition of primates, taphonomists study the way that bones and artifacts become buried, and ethologists study the behavior of social primates. All of these fields have supplied a wealth of information that helps us understand phylogeny within the primate lineage. As you will see in the this section, there are many exciting questions regarding our origins remaining to be answered.

Who Are the Primates?

Primates arose in the late Cretaceous period about 65 mya. Ancestral species were probably insectivores that ran along tree branches and across the ground. Arboreal (tree-dwelling) habits and a shift to diurnal (daytime) activity favored color vision as the primary means for locating food and negotiating uncertain footing. The eyes of most mammals are on the sides of their heads, but the eyes of primates are on the front of the head. This location provides an overlapping field of view for each eye and improved depth perception. Other primate characteristics that can be traced to arboreal origins include a center of gravity that is shifted over hindlimbs, nails that protect the ends of long digits, sensitive foot and hand pads that are used in exploring arboreal environments, and friction ridges that aid in clinging to tree branches. The medial digits of the hands, and usually also the feet, are opposable to allow the hand and foot to close around a branch or other object.

Primates are divided into two suborders (table 22.2). One includes lemurs, the aye-aye, and bush babies. The other includes tarsiers, New and Old World monkeys, gibbons, the gorilla, chimpanzees, the orangutan, and humans (figure 22.18). The latter four are all apes and members of one family, Hominidae. Until recently humans were classified in a family separate from the gorilla, chimpanzees, and the orangutan. DNA and chromosome analysis, however, reveals that chimpanzees and humans are as similar to each other as many sister species. (That is, they are as similar as different species within the same genus.) The classification system in table 22.2 keeps the chimps and humans in different genera, but in the same subfamily (Homininae) and tribe (Hominini). The term "hominin" is used to refer to the chimpanzees and members of this human lineage. (Hominini is sometimes defined to exclude chimpanzees and refer only to modern humans and our human ancestors. When it is used in the more inclusive sense to include chimpanzees and humans, the subtribe Panina designates the chimpanzee lineage and the subtribe Hominina designates the human lineage.)

TABLE 22.2
CLASSIFICATION OF PRIMATES

TAXON	COMMON NAME	DISTRIBUTION
Suborder Strepsirhini		
Seven families	Lemurs, aye-aye, bush babies	Madagascar, Africa, and Asia
Suborder Haplorhini		
Families		
Tarsiidae	Tarsiers	Southeast Asia
Callitrichidae	Marmosets and tamarins	Central and South America
Cebidae	Capuchin-like monkeys	Central and South America
Cercopithecidae	Mandrils, baboons, macaques	Africa and Asia
Hylobatidae	Gibbons and siamang	Asia
Hominidae		
Subfamilies		
Ponginae	Orangutan	Asia
Homininae		
Tribes		
Gorillini	*Gorilla*	Africa
Hominini	*Pan* (chimpanzee), *Sahelanthropus*, *Ardipitheus*, *Australopithecus*, *Homo*	Africa (*Homo*, worldwide)

Evolution of Hominins

The first apes appeared about 25 mya. The fossil record that could document the evolution of the ape lineage to a point of common ancestry of humans and chimpanzees is quite fragmentary. Molecular evidence suggests that divergence between ancestral apes and hominins occurred between 6 and 10 mya, but this evidence cannot help us visualize what this common ancestor looked like. We must avoid the temptation to view living chimpanzees as models for ancestors of the human lineage as the chimpanzee lineage has surely undergone many changes over the time frame that encompasses the human lineage. Unfortunately, changes within the chimpanzee lineage are poorly documented.

Ape evolution was strongly influenced by geography and climate. Continental drift had isolated Asian and African apes. Africa was largely tropical at the time. Climate and geography changed, however, and these changes probably supplied the pressures that fostered evolutionary change. About 20 mya, global temperatures turned sharply cooler. Temperate regions expanded, and seasons became more pronounced. Geological uplift created highlands and dry belts across eastern Africa. Between 5 and 7 mya, global temperatures fell further. The continuous tropical forests of Africa began breaking into a mosaic of forest and vast savannah. Under these circumstances, African apes acquired adaptations that allowed them to move from arboreal habitats to exploit grains, tubers, and dead grazing animals. An upright posture and bipedal locomotion, hallmark characteristics that distinguish the hominins, promoted exploitation of these resources. Earliest hominins were probably not strictly arboreal or ground dwelling, but would have come out of the trees to forage and used trees as refugia. It is very possible that bipedal locomotion evolved in more than one lineage in the bush-like hominin phylogeny. A few of the host of skeletal adaptations that give paleontologists clues to bipedal locomotion are described next.

Bipedal locomotion required adaptations for balancing and adaptations that permit the weight of the body to be supported by two, rather than four, appendages. In humans, the vertebral column is curved in a manner that brings the center of gravity more in line

(a) (b) (c) (d)

FIGURE 22.18

Primates. (*a*) Lemurs are found only on the island of Madagascar, off the eastern coast of Africa. Their eyes are partially directed toward the front, allowing some binocular vision. They have longer hindlimbs than forelimbs and somewhat elongated digits. A black lemur (*Eulemur macaco*) is shown here. (*b*) Old World monkeys are found in Africa, Asia, Japan, and the Philippines. They use their tails as a balancing aid, but they are not prehensile (grasping). A red shanked douc langur (*Pygathrix nemaeus*) is shown here. (*c*) New World monkeys are found in Central and South America. They have prehensile tails. A red howler monkey (*Alouatta seniculus*) is shown here. (*d*) Gorillas (*Gorilla gorilla*), along with chimpanzees, orangutans, and humans, belong to the family Hominidae.

(a) ©byvalet/Alamy Stock Photo (b) ©Terry Whittaker/Science Source (c) ©age fotostock/SuperStock (d) ©Erni/Shutterstock

with the axis of support (figure 22.19). In addition, the vertebrae that make up the vertebral column become larger from the neck to the pelvis as the force of compression increases. Another important skeletal change associated with bipedalism involves a reduction in the size of spinous processes on neck vertebrae. This change is associated with reduced neck musculature required by the positioning of the head on top of the vertebral column, rather than being held horizontally at the end of the column.

Bipedal locomotion is also reflected in the structure of appendages. Unlike the knuckle walking and brachiation of apes, the shorter human forelimbs are not used in bipedal locomotion. The pelvis is short and wide, which transmits weight directly to the legs, maintains the size of the birth canal, and provides surfaces for the attachment of leg muscles. The femur of humans is angled at the knee toward the axis of the body (*see figure 22.19*). Angling the femur places feet under the center of gravity while walking, which results in a smooth stride compared to the "waddle" of other apes.

Although not necessarily directly associated with bipedal locomotion, changes in the skull also accompanied human evolution. The face of humans is less protruding than that of other apes. This change accompanies the expansion of the anterior portion of the skull in association with the enlargement of the brain. Other skull changes include a reduction in the size of jaws, teeth, and the bones that contribute to the ridges above the eyes (supraorbital ridges). The foramen magnum, the opening of the skull for the exit of the spinal cord, is shifted anteriorly in humans. This positioning results in the skull being balanced on top of the vertebral column rather than protruding forward as the skull does in nonhuman apes.

Earliest Hominins

Early hominin evolution occurred between 7 and 5 mya. Recent important discoveries reveal a very bush-like hominin phylogeny that includes multiple species, some of which were contemporaries who lived in relatively close proximity. *Sahelanthropus tchadensis*

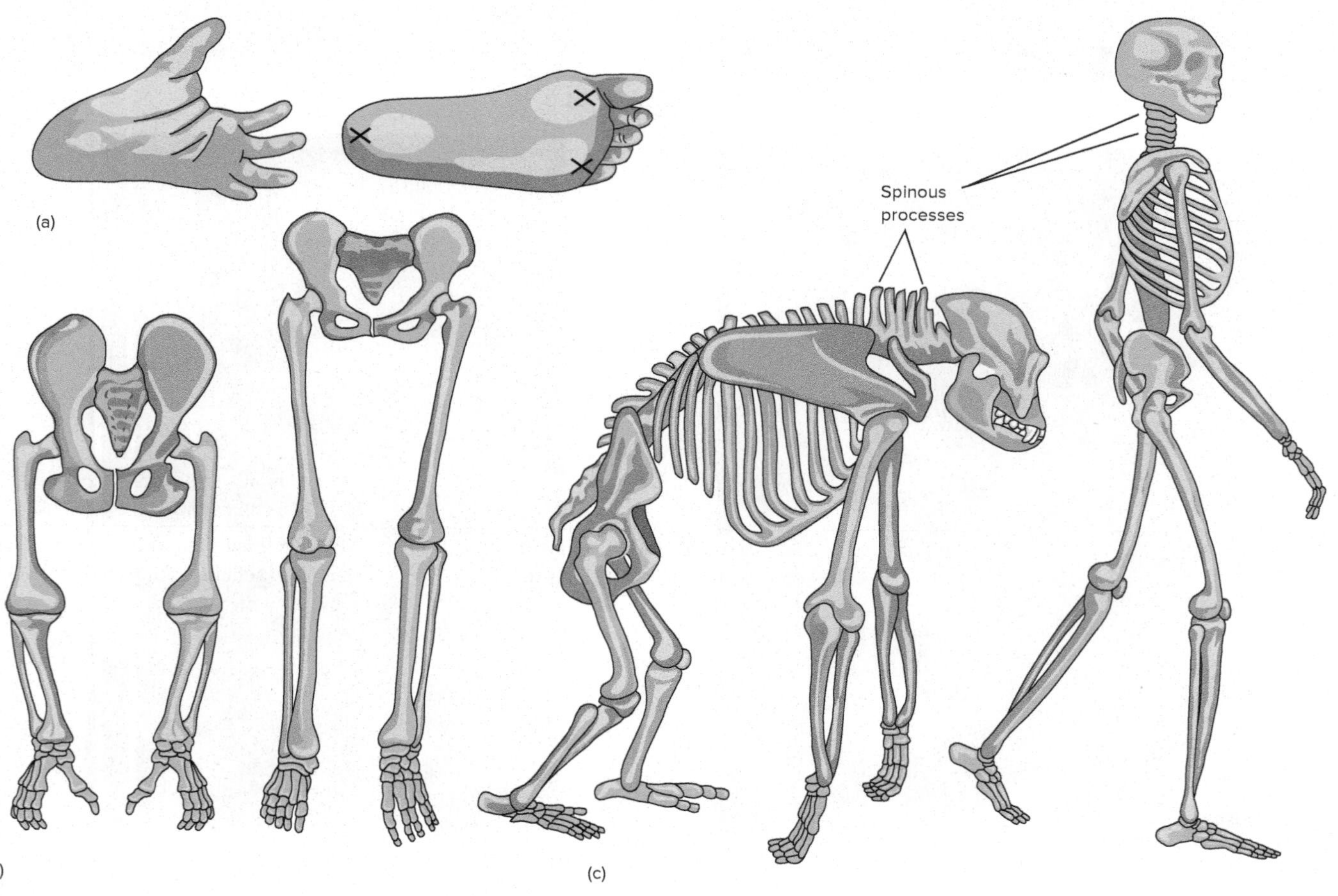

FIGURE 22.19

A Comparison of the Skeletons of Humans and Other Apes. (*a*) The great toe of a human is not opposable; rather, it is parallel to the other toes. Its weight-bearing surfaces (marked by **X**) form a stable tripod. (*b*) The femur of humans is angled toward the body's center of gravity. This makes an upright stride smoother than the "waddle" of other apes. (*c*) The pelvic girdle of humans is relatively short and transfers the weight of the upper body directly to the legs. In other apes more weight is borne by the arms, and the pelvis is more elongate. The greater curvature of the human vertebral column places the vertebrae in line with the body's center of gravity. The vertebral column of other apes is archlike and suspends the body mass below. Large spinous processes of the cervical vertebrae of other apes serve for muscle attachment to support the head at the end of the vertebral column. In humans, the weight of the head is supported by the vertebral column's upright orientation, and the spinous processes of cervical vertebrae are reduced in size.

fossils date to between 6 and 7 mya and show a mixture of ape and hominin features. This species was undoubtedly not a part of the lineage leading to *Homo*, but it has the distinction of being the fossil dating closest to the chimp/human divergence (table 22.3 and figure 22.20). The next oldest hominin fossil is that of *Ardipithecus ramidus,* which has been dated to 5.8 million years. *A. ramidus* shows a mosaic of hominin/ancient ape characteristics, which suggest that this species combined tree climbing with walking on all four limbs—but supporting its weight on the palms of the hand rather than on the knuckles. It also had an anterior position of the foramen magnum and foot-bone structure that suggests intermittent bipedal locomotion or a short-term upright stance, for example, when using hands for holding or carrying objects.

Numerous fossils in the genus *Australopithecus* have been discovered since 1974. These fossils date to more than 4 million years old and provide strong evidence of bipedal locomotion. The discovery of a nearly complete *A. afarensis* in 1974 in East Africa is one of the most famous hominin discoveries of all time (figure 22.21). Dubbed "Lucy," this fossil, and others discovered since 1974 date between 3.9 and 3.0 mya and show pelvis and leg structure that leaves no doubt that this species was bipedal. Females were substantially shorter than males. Height varied between 107 and 152 cm.

TABLE 22.3
SIGNIFICANT EVENTS IN HOMININ EVOLUTION

SPECIES (YEARS BEFORE PRESENT)	BRAIN SIZE AND STATURE	SIGNIFICANT EVENTS	EXTENT OF FOSSIL RECORD
Sahelanthropus tchadensis (7-6 million)	350 cm^3 ? cm Possibly bipedal	Oldest known hominin fossil	Single skull
Ardipithecus ramidus (5.8-4 million)	? cm^3 122 cm Possibly bipedal		Three fossil sites include partial jaw, teeth, and partial arm bones.
Australopithecus anamensis (4.2-3.9 million)	? cm^3 ? cm Probably bipedal		Three fossil sites include partial jaw, humerus, and tibia.
Australopithecus afarensis (3.9-3 million)	375-550 cm^3 107-152 cm Bipedal	Possible divergence point to *Homo* lineage	Multiple fossil sites and numerous individuals, including the 40% complete "Lucy" and another 70% complete specimen.
Australopithecus africanus (3-2 million)	420-500 cm^3 ? cm Bipedal		Multiple fossil sites and numerous individuals. Skull, pelvis, vertebrae, and leg bones. Includes a nearly complete skull of a child about three years old.
Homo habilis (2.4-1.5 million)	500-800 cm^3 127 cm Bipedal	Possibly rudimentary speech. Primitive stone tool use.	Multiple fossil sites with many skeletal remains, including skulls and arm and leg bones.
Homo erectus (1.8 million-300,000)	750-1,225 cm^3 160-180 cm Bipedal	More sophisticated stone tools and fire. Migrated widely out of Africa into Europe and Asia	Multiple fossil sites with many skeletal remains, including skulls and a nearly complete skeleton of "Turkana boy," a 10- or 11-year-old individual discovered near Lake Turkana in Kenya.
Homo heidelbergensis (500,000-200,000)	1,200 cm^3 ? cm Bipedal		Multiple fossil sites with skulls and teeth.
Homo neanderthalensis (230,000-30,000)	1,450 cm^3 170 cm Bipedal	More advanced tools and weapons. Burial rituals. Construction of shelters.	Many fossil sites with nearly complete skeletons.
Homo sapiens (300,000-present)	1,350 cm^3 180 cm Bipedal	More advanced tools and weapons. Developed fine artwork.	Many fossil sites with nearly complete skeletons.

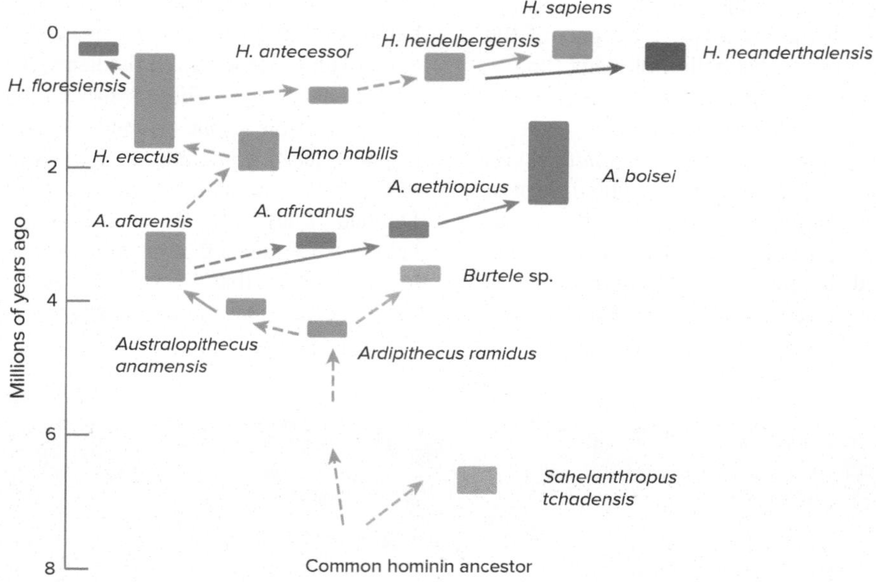

FIGURE 22.20

Human Evolution. This illustration shows the approximate time frames and plausible sequences in human evolution. Solid arrows show fairly certain pathways. Dashed arrows depict uncertain pathways. Colors depict various hominin pathways. Green pathways show the plausible evolutionary sequence leading to modern humans. Numerous species have been omitted and new fossils are being found on a regular basis. Hominin phylogeny is very bush-like. Bipedal adaptations probably arose independently in branches that did not lead to *H. sapiens*. The chimpanzee (*Pan troglodytes*) lineage is poorly documented and is not shown.

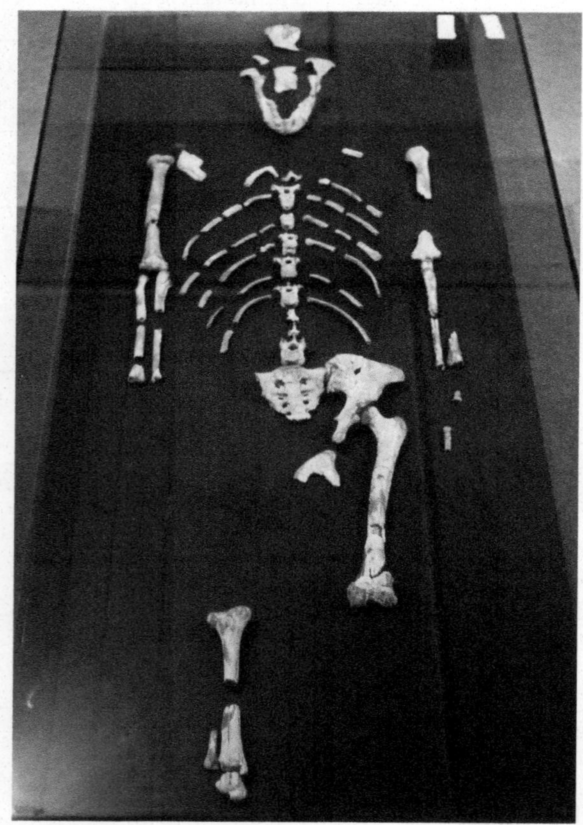

FIGURE 22.21

Australopithecus afarensis. This fossil, named "Lucy," was discovered in 1974. It dates between 3.9 and 3.0 mya and shows pelvis and leg structure that leaves no doubt that this species was bipedal.

A. africanus existed between 3 and 2 mya. Its body and brain were slightly larger than those of *A. afarensis*. The shape of the jaw and sizes of the teeth are more similar to those of *Homo* than are those of *A. afarensis*. Compelling data suggest that *A. afarensis* may be an ancestor of the *Homo* lineage. If that is the case, *A. afarensis* probably represents a point of divergence between the *Homo* lineage and other australopithecines that were contemporaries of early *Homo* species.

Interestingly, a newly described hominin fossil (*Burtele*) has been described from the same time as, and only 50 km distant from, *A. afarensis* fossils. *Burtele* shows fewer bipedal characteristics, for example, it had grasping big toes. As *A. afarensis* foraged for food on the ground, it may have looked into the trees to see *Burtele* looking down at it.

Homo

The criteria for assigning fossils to the genus *Homo* divide paleontologists. There is, however, general acceptance of seven *Homo* species. *H. habilis* is called "handy man" because of the evidence of primitive stone "toolkits" associated with these fossils. The brain size overlapped that of later australopithecines (the low end) and later species of *Homo* (the high end). Casts of their skulls indicate the presence of Broca's area, which is essential for speech. *H. habilis* existed between 2.4 and 1.5 mya. *H. erectus* lived about 1.5 mya (between 1.8 million and 300,000 years ago). *H. erectus* spread widely from its African birthplace. Fossils have been uncovered in Africa, Europe, China (Peking Man), and Indonesia. Artifacts associated with their campsites indicate the use of more sophisticated stone tools, hunting, and fire. One site in China showed ash accumulations 6 m thick!

The following two species were formerly included as subspecies of one species, *H. sapiens.* Taxonomic revisions have resulted in these former subspecies being elevated to species. *H. heidelbergeinsis* appeared about 500,000 years ago and was followed by *H. neanderthalensis.* The latter existed between 230,000 and 30,000 years ago. Neanderthals lived mostly in cold climates, and their body proportions suggest a short (170 cm), solid physique. Their bones were thick and heavy and indicate powerful musculature. They were found throughout Europe and the Middle East and lived in caves and shelters made of wood. They used a diversity of tools made of stone, bone, antler, and ivory and observed burial rituals.

Fossils of a relatively small (adult males weighed around 40 kg and were approximately 150 cm tall) hominid, *Homo naledi*, were found in South Africa in 2015. *H. naledi* is a primitive hominid. It displays skeletal characteristics that are found in both *Homo* and the australopithecines. *H. naledi* possessed an australopithecine-like brain and pelvis, but, among other characteristics, its skull and lower extremities were more *Homo*-like. It is estimated to have lived between 330,000 and 230,000 years ago.

The smallest member of the genus *Homo*, *Homo floresiensis*, is nicknamed "the hobbit." It stood only 106 cm tall and weighed approximately 30 kg. *H. floresiensis* inhabited the island of Flores, Indonesia, between 190,000 and 50,000 years ago. Its small stature may be due to an evolutionary process termed **island dwarfism**. Island dwarfism occurs when a larger species migrates to small, isolated islands and because food resources and predators are limited, successive generations become smaller.

It has long been hypothesized that *H. sapiens* arose in Africa 195,000 years ago. However, recent fossil evidence (from Morocco and Israel, respectively) suggests that *H. sapiens* evolved at least 300,000 years ago, and that they may have migrated outside of Africa as early as 194,000 years ago. In addition to more sophisticated tools for making clothing, sculpting, engraving, and hunting, they produced fine artwork–including spectacular cave paintings like those at Lascaux, France (figure 22.22). Over the 300,000 years of our history, changes toward smaller molars and decreased robustness are evident. Fully modern humans were present about 30,000 years ago.

Cultural Evolution—A Distinctly Human Process of Change

Culture is a system of nongenetic behaviors, symbols, beliefs, institutions, and technology characteristic of a group that is transmitted through generations. Although a rudimentary capacity for culture exists in other animals, culture is considered a predominately human trait. Cultural evolution is said to have occurred when cultures change over time. It is evolution in the sense of change, but it is not organic evolution because it can be based entirely on learning rather than genetic changes. Separating out nongenetic change from genetic change can be very difficult. Our ability to learn from the past, and to pass information to future generations, is linked to the evolution of larger brains on the savannahs of Africa. Genetic and nongenetic evolutionary changes intermingle through most of our evolutionary history. As hunters and gatherers, humans learned to construct tools, shelters, and clothing. They learned to use fire. This knowledge passed rapidly between generations and

FIGURE 22.22

***Homo sapiens* Cave Painting.** Paintings like this one in *Grotte de Niaux*, France, were probably associated with rituals designed to bring good fortune to hunters. These caves show little other evidence of human habitation.

populations–promoted by the development of regions of the brain that supported the capacity for instruction, discussion, and bargaining. As populations grew, humans learned that natural resources are not unlimited. Small campsites became permanent communities as humans learned to cultivate grain crops to supplement hunting and gathering. This was the beginning of the agricultural revolution, which was in full swing in the Fertile Crescent of the Middle East 10,000 years ago. These technologies quickly spread to China about 7,000 years ago and to Central America about 5,000 years ago. The domestication of animals and the advent of metallurgy about 8,000 years ago made both agriculture and warfare more efficient. Bountiful agricultural harvests made it possible for fewer people to supply the needs of large populations. We looked in a new direction, one that has changed the shape of our planet for all times. Industrialization has produced both rewards and scars, such as the imminent extinction of many species. The Industrial Revolution is a reminder that biological evolution and cultural evolution do share a common feature–progress is not guaranteed by either. Finally, we are in the midst of another revolution–the "genetic revolution." We are only beginning to experience its influence. Only the advantage of a historical perspective will reveal its full impact.

Section 22.4 Thinking Beyond the Facts

How were the evolutionary events that occurred early in the human lineage different from the "evolutionary" events of the last 30,000 years?

Summary

22.1 **Evolutionary Perspective**

- Mammalian characteristics evolved from the synapsid lineage over a period of about 200 million years. Mammals evolved from a group of synapsids called therapsids. Extinct cynodont therapsids such as *Cynognathus and Hadrocodium* possessed mammalian-like structures. These animals were critical to the development of hypotheses on the origins of modern mammal groups.

22.2 **Diversity of Mammals**

- Modern mammals include the monotremes, marsupials, and placental mammals.

22.3 **Evolutionary Pressures**

- Hair is uniquely mammalian. It functions in sensory perception, temperature regulation, and communication. Mammals have sebaceous, sudoriferous, scent, and mammary glands. The teeth and digestive tracts of mammals are adapted for different feeding habits. Flat, grinding teeth for mechanically digesting plants are seen in herbivores. Predatory mammals have sharp teeth for killing and tearing prey.
- The feeding habits of mammals are difficult to generalize. A wide variety of food is taken. Food rich in cellulose is difficult for mammals to digest. Gastrointestinal modifications and symbiotic microorganisms make digestions of cellulose possible.
- The mammalian heart has four chambers, and circulatory patterns are adapted for viviparous development. Mammals possess a diaphragm that alters intrathoracic pressure, which helps ventilate the lungs. Mammalian thermoregulation involves metabolic heat production, insulating pelage, and behavior. Mammals react to unfavorable environments by migration, winter sleep, and hibernation.
- The nervous system of mammals is similar to that of other vertebrates. Olfaction and hearing were important for early mammals. Vision, hearing, and smell are the dominant senses in many modern mammals.
- The nitrogenous waste of mammals is urea, and the kidney is adapted for excreting a concentrated urine.
- Mammals have complex behavior to enhance survival. Visual cues, pheromones, and auditory and tactile cues are important in mammalian communication. Most mammals have specific times during the year when reproduction occurs. Female mammals have estrus or menstrual cycles. Monotremes are oviparous. All other mammals nourish young by a placenta.

22.4 **Human Evolution**

- Primates arose about 65 mya. Ancestral primates were likely arboreal and cursorial insectivores. A suite of anatomical modifications including forward-facing eyes for increased depth perception indicate that modern primates had arboreal origins.
- Apes diverged from other primates about 25 mya. Global cooling about 20 mya resulted in the expansion of savannahs in Africa. This change may have caused apes to move from arboreal to terrestrial habitats and selected for bipedalism. The human lineage is traced back about 7 million years to its divergence from that of chimpanzees. The fossil record indicates a very bush-like hominin phylogeny that leaves many unanswered questions.
- The genus *Homo* first appeared about 2.4 mya. Our species arose around 300,000 years ago in Africa and may have migrated from the continent some 194,000 years ago. Hallmarks of *H. sapiens* include complex language and culture, the ability to reason, enhanced logical thinking, and our ability to manipulate the environment to suit our biological needs. Changing human cultures over the last 30,000 years have resulted in agricultural innovations, metallurgy, the Industrial Revolution, and a genetic revolution.

Concept Review Questions

1. Members of this subclass include the monotremes, for example, the spiny anteater and the duck-billed platypus.
   a. Prototheria
   b. Theria

c. Eutheria
d. Metatheria

2. Placental mammals belong to the infraclass
   a. Prototheria.
   b. Theria.
   c. Eutheria.
   d. Metatheria.
3. These glands are associated with hair follicles, and their oily secretion lubricates and waterproofs the skin and hair of mammals.
   a. Sebaceous glands
   b. Sudoriferous glands
   c. Musk glands
   d. Mammary glands
4. Mammals have teeth that are specialized for specific functions. This is called the _____ condition.
   a. homodont
   b. heterodont
   c. odontophore
5. Chimpanzees and humans are all members of the same primate lineage. Which one of the following is the LEAST inclusive shared lineage?
   a. Hominidae
   b. Haplorhini
   c. Hominini
   d. Homininae

Analysis and Application Questions

1. Why is tooth structure important in the study of mammals?
2. What does the evolution of secondary palates have in common with the evolution of completely separated, four-chambered hearts?
3. Why is classifying mammals by feeding habits not particularly useful to phylogenetic studies?
4. Under what circumstances is endothermy disadvantageous for a mammal?
5. What is induced ovulation? Why might it be adaptive for a mammal?
6. Do you think tool use selected for increased intelligence, or did increased intelligence (perhaps selected for by social behaviors) promote tool use? Explain.

23

Protection, Support, and Movement

The chitinous exoskeleton of this flame skimmer dragonfly (*Libellula saturata*), and the muscles associated with the exoskeleton, interact to make flight possible. The exoskeleton also provides support and protection.
Source: Cyndi Souza/USFWS Photo

Chapter Outline

23.1 Integumentary Systems
Invertebrate Integument
Vertebrate Integument
23.2 Skeletal Systems
Invertebrate Skeletons
Vertebrate Skeletons
23.3 Nonmuscular Movement and Muscular Systems
Nonmuscular Movement
Muscular Systems
Skeletal Muscle Structure and Function
Variations in Muscle Structure and Function

In virtually all animal phyla, structures that provide protection, support, and movement are interrelated. Internal bony skeletons of vertebrates provide protection for delicate internal organs. They also support the body and provide points of attachment for muscles, and muscular and skeletal systems act together to promote movement. External chitinous skeletons of arthropods, and their associated muscles, perform all of these functions, and they form the wings that make insect flight possible. These interrelated functions–protection, support, and movement–are discussed in the sections that follow.

23.1 INTEGUMENTARY SYSTEMS

LEARNING OUTCOMES

1. Describe invertebrate integument and integumentary adaptations that provide protection in various invertebrate groups.
2. Contrast integumentary structure and function in fishes, amphibians, reptiles (including birds), and mammals.
3. Describe adaptations of the skin of amphibians, reptiles (including birds), and mammals for life in terrestrial environments.
4. Explain the difference between hair and nails.

The external covering of an animal is called the **integument** (L. *integumentum*, cover). The integument shows adaptations for the animal's environment. It may serve as a surface for the exchange of gases, nutrients, and wastes; protection from injury and invasion by microorganisms; regulation of body temperature; the reception of environmental stimuli; and prevention of desiccation. These important functions occur in what is often the largest organ in an animal's body.

Invertebrate Integument

The integument of many invertebrates consists of a single layer of epithelial cells. This layer is usually called the **epidermis** (Gr. *epi*, upon + *derm*, skin) and is an outer epithelial layer (figure 23.1). The epidermis rests on a thin extracellular layer of collagenous fibers, called the basement membrane, that separates the epidermis from underlying muscles or other tissues.

The outer epidermis of an invertebrate may be modified for many functions described in the previous text. Cilia are present in the epidermis of free-living flatworms (e.g., Platyhelminthes, Turbellaria) that move by ciliary locomotion. Epidermal glands may secrete mucus and noxious chemicals that protect the animal (e.g., Mollusca, Gastropoda). Other

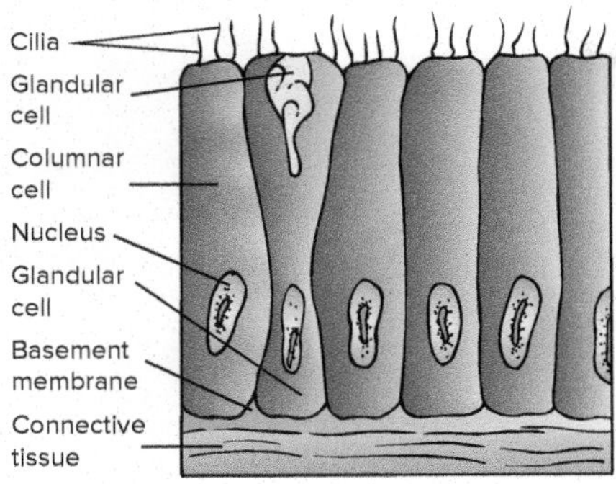

FIGURE 23.1

Integument of Invertebrates. The integument of many invertebrates consists of a simple layer of columnar epithelial cells (epidermis) resting on a basement membrane. A thin layer of connective tissue lies under the basement membrane. Cilia and glandular cells may or may not be present.

glands may secrete skeletal materials like calcium carbonate (e.g., Cnidaria, anthozoan hard corals, *see figure 9.20*). The epidermis of parasitic flatworms (Cestoidea and Trematoda, *see figure 10.11*) is a syncytial tegument that protects against host defenses and aids tapeworms in the absorption of nutrients from the host digestive tract.

All ecdysozoans (*see chapters 13 to 15*) possess a nonliving cuticle that is secreted by the epidermis. These cuticles are either collagenous or chitinous. In the arthropods, cuticles are hardened by the deposition of calcium carbonate or by a tanning process called sclerotization. When the epidermis is overlain by a cuticle, the epidermis is often referred to as a **hypodermis** (Gr. *hypo*, below + *derma*, skin, figure 23.2). (This hypodermis should not be confused with the hypodermis of vertebrate integument, which is described in the next section.) Cuticles provide protection, but they also restrict growth. In order to accommodate growth, new cuticle is periodically formed beneath the old, and then the old cuticle is shed (*see figure 14.5*).

Some invertebrate integuments are comprised of an epidermis and, below the basement membrane, a connective tissue layer called the **dermis**. In the Echinodermata, this dermis secretes the calcium carbonate ossicles that comprise endoskeletal elements.

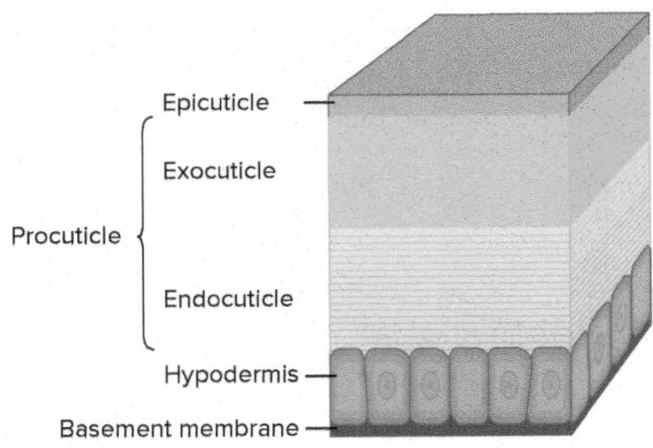

FIGURE 23.2

Cuticles. The cuticle of an arthropod. The underlying hypodermis secretes the cuticle.

Source: Russell-Hunter, WD. 1979. *A life of invertebrates*. New York (NY): Macmillan.

Vertebrate Integument

The vertebrate integument is often simply called **skin**. As with invertebrates, the outer layer is called the epidermis. The epidermis is often comprised of multiple cell layers and is underlain by a basement membrane. Below the basement membrane is the dermis, which is comprised of collagenous and other connective tissue fibers and cells. The vertebrate hypodermis, in contrast to that described for ecdysozoans, consists of connective tissues (including adipose), nerve endings, and blood vessels. It separates the integument from deeper tissues, such as muscle tissue.

The Skin of Fishes

The epidermis of fishes is multilayered and richly supplied with glandular cells. Hagfishes (class Myxini) and lampreys (class Petromyzontida) have a scaleless integument with mucous-secreting cells (figure 23.3). Granular cells contribute to a mucous coat that reduces friction between the water and the surface of the animal. This coat also protects against bacterial, and other, infections. Thread cells in hagfishes produce cords of mucus that are secreted through multicellular slime glands. Slime is secreted when the hagfish is irritated and discourages predators whose gills may be clogged by copious slime discharge. The epidermis of other fish is similarly endowed with mucous-secreting granular cells. Epidermal cells also secrete pheromones (*see chapter 25*) that communicate danger or other chemical messages to members of the same species.

Unlike the hagfishes and lampreys, other fishes have some form of dermally derived **scales** (Fr. *escale*, shell, husk) that protect a fish and are oriented in a fashion that reduces friction as the fish swims. In Chondrichthyes (sharks and their relatives), small placoid scales are in the form of **denticles** (L. *denticulus*, little tooth) arising from, and seated in, the dermis. They are comprised of a bone-like calcified matrix called dentine, and each scale projects through the epidermis. The epidermis secretes a covering of enamel over the dentine core (figure 23.4). Denticles wear and are replaced through the life of the fish.

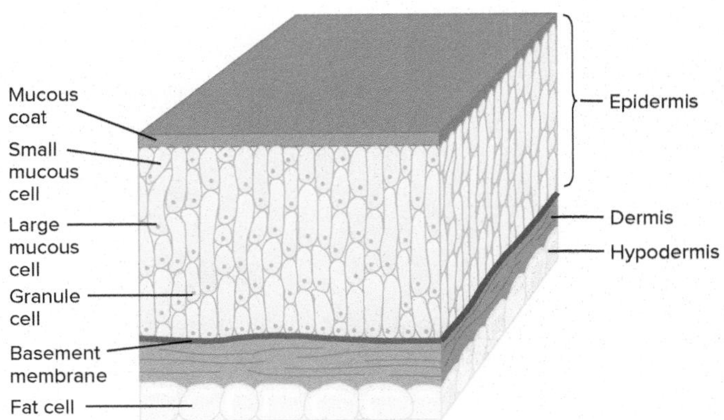

FIGURE 23.3

Skin of Jawless Fishes. The skin of an adult lamprey has a multilayered epidermis with glandular cells and fat storage cells and an underlying hypodermis.

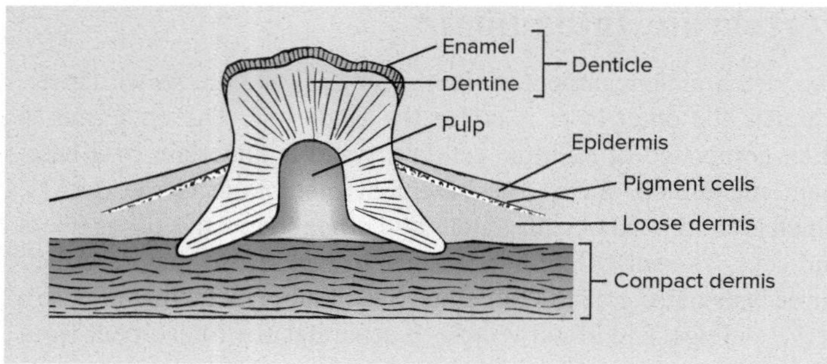

FIGURE 23.4

Skin of Cartilaginous Fishes. Shark skin contains toothlike denticles that become exposed through the loss of the epidermal covering. The skin is otherwise fishlike in structure.

The scales of bony fishes (Actinopterygii and Sarcopterygii) are also dermally derived; however, these scales do not lose their covering of epidermal cells (figure 23.5). Scales are classified based on structure, but they are all comprised of a layering of dermal bone, dentine, and an epidermally derived layer of enamel. Scales of bony fishes are associated with the dermal blood capillaries, and they grow throughout the life of a fish. Growth patterns produce the rings that are useful in determining the age of a fish.

The Skin of Amphibians

The earliest amphibians retained some of the bony dermal elements in the integument of their fish ancestors. During amphibian evolution, the skin lost this bone and became very thin and moist. This thin, moist skin presented three problems associated with a partially terrestrial existence: the danger of desiccation, the damaging effect of ultraviolet radiation, and physical abrasion. A number of adaptations help alleviate these challenges.

The epidermis of amphibian skin is layered (stratified) and gives rise to mucus and granular glands. The base of these glands extends into the dermis, but they are epidermal in origin. Mucus helps prevent desiccation and keeps skin moist, which promotes cutaneous gas exchange. Granular glands (modified serous glands) produce noxious fluids that act as predator deterrents (figure 23.6).

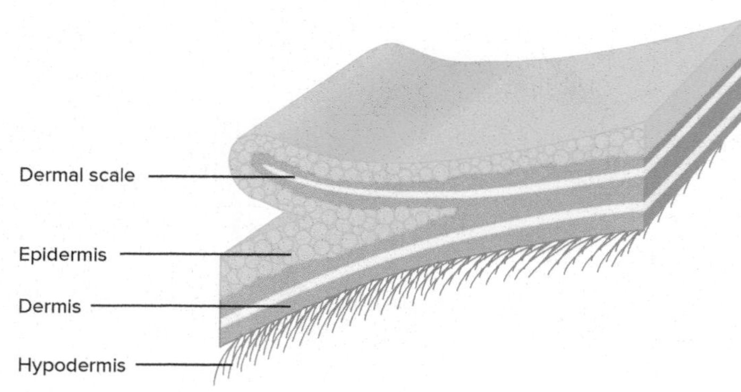

FIGURE 23.5

Skin of Bony Fishes. The skin of a typical bony fish has overlapping scales (two are shown here). The scales are layers of collagenous fibers covered by a thin, flexible layer of bone.

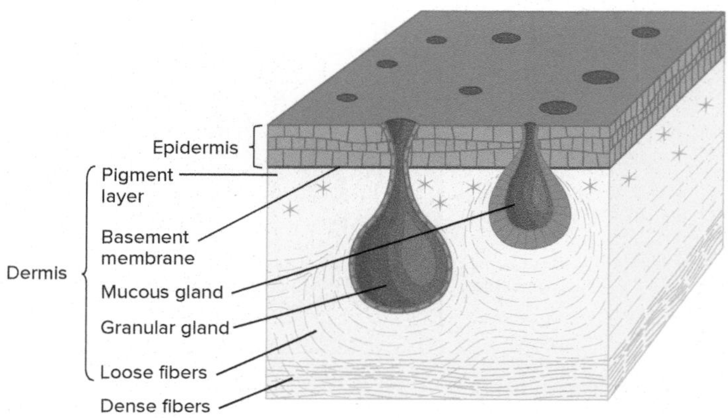

FIGURE 23.6

Skin of Amphibians. Frog skin has a stratified epidermis and several types of glands in the dermis. Notice the pigment layer in the upper part of the dermis.

Keratin is a tough, impermeable protein that is produced by epidermal cells of tetrapods. It is especially prevalent in the skin of amniotes, but it also helps protect amphibians from desiccation and damaging effects of ultraviolet light.

The dermis of amphibians is comprised of fibrous connective tissue, nerves, blood vessels, and pigment cells called chromatophores. Dermal thickenings and aggregations of epithelial glands may produce bumps and sculpturing often referred to as "warts."

The Skin of Reptiles

Many reptiles live their lives away from an aquatic environment, and their integument reflects adaptations to terrestrialism. The outer epidermis (stratum corneum, figure 23.7) of nonavian reptiles is thickened and highly keratinized. The resulting scales are

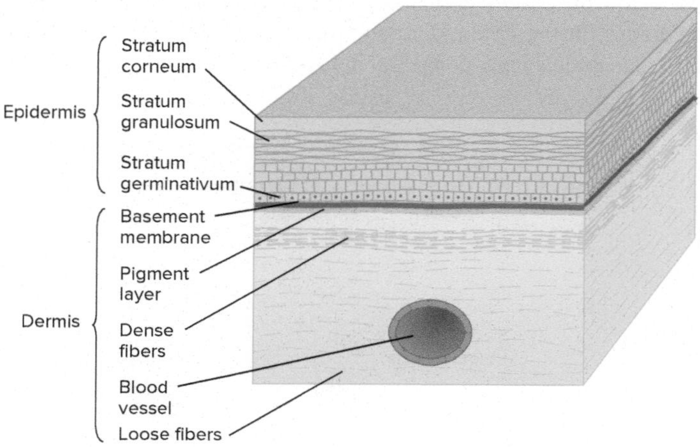

FIGURE 23.7

Skin of Reptiles. Lizard skin has the heavily keratinized outer epidermis (scales) characteristic of reptiles. Notice the absence of integumentary glands, making reptilian skin exceptionally dry.

epidermal in origin (in contrast to the dermal scales of fishes) and protect from physical abrasion and desiccation. Epidermal keratinization also contributes to the formation of scutes (thickened ventral scales) in snakes, beaks in turtles, and claws in most reptiles. The keratinized portion of the epithelium is periodically shed during ecdysis when fluid accumulates between the old stratum corneum and the lower epidermal layers. It may be shed in one piece or in smaller flakes.

The skin of avian reptiles is usually thin and only two or three cell layers thick (figure 23.8). There is less keratinization in the outer epidermis; however, feathers are derived from the stratum corneum and are highly keratinized structures (*see chapter 21 and figure 21.4*).

The dermis of avian reptiles contains blood and lymphatic vessels, nerves, and sensory receptors. The dermis and hypodermis of avian reptiles contain adipose tissue and air spaces that aid in thermoregulation. Smooth muscle fibers in these layers help position feathers during flight and behaviors involved with thermoregulation, courtship, and other functions.

The Skin of Mammals

The epidermis of mammalian skin is stratified. Mitotic cell division in the deepest layer of the epidermis replaces cells as they are sloughed from the skin surface. As cells near the surface of the skin, they become keratinized, die, and form the stratum corneum.

Hair and a variety of glands are epidermal in origin but seated in the dermis. **Hair** is a derived character diagnostic of mammals first seen in Mesozoic (160 mya) mammalian fossils. A mammal's coat of hair (pelage) is important in protection, insulation, sensory perception, and communication. Hair is seated in invaginations of the epidermis, called follicles (figure 23.9). The portion of the hair within a follicle is the root, and the portion above the follicle is the shaft. The hair grows from the base of the follicle, which is supplied with dermal blood capillaries. A band of smooth muscle, the arrector pili muscle, runs between the hair follicle and the lower epidermal layer. When this muscle contracts, the hair is pulled to an erect position, which is used in mammals to increase insulating qualities of pelage and in behavioral displays. Neurons associated with the root of a hair make the hair an important sensory receptor.

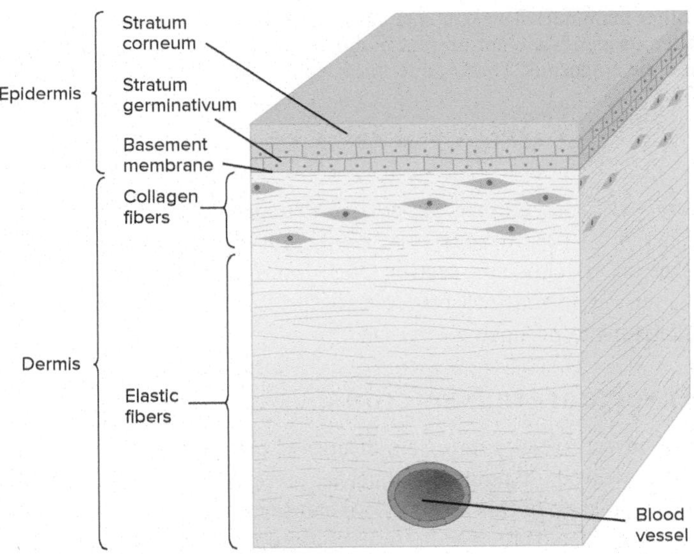

FIGURE 23.8

Skin of Birds. Bird skin has a relatively soft and thin epidermis with no epidermal glands.

Different types of hair are adapted for specific functions. The longer, thick hair of a mammals' pelage is protective and is called **guard hair**. Finer, shorter **under hair** increases the number of air spaces within the pelage and enhances insulating qualities of a mammal's coat. The longest and thickest hair is often found around a mammal's snout. These **vibrissae** are well endowed with nerve endings and have sensory functions.

A variety of glands are also epidermally derived (*see figure 23.9*). **Sebaceous (oil) glands** (L. *sebum*, tallow or fat) release lubricating and waterproofing sebum into hair follicles and ultimately to the surface of the skin. **Sudoriferous (sweat) glands** (L. *sudor*, sweat) are present in many mammals and promote evaporative cooling. **Scent** or **musk glands** secrete pheromones that may be involved with defensive, courtship, and other behaviors. **Mammary glands** are another derived character for Mammalia. They are probably evolutionarily related to a type of sweat gland and supply milk for newborn mammals (see *figure 22.8*).

Other epidermal derivatives of mammalian skin include nails (primates only), claws, and hooves (*see figure 22.7*). These structures form from accumulations of keratin that cover the terminal phalynx (bone) of the digits and provide protection for the tips of the digits and aid in digging and climbing.

The dermis of mammalian skin is comprised of layers of tough fibrous connective tissue. It contains blood vessels that supply the skin (including the avascular epidermis) with nutrients and promote gas exchange. The dermis contains a variety of sensory receptors for temperature, pressure, touch, and pain.

The mammalian hypodermis is comprised of a less fibrous, loose connective tissue and is the location for adipose (fat) tissue. Blood vessels and nerves run through the hypodermis into the dermis (*see figure 23.9*).

Skin and hair color in mammals serves a variety of functions. Pigments in the skin may absorb or reflect light, including potentially harmful ultraviolet wavelengths from the sun. Pigments are produced by chromatophores during the formation of skin and hair cells. Different forms of melanin are the main mammalian skin and hair pigments that are formed in the deepest layers of the epidermis. Skin color is also influenced by blood flow through the dermal layers of the skin, and hair color is influenced by the presence of air pockets in the hair shaft. Skin and hair colors are used in camouflage and other forms of crypsis (*see chapter 6*), aposematic coloration, and social behaviors involved with advertising reproductive status or social rank.

Section 23.1 Thinking Beyond the Facts

What are some of the skin adaptations that occurred in amphibians as this group evolved a life divided between land and water?

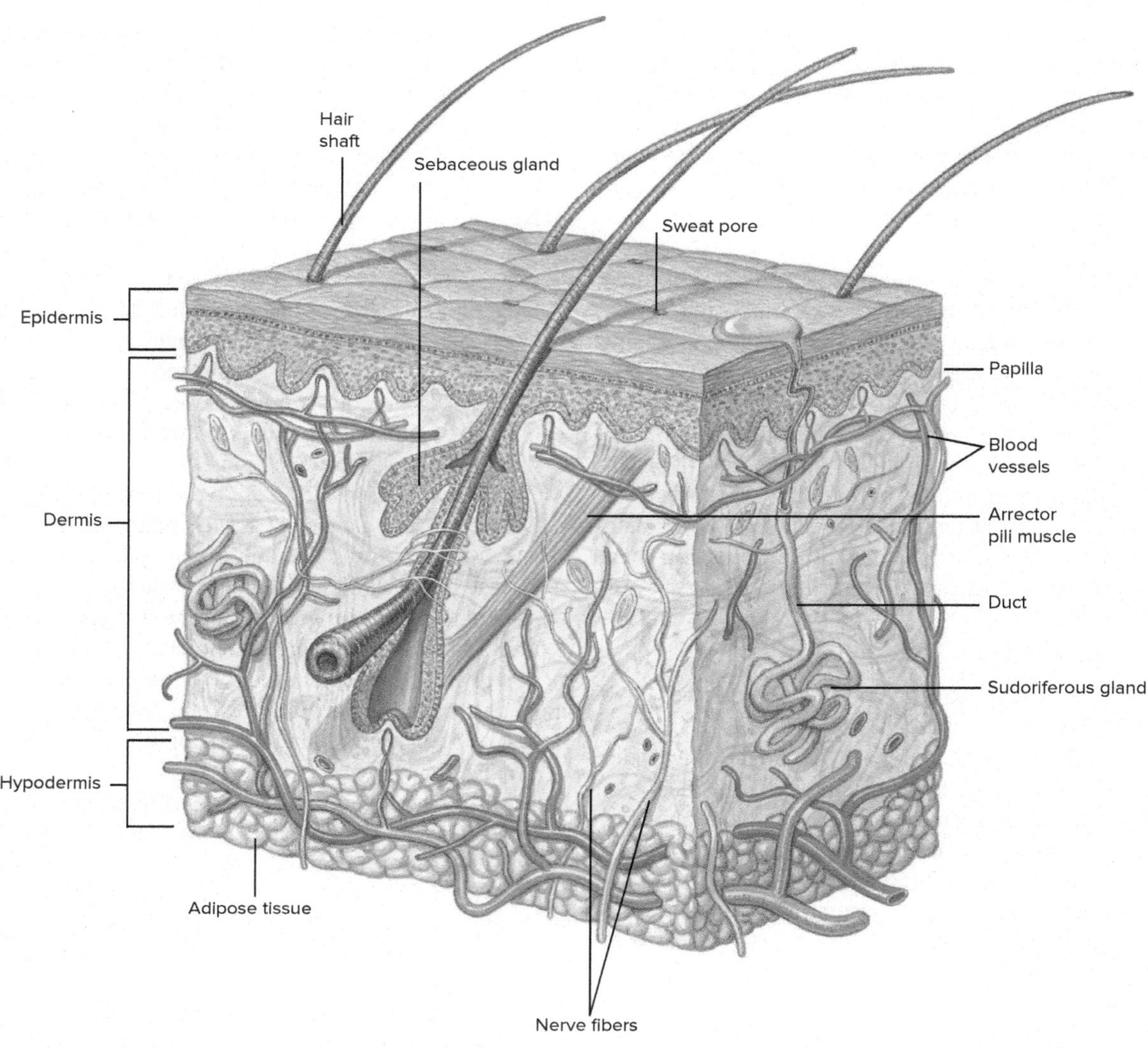

FIGURE 23.9

Skin of Mammals. Human skin has many features in common with the skin of other mammals. The skin is comprised of three regions: epidermis, dermis, and hypodermis. Epidermal derivatives include hair and glands. (Sudoriferous glands are not present in most mammals.) The dermis is comprised of connective tissues and contains the blood supply for the dermis and the epidermis. The hypodermis is comprised of loose connective and adipose tissues.

23.2 SKELETAL SYSTEMS

LEARNING OUTCOMES

1. Compare hydrostatic skeletons, exoskeletons, and endoskeletons.
2. Explain how animals with hydrostatic skeletons achieve movement, support, and locomotion.
3. Contrast the limitations and advantages of an invertebrate's exoskeleton.
4. Describe mineralized tissues of invertebrates.
5. Contrast the structure and function of the endoskeletons of fishes and tetrapods.

The evolution of multicellularity and increased body size was accompanied by a diverse set of skeletal structures that provided support for an animal. Supportive structures did not evolve alone. They must permit body movement. This section discusses animal skeletal systems, and section 23.3 describes how contractile cells (often organized into muscular tissues and organs) interact with skeletons to produce movement.

Invertebrate Skeletons

The invertebrates comprise a diverse assemblage of organisms that have evolved in numerous lineages over more than 600 million years. In spite of this diversity, we can see some common (although often unrelated) themes in the evolution of supportive structures.

Hydrostatic Skeletons

Hydrostatic (Gr. *hydro*, water + *statikos*, to stand) **skeletons** have arisen many times in animal evolution. They are formed when

(a)

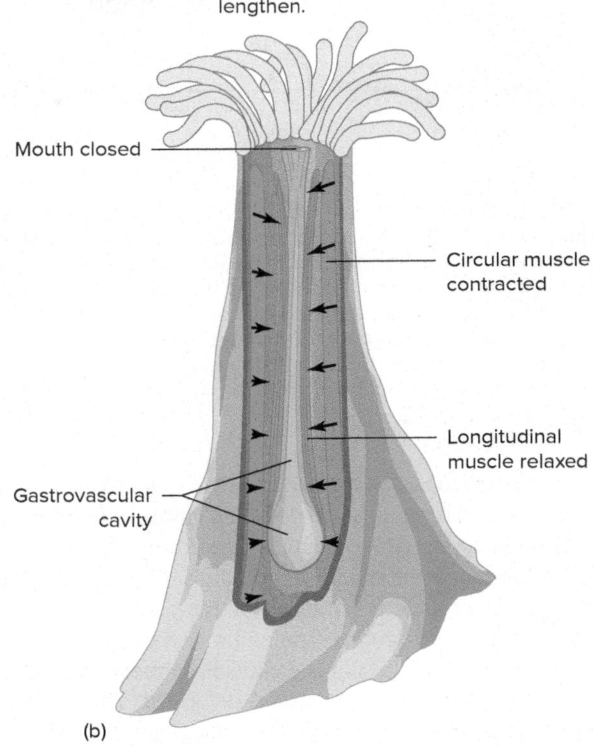

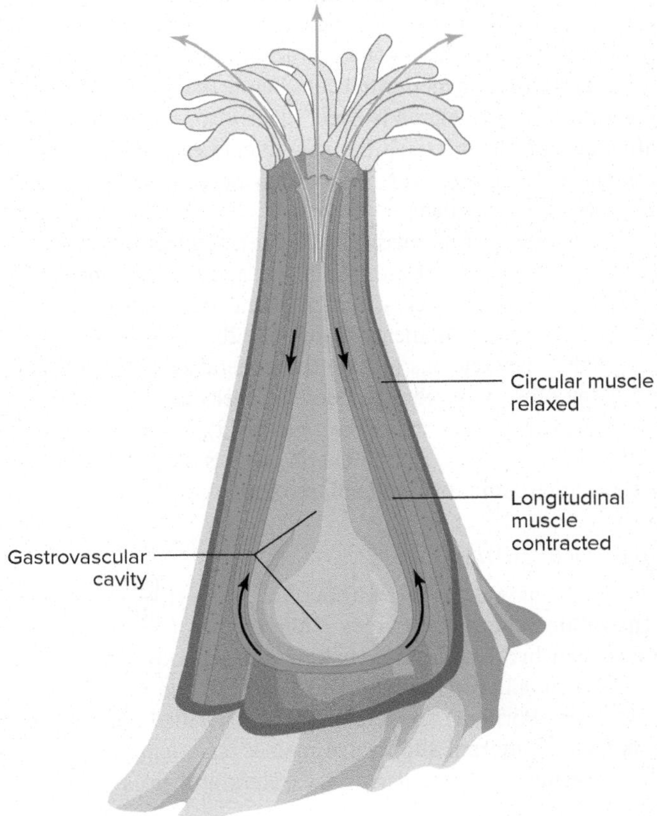

FIGURE 23.10

Hydrostatic Skeletons. (*a*) The hydrostatic skeleton of sea anemones (*Corynactis californica*) allows them to shorten when longitudinal muscles contract, or to lengthen when circular muscles contract. (*b*) Contraction of circular muscles and relaxation of longitudinal muscles compresses water within the hydrostatic gastrovascular cavity. The anemone lengthens. (*c*) Contraction of longitudinal muscles and relaxation of circular muscles causes the anemone to shorten as water escapes through the open mouth.

water or a body fluid is enclosed in a resistant sheath of muscle and/or body wall elements. Contractions of the body wall against the enclosed fluid causes shape changes or movements of the animal. For example, a sea anemone's gastrovascular cavity acts as a hydrostatic chamber. When its mouth is closed to prevent the escape of water, contractions of circular muscles cause the anemone to elongate (figure 23.10*a,b*). If circular muscles relax and longitudinal muscles contract, the anemone shortens and thickens; or if the mouth is open the animal collapses (figure 23.10c), which is a defensive response. Other examples of hydrostatic skeletons can be found throughout chapters 9 through 13. Hydrostatic cavities include the pseudocoelom of rotifers, acanthocephalans (*see chapter 10*), nematodes, and nematomorphs (*see chapter 13*). The coelom acts as a hydrostatic cavity in the annelids (*see chapter 12*). A unique form of hydrostatic skeleton, called a **hydraulic skeleton**, occurs in molluscs (*see chapter 11*). Fluids (blood) are enclosed in, and squeezed through, tissue spaces of their open circulatory system to extend a portion of the head-foot, tentacle, or another body part.

Exoskeletons

Members of the Ecdysozoa have cuticles that are shed to accommodate growth. In arthropods the cuticle is in the form of an **exoskeleton** (Gr. *exo*, outside). It provides structural support, protection, impermeable surfaces for preventing desiccation, and a system of levers for muscle attachment and movement (*see figures 23.2 and 14.3*). The exoskeleton is one of many (but not minor) reasons for the success of the Arthropoda.

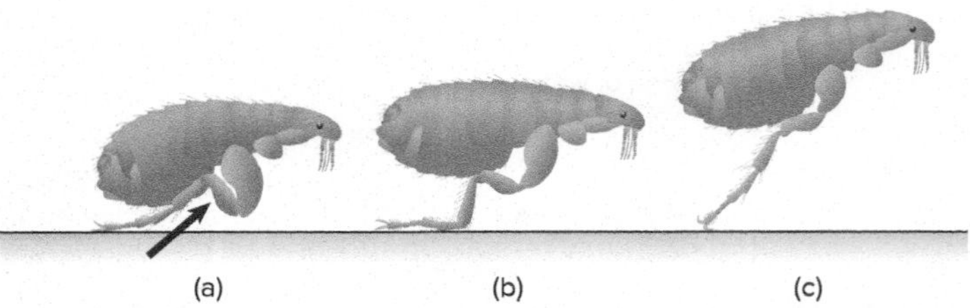

FIGURE 23.11

Jump of Flea. A flea has a jointed exoskeleton. (*a*) When a flea is resting, the femur (black arrow) of the leg (for simplicity, only one leg is shown) is raised, the joints are locked, and energy is stored in the deformed elastic protein ("animal rubber" or resilin) of the cuticle. (*b*) As a flea begins to jump, the relaxation of muscles unlocks the joints. (*c*) The force exerted against the ground by the tibia gives the flea a specific velocity that determines the height of the jump. The jump is the result of the quick release of the energy stored in the resilin of the cuticle.

The exoskeleton is comprised of a waxy outer layer called the epicuticle. It is a barrier to water loss and invasion by microorganisms. The inner layers (the procuticle) are formed from a leathery polysaccharide called chitin and several kinds of proteins. The outer portion of the procuticle is hardened by the deposition of calcium salts (crustaceans) or a protein tanning process called sclerotization (insects and arachnids). Resilin is a cuticular protein that stores and releases energy when the exoskeleton is deformed. This resilience of the cuticle is important in wing movements of insects during flight (*see figure 15.12*) and jumping by some insects (figure 23.11) and other arthropods.

Growth by an arthropod enclosed in an exoskeleton requires that the exoskeleton be periodically shed in a process called ecdysis (figure 23.12*a*, *see figure 14.5*). Preceding ecdysis a new exoskeleton is secreted by hypodermal cells below the existing exoskeleton. When the old exoskeleton is shed, the arthropod engorges its body with water or air to stretch the new pliable exoskeleton to a larger size, and the vulnerable arthropod may become secretive until the new exoskeleton hardens.

Movement of an exoskeletal armor requires joints, which are regions where the exoskeleton is thinner and less hardened. Muscles acting on appendages across a joint create a very effective lever system for arthropod movements (figure 23.12b, *see figure 14.4*).

Other exoskeletons include the mineralized shells of molluscs and many other invertebrates. Mollusc shells are secreted by modified tissue called the mantle (*see figure 11.3*). Mineralization occurs when calcium carbonate or a mineral oxide is deposited within a collagenous matrix to provide a hard, protective covering.

Endoskeletons

Endoskeletons (Gr. *endo*, within) are supportive structures that form within other body tissues. They are less common in invertebrates than hydrostatic skeletons or exoskeletons. The calcareous ($CaCO_3$) or siliceous (SiO_2) spicules, and the protein spongin, of sponges (Porifera) are secreted by cells called archeocytes and comprise an internal supportive matrix (*see figures 9.3 and 9.4*). The skeletons of echinoderms (sea stars and their relatives) are

(a)

(b)

FIGURE 23.12

Exoskeletons. (*a*) A cicada nymph (*Platypedia*) leaves its old exoskeleton as it molts. This exoskeleton provides external support for the body and attachment sites for muscles. (*b*) In an arthropod, muscles attach to the interior of the exoskeleton. In this articulation of an arthropod limb, the cuticle is hardened everywhere except at the joint, where the membrane is flexible. Notice that the extensor muscle is antagonistic to (works in an opposite direction from) the flexor muscle.

(a) ©Steven P. Lynch (b) Source: Russell-Hunter, WD. 1979. *A life of invertebrates*. New York (NY): Macmillan.

calcareous ossicles that are embedded in the body wall. Even the protruding spines of sea urchins are formed internally and exposed when enclosing epithelial layers are worn away. These skeletal elements are highly modified in echinoderms to form rigid tests, jaw structures, pincer-like pedicellariae, and tiny ossicles in the body wall of sea cucumbers (*see chapter 16*).

Vertebrate Skeletons

The endoskeletons of vertebrates are composed of two major regions. The **axial skeleton** includes the skull, vertebral column, sternum, and ribs. The **appendicular skeleton** is composed of the appendages, the pectoral girdle, and the pelvic girdles. These girdles attach the upper and lower appendages to the axial skeleton. This section describes the two major types of supportive tissues that comprise vertebrate skeletons–cartilage and bone–and variations in vertebrate skeletal structure.

Animation Bone Growth in Width

Vertebrate Skeletal Tissues

The two major vertebrate skeletal tissues are connective tissues (*see chapter 2*). They consist of a fiber matrix containing the universal animal protein, collagen, and a solid or semisolid ground substance. Cells are present within the matrix in spaces called lacunae.

Cartilage is the primary skeletal tissue of chondrichthyan fishes. In other vertebrates, cartilage is present at joints where it promotes smooth skeletal movements and serves as a point of attachment for muscles. The matrix of cartilage is usually flexible, but may be hardened by mineralization. The cells that maintain cartilage are called chondrocytes (figure 23.13). Different types of cartilage are determined by the fiber types that predominate, for example, elastic fibers in the elastic cartilage that supports the pinna of your ear as compared to collagenous fibers in the hyaline cartilage of your nose.

Bone is the primary skeletal tissue of bony fishes and all tetrapods. Dermal bone forms within the dermis of the skin. (Recall the discussion of fish scales.) Endochondral bone forms when bone tissue replaces a cartilaginous model during development. Endochondral bone comprises most components of vertebrate endoskeletons. It provides points of attachment for muscles, protection for internal organ systems, reservoirs for storage of calcium and phosphate ions, and sites for the production of blood cells. The matrix of bone is highly mineralized; and, therefore, it is solid.

In a long bone, such as the vertebrate femur, bone develops in a series of concentric rings. One set of these rings is called an osteon, which is supplied with blood by vessels running through a central osteonic canal. Osteocytes maintain bone and are nourished by this blood supply (figure 23.14).

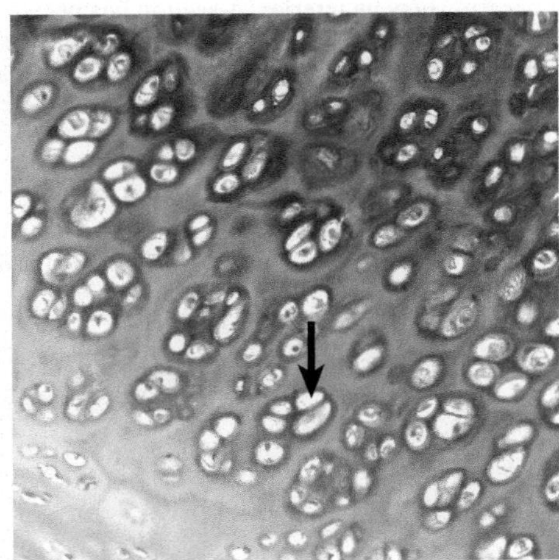

FIGURE 23.13

Hyaline Cartilage. Cartilage cells (chondrocytes) are located in lacunae (arrow) surrounded by intercellular material containing fine collagenous fibers (LM ×160). Hyaline cartilage functions in support and reinforcement.

The Skeleton of Fishes

The axial skeleton, which includes the skull and vertebral column, is the major skeletal feature of fishes. The vertebral column is divided into trunk (including ribs) and the tail regions (figure 23.15). The axial skeleton is comprised primarily of bone, except in the Hyperotreti (hagfishes), Petromyzontida (lampreys) and Chondrichthyes (*see chapter 18*), where cartilage predominates. The appendicular skeleton, pectoral and pelvic fins of Chondrichthyes and bony fishes, is attached to the axial skeleton through tendons and ligaments. Girdle bones are absent. Lampreys and hagfishes lack paired appendages.

The Skeleton of Tetrapods

Tetrapods support themselves on land against the pull of gravity. Amphibians appeared over 400 million years ago (mya) and skeletal adaptations seen in the fossil record include the evolution of a neck (cervical) vertebra from the first trunk vertebra, which facilitated head mobility and feeding on insect prey. A posterior trunk vertebra served for the attachment of larger hind appendages and became the sacral vertebra, which provided support along the water's edge (figure 23.16). Further skeletal changes in the amniote lineages (*see figure 20.3*) included increases in the number of cervical and sacral vertebrae, the division of the trunk into thoracic (with ribs) and lumbar (without ribs) regions, and the evolution of a pectoral girdle for the attachment of fore appendages to the axial skeleton. All of these changes improved support in terrestrial habitats and increased flexibility and freedom of movement. Many other skeletal adaptations occurred during tetrapod evolution and are described in chapters 19 through 22.

Section 23.2 Thinking Beyond the Facts

What limitations does an exoskeleton impose on terrestrial invertebrates?

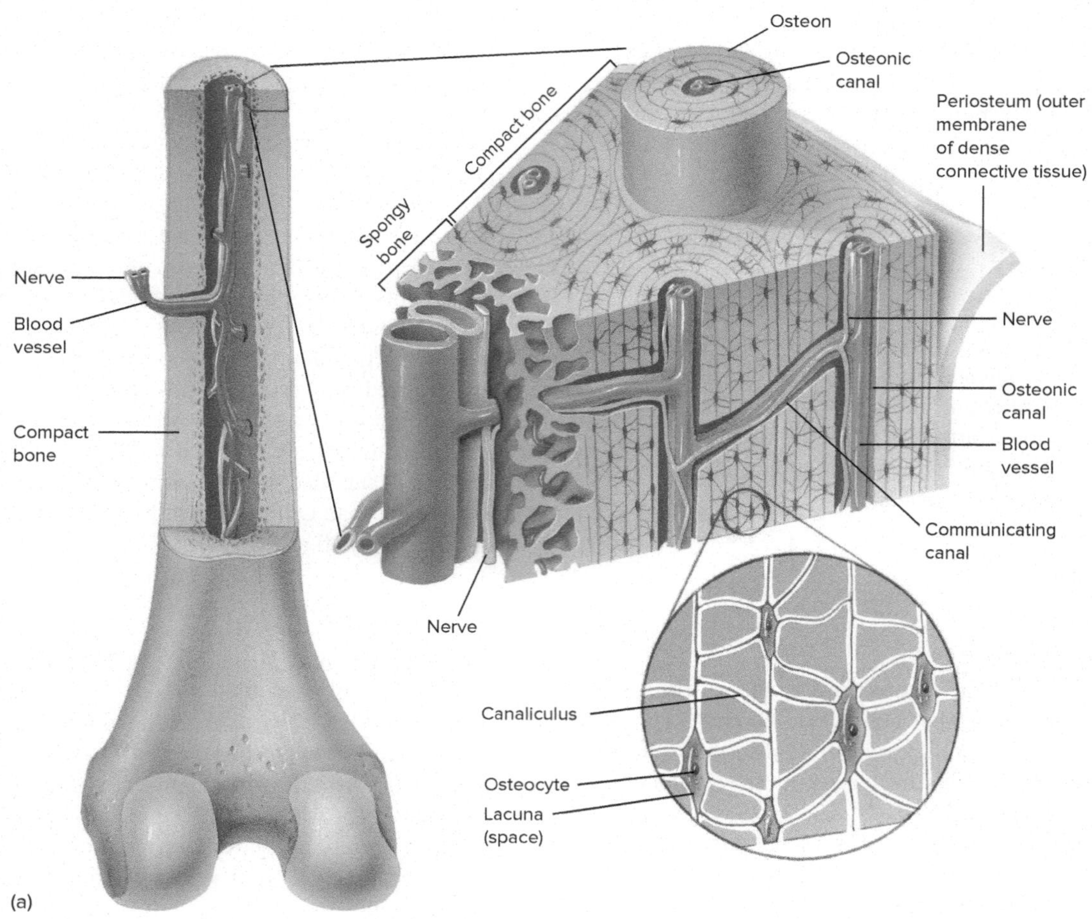

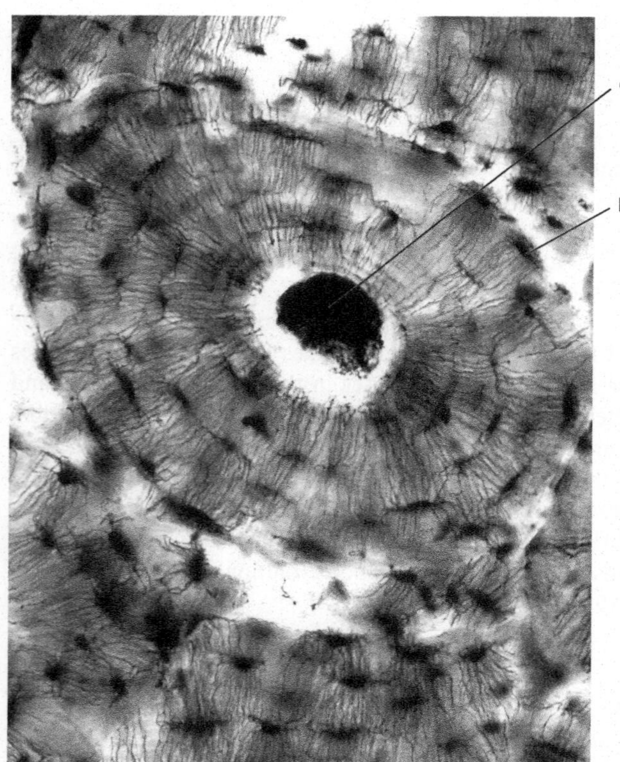

FIGURE 23.14

The Structure of Bone. (*a*) A mammalian humerus (on the left) is partially sliced open to show its interior. A section has been removed and magnified on the right to show the difference between the outer compact bone and the inner spongy bone. Details of the basic layers, osteonic canals, and osteocytes in the lacunae can be seen. (*b*) A single osteon in compact bone (SEM ×450).

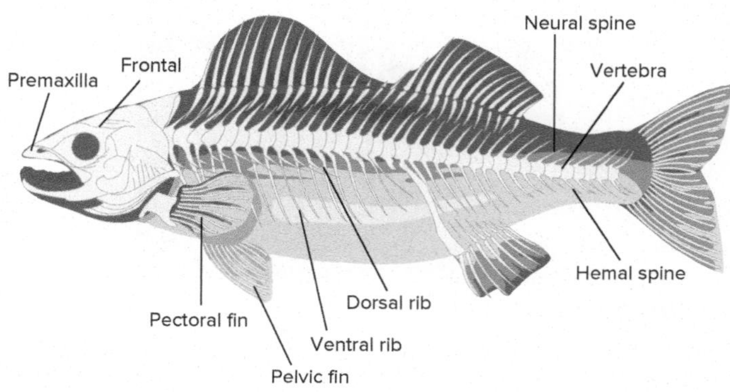

FIGURE 23.15

Fish Endoskeleton. Lateral view of the perch skeleton.

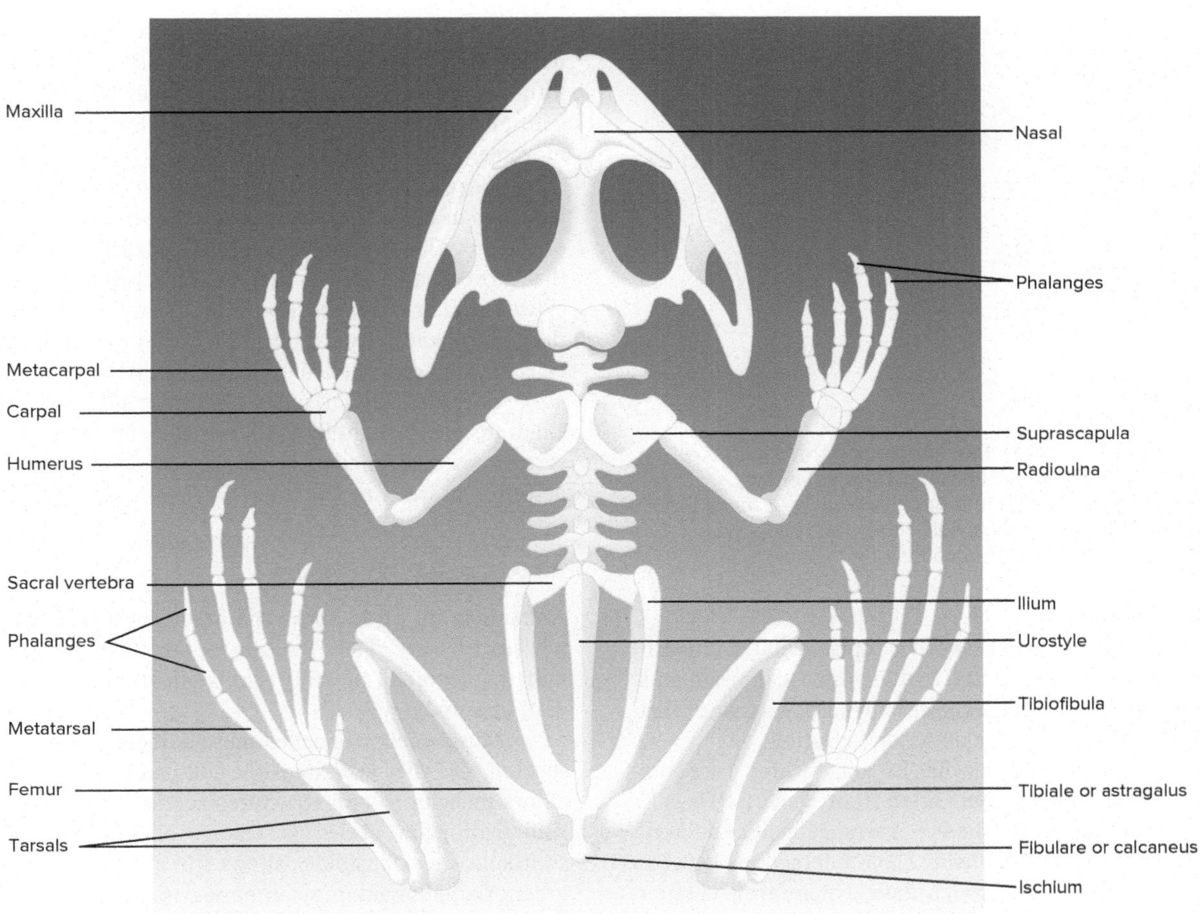

FIGURE 23.16
Tetrapod Endoskeleton. Dorsal view of the frog skeleton.

23.3 NONMUSCULAR MOVEMENT AND MUSCULAR SYSTEMS

LEARNING OUTCOMES

1. Describe three types of nonmuscular movement.
2. Explain the sliding-filament mechanism of muscle contraction.
3. Compare the structure and function of the three types of muscle tissues.

Virtually all animals, and many animal cells, are capable of movement. From the archeocytes of a sponge's mesenchyme to the white blood cells of a mammal's circulatory system, many independent cells have the capacity for ameboid movement. Flagella propel sperm cells, and cilia move materials along cell surfaces. Muscle cells in animals from the Cnidaria through the Chordata work together to produce muscular movements. The common entities in all of these movements are cellular microfilaments that comprise the cytoskeleton (*see figure 2.22*c). These movements are described in this section.

Nonmuscular Movement

Ameboid movement is one of three forms of nonmuscular movement found in animals. Ameboid movement is named based on its function in the amoebozoan protists (*see appendix C*). This same mechanism of movement also occurs in ameboid cells of virtually all animal phyla, including the archeocytes of sponges (phylum Porifera), sperm cells of arthropods and nematodes, and certain white blood cells (neutrophils and monocytes) of mammals (phylum Chordata).

During ameboid movement, a more fluid interior cytoplasm (endoplasm) flows into lobe-like extensions of the cell called pseudopodia (sing., pseudopodium) (Gr. *pseudes*, false + *podium*, little foot) that advance across the substrate. This advancing region, called the fountain zone, adheres to the substrate by adhesion proteins on the plasma membrane. The fluid endoplasm changes into a less fluid ectoplasm as the endoplasm reaches the tip of a pseudopodium. At the opposite end of the cell (the recruitment zone), ectoplasm changes into endoplasm and begins flowing in the direction of advancement. As the cell moves, the plasma membrane appears to be sliding, or rolling, over the ectoplasm (*figure 23.17*). The

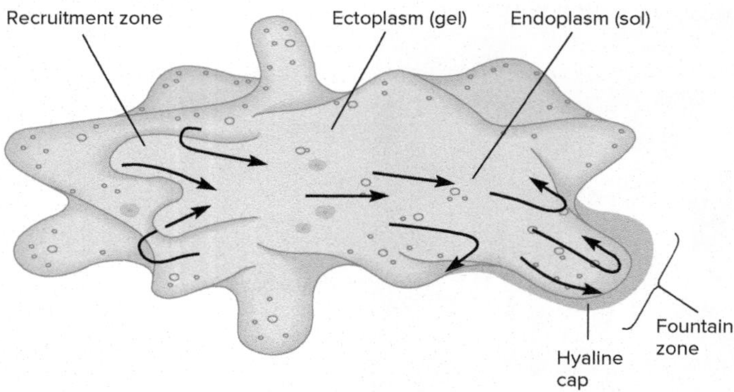

FIGURE 23.17

Mechanism of Ameboid Movement. Endoplasm (sol) flows into an advancing pseudopodium. At the tip (fountain zone) of the pseudopodium, endoplasm changes into ectoplasm (gel). At the opposite end (recruitment zone) of the ameboid cell, ectoplasm changes into endoplasm and begins flowing in the direction of movement.

change of state of the cytoplasm between ectoplasm and endoplasm involves interactions between actin and myosin microfilaments, mediated by calcium ion signaling. Specific hypotheses that explain the details of these interactions are under investigation.

Cilia (L. "eyelashes") and flagella (L. "small whips") are common features of all eukaryotes but differ from bacterial flagella. Eukaryotic cilia and flagella have the 9 + 2 arrangement of microtubules described in chapter 2 (*see figure 2.23*). Interactions between adjacent doublet microtubules through their dynein arms create a sliding between the adjacent doublets and a bending of the cilium or flagellum. Cilia and flagella are essentially the same organelle, the main differences are in their length and their patterns of movement.

Cilia are shorter and more numerous than flagella, and their movements occur in coordinated waves. For example, cilia in mammalian respiratory passages beat slightly out of phase with one another so that ciliary waves move trapped dust particles out of the respiratory passages (figure 23.18). Similarly, bands of cilia in the comb rows of ctenophores (*see figure 9.22*) beat in coordinated waves that move these animals in the water column. Large

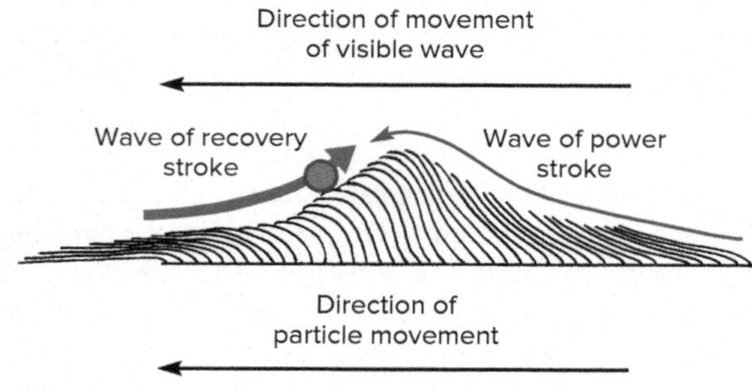

FIGURE 23.18

Ciliary Movement. Cilia move in coordinated (metachronal) waves along the surfaces of animal tissues. In this illustration, the blue particle is being swept across the ciliated surface from right to left.

turbellarian flatworms and nemertines (*see chapter 10*) have a densely ciliated epidermis. Coordinated ciliary waves produce a form of locomotion called **ciliary creeping**, which is often combined with muscular contractions to produce looping and other complex movements.

Flagella usually occur singly in animals, and they are much longer than cilia. They are present in sperm cells and choanocytes of sponges (*see figures 29.6 and 9.3*). Flagella beat in planar waves that move away from the base of the flagellum. In sponge choanocytes, flagellar waves result in water movement through the base of their collar and outward toward the tip of the flagellum. In sperm cells, the cell is propelled opposite the direction of flagellar waves.

Both cilia and flagella also have nonlocomotor functions. Cilia are present within the inner ear and lateral line systems of vertebrates and function in hearing, equilibrium and balance, and detection of water movements (*see figures 24.19 and 24.22*).

Muscular Systems

Animals in all phyla above the basal phyla (*see appendix* B) have muscular tissues. The evolution of muscular tissue in animals permitted a degree of mobility that is unparalleled by any other group of organisms.

Muscle cells are called **muscle fibers**. They can contract and shorten, and they are extensible and can be stretched. Contraction allows muscles to pull structures across hinged joints or, working with hydrostatic skeletons, change body shape. After a muscle contracts, its extensibility allows the muscle to be stretched back to its original position by an opposing muscle without being torn. (Muscle cells cannot lengthen by their own power.) Muscle cells, like nerve cells, are excitable. Their membranes receive and conduct the electrical signals that stimulate contraction. Elasticity of muscles allows them to return to their original shapes after being stretched.

Muscular tissue is divided into three subtypes. Two of these subtypes have alternating light and dark banding patterns when viewed under microscopes. This banding pattern results from a very regular arrangement of contractile microfilaments that are responsible for their contractile mechanisms. They are striated muscles and include skeletal muscle and vertebrate cardiac muscle. The third muscle subtype has similar microfilaments, but the microfilaments are less regularly arranged. This muscle subtype is smooth muscle. The contractile mechanism of skeletal muscle is described in the following paragraphs. It is then followed by a discussion of cardiac and smooth muscles.

Skeletal Muscle Structure and Function

Skeletal muscle is named for its association with animal skeletons. Its structure and function is best known in vertebrates, where it is associated with an endoskeleton. Axial skeletal muscle predominates in fishes (figure 23.19), where it is arranged in segmental blocks called **myomeres**. Muscle fibers of a myomere run between connective tissue sheets creating the familiar zig-zag pattern of fish musculature. This pattern is responsible for the side-to-side bending of a fish

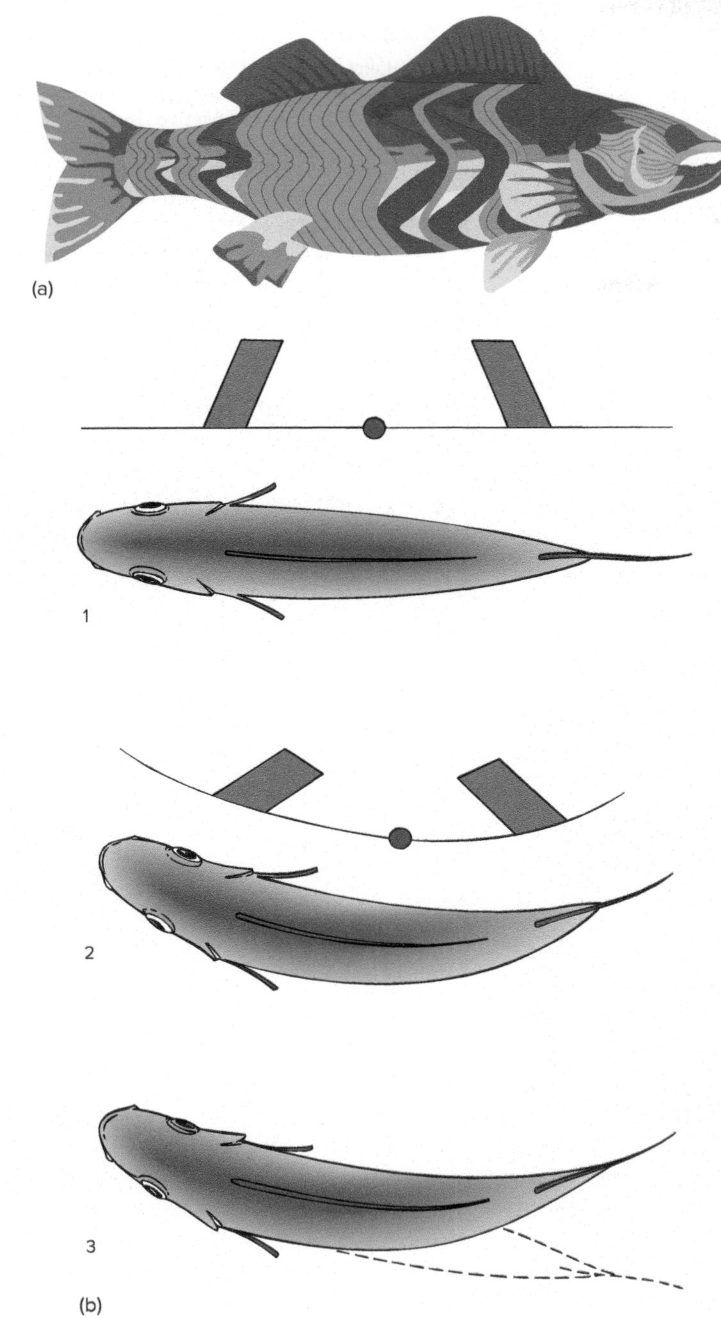

FIGURE 23.19

Fish Musculature. (*a*) Skeletal muscles of a bony fish (perch), showing mainly the large muscles of the trunk and tail. These muscles occur in blocks called myomeres separated by connective tissue sheaths. Notice that the myomeres are flexed so that they resemble the letter W tipped at a 90° angle. The different colors (red, orange, blue) represent different myomeres. (*b*) Fish movements based on myomere contractions. (*1*) Muscular forces cause the myomere segments to rotate rather than constrict. (*2*) The rotation of cranial and caudal myomere segments bends the fish's body about a point midway between the two segments. (*3*) Alternate bends of the caudal end of the body propel the fish forward.

during its locomotion. In tetrapods, most skeletal muscle is usually associated with the appendages (figure 23.20*a*). Fibrous connective tissue bands and sheets, called **tendons**, attach muscle to bone or muscle to muscle to transmit the force of contraction across joints in the skeleton. The elasticity of tendons helps to smooth the contractile force and store energy for recoil of the muscles during their relaxation phase.

Skeletal muscles are comprised of cells (myofibers) bundled within the muscle. Myofibers of skeletal muscle are striated and multinucleate. Because myofibers extend the entire length of a muscle, contraction occurs along the entire length of a skeletal muscle. Myofibers are comprised of collections of subcellular bundles called **myofibrils** (figure 23.20*b, c*). A myofibril is the location of the regularly arranged microfilaments (*see figure 2.22*c) responsible for skeletal muscle's striated histology (figure 23.20*d*) and its contraction. These microfilaments (called myofilaments) are the proteins of muscle contraction—actin and myosin.

Actin and myosin slide past one another to shorten a myofiber and accomplish contraction. **Actin** myofilaments are made of helical strings of globular proteins arranged into a netlike pattern and interconnected across the width of a myofibril at Z lines (figure 23.20*e*). Actin myofilaments are thinner than myosin myofilaments and are often called thin filaments. **Myosin** myofilaments (thick filaments) are arranged between, and overlap, adjacent actin myofilaments in a fashion that results in the striated histology. Dark (A) bands are regions of the myofibril where actin and myosin overlap, and light (I) bands are regions where there is only actin. Myosin molecules have globular projections called cross-bridges (heads*) that extend toward adjacent actin myofilaments. Actin, myosin, and other proteins associated with the Z line-to-Z line distance contain the molecular machinery needed for contraction. This Z line-to-Z line distance is called a **sarcomere**. To set the stage for understanding contraction, notice in figure 23.21 that the sarcoplasmic reticulum (a form of endoplasmic reticulum) is associated with each sarcomere.

The neuromuscular junction is the synaptic connection comprised of a motor neuron ending and a myofiber (figure 23.21). A nerve impulse (action potential, *see chapter 24*) that travels to the myofiber along the neuron causes the release of a chemical neurotransmitter, called acetylcholine, into a 30-nanometer (nm) cleft (the neuromuscular cleft) between the neuron and the muscle cell membrane. Receptors for acetylcholine on the myofiber's plasma membrane (**sarcolemma**) open ion channels. Diffusion of ions creates a sarcolemmal action potential, which is then conducted along the entire length of the myofiber. Invaginations of the sarcolemma into the muscle cell, called **transverse (T) tubules,** conduct the action potential into the interior of the cell. Interactions between the T-tubule action potential and the sarcoplasmic reticulum initiate the release of calcium ions (Ca^{2+}) from the sarcoplasmic reticulum into the cytosol. Calcium ions then interact with sarcomere proteins. This interaction leads to the sliding of actin relative to myosin and is described next (figure 23.22).

*Some authors reserve the term "cross-bridge" for the condition described in the following paragraphs in which a myosin projection (then called "head") is bound to an active site on actin. Other authors use the terms "cross-bridge" and "head" interchangeably, as we have chosen to do here. In this sense, a cross-bridge cycles between free and attached states.

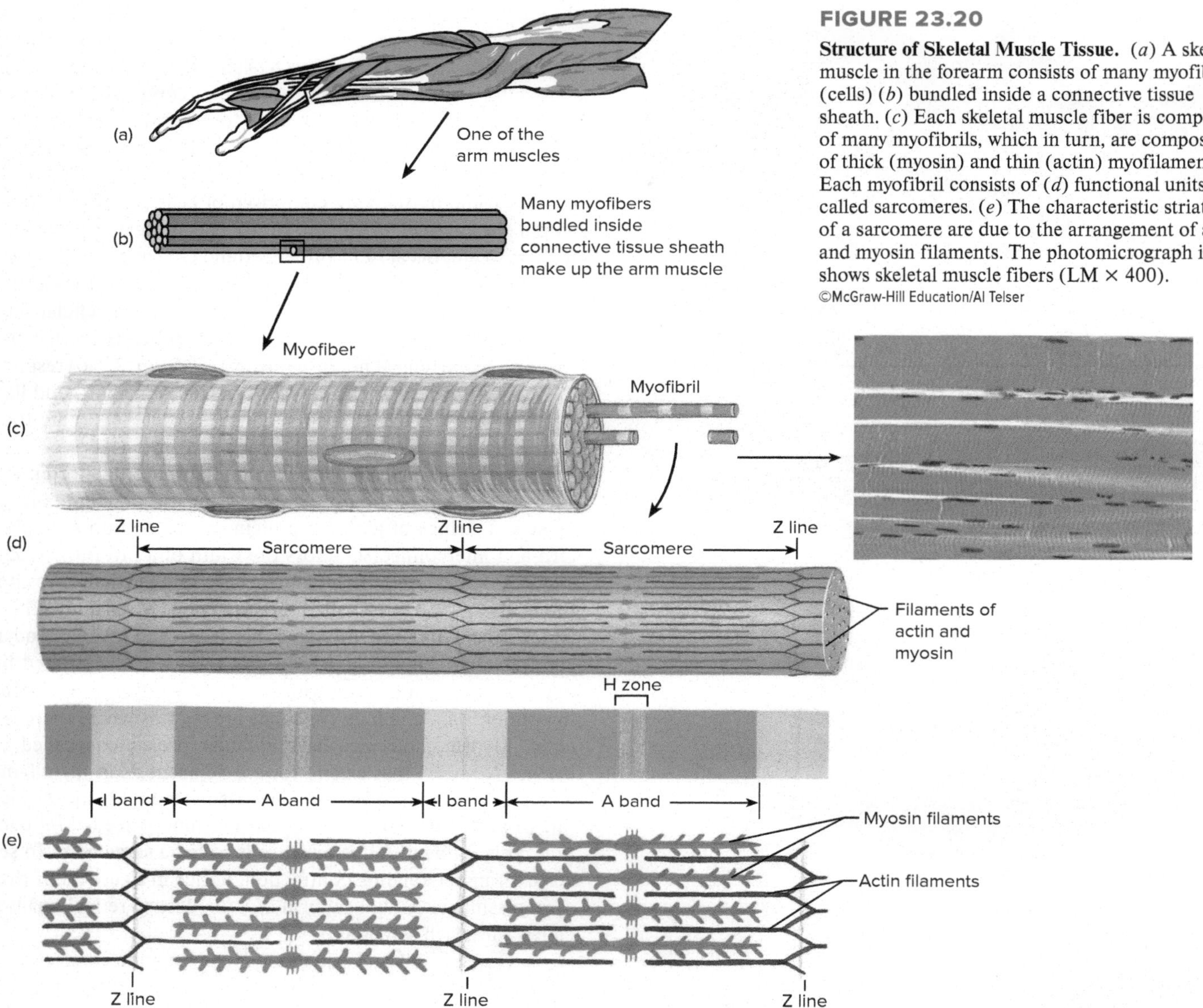

FIGURE 23.20

Structure of Skeletal Muscle Tissue. (*a*) A skeletal muscle in the forearm consists of many myofibers (cells) (*b*) bundled inside a connective tissue sheath. (*c*) Each skeletal muscle fiber is composed of many myofibrils, which in turn, are composed of thick (myosin) and thin (actin) myofilaments. Each myofibril consists of (*d*) functional units called sarcomeres. (*e*) The characteristic striations of a sarcomere are due to the arrangement of actin and myosin filaments. The photomicrograph inset shows skeletal muscle fibers (LM × 400).

Two additional proteins are associated with actin. One is a globular protein called troponin, which binds calcium ions when they are released into the cytosol. Troponin is associated with a second protein, tropomyosin. Tropomyosin is a filamentous protein that winds around actin and covers active sites on actin where myosin cross-bridges will bind. However, the active sites must first be uncovered. Calcium ions binding to troponin cause a conformational change in that molecule, which is then conferred to tropomyosin. It is this conformational change that uncovers the active sites to which myosin cross-bridges bind (figure 23.23).

Actin slides relative to myosin when cross-bridges change conformation (*see figure 23.22*). This conformational change is the cross-bridge power stroke, and it requires an expenditure of ATP (*see figure 2.11*). A cross-bridge that is ready to bind actin and undergo its power stroke has already been energized via hydrolysis of a cross-bridge-bound ATP to ADP and P_i. Energized cross-bridges are ready for their power stroke. A cross-bridge undergoes its power stroke, and actin slides relative to myosin, when (*1*) calcium ions bind troponin, (*2*) tropomyosin uncovers actin's active sites, (*3*) an activated cross-bridge spontaneously binds an active site on actin, (*4*) the stored energy in the cross-bridge is released powering its stroke, and (*5*) ADP is released from the cross-bridge. In order for a cross-bridge to release from actin following a power stroke, another ATP must bind to the cross-bridge. When this ATP is hydrolyzed, the released cross-bridge is activated again for the next cycle of binding and stroking. As long as calcium ions and ATP are present, cross-bridges will continue to be energized, attach to actin, undergo power strokes, release from actin, and be reactivated. These functions result in a decrease in sarcomere lengths and contraction of the entire myofiber. When the impulse at the neuromuscular junction ceases, calcium ion release from the sarcoplasmic reticulum stops, and calcium ions are actively transported from the myofiber's cytosol back into the sarcoplasmic

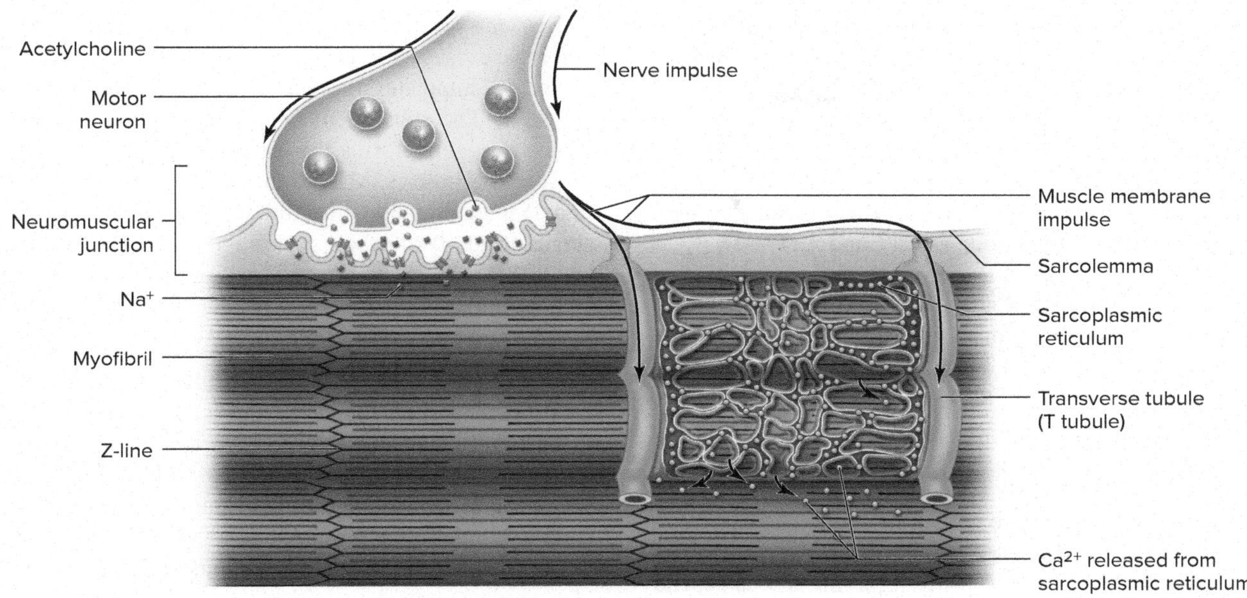

FIGURE 23.21

Conduction across a Neuromuscular Junction. A neuromuscular junction is the synaptic connection between a motor neuron and a myofiber. The neurotransmitter acetylcholine is released into the junction and initiates an impulse on the muscle cell membrane (sarcolemma). The myofiber impulse is conducted along the sarcolemma and into the interior of the fiber through transverse tubules. This transmission initiates the release of Ca^{2+} from sarcoplasmic reticulum into the cytoplasm where it is available to initiate the sliding of actin and myosin filaments.

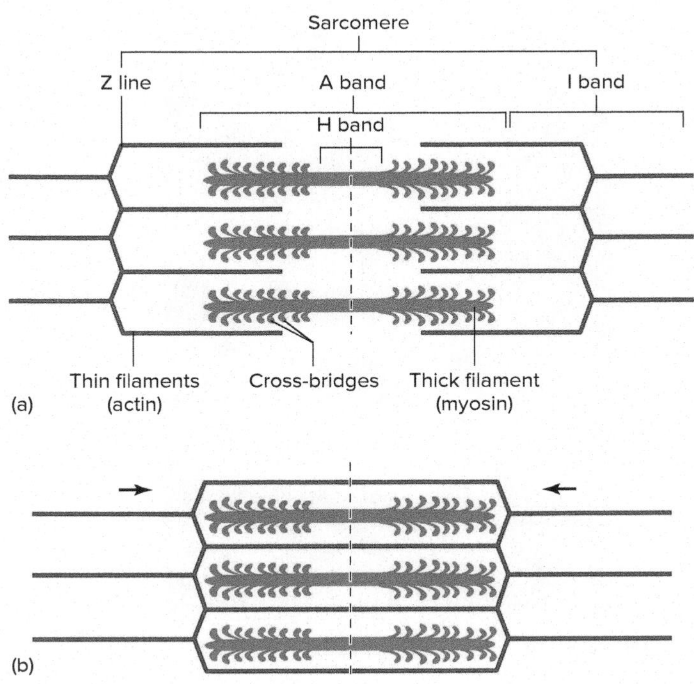

FIGURE 23.22

Sliding-Filament Model of Muscle Contraction. (*a*) A sarcomere in a relaxed position. (*b*) As the sarcomere contracts, the myosin filaments form attachments of its cross-bridges to the actin filaments. Nodding of the cross-bridges pulls the actin filaments so that they slide past the myosin filaments. Compare the length of the sarcomere in (*a*) to that in (*b*).

reticulum. The drop in the cytosol calcium ion concentration causes the release of these ions from troponin, and tropomyosin returns to its original position covering actin's active sites. Cross-bridges can no longer bind actin, and the myofiber relaxes. These events can be summarized as follows:

- Nerve impulse is transferred across a neuromuscular junction to the muscle sarcolemma.
- T-tubules carry the impulse to the interior of the cell and initiate Ca^{2+} release from sarcoplasmic reticulum.

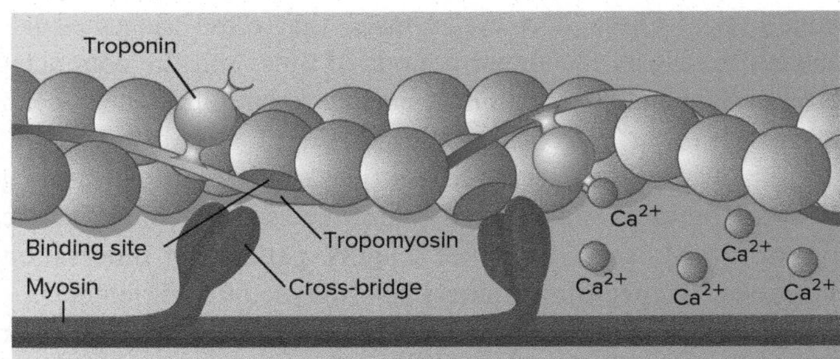

FIGURE 23.23

Model of the Calcium-Induced Changes in Troponin That Allow Cross-Bridges to Form between Actin and Myosin. The attachment of Ca^{2+} to troponin moves the troponin–tropomyosin complex, which exposes a binding site on the actin. The ATP-activated myosin cross-bridge can then attach to actin and undergo a power stroke. The cross-bridge releases from the actin binding site when it is activated by another ATP molecule.

Animation Sarcomere Contraction

Animation Breaking of ATP and Cross-Bridge Movement

- Ca^{2+} binds troponin, causes a conformational change in tropomyosin, and exposes cross-bridge binding sites on actin.
- Cross-bridges, previously activated by ATP binding and hydrolysis, attach to binding sites on actin.
- Cross-bridges undergo their power stroke, ADP is released, and actin slides.
- ATP binds cross-bridges causing cross-bridge release from actin.
- ATP hydrolysis energizes cross-bridges for the subsequent cycle.
- Filament sliding continues as long as Ca^{2+} and ATP are present.
- When nerve impulses cease, Ca^{2+} is actively transported back into the sarcoplasmic reticulum and relaxation occurs.

Variations in Muscle Structure and Function

Skeletal muscles of vertebrates are not all the same. They have contractile characteristics based on aspects of their cellular metabolism. Some fibers, called twitch fibers, contract at varying speeds depending on how quickly they convert ATP to ADP and Pi (inorganic phosphate). They also differ in their ability to sustain contracted states (their tone). Tone is a function of blood supply, the amount of oxygen-storing pigment (myoglobin) that they contain, and the number of mitochondria that they possess. Slow twitch muscle fibers have many mitochondria, slow ATP conversion rates, and abundant blood and myoglobin supplies. Thus, they can sustain contractions for long periods and are prevalent in postural muscles of the vertebrate spinal column. Fast twitch muscle fibers are similar, but they have higher ATP conversion rates. Fast twitch muscle fibers are found in rapidly contracting endurance muscles, like flight muscles of birds. Twitch fibers with high ATP conversion rates but few mitochondria, a poor blood supply, and little myoglobin provide little endurance; however, artificial selection has resulted in a prevalence of these muscles in domestic fowl (i.e., white breast meat). Other fibers, called tonic fibers, are more common in amphibians and reptiles than they are in other vertebrates. They contract very slowly because they have slow ATP conversion rates and very slow cross-bridge cycles. These slow cycles mean that cross-bridges remain attached to myosin for longer periods of time, and the contracted state can be maintained with little energy expenditure. The appendage muscles of turtles contain tonic fibers.

Many invertebrates also have striated muscles whose histological organization is similar to that described for vertebrate muscle. Control mechanisms, sarcomere lengths, patterns of innervation, and the specific neurotransmitters involved are often different.

Insect flight muscle is a unique striated muscle. Powered flight evolved in insects about 350 mya. A direct, or synchronous, flight mechanism uses muscles that are attached to the base of the wing to accomplish an upward wing stroke. Muscles that run between dorsal and ventral aspects of the thorax deform the exoskeleton to accomplish the downward wing stroke (*see figure 15.12*a). The synchronous nature of this mechanism is based on a one-to-one correspondence between nerve impulses and wing beats. Indirect, or asynchronous, flight involves muscles that alter the shape of the exoskeleton to accomplish wing beats (*see figure 15.12*b). Stretching of one set of flight muscles, caused by the deformation of the thoracic exoskeleton during the contraction of a second set of flight muscles, stimulates the subsequent contraction of the stretched muscles. The subsequent contraction of the second set of flight muscles again deforms the exoskeleton, which then stimulates the first set of muscles. Many cycles of contraction are stimulated by a single nerve impulse (thus, the term "asynchronous") and can generate up to 50 wing-beat cycles and an overall wing-beat frequency up to 1,000 cycles per second (e.g., in midges of the order Diptera).

Cardiac Muscles

Vertebrate **cardiac muscle** has a similar sarcomere organization as is present in skeletal muscle and is also striated. Cardiac muscle cells have a single nucleus and are branched (figure 23.24). Cardiac muscle cells are not consciously controlled. They may contract in response to nerve impulses, for example in the hearts of lobsters and other decapod crustaceans. In vertebrates and many invertebrates, cardiac muscle is self-excitatory and needs no stimuli from the nervous system to initiate contractions. In vertebrate cardiac muscle, the single nucleate cells are joined end-to-end by specialized cell (gap) junctions called intercalated discs that strengthen cardiac muscle and allow a muscle membrane impulse to travel quickly between cells (*see figure 23.24*). Thus, an impulse that originates at the pacemaker of the heart (SA node, *see chapter 26*) can be

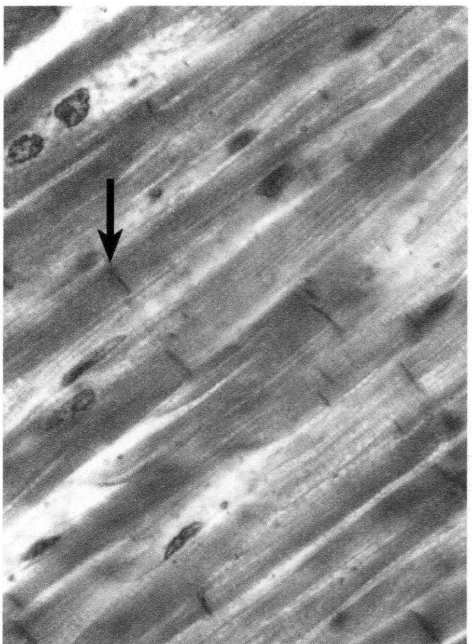

FIGURE 23.24

Cardiac Muscle Tissue. This tissue consists of branched striated cells, each containing a single nucleus and specialized junctions called intercalated disks (arrow) that allow electrical current to flow between cells during an action potential (LM ×500). Cardiac muscle comprises the wall of the heart and helps propel blood into the circulation. It is under involuntary control.

conducted throughout the heart. Branching of cardiac muscle cells allows impulses to radiate in all directions around a heart chamber; the muscle then contracts as a unit, and the contraction squeezes blood from the chamber.

Smooth Muscles

Smooth muscle is associated with the digestive tract, blood vessels, and other internal organs. The ciliary muscle of the eye that controls focusing of the lens is another example of smooth muscle. Smooth muscle cells are spindle shaped and have a single nucleus. They are said to be smooth because they lack striations (figure 23.25). Smooth muscle cells have actin and myosin, but these myofilaments do not have the regular sarcomere organization seen in striated muscles. Actin filaments attach to the inside of a cell's plasma membrane and to structures in the interior of the cell called dense bodies. During contraction, smooth muscle cells shorten and become fatter. They are often linked together in sheets that cause regions of an organ (like an intestine) to constrict or shorten depending on the orientation of the muscle cells (circular vs. longitudinal).

The contractile properties of smooth muscle depend on its innervation and a number of unique metabolic characteristics. Many smooth muscle cells are electrically coupled with gap junctions, in a fashion similar to cardiac muscle, and impulses can pass between cells to excite many cells with a single stimulus. These coordinated (peristaltic) contractions are important when moving materials through a tube, like an animal's intestine. Smooth muscles may be self-excitatory, which enhances motility of the digestive tract. Other smooth muscles require nerve stimulation to initiate contraction, or they may be stimulated to contract by being stretched (e.g., when the intestine receives food from the stomach).

Some smooth muscles remain contracted for very long periods (high muscle tone). The circular smooth muscles of sphincters in animal digestive tracts remain contracted to regulate the passage of food between regions of the digestive tract. During the sustained contraction of these muscles, cross-bridges of myosin remain attached to actin because the ATP conversion that causes cross-bridges to release from actin is extremely slow. Contractions can be maintained with little ATP expenditure. Another example of this "catch" or "latch" state is in the adductor muscles of bivalve molluscs (*see figures 11.8 and 11.10*). These muscles hold the valves of the shell closed for hours or days with little ATP expenditure or O_2 consumption.

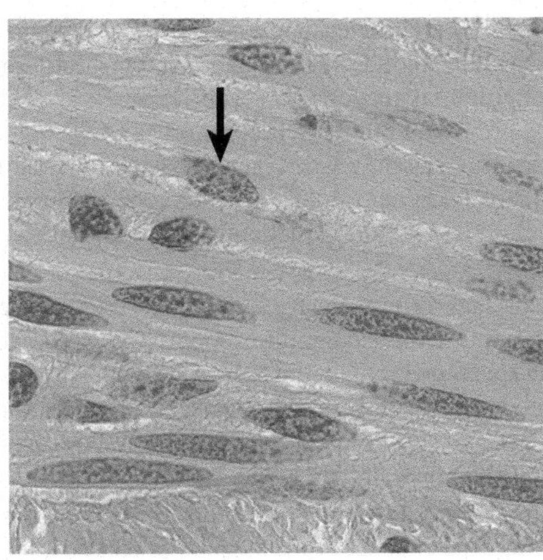

FIGURE 23.25

Smooth Muscle Tissue. This tissue type has spindle-shaped cells, each containing a single centrally located nucleus (arrow) (LM ×1,000). Its cells are arranged to form sheets. Smooth muscle is located in the walls of hollow organs and moves substances along internal passages. Smooth muscle is under involuntary control.

Section 23.3 Thinking Beyond the Facts

Many snake venoms inhibit the enzyme acetylcholinesterase, required to break down acteylcholine. Based on this information, what are the likely effects on a prey?

Summary

23.1 **Integumentary Systems**

- The integumentary system is the external covering of an animal. It primarily protects against mechanical injury and invasion by microorganisms.
- Most invertebrates have an integument consisting of a single layer of columnar epithelial cells called an epidermis. Specializations outside of this epithelial layer may be in the form of cuticles, shells, or teguments.
- Skin is the vertebrate integument. It has two main layers: the epidermis and the dermis. Skin structure varies considerably among vertebrates. Some of these variable structures include scales, hairs, feathers, claws, nails, and baleen plates. The skin of jawless fishes (lampreys and hagfishes) is thick. The skin of cartilaginous fishes (sharks) is multilayered and contains bone in the form of denticles. The skin of bony fishes (teleosts) contains scales. The skin of amphibians is stratified and contains mucous and granular glands plus pigmentation in the form of chromatophores. The skin of reptiles is thick and modified into keratinized scales. The skin of birds is thin and soft and contains feathers. Mammalian skin consists of several layers of a variety of cells.

23.2 **Skeletal Systems**

- Animals have three types of skeletons: hydrostatic skeletons, exoskeletons, and endoskeletons. These skeletons function in animal movement that requires muscles working in opposition (antagonism) to each other.

- The hydrostatic skeleton is a core of liquid (water or a body fluid such as blood) surrounded by a tension-resistant sheath of longitudinal and/or circular muscles. Hydrostatic skeletons are found in invertebrates and can take many forms and shapes, such as the gastrovascular cavity of acoelomates, the rhynchocoel in nemertines, a pseudocoelom in aschelminthes, a coelom in annelids, or a hemocoel in molluscs.
- Rigid exoskeletons also have locomotor functions because they provide sites for muscle attachment and counterforces for muscle movements. Exoskeletons also support and protect the body, but these are secondary functions. In arthropods, the epidermis of the body wall secretes a thick, hard cuticle. In crustaceans (crabs, lobsters, and shrimp), the exoskeleton contains calcium carbonate crystals that make it hard and inflexible, except at the joints.
- Rigid endoskeletons are enclosed by other body tissues. For example, the endoskeletons of sponges consist of mineral spicules, and the endoskeletons of echinoderms (sea stars, sea urchins) are made of calcareous plates called ossicles.
- The most familiar endoskeletons, both cartilaginous and bony, first appeared in the vertebrates. The axial skeleton consists of the skull and vertebral column. The appendicular skeleton consists of appendages and pectoral and pelvic girdles. Endoskeletons consist of two main types of supportive connective tissue: cartilage and bone. Cartilage provides a site for muscle attachment, aids in movement at joints, and provides support. Bone provides a point of attachment for muscles and transmits the force of muscular contraction from one part of the body to another.

23.3 **Nonmuscular Movement and Muscular Systems**

- Movement (locomotion) is characteristic of certain cells, protists, and animals. Ameboid movement and movement by cilia and flagella are examples of locomotion that does not involve muscles.
- The power behind muscular movement in both invertebrates and vertebrates is muscular tissue. The three types of muscular tissue are smooth, cardiac, and skeletal. Muscle tissue exhibits contractility, excitability, extensibility, and elasticity.
- The functional (contractile) unit of a skeletal muscle myofibril is the sarcomere. Contractile proteins, actin and myosin, are arranged in a regular pattern that creates an alternating dark and light banding pattern. During contraction, actin and myosin myofilaments slide past one another, powered by cross-bridges and shortening the sarcomere. Muscles are controlled by neurons that innervate one or more muscle fibers. Contraction involves the expenditure of ATP and the release of calcium ions from sarcoplasmic reticulum.
- Variations in muscle structure and function include metabolic and structural adaptations that promote variations in muscle tone and speed of contraction. Insect asynchronous flight muscle is adapted to generate cycles of contractions that occur at rates up to 1,000 cycles per second. Vertebrate cardiac muscle is self-excitatory striated muscle. It has branching fibers that squeeze blood from heart chambers. Smooth muscle is not striated. It may be self-excitatory, often occurs in sheets, contracts slowly, and has a high degree of tone. Some smooth muscles can enter a "latch" state and maintain contraction for long periods with little energy expenditure or oxygen consumption.

Concept Review Questions

1. The smallest unit of a muscle fiber that is capable of contraction is the
   a. A band.
   b. I band.
   c. M line.
   d. sarcomere.
   e. muscle fiber.
2. In which of the following animals would you find cuticles making up the integumentary system?
   a. Protozoa
   b. Rotifers
   c. *Hydra*
   d. Flukes
   e. Tapeworms
3. Which of the following has denticles?
   a. The skin of jawless fishes
   b. The skin of cartilaginous fishes
   c. The skin of bony fishes
   d. The skin of amphibians
   e. The skin of reptiles
4. Motor neurons stimulate muscle contraction via the release of
   a. calcium ions.
   b. ATP.
   c. hormones.
   d. troponin.
   e. acetylcholine.
5. Stimulation of a muscle fiber by a motor neuron occurs at
   a. the sarcoplasmic reticulum.
   b. the transverse tubules.
   c. the myofibril.
   d. the Z line.
   e. the neuromuscular junction.

Analysis and Application Questions

1. How does the structure of skin relate to its functions of protection, temperature control, waste removal, water conservation, radiation protection, vitamin production, and environmental responsiveness?
2. How does the epidermis of an invertebrate differ from that of a vertebrate?
3. Give an example of an animal with each type of skeleton (hydro-, exo-, endoskeleton), and explain how the contractions of its muscles produce locomotion.
4. Describe how the principle of homology can be illustrated in comparing skeletons of different vertebrates.

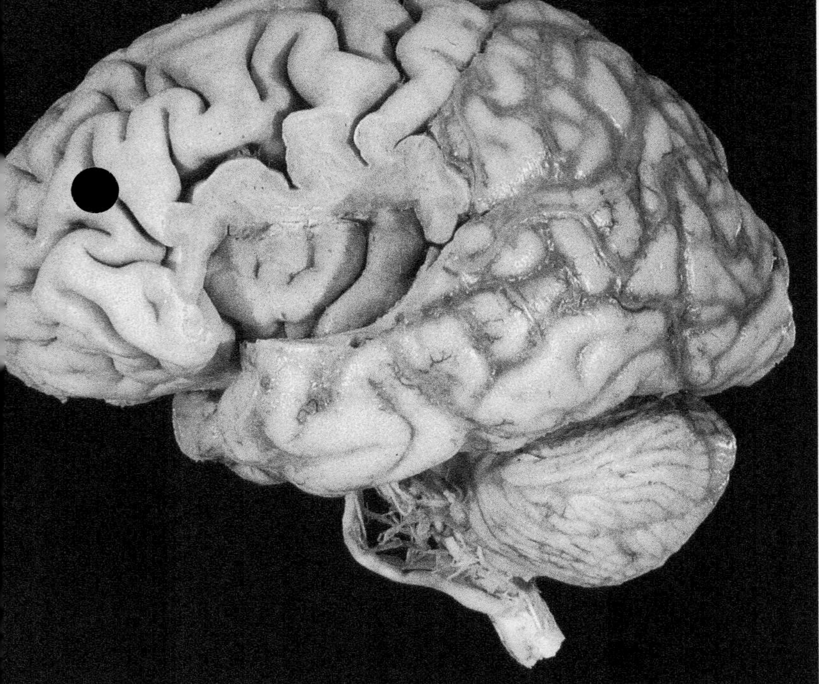

This photograph shows a left lateral view of the external surface of the human brain. The brain is the coordinating center for communication through nervous tissue. Two of the animal communication systems are discussed in this chapter: nervous and sensory.

24

Communication I: Nervous and Sensory Systems

Chapter Outline

24.1 Neurons: The Basic Functional Units of the Nervous System
Neuron Structure: The Key to Function
24.2 Neuron Communication
Resting Membrane Potential
Graded Potentials
Action Potentials
Transmission of Action Potentials
24.3 Invertebrate Nervous Systems
24.4 Vertebrate Nervous Systems
The Spinal Cord
Spinal Nerves
The Brain
Cranial Nerves
The Autonomic Nervous System
24.5 Sensory Reception
24.6 Invertebrate Sensory Receptors
Baroreceptors
Chemoreceptors
Georeceptors
Hygroreceptors
Phonoreceptors
Photoreceptors
Proprioceptors
Tactile Receptors
Thermoreceptors
24.7 Vertebrate Sensory Receptors
Lateral-Line System and Electroreception
Lateral-Line System and Mechanoreception
Hearing and Equilibrium in Air
Hearing and Equilibrium in Water
Sensory Receptors of the Skin
Echolocation
Smell
Taste
Vision
Magnetoreception

The nervous and endocrine systems are intimately associated communicatory systems that integrate complex body functions so that animals can maintain homeostasis in ever-fluctuating environments. Although both systems relay information, they differ in some important ways: The nervous system is fast-acting. It uses cells called neurons to rapidly transmit electrical signals to initiate specific responses. Conversely, the endocrine system uses specialized cells to release chemical messengers for communication. These chemical messengers are typically slower acting and generally trigger prolonged and widespread changes in animal bodies. This chapter focuses on the basic anatomy, physiology, and evolutionary history of animal nervous systems. Chapter 25 examines the endocrine system.

24.1 NEURONS: THE BASIC FUNCTIONAL UNITS OF THE NERVOUS SYSTEM

LEARNING OUTCOMES

1. Describe two properties of neurons.
2. Compare the three functional types of neurons within nervous systems.

The functional unit of the nervous system is the **neuron** (Gr. *neuron*, nerve or cord). Neurons are specialized cells that produce and relay signals from one part of an animal's body to another. Neurons are both excitable (they respond to stimuli) and conductive (they transmit electrical impulses). There are three functional types of neurons. **Sensory** (**receptor** or **afferent**) neurons either act as receptors themselves or are activated by stimuli received from receptors (figure 24.1). Sensory neurons respond to stimuli by sending signals to interneurons (*see figure 24.1*) within integrating centers of the central nervous system. **Interneurons** process and relay signals to motor neurons. **Motor** (**effector** or **efferent**) **neurons** (*see figure 24.1*) send the processed information to effectors, such as muscles or glands, causing them to contract or secrete to effect a change in the animal's physiology (figure 24.2).

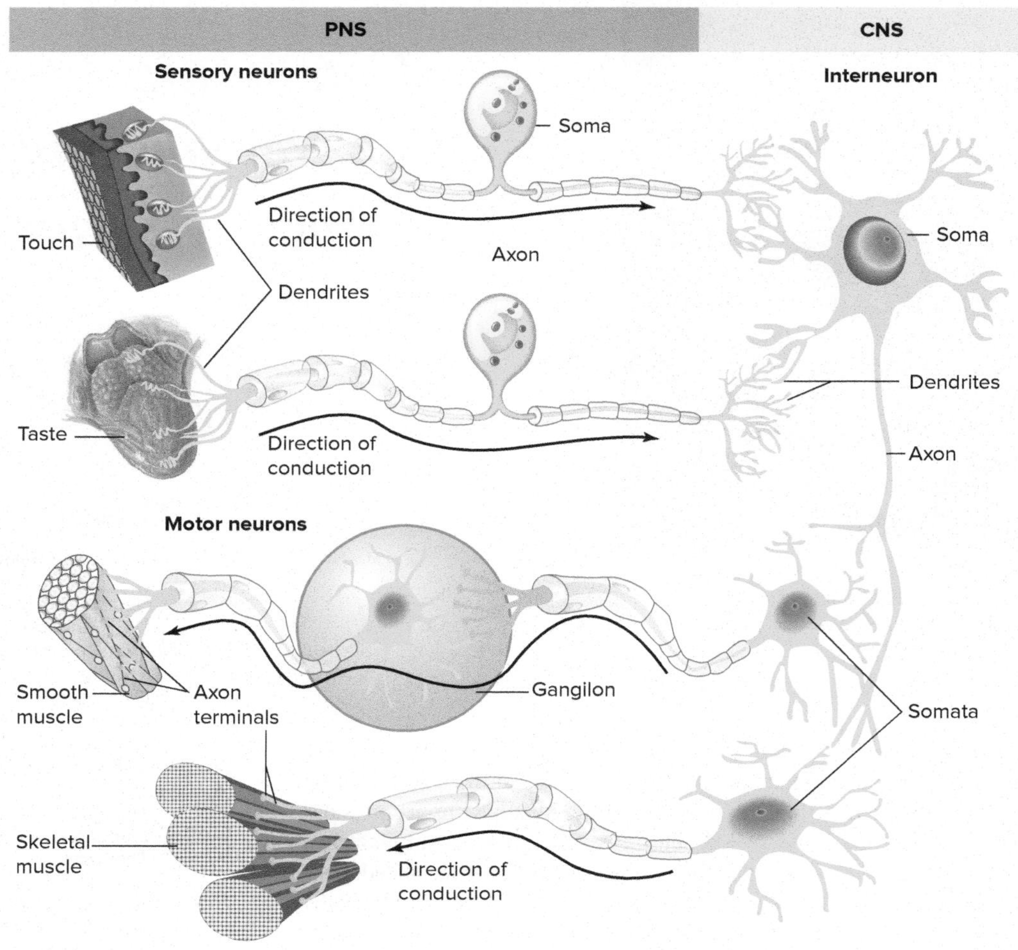

FIGURE 24.1

Types of Vertebrate Neurons. The brain and spinal cord make up the central nervous system (CNS) of vertebrates, and sensory and motor neurons form the peripheral nervous system (PNS). The sensory neurons of the peripheral nervous system carry information about the environment to the CNS. Within the CNS, interneurons provide the links between sensory and motor neurons. Motor neurons of the PNS carry impulses or "commands" to the muscles and glands (effectors) of a vertebrate.

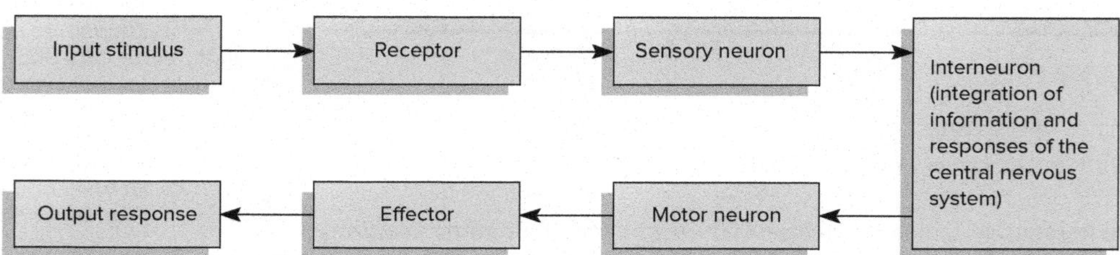

FIGURE 24.2

Generalized Pathway for the Flow of Information within the Nervous System. An input stimulus initiates impulses within some sensory structure (the receptor); the impulses are then transferred via sensory neurons to interneurons. After response selection, nerve impulses are generated and transferred along motor neurons to an effector (e.g., a muscle or gland), which elicits the appropriate output response.

Neuron Structure: The Key to Function

Neurons are usually comprised of three parts: the dendrites, soma (cell body), and axon (*see figure 24.1*). **Dendrites** (Gr. *dendron*, tree) are extensions of the soma. Dendrites receive signals and relay them to the **soma** (Gr. *soma*, body). The soma often receives information from many dendrites, and even other neurons, and combines it into a signal that will be passed to the axon. This process is called integration. **Integration** occurs any time information from one or more sources is combined and a response is initiated. Integration can occur at the level of the neuron (as described above) or within responses involving the entire nervous system of an animal. The latter will be described in subsequent sections.

The **axon** (Gr. *axon*, axis) is a long, cylindrical structure that conducts signals away from the soma to initiate a response. Axons communicate with other neurons or effectors. Although some invertebrates (e.g., Hydrozoa and Anthozoa) possess unsheathed neurons, most animals use sheathed neurons for nervous communication. Sheathed (myelinated) neurons have specialized cells called **neurolemmocytes** (Schwann cells) that wrap around axons in layers to form the laminated, lipid-based **myelin sheath**. Neurolemmocytes insulate neurons, assist with neuronal repair, and speed the rate of signal conduction. Neurolemmocytes cover the majority of the axon in myelinated neurons. However, regularly spaced gaps called **neurofibril nodes** (**nodes of Ranvier**) in myelination are present, and they function to increase the speed of neuron communication. This process is described next.

Section 24.1 Thinking Beyond the Facts

How does the structure of a neuron promote the rapid transfer of information between two distant points in an animal?

24.2 NEURON COMMUNICATION

LEARNING OUTCOMES

1. Identify the ions involved in nerve impulse transmission and their relative concentrations inside and outside the neuron when the neuron is resting.
2. Distinguish between electrical and chemical synapses.
3. Explain the importance of the all-or-none law.
4. Assess the importance of saltatory conduction in the transmission of a nerve impulse.

Neurons communicate via action potentials (nerve impulses). Action potentials are considered the language of the nervous system and result from ion concentration changes that occur in the neuron when a stimulus affects the permeability of its plasma membrane. Action potentials travel along the dendrites, soma, and axon of each neuron.

Resting Membrane Potential

The plasma membrane of a neuron that is resting (not conducting an impulse) is polarized; that is, the intracellular fluid is negatively charged with respect to the positively charged extracellular fluid (figure 24.3). This difference in charge is called the **resting membrane potential**, and it is measured in millivolts (mV, 1/1,000 of a volt). The resting membrane potential is about −70 mV (−40 to −90 mV in vertebrates) due to the unequal distribution of the electrically charged molecules and ions on either side of the neuron's plasma membrane. Large, nondiffusible, negatively charged proteins create a "sink of negative charges" within the cell. The negativity within the cell creates an electrical attraction (gradient) that keeps highly diffusible potassium ions (K^+) at a higher concentration inside the cell than outside. The balance between potassium's diffusion gradient, which draws K^+ out of the cell, and the electrical gradient created that draws K^+ into the cell creates an electrochemical gradient that is responsible for most of the resting membrane potential. Other ions, like sodium (Na^+) and chloride (Cl^-) (higher concentrations outside the cell), contribute to the resting membrane potential to a lesser extent. Potassium and sodium ions constantly leak through ion channels in the neuron's plasma membrane. However, **sodium-potassium ATPase pumps** (*see figure 2.10*) maintain the resting membrane potential by actively transporting three Na^+ out of the neuron for every two K^+ that it moves into the neuron.

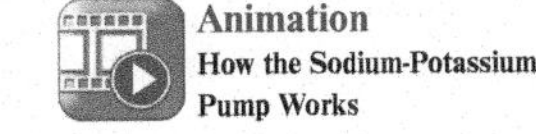

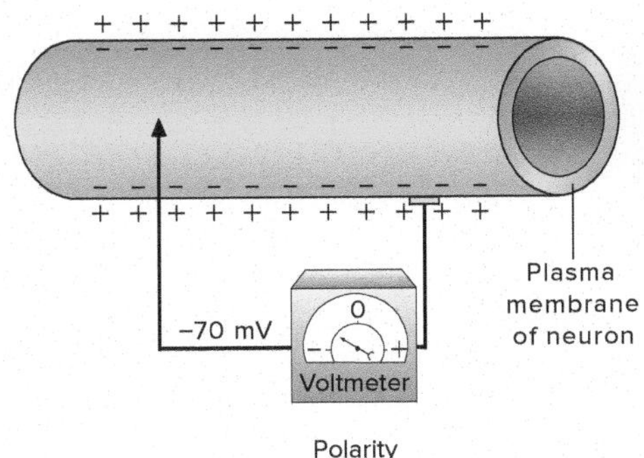

(a)

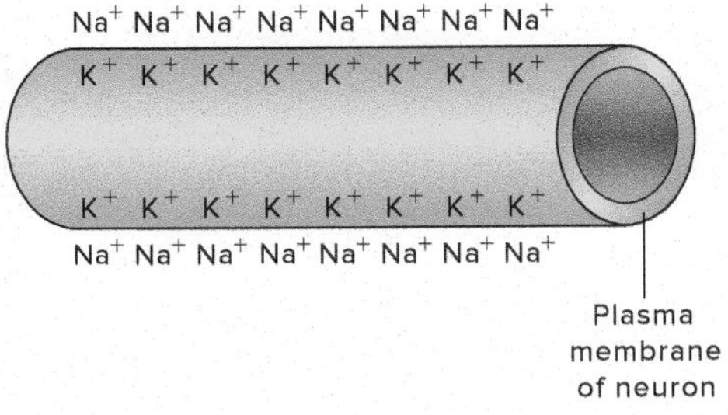

(b)

FIGURE 24.3

Resting Membrane Potential. (*a*) A voltmeter measures the difference in electrical potential between two electrodes. When one microelectrode is placed inside a neuron at rest, and one is placed outside, the electrical potential inside the cell is −70 mV relative to the outside. (*b*) In a neuron at rest, sodium is more concentrated outside the cell and potassium is more concentrated inside the cell. A neuron in this resting condition is said to be polarized.

Graded Potentials

Deviations from a neuron's resting membrane potential produce electrical signals. These signals occur in two forms: graded potentials and action potentials. Graded potentials are small and are propagated over short distances. Unlike action potentials, their strength decreases with increasing distance from their origin, and the magnitude of the graded potential depends upon the strength of the stimulus. Graded potentials can combine and increase in magnitude in a process called summation. Their effects are important to sensory neurons, which must distinguish between strong and weak environmental stimuli. In addition, graded potentials can act as a trigger for an action potential (described in the next section).

Action Potentials

The action potential is the language of the neuron. An action potential is initiated at one point on a membrane and results in an isolated reversal of charges (positive inside and negative outside) on either side of the membrane at that point. An action that brings the membrane potential toward 0 mV is called **depolarization** (figure 24.4). If a stimulus (or summation of graded potentials) depolarizes the membrane potential enough to meet a particular threshold (the threshold potential), then channels that are sensitive to changes in membrane potential (voltage-gated, Na^+) open due to a change in the shape of the gate (proteins). Na^+ from the extracellular fluid then rapidly move down their concentration gradient into the neuron.

During depolarization, the membrane potential at a point on the membrane reverses (positive inside and negative outside). At this reversal potential, Na^+ gates close. Nearly immediately after Na^+ gates close, K^+ channels open, allowing K^+ to rapidly move down its concentration gradient to the extracellular fluid. This process, called **repolarization**, brings the membrane potential toward the resting potential. K^+ gates close slowly in comparison to the rate at which K^+ floods the extracellular fluid. Continued K^+ permeability causes the membrane potential to undergo **hyperpolarization** (i.e., become more negative than the resting potential). During hyperpolarization the membrane is less sensitive to a second stimulus, and it is said to be in its **refractory period**. The sodium-potassium ATPase pumps eventually restore Na^+ and K^+ ions to their resting concentrations outside and inside the membrane. This entire process (from threshold potential to restoration of resting membrane potential) is extremely rapid and occurs in less than 4 milliseconds (ms).

Unlike graded potentials, a threshold potential must be met in order to initiate an action potential. Action potentials are not weak or strong. They either occur to their full extent, or they do not; this is known as the **all-or-none law**. This law makes action potentials well suited for long-distance communication. Conduction over long distances occurs when depolarization at one point on a membrane sets up an electrical field (electron flow) that initiates depolarization in a region of the membrane ahead of the original depolarization. Recall that the membrane at the point of the original action potential hyperpolarizes and is less sensitive to subsequent depolarization. Thus, the action potential can only be conducted in one direction, away from the original depolarization. Because an action potential is all-or-none, it is conducted the entire length of the membrane without dying.

The all-or-none law also means that an increase in stimulus intensity does not increase the strength of the action potential. How then do animals perceive intensity? Stronger stimuli produce more impulses per unit time than weaker stimuli. Additionally, all

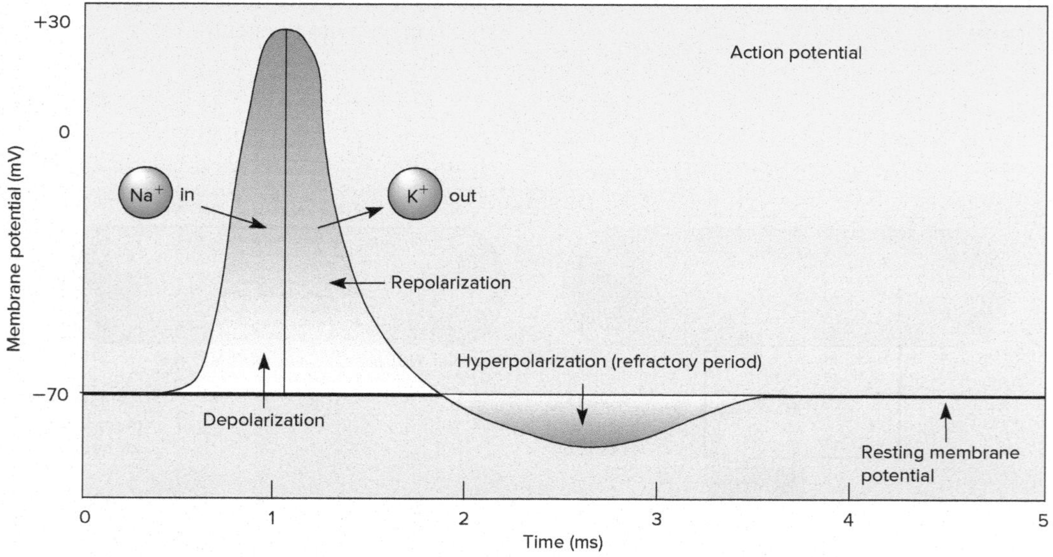

FIGURE 24.4

An Action Potential Recording. In this action potential, the threshold potential occurs at approximately −55 mV to trigger opening of sodium (Na^+) gates that allow ions to rush into the neuron (depolarization). Na^+ channels become inactive, and potassium (K^+) gates open at approximately +30 mV to begin repolarization (K^+ ions exit the neuron). The charges at which these events occur vary and depend upon initial Na^+ and K^+ concentrations in the intra- and extracellular fluid.

action potentials are qualitatively the same. An action potential that is generated in receptors of an ear has no inherent information that distinguishes it from an action potential generated by receptors on the retina of an eye. Animals are able to differentiate action potentials because sensory pathways from different receptors are interpreted by different regions of the brain.

Axons with larger diameters conduct impulses more quickly than smaller diameter axons. However, the myelin sheath greatly increases the speed of conduction regardless of the axon's diameter. Current flow can only initiate depolarization at the uninsulated neurofibril nodes between sheath cells. Action potentials along myelinated axons appear to jump or skip from node to node. This type of transmission presents a pattern of conduction along the axon called **saltatory conduction** (L. *saltare*, to jump). Conduction velocities are greater because depolarization and repolarization events occur at points along an axon rather than along the entire extent of the axon membrane. Thus, myelination permits rapid conduction in small diameter neurons. This configuration facilitated the evolution of rapid and complex nervous systems that did not occupy excessive space in animal bodies.

Transmission of Action Potentials

An action potential traveling along an axon reaches the **end bulb** (end of a branching axon terminal) before encountering a **synapse** (Gr. *synapsis*, connection)—the junction between the axon of a presynaptic neuron and the dendrite of a postsynaptic, or effector, cell. The space (junction) between the end bulb and dendrite of the next neuron is the **synaptic cleft**. All presynaptic cells are neurons, whereas postsynaptic cells can be neurons, muscle cells (*see figure 32.21*), or gland cells.

Synapses are either electrical or chemical. In **electrical synapses**, action potentials are transmitted directly between neurons when current flow generated by depolarization of one cell passes through specialized gap junctions (e.g., intercalated discs of cardiac muscle, *see figure 23.24*) and causes depolarization in the next cell. In chemical synapses, action potentials are relayed via chemicals called neurotransmitters. Presynaptic cells release **neurotransmitters** that alter the permeability of the postsynaptic cell's resting membrane potential in order to depolarize the postsynaptic cell.

Chemical neurotransmission occurs via exocytosis. When an action potential arrives at the end bulb of a presynaptic neuron, Ca^{2+} channels open and allow an influx of Ca^{2+} ions from the extracellular fluid into the neuron. This action causes neurotransmitter-containing synaptic vesicles to fuse with the presynaptic neuron's plasma membrane. During this fusion, the neurotransmitter is released into the synapse where it binds with receptors on the membrane of the postsynaptic cell, causing depolarization in the postsynaptic cell (figure 24.5). Depolarization ceases when the neurotransmitter is either enzymatically degraded or transported back into the presynaptic cell.

There are over 100 types of neurotransmitters. Some may be either excitatory or inhibitory depending on the postsynaptic receptor involved. **Acetylcholine** stimulates skeletal muscle contraction, but slows heart rate. **Norepinephrine** increases heart rate but decreases motility of the gut tract. Gamma-aminobutyric acid (GABA) is an example of an inhibitory neurotransmitter that functions as a Ca^{2+} channel blocker. Each neuron is regularly exposed to both types of neurotransmitters. The soma of a neuron processes signals that arrive from many synapsing neurons in a process called synaptic integration. For instance, if the postsynaptic cell receives more excitatory than inhibitory neurotransmitter, it will generate an action potential.

Section 24.2 Thinking Beyond the Facts

How can the movements of only positive ions result in depolarization and repolarization of the neuron membrane during an action potential?

24.3 Invertebrate Nervous Systems

LEARNING OUTCOMES

1. Describe how coordination occurs in sponges.
2. Compare and contrast the nerve nets of cnidarians and echinoderms.
3. Analyze the evolutionary connection between cephalization and centralization in nervous systems and evolution of bilateral symmetry in animals.

All cells respond to stimuli. Even when no nervous tissues are present, such as in sponges, coordination and reaction to stimuli do occur (*see chapter 9*). In sponges, most reactions result from individual cells responding to a stimulus. For example, water circulating through some sponges is at a maximum just before sunset because light inhibits constriction of dermal pores that serve as entry points for water into a sponge. Cellular responses like these occur slowly and help regulate water filtration rates. In hexactinellid sponges, electrical signals transmitted across the pinacoderm can cause flagellar beating of choanocytes to begin or cease very quickly. These reactions suggest communication across the pinacoderm cell membranes. Furthermore, genomic analyses have found that many genes important for nervous system organization are present in protists, and these genes could have influenced the evolution of neurons in animals.

Virtually all diploblastic and triploblastic animals have nervous tissues. The radially symmetrical cnidarians (hydras, jellyfishes, and sea anemones) and ctenophorans (*see chapter 9*) have nerve nets. In a **nerve net**, the neurons are interconnected into a latticework that conducts impulses in all directions around the animal (figure 24.6*a*). In many nerve nets, impulse conduction is nonpolarized or bidirectional rather than polarized or unidirectional. Two neurons make synaptic contacts at points where they cross, but the neurons cannot be considered either presynaptic or postsynaptic. Nerve nets are involved in swimming movements and responding to stimuli—for example, keeping the body oriented in response to gravity.

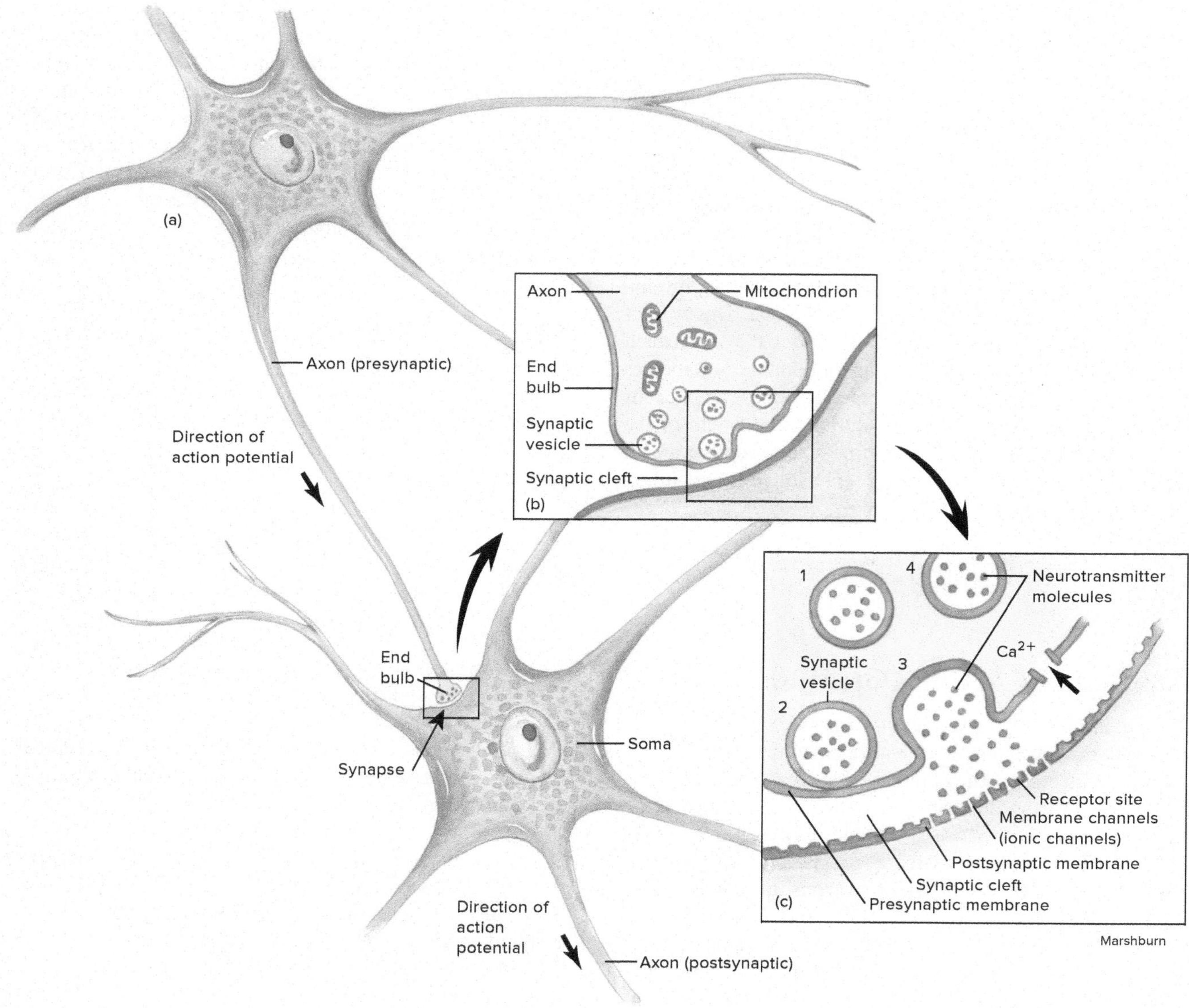

FIGURE 24.5

Chemical Transmission across a Synapse. (*a*) Pre- and postsynaptic neurons with synaptic end bulb. (*b*) Enlarged view of the end bulb containing synaptic vesicles. (*c*) Enlargement of a portion of the end bulb showing exocytosis. The sequence of events in neurotransmitter release is: (1) a synaptic vesicle containing neurotransmitter approaches the plasma membrane; (2) due to the influx of calcium ions, the vesicle fuses with the membrane; (3) exocytosis occurs; and (4) the vesicle re-forms and begins to fill with more neurotransmitter.

Echinoderms (e.g., sea stars, sea urchins, and sea cucumbers) are pentaradially symmetrical, and their nervous systems are also pentaradially structured. For example, a sea star has a circumoral nerve ring that encircles the mouth and five radial nerve cords that run into the animal's arms (*see figures 16.4 and 16.5*). These nerve cords are also interconnected with nerve nets (figure 24.6*f*). Their nerve nets provide for greater variety of responses to stimuli. Most echinoderms have three distinct nerve nets. One nerve net lies under the skin, another net serves the muscles between skeletal ossicles, and the third net connects to the tube feet. This degree of complexity permits locomotion, a variety of reflex responses, and some degree of "central" coordination. For example, when a sea star is flipped over, it can right itself.

Most bilaterally symmetrical animals display cephalization and centralization of their nervous systems (figure 24.6*b–e*). **Centralization** of nervous systems refers to structural organization in which integrating neurons are collected into central integrating areas, usually along the longitudinal axis of the body. **Cephalization** is the concentration of nervous structures and functions at one end of the animal's body—usually the head end. Bilaterally symmetrical animals are active swimming and crawling animals, and they are often hunters. The evolution of a preferred direction of locomotion

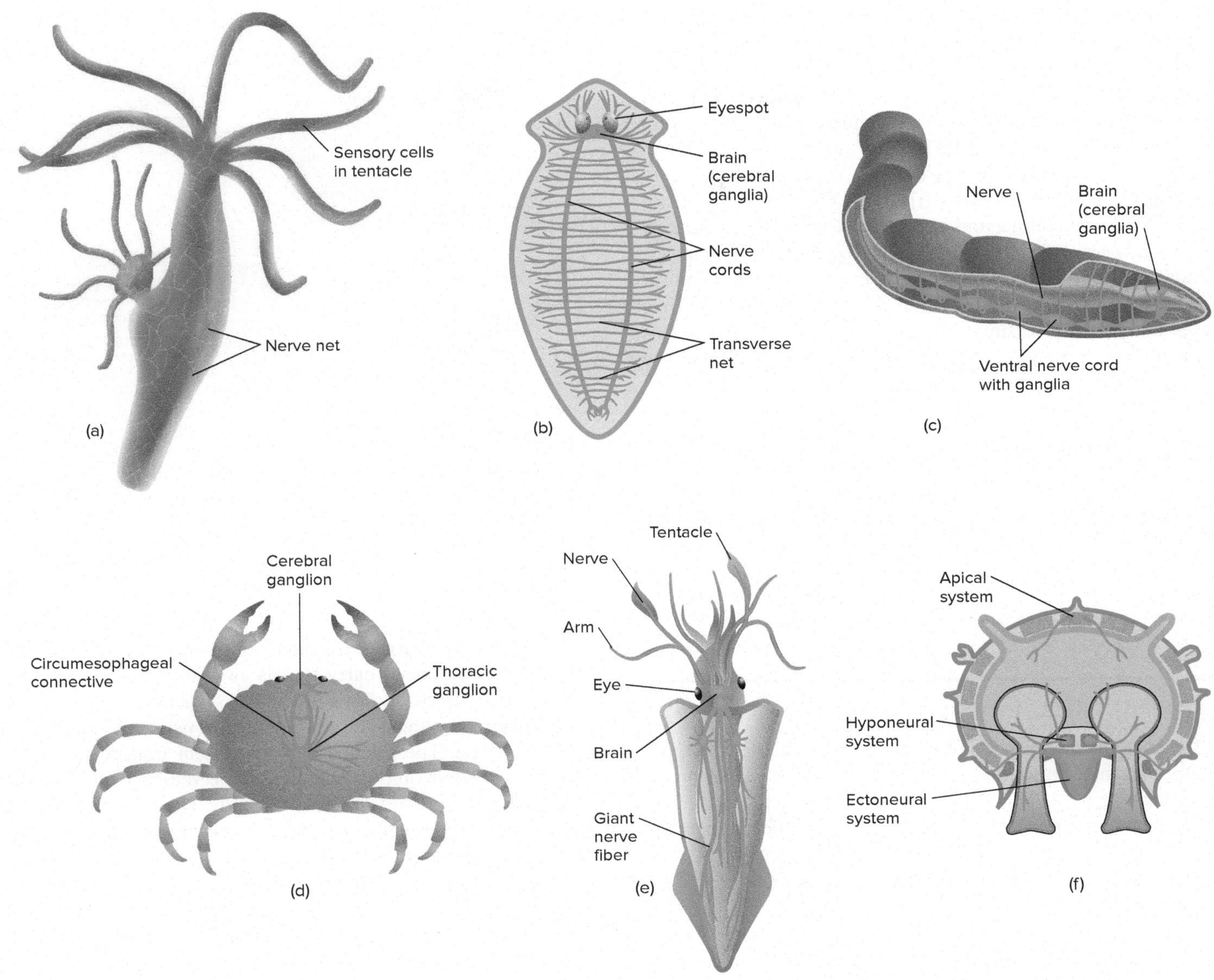

FIGURE 24.6

Some Examples of Invertebrate Nervous Systems. (*a*) Nerve nets are often present in radially symmetrical animals. The cnidarian, *Hydra*, is shown here. (*b*) Brain and paired nerve cords of a planarian flatworm. This nervous system shows differentiation into a peripheral nervous system and a central nervous system. (*c*) Brain, ventral nerve cord, ganglia, and peripheral nerves of the earthworm, an annelid worm. (*d*) A crustacean, showing the principal ganglia and visceral connective nerves. (*e*) Some cephalopods (such as the squid) have nervous systems and behaviors that rival those of fishes. (*f*) Cross section of a sea star arm. Nerves from the ectoneural system terminate on the surface of the hyponeural system, but the two systems have no contact.

apparently included the concentration of sensory receptors at the forward (anterior) end of the body, a concentration of nervous tissue to process sensory information, and centralized nerve cords to coordinate locomotor functions and body responses to stimuli.

Centralization is evidenced by the presence of **nerve cords.** Nerve cords are discrete aggregations of neurons into longitudinally arranged clusters and tracts. Increasing numbers of interneurons (*see figure 24.1*) enhance capacities for centralized integration of sensory stimuli and initiation of motor responses. In metameric animals (e.g., phylum Annelida), segmental ganglia associated with the nerve cord process sensory information from, and control motor responses of, that segment of the body.

Cephalization is evidenced by the concentration of nervous tissue into a head ganglion or brain. In animals like flatworms (e.g., phylum Platyhelminthes), this ganglion is devoted primarily to processing sensory information from receptors around the head. In other bilaterally symmetrical animals (e.g., phyla Arthropoda and Chordata) the head ganglion or brain exerts a degree of domination and control over the other portions of the nervous system. Together, the interconnected longitudinal nerve cords, segmental and other ganglia, and head ganglion or brain comprise the central nervous system (CNS). The peripheral nervous system (PNS) also is increasingly consolidated in bilaterally symmetrical animals. Rather than being netlike, the peripheral sensory and motor processes of neurons are collected into nerves that run between the CNS and peripheral receptors and muscles (figure 24.6*b*–*e*). Large-diameter axons in the peripheral nervous system are common among many invertebrates (e.g., certain decapods and oligochaetes). The largest are those of the squid (*Loligo*), where axon diameter may be over 1 mm. These axons have a rapid conduction velocity (of more than 36 m/s) that causes a maximal contraction of mantle muscles. This contraction forces water out of the mantle, allowing the squid to "jet" away from a predator.

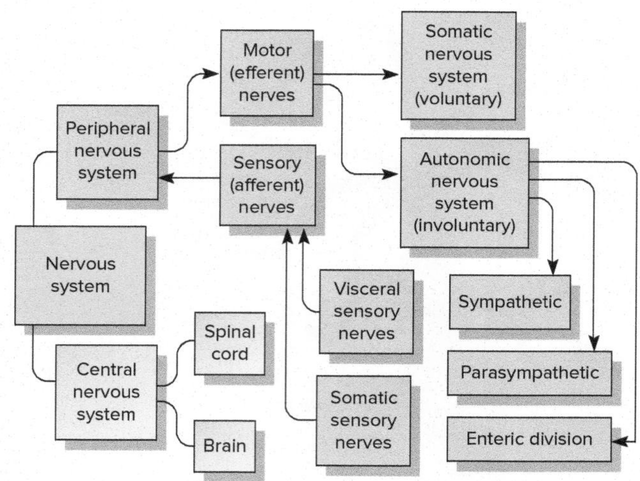

FIGURE 24.7

The Basic Organization of the Nervous System Is Similar in All Vertebrates. This flowchart shows the divisions and nerves of the vertebrate nervous system. Arrows indicate the directional flow of nerve impulses.

Section 24.3 Thinking Beyond the Facts

Explain the usefulness of a large brain, a head, and centralized nerve cords for a relatively sedentary animal like a sea urchin.

24.4 VERTEBRATE NERVOUS SYSTEMS

LEARNING OUTCOMES

1. Describe the organization of the brain in vertebrates.
2. Distinguish between the somatic and autonomic nervous systems.
3. Evaluate the differences between the sympathetic and parasympathetic divisions of the autonomic nervous system.
4. Describe the structure of the spinal cord of several vertebrates.

The basic organization of the nervous system is similar in all vertebrates. The evolution of the vertebrate nervous system responded to the same adaptive pressures associated with a bilaterally symmetrical, active lifestyle described in the previous section. In vertebrates, the nervous system develops in association with one of the hallmark chordate characteristics–the dorsal, tubular nerve cord (*see figure 17.6*).

During vertebrate evolution, the dorsal nerve cord became regionally modified and highly specialized. Its anterior-most end expanded into a brain. Anterior sensory structures also evolved to allow vertebrates to rapidly acquire environmental information and communicate it to the nearby brain. Over time, these receptors evolved into complex paired sensory structures that can detect a wide array of stimuli.

There are two major divisions of the vertebrate nervous system (figure 24.7). The **central nervous system** (CNS) includes the brain and spinal cord and is primarily responsible for processing information and coordinating bodily responses. The **peripheral nervous system** (PNS) includes all nerves outside the CNS. It is subdivided based on the direction of the signals that it transmits. **Sensory (afferent) nerves** transmit information to the CNS and **motor (efferent) nerves** carry signals away from the CNS. Motor nerves are grouped into two functional systems: the **somatic (voluntary) nervous system**, and the **autonomic (visceral or involuntary) nervous system**. The somatic nervous system transmits messages to skeletal muscle, and the autonomic nervous system stimulates involuntary muscle and glands of the body. The nerves of the autonomic nervous system are divided into **sympathetic**, **parasympathetic**, and **enteric** (intestinal) divisions.

Nervous system pathways are composed of individual axons bundled together. In the CNS, these bundles are called **tracts**, whereas in the PNS, they are called **nerves**. The somata from which the axons extend are often clustered together. These clusters are called nuclei if they are in the CNS and **ganglia** if they are part of the PNS.

The Spinal Cord

The spinal cord is the part of the central nervous system that extends from the brain to near or into the tail (figure 24.8). It connects the brain to most of the body and facilitates spinal reflex actions (a reflex is a predictable, involuntary response to a stimulus). Voluntary and involuntary limb movements, and certain organ functions depend on this connection.

The spinal cord is comprised of three major regions. The neural canal is in the center of the spinal cord–it contains cerebrospinal fluid which nourishes, protects, and supports the CNS. Gray matter immediately surrounds the neural canal and is comprised of somata and dendrites. It enables reflex actions to occur at multiple levels of the spinal cord. White matter surrounds the gray matter and is comprised of myelinated axons that facilitate rapid transmission of action potentials.

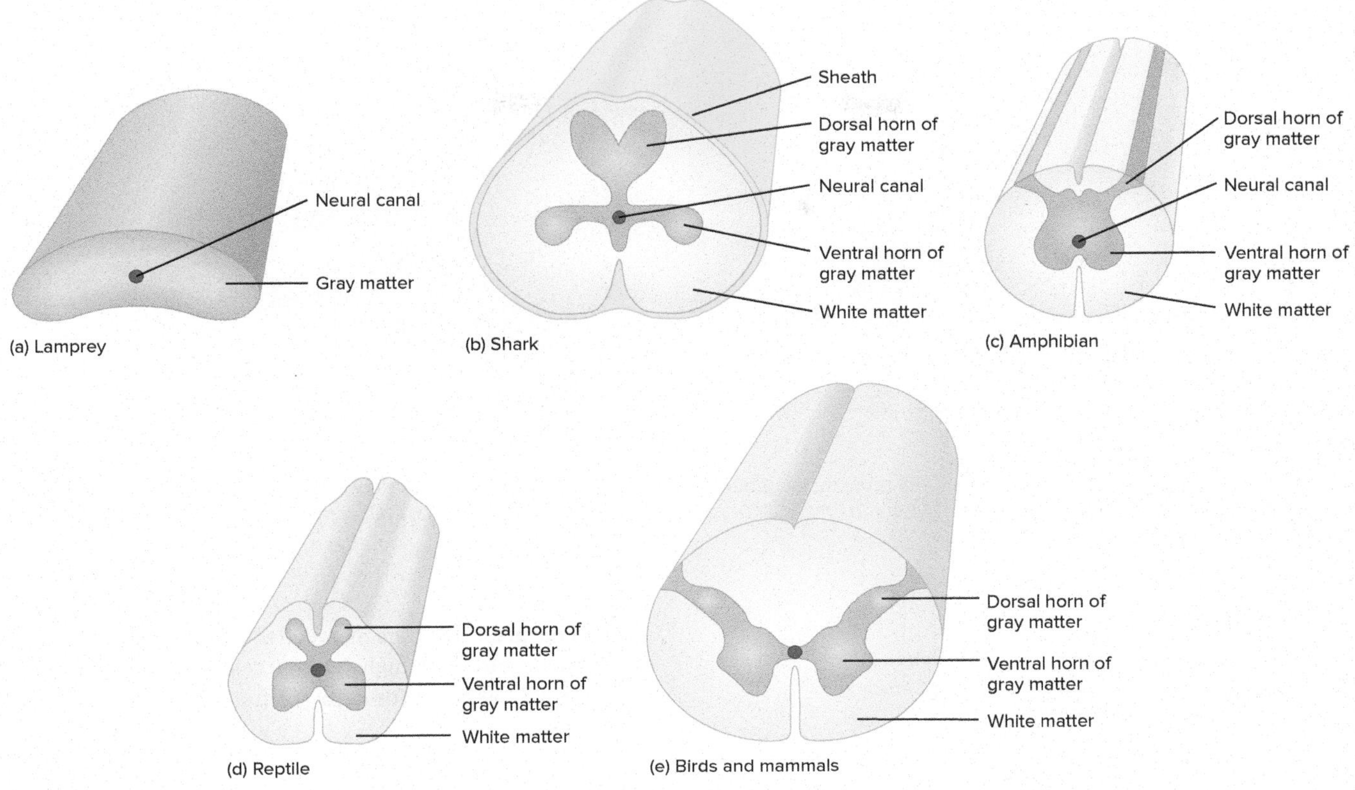

FIGURE 24.8

Spinal Cords of Vertebrates. (*a*) The spinal cord of a typical agnathan (lamprey) is flattened and possesses no differentiated white matter region. Its shape facilitates the diffusion of gases, nutrients, and other products. (*b* and *c*) In fishes and amphibians, the spinal cord is larger, well vascularized, and rounded. With more white matter, the spinal cord bulges outward. The gray matter in the spinal cord of (*d*) a reptile and (*e*) birds and mammals has a characteristic butterfly shape.

Three layers of **meninges** enclose the spinal cord (and brain). The outer layer is called the **dura mater**; it is tough and fibrous. The thinner middle layer, the **arachnoid**, connects to the innermost layer, the **pia mater**. The pia mater is vascular and it nourishes the spinal cord (and brain).

Spinal Nerves

Spinal nerves extend from the spinal cord and are bundles of neurons conducting information to muscles and glands and from sensory receptors along the length of the body. In vertebrates, the number of spinal nerves is generally related to the number of segments in the trunk and tail. Adult anurans, for instance, are tailless and have evolved a compact trunk and strong hind legs for swimming or jumping. Consequently, they only have 10 pairs of spinal nerves. Conversely, snakes have several hundred pairs of spinal nerves that innervate their elongated trunk and tail.

The Brain

During embryonic development, the brain develops from a hollow tube of nervous tissue and expands into three major functional regions: the hindbrain, midbrain, and forebrain (figure 24.9). The spinal cord's neural canal extends into the brain and forms chambers called ventricles that allow cerebrospinal fluid to flow into the brain. The structure and function of the vertebrate brain are discussed next.

Hindbrain

The **hindbrain** is continuous with the spinal cord and includes three major regions necessary for vital body functions. The **medulla oblongata** is an enlarged area of the posterior-most hindbrain. It contains reflex centers for breathing, swallowing, cardiovascular function, and gastric secretion. The medulla oblongata is well developed in gnathostomes and permits an enhanced ability to control visceral functions and screen information that leaves or enters the brain.

The **cerebellum** is an outgrowth of the medulla oblongata that coordinates motor activities and balance. Vertebrates that can execute complex movements have a well-developed cerebellum. The cerebellum in chondrichthyes is lobed, and active teleosts have a comparatively larger and more complex cerebellum than less active fishes. The rather rudimentary cerebellum of amphibians is reflected in these animals' inability to carry out complex movements (figure 24.10). The amniotes, however, have a laterally expanded cerebellum. The avian reptiles and mammals have the largest and most convoluted cerebellum of the amniotes. They are capable of executing correspondingly complex movements.

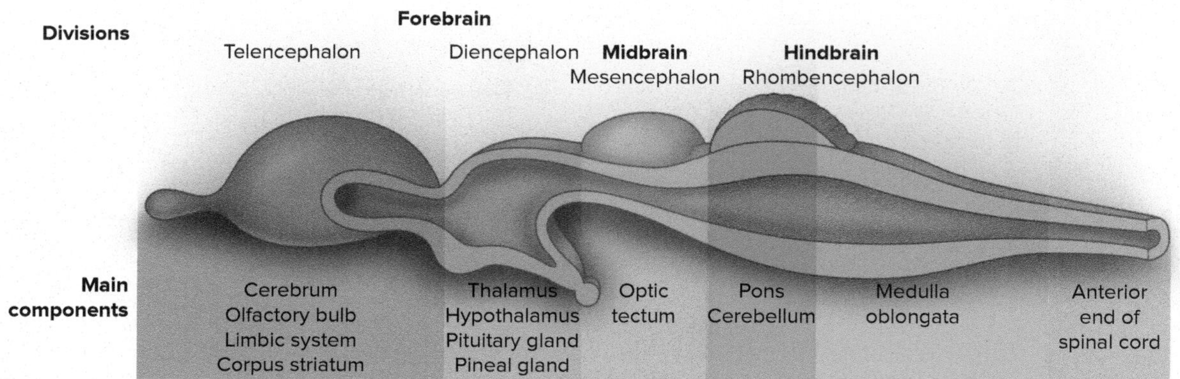

FIGURE 24.9

Development of the Vertebrate Brain. Summary of the three major subdivisions and some of the structures they contain. This drawing is highly simplified and flattened. *Redrawn from Starr/Taggart, BIOLOGY: THE UNITY AND DIVERSITY OF LIFE, 4E. Copyright © 1987 Brooks/Cole, a part of Cengage Learning. Inc. Reproduced by permission. www.cengage.com/permissions.*

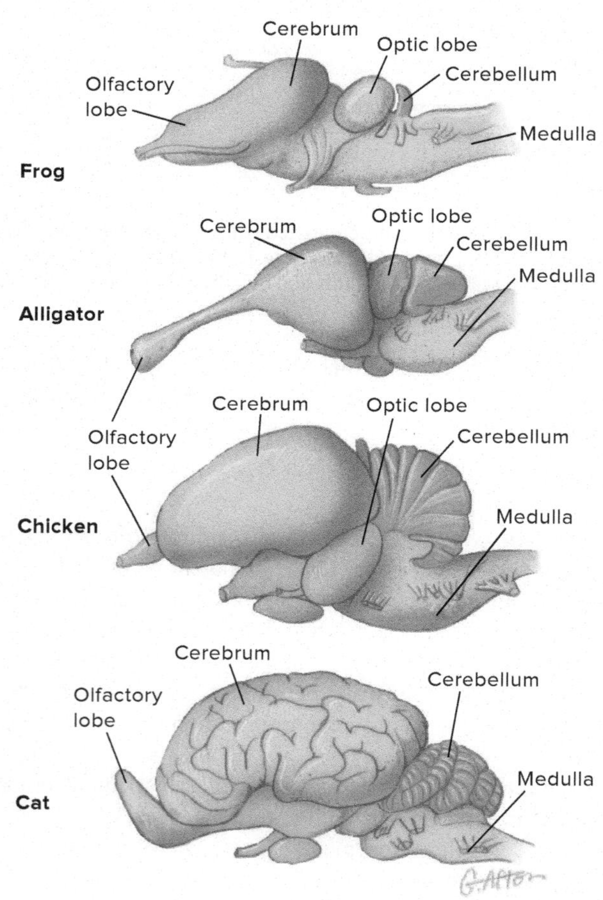

FIGURE 24.10

Vertebrate Brains. Comparison of several vertebrate brains (lateral views). The drawings are not drawn to the same scale. The evolution of vertebrates with complex nervous systems resulted in the enlargement of the cerebrum and folding of the cerebral cortex. Notice the size of the cerebrum of the frog as compared to the remainder of the frog's brain versus the same region of the chicken and the cat.

The **pons** contains nerve tracts that connect the medulla and spinal cord with higher brain centers. It also contains centers involved with a wide array of functions, including control of breathing.

Midbrain

The **midbrain** was originally a center for coordinating reflex responses to visual input. As the brain evolved, it took on added functions relating to tactile (touch) and auditory (hearing) input. The roof of the midbrain, called the tectum, is a thickened region of gray matter that is frequently enlarged in fishes and amphibians due to its function in integrating sensory information and initiating motor responses to this information. In amniotes, many of these functions shift to the forebrain. In all vertebrates, this region is important in the integration of visual and auditory signals. The optic lobes that develop from the tectum are prominent in all vertebrates except the mammals (*see figure 24.10*).

Forebrain

The anterior-most expansion of the vertebrate brain is the **forebrain.** The evolution of the forebrain is reflected in the relative size of this region (figure 24.11) and in key adaptations in vertebrate nervous functions. In basal vertebrate groups (e.g., hagfishes, class Myxini), the sense of smell (olfaction) is a primary sensory function of the forebrain. During vertebrate evolution, forebrain enlargement accompanied increasingly complex behaviors and muscle control. This enlargement is especially evident in amniotes where terrestrial locomotion requires coordination of limb movements. The evolution of the forebrain is also highlighted by its role in regulating many internal functions; coordinating complex behaviors; integrating sensory information from the environment; initiating motor responses to sensory information; and housing the centers for speech, emotions, and intellectual functions. During evolution and development, the forebrain is divided into two regions: the diencephalon and telencephalon (*see figure 24.9*).

EVOLUTIONARY INSIGHTS

Evolution of the Cerebral Cortex in Vertebrates

Cerebral lobes originated as paired outgrowths of the forebrain (*see figure 24.10*). The cerebral lobes were in their most primitive form in fishes, and they functioned primarily in olfaction. Amphibians and nonavian reptiles overlaid this olfactory-oriented cortex with the hippocampus, a part of the forebrain associated with certain innate behaviors, and the corpus striatum, a forebrain structure they used for automatic and instinctual responses. In avian reptiles, the corpus striatum is well developed. In some later nonavian reptiles, the first neurons associated with a structure called the neopallial cortex (a part of the forebrain associated with higher-order brain functions) developed. The growth and expansion of this structure characterizes the mammalian brain and was a hallmark of primate evolution.

In humans, the location of the hippocampus shifted, and became involved with sexual and aggressive behaviors. The corpus striatum evolved into a structure called the corpus callosum, which connects right and left hemispheres of their greatly expanded neopallial (cerebral) cortex. The neopallial cortex is the seat of learning, memory, and intelligence, and it is so expanded that it envelops many other brain structures. In fact, the human cerebral cortex is so large and complex that it (and the rest of the brain) continues to develop at fetal rates even after birth. For humans to achieve the maturity at birth typical of other primates, the human gestation period would have to be about 21 months long, rather than nine. However, if the human cerebral cortex were to develop more fully in utero, the baby's head would not fit through the mother's birth canal. Thus, humans have an abbreviated gestation period when compared to those of other primates.

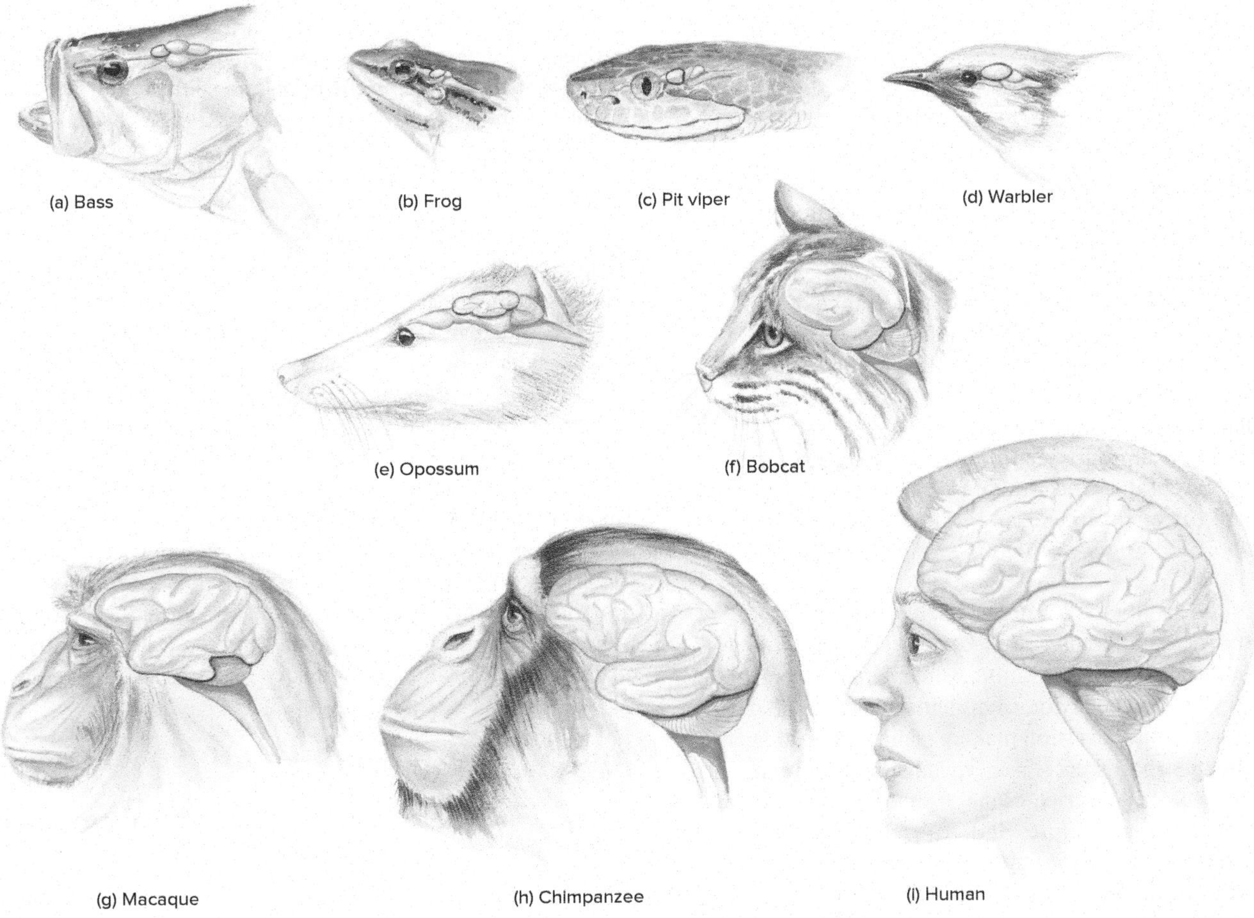

FIGURE 24.11

Cerebrum in Different Vertebrate Species. Increased reliance on the cerebral hemispheres for integrative functions is evidenced in the relative size of this brain region in comparison to other regions of an animal's brain. Overall brain-size differences between species represented here reflect the size of the animals. (*a*) Fishes and (*b*) amphibians lack cerebral cortices, whereas (*c*) reptiles and (*d*) birds have a small amount of gray matter covering their cerebrums. Some mammals, such as (*e*) the opossum, have smooth cortices. Other mammals, such as (*f*) the bobcat, have larger cerebrums, and the cortex has a few convolutions. (*g*-*i*) In the primates, the cerebrum is larger relative to other brain structures, and the cortex is highly convoluted.

The **diencephalon** (Gr. *dia,* through + *enkephalos,* brain) lies just in front of the midbrain and contains the thalamus, hypothalamus, pituitary gland, and pineal gland. The **thalamus** relays all sensory information to higher brain centers. The **hypothalamus** lies below the thalamus and regulates many functions related to homeostasis: temperature regulation, endocrine regulation, behavioral patterns, circadian rhythms, sexual drive, metabolic rate, hunger, and thirst. The pituitary and pineal glands are located in the diencephalon and are endocrine glands that will be discussed in chapter 25.

The diencephalon is also the location for integration of some sensory information from the external environment. In fishes and amphibians, the diencephalon processes olfactory information sent here from the telencephalon, and it initiates motor responses to that information. Other examples include the integration of touch, temperature, vision, hearing, and equilibrium and balance. In reptiles (including birds), the diencephalon contains the corpus striatum (the term describes the striated appearance of gray matter), which plays a role in their complex behavior patterns (*see chapter 21*).

The anterior region of the forebrain, the **telencephalon** (Gr. *telos,* end + *enkephalos* brain, an embryological and evolutionary designation) or **cerebrum** (an adult anatomical reference), rapidly expanded in both size and complexity. The cerebrum is divided into a right and left cerebral hemisphere by a large groove (the longitudinal fissure). In early vertebrates, olfaction was the major function associated with the cerebrum. However, this part of the brain enlarged during vertebrate evolution and became the integrating region for many sensory and motor functions that moved here from the tectum (*see figure 24.11*), and it is the seat of most higher intellectual functions. The increased importance of the cerebrum overshadowed many other brain regions.

In mammals, the outermost part of the cerebrum is a layer of gray matter called the cerebral cortex. This layer folds back on itself to such an extent that it suggests that the evolution of the mammalian cerebrum outpaced the enlargement of the skull housing it. The cerebral cortex integrates virtually all sensory information from the environment and initiates motor responses to it (*see Evolutionary Insights: Evolution of the Cerebral Cortex in Vertebrates*).

The cerebrum is divided into regions or lobes that are often named based on their underlying skull bones. The **frontal lobe** is involved with motor control of voluntary movement, control of emotional expression, and behavior. In humans it is associated with planning, judgement, and decision-making skills. This area also contains **Broca's area** that is critical in vocalizations in mammals and in human speech. The **parietal lobe** is involved with general senses such as touch, temperature, and pain. The **temporal lobe** is involved with hearing, equilibrium, emotion, and memory. The **occipital lobe** is organized for vision and associated forms of expression. The **insula** may be involved with self-awareness, cognitive functions, and gastrointestinal and other visceral activities. Finally, the **limbic lobe** of the cerebrum is combined with structures of the diencephalon and mesencephalon to form the **limbic system**. The limbic system is involved with emotions, behavioral expressions, short-term memory, and smell.

Cranial Nerves

Paired cranial nerves extend from the brain. They are part of the peripheral nervous system, and relay information between the brain and areas of the head, neck, and viscera. Amniotes have 12 pairs of cranial nerves, the first 10 of which are present in fishes and amphibians. Some cranial nerves (e.g., optic nerve) are sensory and relay information to the brain, while others are mixed. The vagus nerve, for instance, has sensory axons leading to the brain and motor axons leading to the heart and smooth muscles of the visceral organs.

The Autonomic Nervous System

The vertebrate autonomic nervous system is composed of parasympathetic and sympathetic divisions. These divisions work continuously, and often antagonistically, to help a vertebrate maintain homeostasis by controlling involuntary muscles and the activity of glands (figure 24.12). The **parasympathetic division** allows an animal to "rest and digest"; heart rate is decreased, digestive secretions and motility are increased, and pupils and respiratory passageways are constricted. It consists of numerous efferent nerves that synapse at ganglia positioned near or within various organs, and shorter efferent nerves that extend from those ganglia to the organs. Parasympathetic nerves originate in the brain and spinal cord. The **sympathetic division** mobilizes body systems necessary for physical activity and is well known for its involvement in the "fight-or-flight" response. The sympathetic division contains efferent thoracic and lumbar spinal nerves that extend to ganglia near the spine and longer efferent nerves that extend from the ganglia to organs. This long-held understanding of autonomic nervous organization has recently been challenged by research that found sacral outflow in mice to be sympathetic, not parasympathetic. More data are needed to confirm these results and to fully understand the implications of this newly observed difference in autonomic nervous arrangement.

The **enteric division** (*see figure 24.7*) of the autonomic nervous system (not shown in figure 24.12) is comprised of many neurons and neurotransmitters. Its functions are not as well understood as the other divisions. However, in humans, it is sometimes referred to as the "brain" of the gut due to its regulation of various gut functions. Enteric neurons control secretions of the pancreas, gall bladder, and digestive tract and peristalsis of gastrointestinal smooth muscle. Although the enteric division can function independently, it is normally regulated by the sympathetic and parasympathetic divisions.

Section 24.4 Thinking Beyond the Facts

Why is it advantageous to have two divisions (sympathetic and parasympathetic) within the autonomic nervous system?

24.5 SENSORY RECEPTION

LEARNING OUTCOMES

1. Describe the basic features of most sensory receptors.
2. Explain how receptors function as transducers.

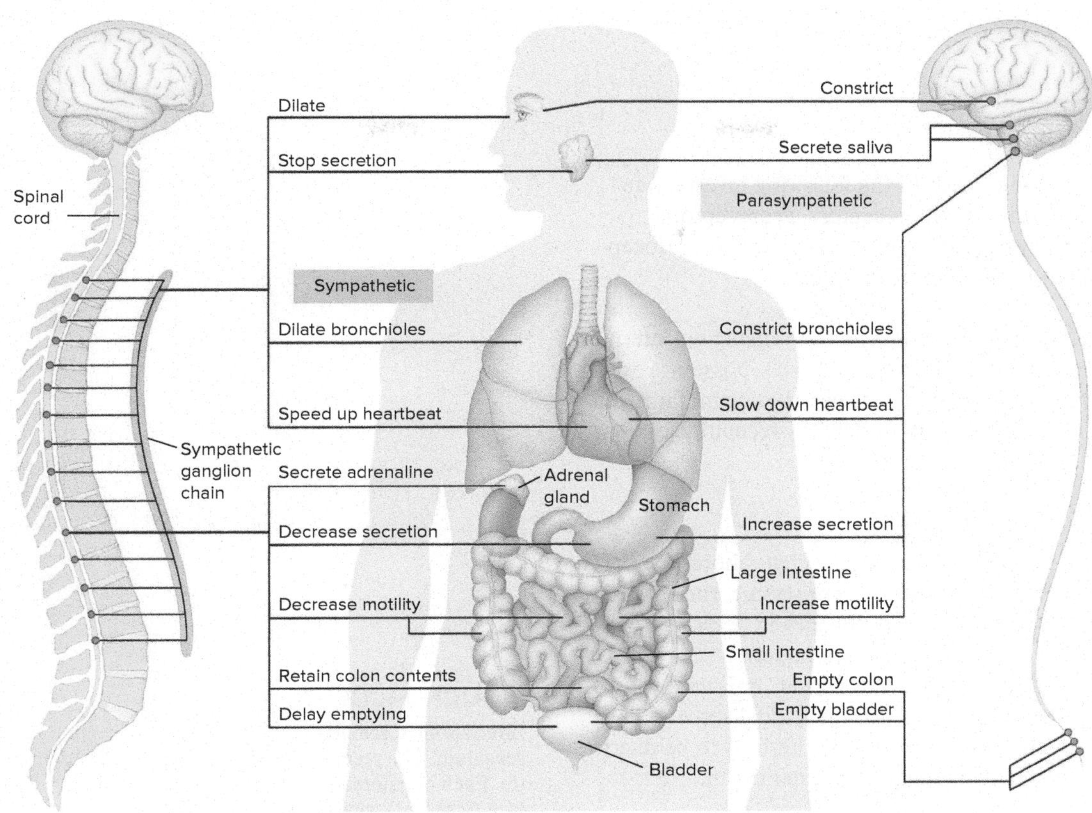

FIGURE 24.12

The Autonomic Nervous System. Preganglionic neurons of the sympathetic division (left) exit thoracic and lumbar areas of the spinal cord. The ganglia are located near the spinal cord. Parasympathetic neurons (right) exit the brain and sacral region of the spinal cord. The ganglia are located near the organs they innervate. Most internal organs are innervated by both divisions. Parasympathetic action includes secretion of saliva, increased intestinal movements, pupil contraction, and sphincter relaxation. Sympathetic action is antagonistic. It inhibits salivation, decreases intestinal movements, dilates pupils, and contracts sphincters of the body.

Animals possess a wide array of sensory receptors that receive and transmit information about their environment. Many of these senses go unnoticed in our everyday lives, but are critical to safely navigating the world and to maintaining homeostasis. Overall, an animal's senses provide an impression of its environment. Different animal groups may perceive similar environments in very different ways. Animals' perceptions are limited by the nature of receptors that each animal uses to capture and convert environmental (both internal and external) stimuli into action potentials and the ability of their central nervous systems to interpret these action potentials. The rest of this chapter examines how animals use specific sensory information to help maintain homeostasis and to navigate their environments.

Specialized cells in **sensory receptors** convert environmental stimuli (sing., **stimulus**) into graded potentials that may lead to action potentials if the threshold potential is met. All sensory receptors convert one form of energy into another; thus they are termed **transducers**. The type of stimulus that is transduced varies. The graded response of a sensory receptor is called a receptor potential, and this receptor potential may result in one or more action potentials being conducted toward the CNS. Recall that the brain can interpret the intensity and nature of a stimulus based on the number of action potentials it receives per unit time and the particular region of the brain that receives the action potential. Sensory receptors have the following properties:

1. They contain receptor cells or small peripheral branches of sensory neurons that respond to stimuli with graded potentials.
2. They are specialized to detect a specific type of stimulus.
3. Their receptor cells synapse with afferent neurons that lead, via specific pathways, to the central nervous system.
4. Action potentials that they generate are translated into a recognizable sensation in the CNS.

Section 24.5 Thinking Beyond the Facts

Explain this statement: "All receptors are transducers."

24.6 INVERTEBRATE SENSORY RECEPTORS

LEARNING OUTCOME

1. Describe one function of each of the following invertebrate receptors: baroreceptors, chemoreceptors, georeceptors, hygroreceptors, phonoreceptors, photoreceptors, proprioceptors, tactile receptors, and thermoreceptors.

Animals change their behaviors in response to changing internal and external environmental stimuli. Invertebrates possess a wide array of receptors through which they receive these stimuli. A discussion of some well-known invertebrate sensory receptors follows.

Baroreceptors

Although it is known that certain arthropods, ctenophores, cnidarians, and molluscs can detect changes in water pressure, the anatomy and physiology of structures that sense changes in pressure (**baroreceptors**, Gr. *baros*, weight + receptor) are largely unknown. An example of baroreception may be seen in certain intertidal crustaceans that coordinate diel movements with daily tidal activity. Their behavioral response to changing tides is hypothesized to be due to pressure changes associated with changes in water depth.

Chemoreceptors

Chemoreceptors (Gr. *chemeia*, pertaining to chemistry) bind chemicals to trigger a response. Chemoreception is among the oldest and most common senses in Animalia, and many different types of chemoreceptors have evolved and detect a wide array of chemical signals. Many aquatic invertebrates have chemoreceptive sensilla (sing., sensillum) located in pits or depressions that receive water containing chemical cues. **Sensilla** are modifications of the exoskeleton that house dendrites of underlying chemosensory neurons (figure 24.13). Sensilla may be modified to detect different stimuli, and they may be positioned on mouthparts, the antennae, and legs.

Invertebrates respond to chemical cues that provide information on humidity and pH changes, prey and food location and recognition, and mate location. For example, the antennae of male silkworm moths (*Bombyx mori*) can detect one female sex pheromone molecule (bombykol) in over a trillion molecules of air. This ability allows males to find receptive females at night from several kilometers downwind.

Georeceptors

Georeceptors (Gr. *geo*, earth + receptor) allow animals to sense their bodily position relative to the force of gravity. This sense is also one of the most ancient in Animalia. Most invertebrate georeceptors are fluid-filled, sensory-epithelium-lined chambers called **statocysts** (Gr. *statos*, standing + *kystis*, bladder)(figure 24.14). These chambers enclose a solid, stonelike granule called a **statolith** (Gr. *lithos*, stone). When an invertebrate moves, so does the statolith

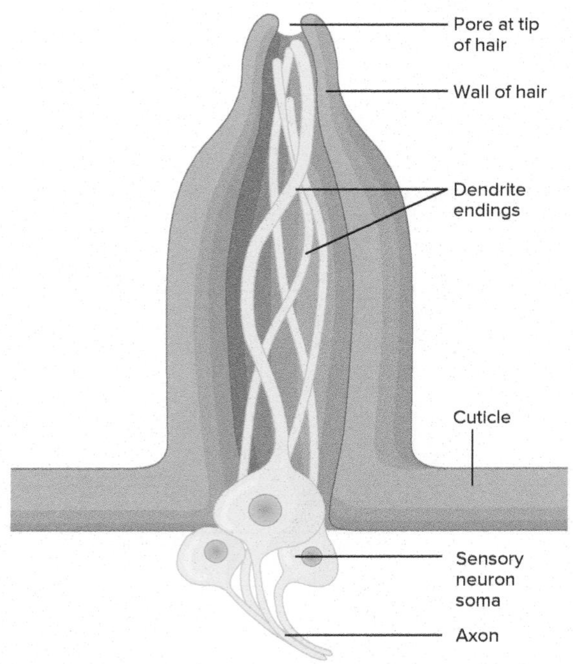

FIGURE 24.13

Invertebrate Chemoreceptor. Longitudinal section through an insect sensillum. The receptor is a projection of the cuticle with a pore at the tip. Each chemoreceptor generally contains four to five dendrites, which lead to sensory neuron somata underneath the cuticle. Each sensory cell has its own spectrum of chemical responses. Thus, a single sensillum with four or five dendrites and somata may be capable of discriminating between many different chemicals.

and fluid within the statocyst–these changes alter the position of cilia that are embedded in sensory epithelium. The altered position of the cilia may lead to an action potential, providing information about an invertebrate's linear and rotational acceleration and position in the environment.

Statocysts are found in animals of various phyla, including Arthropoda (e.g., Crustacea), Annelida (e.g., Errantia), Mollusca (e.g., Gastropoda and Cephalopoda), Nemertea, and Cnidaria (e.g., Scyphazoa), and are especially important to invertebrates that traverse environments devoid of light. Some aquatic insects make use of a different type of georeceptor; they trap air bubbles in tracheal tubes to form a statocyst-like structure. Air bubbles occupy otherwise fluid-filled and receptor-lined tubes and change position according to gravity when the insect moves. The air bubbles stimulate sensory bristles (like the statolith in a statocyst) and provide the insect with information on its position relative to gravity.

Hygroreceptors

Hygroreceptors (Gr. *hygros*, moist) detect the water content of air. Some insects use hygroreceptors to identify environments with a preferred humidity or to modify their physiology or behavior with respect to ambient humidity (e.g., to regulate opening or closing of spiracles). Hygroreceptors are positioned on antennae, palps, the

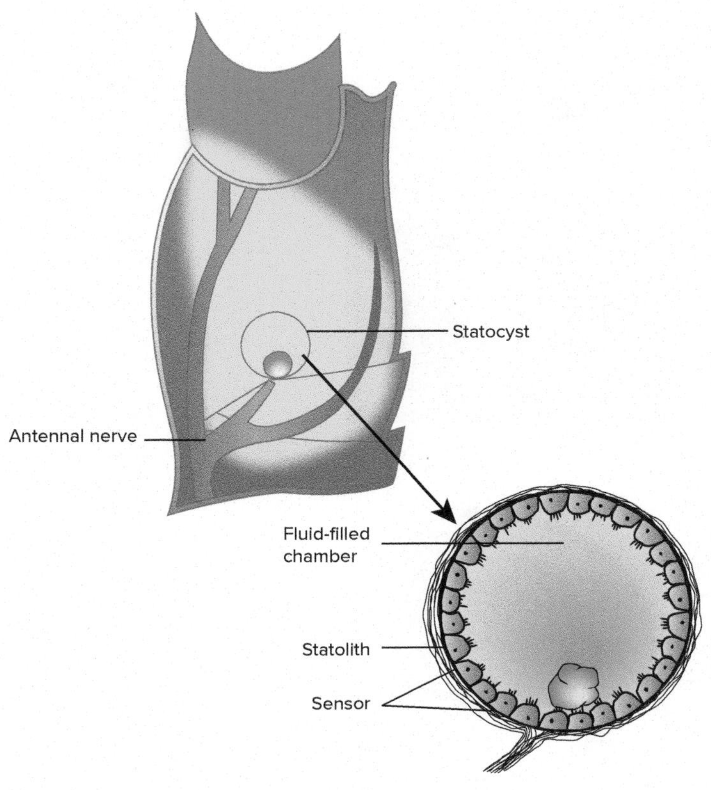

FIGURE 24.14

Invertebrate Georeceptor. A statocyst is present in the base of each first antenna of a crayfish (Crustacea, Decapoda). A statocyst (cross section) consists of a fluid-filled chamber containing a solid granule called the statolith. The inner lining of the chamber contains tactile epithelium from which cilia associated with underlying neurons project. Displacement of cilia by the statolith during bodily movement may cause an action potential that is interpreted in the CNS, thus providing the invertebrate information on its position relative to gravity.

Source: Brusca, RC, Brusca, GJ. 1990. Invertebrates. Sunderland (MA): Sinauer Associates, Inc., Publishers.

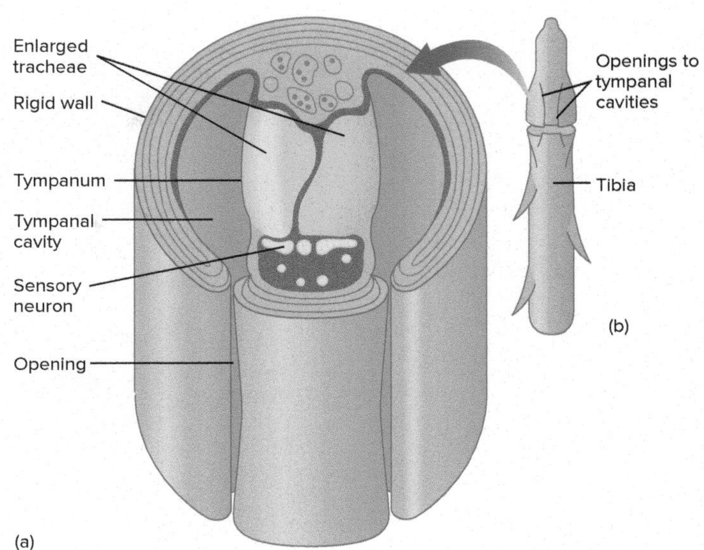

FIGURE 24.15

Invertebrate Phonoreceptor (Tympanal Organ). (*a*) This organ functions on the drumhead principle. The flattened outer wall (tympanum) of each trachea functions as a "drumhead." As the tympanum vibrates in response to sound waves, pressure changes within the tracheae affect the sensory neuron, causing a generator potential. (*b*) The slit openings on the leg (tibia) of a cricket lead to the tympanal cavities.

ventral body, and near the spiracles in insects. Exactly how hygroreceptors transduce humidity into action potentials is unknown.

Phonoreceptors

Although various invertebrates can respond to sound-induced vibrations present in the substrate, Insecta, Arachnida, and Chilopoda are the only known invertebrate groups with species that use **phonoreceptors** (Gr. *phone*, voice + receptor) to detect sound waves transmitted though air or water. Phonoreceptors called **tympanic (tympanal)** organs are present in the legs of Orthoptera and Hemiptera (figure 24.15). These organs consist of a tough, flexible outer tympanum that covers an internal sac. This arrangement permits tympanal vibrations in response to incoming sound waves. Attached sensory neurons are stimulated and depolarized. Noctuid moths (Noctuidae) are equipped with phonoreceptors that permit detection of predatory bats, and water boatmen (Corixidae) and fruit flies (*Drosophila*) use phonoreceptors to identify songs of conspecifics. Most arachnids use cuticular phonoreceptors called slit sense organs to detect sound-induced vibrations. Chilopoda have organs of Tomosvary, which are likely sensitive to sound. The physiology of slit sense organs and organs of Tomosvary is poorly understood.

Photoreceptors

Photoreceptors (Gr. *photos*, light + receptor) contain light-sensitive pigments (e.g., carotenoids and rhodopsin) that absorb and transduce light energy into receptor potentials. The structure and function of photoreceptors vary incredibly among the invertebrates. For instance, some invertebrates, (e.g., the earthworm *Lumbricus*) have simple photoreceptive cells scattered across the epidermis. Other invertebrates have more complex multicellular photoreceptors that can be grouped into three general types: ocelli, compound eyes, and complex eyes.

Ocelli (L. dim. of *oculus*, eye) (sing., ocellus) are integumentary depressions lined with photosensitive (retinular) cells and light-absorbing pigments (figure 24.16*a*; *see also figure 14.10*b). Stimulation of retinular cells leads to chemical changes in the pigment that cause receptor potentials. This visual system provides information about light direction and intensity, but ocelli are not used for image formation. Ocelli are common in various phyla including Annelida, Mollusca, and Arthropoda.

Compound eyes may contain a few or thousands of **ommatidia** (Gr. *ommato*, eye + *ium*, little) (sing., ommatidium), each oriented in a slightly different direction from the others due to the eye's convex shape (figure 24.16*b*, *see figure 15.17*). The field of vision for a given ommatidium overlaps some with adjacent ommatidia; thus, if an object within an invertebrate's total field of vision shifts

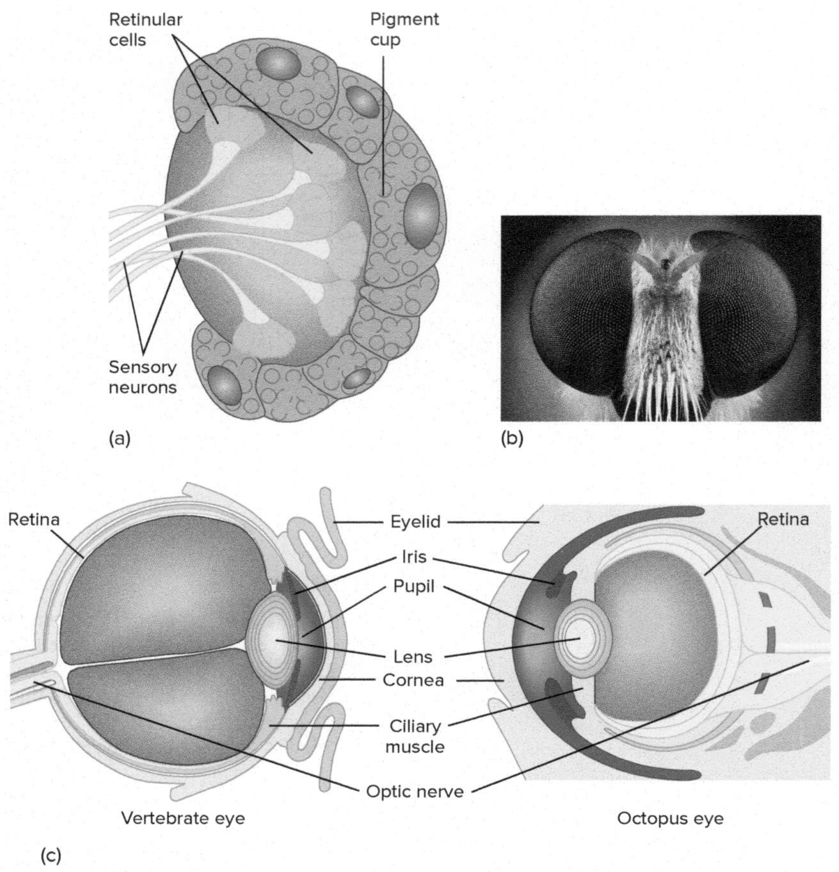

FIGURE 24.16

Invertebrate Photoreceptors. (*a*) Ocellus. The inverted pigment cup ocellus of a flatworm. (*b*) Compound eye. The compound eye of a fly contains thousands of ommatidia. Note the eye's convex shape; no two ommatidia are oriented in precisely the same direction. (*c*) A comparison of the camera eyes of vertebrates and cephalopods (vertical section).

position, the level of stimulation of particular ommatidia changes. This arrangement, coupled with a sophisticated CNS, provides an overall wide field of vision and allows for the detection of fine movements. Invertebrates with compound eyes are capable of forming images and discriminating between some colors, which is particularly important to some jumping spiders (Salticidae) and to diurnal nectar-feeding insects (*see chapter 15*). Compound eyes are most well-developed in the arthropods, but are also present in bivalve molluscs and annelids.

Complex camera eyes of cephalopods are highly modified to form images and are the most sophisticated of the invertebrate camera eyes. The largest complex camera eye in the animal kingdom belongs to a cephalopd, the colossal squid (*Mesonychoteuthis hamiltoni*). The eye exceeds 38 cm in diameter.

The optic anatomy of cephalopods is similar to, and vision is on par with, many vertebrates. The eyes of both groups contain a thin, transparent cornea and a lens that is suspended (and controlled) by ciliary muscles (figure 24.16*c*). Some differences, however, do exist. In cephalopods, photosensitive cells face the direction of light entering the eye, and the optic nerve does not interrupt the distribution of photosensitive cells on the retina (i.e., there is no bind spot). In vertebrates, photosensitive cells face away from incoming light, and their axons traverse the surface of the retina and converge at a point that creates a blind spot. In addition, cephalopods focus by using eye muscles to move the lens toward or away from the retina (like moving a hand lens to focus on different-sized objects). Many vertebrates focus in a similar fashion (e.g., fishes), and others focus by changing the shape of the lens (e.g., mammals).

Proprioceptors

Proprioceptors (L. *proprius*, one's self + receptor) are internal and sense mechanically induced changes caused by stretching, compression, bending, or tension. They provide invertebrates with information about the movement and position of body parts. In arthropods, proprioceptors (stretch receptors) innervate extensor muscles and appendage joints (figure 24.17). Dendrites of proprioceptive neurons associated with cells or tissues in these areas are distorted during movement, causing a receptor potential. Although less well known, proprioception is also an important sense for some soft-bodied invertebrates. For example, the complex feeding behaviors of gastropod molluscs is controlled by proprioceptors.

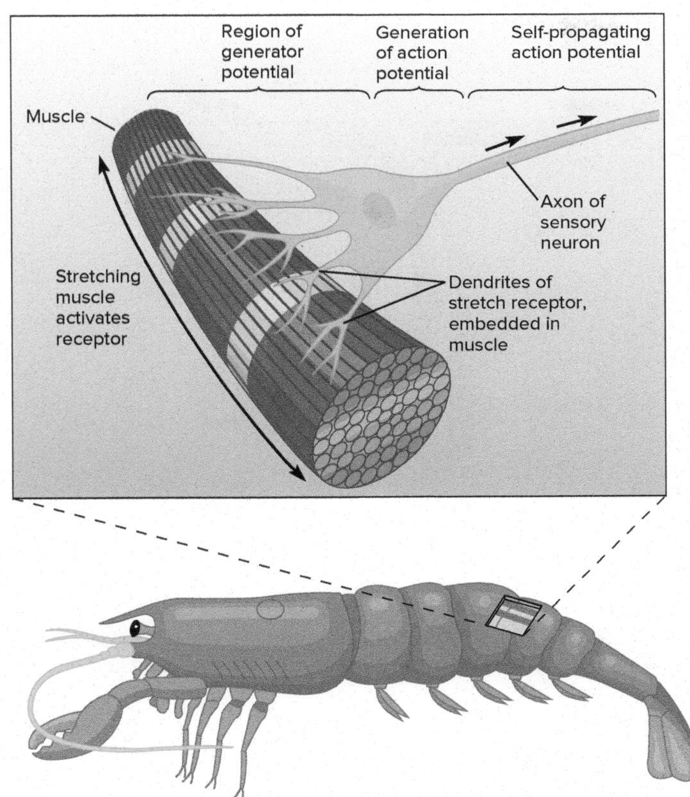

FIGURE 24.17

Invertebrate Proprioceptor. Crayfish stretch receptors are sensory neurons attached to muscles. When crayfish arch their abdomen while swimming, stretch receptors detect changes in muscle length. When muscles are stretched, so are sensory neurons. This action mechanically opens sodium channels in sensory neurons. Inflowing sodium ions may lead to action potentials that travel to the CNS for interpretation.

Tactile Receptors

Tactile (touch) receptors of invertebrates are typically derived from modified epithelial cells associated with sensory neurons. They form body surface projections such as bristles, spines, setae, and tubercles. Displacement or deformation of these receptors through direct contact or vibration activates underlying sensory neurons.

Gastropods use tactile receptors on the tips of retractable tentacles to navigate their environments. Tube-dwelling annelids (e.g., polychaete worms) have tactile receptors that allow them to detect subtle water movements associated with activities of nearby predators. In response, the worms withdraw into their tubes for protection. Web-building spiders (e.g., Araneidae) have tactile receptors that can sense vibrations of struggling prey entangled in webs, and jumping spiders (Salticidae) use their many setae to sense vibrations used in courtship rituals.

Thermoreceptors

Thermoreceptors (Gr. *therme*, heat + receptors) respond to temperature changes. Certain invertebrates can directly sense differences in environmental temperatures. Fruit flies (*Drosophila*) use transient receptor potential ion channels to detect and avoid harmful heat. In addition to other senses, Leeches (Hirudinea) and ticks (Acarina) use specialized heat-sensing mechanisms to locate warm-blooded prey and hosts, respectively. Other insects, some crustaceans, and horseshoe crabs (*Limulus*) are also temperature sensitive. Our understanding of the structure and function of invertebrate thermoreceptors is currently limited.

Section 24.6 Thinking Beyond the Facts

Compare the set of sensory receptors that are most important to a terrestrial insect to the set of receptors that would be most important to a marine crustacean such as the American lobster (*Homarus americanus*).

24.7 Vertebrate Sensory Receptors

Learning Outcomes

1. Compare and contrast the functions of sensory receptors whose physiology depend on the function of ampullary organs or hair cells.
2. Describe the functions of vertebrate tactile, chemical, and visual receptors.

Vertebrate sensory receptors reflect adaptations to the kinds of sensory stimuli present in different external and internal environments. Terrestrial, aquatic, and marine environments have unique sets of challenges to sensory perception. A timber rattlesnake (*Crotalus horridus*) can be found basking on southern-facing rocky outcroppings in Appalachia; piping plovers (*Charadrius melodus*) are shorebirds that scamper the intertidal zone and mudflats, intermittently probing substrates for prey; brook trout (*Salvelinus fontinalis*) may inhabit clear, cool, freshwater streams; and spiny dogfish sharks (*Squalus acanthias*) thrive in seawater. Each of these environments is comprised of media that contain different proportions of specific stimuli. For example, air transmits light and sound waves efficiently, but it carries relatively few molecules detectable via olfaction and transmits little or no electrical energy. In water, however, sound travels both faster and farther than in air, and water carries a wider range of chemicals detectable via olfaction. Water, especially seawater, is also an excellent conductor of electricity, but it absorbs (fails to transmit) many wavelengths of light. Consequently, vertebrates (as in the invertebrates) have evolved receptors (organs) that are sensitive to the different stimuli present in the environments in which they must function. The remainder of this chapter describes the structure and function of selected vertebrate sensory receptors.

Lateral-Line System and Electroreception

Specialized organs for georeception, hearing, and magnetoreception have evolved from the **lateral-line system** of fishes. Lateral-line systems used for electrical sensing are in the head and body areas of most fishes, some amphibians, and the duck-billed platypus (*Ornithorhynchus anatinus*) (figure 24.18*a*). These systems consist

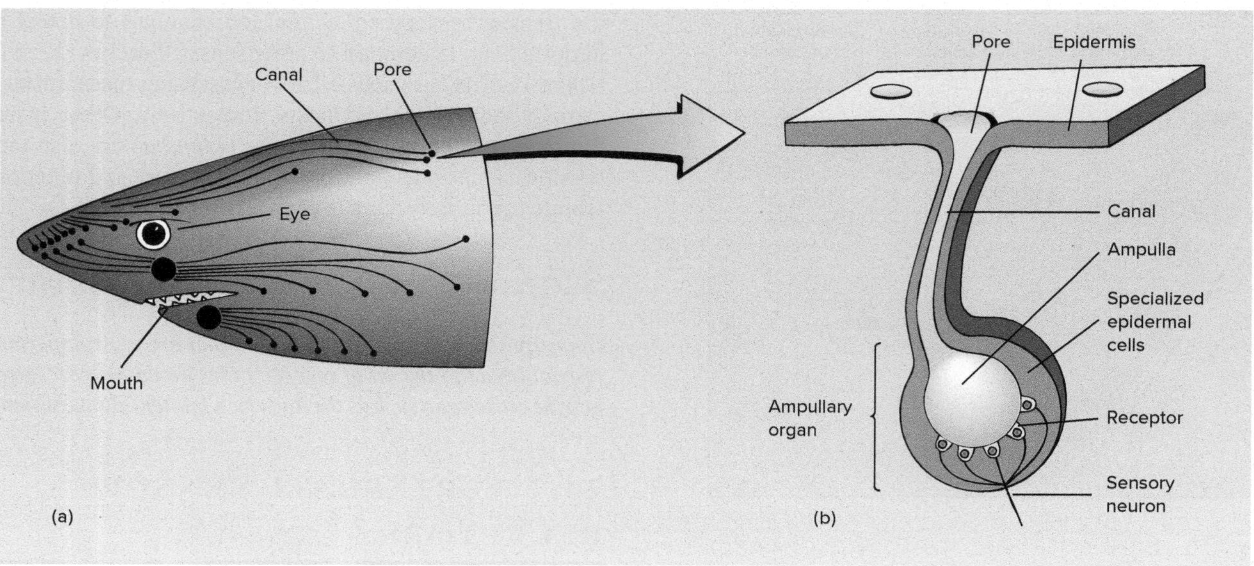

FIGURE 24.18

Lateral-Line System and Electrical Sensing. (*a*) In jawless fishes, jawed fishes, and amphibians, electroreceptors are in the epidermis along the sides of the head and body. (*b*) Pores of the lateral-line system lead into canals that connect to an ampullary organ that functions in electroreception and the production of a generator potential.

of sensory pores in the epidermis that lead, via specialized canals, to electroreceptors called ampullary organs (figure 24.18*b*). Most electroreceptors consist of modified hair cells with bundles of voltage-sensitive protein channels that activate afferent neurons.

Ampullae of Lorenzini are extensions of the lateral-line system that are present in dense clusters over the heads of cartilaginous fishes (e.g., sharks and skates, Elasmobranchii). The ampullae consist of bundles of electrosensory cells inside pores containing a highly conductive hydrogel. Ampullae of Lorenzini permit the detection of electrical fields generated by living organisms. This ability helps fishes find mates, capture prey, or avoid predators and is especially valuable to fishes that inhabit deep, turbulent, or murky waters where vision is limited. In fact, some fishes actually generate electrical fields and then use their electroreceptors to detect how surrounding objects distort the fields (this process is called electrocommunication; *see figure 18.18*). Electroreception is also known in the Guiana dolphin (*Sotalia guianensis*) and the platypus (*Ornithorhynchus anatinus*).

Lateral-Line System and Mechanoreception

Mechanoreceptors are excited by mechanical pressures or distortions (e.g., sound, touch, and muscular contractions). The lateral-line systems of certain fishes and aquatic amphibians include several different kinds of hair-cell mechanoreceptors called **neuromasts**. Neuromasts are confined to pits along the body (figure 24.19*a* and *b*). All neuromasts are responsive to local water displacement or disturbance. When water near the lateral line moves, it moves fluid in the pits and distorts hair cells, causing generator potentials in associated sensory neurons (figure 24.19*c*). This process allows certain aquatic vertebrates to detect the direction and force of water currents. Trout (*S. fontinalis*), for instance, use this sense to orient their heads upstream. Neuromasts also allow the detection of movement of nearby animals in the water, allowing rapid responses to the presence of predators and prey. In some vertebrates (e.g., zebrafish larvae, *Danio rerio*), clusters of hair cells actually project from the surface of skin.

Hearing and Equilibrium in Air

Early terrestrial vertebrates likely used hearing to detect potential danger. It later became used to search for food and mates and in intraspecific communication. Both hearing and equilibrium are sensed in the same vertebrate organ—the ear. The ear has two functional units: the auditory and vestibular apparatus (semicircular canals). The former is used for hearing and the latter for maintenance of posture and equilibrium.

Hearing in air is difficult for vertebrates. Although middle ear structures are sound pressure sensors, pressures generated by airborne sound waves generally contain less than 0.1% of the pressure transmitted through water. The ability to sense airborne sound waves involved the evolution of a structure called the tympanum (the eardrum or tympanic membrane).

Although caudates (salamanders and newts) lack a tympanum and middle ear (*see chapter 19*), the tympanum first evolved in amphibians. Anuran (frogs and toads) ears consist of a tympanum, middle ear, and inner ear (figure 24.20, *see figure 4.12*b). The anuran tympanum is modified integument stretched over a cartilaginous ring that vibrates in response to sound. A middle ear ossicle (collumela or stapes) touches the tympanum at its lateral end and transmits the tympanic vibrations to the oval window membrane at its medial

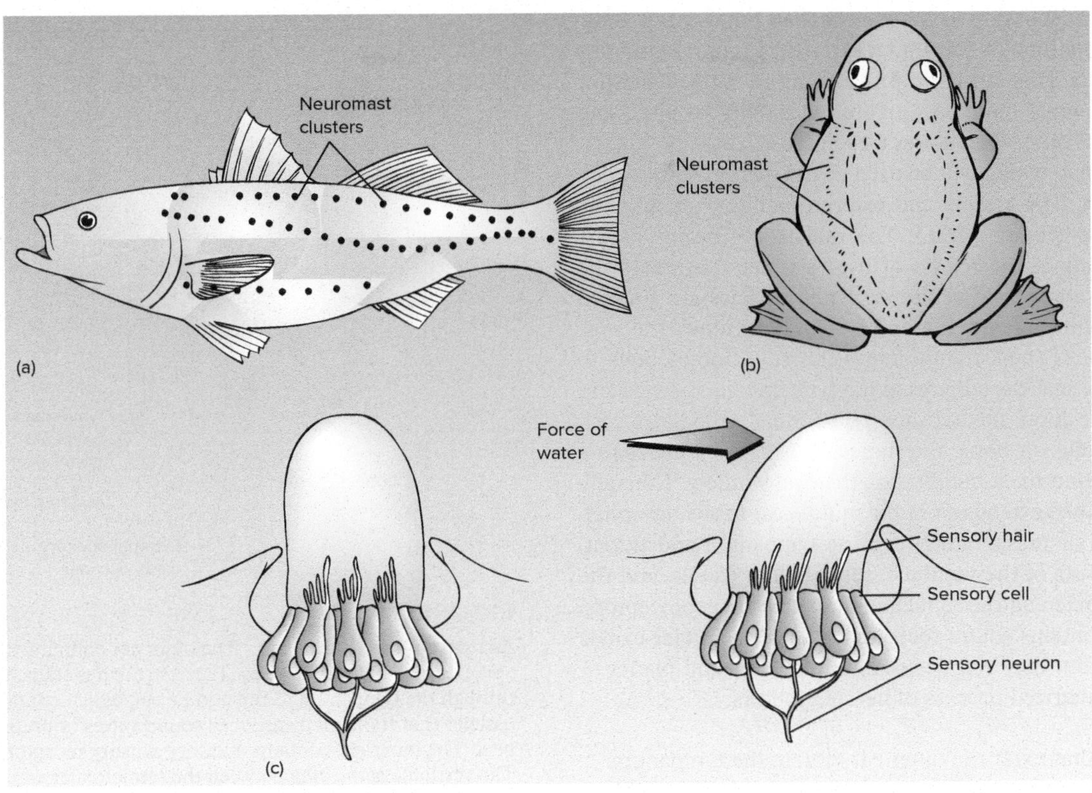

FIGURE 24.19

Lateral-Line System and Mechanoreception. The lateral-line system of (*a*) a bony fish and (*b*) a frog, showing the various neuromast clusters. (*c*) Action of neuromast stimulation. The water movement (blue arrow) forces the caplike structure covering a group of neuromast cells to bend or distort, thereby distorting the small sensory hairs of the neuromast cells, producing a generator potential. The generator potential causes an action potential in the sensory neuron.

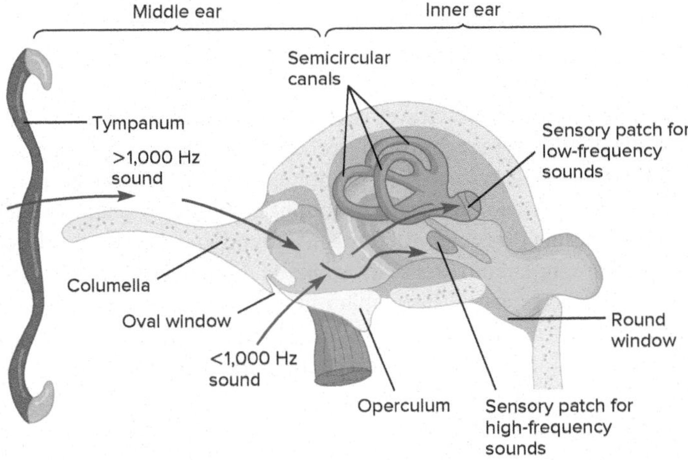

FIGURE 24.20

Ear of an Anuran (Posterior View). Red arrows show the pathway of low-frequency sounds, via the operculum. Dark-blue arrows show the pathway of high-frequency sounds, via the columella (stapes).

end (the oval window membrane stretches between the middle and inner ear). When high-frequency (1,000 to 5,000 Hz) sounds strike the tympanum, they are transmitted through the middle ear via the columella. This action causes pressure waves in inner-ear fluid that stimulate high-frequency receptor cells. Substrate-borne vibrations transmitted through the forelimbs and pectoral girdle cause another ossicle (the operculum) to vibrate. Opercular vibrations generate pressure waves in fluid of the inner ear that stimulate low-frequency receptor cells (100 to 1,000 Hz). Muscles of the operculum and columella can lock either or both of these ossicles, allowing anurans to screen out either high- or low-frequency sounds depending on their need. During the breeding season, many anurans must focus on high-frequency mating calls. Locking the operculum may enhance their focus, but may also make them more vulnerable to predation because they would be less sensitive to low-frequency sounds that may warn of approaching predators.

Nonavian reptile ear structures vary (*see figure 4.12*c). Snakes lack a middle ear cavity and a tympanum, so substrate-borne vibrations are transmitted through a jaw bone to the stapes of the inner ear. In other reptiles, a tympanum may be on the surface of the head, or within small depressions on the head. The nonavian reptile inner ear is similar to that of the amphibians. Hearing is well developed in most avian reptiles. Auriculars (loose, delicate feathers) direct sound into external ear openings.

Auditory senses are important to mammals, both extinct and extant. Mammalian auditory adaptations include an ear flap (the auricle, or pinna), that directs sound into a conductive external auditory canal. The canal conveys sound to the tympanum, which facilitates sound transmission to the mammalian inner ear. A long, coiled, sensory structure of the inner ear (the cochlea) contains receptors for sound. This structure provides more surface area for receptor cells and gives mammals greater sensitivity to pitch and volume than that of other vertebrates (*see figure 4.12*d).

The human ear, which will be used as the mammalian model, has three divisions. The auricle and external auditory canal comprise the **outer ear** (figure 24.21). The **middle ear** begins at the tympanum and ends at the region of the vestibule that bears the membranous oval and round windows. Three ossicles are situated between the tympanic membrane and the oval window: the malleus (hammer), incus (anvil), and stapes (stirrup, homologous to the columella). The malleus adheres to the tympanic membrane and articulates with the more medial incus. The incus articulates with the innermost middle ear bone, the stapes. The stapes adheres to a membrane of the innermost middle ear–the oval window. The auditory (Eustachian) tube extends from the middle ear to the nasopharynx and equalizes air pressure between the tympanum and throat. The **inner ear** consists of the vestibule, semicircular canals, and the cochlea. The vestibule and the semicircular canals are georeceptors, and the cochlea contains sound receptors. The semicircular canals are positioned so that they can detect multiple types and planes of movement. A summarized process of hearing follows:

1. Sound waves that enter the outer ear vibrate the tympanum.
2. The vibrations of the tympanum are transmitted to the malleus and then to the incus and stapes.
3. Vibrations of the stapes are transmitted to the oval window.
4. Movements of the oval window vibrate the fluid in the cochlea, which leads to a rippling of a specialized membrane inside the cochlea.
5. This rippling may stretch attached sound-receptive hair cells and lead to an action potential that travels along the vestibulocochlear nerve to the temporal lobes of the brain for interpretation.
6. Vibrations in the cochlear fluid dissipate as a result of movements of the round window.

Unlike some other terrestrial vertebrates (e.g., African elephants, *Loxodonta africana*), humans cannot hear low-pitched sounds (i.e., sounds below 20 cycles per second [cps]). Children can hear high-pitched sounds up to 20,000 cps, but our ability to detect high-frequency sounds decreases with age, likely due to the degradation of cochlear cells and membranes. Other vertebrates can hear sounds at much higher frequencies. Dogs (*Canis*) can detect sounds of 40,000 cps.

The sense of equilibrium is both static (sensing movement in one plane, either vertical or horizontal) and dynamic (sensing angular and/or rotational movement). Patches of hair cells within the semicircular canals sense dynamic equilibrium. Maculae are sensory patches within the vestibule that detect static equilibrium. Concretions, called otoliths, respond to gravitational force, causing the gelatinous material in which they are embedded to sag and stimulate hair cells (figure 24.22).

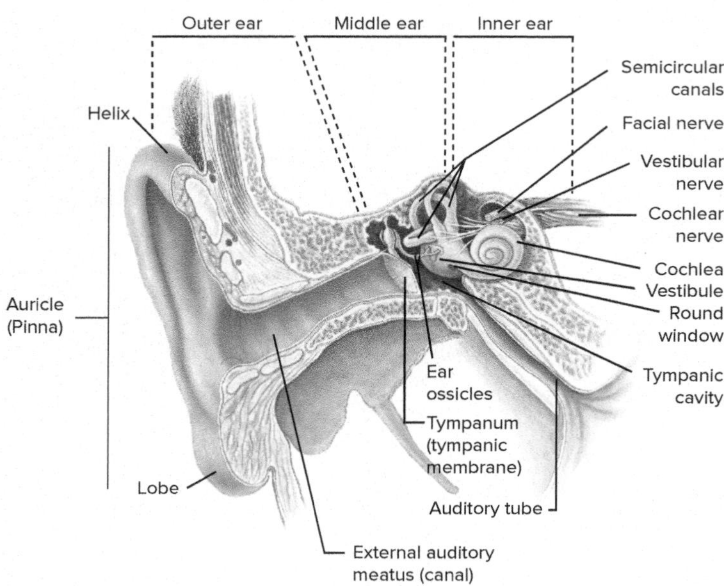

FIGURE 24.21

Anatomy of the Human Ear. The outer ear contains the auricle (pinna) and external auditory meatus. These structures transmit sound waves through the tympanum to the middle ear, which contains specialized ossicles that transmit transduced sound waves (vibrations) to the inner hear. The inner ear contains auditory sensory receptors and georeceptors. The vestibule is the area between the semicircular canals and cochlea. It bears the oval and round windows and is in contact with the vestibular and cochlear branches of the vestibulocochlear nerve.

Hearing and Equilibrium in Water

Hearing is an important sense for most fishes. Although neuromast clusters respond to sound waves, bones associated with the skull also conduct sound waves to the inner ear (there is no middle or outer ear in fishes). The amplitude of sound waves is reduced in water, and sound waves travel more slowly than they do in air. As a result, auditory perception is probably less directional and less sensitive for most fishes when compared to that of terrestrial vertebrates. The **Weberian apparatus** amplifies sound by using specialized (Weberian) ossicles to transmit sound previously amplified in a resonating chamber (the swim bladder) to the inner ear. Consequently, fishes with this apparatus (ostariophysians) probably detect sounds at lower amplitudes than other fishes. In bony fishes, georeceptors (semicircular canals) and auditory receptors are in the inner ear, and they function similarly to those of other vertebrates (figure 24.23).

Sensory Receptors of the Skin

Nociceptors (L. *nocere*, to injure + receptor) (pain receptors) are bare sensory nerve endings present throughout mammalian bodies, except for in the brain. Severe heat, cold, irritating chemicals, strong mechanical stimuli, or the release of chemical signals (e.g., prostaglandins) by injured cells may elicit responses from nociceptors that

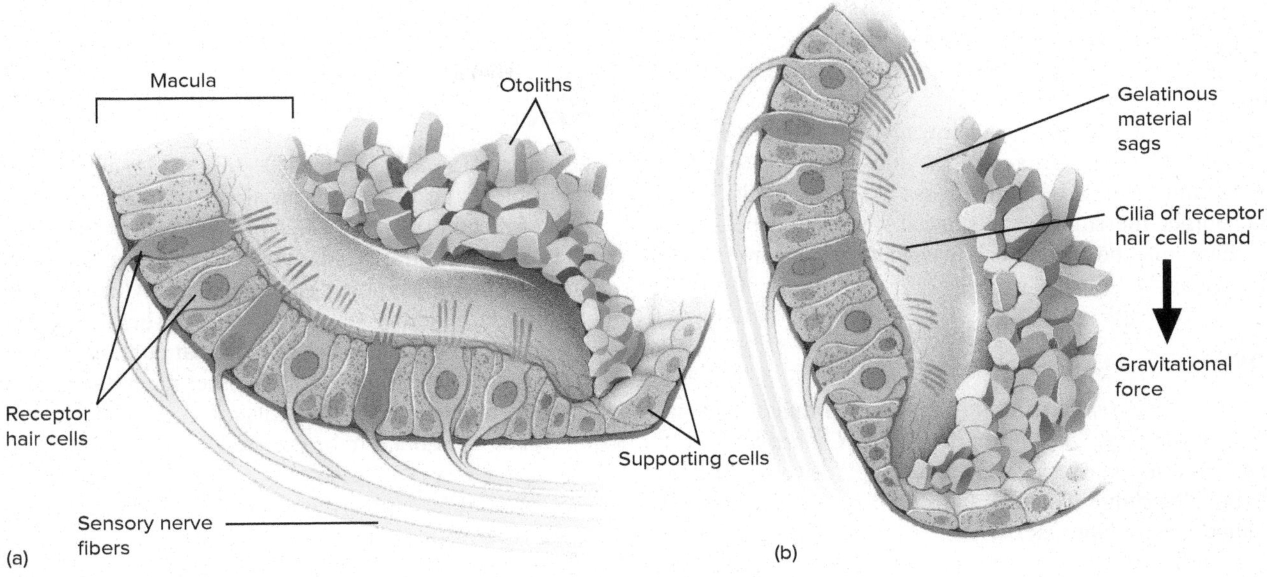

FIGURE 24.22

Static Equilibrium (Balance). Receptor hair cells in the vestibule respond to sideways or up-or-down movement. (*a*) When the head is upright, otoliths are balanced directly over the cilia of receptor hair cells. (*b*) When the head bends forward, the otoliths shift, and the cilia of hair cells bend. This bending of hairs initiates a generator potential.

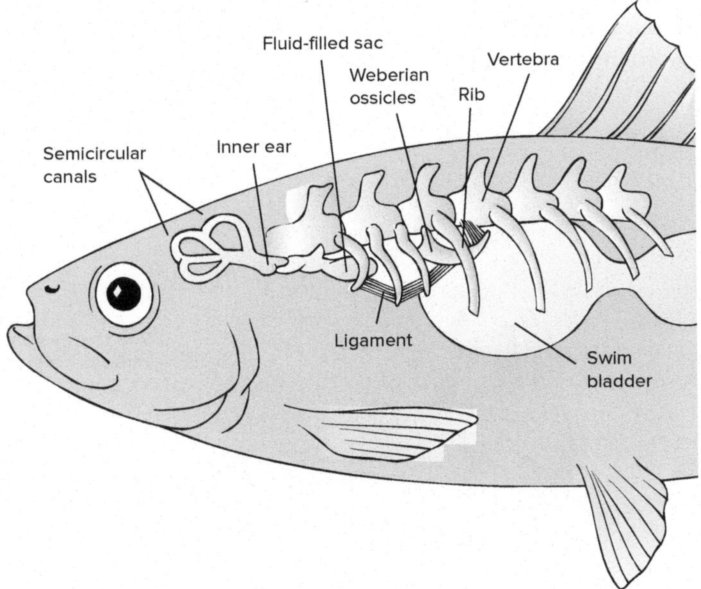

FIGURE 24.23

Inner Ear of a Bony Fish. Sound waves entering the mouth are transmitted to gases in the swim bladder, causing it to expand and contract at frequencies and amplitudes corresponding to the incoming sound waves. Ligament-suspended Weberian ossicles contact the swim bladder and transmit the vibrations to a fluid-filled sac in the inner ear.

the brain interprets as pain, itching, or burning. Certain nociceptor action potentials are carried to the spinal cord, where a reflex response is initiated to reduce damage caused by injurious stimuli. Current knowledge of nociceptor structure and function is limited.

Thermoreceptors (Gr. *therm,* heat + receptor) are bare, temperature-sensitive nerve endings. Thermoreceptors are present in either the epidermis or dermis. Mammals have distinct areas (spots, or small areas of the skin) that are sensitive to either cold or warm. When stimulated, receptors in these areas yield a temperature sensation of warmth or cold. Cold receptors in the skin respond to temperatures below skin temperature, and heat receptors respond to temperatures above skin temperature. Materials that contact skin need not be warm or cold to produce temperature sensations. For example, when metal contacts skin, it absorbs heat, and you feel a sense of coldness. When skin contacts objects that absorb less heat, they would not feel as cold.

The ability to detect temperature changes is well developed in numerous vertebrates. Some snakes are equipped with heat-sensitive **pit organs** located on each side of the face between the eye and the nostril (figure 24.24), or on labial scales. Pit organs contain depressions that are lined with thermoreceptors, which respond to extremely small differences in temperature in the snakes' surroundings. Action potentials generated by these receptors are interpreted by the optic tectum, creating a visual pattern of heat. Pit organs permit location of warm-blooded prey at night. Additionally, vampire bats (Desmodontinae) have highly sensitive rostral thermoreceptors that permit detection of blood-rich veins on mammalian prey.

Many vertebrates use tactile (touch) stimuli to obtain information about their environment. Mechanoreceptors in vertebrate skin detect stimuli that are interpreted as light touch, touch-pressure, and vibration. Light touch is perceived when the skin is not strongly deformed when touched. Light touch receptors include **bare sensory nerve endings** and **Meissner's (tactile) corpuscles**. Meissner's corpuscles are present in areas of mammalian skin that do not contain hair (e.g., fingertips). Other receptors for touch-pressure include **organs of Ruffini** and **Pacinian corpuscles**. These receptors (respectively) detect the duration and extent of touch, and the initiation and cessation of pressure (vibrations).

Many mammals have mechanoreceptors called **vibrissae** (sing., vibrissa) on their wrists, snouts, and eyebrows (e.g., cat whiskers). Around the base of each vibrissa is a blood sinus. When nerves that border the sinus are displaced, they are depolarized. These structures may help mammals navigate their environments, especially at night.

Echolocation

Bats (e.g., *Eptesicus fuscus*), shrews (e.g., *Sorex araneus*), several cave-dwelling birds (e.g., *Steatornis caripensis, Collocalia linchi*), whales (e.g., *Physeter macrocephalus*), and dolphins (e.g., *Tursiops truncatus*) can determine distance and depth by a form of echolocation called **sonar (biosonar)**. These animals emit rapid, high-frequency sounds and then determine how long it takes for the sounds to return after bouncing off objects in the environment. The returning echo created when a moth or other insect flies past a bat can provide enough information for the bat to locate and catch its prey.

The various structures used for echolocation in bats are complex. For instance, a bat's nostrils are widely spaced to produce a megaphone effect when emitting sounds, and its large pinnae and wrinkled rostrum are used to help capture returning echoes. Additionally, the tympanum and middle ear ossicles of bats are very small and light, thus effectively transmitting sound pressures with high fidelity. Furthermore, muscles controlling auditory ossicles contract briefly, reducing the sensitivity of the ear during sound emission, and specialized structures isolate the inner ear from the skull to reduce sound interference during echolocation. Finally, auditory centers that occupy a large portion of the bat's brain are able to construct a sophisticated spatial representation of the external world.

Smell

Olfactory (L. *olere*, to smell + *facere*, to make) receptor cells in the roof of the vertebrate nasal cavity make possible a sense of smell (**olfaction**) (figure 24.25). These cells, which are specialized endings of the first cranial (olfactory) nerve, are densely packed and lie among supporting epithelial cells (e.g., a dog has up to 40 million olfactory receptor cells per square centimeter). Each olfactory cell ends in a tuft of cilia that contains receptor sites for various chemical stimuli. Unlike other neurons, mammalian olfactory receptors are regularly replaced.

For an odorant to be detected, its molecules must diffuse into the air, contact (and dissolve in) olfactory mucus, and bind to specific odorant receptors (there are thousands of different receptors). Each type of odorant receptor only responds to a specific group of molecules. The interaction between an odorant molecule and its receptor alters the membrane permeability of the receptor and leads to depolarization.

In most fishes, external nares in the snout lead to olfactory receptors. Some fishes rely heavily on their sense of smell. Salmon (e.g., *Oncorhynchus nerka*) and lampreys (e.g., *Petromyzon marinus*) return to spawn in their natal streams. Their migrations between the sea and freshwater streams can cover hundreds of miles and are guided by olfactory cues.

FIGURE 24.24

Thermoreception. A subgroup of vipers, the pit vipers, contain a single heat-sensitive pit between each eye and nostril. Pit vipers are distributed throughout Eurasia and the Americas. A representative pit viper, the copperhead (*Agkistrodon contortrix*), is shown here.

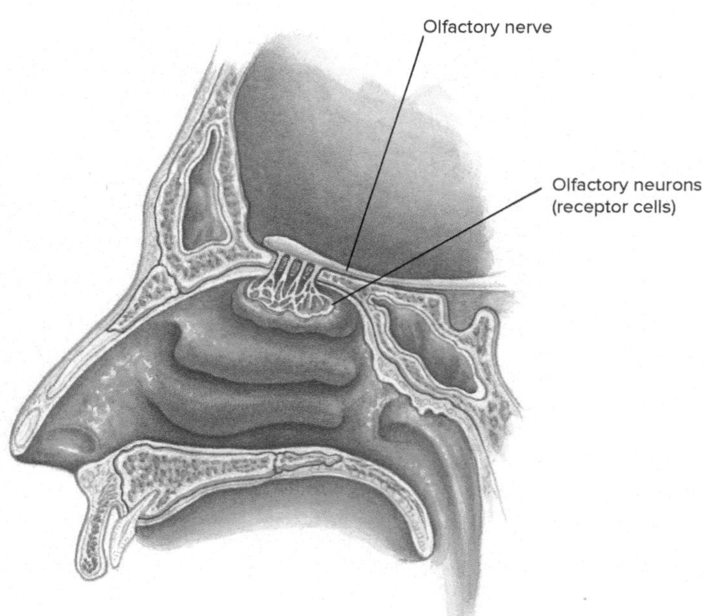

FIGURE 24.25

Smell. Position of olfactory receptors in a human nasal passageway. Columnar epithelial cells support the receptor cells, which have hairlike cilia on dendrites that project into the nasal cavity. When chemicals in the air stimulate these receptor cells, the olfactory nerves conduct nerve impulses to the brain.

Olfaction is important to many amphibians. It is used in mate recognition, detecting noxious chemicals, and locating food.

Olfactory senses are better developed in nonavian reptiles than in amphibians. In addition to having more olfactory epithelium, most nonavian reptiles (except crocodylians) possess blind-ending pouches that open into the mouth. These pouches, called **Jacobson's (vomeronasal) organs,** are best developed in Squamata (figure 24.26). The protrusible, forked tongues of squamates are accessory olfactory organs used to sample airborne chemicals. A snake's tongue flicks out and then moves to the Jacobson's organs, which perceive odor molecules. Turtles (Testudines) and tuataras (Sphenodontidae) use Jacobson's organs to taste objects held in the mouth.

Olfaction plays a minor role in the lives of most avian reptiles. External nares open near the base of the beak, but the olfactory epithelium is poorly developed. Vultures (e.g., Cathartidae) are exceptions; they locate dead and dying prey largely by smell.

Olfactory senses are critical to the survival of many mammals, some of which can perceive olfactory stimuli over long distances (e.g., *Canis*). Olfaction is used for food location, intraspecific recognition, and predator avoidance. **Pheromones** are olfactory signals that are used to maintain social hierarchies and to stimulate reproduction in many animals, including mammals (*see figure 25.1 and table 15.2*).

Taste

Taste (**gustation**, L. *gustus*, taste) receptors are chemoreceptors. The positioning of taste receptors (buds) varies widely among the vertebrates. Taste receptors of avian and nonavian reptiles are mostly in the pharynx. In fishes and amphibians, taste buds are also on the skin. Taste buds on the rostrum of sturgeon (Scaphirhynchinae) permit them to detect food as they search for it while gliding across riverbeds. In other fishes, taste buds occur on the roof, side walls, and floor of the pharynx, where they monitor incoming water. Taste buds of bottom-feeding fishes such as catfish (Siluriformes), and carp and suckers (Cypriniformes), are distributed over the entire body. They are also abundant on catfish barbels ("whiskers").

The surface of the mammalian tongue is covered with many small papillae (sing., papilla) that give the tongue a bumpy appearance (figure 24.27*a*). Thousands of taste buds are situated in crevices between the papillae (figure 24.27*b*). Taste buds are barrel-shaped clusters of gustatory (and supporting) cells arranged like alternating segments of an orange. Gustatory hairs (microvilli) project through tiny openings (taste pores) in gustatory cells. Sensory neurons are associated with the basal ends of gustatory cells.

The five generally recognized taste sensations of vertebrates are sweet (sugars), sour (acids), bitter (alkaloids), salty (electrolytes), and umami (savory; a meaty flavor stimulated by the amino acid glutamate and monosodium glutamate). The exact mechanism(s) that stimulate gustatory cells are not well known. It is likely that different types of gustatory stimuli cause proteins on receptor-cell plasma membranes to change their permeability to cations, resulting in depolarization.

Bitter and sweet taste receptors have been discovered in many organs and tissues that do not contact food. In the airways (and possibly other body parts), these taste receptors play an important role in immunity (see *Evolutionary Insights, Taste-Bud Body Guards*).

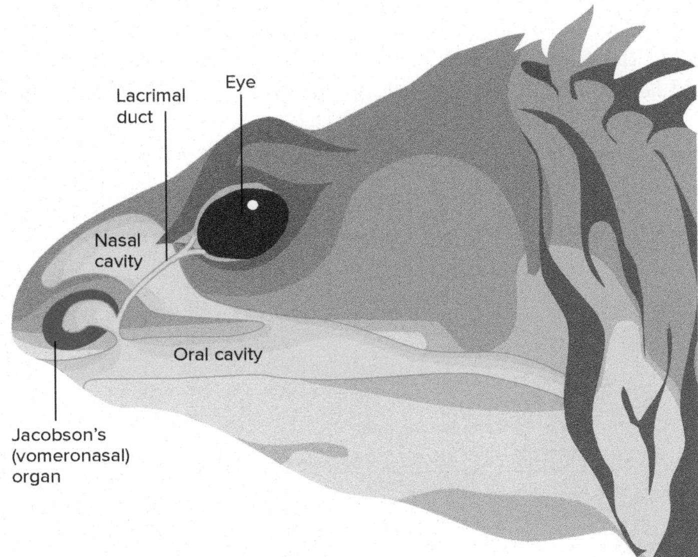

FIGURE 24.26

Smell. Anatomic relationships of Jacobson's (vomeronasal) organ in a generalized lizard. Only the left organ alongside the nasal cavity is shown. Jacobson's organ is a spherical structure, with the ventral side invaginated into the shape of a mushroom. A narrow duct connects the interior of Jacobson's organ to the oral cavity. In many lizards, fluid draining from the eye via the lacrimal duct may bring odoriferous molecules into contact with the sensory epithelium of Jacobson's organ.

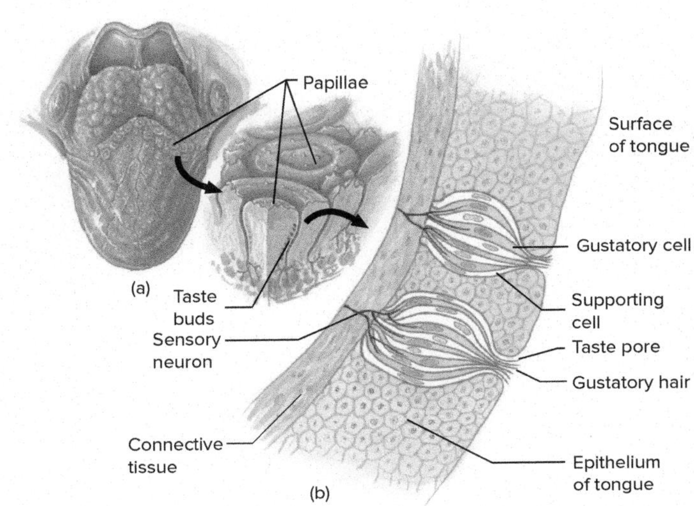

FIGURE 24.27

Taste. (*a*) Surface view of the human tongue, showing the many papillae and the numerous taste buds between papillae. (*b*) Supporting cells encapsulate the gustatory cell and its associated gustatory hair.

EVOLUTIONARY INSIGHTS

Taste-Bud Body Guards

Bitter taste buds are not only on the tongues of vertebrates—they are found throughout the body, where they defend against bacterial invaders. Bitter receptors found in the nose trigger three bacteria-fighting responses. First, they send signals to ciliated cells, which then remove bacteria from the nasal epithelium. Second, the receptors signal epithelial cells to release nitric oxide. (Nitric oxide stimulates ciliary cells to increase ciliary beating and directly kills bacteria). Third, the receptors signal still other cells to release antimicrobial proteins called defensins.

Bitter receptors are found in the respiratory airways, the heart, the lungs, and the intestines. (Most of these organs never come into contact with food.) Physiologists hypothesize that these receptors are part of the vertebrate's innate immune system that reacts rapidly to foreign bacteria. Warning against harm is key to a vertebrate's survival. While sweet, sour, salty, and umami each have only one type of receptor, there are over 25 types of bitter receptors that probably recognize, and protect us from swallowing, a wide variety of poisons. Researchers also suggest that the genetics of these receptors may help us understand why some people get respiratory infections, while others do not.

Vision

Vision (photoreception) is relied upon heavily by vertebrates that inhabit sunlit environments. Most vertebrates have image-forming eyes that contain numerous modifications used to capture and convert light energy into an action potential. The mucous membrane covering the eye is the conjunctiva. It is vascular; and it lubricates, nourishes, and protects the eye. The cornea is a transparent continuation of the toughened outer sclera; it covers and protects the underlying structures. Choroid tissue lies under the sclera. It extends to the front of the eye where it is modified to form the iris, ciliary body, and suspensory ligaments. The iris contains light-screening pigments and smooth muscles that regulate the amount of light entering the pupil. Clear fluid (aqueous humor) fills spaces situated between the lens and cornea (the anterior and posterior chambers). Just behind the iris is the light-focusing apparatus of the eye, the lens. A clear jellylike fluid called vitreous humor fills the large vitreous chamber behind the lens. Suspensory ligaments alter the shape of the lens to focus light energy onto the retina (figure 24.28).

Vertebrates can focus on close-up or distant objects through a process called accommodation. Accommodation mechanisms vary among the vertebrates, but they usually rely on coordinated contraction and relaxation of eye muscles to switch between near and distance vision. In humans, ciliary bodies, suspensory ligaments, and the lens work together to focus light on an area of the retina called the fovea centralis. This area has a high density of cones and is our area of keenest vision. During near vision, the circular

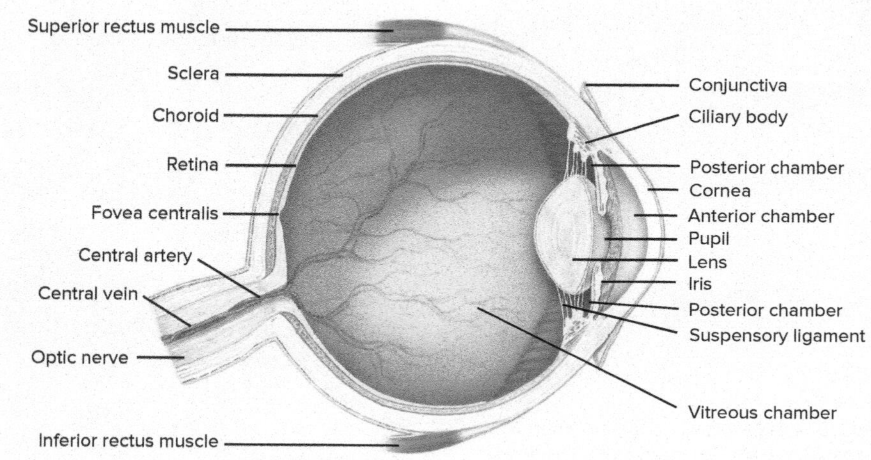

FIGURE 24.28

Internal Anatomy of the Human Eyeball. Light passes through the transparent cornea. The lens focuses the light onto the rear surface of the eye, the retina, at the fovea centralis. The retina is rich in rods and cones.

smooth muscle of ciliary bodies contracts, causing the attached suspensory ligaments to slacken. Because the suspensory ligaments also attach to the lens, this action lessens tension on the elastic lens and causes the previously oval-shaped lens to become more convex. A more convex lens causes rapidly diverging light rays entering the eye to converge at the fovea centralis. Conversely, in distance vision, the circular smooth muscle of ciliary bodies relaxes. This action increases tension on the suspensory ligaments and the lens, and causes the previously rounded lens to become oval shaped. When the lens takes on this shape, it causes parallel light rays that are reflected off of distant objects to more gradually converge onto the fovea centralis.

In vertebrates, the retina is well developed. Its basement layer is composed of pigmented epithelium that covers the choroid layer. Nervous tissue that contains photoreceptors, called rod and cone cells, adheres to this basement layer. **Rod cells** are sensitive to dim light. **Rhodopsin**, a visual pigment in rod cells, absorbs light energy. Absorption can lead to an action potential that travels out of the eye and to the brain via the optic nerve. **Cone cells** respond to high-intensity light and allow for color perception. Three types of cone cells allow primates, birds, reptiles, and fishes to distinguish among light of varying wavelengths. Each type of cone cell contains red-, green-, or blue-absorbing pigments. The pigments are light-absorbing proteins that are sensitive to either long (red), intermediate (green), or short (blue) wavelengths of the visible spectrum. Retinal nerves transduce relative amounts of light absorbed by each type of cone cell into graded potentials that are then transmitted as nerve impulses to the brain. Depending on the extent of stimulation of each type of cone, all the colors of the visible spectrum are perceived.

The eyes of the fishes are typically lidless and positioned laterally on the head. Some fishes (e.g., Carcharhiniformes) are equipped with nictitating membranes that protect the eyes while hunting. The lens in fishes is rounded and close to the cornea; it must be moved forward or backward in order to focus.

Most amphibians are visual hunters. Certain adaptations that are typically not seen in fishes allow amphibian eyes to function on land. For example, the eyes of some anurans and caudates are located close together on the front of the head. This arrangement provides the binocular vision and depth perception necessary for capturing terrestrial prey. Other amphibians with smaller and more laterally positioned eyes lack binocular vision—an arrangement that supports peripheral vision. An amphibian's transparent nictitating membrane functions similar to that of the fishes. It lies underneath the upper and lower eyelids, which can be closed to fully protect the eye (*see chapter 19*).

The eyes of nonavian reptiles are similar to those of amphibians. They are equipped with upper and lower eyelids, and the nictitating membrane (which is also present in avian reptiles) and blood sinus protect and cleanse the eye. In some squamatids, the upper and lower eyelids may fuse embryonically to form a clear protective covering called the spectacle.

A parietal eye that develops from outgrowths of the roof of the midbrain (the optic tectum) is present in some nonavian reptiles (figure 24.29). The parietal eye of tuataras (*Sphenodon*) is particularly well developed; it has a lens, nerve, and retina. Skin covers

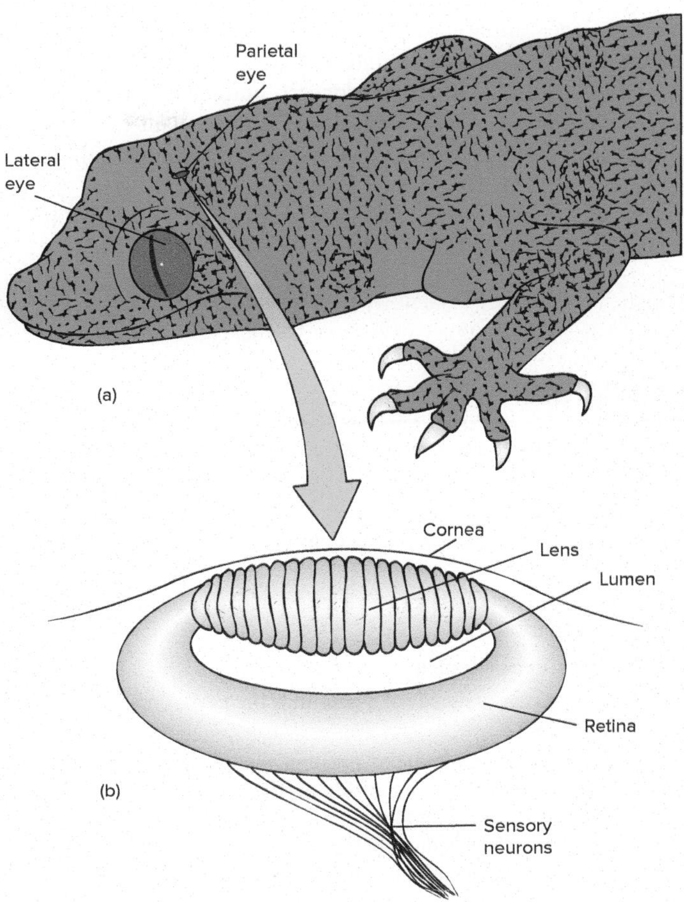

FIGURE 24.29

Median Eye of Reptiles. (*a*) Parietal eye in a reptile and its relationship to the lateral eyes (dorsal view). (*b*) Sagittal section of the median eye.

parietal eyes, so they likely cannot form images. They are involved with regulation of circadian rhythm and orientation to the sun for thermoregulation (a less-developed version of the parietal eye exists in amphibians).

Avian reptile eyes differ somewhat from those of other amniotes. They possess specialized striated musculature and ciliary bodies that can control the curvature of both the lens and cornea. This structure permits nearly instantaneous double focusing so that birds of prey can maintain focus on prey throughout extremely rapid dives (over 320 km/hr in peregrine falcons, *Falco peregrinus*).

Animation
Vision

Magnetoreception

Magnetoreception is the ability to detect magnetic fields. Many animals use the earth's magnetic field to navigate during migration or homing. Magnetoreception has been extensively studied, but the physiology of magnetoreception is not yet thoroughly understood. Many studies are focused on magnetite, an iron oxide found in various animals including avian reptiles (e.g., *Columba livia*), salmon (e.g., *Oncorhynchus tshawytscha*), and sea turtles (e.g., *Caretta*

caretta). Other studies focus on cryptochromes, blue-light photoreceptors that were recently found to enable magnetoreception used in long-distance migrations of monarch butterflies (*Danaus plexippus*). Cryptochromes were first identified in a plant (*Arabidopsis*), and they also mediate various light-related responses in animals.

Section 24.7 Thinking Beyond the Facts

The lateral-line system of primary terrestrial adult amphibians is lost during metamorphosis from aquatic larval to adult stages. Why is the lateral-line system not useful to these adult amphibians?

Summary

24.1 **Neurons: The Basic Functional Units of the Nervous System**

- The functional unit of the nervous system is the neuron. Neurons have two important properties: excitability and conductivity. The three functional types of neurons are sensory neurons, interneurons, and motor neurons.
- A typical neuron has three anatomical parts: dendrites, a soma, and an axon.

24.2 **Neuron Communication**

- The language (signal) of a neuron is the nerve impulse or action potential.
- The plasma membrane of a resting neuron is polarized, meaning that the fluid on the inner side of the membrane is negatively charged with respect to the positively charged fluid outside the membrane. The sodium-potassium ATPase pump and diffusion of ions across membrane channels maintain this polarization.
- Graded potentials are changes in the membrane potential that are confined to a small area of a membrane. Graded potentials are important for the function of sensory neurons, which must distinguish between strong and weak stimuli coming into the animal from the environment.
- Action potentials occur when a threshold stimulus is applied to a resting neuron. An action potential involves a depolarization/repolarization sequence and conduction in an all-or-none fashion. In myelinated neurons, the action potential jumps from one neurofibril node to the next in a process known as saltatory conduction.
- Neuronal activity is transmitted between cells at the synapse. Electrical synapses transmit neuronal activity very rapidly from one neuron to the next. Chemical synapses use chemical neurotransmitter molecules.

24.3 **Invertebrate Nervous Systems**

- Sponges have no true nervous system; nevertheless, coordination and reaction to external and internal stimuli do occur. Cnidarians have radial symmetry and nerve nets. Echinoderms have a pentaradial nervous system and three nerve net systems. This complexity permits locomotion, a variety of useful reflexes, and some degree of "central" coordination. Animals that have bilateral symmetry show centralization and cephalization.

24.4 **Vertebrate Nervous Systems**

- The vertebrate nervous system has two main divisions. The central nervous system is composed of the brain and spinal cord, and the peripheral nervous system is composed of all the nerves (bundles of axons and/or dendrites) outside the brain and spinal cord.
- The spinal cord of a vertebrate serves two important functions. It is the connecting link between the brain of an animal and most of the body, and it is involved in spinal reflex actions. Protective meninges surround the spinal cord.
- The number of spinal nerves is directly related to the number of segments in the trunk and tail of a vertebrate.
- The vertebrate brain develops from three swellings of the anterior region of the spinal cord: hindbrain, midbrain, and forebrain. The hindbrain is continuous with the spinal cord and includes the medulla oblongata, cerebellum, and pons. The midbrain is a thickened region of gray matter that integrates visual and auditory signals. The forebrain contains the pineal gland, pituitary gland, hypothalamus, and thalamus. The anterior part of the forebrain expanded during evolution to give rise to the cerebral cortex.
- The peripheral nervous system of nonavian and avian reptiles and mammals includes 12 pairs of cranial nerves. Fishes and amphibians have only the first 10 pairs.
- The autonomic nervous system consists of two antagonistic parts: the sympathetic and parasympathetic divisions. The enteric division consists of networks of neurons associated with digestive organs.

24.5 **Sensory Reception**

- A stimulus is any form of energy an animal can detect with its receptors. Receptors are nerve endings of sensory neurons or specialized cells that respond to stimuli, such as chemical energy, mechanical energy, light energy, or radiant energy. Receptors transduce energy from one form to another. Stimulation of a receptor initiates a graded potential, which creates an action potential that travels along a nerve pathway to another part of the nervous system, where it is perceived.

24.6 **Invertebrate Sensory Receptors**

- Invertebrates possess an impressive array of receptor structures through which they receive information about their environment.
- Baroreceptors respond to pressure changes in aquatic invertebrates.
- Chemoreceptors are located in pits or depressions through which water or air circulates. They function in humidity detection, pH assessment, food recognition, and mate location.
- Georeceptors respond to gravity and include statocysts.
- Hygroreceptors detect water content of air.

- Phonoreceptors respond to sound. These include tympanic organs.
- Photoreceptors are sensitive to light. These include ocelli, compound eyes, and complex camera eyes.
- Proprioceptors are stretch receptors and give an animal information about the movement of its body parts and their positions relative to each other.
- Tactile or touch receptors include bristles, spines, setae, and tubercles. They are sensitive to mechanically induced vibrations.
- Thermoreceptors respond to temperature changes.

24.7 **Vertebrate Sensory Receptors**

- The lateral-line system for electrical sensing is in the head area of most fishes, some amphibians, and the platypus. This system can sense electrical currents in the surrounding water.
- The lateral-line system of fishes and amphibians also contains neuromasts, which are responsive to local water displacement or disturbances.
- The vertebrate ear has two functional units: the auditory apparatus is concerned with hearing, and the vestibular apparatus is concerned with posture and equilibrium. Hearing in air involves a tympanum, which conducts sound waves to the columella or stapes. This middle ear ossicle transmits pressure waves to the oval window, then to receptors of the inner ear. Mammals possess three middle ear ossicles and specialized structures in the cochlea for the detection of sound.
- Fishes lack outer and middle ears, and sound vibrations are conducted through bones of the skull.
- Pain receptors (nociceptors) are bare sensory nerve endings that produce a painful or itching sensation.
- Sensors of temperature (thermoreceptors) are bare sensory nerve endings and the simplest vertebrate receptors. Some snakes have heat-sensitive pit organs on each side of the face.
- Many vertebrates rely on tactile (pertaining to touch) stimuli to respond to their environment. Receptors include bare sensory nerve endings, tactile (Meissner's) corpuscles, Pacinian corpuscles, organs of Ruffini, and vibrissae.
- Bats, shrews, whales, and dolphins can determine distance and depth by echolocation.
- The sense of smell is due to olfactory neurons in the roof of the vertebrate nasal cavity. Each receptor cell contains one or a few receptor types that responds in unique ways to odorants characterized by specific chemical groups. Vertebrates other than most avian reptiles depend on the sense of smell for mate recognition, detecting home areas, feeding, and other functions.
- The receptors for taste (gustation) are chemoreceptors on the body surface of an animal or in the mouth and throat.
- Most vertebrates have eyes capable of forming visual images. Accommodation (focusing) occurs when light rays fall on the retina of the eye. In most vertebrates, it occurs when muscle of the eye change the shape of the lens. In fish, it occurs as a result of the movement of the lens forward or backward in the eye. Rod and cone cells in the eye respond to light energy.
- Magnetoreception is the ability to detect magnetic fields and is the most elusive of the sensory modalities.

Concept Review Questions

1. Which of the following is NOT a functional type of neuron?
   a. Sensory
   b. Afferent
   c. Interneuron
   d. Efferent
   e. Connecting
2. Graded potentials are important in signaling over long distances, whereas action potentials signal over short distances.
   a. True
   b. False
3. The simplest form of nervous organization is found in
   a. Arthropoda.
   b. Porifera.
   c. *Hydra.*
   d. Echinodermata.
   e. Platyhelminthes.
4. Which of the following is NOT part of the autonomic nervous system?
   a. Sympathetic portion
   b. Parasympathetic portion
   c. Enteric portion
   d. Somatic system
5. Which of the following is NOT part of the vertebrate diencephalon?
   a. Thalamus
   b. Hypothalamus
   c. Pituitary gland
   d. Pineal gland
   e. Cerebrum

Analysis and Application Questions

1. How do inhibitory and excitatory neurotransmitters alter the activity of the human nervous system acting at the synapse?
2. How can the movement of only positive ions result in depolarization and repolarization of the membrane during an action potential?
3. Surveying the functions of the evolutionarily oldest parts of the vertebrate brain gives us some idea of the original functions of the brain. Explain this statement.
4. What are the possible advantages and disadvantages in the evolutionary trend toward cephalization of the nervous system?
5. How is an action potential in one cell converted to an action potential in a second cell?
6. Most mammals lack cones in their retinas. How, then, do such animals view the visual world?
7. Why is the gravitational sense considered a sense of equilibrium?
8. Explain how a lack of vitamin A might affect the vision of a primate.
9. How would you expect your inner ear to behave in zero gravity?

25

Communication II: The Endocrine System and Chemical Messengers

This photograph shows a Monarch butterfly (*Danaus plexippus*) chrysalis (pupa) undergoing a final molt into an adult. This process is highly regulated by hormones, the subject of this chapter.

Chapter Outline

25.1 The Evolution and Diversity of Chemical Messengers
 Diversity of Chemical Messengers
25.2 Hormones and Their Feedback Systems
 Biochemistry of Hormones
 Feedback Control System of Hormone Secretion
25.3 Mechanisms of Hormone Action
 Fixed-Membrane-Receptor Mechanism
 Mobile-Receptor Mechanism
25.4 Endocrine Glands
25.5 Invertebrate Endocrine Control
 Chemical Messengers of Basal and Lophotrochozoan Phyla
 Chemical Messengers of Ecdysozoans
 Chemical Messengers of Deuterostomes
25.6 Endocrine Systems of Fishes and Amphibians
25.7 Endocrine Systems of Amniotes
 Avian and Nonavian Reptiles
 Mammals
25.8 Some Hormones Are Not Produced by Endocrine Glands
25.9 Evolution of Endocrine Systems

Chapter 24 discussed ways that the nervous and sensory systems work together to rapidly communicate information and maintain homeostasis in an animal's body. In addition, many animals have a second, slower form of communication and coordination–the endocrine system with its chemical messengers. This chapter describes the evolution and diversity of chemical messengers, the mechanisms of hormone action, and the functions that specific hormones play in various animal taxa.

25.1 THE EVOLUTION AND DIVERSITY OF CHEMICAL MESSENGERS

LEARNING OUTCOMES

1. Assess the selective pressures that favored the evolution of chemical messenger systems.
2. Compare the sources, transport routes, and destinations of the 5 types of chemical messengers found in animals.

Some scientists suggest that chemical messengers may initially have evolved in single-celled organisms to coordinate feeding or reproduction. As multicellularity evolved, more complex organs also evolved to govern the many individual coordination tasks, but control centers relied on the same kinds of chemical messengers that were present in the simpler organisms. Some of the messengers worked fairly slowly but had long-lasting effects on distant cells; these became the modern hormones. Others worked more quickly but influenced only adjacent cells for short periods; these became the neurotransmitters and local chemical messengers. Clearly, chemical messengers have an ancient origin and must have been conserved for hundreds of millions of years.

One key to the survival of any group of animals is proper timing of activity so that growth, maturation, and reproduction coincide with the times of year when climate and food supply favor survival. It seems likely that the chemical messengers regulating growth and reproduction were among the first to appear. These messengers were probably secretions from neurons. Later, specific hormones developed to play important regulatory roles in molting, growth, metamorphosis, and reproduction in various invertebrates. Chemical messengers and their associated secretory structures became even more complex with the appearance of vertebrates.

We see in the evolution of chemical messengers the idea that when a new function evolves in an animal, the new function often uses preexisting structures and pathways. In the evolution of endocrine functions, "old" messengers were adapted to new purposes. In this chapter, you will see examples of ancient protein messengers functioning in different ways in diverse animal groups.

Diversity of Chemical Messengers

The development of most animals commences with fertilization and the subsequent division of the zygote. Further development then depends on continued cell proliferation, growth, and differentiation. The integration of these events, as well as the communication and coordination of physiological processes, such as metabolism, respiration, excretion, movement, and reproduction, depend on chemical messengers—molecules that specialized cells synthesize and secrete. Chemical messengers can be categorized as follows:

1. **Local chemical messengers.** Many cells secrete chemicals that alter local physiological conditions (figure 25.1*a*). These chemicals predominantly act on the same cell (**autocrine agents**) or adjacent cells (**paracrine agents**) and do not accumulate in blood. Local chemical messengers of vertebrates include the lumones, which are produced in the gut to help regulate digestion. Histamines of vertebrates are released by mast cells (a type of white blood cell) to promote inflammation in response to infection by an external agent.
2. **Neurotransmitters.** Neurons secrete neurotransmitters (e.g., nitric oxide and acetylcholine) to relay action potentials to adjacent target cells (figure 25.1*b*). Neurotransmitters reach high concentrations in synaptic clefts, act quickly, and are actively degraded and recycled (reuptake).
3. **Neuropeptides.** Specialized neurons called neurosecretory cells secrete neuropeptides. Blood or other body fluids transport neuropeptides to nonadjacent target cells, where neuropeptides act (figure 25.1*c*). In mammals, the hypothalamus triggers the posterior pituitary gland to release oxytocin—a neuropeptide produced in the hypothalamus that causes the powerful uterine contractions responsible for childbirth.
4. **Hormones.** Endocrine glands or cells secrete hormones that the bloodstream transports to nonadjacent target cells (figure 25.1*d*). Many examples are given throughout this chapter.
5. **Pheromones.** Pheromones are released externally, and they influence various behaviors of other members of the same species (conspecifics). They have a variety of functions including the indication of reproductive readiness, the location of a food source, or threats to conspecifics (figure 25.1*e*; *see also table 15.2*).

The nervous and endocrine systems work together as a comprehensive communicative and integrative network called the **neuroendocrine system.** This system operates through feedback loops that regulate animal body functions to maintain homeostasis.

Section 25.1 Thinking Beyond the Facts

How might certain chemical messengers function to coordinate the activities of different body systems?

25.2 Hormones and Their Feedback Systems

Learning Outcomes

1. Categorize hormones by their biochemical structures and by their effects on target cells.
2. Contrast the two types of hormonal feedback loops that are involved with regulating biological processes.

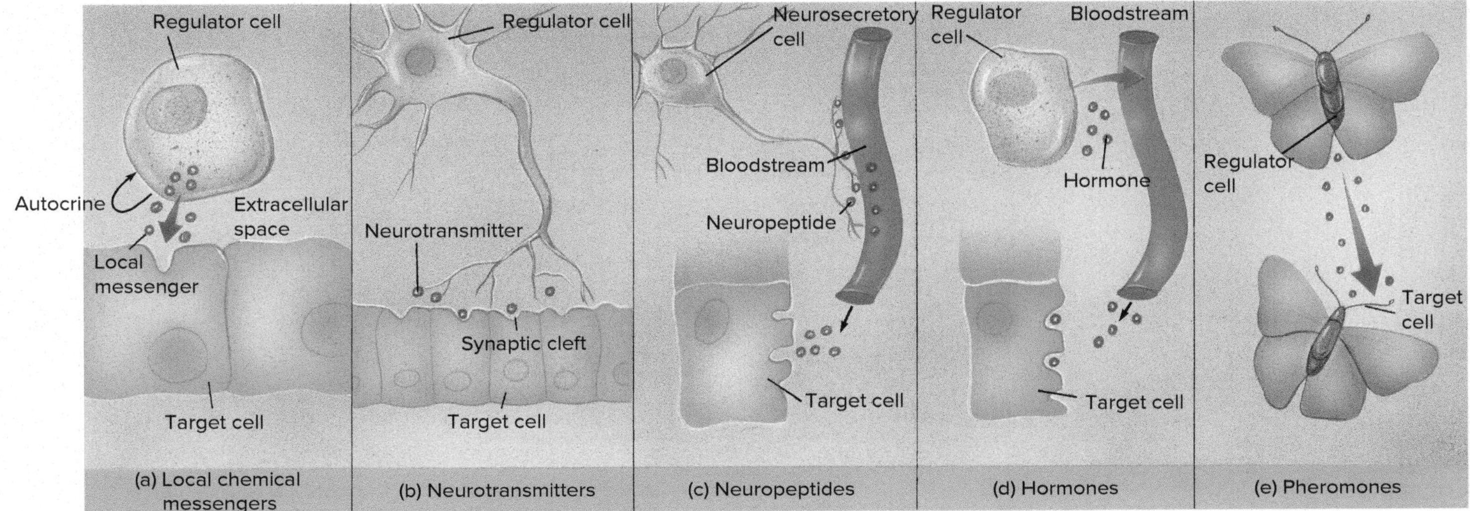

FIGURE 25.1

Chemical Messengers: Targets and Transport. (*a*) Short-distance local messengers act on an adjacent cell (paracrine agents) or the same cell (autocrine agents). (*b*) Individual nerve cells secrete neurotransmitters that cross the synaptic cleft to act on target cells. (*c*) Individual nerve cells can also secrete neuropeptides (neurohormones) that travel some distance in the bloodstream to reach a target cell. (*d*) Regulatory cells, usually in an endocrine gland, secrete hormones, which enter the bloodstream and travel to target cells. (*e*) Regulatory cells in exocrine glands secrete pheromones. They leave the body and stimulate target cells in another animal. Each kind of chemical regulator binds to a specific protein on the surface or within the cells of the target organ.

A **hormone** (Gr. *hormaein*, to set in motion or to spur on) is a specialized chemical messenger that an endocrine gland or tissue produces and secretes. Hormones circulate through body fluids and affect the metabolic activity of a target cell or tissue in a specific way. A **target cell** has receptors to which chemical messengers either selectively bind or affect in some way. Hormones often work in concert with other hormones in feedback networks. In these networks, the concentrations of circulating hormones stimulate, inhibit, or complement the functions of other hormones. The study of endocrine glands and their hormones is called **endocrinology.**

Biochemistry of Hormones

Most hormones are proteins (polypeptides), amino acid derivatives (amines), or steroids. A few (prostaglandins) are fatty acid derivatives. For instance, most invertebrate neurosecretory cells produce polypeptide hormones (neuropeptides). The vertebrate pancreas secretes protein hormones and the thyroid gland secretes amines. The ovaries, testes, and cortex of the adrenal glands secrete steroids.

Generally, only a few hormone molecules are needed to produce dramatic responses in target cells. Hormones help control biochemical reactions in target cells by (1) increasing the rate that other substances enter or leave cells; (2) stimulating target cells to synthesize enzymes, proteins, or other substances; or (3) prompting target cells to activate or suppress existing cellular enzymes. Like enzymes, hormones are not changed by the reactions they regulate.

Feedback Control System of Hormone Secretion

Although hormones are always present in some amount in endocrine cells or glands, they are not secreted continuously. Instead, the glands secrete the amount of the hormone that the animal needs to maintain homeostasis. **Homeostasis** is the maintenance of relatively constant internal conditions that are compatible with life, and it requires a control system. Control systems can involve many animal body systems. The endocrine system of an animal is a control system for many internal variables. For example, endocrine regulation maintains blood glucose levels at or near a particular value or **set point** (70 to 110 mg/dl for blood glucose in humans). Any fluctuation in the variable above or below the set point serves as the stimulus detected by the sensor. Upon receiving a signal from the sensor, an **integrating center** (usually an endocrine gland or the nervous system) generates output that triggers a response by an effector structure, which helps return the variable to the set point. When an effector's response reverses a change detected by a sensor, the control system is operating through **negative feedback.** The blood concentrations of glucose, calcium, and other substances are regulated by negative feedback loops. Steroid hormones that regulate reproductive functions in animals also utilize negative feedback (*see figure 29.8*).

Animation
Hormonal Communication

Two hormones are involved in the negative feedback regulation of blood glucose–insulin and glucagon. After a carbohydrate meal, blood glucose levels rise (figure 25.2*a*) above the set point of 110mg/dl. This rise in glucose (the stimulus) is detected by pancreatic islet cells (the sensor and integrating center) and stimulates insulin secretion from the pancreatic islet cells. Insulin (the effector) causes body cells to absorb glucose from the blood, and blood glucose drops to its normal range (70 to 110 mg/dl). Because a rise in blood glucose stimulates insulin secretion, a lowering of blood glucose caused by the action of insulin inhibits any more insulin secretion–a negative feedback control system. Conversely, blood glucose levels may fall below 70 mg/dl. In this case, insulin secretion decreases and prevents glucose uptake by skeletal muscle, the liver,

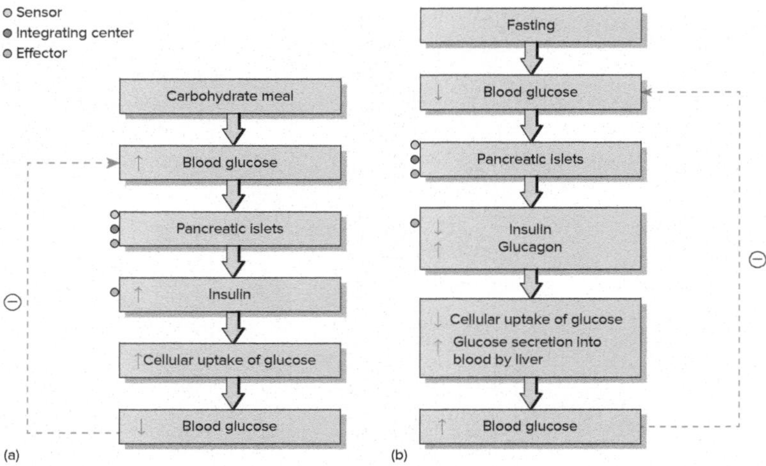

FIGURE 25.2

Negative Feedback Control of Blood Glucose. In flow charts that illustrate control systems, negative feedback is either represented by a dashed line or a negative (−) sign. (*a*) A rise in blood sugar occurs in a mammal after eating a meal high in carbohydrates (e.g., glucose). The pancreatic islets act as both a sensor and an integrating center to detect the rise in blood glucose and release the hormone insulin–the effector. (Insulin is secreted by specialized islet cells called beta cells.) Insulin binds to specific receptors on certain cells allowing them to take-up the excess blood sugar. This uptake allows blood glucose to return to its normal set point. (*b*) When a mammal is fasting, blood sugar drops to low levels and insulin secretion is inhibited. The antagonistic hormone glucagon is then secreted by specialized islet cells called alpha cells. Glucagon decreases the cellular uptake of glucose and simultaneously stimulates the release of glucose from the liver to increase blood glucose levels to normal values.

and adipose tissue. Glucagon is an antagonistic hormone to insulin. Glucagon stimulates the liver to release glucose into the blood stream (figure 25.2*b*). As a result, blood glucose rises back to its normal range (70 to 110 mg/dl).

Unlike negative feedback, **positive feedback** is a control system that amplifies the response to a stimulus. Positive feedback control systems do not play a major role in homeostasis, but instead help drive processes to completion. For example, during mammalian birth, the hormone oxytocin (posterior pituitary gland) and the pressure of the fetal head against the uterine birth canal stimulate uterine contractions. These contractions result in the release of additional oxytocin and even greater pressure against the opening of the uterus, thus heightening the contractions. This positive feedback continues, eventually expelling the fetus from the uterus. Oxytocin is also involved in the positive feedback regulation of milk release from mammary glands during nursing. Positive feedback also ensures that blood clots form rapidly following a wound.

Section 25.2 Thinking Beyond the Facts

Critique the following statement: "In order to be effective, hormones are not continuously secreted."

25.3 Mechanisms of Hormone Action

Learning Outcomes

1. Explain how the signal carried by a peptide hormone induces a change in a target cell.
2. Explain how steroid hormones activate transcription.

Hormones modify the biological activity of target cells via fixed-membrane- or mobile-receptor mechanisms. The fixed-membrane-receptor mechanism applies to protein and amine hormones. These hormones are water soluble (**hydrophilic**) and cannot diffuse across the plasma membrane of target cells. They initiate their effects by first binding to receptors that are fixed on plasma membranes of target cells. Conversely, in the mobile-receptor mechanism, steroid hormones diffuse into the cytoplasm of target cells because they are lipid soluble (**lipophilic**). They initiate their effects by binding to mobile cytoplasmic receptors.

Fixed-Membrane-Receptor Mechanism

Water-soluble hormones that circulate through the bloodstream function via the fixed-membrane-receptor mechanism (figure 25.3*a*). These hormones act as initial (extracellular) messengers, binding to hormone-specific receptor sites on plasma membranes of target cells (figure 25.3*b*). Hormone-receptor complexes activate adenylate cyclase enzymes and convert ATP into nucleotides called cyclic AMP (cAMP). Cyclic AMP nucleotides become second (intracellular) messengers, diffuse throughout the cytoplasm, and activate

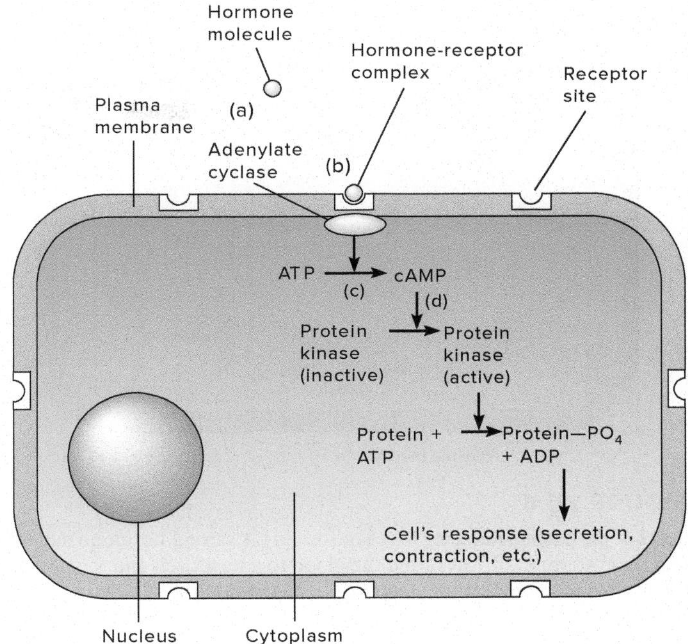

FIGURE 25.3

Steps of the Fixed-Membrane-Receptor Mechanism of Hormonal Action. (*a*) A protein hormone molecule (such as epinephrine) diffuses from the blood to a target cell. (*b*) The binding of the hormone to a specific plasma membrane receptor activates adenylate cyclase (a membrane-bound enzyme system). (*c*) This enzyme system catalyzes cyclic AMP formation (the second messenger) inside the cell. (*d*) Cyclic AMP (cAMP) diffuses throughout the cytoplasm and activates an enzyme called protein kinase, which then phosphorylates specific proteins in the cell, thereby triggering the biochemical reaction, leading ultimately to the cell's response.

protein kinase enzymes that cause a cell to respond in a specific preprogrammed fashion (figure 25.3*c* and *d*). After prompting target cells to function, phosphodiesterases inactivate cAMP. Simultaneously, fixed plasma membrane receptors lose initial messengers and become available for new reactions.

Mobile-Receptor Mechanism

The mobile-receptor mechanism involves steroid hormones that stimulate protein synthesis. Steroids enter target cells by diffusion though plasma membranes. They then bind to specific protein receptors in the cells' cytoplasm where they form steroid–protein complexes (figure 25.4*a* and *b*). These complexes are able to enter a cell's nucleus, bind to hormone response elements on DNA, and regulate transcription of specific genes to form messenger RNA (mRNA) (figure 25.4*c* and *d*). Newly transcribed mRNA leaves the nucleus, moves to the rough endoplasmic reticulum, and begins protein synthesis (figure 25.4*e* and *f*). The metabolic action of selected newly synthesized proteins promotes the functions associated with the specific steroid hormone (figure 25.4*g*).

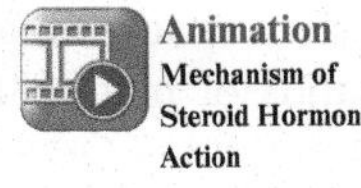

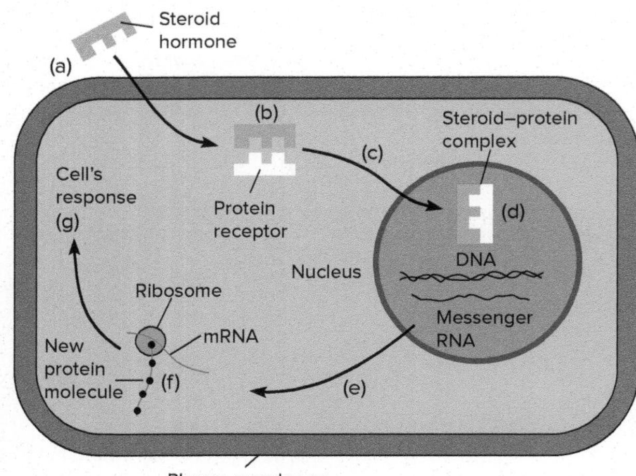

FIGURE 25.4

Steps of the Mobile-Receptor Mechanism. (*a*) A steroid hormone molecule (e.g., testosterone) diffuses from the blood to a target cell and then across the plasma membrane of the target cell. (*b*) Once in the cytoplasm, the hormone binds to a protein receptor that (*c*) carries it into the nucleus. (*d*) This steroid-protein complex triggers the transcription of specific gene regions of DNA. (*e*) The messenger RNA transcript is then translated into a gene product via (*f*) protein synthesis in the cytoplasm. (*g*) The new protein then mediates the cell's response.

Section 25.3 Thinking Beyond the Facts

Explain how a single hormone, such as epinephrine, can have different effects in different cells or tissues.

25.4 ENDOCRINE GLANDS

LEARNING OUTCOMES

1. Distinguish between endocrine and exocrine glands.
2. Compare mechanisms by which the secretion of classical hormones by endocrine glands is controlled.

The types of chemical messengers and the close association of nervous and endocrine systems in animals were described at the beginning of this chapter. Neuropeptides are produced by nerve cells and released directly into body fluids at nerve terminals. As we will see in the next section, neuropeptides evolved very early, and they are found in all animal phyla. Classical hormones are released into body fluids by secretory cells that are organized into discrete organs called **endocrine** (Gr. *endo*, within + *krinein*, to separate) **glands**. Endocrine glands have no ducts, and hormones are released into tissue spaces and body fluids next to each endocrine cell. The hormones are then carried throughout the body to their target cells (figure 25.5).

Endocrine gland secretions are controlled in one of three ways. Endocrine glands are often controlled directly by the nervous system of an animal; for example, the medullary areas of the adrenal glands of mammals are controlled directly by the autonomic nervous system. Other endocrine glands may be controlled by another

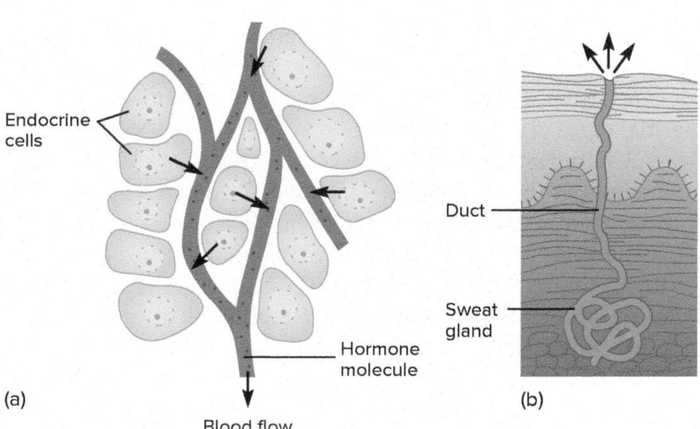

FIGURE 25.5

Glands with and without Ducts. (*a*) An endocrine gland, such as the thyroid, secretes hormones into the extracellular fluid. From there, the hormones pass into blood vessels and travel throughout the body. (*b*) An exocrine gland, such as a sudoriferous (sweat) gland, secretes material (sweat) into a duct that leads to a body surface.

endocrine structure; for example, the Y-organ that induces molting in crustaceans is stimulated to produce its hormone in response to activities of another endocrine structure, the X-organ. Still other endocrine glands are autonomous, responding to conditions within or around the animal. These glands are free from control by the nervous system and other endocrine glands. Insulin secretion by the islets of Langerhans of the pancreas responds to changes in blood glucose in this fashion.

In contrast to endocrine glands, **exocrine** (Gr. *exo*, outside + *krinein*, to separate) **glands** secrete chemicals into ducts that, in turn, empty into body cavities or onto body surfaces (e.g., mammary, salivary, and sweat glands; *see figure 25.5*). You were introduced to examples of these glands in studying the skin (chapter 23) and in many other discussions of animal form and function in chapters 9 through 22.

Section 25.4 Thinking Beyond the Facts

Why is it advantageous to have some endocrine glands directly controlled by the nervous system, some controlled by other glands, and still others that act autonomously?

25.5 INVERTEBRATE ENDOCRINE CONTROL

LEARNING OUTCOMES

1. Explain the roles of neuropeptides in invertebrate chemical regulation.
2. Discuss the range of functions hormones play in regulating invertebrate body systems.
3. Contrast the endocrine regulation of ecdysis in nematodes, crustaceans, and insects.

The survival of any group of animals depends on growth, maturation, and reproduction coinciding with the seasons of the year in which climate and food supply favor the successful development and growth of young. Neurosecretory peptides (neuropeptides) regulating growth, maturation, and reproduction probably were among the first chemical messengers to appear during the course of animal evolution, and neuropeptides are the primary chemical messengers in most invertebrate phyla. Some invertebrates (e.g., molluscs, arthropods, and echinoderms) have classical endocrine glands and hormone messengers.

Chemical Messengers of Basal and Lophotrochozoan Phyla

Members of all animal phyla rely on chemical messengers for cell-to-cell communication. Members of the basal phyla (*see appendix A*) are certainly no exception, but most chemical communication functions in these groups are poorly understood. Among the basal phyla, chemical messengers are best understood in the Cnidaria. Cnidarians possess neurotransmitters, neuropeptides, and classical hormones—many of which are also present in other phyla, including Chordata. Cnidarian chemical messengers facilitate communication within the cnidarian nerve net and help regulate reproduction, development, and planula settling.

Virtually all lophotrochozoans (*see appendix A*) also rely on neuroendocrine control mechanisms. Cerebral ganglia of members of the Platyhelminthes and Nemertea produce neuropeptides that function in regeneration, gonad maturation, and other reproductive functions (e.g., proglottid shedding in some tapeworms). Neuropeptide control of water balance has been demonstrated for some nemerteans.

Lophotrochozoan endocrine functions are best understood in the molluscs and annelids. The ganglia of molluscs produce neurosecretions that regulate heart rate, kidney function, and energy metabolism. The intestines of some bivalves produce a protein, which may have a carbohydrate-regulating role similar to that of insulin in vertebrates (*see figure 25.18*). In certain gastropods (e.g., *Helix*), a specific hormone stimulates spermatogenesis; another egg-laying hormone stimulates egg development; and hormones from the gonads stimulate the function of accessory sex organs. In all snails, a growth hormone controls shell growth. In cephalopods (the octopus and squid), the optic gland in the eyestalk produces hormones that stimulate egg development, proliferation of spermatogonia, and the development of secondary sexual characteristics.

Because annelids have a well-developed and cephalized nervous system, a well-developed circulatory system, and a large coelom, their correspondingly well-developed endocrine control of physiological functions is not surprising. The endocrine functions of annelids are involved with morphogenesis, development, growth, regeneration, and gonadal maturation. For example, in polychaetes, juvenile hormone inhibits the gonads and stimulates growth and regeneration. Gonadotropin stimulates the development of eggs, and annetocin (related to vertebrate oxytocin) elicits egg-laying behavior. In leeches, a neuropeptide stimulates gamete development and triggers color changes. Osmoregulatory hormones have been reported in oligochaetes, and a hyperglycemic hormone that maintains a high concentration of blood glucose has been reported in the oligochaete *Lumbricus*.

Chemical Messengers of Ecdysozoans

Ecdysozoan (*see appendix A*) endocrine functions have been most thoroughly studied in Nematoda and Arthropoda. Not surprisingly, many endocrine functions are associated with the processes of ecdysis that are common to members of this clade.

Nematodes have 41 genes encoding peptide hormones. Most of these peptides are associated with a neuroendocrine network that regulates growth, metabolism, life span, and ecdysis of the old cuticle. There is evidence for the action of an insulin-like peptide and a steroid hormone that initiates ecdysis. Another neuropeptide stimulates the secretion of an enzyme (leucine aminopeptidase) into the space between the old and new cuticle and causes the old cuticle to be split and shed.

Regulation of Arthropod Ecdysis

The endocrine systems of arthropods regulate many functions including ecdysis, color change, embryonic development, growth, and metabolism. The regulation of ecdysis in arthropods is, perhaps, the best-studied endocrine function in invertebrates, and it is the focus of this section.

X-organs are neurosecretory tissues in crustacean eyestalks (figure 25.6*a*). Nearby sinus glands accumulate and release secretions of X-organs. Other glands, called Y-organs, are at the base of maxillae. In the absence of appropriate stimuli, molt-inhibiting hormone (MIH), is produced and released by X-organs and sinus glands, respectively (figure 25.6*b*). Y-organs are the target of MIH. When MIH is present in high concentrations, Y-organs are inactive. Under appropriate internal and external stimuli (e.g., changes in photoperiod and increased soft tissue growth), MIH release is prevented, and Y-organs release the hormone ecdysone, which leads to ecdysis (figure 25.6*c*).

The hormonal regulation of insect ecdysis differs from that of crustaceans. In insects, appropriate stimuli activate certain neurosecretory cells in the optic lobes of the brain (figure 25.7*a*). The optic lobes produce the hormone prothoracicotropic hormone (PTTH), which axons transport to its release site, the corpora cardiaca. External and internal signals (day length, crowding, and others) initiate the release of PTTH from the corpora cardiaca into the hemolymph, and PTTH eventually reaches the prothoracic glands. PTTH causes the prothoracic glands to produce and release ecdysone, which induces ecdysis (figure 25.7*b*).

Another hormone, juvenile hormone (JH), is present during insect ecdysis. Paired corpora allata, positioned just behind the brain, produce JH to allow molting without cell differentiation, which maintains nymphal or larval structures (*see figures 25.7a and 25.7b, juvenile pathway*). When a holometabolous insect nears its last larval molt, the corpora allata decrease JH production. A low concentration of JH in hemolymph promotes molting to the pupal stage (figure 25.7*b*, pupal/adult pathway). A final surge of ecdysone, now in the absence of JH, transforms the pupa into an adult. Other hormones influence other aspects of ecdysis and metamorphosis. For example, bursicon is produced by specialized cells of the brain and nerve cords and influences hardening and darkening (sclerotization) of the outer procuticle (*see figure 14.5*).

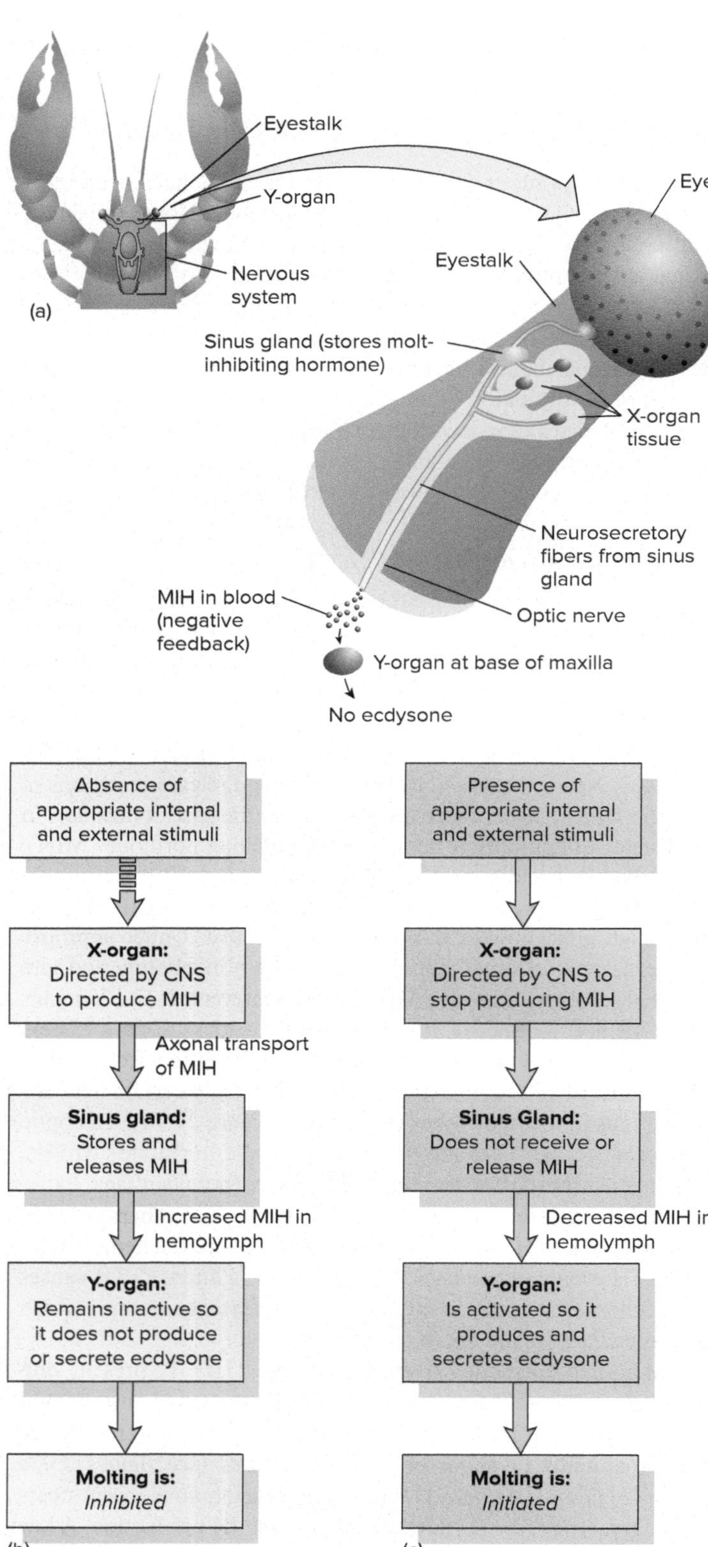

FIGURE 25.6

Control of Ecdysis (Molting) in Crustaceans. (*a*) Neurosecretory apparatus in a crustacean eyestalk. (*b*) Flow diagram of the events inhibiting molting and (*c*) causing molting. (MIH = molt-inhibiting hormone.)

(a) Source: Brusca, RC, Brusca, GJ. 1990. *Invertebrates.* Sunderland (MA): Sinauer Associates, Inc., Publishers.

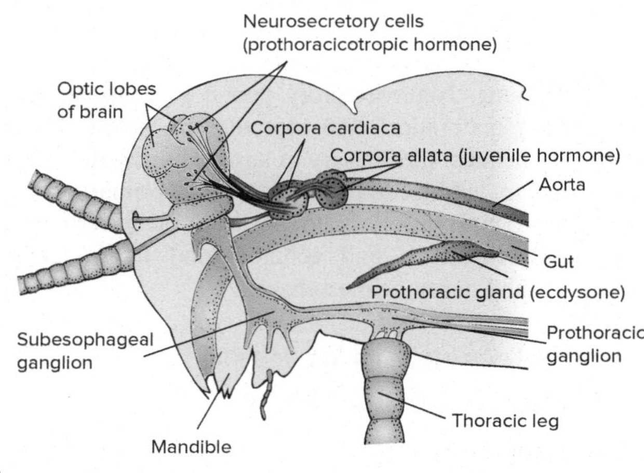

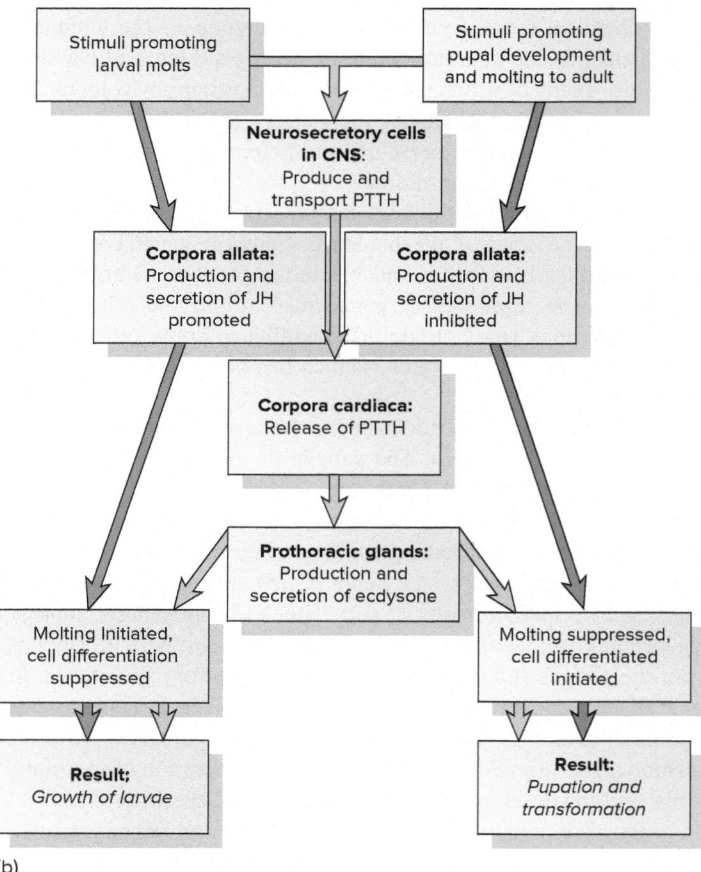

FIGURE 25.7

Hormonal Control of Ecdysis and Development (Metamorphosis) in an Insect. (*a*) Anterior end of an insect showing the location of the insect CNS, secretory centers, and hormones critical to growth and development. (*b*) Simplified flow diagram of hormonal regulation of larval growth and transformation. Prothoracicotropic hormone (PTTH). Juvenile hormone (JH). Removal of juvenile hormone from the hormonal event chain results in drastically different outcomes. The right side of the diagram occurs after the larva has reached an appropriate size.

Chemical Messengers of Deuterostomes

Echinoderms (*see chapter 16*) and invertebrate chordates (*see chapter 17*) use a host of neuropeptides, and many classical hormones, for regulating the maturation and shedding of gametes, reproductive behaviors, growth, and metabolism. In tunicates (e.g., *Ciona*) endocrinologists have found membrane receptors for gonadotropin hormones similar to those found in vertebrates, and there are probably many other similar hormones that will be discovered in the future. The endostyle of tunicates and cephalochordates is the well-established forerunner of the vertebrate thyroid gland, and one of its secretions (T_3 thyroxine) is virtually identical to the same metabolism-regulating hormone in vertebrates. The next section surveys the most thoroughly studied set of endocrine functions–those of the vertebrate deuterostomes.

Section 25.5 Thinking Beyond the Facts

Explain why the concepts described in this section can be highlighted by the term "neuropeptide."

25.6 Endocrine Systems of Fishes and Amphibians

Learning Outcomes

1. Describe how the similar neurosecretory and endocrine actions reflect the shared evolutionary pathways of the vertebrates. .
2. Explain how thyroxine and melatonin are used by fishes and amphibians.

The endocrine systems of all vertebrates are based on a similar set of neurosecretory and endocrine pathways and a similar set of hormones. These similarities reflect the shared evolutionary pathways that comprise the vertebrate clade. The following three principles do not strictly apply to the vertebrates, but they are clearly illustrated by the study of vertebrate endocrine systems:

1. Hormones (or neuropeptides) with the same function in different species may not be chemically identical.
2. Certain hormones are species-specific with respect to their function; conversely, some hormones produced in one species may be completely functional in another species.
3. A hormone from one species may elicit a different response in the same target cell or tissue of a different species.

This section describes the endocrine systems of fishes and amphibians and is followed by a discussion of the endocrine systems of amniotes (nonavian and avian reptiles and mammals).

In fishes, the central nervous system is the most important hormone producer, with other glands being rudimentary (figure 25.8). In jawed fishes, three regions of the central nervous system secrete neuropeptides. The **pineal gland** and **preoptic nuclei** (of the epithalamus and hypothalamus, respectively) are important hormone producers of the brain. Melatonin, a neuropeptide secreted by the pineal gland, affects body metabolism by synchronizing activity patterns with light intensity and day length. Other pineal neuropeptides affect pigmentation and inhibit reproductive development, both of which are influenced by light. The preoptic nuclei produce neuropeptides that regulate, among other processes, growth, sleep, and locomotion. The **urophysis** (Gr. *oura*, tail + *physis*, growth) is a discrete structure of the caudal spinal cord. It produces neuropeptides that regulate water and ion balance, blood pressure, and smooth muscle contraction.

In many fishes and amphibians, melatonin controls variations in skin color (figure 25.9). Prolactin is a pituitary gland-produced hormone that stimulates many different physiological processes among the vertebrates (*see point 2 at the beginning of this section*). For instance, in salamanders, it stimulates seasonal migrations to

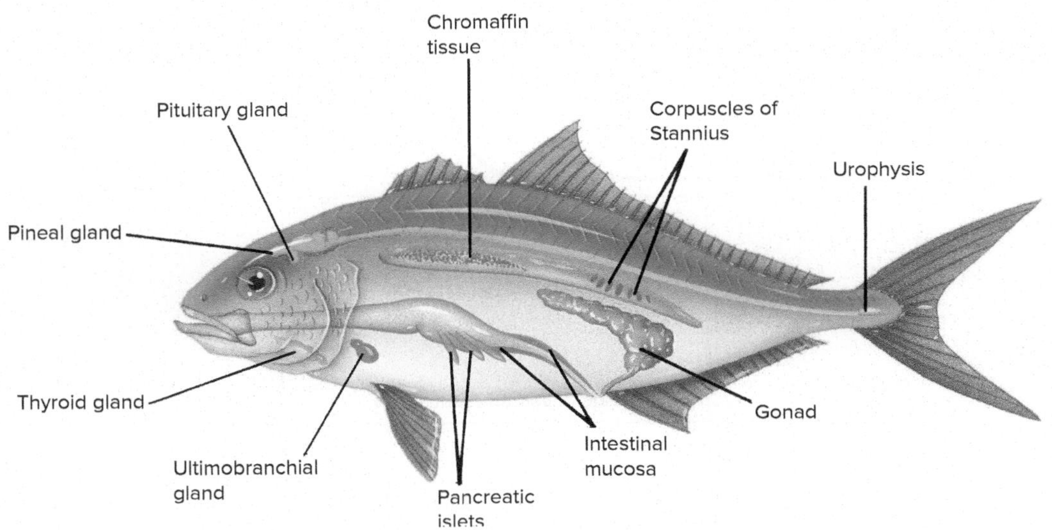

FIGURE 25.8

Major Endocrine Tissues (Glands) in a Bony Fish.

FIGURE 25.9

Hormonal Control of Frog Skin Color. The light-colored frog on the left was immersed in water containing the hormone melatonin. The dark-colored frog on the right received an injection of melanocyte-stimulating hormone.

breeding ponds for reproductive purposes, and in birds it triggers brooding behaviors. Prolactin also helps some fishes osmoregulate, and it enables certain saltwater fishes to enter freshwater during spawning runs. The aforementioned **pituitary gland** is associated with the hypothalamus and will be discussed in detail in the next section.

The thyroid gland likely evolved from a pouchlike structure (the endostyle) that carried food particles in the front end of the digestive tract. (*See chapter 17 box, How Do We Know about the Evolution of the Thyroid Gland from the Endostyle?*) The evolutionary history explains both the position of the thyroid gland (*see figure 25.8*) and the common endocrine functions among chordates. The major thyroid hormones are thyroxine (called T_4 because it contains four iodines) and triiodothyronine (T_3, three iodines). Blood concentrations of T_4 are higher than T_3. However, T_4 is actually a prohormone that is converted to an active form (T_3) by enzymes called deiodinases. T_3 controls the rate of metabolism, growth, and tissue differentiation in vertebrates.

T_4 and T_3 are examples of hormones that are present across the vertebrates, but regulate different processes (that is, they are said to be conserved traits). For example, these hormones regulate overall metabolism in most animals. However, specifically timed changes in T_4 and T_3 also control larval metamorphosis in many amphibians. Low T_4 and T_3 levels in young frog larvae (tadpoles) stimulate larval growth and prevent metamorphosis. As the hypothalamus and pituitary glands develop in growing larvae, the hypothalamus releases thyrotropin-releasing hormone, which causes the pituitary gland to release thyroid-stimulating hormone. As a result, the concentrations of T_4 and T_3 rise, triggering metamorphosis. Tail resorption and other metamorphic changes follow (figure 25.10).

In jawed fishes and primitive tetrapods, several small **ultimobranchial glands** form ventral to the esophagus (*see figure 25.8*). These glands produce calcitonin, a hormone that helps regulate blood calcium and, consequently, the integrity of osseous tissue.

Chromaffin tissue or **adrenal glands** near the kidneys produce the catecholamine hormones epinephrine (adrenaline) and

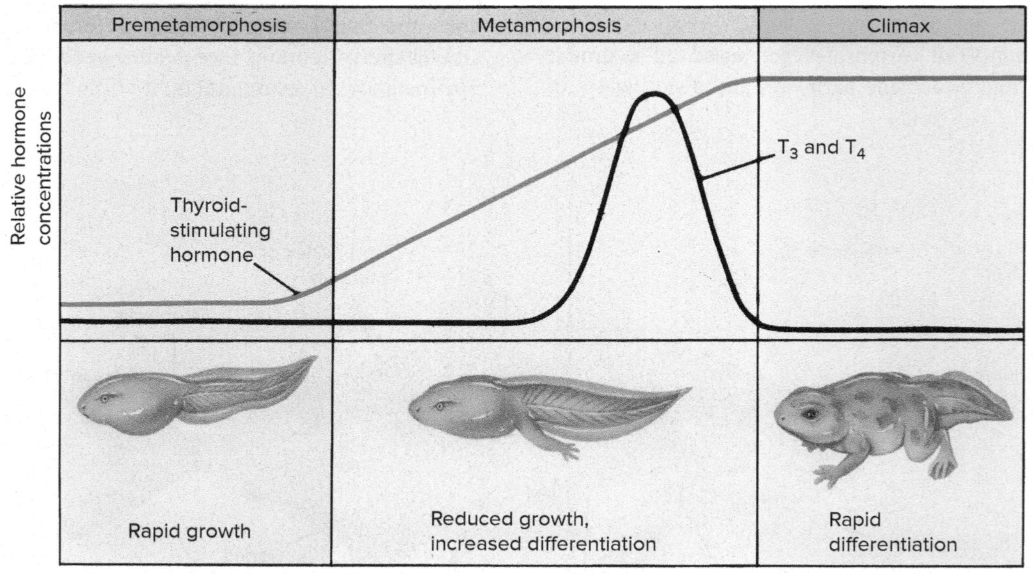

FIGURE 25.10

Frog Tadpole Metamorphosis. The thyroid hormones triiodothyronine (T_3) and thyroxine (T_4) regulate the metamorphosis of an aquatic frog tadpole into a semiterrestrial or terrestrial adult. The anterior pituitary secretes thyroid-stimulating hormone, which regulates thyroid gland activity. During the premetamorphosis (tadpole) stage, the pituitary and thyroid glands are relatively inactive. This keeps the concentration of thyroid-stimulating hormones, T_3 and T_4, at low concentrations. During metamorphosis, the concentrations of the thyroid hormones markedly increase. These (and other) hormonal fluctuations induce rapid differentiation, climaxing in the adult frog.

norepinephrine (noradrenaline). These hormones are involved with maintaining homeostasis and facilitating the stress (fight or flight) response in some vertebrates (figure 25.11). Their effects vary across the vertebrates; however, some of their major effects include regulation of blood vessel diameter, blood pressure, heart rate, and blood glucose levels.

Section 25.6 Thinking Beyond the Facts

Explain the evolutionary origin of the thyroid gland and its functions in vertebrates.

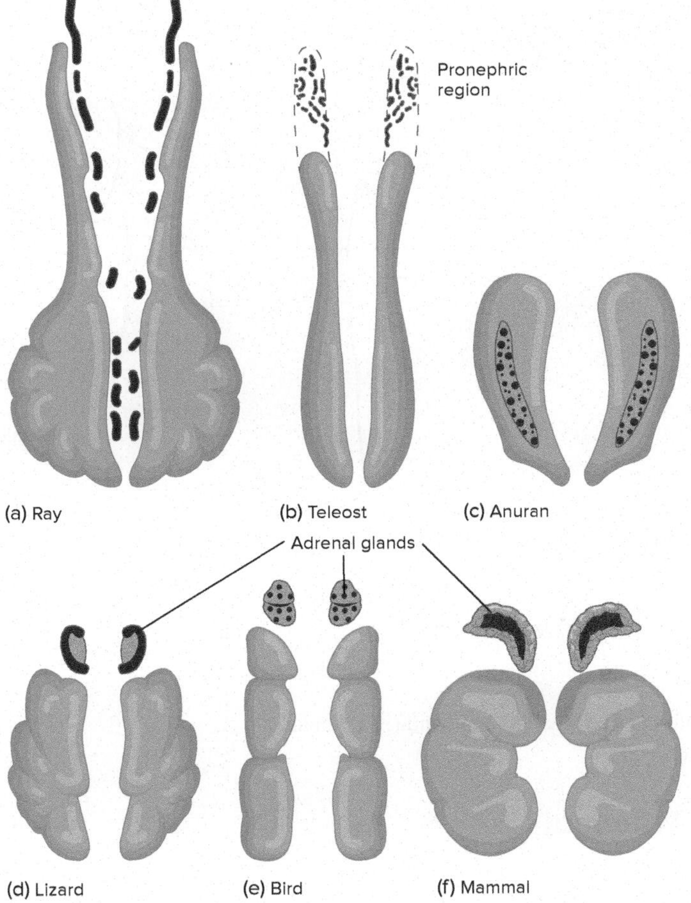

FIGURE 25.11

Chromaffin Tissue and Adrenal Glands in Selected Vertebrates. The chromaffin tissue (steroidogenic) produces steroid hormones and is shown in gray. The aminogenic tissue that produces norepinephrine and epinephrine is shown in black. The kidneys are shown in orange. Note the reversed location of the two components in lizards and mammals. (*a*) In jawless and cartilaginous fishes (elasmobranchs), aminogenic tissue develops as clusters near the kidneys. (*b*) In teleosts, the chromaffin tissue is generally at the anterior end of the kidney (pronephric region). (*c*) In anurans, the chromaffin tissue is interspersed in a diffuse gland on the ventral surface of each kidney. (*d*) In lizards, the chromaffin tissue forms a capsule around the steroidogenic-producing tissue. (*e*) In birds, the chromaffin tissue is interspersed within an adrenal capsule. (*f*) In most mammals, the chromaffin tissue forms an adrenal medulla, and the steroidogenic tissue forms the cortex.

25.7 Endocrine Systems of Amniotes

Learning Outcomes

1. Describe the function of some of the unique endocrine glands found in birds but not mammals.
2. Describe the major amniote endocrine glands and their functions.

With some minor exceptions, amniotes have a similar complement of endocrine glands as described for fishes and amphibians (figure 25.12). Selected amniote endocrine tissues, glands, and hormones are discussed in the following sections and in table 25.1.

Avian and Nonavian Reptiles

The endocrine glands in avian and nonavian reptiles are nearly identical to those found in mammals and are discussed in the next section. Some unique hormonal functions of avian and nonavian reptiles are described in the following paragraphs.

Some avian reptiles use the pituitary hormone prolactin. In pigeons and doves (Columbiformes), emperor penguins (*Aptenodytes forsteri*), and greater flamingos (*Phoenicopterus roseus*), prolactin stimulates the production of a nutrient-rich food source that is fed to chicks. The food source is termed "crop milk" and consists of desquamated cells from the crop and upper digestive tract of both parents. Prolactin also stimulates and regulates broodiness and post-hatching parental behaviors. In the presence of estrogen, prolactin also stimulates the development of a **brood patch** that parents use to incubate eggs (figure 25.13). Prolactin is also used in nonavian reptiles. For instance, certain lizards (Sauria) are capable of caudal autotomy (*see Support and Movement section in chapter 20*) when under duress. Regeneration of the lost tail is influenced by prolactin.

Another pituitary hormone (a neurohormone), arginine vasotocin, is also used by some avian and nonavian reptiles. Interestingly, chickens (*Gallus gallus*) and sea turtles (e.g., *Lepidochelys olivacea* and *Caretta caretta*) use this neurohormone to activate oviduct contraction during oviposition.

In addition to regulating metabolism (and other major functions), thyroid hormones regulate feather development, molting, and migratory behavior in avian reptiles. In nonavian reptiles, thyroid hormones are also involved in metabolic regulation. However, the process of regulation differs between the vertebrate groups. Because nonavian reptiles do not retain body heat as well as avian reptiles, they must bask. The basking, in part, is used to elevate their body temperature so their tissues respond to thyroid hormones. When body temperatures are 20°C or less (in some lizards), thyroid hormones are unable to affect cellular metabolism.

Testes produce the hormone testosterone in both avian and nonavian reptiles. Testosterone controls the secondary sexual characteristics of male avian reptiles, such as bright plumage color, comb (when present), and spurs—all of which strongly influence sexual behavior. It is also important to traits associated

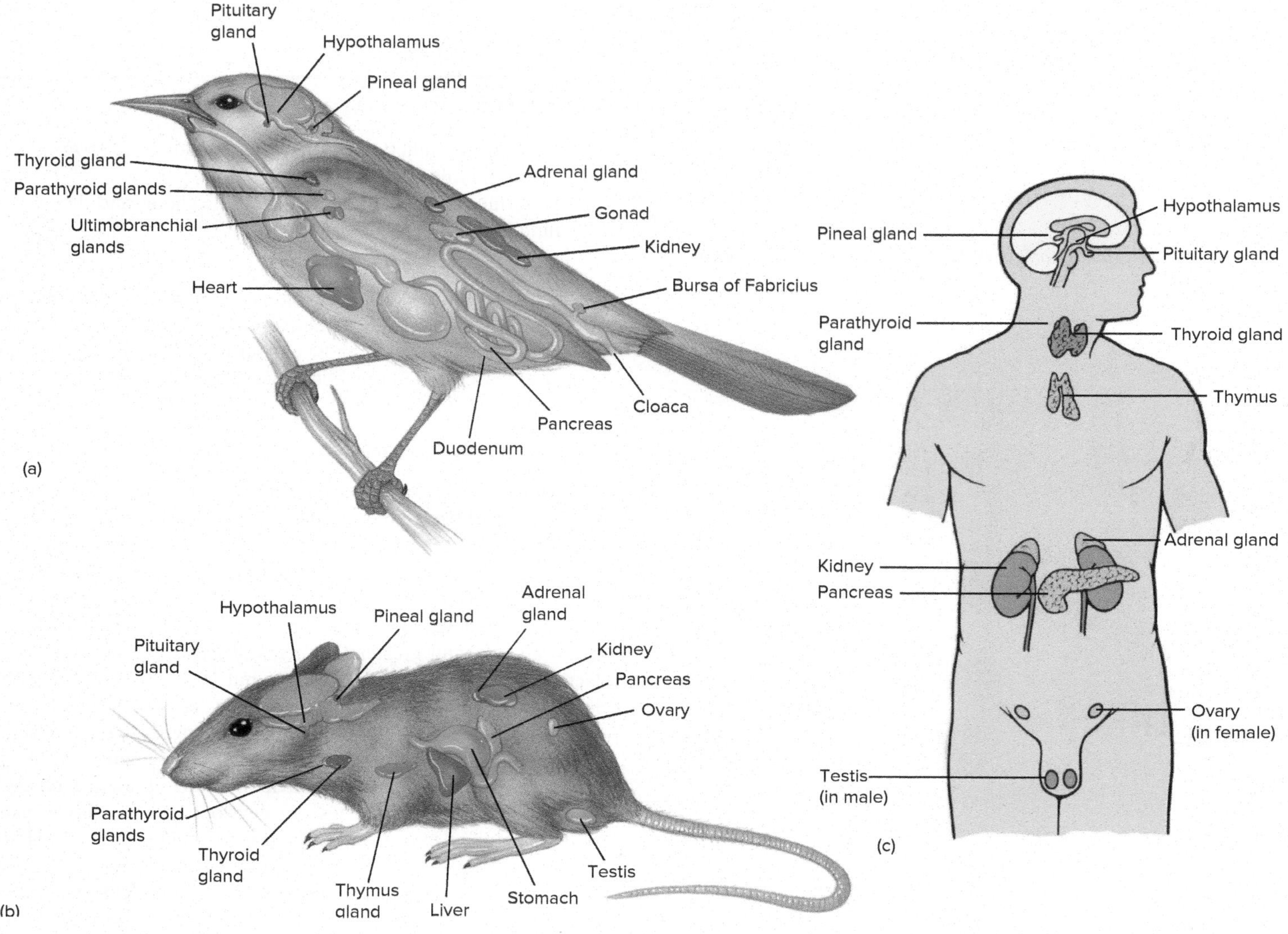

FIGURE 25.12

Endocrine Glands of Birds and Mammals. Locations of the major endocrine glands of (*a*) a bird, (*b*) a rat, and (*c*) a human.

with dominance in nonavian reptiles. For example, in green anoles (*Anolis carolinensis*), blood testosterone levels during the breeding season are positively correlated with enlarged heads, increased bite-force, and larger body dewlap size. Not all lizards, however, are dependent on circulating testosterone to display certain sexual behaviors.

Ultimobranchial glands are small, paired structures in the neck just below the parathyroid glands. They are present in both avian and nonavian amniotes, and their function is similar to that of the anamniotes.

The **bursa of Fabricius** is a saclike lymphoid organ unique to avian reptiles. It is connected to the more ventrally located cloaca, and it produces a hormone called bursin that aids in B-lymphocyte maturation (*see figure 25.12*a). The bursa of Fabricius is well developed during embryonic development, and begins to regress after hatching.

Mammals

Zoologists know more about the endocrine organs, hormones, and target tissues of mammals than of any other animal group. This is especially true for the human body. A brief overview of mammalian endocrinology follows.

Pituitary Gland (Hypophysis)

The pituitary gland (also called the hypophysis) is located directly below the hypothalamus (*see figure 25.12c*). It consists of anterior (adenohypophysis) and posterior lobes (neurohypophysis) (figure 25.14). The adenohypophysis is larger than the neurohypophysis and it contains pituicytes that produce and secrete hormones. The neurohypophysis does not produce the hormones it secretes. Neurosecretory neurons originating in the hypothalamus produce hormones that travel via neurosecretory axonal transport to the neurons' secretory ending in the neurohypophysis. Hormones are

TABLE 25.1
SELECTED AMNIOTE ENDOCRINE TISSUES, GLANDS, AND HORMONES

SOURCE	HORMONES	TARGET CELLS AND PRINCIPAL ACTIONS
Adrenal cortex	Glucocorticoids (e.g., cortisol)	Promote synthesis of carbohydrates and breakdown of proteins; initiate anti-inflammatory and antiallergic actions; mediate response to stress
	Mineralocorticoids (e.g., aldosterone)	Regulate sodium retention and potassium loss through kidneys, and water balance
	Androgens	Small contribution to the development of sex organs in both males and females (*see Testes*)
Adrenal medulla	Epinephrine (adrenaline)	Mobilizes glucose; increases blood flow through skeletal muscle; increases oxygen consumption; increases heart rate
	Norepinephrine (noradrenaline)	Elevates blood pressure; constricts arterioles and venules
Corpus luteum	Progesterone (and estrogen, *see Ovaries entry* for actions)	Maintains pregnancy; stimulates development of mammary glands
Heart	Atrial natriuretic peptide	Sodium secretion by kidneys; blood pressure
Hypothalamus	Thyroid-stimulating hormone (TSH)	Stimulates release of TSH by anterior pituitary
	Corticotropin-releasing hormone (CRH)	Stimulates release of ACTH by anterior pituitary
	Gonadotropin-releasing hormone (GnRH)	Stimulates gonadotropin release by anterior pituitary
Kidneys	Erythropoietin	Erythrocyte production in bone marrow
Ovaries	Estrogens (e.g., estradiol)	Maintain female sexual characteristics; promote oogenesis
Pancreas, islet cells	Insulin (from beta cells)	Promotes glycogen synthesis and glucose utilization and uptake from blood
	Glucagon (from alpha cells)	Raises blood glucose concentration; stimulates breakdown of glycogen in liver
Parathyroid glands	Parathormone	Regulates calcium concentration; activates vitamin D
Pineal gland	Melatonin	Sexual maturity; body rhythms
Pituitary gland, anterior lobe (adenohypophysis)	Somatotropin (STH, or growth hormone [GH])	Stimulates growth of bone and muscle; promotes protein synthesis; affects lipid and carbohydrate metabolism; increases cell division
	Adrenocorticotropic hormone (ACTH)	Stimulates secretion of adrenocortical steroids such as cortisol; is involved in stress response
	Thyrotropin (TSH) or thyroid-stimulating hormone	Stimulates thyroid gland to synthesize and release thyroid hormones (T_3, T_4) concerned with growth, development, metabolic rate
	Gonadotropins: luteinizing hormone (LH)	In ovary: Forms corpora lutea; secretes progesterone; probably acts in conjunction with FSH
		In testis: Stimulates the interstitial cells, thus promoting the secretion of testosterone
	Follicle-stimulating hormone (FSH)	In ovary: Stimulates growth of follicles; functions with LH to cause estrogen secretion and ovulation
		In testis: Acts on seminiferous tubules to promote spermatogenesis
	Prolactin (PRL)	Initiates milk production by mammary glands; acts on crop sacs of some birds; stimulates maternal behavior in birds
Pituitary gland, intermediate or posterior lobe	Melanocyte-stimulating hormone (MSH)	Expands amphibian melanophores; contracts iridophores and xanthophores; promotes melanin synthesis; darkens the skin; responds to external stimuli
Pituitary gland, posterior lobe (neurohypophysis)	Antidiuretic hormone (ADH or vasopressin)	Elevates blood pressure by acting on arterioles; promotes reabsorption of water by kidney tubules
	Oxytocin	Affects postpartum mammary gland, causing ejection of milk; promotes contraction of uterus; has possible action in parturition and in sperm transport in female reproductive tract

(*Continued*)

TABLE 25.1 Continued

Placenta	Estrogens	*See Ovaries*
	Progesterone	*See Ovaries*
Testes	Androgens (e.g., testosterone and dihydrotestosterone)	Maintain male sexual characteristics; promote spermatogenesis
Thymus	Thymopoietin	T-lymphocyte function
	Thymosins	
Thyroid gland	Thyroxine (T_4), triiodothyronine (T_3)	Affect growth, amphibian metamorphosis, molting in birds, metabolic rate in birds and mammals, development
	Calcitonin	Lowers blood calcium level by inhibiting calcium reabsorption from bone

stored there and are released under control of the hypothalamus. Neurosecretory neurons and blood vessels carrying other hypothalamic hormones are collected into a stalklike connection between the hypothalamus and pituitary gland called the infundibulum or pituitary stalk. These structures are conserved traits, suggesting ancient and important functions in vertebrates.

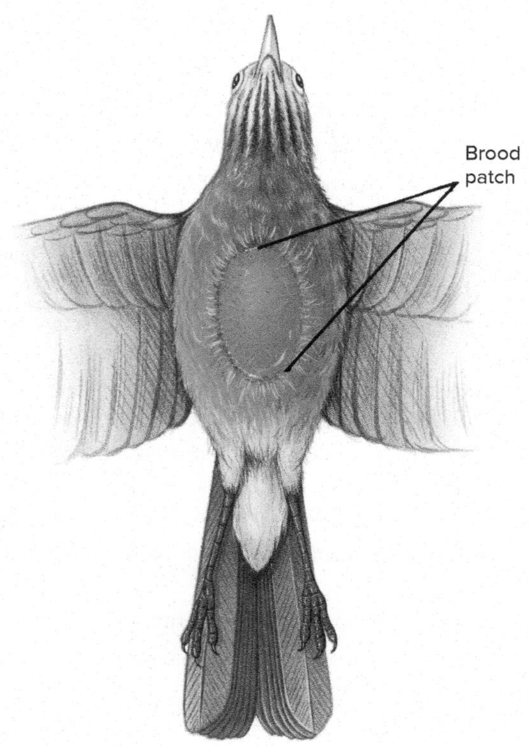

FIGURE 25.13

A Bird's Brood Patch. In this example, a robin's (*Turdus migratorius*) single brood patch appears (due to the effect of the hormone prolactin) a few days before eggs are laid. Prolactin causes down feathers to drop from the incubating robin's abdomen, and the bare patch becomes swollen and richly supplied with blood vessels. After laying the eggs, the robin settles on its nest and brings this warm patch in contact with its eggs, thereby transferring heat to the developing embryos. Incubation temperatures are between 33 and 37°C.

The pituitary gland of many vertebrates (except humans, avian reptiles, and cetaceans) also has an additional functional **intermediate lobe** (or **pars intermedia**). It consists of glandular tissues that secrete melanocyte-stimulating hormones. These hormones induce body-surface color changes in response to various external stimuli.

Hormones of the Neurohypophysis The neurohypophysis secretes two neurohormones, antidiuretic hormone (ADH or vasopressin) and oxytocin. These hormones are structurally similar and data suggest that they evolved from an ancestral chemical messenger involved with osmoregulation.

ADH helps facilitate osmoregulation in mammals by decreasing urine excretion. As a mammal becomes dehydrated, osmoreceptors in the hypothalamus lose water and shrivel. This response initiates the release of ADH from the neurohypophysis. ADH stimulates water reabsorption from urine, and water is conserved. When a mammal is sufficiently hydrated, ADH production decreases and urine becomes more voluminous and dilute. Oxytocin is involved in reproduction. It stimulates the contraction of smooth muscle of the uterus (uteri) during parturition, and it promotes let-down of milk from mammary glands (*see chapter 29*).

Hormones of the Adenohypophysis The adenohypophysis synthesizes six different polypeptide hormones, two of which have direct effects on certain cells and tissues. In mammals, these **nontropic hormones** are growth hormone and prolactin. Growth hormone (GH), or somatotropin (STH), affects all body cells concerned with growth by inducing mitosis and protein synthesis. Prolactin (PRL) is an adenohypophyseal hormone that is essential to many aspects of mammalian reproduction. For example, it stimulates reproductive migrations in many mammals, including elk (*Cervus canadensis*) and caribou (*Rangifer tarandus*), and enhances mammary gland development and milk production in female mammals (*see chapter 29*).

Tropic hormones target endocrine glands and cause them to release additional hormones (figure 25.14). The release of hormones of the adenohypophysis is nearly all controlled by tropic (releasing or inhibiting) hormones from the hypothalamus. A number of hormones thus released are also tropic in function. Thyrotropin, or thyroid-stimulating hormone (TSH), causes the thyroid gland to synthesize and secrete thyroid hormones (*see discussion in*

EVOLUTIONARY INSIGHTS

The Evolution of New Receptors for the Hormone Prolactin Accounts for Its Diverse Functions

A particular hormone can have multiple functions due to the effects of different receptors on different target cells. Thus, hormones typically acquire new functions, rather than undergo structural changes. Prolactin, for instance, regulates osmotic balance in fishes that adapt to various salinities. It also stimulates aspects of skin development, including molting in nonavian reptiles, defeathering of brood patches in avian reptiles, and production of skin secretions that nourish hatchlings in some teleosts. Additionally, prolactin induces cyclical fat deposition in many vertebrates, influences carbohydrate metabolism, and stimulates production of avian reptile crop milk and mammalian milk. More actions include the stimulation of breeding migrations in some salamanders, activation of estrus cycles in rats, and the initiation of nest building, brooding behaviors, and protection of nestlings in avian reptiles. In humans, prolactin interacts with dopamine to motivate parents to nurture, bond with, and protect their offspring. These behaviors shape neural development of the infant social brain.

Prolactin is an ancient hormone found in all vertebrates. Its genetic code shows that it probably evolved with growth hormone from an even more ancient ancestral molecule. Although the structure of prolactin has not changed, different receptors have evolved for the same hormone, allowing it to function differently in different vertebrates.

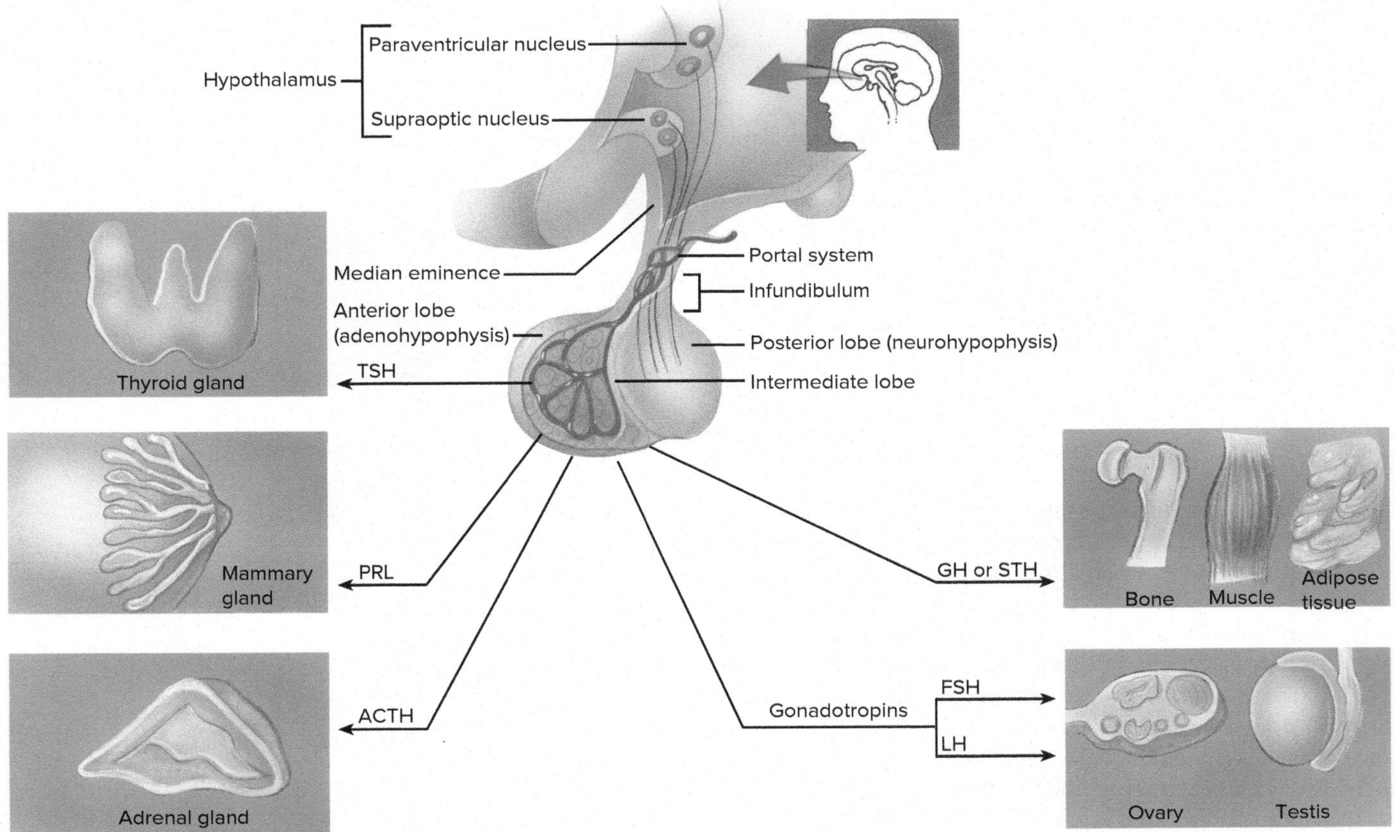

FIGURE 25.14

Functional Links between the Pituitary Gland and the Hypothalamus. Target areas for each hormone are shown in the corresponding box. The blood vessels that make up the hypothalamic-hypophyseal portal system provide the functional link between the hypothalamus and the adenohypophysis, and the axons of the hypothalamic neurosecretory cells provide the link between the hypothalamus and the neurohypophysis. (TSH = thyroid-stimulating hormone; PRL = prolactin; ACTH = adrenocorticotropic hormone; GH = growth hormone; STH = somatotropin; FSH = follicle-stimulating hormone; LH = luteinizing hormone.)

the following text). Adrenocorticotropic hormone (ACTH) causes adrenal glands to produce and secrete steroid hormones called glucocorticoids (cortisol). Glucocorticoids have wide-ranging effects, but they are most well known for their involvement in regulating glucose metabolism and inflammatory and immune responses. ACTH secretion is regulated by the secretion of corticotropin-releasing hormone (CRH) from the hypothalamus, which, in turn, is regulated by a feedback system involving stress, insulin, ADH, and other hormones.

The adenohypophysis produces two gonad-stimulating hormones (gonadotropins): luteinizing hormone (LH) and follicle-stimulating hormone (FSH). LH receives its name from the corpus luteum, a temporary endocrine tissue in the ovaries that secretes the female sex hormones estrogen and progesterone. In females, a surge of LH in the blood triggers ovulation. In males, LH causes interstitial cells in the testes to secrete the male hormone testosterone. In females, follicle-stimulating hormone (FSH) stimulates the development of the ovarian follicle, the production of estrogen, and development of a mature egg. In the male, FSH causes spermatogenic cells to produce sperm (*see chapter 29*). Release of gonadotropins is triggered by hypothalamic secretion of gonadotropin-releasing hormone (GnRH).

Pineal Gland

The pineal gland (body) is pine cone shaped, hence its name. Photoreceptors of ancestral vertebrates gave rise to the modern pineal gland. Its cells synthesize melatonin, and because light inhibits enzymes necessary for melatonin production, pineal cells are most active in the dark. Melatonin can affect numerous physiological processes. In mammals (and other vertebrates), its functions are associated with ensuring that certain activities occur when environmental conditions are most appropriate for those activities–in other words, it regulates diel and seasonal cycles. In humans, decreased melatonin secretion is also related to the onset of puberty.

Thyroid Gland

The **thyroid gland** varies in shape among the different vertebrates, but it is always found in the pharyngeal area (*see figure 25.12*). As in other vertebrates, two of its secretions, T_3 and T_4, are under the control of thyroid-stimulating hormone (TSH) from the adenohypophysis and influence growth, development, and metabolic rates. Another thyroid hormone, calcitonin, facilitates deposition of calcium ions (Ca^{2+}) into osseus tissue when extracellular concentrations of calcium rise. Thyroid cells decrease secretion of calcitonin when extracellular calcium levels return to a homeostatic optimum. Recall that in other vertebrates, ultimobranchial glands perform this function.

Animation Mechanism of Thyroxine Action

Parathyroid Glands

Parathyroid glands are small glands embedded in the thyroid lobes. There are typically two glands in each lobe (*see figure 25.12*), and they secrete a hormone called parathormone (PTH), which regulates the concentrations of calcium (Ca^{2+}) and phosphate ($HPO_2{}^{-4}$) ions in the blood. In contrast to the function of thyroxine, when calcium concentrations in the blood are low PTH secretion causes osteoclasts (cells that break down osseous tissue) to release calcium ions into the blood and increase calcium absorption from the small intestine and reabsorption from urine by the kidneys. These functions increase blood calcium levels and are regulated by negative feedback (figure 25.15).

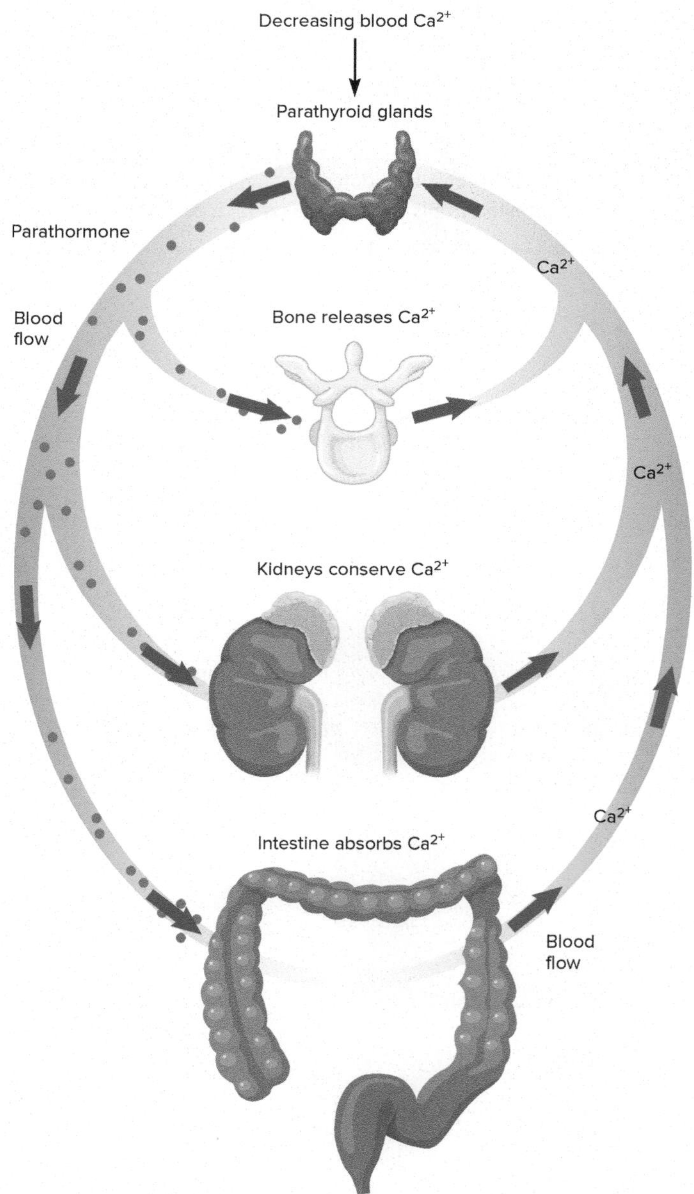

FIGURE 25.15

Hormonal Feedback. The negative feedback mechanism of the parathyroid glands (parathormone). Parathormone stimulates bones to release calcium and the kidneys to conserve calcium. It indirectly stimulates the intestine to absorb calcium. The result increases blood calcium, which then inhibits parathormone secretion.

Adrenal Glands

In mammals, two adrenal glands are immediately anterior (or superior) to the kidneys. Each gland contains two separate glandular tissues; the outer cortex and inner medulla (figure 25.16). The **adrenal cortex** secretes three classes of steroid hormones: glucocorticoids (cortisol), mineralocorticoids (aldosterone), and some sex hormones (androgens). Glucocorticoids (e.g., cortisol) help regulate metabolism and blood sugar levels, and function in defense responses to infection or injury. Aldosterone helps maintain extracellular fluid solute (i.e., sodium) concentrations in response to food intake or changes in metabolic activity. Aldosterone also promotes sodium reabsorption by the kidneys and, thus, it is necessary for hydrating the body (see chapter 28). Sex hormones secreted by the adrenal cortex are secreted in smaller amounts than those secreted by the adenohypophysis. They have minor effects on gonadal functions and consist mainly of androgens and estrogens.

Animation
Glucocorticoid Hormones

The **adrenal medulla** is under neural control. It contains neurosecretory cells that secrete epinephrine (adrenaline) and norepinephrine (noradrenaline), both of which influence heart rate and carbohydrate metabolism. During periods of excitement (including stress responses), the adrenal medulla contributes to body system mobilization through the release of epinephrine and norepinephrine. Their release is controlled, in part, by the hypothalamus and sympathetic nerves. Epinephrine and norepinephrine cause dilation of pulmonary airways, and increased heart rate and blood flow to vital organs. Consequently, more oxygen is delivered to all body cells. These events are often called the fight-or-flight response and they allow rapid and strong responses to emergencies.

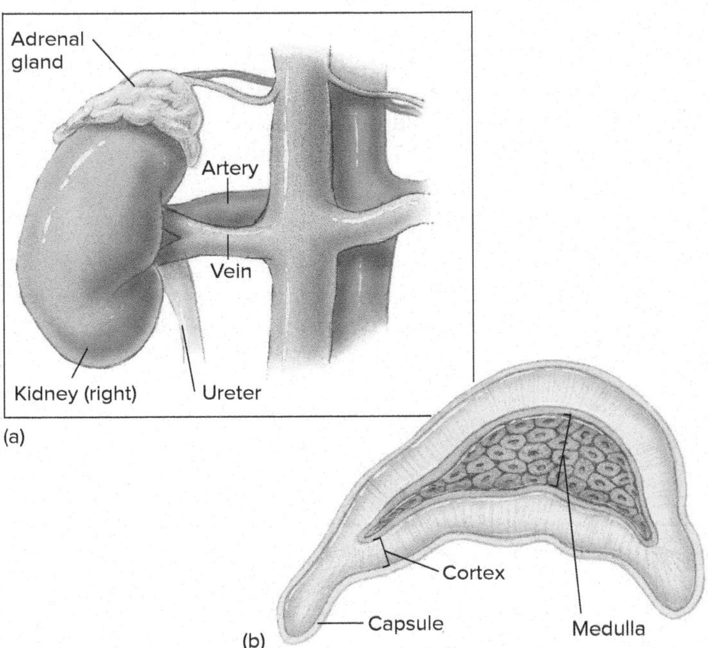

FIGURE 25.16

Adrenal Gland of a Mammal. (*a*) An adrenal gland sits on top of each kidney. (*b*) Each gland contains two structurally, functionally, and developmentally distinct regions. The outer cortex is endocrine and produces glucocorticoids (cortisol), mineralocorticoids (aldosterone), and androgens (sex hormones). The inner medulla is nervous tissue that produces epinephrine (adrenaline) and norepinephrine (noradrenaline).

Pancreas

The **pancreas** is an elongated, glandular organ adjacent to the proximal small intestine that has both exocrine (with ducts) and endocrine (ductless) regions (figure 25.17). The exocrine region secretes digestive enzymes, and the endocrine region (which makes up about 1% of the gland) synthesizes, stores, and secretes hormones from groups of cells called **pancreatic islets.** Each islet contains groups of four types of specialized cells. Alpha (α) and beta (β) cells produce the hormones glucagon and insulin, respectively. These hormones are critical to blood sugar homeostasis (figure 25.18; *see figure 25.2*). When blood glucose concentrations are high (i.e., after a meal), beta cells secrete insulin. Insulin promotes cellular uptake of glucose. When blood glucose concentrations are low, alpha cells secrete glucagon. Glucagon stimulates the breakdown of glycogen (a stored polysaccharide) into glucose units, which are released into the bloodstream to raise blood glucose concentrations. The pancreas works in conjunction with the liver, which when stimulated by insulin, converts excess glucose to glycogen.

Delta (δ) cells secrete the hormone somatostatin, which inhibits glucagon and insulin secretion. F cells secrete pancreatic polypeptide (PP), which inhibits somatostatin secretion, gallbladder contraction, and the secretion of pancreatic enzymes when they are no longer needed to break down ingested food.

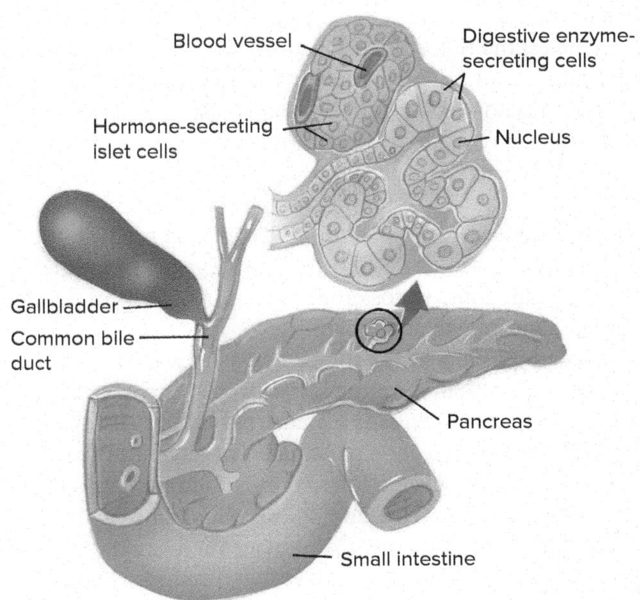

FIGURE 25.17

Pancreas. The hormone-secreting cells of the pancreas are arranged in clusters or islets closely associated with blood vessels. Other pancreatic cells secrete digestive enzymes into ducts.

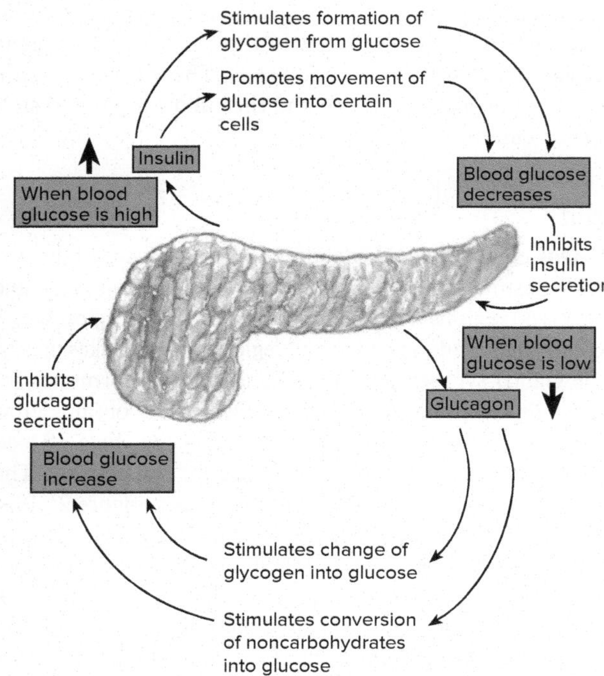

FIGURE 25.18

Two Pancreatic Hormones (Insulin and Glucagon) Regulate the Concentration of Blood Glucose. The negative feedback mechanism for regulating glucagon and insulin secretion helps maintain a homeostatic blood glucose concentration.

Gonads

Hormones that aid in regulation of reproductive functions are secreted by male and female **gonads** called testes and ovaries, respectively. In males, the testes secrete testosterone. Testosterone, luteinizing hormone, and follicle-stimulating hormone work together to stimulate spermatogenesis. Testosterone is also needed for growth and maintenance of male sex organs, the development and maintenance of sexual behavior, and the development of secondary sexual characteristics (in humans this includes deepening of the voice and growth of facial and pubic hair). The testes also produce inhibin, which inhibits the secretion of FSH (*see chapter 29*). Aside from reproductive functions, testosterone is also required for adequate muscle and bone growth and repair.

Female reproductive functions are partially regulated by four major classes of ovarian hormones. Estrogens (estrin, estrone, and estradiol) take part in regulating menstrual and estrus cycles and development of female secondary sexual characteristics (e.g., mammary glands). Progestins (primarily progesterone) also regulate menstrual and estrus cycles and the development of mammary glands. In addition, progestins aid in placenta formation. Relaxin, which is produced in smaller quantities than estrogens and progestins, softens the cervix for delivery. Inhibin prevents maturation of an ovarian follicle by suppressing FSH production (*see chapter 29*).

Thymus

The **thymus gland** is positioned near the heart (*see figure 25.12*). It is large and conspicuous in young mammals (and birds), but gets smaller with age. In humans, this gland is bilobed and consists of numerous septae that subdivide the lobes into lobules. The thymus gland's major function is to receive immature T-cells from bone marrow, and through specific selection methods, train them to detect antigens. The major thymic hormones are a family of peptide hormones, some of which are involved in this process (thymopoietin and thymosins; *see table. 25.1*).

Section 25.7 Thinking Beyond the Facts

Compare the function and regulation of hormones that are under nervous system control with those that are free of nervous system control.

25.8 SOME HORMONES ARE NOT PRODUCED BY ENDOCRINE GLANDS

LEARNING OUTCOME

1. Other than the major endocrine glands, describe some other organs/tissues of mammals that produce hormones.

In mammals, some hormones are secreted by tissues and organs that are not exclusively endocrine glands. The heart's right atrium secretes atrial natriuretic hormone (ANP). ANP influences blood pressure by causing kidneys to move salts and water to urine. It works antagonistically with aldosterone to maintain osmotic homeostasis, and thus adequate blood pressure. The kidneys secrete erythropoietin (EPO), a hormone that stimulates bone marrow to produce red blood cells, which are necessary for gas exchange (*see chapter 26*). Other tissues and organs, such as adipose tissue, the liver, gastrointestinal tract, and placenta, also secrete hormones.

Section 25.8 Thinking Beyond the Facts

How are the actions of atrial natriuretic hormone and erythropoietin critical to both the function of the endocrine system itself and to homeostasis of the animal?

25.9 EVOLUTION OF ENDOCRINE SYSTEMS

LEARNING OUTCOME

1. Compare the diversity of structure and function of animal endocrine systems to the diversity of animal nervous systems.

As noted throughout this book, cell-to-cell signaling plays an important role in the maintenance of homeostasis and the coordination of reproduction, growth, and development in all animals. Cell-to-cell signaling involves both the transmission of electrical signals and chemical communication. The advent of nervous tissues and neuropeptides early in the Animalia means that some form of both kinds of communication are vital in virtually all animal phyla. Since all animals share a common ancestry, it is not surprising that all animals share a basic set of signal transduction mechanisms involved with chemical messaging systems.

Over time, animal cell-to-cell communication mechanisms have diverged into complex endocrine systems—but the evolutionary connections between these diverse endocrine systems are still evidenced by the common molecules involved. Endocrine systems evolved in conjunction with circulatory systems, discussed in chapter 26, that allow hormones to be transported across long distances.

Section 25.8 Thinking Beyond the Facts

Explain the relationships between nervous, endocrine, and circulatory systems in chemical messaging in animals.

Summary

25.1 **The Evolution and Diversity of Chemical Messengers**

- Chemical messengers evolved in single-celled organisms. The same kinds of chemical messengers are adapted as neurotransmitters and hormones in animals where they are used in timing growth, maturation, reproduction, and many other functions.
- Specialized cells secrete chemical messenger molecules. These chemical messengers can be categorized as local chemical messengers (autocrines, paracrines), neurotransmitters (e.g., acetylcholine), neuropeptides, hormones, and pheromones (e.g., sex attractants).

25.2 **Hormones and Their Feedback Systems**

- A hormone is a chemical messenger that is secreted by an endocrine gland or tissues and has an effect at a target cell or tissue.
- Hormones control biochemical reactions by increasing the rate at which other substances enter or leave a cell, stimulating a target cell to synthesize a cellular product, or by activating or suppressing a cellular activity.
- Hormones usually act through negative feedback systems. An increase in some cellular function stimulates the release of a hormone to reverse the change and keep the level of that cellular function relatively stable.

25.3 **Mechanisms of Hormone Action**

- Hydrophilic hormones, such as peptides and amino acids, bind externally to the plasma membrane receptors that activate protein kinases directly or that operate through a second-messenger system such as cAMP.
- Lipophilic hormones, such as steroids, pass through a target cell's plasma membrane and bind to intracellular receptor proteins. The hormone-receptor complex then binds to the hormone response element in the target gene.

25.4 **Endocrine Glands**

- Chemical messengers in the form of neuropeptides function in all animal phyla. Classical hormones are secreted by endocrine glands. Endocrine glands have no ducts and secrete their hormones into body fluids. Endocrine glands can be controlled by the nervous system or other glands, or they may be autonomous. Exocrine glands have ducts that lead to the body surface or a body cavity.

25.5 **Invertebrate Endocrine Control**

- Members of the basal phyla and Lophotrochozoa rely on neurotransmitters, neuropeptides, and classical hormones for chemical signaling. Chemical regulation is best known in the molluscs and annelids.
- Members of the Ecdysozoa have endocrine functions related to ecdysis. These functions are best known in crustaceans and insects.
- Deuterostomes utilize neuropeptides and many classical hormones for regulating the maturation and shedding of gametes, reproductive behaviors, growth, and metabolism. The endostyle of tunicates and cephalochordates is the forerunner of the vertebrate thyroid gland.

25.6 **Endocrine Systems of Fishes and Amphibians**

- Endocrine structures of fish include the pineal gland and preoptic nuclei of the brain and the urophysis of the spinal cord. The pituitary gland of amphibians stimulates reproductive migrations in salamanders, and thyroxine controls metamorphosis in amphibians. Ultimobranchial glands help regulate blood calcium, and chromaffin tissues help vertebrates respond to stress.

25.7 **Endocrine Systems of Amniotes**

- Birds have unique endocrine functions associated with courtship, mating, and nesting activities. The bursa of Fabricius is unique to birds and functions in immunological reactions.
- Major endocrine glands of amniotes include the pituitary gland, thyroid gland, parathyroid glands, adrenal glands, pancreas, gonads, and thymus.

25.8 **Some Hormones Are Not Produced by Endocrine Glands**

- In mammals, various hormones are produced and secreted by tissues and/or organs that are not exclusively endocrine glands. The heart secretes atrial natriuretic hormone, and the kidneys secrete erythropoietin.

25.9 **Evolution of Endocrine Systems**

- All animals possess some form of chemical regulation. Chemical regulation evolved in concert with nervous regulation as evidenced by the neuropeptides that the two systems share. Endocrine systems became important following the evolution of circulatory systems because most hormones require the circulatory system for distribution in an animal.

Concept Review Questions

1. Short-distance local messengers that act on adjacent cells are called
   a. neurotransmitters.
   b. neuropeptides.
   c. hormones.
   d. pheromones.
   e. paracrine agents.

2. Which of the following is NOT a biochemical category of hormones?
   a. Proteins
   b. Amines
   c. Steroids
   d. Prostaglandins
   e. Nucleic acids
3. Which of the following hormones increase when a metamorphosing anuran tadpole is transitioning into an adult?
   a. Testosterone
   b. Estrogen and progesterone
   c. Thyroid stimulating hormone, thyroxine and triiodothyronine
   d. Epinephrine and thyroxin
4. In birds, the bursa of Fabricius produces
   a. thyroxine.
   b. bursin.
   c. testosterone.
   d. calcitonin.
5. Which of the following are nontropic hormones produced by the adenohypophysis?
   a. Growth hormone and estrogen
   b. Growth hormone and prolactin
   c. Prolactin and thyroxin
   d. ADH and oxytocin

Analysis and Application Questions

1. How do hormones encode information? How do cells "know what to do" in response to hormonal information?
2. Summarize your knowledge of how endocrine systems work by describing the "life" of a hormone molecule from the time it is secreted until it is degraded or used up.
3. All cells secrete or excrete molecules, and all cells respond to certain biochemical factors in their external environments. Could the origin of endocrine control systems lie in such ordinary cellular events? How might the earliest multicellular organisms have evolved some sort of endocrine coordination?
4. Mental states strongly affect the function of many endocrine glands. This mind–body link occurs through the hypothalamus. Can you describe how thoughts are transformed into physiological responses in the hypothalamus?
5. Compared to enzymes and genes, hormones are remarkably small molecules. Would larger molecules be able to carry more information? Explain.

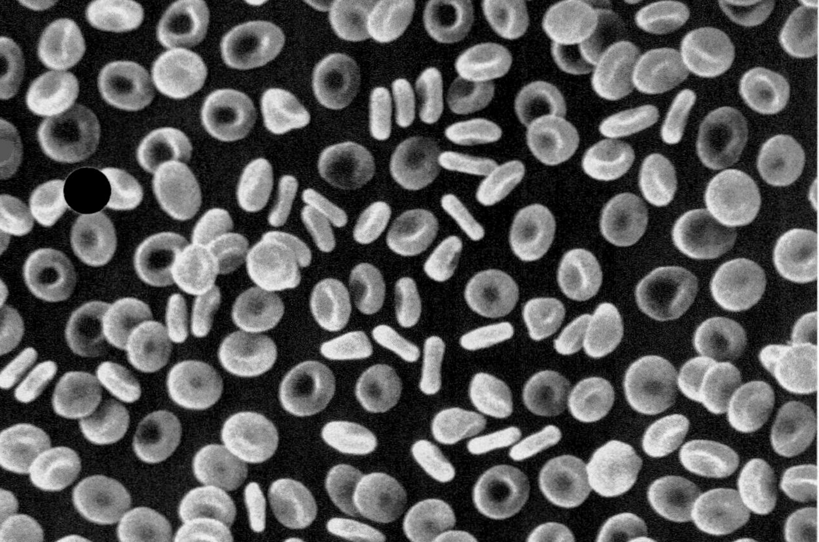

Blood cells, like these red blood cells (erythrocytes), participate in internal transport processes occurring in many animals. In their roles in oxygen and carbon dioxide transport and exchange, red blood cells form an essential link between the two systems discussed in this chapter: circulation and gas exchange.

26

Circulation and Gas Exchange

Chapter Outline

26.1 Transport Systems in Invertebrates
Characteristics of Invertebrate Hemolymph and Blood
26.2 Transport Systems in Vertebrates
Characteristics of Vertebrate Blood and Blood Cells
Vertebrate Blood Vessels
26.3 Vertebrate Circulatory Systems
Circulatory Systems of Fishes and Amphibians
Circulatory Systems of Amniotes
26.4 The Mammalian Circulatory System
The Mammalian Heart
Blood Pressure
26.5 Lymphatic Systems
26.6 Gas Exchange
Invertebrate Respiratory Systems
26.7 Vertebrate Respiratory Systems
Cutaneous Exchange
Gills
Lungs
Lung Ventilation
26.8 The Mammalian System
Air-Conduction Portion
Gas-Exchange Portion
Ventilation
Gas Transport

Maintaining homeostasis requires that animals exchange materials with their surroundings. Virtually all animals have mechanisms and structures that reduce the distances that nutrients, wastes, and gases must diffuse. Structures that reduce this diffusion distance by increasing exchange surface areas or by transporting substances aid in these exchanges. These structures include cavities and chambers through which water or body fluids circulate and complex organ systems involving hearts to pump blood through blood vessels and body spaces. Similarly, large animals require systems that bring oxygen into contact with their circulatory systems and help remove carbon dioxide from their bodies. Thus, most animals rely on the specialized systems for transport and exchange that are described in this chapter: circulatory and respiratory systems.

26.1 TRANSPORT SYSTEMS IN INVERTEBRATES

LEARNING OUTCOMES

1. Compare and contrast the transport processes of invertebrates.
2. Describe hemolymph and give several of its functions.

Many invertebrates exchange nutrients, wastes, and gases without using a circulatory system. For example, sponges circulate water from the external environment through their bodies, instead of circulating an internal fluid (figure 26.1*a; see also figure 9.3*). Cnidarians, such as *Hydra,* have a fluid-filled internal **gastrovascular cavity** (figure 26.1*b; see also figure 9.7*). This cavity supplies nutrients for all body cells lining the cavity, provides oxygen from the water in the cavity, and is a reservoir for carbon dioxide and other wastes. Simple body movement moves the fluid.

The gastrovascular cavity of flatworms, such as the planarian *Dugesia,* has branches that penetrate to all parts of the body (figure 26.1*c; see also figure 10.4*). Because this branched gastrovascular cavity runs close to all body cells, diffusion distances for nutrients, gases, and wastes are short. Body movement helps distribute materials to various parts of the body. One disadvantage of this system is that it limits these animals to relatively small sizes or to shapes that maintain small diffusion distances.

Invertebrates that have a pseudocoelom, such as rotifers, gastrotrichs, and nematodes, use the coelomic fluid of their body cavity for transport (figure 26.1*d; see also figure 13.4*). Coelomic fluid may be identical in composition to interstitial fluids or may differ, particularly with respect to specific proteins and cells. Coelomic fluid transports gases, nutrients, and

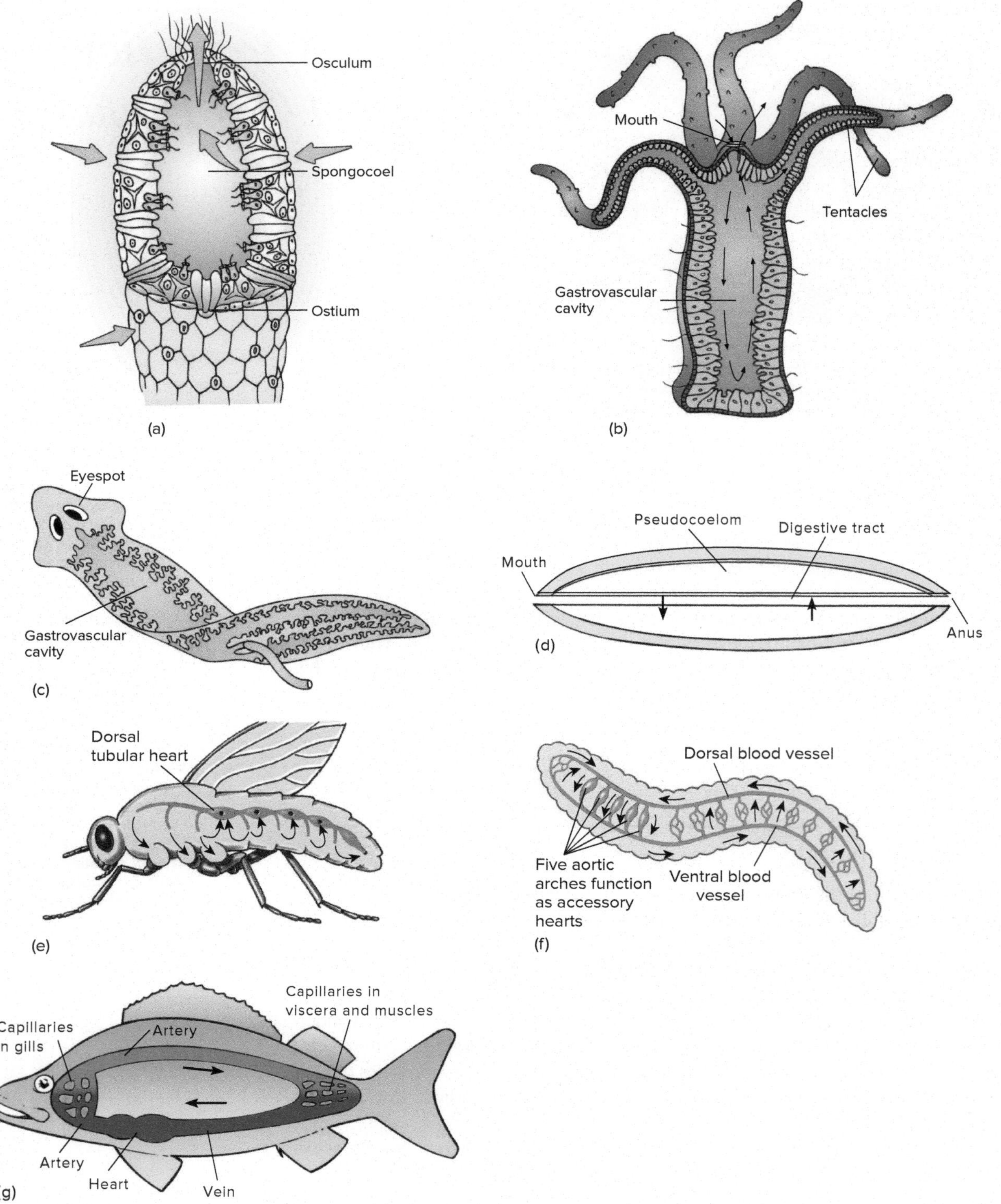

FIGURE 26.1

Some Transport and Circulatory Systems. (*a*) Sponges use water from the environment as a circulatory fluid by passing it through their bodies (blue arrows) using many incurrent pores and one or more excurrent openings (oscula). (*b*) Cnidarians, such as this *Hydra,* also use water from the environment and circulate it (black arrows) through the gastrovascular cavity by way of muscular contraction. Cells lining the cavity exchange gases and nutrients from the water and release waste into it. (*c*) The planarian's gastrovascular cavity is branched, allowing for more effective distribution of materials. (*d*) Invertebrates with a pseudocoelom, such as this nematode, use their body cavity fluid for internal transport from and to the digestive tract, as the black arrows indicate. (*e*) The dorsal tubular heart of an insect pumps hemolymph through an open circulatory system. In this example, hemolymph and body cavity (hemocoelic) fluid are one and the same.(*f*) The circulatory system of an earthworm contains blood that is kept separate from the coelomic fluid. This is an example of a closed circulatory system. (*g*) Octopuses, other cephalopod molluscs, annelids, and vertebrates, such as this fish, have closed circulatory systems. In a closed system, the walls of the heart and blood vessels are continuously connected, and blood never leaves the vessels. Black arrows indicate the direction of blood flow.

waste products. It also may function as a hydrostatic skeleton (*see figure 23.10*). Some of these animals are small, and movements of the body against the coelomic fluids, which are in direct contact with the internal tissues and organs, produce adequate transport. A few other invertebrates (e.g., ectoprocts, sipunculans, and echinoderms) also depend largely on the body cavity as a coelomic transport chamber.

A **circulatory** or **cardiovascular system** (Gr. *kardia,* heart + L. *vascular,* vessel) is a specialized system in which a muscular, pumping heart moves the fluid medium, called either hemolymph or blood, in a specific direction determined by the presence of unidirectional blood vessels.

The animal kingdom has two basic types of circulatory systems: open and closed. In an **open circulatory system,** the heart pumps hemolymph out into the body cavity or at least through parts of the cavity, where the hemolymph bathes the cells, tissues, and organs. Most molluscs and arthropods have open circulatory systems in which hemolymph directly bathes the cells and tissues (figure 26.1*e*). For example, an insect's heart pumps hemolymph through vessels that open into a body cavity (hemocoel).

In a **closed circulatory system**, blood circulates in the confines of tubular vessels. The annelids, such as the earthworm, have a closed circulatory system in which blood travels through vessels delivering nutrients to cells and removing wastes (figure 26.1*f*; *see figure 12.6*).

Characteristics of Invertebrate Hemolymph and Blood

Hemolymph (Gr. *haima,* blood + *lympha,* water) is the circulating fluid of animals with an open circulatory system. Most arthropods and ascidians have hemolymph. In these animals, a heart pumps hemolymph at low pressures through vessels to tissue spaces (hemocoel) and sinuses. Generally, the hemolymph volume is high and the circulation slow. In the process of movement, essential gases, nutrients, and wastes are transported.

Many times, hemolymph has noncirculatory functions. For example, in insects, hemolymph pressure assists in molting of the old cuticle and in inflation of the wings. In certain jumping spiders, hydrostatic pressure of the hemolymph provides a hydraulic mechanism for limb extension.

The hemolymph of most animals contains circulating cells called **hemocytes.** Some cells contain a respiratory pigment, such as hemoglobin, and are called erythrocytes. These cells are usually present in high numbers to facilitate oxygen transport. Cells that do not contain respiratory pigments have other functions, such as blood clotting. The hemolymph of molluscs has two general types of hemocytes (amoebocytes and granulocytes) that have most of the aforementioned functions as well as nacrezation (pearl formation) in some bivalves. Insect hemolymph contains large numbers of various hemocyte types that function in phagocytosis, encapsulation, and clotting.

Blood is the circulating fluid of animals with a closed circulatory system. The number and types of blood cells vary dramatically in different invertebrates. For example, annelid blood contains blood cells that are phagocytic. The coelomic fluid contains a variety of coelomocytes (amebocytes, eleocytes, lampocytes, and linocytes) that function in phagocytosis, glycogen storage, encapsulation, defense responses, and excretion.

Section 26.1 Thinking Beyond the Facts

Why are circulatory systems characteristically found in most larger animals?

26.2 TRANSPORT SYSTEMS IN VERTEBRATES

LEARNING OUTCOMES

1. Compare the functions of plasma with the function of serum.
2. Relate the structures of mature formed elements to their functions.

All vertebrates have a closed circulatory system (figure 26.1g). Blood moves from the heart, through arteries, arterioles, capillaries, venules, veins, and back to the heart. Exchange between the blood and extracellular fluid only occurs at the capillary level.

Characteristics of Vertebrate Blood and Blood Cells

Overall, vertebrate blood transports oxygen, carbon dioxide, and nutrients; defends against harmful microorganisms, cells, and viruses; prevents blood loss through coagulation (clotting); and helps regulate body temperature and pH. Because it is a liquid, vertebrate blood is classified as a specialized type of connective tissue. Blood contains a fluid matrix called plasma and cellular elements called formed elements.

Plasma

Plasma (Gr., anything formed or molded) is the straw-colored, liquid part of blood. In mammals, plasma is mostly water and provides the solvent for dissolving and transporting nutrients. A small percentage of plasma is composed of electrolytes, amino acids, glucose and other nutrients, various enzymes, hormones, and metabolic wastes.

A group of proteins (albumin, fibrinogen, and globulins) influences the distribution of water between the blood and extracellular fluid. Because most of the protein is albumin, it plays important roles with respect to water movement. Fibrinogen is necessary for blood coagulation (clotting), and the globulins include the immunoglobulins and various metal-binding proteins. **Serum** is plasma from which the proteins involved in blood clotting have been removed. The gamma globulin portion functions in the immune response because it consists mostly of antibodies.

Formed Elements

The **formed-element fraction** (cellular component) of vertebrate blood consists of erythrocytes (red blood cells; RBCs), leukocytes (white blood cells; WBCs), and platelets (thrombocytes). These cells function in the transport of respiratory gases, body defenses, and blood clotting.

Red Blood Cells **Red blood cells** (erythrocytes; Gr. *erythros,* red + cells) vary dramatically in size, shape, and number in the different vertebrates (figures 26.2 and 26.3*a*). For example, the mature RBCs of most vertebrates are nucleated, but mammalian RBCs are enucleated (without a nucleus). The loss of the nucleus from mammalian RBCs usually leaves the cells in the shape of biconcave disks (figure 26.3*a*). The shape of a biconcave disk provides a larger surface area for gas diffusion than a flat disk or sphere.

Almost the entire mass of an RBC consists of **hemoglobin** (Gr. *haima,* blood + L. *globulus,* little globe), an iron-containing protein. The major function of an erythrocyte is to pick up oxygen from the environment, bind it to hemoglobin to form **oxyhemoglobin,** and transport it to body tissues. Hemoglobin also carries waste carbon dioxide (in the form of carbaminohemoglobin) from the tissues to the lungs (or gills) for removal from the body.

White Blood Cells **White blood cells (leukocytes)** (Gr. *leukos,* white + cells) phagocytize microorganisms at infection sites, remove foreign chemicals, and remove debris that results from dead or injured cells. All WBCs are derived from immature cells (called stem cells) in bone marrow.

Eosinophils are phagocytic and ingest foreign proteins and immune complexes (figures 26.3*b*). In mammals, eosinophils also release chemicals that counteract the effects of certain inflammatory chemicals released during allergic reactions. **Basophils** are the least numerous WBC (figures 26.3*c*). When they react with a foreign substance, their granules release histamine (a local chemical messenger) and heparin. Histamine causes blood vessels to dilate and leak fluid at a site of inflammation, and heparin prevents blood clotting. **Neutrophils** are the most numerous of the white blood cells (figures 26.3*d*). They are chemically attracted to sites of inflammation and are active phagocytes.

The two types of agranulocytes are the **monocytes** and **lymphocytes** (figures 26.3*e* and *f*). Monocytes have a variety of immune functions, including phagocytosis of pathogens and the presentation of foreign materials (antigens) to lymphocytes. Two distinct types of lymphocytes are B cells and T cells, both of which are central to the immune response. **B cells** originate in the bone marrow (or bursa of Fabricius in birds) and colonize the lymphoid tissue, where they mature, develop receptors for foreign substances (antigens), and function in the formation of antibodies. In contrast, **T cells** are associated with and influenced by the thymus gland before they colonize lymphoid tissue and play their role in a cellular immune response. Some T cells (cytotoxic T cells) recognize and destroy virus-infected and tumor cells through the release of cytotoxic enzymes. Other T cells help to activate cytotoxic T cells, macrophages, and help stimulate antibody production.

Platelets (Thrombocytes) Platelets, or **thrombocytes** (Gr. *thrombus,* clot + cells), are disk-shaped cell fragments that initiate blood clotting. When a blood vessel is injured, platelets immediately move to the site, attach to exposed collagen fibers of the severed vessel, clump with other platelets, and form a temporary plug that reduces blood loss, which is the beginning of the blood-clotting process.

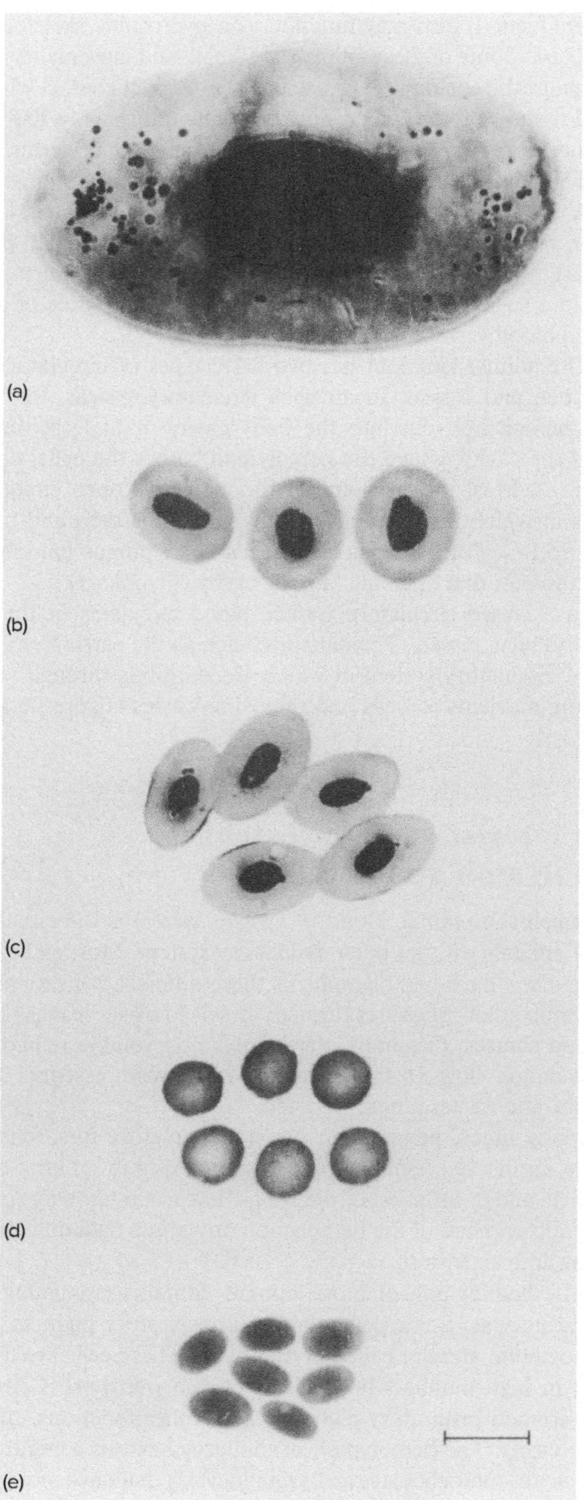

FIGURE 26.2

Comparison of Red Blood Cells from a Variety of Vertebrates. Light micrographs of (*a*) a nucleated cell from a salamander; (*b*) nucleated cells from a snake; (*c*) nucleated cells from an ostrich; (*d*) enucleated cells (biconcave disk) from a red kangaroo; and (*e*) enucleated cells (ellipsoid) from a camel (bar = 10 μm). Animals with lower metabolic demands tend to have larger (and often fewer) nucleated RBCs.

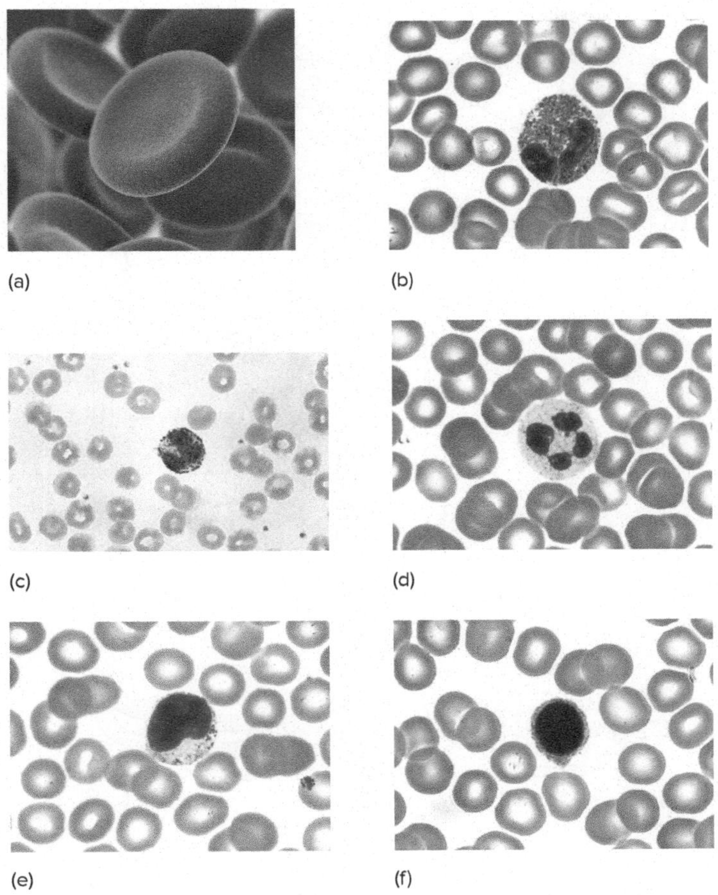

FIGURE 26.3

Blood Cells. (*a*) The biconcave shape of human erythrocytes (SEM × 1,500). (*b*) Red-staining cytoplasmic granules characterize an eosinophil. (*c*) Blue-staining granules characterize a basophil. (*d*) Light-pink granules and a multilobed nucleus characterize a neutrophil. Cells in *b–d* are also known as granulocytes. The agranulocytes consist of large monocytes (*e*) and lymphocytes (*f*). (*b–f* are LM × 1,600)

Vertebrate Blood Vessels

Arteries carry blood away from the heart. The walls of arteries are composed of an outer layer of connective tissue, a middle layer(s) of smooth muscle and elastic connective tissue, and an inner epithelium called an endothelium. The endothelium is a single cell in thickness and lines the lumen of this vessel (figure 26.4*a*). The relatively thick muscle and elastic middle layer is important to an artery's function in maintaining blood pressure within the arterial circulation.

Veins are larger in diameter, but thinner-walled vessels, in comparison to arteries. They carry blood toward the heart, and they are also comprised of three histological layers (figure 26.4*b*). Larger veins have valves that allow one-way flow toward the heart. This is important because veins function under lower pressure. Flow through veins is aided by skeletal muscle contractions that compress the sides of veins, squeezing blood from valve-to-valve through the vessels. Veins have a larger total cross-sectional area than do arteries. As a result, they hold a reserve of blood that can be quickly returned to the heart when the heart rate increases in response to various physiological events.

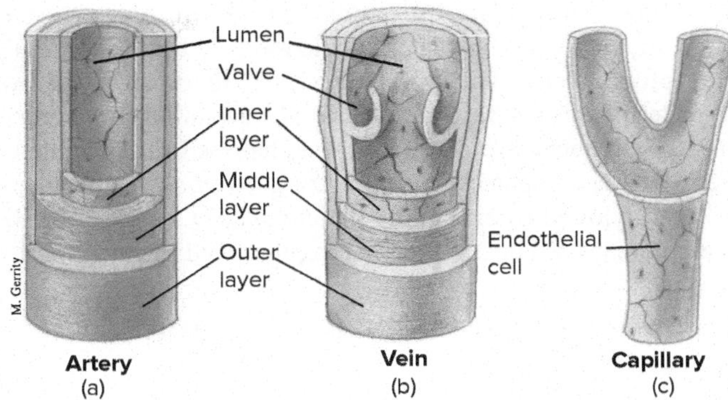

FIGURE 26.4

Structure of Blood Vessels. (*a and b*) The walls of arteries and veins have three layers (tunicae). The outermost layer consists of connective tissue, the middle layer has elastic connective and smooth muscle tissue, and the inner layer consists of a single layer of smooth endothelial cells (endothelium). Notice that the wall of an artery is much thicker than the wall of a vein. The middle layer is greatly reduced in a vein. (*c*) A capillary consists of a single layer of endothelial cells.

Arteries lead to **arterioles**. Arterioles constrict or dilate involuntarily in response to local metabolic conditions and nerve stimulation, which functions in regulating blood flow to various tissues. The arterioles branch to form **capillaries** (L. *capillus*, hair), which connect to **venules** and then to veins. Capillaries are generally composed of a single layer of endothelial cells and are the most numerous blood vessels in an animal's body (figures 26.4*c*). An abundance of capillaries makes an enormous surface area available for the exchange of gases, fluids, nutrients, and wastes between the blood and nearby cells.

Section 26.2 Thinking Beyond the Facts

Although white blood cells have many different and specific functions, they all have one common function. What is one common collective function of white blood cells?

26.3 VERTEBRATE CIRCULATORY SYSTEMS

LEARNING OUTCOMES

1. Compare the circulatory requirements of fishes and amphibians.
2. Assess the role of sarcopterygian fishes in the evolution of vertebrate circulatory systems.

The heart and its associated blood vessels of vertebrates is a closed system that developed evolutionarily from swellings of a major vessel, the ventral aorta, of chordate ancestors. Ancestral chordates did not have a true heart. This condition is seen in cephalochordates (*see chapter 17*), where contractile waves in the walls of major vessels propel blood. The evolution of gills, followed by the invasion of terrestrial habitats, influenced the evolution of vertebrate circulatory systems. We can follow the changes in the circulatory systems that accompanied this terrestrial invasion and the adaptations that followed by examining the circulatory systems of fishes and amphibians, avian and nonavian reptiles, and the mammals.

Circulatory Systems of Fishes and Amphibians

The distribution of blood into capillaries of gills was accompanied in fishes by the evolution of muscular pumping structures associated with the ventral aorta (figure 26.5*a*). The heart of fishes is includes a sinus venosus that receives blood from the **systemic circuit** (vessels associated with structures other than gills or lungs). The sinus venosus also serves as the pacemaker of the heart, initiating the heartbeat. It is followed by two relatively muscular chambers, an atrium and a ventricle. The conus arteriosus directs blood to the anterior ventral aorta and gill capillaries. After leaving the gill capillaries, blood flows through the dorsal aorta to systemic capillaries. Circulation through two sets of capillaries offers resistance to flow, and blood pressure and rates of flow are relatively low in fishes.

One group of fishes, the ancestral sarcopterygians (*see chapter 18 and figure 18.15*b), are the ancestors of the tetrapods, and they set the stage for the evolution of the heart of terrestrial vertebrates. The evolution of the sarcopterygian lungs involved the development of a **pulmonary circuit** that included a heart in which blood pumped from the heart to the lungs for oxygenation is returned to the heart to be pumped anew to the rest of the body. The systemic circuit supplies all body cells with this oxygen-rich blood. This circuit then returns blood to the heart after releasing oxygen and picking up carbon dioxide and other wastes. In addition to the sinus venosus (still the pacemaker) and conus arteriosus, the sarcopterygian heart included two atria and a partially divided ventricle to separate these two circuits.

In amphibians, the sinus venosus (the pacemaker) delivers blood to the right atrium, which connects to a single ventricle. The ventricle pumps blood into the conus arteriosus and to the lungs and the rest of the body (figure 26.5*b*). Blood destined for the lungs and skin comprises the **pulmocutaneous circuit**, as oxygenation of the blood occurs in both locations. Blood returning from the lungs returns to the left atrium and the ventricle. Blood returning from the skin and body enters the right atrium. Both pathways usually contribute oxygenated blood to the ventricle. Depending on the time of year, activity states, and the amphibian species, oxygenation of the blood may primarily involve either cutaneous or lung surfaces. Although the mechanism is not fully understood, under these conditions some amphibians are able to maintain nearly complete separation of pulmonary and systemic circuits in spite of having an undivided ventricle. Internal ventricular channeling may aid in this separation.

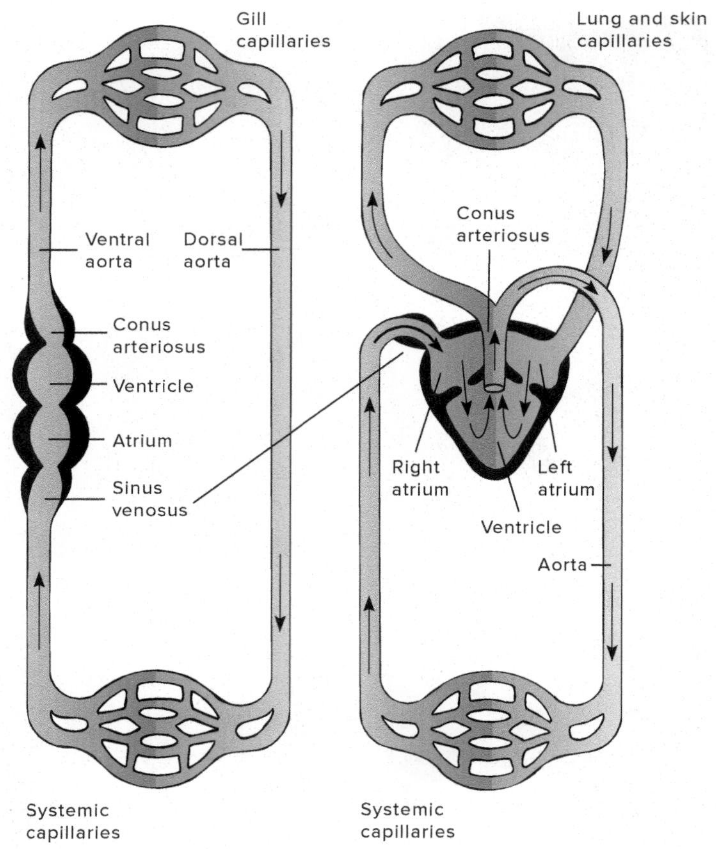

FIGURE 26.5

Heart and Circulatory Systems of Fishes and Amphibians. Oxygenated blood is red; less oxygenated blood is blue; a mixture of oxygenated and less oxygenated blood is purple. (*a*) In bony fishes, the heart's chambers (sinus venosus, atrium, ventricle, and conus arteriosus) pump in series. Gill and systemic capillary beds create resistance to flow in this relatively low-pressure system. (*b*) The amphibian heart's atrium is divided, and this sets up circulation through pulmonary and systemic circuits. Blood from the lungs enters the left atrium, and blood from the body enters the right atrium. The blood from both atria empties into one ventricle, which then pumps it into the respiratory and systemic circulations. Black arrows indicate the direction of blood flow.

Circulatory Systems of Amniotes

In the heart of most reptiles, (Lepidosauria and Testudines, *see figure 20.3*), the ventricle is partially divided into a right and left side (figure 26.6*a*). This division was established in the sarcopterygian ancestors of all tetrapods (*see figure 18.15b*). In addition to having two atria and a partially divided ventricle, the conus arteriosus of reptiles is now incorporated into the bases of pulmonary arteries that carry blood to the lungs from the right ventricle and into systemic arteries that carry blood to the rest of the body. Oxygenated blood from the lungs returns to the left side of the heart via the pulmonary vein without mixing with less oxygenated blood in the right side of the heart. When the ventricles contract, blood is pumped out of the two systemic arteries for distribution throughout the body, as well as to the lungs through the pulmonary arteries. The incomplete separation of the ventricles is an important adaptation for reptiles, such as turtles, because it allows blood to be diverted away from the pulmonary circulation during diving and when the turtle is withdrawn into its shell. This conserves energy and diverts blood to vital organs during the time when the lungs cannot be ventilated. Notice

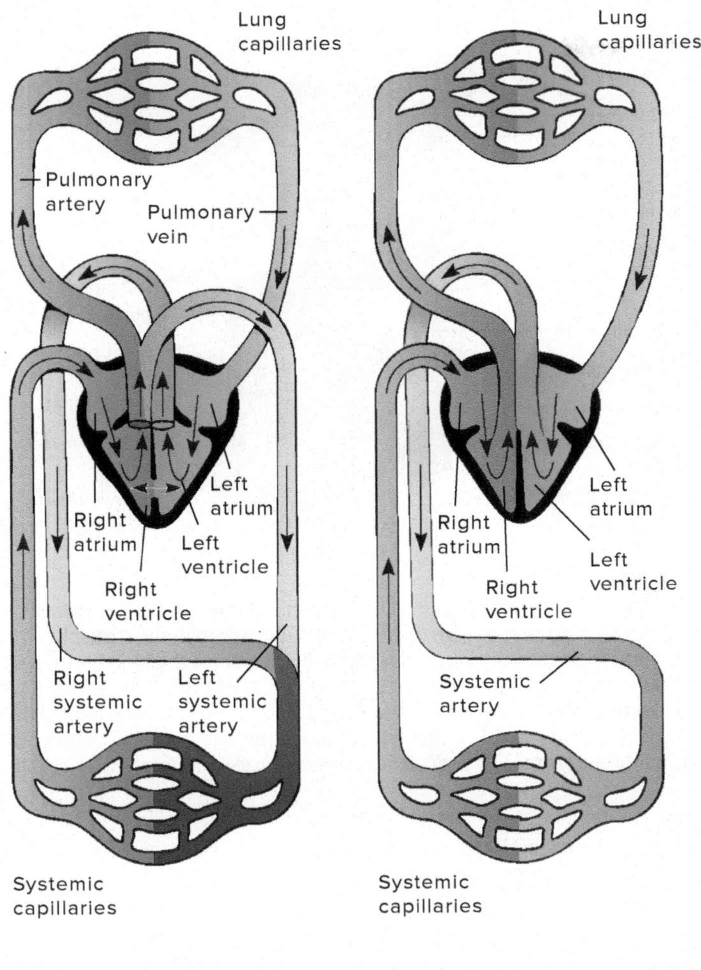

FIGURE 26.6

Heart and Circulatory System of Reptiles and Birds. (*a*) Reptiles have two atria and an incompletely divided ventricle. A pulmonary artery exits the right ventricle to take blood to the lungs. Systemic arteries exit both right and left ventricles. These adaptations permit the shunting of blood away from the lungs when a reptile is submerged and not breathing or when a turtle is withdrawn into its shell. Left and right systemic arteries are named based on the direction that they arch, not on the ventricle from which they originate. (*b*) Archosaurs, including the birds (shown here), have completely divided ventricles. A bird's systemic artery is derived from the right systemic artery of its reptilian ancestors and originates from the left ventricle. These adaptations result in a complete separation of pulmonary and systemic circuits. (As explained in the text, crocodylians retain two systemic arteries.)

in figure 26.6a that the left systemic artery originates with the pulmonary artery from the right ventricle and is the route for systemic blood leaving the heart when flow through the pulmonary circuit is reduced (as occurs in the submerged turtle). With the exception of the turtles, the sinus venosus is reduced to a patch of pacemaker tissue in the wall of the right atrium of reptiles.

Circulatory Systems of Archosaurs

Even though the physiological separation is almost complete in most reptiles, the complete anatomical separation of ventricles occurs only in the archosaur (crocodylians, dinosaurs, and birds) and synapsid (mammals) vertebrate lineages (*see figures 20.3 and 22.12*). This anatomical separation evolved separately in the two lineages, a testament to the importance of its function in active, mostly endothermic animals.

The crocodylian heart and major arteries are similar to those seen in other reptiles, but with two exceptions that allow blood to be diverted away from the lungs when a crocodile or alligator dives. First, the ventricle is completely divided (as it is in all archosaurs). Second, a narrow channel called the **foramen of Panizza** (a cardiac shunt not shown in figure 26.6*a*) connects the left and right systemic arches. When a crocodylian is breathing, blood pressure differences between the systemic arteries shunts blood from the right systemic artery into the left systemic artery and keeps both systemic arteries filled with oxygenated blood from the left ventricle. When a crocodylian is diving, resistance to pulmonary flow increases because the animal is breath-holding, and blood from the right ventricle is diverted away from the pulmonary artery and into the left systemic artery.

Avian reptiles are archosaurs, and their heart and major arteries are similar to the crocodylian pattern. In the birds, the ventricles are again completely divided. Unlike crocodylians, only the right systemic artery remains. Since this vessel originates from the left ventricle, it carries only oxygenated blood and delivers it to the systemic circulation (figure 26.6*b*). In addition, a bird's heart is much larger and more muscular (in comparison to overall body size) than that of other vertebrates. These features ensure efficient delivery of oxygenated blood to the tissues of these active, endothermic animals.

The structure and function of a bird's heart looks very similar to that seen in mammals. Both avian and mammalian hearts have four completely separated chambers and single pulmonary and systemic arteries that keep pulmonary and systemic circuits fully separated. This similarity is evolutionarily superficial. It reflects convergent metabolic adaptions for an active, endothermic lifestyle. In order to trace the origins of the mammalian circulatory system, we must return to the sarcopterygian circulatory pattern—which we do in the next section.

Section 26.3 Thinking Beyond the Facts

Vertebrate heart evolution is often portrayed as a progression from a two-chambered heart (fishes) to three chambers (amphibians) to four chambers (reptiles, birds, and mammals). Assess the validity of this portrayal.

26.4 THE MAMMALIAN CIRCULATORY SYSTEM

LEARNING OUTCOMES

1. Compare the pattern of circulation through mammalian and avian hearts.
2. Explain the origin and conduction of action potentials that initiate the beat of the mammalian heart.
3. Describe the events resulting in the generation of systolic and diastolic pressures of the heart.

Mammals are synapsid amniotes and have their ancestry deep within the sarcopterygian lineage (*see figure 20.3*). The sarcopterygian heart and arteries consisted of a sinus venosus, two atria, a partially divided ventricle, and a conus arteriosus (*see figure 18.15*). The advent of lungs in the sarcopterygian lineage established the separate pulmonary and systemic circuits. Terrestrial habitats and active endothermic lifestyles of ancestral synapsids selected for a complete separation of these circuits. The result is a circulatory system that looks and functions much like that seen in avian reptiles. The ventricles are completely divided, and single pulmonary and systemic arteries arise from the right and left ventricles, respectively. The conus arteriosus is gone as a chamber, and the sinus venosus is reduced to a patch of pacemaker tissue in the wall of the right atrium. This section describes the structure and function of the modern mammalian circulatory system.

The Mammalian Heart

Like the bird heart, the mammalian heart is an efficient, high-pressure pump. Heart rates of mammals vary depending on body sizes and metabolic rates. An elephant's (order Proboscidea) heart beats about 25 times per minute (bpm). The human (order Primates) heart rate is about 70 bpm. Smaller animals have higher metabolic rates and higher heart rates (rabbits [Lagomorpha] 200 bpm; domestic cats [Carnivora], 220 bpm; mice [Rodentia], 400 bpm; and shrews [Eulipotyphla], 800 bpm).

The outer layer of the heart is comprised of fibrous connective tissue and is called the epicardium (Gr. *epi*, upon). The middle myocardium is cardiac muscle, and the inner endocardium (Gr. *endo*, inside) that lines the heart chambers is similar to, and continuous with, the endothelium of the arteries and veins associated with the heart (*see figure 26.4*). Connective tissues within the heart chambers form tendinous cords and valves that will be described next.

The left and right halves of the heart are two separate pumps, each containing two chambers (figure 26.7). In each half, blood first flows into a thin-walled atrium (L. *antichamber,* waiting room) (pl., atria), then into a thick-walled ventricle. Valves are between the upper (atria) and lower (ventricles) chambers. The tricuspid valve is between the right atrium and right ventricle, and the bicuspid valve is between the left atrium and left ventricle. These valves are anchored to the ventricular walls by tendinous cords, which prevent the valves from everting into the atria when the ventricles contract. (Collectively, these are referred to as the AV valves—atrioventricular valves.) The pulmonary semilunar valve is at the exit of the right ventricle, and the aortic semilunar valve is at the exit of the left ventricle. All of these valves open and close due to blood pressure changes when the heart contracts during each heartbeat. Heart valves keep blood moving in one direction, preventing backflow from the ventricles into the atria (the AV valves) or from the pulmonary artery and aorta into the ventricles (semilunar valves).

The heartbeat is a sequence of muscle contractions and relaxations called the cardiac cycle. The cardiac cycle is initiated by the pacemaker remnant from the sarcopterygian sinus venosus, a small mass of tissue called the sinoatrial node (SA node). It is positioned at the entrance to the right atrium. The pacemaker action potential travels over specialized cardiac muscle cells organized into a conducting system (figure 26.8). The action potential quickly spreads over both atria, causing them to contract simultaneously. The action potential then passes into a patch of conducting fibers called the atrioventricular node (AV node), which is located at the base of the heart (near the top of the interventricular septum). A slight (0.1 s in

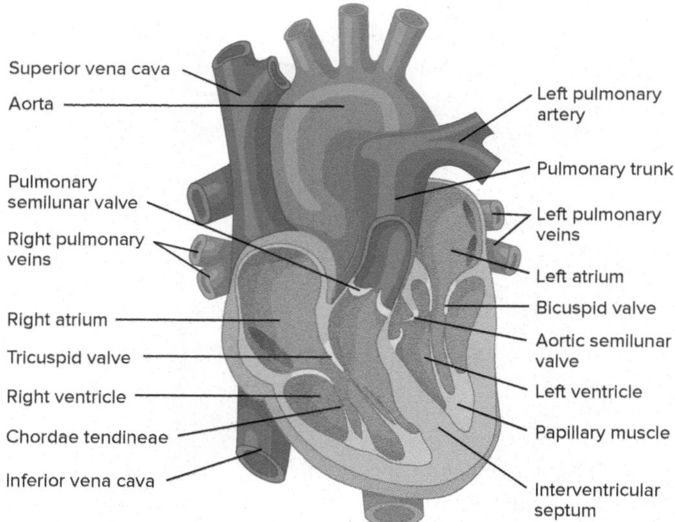

FIGURE 26.7

Structures of the Human Heart. Less oxygenated blood from the tissues of the body returns to the right atrium and flows through the tricuspid valve into the right ventricle. The right ventricle pumps the blood through the pulmonary semilunar valve into the pulmonary circuit, from which it returns to the left atrium and flows through the bicuspid valve into the left ventricle. The left ventricle then pumps blood through the aortic semilunar valve into the aorta. The various heart valves are shown in yellow.

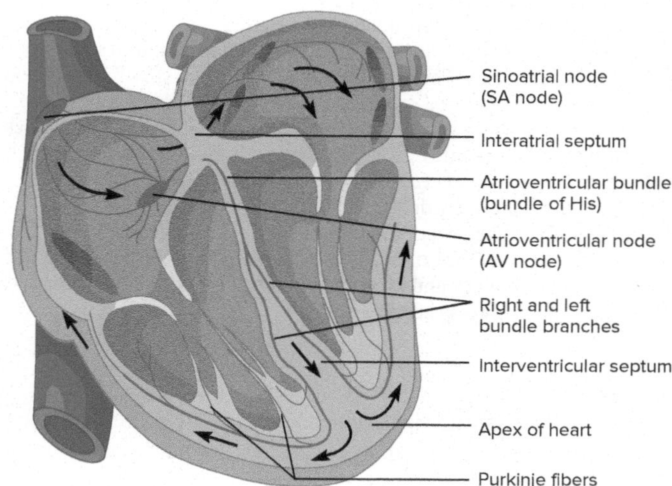

FIGURE 26.8

Electrical Conduction System of the Human Heart. The SA node initiates the depolarization wave, which passes successively through the atrial myocardium to the AV node, the atrioventricular bundle, the right and left bundle branches, and the Purkinje fibers in the ventricular myocardium. Black arrows indicate the direction of the electrical current flow.

the human heart) delay gives blood time to pass into the ventricles during atrial contraction. From here, the action potential continues through the atrioventricular bundle, at the lower apex of the interventricular septum. The atrioventricular bundle divides into right and left branches, which are continuous with the Purkinje fibers in the ventricular walls. Stimulation of these fibers causes the ventricles to contract almost simultaneously and eject blood into the pulmonary and systemic circulations.

Animation Cardiac Cycle

The action potential moving over the surface of the heart causes current flow, which can be recorded at the surface of the body as an electrocardiogram (ECG or EKG).

Animation Conducting System of the Heart

During each cycle, the atria and ventricles go through a phase of contraction called **systole** and a phase of relaxation called **diastole.** During diastole, blood is returning to the heart from the systemic and pulmonary veins to the right and left atria (respectively). AV valves are open during diastole, so blood can flow into the ventricles from the atria. Semilunar valves are closed, preventing backflow of blood from the pulmonary artery and aorta into the ventricles. During systole, atrial contraction forces the remaining atrial blood into the ventricles. Ventricular contractions follow and push blood against the AV valves, forcing them to close. Pressure builds in the ventricles, and it eventually exceeds the pressure of blood in the pulmonary artery and aorta. This increased ventricular pressure causes the semilunar valves to open and blood is forced into the two arteries, and blood then flows away from the heart. Relaxation of ventricular muscle leads to the next diastolic phase.

Blood Pressure

Ventricular contraction generates the fluid pressure, called **blood pressure,** that forces blood through the pulmonary and systemic circuits. All fluids flow from areas of higher pressure to areas of lower pressure. Pressure is highest near the heart, so blood flows away from the heart toward body tissues. As blood flows, it is distributed into more vessels with larger total cross-sectional area, and it also encounters resistance as individual vessels become smaller in diameter. Blood pressure thus decreases as blood moves away from the heart. The high-pressure circulatory systems of birds and mammals have been instrumental in meeting the aerobic metabolic requirements of these endotherms.

Arterial blood pressure rises and falls in a pattern corresponding to the phases of the cardiac cycle. When the ventricles contract (ventricular systole), blood pressure in the pulmonary arteries and aorta rises sharply. The maximum pressure achieved during ventricular contraction is called the **systolic pressure.** When the ventricles relax (ventricular diastole), the arterial pressure drops, and the lowest pressure that remains in the arteries before the next ventricular contraction is called the **diastolic pressure.**

In humans, normal systolic pressure for a young adult is about 120 mm Hg. Diastolic pressure is approximately 80 mm Hg. Conventionally, these readings are expressed as 120/80. Blood pressure is higher in birds and mammals than it is in other vertebrates. Blood pressure in mammals often reaches values similar to that of the bottlenose dolphin (*Tursiops truncatus*, 150/121). In birds, it is often higher (220/154 in the canary, *Serinus*). These are high compared to low blood pressures of 25/10 in turtles (class Testudines) and 40/30 in catfishes (order Siluriformes).

Section 26.4 Thinking Beyond the Facts

It is not necessary to reconnect nerves to a newly transplanted heart in order for the heart to maintain a regular rhythm. Explain.

26.5 LYMPHATIC SYSTEMS

LEARNING OUTCOME

1. Describe how the lymphatic system functions.

Capillaries of the circulatory system are leaky. Fluids at the arteriolar end of capillary beds are under pressure (blood pressure), which force the fluids out of capillary beds between cellular junctions. Filtered fluids carry small organic molecules, including nutrients used by cells in surrounding tissues and some larger proteins. Fluids and many organic molecules that are filtered must be returned to the circulatory system. If they accumulate in tissue spaces, edema results. Some fluids return to the blood stream at the capillary's venous end through osmotic forces. Other fluids and organic molecules move back into the circulatory system through the **lymphatic system.** The lymphatic system is a system of blind-ending lymphatic capillaries that originate in tissue spaces, coalesce into larger vessels, and empty into the venous circulation near the heart (figure 26.9). The fluid within lymphatic vessels is called **lymph** (L. *lympha*, clear water). Lymphatic vessels have valves, similar to those of veins; and, in mammals, lymph moves as a result of skeletal muscles compressing vessels and moving fluids from valve-to-valve. In all other vertebrate groups, lymphatic vessels are often contractile, and **lymph hearts** may propel lymph.

The lymphatic system also functions in transporting lipids absorbed from the digestive system and in body defenses (table 26.1). Immune cells emigrate from lymph nodes to sites of infection. Other lymphatic organs include the spleen, tonsils, bursa of Fabricius (birds), and the thymus gland.

Section 26.5 Thinking Beyond the Facts

What is the physiological relationship between the lymphatic and circulatory systems?

26.6 GAS EXCHANGE

LEARNING OUTCOMES

1. Compare and contrast the variety of respiratory systems found in animals.
2. Describe a book lung.

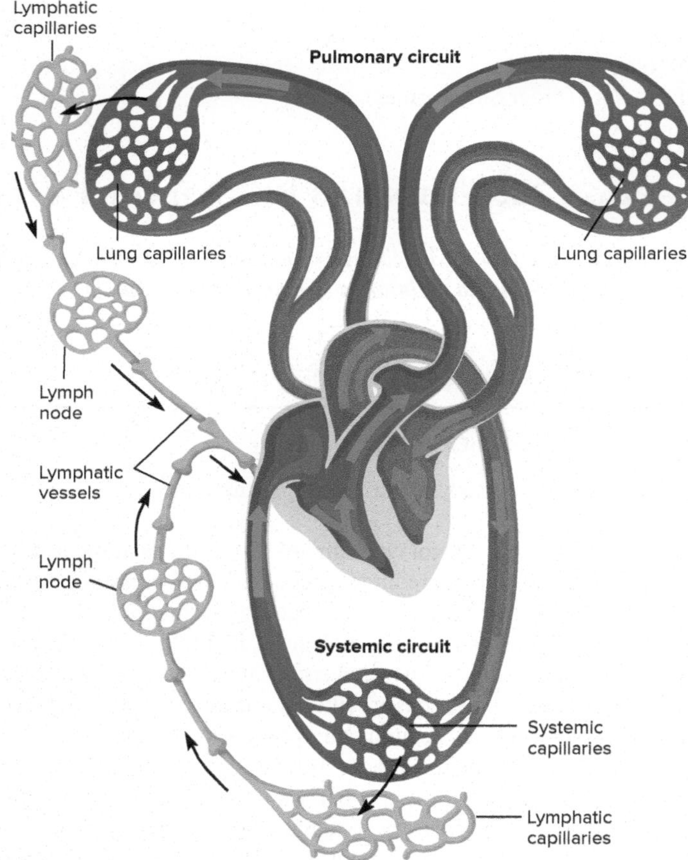

FIGURE 26.9

The Lymphatic System. The lymphatic system consists of one-way vessels that return tissue fluid, called lymph, to the major veins that drain into the right atrium. The black arrows indicate the direction of lymph flow, and the colored (blue and pink) arrows indicate the direction of blood flow.

TABLE 26.1

MAJOR STRUCTURAL AND FUNCTIONAL COMPONENTS OF THE LYMPHATIC SYSTEM IN VERTEBRATES

STRUCTURE	FUNCTION
Lymphatic capillaries	Collect excess extracellular fluid in tissues
Lymphatics	Carry lymph from lymphatic capillaries to veins in the neck, where lymph returns to the bloodstream
Lymph nodes	House the white blood cells that destroy foreign substances; play a role in antibody formation
Spleen	Filters foreign substances from blood; manufactures phagocytic lymphocytes; stores red blood cells; releases blood to the body when blood is lost
Thymus gland (in mammals)	Site of antibodies in the newborn; is involved in the initial development of the immune system; site of T cell differentiation
Bursa of Fabricius (in birds)	A lymphoid organ at the lower end of the alimentary canal in birds; the site of B cell maturation

Aerobic metabolism evolved more than 2 billion years ago (*see chapter 8*). For the next 1.5 billion years, the oxygen that was required to complete the steps of electron transport and chemiosmosis (*see chapter 2*) was acquired by simple diffusion across plasma membranes of single-celled protists (*see appendix C*) and very small animals. All gases move into cells by simple diffusion; however, diffusion can supply oxygen sufficient to support aerobic metabolism only when cells are no more than 0.5 mm from the oxygen source. Evolutionary events that began over 600 million years ago saw an amazing radiation of animals that included the evolution of a diversity of body shapes and sizes. An integral part of this radiation was the evolution of mechanisms that carried air or water into contact with respiratory surfaces where oxygen and carbon dioxide could be exchanged and carried to and from the animal's deepest tissues. These mechanisms include tracheae and various forms of gills and lungs that function in concert with circulatory systems.

Invertebrate Respiratory Systems

Although sometimes aided by countercurrent exchange systems and respiratory pigments, all animals use diffusion to move respiratory gases across plasma membranes (figure 26.10*a*). Some multicellular invertebrates either have very flat bodies (e.g., Platyhelminthes) in which all body cells are relatively close to the body surface, or they have a cavity (e.g., Cnidaria) (figure 26.10*b and c*). Gases diffuse into and out of the animal. Diffusion of gases across body surfaces requires a thin, moist epithelium. In triploblastic animals like the Annelida (*see chapter 12*), cutaneous capillary networks provide a means for respiratory gases to be exchanged across the integument and transported to and from deeper tissues. This exchange is called **integumentary exchange**.

Many aquatic invertebrates possess modifications of the body wall that increase surfaces for gas exchange. These structures are called **gills** and have arisen independently multiple times in animal evolution. The relatively simple folds of the body wall of echinoderms, called dermal branchiae or papulae, protrude between skeletal ossicles to provide gas-exchange surfaces (*see figure 16.4*). Other gills, like those of crustaceans (*see chapter 15*) and most molluscs (*see figure 11.10*) are complexly branched and folded structures that maximize gas-exchange surfaces between water and body fluids.

Some terrestrial invertebrates including insects (Insecta), centipedes (Chilopoda), and some ticks, chiggers (Acarina), and spiders (Araneae) have **tracheal systems**. Tracheal systems consist of highly branched chitin-lined tubes called tracheae (figure 26.11*a*), which exchange gases directly with body cells rather than with hemolymph or blood. Tracheae open to the outside of the body through spiracles, which usually have some kind of closure device to prevent excessive water loss. Spiracles lead to branching tracheal trunks that eventually give rise to smaller branches called tracheoles, whose

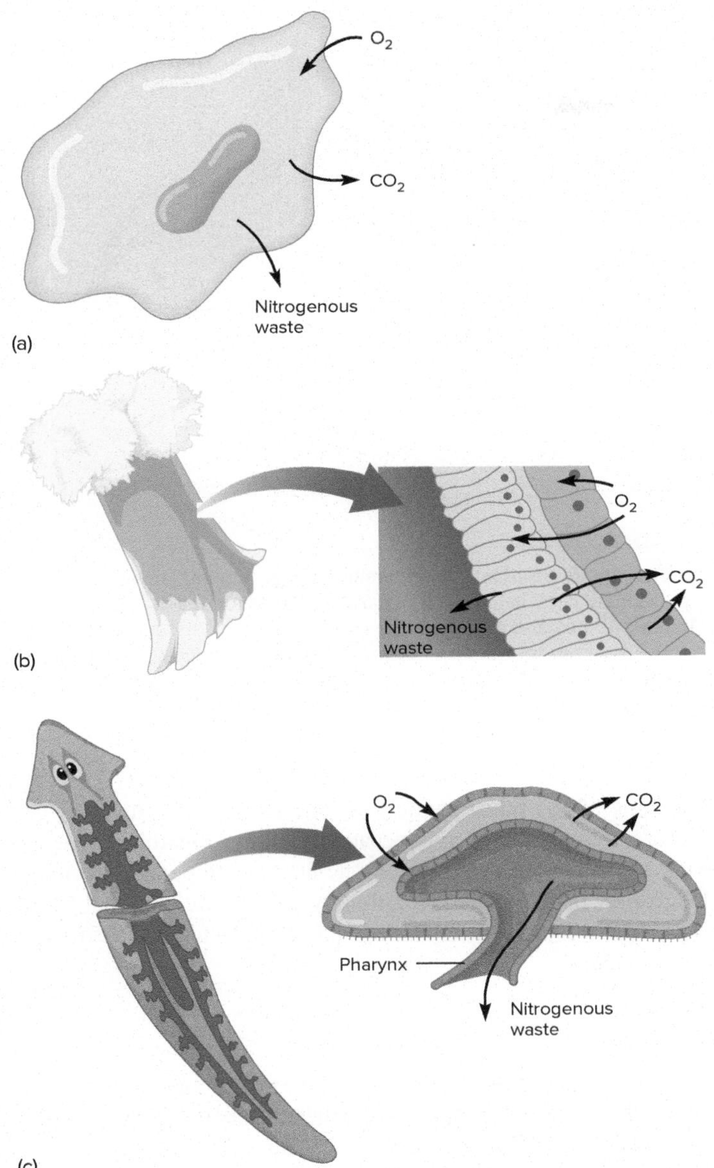

FIGURE 26.10

Invertebrate Respiration: Diffusion through Body Surfaces. Gases are exchanged across the plasma membranes of all animal cells by diffusion (*a*). Some small, or very flat, animals such as cnidarians (*b*) and flatworms (*c*) maintain close enough contact with the environment that they have no need for a respiratory system. Diffusion moves gases, as well as waste products, into and out of these organisms.

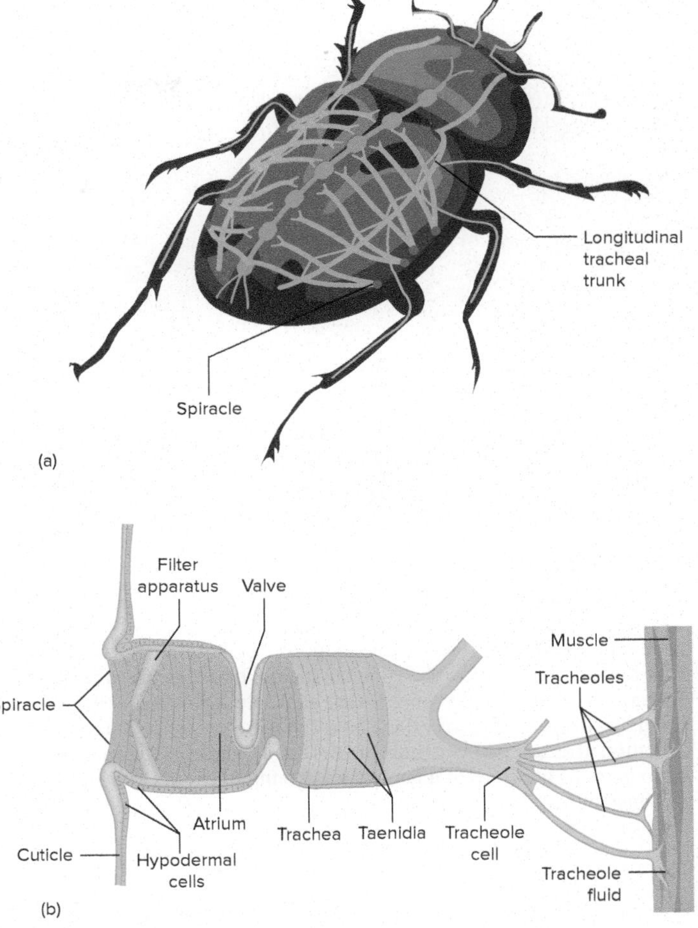

FIGURE 26.11

Invertebrate Respiration: A Tracheal System. (*a*) Tracheal system of an insect, showing the major tracheal trunks. (*b*) Tracheoles end at cells, and the terminal portions of tracheoles are fluid filled. The fluid acts as a solvent for gases.

blind ends lie close to all cells of the body. Because no cells are more than 2 or 3 μm from a tracheole, gases move between the tracheole and the tissues of the body by diffusion (figure 26.11*b*). Most insects have ventilating mechanisms that move air into and out of the trachea. For example, contracting flight muscles of insects alternately compress and expand the large tracheal trunks and thereby ventilate the tracheae.

Lungs are structures that exchange respiratory gases between the air and body fluids. Arachnids possess tracheae, book lungs, or both. **Book lungs** are paired invaginations of the ventral body wall that are folded into a series of leaflike lamellae (figure 26.12; *see also figures 14.9 and 14.12*). Air enters the book lung through a slitlike opening called a spiracle and circulates between lamellae. Respiratory gases diffuse between the hemolymph moving along the lamellae and the air in the air chamber. Some ventilation also results from the contraction of a muscle attached to the dorsal side of the air chamber. This contraction dilates the chamber and opens the spiracle, but most gas movement is still by diffusion.

Another form of invertebrate lung evolved in the pulmonate molluscs (slugs, most freshwater, and all terrestrial gastropods). Their subclass name, Pulmonata (L. *pulmo*, lung), is derived from the presence of a **pulmonate lung** that develops from their mantle (*see figure 11.2*) and which secretes the shell in all molluscs. A mantle cavity is enclosed by the mantle and shell and opens to the outside through a **pneumostome** (*Gr. pneumo*, breath + *stoma*, mouth) (figure 26.13, *see figure 11.7*c). Gas exchange with air occurs by diffusion through the pneumostome, or it is aided by body wall muscles that alternately compress and expand the mantle cavity.

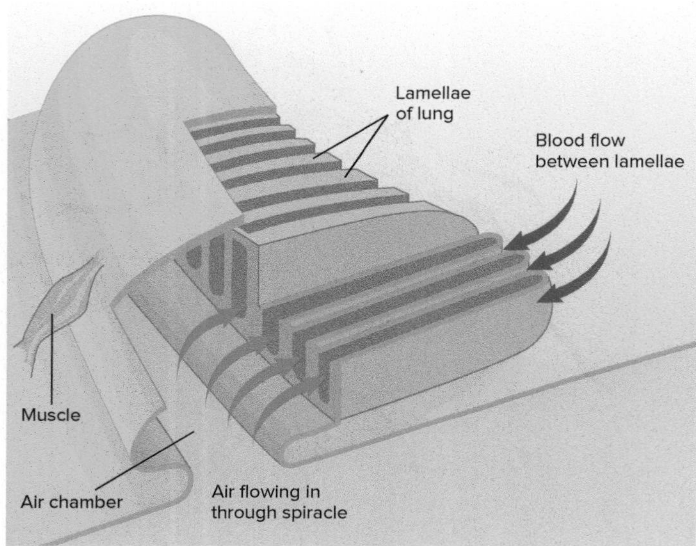

FIGURE 26.12

Invertebrate Respiration: A Book Lung. Structure of an arachnid (spider) book lung. Air enters through a spiracle into the air chamber by diffusion and by ventilation due to muscle contraction. Air diffuses from the air chamber into the lamellar spaces; hemolymph circulates through the blood lamellar spaces that alternate with air lamellar spaces. Small, peglike surface projections hold the lamellae apart. Due to this structural arrangement, air (blue arrows) and blood (purple arrows) move on opposite sides of a lamella in a countercurrent flow, allowing the exchange of respiratory gases by diffusion.

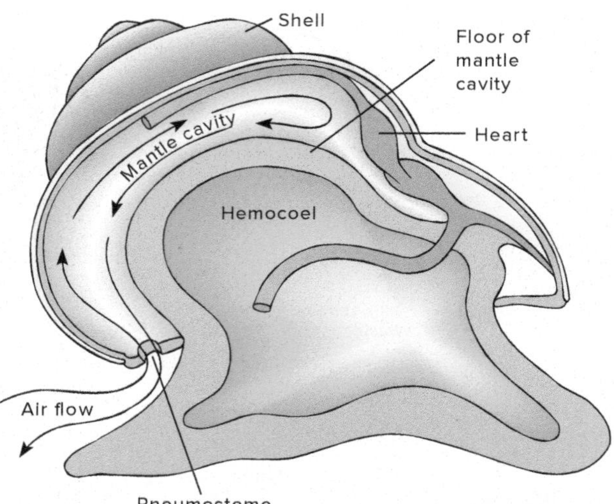

FIGURE 26.13

Invertebrate Respiration: The Pulmonate Lung. The mantle cavity of a pulmonate snail is highly vascularized and functions as a lung. Downward movement of the floor of the mantle cavity increases the cavity's volume, so that air is drawn into the mantle cavity through the pneumostome. Decreasing the volume of the mantle cavity expels the air. Black arrows indicate the direction of air flow.

Section 26.6 Thinking Beyond the Facts

How has the evolution of respiratory systems in invertebrates been influenced by the evolution of increased body size, the presence or absence of exoskeletons, and aquatic versus terrestrial habitats?

26.7 VERTEBRATE RESPIRATORY SYSTEMS

LEARNING OUTCOMES

1. Analyze the importance of bimodal breathing in the evolutionary transition between aquatic and terrestrial environments.
2. Compare and contrast the mechanisms of gas exchange over cutaneous surfaces, gill surfaces, and lung surfaces.

Aquatic vertebrates (fish, amphibians, and some reptiles) rely on cutaneous surfaces or gills for gas exchange. Lungfishes also possess lungs in addition to gills. Other vertebrates, like many amphibians, use a combination of cutaneous surfaces and lungs (adult anurans), or cutaneous surfaces and gills (many salamanders), for gas exchange in either air or water. Similar adaptations occur in many invertebrates. **Bimodal breathing** is the use of respiratory surfaces to exchange gases with either air or water. The evolution of bimodal breathing allows fundamentally aquatic animals to withstand periods of water stagnation or desiccation. Bimodal breathing arose very early in the sarcopterygian vertebrate lineage and persists in modern lungfishes. In the longer evolutionary perspective, bimodal breathing played a significant role in the transition to terrestrialism in the vertebrates.

Cutaneous Exchange

Cutaneous respiration is more important to the survival of amphibians than it is to members of any other vertebrate group. The thin, moist skin of amphibians contains a subepidermal capillary network and is a surface for exchanges of oxygen and carbon dioxide (*see figure 23.6*). In some salamanders (Caudata), cutaneous respiration accounts for 30 to 90% of total gas exchange. A relatively small percentage of total gas exchange can occur across moist surface of the mouth and pharynx of amphibians (**buccopharyngeal respiration**).

Gills

Gills are respiratory organs that have either a thin, moist, vascularized layer of epidermis to permit gas exchange across thin gill membranes, or a very thin layer of epidermis over highly vascularized dermis. Larval forms of a few fishes and amphibians have external gills projecting from their bodies (figure 26.14). Adult fishes have internal gills.

Gas exchange across internal gill surfaces is extremely efficient (figure 26.15; *see also figure 18.16*). It occurs as blood and water move in opposite directions on either side of the lamellar epithelium. Water passing between gill lamellae first passes the

FIGURE 26.14

Vertebrate Respiration: External Gills. This axolotl (*Ambystoma mexicanum*) has elaborate external gills with a large surface for gas exchange with the water.

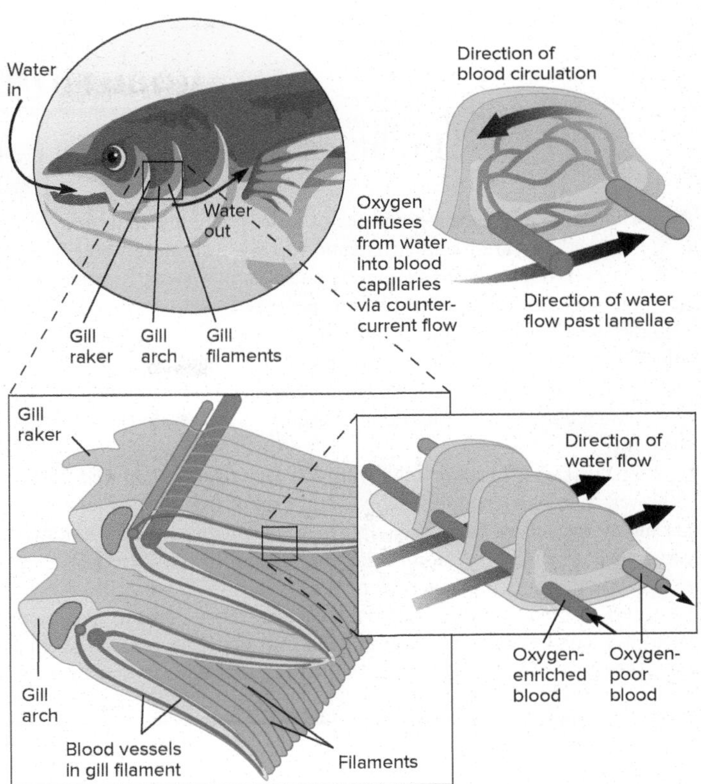

FIGURE 26.15

Vertebrate Respiration: Internal Gills. Removing the protective operculum exposes the feathery internal gills of this bony fish. Each side of the head has four gill arches, and each arch consists of many filaments. A filament houses capillaries within lamellae. Note that the direction of water flow opposes that of blood flow. This countercurrent flow allows the fish to extract the maximal amount of oxygen from the water.

portion of a gill lamella containing blood that is about to leave the lamella. This blood has a relatively high oxygen concentration, having picked up oxygen earlier in the lamella. Because water at this point has lost none of its oxygen, a diffusion gradient still favors movement of more oxygen into the blood. Water then passes by the portion of the vessels bringing blood from deep within the body. This blood is lower in oxygen. Even though the water has already lost some oxygen to the blood earlier in its movement through the gills, there is still a higher concentration of oxygen in the water than in the blood. Thus, a diffusion gradient still favors the movement of oxygen into the blood. Carbon dioxide also diffuses into the water because its concentration is higher in the blood than in the water. This countercurrent exchange mechanism provides efficient gas exchange by maintaining a concentration gradient between the blood and water over the length of the capillary bed.

Lungs

A **lung** is an internal sac-shaped respiratory organ. The lung of a terrestrial vertebrate comprises one or more internal blind pouches into which air is either drawn or forced. The respiratory epithelium of lungs is thin, well vascularized, and divided into a large number of small units, which greatly increase the surface area for gaseous exchange between the lung air and the blood. This blind-pouch construction, however, limits the efficiency with which oxygen and carbon dioxide are exchanged with the atmosphere because only a portion of the lung air is ever replaced with any one breath. Birds are an exception in that they have very efficient lungs with a one-way pass-through system (*see figure 21.11*). For example, a mammal removes approximately 25% of the oxygen from air with each breath, whereas a bird removes approximately 90%.

The evolution of the vertebrate lung is related to the evolution of the swim bladder. The swim bladder is an air sac located dorsal to the digestive tract in the body of many modern fishes. Both lungs and swim bladders evolved from pneumatic sacs present in primitive fishes that were ancestors of both present-day fishes and tetrapods (amphibians, reptiles, birds, and mammals). These ancestral fishes probably had a ventral sac attached to the esophagus (*see figure 18.17*). This sac may have served as a supplementary gas-exchange organ when the fishes could not obtain enough oxygen through their gills (e.g., in stagnant or oxygen-depleted water). By swimming to the surface and gulping air into this sac, ancestral fishes could exchange gas through its wall.

Further evolution of this blind sac proceeded in two different directions (*see figure 18.17*). One adaptation is in the majority of modern bony fishes, where the swim bladder lies dorsal to the digestive tract. The other adaptation is in the form of the lungs, which are ventral to the digestive tract. A few present-day fishes and the tetrapods have ventral lungs. The evolution of the structurally complex lung paralleled the evolution of the larger body sizes and higher metabolic rates of endothermic vertebrates (birds and mammals), which necessitated an increase in lung surface area for gas exchange, compared to the smaller body size and lower metabolic rates of ectothermic vertebrates (figure 26.16).

EVOLUTIONARY INSIGHTS

Evolutionary Adaptations in the Gills of Tunas (Scombridae)

Tunas are among the pinnacles of water-breathing endurance athletes. They are highly active and rapidly swimming water predators. Using their red swimming muscles, they swim continuously, day and night, covering over 100 km per day in search of prey. As a result, these endurance athletes must be able to acquire oxygen very rapidly. Water is not a rich source of oxygen, so tunas have evolved a respiratory system (gills) that enables them to take up oxygen rapidly from the water and a circulatory system that delivers this oxygen to the tissues of the body.

Like most fish, tunas breathe with gills (*see figures 18.16 and 26.15*). However, their gills are not like the average set of fish gills. Instead, they are highly specialized for oxygen uptake. This illustrates the evolutionary principle that a single type of breathing system, in this case gills, may exhibit a wide range of evolutionary adaptations. First, the gills of tuna have about nine times more surface area than those of an average fish, such as a bass or bluegill. (Centrarchidae) Second, in a bass or bluegill, the distance between the water and blood is about 6 μm, whereas in the tuna it is only 0.5 μm. From these two physiological variables, tunas have evolved gills that present a very large surface area and a very thin membranous surface between the water and the blood in order to allow gas exchange to occur rapidly.

Bass and bluegills have a muscular pumping mechanism to move the water into the mouth and pharynx, over the gills, and out of the fish through gill openings. Muscles surrounding the pharynx and the opercular cavity power the pump. During their evolution, tunas lost this pumping mechanism and became ram ventilators. In ram ventilation, tunas hold their mouths open while swimming, and there are no visible breathing movements. This forces (rams) large amounts of water per unit time into the buccal cavity and across the gills; more than 10 times the movement of water across the gills of a bass or bluegill! In this way, the white and red swimming muscles take over the responsibility for powering the flow of water across the gills. As a result, tunas have no choice regarding how much time they spend swimming. They must swim continuously forward or die from a lack of oxygen.

Lung Ventilation

Ventilation is based on several physiological principles that apply to all air-breathing animals with lungs:

1. Air moves by bulk flow into and out of the lungs in the process called ventilation.
2. Carbon dioxide diffuses across the respiratory surface of the lung tissue from pulmonary capillaries and oxygen diffuses from the lung spaces into the pulmonary capillaries.
3. At systemic capillaries, oxygen and carbon dioxide diffuse between the blood and interstitial fluid in response to concentration gradients.
4. Oxygen and carbon dioxide diffuse between the interstitial fluid and body cells.

Vertebrates exhibit two different mechanisms for lung ventilation based on these physiological principles. Amphibians and some reptiles use a positive pressure pumping mechanism. They push air into their lungs. Figure 26.17 shows the positive pressure pumping mechanism of an amphibian. The muscles of the mouth and pharynx create a positive pressure to force air into the lungs.

Most reptiles (including birds) and all mammals use negative pressure to draw air into the lungs. Expansion of the body

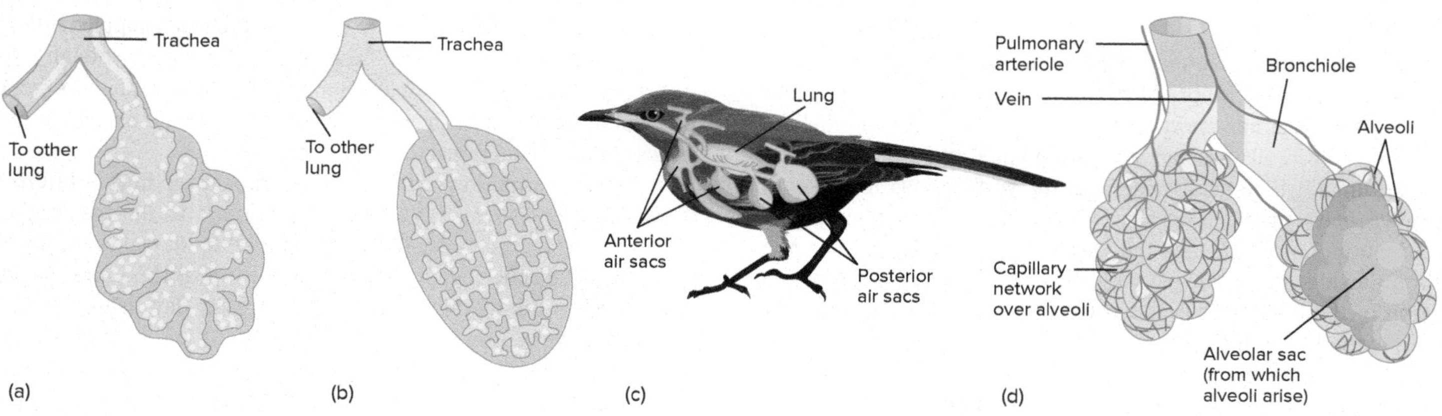

FIGURE 26.16

Vertebrate Respiration: Lungs. Evolution of the vertebrate lung, showing the increased surface area from (*a*) amphibians and (*b*) nonavian reptiles to (*c*) birds and (*d*) mammals. This evolution has paralleled the evolution of larger body size and higher metabolic rates.

FIGURE 26.17

Ventilation in Amphibians. The positive pressure pumping mechanism in an anuran. The breathing cycle has several stages. (*a*) Air is taken into the mouth and pharynx by lowering the floor of the mouth. Notice that the glottis is closed. (*b*) The glottis is then opened, and air is permitted to escape from the lungs, passing over the air just taken in. (*c*) With the nostrils and mouth firmly shut, the floor of the mouth is raised. This positive pressure forces air into the lungs. (*d*) With the glottis closed, fresh oxygenated air can again be brought into the mouth and pharynx. Red arrows indicate body wall movement, and blue arrows indicate air flow.

cavity surrounding the lungs creates pressure lower than atmospheric pressure. Air flows from the higher atmospheric pressure to the lower pressure in the lungs. Most nonavian reptiles expand the body cavity with a posterior movement of the ribs. This expansion decreases pressure in the lungs and draws air into the lungs. Elastic recoil of the lungs and the movement of the ribs and body wall, which compress the lungs, expel air. The ribs of turtles are a part of the carapace (*see figure 20.5*); thus, movements of the body wall to which they attach are impossible. Turtles exhale by contracting muscles that force the viscera upward, compressing the lungs. They inhale by contracting muscles that increase the volume of the visceral cavity, creating negative pressure to draw air into the lungs.

Because of the high metabolic rates associated with flight, birds have a greater rate of oxygen consumption than any other vertebrate. Mammals have small, blind-ending air sacs, called **alveoli**, where gas exchange occurs (figure 26.18*a*). Air brought into the lungs must be expelled by the same pathways that it entered. In contrast, birds have a lung structure that allows a one-way flow through parabronchi (figure 26.18*b*). Small air capillaries branch from parabronchi and are the location of gas exchange with blood capillaries. In addition to the lungs, the respiratory system of a bird includes air sacs that ramify through the body. Two respiratory cycles are required to move a single volume of air through the bird respiratory system (*see chapter 21 and figure 21.11*). Air sacs are collapsible, and open and close as a result of muscle contractions around them. Inhaled air bypasses the lungs and enters the abdominal (posterior) air sacs. It then passes through the lungs into the thoracic (anterior) air sacs. Finally, air is exhaled from the thoracic air sacs (figure 26.19).

SECTION 26.7 THINKING BEYOND THE FACTS

What environmental selection pressure might play a role in the evolution of a bird's highly efficient lungs?

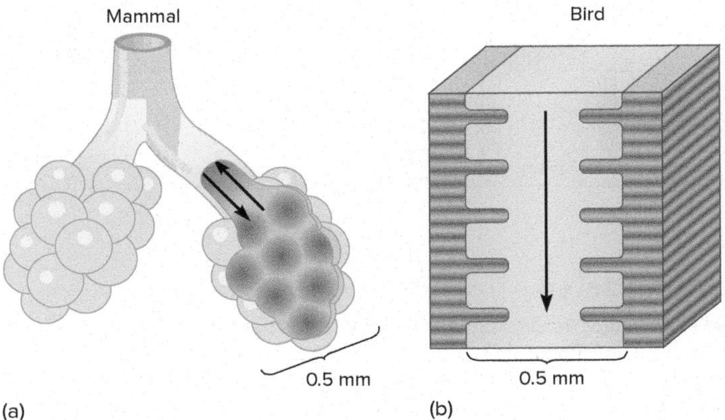

FIGURE 26.18

Gas-Exchange Surfaces in Mammals and Birds. (*a*) The gas-exchange surfaces in a mammal's lung are in saclike alveoli. Ventilation is by an ebb-and-flow mechanism (arrows), and the air inside the alveoli can never be completely replaced. (*b*) The parabronchi in a bird lung are tubes that are open at both ends. Ventilation is by one-way flow (arrow), and complete replacement of air in the tubes is continuous.

26.8 THE MAMMALIAN RESPIRATORY SYSTEM

LEARNING OUTCOMES

1. Relate the mechanisms of gas exchange to structures of the human respiratory system.
2. Differentiate the mechanism of inhalation and the mechanism of exhalation.
3. Compare and contrast the four common respiratory pigments found in animals.

The structure and function of external respiration in humans are typical of mammals. Thus the human respiratory system is used here to describe those principles that apply to all air-breathing mammals.

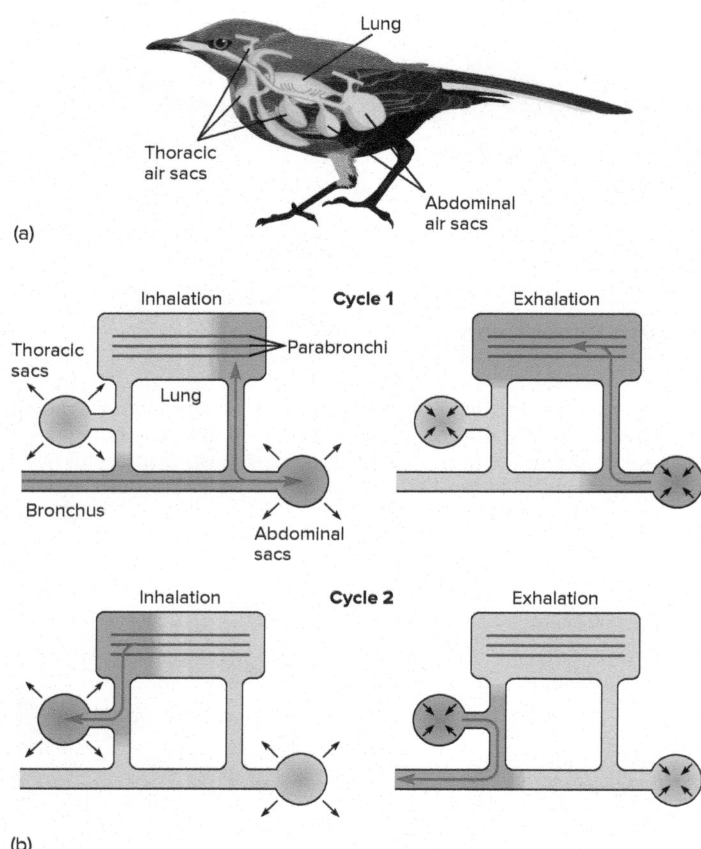

FIGURE 26.19

Gas-Exchange Mechanism in Birds. (*a*) Birds have a number of large air sacs. Some of them (abdominal) are posterior to the small pair of lungs, and others (thoracic) are anterior to the lungs. The main bronchus (air passageway) that runs through each lung has connections to air sacs, as well as to the lung. In (*b*), abdominal and thoracic air sacs are sketched as single functional units to clarify their relationship to the lung and bronchus. (*b*) Air flow through the bird respiratory system. The darker blue portion in each diagram represents the volume of a single inhalation and distinguishes it from the remainder of the air in the system. Two full breathing cycles are needed to move the volume of gas taken in during a single inhalation through the entire system and out of the body. This system is associated with one-way flow through the gas-exchange surfaces in the lungs. Black arrows indicate expansion and contraction of air sacs. Red arrows indicate movement of air.

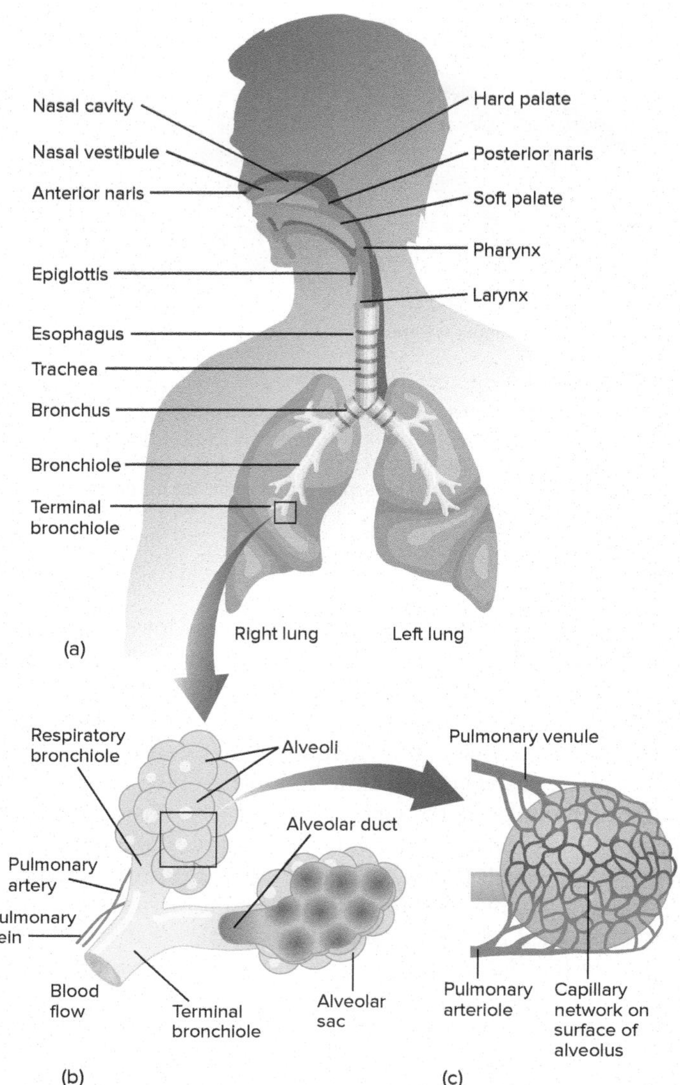

FIGURE 26.20

Organs of the Human Respiratory System. (*a*) Basic anatomy of the respiratory system. (*b, c*) The respiratory tubes end in minute alveoli, each of which is surrounded by an extensive capillary network.

Air-Conduction Portion

Figure 26.20 shows the organs of the human respiratory system. Air normally enters and leaves this system through either nasal or oral cavities. From these cavities, air moves into the pharynx, which is a common area for both the respiratory and digestive tracts. Air-conducting structures leading from the pharynx into the lungs divide in a treelike fashion into ever smaller branches. This respiratory tree begins with the cartilage-lined larynx and trachea. The opening of the larynx is the glottis. A flap of cartilage, the epiglottis, covers this opening when one swallows. The epiglottis prevents food or water from entering the trachea and shunts it to the esophagus and stomach.

During inhalation, air passes through the glottis and larynx and moves into the trachea, which branches into a right and left bronchus (pl., bronchi). After each bronchus enters the lungs, it branches into smaller tubes called bronchioles, and eventually into the gas-exchange portion of the respiratory system.

Gas-Exchange Portion

Small tubes called alveolar ducts connect the respiratory bronchioles to grapelike clusters of alveoli (sing., alveolus) (L. *alveus,* hollow) (figure 26.20*b*). The alveoli cluster to form an alveolar sac. Surrounding the alveoli are many capillaries (figure 26.20*c*). Alveoli are the functional units of the lungs (gas-exchange portion). As with gas exchange in all other animals, respiratory gases move by simple diffusion. Concentration gradients for oxygen and carbon dioxide are compared to the composition of these gases in air at sea level

How Do We Know Diving Leatherback Sea Turtles (*Dermochelys coriacea*) Can Survive Anoxia at 1,000 m?

Leatherback sea turtles can do something no other reptile on earth can do—they can dive as deep as 1,000 m. This is truly remarkable considering that the female of this species weighs about 400 kg and males nearly twice this amount. Prior to tracking experiments in the 1980s, Weddell seals, fin whales, and some other marine mammals were the only natural deep-sea divers known. Because leatherbacks only leave the water to breed and lay their eggs, how are scientists gathering information on their physiology?

During the month of May in the Caribbean, scientists watch the beach for leatherbacks coming out of the sea. When the female gets on her nest in the sand, the scientists put a collection device over her head to collect and later measure the blood gases. At the same time, another scientist draws blood from the turtle for later blood analysis to determine hemoglobin content, red blood cell count, blood gases (carbon dioxide, oxygen concentrations), and blood pH. Any turtles that fall prey to predators or die on the beach have skeletal muscle, heart, and brain tissue removed.

Scientists have estimated that this turtle would have to swim for nearly 40 minutes to reach depths of 1,000 m—all without taking a breath. How can this turtle dive so deep, for so long, and still have enough oxygen for normal aerobic metabolism in the brain and heart? One of the answers is that this turtle has a very large amount of myoglobin (the oxygen-binding protein) in its muscle cells and can store a large amount of oxygen. Compared to other mammals, this turtle also has more blood per unit of body weight and more red blood cells, and hence more hemoglobin. The hemoglobin has a very high affinity to binding oxygen and as a result, the turtle's oxygen-carrying capacity is the highest recorded for any reptile. Thus, during the dive, oxygen is preferentially delivered to the brain and heart, and these organs remain aerobic and do not produce lactic acid. During the dive, lactic acid also remains sequestered in the skeletal muscles and other vasoconstricted tissues, rather than entering the circulatory system. Blood flow to other organs is also reduced, and these tissues adopt anaerobic metabolic pathways. Add to these physiological adaptations a streamlined body and massive front flippers, and you have a reptile uniquely adapted for deep-sea diving without any deleterious side effects. These adaptations allow the turtle to engage in aerobic respiration without having to come up for air.

---

and are measured as partial pressures of these gases in units of mmHg. (figure 26.21). Collectively, the alveoli provide a large surface area for gas exchange. If the alveolar epithelium of a human was removed from the lungs and put into a single layer of cells side by side, the cells would cover the area of a tennis court.

Ventilation

Breathing (also called pulmonary ventilation) has two phases: (1) inhalation, the intake of air and (2) exhalation, the outflow of air. These air movements result from the rhythmic increases and decreases in thoracic cavity volume. Changes in thoracic volume lead to reversals in the pressure gradients between the lungs and the atmosphere; gases in the respiratory system follow these gradients. The mechanism of inhalation operates in the following way (figure 26.22):

1. Several sets of muscles, the main ones being the diaphragm and intercostal muscles, contract. The intercostal muscles stretch from rib to rib, and when they contract, they pull the ribs closer together, causing the entire rib cage to move upward and outward. This increases the volume of the thoracic cavity.
2. The thoracic cavity further enlarges when the diaphragm contracts and flattens.
3. The increased size of the thoracic cavity causes pressure in the cavity to drop below the atmospheric pressure. Air rushes into the lungs, and the lungs inflate.

During ordinary exhalation, air is expelled from the lungs in the following way:

1. The intercostal muscles and the diaphragm relax, allowing the thoracic cavity to return to its original, smaller size and increasing the pressure in the thoracic cavity.
2. The action in step 1 causes the elastic lungs to recoil and compress the air in the alveoli, and surface-tension forces in the alveoli further the movement of gases into larger air passageways. With this compression, alveolar pressure becomes greater than atmospheric pressure, causing air to be expelled (exhaled) from the lungs.

Gas Transport

The evolution of larger body sizes and higher metabolic rates required more oxygen than could be obtained by simple diffusion

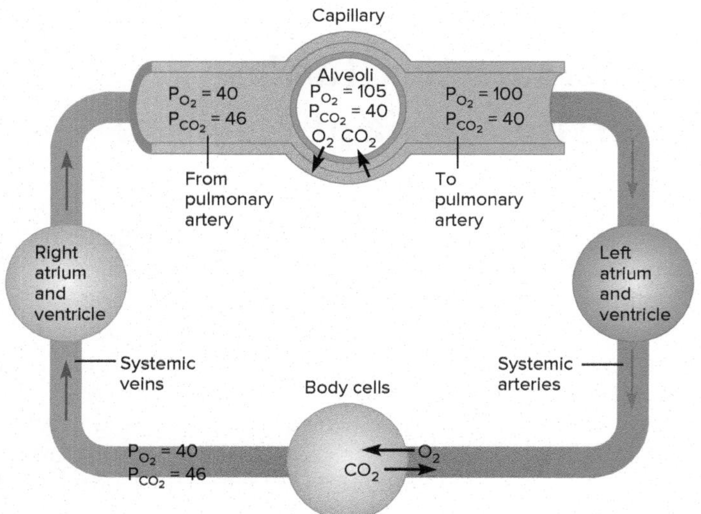

FIGURE 26.21

Gas-Exchange between the Lungs and Tissues. Concentrations of blood and alveolar gases are indicated in partial pressures of the gas (e.g., P_{O_2}) expressed in millimeters of mercury (mmHg). Partial pressures are based on the relative composition of the gas in alveolar air or blood in comparison to total atmospheric pressure (760 mmHg). Comparing the partial pressure differences at the alveoli versus pulmonary blood reveals diffusion gradients that favor the movement of oxygen into blood at the alveoli and diffusion of carbon dioxide from the blood into alveoli. Similar comparisons at body cells explain the diffusion of oxygen from the blood into body cells and the diffusion of carbon dioxide from body cells into the blood.

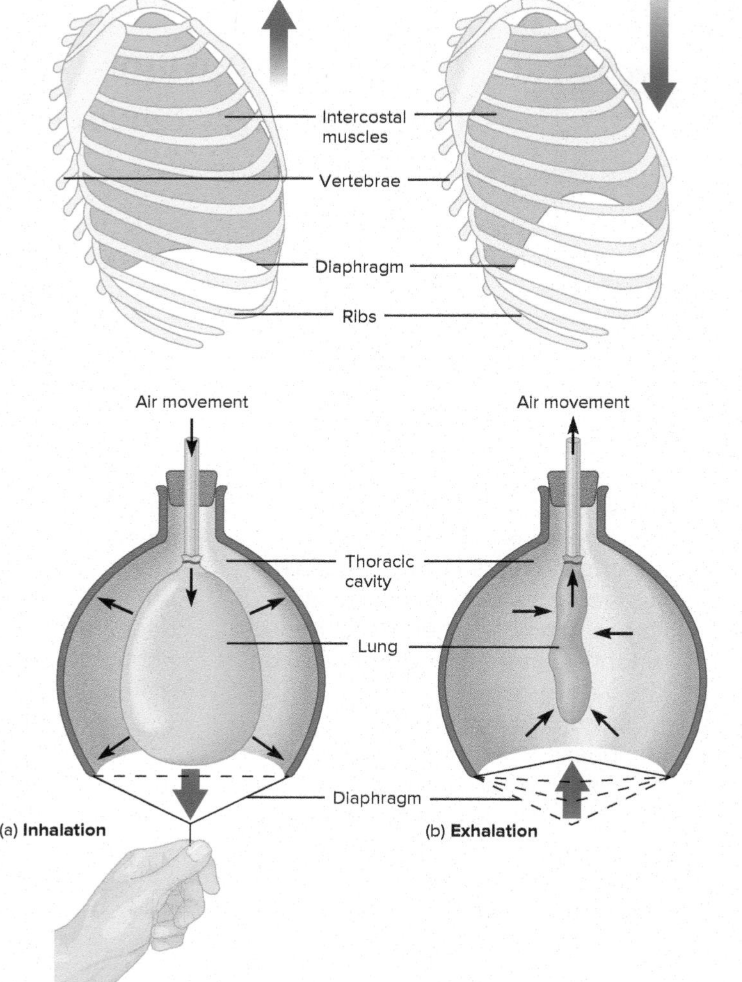

FIGURE 26.22

Ventilation of Human Lungs as an Example of Breathing in Mammals. (*a*) During inhalation, muscle contractions lift the ribs up and out (upper diagram arrows) and lower the diaphragm. These movements increase the size of the thoracic cavity and decrease the pressure around the lungs. This negative pressure causes more air to enter the lungs. (*b*) Exhalation follows the relaxation of the rib cage and diaphragm muscles, as the increased pressure forces the air out of the lungs. Arrows indicate the direction pressure changes take in the thoracic (lower diagrams) cavity during inhalation and exhalation.

between the environment and body cells. Respiratory gases must be carried within an animal's body and released at the locations where the gases are needed. Circulatory systems are the vehicles for carrying these gases. Circulatory systems, like all life, are water-based systems. Respiratory gases (O_2 and CO_2) are relatively insoluble in water. Active homeotherms require about 200 ml of oxygen per liter of blood to support their metabolism. Blood plasma can only carry about 3 ml of oxygen per liter dissolved in the plasma. To make up this difference, animals in phyla from Mollusca through Chordata employ pigments, called respiratory pigments, to aid in the transport of blood gases.

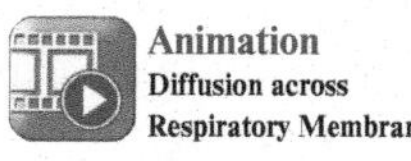

Respiratory pigments are organic compounds that have either metallic copper or iron that binds oxygen. These pigments may be in solution within the blood or body fluids, or they may be in specific blood cells. The most common iron-containing pigment is **hemoglobin.** It is found in protists and many animal phyla where it may occur dissolved in blood or coelomic fluid or in coelomic or blood cells. Other iron-containing pigments include **hemerythrin** (Priapulida, Annelida) and chlorocruorin (a green-colored pigment found in some annelids). **Hemocyanin** is a copper-containing pigment that gives the blood of some molluscs and crustaceans a bluish color.

Respiratory pigments reversibly bind oxygen. In a mammal, for example, blood in alveolar capillaries encounters air with a partial pressure of oxygen (P_{O_2}) of about 100 mmHg. Hemoglobin is saturated with oxygen at this P_{O_2}. As blood circulates into tissue capillaries, it encounters tissue P_{O_2} levels of about 40 mmHg. At this P_{O_2}, the affinity of hemoglobin for oxygen decreases, and hemoglobin releases about 25% of the oxygen it is carrying. This oxygen diffuses from the red blood cells and into the tissues, where it is used to support cellular metabolism. Blood is then transported back to the lungs, where it loads oxygen again. These loading and unloading relationships are depicted in the **oxyhemoglobin dissociation curve** shown in figure 26.23. The hemoglobin molecule is responsive to metabolic conditions at the tissues. The sigmoid shape of the curve demonstrates that more oxygen will be released to the tissues when metabolism increases and the P_{O_2} of tissues drops. Other conditions, for example, a decrease in body pH due to carbon dioxide accumulation or an increase in body temperature, also promote additional oxygen unloading to help recover from these stressful conditions.

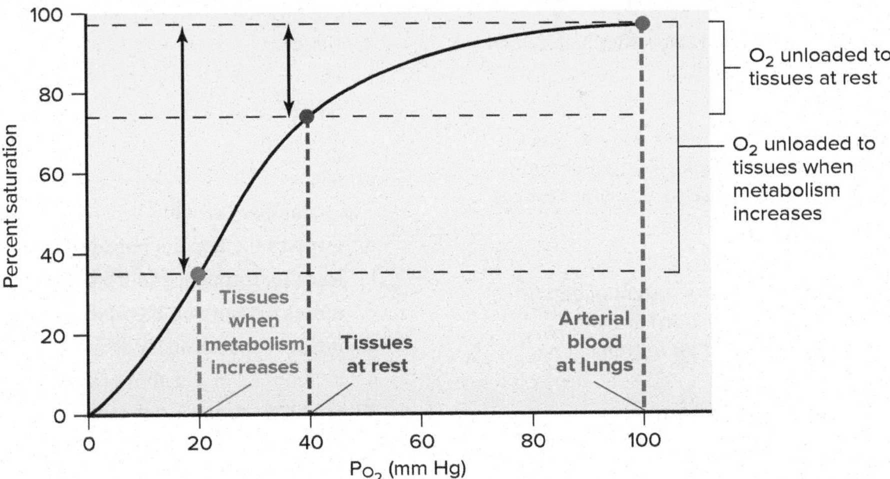

FIGURE 26.23

Oxyhemoglobin Dissociation Curve. Hemoglobin is saturated with oxygen at the lungs. At rest, hemoglobin in tissue capillaries releases about 25% of the oxygen it is carrying. Blood is then transported back to the lungs, where it loads oxygen again. When metabolism increases and the P_{O_2} of tissues drops, more oxygen will be released to the tissues.

Carbon dioxide must also be transported. It is in a higher concentration at the tissues and diffuses into the blood. Some of the carbon dioxide is transported to the lungs by combining with hemoglobin using a different binding mechanism than what is used by oxygen. Most carbon dioxide is transported in the plasma in combination with water in the form of bicarbonate ions. This conversion occurs in blood cells, but the bicarbonate then leaves the blood cells to be carried in plasma. At the lungs, the reactions are reversed and carbon dioxide is released to alveoli.

$$CO_2 + H_2O \rightleftharpoons H_2CO_3 \rightleftharpoons H^+ + HCO_3^-$$

Respiratory pigments have had a complex evolutionary history. The prevalence of hemoglobin in animals can be traced back about 800 million years. Its relationship to an oxygen-storing pigment in muscle, called myoglobin, is well documented. Multiple gene duplications in the vertebrate lineage have resulted in many different forms of hemoglobin that exist today (*see figure 4.14*). Other respiratory pigments have very sporadic distributions among the animal phyla, appearing in very distantly related groups. This distribution pattern suggests that some pigments may have evolved independently in different animal groups.

Section 26.8 Thinking Beyond the Facts

What are the differences in the ways oxygen and carbon dioxide are transported in the blood in a mammal?

Summary

26.1 Transport Systems in Invertebrates

- A circulatory system is comprised of a muscular pumping heart that moves a fluid medium through blood vessels. The two basic types of circulatory systems are open and closed.
- Open systems generally circulate hemolymph, and closed systems circulate blood.

26.2 Transport Systems in Vertebrates

- Blood is a type of connective tissue made up of blood cells (red blood cells and white blood cells), plasma, and platelets.
- Vertebrates use blood vessels and blood for the transport of gases, nutrients, and wastes.

26.3 Vertebrate Circulatory Systems

- The circulatory system of fishes is comprised of a heart consisting of a sinus venosus, atrium, ventricle, and conus arteriosus. Circulation in fishes is a low-pressure system involving gill and systemic capillary beds. Amphibian hearts possess two atria and pulmonary and systemic circuits. Gas exchange occurs across both lung and skin surfaces.
- The hearts of reptiles include partially divided ventricles and systemic and pulmonary arteries. Adaptations are present to keep oxygenated blood separate from less oxygenated blood. The separation of pulmonary and systemic circuits is completed in archosaurs and mammals.

26.4 The Mammalian Circulatory System

- The mammalian heart evolved in the synapsid amniote lineage and includes separate atria and ventricles, single pulmonary and systemic arteries, and a pacemaker derived from the sinus venosus.
- The action of the heart consists of cyclic contraction (systole) and relaxation (diastole). The electrical conduction system is comprised of the SA node, the AV bundle, and Purkinje fibers.
- Systolic contraction generates blood pressure that forces blood through the closed system of vessels.

26.5 **Lymphatic Systems**

- The lymphatic system consists of one-way vessels that help return fluids and proteins to the circulatory system.

26.6 **Gas Exchange**

- Very small invertebrates rely on diffusion across body surfaces for gas exchange. Tracheae, book lungs, book gills, molluscan gills, and lungs are adaptations that increase surface areas for gas exchange.

26.7 **Vertebrate Respiratory Systems**

- The air-conducting portion of the respiratory system of air-breathing vertebrates moves air into (inhalation) and out of (exhalation) this system. This process of air movement is called bimodal breathing.
- Cutaneous respiration is common in amphibians.
- Gills are respiratory organs with a thin, moist epidermis that exchanges gases between water and blood.
- Lungs are internal sac-shaped respiratory organs that exchange gas between air and blood.
- Lung ventilation can occur through positive or negative pressure mechanisms.

26.8 **The Mammalian Respiratory System**

- The mammalian respiratory system is comprised of air-conducting portions and gas-exchange portions.
- Oxygen and carbon dioxide diffuse from areas of higher concentration to areas of lower concentration.
- Once in the blood, oxygen diffuses into red blood cells and binds to hemoglobin for transport to the tissues. Carbon dioxide is transported bound to hemoglobin, as well as in the form of the bicarbonate ion and carbonic acid.
- Respiratory pigments are organic compounds that have either metallic copper or iron that binds oxygen. Examples include hemoglobin, hemocyanin, hemerythrin, and chlorocruorin. Respiratory pigments are adapted to bind with oxygen at a respiratory surface and deliver the oxygen to tissues. Respiratory pigments respond to local metabolic needs of tissues by delivering more oxygen when metabolism increases. Respiratory pigments have a complex evolutionary history.

Concept Review Questions

1. Which one of the following would NOT have an open circulatory system?
   a. Grasshopper
   b. Bivalve
   c. Bird
   d. Earthworm
   e. Both c and d have closed circulatory systems.
2. Which of the following animals has a closed circulatory system?
   a. Insect
   b. Human
   c. Spider
   d. Snail
   e. Clam
3. The pulmonary circuit
   a. involves the hepatic portal vein.
   b. leads to, through, and from the lungs.
   c. moves oxygen-rich blood to the lungs.
   d. includes the coronary arteries.
   e. includes all of the above (a–d).
4. Which of the following statements is FALSE?
   a. Blood pressure is higher in humans compared to fishes.
   b. Blood pressure is lower in the capillary beds than in the major blood vessels.
   c. Blood flows slowly through fishes because it passes through two capillary beds during a single circuit.
   d. Blood circulates faster through a human than through a fish.
   e. Fishes have only one heart pump, but humans have a heart divided into two pumps.
5. One difference between the amphibian and mammalian heart is that
   a. in the amphibian heart, oxygenated blood and blood low in oxygen mix completely in a single ventricle.
   b. in the amphibian heart, there are two SA nodes so that contractions occur simultaneously throughout the heart.
   c. in the amphibian ventricle, internal channels reduce mixing of oxygenated and less oxygenated blood.
   d. in the amphibian heart, the left atrium receives blood oxygenated by diffusion through the skin.

Analysis and Application Questions

1. Many invertebrates utilize the body cavity as a circulatory system. However, in humans, the body cavity plays no role whatsoever in circulation. Why?
2. Describe the homeostatic functions of the vertebrate circulatory system. What functions are maintained at relative stability?
3. The area of an animal's respiratory surface is usually directly related to the animal's body weight. What does this tell you about the mechanism of gas exchange?
4. How can seals, whales, and sea turtles stay under water for long periods?
5. With respect to respiration, why were arthropods able to invade terrestrial environments?

A broad-billed humming bird (*Cynanthus latirostris*) at a flower feeding. Unlike most birds, hummingbirds have retained vertebrate receptors that allow them to taste sweet nectars. Unlike vertebrates that do not feed on sucrose, they also produce the enzyme sucrase, which is needed for digesting the sucrose found in flower nectars. Feeding adaptations like these abound in the animal kingdom.

27

Nutrition and Digestion

Chapter Outline

27.1 Evolution of Heterotrophy
27.2 The Metabolic Fates of Nutrients in Heterotrophs
Macronutrients
Micronutrients
27.3 Digestion
27.4 Animal Strategies for Getting and Using Food
Continuous versus Discontinuous Feeders
Suspension Feeders
Deposit Feeders
Herbivory
Predation
Surface Nutrient Absorption
Fluid Feeders
27.5 Diversity in Digestive Structures: Invertebrates
27.6 Diversity in Digestive Structures: Vertebrates
The Oral Cavity
The Alimentary Tract
Digestive Glandular Systems
27.7 The Mammalian Digestive System
Gastrointestinal Motility and Its Control
Oral Cavity
Pharynx and Esophagus
Stomach
Small Intestine: Main Site of Digestion
Large Intestine
Role of the Pancreas in Digestion
Role of the Liver and Gallbladder in Digestion

Animals feed on other organisms (living, dying, dead, decomposing, or tissue fragments) or on products produced by other organisms (nectar, honey, nuts, and berries). There are about as many animal diets and feeding habits as there are species of animals, but the goals are always the same. Animals strive to obtain nutrients to meet their own daily maintenance requirements, to grow to reproductive maturity, and to reproduce (and in some cases nurture offspring) so their species can continue for another generation. The common themes uniting animal feeding strategies are the subjects of this chapter.

27.1 EVOLUTION OF HETEROTROPHY

LEARNING OUTCOMES

1. Categorize the different types of nutrition found in animals.
2. Explain how the loss of biosynthetic abilities can provide a selective advantage in animal evolution.

Nutrition (L. *nutrio*, nourishing) includes all of those processes by which an animal takes in, digests, absorbs, stores, and uses food (nutrients). **Digestion** (L. *digestio*, from *dis*, apart + *gerere*, to carry) is the mechanical breakdown of food particles followed by the chemical hydrolysis of large organic molecules into a form that an animal can absorb into its cells.

Animals are **heterotrophs** (Gr. *heteros*, another + *trophe*, feeding). They cannot synthesize all of the organic molecules they require to sustain life processes, so they must consume these molecules. Animals may feed on **autotrophs** (Gr. *auto*, self) that carry out photosynthesis. These animals are called **herbivores** (L. *herba*, plant + *vorare*, to eat). Others feed on other animals and are called **carnivores** (L. *carno*, flesh). Still others combine these two feeding strategies and are called **omnivores** (L. *omnius*, all). Within each of these feeding strategies there are many feeding specializations that have been described in chapters 9 through 22.

Heterotrophic nutritional methods are very old. In fact, the first life-forms (3.8 billion years ago [bya]) were heterotrophic bacteria that survived by consuming organic matter that formed spontaneously in primordial seas (*see chapter 8*). Cyanobacteria were the first autotrophs, and their success eventually allowed heterotrophic organisms to flourish. Animals arose 800 million years ago (mya). The selective advantages of heterotrophy are reflected in the success of animals. Obtaining essential, complex organic molecules in the diet, rather than synthesizing them, means less energy and fewer resources are expended in supporting an animal's existence. The energy and resources saved can be devoted to the most important productive function–reproduction.

SECTION 27.1 THINKING BEYOND THE FACTS

What evolutionary "trade-offs" accompany autotrophy versus heterotrophy?

27.2 THE METABOLIC FATES OF NUTRIENTS IN HETEROTROPHS

LEARNING OUTCOME

1. Justify the statement that a "mammal must have both micro- and macronutrients as well as vitamins in its diet."

The nutrients that a heterotroph ingests can be divided into macronutrients and micronutrients. **Macronutrients** are needed in large quantities and include the carbohydrates, lipids, and proteins. **Micronutrients** are needed in small quantities and include organic vitamins and inorganic minerals. Along with water, these comprise an animal's dietary requirements.

One of the most important functions of an animal's diet is to provide energy to drive life's reactions. The metabolic pathways involved in processing organic molecules to derive energy in the form of ATP and, in some cases, converting one organic molecule to another are described in chapter 2, and frequent references to that chapter will be made in the discussions that follow.

Macronutrients

All macronutrients can be used in the production of ATP in the steps of aerobic cellular respiration (*see figures 2.15 and 2.16*). In addition, these molecules provide essential building blocks for the synthesis of cells and tissues and the regulation of chemical reactions. Metabolic pathways are also capable of interconverting macronutrients. Thus, one can accumulate lipids in fat deposits with high-carbohydrate diets.

Carbohydrates are a common energy source for heterotrophs. Simple sugars, like glucose, can be metabolized in the steps of glycolysis and the citric acid cycle (*see figures 2.13, 2.15, and 2.16*) to provide ATP to support the energy needs of animals. Starches and more complex sugars are broken down to provide these simple sugars. Cellulose can be used by many herbivores, and it provides fiber that helps maintain healthy digestive systems of many animals.

Lipids include sterols, triacylglycerols, and fatty acids. Most lipids, in the form of triacylglycerols, are energy storage molecules. Their fatty acids and glycerol can be broken down and used at various locations in the citric acid cycle (*see figure 2.15*) to derive ATP. Deposits of these lipids help insulate the bodies of some vertebrates. Some fatty acids are precursor molecules for hormones and form structural components of cellular membranes. Some fatty acids can be synthesized by animals, but some cannot and are required in the diet. These fatty acids are called essential fatty acids. Sterols are required for the synthesis of steroid hormones and are incorporated into cellular membranes.

Proteins are the building blocks of animals and contribute to all animal tissues. They are comprised of amino acids, which are assembled into very large structural proteins, enzymes, and hormones (*see chapters 3 and 25*). Proteins can also be metabolized in cellular respiration to provide ATP.

Micronutrients

Micronutrients are ions (minerals) or organic molecules (vitamins) that are used in small amounts in enzymatic reactions or as parts of certain proteins (e.g., copper in hemocyanin and iron in hemoglobin). Animals cannot synthesize them rapidly (if at all); thus, they must be obtained from the diet.

Minerals are ions that are essential to animal functions. You have encountered many examples of these minerals in previous chapters. Sodium and potassium ions are vital in action potentials. Calcium is critical in bone formation and muscle contraction. Iron and copper are incorporated into the respiratory pigments hemoglobin and hemocyanin. Phosphorus is found in DNA and RNA. Other minerals, like zinc, cobalt, iodine, and manganese, are cofactors that are essential for the function of enzymes and hormones. They are required in very small amounts, but they are still essential components of an animal's diet (table 27.1).

Vitamins are organic molecules that are required in small amounts and participate in various metabolic reactions (table 27.2). For example, niacin (nicotinic acid or vitamin B_3) is a component of the NADH described in chapter 2. It is vital in the processing

TABLE 27.1
PHYSIOLOGICAL ROLES OF SELECTED ANIMAL MINERALS

MINERAL	MAJOR PHYSIOLOGICAL ROLES
Calcium (Ca)	Component of bone and teeth; blood clotting; muscle, neuron, and cellular functions
Phosphorus (P)	Major constituent of bones, blood plasma; DNA, RNA, ATP, energy metabolism
Potassium (K)	Major positive ion in cells; muscle contraction and neuron excitability
Sodium (Na)	Fluid important in fluid balance; conduction of action potentials
Sulfur (S)	Protein structure; detoxification reactions
Copper (Cu)	Enzyme cofactor; melanin and hemoglobin synthesis; component of hemocyanin and cytochromes
Fluorine (F)	Bone and tooth structure
Iodine (I)	Thyroid hormones
Iron (Fe)	Component of hemoglobin, myoglobin, and cytochromes

TABLE 27.2

SELECTED ANIMAL VITAMINS AND THEIR FUNCTIONS (SOME VITAMINS LISTED ARE NOT SINGLE COMPOUNDS BUT GROUPS OF RELATED COMPOUNDS.)

VITAMIN	FUNCTIONS	SOURCES
Water-Soluble Vitamins		
Vitamin B_1 (thiamin)	Oxidation of carbohydrates and synthesis of ribose	Lean meats, liver, eggs, whole-grain cereals, leafy green vegetables, legumes
Vitamin B_2 (riboflavin)	Oxidation of glucose and fatty acids	Meats, dairy products, leafy green vegetables, whole-grain cereals
Vitamin B_3 (niacin [nicotinic acid])	Oxidation of glucose and synthesis of proteins, fats, and nucleic acids	Liver, lean meats, poultry, peanuts, legumes
Vitamin B_7 (biotin)	Metabolism of amino acids and fatty acids; DNA synthesis	Liver, egg yolk, nuts, legumes, mushrooms
Vitamin B_9 (folate [folic acid])	Metabolism of amino acids, DNA synthesis, and red blood cell production	Liver, leafy green vegetables, whole-grain cereals, legumes
Vitamin C (ascorbic acid)	Collagen synthesis, metabolism of amino acids, absorption of iron, synthesis of steroid hormones	Citrus fruits, citrus juices, tomatoes, cabbage, potatoes, leafy green vegetables, fresh fruits
Fat-Soluble Vitamins		
Vitamin A (retinol)	Synthesis of visual pigments, bone and tooth development, maintenance of epithelial tissues	Liver, fish, dairy products, leafy green vegetables, and yellow and orange vegetables and fruits
Vitamin D	Absorption of calcium and phosphorus, tooth and bone development	Produced in skin exposed to ultraviolet light; milk, egg yolk, fish-liver oils
Vitamin E (tocopherol)	Antioxidant (disables free radicals), smooth muscle growth	Oils from cereal seeds, fruits, nuts, and vegetables
Vitamin K (phylloquinone)	Synthesis of blood-clotting proteins	Leafy green vegetables, egg yolk, pork liver, soy oil, tomatoes, cauliflower

of macronutrients to derive ATP during the steps of aerobic cellular respiration. Some vitamins are water soluble and often function with enzymes in regulating metabolic reactions. These are called coenzymes. Others are fat soluble and are incorporated into the structure of other molecules. For example, retinol (vitamin A) is incorporated into the structure of visual pigments. Other fat-soluble vitamins are involved in promoting the formation of proteins. For example, phylloquinone is an intermediary in the production of prothrombin and other blood-clotting proteins.

SECTION 27.2 THINKING BEYOND THE FACTS

Before migration begins, the ruby-throated hummingbird* (Archilochus colubris) *undertakes a fall feeding frenzy in order to accumulate fat reserves for its trip from North America to Mexico and Central America. How can feeding on high carbohydrate flower nectar promote accumulation of fat?

27.3 DIGESTION

LEARNING OUTCOME

1. Compare and contrast extracellular and intracellular digestion.

Sponges take in whole food particles trapped by choanocyte collars. Food particles are engulfed by endocytosis and broken down with enzymes to obtain nutrients. This is called **intracellular** ("within the cell") **digestion** (figure 27.1*a*; *see also figure 9.3*b). Intracellular digestion circumvents the need for the mechanical breakdown of food, and some animals that rely on intracellular digestion lack gut cavities (e.g., Porifera and some Platyhelminthes). Many animals use intracellular digestion to supply some of their nutritional requirements (e.g., Cnidaria, Platyhelminthes, Rotifera, bivalve molluscs, and invertebrate chordates). Few animals rely on it for all of their nutritional requirements (e.g., Porifera). Total reliance on intracellular digestion limits an animal's size and complexity.

Most animals have adaptations for **extracellular digestion:** the enzymatic breakdown of larger pieces of food into constituent molecules, usually in a special organ or cavity (figure 27.1*b*). Nutrients from the food then pass into body cells lining the organ or cavity and can take part in energy metabolism or biosynthesis.

SECTION 27.3 THINKING BEYOND THE FACTS

Why do you think that very few animals rely exclusively on intracellular digestion to supply nutrients?

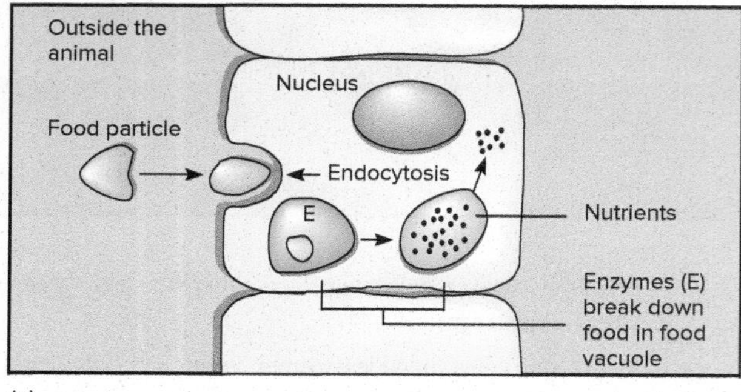

(a)

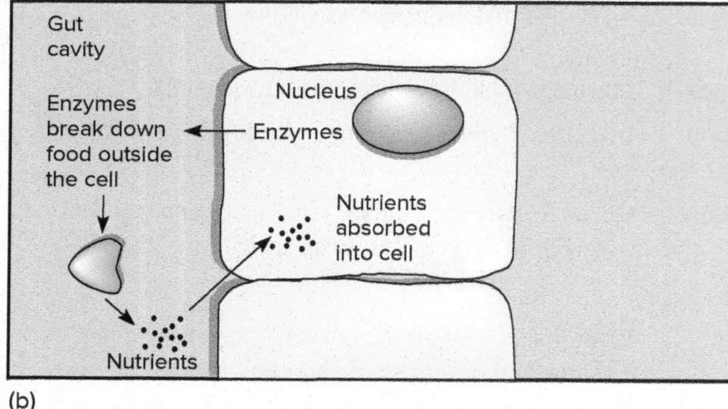

(b)

FIGURE 27.1

Intracellular and Extracellular Digestion. (*a*) A sponge has no gut and thus carries out intracellular digestion. Tiny food particles are taken into the body wall cells by endocytosis. Digestive enzymes in the vacuole then break the small particles into constituent molecules. (*b*) Most animals have a gut and take in and digest (extracellularly) relatively large food particles. Cells lining the gut cavity secrete enzymes into the cavity. There, the enzymes break down food materials into constituent nutrients, and the nearby cells absorb these nutrients.

27.4 ANIMAL STRATEGIES FOR GETTING AND USING FOOD

LEARNING OUTCOMES

1. Contrast the following pairs of feeding strategies: continuous vs. discontinuous feeding, suspension vs. deposit feeding, and herbivory vs. predation.
2. Compare methods of food capture by predators.
3. Hypothesize on why some animals are adapted for feeding exclusively on fluid.

The methods employed by animals for obtaining food are almost as numerous and varied as species are numerous and varied. The adaptations often have little phylogenetic significance. For example within the molluscs, we can see adaptations for filter feeding on suspended particulate matter by bivalves (Bivalvia) and hunting young sperm whales by giant squid (Cephalopoda, *Architeuthis*). One can observe the following feeding strategies across the animal kingdom.

Continuous versus Discontinuous Feeders

Animals may feed continuously or at irregular intervals. Many **continuous feeders** are slow-moving or completely sessile animals. For example, aquatic **suspension feeders,** such as tube worms (Annelida) and bivalves (Mollusca), remain in one place and continuously filter small food particles from the water.

Discontinuous feeders tend to be active, sometimes highly mobile, animals. Typically, discontinuous feeders have more digestive specializations than continuous feeders because discontinuous feeders take in large meals that must be temporarily stored or immediately digested. Many carnivores, for example, pursue and capture relatively large prey. Their digestive systems permit storage and digestion of infrequent, large meals. Other time is devoted to social interactions, reproduction, migration, or other activities.

Herbivores spend more time eating than carnivores do, but they are also discontinuous feeders. They may migrate as seasons change or when food is exhausted and must limit their grazing time to avoid excessive exposure to predators. Thus, their digestive systems are adapted for storing food that is eaten rapidly, followed by digestion in sheltered habitats.

Suspension Feeders

Suspension feeding is the removal of suspended food particles from the surrounding water by a filtering structure. Suspension feeders circulate water past a filtering structure, filter particles from the water, and move the nutrients to the mouth. Suspension feeding is found in phyla across the animal kingdom from Porifera through Chordata (*see figures 9.3, 9.20, 10.27, 11.10, 12.12, 15.10, 16.14, and 17.6*).

Deposit Feeders

Deposit feeding animals obtain their nutrients from the sediments of soft-bottom habitats (muds and sands) or terrestrial soils. Deposit feeders may swallow large quantities of sediment (mud, soil, sand, and organic matter). The usable nutrients are digested, and the indigestible waste passes out the anus. Some gastropod molluscs and many annelids deposit feed in this fashion (*see figure 12.17*). Other deposit feeders use tentaclelike structures to ingest sediment (sea cucumbers, Echinodermata; Sipuncula; and some annelids).

Herbivory

Herbivory (L. *herba,* herb + *vorare,* to eat) is the consumption of macroscopic plants. This common feeding strategy requires the ability to process large pieces of plant matter. Biting and chewing mechanisms are often characterized by the development of hard surfaces (e.g., teeth) and powerful jaw muscles.

Herbivory is commonplace in nearly all animal phyla. Molluscs that are herbivorous use their radula to scrape algae off rocks or tear leaves off terrestrial plants. Herbivory is very common in the arthropods, and their mouthparts are often adapted for cutting

How Do We Know That Pythons Do Not Starve between Meals?

Pythons (Squamata, Pythonidae) utilize bulk feeding because they cannot chew their food into small pieces and must therefore swallow it whole. One of the most extraordinary scientific findings about the digestive system involves pythons. These snakes obtain their food by simply waiting for a prey animal to pass by. With this strategy, many days or weeks may pass between meals. During this interval, why don't the pythons starve?

One feeding adaptation of pythons is that they are equipped to ingest very large animals (box figure 27.1 shows a rock python beginning to ingest a gazelle *(Gazella)* it has captured and killed). An adult snake can weigh more than a human and can eat animals that weigh 70% as much as it does. Thus, reports of sizable artiodactylids (goats, gazelles, and antelopes) being eaten are true.

Unlike mammals that eat on a daily basis and maintain their digestive systems in a state of readiness between meals, pythons immobilize their digestive systems between meals. For example, Burmese pythons (*Python molurus*) undergo extensive immobilization of their digestive systems if they go without food for a month or more. Then, when they obtain food, their metabolic rate increases dramatically in order to begin reconstructing their digestive system. In the first 24 hours after feeding, these pythons double the mass of their intestines by the growth of new gut epithelium (critical to digestion and absorption). At the same time, they produce large amounts of intestinal transport proteins (e.g., the total numbers of glucose transporters may increase by more than 20 times).

BOX FIGURE 27.1 **Python Feeding.** A rock python (*Python sebae*) beginning to ingest its prey. After swallowing its prey, which may take several hours, the python will spend two or more weeks digesting its meal.

and chewing plant material (*see figures 15.5 and 15.13*). The Aristotle's lantern (*see figure 16.11*) of sea urchins (Echinoidea) is used for cutting kelp and other macroalgae. All classes of vertebrates have herbivores with tooth and digestive specializations for feeding on, and digesting, plant matter. Digestion of cellulose found in plants also requires adaptations that will be described in subsequent sections.

Predation

Predation (L. *praedator,* a plunderer, pillager) is an interaction in which one animal (the predator) captures, kills, and consumes another animal (the prey). Predation can be an energetically costly method of feeding. It often takes large amounts of energy to hunt and kill another animal. To be a successful predator, an animal must derive more energy from the hunt than is expended during the hunt. Some predators are generalists, feeding on a wide variety of prey items. Crocodylians, for example, are very opportunistic in their prey choices. Other predators are specialized for feeding on specific prey, which may lead to a coevolutionary "arms race" between predator and prey. For example, echolocation allows bats (order Chiroptera) to find and capture their moth prey in flight. However, some moths (family Noctuidae) can detect the ultrasonic emissions from bats, initiate evasive maneuvers, and even return their own ultrasonic clicks to confuse a bat's echolocating mechanisms.

Predators may stalk their prey, sit and wait for prey (lurking and opportunistic contact), or graze around their environment picking up smaller (often sessile) prey items. These predatory strategies are widely distributed throughout the animal kingdom, and many examples are described in chapters 9 through 22.

Surface Nutrient Absorption

Some animals absorb nutrients from the external medium across their body surfaces. This medium may be nutrient-rich seawater, fluid in other animals' digestive tracts, or the body fluids of other animals. For example, cestode worms, endoparasitic gastropods, and crustaceans (all of which lack mouths and digestive systems) absorb all of their nutrients across their body surface.

A few nonparasitic animals also lack a mouth and digestive system and absorb nutrients across their body surface. Examples include the gutless bivalves and siboglinid worms. Interestingly, many siboglinid worms absorb some nutrients from seawater across their body surface and also supplement their nutrition with organic carbon that symbiotic bacteria fix within the siboglinid's tissues.

Fluid Feeders

The biological fluids of animals and plants are a rich source of nutrients. **Fluid feeding** occurs in some parasites, such as the intestinal nematodes that attach to intestinal epithelia and feed on host blood. External parasites (ectoparasites), such as leeches, ticks, mites, lampreys, and certain crustaceans, use a wide variety of mouthparts to feed on body fluids. For example, the sea lamprey has a funnel structure surrounding its mouth (*see figure 27.3*a). The funnel is lined with over 200 rasping teeth and, it has a rasplike tongue. The lamprey uses the funnel like a suction cup to grip its fish host, and then with its tongue, rasps a wound in the fish's body wall. The lamprey then feeds on body fluids from the wound.

Many insects are fluid feeders and have mouthparts modified for sucking plant fluids (e.g., Lepidoptera, *see figure 15.14*). Others have mouthparts adapted for piercing and sucking plant or animal fluids (e.g., many hemipterans, like aphids, and dipterans, like female mosquitoes).

Most pollen- and nectar-feeding birds have long bills and tongues. Bills are often specialized (in shape, length, and curvature) for particular types of flowers (*see figure 21.8 and chapter-opener figure, page 501*). In other cases, bills are short and are used to pierce the base of a flower. The tongue is then used to obtain nectar through the opening. The tongues of some birds have a brushlike tip or are hollow, or both, to collect the nectar from flowers.

Relatively few mammals feed exclusively on fluids. The only mammals that feed exclusively on blood are the vampire bats (e.g., *Desmodus rotundus*) of tropical South and Central America. These bats prey upon birds, cattle, and horses, using teeth to pierce the surface blood vessels, and then lap at the oozing wound. Nectar-feeding bats (e.g., *Choeronycteris mexicana*) have a long tongue to extract the nectar from flowering plants, and compared to the blood-feeding bats, have reduced dentition. The nectar-feeding honey possum (*Tarsipes rostratus*) is an Australian marsupial that is an important pollinator of the plants that supply its food. It uses grasping feet and a prehensile tail to hold to branches as it feeds with its long, brush-tipped tongue and reduced dentition.

Section 27.4 Thinking Beyond the Facts

Many birds, like hummingbirds, that we think of as nectar feeding, derive some of their nutrients from other food sources, like insects. Why is nectar feeding often supplemented with other food sources?

27.5 DIVERSITY IN DIGESTIVE STRUCTURES: INVERTEBRATES

LEARNING OUTCOMES

1. Contrast feeding and digestion as it occurs in sponges with the pattern found in virtually all other animals.
2. Contrast incomplete and complete digestive tracts.
3. Describe the digestive processes occurring in chewing, herbivorous insects.

Digestive structures of invertebrates are diverse. It is not possible to review all of the variations already discussed in chapters 9 through 22. A few of the basic digestive strategies are described in the following paragraphs.

In the sponges (Porifera), digestion is completely intracellular. All sponges are sedentary filter feeders. Water currents are established by flagellated choanocytes (*see figure 9.3*) that line the inner chambers of the sponge. The choanocytes' collars trap microscopic food particles. Trapped food is moved along collar microvilli to the base of the collar where it is incorporated into an endocytic (food) vacuole. Intracellular digestion begins in the food vacuole by lysosomal enzymes and pH changes (*see figure 27.1*). Partially digested food passes into ameboid cells (archeocytes), which distribute the food to other cells. Placozoans (*see table 9.4*) also lack a digestive tract. They secrete digestive enzymes onto organic detritus and absorb nutrients through their ventral epithelium.

Virtually all other animals have some type of digestive tract. In cnidarians (Cnidaria), the gut is a closed sac called a gastrovascular cavity (figure 27.2*a*). In flatworms (phylum Platyhelminthes) the digestive tract is often highly branched, but has a single opening for ingestion of food and the elimination of indigestible wastes (figure 27.2*b*). A similar arrangement is present in the acoelomorphs (*see table 9.4*). This type of digestive tract that has a single opening (the mouth) for both ingesting food and the elimination of wastes is an **incomplete digestive tract**. Some specialized cells in the digestive cavity secrete digestive enzymes that begin the process of extracellular digestion. Other phagocytic cells that line the cavity engulf food material and continue intracellular digestion inside food vacuoles (*see figure 9.7*).

The evolution of a second opening (the anus) from the digestive tract to the outside of the animal permitted ingestion, digestion, and elimination of wastes to occur simultaneously or in a sequence. A digestive system with both mouth and anus is called a **complete**

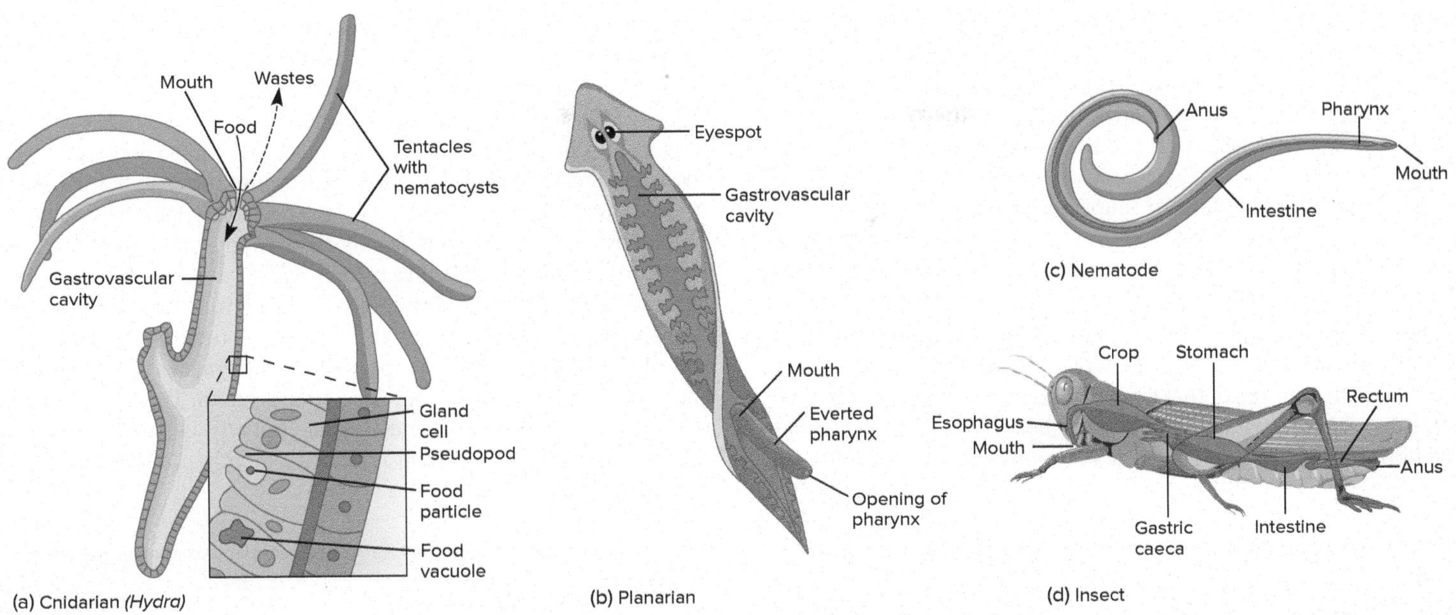

FIGURE 27.2

Digestive Structures in Invertebrates. (*a*) The gastrovascular cavity of a cnidarian (*Hydra*) is an incomplete digestive tract because its one opening, a mouth, must serve as the entry and exit point for food and waste. Extracellular digestion occurs in the gastrovascular cavity, and intracellular digestion occurs inside food vacuoles formed when phagocytic cells engulf food particles. (*b*) Even though the gastrovascular cavity in a platyhelminth (*Planaria*) branches extensively, it is also an incomplete digestive tract with only one opening. When a planarian feeds, it protrudes its muscular pharynx from its mouth and sucks in food. (*c*) A nematode (*Ascaris*) has a complete digestive tract with a mouth, pharynx, intestine, and anus. (*d*) The complete digestive tract of an insect (*Melanoplus*) has an expanded region called a crop that functions as a food storage organ.

digestive tract (**alimentary tract**). A complete digestive tract permits the one-way flow of ingested food without mixing it with previously ingested food or waste (figure 27.2*c*). Complete digestive tracts also have the advantage of progressive digestive processing in specialized regions, resulting in a digestive system. Food can be digested efficiently in a series of steps by specialized digestive structures. Virtually all bilaterians other than the flatworms and acoelomorphs have complete digestive tracts. Many variations exist that correlate with different food-gathering mechanisms and diets. Even within a single phylum, like Mollusca, variations range from digestive system specializations in the sessile filter-feeding bivalves to those present in predatory cephalopods. These accounts will not be revisited here. Instead, one example follows that will provide useful comparisons to the description of a vertebrate's complete digestive system in the next section.

As with other animal phyla, there is tremendous diversity of digestive structures within the insects (*see figures 15.13, 15.14, and 15.15*). Grasshoppers (Orthoptera) are often studied in zoology laboratories and are insects with a complete digestive tract and extracellular digestion (figure 27.2*d*). During feeding, the mandibles and maxillae first break up (masticate) the food, which is then taken into the mouth and passed to the crop via the esophagus. During mastication, the salivary glands add saliva to the food to lubricate it for passage through the digestive tract. Saliva also contains the enzyme amylase, which begins the enzymatic digestion of carbohydrates. This digestion continues during food storage in the crop. The midgut secretes other enzymes (carbohydrases, lipases, and proteases) that enter the crop. Food passes slowly from the crop to the stomach, where it is mechanically reduced and the nutrient particles sorted. Large particles are returned to the crop for further processing; the small particles enter the gastric cecae, where extracellular digestion is completed. Most nutrient absorption then occurs in the intestine. Wastes are moved along the intestine and into the rectum, where water and ions are absorbed. The solid fecal pellets that form then pass out of the animal via the anus. During this entire feeding process the nervous system, the endocrine system, and the presence of food exert considerable control over enzyme production and gut tract motility.

SECTION 27.5 THINKING BEYOND THE FACTS

Why are regionally specialized digestive tracts primarily found in animals with complete digestive tracts?

27.6 DIVERSITY IN DIGESTIVE STRUCTURES: VERTEBRATES

LEARNING OUTCOMES

1. Explain how the structure and function of tongues, teeth, and salivary glands reflect feeding habits of vertebrates.
2. Explain the functions of the component parts of the vertebrate digestive tract.
3. Hypothesize why the ruminant lifestyle evolved.

The vertebrate digestive tract is a complete digestive tract and often highly specialized for the digestion of a wide variety of foods. The oral cavity, teeth, intestines, and other major digestive structures usually reflect the way an animal gathers food, the type of food it eats, and the way it digests that food. There is an amazing diversity in the form and function of vertebrate digestive structures. Some of this diversity is discussed in the following paragraphs.

The Oral Cavity

A tongue or tonguelike structure develops in the floor of the oral cavity in many vertebrates. For example, a lamprey has a protrusible tongue with horny teeth that rasp its prey's flesh (figure 27.3*a*). Fishes may have a primary tongue that bears teeth that help hold prey; however, this type of tongue is not muscular (figure 27.3*b*).

Tetrapods have evolved mobile tongues for gathering food. Frogs and salamanders and some lizards can rapidly project part of the tongue from the mouth to capture an insect (figure 27.3*c; see also figure 20.12*). For example, a frog (Anura) uses its whiplike tongue to snag its prey faster than a human can blink, hitting the prey with force five times greater than gravity. But the frog also needs to hold onto its prey as the food rockets back into its mouth. Prey capture is accomplished by unique saliva that quickly changes state and a very soft and elastic tongue. The frog's saliva is thick and sticky as the tongue flicks out and back in during prey capture. The saliva then quickly turns into a thin and watery fluid as the food enters the mouth. The frog's soft and elastic tongue is able to stretch and store energy much like a spring that snaps captured food back to the mouth. This combination of salivary transformation and tongue elasticity is so effective that prey is secured as it rockets into the frog's mouth.

(a)

(c)

(b)

(d)

FIGURE 27.3

Tongues. (*a*) Rasping tongue and mouth of a lamprey (*Petromyzon marinus*). (*b*) Fish (Serranidae) tongue. (*c*) A panther chameleon (*Furcifer pardalis*) using its tongue to capture prey. (*d*) A piliated woodpecker (*Dryocopus pileatus*) uses its tongue to extract insects from a tree.
(a)Source: USEPA/Great Lakes National Program office (b)©DebbiSmirnoff/Getty Images (c)©Svoboda Pavel/Shutterstock (d)©MARSHFIELD/Getty Images

The tongues of other tetrapods have also become specialized for their specific feeding styles. A woodpecker has a long, spiny tongue for gathering insects and grubs (figure 27.3*d*). Ant- and termite-eating mammals also gather food with long, sticky tongues. Spiny papillae on the tongues of cats and other carnivores help these animals rasp flesh from a bone.

With the exception of avian reptiles, turtles, and baleen whales, the oral cavities of most vertebrates have teeth (figure 27.4). Avian reptiles lack teeth, probably to reduce body weight for flight. Tooth structure and arrangement reflect an animal's diet (herbivore, carnivore, or omnivore) and feeding strategy (e.g., grazing predator versus stalker). The teeth of snakes slope backward to aid in the retention of prey while swallowing (*see figure 20.13*), and the canine teeth of wolves are specialized for ripping food. Herbivores, such as deer, have predominantly grinding teeth, the front teeth of a beaver are used for chiseling trees and branches, and the elephant has its two upper incisors specialized as tusks for defense, social displays, and digging. Because humans, pigs (Suidae), bears (Ursidae), raccoons (Procyonidae), and a few other mammals are omnivores, they have teeth that can perform a number of tasks—tearing, ripping, chiseling, and grinding.

The Alimentary Tract

The alimentary tract runs from the mouth to the anus. Adaptations reflect feeding strategies. Alimentary tracts of carnivores tend to be shorter and less complex than those of herbivores, because cellulose diets often require specializations where bacterial and protist fermentation of plant fibers can occur.

The oral cavity leads to the **esophagus**. The esophagus is a muscular (skeletal and smooth muscle) food-conducting tube that leads to the stomach. Grain- and seed-eating birds have a crop that develops from the caudal portion of the esophagus (figure 27.5*a*). Storing food in the crop ensures an almost continuous supply of food to the stomach and intestine for digestion. This structure allows these birds to reduce the frequency of feeding and still maintain a high metabolic rate.

The stomachs of vertebrates store, and begin the digestion of, large masses of food that are ingested at less frequent intervals. The stomachs of carnivores are often relatively small and simple. Most vertebrates have gastric glands that secrete hydrochloric acid (HCl). HCl facilitates the function of the primary stomach enzyme, pepsin, which is involved with the digestion of proteins. Stomachs of carnivores, and especially carrion feeders, are much more acidic than those of herbivores (pH 1 vs. 4). This difference suggests that the primary (and perhaps original) function of stomach acid is to serve as a bacterial filter that protects the animal from pathogens in decaying animal flesh, and it probably protects the integrity of the gut microbiome of all vertebrates (*see* Evolutionary Insights: *Gut Microbiomes—The Hidden World Within*).

Herbivores often have complex chambered stomachs and intestines. Ruminant (L. *ruminare*, to chew the cud) mammals—animals that "chew their cud," such as cows, sheep, elk, bison, water buffalo, goats, giraffes, caribou, and deer (all order Artiodactyla)—have multichambered stomachs. Ruminant stomachs provide fermentation chambers for large numbers of microorganisms to digest the cellulose walls of grass and other vegetation. Cellulose contains a large amount of energy; however, animals generally lack the ability to produce the enzyme cellulase for digesting cellulose and obtaining its energy. The diverse microbial communities in these stomachs break down cellulose through fermentation processes, and some of the microorganisms produce cellulase.

In ruminants, the upper portion of the stomach is comprised of a large storage and fermentation chamber called the **rumen** and a smaller fermentation chamber called the reticulum (figure 27.6). Fermentation by bacteria and protists begin to break down complex carbohydrates, principally into short-chain fatty acids, which can be absorbed into the bloodstream. Fluid secretion, body heat, and churning movements reduce the food to a pulpy mass. This partially digested food can be regurgitated, chewed as "cud," and swallowed a second time. The contents of the "reticulorumen" then moves to the omasum (absorption of minerals) and abomasum. The latter is often called the "true stomach" as gastric enzymes are first secreted here.

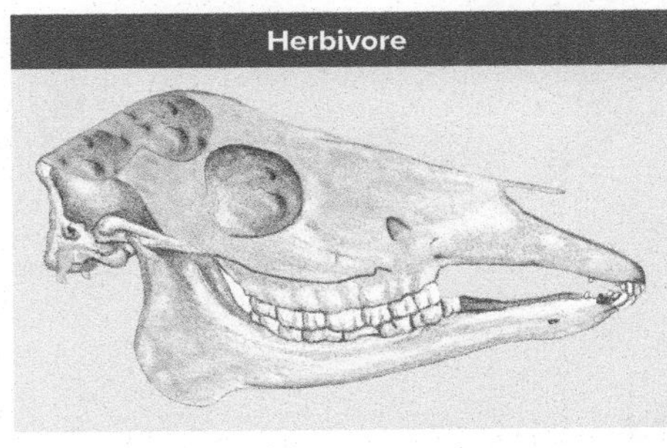

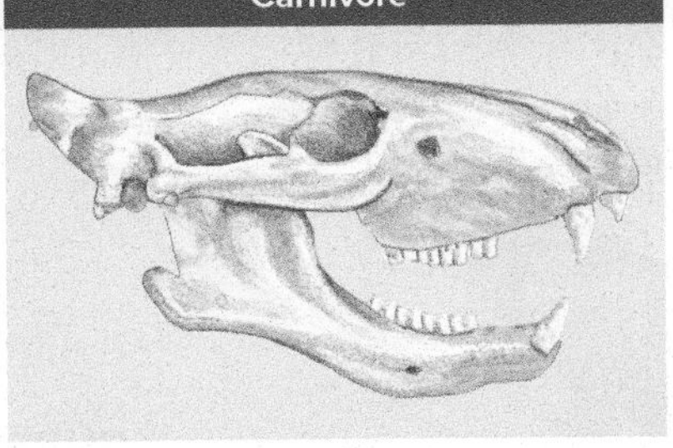

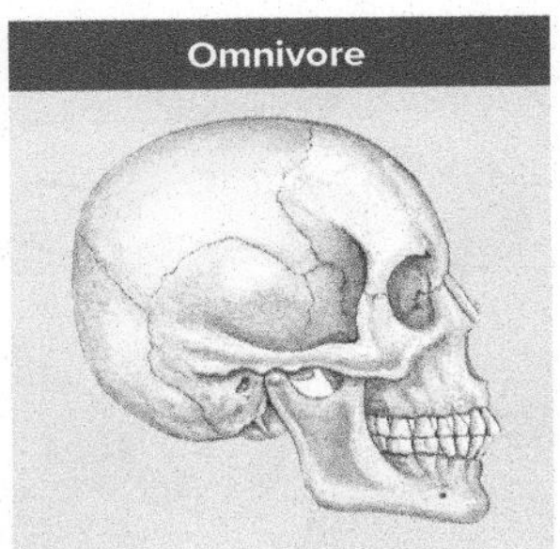

FIGURE 27.4

Arrangement of Teeth in a Variety of Mammals. The various patterns of dentition (an animal's assortment of teeth) shown here depend on the animals' diet. Different vertebrates (herbivore, carnivore, or omnivore) have evolved specific patterns and variations from a generalized pattern of dentition depending on their nutritional source. For example, herbivores, such as this horse, have sharp incisors for clipping grasses and large, flat premolars and molars for grinding. Carnivores, such as this lion, have enlarged canines for killing their prey, short incisors for scrapping bones, and jagged molars for tearing apart flesh. The front teeth of humans, an omnivore, resemble those of carnivores and the back teeth are like herbivores.

Some fishes, crocodylians, and all avian reptiles have a muscular **gizzard** (L. *gigeria*, cooked entrails of poultry) for grinding up food (*see figure 27.5*a). The bird's gizzard develops from the posterior part of the stomach called the ventriculus. Pebbles that have been swallowed are often retained in the gizzard of grain-eating birds and facilitate the digestive process.

The small and large intestines continue the digestive and absorptive processes in vertebrates. The configuration and divisions of intestines vary greatly among vertebrates. These variations are related to the animal's type of food, body size, and levels of activity. Intestines generally increase in length and complexity as one proceeds through fish and tetrapod taxa (*see figure 27.5*b–d). Specializations to increase the surface area for digestion and absorption, like the spiral valve of chondrichthians (*see figure 27.5*b), are common. The small intestine is usually where the final stages of digestion and nutrient absorption occur, and water and ions are often absorbed in the large intestine.

The cellulose digestion (previously described) by ruminants occurs in modifications of the stomach. Therefore, ruminants are

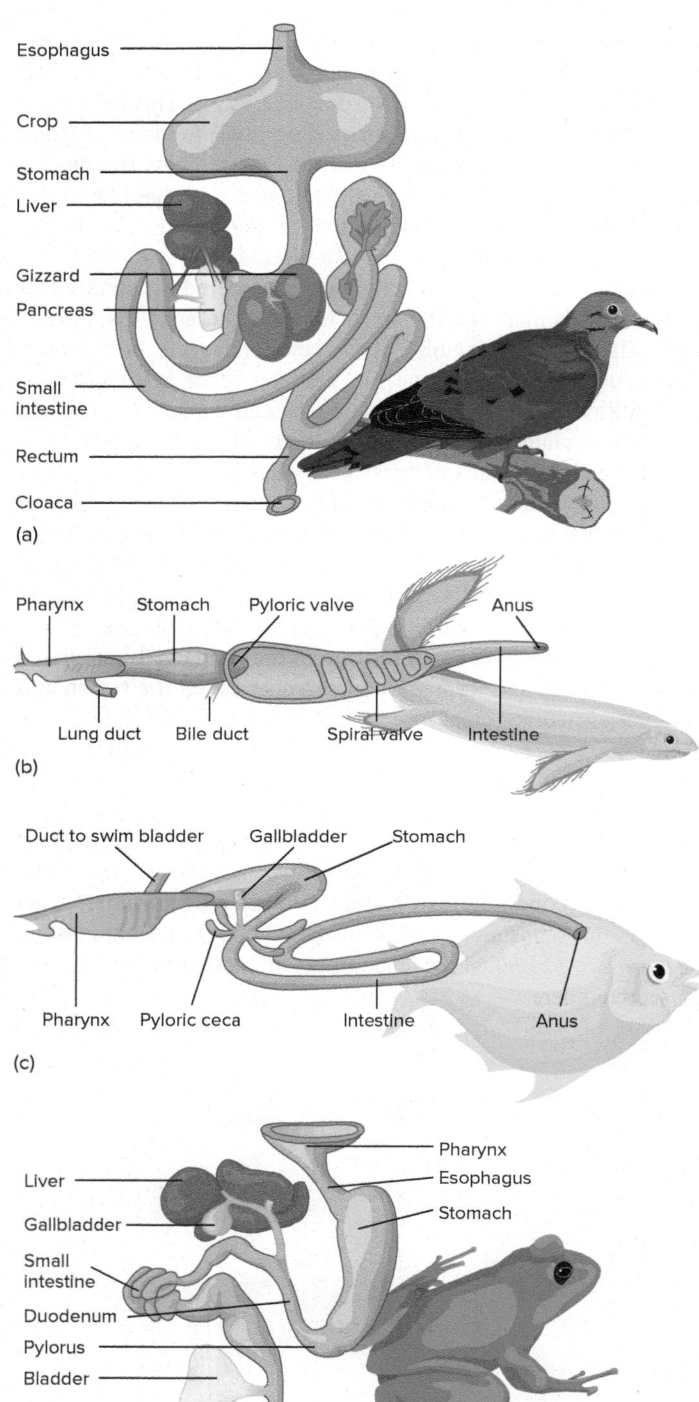

FIGURE 27.5

Generalized Arrangement of Stomachs and Intestines in a Variety of Vertebrates. (*a*) Pigeon (avian reptile). (*b*) Lungfish (Sarcopterygii). (*c*) Teleost fish. (*d*) Frog (Anura).

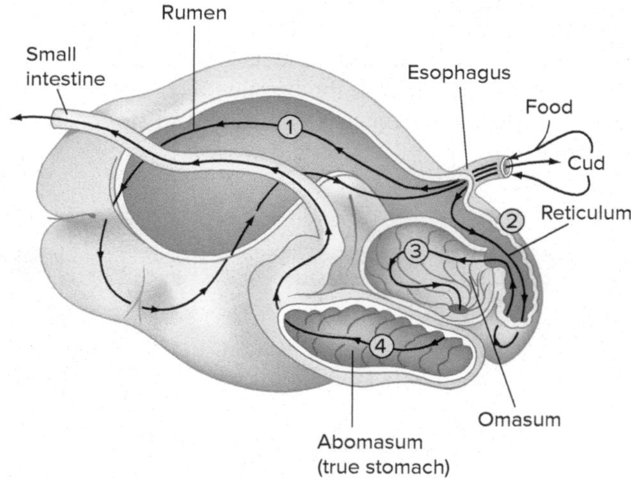

FIGURE 27.6

Ruminant Mammal. Ruminant mammals have four-chambered stomachs where fermentation and enzymatic digestion is facilitated by symbiotic microorganisms. Vegetation eaten by a ruminant enters the rumen (1) and the reticulum (2) from which food may be regurgitated and rechewed (the cud). The food is then swallowed a second time and eventually transported to the posterior two chambers, the omasum (3) and abomasum (4). Only the abomasum secretes gastric juice.

EVOLUTIONARY INSIGHTS

Gut Microbiomes–The Hidden World Within

Virtually all animal alimentary tracts are inhabited by Bacteria, Archaea, viruses, protists, and Fungi. These microorganisms comprise animals' gut biomes. Most of us are familiar with the role that microorganism symbionts play in aiding the digestion of cellulose by termites (order Isoptera, *see chapter 15*). One termite, *Reticulitermes flavipes*, is the most important wood-destroying termite pest in the United States. The hindgut of this species is inhabited by 12 species of flagellated protists (*see appendix C*) and many species of Bacteria, Archaea, and Fungi. These symbionts produce cellulose-digesting enzymes and carry out fermentation, which makes wood-consumption a viable nutritional mode for termites. Foregut-fermenting ruminants and hindgut-fermenting herbivores are described in this chapter. Virtually every animal that has a digestive tract benefits from symbiotic relationships with microorganisms.

Most data indicate that animals are born (or hatch) with an essentially sterile alimentary tract. Humans and other mammals initially acquire gut microbes through nursing, and that source is later supplemented through food and exposure to environmental microorganisms. Termite nymphs acquire microbes through grooming by adults. Ruminants acquire microorganisms through grooming and shared water sources, and other animals acquire microbes through trophallaxis (mouth-to-mouth transfer or anus-to-mouth) transfer or contact with feces.

Over 1,000 bacterial species live in the alimentary tract of healthy adult humans. About two-thirds of these species are unique to each human, and this composition is relatively stable over decades. As in many other animals, gut microbes ferment otherwise indigestible plant fibers to short-chain fatty acids that can be metabolized by our cells. They aid in the production of B and K vitamins (*see table 27.2*). Gut microbes also promote the development of our immune systems. T cells (a type of lymphocyte) engulf bacterial proteins and transport them to lymph nodes, where these proteins are identified as "friendly." Thus, these mutualistic bacteria coexist with us and do not elicit an immune response.

The human gut biome has evolved within the primate lineage. A comparative analysis of gut biomes of chimpanzees (*Pan troglodytes*), bonobos (*Pan paniscus*), and western gorillas (*Gorilla gorilla*) suggests that gut biomes evolve at a relatively constant rate in wild populations as species within a lineage diverge from a common ancestor.* Wild primates have unique, but very diverse, gut biomes. This gut biome diversity is probably linked to the diverse diets of wild primates. On the other hand, the human gut biome has experienced a very rapid evolution. It has become much less diverse than that of our primate relatives. This decreased gut biome diversity is pronounced in the U.S. population. It is probably associated with an increasing reliance on meat in modern diets.

Is there a consequence to lowered microbial diversity in humans? Low biome diversity is associated with inflammatory bowel disease, increased risk of type I diabetes, obesity, and some forms of cancer. *Clostridium difficile* (*C. diff*) is a very resistant bacterium that proliferates after rounds of antibiotic treatment have reduced the diversity of the healthy bacterial symbionts that are normally present in our system. When *C. difficile* dominates gut bacteria, it alters gut-wall permeability, creating a harmful, and sometimes deadly, inflammation of the bowel. It is likely that lowered gut microbial diversity is equally detrimental to other mammals, and it is linked to human-induced habitat degradation. For instance, habitat degradation has been shown to reduce the diversity of food resources available to black howler monkeys (*Alouatta pigra*).** This reduction in resource diversity, in turn, reduces the microbial diversity of their gastrointestinal tracts and the overall health of the animal.

To stay healthy we must encourage and protect a healthy gut microbiome. Humans can accomplish this through a diverse diet, exercise, and the reduced use of nonessential antibiotics. Other animals are not so lucky. Landscape conservation is necessary for the maintenance of gastrointestinal health and the long-term survival of other animals that are also dependent on a healthy diversity of intestinal flora.

*Moeller AH, Li Y, Ngole EM, Ahuka-Mundeke S, Lonsdorf EV, Pusey AE, Peeters M, Hahn BH, and Ochman H. 2014. Rapid changes in the gut microbiome during human evolution. Proc. Natl. Acad. Sci. U.S.A. 111(46): 15431-16435.

**Amato KR, Yeoman CJ, Kent A, Righini N, Carbonero F, Estrada A, Gaskins HR, Stumpf RM, Yildirim S, Torralba M, Gillis M, Wilson BA, Nelson KE, White BA, Leigh SR. 2013. Habitat degradation impacts black howler monkey (*Alouatta pigra*) gastrointestinal microbiomes. The ISME Journal 7(2013): 1344-53; doi:10.1038/ismej.2013.16; published online 14 March 2013.

sometimes called "foregut fermenters." Many other herbivores carry out microbial fermentation and digestion of cellulose in a pouch at the junction of the small and large intestines called the **cecum** (L. *caecum*, blind gut; pl., ceca). The ceca of rabbits (Lagomorpha, figure 27.7), horses (Perissodactyla), and rats (Rodentia) are large fermentation chambers, and these animals are often called "hindgut fermenters." Some hindgut fermenters (e.g., rabbits) consume special "soft fecal pellets" for a second trip through the digestive tract. Microorganisms in these feces aid in this additional digestion. Consumption of fecal materials is not uncommon in animals and is called coprophagia.

Digestive Glandular Systems

There is a host of single-celled glands that line most regions of the alimentary canal. These glands include mucus-secreting goblet cells and cells that secrete digestive enzymes in the small intestine. Small multicellular glands, like those of the stomach that secrete protein-digesting pepsin and hydrochloric acid, are also present in many regions of the alimentary tract. In addition to those glands, larger glandular organs assist with food processing. These include salivary glands, livers, gallbladders, and pancreata.

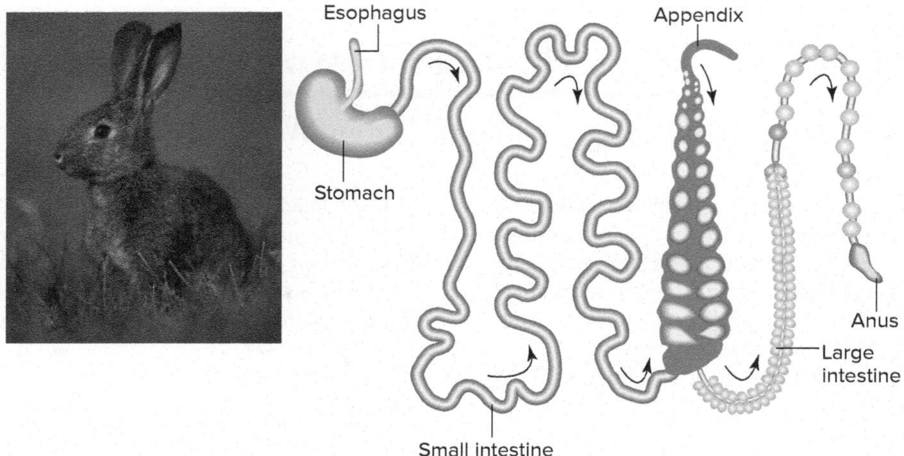

FIGURE 27.7

Extensive Cecum of a Nonruminant Herbivore, Such as a Rabbit (Lagomorpha). The cecum contains microorganisms that produce digestive enzymes (e.g., cellulase that helps break down cellulose). Black arrows indicate the direction of food movement.

Most fishes lack salivary glands in the head region. Lampreys (Petromyzontida) are an exception because they have a pair of glands that secrete an anticoagulant needed to keep their prey's blood flowing as they feed. Modified salivary glands of some snakes (Squamata) produce venom that is injected through fangs to immobilize prey. Salivary glands are absent in most amphibians and avian and nonavian reptiles. All mammals have them.

In those vertebrates with a gallbladder, it is closely associated with the liver. The **liver** manufactures bile, which the gallbladder then stores. **Bile** is a fluid containing bile salts and bile pigments. Bile salts play an important role in the digestion of fats, although they are not digestive enzymes. They emulsify dietary fat, breaking it into small globules (emulsification), increasing the surface area for the fat-digesting enzyme lipase. Bile pigments result from phagocytosis of red blood cells in the spleen, liver, and red bone marrow. Phagocytosis cleaves the hemoglobin molecule, releasing iron, and the remainder of the molecule is converted into pigments that enter the circulation. These pigments are subsequently extracted from the circulation in the liver and excreted in the bile as bilirubin ("red bile") and biliverdin ("green bile").

Because of the importance of bile in fat digestion, the gallbladder is relatively large in carnivores and vertebrates in which fat is an important part of the diet. It is much reduced or absent in blood-feeding vertebrates, such as the lamprey, and in herbivores.

Every vertebrate has a **pancreas** (pl., pancreata); however, in lampreys and lungfishes it is embedded in the wall of the intestine and is not a visible organ. Both endocrine and exocrine tissues are present, but the cell composition varies. Pancreatic fluid containing many enzymes empties into the small intestine, and these digestive processes will be described in the following section.

Section 27.6 Thinking Beyond the Facts

How would you explain the observation that the alimentary tracts of carnivores tend to be shorter and simpler than those of herbivores?

27.7 THE MAMMALIAN DIGESTIVE SYSTEM

LEARNING OUTCOMES

1. Categorize the key processes that are involved in digesting and absorbing nutrients in mammals.
2. Describe the roles of the accessory organs (glands) of digestion in a mammal.

Humans (*Homo*), pigs (*Sus*), most bears (Ursidae), and raccoons (*Procyon*) are omnivores. The digestive system of an omnivore has the mechanical and chemical ability to process many kinds of foods. The sections that follow examine the control of gastrointestinal motility, the major parts of the alimentary canal, and the accessory organs of digestion (figure 27.8).

The process of digesting and absorbing nutrients in a mammal includes:

1. Ingestion—eating
2. Peristalsis—the involuntary, sequential muscular contractions that move ingested nutrients along the digestive tract
3. Segmentation—mixing the contents in the digestive tract
4. Secretion—the release of hormones, enzymes, and specific ions and chemicals that take part in digestion
5. Digestion—the conversion of large nutrient particles or molecules into small particles or molecules
6. Absorption—the passage of usable nutrient molecules from the small intestine into the bloodstream and lymphatic system for the final passage to body cells
7. Defecation—the elimination from the body of undigested and unabsorbed material as waste

Animation Organs of Digestion

Animation Digestion and Metabolism Overview

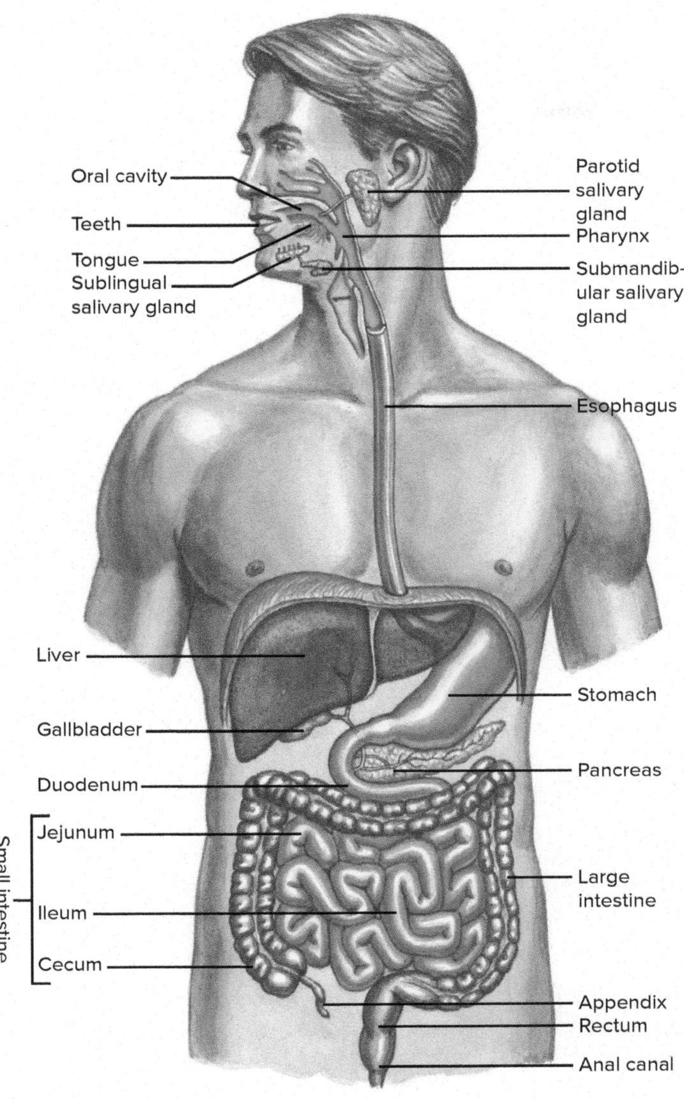

FIGURE 27.8

Major Organs and Parts of the Human Digestive System. Food passes from the mouth through the pharynx and esophagus to the stomach. From the stomach, it passes to the small intestine, where nutrients are broken down and absorbed into the circulatory and lymphatic systems. Nutrients then move to the large intestine, where water is reabsorbed, and feces form. Feces exit the body via the anal canal. Digestion is also aided by the accessory organs of digestion: the liver, pancreas, and gallbladder.

Gastrointestinal Motility and Its Control

The histology of the gastrointestinal tract is based upon a common layering of tissues throughout its length (figure 27.9). Knowing this structure helps one understand the functions occurring in each of its regions. The outermost layer of a section of the gastrointestinal tract is an epithelial-covered connective tissue layer called the serosa. This layer is actually continuous with the **mesentery**

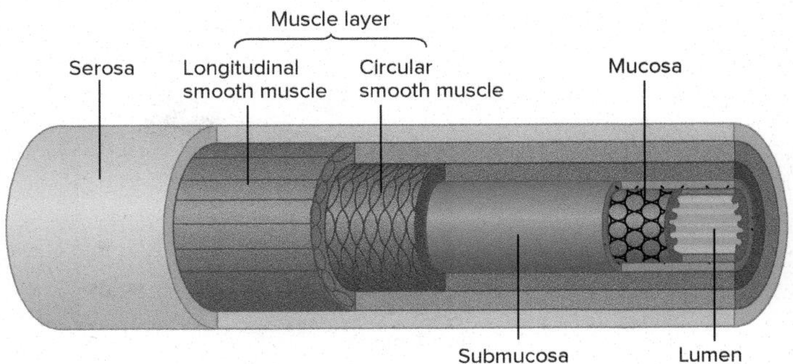

FIGURE 27.9

Mammalian Gastrointestinal Tract. Common structural layers of the gastrointestinal tract. The central lumen extends from the mouth to the anus.

that suspends the gastrointestinal tract in the body cavity and the **peritoneum** that lines the abdominal coelom (*see figure 7.11*c). Inside the serosa are muscular layers. For most of the length of the gastrointestinal tract, there are two layers of outer-longitudinal smooth muscle and inner-circular smooth muscle. The anterior (superior) end of the esophagus has skeletal muscle that initiates voluntary swallowing movements. The submucosa underlies the muscle layers. It is comprised of connective tissues and is rich with blood and lymphatic vessels and nerves. The mucosa is an epithelial layer that lines the lumen of the gastrointestinal tract. The mucosa usually carries out secretory and absorptive functions. (Again, the esophagus is the exception.)

The coordinated contractions of the muscle layers of the gastrointestinal tract mix the food material with various secretions and move the food from the oral cavity to the rectum. The two types of movement involved are peristalsis and segmentation.

During **peristalsis** (Gr. *peri,* around + *stalsis,* contraction), food advances through the gastrointestinal tract when the rings of circular smooth muscle contract behind it and relax in front of it (figure 27.10*a*). Peristalsis is analogous to squeezing icing from a pastry tube. The small and large intestines also have rings of smooth muscles that alternately contract and relax but do not move along the intestine. These contractions mix the intestinal contents, exposing the contents to mucosal surfaces. These contractions are called **segmentation** (figure 27.10*b*).

Sphincters also influence the flow of material through the gastrointestinal tract and prevent backflow. Sphincters are rings of smooth or skeletal muscle at the beginning or ends of specific regions of the gut tract. For example, the cardiac sphincter is between the esophagus and stomach, and the pyloric sphincter is between the stomach and small intestine.

Control of gastrointestinal activity is based on the volume and composition of food in the lumen of the gut. For example, ingested food distends the gut and stimulates stretch receptors in the gut wall. In addition, the digestion of carbohydrates, lipids, and proteins stimulates various chemical receptors in the gut wall. Nerve impulses from these stretch and chemical stimuli travel

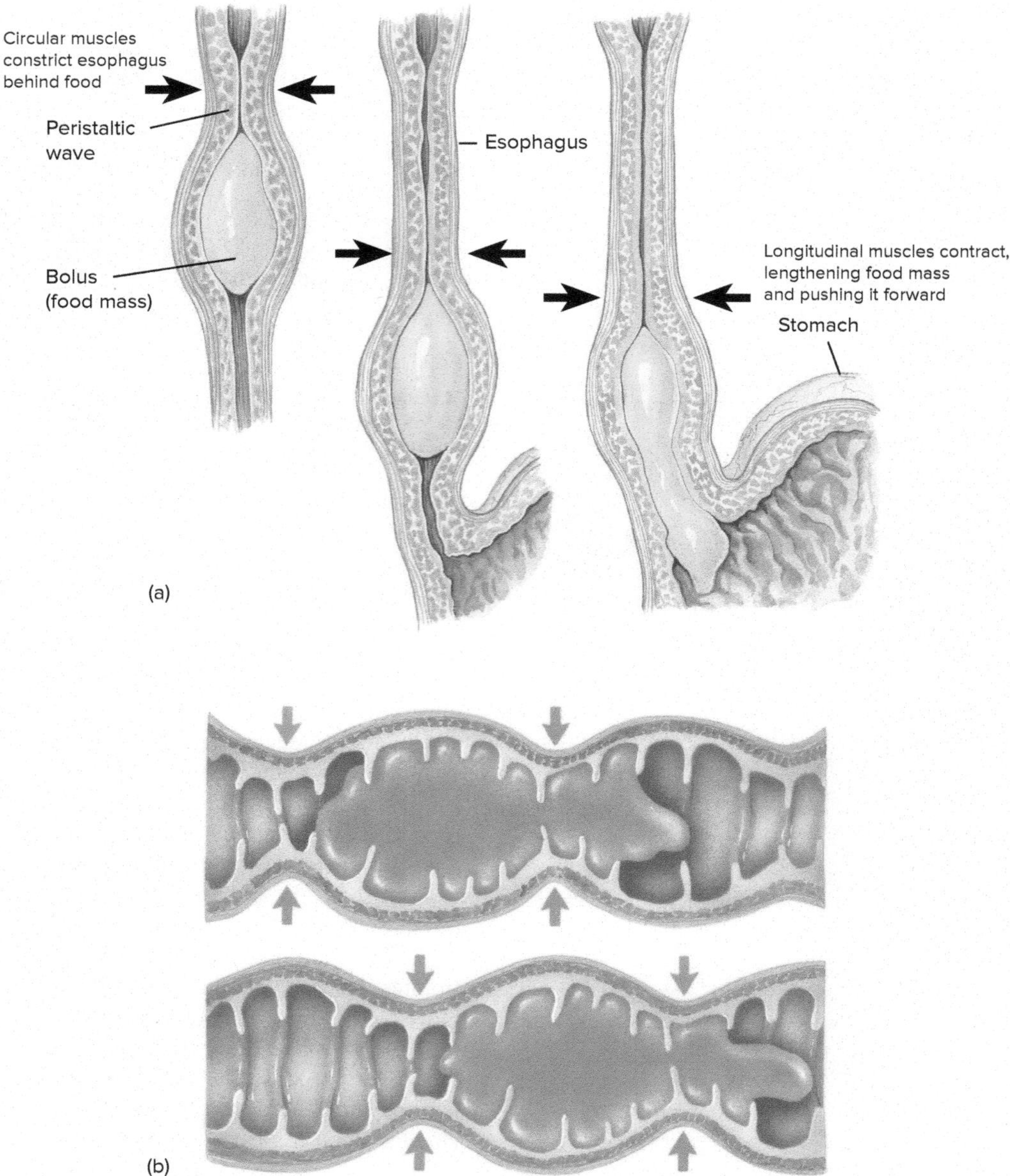

FIGURE 27.10

Peristalsis and Segmentation. (*a*) Peristaltic waves (black arrows) move food through the esophagus to the stomach. (*b*) In segmentation, simultaneous muscular contractions of many sections of the intestine (blue arrows) help mix nutrients with digestive secretions.

through nerve plexuses in the gut wall to control the muscular contraction that leads to peristalsis; segmentation; and secretion of mucus, enzymes, and other substances into the gut lumen. In addition to this local control, nerve pathways connect the receptors and effectors with the central nervous system. The central nervous system's control of gastrointestinal motility and secretion is largely through the parasympathetic component of the autonomic nervous system.

In addition to control through the nervous system, local hormones, produced by mucosal cells, aid in gastrointestinal control. Gastrin is a hormone secreted by the stomach mucosa, and cholecystokinin and secretin are secreted by the small intestine. These, and other, local hormones help control gastrointestinal motility and secretion.

Oral Cavity

A pair of lips protects the **oral cavity** (mouth). The lips are highly vascularized, skeletal muscle tissue with an abundance of sensory nerve endings. Lips help retain food as it is being chewed and play a role in phonation (the modification of sound).

The oral cavity is continuously bathed by **saliva**, a watery fluid that at least three pairs of salivary glands secrete. Saliva moistens food, binds it with mucins (glycoproteins). The tongue helps form the ingested food into a moist mass called a bolus. Saliva also contains bicarbonate ions (HCO_3^-), which buffer chemicals in the mouth, and thiocyanate ions (SCN^-) and the enzyme lysozyme, which kill microorganisms. It also contributes an enzyme (amylase), which initiates carbohydrate digestion.

Pharynx and Esophagus

Chapter 26 discussed how air, swallowed food, and liquids pass from the mouth into the **pharynx** (Gr. "the throat")–the common passageway for both the digestive and respiratory tracts. The epiglottis temporarily seals off the opening (glottis) to the trachea so that swallowed food does not enter the trachea. The upper (anterior) third of the esophagus is comprised of skeletal muscle, and the initiation of swallowing can be voluntary. After swallowing is initiated, food moves into the lower (posterior) portion of the esophagus and an involuntary peristaltic wave propels the bolus or liquid to the stomach. Neither the pharynx nor the esophagus contributes to digestion.

Stomach

The mammalian stomach is a muscular, distensible sac with three main functions. It (1) stores and mixes the food bolus received from the esophagus, (2) secretes enzymes, mucus, and hydrochloric acid (HCl) that start the digestion of proteins, and (3) helps control the rate at which food moves into the small intestine via the pyloric sphincter (figure 27.11*a*).

The mucosa of the stomach is comprised of thousands of gastric glands (figure 27.11*b*). Gastric glands are comprised of three types of secretory cells. **Parietal cells** secrete a solution containing HCl, and **chief cells** secrete pepsinogen, the precursor of the enzyme pepsin. These secretory cells are in the pits of the gastric glands. The third cell type, **mucous cells**, are located near the openings of gastric glands. The mucus they secrete coats the mucosa of the stomach, protecting it from HCl and digestive enzymes. Endocrine cells in one part of the stomach mucosa release the hormone gastrin, which travels to target cells in the gastric glands, further stimulating them.

Animation
Three Phases of Gastric Secretion

When the bolus of food enters the stomach, it distends the walls of the stomach. This distention, as well as the act of eating, causes the gastric pits to secrete HCl (as H^+ and Cl^-) and pepsinogen. The H^+ ions cause pepsinogen to be converted into the active enzyme pepsin. As pepsin, mucus, and HCl mix with and begin to break down proteins, smooth muscles contract and vigorously churn and mix the food bolus. About three to four hours after a meal, the stomach contents are reduced to a semiliquid mass called **chyme** (Gr. *chymos,* juice). The pyloric sphincter regulates the release of the chyme into the small intestine.

When the stomach is empty, peristaltic waves cease; however, after about 10 hours of fasting, new waves may occur in the upper region of the stomach. These waves can cause "hunger pangs" as sensory nerve fibers carry impulses to the brain.

Small Intestine: Main Site of Digestion

Most of the food a mammal ingests is digested and absorbed in the **small intestine.** The human small intestine is about 4 cm in diameter and 7 to 8 m in length (*see figure 27.8*). The length of the small intestine; the many circular folds; and thousands of fingerlike projections, called **villi** (L. *villus*, tuft of hair; sing., villus), all contribute to a very large surface area for digestion and absorption functions (figure 27.12*a–c*). The luminal plasma membranes of these epithelial cells are folded into microvilli (figure 27.12*d*). These minute projections are so dense that the inner wall of the human small intestine has a total surface area of approximately 300 m^2–the size of a tennis court.

The superior (anterior) portion of the small intestine, called the duodenum, functions primarily in digestion. The next region is the jejunum, and the inferior (posterior) small intestine is the ileum. The jejunum and ileum function in nutrient absorption.

The duodenal mucosa contains many unicellular glands that secrete mucus and digestive enzymes. The pancreas secretes other enzymes. In the duodenum, digestion of carbohydrates and proteins is completed, and most lipids are digested. The jejunum and ileum absorb the end products of digestion (amino acids, simple sugars, fatty acids, glycerol, nucleotides, and water). Much of this absorption involves active transport and the sodium-dependent ATPase pump. Sugars and amino acids are absorbed into the capillaries of the villi, whereas free fatty acids enter the epithelial cells of the villi and recombine with glycerol to form triglycerides. The triglycerides are coated with proteins to form small droplets called **chylomicrons**, which enter the lacteals of the villi (*see figure 27.12*d). From the lacteals, the chylomicrons move into the lymphatics and eventually into the bloodstream for transport throughout the body.

Besides absorbing organic molecules, the small intestine absorbs water and dissolved mineral ions. The small intestine absorbs about 9 L of water per day, and water absorption continues in the large intestine.

Animation
Enzymatic Action and Hydrolysis of Sucrose

Large Intestine

Unlike the small intestine, the **large intestine** has no circular folds, villi, or microvilli; thus, the surface area is much smaller. The small intestine joins the large intestine at the cecum (L. *caecum,* blind gut) (*see figure 27.8*). The human cecum and its extension, the **appendix** (L. *appendere,* to hang upon), are storage sites and evolutionary remnants of a larger, functional cecum, such as those found in herbivores (*see figure 27.7*). The appendix contains an abundance of lymphoid tissue and may function as part of the immune system.

The major functions of the large intestine include the reabsorption of water and minerals and the formation and storage of feces. As peristaltic waves move food residue along, minerals diffuse or are actively transported from the residue across the epithelial surface of the large intestine into the bloodstream. Water follows osmotically and returns to the lymphatic system and bloodstream. When water reabsorption is insufficient, diarrhea (Gr. *rhein,* to flow)

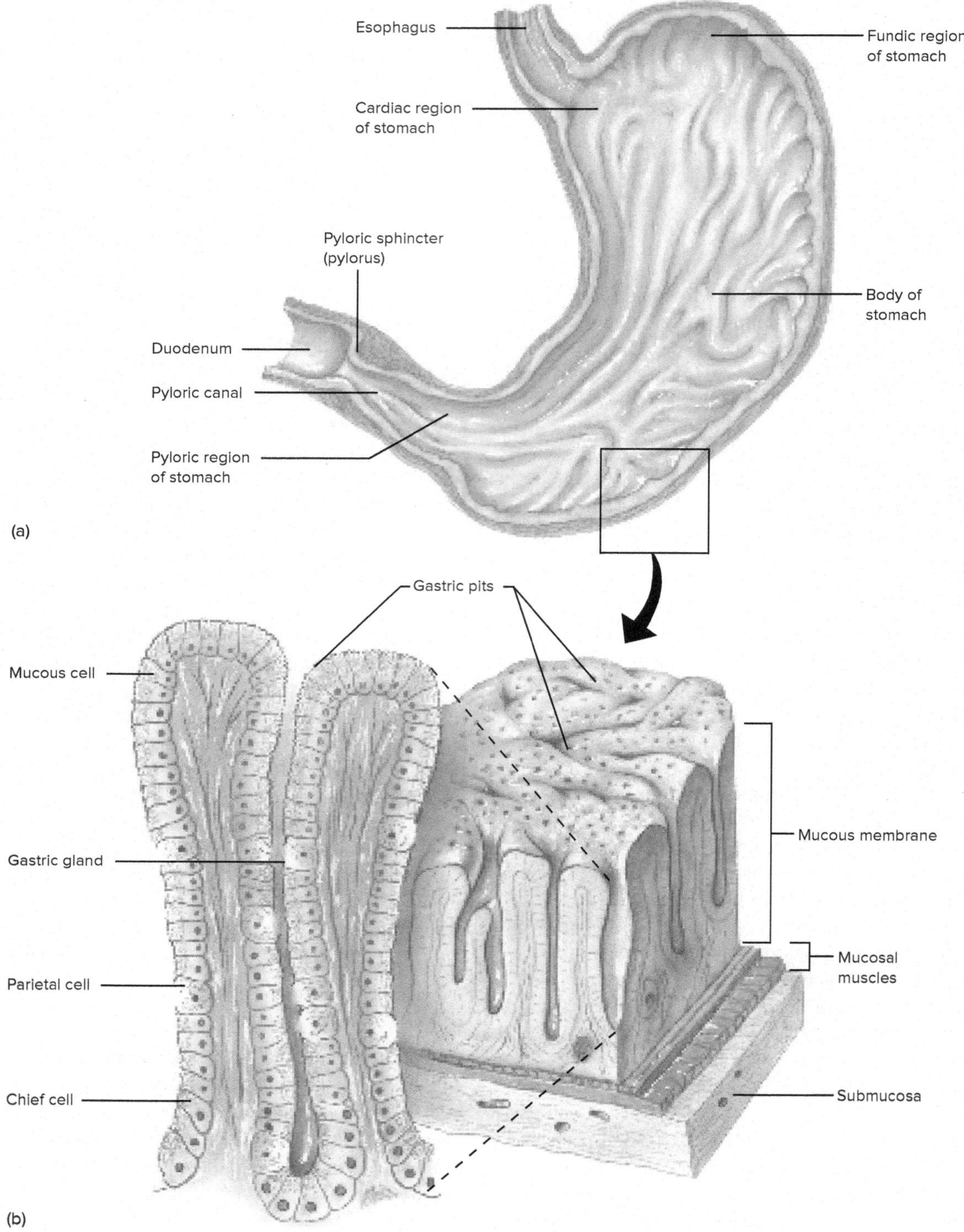

FIGURE 27.11

Stomach. (*a*) Food enters the stomach from the esophagus. (*b*) Gastric glands cover the mucosa of the stomach and include mucous cells, parietal cells, and chief cells. Each type produces a different secretion.

results. If too much water is reabsorbed, fecal matter becomes too thick, resulting in constipation.

Many bacteria and fungi exist symbiotically in the large intestine (*see Evolutionary Insights, page 511*). They feed on the food residue and further break down its organic molecules to waste products. In turn, they secrete amino acids and vitamin K, which the host's gut absorbs. What remains—feces—is a mixture of bacteria, fungi, undigested plant fiber, sloughed-off intestinal cells, and other waste products.

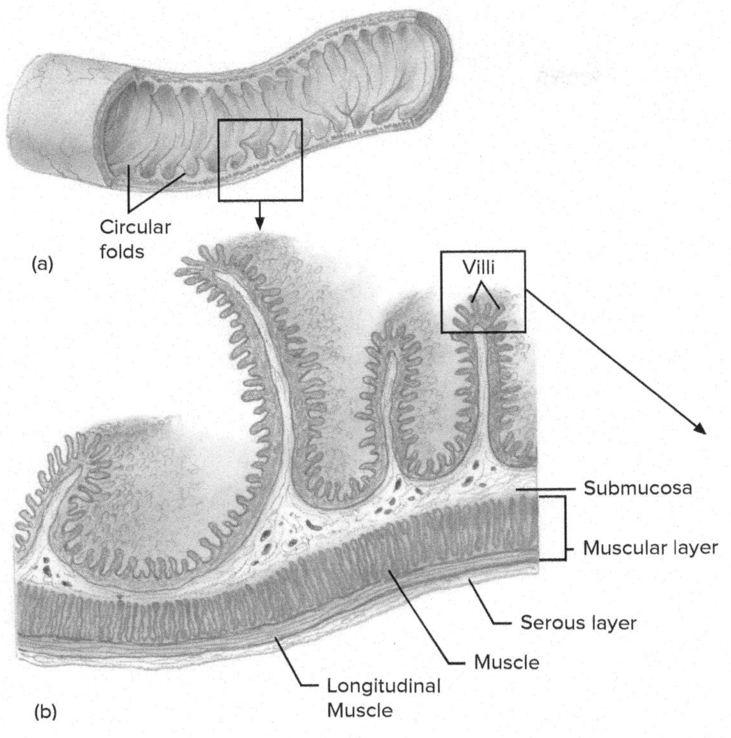

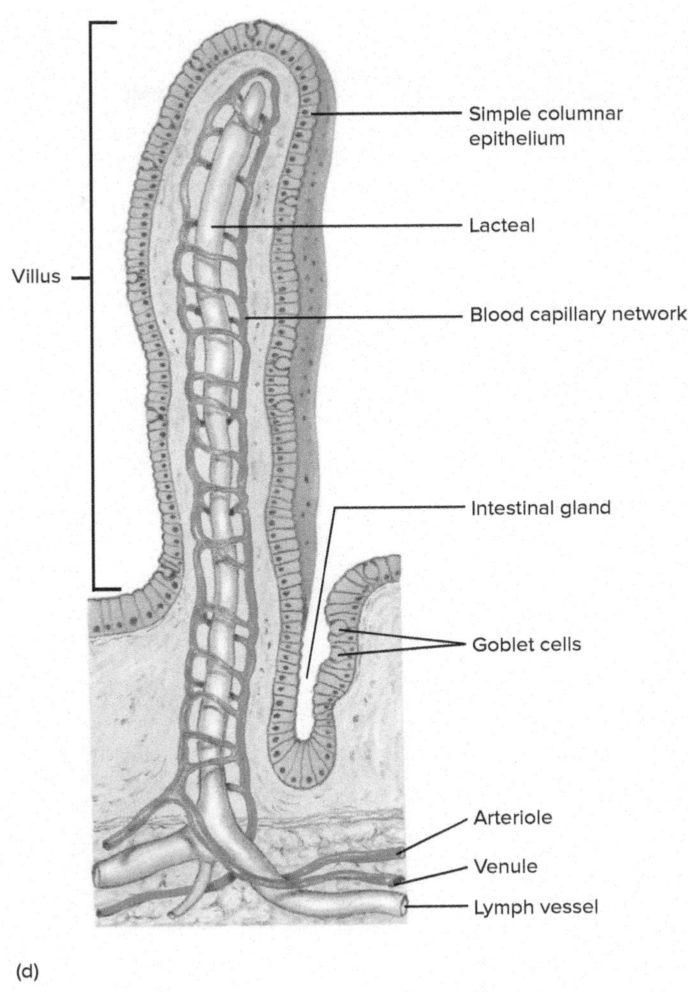

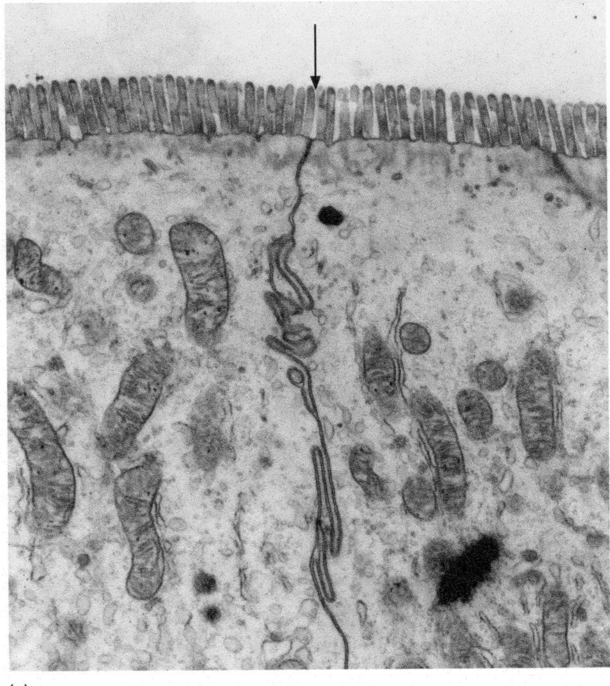

FIGURE 27.12

Small Intestine. The small intestine absorbs food over a large surface area. (*a*) The lining of the intestine has many circular folds. (*b* and *d*) Fingerlike villi line the intestine. A single villus contains a central capillary network and a lymphatic lacteal, both of which transport nutrients absorbed from the lumen of the intestine. (*c*) The plasma membrane of the simple columnar epithelial cells covering the villi fold into microvilli (arrow), which further increase the surface area facing the lumen.

Role of the Pancreas in Digestion

The pancreas (Gr. *pan,* all + *kreas,* flesh) is an organ that lies just ventral to the stomach and has both endocrine and exocrine functions (table 27.3). Exocrine cells in the pancreas secrete digestive enzymes into the pancreatic duct, which merges with the hepatic duct from the liver to form a common bile duct that enters the duodenum. Pancreatic enzymes complete the digestion of carbohydrates and proteins and initiate the digestion of lipids. Trypsin, carboxypeptidase, and chymotrypsin digest proteins into small peptides and individual amino acids. Pancreatic lipases split triglycerides into smaller, absorbable glycerol and free fatty acids. Pancreatic amylase converts polysaccharides into disaccharides and monosaccharides.

TABLE 27.3

MAJOR DIGESTIVE GLANDS, SECRETIONS, AND ENZYMES IN MAMMALS

PLACE OF DIGESTION	SOURCE	SECRETION	ENZYME	DIGESTIVE FUNCTION(S)
Mouth (oral cavity)	Salivary glands	Saliva	Salivary amylase	Begins the digestion of carbohydrates; inactivated by stomach HCl
	Mucous glands	Mucus	–	Lubricates food bolus
Esophagus	Mucous glands	Mucus	–	Lubricates food bolus
Stomach	Gastric glands	Gastric juice	Lipase Pepsin	Digests lipids into fatty acids and glycerol Digests proteins into polypeptides
	Gastric mucosa	HCl	–	Converts pepsinogen into active pepsin; kills microorganisms
	Mucous glands	Mucus	–	Lubricates and protects
Small intestine	Liver	Bile	–	Emulsifies lipids; activates lipase
	Pancreas	Pancreatic juice	Amylase	Digests starch into maltose
			Chymotrypsin	Digests proteins into peptides and amino acids
			Lipase	Digests lipids into fatty acids and glycerol (requires bile salts)
			Nuclease	Digests nucleic acids into mononucleotides
			Trypsin	Digests proteins into peptides and amino acids
	Intestinal glands	Intestinal juice	Enterokinase	Digests inactive trypsinogen into active trypsin
			Lactase	Digests lactose into glucose and galactose
			Maltase	Digests maltose into glucose
			Peptidase	Digests polypeptides into amino acids
			Sucrase	Digests sucrose into glucose and fructose
	Mucous glands	Mucus	–	Lubricates
Large intestine	Mucous glands	Mucus	–	Lubricates

The pancreas also secretes bicarbonate (HCO_3^-) ions that help neutralize the acidic food residue coming from the stomach. Bicarbonate raises the pH from 2 to 7, the optimal pH for pancreatic enzymes.

Role of the Liver and Gallbladder in Digestion

The liver, the largest internal organ a mammal, is just under the diaphragm (*see figure 27.8*). In the liver, millions of specialized cells called hepatocytes take up nutrients absorbed from the intestines and release them into the bloodstream. Hepatocytes also manufacture the blood proteins prothrombin and albumin.

In addition, some major metabolic functions of the liver include:

1. Removal of amino acids from organic compounds.
2. Urea formation from protein and amino acid metabolism.
3. Manufacture of most of the plasma proteins, formation of fetal erythrocytes, destruction of worn-out erythrocytes, and synthesis of the blood-clotting agents prothrombin and fibrinogen from amino acids.
4. Synthesis of nonessential amino acids.
5. Conversion of galactose and fructose to glucose.
6. Oxidation of fatty acids.
7. Formation of lipoproteins, cholesterol, and phospholipids (essential cell membrane components).
8. Conversion of carbohydrates and proteins into fat.
9. Modification of waste products, toxic drugs, and poisons (detoxification).
10. Synthesis of vitamin A from carotene, and with the kidneys, participation in the activation of vitamin D.
11. Maintenance of a stable body temperature by raising the temperature of the blood passing through it. Its many metabolic activities make the liver the major heat producer in a mammal's body.
12. Manufacture of bile salts.
13. Storage of glucose in the form of glycogen, and with the help of insulin and enzymes, conversion of glycogen back into

glucose as the body needs it. Storage of fat-soluble vitamins (A, D, E, and K), and minerals. Storage of fats and amino acids, and conversion of them into usable glucose as required.

The **gallbladder** (L. *galbinus,* greenish yellow) is a small organ near the liver (*see figure 27.8*). The gallbladder stores the greenish fluid called bile that the liver cells continuously produce. Bile is very alkaline and contains pigments, cholesterol, lecithin, mucin, bilirubin, and bile salts that act as detergents to emulsify fats (form them into droplets suspended in water) and aid in fat digestion and absorption. (Recall that fats are insoluble in water.) Bile salts also combine with the end products of fat digestion to form micelles. **Micelles** are lipid aggregates (fatty acids and glycerol) with a surface coat of bile salts. Because they are so small, they can cross the microvilli of the intestinal epithelium.

Section 27.6 Thinking Beyond the Facts

Why does fat not require carrier proteins to cross the intestinal epithelium?

Summary

27.1 **Evolution of Heterotrophy**

- Nutrition describes all of those processes by which an animal takes in, digests, absorbs, stores, and uses food (nutrients) to meet its metabolic needs. Digestion is the mechanical and chemical breakdown of food into smaller particles that the individual cells of an animal can absorb. Animals are heterotrophs that consume other organisms for their nutrients. Heterotrophy allows animals to devote energy to growth and reproduction.

27.2 **The Metabolic Fates of Nutrients in Heterotrophs**

- Macronutrients are large organic molecules that are needed in large quantities and include the carbohydrates, lipids, and proteins.
- Micronutrients are needed in small quantities and include the vitamins and minerals.

27.3 **Digestion**

- Two types of digestion exist: intracellular and extracellular. Large animals have special organs for digestion.

27.4 **Animal Strategies for Getting and Using Food**

- Only a few animals can absorb nutrients directly from their external environment. Most animals must work for their nutrients. Various specializations have evolved for food procurement (feeding) in animals.
- Continuous feeders include filter feeders such as tube worms. Discontinuous feeders have specializations for procuring, storing, and digesting food.
- Suspension feeders remove food particles from surrounding water.
- Deposit feeders obtain nutrients from sediments.
- Herbivores consume plant material.
- Predators capture and kill other animals for food.
- Some parasites and nonparasites lack a mouth and digestive system and absorb nutrients across their body surface.
- Some animals are fluid feeders. These include animals that feed on nectar and ectoparasites and predators that feed on body fluids, including blood.

27.5 **Diversity in Digestive Structures: Invertebrates**

- A few invertebrates, like sponges, lack a digestive tract and digest food intracellularly.
- Extracellular digestion occurs in most animals. Animals with incomplete digestive tracts have a mouth that serves as both the point of entrance for food into the digestive tract and the point for elimination of indigestible wastes. Animals with a complete digestive tract have an anus; and food procurement, digestion, and waste elimination occur sequentially. Chewing insects have a complete digestive tract.

27.6 **Diversity in Digestive Structures: Vertebrates**

- Vertebrates have a complete digestive tract.
- Tongues are specialized for procuring and manipulating food. Teeth are specialized and reflect the nature of an animal's diet.
- Esophagi move food to the stomach. Stomachs store food ingested at infrequent intervals. Gizzards are present in some fishes, crocodylians, and all avian reptiles. Rumens aid in the digestion of cellulose-containing food. Intestines are relatively short in fishes and longer in birds and mammals to increase surface area for digestion. Ceca are blind pouches that house microorganisms that aid in cellulose digestion.
- Unicellular and multicellular glands associated with the digestive tract aid in digestion. Salivary glands are present in mammals, but most other vertebrates lack them. Secretions of the liver aid in fat digestion. The pancreas secretes a variety of enzymes into the small intestine.

27.7 **The Mammalian Digestive System**

- The coordinated contractions of the muscle layer of the gastrointestinal tract mix food material with various secretions and move the food from the oral cavity to the rectum. The two types of movement involved are segmentation and peristalsis.
- The oral cavity functions in chewing food and mixing food with saliva.
- The pharynx and esophagus conduct food to the stomach.
- The stomach is a muscular sac that stores and mixes food, secretes enzymes and HCl, and helps control the rate at which food moves into the small intestine.
- Most digestion occurs in the duodenal portion of the small intestine. The products of digestion are absorbed in the walls of the jejunum and ileum.
- The large intestine functions in the absorption of water and minerals and the formation of feces.
- The pancreas secretes digestive enzymes into the small intestine for the digestion of all macronutrients.
- In the process of digestion, bile that the liver secretes makes fats soluble. The liver has many diverse functions. It controls the fate of newly synthesized food molecules, stores excess glucose as glycogen, synthesizes many blood proteins, and converts nitrogenous and other wastes into a form that the kidneys can excrete.

Concept Review Questions

1. Which of the following animals possess an incomplete digestive system?
   a. echinoderms
   b. annelids
   c. flatworms
   d. arthropods
   e. molluscs
2. Ruminants need multiple stomach chambers in order to digest
   a. starch.
   b. proteins.
   c. cellulose.
   d. lignin.
   e. enzymes.
3. In a robin, the muscular digestive organ in which food is crushed into small bits is the
   a. cloaca.
   b. lumen.
   c. stomach.
   d. crop.
   e. gizzard.
4. Which of the following organs of the digestive system is different from the four others because it does NOT produce any secretions that aid in the digestive process?
   a. Salivary gland
   b. Pancreas
   c. Liver
   d. Stomach
   e. Esophagus
5. Which of the following vitamins functions in the formation of a blood clot?
   a. A
   b. E
   c. K
   d. D
   e. C

Analysis and Application Questions

1. What advantages are there to digestion? Would it not be simpler for a vertebrate to simply absorb carbohydrates, lipids, and proteins from its food and to use these molecules without breaking them down?
2. What might have been some evolutionary pressures acting on animals that led to the internalization of digestive systems?
3. Many digestive enzymes are produced in the pancreas and released into the duodenum. Why, then, has the mammalian stomach evolved the ability to produce pepsinogen?
4. Human vegetarians, unlike true herbivores, have no highly specialized fermentation chambers. Explain how this is reflected in the structure of the digestive systems of a human and a deer.
5. Trace the fate of a hamburger from the mouth to the anus, identifying sites and mechanisms of digestion and absorption.

28

Temperature and Body Fluid Regulation

The Heermanni kangaroo rat (*Dipodomys heermanni*) is native to grasslands and chaparral of southern California. It survives in hot, dry environments without drinking water. The amazing adaptations for water conservation, despite needs for evaporative cooling and excretion of nitrogenous wastes, are hallmarks of this species. Amazing as they are, these adaptations are not atypical of the temperature and fluid regulation strategies of animals living in extreme environments.
Source: Moose Peterson/USFWS photo

Chapter Outline

28.1 Homeostasis and Temperature Regulation
- *Heat Gains and Losses*
- *Some Solutions to Temperature Fluctuations*
- *Temperature Regulation in Invertebrates*
- *Temperature Regulation in Fishes*
- *Temperature Regulation in Terrestrial Ectotherms*
- *Temperature Regulation in Endotherms*

28.2 Control of Water and Electrolytes

28.3 Invertebrate Excretory Systems
- *Contractile Vacuoles*
- *Protonephridia*
- *Metanephridia*
- *Antennal, Maxillary, and Coxal Glands*
- *Malpighian Tubules*

28.4 Vertebrate Excretory Systems
- *How Vertebrates Regulate Ions and Water*
- *Evolution of the Vertebrate Kidney*
- *How the Metanephric Kidney Functions*

The earth's environments vary dramatically in temperature and amount of water present. In the polar regions, high mountain ranges, and deep oceans, the temperature remains near or below 0 °C throughout the year. Temperatures exceeding 40 °C are common in equatorial deserts. Similarly, water is the solvent that supports all life, but water availability varies from an average rainfall of 15 mm per year in the Atacama Desert on the Pacific coast of South America to over 12,500 mm per year in the northeastern India state of Meghalaya. Some water is essentially ion free, but other bodies of water are very salty—brine shrimp (Crustacea, *Artemia*) survive in the Great Salt Lake of Utah, whose salinity approaches 27%. (Salinity of seawater is approximately 3.5%.)

Animal species have adapted to living at these environmental extremes and to most conditions in between—even when the conditions fluctuate widely. Often, animal species must balance competing challenges. The kangaroo rat (*Dipodomys heermanni*) lives in the chaparral of southern California and must balance demands of water conservation with the need for evaporative cooling and eliminating nitrogenous waste with watery urine. Life in other environments faces similar competing challenges. Marine animals must excrete nitrogenous wastes, counter ion accumulation from their seawater surroundings, and conserve water that tends to be lost osmotically. These, and other, evolutionary challenges are met in diverse ways and are the subject of this chapter.

28.1 HOMEOSTASIS AND TEMPERATURE REGULATION

LEARNING OUTCOMES

1. Describe why an animal's physiological functions are linked to its body temperature.
2. Define how animals can be classified with respect to temperature regulation.
3. Describe several different mechanisms for temperature homeostasis in animals.

Metabolic reactions of all cells are influenced by temperature. There is always a temperature range that promotes optimal cellular functions, and that range often varies from species to species. For example, muscle cells of an insect contract forcefully and fast enough for flight when the insect's body musculature is within an appropriate temperature range. (This range varies with species, but is usually between 40 and 45 °C.) Enzyme systems of a rainbow trout (*Oncorhynchus mykiss*) function best between 10 and 15 °C. Enzyme systems of most mammals function within a 36 to 39 °C range.

When body temperatures fall outside their optimal ranges, physiological functions are compromised. Rates of enzymatic reactions decline at both temperature extremes. Extreme

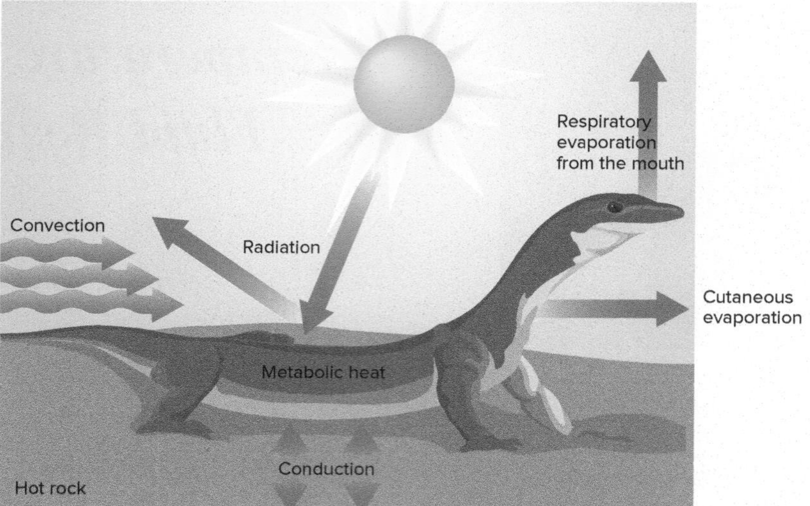

FIGURE 28.1

Heat Gain and Loss for a Terrestrial Reptile in a Typical Terrestrial Environment. Heat is either gained or lost by objects in direct contact with the animal (conduction), by air currents (convection), by exhaled air or water on the surface of an animal (evaporative cooling), and by electromagnetic waves (radiation).

temperatures affect the fluidity of lipids in cellular membranes, which inhibits many cellular transport functions. High temperatures cause nucleic acids to denature. Fortunately, evolutionary adaptations provide physiological and behavioral mechanisms that provide most animals with limited ability for **thermoregulation**—that is, most animals can influence their body temperature. The ability to thermoregulate is strongly influenced by heat gains and losses from the environment.

Heat Gains and Losses

Animals produce heat as a by-product of metabolism and either gain heat from, or lose it to, the environment. An animal's body temperature is a result of an interaction of these factors and can be expressed as:

Body temperature = heat produced metabolically
+ heat gained from the environment
− heat lost to the environment

Four physical processes influence heat exchanges with the environment: conduction, convection, evaporation, and radiation (figure 28.1). **Conduction** is the direct transfer of thermal motion (heat) between molecules of the environment and those on the body surface of an animal. The transfer of heat moves along thermal gradients—from warmer to cooler surfaces. Thus, animals can absorb heat from warm surfaces or lose heat to cold surfaces.

Convection is heat transfer as air or a liquid moves over a warm (or cool) surface. Arctic animals may seek shelter from cold winds because their warmer skin loses heat to cold air moving over their skin.

Evaporation is a change of state of water from a liquid to a gas. Evaporation involves breaking hydrogen bonds of water molecules to free individual molecules to depart from a watery surface. Breaking these hydrogen bonds requires energy. Evaporation from body surfaces (in the form of sweating in primates and horses [*Equus*], and panting in dogs [*Canis*] and avian and nonavian reptiles) uses body heat to break these hydrogen bonds. Thus, heat is lost, and cooling occurs.

Radiation is the emission of electromagnetic waves that objects, such as another animal's body or the sun, produce. Radiation can transfer heat between objects that are not in direct contact with each other, as happens when an animal suns itself (figure 28.2).

FIGURE 28.2

Radiation Warms an Animal. After a cold night in its den on the Kalahari Desert (southwest Africa), a meerkat (*Suricata suricatta*) stands at attention, allowing the large surface area of its body to absorb radiation from the sun. The meerkat, like many endotherms, uses behaviors to supplement metabolic thermoregulatory mechanisms.

Other factors related to the structure of an animal and its environment influence heat gains and losses. Small animals that have a high surface-to-volume ratio lose or gain heat more rapidly from external sources. Animals that have insulating fat deposits, fur, or feathers have a lower heat conductance. Animals that live in environments where the ambient temperature is close to their preferred body temperature have lower rates of heat gain or loss.

Some Solutions to Temperature Fluctuations

Animals have adapted to temperature fluctuations in one or more of the following ways: (1) they may occupy a place in the environment where the temperature remains constant and compatible with their physiological processes; (2) their physiological processes may have adapted to the range of temperatures in a specific environment; and (3) they may generate and trap heat internally to maintain a relatively stable body temperature, despite fluctuations in the temperature of the external environment.

Zoologists classify thermoregulation by animals in one of two ways. Animals can be classified according to their source of body heat and their ability to maintain body temperature. **Heterotherms** have a body temperature that may vary with environmental temperature [quadrants (1), (2), and (3) of figure 28.3*a* and figure 28.3*b*]. **Homeotherms** maintain a relatively constant body temperature despite swings in the environmental temperature by variable production and retention of metabolic heat [quadrant (4) of figure 28.3*a* and *see figure 28.3*b]. This dichotomy, based on relative constancy of body temperature, does not tell the whole temperature regulation story.

The second way of classifying animals is based on whether or not body heat is metabolically generated. Animals that use metabolism to generate body heat are called **endotherms** (Gr. *endos*, within) [quadrants (2) and (4) of figure 28.3*a*]. Most endotherms have a lower thermoconductivity (the flow of heat from warm to cold) due to fur, feathers, and/or fat insulating mechanisms. These animals usually maintain a relatively stable body temperature and are also homeotherms [quadrant (4) of figure 28.3*a*]. A few endotherms do not retain body heat and are heterothermic [quadrant (2) of figure 28.3*a*; *see How Do We Know That Not All Mammals Are Homeotherms, page 524*]. Animals that have a relatively low metabolic rate and do not generate heat metabolically are called **ectotherms** (Gr. *ectos*, outside). Ectotherms tend to have a higher thermoconductivity and lack insulation [quadrants (1) and (3) of figure 28.3*a*], but many ectotherms do thermoregulate through specific behaviors that warm or cool their bodies [quadrant (3) of figure 28.3*a*). Cooling behaviors include retreating to cooler burrows during the hottest part of the day and raising their body off the ground to reduce conduction from warm surfaces. Warming

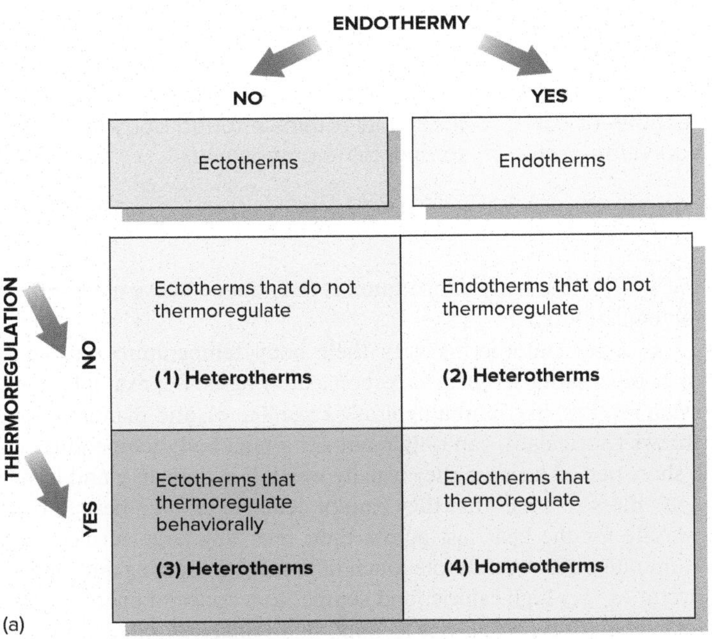

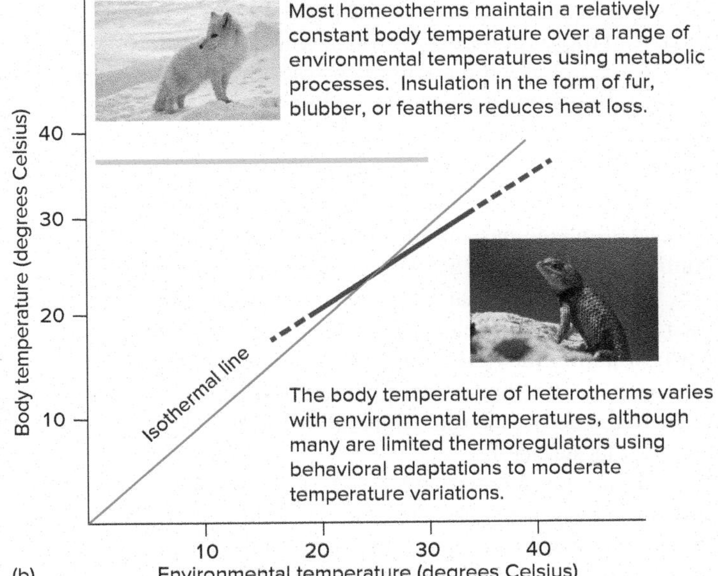

FIGURE 28.3

Thermoregulation. (*a*, 1–3 and *b*) Heterotherms have body temperatures that can vary with their environment. (*a*, 4 and *b*) Homeotherms produce heat metabolically and have relatively stable body temperatures because insulation reduces heat loss to the environment. (*a*, 1 and 3) Ectotherms derive body heat from their environment. (*a*, 3) Even though their body temperatures vary, ectotherms may thermoregulate using behavioral mechanisms. (*a*, 2 and 4) Endotherms generate heat metabolically. Most endotherms are also homeotherms (*a*, 4), but a few lack insulation in the form of fur or feathers and cannot maintain stable body temperatures over a range of environmental temperatures (*a*, 2). (b) Mammals like the Arctic fox (*Vulpes lagopus*) shown here are usually homeotherms (shown in yellow). Lizards like this male spiny lizard (*Sceloporus magister*) are heterotherms (shown in blue). This lizard is cooling itself by raising its body off the warm rock to decrease thermal conduction from the rock. Thermal responses of heterotherms vary at temperature extremes (broken blue lines). The isothermal line (shown in red) represents a body temperature that varies exactly with environmental temperature.

How Do We Know That Not All Mammals Are Homeotherms

Naked mole rats (*Heterocephalus glaber*) live in complex tunnels in equatorial African grasslands (box figure 28.1). They have no body fur, subcutaneous fat, or insulation. Consequently, they cannot regulate their body temperature like other endothermic mammals. Their body temperature fluctuates around 31°C, but some stasis is achieved by employing behaviors similar to that of certain social insects.

To keep warm, naked mole rats often bask in their sun-exposed, shallowly constructed tunnels and also huddle together while sleeping. These thermoregulatory mechanisms are similar to those used by termites (Isoptera) in their mounds. In addition, termites and mole rats have a social structure (called eusocial) that is similar to hive caste systems of social hymenopterans. There is a single queen mole rat that only mates with a few dominant males. Workers dig the tunnels and gather food, while soldiers protect the tunnels from predators. Overall, the entire mole rat colony must cooperate to survive—part of this cooperation ensures that each individual retains enough body heat to maintain homeostasis.

BOX FIGURE 28.1 **Naked African Mole Rat.** *Heterocephalus glaber* in their tunnel.

behaviors include basking in the sun and increasing body surface contact with rocks or other surfaces warmed by the sun.

Most birds and mammals are endotherms because they obtain their body heat from cellular processes. A constant source of internal heat allows them to maintain a nearly constant core temperature, despite the fluctuating environmental temperature. ("Core" refers to the body's internal temperature as opposed to the temperature near its surface.)

Most endotherms have bodies insulated by fur or feathers and a relatively large amount of fat. This insulation enables them to retain heat more efficiently and to maintain a high core temperature. Endothermy allows animals to stabilize their core temperature so that biochemical processes can proceed at homeostatic levels. Endothermy allows some animals to colonize habitats denied to ectotherms.

The advantages of endothermy are achieved at a high energy cost. Efficient circulatory systems, high respiratory rates, increased food requirements, maintenance of insulation, and greater excretory demands all accompany endothermy. Metabolic rates and energy expenditures related to thermoregulation approximately double for every 10°C that body temperature is maintained above environmental temperature.

Some endotherms vary their body temperature seasonally (e.g., hibernation); others vary it on a daily basis. For example, some avian reptiles, like hummingbirds (Trochilidae), and mammals, like shrews (Soricidae), can only maintain a high body temperature for a short period because they usually weigh less than 10 g and have a body mass so small that they cannot generate enough heat to compensate for the heat lost across their relatively large surface area. Hummingbirds must devote much of the day to locating and sipping nectar (a very-high-calorie food source) as a constant energy source for metabolism. When not feeding, hummingbirds rapidly run out of energy unless their metabolic rates decrease. At night, hummingbirds enter a sleep-like state, called daily torpor, and their body temperature approaches that of the cooler surroundings. Some bats also undergo daily torpor to conserve energy.

In general, ectotherms are more common in temperate and tropical regions because they are dependent on, and limited by, their environment to maintain body temperatures. Their success in these regions is, in part, due to a greater portion of their energy expenditures being devoted to productive functions (growth and reproduction) than is the case for endotherms. Indeed, in the tropics, amphibians are far more abundant than mammals.

Temperature Regulation in Invertebrates

Many invertebrates have no thermoregulatory mechanisms and passively conform to environmental temperatures. These are said to be **thermoconformers**. Others can detect temperature differences and have physiological and behavioral mechanisms that influence body temperature.

Many arthropods have unique mechanisms for surviving temperature extremes. For example, temperate-zone insects avoid freezing by reducing the water content in their tissues as winter approaches. Other insects can produce glycerol or other glycoproteins that act as an antifreeze. Some moths and bumblebees warm up prior to flight by shivering contractions of their thoracic flight muscles. Most large, flying insects have evolved a mechanism to prevent overheating during flight; hemolymph circulating through the flight muscles carries heat from the thorax to the abdomen, which gets rid of the heat through radiation and convection. Citrus cicadas (*Diceroprocta apache*) have adapted to conditions in the Sonoran Desert (northwest Mexico) using an evaporative cooling mechanism. When threatened with overheating, these cicadas extract water from their hemolymph and transport it through large ducts to the surface of the body, where it passes through sweat pores and evaporates. In other words, these insects can sweat.

Body posture and orientation of the wings to the sun can markedly affect the body temperature of basking insects. For example, perching dragonflies *(see cover photograph)* and butterflies can regulate their radiation heat gain by postural adjustments.

Many endothermic insects (such as bumblebees, honeybees, and some moths) have a countercurrent heat exchanger (*see figures 28.4 and 28.5*) that helps maintain a high temperature in the thorax, where the insect's flight muscles are located. This mechanism allows the insect to control heat gain and loss by regulating the flow of hemolymph through the heat exchanger. Hemolymph flowing through the heat exchanger allows body heat to be retained in the flight muscles. Diverting hemolymph from the countercurrent heat exchanger to the abdomen promotes heat loss from that region of the body.

To prevent overheating, many ground-dwelling arthropods (*Tenebrio* beetles, locusts, and scorpions) raise their bodies as high off the ground as possible to minimize heat gain from the ground. Some caterpillars and locusts orient with reference to both the sun and wind, thus regulating heat gain or loss from radiation and convection. Some desert-dwelling beetles can exude waxes from

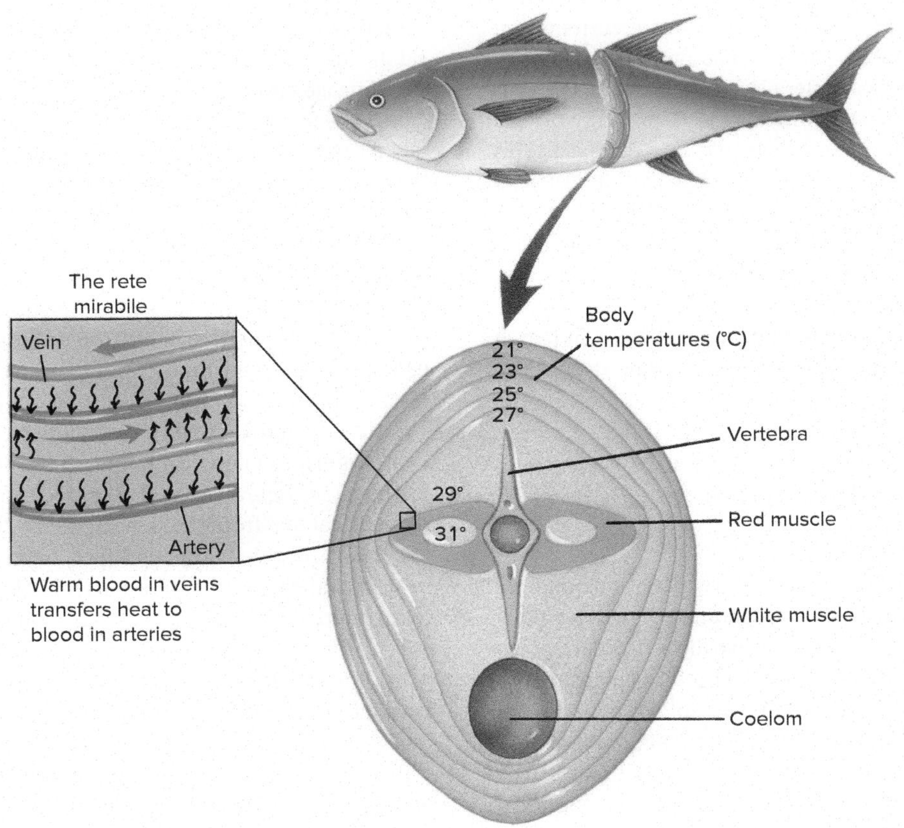

FIGURE 28.4

Thermoregulation in Large, Active Fishes. In the bluefin tuna, the rete mirabile of arteries and veins acts as a countercurrent exchange system that helps reduce the loss of body heat. The cross section through the body shows that the temperature is the highest around the red swimming muscles.

thousands of tiny pores on the cuticle. These "wax blooms" prevent dehydration and also are an extra barrier against the desert sun.

Color has a significant effect on thermoregulation, as 50% of the radiant energy from the sun is in the visible spectrum. A black surface reflects less radiant energy than a white surface. Thus, many black beetles may be more active earlier in the day because they absorb more radiation and heat faster. Conversely, white beetles are more active in the hotter parts of the day because they absorb less heat.

Temperature Regulation in Fishes

The temperature of the surrounding water determines the body temperature of most fishes. Fishes that live in extremely cold water have "antifreeze" materials in their blood. Polyalcohols (e.g., sorbitol and glycerol) or water-soluble peptides and glycopeptides lower the freezing point of blood plasma and other body fluids. These fishes also have proteins or protein-sugar compounds that stunt the growth of ice crystals that begin to form. These adaptations enable these fishes to stay flexible and swim freely in a supercooled state (i.e., at a temperature below the normal freezing temperature of a solution; −2 °C.

Some active fishes maintain a core temperature significantly above the temperature of the water. Bluefin tuna, swordfish, and the great white shark have major blood vessels just under the skin. Branches deliver blood to the deeper, powerful, red swimming muscles, where smaller vessels are arranged in a countercurrent heat exchanger called the **rete mirabile** ("miraculous net") (figure 28.4). The heat that these red muscles generate is not lost because it is transferred in the rete mirabile from venous blood passing outward to cold arterial blood passing inward from the body surface. This arrangement of blood vessels enhances vigorous activity by keeping the swimming muscles several degrees warmer than the tissue near the surface of the fish. This system has been adaptive for these fishes. Their muscular contractions can have four times as much power as those of similar muscles in fishes with cooler bodies. Thus, they can swim faster and range more widely through various depths in search of prey than can other predatory fishes more limited to given water depths and temperatures.

Temperature Regulation in Terrestrial Ectotherms

Terrestrial ectotherms (i.e., most nonavian reptiles and amphibians on land) are in a medium whose temperature fluctuates daily and seasonally. Most of these animals have little insulation, so metabolic heat that is generated is rapidly lost to the environment. Similarly, in hot environments, heat may be rapidly gained from the environment. Behavioral thermoregulatory adaptations are common in these animals. They may select favorable times of the day for activities and seek shelter when temperatures are not favorable (nocturnal versus diurnal). They orient their bodies to absorb, or avoid, radiant energy. They may bask to absorb heat (conduction) or lift their bodies from warm surfaces to cool themselves (convection).

Most amphibians exchange at least some of their respiratory gases across the skin surface. Their moist skin acts an evaporative cooling system, and it may contribute to excessive water loss. These considerations help explain the abundance of amphibians in warm, moist environments. Some amphibians, such as American bullfrogs (*Lithobates catesbeianus*), can vary the amount of mucus they secrete from the body surface—a physiological response that helps regulate evaporative cooling.

Nonavian reptiles have dry skin, which reduces the loss of body heat through evaporative cooling of the skin. Their expandable rib cage allows for efficient ventilation and promotes evaporative cooling through the respiratory system. In addition to the behavioral mechanisms described earlier, many nonavian reptiles have physiological specializations that aid in thermoregulation. For example, diving reptiles (e.g., sea turtles and sea snakes) conserve body heat by routing blood through circulatory shunts into the center of the body. These animals can also increase heat production in response to the hormones thyroxine and epinephrine. In addition, tortoises and land turtles can cool themselves through salivating and frothing at the mouth, urinating on the back legs, moistening the eyes, and panting.

Temperature Regulation in Endotherms

Endothermy evolved twice in the amniote lineage—once in the synapsid lineage (i.e., leading to modern mammals) and once in the archosaur lineage (leading to dinosaurs, crocodylians, and avian reptiles) (*see figure 20.3*). Avian reptiles and mammals are able to live in habitats all over the earth because they are homeothermic endotherms; they can maintain body temperatures between 35 and 42 °C with metabolic heat.

Various cooling mechanisms prevent excessive warming in birds. Because they have no sweat glands, birds pant to lose heat through evaporative cooling. Some species have a highly vascularized pouch (gular pouch) in their throat that they can flutter (a process called **gular flutter**) to increase evaporation from the respiratory system.

Some birds possess mechanisms for preventing heat loss. Feathers are excellent insulators for the body, especially downy-type feathers (*see figure 21.4*b) that trap a layer of air next to the body to reduce heat loss from the skin (figure 28.5*a*). Aquatic species, which lose heat from their legs and feet, have peripheral countercurrent heat exchange vessels (rete mirabile) in their legs to reduce heat loss (figure 28.5*b*).

Mammals have similar heat-conservation mechanisms. Pelage provides insulation in nearly all mammals. Some other mammals have a thick layer of insulating fat called **blubber** just under the skin that helps marine animals, such as seals (Carnivora, Otariidae and Phocidae) and whales (Cetacea), maintain a body temperature of around 36 to 38 °C. In the tail and flippers, which have no blubber, a countercurrent system of arteries and veins helps minimize heat loss. Arctic mammals, such as the arctic fox (*Vulpes lagopus*) and barren-ground caribou (*Rangifer tarandus groenlandicus, see figure 22.14*), also have similar exchange vessels in other extremities

(a)

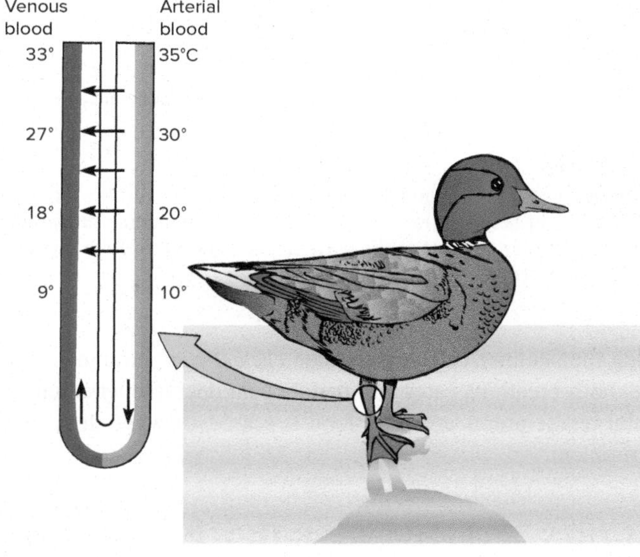

(b)

FIGURE 28.5

Insulation and Countercurrent Heat Exchange. (*a*) A thick layer of down feathers keeps these Chinstrap penguins (*Pygoscelis antarcticus*) warm. Their covering of short, stiff feathers interlocks to trap air, forming the ornithological equivalent of a diver's suit. (*b*) The countercurrent heat exchanger in a bird foot. Some aquatic birds, such as this duck, possess countercurrent systems of arteries and veins (rete mirabile) in their legs that reduce heat loss. The arteries carry warm blood down the legs to warm (arrows) the cooler blood in the veins, so the heat is carried back to the body rather than lost through the feet that are in contact with a cold surface.

(e.g., legs, tails, ears, and nose). Animals in hot climates, such as jackrabbits (*Lepus*), have mechanisms (e.g., large ears) to rid the body of excess heat (figure 28.6).

Birds and mammals also use behavioral mechanisms to cope with external temperature changes. Like ectotherms, they sun themselves or seek shade as the temperature fluctuates (*see figure 28.2*).

FIGURE 28.6

Temperature Regulation. This antelope jackrabbit (*Lepus alleni*) must get rid of excess body heat. Its huge, thin, highly vascularized ears that are perfused with warm blood have a large surface area for heat exchange.

Many avian reptiles and mammals huddle to keep warm; others share burrows for protection from temperature extremes. Migration to warm climates and hibernation enable many birds and mammals to survive the harsh winter months.

Heat Production in Endotherms

In endotherms, heat is generated **(thermogenesis)** as a by-product of daily metabolism and muscle contraction. Other physiological specializations also contribute to thermogenesis.

Every time a muscle cell contracts, the actin and myosin filaments sliding over each other and the hydrolysis of ATP molecules generate heat. Both voluntary muscular work (e.g., running, flying, and jumping) and involuntary muscular work (e.g., shivering) generate heat. Heat generation by shivering is called **shivering thermogenesis**.

Birds and mammals have a unique capacity to generate heat by using specific enzymes of ancient evolutionary origin–the ATPase pump enzymes in the plasma membranes of most cells. When the body cools, the thyroid gland releases the hormone thyroxine. Thyroxine increases the permeability of many cells to sodium (Na^+) ions, which leak into the cells. The ATPase pump quickly pumps these ions out. In the process, ATP is hydrolyzed, releasing heat energy. The hormonal triggering of heat production is called **nonshivering thermogenesis**.

Brown fat is a specialized type of fat found in newborn mammals, in mammals that live in cold climates, and in mammals that hibernate (figure 28.7). The brown color of this fat comes from the large number of mitochondria with their iron-containing cytochromes. Deposits of brown fat are beneath the ribs and in the shoulders. A large amount of heat is produced when brown fat cells oxidize fatty acids because mitochondrial membrane transport systems are modified for heat production rather than phosphorylating ADP to produce ATP. Blood flowing past brown fat is heated and contributes to warming the body.

In all tetrapods, the hypothalamus of the brain contributes to thermoregulation; however, in avian reptiles and mammals, the hypothalamus is especially important in thermoregulatory functions. The two hypothalamic thermoregulatory areas are the heating center and the cooling center. The heating center controls vasoconstriction of superficial blood vessels, erection of hair and fur, and shivering or nonshivering thermogenesis. The cooling center controls vasodilation of blood vessels, sweating, and panting. Negative feedback mechanisms (with the hypothalamus acting as a thermostat) trigger either the heating or cooling of the body and thereby control body temperature (figure 28.8). Specialized neuronal receptors in the skin and other parts of the body sense temperature changes. Warm neuronal receptors excite the cooling center and inhibit the heating center. Cold neuronal receptors have the opposite effects.

During the winter some endotherms, including bats (e.g., *Myotis*), woodchucks (*Marmota*), chipmunks (*Tamias*), and ground squirrels (Sciuridae), go into **hibernation** (L. *hiberna*, winter). During hibernation, the metabolic rate slows, as do the heart and breathing rates. Mammals prepare for hibernation by building up fat reserves and growing long winter pelts. All hibernating animals have brown fat. Decreasing day length stimulates both increased fat deposition and fur growth. Other mammals such as black bears (*Ursus americanus*) and skunks (Mephitidae) enter a state of prolonged winter sleep in which body temperatures and metabolic rates decrease somewhat, but they do not necessarily remain inactive all winter (*see chapters 6 and 22*).

Some very small endotherms like bats, chickadees (*Poecile*), and hummingbirds (Trochilidae) can also reduce both metabolic rate and body temperature to produce a state of dormancy called daily torpor. **Daily torpor** (*see chapter 6*) (L. *torpore*, to be sluggish) allows an animal to reduce the need for food by reducing metabolism. For example, hummingbirds allow their body temperature to drop as much as 25 °C at night when food supplies are low. This strategy is only found in small endotherms, as larger ones have too large of a body mass to allow rapid cooling.

Other vertebrates (desert tortoise [*Gopherus*], pygmy mouse [*Baiomys*], and ground squirrels) will enter a state of dormancy during the summer called **aestivation** (L. *aestivus*, summer). In this state, both breathing rates and metabolism decrease when environmental temperatures are high, food is scarce, or when dehydration is a problem.

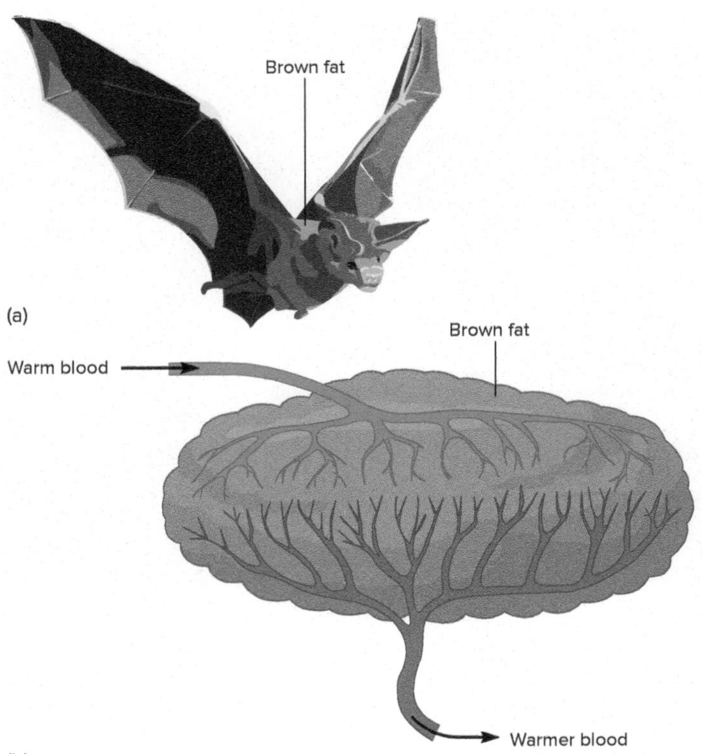

FIGURE 28.7

Brown Fat. (*a*) Many mammals, such as this bat, have adipose tissue called brown fat between the shoulder blades. (*b*) The area of brown fat is much warmer than the rest of the body. Blood flowing through the brown fat is further warmed.

SECTION 28.1 THINKING BEYOND THE FACTS

Based on what you have learned in this section, why are the terms "cold blooded" and "warm blooded" outmoded and inaccurate with respect to describing temperature regulation in animals?

28.2 CONTROL OF WATER AND ELECTROLYTES

LEARNING OUTCOMES

1. Explain the importance of an animal's maintenance of osmotic balance.
2. Describe how animals are classified based on their method of osmoregulation.

Excretion (L. *excretio*, to eliminate) is the elimination of metabolic wastes from an animal's body. These wastes include. carbon dioxide

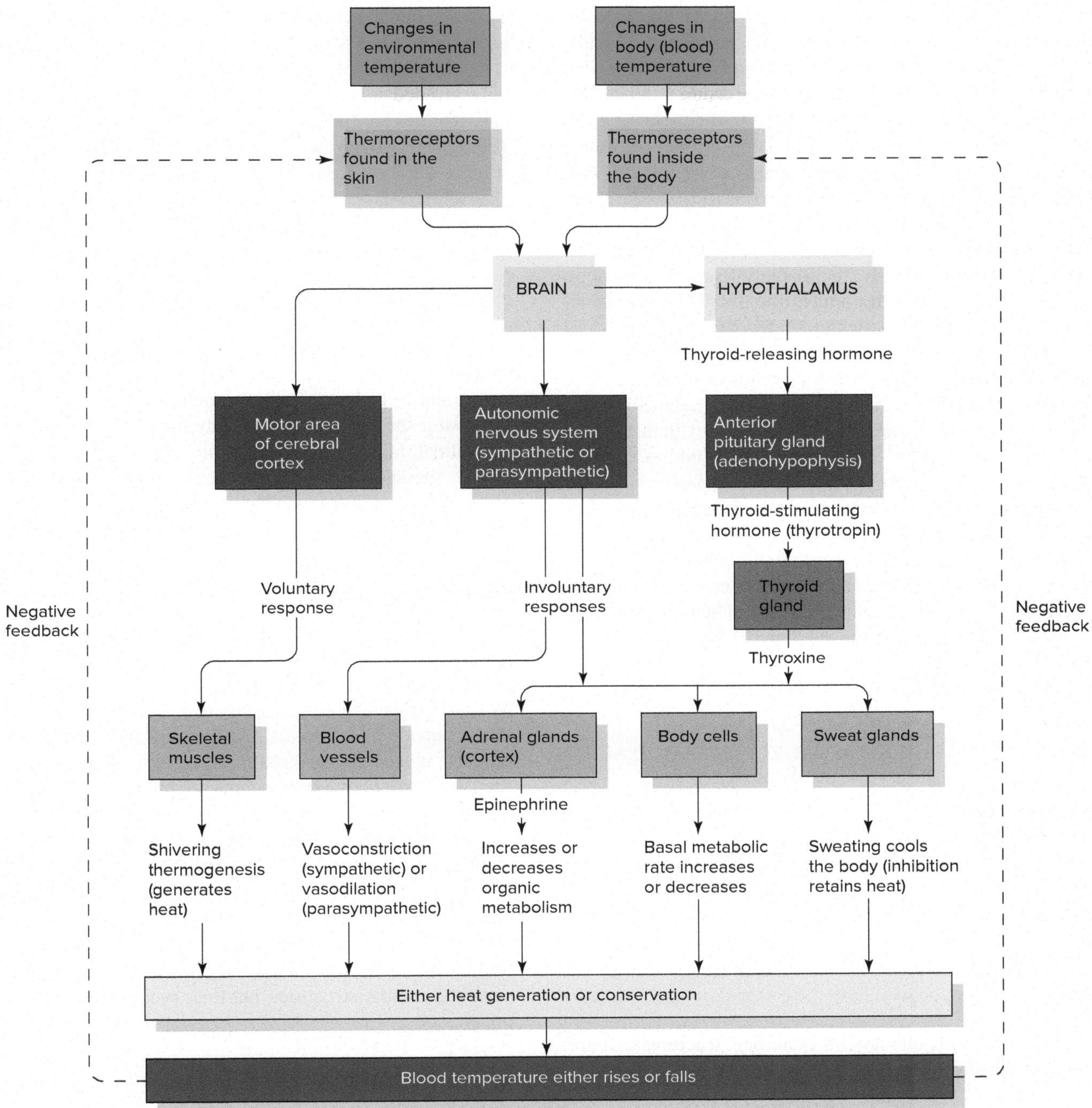

FIGURE 28.8

Thermoregulation. Overview of the feedback pathways that control the core body temperature of a mammal. Arrows show the major control pathways.

(see chapter 26) and excess nitrogen released during the deamination of amino acids. Nitrogen is excreted as ammonia, urea, or uric acid. Excretory systems are also involved in regulating water and electrolyte balances. These functions are discussed in the remainder of this chapter.

No animal can completely conform to the concentration of water and electrolytes in its external environment. Complete conformity would mean that conditions inside cells would be identical to conditions in the environment. The concentrations of specific ions (e.g., Na^+ and K^+) must be maintained at intracellular concentrations that support functions like nerve impulse conduction and muscle contraction. Even though the sum of all ions inside an animal's cells might equal that of the environment, the concentrations of physiologically important ions are always homeostatically regulated. Most animals actively regulate the concentrations of osmotically active solutes (e.g., electrolytes and urea) at levels different from their environment. Water moves between the animal and its environment by osmosis in response to solute concentrations. Regulation of these exchanges is called osmoregulation.

Osmolarity is a measure of the osmotic pressure (strength) of a solution measured in osmoles and is related to the molar concentration of solutes in a solution. If the osmolarity of the body fluids of an animal varies depending on the osmolarity of its environment, the animal is an **osmoconformer**. Even though specific ion concentrations will be regulated, the total ion concentration of body tissues approximates that of the environment, and the animal is nearly isosmotic* to its medium. Obviously, the inability to regulate osmotic concentrations of body fluids has limited the distribution of osmoconformers. Many marine invertebrates are osmoconformers. In contrast, an animal that maintains its body fluids at a different osmolarity from that of its surrounding environment is an **osmoregulator**.

Most vertebrates living in seawater have body fluids with an osmolarity that is about a third less (hypoosmotic) than the surrounding seawater, and water tends to leave their bodies continually. To compensate for this problem, mechanisms evolved in these animals to conserve water and prevent dehydration. Freshwater animals have body fluids that are hyperosmotic with respect to their environment, and water tends to continually enter their bodies. Mechanisms evolved in these animals that excrete water and prevent fluid accumulation. Land animals have a higher concentration of water in their fluids than in the surrounding air. They tend to lose water to the air through evaporation and may use considerable amounts of water to dispose of wastes.

Section 28.2 Thinking Beyond the Facts

How are the osmoregulatory requirements of an animal are related to the animal's energy budget?

28.3 INVERTEBRATE EXCRETORY SYSTEMS

LEARNING OUTCOME

1. Describe the osmoregulatory and excretory structures that are found in major invertebrate taxa.

Aquatic invertebrates occur in a wide range of media, from freshwater to markedly hypersaline water (e.g., salt lakes). Generally, marine invertebrates have about the same osmotic concentration as seawater. Even though they are osmoconformers, marine invertebrates usually regulate the concentrations of specific ions used in achieving this conformity. Electrolytes are transported across membrane surfaces of the integument, digestive tract, and gills. They are lost by diffusion or transport cross these membranes or in the urine.

Freshwater invertebrates are strong osmoregulators because it is impossible to be isosmotic with dilute media. Any water gain is usually eliminated as urine. Electrolytes must be constantly conserved, and losses are replaced through active transport and feeding.

A number of invertebrate taxa have more or less successfully invaded terrestrial habitats. The most successful terrestrial invertebrates are the arthropods, particularly arachnids and myriapods. Overall, the water and ion balance of terrestrial invertebrates is quite different from that of aquatic animals because terrestrial invertebrates face limited water supplies and water loss by evaporation from their integument. Some of the invertebrate excretory mechanisms and systems are now discussed.

Contractile Vacuoles

Some marine invertebrates (e.g., cnidarians, echinoderms, and sponges) do not have specialized excretory structures because wastes simply diffuse into the surrounding isosmotic water. In some freshwater species, cells on the body surface actively pump ions into the animal. Many freshwater species (Porifera), however, have contractile vacuoles that pump out excess water. **Contractile vacuoles** are energy-requiring organelles that expel excess water from individual cells exposed to hypoosmotic environments .

Protonephridia

Nephridia (Gr. *nephros*, kidney) (sing., nephridium) are excretory structures that are common to many invertebrates. Two types of nephridia function in animals.

The **protonephridium** consists of a tubular system that opens to the outside of the animal. At the opposite (internal) end of the tubule a closed cell or group of cells draws body fluids into the tubule system. Flame cell systems are protonephridia that are present in rotifers, some annelids, larval molluscs, and some flatworms (figure 28.9) that live in freshwater. Ciliated **flame cells** are located along the tubule system. Beating of the cilia drive fluids from tissues into the tubules and eventually out of the animal through an excretory pore. Flame-cell systems function primarily in eliminating excess water. Nitrogenous waste simply diffuses across the body surface into the surrounding water. Protonephridia are very ancient excretory structures, but their evolutionary history is poorly understood.

Metanephridia

A second, and more common, invertebrate nephridium is the **metanephridium** (Gr. *meta*, beyond) (pl., metanephridia). Like protoneprhridia, metanephridia open to the outside of the animal. Their tubule system also has a second opening within the coelom of the invertebrate, usually in the form of a ciliated funnel, called the nephrostome. Cilia of the nephrostome drive fluids into, and through, the metanephridium. The tubule system may be modified with bladder-like structures where ions or molecules can be conserved through selective reabsorption into blood or other body fluids.

*In this chapter we use the terms "isosmotic," "hypoosmotic," and "hyperosmotic," which refer specifically to osmolarity. The terms "isotonic," "hypotonic," and "hypertonic" are more limited because they apply only to the response of animal cells—whether they swell or shrink—in solutions of known solute concentrations.

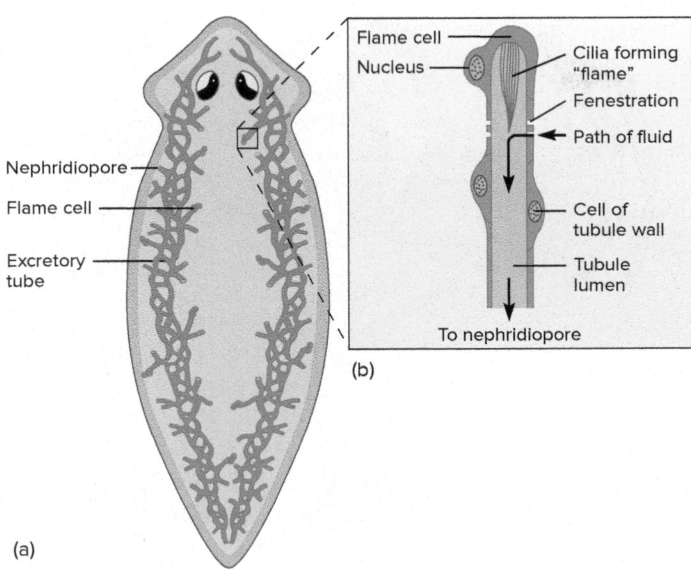

FIGURE 28.9

Protonephridial (Excretory) System in a Turbellarian. (*a*) The system lies in the mesenchyme and consists of a network of fine tubules that run the length of the animal on each side and open to the surface by minute excretory pores called nephridiopores. (*b*) Numerous fine side branches from the tubules originate in the mesenchyme in enlargements called flame cells.

Most annelids (e.g., the earthworm, *Lumbricus*) and a variety of other invertebrates have a metanephridial excretory system. Recall that the earthworm's body is divided into segments and that each segment has a pair of metanephridia. The nephrostome is open to the coelom of one segment and anchored to a septum that separates two body segments. The tubule passes through the septum into the next posterior segment. (figure 28.10; *see also figure 12.8*). As beating cilia move the fluid through the tubule, a network of capillaries surrounding the tubule and bladder reabsorb and carry away ions. Each tubule empties urine to the outside of the body through the nephridiopore. Each day an earthworm may produce a volume of urine that is equal to 60% of its body weight.

Some research suggests homologies between metanephridia and mollusc kidneys (covered next).

Mollusc Kidneys (Nephridia)

The renal organs of adult molluscs are tubular (saccular) structures called nephridia (Gr. *nephra*, kidney) or kidneys. These nephridia either empty into the mantle cavity or directly to the outside of the mollusc. Bivalves, most cephalopods, and some gastropods have two nephridia. Most gastropods only have one nephridium, since the right nephridium has disappeared as a result of embryonic torsion (*see figure 11.5*). Each nephridium consists of a sac with highly folded walls and connects to the reduced coelom (the pericardial cavity) that surrounds the heart (*see figures 11.11 and 11.17*). Excretory wastes are derived largely from fluids filtered and secreted into the coelom from the blood. The nephridum modifies this waste by selectively reabsorbing certain ions and molecules. (Reabsorption

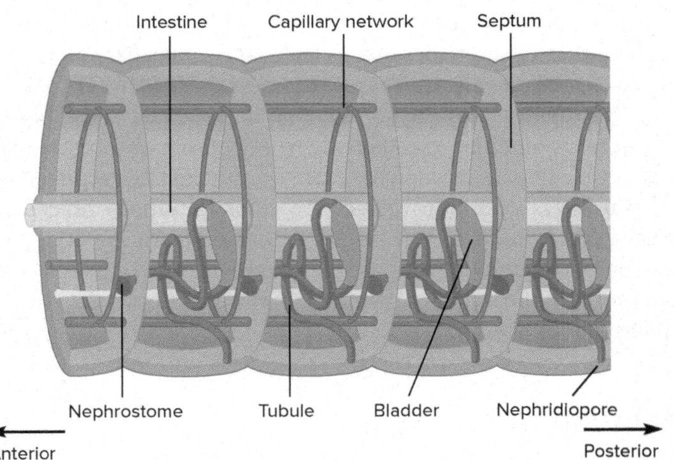

FIGURE 28.10

Earthworm (*Lumbricus*) Metanephridium. The metanephridium opens by a ciliated nephrostome into the cavity of one segment, and the next segment contains the nephridiopore. The main tubular portion of the metanephridium is coiled and is surrounded by a capillary network. Waste can be stored in a bladder before being expelled to the outside. Most segments contain two metanephridia.

of nonwaste materials from urine is an important function of most animal excretory systems.) Aquatic mollusc species excrete ammonia because they have access to water in which the toxic ammonia is diluted. In contrast, terrestrial snails must convert ammonia to a less-toxic form–uric acid. Because uric acid is less soluble in water and less toxic, it can be excreted in a semisolid form, which helps conserve water.

Antennal, Maxillary, and Coxal Glands

Three types of excretory structures found in arthropods probably originated from nephridia and are homologous with each other. These structures lack an open nephrostome and function by filtration through a closed filtration membrane, probably because they are bathed in hemolymph rather than being associated with a coelomic cavity. Hemolymph pressure forces fluids, ions, and small molecules through the filtration membrane and into the tubule system. Modifications of their tubule systems promote reabsorption functions. The paired occurrence of these organs in various locations suggests that they originated from segmentally arranged nephridia. They are antennal, maxillary, and coxal glands.

In those crustaceans that have gills, nitrogenous wastes are removed by simple diffusion across the gills. Most crustaceans are aquatic and release ammonia, although they also produce some urea and uric acid as waste products. Thus, the excretory organs of freshwater species may be more involved with the reabsorption of ions and elimination of water than with the discharge of nitrogenous wastes. The excretory organs in some crustaceans (crayfish and crabs) are called **antennal glands** or **green glands** because of their location near the antennae and their green color (figure 28.11). Marine crustaceans have a short nephridial canal and produce urine that is isosmotic to their hemolymph. The nephridial canal is longer in freshwater crustaceans; permits efficient selective reabsorption of conserved ions.

In other crustaceans (some malacostracans [crabs, shrimp, and pillbugs]), the excretory organs are near the maxillary segments and are termed **maxillary glands**. In maxillary glands, fluid collects within the tubules from the surrounding hemolymph, and this primary urine is modified substantially by selective reabsorption and secretion as it moves through the excretory system and rectum.

Coxal (L. *coxa*, hip) **glands** are common among arachnids (spiders, scorpions, ticks, and mites). Wastes are discharged through pores on one to several pairs of appendages near the proximal segment (coxa) of the leg. Coxal glands may also function in the release of pheromones (figure 28.12).

Malpighian Tubules

Some arthropods have excretory structures that are not nephridial in origin. Insects, arachnids, and myriapods (in addition to, or instead of, coxal glands) have an excretory system made up of the gut and **Malpighian tubules** (named after Marcello Malpighi, Italian anatomist, 1628–1694) attached to the gut (figure 28.13). Excretion involves the active transport of potassium ions into the tubules from the surrounding hemolymph and the osmotic movement of water that follows. Nitrogenous waste (uric acid) also enters the tubules. As fluid moves through the Malpighian tubules, some of the water and certain ions are recovered. All of the uric acid passes into the gut and out of the body in the feces. Because insects are capable of conserving water very effectively, the insect's excretory system is a key adaptation contributing to these animals' tremendous success on land.

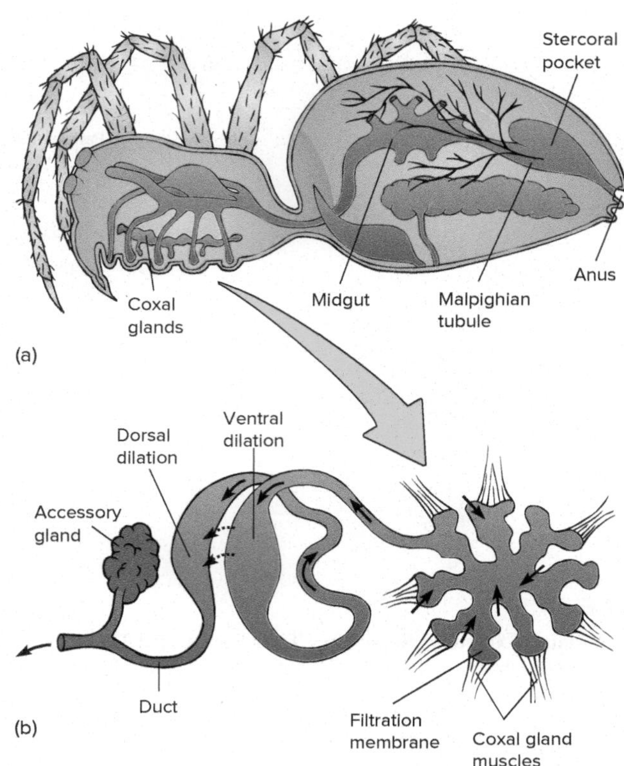

FIGURE 28.12

Coxal Glands in Arachnids. (*a*) The gut and excretory systems of a spider. The stercoral pocket is a diverticulum off the hindgut that stores waste prior to waste elimination. (*b*) Coxal gland muscles attach to the thin saccular filtration membrane. These muscles promote filtration and fluid flow (black arrows) by contracting and relaxing along the tubular duct. Water and solutes are reabsorbed along the tubular duct.

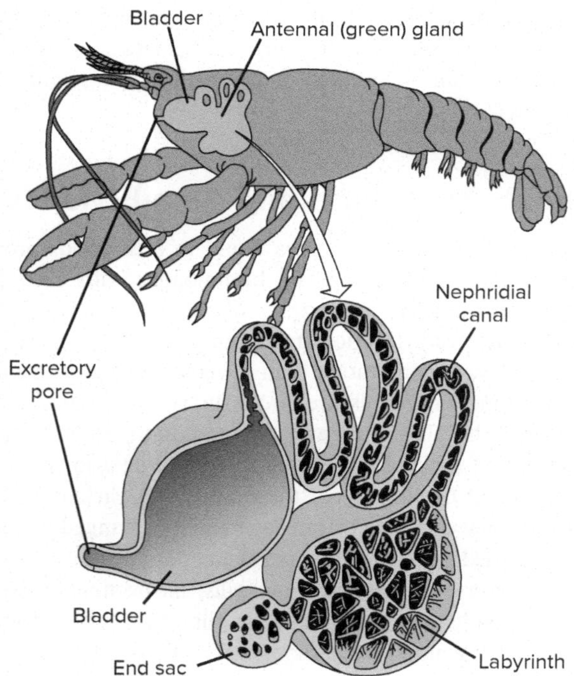

FIGURE 28.11

Antennal (Green) Gland of the Crayfish. The antennal gland, which lies in front of and to both sides of the esophagus, is divided into an end sac, where fluid collects by filtration, and a labyrinth. The labyrinth walls are greatly folded and glandular and appear to be an important site for reabsorption. The labyrinth leads via a nephridial canal into a bladder. From the bladder, a short duct leads to an excretory pore.

Soruce: Purves WK and Orlans GH. 1987 Life: The Science of Biology, 2e. Sunderland (MA). Sinauer Associates, Inc.

Section 28.3 Thinking Beyond the Facts

How are the functions of the Malpighian tubules different from the functions of other invertebrate excretory structures?

28.4 VERTEBRATE EXCRETORY SYSTEMS

LEARNING OUTCOMES

1. Explain the three key physiological functions vertebrates use to achieve osmoregulation.
2. Compare and contrast osmoregulation by freshwater fishes and osmoregulation by marine fishes.
3. Compare the functions of antennal glands and vertebrate kidneys.

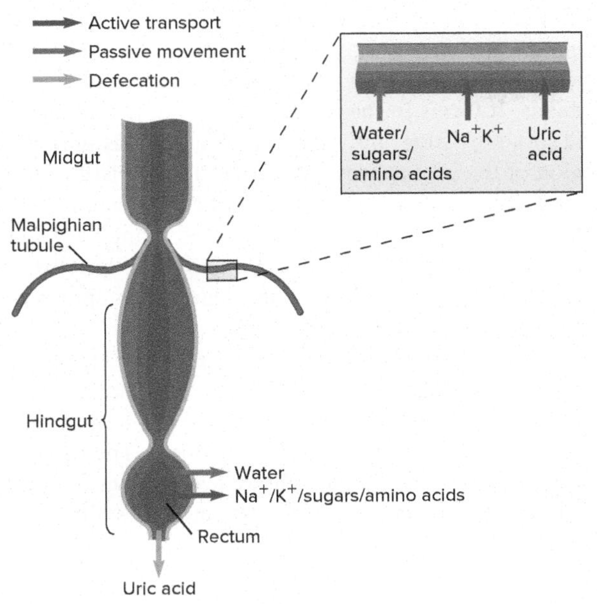

FIGURE 28.13

Malpighian Tubules. Malpighian tubules remove nitrogenous wastes (uric acid) from the hemocoel. Various ions are actively transported across the outer membrane of the tubule. Water follows these ions into the tubule and carries amino acids, sugars, and some nitrogenous wastes along passively. Some water, ions, and organic compounds are reabsorbed in the basal portion of the Malpighian tubules and the hindgut; the rest are reabsorbed in the rectum. Uric acid moves into the hindgut and is excreted in the feces.

Vertebrates face the same problems as invertebrates in controlling water and ion balance. Generally, water losses are balanced precisely by water gains (table 28.1). Vertebrates gain water by absorption from liquids and solid foods in the small and large intestines and by metabolic reactions that yield water as an end product. They lose water by evaporation from respiratory surfaces, evaporation from the integument, sweating or panting, elimination in feces, and excretion by the urinary system.

Solute losses also must be balanced by solute gains. Vertebrates take in solutes by the absorption of minerals from the small and large intestines, through the integument or gills, from secretions of various glands or gills, and by metabolism (e.g., the waste products of degradative reactions). They lose solutes in sweat, feces, urine, and gill secretions, and as metabolic wastes. The major metabolic wastes that must be eliminated are ammonia, urea, or uric acid.

Vertebrates live in saltwater, freshwater, and on land; each of these environments presents different water and solute problems. The next section discusses how vertebrates avoid losing or gaining too much water and, in turn, how they maintain a homeostatic solute concentration in their body fluids. The excretion of certain metabolic waste products is also coupled with osmotic balance and is discussed with the urinary system.

How Vertebrates Regulate Ions and Water

Various mechanisms have evolved in vertebrates to cope with their osmoregulatory problems, and most of them are adaptations of the urinary system. As presented in chapter 26, vertebrates have a closed circulatory system containing blood that is under pressure. As blood passes through kidney capillary beds, blood pressure forces plasma fluids, ions, and small molecules through capillary walls and into a tubule system where this filtrate will be processed into urine. The following functions occur in the kidney:

1. Filtration, in which blood passes through a capillary bed that retains blood cells, proteins, and other large solutes but lets small molecules, ions, and urea pass into the kidney tubule system
2. Reabsorption, in which selective ions and molecules are taken back into the bloodstream from the filtrate
3. Secretion, whereby selective ions and end products of metabolism (e.g., K^+, H^+, and NH_3) that are in the blood are added to the filtrate for removal from the body

TABLE 28.1
AVERAGE WATER GAIN AND LOSS IN A HUMAN AND A KANGAROO RAT

VERTEBRATE	WATER GAIN (ML)		WATER LOSS (ML)	
Human (daily)	Ingested in solid food	1,200	Feces	100
	Ingested as liquids	1,000	Urine	1,500
	Metabolically produced	350	Skin and lungs	950
	Total	2,550		2,550
Kangaroo rat (over four weeks)	Ingested in solid food	6	Feces	3
	Ingested in liquids	0	Urine	13
	Metabolically produced	54	Skin and lungs	44
	Total	60		60

Evolution of the Vertebrate Kidney

Vertebrates have two kidneys that are positioned dorsally in the abdominal cavity between the abdominal peritoneum and body wall muscle. (They are said to be retroperitoneal.) Kidneys form embryologically from tissue that extends the length of the body cavity. This embryological position is retained only in adult hagfishes (Hyperotreti, *see figure 28.14*a). In Vertebrata, various regions of the embryological tissue contribute to the kidneys of embryos and adults. There are three kinds of vertebrate kidneys: the pronephros, mesonephros, and metanephros. The **pronephros** (L. *pro*, before + *nephros*, kidney) forms in the anterior body cavity; and it appears only briefly in many vertebrate embryos, and not at all in mammalian embryos (figure 28.14*a*). The pronephros is the first osmoregulatory and excretory organ of some embryos and amphibian larvae. During the embryonic development of amniotes, or during metamorphosis in amphibians, the mesonephros replaces the pronephros (figure 28.14*b*). The **mesonephros** (Gr. *mesos*, middle + L. *nephros*, kidney) is the functioning embryonic kidney of many vertebrates and also adult fishes and amphibians. It forms in the middle portion of the abdominal cavity, and in adult fishes and amphibians it extends anteriorly into the region of the pronephros and is called the opisthonephros. Its embryological origin is similar to that of the pronephros and uses the pronephric duct system. The functional kidney of adult amniotes is the **metanephros** (Gr. *meta*, beyond + L. *nephros*, kidney) (figure 28.14*c*). It replaces the embryonic mesonephros in these animals and uses a new duct system.

The physiological differences between these kidney types are primarily related to the number of blood-filtering units they contain. The pronephric kidney contains fewer blood-filtering units than either the mesonephric or metanephric kidneys. The larger number of filtering units in the latter has allowed vertebrates to face the rigorous osmoregulatory and excretory demands of freshwater and terrestrial environments.

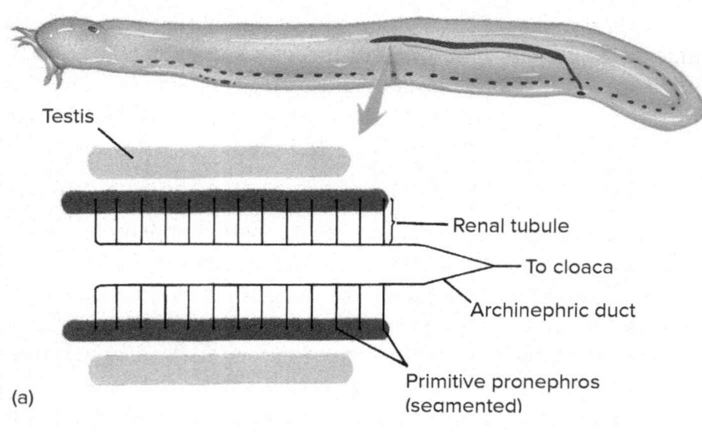

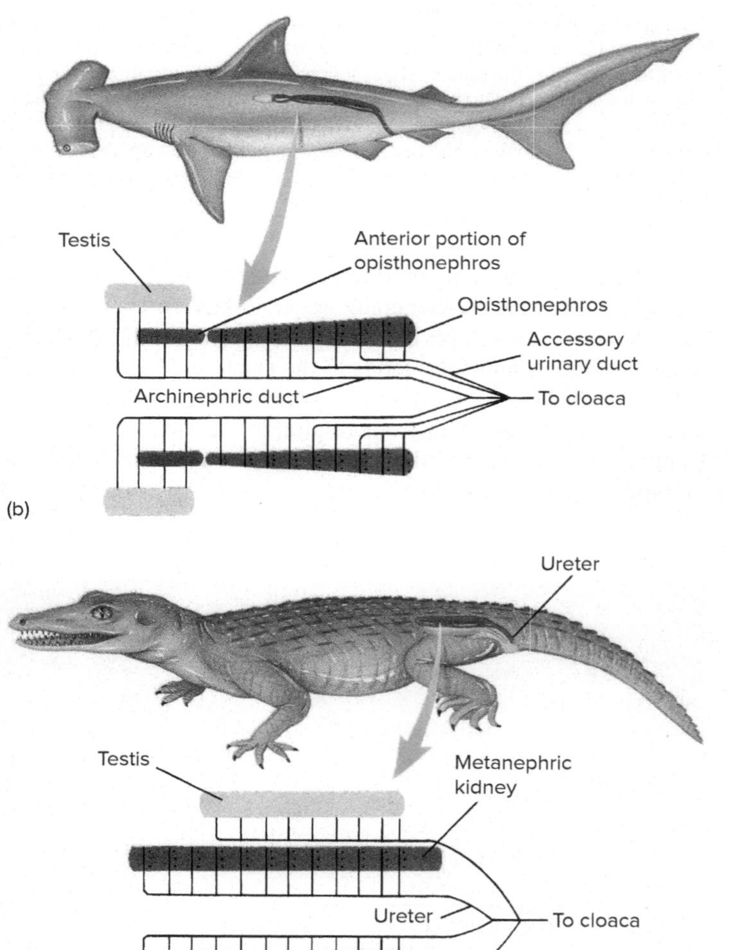

FIGURE 28.14

Types of Kidneys in Chordates and Their Association with the Male Reproductive System. The colored portions of the drawings represent the mesoderm that forms both the kidneys and gonads. Notice that it extends much of the length of the coelom during development. (*a*) The kidney of a hagfish (Hyperotreti) extends most of the length of the body cavity. In vertebrate fish embryos, the pronephros develops in the anterior portion of the body cavity. Segmental renal tubules lead from the kidney to the archinephric duct, which carries urine to a common opening with the anus (the cloaca). (*b*) The mesonephros is the functional kidney in the amniote embryo, adult fishes, and amphibians. It occupies the region of the body cavity posterior to the pronephros. In elasmobranchs, the kidney (opisthonephros) extends into the region of the pronephros. The anterior portion of the opisthonephros functions in blood cell formation and secretion of sex hormones. Notice that the testes occupy the position of the anterior opisthonephros, and the archinephric duct carries both sperm and urine. (*c*) The metanephric kidney of adult amniotes (nonavian and avian reptiles and mammals) forms in the posterior region of the body cavity. The kidney is compact and has many more filtering units than either the pronephros or mesonephros. It is displaced laterally. Notice the separate ureters (new ducts) for carrying urine. The archinephric duct becomes the ductus deferens for carrying sperm.

Cartilaginous Fishes (Elasmobranchs) Retain Urea and Pump Out Electrolytes

Elasmobranchs (sharks, skates, and rays), like other fishes, have mesonephric kidneys. (*see figure 28.14*b). Instead of actively pumping ions out of their bodies through the kidneys, they have a **rectal gland** that secretes a highly concentrated salt (NaCl) solution (table 28.2). To reduce water loss, they use two organic molecules–urea and trimethylamine oxide (TMAO)–in their body fluids to raise tissue osmolarity equal to or higher than that of the seawater.

Urea denatures proteins and inhibits enzymes, whereas TMAO stabilizes proteins and activates enzymes. Together in the proper ratio, they counteract each other, raise the osmotic pressure, and do not interfere with enzymes or other proteins. This reciprocity is termed the **counteracting osmolyte strategy.**

A number of other fishes and invertebrates have evolved the same mechanism and employ pairs of counteracting osmolytes to raise the osmotic pressure of their body fluids.

Freshwater Teleosts Must Keep Water Out and Retain Electrolytes

Most teleost fishes (a subgroup of Neopterygii, *see chapter 18*) have mesonephric kidneys. Because the body fluids of freshwater fishes are hyperosmotic relative to freshwater (*see table 28.2*), water tends to enter the fishes, causing excessive hydration or bloating (figure 28.15*a*). At the same time, body ions tend to move outward into the water. To solve this problem, freshwater fishes usually do not drink much water. Their bodies are coated with mucus, which helps stem inward water movement. They absorb salts and ions by active transport across their gills. They also excrete a large volume of water as dilute urine.

Marine Teleosts Must Keep Water in and Excrete Electrolytes

Marine teleosts face a different problem of water balance–their body fluids are hypoosmotic with respect to seawater (*see table 28.2*), and water tends to leave their bodies, resulting in dehydration (figure 28.15*b*). To compensate, marine teleosts drink large quantities of seawater, and they secrete Na^+, Cl^-, and K^+ ions through secretory cells in their gills. Channels in plasma membranes of their kidneys actively transport the ions that are abundant in seawater (e.g., Ca^{2+}, Mg^{2+}, SO_4^{2-}, and PO_4^{3-}) out of the extracellular fluid and into the nephron tubes. The ions are then excreted in a concentrated urine.

Diadromous Fishes

Some fishes encounter both fresh- and saltwater during their lives. **Diadromous** fish migrate between freshwater and saltwater. For example, Atlantic salmon (*Salmo salar*) swim downstream from their natal freshwater streams and enter the sea. Instead of continuing to pump ions in, as they have done in freshwater, the salmon must now rid their bodies of salt. Years later, these same salmon migrate from the sea to their freshwater home to spawn. As they do, the pumping mechanisms reverse themselves.

Amphibians Adapt to Two Environments

The amphibian kidney is identical to that of freshwater fishes (*see figure 28.15*), which is not surprising, because amphibians spend a large portion of their time in freshwater, and when on land, they tend to seek out moist places. Amphibians take up water and ions in their food and drink, through the skin that is in contact with moist substrates, and through the urinary bladder (figure 28.16). This

TABLE 28.2
SALT AND WATER BALANCE IN SELECTED VERTEBRATES

ORGANISM	ENVIRONMENTAL CONCENTRATION RELATIVE TO BODY FLUIDS	URINE CONCENTRATION RELATIVE TO BLOOD	MAJOR NITROGENOUS WASTE(S)	KEY ADAPTATION
Freshwater fishes	Hypoosmotic	Hypoosmotic	Ammonia	Absorb ions through gills
Saltwater fishes	Hyperosmotic	Isosmotic	Ammonia	Secrete ions through gills
Sharks	Isosmotic	Isosmotic	Ammonia	Secrete ions through rectal gland
Amphibians	Hypoosmotic	Very hypoosmotic	Ammonia and urea	Absorb ions through skin
Marine nonavian reptiles	Hyperosmotic	Isosmotic	Ammonia and urea	Secrete ions through salt gland
Marine mammals	Hyperosmotic	Very hyperosmotic	Urea	Drink some water
Desert mammals	No comparison	Very hyperosmotic	Urea	Produce metabolic water
Marine avian reptiles	No comparison	Weakly hyperosmotic	Uric acid	Drink seawater and use salt glands
Terrestrial avian reptiles	No comparison	Weakly hyperosmotic	Uric acid	Drink freshwater

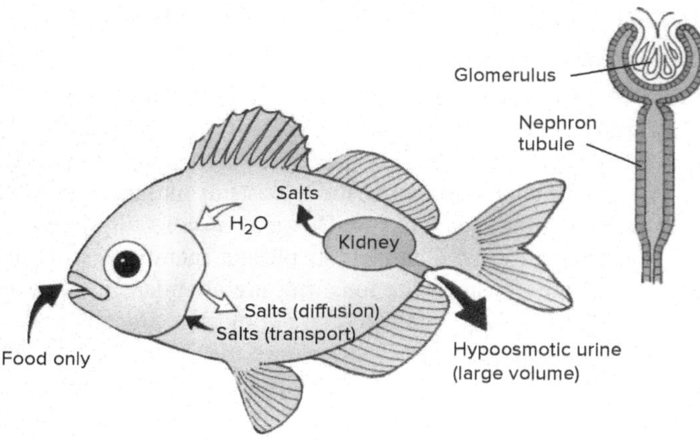

(a) **Freshwater teleosts** (hypertonic blood)

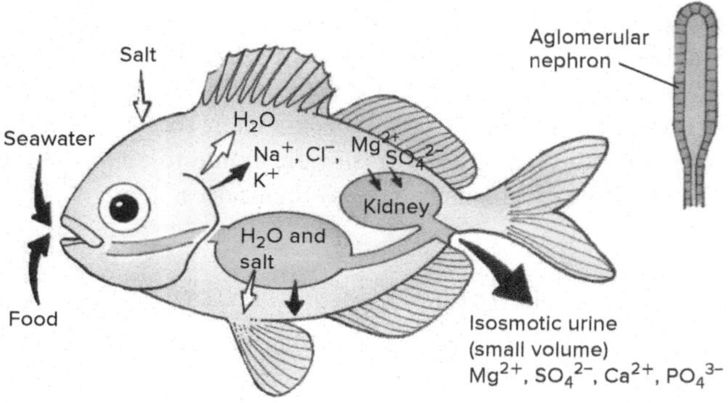

(b) **Marine teleosts** (hypotonic blood)

FIGURE 28.15

Osmoregulation. Osmoregulation by (*a*) freshwater and (*b*) marine fishes. Large black arrows indicate passive uptake or loss of water or ions. Small black and white arrows indicate active transport processes at gill membranes and kidney tubules. Insets of kidney nephrons depict adaptations within the kidney. Water, ions, and small organic molecules are filtered from the blood at the glomerulus of the nephron. Essential components of the filtrate can be reabsorbed within the tubule system of the nephron. Marine fishes conserve water by reducing the size of the glomerulus of the nephron, and thus reducing the quantity of water and ions filtered from the blood. Ions can be secreted from the blood into the kidney tubules. Marine fishes can produce urine that is isosmotic with the blood. Freshwater fishes have enlarged glomeruli and short tubule systems. They filter large quantities of water from the blood, and tubules reabsorb some ions from the filtrate. Freshwater fishes produce a hypoosmotic urine.

uptake counteracts what is lost through evaporation and prevents osmotic imbalance (*see table 28.2*).

The urinary bladder of anurans and caudates is an important water and ion reservoir. For example, when the environment becomes dry, the bladder enlarges for storing more urine. If an amphibian becomes dehydrated, a brain hormone causes water to leave the bladder and enter the body fluid.

Amniotes Are Able to Retain Water and Excrete Concentrated Urine

Avian and nonavian reptiles and mammals all possess metanephric kidneys (*see figure 28.14*c). Their kidneys are by far the most complex animal kidneys, well suited for these animals' high rates of metabolism.

In most amniotes the kidneys can remove far more water than can those in amphibians, and the kidneys are the primary regulatory organs for controlling the osmotic balance of the body fluids. Some desert and marine reptiles and birds build up high salt (NaCl) concentrations in their bodies because they consume salty foods or seawater, and they lose water through evaporation and in their urine and feces. To rid themselves of excess salt, these animals also have salt glands near the eye or in the tongue that remove excess salt from the blood and secrete it as tear-like droplets (figure 28.17). Examples include green sea turtles (*Chelonia mydas*), albatrosses (Diomedeidae), marine iguanas (Amblyrhynchus cristatus), and gulls (e.g., *Larus marinus*)

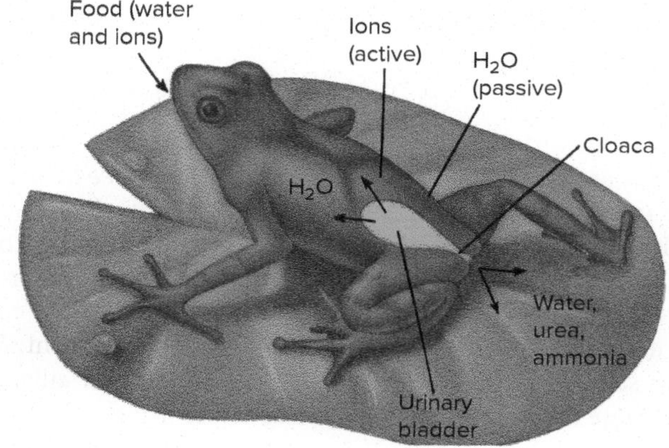

FIGURE 28.16

Water and Ion Uptake in an Amphibian. Water can enter this frog via food, through its highly permeable skin, or from the urinary bladder. The skin also actively transports ions such as Na^+ and Cl^- from the environment. The kidney forms a dilute urine by reabsorbing Na^+ and Cl^- ions. Urine then flows into the urinary bladder, where most of the remaining ions are reabsorbed.

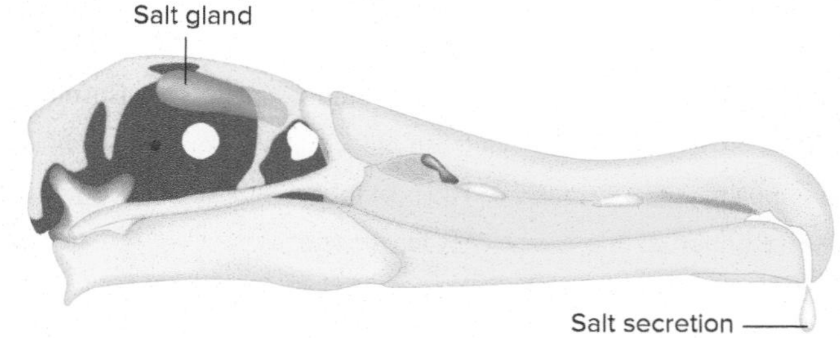

FIGURE 28.17

How Marine Birds Cope with Excess Salt in Their Diets. Because marine birds drink seawater (saltwater), they excrete the excess salt from salt glands near the eyes. The extremely salty fluid produced by these glands can then dribble down the beak into the environment.

A major site of water loss in mammals is the lungs. To reduce this evaporative loss, many mammals have nasal cavities that act as countercurrent exchange systems (figure 28.18). When the animal inhales, air passes through the nasal cavities and is warmed by the surrounding tissues. In the process, the temperature of this tissue drops. When the air gets deep into the lungs, it is further warmed and humidified. During exhalation, as the warm, moist air passes up the respiratory tree, it gives up its heat to the nasal cavity. As the air cools, much of the water condenses on the nasal surfaces and does not leave the body. This mechanism explains why a dog's nose is usually cold and moist.

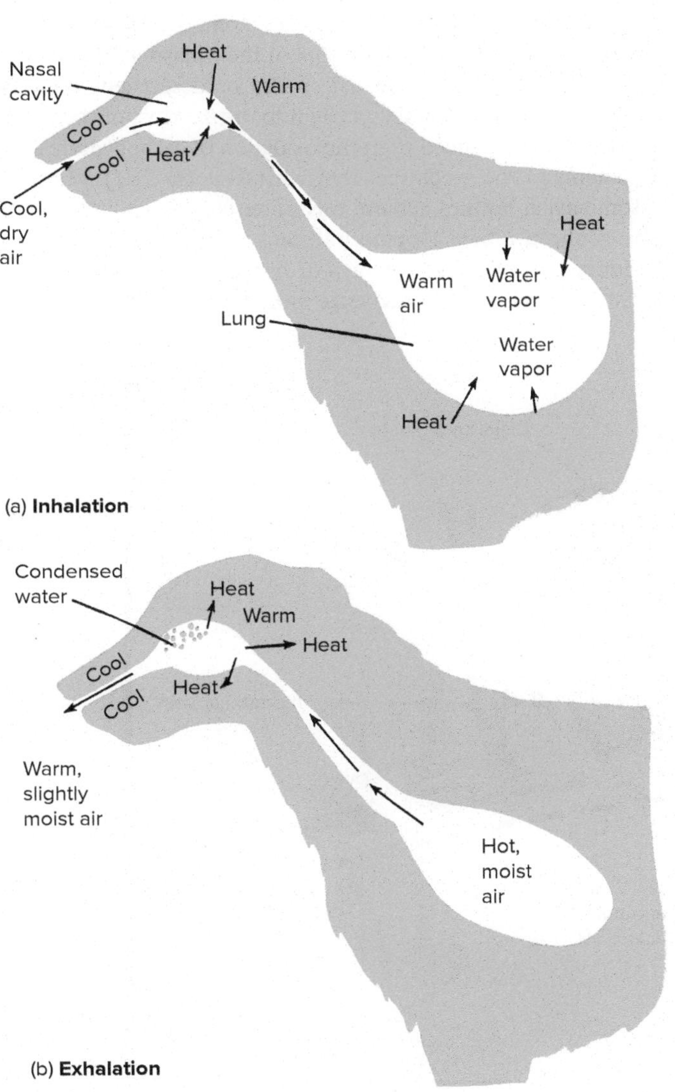

FIGURE 28.18

Water Retention in the Respiratory System of a Mammal. (*a*) When this animal inhales, the cool, dry air passing through its nose is heated and humidified. At the same time, its nasal tissues are cooled. (*b*) When the animal exhales, it gives up heat to the previously cooled nasal tissue. The air carries less water vapor, and condensation occurs in the animal's nose. Black arrows indicate the direction of air movement.

How the Metanephric Kidney Functions

The inner portion of the kidney is called the medulla, and the outer portion is called the cortex. The cortex is surrounded by a coat of connective tissue called the renal (L. *renes*, kidney) capsule (figure 28.19*a*).

The functional units of the metanephric kidney are the **nephrons** (Gr. *nephros*, kidney), and over 1 million are present in each kidney. Each nephron filters blood, secretes excess ions, and reabsorbs required ions and nutrients while cleansing the blood

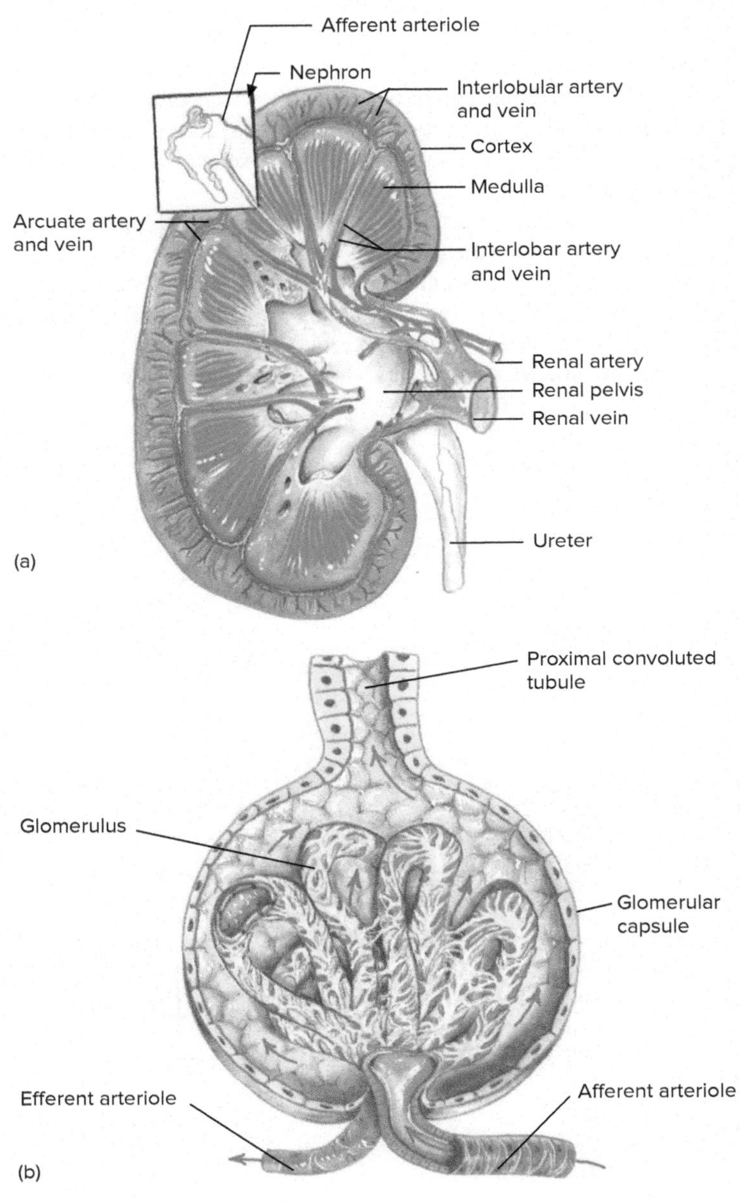

FIGURE 28.19

Filtration Device of the Metanephric Kidney. (*a*) Interior of a kidney, showing the positioning of the nephron and the blood supply to and from the kidney. (*b*) Glomerular capsule. Red arrows show that high blood pressure forces water and ions through small perforations in the walls of the glomerular capillaries to form the glomerular filtrate.

of metabolic waste (*see figure 28.19*a). The filtration apparatus of the nephron is comprised of a glomerular capsule and a glomerulus (figure 28.19*b*). The capsules are in the cortical (outermost) region of the kidney. In each capsule, an afferent ("going to") arteriole enters and branches into a fine network of capillaries, the **glomerulus** (L. *glomeratus*, wound into a ball). The walls of these glomerular capillaries contain small perforations called filtration slits. Blood pressure forces fluid from the blood through these slits. The fluid is now known as glomerular filtrate and contains small molecules, such as glucose, ions, and the primary nitrogenous waste product of metabolism–urea or uric acid. Because the filtration slits are small, large proteins and blood cells remain in the blood and leave the glomerulus via the efferent ("outgoing") arteriole. The efferent arteriole then divides into a set of capillaries called the peritubular capillaries that follow the tubular portions of the nephron into, and then back out of, the kidney medulla (figure 28.20). Eventually, they merge to form veins that carry blood out of the kidney.

The tubule system of the nephron begins at the glomerular capsule and includes the proximal convoluted tubule, the loop of the nephron, and the distal convoluted tubule (*see figure 28.20*). At various places along these structures, the glomerular filtrate is selectively reabsorbed, returning certain ions (e.g., Na^+, K^+, and Cl^-) to the bloodstream. Both active (ATP-requiring) and passive transport mechanisms are involved in the recovery of these substances. Potentially harmful compounds, such as hydrogen (H^+) and ammonium (NH_4^+) ions, drugs, and various other foreign materials are secreted into the nephron lumen from the peritubular capillaries. The last portion of the nephron is called the collecting duct. As described in the following section, a hypoosmotic urine may be eliminated from the kidney through this collecting duct, thus ridding the animal of excess water. Alternatively, water reabsorption, under the influence of a hypothalamic hormone, can promote water reabsorption to form a hyperosmotic urine. Water is thus conserved.

Mammalian, and to a lesser extent avian, kidneys can remove far more water from the glomerular filtrate than can the kidneys of amphibians. For example, human urine can be concentrated fourfold as compared to blood plasma, and some desert rodents can concentrate urine up to 20 times that of their plasma. This concentrated waste enables them to live in dry or desert environments, where little water is available for them to drink. Most of their water is metabolically produced from the oxidation of carbohydrates, fats, and proteins in the seeds that they eat (*see table 28.1*). Mammals and some avian reptiles achieve this water conservation as a result of the nephron tubule looping into and out of the kidney medulla (*see figure 28.20*) and an associated countercurrent exchange mechanism.

Animation
Kidney Function

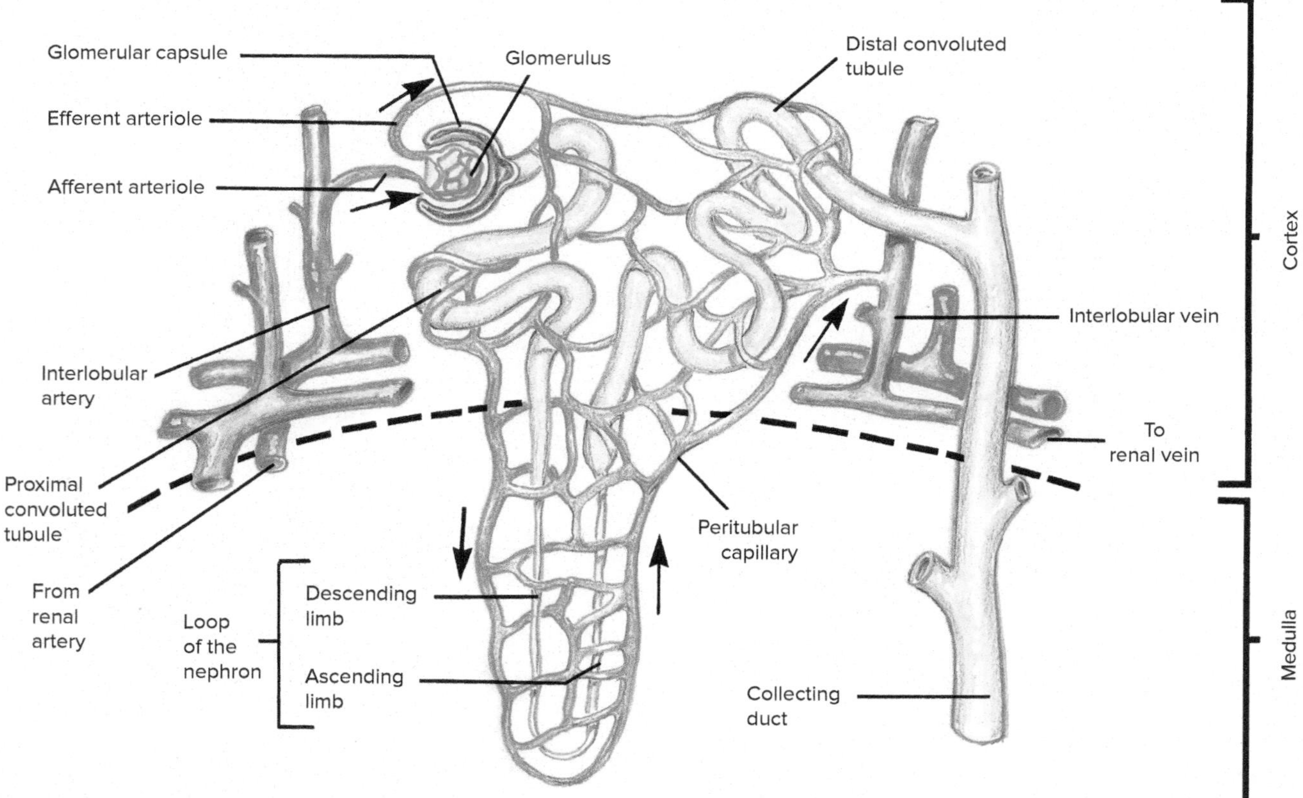

FIGURE 28.20

Metanephric Nephron. The proximal convoluted tubule reabsorbs glucose and some ions. The distal convoluted tubule reabsorbs other ions and water. Final water reabsorption takes place in the collecting duct. Black arrows indicate the direction of movement of materials in the nephron.

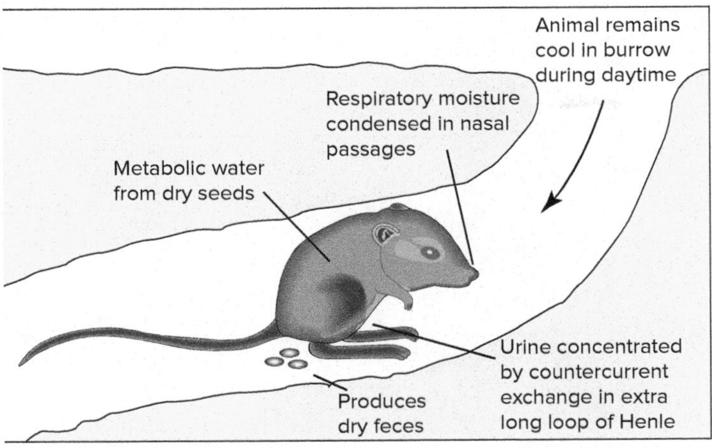

FIGURE 28.21

Kangaroo Rat (*Dipodomys heermanni*), a Master of Water Conservation. Its efficient kidneys can produce an ion concentration in its urine that is 20 times that of its blood plasma. As a result, these kidneys, as well as other adaptations, prevent unnecessary water loss to the environment (*see table 28.1 and chapter opener photo on page 521*).

Countercurrent Exchange

The loop of the nephron increases the efficiency of reabsorption by a countercurrent flow similar to that in the gills of fishes or in the legs of birds, but with water and ions being reabsorbed instead of oxygen or heat (*see figure 28.5*). Generally, the longer the loop of the nephron, the more water and ions that can be reabsorbed. It follows that desert rodents (e.g., *Dipodomys heermanni*) that form highly concentrated urine have very long nephron loops (figure 28.21).

Figure 28.22 shows the countercurrent flow mechanism for concentrating urine. Reabsorption in the proximal convoluted tubule removes some salt (NaCl) and water from the glomerular filtrate and reduces its volume by approximately 25%. However, the concentrations of salt and urea are still isosmotic with the extracellular fluid.

As the filtrate moves to the descending limb of the loop of the nephron, it becomes further reduced in volume and more concentrated. Water moves out of the tubule by osmosis due to the high urea and salt concentrations (the "urea-brine bath") in the surrounding medullary tissue.

Notice in figure 28.22 that the highest urea-brine bath concentration is around the lower portion of the loop of the nephron. As the filtrate passes into the ascending limb, sodium (Na^+) ions are actively transported out of the filtrate into the extracellular fluid, with chloride (Cl^-) ions following passively. Water cannot flow out of the ascending limb because the cells of the ascending limb are impermeable to water. Thus, the salt concentration of the extracellular fluid becomes very high. The salt flows passively into the descending loop, only to move out again in the ascending loop, creating a recycling of salt through the loop and the extracellular fluid. Because the flows in the descending and ascending limbs are

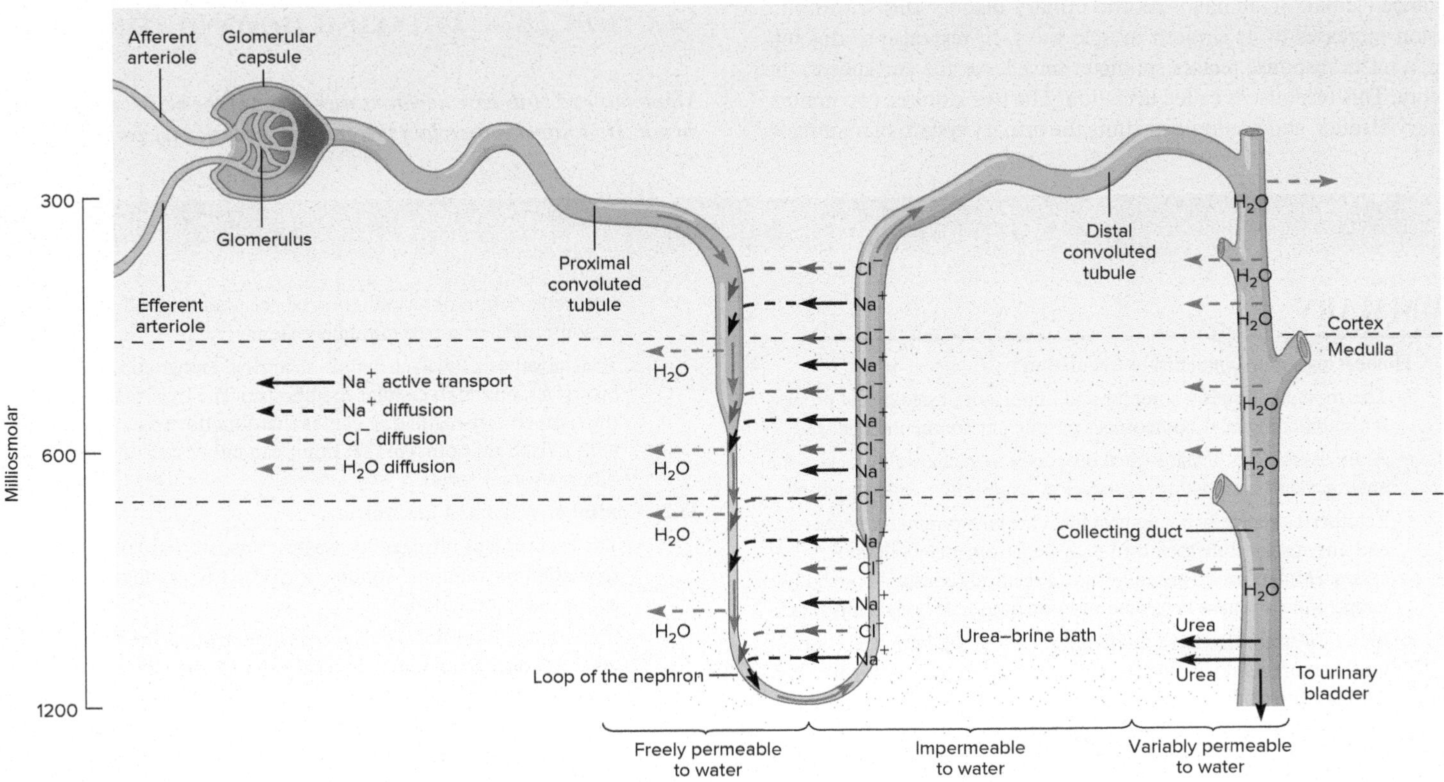

FIGURE 28.22

Countercurrent Exchange. Movement of materials in the nephron and collecting duct. Solid arrows indicate active transport; dashed arrows indicate passive transport. The stippling at intervals along the tubules illustrates the relative concentration of sodium chloride and urea in the medullary tissues.

in opposite directions, a countercurrent gradient in salt is set up. As described next, the permeability of the collecting duct to urea also contributes to the very high osmolarity of the medullary tissues.

Finally, the distal convoluted tubule empties into the collecting duct, which carries urine through the kidney medulla. Amniotes can excrete a very dilute urine if urine moves out of the body with no further modification. Thus, excess water is eliminated. If water conservation is required, osmoreceptors of the hypothalamus initiate the release of antidiuretic hormone (ADH), produced in the hypothalamus but released from the posterior pituitary (neurohypophysis). Under the influence of ADH, the permeability of the collecting duct wall to water increases. Water is osmotically reabsorbed from the collecting duct into the kidney medulla and returned to the peritubular capillaries and the blood. A hyperosmotic urine is then excreted.

Many avian (and all nonavian) reptiles lack a long loop of the nephron. In spite of this absence, their excretory systems are able to conserve water. Reptiles primarily excrete nitrogenous wastes in the form of uric acid. Uric acid is less water soluble than is urea and requires less water for its excretion. Reabsorption of water occurs in the cloaca of a reptile prior to semisolid uric acid being released from the body.

The renal pelvis of the mammalian kidney is continuous with a tube called the **ureter** that carries urine to a storage organ called the **urinary bladder** (figure 28.23). Urine from two ureters (one from each kidney) accumulates in the urinary bladder. The urine leaves the body through a single tube, the **urethra,** which opens at the cloaca of reptiles, at the end of the penis (male mammals), or in front of the vaginal entrance (female mammals). As the urinary bladder fills with urine, tension increases in its smooth muscle walls. In response to this tension, a reflex response relaxes sphincter muscles at the entrance to the urethra. This response is called urination. The two kidneys, two ureters, urinary bladder, and urethra constitute the urinary system of mammals.

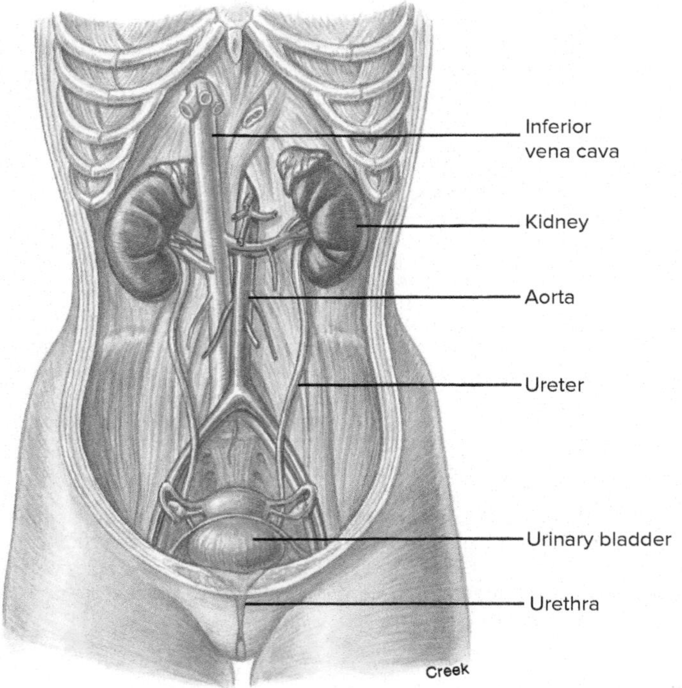

FIGURE 28.23

Component Parts of the Human Urinary (Renal) System. The positions of the kidneys, ureters, urinary bladder, and urethra.

Section 28.4 Thinking Beyond the Facts

Mammals and birds have nephrons with a loop of the nephron but reptiles do not. How would you explain this from an evolutionary point of view?

Summary

28.1 **Homeostasis and Temperature Regulation**

- Thermoregulation is a complex and important physiological process for maintaining heat homeostasis despite environmental changes.
- Animals use four physiological processes to exchange heat with the environment: conduction, convection, evaporation, and radiation.
- Ectotherms generally obtain heat from the environment, whereas endotherms generate their own body heat from metabolic processes. Homeotherms generally have a relatively constant core body temperature, whereas heterotherms have a variable body temperature.
- Many invertebrates have relatively low metabolic rates and have no thermoregulatory mechanisms. They passively conform to the temperature of the external environment.
- The temperature of the surrounding water determines the body temperature of most fishes.
- Most amphibians and reptiles are ectotherms and derive heat from their environment, and their body temperatures vary with external temperatures.
- The high, constant body temperature of birds and mammals also depends on insulation, panting, sweating, specific behaviors, vasoconstriction or vasodilation of peripheral blood vessels, and in some species, a rete mirabile system.
- Thermogenesis involves mainly shivering, enzymatic activity, brown fat, and high cellular metabolism The hypothalamus is the temperature-regulating center that functions as a thermostat with a fixed set pointThis set point can either rise or fall during hibernation or torpor.

28.2 **Control of Water and Electrolytes**

- The excretion of nitrogenous wastes is usually associated with the regulation of water and solutes (ions) by a physiological process called osmoregulation.
- If an animal does not regulate the osmolarity of the body fluids when environmental osmolarity changes, the animal is an osmoconformer.
- An animal that maintains its body fluids at different osmotic concentrations from that of its surrounding environment is an osmoregulator.

28.3 **Invertebrate Excretory Systems**

- Some invertebrates have contractile vacuoles, flame-cell systems, antennal (green) glands, maxillary glands, coxal glands, nephridia, or Malpighian tubules for osmoregulation.

- Some freshwater invertebrates, such as sponges, have contractile vacuoles that pump out excess water.
- The earliest type of nephridium to appear in the evolution of animals was the protonephridium.
- The most common type of excretory structure among invertebrates is the metanephridium.
- The renal organs of adult molluscs are tubular nephridia.
- The excretory organs in some crustaceans are called antennal glands or green glands. Coxal glands are common among the arachnids.
- Insects have an excretory system made up of the gut and Malpighian tubules.

28.4 **Vertebrate Excretory Systems**

- The osmoregulatory system of vertebrates governs the concentration of water and ions; the excretory system eliminates metabolic wastes, water, and ions from the body. Processes occurring at the kidney include filtration of the blood, reabsorption of ions and water, and secretion of other ions and end products of metabolism into the urine.
- There are three kinds of vertebrate kidneys: the pronephros, mesonephros, and metanephros. Freshwater animals tend to lose ions and take in water. To avoid hydration, freshwater fishes rarely drink much water, have impermeable body surfaces covered with mucus, excrete a dilute urine, and take up ions through their gills. Marine animals tend to take in ions from the seawater and to lose water. To avoid dehydration, they frequently drink water, have relatively permeable body surfaces, excrete a small volume of concentrated urine, and secrete ions from their gills. Amphibians can absorb water across the skin and urinary bladder wall. Desert and marine reptiles and birds have salt glands to remove and secrete excess salt (NaCl). In reptiles, birds, and mammals, the kidneys are important osmoregulatory structures
- The functional unit of the kidney is the nephron, composed of the glomerular capsule, proximal convoluted tubule, loop of the nephron, distal convoluted tubule, and collecting duct. The loop of the nephron and the collecting duct are in the kidney's medulla; the other nephron parts lie in the kidney's cortex. Urine passes from the pelvis of the kidney to the urinary bladder. To make urine, kidneys produce a filtrate of the blood and reabsorb most of the water, glucose, and needed ions, while allowing wastes to pass from the body. Three physiological mechanisms are involved: filtration of the blood through the glomerulus, reabsorption of the useful substances, and secretion of toxic substances. In those animals with a loop of the nephron, salt (NaCl) and urea are concentrated in the extracellular fluid around the loop, allowing water to move by osmosis out of the loop and into the peritubular capillaries.

Concept Review Questions

1. Which of the following represents heat loss from an animal's body due to the movement of air over the animal's body?
   a. Conduction
   b. Convection
   c. Evaporation
   d. Radiation
   e. None of the above (a–d)
2. Most birds and mammals are called
   a. ectotherms.
   b. endotherms.
   c. homeotherms.
   d. heterotherms.
   e. both b and c.
3. Most amphibians have difficulty controlling body heat because they produce little of it metabolically and rapidly lose most of it from their body surfaces.
   a. True
   b. False
4. The hormonal triggering of heat production is called
   a. shivering thermogenesis.
   b. gular flutter.
   c. thermogenesis.
   d. nonshivering thermogenesis.
   e. hibernation.
5. Which of the following is a function of the kidneys?
   a. The kidneys remove harmful substances from the body.
   b. The kidneys recapture water for use by the body.
   c. The kidneys regulate the concentration of ions in the blood.
   d. All of these (a–c) are functions of the kidney.
6. Humans excrete their excess nitrogenous waste as
   a. uric acid crystals.
   b. molecules containing proteins.
   c. ammonia.
   d. urea.
7. An osmoregulator would maintain its internal fluids at a concentration that is __________ relative to its specific surroundings (environment).
   a. isosmotic
   b. hyperosmotic
   c. hypoosmotic
   d. All of these (a–c).

Analysis and Application Questions

1. Reptiles are said to be behavioral homeotherms. Explain what this means.
2. Why do very small birds and mammals go into a state of torpor at night?
3. How does the countercurrent mechanism help regulate heat loss?
4. In endotherms, what controls the balance between the amount of heat lost and the amount gained?
5. If marooned on a desert isle, do not drink seawater; it is better to be thirsty. Why is this true?

29

Reproduction and Development

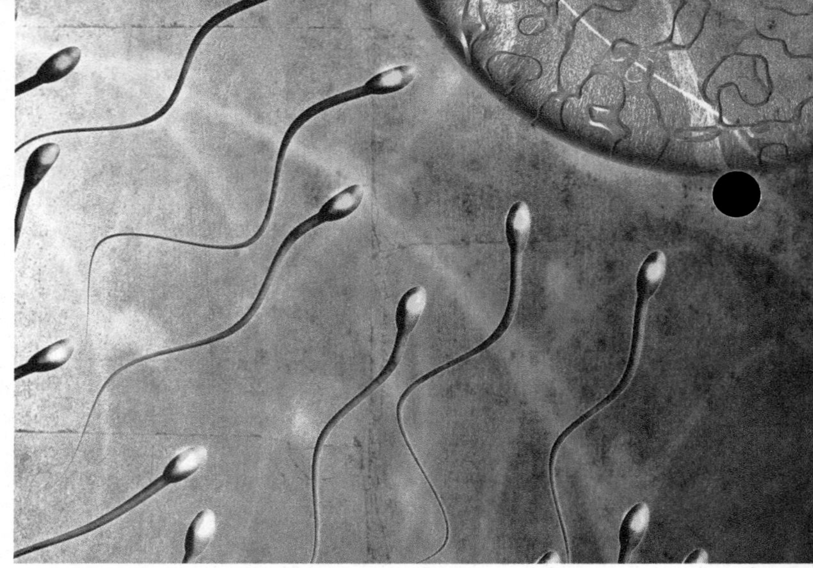

Reproduction is as vital for a species as breathing is for an individual. This chapter describes reproductive strategies found throughout the animal kingdom. Fertilization of a human egg is shown here.
©Colin Anderson/Stockbyte/Getty Images

Chapter Outline

29.1 Asexual Reproduction in Invertebrates
Fission
Budding
Fragmentation
Parthenogenesis
Advantages and Disadvantages of Asexual Reproduction
29.2 Sexual Reproduction in Invertebrates
External Fertilization
Internal Fertilization
Advantages and Disadvantages of Sexual Reproduction
29.3 Sexual Reproduction in Vertebrates
Vertebrate Reproductive Strategies
Fishes
Amphibians
Amniotes
29.4 The Human Male Reproductive System
Production and Transport of Sperm
Hormonal Control of Male Reproductive Function
29.5 The Human Female Reproductive System
Production and Transport of the Egg
Hormonal Control of Female Reproductive Function
Hormonal Regulation in the Pregnant Female
Mammals with Estrous Cycles
29.6 Prenatal Development and Birth in a Human
Events of Prenatal Development: From Zygote to Newborn
The Placenta: Exchange Site and Hormone Producer
Birth: An End and a Beginning
Milk Production and Lactation

Reproduction is a basic attribute of all forms of life. Chapter 3 describes the general features of animal development and the control processes that allow a genotype to be translated into its phenotype. Although in modern zoology, development is "the center stage" in reproduction, the whole process includes the behavior, anatomy, and physiology of adults—whether protists, invertebrates, or vertebrates. This chapter begins with a comparative focus on the different reproductive strategies observed in protists and major groups of animals. The chapter concludes with a discussion of human reproduction, not only because of the subject's basic interest to everyone, but because scientists know more about the biochemistry, hormones, anatomy, and physiology of human reproduction than they do about those of any other species.

29.1 ASEXUAL REPRODUCTION IN INVERTEBRATES

LEARNING OUTCOMES

1. Hypothesize why those organisms that reproduce asexually evolve very slowly.
2. Explain one advantage and one disadvantage to asexual reproduction.

In the biological sense, reproduction means producing offspring that may (or may not) be exact copies of the parents. Reproduction is part of a life cycle, a recurring frame of events in which animals grow, develop, and reproduce according to a program of instruction encoded in the DNA they inherit from their parents. One of the two major types of reproduction in the biological world is asexual reproduction.

The first organisms to evolve probably reproduced by pinching in two, much like the simplest organisms that exist today do. This is a form of **asexual reproduction,** which is reproduction without the union of gametes or sex cells. In the first 2 billion years or more of evolution, forms of asexual reproduction were probably the only means by which the primitive organisms could increase their numbers. Although asexual reproduction effectively increases the numbers of a species, those species reproducing asexually tend to evolve very slowly, because all offspring of any one individual are alike, providing less genetic diversity for evolutionary selection.

Asexual reproduction is common among poriferans, cnidarians, platyhelminths, and many annelids. Asexual reproduction is rare among the other invertebrates. The ability to reproduce asexually often correlates with a marked capacity for regeneration. Asexual reproduction is usually complemented by sexual reproduction at some stage in an invertebrate's life history.

The most common forms of asexual reproduction by invertebrates are fission, budding (both internal and external), and fragmentation. Parthenogenesis is an important asexual reproductive adaptation in a few invertebrates.

Fission

Protists and certain invertebrate animals (e.g., cnidarians and annelids) may reproduce by **fission** (L. *fissio*, the act of splitting). Fission occurs when one cell body or body part divides into two (figure 29.1*a*). In protists the plasma membrane invaginates, divides the protoplasm, and results in two (in binary fission) or more individuals (in multiple fission). Binary fission is a common mode of reproduction in protozoa.

Animation
Binary Fission

The plane of division varies among the species that reproduce by fission. Certain turbellarians employ longitudinal fission, where the plane of division is along the longitudinal axis of the animal's body (figure 29.1*b*; *see also figure 10.8*). Some annelids use a form of multiple fission for reproduction. These animals form transverse constrictions along the length of the body that produce a chain of daughter cells (figure 29.1*c*).

Budding

Another mode of asexual reproduction found in some invertebrates is **budding** (L. *bud*, a small protuberance). In some cnidarians and poriferans, certain cells divide rapidly and proliferate at the animal's

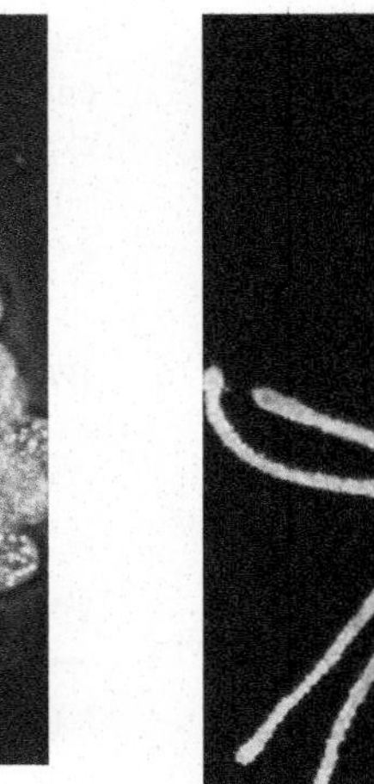
(a)

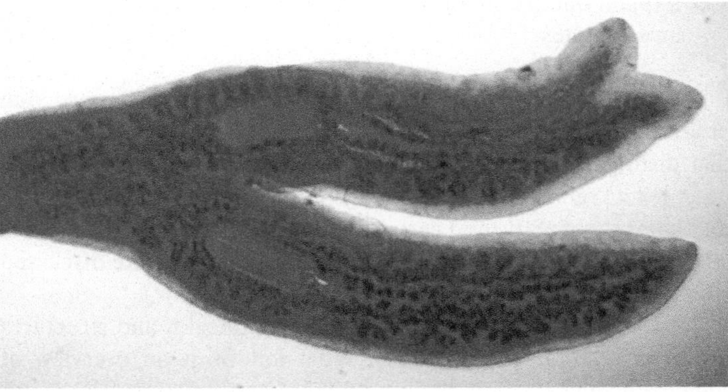
(b)

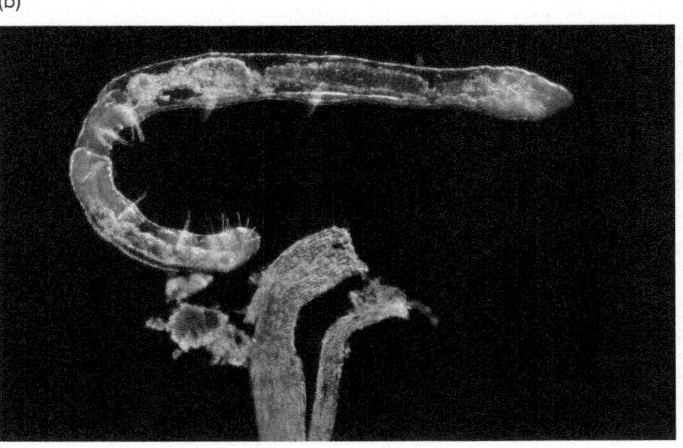
(c)

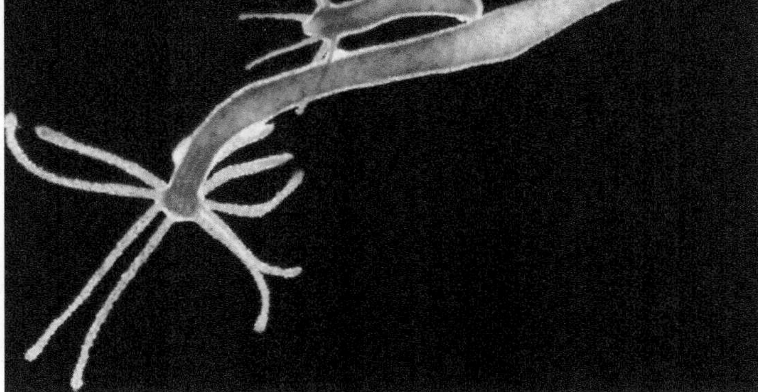
(d)

(e)

FIGURE 29.1

Asexual Reproduction. (*a*) An amoeba (a protist) undergoes fission to form two individual organisms. (*b*) Planarian worms undergoing longitudinal fission. (*c*) An annelid, Naididae, with a single transverse constriction that will eventually yield two individuals. (*d*) Full-length view of Hydra (*Chlorohydra viridissima*) showing asexual budding on a black background. (*e*) Northern red anemone (*Urticina felina*) regenerating a new mouth and set of tentacles by fragmentation.

(a) ©Biophoto Associates/Science Source (b) ©Science Source (c) ©Rubén Duro/BioMEDIA ASSOCIATES LLC/Science Source (d) ©NHPA/M. I. Walker (e) ©Andrew J. Martinez/Science Source

surface to form an external bud (figure 29.1*d*, *see also figures 9.10 and 9.16*). Buds develop into new individuals that usually break away from the parent. If buds remain attached to the parent, they form a **colony** (a group of closely associated or interconnected individuals of one species). **Gemmulation** is a specific type of internal budding used by some freshwater poriferans. The process of gemmulation produces **gemmules**–tough cyst-like structures that contain aggregates of stem cells. When the parent dies and degenerates (and presumably when conditions are favorable), stem cells exit the gemmule and give rise to a new individual (*see figure 9.6*d).

Fragmentation

A type of asexual reproduction that results when a lost body part regenerates into a new organism is termed **fragmentation**. Some cnidarians, platyhelminths, nemerteans, and echinoderms use fragmentation (along with sexual reproduction) to perpetuate the species. Adult sea anemones are primarily sessile animals. However, when anemones do detach or creep along their substrate, small pieces may break off from the adult and develop into new individuals (figure 29.1*e*).

Parthenogenesis

Parthenogenesis is an uncommon mode of reproduction that some invertebrates (and a few chordates including certain snakes, lizards, and fish) use as part of a larger reproductive strategy. In **parthenogenesis** (Gr. *parthenos*, virgin + *genesis*, production), animals are capable of producing diploid or haploid ova that develop into diploid females or haploid males without being fertilized. Because offspring produced parthenogenetically lack sets of chromosomes from two parents, parthenogenetic populations exhibit less genetic diversity than those produced in a sexual cycle. Some parthenogenetic animals can quickly, and efficiently, populate relatively stable environments when conditions are most favorable to the particular species. However, when some parthenogenetic species encounter environmental instability, they may revert to a form of sexual reproduction that produces, for instance, resistant eggs and increased genetic diversity in the population. Parthenogenesis is also integral to social organization in colonies of certain hymenopterans. In these insects, males (drones) are produced parthenogenetically, whereas sterile females (workers) and reproductive females (queens) are produced sexually (*see chapter 15*).

Advantages and Disadvantages of Asexual Reproduction

Most asexually reproducing invertebrates are marine. As a whole, marine environments are comparatively more stable than freshwater and terrestrial habitats. In this environment, a species' genotype is likely to match the unchanging system in which they live. Independent assortment, crossing over, nonlethal mutation, and random selection of gametes in sexual reproduction (*see chapter 3*) may produce offspring with maladaptive genotypes. In stable environments, the production of variable offspring with maladaptive traits would be energetically inefficient and detrimental to the species. Therefore, in certain stable environments, asexual reproduction–the ability of fit parents to bequeath their exact genetic makeup to offspring–may have a selective advantage over sexual reproduction. Some freshwater invertebrates also make use of asexual reproduction, even though the environment is more variable. In such cases, asexual reproduction coincides with predictably favorable environmental periods.

The obvious disadvantage to asexual reproduction is that a population of genetically identical individuals may not have enough genetic variability to withstand rapid and detrimental environmental changes or emerging infectious diseases. Although nonlethal mutation may enhance genetic variability in asexually reproducing populations, most mutations are maladaptive or lethal. Given that most mutations are detrimental, their negative effects would be present in every offspring of an asexually reproducing individual that experienced a mutation.

Section 29.1 Thinking Beyond the Facts

What are some advantages and disadvantages to parthenogenesis?

29.2 SEXUAL REPRODUCTION IN INVERTEBRATES

LEARNING OUTCOMES

1. Describe several forms of sexual reproduction in invertebrates.
2. Explain one advantage of sexual reproduction in invertebrates.

If successful, **sexual** (L. *sexualis*, pertaining to sex) **reproduction** produces offspring that contain a unique combination of genes that distinguishes them from both the parents and siblings. Over time, sexual reproduction leads to genetic variability in populations that helps them withstand certain environmental disturbances. The advantages and disadvantages of sexual reproduction are discussed later in the chapter.

The diversity of sexual reproductive strategies and structures in the invertebrates is immense. What follows is an overview of some principles of reproductive structure and function. The coverage of each invertebrate phylum in chapters 9 through 17 provides more details.

External Fertilization

Many invertebrates, particularly those that live in marine environments, release gametes into the surrounding environment and rely on water currents for gamete dispersal and fertilization. In these invertebrates, reproduction is often seasonal and is coordinated by a suite of environmental cues and the release of pheromones. Structures that make this type of external fertilization possible (**broadcast spawning**) include various arrangements of coelomic ducts, sperm ducts, oviducts, and metanephridia that permit the release of gametes into the surrounding water. Oogenesis and spermatogenesis occur in simple, ephemeral gonads that are greatly reduced in the nonreproductive season.

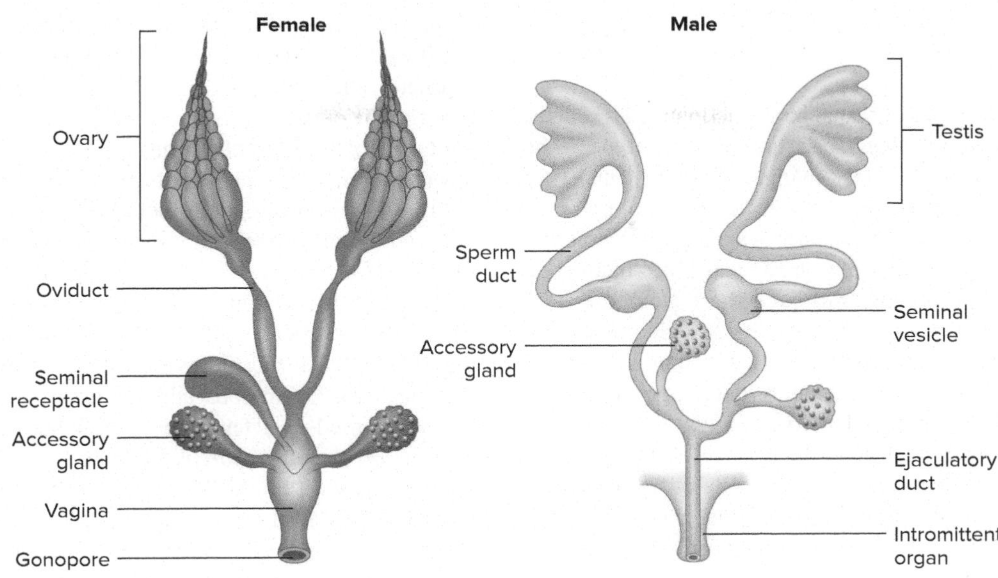

FIGURE 29.2

Stylized Male and Female Reproductive Systems in Invertebrates. Sexual reproduction is possible via these systems.

Internal Fertilization

Other invertebrates (e.g., certain gastropods, cephalopods, and hexapods) use specialized copulatory organs to directly facilitate sperm transfer between reproductive individuals, allowing internal fertilization (figure 29.2).

In males, sperm are produced by spermatogenic cells in the testes and move through sperm ducts to storage areas called seminal vesicles. Prior to mating, some invertebrates (e.g., Cephalopoda, Scorpionida, and some Hirudinea) incorporate many sperm into protective, and sometimes motile, casings called **spermatophores**. Sperm or the spermatophores are then passed into an ejaculatory duct to a copulatory organ (e.g., penis, cirrus, and gonopore) that is used to introduce sperm into the female's system. Seminal vesicles or other accessory glands may be used to produce seminal fluid or spermatophores.

In females, ovaries produce ova (eggs) that are transported to oviducts where fertilization may occur. Accessory glands, some of which produce egg capsules or shells to protect the zygotes, may also be present in the female's reproductive tract. Specialized ducts connect seminal receptacle and accessory glands to oviducts.

The most familiar form of sexual reproduction involves the union of male and female gametes from sexually dioecious parents—that is male and female reproductive structures exist in separate individuals. However, variations from this mode of reproduction exist.

Hermaphroditism (Gr. *hermaphroditos*, an organism with the attributes of both sexes) is a reproductive condition where an animal is equipped with both male and female reproductive systems. This condition is beneficial in slow-moving or sessile animals (e.g., certain pulmonates, oligochaetes and maxillopods) because contact between individuals of the same species may be fortuitous, or difficult. When contact occurs, the chances of successful fertilization are higher if individuals can serve as both male and female, donating and receiving sperm (figure 29.3). Although some hermaphrodites self-fertilize, most have mechanisms in place to prevent it from occurring. Hermaphroditism is also referred to as the monoecious condition (Gr. *monos*, single + *oikos*, house).

A variation of hermaphroditism—**sequential hermaphroditism**—occurs when an animal switches sexes during different phases of its life cycle. Hermaphrodites can be either **protogynous** (Gr. *protos*, first + *gyne*, women) or **protandrous** (Gr. *protos*, first + *andros*, man). Protandrous animals either start out as male and become female later in the life cycle, or male reproductive structures and gametes simply mature before female structures do. The reverse is true for protogynous animals. A sex ratio change in a population can induce sequential hermaphroditism, which is common in certain oysters (e.g., *Ostrea edulis*).

FIGURE 29.3

Mating in Hermaphroditic Earthworms (*Lumbricus terrestris*). During mating, each partner passes sperm from genital pores along grooves to seminal receptacles of its mate (*see also figure 12.19*). Mucous secretions hold mating earthworms together during this process.

Advantages and Disadvantages of Sexual Reproduction

The most significant advantage of sexual reproduction is that genetic recombination (*see figure 3.7*) can lead to novel genotypes that improve a species' fitness in randomly changing environments. This change in the frequency of alleles expressed is the foundation of evolution and, in part, accounts for the diversity of species on the planet.

Genetic recombination can also lead to the production of deleterious genotypes in offspring. Although these genotypes will be quickly eliminated from a population in sexually reproducing individuals, their production represents a substantial waste of metabolic effort. Additionally, many gametes that are produced and released are not fertilized; this also depletes energy.

The advantages of sexual reproduction seem to outweigh the disadvantages, as it is a more common mode of reproduction in Animalia. Although many animals use asexual methods, these methods are typically incorporated into broader reproductive strategies that also include sexual phases.

Section 29.2 Thinking Beyond the Facts

Describe the different types of hermaphroditism and give an example of each.

29.3 SEXUAL REPRODUCTION IN VERTEBRATES

LEARNING OUTCOMES

1. Contrast the strategies used by vertebrates to promote successful fertilization of eggs and early development of embryos.
2. Compare reproductive habits of fishes and amphibians.
3. Hypothesize on the selective pressures that promoted the evolution of, and the evolutionary conservation of, internal fertilization and the amniotic egg in the amniote lineage.

Since the evolution of the first animals, the basic use of male and female gametes has been preserved. Vertebrate evolution has also given rise to the close link between reproductive biology and sexual behavior. The strong drive to mate or reproduce dominates the lives of many vertebrates, as illustrated by the salmon's (Salmonidae) fateful spawning run or the rutting of bull elk (*Cervus canadensis*). This section describes reproductive strategies seen in vertebrates and variations to those strategies in the major vertebrate groups: fishes, amphibians, and amniotes.

Vertebrate Reproductive Strategies

Some of the most important differences in vertebrate reproductive adaptations reflect whether reproduction is occurring in either aquatic or terrestrial environments. Many vertebrates that reproduce in aquatic environments release their gametes or fertilized eggs into the environment without risk of desiccation for the gametes and embryos. Animals that release gametes or fertilized eggs to the outside of the body are **oviparous** (L. *ovum*, egg + *parere*, to bring forth). Eggs of oviparous animals often develop unattended by the parents, and the parents produce thousands of eggs–although only a few survive to adulthood. Fewer animals in either aquatic or terrestrial environments are ovoviviparous. **Ovoviviparous** (L. *ovum*, egg + *vivere*, to live + *parere*, to bring forth) animals form eggs that are fertilized–and develop from stored nutrients (yolk)–within the reproductive tract of the female. Ovoviviparity is one way to protect embryos during development. It occurs in a wide variety of animals (e.g., Nematoda and Arthropoda). Within the vertebrates it is observed in a few fish, amphibians, and reptiles. **Viviparous** (L. *vivere*, to live + *parere*, to bring forth) animals are those that retain young within the female reproductive tract, and the young are nourished by food supplied by the mother. **Gestation** (L. *gestare*, to bear) is the time during which the embryo is nourished with nutrients and oxygen by the mother's uterus or placenta. Viviparity is relatively common within the vertebrates. It is present in many fishes, some amphibians and snakes, and most mammals.

Internal fertilization evolved many times within virtually all vertebrate lineages. Its single origin in amniote ancestors, and its retention within that lineage, is a reflection of the importance of internal fertilization in fully terrestrial vertebrates.

Fishes

All fishes reproduce in aquatic environments. Members of the classes Sarcopterygii and Actinopterygii are comprised mostly of oviparous species. Fertilization is usually external, and eggs contain only enough yolk to sustain the developing fish for a short time. After this yolk is consumed, the growing fish must seek food. Although many thousands of eggs are produced and fertilized, few survive and grow to maturity. Some succumb to fungal and bacterial infections, others to siltation, and still others to predation. Thus, for reproduction to be successful, the fertilized egg must develop rapidly, and the young must achieve maturity within a short time. Other bony fish are ovoviviparous or viviparous. In contrast, fertilization by members of the class Chondrichthyes is internal. The male introduces sperm into the female through a modified pelvic fin. Many chondrichthians are ovoviviparous or viviparous. In some oviparous species, the fertilized eggs are enclosed in an egg case that is deposited outside the female and from which young hatch.

Amphibians

The amphibians were the first vertebrates to invade the land. Most amphibians have not, however, become adapted to completely terrestrial environments. Their life cycle is still usually linked to water because most amphibians have thin, permeable skin that makes them susceptible to drying. In addition, amphibians do not produce dessication-resistant eggs.. Although internal fertilization occurs in some amphibians, it is the exception–fertilization is usually external.

EVOLUTIONARY INSIGHTS

The Evolution of Mechanisms to Delay Birth

To enhance reproductive success, animals must not only produce as many offspring as possible, but also produce them at a time that ensures the greatest likelihood of their survival to reproductive adulthood. Environmental and development constraints often influence the timing of reproductive functions. For example, mammals living in tropical regions must time the birth of their young so that they become independent during the least harsh time of the year and not during the rainy season. Large mammals that have long pregnancies, such as elephants (Elephantidae), must pace their breeding to allow for the development and suckling of their young. Because elephants have a gestation period of 22 months and nurse their young for three years, the time between the birth of newborns ranges from four to nine years.

In some mammalian species, the time required for embryonic development is shorter than the interval between the best time to mate and the best time to give birth. For these species, three mechanisms evolved to delay birth until a time is favorable for the newborns: sperm storage, delayed embryonic development, and delayed implantation.

For example, several species of female bats (Vespertilionidae) in northern zones either store sperm or slow the development of the embryo. These bats can store sperm in their uteri for up to six months. By increasing the interval between mating and fertilization, they delay the time of birth. Slowing embryonic development also delays birth. The latter is thought to be a response to low ($<$30°C) environmental temperatures.

Mammals such as white-footed mice (*Peromyscus leucopus*), certain seals, (Pinnipedia) and wallabies (Macropodidae) all delay the implantation of the blastocyst in order to delay birth. Large, long-lived mammals nurse their young for long periods. This spaces the birth of their offspring over several years. During lactation, ovulation is suppressed due to high levels of the hormone prolactin (*see table 25.1 and Evolutionary Insights, chapter 25*).

Failure of a species to reproduce ultimately leads to extinction. The reproductive variations described in this reading, and in this chapter as a whole, helps us understand that reproduction is the most important evolutionary function for all species.

Among the anurans, the male grasps the female and discharges fluid containing sperm onto the eggs as she releases them into the water (figure 29.4*a, see figure 19.15*).

The developmental period is much longer in amphibians than in fishes, although the eggs do not contain appreciably more yolk. An evolutionary adaptation of amphibians is the presence of two periods of development: larval and adult stages. The aquatic larval stage (tadpole) develops rapidly, and the animal spends much time eating and growing. After reaching a sufficient size, the tadpole undergoes a developmental transition called metamorphosis into the adult (often terrestrial) form (*see figure 19.18*).

Amniotes

Ancestral amniotes acquired independence from the aquatic habitat, in part, because of adaptations that permitted sexual reproduction and embryonic development on land. These adaptations include the evolution of internal fertilization, which protects gametes from desiccation in terrestrial environments, and the evolution of the amniotic egg.

The amniotic egg permits development on land (*see figure 20.2*). The extra embryonic membranes of this egg allow amniote embryos to develop on land without the danger of desiccation. The amnion encloses the embryo in a fluid-filled sac and protects against shock and desiccation. The chorion and allantois permit gas exchange and store excretory products. Thus, complete development can occur within these three membranes. The evolution of a shell surrounding

(a) (b)

(c) (d)

FIGURE 29.4

Vertebrate Reproductive Strategies. (*a*) A male anuran clasping the female in amplexus, a form of external fertilization. As the female releases eggs into the water, the male releases sperm over them. (*b*) Reptiles, such as these tortoises, use internal fertilization. (*c*) Avian reptiles, like most reptiles, are oviparous. They have shelled, amniotic eggs and often provide parental care for eggs and hatchlings. (*d*) A placental mammal. This female dog is nursing her puppies.

(a) ©Steve Knight/Getty Images (b) ©Rob Covert/Getty Images (c) ©Dave Cole/Alamy (d) ©Comstock Images/Alamy

How Do We Know That Sperm May Serve as Competitors of Sperm from Other Males?

In those female mammals, such as chimpanzees (*Pan troglodytes*), that routinely copulate with many males, males are said to engage in sperm competition. These males have testes much larger than the testes of other mammalian species. The result is the production of larger numbers of sperm to outcompete, and possibly to block and kill, rival sperm in the female reproductive tract.

In the past, sperm competition has been regarded as a matter of mere numbers. The male that deposits the most sperm in the female is more likely to fertilize the female. Scientists hypothesize that many sperm are not designed for fertilization. Instead, their role may be to prevent fertilizing sperm from other males from reaching the ovum (ova) in the female reproductive tract. Evidence suggests that there are two types of these nonfertilizing sperm: (1) Blocker sperm have hooked flagella that allow them to join together in large numbers (masses) and form a physical barrier to the fertilizing sperm from other individuals. (2) Killer sperm are very active and bind to the immunologically different competing sperm and kill them via their acrosomal enzymes. Because the evidence for blocker sperm and killer sperm is still circumstantial and more research is needed, the issue remains unresolved.

an amniotic egg in reptiles (including birds) adds additional protection from mechanical injury for the embryo. When the animal hatches or is born, it has developed to a point that it can survive on its own with only some parental care (*see figures 20.15 and 20.16*).

Nonavian and Avian Reptiles

Eggs of all reptiles possess a leathery or calcareous shell, which necessitates that the eggs are fertilized prior to shell formation. Courtship and copulation behaviors frequently precede fertilization (figure 29.4*b*). Most nonavian reptiles and all avian reptiles are oviparous. Eggs of most nonavian reptiles are usually laid and then left to develop and hatch unattended (crocodylians are an exception, *see figure 20.16*).

With the exception of most waterfowl (Anseriformes) and ostriches (*Struthio camelus*), birds lack a penis. Males simply deposit semen against the cloaca for internal fertilization. Sperm then migrate up the cloaca and fertilize the eggs before hard shells form. This method of mating occurs more quickly than the internal fertilization that nonavian reptiles practice. The eggshells of birds are much thicker than those of nonavian reptiles. Thicker shells permit birds to sit on their eggs and warm them. This brooding, or incubation, hastens embryo development. Eggs often require significant parental care involving incubating, maintaining humidity, and turning eggs. When many young birds hatch, they are incapable of surviving on their own, and they must be fed and protected (figure 29.4*c*, *see figure 21.16*).

Mammals

The Prototheria (e.g., the duckbilled platypus and spiny anteater, *see table 22.1*) are oviparous. Oviparity is retained from the synapsid ancestors of the mammalian lineage. All other mammals (Theria, *see table 22.1*) are viviparous. Mammalian viviparity takes on two forms. The marsupials (infraclass Metatheria, opossums and kangaroos) nourish their young in a marsupial pouch after a short gestation inside the female. The other, much larger, group is comprised of the placental mammals (Eutheria) that retain their young inside the female, where the mother nourishes them by means of a placenta. During gestation the embryo is nourished with nutrients and oxygen, and it is protected from attack by the female's immune system.

Mammary glands are a unique mammalian adaptation that permits the female to nourish the young with milk that she produces (figure 29.4*d*). Young mammals are nursed until they are able to survive on adult food, which is usually correlated to tooth eruption. Some mammals nurture their young until adulthood, when they are able to mate and fend for themselves.

Certain female primates, including apes and monkeys, are asynchronous breeders. Mating and births can take place over much of the year. Females mate only when in estrus, increasing the probability of fertilization. Human females show a less distinctive estrus phase and can reproduce throughout the year. They can also engage in sexual activity without reproductive purpose; no longer is sexual behavior precariously tied to ovulation. The source of this important reproductive adaptation may be physiological or a result of concomitant evolution of the brain—a process that gave humans some conscious control over their emotions and behaviors that hormones, instincts, and the environment control in other animals. This separation of sex from a purely reproductive function has evolved into the long-lasting pair bonds between human males and females that further support the offspring. This type of behavior has also resulted in the transmission of culture—a key to the evolution and success of the human species.

Section 29.3 Thinking Beyond the Facts

How did the evolution of terrestrial lifestyles influence reproductive physiology and behaviors of vertebrates?

29.4 THE HUMAN MALE REPRODUCTIVE SYSTEM

LEARNING OUTCOMES

1. Describe semen and how it is released during mating.
2. Explain how hormones regulate human male reproductive function.

The reproductive role of the human male is to produce sperm and deliver them to the vagina of the female. This function requires the following structures:

1. Two testes that produce sperm and the male sex hormone, testosterone.
2. Accessory glands and tubes that furnish a fluid for carrying the sperm to the penis. This fluid, together with the sperm, is called semen.
3. Accessory ducts that store and carry secretions from the testes and accessory glands to the penis.
4. A penis that deposits semen in the vagina during sexual intercourse.

Animation
Spermatogenesis

Production and Transport of Sperm

The paired **testes** (sing., testis) (L. *testis*, witness; to bear witness to a man's virility) are male reproductive organs that house spermatogenic tissues (figure 29.5). Shortly before birth, the testes descend from the abdominal cavity into a specialized pouch called the **scrotum** (L. *scrautum*, a leather pouch for arrows), which hangs outside the body between the thighs. The temperature required for spermatogenesis is a few degrees cooler than core body temperature (37°C). The extracoelomic positioning of the scrotum allows for testes to remain within an optimal temperature for spermatogenesis. Cremaster muscles elevate or lower the testes, depending on the outside air temperature. The testes contain over 1,600 tightly coiled **seminiferous tubules** (figure 29.6*a* and *b*). These tubules are lined with spermatogenic cells, which are capable of producing thousands of sperm every second, and sustentacular (Sertoli) cells, which secrete fluids and a hormone (inhibin) necessary for sperm survival and efficient spermatogenesis. Between the seminiferous tubules are clusters of interstitial (Leydig) cells that secrete the male sex hormone testosterone.

A system of tubes carries sperm from the testes to the penis. Seminiferous tubules come together to form the rete testis (L. *rete*, net). This network of tubules connects the seminiferous tubules

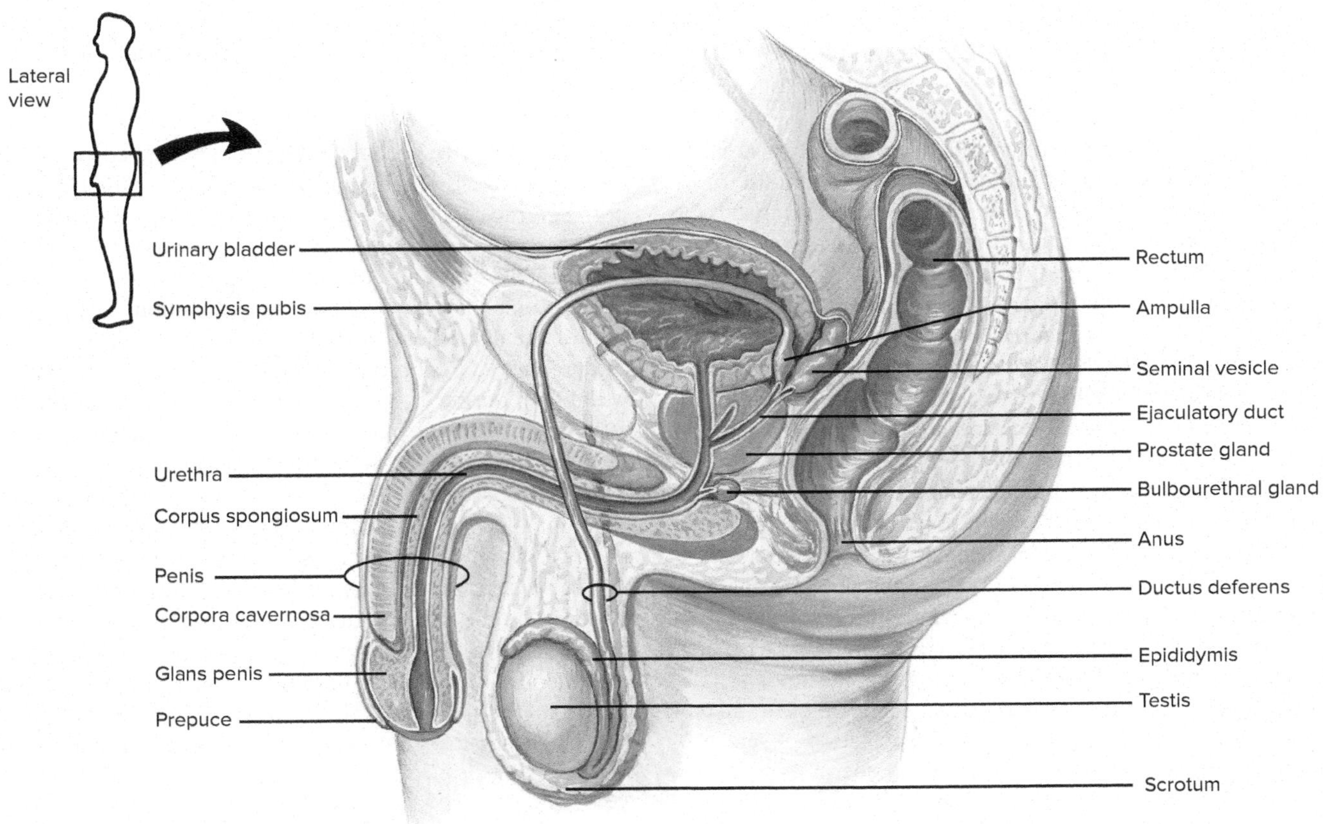

FIGURE 29.5

Lateral View of the Human Male Reproductive System. There are two each of the following structures: testis, epididymis, ductus deferens, seminal vesicle, ejaculatory duct, and bulbourethral gland. The penis and scrotum are the external genitalia.

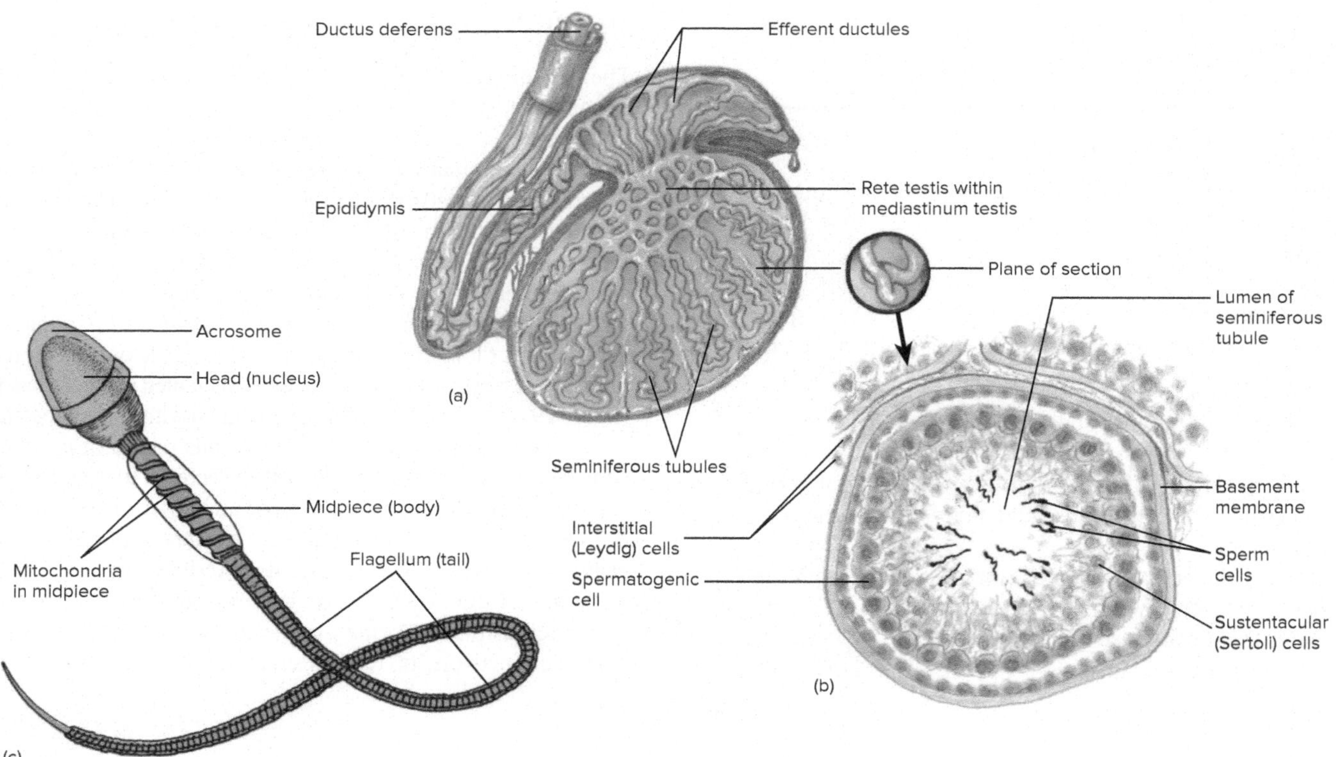

FIGURE 29.6

Human Male Testis. (*a*) Sagittal section through a testis. (*b*) Cross section of a seminiferous tubule, showing the location of spermatogenesis. (*c*) A mature sperm.

to an adjacent sperm storage center called the epididymis. The main functions of the epididymis are to: (1) store sperm until they are mature, (2) propel the sperm toward the penis via peristalsis of associated smooth muscle, and (3) serve as a passageway from the testis to the ductus (vas) deferens. The ductus deferens is an expanded continuation of the epididymis that passes through the lower abdominal wall via the inguinal canal. The ductus deferens loops around the urinary bladder and enlarges to form the ampulla, which stores some sperm until they are ejaculated (*see figure 29.5*). Distal to the ampulla, the ductus deferens becomes the ejaculatory duct. The urethra is the final section of the reproductive ductwork.

After the ductus deferens passes around the urinary bladder, several accessory glands add secretions to sperm as they are propelled toward the urethra. The paired **seminal vesicles** secrete water, fructose, prostaglandins, and vitamin C. This fluid provides energy to motile sperm and helps neutralize the natural protective acidity of the vagina (the pH increases from as low as 3 to around 6). The **prostate gland** secretes water, enzymes, cholesterol, buffering salts, and phospholipids. The **bulbourethral gland** secretes a clear, alkaline fluid that lubricates the urethra prior to ejaculation of semen. The combination of sperm (300 to 400 million sperm cells) and these glandular secretions is 3 to 4 ml of **semen** (L. *seminis*, seed).

The penis has both excretory and reproductive functions. During urination, the penis carries urine outside of the body via the urethra. The penis also transports semen through the urethra during ejaculation. The penis contains three cylindrical strands of erectile tissue: two corpora cavernosa and the corpus spongiosum (*see figure 29.5*). The corpus spongiosum extends beyond the corpora cavernosa and becomes expanded distally to form the neuron-rich glans penis. During erection, the corpus spongiosum (which immediately surrounds the urethra) also swells with blood, but it retains some elasticity to allow for ejaculation. The loosely fitting skin of the distal penis folds forward over the glans penis to form the prepuce (foreskin). The tissues of the corpora cavernosa fill with blood during erection and are largely responsible for erection, allowing the penis to function as an intromittent organ. (Most non-human mammals use a bone called a baculum to aid in copulation.)

A mature human sperm cell (spermatozoan) consists of a head, midpiece, and tail (figure 29.6*c*). The head contains a haploid nucleus, which is mostly DNA. The acrosome covers most of the head and contains the enzyme acrosin, which allows spermatozoa to penetrate the zona pellucida (outer layer of a secondary oocyte) during fertilization. The midpiece is equipped with spiral-shaped mitochondria that supply the ATP necessary for movement of spermatozoa. The tail consists of a bundle of flexible microtubules that beat and generate a propulsive force, allowing locomotion of the cell.

Hormonal Control of Male Reproductive Function

Special regulatory hormones guide the transition from prepubescence to sexual maturity (table 29.1). Collectively these hormones

TABLE 29.1
MAJOR HUMAN MALE REPRODUCTIVE HORMONES IN AN ADULT

HORMONE	FUNCTION(S)	SOURCE
FSH (follicle-stimulating hormone)	Aids sperm maturation; increases testosterone production	Pituitary gland
GnRH (gonadotropin-releasing hormone)	Controls pituitary secretion	Hypothalamus
Inhibin	Inhibits FSH secretion	Sustentacular cells in testes
LH (luteinizing hormone) also called ICSH (interstitial cell–stimulating hormone)	Causes interstitial (Leydig) cells to secrete testosterone	Pituitary gland
Testosterone	Increases sperm production; stimulates development of male primary and secondary sexual characteristics; inhibits LH secretion	Interstitial cells in testes

are called **androgens** (Gr. *andros*, man + gennan, to produce) and **gonadotropins**. **Testosterone** is an important androgen that is involved directly with the development of male secondary sex characteristics and spermatogenesis. Gonadotropins travel from the brain and pituitary gland to the testes and regulate androgen production. The role of these hormones in spermatogenesis is described next.

Spermatogenesis is initiated when the hypothalamus is stimulated to secrete GnRH (gonadotropin-releasing hormone). GnRH stimulates the secretion of the gonadotropins, FSH (follicle-stimulating hormone) and LH (luteinizing hormone), into the bloodstream. FSH causes spermatogenic cells in seminiferous tubules to initiate spermatogenesis, and LH stimulates interstitial cells (also called Leydig cells) to secrete testosterone. The cycle is completed when testosterone inhibits the secretion of LH, and another hormone, inhibin, is secreted by sustentacular cells. Inhibin inhibits FSH secretion from the anterior pituitary (stopping spermatogenesis) but does not affect testosterone production, as this androgen is essential for other biological functions (figure 29.7). Interestingly, FSH and LH were first named for their functions in females, but their molecular structure is exactly the same in males. Consequently, LH is referred to as ICSH (interstitial cell–stimulating hormone) in males.

Section 29.4 Thinking Beyond the Facts

Would natural selection favor those males that produce more sperm over those males that produce fewer sperm? Explain your answer.

29.5 THE HUMAN FEMALE REPRODUCTIVE SYSTEM

LEARNING OUTCOMES

1. Describe the sequence of events in the production of an oocyte.
2. Explain how hormones regulate female reproductive functions.

The reproductive role of human females is complex. In addition to producing gametes, their intricate reproductive anatomy (figure 29.8) undergoes hormone-regulated and cyclical structural changes that facilitate implantation and embryonic and fetal development. After the offspring is born, additional hormone-regulated physical changes in the mother allow her to produce antibody and calorie-rich nourishing foods for the baby—colostrum and milk. Aspects of female reproductive anatomy and physiology and the hormonal changes associated with ovulation, menstruation, childbirth, and milk production are described next.

Production and Transport of the Egg

Paired female gonads, or **ovaries** (L. *ovum*, egg), produce eggs (secondary oocytes) and female hormones (estrogen and progesterone). Ovaries are located in the lower abdomen, just lateral to the uterus, and contain vesicles called follicles, which are the actual centers of oogenesis (figure 29.9). Follicles contain primary oocytes (immature eggs) and undergo several stages of development. After **ovulation** (release of a secondary oocyte), the follicle lining grows inward and forms an endocrine tissue called the corpus luteum, which temporarily secretes estrogen and progesterone.

Uterine tubes allow the secondary oocyte to move from the ovary to the uterus (*see figure 29.8*). Fingerlike fimbriae fringe the area where the uterine tube encircles the ovary. During ovulation, fimbriae sweep across the ovary and allow the secondary oocyte to move into the outermost portion of the uterine tube (the infundibulum).

Because the secondary oocyte is not motile, peristaltic contractions and ciliary action in the uterine tube carry the secondary oocyte toward the uterus (figure 29.10). If fertilization occurs (usually in the upper third of the uterine tube), the zygote will implant in the endometrium of the uterus. This process takes four to seven days. If not fertilized, the secondary oocyte degenerates in the uterine tube.

Animation
Maturation of the Follicle and Oocyte

The uterine tubes terminate at the **uterus**, a hollow, muscular organ positioned between the rectum and urinary bladder

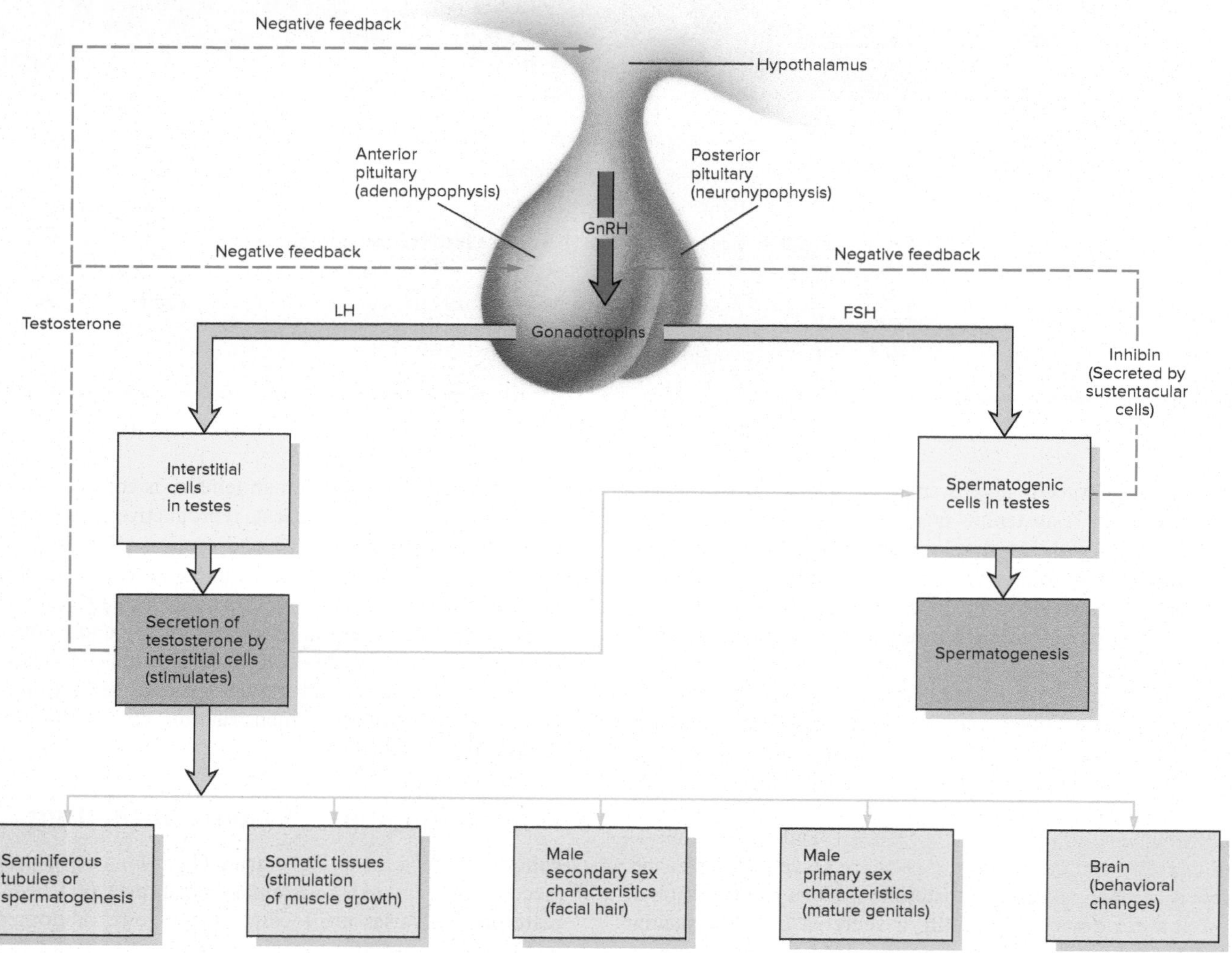

FIGURE 29.7

Hormonal Control of Reproductive Function in Adult Human Males. Negative feedback mechanisms (red dashed pathways) by which the hypothalamus controls sperm maturation and the development of male secondary sexual characteristics. (GnRH = gonadotropin-releasing hormone; LH = luteinizing hormone; FSH = follicle-stimulating hormone.) In response to LH, interstitial (Leydig) cells in the testes secrete testosterone that, along with FSH, facilitates spermatogenesis.

(figure 29.11). The lower uterus narrows into the cervix, which attaches the uterus to the **vagina**, an 8–10 cm long tube composed of smooth muscle and elastic tissue. The perimetrium is the outer layer of the uterus. It extends to form broad ligaments that affix the uterus to the pelvis. The myometrium is the muscular middle layer of the uterus. It comprises most of the uterine wall. The endometrium is the inner layer of the uterus. This layer bears a specialized mucous membrane that contains many blood vessels and glands.

The external genital organs, or genitalia, include the mons pubis, labia majora, labia minora, glans clitoris, prepuce of clitoris, and urethral and vaginal openings (*see figure 29.8 for major female organs*). As a group, these are called the **vulva**. They function, in part, in sexual stimulation. The hymen is a thin membrane that partially covers the vaginal opening in young women. The hymen's function is largely unknown, but it is thought to be a result of embryonic development of the reproductive tract. The hymen can be ruptured during strenuous physical activity or sexual intercourse.

Mammary glands (L. *mammae*, breasts) are modified sweat glands that produce and secrete colostrum and milk. They contain mammary and adipose tissue and are suspended by bands of connective tissue called suspensory ligaments. A network of ducts terminate at the nipple which, through a positive feedback mechanism, allows release of mammary secretions to the infant while suckling. The surrounding areola contains sebaceous glands.

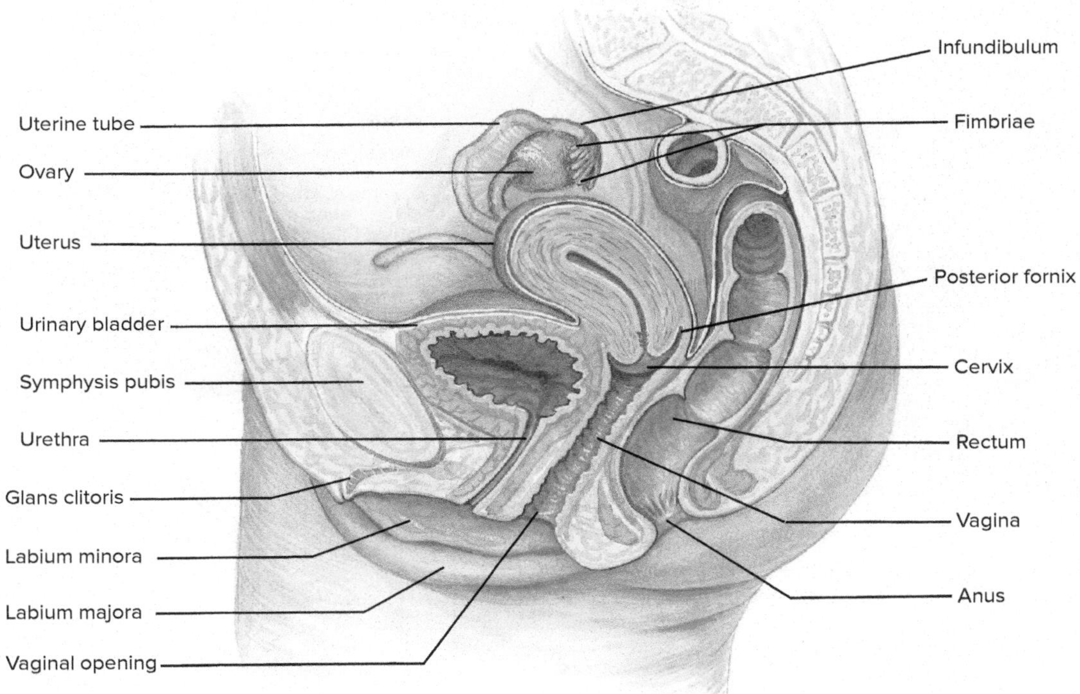

FIGURE 29.8

Lateral View of the Human Female Reproductive System. Major female reproductive organs are illustrated here. Ovaries produce eggs, estrogen, and progesterone. Uterine tubes carry eggs from the ovary to the uterus. The uterus receives the blastocyst and houses the developing embryo/fetus. Insemination occurs in the vagina; it is also the exit point for menstrual flow, and the baby during childbirth. External sexual organs facilitate sexual arousal and are collectively called the vulva. The mons pubis (not labeled here) is the protective layer of adipose tissue (illustrated in yellow) just in front of the symphysis pubis. The prepuce of clitoris (also not labeled here) is an extension of the labia minora that cover the glans clitoris.

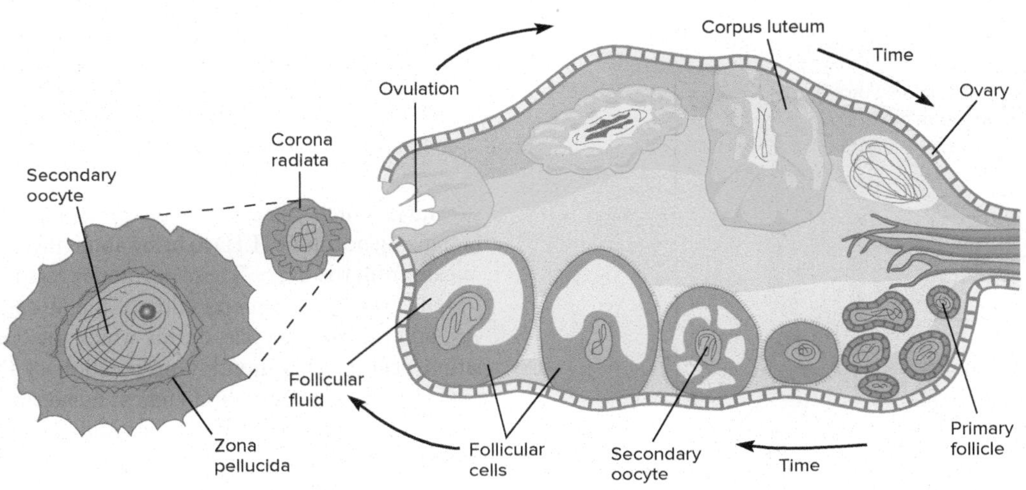

FIGURE 29.9

Cross Section through a Human Ovary. Events in the ovarian cycle proceed from growth and maturation of the primary follicle, through ovulation, and through the formation and maintenance (or degeneration) of the corpus luteum. The positions of the oocyte and corpus luteum are varied for illustrative purposes only. An oocyte matures at the same site, from development through ovulation.

FIGURE 29.10

Cilia Lining the Uterine Tubes. The cilia on the inner surface of the uterine tube propel the secondary oocyte downward and perhaps the sperm upward (EM ×1,000).

Hormonal Control of Female Reproductive Function

Unlike males, females, are only fertile during a few days each month. Cyclical hormone production controls development of a secondary oocyte in a follicle (figure 29.12; table 29.2). Gonadotropin-releasing hormone (GnRH) from the hypothalamus acts on the anterior pituitary, which releases follicle-stimulating hormone (FSH) and luteinizing hormone (LH) to bring about the oocyte's maturation and release from the ovary. These hormones regulate the **menstrual cycle**, which is the cyclical preparation of the uterus to receive a fertilized egg, and the **ovarian cycle**, during which the oocyte matures and ovulation occurs. This monthly preparation of the uterine lining normally begins at puberty and ends at **menopause** (Gr. *men*, month + *pausis*, cessation). At 45 to 55 years old, a woman's ovaries lose their sensitivity to FSH and LH and stop making normal amounts of progesterone and estrogen.

One way to understand the hormonal pattern in the normal monthly cycle is to follow the development of the oocyte and the physical events in the menstrual cycle (figure 29.13; table 29.3). On average, it takes 28 days to complete one menstrual cycle, although the range may be from 22 to 45 days. During this time, the following events take place:

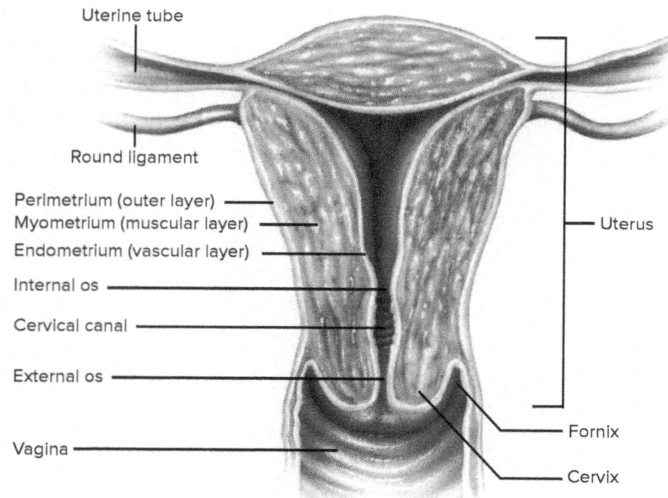

FIGURE 29.11

Human Female Uterus. The outer layer is the perimetrium. The middle myometrium makes up the bulk of the uterine wall. It is composed of smooth muscle fibers. The innermost layer is composed of the endometrium, which is deep and velvety in texture. Breakdown of the endometrium comprises part of the menstrual flow.

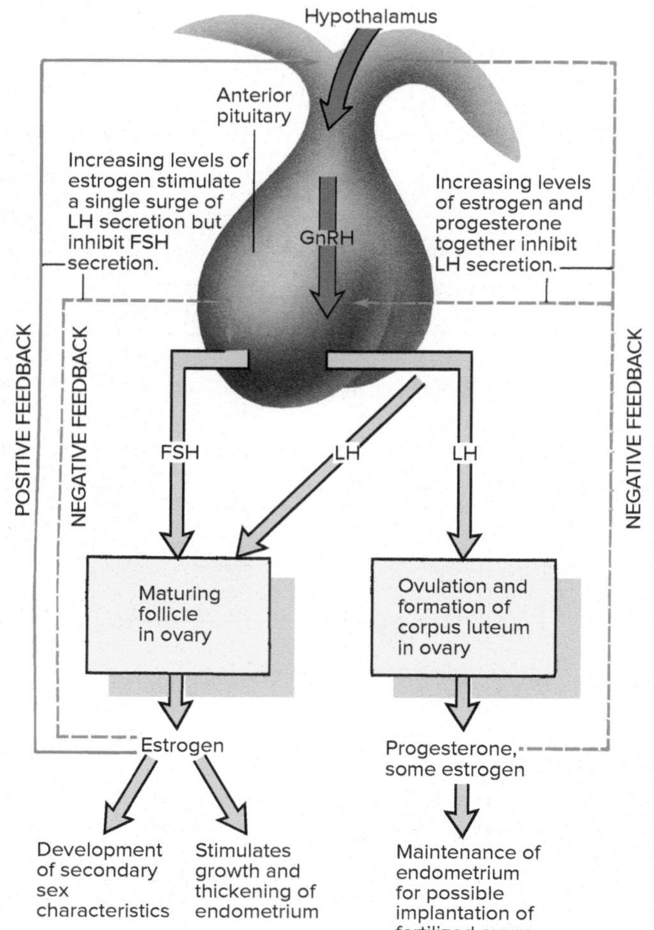

FIGURE 29.12

Hormonal Control of Reproductive Functions in an Adult Human Female. Feedback loops involve the hypothalamus, anterior pituitary, and ovaries. Gonadotropin-releasing hormone (GnRH) stimulates the release of both follicle-stimulating hormone (FSH) and luteinizing hormone (LH). Two negative feedback systems (red dashed pathways) and a positive feedback system (green pathway) control the ovarian cycle.

TABLE 29.2
MAJOR HUMAN FEMALE REPRODUCTIVE HORMONES

HORMONE	FUNCTION(S)	SOURCE(S)
Estrogen (estradiol)	Stimulates thickening of uterine wall, maturation of oocyte, and development of female sexual characteristics; inhibits FSH secretion; increases LH secretion	Ovarian follicle, corpus luteum
FSH (follicle-stimulating hormone)	Causes immature oocyte and follicle to develop; increases estrogen secretion; stimulates new gamete formation and development of uterine wall after menstruation	Pituitary gland
GnRH (gonadotropin-releasing hormone)	Controls pituitary secretion	Hypothalamus
hCG (human chorionic gonadotropin)	Prevents corpus luteum from disintegrating; stimulates corpus luteum to secrete estrogen and progesterone	Embryonic membranes and placenta
Inhibin	Inhibits secretion of FSH from the anterior pituitary gland	Ovaries
LH (luteinizing hormone)	Stimulates further development of oocyte and follicle; stimulates ovulation; increases progesterone secretion; aids in development of corpus luteum	Pituitary gland
Oxytocin	Stimulates uterine contractions during labor and milk release during nursing	Pituitary gland
Prolactin	Promotes milk production by mammary glands	Pituitary gland
Progesterone	Stimulates thickening of uterine wall	Corpus luteum
Relaxin	Increases flexibility of pubic symphysis during pregnancy and helps dilate uterine cervix during labor and delivery	Placenta and ovaries

1. The controlling center for ovulation and menstruation is the hypothalamus. It releases, on a regular cycle, GnRH, which stimulates the anterior pituitary to secrete FSH and LH (*see figure 29.12*).
2. FSH promotes the development of the oocyte in one of the immature ovarian follicles.
3. The follicles produce estrogen, causing a build-up and proliferation of the endometrium, as well as the inhibition of FSH production.
4. The elevated estrogen level about midway in the cycle triggers the anterior pituitary (via the hypothalamus) to secrete LH. This positive feedback causes the mature follicle to enlarge rapidly and release the secondary oocyte (ovulation). LH also causes the collapsed follicle to become another endocrine tissue, the corpus luteum.
5. The corpus luteum secretes estrogen and progesterone, which act to complete the development of the endometrium and maintain it for 10 to 14 days.
6. If the oocyte is not fertilized, the corpus luteum disintegrates into a corpus albicans, and estrogen and progesterone secretion cease.
7. Without estrogen and progesterone, the endometrium breaks down, and **menstruation** occurs. The menstrual flow is composed mainly of sloughed-off endometrial cells, mucus, and blood.
8. As progesterone and estrogen levels decrease further, the pituitary renews active secretion of FSH, which stimulates the development of another follicle, and the monthly cycle begins again.

Hormonal Regulation in the Pregnant Female

Pregnancy sets into motion a new series of physiological events. The ovaries are directly affected because, as the embryo develops, the cells of the embryo and placenta release the hormone human chorionic gonadotropin (hCG), which prevents degeneration of the corpus luteum. The progesterone that it secretes maintains the uterine lining. Eventually the placenta takes over estrogen and progesterone production, and the corpus luteum degenerates. Blood and urine concentrations of hCG will be detectable at around two weeks after implantation. Other hormones are secreted during and after pregnancy. For example, prolactin induces milk production during (and after) pregnancy, and oxytocin causes letdown after childbirth. Additionally, oxytocin and prostaglandins cause uterine contractions that expel the baby and placenta from the uterus during childbirth.

Mammals with Estrous Cycles

Mammals with an estrous cycle regularly shed endometrial cells without bleeding. The estrous cycle is comprised of four phases: proestrus (follicular development), estrus (sexual receptivity), metestrus (a decline in estrogen), and diestrus (a period of sexual inactivity between phases). These phases correspond with the four phases of the endometrium menstrual cycle

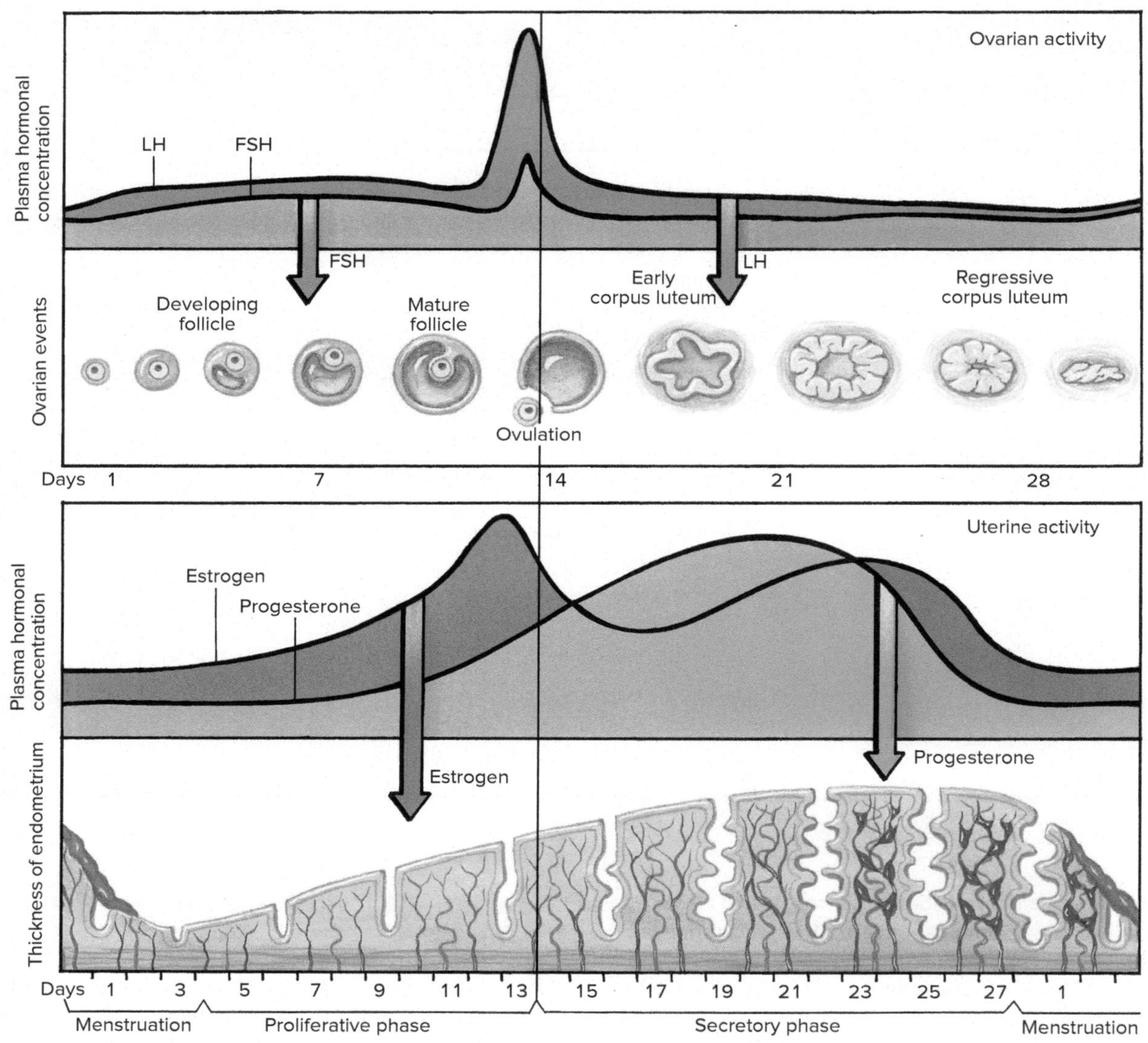

FIGURE 29.13

Major Events in the Female Ovarian and Menstrual Cycles. The two charts correlate the gonadotropins, ovarian hormones, follicle development, ovulation, and changes in uterine anatomy during the cycles.

TABLE 29.3
SUMMARY OF THE MENSTRUAL CYCLE EVENTS

PHASE	EVENT(S)	DURATION IN DAYS*
Follicular	Follicle matures in the ovary; menstruation (endometrium breaks down); endometrium rebuilds	1–5
Ovulation	Ovary releases secondary oocyte	6–14
Luteal	Corpus luteum forms; endometrium thickens and becomes glandular	15–28

*Using a 28-day menstrual cycle as an example.

(*see figure 29.13; see also chapter 22 for more details on mammalian reproductive cycles.*)

SECTION 29.5 THINKING BEYOND THE FACTS

How are the functions of FSH and LH similar in human males and females? How do they differ?

29.6 PRENATAL DEVELOPMENT AND BIRTH IN A HUMAN

LEARNING OUTCOMES

1. Describe the major developmental events that occur during each of the three trimesters of a human pregnancy.
2. Explain the role of the placenta.

This section covers the main event in reproduction—the nine-month pregnancy period, during which time the human female's body carries, nourishes, and protects the embryo as it grows to a full-term baby.

Events of Prenatal Development: From Zygote to Newborn

The development of a human can be divided into prenatal ("before birth") and postnatal ("after birth") periods. The prenatal period begins with a zygote that undergoes cleavage to become a **morula** (solid ball of embryonic stem cells), and later, a **blastocyst** (a human blastula in which some differentiation has occurred) that implants in the endometrium. From two weeks after fertilization until the end of the eighth week, the developing individual is called an embryo. From nine weeks until birth, it is termed a fetus. During or after birth, it is called a newborn, or baby.

Pregnancy is divided into three-month periods called trimesters. The divisions are somewhat arbitrary, but they are framed by specific fetal developments. The first trimester begins at fertilization and is when the bulk of organogenesis takes place (0–13 weeks). The second (14–26 weeks) and third (27–40 weeks) trimesters are mainly periods of growth for the fetus.

The First Trimester

After fertilization, usually in the upper third of the uterine tube, the zygote goes through several cleavages as it moves down the tube (figure 29.14). It eventually becomes the morula, and by the fourth day, it develops into the blastocyst. The blastocyst then adheres to the uterine wall and implants. During implantation, the outer blastocyst (the trophoblast), invades the endometrium. This process is typically completed 11 to 12 days after fertilization; from then on, the female is considered to be pregnant.

In mammalian development, most cells of the early embryo do not contribute to the development of the embryo's body. They eventually form essential supportive and protective membranes. Only the inner cell mass gives rise to the embryonic body. Eventually, these cells arrange in a flat sheet and undergo a gastrulation similar to that of the avian and nonavian reptiles.

After gastrulation, most of the first trimester is devoted to organogenesis and fetal growth (figure 29.15). Complex developmental processes called primary and secondary **inductions** shape most of the organ systems.

The Second Trimester

In the second trimester, fetal growth is spectacular. In month four, the pregnant mother is aware of fetal movements, and fetal bones enlarge. In month five, the fetal heartbeat can be heard with a stethoscope. Later in this trimester, the upper and lower eyelids separate, eyelashes form, and the eyes open.

The Third Trimester

During the third trimester (month seven to birth), fetal circulatory and respiratory systems develop sufficiently so that if the baby were born prematurely, it could potentially survive. Increased development of the central and peripheral nervous systems also occurs during this period. During the last month, fetal weight doubles.

The Placenta: Exchange Site and Hormone Producer

The lengthy gestation of mammals is possible, in part, because of evolutionary modifications of extraembryonic membranes present in early amniotes. For instance, the chorion and allantois gave rise to embryonic components of the mammalian **placenta**. This organ sustains the developing embryo and fetus during pregnancy by allowing the exchange of gases, nutrients, and wastes between maternal and fetal systems (figure 29.16). The small projections that were initially sent out from the blastocyst during implantation develop into structures called chorionic villi. Chorionic villi are specialized areas of the placenta where fetal and maternal capillaries form an integral countercurrent exchange system that permits efficient exchange. Maternal and fetal blood vessels do not merge so the two bloodstreams remain separate. The fetal capillaries come together and become two umbilical arteries and one umbilical vein that spiral about each other to form a structure called the **umbilical cord**. The umbilical cord connects the fetal abdomen to the placenta.

The placenta also secretes a suite of hormones essential for successful embryonic and fetal development, and parturition. Some of these hormones were previously described (*see Hormonal Regulation in the Pregnant Female*) and others are discussed in the next section.

Birth: An End and a Beginning

Approximately 266 days after fertilization, the human infant is born. Birthing is called **parturition** (L. *parturire*, to be in labor), and changing hormone levels initiate parturition. The hormone relaxin, produced by the placenta and ovaries, causes cervical dilation and the mother's pubic symphysis to soften so that the baby can successfully exit the birth canal.

When parturition begins, the baby's pituitary gland secretes adrenocorticotropic hormone (ACTH), which stimulates its adrenal glands to secrete steroid hormones. These steroids contact the placenta and stimulate it to produce prostaglandins and placental estrogen. Prostaglandins, increased estrogen, and stretching of the cervix initiate uterine contractions and cause oxytocin to be released from the mother's posterior pituitary. This hormone causes major uterine contractions that build in length and increase in frequency over a period that most often lasts between 2 and 18 hours. During that time, the cervix becomes fully dilated, and the amniotic sac ruptures (this is termed "water breaking").

Usually within an hour of these events, the baby is born (figure 29.17*a–c*). After the baby emerges, uterine contractions continue to expel the placenta and extraembryonic membranes (commonly called the **afterbirth**; figure 29.17*d*). The umbilical cord is cut (or bitten in nonhuman mammals), and the newborn embarks on its nurtured existence in the outside world. Because

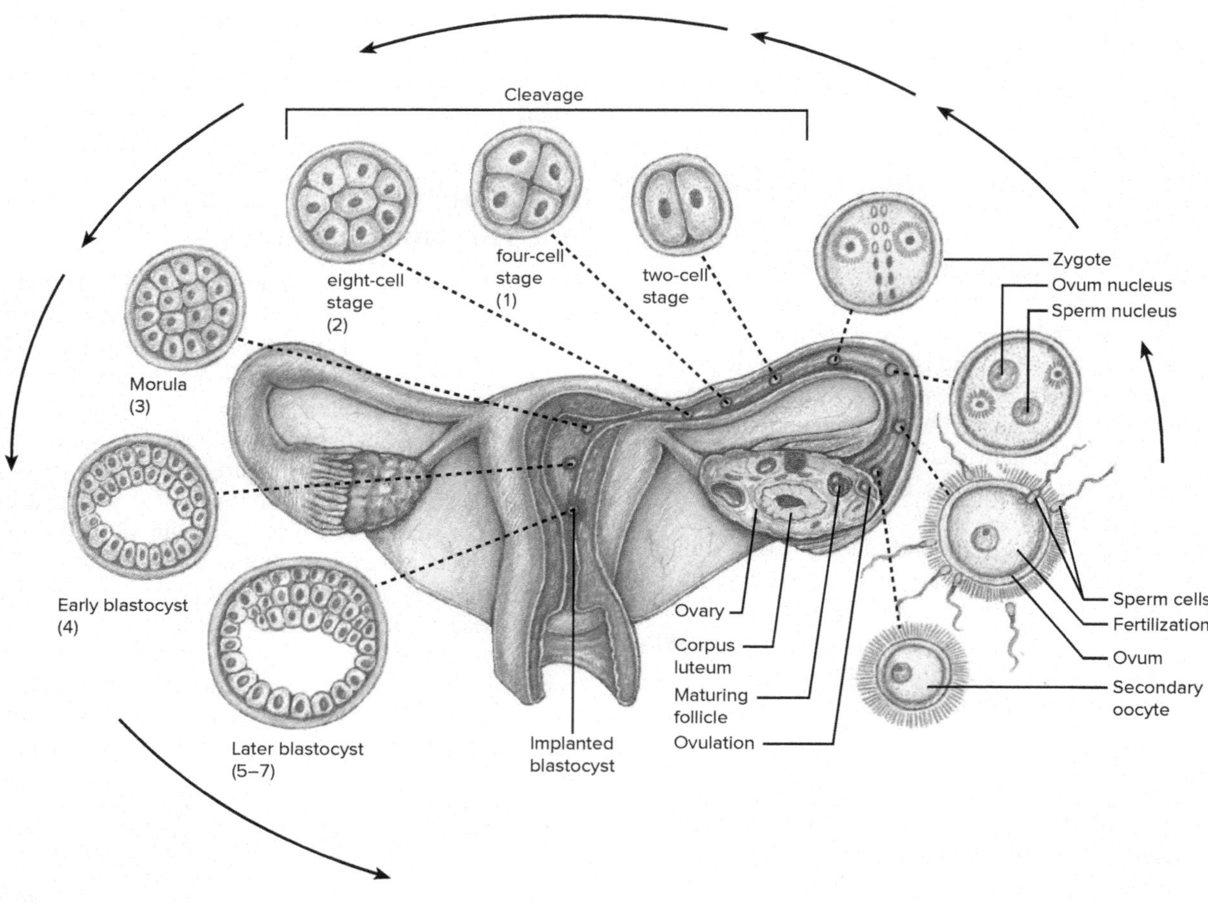

FIGURE 29.14

Early Stages in Human Development. The numbers in parentheses indicate the days after fertilization. The secondary oocyte is fertilized in the upper third of the uterine tube, undergoes cleavage while traveling down the tube, and finally implants in the endometrium of the uterus.

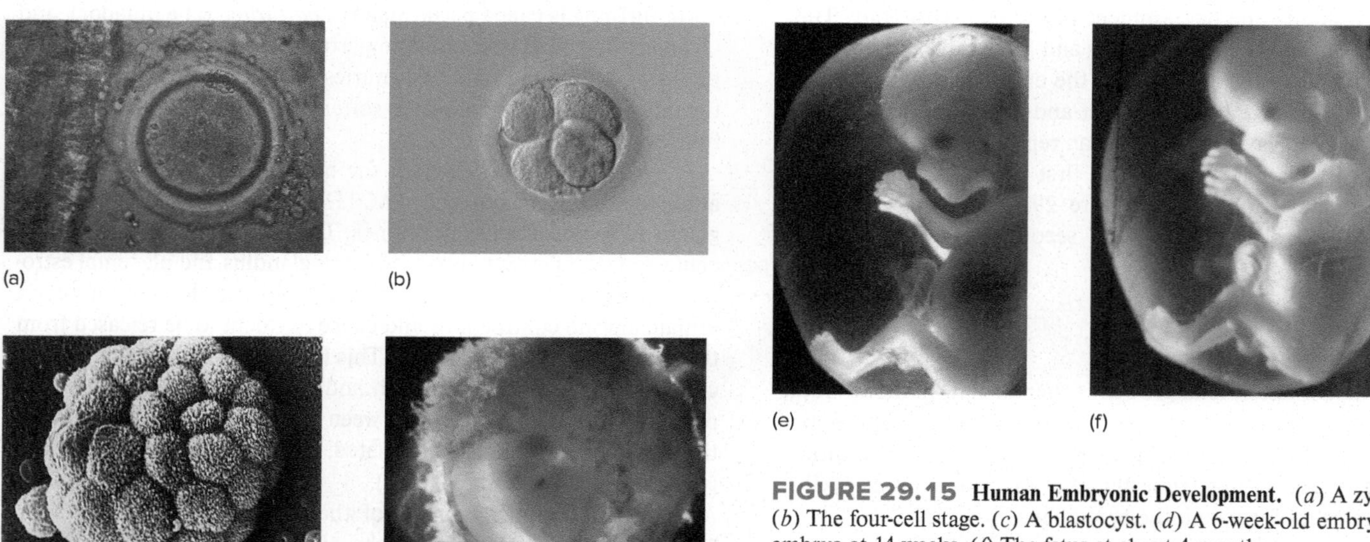

FIGURE 29.15 Human Embryonic Development. (*a*) A zygote. (*b*) The four-cell stage. (*c*) A blastocyst. (*d*) A 6-week-old embryo. (*e*) The embryo at 14 weeks. (*f*) The fetus at about 4 months.

(a) ©Petit Format/Science Source (b) ©Red_Hayabusa/Getty Images (c-d) ©Petit Format/Science Source (e) ©Claudia Edelmann/Science Source (f) ©Petit Format/Science Source

How Do We Know That Breast Milk Is Nature's First Functional Food?

Human breast milk is considered the ideal source of nutrition for infants, and it probably played a critical role in the biological and cultural evolution of human beings. The study of mother's breast milk is now booming, thanks to new analytical techniques, a growing interest in milk-microbiome connections, and the resurgence of breastfeeding. Building on a century-old study that first indicated that milk nourished certain bacteria in infants, the new work has characterized the complexity of breast milk carbohydrates that nourishes a multitude of beneficial gut bacteria *(see Gut Biomes p. xxx)* and promotes a healthy gut immune system. These carbohydrates also protect the gut against pathogens, as do peptides released from breast milk proteins by the enzymes also supplied in breast milk. Because the benefits of bacteria seem to help protect against infection that can be deadly in premature infants, researchers are busy developing probiotics (microorganisms that can be added to the diet in order to provide health benefits) to encourage the microbial health of these vulnerable babies.

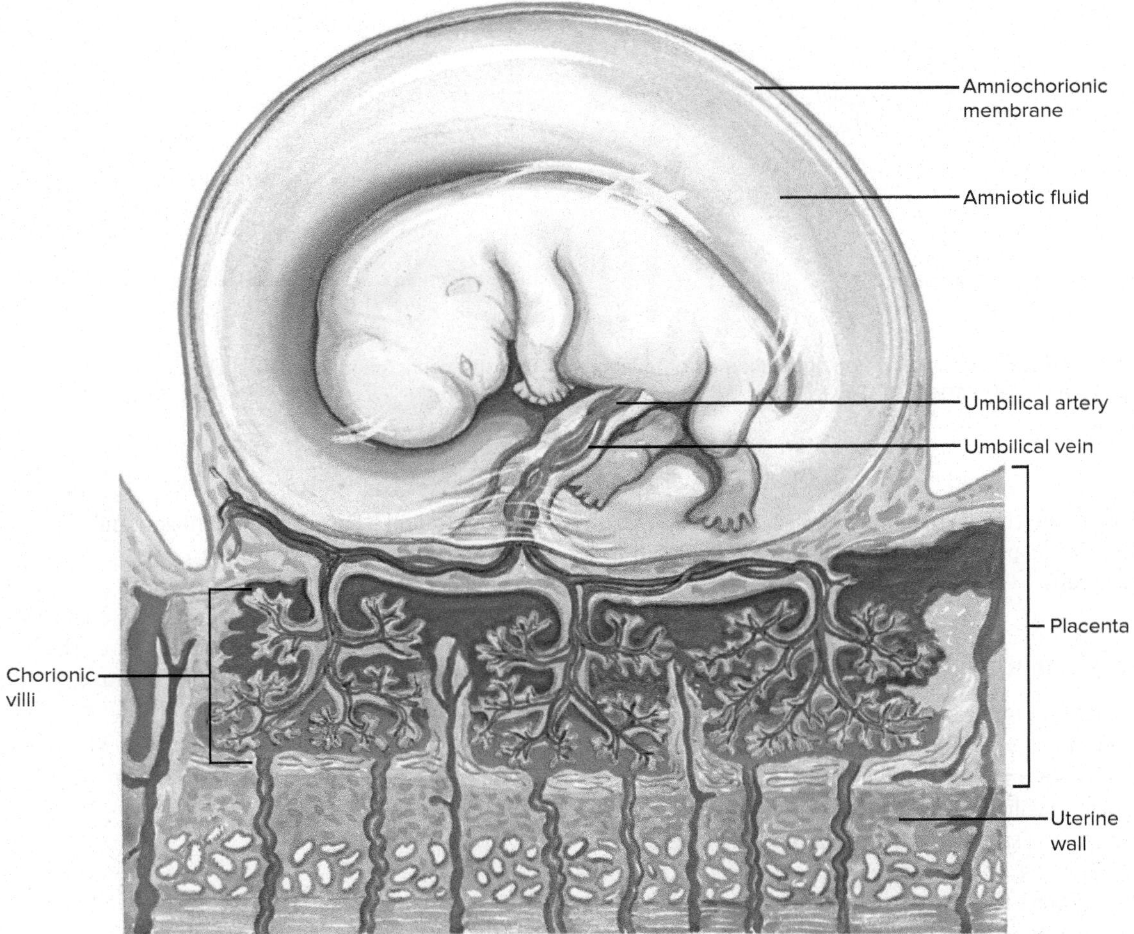

FIGURE 29.16

Fetus and Placenta at Seven Weeks. The circulations of mother and fetus are in close contact at the site of the chorionic villi, but they do not actually mix. Branches of the mother's arteries in the wall of her uterus open into pools near the chorionic villi. Oxygen and nutrients from the mother's blood diffuse into the fetal capillaries of the placenta. The fetal capillaries lead into the umbilical vein, which is enclosed within the umbilical cord. From here, the fresh blood circulates through the fetus's body. Blood that the fetus has depleted of nutrients and oxygen returns to the placenta in the umbilical arteries, which branch into capillaries, from which waste products diffuse to the maternal side.

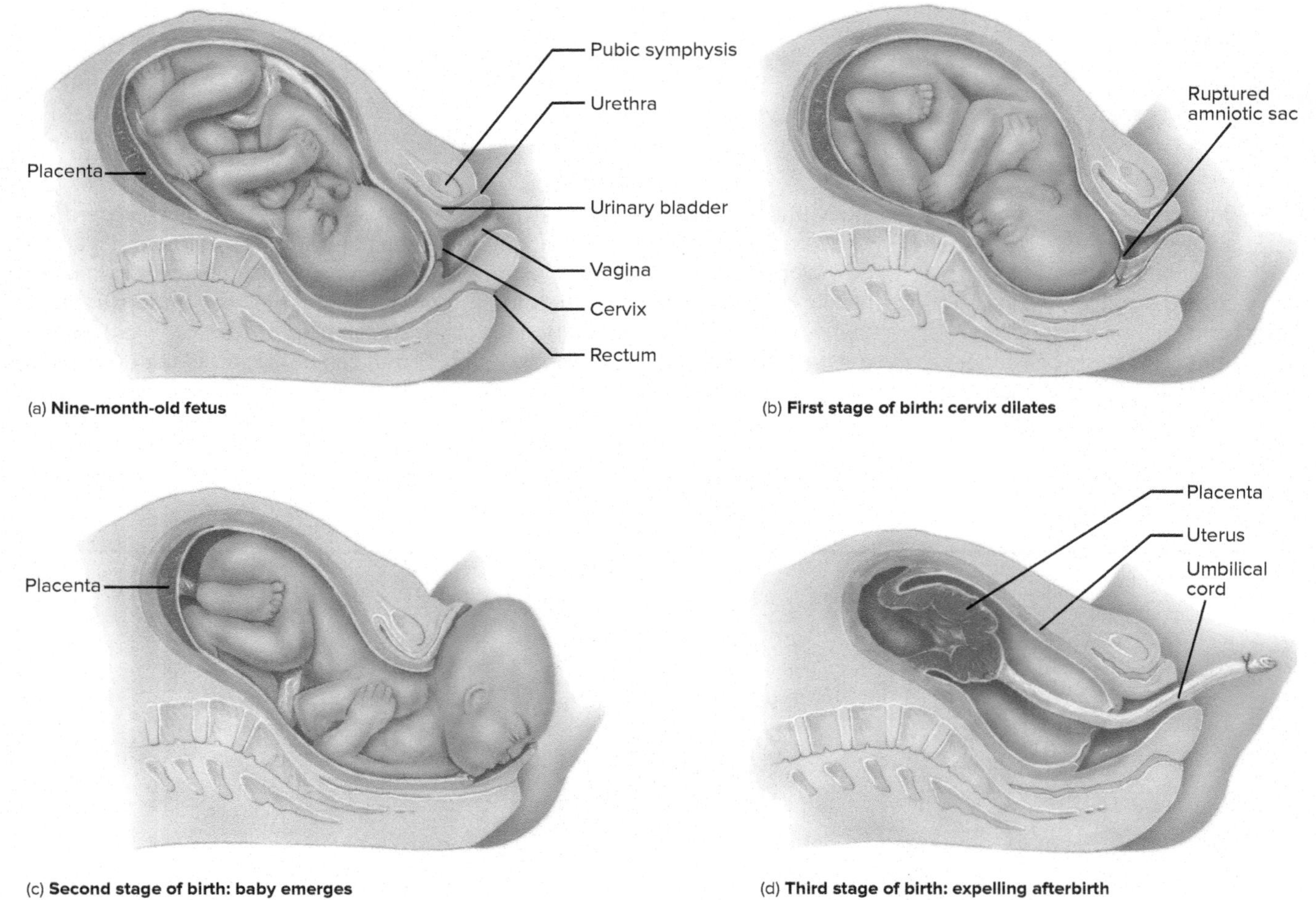

FIGURE 29.17

Stages of Labor and Parturition. (*a*) The position of the fetus prior to labor. (*b*) The ruptured amniotic sac and early dilation of the cervix. (*c*) The expulsion stage of parturition. (*d*) Passage of the afterbirth.

birth is painful and lengthy, beta-endorphins and cortisol are released to help alleviate pain, and increase energy available to the mother, respectively.

Milk Production and Lactation

Like the other reproductive processes, **lactation** (L. *lactare*, to suckle) is also regulated hormonally. During pregnancy, estrogen prepares the mammary glands for milk production. Two other hormones, prolactin and oxytocin, facilitate the two aspects of lactation. After birth, prolactin secretion is stimulated by regulatory hormones from the hypothalamus. Suckling by the baby results in a specific neuronal stimulation that triggers the further release of prolactin and the release of oxytocin from the pituitary gland. Prolactin causes mammary glands to produce milk. Oxytocin has two roles: It causes cells associated with mammary alveoli to contract and move milk toward the nipple, and it stimulates widening of milk ducts to allow adequate flow to the suckling newborn.

The initial product of lactation and letdown is not actually milk, it is **colostrum**, a thick and concentrated protein and antibody-rich fluid that helps to strengthen the baby's immune system. Colostrum also functions as a laxative, facilitating defecation of material ingested by the baby while in utero. This material, called **meconium**, contains bile, amniotic fluid, epithelial cells, fetal body hair, mucus, and water. Colostrum is produced for about three days after birth.

Section 29.6 Thinking Beyond the Facts

Why is the circulation through the placenta like that of the lungs* (see figure 29.16)*?

SUMMARY

29.1 **Asexual Reproduction in Invertebrates**

- Asexual reproductive processes do not involve the production and subsequent fusion of haploid cells, but rely solely on vegetative growth through mitotic cell division.
- Protists and some animals (cnidarians and annelids) may reproduce by fission. Fission is the division of one cell, body, or body part into two.
- Another method of asexual reproduction found in some invertebrates is budding. Bud cells proliferate and break away from the parent.
- Fragmentation is a type of asexual reproduction whereby a body part is lost and then regenerates into a new organism.
- Certain invertebrates can reproduce without sperm and normal fertilization in a process called parthenogenesis.
- The predominance of asexual reproduction in some invertebrates can be partially explained by the environment in which they live.

29.2 **Sexual Reproduction in Invertebrates**

- Sexual reproduction involves the formation of haploid cells through meiosis and the subsequent fusion of pairs of those cells to produce a diploid zygote.
- Many invertebrates simply release their gametes into the water in which they live via broadcast spawning.
- Other invertebrates utilize internal fertilization to transfer sperm from male to female and have structures that facilitate such transfer.
- In sexual reproduction, new combinations of traits can arise rapidly because of genetic recombinations. On the other hand, a sexually reproducing animal can never bequeath its own exact set of genetic material to its progeny.

29.3 **Sexual Reproduction in Vertebrates**

- Many vertebrates are oviparous. These animals often produce many eggs that are fertilized externally and develop with little parental care. Fewer vertebrates are ovoviviparous. Internal fertilization and development protects embryos until birth. Viviparous vertebrates retain young within the female reproductive tract, and young are nourished by food supplied by the mother during gestation.
- Most fish are oviparous. Others are ovoviviparous or viviparous. Fertilization is internal in the Chondrichthyes.
- Amphibians usually rely on aquatic habitats for reproduction. Fertilization is usually external, and development involves aquatic larval stages.
- Internal fertilization and the evolution of the amniotic egg permits terrestrial development by avian and nonavian reptiles and mammals. All reptiles possess a leathery or calcareous shell. Eggs of nonavian reptiles usually develop unattended by the parents. Eggs of birds often require parental care. Primitive mammals, the Prototheria, are oviparous. The Theria are viviparous. Eutherian mammals retain young inside the female, where young are nourished through a placenta. After birth, mammalian young are nourished with milk from mammary glands. Primates are asynchronous breeders.

29.4 **The Human Male Reproductive System**

- The reproductive role of the human male is to produce sperm and deliver them to the vagina of the female. This function requires different structures. The testes produce sperm and the male sex hormone, testosterone. Accessory glands furnish a fluid, called semen, for carrying the sperm to the penis. Accessory ducts store and carry secretions from the testes and accessory glands to the penis. The penis deposits semen in the vagina during sexual intercourse.
- Before a human male can mature and function sexually, special regulatory hormones (FSH, GnRH, inhibin, LH, and testosterone) must function.

29.5 **The Human Female Reproductive System**

- The reproductive roles of the human female are more complex than those of the male. Not only do females produce eggs, but after fertilization, they also nourish, carry, and protect the developing embryo. They may also nourish the infant for a time after it is born.
- The female reproductive system consists of two ovaries, two uterine tubes, the uterus, the vagina, and external genitalia. The mammary glands contained in the paired breasts produce milk for the newborn baby.
- The human female is fertile for only a few days each month, and the pattern of hormone secretion is intricately related to the cyclical release of a secondary oocyte from the ovary. Various hormones regulate the menstrual and ovarian cycles.
- Pregnancy sets into motion a new series of physiological events involving human chorionic gonadotropin hormone and progesterone. Oxytocin and prostaglandins also stimulate uterine contractions that expel the baby from the uterus during childbirth.

29.6 **Prenatal Development and Birth in a Human**

- Pregnancy sets a new series of physiological events into motion that are directed to housing, protecting, and nourishing the embryo.
- The development of a human may be divided into prenatal and postnatal periods. Pregnancy is arbitrarily divided into trimesters.
- The placenta is the organ that sustains the embryo and fetus throughout the pregnancy. The birth process is called parturition and occurs about 266 days after fertilization.
- Lactation includes both milk secretion (production) by the mammary glands and milk release from the breasts. The initial product of lactation is colostrum, which helps to strengthen the baby's immune system and functions as a laxative. Milk production and release begins three days after parturition.

CONCEPT REVIEW QUESTIONS

1. The cnidarian, *Hydra,* reproduces
   a. only sexually.
   b. asexually by budding.
   c. by parthenogenesis.
   d. by fragmentation.
   e. by both b and c.
2. Asexual reproduction is common in
   a. the protozoa.
   b. sponges.
   c. jellyfish.
   d. flatworms.
   e. all of the above (a–d).

3. Broadcast spawning is an example of internal fertilization.
   a. True
   b. False
4. In protandry, an animal is a male during its early life history and a female later in the life history.
   a. True
   b. False
5. The three basic embryonic membranes that characterize the mammalian embryo are also found in
   a. fishes.
   b. amphibians.
   c. reptiles.
   d. birds.
   e. Both c and d are correct.
6. Which of the following structures is the site of spermatogenesis?
   a. Prostate
   b. Bulbourethral gland
   c. Urethra
   d. Seminiferous tubule
   e. Both a and b
7. FSH and LH are produced by the
   a. ovaries.
   b. testes.
   c. anterior pituitary gland.
   d. adrenal glands.

Analysis and Application Questions

1. Is the fertility of a woman affected by the length of a given menstrual cycle or whether the cycles are regular or irregular? Explain.
2. Looking at a variety of animals, what are the advantages to restricting reproduction to a limited time period? Why do so many animals have a sharply defined reproductive season during the year?
3. In most sexual species, males produce far more gametes than do females. Why does this occur when, in most cases, only one male gamete can fertilize one female gamete?
4. Why are the accessory glands of the male so important in reproduction?
5. Why doesn't a women menstruate while she is pregnant?

APPENDIX A
ANIMAL PHYLOGENY

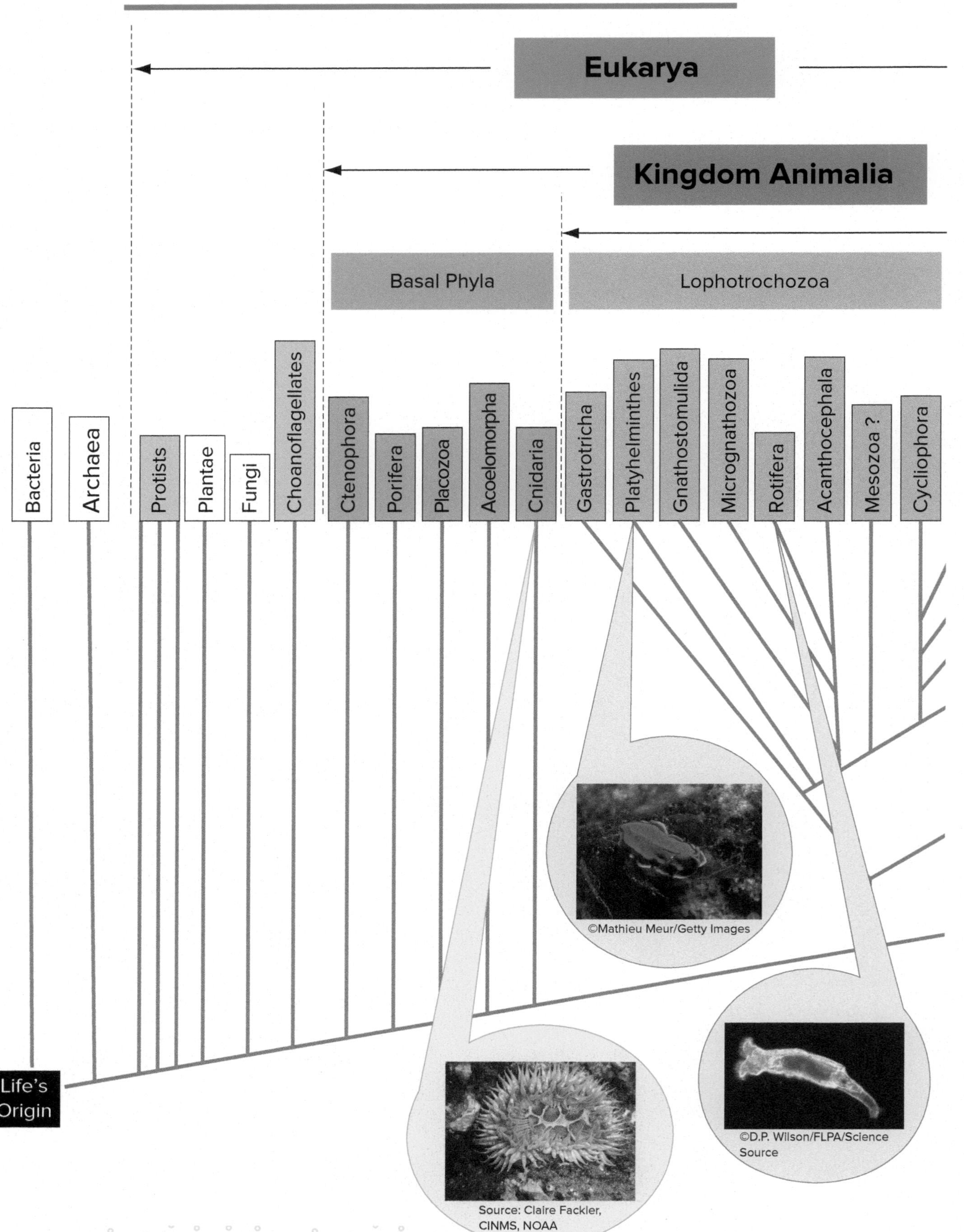

©Mathieu Meur/Getty Images

Source: Claire Fackler, CINMS, NOAA

©D.P. Wilson/FLPA/Science Source

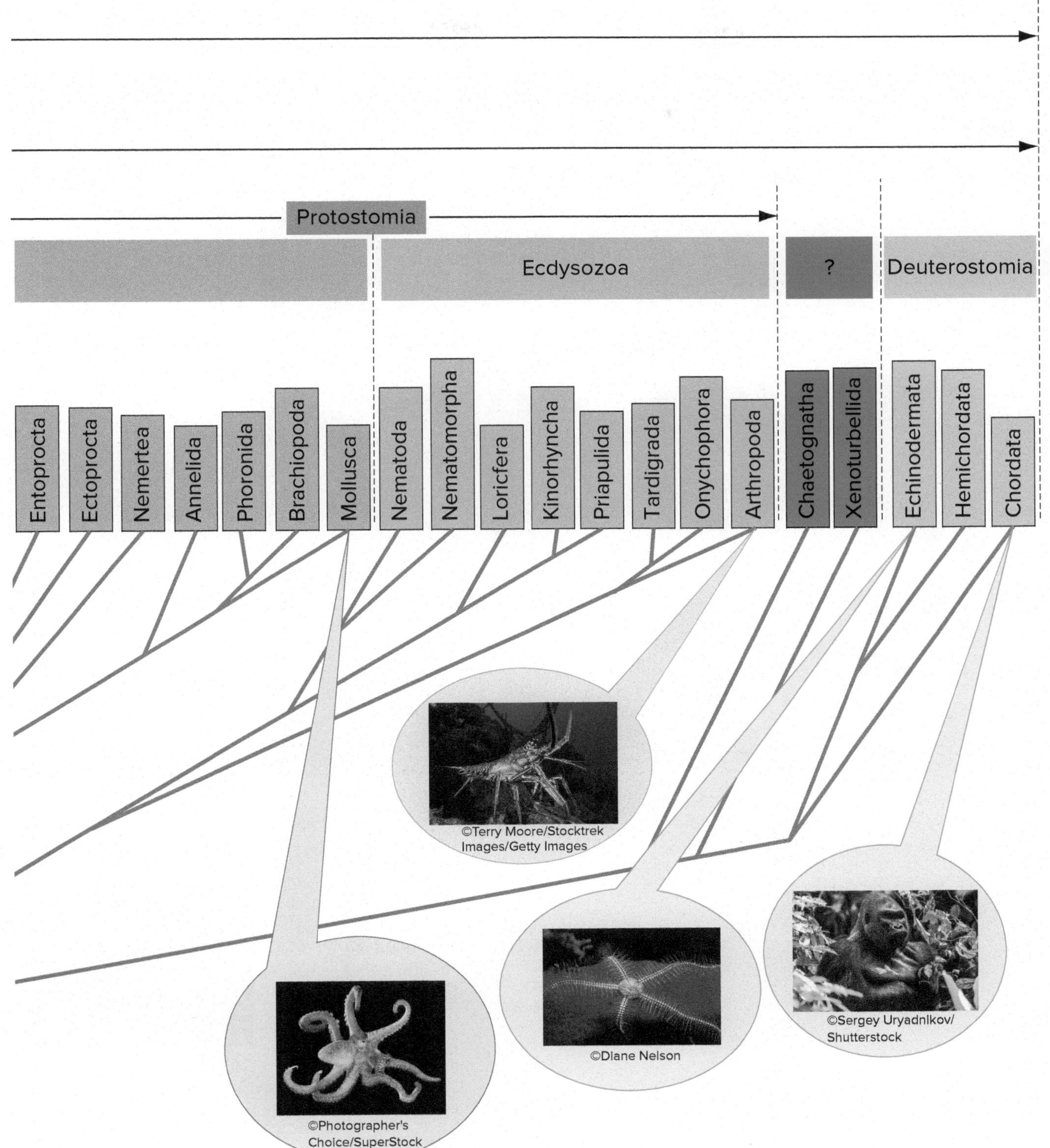
Protostomia
Ecdysozoa
?
Deuterostomia
Entoprocta
Ectoprocta
Nemertea
Annelida
Phoronida
Brachiopoda
Mollusca
Nematoda
Nematomorpha
Loricfera
Kinorhyncha
Priapulida
Tardigrada
Onychophora
Arthropoda
Chaetognatha
Xenoturbellida
Echinodermata
Hemichordata
Chordata
©Terry Moore/Stocktrek Images/Getty Images
©Photographer's Choice/SuperStock
©Diane Nelson
©Sergey Uryadnikov/ Shutterstock

APPENDIX B
THE HISTORY OF THE EARTH: GEOLOGICAL EONS, ERAS, PERIODS, AND MAJOR BIOLOGICAL EVENTS

THE HISTORY OF THE EARTH: GEOLOGICAL EONS, ERAS, PERIODS, AND MAJOR BIOLOGICAL EVENTS

EON	ERA	PERIOD	AGE AT BEGINNING OF THE PERIOD (MILLIONS OF YEARS)	MAJOR EVENTS
PHANEROZOIC			542	*See facing page.*
PROTEROZOIC	EDIACARAN		635	Virtually all modern animal phyla were present by the end of this period.
			2,500	Animals originated in Opisthokonta (802 million years ago [mya]). Multicellular organisms appeared (1,200 mya). Eukaryotic organisms appeared (2,100 mya). Oxygen accumulated in the atmosphere and oceans.
ARCHEAN			4,000	Photosynthetic cyanobacteria began releasing oxygen to the oceans and atmosphere (2,500 mya). First bacterial life formed (3,800 mya).
HADEAN			4,600	Organic molecules. Cooling of molten Earth. Formation of the moon. Heavy bombardment from the solar system.

EON	ERA	PERIOD	AGE AT BEGINNING OF THE PERIOD (MILLIONS OF YEARS)	MAJOR BIOLOGICAL EVENTS
PHANEROZOIC	CENOZOIC	Quaternary	2.6	Subtropical forests gave way to cooler forests and grassland areas. Evolution of *Homo*.
		Tertiary	66	Modern orders of mammals expanded. Hominins evolved in the last 5–8 million years.
	MESOZOIC	Cretaceous	144	Continental seas and swamps spread. Extinction of ancient birds and reptiles. Rise of flowering plants and expansion of insect diversity.
		Jurassic	201	Climate warm and stable. High reptilian diversity. Avian reptiles and stem mammalian forms appeared.
		Triassic	252	Climate warm. Extensive deserts. Diapsid reptiles diversify. Archosaur lineage appeared.
	PALEOZOIC	Permian	299	Climate cold early, but then warmed. Diapsid and synapsid amniote lineages diversify; first mammals. Widespread extinction of amphibians.
		Carboniferous	356	Warm and humid with extensive coal-producing swamps. Arthropods and amphibians common. First reptiles appeared.
		Devonian	419	Land high and climate cool. Freshwater basins developed. Fishes diversified. Early amphibians appeared. First insects.
		Silurian	444	Extensive shallow seas. Warm climate. First terrestrial arthropods: Chelicerata then Myriapoda. First jawed Fishes.
		Ordovician	485	Shallow extensive seas. Climate warmed. Many marine invertebrates. Jawless fishes widespread.
		Cambrian	541	Extensive shallow seas and warm climate. Trilobites and brachiopods common. Earliest chordates found late in the Cambrian.

©The Natural History Museum/Alamy Stock Photo

©Elenarts/Shutterstock

©Nicolas Primola/Shutterstock

©Publiphoto/Science Source

APPENDIX C
ANIMAL-LIKE PROTISTS: THE PROTOZOA

EVOLUTIONARY PERSPECTIVE

The origin of life and its diversification into three domains were described in chapter 8. This textbook has been primarily concerned with one domain (Eukarya) and a single lineage within Eukarya, the supergroup Opisthokonta (*see figure 8.2*). The single-celled opisthokonts are important because they include the choanoflagellate ancestors of the Animalia (*see figures 8.5 and 8.6*). Other protists are found within the remaining four eukaryotic supergroup lineages, and many of these are important in the study of zoology. Some of these are important because of their heterotrophic (feeding on other organisms) characteristic. Very early classification schemes treated these protists as animals and grouped them within an animal taxon called "Protozoa." Protozoans have been traditionally covered in zoology courses. More significantly, the protozoa (now a nontaxonomic designation) play important roles in Earth's ecosystems, and they interact with animals in both beneficial and harmful ways. In addition to the opisthokonts described in chapter 8, most of the heterotrophic protozoans are members of three of the five eukaryotic supergroups. A few of these protozoans are described in the following paragraphs.

PROTOZOANS: MAJOR ECOLOGICAL ROLES

Protozoans feed primarily on bacteria, other protists, soluble organic matter, and sometimes fungi; and they range in size from 1 to 150 μm. They are an important constituent of ecosystems worldwide. Photosynthetic protists and cyanobacteria are the major primary producers in oceanic ecosystems, and heterotrophic protozoans–especially flagellated protozoans–convert much of this primary production into food for zooplankton, like copepods, and krill (*see chapter 15*). Zooplankton and krill then support higher trophic levels of marine ecosystems including animals as large as blue whales (*Balaenoptera musculus, see figure 6.10*). Benthic protozoans that inhabit anoxic marine sediments are the only eukaryotes that are known to be capable of denitrification, reducing nitrate to nitrogen gas within Earth's nitrogen cycles. (This role is often played by denitrifying bacteria in all Earth's ecosystems [*see chapter 6*]). Protozoans are important decomposers in virtually all of Earth's ecosystems. Up to 90% of carbon mineralization and decomposition, which makes components of organic matter available to other organisms, is carried out by microorganisms including the protozoans.

Some protozoans are important symbionts of animals. Many of these live in mutualistic relationships with their hosts. For example, termites (order Isoptera, *see chapter 15*) have protozoans (*Mixotricha paradoxa*) living in their gut that help the termites digest the cellulose in the wood they eat. All ruminant mammals (*see figure 27.6*) have similar mutualistic relationships with their symbiotic protozoan partners. Many other protozoans live in commensalistic or parasitic relationships with their hosts, and may be pathogenic. Protozoans may also cause human diseases and crop failures. Selected human parasites are described next.

SUPERGROUP EXCAVATA
Underlying Features Suspension feeding groove (cytosome) present or presumed to have been lost; feed by flagella-generated currents
Examples *Giardia, Trichomonas, Euglena, Leishmania, Trypanosoma*

Giardia intestinalis

Giardiasis is worldwide in distribution and occurs in most mammals, including humans. In the United States, *Giardia intestinalis* (figure 1) is the most common cause of epidemic waterborne diarrheal disease. Transmission commonly occurs by cyst-contaminated water supplies, where cysts remain viable for one to two months. Following ingestion, the cysts undergo excystment in the duodenum and form **trophozoites** (the active feeding stage of a protozoan). The trophozoites inhabit the upper portions of the small intestine, where they attach to the intestinal mucosa by means of their suction disks. The trophozoites feed on mucous secretions and reproduce to form such a large population that they may interfere with nutrient absorption by the intestinal epithelium. The trophozoites form cysts that pass out of the host with the feces and are

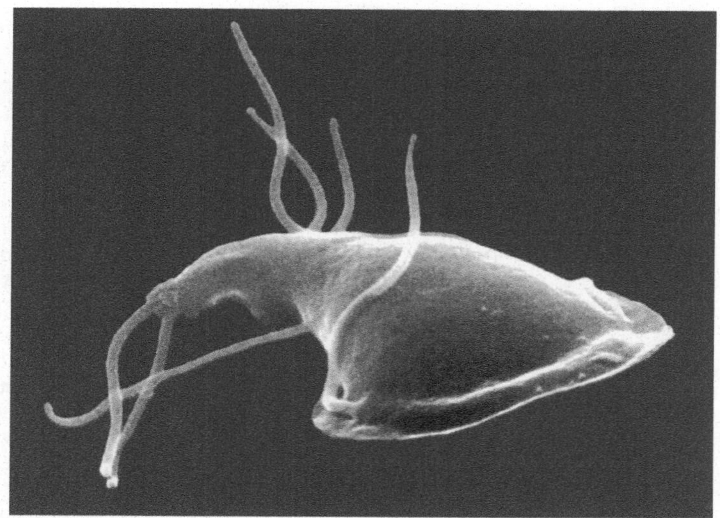

FIGURE 1

The Protozoan *Giardia intestinalis*. This parasite causes the waterborne disease giardiasis in humans. Related species infect other mammals. This digitally colorized SEM depicts the parasite's upper surface. Pairs of flagella seen here include anterior, posterior-lateral, and caudal.
Source: CDC/Dr. Stan Erlandsen and Dr. Dennis Feely

diagnostic in medical examinations. Acute giardiasis is characterized by severe diarrhea, epigastric pain, cramps, and flatulence. It affects children more seriously than adults. Infection rates range from 2% to 4% in developed countries to 40% in developing countries. Several drugs are available for treatment. Prevention and control involve proper treatment of community water supplies.

Trichomonas vaginalis

Trichomoniasis is a sexually transmitted infection of the urogenital system caused by the flagellate *Trichomonas vaginalis* (figure 2). It is one of the most common sexually transmitted diseases in the United States and the most common protozoan parasite in developed countries. There is no cyst in the *T. vaginalis* life cycle; therefore, transmission occurs through the trophozoite stage, usually during sexual intercourse. Infected males are generally asymptomatic because of the trichomonacidal secretions from the prostate gland. A few days following infection in a female, the proliferating flagellates cause degeneration of the vaginal epithelium. The cervical mucosa becomes covered with papules, vesicles, and small hemorrhages leading to the accumulation of leukocytes at the site of infection. Vaginal symptoms include a profuse, purulent (yellow to cream color pus-like), and odiferous discharge that is accompanied by itching. Diagnosis is made in females by identification of the protozoan through microscopic examination of the discharge. Treatment is with antibiotics.

Trypanosoma brucei

Trypanosoma brucei causes the aggregate of diseases called trypanosomiasis. This species is divided into three subspecies collectively referred to as the *Trypanosoma brucei* complex. *T. b. brucei* is a parasite of nonhuman mammals of Africa. *T. b. gambiense,* and *T. b. rhodesiense* are morphologically indistinguishable parasites of humans (figure 3). The former causes West African sleeping sickness, and the latter causes East African sleeping sickness.

The life cycle of *T. brucei* involves an infected intermediate host, the tsetse fly (*Glossina*). When the fly bites, trypanosomes are introduced into the definitive mammalian host with salivary secretions (figure 4). Trypanosomes enter the mammal's lymphatic and circulatory systems where they replicate by binary fission. They eventually infect other sites. These sites include cerebrospinal fluid, where they cause the nervous system complications described in the next paragraph. Infections to new hosts occur when another tsetse fly bites an infected mammal. Trypanosomes ingested by the fly replicate by binary fission in the gut of the fly. Trypanosomes then migrate to the fly's salivary glands where they are available to infect a new host (*see figure 4*).

Symptoms progress rapidly in *T. b. rhodesiense* (death within months if untreated) and more slowly in *T. b. gambiense* (death within three years if untreated). Nervous system symptoms include general apathy, headaches, mental dullness, and lack of coordination. "Sleepiness" develops, and the infected individual may fall asleep during normal daytime activities. Coma and death result from the pathology occurring in the nervous system as well as from heart failure, malnutrition, and other weakened conditions. Treatment is available for trypanosomiasis, and the treatment varies depending on the stage of the disease. There is no way to confirm a cure; thus, asymptomatic treated individuals must be retested for parasites every two years. Prevention and control of *T. b. rhodesiense* is difficult because a variety of nonhuman mammals can serve as reservoir hosts for this subspecies.

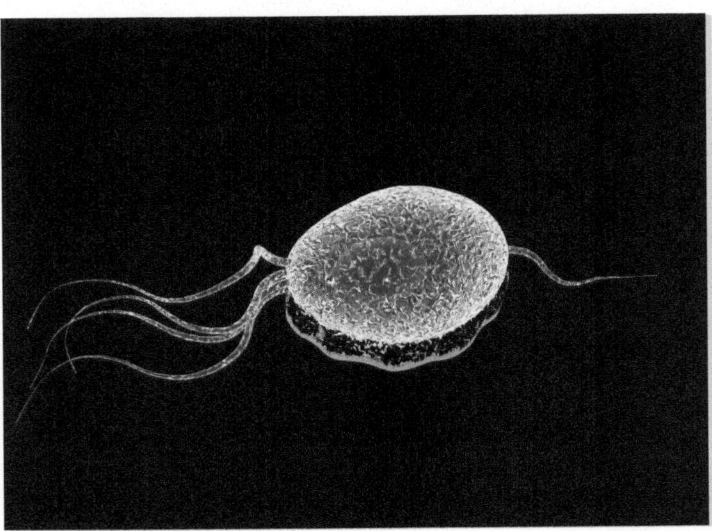

FIGURE 2

The Protozoan *Trichomonas vaginalis*. This parasite causes the sexually transmitted disease trichomoniasis in humans.

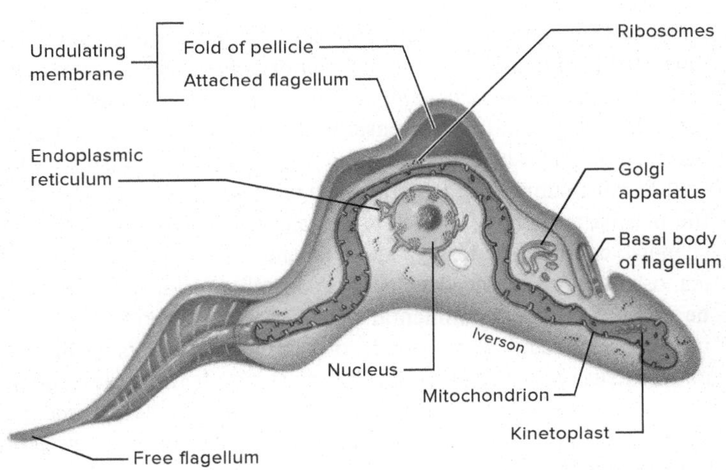

FIGURE 3

***Trypanosoma brucei* Structure.** The large single mitochondrion is characteristic of flagellated protozoans. It contains an organized mass of DNA called the kinetoplast. The undulating membrane helps propel the trypanosome through its host. This flagellate is about 25 μm long.

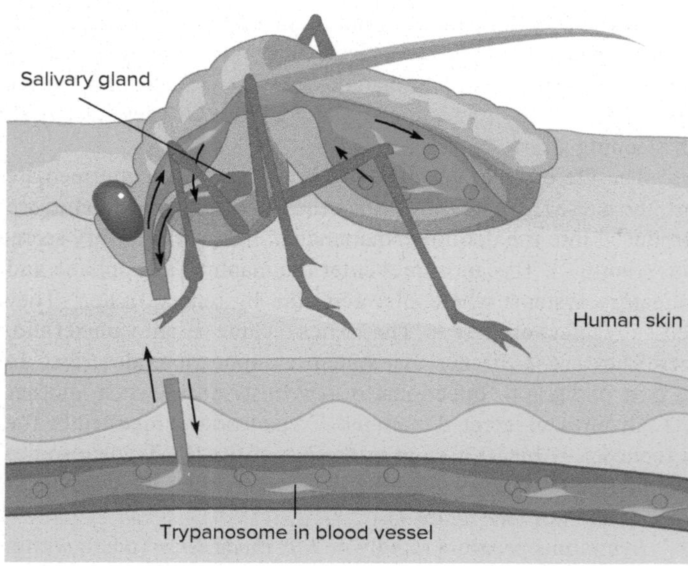

FIGURE 4

***Trypanosoma* Infection Cycle.** When a tsetse fly (*Glossina*) feeds on a mammalian host, trypanosomes enter the host's circulatory system with the fly's saliva. Trypanosomes multiply in the host's circulatory and lymphatic systems by binary fission. When another tsetse fly bites this vertebrate host again, trypanosomes move into the gut of the fly and undergo binary fission. Trypanosomes then migrate to the fly's salivary glands, where they are available to infect a new host.

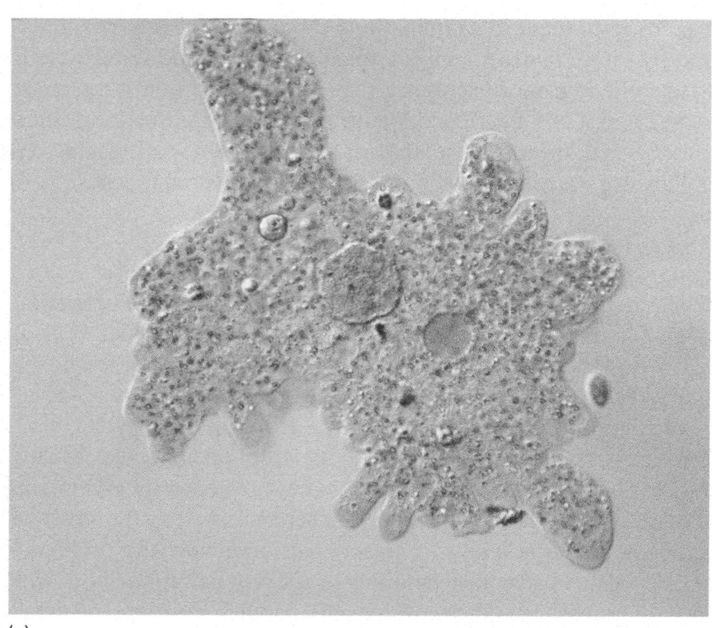

(a)

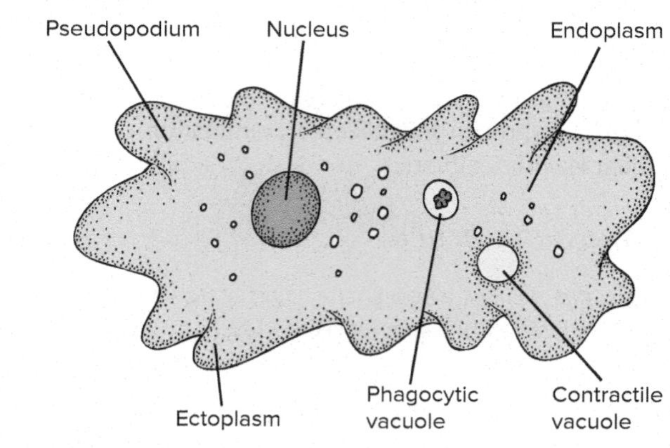

(b)

FIGURE 5

An Amoebozoan. (a) *Amoeba proteus* showing blunt pseudopodia. (b) Structure of an ameba.

(a) ©Stephen Durr

SUPERGROUP AMOEBOZOA
Underlying Features Ameboid motility with pseudopodia; naked or with test (testate); uninucleate or multinucleate; cysts common
Examples *Amoeba, Acanthamoeba, Naegleria, Entamoeba*

Amebae share many common features of structure and function. One of the hallmarks of ameboid motility is the use of structures called pseudopodia (L. *pseudo*, false + Gr. *pous*, foot). These structures are cytoplasm-filled extensions that are used for locomotion and feeding (figure 5 and *see figure 23.17*). Amebae that lack a cell wall or other supporting structures are termed "naked." In contrast, the plasma membrane of testate amebae may secrete a protective shell-like coating (the test) made of calcium carbonate, protein, silica, or chitin (a polysaccharide). Tests of other testate amebae may be composed of environmental debris that is cemented together into a specialized matrix. Usually, one or more openings in the test allow extrusion of pseudopodia. Binary fission is the typical mode of asexual reproduction in Amoebozoa; however, some amebae form cysts and undergo multiple fission, which result in a new generation of many individuals.

Naegleria and *Acanthamoeba*

Acanthamoeba and *Naegleria* are free-living, naked, aerobic inhabitants of soil and water. They possess flagellated and ameboid stages. Certain species within these genera can become facultative parasites of humans when ingested from water that contains the otherwise free-living forms. Parasitic species can cross the blood–brain barrier (a semipermeable membrane that keeps circulating blood from entering the brain and the central nervous system's extracellular fluid), and the resulting immune response causes dangerous inflammation of brain tissues (meningoencephalitis). Neuronal damage rapidly leads to death of the host. Some *Acanthamoeba* species can also infect the keratin layers of the cornea of the eye, leading to inflammation, opacity, ulceration, and in extreme cases, blindness. Infection has been associated with contaminated contact lenses.

Entamoeba histolytica

Entamoeba histolytica is an anaerobic, naked, parasitic amoebozoan. Infection primarily occurs when mature cysts are ingested due to fecal contamination of food or water. Excysted phases released in the lower small intestine divide rapidly to form phagocytic amebas called trophozoites. Trophozoites move to the large intestine where they feed on erythrocytes and release hydrolytic enzymes that result in ulceration of the large intestine's epithelial lining. The damaged intestine and associated immune response produce the disease, amebic dysentery (a debilitating diarrhea that includes blood and mucus). In severe cases, the trophozoites may reach other organs, including the liver and lungs. Amebic dysentery is most frequent in crowded, unsanitary conditions. It is distributed globally.

SUPERGROUP CHROMALVEOLATA

Underlying Features
Members can be autotrophic, heterotrophic, or use a combination of both modes of nutrition (mixotrophic). All Chromalveolata have a single endosymbiotic plastid.

Examples
Apicomplexa (e.g., *Plasmodium, Toxoplasma, Eimeria, Cryptosporidium*)

Most apicomplexans are parasites of animals and some are highly pathogenic. Apicomplexans are named after an organelle called the apical complex that is used to penetrate host cells during a specific stage in the life cycle of the parasite. The life cycle of *Plasmodium vivax*, one of the protozoans that causes the disease malaria, is widely used to demonstrate a complex life cycle involving asexual and sexual phases that occur in humans and *Anopheles* mosquitos, respectively.

When an infected *Anopheles* mosquito bites a human, it injects sporozoites (motile infectious forms of *P. vivax* that are produced asexually) into the human's circulatory system (figure 6). Once the sporozoite reaches the liver, it will use its apical organelle to penetrate and infect hepatocytes, where an asexual cycle called schizogony occurs. Schizogony is a type of multiple fission that eventually produces a generation of merozoites, egg-shaped parasites within hepatocytes that are released into the bloodstream when infected hepatocytes rupture.

Once in the bloodstream, merozoites infect erythrocytes and complete additional cycles of schizogony. This additional schizogony leads to the production of trophozoites. Trophozoites undergo multiple fission to produce merozoites that reinfect more red blood cells, or they divide to produce two types of gametocytes (male and female). Erythrocytes rupture due to the infection by merozoites and release gametocytes (some multiply asexually in the human host), which are ingested by female *Anopheles* as they feed. Gametocytes fuse in the gut of *Anopheles* creating a motile zygote that penetrates the gut lining and forms a cyst elsewhere in the mosquito's abdomen. Multiple sporozoites develop in this cyst (sporogony), migrate to the salivary glands of *Anopheles*, and get injected into a human while the mosquito is feeding.

Symptoms of malaria recur periodically and are called paroxysms. Chills and fever correlate with parasite maturation, the rupture of erythrocytes, and the release of toxic metabolites.

Five *Plasmodium* species affect humans. *P. falciparum* causes the most virulent form of malaria in humans, and its paroxysms are irregular. *P. falciparum* once had a worldwide distribution, but it is now mainly confined to the tropics and subtropics. It remains one of the greatest protozoan-induced sources of human mortality, especially in Africa. *P. malariae* is worldwide in distribution and causes malaria with paroxysms that recur every 72 hours. *P. vivax* occurs in temperate regions and has been eradicated in many parts of the world. Infection leads to paroxysms that recur every 48 hours. *P. vivax* and the relatively rare and tropically distributed *P. ovale* can remain dormant in hepatocytes for several years after initial infection. Malaria in humans due to *P. knowlesi* is also rare. It primarily affects long-tailed (*Macaca fascicularis*) and pig-tailed macaques (*Macaca nemestrina*), but human fatalities have been reported.

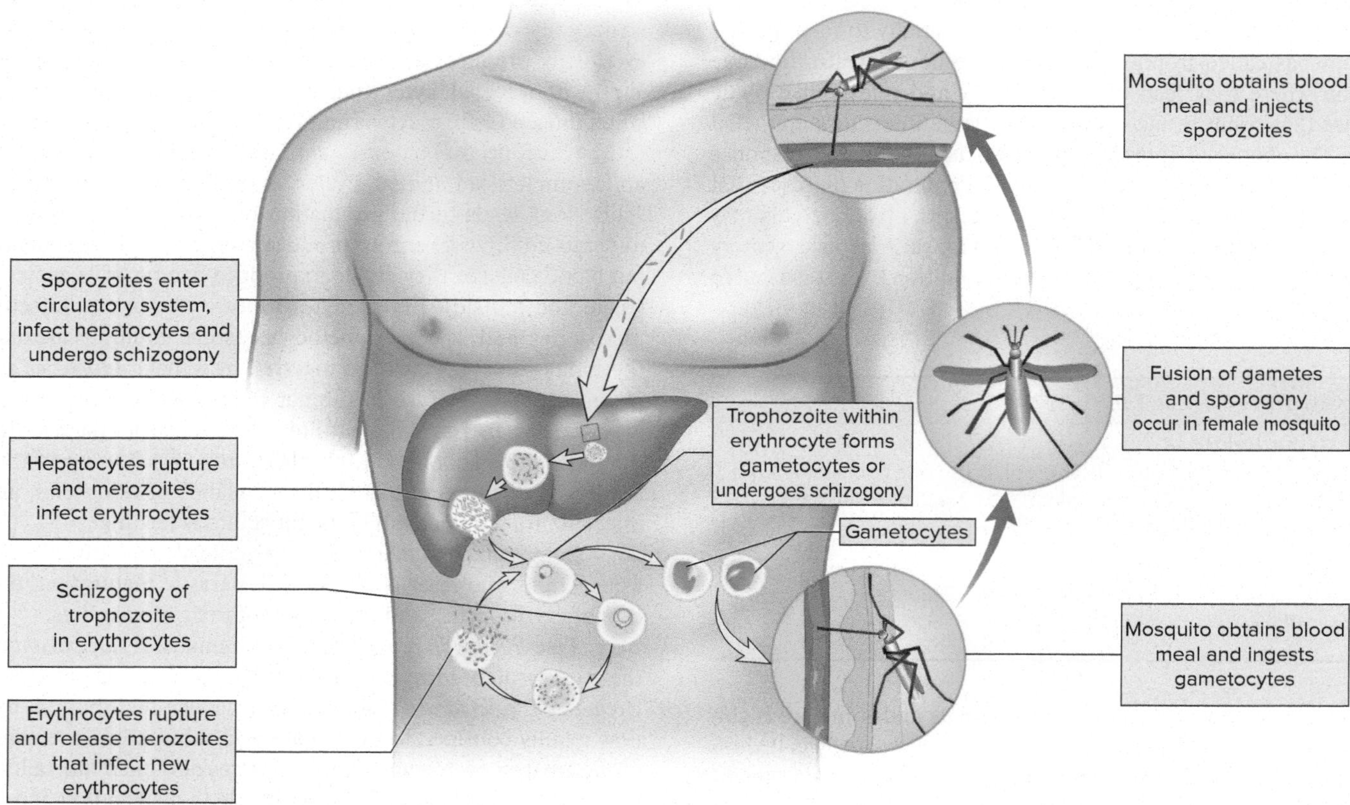

FIGURE 6

The Life Cycle of *Plasmodium*. Humans are infected when sporozoites are introduced into the blood stream with the saliva of a female mosquito. Sporozoites enter the liver cells and undergo schizogony. Merozoites are released from liver cells and infect red blood cells to produce trophozoites. Schizogony of trophozoites produces additional merozoites that repeat red blood cell stages. Gametogony in red blood cells produces male and female gametocytes that are taken up into another female mosquito, where gametocyte fusion and sporogony occur. Sporozoites produced within the female mosquito may pass the infection to a new human host.

GLOSSARY

In this pronunciation system, long vowels that stand alone or occur at the end of a syllable are unmarked. When a long vowel occurs in the middle of a syllable, it is marked with a macron (¯). Short vowels that occur in the middle of a syllable are unmarked. When a short vowel occurs at the end of a syllable, or when it stands alone, it is marked with a breve (˘). The following are examples: cape (cā p), cell (sel), vacation (va-ka′shun), and vegetal (vej″ĕ-tal″).

The vowel heard in "foot" and "book" is indicated as "fŏot" and "bŏok." The vowel heard in "loose" and "too" is unmarked and is indicated as "loos" and "too."

A

aboral (ab-or′al) The end of a radially symmetrical animal opposite the mouth.

acanthella (a-kan′thel-a) Developing acanthocephalan larva between an acanthor and a cystacanth, in which the definitive organ systems are developed. Develops in the intermediate host.

Acanthocephala (a-kan″tho-sef′a-lah) The phylum of lophotrochozoans commonly called the spiny-headed worms.

acanthor (a-kan′thor) Acanthocephalan larva (first larval stage) that hatches from the egg. The larva has a rostellum with hooks that penetrate host tissues.

acetabulum (as″e-tab′u-lum) Sucker. The ventral sucker of a fluke. A sucker on the scolex of a tapeworm.

acetylcholine (as-e-tel-ko′lē n) A neurotransmitter that certain neurons liberate. It is excitatory at neuromuscular junctions and inhibitory at other synapses.

acoelomate (a-se′lah-mā t) Without a body cavity.

Acoelomorpha (a-se′lă-mor″fah) A phylum of acoelomate, triploblastic marine animals. They are free-living and probably occupy a basal position in bilateral animal phylogeny.

acrodont (ak′ro-dont) Tooth condition in which teeth are affixed, sometimes in a shallow depression, to the summit of the jawbone. Occurs in fish, amphibians, snakes, some lizards, and tuataras.

actin (ak′tin) A protein in a muscle fiber that, together with myosin, is responsible for contraction and relaxation.

Actinopterygii (ak″tin-op′te-rig-e-i) The class of bony fishes characterized by paired fins supported by dermal rays, fin bases not especially muscular, tail fin with approximately equal upper and lower lobes, pneumatic sacs used as swim bladders, and blind olfactory sacs. The ray-finned fishes.

action potential The sequence of electrical changes when a nerve cell membrane is exposed to a stimulus that exceeds its threshold.

active transport A process that requires an expenditure of ATP energy to move molecules across a cell membrane. Usually moved against the concentration gradient with the aid of specific carrier (transport) proteins.

adaptation Structures or processes that increase an organism's potential to successfully reproduce in a specified environment.

adaptive radiation Evolutionary change that results in the formation of a number of new characteristics from an ancestral form, usually in response to the opening of new habitats.

adenosine triphosphate (ATP) (ah-dē n′o-sē n tri-fos′fā t) Molecule that stores and releases energy for use in cellular processes. A nucleotide composed of adenine, ribose, and three phosphate groups.

adhesive gland Attachment glands in turbellarians that produce a chemical that attaches part of the turbellarian to a substrate.

adrenal cortex (ah-dre′nal kor′tex) The outer glandular portion of the adrenal gland that secretes three classes of steroid hormones: glucocorticoids (cortisol), mineralocorticoids (aldosterone), and some sex hormones (androgens).

adrenal gland (ah-dre′nal) The endocrine gland on top of the kidney.

adrenal medulla (ah-dre′nal me-dul′ah) The inner neurosecretory portion of the adrenal gland that secretes epinephrine (adrenaline) and norepinephrine (noradrenaline), both of which influence heart rate and carbohydrate metabolism.

aestivation (es′te-va′shen) A period of inactivity in some animals that must withstand periods of heat and drying.

afterbirth (af′ter-berth) The placental and fetal membranes expelled from the uterus after the birth of a mammal.

age structure The proportion of a population that is in prereproductive, reproductive, and postreproductive classes.

airfoil A surface, such as a wing, that provides lift by using currents of air it moves through.

akinetic skull (a-kin-net″ik) A skull in which the upper jaw is firmly attached to the upper skull. Akinetic skulls are common in most vertebrates other than squamate reptiles.

alleles (ah-lē lz′) Alternate forms of a gene that occur at the same locus of a chromosome. For example, in humans, three alleles (I^A, I^B, and i) that occur at the same locus encode the A, B, and O blood types.

allopatric speciation (al″o-pat′rik spe″se-a′shun) Speciation that occurs in populations separated by geographical barriers.

all-or-none law The phenomenon in which a muscle fiber contracts completely when it is exposed to a stimulus of threshold strength. Also, the principle that states that an axon will "fire" at full power or not at all.

altricial (al-trish′al) An animal that is helpless at hatching or birth.

alula (al′u-lah) A group of feathers on the wing of a bird that is supported by the bones of the medial digit. The alula reduces turbulent air flow over the upper surface of the wing.

alveolus (al-ve′o-lus; pl. alveoli) An air sac of a lung. A saclike structure.

ambulacral groove (am″byū l-ac′ral) The groove along the length of the oral surface of a sea star arm. Ambulacral grooves contain tube feet.

Ambulacraria (am″byū l-ac-rar′eah) The deuterostome clade that includes the phyla Echinodermata and Hemichordata.

ametabolous metamorphosis (a″met-ab′ol-us met″ah-mor′fe-sis) Development in which the number of molts is variable. Immature stages resemble adults and molting continues into adulthood.

amictic eggs (e-mik′tic) Pertaining to female rotifers that produce only diploid eggs that cannot be fertilized. The eggs develop directly into amictic females.

amnion (am′ne-on) One of the extraembryonic membranes of the embryos of reptiles, birds, and mammals. The amnion encloses the embryo in a fluid-filled sac.

Amniota (am′ne-ō t-ah) A monophyletic lineage that includes the animals in classes traditionally designated as Reptilia (the reptiles), Aves (the birds), and Mammalia (the mammals). This lineage is characterized by the presence of amniotic eggs.

amniotic egg (am″ne-ot′ik) The egg of reptiles, birds, and mammals. It has extraembryonic membranes that help prevent desiccation, store wastes, and promote gas exchange. These adaptations allowed vertebrates to invade terrestrial habitats.

ameboid movement (ah-me′boid) A form of movement similar to that found in some protists, some white blood cells, arthropod and nematode sperm cells, and sponge archeocytes. Fluid endoplasm (plasmasol) flows forward inside the cell and changes state to viscous ectoplasm (plasmagel) on reaching the tip of a pseudopodium. At the opposite end of the cell, ectoplasm changes into endoplasm.

Amphibia (am-fib′e-ah) The class of vertebrates whose members are characterized by skin with mucoid secretions, which serves as a respiratory organ. Developmental stages are aquatic and are usually followed by metamorphosis to an amphibious adult. Frogs, toads, and salamanders.

amphid (am′fed) One of a pair of chemosensory organs on the anterior end of certain nematodes.

amplexus (am-plek′sus) The positioning of a male amphibian dorsal to a female amphibian, his forelimbs around her body, either just anterior to her hind limbs, or just posterior to her forelimbs. The male releases sperm as the female releases eggs.

ampullae of Lorenzini (am′pu-la of loren′ze-ne) Extensions of the lateral-line system that are present in dense clusters over the heads of cartilaginous fishes. Comprised of electrosensory cells that permit the detection of electrical fields generated by living organisms.

ampullary organ (am′pu-la″re) A receptor that can detect electrical currents. These electroreceptors are found in most fishes, some amphibians, and the platypus. *See also* **electroreceptor.**

anabolic pathways (an-a-bal′ik) Constructive metabolic processes by which organisms convert substances into other components of the organism's chemical architecture. System of biosynthetic reactions in a cell by which large molecules are made from smaller ones.

analogous (a-nal′a-ges) Describes structures that have similar functions in two organisms but that have not evolved from a common ancestral form. Analogous structures often arise through convergent evolution.

anaphase (an′ah-fā z) The stage in mitosis and meiosis, following metaphase, in which the centromeres divide and the chromatids, lined up on the mitotic spindle, begin to move apart toward the poles of the spindle to form the daughter chromosomes.

ancestral character states Attributes of species that are old and have been retained from a common ancestor. Also called **apomorphies.**

androgen (an′dro-jen) Any substance that contributes to masculinization, such as the hormone testosterone.

aneuploidy (an′u-ploid″e) An addition or deletion of one or more chromosomes. Aneuploidy may be represented by $2N + 1$, $2N - 1$, etc.

Animalia (an″i-mal′e-ah) The kingdom of organisms whose members are multicellular, eukaryotic, and heterotrophic. The animals.

Annelida (ah-nel′i-dah) The phylum of triploblastic, coelomate animals whose members are metameric (segmented) and wormlike. Annelids have a complete digestive tract and a ventral nerve cord.

annuli (an′u-li) Secondary divisions of each body segment of a leech (phylum Annelida, subclass Hirudinea).

antennal gland (an-ten′al) The excretory organ in some crustaceans (crayfish). Called antennal gland because of its location near the base of each second antenna. Also called *green gland* because of its green color.

anterior The head end. Usually, the end of a bilateral animal that meets its environment.

Anthozoa (an″tho-zo′ah) The class of cnidarians whose members are solitary or colonial polyps. Medusae are absent. Gametes originate in the gastrodermis. Mesenteries divide the gastrovascular cavity. Sea anemones and corals.

anticodon (an′ti-ko″don) A sequence of three bases on transfer RNA that pairs with codons of messenger RNA to position amino acids during protein synthesis.

antiparallel (an″ti-par′ah-lel″) Refers to opposing strands of DNA that are oriented in opposite directions.

Aplacophora (a″pla-kof′o-rah) The class of molluscs whose members lack a shell, mantle, and foot. Wormlike, burrowing animals with poorly developed heads. Some zoologists divide this group into two classes: Caudofoveata and Solenogastres.

Apoikozoa (a-poi″ko-zo′ah) The monophyletic clade of opisthokonts that includes the choanoflagellates and animals.

apomorphies (ah″po-mor′fez) *See* **derived character states.**

aposematic coloration (ah″pos-mat′ik) Sharply contrasting colors of an animal that warn other animals of unpleasant or dangerous effects.

appendicular skeleton (ap″en-dik′u-lar) The bones of the upper and lower extremities. Includes the shoulder and pelvic girdles.

appendix (a-pen′diks) Refers to the appendix vermiformis of the colon.

aquaporin (ah-qua″pore-in) A membrane channel that allows water to cross the membrane more easily than by diffusion through the membrane.

Arachnida (ah-rak′nĭ-dah) The class of chelicerate arthropods whose members are mostly terrestrial, possess book lungs or tracheae, and usually have four pairs of walking legs as adults. Spiders, scorpions, ticks, mites, and harvestmen.

arachnoid (ah-rak′noid) The weblike middle covering (meninx) of the central nervous system.

Archaea (ahr-kay′e-ah) A domain of organisms that includes prokaryotic microbes that live in harsh anaerobic environments.

archeocytes (ar-ke′o-cit) Ameboid stem cells present in the sponge mesohyl that can differentiate into any sponge cell type.

Aristotle's lantern The series of ossicles making up the jawlike structure of echinoid echinoderms.

arteriole (ar-te′re-ō l) A minute arterial branch, especially just proximal to a capillary.

artery A vessel that transports blood away from the heart.

Arthropoda (ar″thrah-po′dah) The phylum of animals whose members possess metamerism with tagmatization, a jointed exoskeleton, and a ventral nervous system. Includes insects, crustaceans, spiders, and related animals.

Ascidiacea (as-id″eas′e-ah) A class of urochordates whose members are sessile as adults, and solitary or colonial.

ascon (as′kon) The simplest of the three sponge body forms. Ascon sponges are vaselike, with choanocytes directly lining the spongocoel.

asexual reproduction Having no sex; not sexual; not pertaining to sex. Reproduction of an organism without fusion of gametes by fission, budding, or some other method not involving the fertilization of gametes.

aster (as′ter) The star-shaped structure in a cell during prophase of mitosis. Composed of a system of microtubules arranged in astral rays around the centrosome. May emanate from a centrosome or from a pole of a mitotic spindle.

Asteroidea (as″te-roi′de-ah) The class of echinoderms whose members have rays that are not sharply set off from the central disk, ambulacral grooves with tube feet, and suction disks on tube feet. Sea stars and sea daises.

asymmetry (a-sim′i-tre) Without a balanced arrangement of similar parts on either side of a point or axis.

asynchronous flight *See* **indirect (asynchronous) flight.**

ATP *See* **adenosine triphosphate.**

auricle (aw′re-kl) The portion of the external ear not connected within the head. Also used to designate an atrium of a heart. In the class Turbellaria, the sensory lobes that project from the side of the head.

autapomorphies (au′ta-po-mor′fez) Characters that define a particular taxon. Autapomorphy is a relative term that depends on the taxonomic rank being considered.

autocrine agent (aw′to-krin) Denoting self-stimulation through cellular production of a factor and a specific receptor for it.

autonomic (visceral or involuntary) nervous system Regulates smooth and cardiac muscle and glands of the body.

autosome (au′te-sō m″) Chromosome other than sex chromosomes.

autotomy (au-tot′o-me) The self-amputation of an appendage. For example, the casting off of a section of a lizard's tail caught in the grasp of a predator. The autotomized appendage is usually regenerated.

autotroph (aw′to-trō f) An organism that uses carbon dioxide as its sole or principal source of carbon and makes its organic nutrients from inorganic raw materials.

autotrophic (au″to-trō f′ic) Having the ability to synthesize food from inorganic compounds.

Aves (a′vez) A class of vertebrates whose members are characterized by scales modified into feathers for flight, endothermy, and amniotic eggs. The birds.

axial skeleton (ak′se-al) Portion of the skeleton that supports and protects the organs of the head, neck, and trunk.

axon (ak′son) A fiber that conducts a nerve impulse away from a neuron cell body. Action potentials move rapidly, without alteration, along an axon. Their arrival at an axon ending may trigger the release of neurotransmitter molecules that influence an adjacent cell.

B

Bacteria (bak′te-re-ah) Domain of life characterized by cells lacking a nucleus and other membranous organelles. The most abundant organisms on Earth. Distinguished from Archaea based on aspects of chromatin structure and gene regulation.

balanced polymorphism (pol″e-morf′izm) Occurs when different phenotypic expressions are maintained at a relatively stable frequency in a population.

bare sensory nerve endings Nerve endings that are sensitive to pain.

baroreceptor (bar″o-re-sep′tor) A specialized nerve ending that is stimulated by changes in pressure.

basal body (ba′sel) A centriole that has given rise to the microtubular system of a cilium or flagellum and is located just beneath the plasma membrane. Serves as a nucleation site for the growth of the axoneme.

basal phyla A group of five animal phyla that probably originated independently from other animal groups. These include the Ctenophora, Porifera, Placozoa, Acoelomorpha, and Cnidaria.

basement membrane A thin fibrous matrix upon which epithelia rest. Basement membranes are comprised of collagenous fibers and separate epithelia from underlying tissues.

basophil (ba′so-fil) White blood cell characterized by the presence of cytoplasmic granules that become stained blue-purple by a basophilic dye.

B cell A type of lymphocyte derived from bone marrow stem cells that matures into an immunologically competent cell under the influence of the bursa of Fabricius in the chicken, and the bone marrow in nonavian species. Following interaction with antigen, it becomes a plasma cell, which synthesizes and secretes antibody molecules involved in humoral immunity. Also called *B lymphocyte.*

Bdelloidea (del-oid′e-ah) A class of rotifers without males. Anterior end is retractile and bears two disks. Mastax is adapted for grinding. Paired ovaries. Cylindrical body. Example: *Rotaria.*

bilateral symmetry (bi-lat′er-al sim′i-tre) A form of symmetry in which only the midsagittal plane divides an organism into mirror images. Bilateral symmetry is characteristic of actively moving organisms that have definite anterior (head) and posterior (tail) ends.

bile (bī l) A fluid secreted by the liver, stored in the gallbladder, and released into the small intestine via the bile duct. Emulsifies fats.

bimodal breathing (bi-mo′del) The ability of an organism to exchange respiratory gases simultaneously with both air and water, usually using gills for water breathing and lungs for air breathing.

binomial nomenclature (bi-no′me-al no′men-kla″cher) A system for naming in which each kind of organism (a species) has a two-part name: the genus and the species epithet.

biodiversity The variety of organisms in an ecosystem.

biogeochemical cycles The cycling of elements between reservoirs of inorganic compounds and living matter in an ecosystem.

biogeography The study of the geographic distribution of life on Earth. Biogeographers attempt to explain the factors that influence where species of plants and animals live on the earth.

biological species concept In the biological species concept, a species is defined as a group of populations in which genes are actually, or potentially, exchanged through interbreeding.

biomagnification The concentration of substances in animal tissues as the substances pass through ecosystem food webs.

biomass The total mass of all organisms in an ecosystem.

biosonar *See* **sonar.**

biramous appendages (bi-ra′mus ah-pen′ dij-ez) Appendages having two distal processes that a single proximal process connects to the body.

Bivalvia (bi′val ve″ah) The class of molluscs whose members are enclosed in a shell consisting of two dorsally hinged valves, lack a radula, and possess a wedge-shaped foot. Clams, mussels, and oysters.

bladder worm The unilocular hydatid cyst of a tapeworm. *See also* **cysticercus.**

blastocyst (blas′to-sist) An early stage of embryonic development consisting of a hollow ball of cells. The product of cleavage.

blood pressure The force (energy) with which blood pushes against the walls of blood vessels and circulates throughout the body when the heart contracts.

blubber (blub′er) The fat found between the skin and muscle of whales and other cetaceans, from which oil is made. Functions in insulation.

bone The hard, rigid form of connective tissue constituting most of the skeleton of vertebrates. Composed chiefly of calcium salts. Also called *bone (osseous) tissue.*

book gill Modifications of a horseshoe crab's exoskeleton into a series of leaflike plates that are a surface for gas exchange between the arthropod and the water (phylum Arthropoda, class Merostomata).

book lung Modification of the arthropod exoskeleton into a series of internal plates that provide surfaces for gas exchange between the blood and air. Found in spiders.

bottleneck effect Changes in gene frequency that result when numbers in a population are drastically reduced, and genetic variability is reduced as a result of the population being built up again from relatively few surviving individuals.

Brachiopoda (bra-ke-op′o-dah) A phylum of marine animals whose members possess a bivalved calcareous and/or chitinous shell that a mantle secretes and that encloses nearly all of the body. Unlike in the molluscs, the valves are dorsal and ventral. Possess a lophophore. Lampshells.

branch An evolutionary connection between molecules, individuals, populations, or species in a phylogenetic tree. The longer the branch, the greater the variation and the more distant the evolutionary relationship between molecules (individuals, populations, or species).

broadcast spawning The release of gametes into the water for external fertilization.

Broca's area (bro′kez) The region of the frontal lobe of mammals associated with vocalizations and, in humans, speech.

brood patch The patch of feathers birds use to incubate eggs. Also known as *incubation patch.*

brown fat Fat deposits with extensive vascularization, mitochondria, and enzyme systems for oxidation. Found in small, specific deposits in a few mammals and used for nonshivering thermogenesis.

brumation (bru′ma-shen) A period of uncontrolled hypothermia in amphibians and reptiles.

buccal pump (buk′el) The mechanism by which lung ventilation occurs in amphibians. Muscles of the mouth and pharynx create positive pressure to force air into the lungs.

buccopharyngeal respiration (buk″o-fah-rin′je-al res″pah-ra′shun) The diffusion of gases across moist linings of the mouth and pharynx of amphibians.

budding (bud′ing) The process of forming new individuals asexually in many different invertebrates. Offspring arise as outgrowths from the parent and are smaller than the parent. When the offspring do not separate from the parent, a colony forms.

bulbourethral gland (bul″bo-u-re′thral) Gland that secretes a viscous fluid into the male urethra during sexual excitement.

bursa (bur′sah) A membranous sac that invaginates from the oral surface of ophiuroid echinoderms. Functions in diffusion of gases and waste material.

bursa of Fabricius (bur′sah of fah-bris′e-us) The lymphoid organ of birds that, like the thymus, develops as an outpouching of the gut near the cloaca rather than the foregut. It is responsible for the maturation of lymphocytes.

C

Calcarea (kal-kar′e-ah) The class of sponges whose members are small and possess monaxon, triaxon, or tetraaxon calcium carbonate spicules.

calcitonin (kal″si-to′nin) A thyroid hormone that lowers calcium and phosphate levels in the blood. Also called *thyrocalcitonin.*

calyx (ka′liks) 1. A boat-shaped or cuplike central body of an entoproct. The body and tentacles of an entoproct. 2. A cuplike set of ossicles that support the crown of a sea lily or feather star (class Crinoidea, phylum Echinodermata).

capillary (kap′i-lar-e) A small blood vessel that connects an arteriole and a venule. The functional unit of the circulatory system. The site of exchange between the blood and tissue cells.

carapace (kar′ah-pā s) The dorsal portion of the shell of a turtle. Formed from a fusion of vertebrae, ribs, and dermal bone.

cardiac muscle Specialized type of muscle tissue found only in the heart.

cardiovascular system *See* **circulatory system.**

carnivore (kar′ne-vor) One of the flesh-eating animals of the order Carnivora. Also, any organism that eats flesh.

carrying capacity The maximum population size that an environment can support.

cartilage (kar′ti-lij) Type of connective tissue in which cells are within lacunae and separated by a semisolid matrix. Provides a site for muscle attachment, aids in movement of joints, and provides support.

caste (kast) One of the distinct kinds of individuals in a colony of social insects (e.g., queens, drones, and workers in a honeybee colony).

catabolic pathways (ka-ta′ba-lik) Metabolic process by which large molecules are broken down into smaller ones. Intermediates in these reactions are called catabolites.

Caudofoveata (kaw′do-fo″ve-ah″ta) The class of molluscs characterized by a wormlike, shell-less body and scalelike, calcareous spicules. Lack eyes, tentacles, statocysts, crystalline style, foot, and nephridia. Deep-water marine burrowers. *Chaetoderma.*

cecum (se′kum) 1. Each arm of the blind-ending, Y-shaped digestive tract of trematodes (phylum Platyhelminthes). 2. A region of the vertebrate digestive tract where fermentation can occur. It is at the proximal end of the large intestine.

cell cycle The regular sequence of events (including interphase and the mitotic phase) during which a eukaryotic cell grows, prepares for division, duplicates its contents, and divides to form two daughter cells.

cellular respiration Process by which energy is released from organic compounds within cells. The aerobic process that involves glycolysis, the citric acid cycle, the electron transport chain, and chemiosmosis.

centralization The collection of integrating neurons into regions, usually along the longitudinal axis of an animal.

central nervous system The brain and spinal cord.

centriole (sen′tre-ol) A cellular organelle (composed of microtubules) that organizes the mitotic spindle during mitosis. A pair of centrioles is usually in the center of a microtubule-organizing center in animal cells.

centromere (sen′tro-mer) Constricted region of a mitotic chromosome that holds sister chromatids together. Also, the site on the DNA where the kinetochore forms and then captures microtubules from the mitotic spindle.

centrosome (sen′tro-som) A microtubule-organizing center found near the nucleus of animal cells. It is associated with centrioles, which function in the assembly of microtubules involved with cell division.

cephalic (se-fal′ik) Having to do with, or toward, the head of an animal.

cephalization (sef″al-iz-a′shun) The development of a head with an accumulation of nervous tissue into a brain.

Cephalocarida (sef″ah-lo-kar′ĭ-dah) A class of marine crustaceans characterized by uniform, leaflike, triramous appendages.

Cephalochordata (sef″a-lo-kor-dat′ah) The subphylum of chordates whose members possess a laterally compressed, transparent body. They are fishlike and possess all four chordate characteristics throughout life. Amphioxus.

Cephalopoda (sef″ah-lop′o-dah) The class of molluscs whose members have a foot modified into a circle of tentacles and a siphon. A shell is reduced or absent, and the head is in line with the elongate visceral mass. Octopuses, squid, and cuttlefish.

cephalothorax (sef″al-o-thor′aks) The fused head and thoracic regions of crustaceans and some arachnids.

cercaria (ser-kar′e-a) Juvenile digenetic trematode produced by asexual reproduction within a sporocyst or redia. Cercariae are free swimming and have a digestive tract, suckers, and a tail. They develop into metacercariae.

cerebellum (ser″e-bel′um) Portion of the brain that coordinates skeletal muscle movement. Part of the metencephalon, it consists of two hemispheres and a central vermis.

cerebrum (sah-re′brum) The main portion of the brain, occupying the upper part of the cranial cavity. Its two hemispheres are united by the corpus callosum. It forms the largest part of the central nervous system in mammals.

Cestoidea (ses-toid′e-ah) The class of Platyhelminthes with members that are all parasitic with no digestive tract. Have great reproductive potentials. Tapeworms.

chaetae (ke′tah) *See* **setae.**

Chaetognatha (ke″tog-nath′ah) A phylum of planktonic, marine animals whose body consists of a head, trunk, and tail. Their streamlined shape and darting locomotion are reflected in their common name—arrowworms. Chaetognaths are of uncertain phylogenetic affinity.

character Any animal attribute that has a genetic basis and that can be measured. Used by systematists in investigating relatedness among animal groups.

chelicerae (ke-lis′er-ae) One of the two pairs of anterior appendages of arachnids. May be pincerlike or modified for piercing and sucking or other functions.

Chelicerata (ke-lis″er-ah′tah) The subphylum of arthropods whose members have a body that is divided into prosoma and opisthosoma. The first pair consists of feeding appendages called chelicerae. Spiders, scorpions, mites, and ticks.

chemical synapse (sin′apse) A synapse at which neurotransmitters that one neuron releases diffuse across an extracellular gap to influence a second neuron's activity.

chemiosmosis (kem″e-os-mo′sis) The process whereby electron transport along the inner mitochondrial membranes generates a proton gradient between the outer and inner mitochondrial membranes. Electrical and concentration gradients cause protons to diffuse through ATP synthase enzymes, powering the generation of ATP.

chemoreceptor (ke″mo-re-sep′tor) A receptor that is stimulated by the presence of certain chemical substances.

chief cell Cell of a gastric gland that secretes various digestive enzymes, including pepsinogen.

Chilopoda (ki″lah-pod′ah) The class of arthropods whose members have one pair of legs per segment and whose body is oval in cross section. Centipedes.

chitin (ki′tin) The polysaccharide in the exoskeleton of arthropods.

chlorocruorin (klo″ro-kroo′e-ren) The greenish iron-containing respiratory pigment dissolved in the blood plasma of certain marine annelids.

choanocytes (ko-an′o-sī tz) Cells of sponges that create water currents and filter food.

choanoderm (ko-an′o-derm) The inner epithelial lining of flagellated chambers of a sponge (phylum Porifera). *See* **choanocytes**.

Chondrichthyes (kon-drik′thi-es) The class of vertebrates whose members are fishlike, possess paired appendages and a cartilaginous skeleton, and lack a swim bladder. Skates, rays, and sharks.

Chordata (kor-dat′ah) A phylum of animals whose members are characterized by a notochord, pharyngeal slits, a dorsal tubular nerve cord, and a postanal tail.

chromaffin tissue (kro′maf-in) Specialized endocrine cells near the kidneys of various vertebrates. Produces various steroid hormones.

chromatid (kro′mah-tid) One copy of a eukaryotic chromosome formed by DNA replication.

chromatin (kro′mah-tin) Nuclear material that gives rise to chromosomes during mitosis. Complex of DNA, histones, and nonhistone proteins.

chromatophores (kro-mah′tah-forz) Cells containing pigment that, through contraction and expansion, produce temporary color changes.

chromosome (kro′mo-sō m) Rodlike structure that appears in the nucleus of a cell during mitosis. Contains the genes responsible for heredity. Structure composed of a long DNA molecule and associated proteins that carries part (or all) of the hereditary information of an organism.

chrysalis (kris′ah-lis) The pupal case of a butterfly that forms from the exoskeleton of the last larval instar.

chylomicron (ki″lo-mi′kron) A particle of the class of lipoproteins responsible for the transport of cholesterol and triglycerides from the small intestine to tissues after meals.

chyme (kī m) Semifluid mass of food materials that passes from the stomach to the small intestine. Consists of partially digested food and gastric juice.

cilia (sil′e-ah) Microscopic, hairlike processes on the exposed surfaces of certain eukaryotic cells. Cilia contain a core bundle of microtubules and can perform repeated beating movements. They are also responsible for the swimming of many single-celled organisms.

ciliary creeping The principal means of nemertine locomotion.

circulatory system (ser′ku-lah-to″re) Pertaining to the circulation.

Cirripedia (sir″ĭ-pē d′e-ah) The class of crustaceans whose members are sessile and highly modified as adults. Enclosed by calcium carbonate valves. Barnacles.

citric acid cycle (sit′rik) A series of chemical reactions in the mitochondrion by which various molecules are oxidized and energy is released in the form of ATP, NADH, and $FADH_2$.

clade (klā d) A subset of organisms in a phylogenetic group that share a certain synapomorphy (derived character).

cladistics (klad-is′tiks) *See* **phylogenetic systematics.**

cladogram (klad′o-gram) Diagram depicting the evolutionary history of taxa. Derived from phylogenetic systematics (cladistics).

class A level of classification between phylum and order.

claw The sharp, usually curved, nail on the foot of a tetrapod vertebrate or an arthropod. The pincerlike extremity of specific limbs of certain arthropods (e.g., lobster claws).

climax community A final, relatively stable stage in an ecological succession.

Clitellata (kli-te′lah-tah) The class of Annelida characterized by no parapodia, few or no setae, a clitellum that functions in cocoon formation, and monoecious reproduction. Earthworms and leeches.

clitellum (kli-te′lum) The region of an annelid responsible for secreting mucus around two worms in copula and for secreting a cocoon to protect developmental stages.

cloaca (klo-a′kah) A common opening for excretory, digestive, and reproductive systems.

closed circulatory system A circulatory system in an animal (vertebrates) in which blood is confined to vessels throughout its circuit.

clutch The number of eggs laid and chicks produced by a female bird.

cnida (ni′dah) An organelle characteristic of the cnidaria that is used in defense, food gathering, and attachment.

Cnidaria (ni-dar′e-ah) The phylum of animals whose members are characterized by radial or biradial symmetry, diploblastic organization, a gastrovascular cavity, and nematocysts. Jellyfish, sea anemones, and their relatives.

cnidocytes (ni′do-sī tz) The cells that produce and discharge the cnidae in members of the phylum Cnidaria.

cocoon The protective covering of a resting or developmental stage. Sometimes refers to both the covering and the contents.

codominance (ko-dom′ah-nens) An interaction of two alleles such that both alleles are expressed in a phenotype.

codon (ko′don) A sequence of three bases on messenger RNA that specifies the position of an amino acid in a protein.

coelom (se′lom) A fluid-filled body cavity lined by mesoderm.

coevolution (ko″ev″ah-lu′shun) The evolution of ecologically related species such that each species exerts a strong selective influence on the other.

colloblasts (kol′ah-blasts) Adhesive cells on ctenophoran tentacles for capturing prey.

colony An aggregation of organisms. Usually a group of closely associated individuals of one species.

colostrum (ko-los′trum) The first secretion of the mammary glands following the birth of a mammal.

comb rows Rows of cilia that are the locomotor organs of ctenophorans.

commensalism (kah-men′sal-izm) Living within or on an individual of another species without harm. The commensal benefits, but the host is unharmed.

community All populations living in an area.

community diversity The number of different kinds of organisms living in an area.

comparative anatomy The study of animal structure to deduce evolutionary pathways in particular animal groups.

competitive exclusion principle The idea that two species with identical niches cannot coexist.

complete digestive tract A digestive system with both mouth and anus. A complete digestive tract (alimentary tract) permits the one-way flow of ingested food without mixing it with previously ingested food or waste and allows progressive digestive processing in specialized regions.

complex camera eye The type of image-forming eye found in squids and octopuses.

compound eye An eye consisting of many individual lens systems (ommatidia). Present in many members of the phylum Arthropoda.

concentration gradient The difference in concentration of a substance between two points of reference. Substances diffuse down concentration gradients (from an area of higher concentration to an area of lower concentration).

conduction (ken-duk′shun) The conveyance of energy, such as heat, sound, or electricity. The direct transfer of thermal motion (heat) between molecules of the environment and those on the body surface of an animal.

cone cell A color-sensitive photoreceptor cell concentrated in the retina.

connective tissue A basic type of tissue that includes bone, cartilage, and various fibrous tissues. Connective tissue supports and binds.

continuous feeder Usually slow-moving, sessile animals that feed all of the time.

contour feathers Feathers that cover the body, wings, and tail of a bird. Contour feathers provide flight surfaces and are responsible for plumage colors.

contractile vacuole (kon-trak′til vak′u- ōl″) An organelle that collects and discharges water in protists and a few lower metazoa in a cyclical manner to accomplish osmoregulation and some excretion.

controlled hypothermia (hi″po-ther′me-ah) Processes involved with maintaining unusually low body temperatures to withstand periods of unfavorable environmental conditions. *See also* daily **torpor**, **hibernation**, and **aestivation**.

convection (ken-vek′shun) The act of conveying or transmission. The movement of air (or a liquid) over the surface of a body that contributes to heat loss (if the air is cooler than the body) or heat gain (if air is warmer than the body).

convergent evolution Evolutionary changes that result in members of one species resembling members of a second, unrelated (or distantly related) species.

Copepoda (ko″pe-pod′ah) A class of crustaceans characterized by maxillipeds modified for feeding and antennae modified for swimming. The copepods.

coracidium (kor″ah-sid′e-um) Larva with a ciliated epithelium hatching from the egg of certain cestodes. A ciliated, free-swimming oncosphere.

corona (ko-ro′nah) A crown. An encircling structure. The ciliated organ at the anterior end of rotifers used for swimming or feeding.

cortisol (kor′ti-sol) A glucocorticoid that the adrenal cortex secretes that helps regulate overall metabolism and the concentration of blood sugar.

counteracting osmolyte strategy (os-mo-lyt) An osmolyte (ion) that counteracts another ion.

countercurrent exchange mechanism The passive exchange of something between fluids moving in opposite directions past each other.

coxal glands (koks′el) An organ of excretion in some arthropods (spiders) that empties through a pore near the proximal joint (coxa) of the leg.

Craniata (kra″ne-ah′tah) A subphylum of chordates characterized by a skull that surrounds the brain, olfactory organs, ears, and eyes. They possess a unique neural crest tissue. Hagfishes and vertebrates.

Crinoidea (krin-oi′de-ah) The class of echinoderms whose members are attached by a stalk of ossicles or are free living. Possess a reduced central disk. Sea lilies and feather stars.

crossing-over The exchange of material between homologous chromosomes, during the first meiotic division, resulting in a new combination of genes.

Crustacea (krus-tā s′e-ah) A traditional subphylum comprised of mandibulate arthropods whose members are characterized by having two pairs of antennae, one pair of mandibles, two pairs of maxillae, and biramous appendages. "Crustacea" is probably paraphyletic. Crabs, crayfish, and lobsters. *See also* **Pancrustacea.**

crypsis (krip′sis) All instances of an animal avoiding detection. Crypsis includes visual, auditory, and chemical crypsis.

crystalline style A proteinaceous, rodlike structure in the digestive tract of a bivalve (Mollusca) that rotates against a gastric shield and releases digestive enzymes.

Ctenophora (te-nof′er-ah) The phylum of animals whose members are characterized by biradial symmetry, diploblastic organization, colloblasts, and meridionally arranged comb rows. Comb jellies.

Cubozoa (ku″bo-zo′ah) The class of cnidarians whose members have prominent cuboidal medusae with tentacles that hang from the corner of the medusa. Small polyp, gametes gastrodermal in origin. *Chironex.*

cutaneous respiration (kyoo-ta′ne-us) Gas exchange across thin, moist surfaces of the skin. Also called *cutaneous exchange* or *integumentary exchange.*

cuticle (kyoo′tĭ-kel) A noncellular, protective, organic layer secreted by the external epithelium (hypodermis) of many invertebrates. Refers to the epidermis or skin in higher animals and usually consists of cornified dead cells.

Cycliophora (si″kle-o-for′ah) A phylum of animals whose members are marine, acoelomate, and bilaterally symmetrical. They live in association with the mouthparts of lobsters.

cystacanth (sis′ta-kanth) Juvenile acanthocephalan that is infective to its definitive host.

cysticercosis (sis″ti-ser-ko′-sis) Disease in humans that results from cysticercus development in body tissues following infection by *Taenia solium.*

cysticercus (sis″ti-ser′kus) Metacestode developing from the oncosphere in most Cyclophyllidea. Usually has a tail and a well-formed scolex and is characterized by a fluid-filled oval body with an invaginated scolex. Cysticercoid.

cytokinesis (si″ta-kin-e′sis) The division of the cytoplasm of a cell into two parts, as distinct from the division of the nucleus (which is mitosis).

cytoplasm (si′to-plazm) The contents of a cell surrounding the nucleus. Consists of a semifluid medium and organelles.

cytoskeleton (si″to-skel′e-ton) In the cytoplasm of eukaryotic cells, an internal protein framework of microtubules, microfilaments, and intermediate filaments by which organelles and other structures are anchored, organized, and moved about.

D

daily torpor (tor′por) Daily sluggishness that some animals experience. A period of inactivity that is normally induced by cold. Observed in numerous ectotherms and also endotherms. Often used to describe the specific physiological state of endotherms with a circadian cycle of lowered body temperature and depressed metabolic rate.

daughter sporocysts In digenetic trematodes, the embryonic cells that develop from sporocysts and give rise to rediae.

decomposer Mostly heterotrophic bacteria and fungi that obtain organic nutrients by breaking down the remains or products of other organic compounds. Their activities help cycle the simple compounds back to the autotrophs.

degeneracy (de-jen′er-ah-se) The genetic code is said to be degenerate because more than one three-base sequence in DNA can code for one kind of amino acid.

delayed fertilization When fertilization of an egg does not occur immediately following coitus but is delayed for weeks or months.

deme (deem) A small, local subpopulation. Isolated subpopulations sometimes display genetic changes that may contribute to evolutionary change in the subpopulation.

Demospongiae (de-mo-spun′je-a) The class of poriferans whose members have monaxon or tetraaxon siliceous spicules or spongin. Leucon body forms are present and vary in size from a few centimeters to 1 m in height.

dendrite (den′drī t) Nerve fiber that transmits impulses toward a neuron cell body. Dendrites compose most of the receptive surface of a neuron.

density-dependent factors Environmental parameters that are more severe when population density is high (or sometimes very low) than they are at other densities. Disease, predation, and parasitism are density-dependent factors.

density-independent factors Environmental parameters that influence the number of animals in a population without regard to the number of animals per unit space (density). Weather conditions and human influences often have density-independent effects on animal populations.

dental formula A notation that indicates the number of incisors, canines, premolars, and molars in the upper and lower jaws of a mammal.

denticle (den′ti-kl) A small, toothlike process.

deoxyribonucleic acid (DNA) (de-oks′e-ri″bo-nu-kla′ik) A polymer of deoxyribonucleotides in the form of a double helix. DNA is the genetic molecule of life in that it codes for the sequence of amino acids in proteins. Contains nucleotide monomers with deoxyribose sugar and nitrogenous bases adenine (A), cytosine (C), guanine (G), and thymine (T). *See also* **gene.**

depolarization (de-po″lar-i-za′shun) The change in electrical polarity across a membrane to a more positive intracellular state.

deposit feeding The type of feeding whereby an animal obtains its nutrients from the sediments of soft-bottom habitats (mud or sands) or terrestrial soils. Examples include annelids, some snails, and some sea urchins.

derived character states Characters that have arisen since common ancestry with an outgroup. Also called **apomorphies**.

dermal branchiae (der′mal branch′e-a) Thin folds of the body wall of a sea star that extend between ossicles and function in gas exchange and other exchange processes.

dermal pores (der′mal porz) Inlets for water into a sponge. They are either formed by rolling of an embryonic cell to form a pore through the cell or an opening created between a group of surrounding cells. Also called **ostia**.

dermis (der′mis) The layer of the skin deep to the epidermis, consisting of a dense bed of vascular connective tissue.

Deuterostomia (du′te-ro-stom″e-ah) A subkingdom level of classification that includes bilateral animals in which the anus forms from, or in the region of, the blastopore. Deuterostomes are often characterized by enterocoelous coelom formation, and radial cleavage. It includes the phyla Echinodermata, Hemichordata, and Chordata.

diabetes (di″ah-be′tez) Condition characterized by a high blood glucose level and the appearance of glucose in the urine due to a deficiency of insulin or the inability of body cells to respond to insulin. Diabetes mellitus.

diadromous (di-ad′ro-mus) Fishes or other animals that migrate between freshwater and saltwater. For example, Atlantic salmon juveniles migrate from natal streams to the ocean and return to the streams of their hatching to spawn.

diapause (di′ah-pauz) The physiological state of dormancy triggered by environmental conditions.

diaphragm (di′ah-fram) The domed respiratory muscle between thoracic and abdominal compartments of mammals.

diastole (di-as′to-le) Phase of the cardiac cycle during which a heart chamber wall relaxes. *See also* **diastolic pressure.**

diastolic pressure (di″ah-stol′ik) The blood pressure measurement during the interval between heartbeats. It is the second number in a blood-pressure reading.

diencephalon (di-en sef′a-lon) The portion of the vertebrate forebrain that contains the thalamus, hypothalamus, pituitary gland, and pineal gland. It is an integration center for some sensory information.

diffusion (dĭ-fu′zhun) The random movement of molecules from one location to another because of random thermal molecular motion. Net diffusion always occurs from a region of higher concentration to a region of lower concentration.

digestion (di-jest′yun) The process by which mechanical and chemical means break down larger molecules of food into smaller molecules that the digestive system can take up. Hydrolysis.

dihybrid cross (di-hi′brid) A mating between individuals heterozygous for two traits. Dihybrid crosses usually result in a phenotypic ratio of 9 dominant, dominant: 3 dominant, recessive: 3 recessive, dominant: 1 recessive, recessive in the offspring.

dioecious (di-e′shus) Having separate (male and female) sexes. Male and female organs are in separate individuals.

diphyodont (dĭ-fi′o-dont) Single replacement teeth. Deciduous teeth develop first and are replaced by a second set of permanent teeth. Teeth of mammals are diphyodont.

diploblastic (dip″lo-blas′tik) Animals whose body parts are organized into layers that are derived embryologically from two tissue layers: ectoderm and endoderm. Animals in the phyla Cnidaria and Ctenophora are diploblastic.

diploid (di′ploid) Having two sets of chromosomes. The $2N$ chromosome number.

Diplopoda (dip″lah-pod′ah) The class of arthropods whose members are characterized by two pairs of legs per apparent segment and a body that is round in cross section. Millipedes.

direct (synchronous) flight Insect flight that is accomplished by flight muscles acting on wing bases and in which a single nerve impulse results in a single wing cycle. *See also* **indirect (asynchronous) flight.**

directional selection Natural selection that occurs when individuals at one phenotypic extreme have an advantage over individuals with more common phenotypes.

discontinuous feeder An animal that does not feed all the time. Instead, it generally eats large meals sporadically and does not spend time in the continuous pursuit of prey. Most carnivores are discontinuous feeders.

disruptive selection Natural selection that occurs when individuals of the most common phenotypes are at a disadvantage. Produces contrasting subpopulations. Also called **diversifying selection.**

distal Away from the point of attachment of a structure on the body (e.g., the toes are distal to the knee).

domain (do-mān′) The broadest taxonomic grouping. Recent evidence from molecular biology indicates that there are three domains: Archaea, Bacteria, and Eukarya.

dominant Describes a gene that masks one or more of its alleles. For a dominant trait to be expressed, at least one member of the gene pair must be the dominant allele. *See also* **recessive.**

dorsal (dor′sal) The back of an animal. Usually the upper surface. Synonymous with *posterior* for animals that walk upright.

down feathers Feathers that provide insulation for adult and immature birds.

dura mater (du′rah ma′ter) The outermost and toughest meninx covering the brain and spinal cord.

E

ecdysis (ek-dis′is) 1. The shedding of the arthropod exoskeleton to accommodate increased body size or a change in morphology (as may occur in molting from immature to adult). 2. The shedding of the cuticle in any ecdysozoan in order to grow. 3. The shedding of the epidermal layers of the skin of a reptile. Also called *molting.*

Ecdysozoa (ek-dis″o-zo′ah) A subkingdom level of classification that includes animals such as the arthropods (insects and their relatives) and nematodes (roundworms). These animals possess an outer covering called a cuticle that is shed or molted periodically during growth (not homologous to vertebrate molting).

Echinodermata (i-ki″na-dur′ma-tah) The phylum of coelomate animals whose members are pentaradially symmetrical as adults and possess a water-vascular system and an endoskeleton covered by epithelium. Sea stars, sea urchins, sea cucumbers, and sea lilies.

Echinoidea (ek″i-noi′de-ah) The class of echinoderms whose members are globular or disk shaped, possess movable spines, and have a skeleton of closely fitting plates. Sea urchins and sand dollars.

Echiura (ek-e-yur′e-ah) A group of annelids whose members burrow in mud or sand or live in rock crevices. They possess a spatula-shaped proboscis and are 15 to 50 cm in length. Spoon worms.

ecological niche (ek′o-loj′i-kal nich) The role of an organism in a community.

ecology (e-kol′ah-je) The study of the relationships between organisms and their environment.

ecosystem (ek′o-sis″tem) Biological communities coupled with their abiotic environment.

ectoderm (ek′ta-durm) The outer embryological tissue layer. Gives rise to skin epidermis and glands, also hair and nervous tissues in some animals.

Ectoprocta (ek-to-prok′tah) A phylum of animals whose members are colonial and freshwater or marine. Anus ends outside a ring of tentacles. Lophophore used in feeding. Moss animals or bryozoans.

ectotherm (ek′to-therm) Having a variable body temperature derived from heat acquired from the environment. Contrasts with **endotherm.**

ectothermic (ek′to-therm-ik) *See* **ectotherm.**

electrical synapse (sin′aps) A synapse at which local currents resulting from electrical activity flow between two neurons through gap junctions joining them.

electron transport system Electrons from NADH and $FADH_2$ are passed through electron carrying molecules within the inner mitochondrial membrane. Molecular oxygen is the final electron acceptor, producing water. In the process, ATP is generated through chemiosmosis. *See* **chemiosmosis.**

electroreception (e-lek″tro-re-sep′shun) The ability to detect weak electrical fields in the environment.

electroreceptor (e-lek″tro-re-sep′tor) A receptor that senses changes in an electrical current, usually in the surrounding water. Also called an *ampullary organ.*

elephantiasis (el″e-fan-ti′ah-sis) A chronic filarial disease most commonly occurring in the tropics due to infection of the lymphatic vessels with the nematode *Wuchereria* spp.

embryonic diapause (em″bre-on′ik di′ah-pauz) The arresting of early development to allow young to hatch, or be born, when environmental conditions favor survival.

endangered species A species that is in imminent danger of extinction throughout its range.

end bulb A tiny swelling on the terminal end of telodendria at the distal end of an axon. Also called the *synaptic bouton.*

endocrine gland (en′do-krin) Ductless, hormone-producing gland (e.g., pituitary, hypothalamus, thyroid) that is part of the endocrine system.

endocrinology (en″do-kri-nol′o-je) The study of the endocrine system and its role in the physiology of an animal.

endocytosis (en″do-si-to′sis) Physiological process by which substances move across a plasma membrane into the cell via membranous vesicles or vacuoles.

endoderm (en′do-durm) The innermost embryological tissue layer. Gives rise to the inner lining of the gut tract, digestive gland, and the inner lining of the respiratory system.

endomembrane system (en″do-mem′brān) An interconnected system of membranes that includes the endoplasmic reticulum, the Golgi apparatus, various types of vesicles, and the nuclear envelope.

endoplasmic reticulum (ER) (en-do-plaz′mic re-tik′u-lum) Cytoplasmic organelle composed of a system of interconnected membranous tubules and vesicles. Rough ER has ribosomes attached to the side of the membrane facing the cytoplasm, and smooth ER does not. Rough ER functions in protein synthesis, while smooth ER functions in lipid synthesis.

endopodite (end-op′o-dīt) The medial ramus of the biramous appendages of crustaceans and trilobites (phylum Arthropoda).

endoskeleton (end″o-skel′ĕ-ton) A skeleton that lies beneath the surface of the body (e.g., the bony skeleton of vertebrates and the calcium carbonate skeleton of echinoderms).

endosome (en′do-sōm) A cellular vesicle created when the plasma membrane invaginates to engulf materials from the outside of the cell. A food vacuole is an example of an endosome.

endostyle (en′do-stīl″) A ciliated tract within the pharynx of some chordates that is used in forming mucus for filter feeding.

endosymbiont theory (en′do-sim″be-ont ther-e) The idea that eukaryotic cells evolved from large archaean cells that engulfed once-free-living bacteria.

endotherm (en′do-therm) Having a body temperature determined by heat derived from the animal's own metabolism. Contrasts with **ectotherm.**

endothermic (en′do-therm-ik) *See* **endotherm.**

end-product (feedback) inhibition The inhibition of the first enzyme in a pathway by the end product of that pathway.

energy The capacity to do work. Energy in living systems is supplied in the form of adenosine triphosphate (ATP).

energy budget An accounting of the way in which organisms of an ecosystem process and lose energy from the sun.

energy of activation The energy input required for a reaction to occur. Enzymes lower the energy of activation to allow reactions to occur under conditions compatible with life.

enteric division (en-ter′ik) That division of the autonomic nervous system that regulates the digestive tract, pancreas, and gallbladder.

Enteropneusta (ent″er-op-nus′tah) A class of hemichordates whose members live in burrows in shallow marine water. Their bodies are divided into three regions: proboscis, collar, and trunk. Acorn worms.

Entognatha (en″to-na′tha) The class of Hexapoda whose members possess mouth appendages hidden within the head, mandibles with a single articulation, and legs with one undivided tarsus. Collembolans, proturans, and diplurans.

Entoprocta (en-to-prok′tah) Phylum of mostly marine, goblet-shaped lophotrochozoans. They are colonial animals whose mouth and anus lie within the ciliated tentacles of their lophophore.

environmental resistance The constraints that climate, food, space, and other environmental factors place on a population.

enzymes (en′zīmz) Proteins that are synthesized by a cell and act as catalysts in specific cellular reactions. The substances that each type of enzyme acts upon are called its substrates.

eosinophil (e″o-sin′o-fil) White blood cell characterized by the presence of cytoplasmic granules that become stained red by an acid dye (eosin).

ephyra (e-fi′rah; ephyrae) Miniature medusa produced by asexual budding of a scyphistoma (class Scyphoza, phylum Cnidaria). Ephyrae mature into sexually mature medusae.

epidermis (ep′ĭ-dur-mis) A sheet of cells covering the surface of an animal's body. In invertebrates, a single layer of ectodermal epithelium.

epigenetics (ep″ĭ-je-net′iks) The inheritance of environmentally induced changes in a species.

epithelial tissue (ep″ĭ-the′le-al) The cellular covering of internal and external surfaces of the body. Consists of cells joined by small amounts of cementing substances. Epithelium is classified into types based on the number of layers deep and the shape of the superficial cells.

epitoky (ep′ĭ-to″ke) The formation of a reproductive individual (epitoke) that differs from the nonreproductive (atoke) form of that species.

Errantia (er-ran′tiah) A clade of marine annelids; parapodia with prominent lobes supported by internalized chaetae and ventral cirri; palps well developed. *Nereis, Eunice,* and *Glycera*.

esophagus (e-sof′ah-gus) The passage extending from the pharynx to the stomach.

estrus (es′trus) The recurrent, restricted period of sexual receptivity in female mammals (other than primates) marked by intense sexual urges. Also known as "being in heat."

estrus cycle (es′trus) A recurrent series of changes in the reproductive physiology of female mammals other than primates. Females are receptive, physiologically and behaviorally, to the male only at certain times in this cycle.

euchromatic regions (u″kro-mat′ik) Less densely staining regions of chromosomes that contain active genes.

Eukarya (u″kar′e-ah) The domain that includes all eukaryotic organisms: protists, fungi, plants, and animals.

eukaryote (u-kar′e-ōt) Having a true nucleus. A cell that has membranous organelles, most notably the nucleus. All organisms except bacteria are composed of eukaryotic cells.

eukaryotic cell *See* **eukaryote**.

evaporation (e-vap″o-ra′shun) The change of state of water from a liquid to a gas. Evaporative cooling occurs when heat is lost from a surface as water molecules escape in the form of a gas.

evolution (ev″o-lu′shun) Change over time. Organic or biological evolution is a series of changes in the genetic composition of a population over time. *See also* **natural selection** and **punctuated equilibrium model.**

evolutionary conservation The slowness of change in a characteristic of an animal over time. Evolutionary conservation usually indicates that the characteristic is vital for normal functions and that change is not tolerated.

evolutionary systematics The study of the classification of, and evolutionary relationships among, animals. Evolutionary systematists attempt to reconstruct evolutionary pathways based on resemblances between animals that result from common ancestry.

excretion (eks-kre′shun) The act, process, or function of excreting. The elimination of metabolic waste products from an animal's body.

exocrine gland (ek′so-krin) A gland (e.g., mammary, salivary, sweat) that secretes its product to an epithelial surface or body cavity directly or through ducts.

exocytosis (eks″o-si-to′sis) The process by which substances move out of a cell. The substances are transported in the cytoplasmic vesicles, the surrounding membrane of which merges with the plasma membrane in such a way that the substances are dumped outside.

exopodite (eks-op′o-dīt) The lateral ramus of the biramous appendages of crustaceans and trilobites (phylum Arthropoda).

exoskeleton (eks″o-skel′ĕ-ton) A skeleton that forms on the outside of the body (e.g., the exoskeleton of an arthropod).

exponential growth (ek″spo-nen′shal) Population growth in which the number of individuals increases by a constant multiple in each generation.

extracellular digestion (eks″tra-sel′u-ler) Digestion that occurs outside the cell, usually in a special organ or cavity.

F

facilitated diffusion (fah-sil′ĭ-tāt″id di-fu′zhun) Diffusion in which a substance is moved across a membrane from a region of higher concentration to a region of lower concentration (down its concentration gradient) by carrier protein molecules.

family The level of classification between order and genus.

fibrillar flight muscle (fi′bra-lar) Insect flight muscle responsible for indirect flight. A single nerve impulse results in many cycles of flight muscle contraction and relaxation.

filtration (fil-tra′shun) Movement of material across a membrane as a result of hydrostatic pressure.

fission (fish′un) Asexual reproduction in which the cell divides into two (binary fission) or more (multiple fission) daughter parts, each of which becomes an individual organism.

fixed action pattern A genetically determined and unchangeable stereotypic animal behavior that occurs in response to a given stimulus. Once initiated, this stereotypic behavior is usually continued until completion.

flagella (flah-jel′ah) Relatively long, motile processes that extend from the surface of a cell. Eukaryotic flagella are longer versions of cilia. Flagellar undulations drive a cell through a fluid medium. Like cilia, flagella have a 9 + 2 arrangement of microtubules covered by the cell's plasma membrane.

flame cell Specialized, hollow excretory or osmoregulatory structure consisting of one to several cells containing a tuft of cilia (the "flame") and located at the end of a minute tubule. Flame bulb.

flight feathers Feathers that line the tip and trailing edge of the wing of a bird. They are asymmetrical, with longer barbs on one side of the shaft.

fluid feeding The process by which an animal feeds on fluid. Examples include some parasites, leeches, ticks, mites, lampreys, and certain crustaceans.

fluke Any trematode worm. A member of the class Trematoda or class Monogenea.

food chain A linear sequence of organisms through which energy is transferred in an ecosystem from producers through several levels of consumers.

food groves Ciliated tracts along the dorsal and ventral margins of the gills of bivalve molluscs. These tracts transport food filtered by the gills toward the labial palps and the mouth.

food web A sequence of organisms through which energy is transferred in an ecosystem. Rather than being linear, a food web has highly branched energy pathways.

foramen of Panizza (fah′ra-men of pah-ne′zah) A narrow shunt that connects the left and right systemic arteries in archosaurs. This blood pathway is used to equalize flow of oxygenated blood within these vessels when a crocodylian is breathing air.

forebrain (for′brān) Consists of the diencephalon and telencephalon. Also known as the *prosencephalon*.

formed-element fraction The cellular component of vertebrate blood.

fossil Any remains, impressions, or traces of organisms of a former geological age.

founder effect Changes in gene frequency that occur when a few individuals from a parental population colonize new habitats. The change is a result of founding individuals not having a representative sample of the parental population's genes.
fragmentation (frag″men-ta′shun) Division into smaller units. A type of asexual reproduction whereby a body part is lost and then regenerates into a new organism.
frontal lobe The portion of a vertebrate's cerebral hemispheres involved with motor control of voluntary movement, control of emotional expression, and behavior.
Fungi (fun′ji) The kingdom of life whose members are eukaryotic, multicellular, and saprophytic (mushrooms, molds).
furcula (fur′kyah-lah) The fused clavicles of a bird. The furcula is one point of attachment for flight muscles. The wishbone.

G

Galápagos Islands (gah-lah′pĕ-gō s″) An archipelago on the equator in the Pacific Ocean about 900 km west of Ecuador. Charles Darwin's observations of the plant and animal life of these islands were important in the formulation of the theory of evolution by natural selection.
gallbladder (gawl′blad-der) The pear-shaped reservoir for bile under the right lobe of the liver.
gamete (gam′ēt) Mature haploid cell (sperm or egg) that functions in sexual reproduction. The union of two gametes of opposite sex (fertilization) produces a zygote.
ganglia (gang-glee-ah) Found in peripheral nerves; regions of gray matter made up of unmyelinated fibers.
gastric shield A chitinized plate in the stomach of a bivalve (phylum Mollusca) on which the crystalline style rotates.
gastrodermis (gas-tro-derm′is) The endodermally derived lining of the gastrovascular cavity of Cnidaria.
Gastropoda (gas-trop′o-dah) The class of molluscs characterized by torsion. A shell, when present, is usually coiled. Snails.
Gastrotricha (gas-tro-tri′kah) A small phylum of marine and freshwater species of gastrotrichs that inhabit the spaces between bottom sediments.
gastrovascular cavity (gas″tro-vas′ku-lar kav′ĭ-te) The large central cavity of cnidarians and flatworms that receives and digests food. Has a single opening serving as both mouth and anus.
gastrozooid (gas″tro-zo′oid) A feeding polyp in a colonial hydrozoan (phylum Cnidaria). Also called **hydranth**.
gemmule (jem′yool) Resistant, overwintering capsule formed by freshwater, and some marine, sponges that contains masses of mesenchyme cells. Ameboid mesenchyme cells are released and organize into a sponge.
gemmulation (jem-yoo-la′shun) The formation of gemmules.
gene A heritable unit in a chromosome. A series of nucleotide bases on the DNA molecule that codes for a single polypeptide.
gene flow Changes in gene frequency in a population that result from emigration or immigration.
gene pool The sum of all genes in a population.
genetic drift (je-net′ik) Occurs when chance events influence the frequency of genes (evolutionary change). Genetic drift often leads to loss of alleles from a population.
genetic recombination (je-net′ik re-kom-b ē-na′shun) Crossing-over. A major source of genetic variation in a population or a given species.
genetics (je-net′iks) The study of the mechanisms of transmission of genes from parents to offspring.
genotype (je′no-t īp) The specific gene combinations that characterize a cell or an individual.
genotypic ratio (je″no-tip′-ik) The relative numbers of progeny in each genotypic category that a genetic cross produces.
genus (je′nus) The level of classification between species and family.
georeceptor (je″o-re-cep′tor) A specialized nerve ending that responds to the force of gravity.
germ-line cells A population of cells that develops the ability to pass genes on to progeny, usually through sexual reproduction.
gestation (jes-ta′shun) Period of development of the young in viviparous animals, from the time of fertilization of the ovum until birth. Also called *gestation period.*
gill An aquatic respiratory organ for obtaining oxygen and getting rid of carbon dioxide. Found in fishes, molluscs, and many arthropods.
gill arches Bony or cartilaginous gill support of some vertebrates. Also called *visceral arches.*
gill filament A thin-walled, fleshy extension of a gill arch that contains vessels carrying blood to and from gas-exchange surfaces.
gill rakers Processes extending from pharyngeal arches of some fishes that trap food and protect gill filaments from mechanical damage.
gizzard (giz′erd) In birds, it is the highly modified posterior portion of the stomach in which, with aid of gravel, food is broken up. In some invertebrates, such as insects, a gizzard is present and has the same function.
glochidium (glo-kid′e-um) A larval stage of freshwater bivalves in the family Unionidae. It lives as a parasite on the gills or fins of fishes.
glomerulus (glo-mer′u-lus) A capillary tuft (coiled mass of capillaries) within the capsule (Bowman's) of a nephron. Forms the glomerular filtrate.
glycocalyx (gli′ko-kal′iks) The glycoprotein and glycolipid covering (cell coat) that surrounds many eukaryotic cells.
glycolysis (gli′kol-ah-sis) A metabolic sequence occurring in the cytoplasm of a cell. Glucose is broken down to form pyruvate (aerobic conditions) or lactic acid (anaerobic conditions). ATP and NADH (aerobic only) are generated in the process.
Gnathostomata (na″tho-sto′ma-tah) A superclass of vertebrates whose members possess hinged jaws and paired appendages. A vertebral column may have replaced the notochord.
Gnatostomulida (na″tho-sto-mū l′ida) Marine, threadlike lophotrochozoans. Possess a unique jawed pharyngeal apparatus. Body divided into head, trunk, and tail regions. Jaw worms.
Golgi apparatus (gol′je ap″ah-r ă′tus) The membrane-bound cytoplasmic organelle where the proteins and lipids made in the endoplasmic reticulum are modified and stored.
gonad (go′nad) A gamete-producing gland. Ovary or testis.
gonadotropin (go-nad″o-tr ōp′in) A hormone that stimulates activity in the gonads.
gonozooid (gon′o-zo″id) A polyp of a hydrozoan cnidarian that produces medusae. Also called *gonangium.*
graded potential A membrane potential change of variable amplitude and duration that is conducted decrementally; has no threshold or refractory period.
Gordian worm *See* **horsehair worms** *and* **Nematomorpha.**
green gland *See* **antennal gland.**
guard hair The protective, longer hair of a mammal's pelage.
gular flutter (gu′lar) A cooling mechanism used by some birds. Rapid movement of the throat region promotes evaporative water loss.
gustation (gus-ta′shun) The act of tasting or the sense of taste.

H

habitat The native environment of an organism.
hair A long, slender filament. Applied especially to such filamentous appendages of the skin.
haploid (hap′loid) Having one member of each pair of homologous chromosomes. Haploid cells are the product of meiosis and are often gametes.
Hardy-Weinberg theorem (har′de-w īn′berg) A theorem stating that the frequency of genes

in a population does not change from one generation to another if specified conditions are met.

head-foot The body region of a mollusc that contains the head and is responsible for locomotion as well as retracting the visceral mass into the shell.

heartworm disease A parasitic infection in dogs caused by the nematode *Dirofilaria immitis.*

hectocotylus (hek″to-kot′ĭ-lus) A modified arm of some male cephalopods that is used in sperm transfer.

hemal system (he′mal) Strands of tissue found in echinoderms. The hemal system is of uncertain function. It may aid in the transport of large molecules or coelomocytes, which engulf and transport waste particles within the body.

hemerythrin (hem″e-rith′rin) The red iron-containing respiratory pigment in the blood plasma of some annelids priapulids, and brachiopods.

Hemichordata (hem″e-kor-da′tah) The phylum of marine, wormlike animals whose members have an epidermal nervous system and pharyngeal slits. Acorn worms and pterobranchs.

hemimetabolous metamorphosis (hem″e-met-ab′ol-us met-ah-mor′fe-sis) A type of insect metamorphosis involving a species-specific number of molts between egg and adult stages, during which immatures gradually take on the adult form.

hemocoel (he′mo-s ēl) Large tissue space within an arthropod that contains blood. Derived from the blastocoel of the embryo.

hemocyanin (he″mo-si′ah-nin) A nonheme, blue respiratory pigment in the plasma of many molluscs and arthropods. Composed of monomers, each of which contains two atoms of copper and can bind one molecule of oxygen.

hemocyte (he′mo-s īt) Any blood corpuscle or formed element of the blood in animals with an open circulatory system.

hemoglobin (he′mo-glo″bin) An iron-containing respiratory pigment of red blood cells responsible for the transport of oxygen and carbon dioxide. Occurs in vertebrate red blood cells and in the plasma of many invertebrates.

hemolymph (he′mo-limf) The fluid in the coelom or hemocoel of some invertebrates that represents the blood and lymph of higher animals. Found in animals with an open circulatory system.

herbivore (her′b ĭ-vor) A plant-eating animal. Any organism that subsists on plants.

herbivory (her-b ĭ′vor-e) The process of existing by eating macroscopic plants. Examples include molluscs, annelid worms, arthropods, and sea urchins.

hermaphroditism (her-maf′ro-di″tizm) A state characterized by the presence of both male and female reproductive organs in the same animal. Monoecious.

heterochromatic regions (het″er-o-kr ōm-ă′tik) Having inactive genes. Inactive regions of chromosomes are heterochromatic.

heterodont (het′er-o-dont) Having a series of teeth specialized for different functions.

heterotherm (het′e-ro-therm) An animal whose body temperature fluctuates markedly. "Cold-blooded." An animal that derives essentially all of its body heat from the environment.

heterotrophic (het″er-o-tro′fik) The type of nutrition in which organisms derive energy from the oxidation of organic compounds either by consumption or absorption of other organisms.

heterotrophs (het′er-o-trofs) Organisms that obtain both inorganic and organic raw materials from the environment to live. Animals, fungi, many protists, and most bacteria are heterotrophs.

heterozygous (het″er-o-zi′ges) Having different expressions of a gene on homologous chromosomes.

Hexactinellida (hex-act″in-el′id-ah) The class of sponges whose members are characterized by triaxon siliceous spicules, which sometimes form an intricate lattice. Cup or vase shaped. Sycon body form. Glass sponges.

Hexapoda (hex″sah-pod′ah) The subphylum of mandibulate arthropods whose members are characterized by a body consisting of a head, thorax, and abdomen; five pairs of head appendages; and three pairs of uniramous appendages on the thorax.

hibernation (hi″ber-na′shun) Condition of mammals that involves passing the winter in a torpid state in which the body temperature drops to nearly freezing and the metabolism drops close to zero. May last weeks or months.

hierarchical nesting (hi′ah-rar″ki-kel nes′ting) A grouping of organisms by derived characters in a cladogram showing degrees of relatedness between related clades.

hindbrain (h īnd′br ān) Includes the medulla oblongata, cerebellum, and pons. Also known as the *rhombencephalon.*

Hirudinea (hi″ru-din′e-ah) Sedentarian annelids whose members are characterized by bodies with 34 segments, each of which is subdivided into annuli. Anterior and posterior suckers are present. Leeches.

histology (his-tol′e-je) The study of tissues; also known as *microanatomy* or *microscopic anatomy.*

holometabolous metamorphosis (hol″o-met-ab′ol-us met-ah-mor′fe-sis) A type of insect metamorphosis in which immatures, called larvae, are different in form and habitats from the adult. The last larval molt results in the formation of a pupa. Radical cellular changes in the pupal stage end in adult emergence.

homeostasis (ho″me-o-sta′sis) A state of equilibrium in which the internal environment of the body of an animal remains relatively constant with respect to the external environment.

homeotherm (ho′me-o-therm) Having nearly uniform body temperature, regulated independently of the environmental temperature. "Warm-blooded."

homeobox (*Hox*) genes (ho'me-o-box) Sets of genes that direct development of body regions in embryos. Diverse animal groups share these genes.

homodont (ho′mo-dont) Having a series of similar, unspecialized teeth.

homologous (ho-mol′o-ges) Describes structures, processes, or molecules that have a common evolutionary origin. The wing of a bat and the arm of a human are homologous, as each can be traced back to a common ancestral appendage.

homologous chromosomes (ho-mol′o-ges kro′mo-s ōmz) Chromosomes that carry genes for the same traits. One of two copies of a particular chromosome in a diploid cell, each copy being derived from a different parent.

homoplasy (ho′mo-plas″e) Analogous resemblances between animals that result from animals adapting under similar evolutionary pressures. Homoplasies result from convergent evolution and are not useful for establishing evolutionary relationships.

homozygous (ho″mo-zi′ges) Having the same expression of a gene on homologous chromosomes.

horizontal gene transfer (HGT) The movement of genes between species, for example, through phagocytosis of one organism by a second organism and the incorporation of DNA from the first into the genome of the second. Also called **lateral gene transfer (LGT)**.

hormone (hor′m ōn) A chemical secreted by an endocrine gland that is transmitted by the bloodstream or body fluids to a target cell or tissue.

horsehair worms Ecdysozoan animals that belong to the phylum Nematomorpha. Also known as "*Gordian worms* or *hairworms*". (Gordius is the name of an ancient king who tied an intricate knot.)

hydraulic skeleton (hi-drol-ĭk) The use of body fluids in open circulatory systems to give support and facilitate movement. Muscles contracting in one part of the body force body fluids into some distant tissue space, thus causing a part of the body to extend or become turgid.

hydrophilic (hi″dro-fil′ic) Literally translates as "water-loving" and describes substances that are soluble in water. These are either polar or charged molecules such as ions or amine hormones.

hydrostatic skeleton (hi″dro-stat′ik) The use of body cavity fluids, confined by the body wall, to give support (e.g., the hydrostatic skeleton of nematodes and annelids). Also called *hydroskeleton*.

hydrothermal vents (hi″dro-thur′mal) Deep, oceanic regions where the tectonic plates in the earth's crust are moving apart. They are characterized by occasional lava flows and hot water springs. These vents support a rich community by chemolithotrophy.

Hydrozoa (hi″dro-zo′ah) The class of cnidarians whose members have epidermally derived gametes, mesoglea without wandering ameboid cells, and gastrodermis without nematocysts. Medusae, when present, with a velum. *Hydra, Obelia,* and *Physalia.*

hygroreceptor (hi″gro-re-sep′tor) A receptor in insects that detects the water content of air.

Hyperotreti (hi″per-ot′re-te) The infraphylum of craniates whose skull consists of cartilaginous bars, are jawless, and have no paired appendages, four head tentacles, and olfactory sacs that open to the mouth cavity. Hagfishes.

hyperpolarization (hi′per-po″lar-i-za′shun) Condition in which a membrane is more negative than the resting membrane potential. Hyperpolarization of a neuron or muscle fiber following a stimulus makes the membrane less sensitive to a second stimulus. *See* **refractory period**.

hypertonic (hi″per-ton′ik) A solution having a greater number of solute particles than another solution to which it is compared.

hypodermis (hi″po-der′mis) The layer of integument below the cuticle. The outer cellular layer of the body of invertebrates that secretes the cuticular exoskeleton.

hypothalamus (hi″po-thal′ah-mus) A structure within the diencephalon and below the thalamus that functions as an autonomic center and regulates the pituitary gland.

hypotonic (hi″po-ton′ik) A solution having a lesser number of solute particles than another solution to which it is compared.

I

incomplete digestive tract A digestive tract that has a single opening (the mouth) for both ingesting food and the elimination of wastes. For example, the digestive tract in members of the phylum Platyhelminthes.

incomplete dominance An interaction between alleles in which both alleles are expressed more or less equally, and the phenotype of the heterozygote is different from either homozygote.

indirect (asynchronous) flight Insect flight accomplished by flight muscles acting on the body wall. Changes in shape of the thorax cause wing movements. A single nerve impulse results in many cycles of the wings. *See also* **direct (synchronous) flight.**

induction The process by which the presence of one cell or tissue influences the development of adjacent cells or tissues.

inferior Below a point of reference (e.g., the mouth is inferior to the nose in humans).

inner ear Region of the vertebrate ear that contains auditory receptive cells and georeceptors. In humans, these structures are found within the cochlea and semicircular canals, respectively. The inner ear's vestibulocochlear nerve relays auditory signals to the brain.

Insecta (in-sekt′ah) Class of Hexapoda with exposed mouth appendages, mandibles with two points of articulation, and well-developed Malpighian tubules. Insects.

instar (in′star) Any one of the different immature feeding stages of an insect.

insula (in′su-lah) The region of the vertebrate cerebral hemispheres involved with self-awareness, cognitive functions, and gastrointestinal and other visceral activities.

integrating center A portion of the nervous system, or an endocrine gland, that receives and processes sensory information and elicits a response in an effector structure.

integration The combination of sensory information from one or more sources, and the initiation of a response to that information. Integration can occur at the level of the neuron or within responses involving the entire nervous or endocrine system of an animal.

integument (in-teg′u-ment) A covering (e.g., the skin).

integumentary exchange (in-teg″u-men′ tahre) Gas exchange through the integument. Also called *cutaneous exchange.*

intermediate filament The chemically heterogeneous group of protein fibers, the specific proteins of which vary with cell type. One of the three most prominent types of cytoskeletal filaments. Made of fibrous proteins.

intermediate lobe The area in the pituitary gland between the anterior and posterior lobes. Also called the *pars intermedia.* Produces melanophore-stimulating hormone.

interneuron (in″ter-nu′ron) A neuron between a sensory neuron and a motor neuron. Interneurons function as integrating centers.

interphase (in′ter-fāz) Period between two cell divisions when a eukaryotic cell is carrying on its normal functions. Long period of the cell cycle between one mitosis and the next. Includes G_1 phase, G_0 phase, S phase, and G_2 phase. DNA replication occurs during interphase.

intracellular digestion Digestion that occurs inside a cell, as in many protozoans, sponges, cnidarians, flatworms, rotifers, bivalve molluscs, and primitive chordates.

intraspecific competition Competition among members of the same species for environmental resources.

introvert (in′tro-vert) The anterior narrow portion that can be withdrawn (introverted) into the trunk of a sipunculid worm, a loriciferan, or a bryozoan.

island dwarfism An evolutionary process that leads to successively smaller sized individuals in successive generations. It is thought to be due to a lack of predators that would normally exert a selective pressure for larger individuals and fewer island resources available for growth.

isosmotic (i″sos-mot′ik) Having the same osmotic characteristics (pressure).

isotonic (i″so-ton′ik) In a comparison of two solutions, both have equal concentrations of solutes.

J

Jacobson's (vomeronasal) organ Olfactory receptor present in most reptiles. Blind-ending sacs that open through the secondary palate into the mouth cavity. Used to sample airborne chemicals.

Johnston's organ Mechanoreceptor (auditory receptor) found at the base of the antennae of male mosquitoes, midges, and most other insects.

K

keratin (ker′ă-tin) A tough, water-resistant protein found in the epidermal layers of the skin of reptiles, birds, and mammals. Found in hair, feathers, hooves, nails, claws, bills, etc.

keystone species A species that is of overriding importance in a community and controls the community characteristics.

kinetic skull (ki-nĕ′tic skul) Skull modifications, including moveable quadrate bones, that increase skull flexibility. Aid in swallowing large prey. Characteristic of squamate reptiles.

kinetochore (kĭ-nĕ′to-kor) A specialized group of proteins and DNA at the centromere of a chromosome. Is an attachment site for the microtubules of the mitotic spindle and plays an active part in the movement of chromosomes to the pole.

kingdom (king′dom) The level of classification above phylum. Three monophyletic kingdom-level lineages include Fungi, Plantae, and Animalia.

Kinorhyncha (kin″o-rink′ah) The phylum of ecdysozoans that contains members called kinorhynchs. Small, elongate worms found exclusively in marine environments, where they live in mud and sand.

kin selection The idea that natural selection acting on related animals can affect the fitness of an individual. When genes are common to related animals, an individual's fitness is based on the genes the individual passes on, and on those common genes that relatives pass on. Kin selection is thought to explain how altruism could evolve in a population.

L

labial palp (la′be-al palp) 1. Chemosensory appendage found on the labium of insects (Arthropoda). 2. Flaplike lobe surrounding the mouth of bivalve molluscs that directs food toward the mouth.

labium (la′be-um) The posterior mouthpart of insects. It is often referred to as the "lower lip," is chemosensory, and was derived evolutionarily from paired head appendages (Hexapoda, Arthropoda).

lactate fermentation (lak′tat) Glycolysis occurring under anaerobic conditions in animals and some bacteria produces lactic acid and yields a small amount of ATP. *See* **glycolysis**.

lactation (lak-ta′shun) The production of milk by the mammary glands.

large intestine That part of the digestive system between the ileocecal valve of the small intestine and the anus. Removes salt and water from undigested food and releases feces through the anus.

larva (lar′vah) The immature feeding stage of an insect that undergoes holometabolous metamorphosis. The immature stage of any animal species in which adults and immatures are different in body form and habitat.

Larvacea (lar-vas′e-ah) The class of urochordates whose members are planktonic and whose adults retain a tail and notochord with a gelatinous covering of the body.

lateral (lat′er-al) Away from the plane that divides a bilateral animal into mirror images.

lateral gene transfer (LGT) *See* **horizontal gene transfer (HGT).**

lateral-line system (lat′er-al) 1. A line of sensory receptors along the side of some fishes and amphibians used to detect water movement (phylum Chordata). 2. The external manifestation of a lateral excretory canal of nematodes (phylum Nematoda).

leucon (lu′kon) The sponge body form that has an extensively branched canal system. The canals lead to chambers lined by choanocytes.

limbic lobe (lim′bik) The portion of the vertebrate cerebrum that functions with regions of the diencephalon and mesencephalon to form the limbic system.

limbic system (lim′bik) A portion of the vertebrate brain involved with emotions, behavioral expressions, short-term memory, and smell.

limiting factor A nutrient or other component of an organism's environment that is in relatively short supply and therefore restricts the organism's ability to reproduce successfully.

lipophilic (lipo-phil′ic) Nonpolar, fat-soluble substances such as steroid and thyroid hormones.

Lissamphibia (lis′am-fib′e-ah) The group of amphibians comprised of living species. Members of the orders Anura, Caudata, and Gymnophiona.

liver (liv′er) A large, dark-red gland that carries out many vital functions, such as the formation of urea, manufacture of plasma proteins, synthesis of amino acids, synthesis and storage of glycogen, and many others.

local chemical messenger A chemical that acts on nearby cells.

locus (lo′kus) The position of a gene in a chromosome.

logistic population growth The population growth pattern that occurs when environmental resistance limits exponential growth.

lophophore (lof′a-for) Ciliated, tentacle-bearing ridge or arm within which is an extension of the coelomic cavity in some lophotrochozoans (e.g., brachiopods, ectoprocts, phoronids).

Lophotrochozoa (lo-fo-tro″ko-zo′ah) A subkingdom level of classification that includes animals such as the annelids (segmented worms) and molluscs (bivalves, snails, and their relatives). The name "Lophotrochozoa" is derived from the presence of a lophophore or a trochophore larval stage in some members of this group. Also known as *Spiralia*.

lorica (lo-ri′kah) The protective external case in rotifers and some protozoa. It is formed by a thickened cuticle.

Loricifera (lor″a-sif′er-ah) A phylum of lophotrochozoans. A recently described animal phylum. Members are commonly called loriciferans.

lung An organ of the respiratory system in which gas exchange occurs between body fluids (e.g., blood) and air.

lymph (limf) Fluid that the lymphatic vessels transport.

lymph heart Lymph hearts propel the movement of lymph in many fishes, all amphibians and reptiles, bird embryos, and in some adult birds.

lymphatic system (lim-fat′ik) The one-way system of lymphatic vessels, lymph nodes, and other lymphoid organs and tissues. Drains excess tissue fluid from the extracellular space and provides a site for immune surveillance.

lymphocyte (lim′fo-s īt) A type of white blood cell that provides protection to an animal.

lysosome (li′so-s ōm) Cytoplasmic, membrane-bound organelle that contains digestive and hydrolytic enzymes, which are typically most active at the acid pH found in the lumen of lysosomes.

M

M (mitotic) phase The stages of the cell cycle associated with partitioning chromosomes between two daughter cells and the division of the cytoplasm.

macroevolution (mak″ro-ev″o-lu′shun) Large-scale evolutionary changes that result in extinction and the formation of new species.

macronutrient (mak″ro-noo′tre-ent) An essential nutrient for which an animal has a large minimal daily requirement (greater than 100 mg). For example, calcium, phosphorus, magnesium, potassium, sodium, and chloride, along with carbohydrates, lipids, and proteins.

magnetoreception The ability to detect magnetic fields; the least understood of all the sensory modalities.

Malacostraca (mal-ah-kos′trah-kah) The class of crustaceans whose members are characterized by having appendages modified for crawling along the substrate, as in lobsters, crayfish, and crabs. Alternatively, the abdomen and body appendages may be used in swimming, as in shrimp.

Malpighian tubules (mal-pig′e-an tu′b ūlz) The blind-ending excretory and osmoregulatory tubules that join the midgut of insects and some other arthropods. Secrete waste products and form urine.

Mammalia (mah-ma′le-ah) The class of vertebrates whose members are at least partially covered by hair, have specialized teeth, and are endothermic. Young are nursed from mammary glands. The mammals.

mammary gland (mam′ar-e) The breast. In female mammals, the mammary glands produce and secrete milk to nourish developing young.

mandible (man′dib-el) 1. The lower jaw of vertebrates. 2. The paired, grinding and tearing arthropod mouthparts, derived from anterior head appendages.

mantle (man′tel) The outer fleshy tissue of molluscs that secretes the shell. The mantle of cephalopods may be modified for locomotion.

mantle cavity (man′tel kav′ĭ-te) The space between the mantle and the visceral mass of molluscs.

manubrium (mah-nu′bre-um) A structure that hangs from the oral surface of a cnidarian medusa and surrounds the mouth.

mastax (mas′tax) The pharyngeal apparatus of rotifers used for grinding ingested food.

maxilla (maks′il-ah) One member of a pair of mouthparts just posterior to the mandibles of many arthropods.

maxillary gland (mak′si-ler″e) In malacostracan crustaceans, the excretory organ located near the maxillary segments. Helps regulate ion concentrations.

Maxillopoda (maks″il-ah-pod′ah) The class of Crustacea characterized by five head, six thoracic, and four abdominal somites plus a telson. Abdominal segments lack typical appendages. Abdomen often reduced. Barnacles and copepods.

mechanoreceptor (mek″ah-no-re-sep′tor) A sensory receptor that is sensitive to mechanical stimulation, such as changes in pressure or tension.

meconium (mĭ-ko′ne-um) A dark green mucilaginous material in the intestine of the full-term fetus, being a mixture of the secretions of the intestinal glands and some amniotic fluid.

medial (me′de-al) On or near the plane that divides a bilateral animal into mirror images. Also median.

median (parietal) eye (me′de-an) A photoreceptor located middorsally on the head of certain fishes, amphibians and non-Archosaur reptiles. It is associated with the vertebrate epithalamus.

medulla oblongata (me-dul′ah ob″long-gah′tah) Inferior-most portion of the brain stem between the pons and the spinal cord.

medusa (me-du′sah) Usually, the sexual stage in the life cycle of cnidarians. The jellyfish body form.

meiosis (mi-o′sis) Process of cell division by which egg and sperm cells form, involving a diminution in the amount of genetic material. Comprises two successive nuclear divisions with only one round of DNA replication, which produces four haploid daughter cells from an initial diploid cell.

Meissner's (tactile) corpuscles (mis′nerz kor′pus-celz) Light touch receptors in vertebrate skin.

melatonin (mel″ah-to′nin) The hormone that the pineal gland secretes. Regulates photoperiodicity.

meninges (me-nin′jēz) A group of three membranes that covers the brain and spinal cord.

menopause (men′o-pawz) Termination of the menstrual cycle. Period of life in a female during which hormonal changes cause ovulation and menstruation to cease.

menstrual cycle (men′stroo-al) The period of regularly recurring physiologic changes in the endometrium that culminates in its shedding (menstruation).

menstruation (men″stroo-a′shun) Loss of blood and tissue from the uterus at the end of a female primate's reproductive cycle.

Merostomata (mer″o-sto′mah-tah) The class of arthropods whose members are aquatic and possess book gills on the opisthosoma. Eurypterids (extinct) and horseshoe crabs.

mesentery (mez′en ter-e) The double-layered membrane that suspends the alimentary tract of a triploblastic coelomate animal within the abdominal cavity. The mesentery is continuous with the serosal and peritoneal layers.

mesoderm (mez′ah-durm) The embryonic tissue that gives rise to tissues between the ectoderm and endoderm (e.g., muscle, skeletal tissues, and excretory structures).

mesoglea (mez-o-gle′ah) A gel-like matrix between the epidermis and gastrodermis of cnidarians.

mesohyl (mez-o-hīl′) A jellylike layer between the outer (pinacocyte) and inner (choanocyte) layers of a sponge. Contains wandering ameboid cells.

mesonephros (mez″o-nef′rōs) The middle of three pairs of embryonic renal organs in vertebrates. The functional kidney of fishes and amphibians. Its collecting duct is a wolffian duct.

mesothorax (mez″o-thor′aks) The middle of the three thoracic segments of an insect. Usually contains the second pair of legs and the first pair of wings.

Mesozoa (mes″o-zo′ah) A phylum of lophotrochozoans whose members are parasites of marine invertebrates. Two-layered body organization. Dioecious, complex life histories. Orthonectids and dicyemids.

messenger RNA (mRNA) A single-stranded polyribonucleotide. Formed in the nucleus from a DNA template and carries the transcribed genetic code to the ribosome, where the genetic code is translated into protein.

metabolism (me-tab′o-lizm) All of the chemical changes and processes within cells. All controlled, enzyme-mediated chemical reactions by which cells acquire and use energy.

metacercaria (mĕ″tă-ser-ka′re-ah) Stage between the cercaria and adult in the life cycle of most digenetic trematodes. Usually encysted and quiescent.

metamerism (mĕt-tam′ă-riz″em) A segmental organization of body parts. Metamerism occurs in the Annelida, Arthropoda, and other smaller phyla.

metamorphosis (met″ah-mor′fe-sis) Change of shape or structure, particularly a transition from one developmental stage to another, as from larva to adult form.

metanephridium (met″ah-nē-frid′e-um) An excretory organ in many invertebrates. It consists of a tubule that has one end opening at the body wall and the opposite end in the form of a funnel-like structure that opens to the body cavity.

metanephros (me″tah-nef′rōs) The renal organs of some vertebrates arising behind the mesonephros. The functional kidney of nonavian and avian reptiles, and mammals. It is drained by a ureter.

metaphase (met′ah-fāz) Stage in mitosis when chromosomes align in the middle of the cell and firmly attach to the mitotic spindle but have not yet segregated toward opposite poles.

metathorax (met″ah-thor′aks) The posterior of the three segments of an insect thorax. It usually contains the third pair of walking legs and the second pair of wings (Arthropoda).

micelle (mi-sĕl′) Lipid aggregates with a surface coat of bile salts. A stage in the digestion of lipids in the small intestine.

microevolution (mi″kro-ev′o-lu′shun) A change in the frequency of alleles in a population over time.

microfilament (mi″kro-fil′ah-ment) Component of the cytoskeleton. Involved in cell shape, motion, and growth. Helical protein filament formed by the polymerization of globular actin molecules.

microfilaria (mi″kro-fi-lar′e-ah) The prelarval stage of filarial worms. Found in the blood of humans and the tissues of the vector.

Micrognathozoa (mi″kro-nath-o-zo′ah) A recently discovered phylum of lophotrochozoans that inhabit homeothermic springs in Greenland. Body divided into head, thorax, and abdomen. Head contains a complex jaw structure.

micronutrient (mi″kro-nu′tre-ent) A dietary element essential in only small quantities. For example, iron, chlorine, copper, and vitamins.

microtubule (mi″kro-tu′bŭl) A hollow cylinder of tubulin subunits. Involved in cell shape, motion, and growth. Functional unit of cilia and flagella. It is one of three major classes of filaments of the cytoskeleton.

microtubule-organizing center (MTOC) (mi″kro-tu′bŭl) Region in a cell, such as a centrosome or a basal body, from which microtubules grow.

mictic eggs (mik′tik) Pertaining to the haploid eggs of rotifers. If not fertilized, the egg develops parthenogenetically into a male. If fertilized, mictic eggs secrete a heavy shell and become dormant, hatching in the spring into amictic females.

midbrain The portion of the brain between the pons and forebrain (diencephalon).

middle ear Region of the vertebrate ear situated between the outer and inner ear. It contains the tympanum (tympanic membrane) and one or more ossicles, which transmit transduced sound energy to the inner ear. Auditory tubes, also present in the middle ear, function to equalize pressure on each side of the tympanum.

migration Periodic round trips of animals between breeding and nonbreeding areas or to and from feeding areas.

mimicry (mim′ik-re) When one species resembles one or more other species. Often, protection is afforded to the mimic species.

minerals Ions essential in the diet of animals. Sodium and potassium ions are vital in action potentials. Calcium is critical in bone formation and muscle contraction. Zinc, cobalt, iodine, and manganese are cofactors that are essential for the function of enzymes and hormones.

miracidium (mi-rah-sid′e-um) The ciliated, free-swimming, first-stage larva of a digenean trematode that undergoes further development in the body of a snail.

mitochondria (mi″to-kon′dre-ah) Membrane-bound organelles that specialize in aerobic respiration (oxidative phosphorylation) and produce most of the ATP in eukaryotic cells.

mitosis (mi-to′sis) Nuclear division occurring during the M phase of the cell cycle. *See also* **mitotic cell division.**

mitotic cell division (mi-tŏ′tic) Cell division in which the parental number of chromosomes is maintained from one cell generation to the next. Basis of reproduction of single-cell eukaryotes. Basis of physical growth in multicellular eukaryotes.

mitotic spindle (mi-t ŏ′tic spin′dl) Collectively, the asters, spindle, centrioles, and microtubules of a dividing cell.

modern synthesis The combination of the principles of population genetics and Darwinian evolutionary theory.

molecular clock An evolutionary dating technique based on the assumption that the rate of change in a particular region of DNA is relatively constant over time. The amount of genetic change in that region of DNA can be used to assign a date to evolutionary events.

molecular genetics The study of the biochemical structure and function of DNA.

Mollusca (mol-lus′kah) The phylum of coelomate animals whose members possess a head-foot, visceral mass, mantle, and mantle cavity. Most molluscs also possess a radula and a shell. The molluscs. Bivalves, snails, octopuses, and related animals.

molting The periodic renewal of feathers of birds by shedding and replacement. In arthropods and other invertebrates, the shedding of the exoskeleton or other body covering is called molting or ecdysis. *See also* **ecdysis**.

monocyte (mon′o-s ī t) A type of white blood cell that functions as a phagocyte and develops into macrophages in tissues. A monocyte has a kidney-shaped nucleus and gray-blue cytoplasm.

monoecious (mah-ne′shus) An organism in which both male and female sex organs occur in the same individual. Hermaphroditic.

monogamous (mah-nog′ah-mus) Having one mate at a time.

Monogenea (mon″oh-gen′e-ah) The class of Platyhelminthes with members that are called monogenetic flukes. Most are ectoparasites on vertebrates (usually on fishes, occasionally on turtles, frogs, copepods, squids). One life-cycle form in only one host. Bear an opisthaptor. Examples: *Discocotyle, Gyrodactylus,* and *Polystoma.*

Monogononta (mon″o-go-non′tah) A class of rotifers with members that possess one ovary. Mastax not designed for grinding. Produce mictic and amictic eggs. Example: *Notommata.*

monohybrid cross (mon-o-hi′brid) A mating between two individuals heterozygous for one particular trait. Monohybrid crosses usually result in a phenotypic ratio of 3 dominant: 1 recessive in the offspring.

monophyletic group (mon″o-fi-let′ik) A group of organisms descended from a single ancestor.

Monoplacophora (mon″o-pla-kof′o-rah) The class of molluscs whose members have a single, arched shell; a broad, flat foot; and certain serially repeated structures. Example: *Neopilina.*

morula (mor′u-lah) A solid ball of embryonic stem cells that develops from mitotic divisions of a zygote.

mosaic evolution (mo-za′ik ev″ah-loo′shun) A change in a portion of an organism (e.g., a bird wing) while the basic form of the organism is retained.

motor (effector or **efferent) neuron** or **nerve** A neuron or nerve that transmits impulses from the central nervous system to an effector, such as a muscle or gland.

mucous cell A glandular cell that secretes mucus.

Müller's larva A free-swimming, ciliated larva that resembles a modified ctenophore. Characteristic of many marine polyclad turbellarians.

multiple alleles (mul′t ĭ-pl ah-lēlz′) The presence of more than two alleles in a population.

muscle fiber The contractile cell within muscle tissue. Also known as a myofiber in skeletal muscle.

muscle tissue The type of tissue that allows movement. The three kinds are skeletal, smooth, and cardiac. Tissue made of bundles of long cells called muscle fibers.

musk gland *See* **scent gland**.

mutation pressure (myoo-ta′shun presh′er) A measure of the tendency for gene frequencies to change through mutation.

mutualism (myoo′choo-ah-liz-em) A relationship between two species that benefits both members. The association is necessary to both species.

myelin (mi′ĕ-lin) Lipoprotein material that forms a sheathlike covering around some nerve fibers.

myocytes (mi′ŏ-sits) Contractile cells present in sponges (phylum Porifera) that are responsible for body movements.

myofibril (mi″o-fi′bril) Subcellular bundles of microfilaments (actin and myosin) in skeletal muscle.

myomere (mi′o-m ēr) The muscle plate or portion of a somite that develops into voluntary muscle.

myosin (mi′ah-sin) A protein that, together with actin, is responsible for muscular contraction and relaxation.

Myriapoda (mir″e- ă-pod′ah) The subphylum of arthropods characterized by a body consisting of two tagmata and uniramous appendages.

N

naiad (na′ad) The aquatic immature stage of any hemimetabolous insect.

nail The horny, cutaneous plate of the dorsal surface of the distal end of a finger or toe. Homologous to a claw, but broader, flatter, and does not come to a point.

natural selection A theory, conceived by Charles Darwin and Alfred Wallace, of how some evolutionary changes occur. The idea that some individuals in a population possess variations that make them less able to survive and/or reproduce. These individuals die or fail to reproduce. Their genes are less likely to be passed into the next generation. Therefore, the genetic composition of the population in the next generation changes (i.e., evolution occurs).

negative feedback An endocrine response that reverses or opposes the effect of a particular stimulus. Plays a large role in the maintenance of homeostasis.

nematocyst (n ĕ-mat′ah-sist) A cnidarian cnida usually armed with spines or barbs and containing a venom that is injected into a prey's flesh.

Nematoda (nem- ă-to′dah) The phylum of ecdysozoans that contains members commonly called either roundworms or nematodes. Triploblastic, bilateral, vermiform, unsegmented, and pseudocoelomate.

Nematomorpha (nem″ă-to-mor′fah) The phylum of ecdysozoans commonly called horsehair worms.

Nemertea (nem-er′te-ah) The phylum that has members commonly called the proboscis worms. Elongate, flattened worms in marine mud and sand. Triploblastic. Complete digestive tract with anus. Closed circulatory system.

nephridiopore (nĕ-frid′e-o-por) The opening to the outside of a nephridium. An excretory opening in invertebrates.

nephridium (nĕ-frid′e-um; pl. nephridia) The excretory organ of the embryo. The embryonic tube from which the kidney develops. Tubular osmoregulatory and excretory organ of many invertebrates. Functions in excretion, osmoregulation, or both.

nephron (nef′ron) The functional unit of a vertebrate kidney, consisting of a renal corpuscle and a renal tubule.

nerve A bundle of neurons or nerve cells outside the central nervous system.

nerve cords Aggregations of neurons into longitudinally arranged clusters and tracts in animals.

nerve net A diffuse, two-dimensional plexus of bi- or multipolar neurons found in cnidarians. The simplest pattern of invertebrate nervous systems.

nerve tract A bundle of nerve fibers within the central nervous system.

nervous tissue The type of tissue composed of individual cells called neurons and supporting neuroglial cells.

neuroendocrine system (nur″o-en′do-krin) The combination of the nervous and endocrine systems.

neurofibril node (nur″o-fib′ril nōd) Regular gaps in a myelin sheath around a nerve fiber. Formerly called **node of Ranvier.**

neurohormone (nur″o-hor′m ōn) A chemical transmitter that nervous tissue produces. Uses the bloodstream or other body fluids for distribution to its target site.

neurolemmocyte (nur″o-lem′o-s īt) The cell that surrounds a fiber of a peripheral nerve and forms the neurolemmal sheath and myelin. Formerly called **Schwann cell.**

neuromast (nur′o-mast) The hair-cell mechanoreceptor within pits of the lateral-line system. Used to detect water currents and the movements of other animals.

neuron (nu′ron) A nerve cell that consists of a cell body and its processes. The basic functional unit of communication in nervous systems.

neuropeptide (nur″o-pep′t īd) A hormone (neurohormone) that secretory nervous tissue produces.

neurotransmitter (nur″o-trans-mit′er) Chemical substance that the terminal end of an axon secretes that either stimulates or inhibits a muscle fiber contraction or an impulse in another neuron.

neutral theory The description of how allelic frequencies change in populations when allelic variation is selectively neutral. That is, two alleles provide neither selective advantage nor disadvantage in a population.

neutrophil (nu′tro-fil) A type of phagocytic white blood cell with a multilobed nucleus and inconspicuous cytoplasmic granules.

nictitating membrane (nik′t ĭ-t āt-ing mem′br ān) The thin, transparent lower eyelid of amphibians and reptiles.

nociceptor (no″se-sep′tor) A sensory receptor responding to potentially harmful stimuli. Produces a sensation of pain.

node of Ranvier (n ōd ov Ran′ve-a) A constriction of myelinated nerve fibers at regular intervals at which the myelin sheath is absent and the axon is enclosed only by sheath cell processes. Also known as a **neurofibril node**.

nodes Branch points in phylogenetic trees that represent ancestral molecules, individuals, populations, or species.

nomenclature (no′men-kla-cher) The study of the naming of organisms in the fashion that reflects their evolutionary relationships.

nondisjunction (non″dis-junk′shun) The failure of homologous chromosomes to separate during meiosis. Nondisjunction results in gametes with a deficiency of one chromosome or an extra chromosome.

nonrenewable resources Resources that cannot replenish themselves over meaningful time frames when used within natural ecosystems or by humans. Contrasts with **renewable resources**.

nonshivering thermogenesis (non-shiv′ er-ing ther-mo-gen′esis) The hormonal triggering of heat production. A thermogenic process in which enzyme systems for fat metabolism are activated, breaking down and oxidizing conventional fats to produce heat.

nontropic hormones (non″tro-pik) Poplypeptide hormones released from the anterior pituitary that have direct effects on certain cells and tissues. Growth hormone and prolactin.

norepinephrine (nor″ep-i-nef′rin) A catecholamine neurotransmitter released from the axon ends of some nerve fibers. Released as a hormone from the adrenal medula. Also known as *noradrenaline*.

notochord (no′tah-kord) A rodlike, supportive structure that runs along the dorsal midline of all larval chordates and many adult chordates.

nuclear envelope Double membrane forming the surface boundary of a eukaryotic nucleus. Consists of outer and inner membranes perforated by nuclear pores.

nucleolus (nu-kle′o-lus) A small structure within the nucleus of a cell that transcribes ribosomal RNA and assembles ribosomal subunits.

nucleosome (nu-kle′ah-s ōm) An association of DNA and histone proteins that makes up chromatin. Many nucleosomes are linked in a chromatin strand.

nucleotide (nu′kle-o-t īd) A component of a nucleic acid molecule consisting of a sugar, a nitrogenous base, and a phosphate group. Nucleotides are the building blocks of nucleic acids.

nucleus (nu′kle-us) 1. Cell nucleus. A spheroid body within a cell, contained in a double membrane, the nuclear envelope, and containing chromosomes and one or more nucleoli. The genetic control center of a eukaryotic cell. 2. The cell bodies of nerves within the central nervous system. 3. An atom's central core, containing protons and neutrons.

Nuda (nu′dah) The class of ctenophorans whose members lack tentacles and have a flattened body with a highly branched gastrovascular cavity.

nutrition (nu-tr ĭ ′shun) The study of the sources, actions, and interactions of nutrients. The study of foods and their use in diet and therapy.

nymph (nimf) The immature stage of a hemimetabolous insect. Resembles the adult but is sexually immature and lacks wings (Arthropoda).

O

occipital lobe (oc-cip′i-tal) The region of the vertebrate cerebral hemispheres organized for vision and associated forms of expression.

ocellus (o-sel′as; ocelli) A simple eye or eyespot in many invertebrates. A small cluster of photoreceptors.

odontophore (o-dont′o-for″) The cartilaginous structure that supports the radula of molluscs.

olfaction (ol-fak′shun) The act of smelling. The sense of smell.

ommatidia (om″ah-tid′e-ah) The sensory unit of the arthropod compound eye.

omnivore (om′n ĭ-vor) Subsisting upon both plants and animals. An animal that obtains its nutrients by consuming plants and other animals.

oncosphere (ong′ko-sf ēr) The larva of the tapeworm contained within the external embryonic envelope and armed with six hooks and cilia. Typically referred to as a coracidium when released into the water.

oncomiracidium (on″ko-mir-ă-sid′e-um) Ciliated larva of a monogenetic trematode.

Onychophora (on-e-kof′o-rah) A phylum of terrestrial animals with 14 to 43 pairs of unjointed legs, oral papillae, and two large antennae. Onycophorans live in humid, tropical areas of the world. Along with Tardigrada and Arthropoda, Onycophora comprises the clade Panarthropoda.

oogenesis (o″o-jen′ĕ-sis) The process by which an egg cell forms from an oocyte.

open circulatory system A circulatory system found in insects and some other invertebrates in which blood is not confined to vessels in part of its circuit. Blood bathes tissues in blood sinuses.

operculum (o-per′ku-lum) A cover. 1. The cover of a gill chamber of a bony fish (Chordata). 2. The cover of the genital pore of a horseshoe crab (Meristomata, Arthropoda). 3. The cover of the aperture of a snail shell (Gastropoda, Mollusca).

Ophiuroidea (o-fe-u-roi′de-ah) The class of echinoderms whose members have arms sharply set off from the central disk and tube feet without suction disks. Brittle stars.

opisthaptor (ah″pis-thap′ter) Posterior attachment organ of a monogenetic trematode.

opisthosoma (ah″pis-tho-so′mah) The portion of the body of a chelicerate arthropod that contains digestive, reproductive, excretory, and respiratory organs.

oral Having to do with the mouth. The end of an animal containing the mouth.

oral cavity The cavity within the mouth.

oral sucker The sucker on the anterior end of a tapeworm, fluke, or leech.

order The level of classification between class and family.

organ A structure consisting of a group of specialized tissues that performs a specialized function.

organelles (or″gah-nelz′) Small structures within cells that perform specific functions.

organic evolution The change in an organism over time. A change in the sum of all genes in a population.

organ of Ruffini Sensory receptor in the skin believed to be a sensor for touch-pressure, position sense of a body part, and movement. Also known as *corpuscle of Ruffini.*

organ system A set of interconnected or interdependent parts that function together in a common purpose or produce results that cannot be achieved by one of them acting alone.

orifice (or′ĭ-fis) The entrance or outlet of any cavity in an animal's body.

oscula (os-cu-lah, sing. *osculum*) Openings in the body of a sponge (phylum Porifera) through which water exits the sponge.

osmoconformer (oz″mo-con-form′er) An organism whose body fluids have the same or very similar osmotic pressure as that of its aquatic environment. A marine organism that does not utilize energy in osmoregulation.

osmolarity (oz″mo-lar′i-te) A measure of the osmotic properties of a solution, as measured in osmoles.

osmoregulation (oz″mo-reg″u-la′shun) The maintenance of osmolarity by an organism or body cell with respect to the surrounding medium.

osmoregulator (oz″mo-reg′u-la-ter) An organism that regulates its internal osmolarity with respect to the environment.

osmosis (oz-mo′sis) Net movement of water across a selectively permeable membrane driven by a difference in concentration of solute on either side. The membrane must be permeable to water but not to the solute molecules.

osphradia (os′fra-de-ah) Chemoreceptors in the anterior wall of the mantle cavity of a gastropod that detect sediment and chemicals in inhalant water or air.

ostia (os′te-ah) *See* **dermal pores**.

outer ear Region of the vertebrate ear that is responsible for directing sound energy into more interior regions of the ear. It contains the external auditory meatus and, in mammals, the exterior pinnae (auricles).

outgroup In cladistic studies, an outgroup is a group outside of a study group that shares an ancestral characteristic with the study group.

ovarian cycle (o-va′re-an) The cycle in the ovary during which the oocyte matures and ovulation occurs.

ovary (o′var-e pl.ovaries) The primary reproductive organ of a female. Where eggs (ova) are produced.

oviparous (o-vip′er-us) Organisms that lay eggs that develop outside the body of the female.

ovipositor (ov-ĭ-poz′it-or) A modification of the abdominal appendages of some female insects that is used for depositing eggs in or on some substrate (Arthropoda, Hexapoda).

ovoviviparous (o′vo-vi-vip′er-us) Organisms with eggs that develop within the female reproductive tract and that are nourished by food stored in the egg.

ovulation (ov″vu-la′shun) The release of an egg (female gamete) from a mature ovarian follicle or ovary.

oxyhemoglobin (ok″s ĭ-he′mo-glo″bin) Compound formed when oxygen combines with hemoglobin.

oxyhemoglobin dissociation curve A graphic description of the loading of the hemoglobin molecule with oxygen at a gas-exchange surface (e.g., skin, lung, or gill) and the release of oxygen from the hemoglobin molecule at body tissues.

P

Pacinian corpuscle (pah-sin′e-an kor′pus-l) A sensory receptor in skin, muscles, body joints, body organs, and tendons that is involved with the vibratory sense and firm pressure on the skin. Also called a *lamellated corpuscle.*

pain receptor A modified nerve ending that, when stimulated, gives rise to the sense of pain.

paleontology (pa″le-on-tol′o-je) The study of early life-forms on the earth.

Panarthropoda (pan-ar″thra-po′dah) The ecdysozoan clade that includes the phyla Onychophora, Tardigrada, and Arthropoda.

pancreas (pan′kre-as) Glandular organ in the abdominal cavity and behind the stomach that secretes hormones and digestive enzymes.

pancreatic islet (pan″kre-at′ic i′let) An island of special tissue in the pancreas. Secretes insulin or glucagon.

Pancrustacea (pan′krus-tas″e-ah) The arthropod clade including crustaceans and hexapods. The traditional subphylum "Crustacea" is a paraphyletic group.

papulae (pap′yoo-lah) *See* **dermal branchiae.**

parabronchi (par″ah-brong′ke) The tiny air tubes within the lung of a bird that lead to air capillaries where gases are exchanged between blood and air.

paracrine agent (par′a-krin) Relating to a type of hormone function in which the effects of the hormone are restricted to the local environment.

parapatric speciation (par″ah-pat′rik spe″she-a′shun) Speciation that occurs in small, local populations, called **demes.**

paraphyletic group (par″ah-fi-let′ik) A group that includes some, but not all, members of a lineage. Paraphyletic groups result from insufficient knowledge of the group.

parapodia (par″ah-pod′e-ah) Paired lateral extensions on each segment of some annelids. May be used in swimming, crawling, and burrowing.

parasitism (par′ah-si″tizm) A relationship between two species in which one member (the parasite) lives at the expense of the second (the host).

parasympathetic division (par″ah-sim″pah-thet′ik) Portion of the autonomic nervous system that arises from the brain and sacral region of the spinal cord.

parathyroid gland (par″ah-thi′roid) One of the small glands within a lobe of the thyroid gland. Produces the hormone parathormone.

parenchyma (p ă-ren′k ă-m ă) A spongy mass of mesenchyme cells filling spaces around viscera, muscles, or epithelia in acoelomate animals. Depending on the species, parenchyma may provide skeletal support, nutrient storage, motility, reserves of regenerative cells, transport of materials, structural interactions with other tissues, modifiable tissue for morphogenesis, oxygen storage, and perhaps other functions that have yet to be determined.

parietal cell (pah-ri′ĕ-tal) Cell of a gastric gland that secretes hydrochloric acid and intrinsic factor.

parietal eye (pah-ri′ĕ-tal) *See* **median (parietal) eye.**

parietal lobe (pah-ri-tal) The region of the vertebrate cerebral hemispheres involved with general senses such as touch, temperature, and pain.

pars intermedia *See* **intermediate lobe.**

parthenogenesis (par″th ĕ-no-jen′ĕ-sis) A modified form of sexual reproduction by the development of a gamete without fertilization, as occurs in some bees, wasps, certain lizards, and a few other animals.

parturition (par″tu-rish′un) The process of childbirth.

Pauropoda (por″o-pod′ah) A class of arthropods whose bodies are small, soft, and divided into 11 segments, and who have 9 pairs of legs.

pedicel (ped′ĕ-sel) A footlike part, especially any of the secondary structures of a podocyte; also called foot process.

pedicellariae (ped″ĕ-sel-ar′i-a) Pincerlike structures found on the body wall of many echinoderms. They are used in cleaning and defense.

pedipalps (ped′ĕ-palps) The second pair of appendages of chelicerate arthropods. These appendages are sensory in function.

pennaceous feathers (pen″a-ce′us) Tightly closed feathers that create aerodynamic surfaces to birds—such as those found in the wings and tails. These feathers have a prominent shaft or rachis from which barbs branch. Pennaceous feathers include flight and contour feathers.

pentaradial symmetry (pen″tah-ra′de-al sim′ĭ-tre) A form of radial symmetry in the echinoderms in which body parts are arranged in fives around an oral-aboral axis.

Pentastomida (pen-tah-stom′id-ah) A phylum of worms that are all endoparasites in the lungs or nasal passageways of carnivorous vertebrates. Tongue worms.

peripheral nervous system (p ĕ-rif′er-al) The nerves and ganglia of the nervous system that lie outside of the brain and spinal cord.

peristalsis (per″ĭ-stal′sis) Rhythmic waves of muscular contraction in the walls of various tubular organs that move material through the organs.

peristomium (per″i-st ōm′e-um) The segment of an annelid body that surrounds the mouth.

peritoneum (per″i-to-ne′um) The serous membrane lining the abdominal cavity.

peroxisome (per-ok′si-som) A microbody that plays an important role in the breakdown of hydrogen peroxide by the enzyme catalase.

Petromyzontida (pet′tro-mi-zon″tid-ah) The class of Petromyzontomorphi characterized by a sucking mouth with teeth and rasping tongue, seven pairs of pharyngeal slits, and blind olfactory sacs. The lampreys.

Petromyzontomorphi (pet′tro-mi-zon″to-morf′e) The superclass of fishes characterized by a large, sucker-like mouth reinforced by cartilage. Gill arches with spine-shaped processes. *See also* **Petromyzontida**.

phagocytosis (fag″o-si-to′sis) Process (endocytosis) by which a cell engulfs bacteria, foreign proteins, macromolecules, and other cells and digests their substances. Cellular eating.

pharyngeal lamellae Thin plates of tissue on gill filaments that contain the capillary beds across which gases are exchanged.

pharyngeal slit (far-in′je-al) One of several openings in the pharyngeal region of chordates. Pharyngeal slits allow water to pass from the pharynx to the outside of the body. In the process, water passes over gills, or suspended food is removed in a filter-feeding mechanism.

pharynx (far′inks) The passageway posterior to the mouth that is common to respiratory and digestive systems.

phasmid (faz′mid) Sensory pit on each side near the posterior end of some nematodes.

phenotype (fe′no-t īp) The expression that results from an interaction of one or more gene pairs and the environment.

phenotypic ratio (fe″no-tip′ik) The relative numbers of progeny in each phenotypic category that a genetic cross produces.

pheromone (fer′o-m ōn) A chemical that is synthesized and secreted to the outside of the body by one organism and that is perceived (as by smell) by a second organism of the same species, releasing a specific behavior in the recipient.

phonoreceptor (fo″no-re-sep′tor) A specialized nerve ending that responds to sound.

Phoronida (fo-ron′ĭ-dah) A phylum of marine animals whose members live in permanent, chitinous tubes in muddy, sandy, or solid substrates. They feed via an anterior lophophore with two parallel rings of long tentacles.

photoreceptor (fo″to-re-sep′tor) A nerve ending that is sensitive to light energy.

phyletic gradualism (fi-let′ik graj′oo-el-izm) The idea that evolutionary change occurs at a slow, constant pace over millions of years.

phylogeny (fi-log′e-ne) A description of the evolutionary history of a group of taxa, usually depicted by tree diagrams.

phylogenetic species concept In the phylogenetic species concept, a species is defined in terms of monophyly. A species is a group of populations that has evolved independently of other groups of populations. A species shares one or more synapomorphies (shared derived characteristics).

phylogenetic systematics (fi-lo-je-net′ik sis-tem-at′iks) The study of the phylogenetic relationships among organisms in which true and false similarities are differentiated. Cladistics.

phylogenetic trees Branching diagrams that depict lines of descent within a taxonomic group.

phylogeny (fi′loj′en-e) The evolutionary relationships among species.

phylum (fi′lum) The level of classification between kingdom and class. Members are considered a monophyletic assemblage derived from a single ancestor.

pia mater (pi′ah m ā′ter) The innermost meninx that is in direct contact with the brain and spinal cord.

pilidium larva (pi-lid-e-um lar′va) Free-swimming, hat-shaped larva of nemertean worms characterized by an apical tuft of cilia.

pinacocyte (pin″ah-ko′s īt) Thin, flat cell covering the outer surface, and some of the inner surface, of poriferans.

pinacoderm (pin″ah-ko′derm) The outer epithelium of a sponge (phylum Porifera) body wall.

pineal gland (pin′e-al) A small gland in the midbrain of vertebrates that converts a signal from the nervous system into an endocrine signal. Also called the *pineal body*.

pinocytosis (pin″o-si-to′sis) Cell drinking. The engulfment into the cell of liquid and of dissolved solutes by way of small membranous vesicles. A type of endocytosis.

pioneer community The first community to become established in an area.

pit organ Receptor of infrared radiation (heat) on the heads of some snakes (pit vipers).

pituitary gland (p ĭ-tu′ ĭ-tar″e) Endocrine gland that attaches to the base of the brain and consists of anterior and posterior lobes. Hypophysis.

placenta (plah-sen′tah) Structure by which an unborn mammal attaches to its mother's uterine wall and through which it is nourished.

placid (plas′id) Plates on the Kinorhyncha. Modified scalid that grips the substrate during burrowing.

Placozoa (plak″o-zo′ah) A phylum of small, flattened, marine animals that feed by forming a temporary digestive cavity. Example: *Tricoplax adherans*.

Plantae (plant′a) One of the three monophyletic kingdoms of life. Characterized by being eukaryotic and multicellular, and having rigid cell walls and chloroplasts.

planula (plan′u-lah) A ciliated, free- swimming larva of most cnidarians. The planula develops following sexual reproduction and metamorphoses into a polyp.

plasma (plaz′mah) The fluid or liquid portion of circulating blood within which formed elements and various solutes are suspended.

plasma membrane (plaz′mah mem′br ān) Outermost membrane of a cell. Its surface has molecular regions that detect changes in external conditions and act as a selective barrier to ions and molecules passing between the cell and its environment. Consists of a phospholipid bilayer in which are embedded molecules of protein and cholesterol.

plastron (plas′tron) The ventral portion of the shell of a turtle. Formed from bones of the pectoral girdle and dermal bone.

platelet (pl āt′let) Cytoplasmic fragment formed in the bone marrow that functions in blood coagulation. Also called *thrombocyte*.

Platyhelminthes (plat″e-hel-min′th ēz) The phylum of flatworms. Bilateral acoelomates.

plerocercoid larva (ple″ro-ser′koid) Metacestode that develops from a procercoid larva. It usually shows little differentiation.

pleurodont (plu″ro-dont) Teeth that are not set into sockets, but attached along the medial edge of the jaw. Most lizards have pleurodont teeth.

plesiomorphies (ples″e-o-mor′fez) *See* **ancestral character state.**

plumulaceous feathers (plu′mu-la″ceous) Feathers that have a rudimentary shaft to which a wispy tuft of barbs and barbules are attached. Plumulaceous feathers include insulating down feathers.

pneumatic sacs (nu-mat′ik) Gas-filled sacs that arise from the esophagus, or another part of the digestive tract, of fishes. Pneumatic sacs are used in buoyancy regulation (swim bladders) or gas exchange (lungs).

pneumostome (nu′mo-st ōm) The outside opening of the gas-exchange structure (lung) in land snails and slugs (Pulmonata).

point mutations A change in the structure of a gene that usually arises from the addition, deletion, or substitution of one or more nucleotides.

polyandrous (pol″e-an′drus) Having more than one male mate. Polyandry is advantageous when food is plentiful, but because of predation or other factors, the chances of successfully rearing young are low.

polygynous (pah′lij′ah-nus) Having more than one female mate. Polygyny tends to occur in species whose young are relatively independent at birth or hatching.

polyp (pol′ip) The attached, usually asexual, stage of a cnidarian.

polypide (pol′li-pid) An individual (zooid) in an ectoproct colony. It is comprised of a lophophore, a digestive tract, muscles, and nerve centers. A polypide is surrounded by the colony exoskeleton.

polyphyletic *See* **polyphyletic group.**

polyphyletic group (pol-e-fi-let′ik) An assemblage of organisms that includes multiple evolutionary lineages. Polyphyletic assemblages usually reflect insufficient knowledge regarding the phylogeny of a group of organisms.

polyphyodont (pol-if′e-o-dont) A tooth condition in which teeth can be replaced multiple times. Most nonavian reptiles have polyphyodont teeth.

Polyplacophora (pol″e-pla-kof′o-rah) The class of molluscs whose members are elongate and dorsoventrally flattened, and have a shell consisting of eight dorsal plates.

polyploidy (pol′e-ploi-de) Having more than two sets of chromosomes (3*N*, 4*N*, and so forth).

pons (ponz) A portion of the brain stem above the medulla oblongata and below the midbrain.

population A group of individuals of the same species that occupy a given area at the same time and share a unique set of genes.

population genetics The study of genetic events in gene pools.

Porifera (po-rif′er-ah) The animal phylum whose members are sessile and either asymmetrical or radially symmetrical. Body is organized around a system of water canals and chambers. Cells are not organized into organs. Sponges.

porocytes (por′o-s ītz) Tubular cells in sponge body wall that create a water channel to an interior chamber.

positive feedback An endocrine response that increases or amplifies the effect of a particular stimulus. Used largely to drive biological processes to completion and does not play a major role in maintaining homeostasis.

postanal tail (post-an′al) A tail that extends posterior to the anus. One of the five distinguishing characteristics of chordates.

preadaptation (pre-ă-dap-ta′shun) Occurs when a structure or a process present in members of a species proves useful in promoting reproductive success when an individual encounters new environmental situations.

precocial (pre-ko′shel) Having developed to a high degree of independence at the time of hatching or birth.

predation (pre-da′shun) The derivation of nutrients essential for an animal's existence through the capture and consumption of individuals of another species. The ingestion of prey by a predator for energy and nutrients.

preoptic nuclei Neurons in the hypothalamus that produce various neuropeptides (hormones).

Priapulida (pri″ă-pyu′-li-da) A phylum of lophotrochozoans commonly called priapulids.

primary production The total amount of energy converted into organic compounds in a given area per unit time.

Primates The order of mammals whose members include humans, monkeys, apes, lemurs, and tarsiers.

principle of independent assortment One of Mendel's observations on the behavior of hereditary units (genes) during gamete formation. A modern interpretation of this principle is that genes carried in one chromosome are distributed to gametes without regard to the distribution of genes in nonhomologous chromosomes.

principle of segregation One of Mendel's observations on the behavior of hereditary units (genes) during gamete formation. A modern interpretation of this principle is that genes exist in pairs, and during gamete formation, members of a pair of genes are distributed into separate gametes.

probiotics (pro-bi′ot-ics) Living organisms that may provide health benefits beyond their nutritional value when ingested.

procercoid larva (pro-ser′koid lar′v ă) Cestode developing from a coracidium in some orders. It usually has a posterior cercomer. Developmental stage between oncosphere and plerocercoid.

proglottid (pro-glot′id) One set of reproductive organs in a tapeworm strobila. Usually corresponds to a segment. One of the linearly arranged segmentlike sections that make up the strobila of a tapeworm.

prometaphase (pro-met′ah-fāz) The stage of mitosis that follows the breakup of the nuclear envelope. A microtubule attaches to the kinetochore of one chromatid and to one of the poles of the cell at the other end of the microtubule.

pronephros (pro-nef′res) Most anterior of three pairs of embryonic renal organs of vertebrates. Functional only in larval amphibians and fishes, and in adult hagfishes. Vestigial in mammalian embryos.

prophase (pro′f āz) The stage of mitosis during which the chromosomes become visible under a light microscope. The first stage of mitosis during which the chromosomes are condensed but not yet attached to a mitotic spindle.

proportion of polymorphic loci A calculation used to quantify the amount of genetic variation in a population. Populations with more loci containing more than one allele display higher levels of variation.

proprioceptor (pro″pre-o-sep′tor) A sensory nerve terminal that gives information concerning movements and positions of the body. They occur chiefly in the muscles, tendons, and the labyrinth of the mammalian ear and in the joints of arthropods.

prosoma (pro-so′mah) A sensory, feeding, and locomotor tagma of chelicerate arthropods.

prostate gland (pros′t āt) Gland around the male urethra below the urinary bladder that adds its secretions to seminal fluid during ejaculation.

prostomium (pro-st ōm′e-um) A lobe lying in front of the mouth, as found in the Annelida.

protandrous (pro-tan′drus) The condition in a monoecious (hermaphroditic) organism in which male gonads mature before female gametes. Prevents self-fertilization.

protandry (prot-an′dre) The monoecious (hermaphroditic) condition in which male gametes mature before female gametes in order to prevent self-fertilization.

prothorax (pro-thor′aks) The first of the three thoracic segments of an insect. Usually contains the first pair of walking appendages.

protists (pro-tists′) Eukaryotes that are comprised of unicellular or colonial organisms. A polyphyletic assemblage.

protogynous (pro-toj′ĭ-nus) Hermaphroditism in which the female gonads mature before the male gonads.

protonephridium (pro-to-n ĕ-frid′e-um) Primitive osmoregulatory or excretory organ composed of a tubule terminating internally with a flame cell or solenocyte. The unit of a flame-cell system. Protonephridia are specialized for ultrafiltration.

protopodite (pro″to-po′d īt) The basal segment of a biramous appendage of a crustacean.

Protostomia (pro′to-stom″e-ah) Animals in the phyla Platyhelminthes, Nematoda, Mollusca, Annelida, Arthropoda, and others. Protostomia does not represent a formal taxonomic rank, but its members share certain developmental stages. The embryonic blastopore becomes the mouth. Often possesses a trochophore larva, schizocoelous coelom formation, and spiral embryonic cleavage.

protostyle (pro′to-st īl″) A rotating mucoid mass into which food is incorporated in the gut of a gastropod (phylum Mollusca).

proximal (proks′em-al) Toward the point of attachment of a structure on an animal (e.g., the hip is proximal to the knee).

pseudocoelom (soo″do-se′lom) A body cavity between the mesoderm and endoderm. A persistent blastocoel that is not lined with peritoneum. Also called *pseudocoel.*

Pterobranchia (ter″o-brang′ke-ah) The class of hemichordates whose members lack gill slits and have two or more arms. Colonial, living in externally secreted encasements.

pulmonary circuit (pul′mo-ner″e) The system of blood vessels from the right ventricle of the heart to the lungs, transporting deoxygenated blood and returning oxygenated blood from the lungs to the left atrium of the heart.

pulmonate lung (pul′mon- āt) The gas-exchange structure in the Pulmonata—the land snails and slugs.

pulmocutaneous circuit (pul′mo-ku-ta′ne-es) Because most amphibians absorb oxygen through their skin as well as through their lungs or gills, blood returning from the skin also contributes oxygenated blood to the ventricle. *See also* **cutaneous respiration.**

punctuated equilibrium model (pungk′choo-āt″ed e″kw ĕ-lib′re-ahm) The idea that evolutionary change can occur rapidly over thousands of years and that these periods of rapid change are interrupted by periods of constancy (stasis).

Punnett square A tool geneticists use to predict the results of a genetic cross. Different kinds of gametes that each parent produces are placed on each axis of the square. Combining gametes in the interior of the square gives the results of random fertilization.

pupa (pu′pah) A nonfeeding immature stage in the life cycle of holometabolous insects. It is a time of radical cellular changes that result in a change from the larval to the adult body form.

puparium (pu-par′e-um) A pupal case formed from the last larval exoskeleton. *See also* **pupa.**

purine (pyoor′ēn) A nitrogen-containing organic compound that contributes to the structure of a DNA or an RNA nucleotide. Uric acid is also derived from purines.

Pycnogonida (pik″no-gon′ĭ-dah) The class of chelicerate arthropods whose members have a reduced abdomen and four to six pairs of walking legs. Without special respiratory or excretory structures. Sea spiders.

pygostyle (pig′o-st īl) The fused posterior caudal vertebrae of a bird. Helps support tail feathers that are important in steering.

pyrimidine (pi-rim′i-d ēn) A nitrogen-containing organic compound that is a component of the nucleotides making up DNA and RNA.

R

radial symmetry A form of symmetry in which any plane passing through the oral-aboral axis divides an organism into mirror images.

radiation (ra″de-a′shun) A form of energy that includes visible light, ultraviolet light, and X rays. Means by which body heat is lost in the form of infrared rays.

radula (raj′oo-lah) The rasping, tonguelike structure of most molluscs that is used for scraping food. Composed of minute chitinous teeth that move over a cartilaginous odontophore.

ram ventilation The movement of water across gills as a fish swims through the water with its mouth open.

range of optimum The range of values for a condition in the environment that is best able to support survival and reproduction of an organism.

receptor-mediated endocytosis (re-cep′tor-me′de-āt-ed en″do-si-to′sis) The type of endocytosis that involves a specific receptor on the plasma membrane that recognizes an extracellular molecule and binds with it.

recessive A gene whose expression is masked when it is present in combination with a dominant allele. For a recessive gene to be expressed, both members of the gene pair must be the recessive form of the gene.

rectal gland (rek′tel) The excretory organ of elasmobranchs and the coelacanth near the rectum. It excretes a hyperosmotic salt (NaCl) solution.

red blood cell (erythrocyte) The type of blood cell that contains hemoglobin and no nucleus. During their formation in mammals, erythrocytes lose their nuclei. Those of other vertebrates retain the nuclei.

redia (re′de-ah) A larval, digenetic trematode produced by asexual reproduction within a miracidium, sporocyst, or mother redia.

refractory period (re-frak′ter-e) Time period following stimulation during which a neuron or muscle fiber cannot respond to a stimulus.

releaser gland A gland in turbellarians that secretes a chemical that dissolves the organism's attachment to a substrate.

Remipedia (re-mi-pe′de-ah) A class of crustaceans whose members possess about 30 body segments and uniform, biramous appendages. This class contains a single species of cave-dwelling crustaceans from the Bahamas.

renette (re′net) An excretory structure in some worms.

renewable resources Resources that can replenish themselves when used within natural ecosystems or by humans. Contrasts with **nonrenewable resources.**

repolarization (re-po′lar-i-za′shun) The reestablishment of polarity, especially the return of plasma membrane potential to resting potential after depolarization.

reproductive isolation Occurs when individuals are prevented from mating, even though they may occupy overlapping ranges.

Reptilia (rep-til′e-ah) The class of vertebrates whose members have dry skin with epidermal scales and amniotic eggs that develop in terrestrial environments. Snakes, lizards, and alligators.

respiratory pigment Organic compounds that have either metallic copper or iron with which oxygen combines.

respiratory tree A pair of tubules attached to the rectum of a sea cucumber that branch through the body cavity and function in gas exchange.

resting membrane potential The potential difference that results from the separation of charges along the plasma membrane of a neuron or other excitable cell. The steady voltage difference across the plasma membrane.

rete mirabile (re′te mă-rab′ă-le) A network of small blood vessels arranged so that the incoming blood runs countercurrent to the outgoing blood, and thus makes possible efficient exchange of heat or gases between the two bloodstreams. An extensive countercurrent arrangement of arterial and venous capillaries.

rhabdite (rab′d ī t) A rodlike structure in the cells of the epidermis or underlying parenchyma in certain turbellarians that is discharged in mucous secretions, possibly in response to attempted predation or desiccation.

rhodopsin (ro-dop′sin) Light-sensitive substance in the rods of the retina. Visual purple.

rhopalium (ro-pal′e-um) A sensory structure at the margin of the scyphozoan medusa. It consists of a statocyst and a photoreceptor (phylum Cnidaria).

rhynchocoel (ring′ko-s ēl) In nemerteans, the fluid-filled coelomic cavity that contains the inverted proboscis.

ribonucleic acid (RNA) (ri″bo-nu-kle′ik) A single-stranded polymer of ribonucleotides. RNA forms from DNA in the nucleus and carries a code for proteins to ribosomes, where proteins are synthesized. Contains the nitrogenous bases adenine (A), cytosine (C), guanine (G), and uracil (U).

ribosomal RNA (rRNA) (ri″bo-s ōm′al) A form of ribonucleic acid that makes up a portion of ribosomes.

ribosome (ri′-bo-s ōm) Cytoplasmic organelle that consists of protein and RNA, and functions in protein synthesis.

rod cell A type of light receptor that is sensitive to dim light.

Rotifera (ro-tif′er-ah) The phylum of lophotrochozoans that has members with a ciliated corona surrounding a mouth. Muscular pharynx (mastax) present with jawlike features. Nonchitinous cuticle. Parthenogenesis common. Both freshwater and marine species.

rumen (roo′men) The first stomach of a ruminant, or cud-chewing, animal.

S

saliva (sah-li′vah) The enzyme-containing secretion of the salivary glands.

saltatory conduction (sal′tah-tor-e kon-duk′shun) A type of nerve impulse conduction in which the impulse seems to jump from one neurofibril node to the next.

sarcolemma (sar″ko-lem′ah) The plasma membrane of a muscle fiber.

sarcomere (sar′ko-mer) The contractile unit of a myofibril. The repeating units, delimited by the Z bands along the length of the myofibril.

Sarcopterygii (sar-kop-te-rij′e-i) The class of bony fishes characterized by paired fins with muscular lobes, pneumatic sacs that function as lungs, and atria and ventricles that are at least partly divided. The lungfishes and coelacanths (lobe-finned fishes).

scale A thin, compacted, flaky fragment. One of the thin, flat, horny plates forming the covering of certain animals (snakes, fishes, and lizards).

scalid (sca′lid) A set of complex spines found on the kinorhynchs, loriciferans, priapulans, and larval nematomorphs, with sensory, locomotor, food capture, or penetrant function.

Scaphopoda (ska-fop′o-dah) A class of molluscs whose members have a tubular shell that is open at both ends. Possess tentacles but no head. Example: *Dentalium.*

scent gland A gland around the feet, face, or anus of many mammals. Secretes pheromones, which may be involved with defense, species and sex recognition, and territorial behavior. Musk gland.

Schwann cell *See* **neurolemmocyte.**

scolex (sko′leks) The attachment or holdfast organ of a tapeworm, generally considered the anterior end. It is used to adhere to the host.

scrotum (skro′tum) A pouch of skin that encloses the testes.

scyphistoma (si-fis′to-mah) The polyp stage of a scyphozoan (phylum Cnidaria). Develops from a planula and produces ephyrae by budding.

Scyphozoa (si″fo-zo′ah) A class of cnidarians whose members have prominent medusae. Gametes are gastrodermal in origin and are released to the gastrovascular cavity. Nematocysts are present in the gastrodermis. Polyps are small. Example: *Aurelia.*

sebaceous (oil) gland (se-ba′shus) Gland of the skin that secretes sebum.

secondary palate (pal′et) A plate of bone that separates the nasal and oral cavities of mammals and some reptiles.

Sedentaria (sed-en-ter′iah) Marine, freshwater, and terrestrial annelids; parapodia with reduced lobes or parapodia lacking; setae associated with the stiff body wall to facilitate anchoring in tubes and burrows; palps reduced. Includes the clade Clitellata, echiurans, and siboglinids. *Arenicola, Riftia, Lumbricus,* and *Hirudo.*

segmentation (seg″men-ta′shun) 1. In many animal species, a series of body units that may be externally similar to, or quite different from, one another. 2. The oscillating back-and-forth movement in the small intestine that mixes food with digestive secretions and increases the efficiency of absorption.

Seisonidea (sy″son-id′e-ah) A class of rotifers with members that are commensals of crustaceans. Large and elongate body with rounded corona. Example: *Seison.*

selection pressure The tendency for natural selection to occur. Natural selection occurs whenever some genotypes are more fit than other genotypes.

selective permeability (s ĭ-lek′tiv per″me-ah-bil′ĭ-te) The ability of the plasma membrane to let some substances in and keep others out.

semen (se′men) The thick, whitish secretion of the reproductive organs in the male; composed of sperm and secretions from the prostate, seminal vesicles, and various other glands and ducts. Fluid containing sperm.

seminal receptacle (sem′ĭ-nal r ĭ-sep′tah-kl) A structure in the female reproductive system that stores sperm received during copulation (e.g., in many insects and annelids).

seminal vesicle (sem′ĭ-nal ves′ĭ-kl) 1. One of the paired accessory glands of the reproductive tract of male mammals. It secretes the fluid medium for sperm ejaculation (phylum Chordata). 2. A structure associated with the male reproductive tract that stores sperm prior to its release (e.g., in earthworms—phylum Annelida).

seminiferous tubule (se″m ĭ-nif′er-us tu′b ūl) The male duct that produces sperm. Highly convoluted tube in testis.

sensilla (sen-sil′ah) Modifications of the arthropod exoskeleton that, along with nerve cells, form sensory receptors.

sensory (afferent) neuron or **nerve** A neuron or nerve that conducts an impulse from a receptor organ to the central nervous system. *See also* **sensory receptors.**

sequential hermaphroditism (s ĭ-kwen′shal her-maf′ro-di-tizm) The type of hermaphroditism that occurs when an animal is one sex during one phase of its life cycle and an opposite sex during another phase.

sensory receptors Structures modified to detect physical stimuli in internal or external environments. They contain, or are themselves, sensory (afferent) neurons.

seral stage (ser′al) A successional stage in an ecosystem.

sere *See* **seral stage**.

serially homologous (ser′e-al-e ho-mol′o-ges) Metameric structures that have evolved from a common form. The biramous appendages of crustaceans are serially homologous.

serum (se′rum) The fluid portion of coagulated blood. The protein component of plasma.

set point The value of an internal parameter (e.g., blood glucose) that is maintained by an animal's homeostatic mechanisms.

setae (se′tah) 1. Hairlike modifications of an arthropod's exoskeleton that may be set into a membranous socket. Displacement of a seta initiates a nerve impulse in an associated cell. 2. Hairlike bristles found on parapodia of annelids and on the body wall of oligochaete annelids. Also called **chaetae.**

sex chromosome A chromosome that carries genes determining the genetic sex of an individual.

sexual reproduction The generation of a new cell or organism by the fusion of two haploid cells so that genes are inherited from each parent.

sexual selection Variation in reproductive success that occurs when individuals have varying success in obtaining mates. It often results in the evolution of structures used in combat between males for mates, such as antlers and horns, or ornamentation that attracts individuals of the opposite sex, such as the brightly colored tail feathers of a peacock.

shivering thermogenesis (ther″mo-jen′e-sis) The generation of heat by shivering, especially within the animal body.

Siboglinidae (si′bo-glin″ida) A group of annelids that is distributed throughout the world's oceans. Live in secreted, chitinous tubes in cold water at depths exceeding 100 m. Lack a mouth and digestive tract. Nutrients absorbed across the body wall and from endosymbiotic bacteria that they harbor. Beard worms. Formerly the phylum Pogonophora.

Simphyla (sim-fi′lah) A class of arthropods whose members are characterized by having long antennae, 10 to 12 pairs of legs, and centipede-like bodies. Occupy soil and leaf mold.

simple diffusion (dĭ-fu′zhun) The process of molecules spreading out randomly from where they are more concentrated to where they are less concentrated until they are evenly distributed.

sinus venosus (si′nes vĕ-no′sus) The common venous receptacle in the embryonic heart, attached to the wall of the primitive atrium; it receives the umbilical and vitelline veins and the common cardinal veins.

siphon (si′fon) A tubular structure through which fluid flows. Siphons of some molluscs allow water to enter and leave the mantle cavity.

Sipuncula (si-pun′kyu-lah) A group of protostomate worms whose members burrow in soft marine substrates throughout the world's oceans. Range in length from 2 mm to 75 cm. Peanut worms. Likely basal annelids.

sister chromatid (kro′mah-tid) One of the two identical parts of a duplicated chromosome in a eukaryotic cell. Sister chromatids are exact copies of a long, coiled DNA molecule with associated proteins. Sister chromatids join at the centromere of a duplicated chromosome.

sister groups Two groups that share a most recent common ancestor in a phylogeny.

skeletal muscle Muscle consisting of cylindrical multinucleate cells with obvious striations; the muscle(s) attached to the animal's skeleton; voluntary muscle.

skin The outer integument or covering of an animal body, consisting of the dermis and the epidermis and resting on the subcutaneous tissues.

small intestine The part of the digestive system consisting of the duodenum, jejunum, and ileum.

smooth muscle Type of muscle tissue in the walls of hollow organs. Visceral muscle.

sodium-potassium ATPase pump The active transport mechanism that concentrates sodium ions on the outside of a plasma membrane and potassium ions on the inside of the membrane.

soma (so-mă) Portion of a nerve cell that includes a cytoplasmic mass and a nucleolus, and from which the nerve fibers extend. Also called *cell body*.

somatic cell (so-mat′ik) Ordinary body cell. Pertaining to or characteristic of a body cell. Any cell other than a germ cell or germ-cell precursor.

somatic (voluntary) nervous system The part of the nervous system that relays commands to skeletal muscles.

sonar (so′nar) or **biosonar** A system that uses sound at sonic or ultrasonic frequencies to detect and locate objects.

speciation (spe″she-a′shun) The process by which two or more species form from a single ancestral stock.

species A group of populations in which genes are actually, or potentially, exchanged through multiple generations. Numerous problems with this definition make it difficult to apply in all circumstances.

species diversity *See* **community diversity**.

spermatogenesis (sper″mah-to-jen′ĕ-sis) The production of sperm cells.

spermatophores (sper-mat′ah-fors) Encapsulated sperm that a male can deposit on a substrate for a female to pick up or that a male can transfer directly to a female.

sphincter (sfingk′ter) A ringlike band of muscle fibers that constricts a passage or closes a natural orifice. Acts as a valve.

spicules (spik′ūlz) Skeletal elements that some mesenchyme cells of a sponge body wall secrete. May be made of calcium carbonate or silica.

spiracle (spi′rah-kel) An opening for ventilation. The opening(s) of the tracheal system of an arthropod or an opening posterior to the eye of a shark, skate, or ray.

spongin (spun′jin) A fibrous protein that makes up the supportive framework of some sponges.

sporocyst (spor′o-sist) An asexual stage of development in some digenean trematodes that arises from a miracidium and gives rise to rediae.

stabilizing selection Natural selection that results in the decline of both extremes in a phenotypic range. Results in a narrowing of the phenotypic range.

statocyst (stat′o-sist) An organ of equilibrium and balance in many invertebrates. Statocysts usually consist of a fluid-filled cavity containing sensory hairs and a mineral mass called a statolith. The statolith stimulates the sensory hairs, which helps orient the animal to the pull of gravity.

statolith *See* **statocyst**.

stimulus (stim′u-lus) Any form of energy an animal can detect with its receptors.

strobila (stro-bi′lah) The chain of proglottids constituting the bulk of the body of adult tapeworms.

substrates The reactants in enzyme-mediated reactions. *See* **enzymes**.

succession The sequence of community types during the maturation of an ecosystem.

sudoriferous gland (su″do-rif′er-us) A sweat gland.

superior Above a point of reference (e.g., the neck is superior to the chest of humans).

suspension feeder The type of feeding whereby an animal obtains its nutrients by removing suspended food particles from the surrounding water by some sort of capturing, trapping, or filtering structure. Examples include tube worms and barnacles. Also, **suspension feeding.**

swim bladder A gas-filled sac, usually along the dorsal body wall of bony fishes. It is an outgrowth of the digestive tract and regulates the buoyancy of a fish.

sycon (si′kon) A sponge body form characterized by choanocytes lining radial canals.

symbiosis (sim′bi-os′es) The living together of two different species in an intimate relationship. In this relationship, the symbiont always benefits; the host may benefit, may be unaffected, or may be harmed, as in mutualism, commensalism, and parasitism, respectively.

symmetry (sim′ĭ-tre) A balanced arrangement of similar parts on either side of a common point or axis.

sympathetic division (sim″pah-thet′ik) Portion of the autonomic nervous system that arises from the thoracic and lumbar regions of the spinal cord. Also called *thoracolumbar division.*

sympatric speciation (sim-pat′rik spe″she-a′shun) Speciation that occurs in populations that have overlapping ranges.

symplesiomorphies (sim-ples″e-o-mor′f ēz) Taxonomic characters that are common to all members of a group of organisms. These characters indicate common ancestry but cannot be used to describe relationships within the group.

synapomorphies (sin-ap′o-mor″f ēz) Characters that have arisen within a group since it diverged from a common ancestor. Synapomorphies indicate degrees of relatedness within a group. Also called *shared, derived characters.*

synapse (sin′aps) The junction between the axon end of one neuron and the dendrite or cell body of another neuron or effector cell.

synapsis (s ĭ-nap′sis) The time in reduction division when the pairs of homologous chromosomes lie alongside each other in the first meiotic division.

synaptic cleft (s ĭ-nap′tik) The narrow space between the terminal ending of a neuron and the receptor site of the postsynaptic cell.

synchronous flight *See* **direct (synchronous) flight.**

synsacrum (sin-sak′rum) The fused posterior thoracic vertebrae, all lumbar and sacral vertebrae, and anterior caudal vertebrae of a bird. Helps maintain proper flight posture.

syrinx (sir′ingks) The vocal apparatus of a bird. Located near the point where the trachea divides into two bronchi.

systematics (sis-tem-at′iks) The study of the classification and phylogeny of organisms. *See also* **taxonomy.**

systemic circuit (sis-tem′ik) The portion of the circulatory system concerned with blood flow from the left ventricle of the heart to the entire body and back to the heart via the right atrium.

systole (sis′to-le) Phase of the cardiac cycle during which a heart chamber wall contracts. *See also* **systolic pressure.**

systolic pressure (sis-tol′ik) The portion of blood pressure measurement that represents the highest pressure reached during ventricular ejection. It is the first number shown in a blood-pressure reading.

T

tactile (touch) receptor (tak′t ĭl r ĭ-sep′ter) A sensory receptor in the skin that detects light pressure. Formerly called **Meissner's corpuscle.**

tagmatization (tag″mah-ti-za′shun) The specialization of body regions of a metameric animal for specific functions. The head of an arthropod is specialized for feeding and sensory functions, the thorax is specialized for locomotion, and the abdomen is specialized for visceral functions.

Tardigrada (tar-di-gra′dah) A phylum of animals whose members live in marine and freshwater sediments and in water films on terrestrial lichens and mosses. Possess four pairs of unsegmented legs and a proteinaceous cuticle. Water bears. Along with the Onycophora and the Arthropoda, Tardigrada comprises the clade Panarthropoda.

target cell The cell a specific hormone influences.

taste buds The receptor organs in the tongue that are stimulated and give rise to the sense of taste.

taxis (tak′sis) The movement of an organism in a particular direction in response to an environmental stimulus.

taxon (tak′son) A group of organisms that are genetically (evolutionarily) related.

taxonomy (tak″son′ah-me) The description of species and the grouping of organisms to reflect evolutionary relationships. *See also* **phylogenetic systematics, evolutionary systematics,** and **systematics.**

T cell A type of lymphocyte, derived from bone marrow stem cells, that matures into an immunologically competent cell under the influence of the thymus. T cells are involved in a variety of cell-mediated immune reactions. Also known as a *T lymphocyte.*

tegument (teg′u-ment) The external epithelial covering in cestodes, trematodes, and acanthocephalans. Once called a cuticle.

telencephalon (tel′en-cef″ah-lon) An embryological term designating the anterior region of the vertebrate brain. It is the cerebrum of the adult vertebrate brain.

telophase (tel′o-f āz) Stage in mitosis during which daughter cells become separate structures. The two sets of separated chromosomes decondense, and nuclear envelopes enclose them.

temporal lobe (tem′po-ral) The portion of the vertebrate cerebral hemispheres involved with hearing, equilibrium, emotion, and memory.

tendon (ten′den) A cord- or bandlike mass of white, fibrous connective tissue that connects a muscle to a bone or to another muscle.

Tentaculata (ten-tak′u-lah-tah) The class of ctenophorans with tentacles that may or may not be associated with sheaths into which tentacles can be retracted. Example: *Pleurobranchia.*

testis (tes′tis pl.**testes**) Primary reproductive organ of a male. A sperm-cell-producing organ.

testosterone (tes-tos′t ĕ-r ōn) Male sex hormone that the interstitial cells of the testes secrete.

tetrad (tet′rad) A pair of homologous chromosomes during synapsis (prophase I of meiosis). A tetrad consists of four chromatids.

tetrapod (t ĕ′trah-pod) A taxonomic designation that refers to extant amphibians, nonavian and avian reptiles, and mammals and their closest common ancestor.

thalamus (thal′ah-mus) An oval mass of gray matter within the diencephalon that is a sensory relay area.

Thaliacea (tal″e-as′e-ah) A class of urochordates whose members are planktonic. Adults are tailless and barrel shaped. Oral and atrial openings are at opposite ends of the tunicate. Water currents are produced by muscular contractions of the body wall and result in a weak form of jet propulsion.

Thecostracea (the′ko-stra′se-ah) The subclass of maxillopod crustaceans that includes the barnacles.

thecodont (the′ko-dont) The tooth condition in which teeth are set into deep sockets.

thermoconformer (ther″mo-con-form′er) To conform to the temperature of the external environment.

thermogenesis (ther″mo-jen′ah-sis) The metabolic generation of heat by muscle contraction or brown fat metabolism.

thermoreceptor (ther″mo-re-sep′tor) A sensory receptor that is sensitive to changes in temperature. A heat receptor.

thermoregulation (ther″mo-reg″u-la′shun) Heat regulation. Involves the nervous, endocrine, respiratory, and circulatory systems in higher animals.

threatened species A species that is likely to become endangered in the near future. *See also* **endangered species.**

thrombocyte *See* **platelet.**

thymus gland (thi′mus) A ductless mass of flattened lymphoid tissue behind the top of the sternum. It forms antibodies in the newborn and is involved in the development of the immune system.

thyroid gland (thi′roid) An endocrine gland in the neck of vertebrates and involved with the metabolic functions of the body. Produces thyroxins.

tissue (tish′u) A group of similar cells bound by an extracellular matrix that performs a specialized function.

tolerance range The range of variation in an environmental parameter that is compatible with the life of an organism.

tonicity (to-nis′ ĭ-te) The state of tissue tone or tension. In body fluid physiology, the effective osmotic pressure equivalent. Tonicity refers to the relative concentration of solutes in the water inside and outside a cell.

tornaria (tor-nar′e-ah) The ciliated larval stage of an acorn worm (class Enteropneusta, phylum Hemichordata).

torsion (tor′shun) A developmental twisting of the visceral mass of a gastropod mollusc that results in an anterior opening of the mantle cavity and a twisting of nerve cords and the digestive tract.

tracheae (tra′che-e) The small tubes that carry air from spiracles through the body cavity of an arthropod. Arthropod tracheae are modifications of the exoskeleton.

tracheal system *See* **tracheae.**

tract A bundle of nerve fibers within the central nervous system.

transcription (tran-skrip′shun) The formation of an RNA molecule from DNA.

transducer (trans-du′ser) A receptor that converts one form of energy into another.

transfer RNA (tRNA) (trans′fer) A single-stranded polyribonucleotide that carries amino acids to a ribosome and positions those amino acids by matching the tRNA anticodon with the messenger RNA codon.

translation (trans-la′shun) The production of a protein based on the code in messenger RNA.

transverse (T) tubules Invaginations of the sarcolemma of a muscle cell that carry an action potential into the interior of the cell.

Trematoda (trem″ah-to′dah) The class of Platyhelminthes with members that are all parasitic. Several holdfast devices are present. Complicated life cycles involve both sexual and asexual reproduction.

trichinosis (trik″ĭ-no′sis) A disease resulting from infection by *Trichinella spiralis* (Nematoda) larvae from eating undercooked meat. Characterized by muscular pain, fever, edema, and other symptoms.

trichomoniasis (trik″o-mo-ni′ah-sis) A sexually transmitted disease caused by the parasitic protozoan *Trichomonas vaginalis.*

Trilobitamorpha (tri″lo-bit″a-mor′fah) The subphylum of arthropods whose members had bodies divided into three longitudinal lobes. Head, thorax, and abdomen were present. One pair of antennae and biramous appendages. Entirely extinct.

triploblastic (trip″lo-blas′tik) Animals whose body parts are organized into layers that are derived embryologically from three tissue layers: ectoderm, mesoderm, and endoderm. Platyhelminthes and all coelomate animals are triploblastic.

trochophore larva (trok′o-for lar′va) A larval stage characteristic of many molluscs, annelids, and some other protostomate animals.

trophic level (tr ōf′ik) The feeding level of an organism in an ecosystem. Green plants and other autotrophs function at producer trophic levels. Animals function at the consumer trophic levels.

trophozoites (tro′fah-zo″ītz) The active feeding stage of a protozoan.

tropic (tro′-pik) **hormones** Polypeptide hormones released by the anterior pituitary that target endocrine glands and cause them to release additional hormones.

tube feet Muscular projections from the water-vascular systems of echinoderms that are used in locomotion, gas exchange, feeding, and attachment.

tubular nerve cord A hollow nerve cord that runs middorsally along the back of chordates. One of five distinguishing chordate characteristics. Also called the *neural tube* and, in vertebrates, the *spinal cord.*

Turbellaria (tur″bel-lar′e-ah) The class of Platyhelminthes with members that are mostly free living and aquatic. External surface usually ciliated. Predaceous. Possess rhabdites. Protrusable proboscis. Mostly hermaphroditic. Examples: *Convoluta, Notoplana, Dugesia*. "Turbellaria" is a paraphyletic grouping that should be abandoned as a taxonomic designation.

tympanic (tympanal) organs (tim-pan′ik) Auditory receptors present on the abdomen or legs of some insects.

U

ultimobranchial gland (ul″ti-mo-bronk′e-el) In jawed fishes, primitive tetrapods, and birds, the small gland(s) that forms ventral to the esophagus. Produces the hormone calcitonin that helps regulate calcium concentrations.

umbilical cord (um-bil′ĭ-kal) Cordlike structure that connects the fetus to the placenta.

umbo (um′bo) The rounded prominence at the anterior margin of the hinge of a bivalve (Mollusca) shell. It is the oldest part of the shell.

under hair Fine, short hair of a mammal's pelage. Provides insulation for a mammal.

uniformitarianism (u″nah-for″m ĭ-tar′e-an-ism) The idea that forces of wind, rain, rivers, volcanoes, and geological uplift shape the earth today, just as they have in the past.

ureter (u-re′ter) The tube that conveys urine from the kidney to the bladder.

urethra (u-re′thrah) The tube that conveys urine from the bladder to the exterior of the body.

urinary bladder (u′r ĭ-ner″e blad′er) The storage organ for urine.

Urochordata (u″ro-kor-da′tah) The subphylum of chordates whose members have all four chordate characteristics as larvae. Adults are sessile or planktonic and enclosed in a tunic that usually contains cellulose. Sea squirts or tunicates.

urophysis (u″ro-fi′sis) A discrete structure in the spinal cord of the fish tail that produces neuropeptides (hormones) that help control water and ion balance, blood pressure, and smooth muscle contractions.

uterine tube (u′ter-in) The tube that leads from the ovary to the uterus and transports the ovum or egg(s).

uterus (u′ter-us) The hollow, muscular organ in female mammals in which the fertilized ovum normally becomes embedded and in which the developing embryo/fetus is nourished.

V

vacuole (vak′u- ōl) Any small membrane-bound space or cavity formed in the cytoplasm of a cell. Functions in either food storage or water expulsion.

vagina (vah-ji′nah) Tubular organ that leads from the uterus to the vestibule of the female reproductive tract. Receptacle for the penis during copulation.

valves 1. Devices that permit a one-way flow of fluids through a vessel or chamber. 2. The halves of a bivalve (Mollusca) shell.

vault Cytoplasmic ribonucleoproteins shaped like octagonal barrels. Aid in the transport of materials, including RNA, between the nucleus and cytoplasm.

vein A vessel that carries blood toward the heart from the various organs.

veliger larva (vel′ĭ-jer lar′va) The second free-swimming larval stage of many molluscs. Develops from the trochophore and forms rudiments of the shell, visceral mass, and head-foot before settling to the substrate and undergoing metamorphosis.

ventral The belly of an animal. Usually the lower surface. Synonymous with *anterior* for animals that walk upright.

venule (ven′y ūl) A small blood vessel that collects blood from a capillary bed and joins a vein.

Vertebrata (ver″te-bra′tah) The infraphylum of chordates whose members are characterized by cartilaginous or bony vertebrae surrounding a nerve cord. The skeleton is modified anteriorly into a skull for protection of the brain.

vertical gene transfer The passing of genes between generations in members of a population, for example, through sexual or asexual reproduction.

vessicle (ves′i-kel) Small membrane-bound structures used in transporting materials within cells. *See* **endomembrane system**.

vestigial structures (ve-stij'e-al) Structures that have no apparent function in modern animals but clearly evolved from functioning structures in ancestors. Provide a source of evidence of macroevolutionary change.

vibrissa (vi-bris'ah) A long, coarse hair, such as those occurring about the nose (muzzle) of a dog or cat. A sensor for mechanical stimuli.

villus (vil'us pl.villi) Tiny, fingerlike projection that extends outward from the inner lining of the small intestine and increases the surface area for absorption.

visceral arches *See* **gill arches**.

visceral mass (vis'er-al) The region of a mollusc's body that contains visceral organs.

vitamin An organic substance other than a carbohydrate, lipid, or protein that is needed for normal metabolism but that the body cannot synthesize in adequate amounts.

viviparous (vi-vip'er-us) Organisms with eggs that develop within the female reproductive tract and are nourished by the female.

vomeronasal organ *See* **Jacobson's (vomeronasal) organ.**

vulva (vul'vah) The external genital organs in the female.

W

water-vascular system (wah'ter vas'ku-ler) A series of water-filled canals and muscular tube feet in echinoderms. Provides the basis for locomotion, food gathering, and attachment.

Weberian apparatus (Weberian ossicles) (web-ber'e-an) Specialized ossicles present in ostariophysian fishes used to transmit sound previously amplified in a resonating chamber (the swim bladder) to the inner ear.

white blood cell (leukocyte) A type of blood cell involved with body defenses.

winter sleep A period of inactivity in which a mammal's body temperature remains near normal and the mammal is easily aroused.

X

Xenoturbellida (zen'ah-tur-bel"i-dah) Phylum of animals with uncertain phylogenetic affinities whose members live in muddy sediment of the North Sea. Larvae are parasites of molluscs. Adults feed on bivalves and bivalve eggs.

zoarium (zo-ar'ium) The colony of an ectoproct.

zoecium (zo-e'she-em) The cuticular sheath, secreted exoskeleton, or shell of the Ectoprocta.

zonite (zo'n ĭt) The individual body unit of a member of the phylum Kinorhyncha.

zooid (zo'oid) An individual member of a colony of animals, such as colonial cnidarians and ectoprocts, produced by incomplete budding or fission.

zoology (zo-ol'o-je) The study of animals.

zygote (zi'g ōt) Diploid cell produced by the fusion of an egg and sperm. Fertilized egg cell.

INDEX

A

abiotic environment
- energy, 88
- other abiotic factors, 89
- temperature, 88–89

acanthella, 175
Acanthocephala/acanthocephalans, 174–176
acanthor, 175
Acarina, 251
acetabulum, 165
acetylcholine, 429, 439
Acoelmorpha, 155
acoelomate, 118
acorn worms, 297–298
actin, 429
Actinopterygii, 322–324
action potentials
- neuron communication, 438–439
- recording, 438
- transmission of, 439

active transport, 17
adaptation, 59–60
adaptive mutation. *See* mutation(s)
adaptive radiation, 57
- of Galápagos finches, 58

adenosine diphosphate (ADP), 528
adenosine triphosphate (ATP), 13, 18, 84, 502
adhesive cells, 154
adhesive glands, 161
adrenal cortex, 473, 477
adrenal gland, 470, 477
adrenal medulla, 473, 477
adrenocorticotropic hormone (ACTH), 473, 476, 557
aerobic cellular respiration, 502
aestivation, 89, 528
afferent neuron, 435
afterbirth, 560
age structure, of population, 99
agnatha, 314
air-conduction portion, 496
alimentary tract, 507, 509–510
Alitta, 214
alleles, 45, 72
allopatric speciation, 81
all-or-none law, 438
alpha (α) cells, 464, 477
altricial chicks, 388
alula, 382
alveolar ducts, 496
alveoli, 495
Ambulacraria
- Echinodermata (*See* Echinodermata/echinoderms)
- Hemichordata, 295–299

Ambystoma maculatum, 340
ameboid movement, mechanism of, 428
American hookworm, 232
Amia calva, 323
amictic, 174
amino acids. *See* DNA; proteins
Amniota/amniotes
- able to retain water, 536–537
- of circulatory systems, 486–487
- cladistic interpretation of, 355–356
- endocrine tissues, 473–474
- evolution, 356–357
- excrete concentrated urine, 536–537
- mammals, 548
- nonavian and avian reptiles, 548
- sexual reproduction, in vertebrates, 547–548
- skull structure, 356–357

Amoeba, 13
Amphibia/amphibians
- adapt to two environments, 535
- amniote lineage, 352
- Anura, 340–341
- body systems, 341
- Caudata, 339–340
- circulatory system, 345, 486
- excretion and osmoregulation, 347
- external structure and locomotion, 341–342
- gas exchange, 345–346
- Gymnophiona, 338–339
- heart of, 486
- locomotion, 341–342
- metamorphosis, 350
- nervous system, 346–347
- nutrition and digestive system, 344, 345
- parental care, 349
- in Peril, 350–352
- phylogenetic relationships, 336–338
- reproduction, 347–350
- sensory functions, 346–347
- sexual reproduction, in vertebrates, 546–547
- of skin, 420
- support and movement, 342–344
- temperature regulation, 346
- ventilation in, 495
- vocalization, 349
- water and ion uptake in, 536

Amphisbaenia, 361
amplified fragment length polymorphisms (AFLP), 108
ampullae of lorenzini, 452
ampullary organs, 452
anabolic pathways, 18
anaphase, 36
ancestral character states, 109
androgens, 473, 551
Anguimorpha, 361
Animalia, 123
animal origins
- animal radiation and Cambrian explosion, 128–129
- earth's beginning and evidence of early life, 121–122
- endosymbiosis and eukarya, 125–126
- life's beginning and first 3 billion years, 122–126
- life's origins and early evolution, 122
- and multicellularity, 126–129
- phylogenetic highlights of, 129–132
- protist/animal crossroads, 126–128
- three domains, 122–125

animal radiation, 128–129
animal's diet, 502
animal strategies, for food
- continuous *vs.* discontinuous feeders, 504
- deposit feeders, 504
- fluid feeders, 506
- herbivory, 504–505
- predation, 505–506
- surface nutrient absorption, 506
- suspension feeders, 504

animal systematics
- cladistics/evolutionary systematics, 112–115
- evolutionary systematics, 111–112
- phylogenetic systematics/cladistics, 108–111

animal taxonomy, 130
Annelida/annelids. *See also* segmented worms
- basal annelid groups, 221–222
- body cavity of, 208
- characteristics of, 206
- classification of, 208
- development of metameric, coelomic spaces, 209
- evolutionary perspective, 206–209
- excretion, 212–213
- external structure, 210

Annelida/annelids–*Cont.*
feeding and digestive system, 210–211
gas exchange and circulation, 211
locomotion, 210, 211
Lophotrochozoans, 223–224
metamerism and tagmatization, 208–209
nephridia, 213
nervous and sensory functions, 211–212
parapodia, 210
phylogenetic considerations, 222–224
phylogeny, 222
Polychaeta and Clitellata, 208
regeneration, 213–214
relationship of, 207
reproduction, and development, 213–214
structure and function, 209–214
Anomalocaris, 129
Anostraca, 265
antennal glands, 531, 532
Anthozoa/anthozoans, 147–151. *See also* sea anemones
Anthracosaurs, 353
anthropogenic greenhouse gas emission, 7, 9
anticodon, 40
antidiuretic hormone (ADH), 473, 474, 540
ants, 279
Anura/anurans, 340–341. *See also* frogs; toads
vocalization, 349
water conservation, 348
Apoikozoa, 127
apomorphies, 109
aposematic coloration, 93
appendicular skeleton, 425
Arachnida/arachnids
Acarina, 251
Araneae, 248–251
coxal glands in, 532
form and function, 246–247
Opiliones, 251
Scorpionida, 247–248
arachnoid, 443
aragonite, 101
Araneae, 248–251. *See also* spiders
Archaea, 13, 123
Archaean eon, 121
Archaeognathans, 282
Archaeopteryx, 375
archeocytes, 137, 424
archosaurs, 359–360
of circulatory systems, 487
Aristotle's lantern, 293
arrector pili muscle, 397
arteries, 485
arterioles, 485
Arthropoda/arthropods
characteristics of, 239–240
Chelicerata, 244–252
classification of, 240–241
Ediacaran fauna, 280
exoskeleton, 241–243
hemocoel, 242–243
metamerism and tagmatization, 241
metamorphosis, 243–244
Myriapoda, 253–254
Trilobitomorpha, 244
artificial selection, 58
Ascaris lumbricoides, 231
ascon body form, 137
asexual reproduction, 542, 543
colony, 544
gemmulation, 544
gemmules, 544
in invertebrates
advantages and disadvantages of, 544
budding, 543–544
fission, 543
fragmentation, 544
parthenogenesis, 544
in Turbellaria, 164
Asterias, 287
Asteroidea
maintenance functions, 288–289
regeneration, reproduction, and development, 289
sea daisies, 289–290
asters, 35
asymmetry, 116
ATPase, 528
atrial natriuretic hormone (ANP), 478
atrial natriuretic peptide, 473
atrioventricular (AV) valves, 488
audition. *See* ears; hearing
Aurelia, 147
life history, 148
auricles, 163
Australian marsupial, 506
Australopithecus afarensis, 414
autapomorphies, 110
autocrine agents, 463
autonomic nervous system, 446, 447
autonomic (visceral or involuntary) nervous system, 442
autosomes, 32–33
autotrophic, 88
autotrophs, 501
Aves/avians. *See* birds
axial skeleton, 425
axon, 437

B

bacteria, 13, 122
baculum, 550
balance. *See* equilibrium
balanced polymorphism, 79–80
bare sensory nerve endings, 455
barn owl, 386
baroreceptors, 448
basal animal
Cnidaria, 141–151
Ctenophora, 151–154
evolutionary perspective, 134–136
phylogenetic considerations, 154–156
Porifera, 136–141
basal phyla, 129
basement membrane, 28, 418
basket stars, 290
basophils, 484
B cells, 484
bees, 279
beetles, 277–278
beta (β) cells, 464, 477
bicuspid valve, 488
bilateral symmetry, 117
bile, 512, 519
bimodal breathing, 492
biodiversity, 100
biogeochemical cycles, 98
biogeography, 62, 63
biological species concept, 80
biomagnification, 97
biomass, 95
biotic factors
interspecific interactions
coevolution, 91
herbivory and predation, 91
interspecific competition, 91
other interspecific adaptations, 93
symbiosis, 92–93
populations
intraspecific competition, 91
population density, 90–91
population growth, 89–90
population regulation, 90–91
biramous, 244
birds
ancient, 375–376
brood patch, 474
circulation, 384
of circulatory systems, 487
diversity of, 376–378
of endocrine gland, 472
excretion and osmoregulation, 387
external structure and locomotion, 376, 378
feathers, 376
flight, 381–382
gas-exchange, 384–385, 495, 497
locomotion, 376, 378
migration, 389
muscles, 381
navigation, 389

nervous and sensory systems, 386, 387
nesting behavior, 388–389
nonavian theropods, 374–375
nutrition and the digestive system, 383–384
phylogenetic relationships, 373–374
reproduction and development, 387–388
skeleton, 378–381
of skin, 421
thermoregulation, 385–386
Bivalvia/bivalves
circulation, 192
crystalline style, 192
digestion, 191–192
diversity, 194, 195
filter feeding, 191–192
gas exchange, 191–192
gastric shield, 192
Lamellibranch Gill of, 191
larval stages of, 193
other maintenance functions, 192–193
reproduction and development, 193–194
shell and associated structures, 190–191
suprabranchial chamber, 191
umbo, 190
valves, 190
water tubes, 191
bladder worm, 169
blastocoele, 118
blastocyst, 557
Blattaria, 269
blindsnakes, 361, 363
blood cells, 485. *See also* red blood cells; white blood cells
blood flows, 489
blood glucose
concentration of, 478
negative feedback control of, 464
blood pressure, 489. *See also* circulatory systems
blood, systemic circuit, 486
blood vessels, 485
bloodworm, 215
blubber, 526
blue sheep (*Pseudois nayaur*), 50
body fluid regulation. *See* temperature regulation
body heat, 523
body plans
diploblastic, 118
triploblastic, 119
body temperature. *See* temperature regulation; thermoregulation
body wall
Cnidaria, 141–142
of *Hydra*, 142
Porifera, 136–137
Turbellaria/turbellarian, 161
bolus, 515
bone, structure of, 426
bony fishes
Actinopterygii, 322–324
endocrine tissues (glands) in, 469
fossils of, 321
Sarcopterygii, 322
of skin, 420
book lungs, 491, 492
bottleneck effect, 75–76
box jellyfish, 147
box turtle (Wildlife Alert). *See* turtle(s)
Brachiopoda, 179
Brachiosaurus, 366
brain
forebrain, 444–446
hindbrain, 443–444
midbrain, 444
vertebrate nervous systems, 443–446
branchial hearts, 197
Branchiopoda, 265–267
breast milk, 559
breeding, for speed, 49
brine shrimp, 265
brittle stars, 290
broadcast spawning, 544
Broca's area, 446
bronchioles, 496
brood parasites, 81
brood patch, 471, 474
brook lampreys, 318
brown fat, 402, 528
brumation, 89
Bryozoa, 177–179
buccopharyngeal respiration, 492
budding, 543–544. *See also* reproduction
Buffon, Georges-Louis, 53
Burgess Shale, 128, 129
burrowing, worm. *See* locomotion
bursa of fabricius, 472
butterfly kisses, 278

C

caecilian, 339
calcitonin, 474
calcium carbonate skeleton, 150
calcium-induced changes, in troponin, 431
Cambarus aculabrum, 266
Cambrian explosion, 128–129. *See also* fossils/fossil records
camouflage, 93
capillaries, 485
carbon cycle, 98
cardiac cycle, 488
cardiac muscle, 432–433
cardiovascular system, 483. *See also* circulatory systems; heart
carnassial apparatus, 400
Carnivora, 398, 407
carnivores, 501
carrier proteins. *See* transport proteins
carrying capacity, 90
cartilage, 425
cartilaginous fishes, 535
of skin, 420
catabolic pathways, 18
cattle, 168
Caudata, 339–340
Caudipteryx, 375
Caudofoveata, 201, 202
cecum, 511
cell(s)
alternative pathways, 22–23
cytoskeleton and cellular movement, 27–28
energy metabolism, 18
limits on size, 12
membrane transport, 15–17
mitochondrion, 19–22
nuclear envelope, 23
nucleolus, 23
origin, 12–13
peroxisomes, 26
plasma membrane, 14–15
properties and varieties, 11–13
surface area *vs.* volume, 13
types, 13
vacuoles, 26
cell adhesion, 134
cell cycle
control, 36
interphase, hereditary material replication, 34–35
mitosis, 34
cell-to-cell signaling, 478
cellular membranes, 502
cellular movement
cilia and flagella, 27–28
intermediate filaments, 27
microtubules, 27
cellular respiration, 18
centipedes, 253, 254
centralization, 440
central nervous system (CNS), 442
in fishes, 327–328, 469
centrioles, 27
centromere, 35
centrosome, 27. *See also* microtubule-organizing center (MTOC)
cephalization, 117, 440
Cephalopoda
arms and tentacles, 198
digestion, 196–197

Cephalopoda—*Cont.*
 eye, 198
 feeding, 196–197
 learning, 199
 locomotion, 196
 other maintenance functions, 197–198
 reproduction and development, 199
 shells, 196
cercariae, 166
cerebellum, 443
cerebral cortex, 446
cerebrum, 446
 in different vertebrate species, 445
Cestoidea/cestodarians/cestodes, 168. *See also* tapeworms
 Eucestoda, 169
 tapeworm of humans, 169–170
chaetae, 210
Chaetognatha, 300
Chaetopteridae, 221
Chaetopterus, 221
chameleon, 365
Chelicerata
 Arachnida, 245–252
 Merostomata, 245
 Pycnogonida, 252–253
chemical messengers
 of basal and lophotrochozoan phyla, 467
 of ecdysozoans, 467–469
 of evolution and diversity, 462–463
 targets and transport, 463
chemiosmosis, 22, 490
chemoreceptors, 448
chief cells, 515
Chilopoda, 254. *See also* centipedes
Chimaera, 321
Chinlestegophis jenkinsi, 339
chitin, 424
choanocytes, 127
choanoderm, 137
choanoflagellates, 126, 127
Chondrichthyes, 320–321
chondrocytes, 425
chorionic villi, 556
choriovitelline (yolk sac) placenta, 409
chromaffin tissue, 470
chromatid, 34
chromatin, 32
chromatophores, 198, 342, 420
chromosomes, 72
 aneuploidy, 44
 eukaryotic (*See* eukaryotic chromosomes)
 replication, 34
 sex, 32–33
 structure, 44
chylomicrons, 515
chyme, 515
cichlids, 2
 scale-eating, 3
cilia, 27–28
ciliary creeping, 428
ciliary movement, 428
Cingulata, 397
circulation and gas exchange
 circulatory systems
  mammalian, 487–489
  vertebrate, 485–487
 gas exchange, 489–492
 lymphatic systems, 489
 transport systems
  in invertebrates, 481–483
  in vertebrates, 483–485
circulatory systems, 482, 483
 of amniotes, 486–487
 of amphibians, 345, 486
 of archosaurs, 487
 of birds, 487
 of fishes, 486
 in mammalian, 487–489
 of polychaete, 212
 of reptiles, 487
 in vertebrate, 485–487
citric acid cycle, 20, 502
clade, 110
cladistics
 evolutionary systematics, 112–115
 phylogenetic systematics, 108–111
Cladocera, 266–267
cladograms/cladistics/cladists, 109. *See also* phylogeny
 vertebrate phylogeny, 112
class mammalia, 393
claws, structure of, 398
climate change, 101
climax community, 95
Clitellata, 216–221
 external structure and locomotion, 217–218
 maintenance functions, 218–221
 oligochaetes, 216–217
 reproduction and development, 218–219, 221
clitellum, 217
Clonorchis sinensis, 167
closed circulatory system, 197, 483
cnida, 141
Cnidaria/Cnidarian
 alternation of generations, 142
 body wall, 141–142
 characteristics of, 141
 classification of, 141
 gastrovascular cavity, 142
 life cycle, 143
 maintenance functions, 142–143
 and nematocysts, 141–142
 nerve cells, 143
 reproduction, 143
 taxonomy, 156
cnidocil, 142
cnidocyte, 141
 structure, 142
cochlea, 405
cocoon, 164
codominance, 48
codon, 39
coelomate pattern, 119
coelomic fluid, 481
coenzymes, 503
coevolution, 91
coexistence, of competing species, 92
Coleoptera, 277–278
Collembola, 274
colloblasts, 152, 154
Colony Collapse Disorder (CCD), 277
colostrum, 560
comb jellies, 151
comb rows, 152
commensalism, 92
community, 94
community (species) diversity, 94
community stability, 94–95
community succession, 94
competitive exclusion principle, 91
complete digestive tract, 506–507
complex camera eyes, 450
compound eyes, 449
conduction, 522
cone cells, 459
connective tissues, 28
conodont, 317
conservation program, animals, 50
continuous feeders, 504
contractile vacuoles, 530
controlled hypothermia, 88
convection, 522
convergent evolution, 64, 109, 402
Copepoda/copepods, 170, 267–268
 crustaceans, 257, 258
coprophagia, 511
coracidia, 170
coral bleaching, 153
coral reefs ecosystem, 153–154
corpus callosum, 445
corpus luteum, 473, 551
corticotropin-releasing hormone (CRH), 473, 476
counteracting osmolyte strategy, 535
countercurrent exchange, 405, 525, 526, 539–540
 mechanism, 326
cow, stomach, 13. *See also* cattle
coxal glands, 532
cranial nerves, 446

Craniata, 314
crayfish, 261-264, 266
antennal gland of, 532
appendages, 261, 262
endocrine system of, 264
excretory organs of, 264
external structure of, 261, 262
internal structure of, 263
Crinoidea, 294-295
Crocodylia, 359-360
cross-bridges, 429
Crotalus adamanteus, 362-363
Crustacea/crustaceans. *See also* crayfish; gills
Branchiopoda, 265-267
classification of, 259
control of ecdysis (molting) in, 468
Copepods, 257, 258
Malacostraca, 261-265
Maxillopoda, 267-268
Ctenophora, 151-154
characteristics of, 152
classification of, 152
Cubozoa, 147
cutaneous respiration, 492
cuticles, 130, 419
cuttlebone, 196
cyanobacteria, 501
cyclic AMP (cAMP), 465
Cycliophora, 107, 177, 178
Cynotilapia afra, 2
Cyphontilapia fontosa, 2
cystacanth, 175
cysticercosis, 170
cysticercus, 169
cytokinesis, 34
cytoplasm, 12
cytoplasmic, 117-118
bridges, 165
cytoskeleton
cilia and flagella, 27-28
intermediate filaments, 27
microtubules, 27

D

dactylozooids, 145
daily torpor, 88, 528. *See also* torpor
Darwin, Charles Robert, 54
Darwin's finches, 82
Darwin's ideas of evolution
adaptation, 59-60
analogous, 64
fossil evidence, 56
Galápagos Islands, 56-57
geology, 55-56
homologous, 64
natural selection, 59
voyage of HMS *Beagle,* 54-55
Wallace, Alfred Russel, 60
Decapoda, 261-264
deciduous/milk teeth, 399
defensins, 458
degeneracy, 40
deiodinases, 470
delay birth, evolution of, 547
delayed fertilization, 407
delta (δ) cells, 477
demes, 81
Demodex folliculorum, 252
dendrites, 436
dense bodies, 433
density-dependent factors, 91
density-independent factors, 90
dental formula, 399
denticles, 419
dentine, 419
deoxyribonucleic acid (DNA), 13
depolarization, 438
deposit feeding, 504
derived character states, 109
Dermacentor andersoni, 252
dermal branchiae, 490
dermal pores or ostia, 137
dermis, 419
Dermochelys coriacea, 496
Deuterostomia/deuterostomes, 130, 284
phylogeny, 299
Diadectomorphs, 353
diadromous fishes, 330, 535
diastema, 392, 400
diastole, 489
diastolic pressure, 489
diencephalon, 446
digestion, 501, 503-504
Bivalvia/bivalves, 191-192
Cephalopoda, 196-197
extracellular, 503, 504
fat, 512
Gastropoda, 188
heterotrophy, 503-504
intracellular, 503, 504
Turbellaria/turbellarian, 162
digestive glandular systems, 511-512
digestive glands, 518
digestive structures
in invertebrates, 506-507
in vertebrates
alimentary tract, 509-510
digestive glandular systems, 511-512
esophagus, 509
evolutionary insights gut microbiomes, 510-511
oral cavity, 508-509
tongues, 508
digestive systems
in Annelida, 210-211
of human organs, 513
mammalian (*See* mammalian digestive system)
in Rotifera, 174
in Turbellarians, 163
dihybrid cross, 46
Dimetridon, 356
dinosaurs. *See* archosaurs
dioecious, 142
Diphyllobothrium latum, life cycle of, 171
diphyodont condition, 399
diploblastic, 118
diploid, 33
Diplopoda, 253-254. *See also* millipedes
Diplura/diplurans, 268
Diptera, 278-279
directional selection, 78
Dirofilaria immitis, 233
discontinuous feeders, 504
disruptive selection, 79
diversifying selection, 79
DNA, 108, 122, 502
antiparallel, 39
double helix model, 38-39
molecular genetics, 38
point mutation, 42-44
replication, 39, 41
sequencers, 67
structure, 40
transcription, 39, 40
translation, 39, 40
dominance, molecular basis, 49
dominant alleles, 45
Draco, 361
Drosophila melanogaster, 45
duodenum, 515
dura mater, 443

E

ears. *See also* hearing
anatomy of, 454
of anuran, 453
ossicles, evolution of, 66
earth's crust, 122
earth's oceans, 122
earth's resources and global inequality
biodiversity, 100
climate change and ocean acidification, 101
habitat loss, 100-101
invasive species, 104
nutrient load and pollution, 101-103
overexploitation of resources, 103-104

earthworm
locomotion, 217
metanephridium, 531
reproduction, 219
structure, 218
ecdysis, 130, 227, 424
Ecdysozoa/ecdysozoans, 130, 419
of chemical messengers, 467–469
interpretation of, 237
Kinorhyncha, 234–235
Loricifera, 236
morphological character, 226, 227
Nematoda/nematodes (*See* Nematoda/nematodes)
Nematomorphs, 234
Priapulida/priapulids, 235–236
Echinodermata/echinoderms
Asteroidea, 287–290
characteristics, 286–287
Crinoidea, 294–295
Echinoidea, 291–293
Holothuroidea, 293–294
integument, 419
Ophiuroidea, 290–291
Echinoidea, 291–293
Echiura, 216
echolocation, 456
ecological niche, 94
ecological problems
earth's resources and global inequality, 99–104
human population growth, 99
ecology, 87
animals and their abiotic environment, 87–89
biotic factors, 89–93
communities, 94–95
cycling within ecosystems, 97–99
ecological problems, 99–104
endangered species, 8
hierarchy of relatedness, 5
threatened species, 8
trophic structure, of ecosystems, 95–97
world population, 6
world resources, 7–9
ecosystems
coral reefs, 153–154
cycling within, 97–99
energy flow through, 97
primary production, 95
trophic structure of, 95–97
ectoderm, 118
ectodermal cells, 118
ectoparasites, 219
Ectoprocta, 177–179
ectothermic, 89
ectotherms, 368, 523
homeostasis and temperature regulation in, 526
Ediacaran
fauna, 280
fossil, 128
eggs, in vertebrate reproduction. *See* reproduction
Elasmobranchii, 320
electrical conduction system, 488
electrical synapses, 439
electric eel, 328
electric fishes, 328–329
electrocardiogram (ECG), 489
electrocommunication, 452
electrolytes
excrete, 525
temperature control in, 528–530
electron transport, 22, 126, 490
electroreception, 451–452
embryo, 546–547, 555
embryonic diapause, 407
endangered species, 8
end bulb, 439
endocrine gland, 466
of birds, 472
hormones, 478
of mammals, 472
endocrine system
of amniotes, 471–478
avian and nonavian reptiles, 471–472
and diversity, 462–463
endocrine glands, 466
evolution, 462–463, 478–479
feedback systems, 463–465
of fishes and amphibians, 469–471
hormone action mechanisms, 465–466
hormones are not produced by endocrine glands, 478
invertebrate endocrine control, 466–469
mammals, 472–478
endocrinology, 464
endocytosis, 26
endoderm, 118
endomembrane system
endocytosis, 26
ER, 24–25
exocytosis, 26
Golgi apparatus, 24–25
vesicles and cellular transport, 25–26
endoplasmic reticulum (ER), 24–25
endoskeletons, 426
endosome, 26
endosymbiont theory, 125
endosymbiosis, origins of, 125–126
endothelium, 485
endothermic, 88
endotherms, 523, 524
heat production in, 527–528
homeostasis and temperature regulation in, 526–528
energy
abiotic environment, 88
of activation, 18, 19
budget, 88
metabolism, 18
Enhydra lutris nereis, 408
enteric division, 446
enteric, nervous system, 442
Enterobius vermicularis, 231, 232
Enteropneusta, 295, 297–298
Entognatha, 268
environmental resistance, 90
enzymes, 18, 19
in mammals, 518
Eoalulavis, 376
eosinophils, 484
Ephemeroptera, 275
epicardium, 488
epicuticle, 424
epidermal glands, 418
epidermis, 141, 418, 419. *See also* skin
epididymis, 550
epigenetics, 54
epinephrine, 473
epithelial cells, 515
epithelial tissues, 28
epitheliomuscular cells, 142
Epitoky, 213
equilibrium
in air, 452–454
in water, 454
ER. *See* endoplasmic reticulum (ER)
Errantia, 210, 215
fireworms, 215
Glycera, 214–215
Nereis, 214
erythrocytes, 483, 485
erythropoietin (EPO), 473, 478
Escherichia coli, 13
esophagus, 509, 515
essential fatty acids, 502
estrogens, 473, 474
estrous cycles, 555
estrus cycle, 407
euchromatic regions, 32
Eukarya, 13, 123
origins of, 125–126
eukaryotic cell, 13
origin of, 125
eukaryotic chromosomes
chromosomes, 32
genetics, 31
number of, 33

organization, 32
sex chromosomes, 32–33
Euphausiacea, 264
evaporation, 522
evo-devo, 68
evolution
Darwin's ideas of (*See* Darwin's ideas of evolution)
of feathers, 381
hominin, 413
human, 414
tetrapod limbs, 332, 333
evolutionary change
biogeography evidence of, 62
paleontological evidence of, 64
evolutionary conservation, 108
evolutionary groups, 109
evolutionary mechanisms
balanced polymorphism, 79–80
gene flow, 76
heterozygote superiority, 79–80
mutation, 76–77
natural selection reexamined, 77–79
neutral theory and genetic drift, 74–76
evolutionary systematics, 111–112
evolution rates, 81–83
excretion/excretory systems, 528
amphibians, 347
birds, 387
fishes, 329
insects, 273, 274
in invertebrate
antennal, maxillary, and coxal glands, 531–532
contractile vacuoles, 530
malpighian tubules, 532
metanephridia, 530–531
protonephridia, 530
in vertebrate
evolution of vertebrate kidney, 534–537
metanephric kidney functions, 537–540
regulate ions and water, 533
excurrent canals, 137
exocrine glands, 466
exocytosis, 26
exoskeletons, 423, 424
arthropods, 241–243
exponential growth, 90
external fertilization, 544
external gills, 493
extracellular digestion, 503, 504
extracellular matrix, 134, 135
eyes
Cephalopoda, 198
complex camera, 450
compound, 449

F

facilitated diffusion, 17
fairy shrimp, 265
Fasciola hepatica, 165
fat digestion, 512
feathers. *See* birds
feather stars, 285, 286, 295
feedback control system, 464–465
feeding. *See* digestion
fermentation
autotrophs, 18
heterotrophs, 18
lactate, 19, 20
fertilization. *See also* asexual reproduction; reproduction
delayed, 407
external, 544
internal, 545
fetus, 559
fight-or-flight response, 477
filarial worms, 233–234
filter feeders, 284, 286, 324
filtration, 16
fireworms, 215
fish(es). *See also* bony fishes
bony, 321–324, 420
buoyancy regulation, 326–327
cartilaginous, 420
in central nervous system, 327–328, 469
Chondrichthyes, 320–321
circulation, 324–325, 486
electric fishes, 328–329
electroreception, 328
endoskeleton, 426
excretion, 329
gas exchange, 325–326
gnathostome, 319–320
heart of, 486
homeostasis and temperature regulation in, 526
jawless, 419
lampreys, 318–319
locomotion, 324
musculature, 429
Myxini, 317–318
osmoregulation, 329, 330
phylogenetic relationships, 314–317
reproduction and development, 330–331
sensory receptors, 327–328
sexual reproduction, in vertebrates, 546
swim bladders and lungs, 326
thermoregulation in large, active, 525
fission, 543
fixed-membrane-receptor mechanism, 465
flagella, 27–28
flame cells, 162, 163, 530
flatworms, 161. *See also* Platyhelminthes; Turbellaria/turbellarian
flea, 424
flies. *See* Diptera
flight. *See* birds
fluid feeding, 506
flukes, 165. *See also* Trematoda/trematodes
follicles, 421, 551
hair, 397
follicle-stimulating hormone (FSH), 473, 476, 551, 554, 555
food chain, 95
food grooves, 191
food storage organ, 507
food webs, 95, 96
foramen of Panizza, 487
forebrain, 444–446
foregut fermenters, 511
formed-element fraction, 483
fossil evidence, 56
fossils/fossil records, 64
birds, ancient, 375–376
caecilian, 339
Seymouria baylorensis, 353
Tiktaalik, 332
founder effect, 74
fountain zone, 427
fovea centralis, 458
fragmentation, 544
freshwater bivalves, 202–203
freshwater teleosts, 535, 536
frogs
hormonal control of, 470
saliva, 508
tadpole metamorphosis, 470
frontal lobe, 446
fungi, 123

G

Galápagos finches, 58
Galápagos Islands, 54
Galápagos tortoises, 57
gallbladder, 518–519
gamete, 36, 542
primary and secondary nondisjunction in, 44
ganglia, 442
gas-conduction portion, 496
gas-exchange
amphibians, 345–346
birds, 384–385
fishes, 325–326
insects, 270–271
invertebrate respiratory systems, 490–492
lungs *vs.* tissues, 498
mechanism in birds, 497
nonavian reptiles, 368

gas transport
  mammalian respiratory system, 497–499
  oxyhemoglobin dissociation curve, 499
gastric enzymes, 509
gastrodermal cells, 141
gastrodermis, 141
gastrointestinal motility, 513–514
gastrointestinal tract, mammalian, 513
Gastropoda
  body form, 189
  digestion, 188
  diversity, 189–190
  feeding, 188
  locomotion, 187
  other maintenance functions, 188–189
  reproduction and development, 189
  shells, 187
  structure, 188
  torsion, 186, 187
Gastrotricha, 171–172
gastrovascular cavity, 142, 481, 506
geckos, 361
gemmulation, 544
gemmules, 141, 544
gene(s), 13
  duplication, 84
  flow, 76, 77
  genetic code, 39–41
  pools, 72–73
  RNA, 39
  transcription, 40
  translation, 40
generalized animal cell, 12
genetics, 31
  diversity, preservation, 50
  drift, 74–76
  recombination, 37
genotype, 46
geological time, 60–61
georeceptors, 448
germ-line cells, 36
gestation, 546
  period, 409
giant sloth, 56
gills, 490
  rakers, 324
  of tunas, 494
glands, 466
  with and without Ducts, 466
global warming, 154, 292
glochidium, 193
glomerulus, 405, 538
glucagon, 473, 478
glucocorticoids, 473, 476, 477
glucose metabolism, 23
Glycera, 214–215
glycocalyx, 165
glycolysis, 18
  autotrophs, 18
  heterotrophs, 18
  lactate fermentation, 19, 20
Gnathostomata/gnathostomes, 317, 319–320
Gnathostomulida, 172–173
golden toad (Wildlife Alert), 351
gonadotropin-releasing hormone (GnRH), 473, 476, 551, 554, 555
gonadotropins, 551
gonads, 478
gonangium, 144
Gonionemus, 144
gonozooid, 144
Gordian worm, 234, 235
green glands, 531
growth hormone (GH), 473, 474
guanine triphosphate (GTP), 41
guard hair, 421
gular flutter, 526
gustation, 457. *See also* taste
gut microbiomes, 510–511
Gymnophiona, 338–339

H

habitat, 87
  loss, 100–101
Hadean eon, 121
hagfish, 318
hair, 421
  follicle, 397
Haplochromis pyrrhocephalus, 5
haploid, 33
Hardy-Weinberg theorem, 73
hearing. *See also* ears
  in air, 452–454
  in water, 454
heart
  of amphibians, 486
  branchial, 197
  electrical conduction system, in human, 488
  of fishes, 486
  lymph, 489
  mammalian circulatory system, 488–489
  structures of human, 488
heartworm disease, 233–234
heat exchange
  countercurrent, 525–527
  hemolymph flow, 525
  insulation, 527
heat gains, 522–523
heat losses, 522–523
heat production, in endotherms, 527–528
hectocotylus, 199
hemerythrin, 498
Hemichordata
  characteristics of, 295
  Enteropneusta, 295, 297
  Pterobranchia, 298–299
hemocyanin, 498
hemocytes, 483
hemoglobin, 498
  phylogenetic trees of, 68
hemolymph, 483
hepatocytes, 518
herbivory, 501, 504–505
  and predation, 91
hereditary information. *See* chromosomes; DNA
hermaphroditic earthworms, mating in, 545
hermaphroditism, 545
herring, 324
Heterocephalus glaber, 524
heterochromatic regions, 32
heterodont, 398, 399
heterotherms, 523
heterotrophy, 88, 501
  animal strategies, for food, 504–506
  digestion, 503–504
  diversity in digestive structures
    invertebrates, 506–507
    vertebrates, 507–512
  evolution of, 501–502
  mammalian digestive system, 512–519
  metabolic fates of nutrients, 502–503
heterozygote superiority, 79–80
Hexapoda
  classification of, 259–260
  Insecta/insects, 268–279
HGT. *See* horizontal gene transfer (HGT)
hibernation, 89, 404, 528
hierarchical nesting, 111
hindbrain, 443–444
hindgut fermenters, 511
Hirudinea
  annuli, 219
  leech, 219
  maintenance functions, 219–221
  reproduction and development, 221
Hirudo medicinalis, 220
histology, 28. *See also* tissue(s)
Holocephali, 321
Holothuroidea/holothuroieds. *See* sea cucumbers
Homarus americanus, 261
homeobox (Hox) genes, 68, 129
homeostasis, 464
  in endotherms, 526–528
  in fishes, 526
  heat gains and losses, 522–523
  in invertebrates, 525–526
  temperature fluctuations, 523–525
  in terrestrial ectotherms, 526

homeotherms, 523
homing pigeons, 389
hominin evolution, 413
hominins arose, 394
homologous chromosomes, 34, 36
homology, 66
Homo sapiens cave painting, 415
hookworm, 232
hooves, structure of, 398
horizontal gene transfer (HGT), 123, 125
hormonal control
    of ecdysis and development (metamorphosis), 468
    of female reproductive function, 554
    of frog skin color, 470
    of male reproductive function, 552
hormonal feedback, 476
hormonal regulation, in pregnant female, 555
hormone(s), 464. *See also* endocrine system
    biochemistry of, 464
    diversity of chemical messengers, 463
    endocrine system mechanisms of, 465–466
    feedback control system of, 464–465
    fixed-membrane-receptor mechanism, 465
    mobile-receptor mechanism, 465–466
    producer, 556
    prolactin, 475
horsehair worms, 234, 235
host, 92
human age pyramids, 100
human development, early stages in, 558
human embryonic development, 558
human evolution, 414
human eyeball, internal anatomy of, 458
human female reproductive system
    hormonal control of, 554
    hormonal regulation, in pregnant female, 555
    in invertebrates, 545
    major hormones in, 556
    production and transport, of sperm, 549–550
human male reproductive system, 534
    hormonal control of, 550–551
    in invertebrates, 545
    production and transport, of egg, 551–553
hyaline cartilage cells, 425
Hydra/Hydrozoa/hydrozoans, 143–145
    body wall of, 142
    freshwater hydrozoan, 145
    gastrovascular cavity, 481
    medusa, 145
hydraulic skeleton, 188, 423
hydrological cycle, 98
hydrophilic, 465
hydrostatic skeletons, 142, 422–423
hygroreceptors, 448–449
Hymenoptera, 279
hyomandibular, 66
Hyperotreti, 314, 315, 317–318
hyperpolarization, 438
hypertonic solution, 16
hypodermis, 419
hypophysis, 472–476. *See also* pituitary gland
hypothalamus, 446, 473, 475
hypotonic solution, 16

I

ibex *(Capra sibirica),* 50
Ichthyostega, 337
Iguania, 361
imperiled sea cucumbers (Wildlife Alerts), 296–297
Incilius periglenes, 351
incomplete digestive tract, 506
incomplete dominance, 48
incurrent canals, 137
independent assortment, principle of, 48
infraphylum vertebrata, 314, 318–324
infundibulum, 474
inheritance. *See* genetics
inheritance pattern, animal
    independent assortment, 46–48
    molecular basis, 49
    multiple alleles, 48
    other patterns, 48–49
    segregation, 45
inner ear, 454
    of bony fish, 455
Insecta/insects
    chemical regulation, 273–274
    circulation and temperature regulation, 271
    Coleoptera, 277–278
    digestive system, 270
    Diptera, 278–279
    effects on humans, 277
    excretion, 273, 274
    flight, 268–269
    gas exchange, 270–271
    Lepidoptera, 278
    locomotion, 269
    metamorphosis, 275–276
    nervous and sensory functions, 271–272
    nutrition, 269–270
    reproduction and development, 274–275
    social behavior of, 276–277
insect protenor, 33
insula, 446
insulin, 473, 478
integrating center, 464
integration, neuron structure, 436
integument, 418
integumentary exchange, 490
integumentary systems. *See also* skin
    invertebrate, 418–419
    vertebrate, 419–422
intercalated discs, 432
intermediate filaments, 27
intermediate host, 92
internal fertilization, 545
internal gills, 493
interneurons, 435
interpreting cladograms, 111
interspecific competition, 91, 94
interstitial cells, 145
interstitial cell-stimulating hormone (ICSH), 551
intestines
    generalized arrangement of, 510
    large, 515–517
    mammalian digestive system, 515
    small, 515
intracellular digestion, 503, 504
intraspecific competition, 91
introvert, 234, 236
invasive species, 104
invertebrate(s)
    digestive structures, 506–507
    homeostasis and temperature regulation in, 525–526
    integument of, 418–419
    respiratory systems, 490–492
    skeletons, 422–425
invertebrate chemoreceptor, 448
invertebrate endocrine control
    of basal and lophotrochozoan phyla, 467
    of ecdysozoans, 467–469
invertebrate georeceptor, 449
invertebrate nervous systems, 439–442
invertebrate phonoreceptors, 449
invertebrate photoreceptors, 450
invertebrate proprioceptor, 451
invertebrate respiration
    book lung, 492
    diffusion through body surfaces, 491
    pulmonate lung, 492
    tracheal system, 491
invertebrate sensory receptors
    baroreceptors, 448
    chemoreceptors, 448
    georeceptors, 448
    hygroreceptors, 448–449
    phonoreceptors, 449
    photoreceptors, 449–450
    proprioceptors, 450–451
    tactile receptors, 451
    thermoreceptors, 451

invertebrate skeletons
endoskeletons, 424-425
exoskeletons, 423-424
hydrostatic skeletons, 422-423
island dwarfism, 415
isotonic solution, 16
Ixodes scapularis, 252

J

Jacobson's organs, 369, 457
jawless fishes, skin, 419
jaw worms, 172
jellyfish, 134, 146, 147, 542
Johnston's organs, 272
jumping bristletails, 259
juvenile hormone (JH), 467, 468

K

kangaroo rat, 539
keystone species, 94
kidneys, 473
in chordates, 534
metanephric kidney functions, 537-540
mollusc, 531
vertebrate excretory systems, 534-537
kinetochore, 35
Kinorhyncha/kinorhynchs, 234-235
Krebs cycle, 126
krill, 96, 264

L

labial palps, 191
labor, stages of, 560
lactate fermentation, 19
lactation, 560
lacunae, 425
Lake kivu, 3, 4
Lake Malawi, 2
Lake Tanganyika, 2-4
Lake Victoria, 1-6
Lamarck, Jean Baptiste, 54
Lamellibranch Gill, 191
Lampetra, 318
lampreys, 318, 319
lamp shells, 179
large intestine, 515-517
larva, 140
lateral gene transfer (LGT), 123, 125
lateral-line system, 451
and electrical sensing, 452
and electroreception, 451-452
and mechanoreception, 452, 453
Latimeria chalumnae, 322
Latimeria menadoensis, 322
leeches
Hirudinea leech, 219
internal structure of, 219
locomotion, 220
Lepidoptera, 278
Lepidosauria, 360-363
Lepisosteus, 323
Leptodactylus pentadactylus, 349
Lepus americanus, 59
leucon sponges, 137
leukocytes, 484
level of organization
organ and organ system, 29
tissues, types, 28-29
unicellular (cytoplasmic), 117-118
Leydig cells, 551
life cycle
of Cnidaria, 143
of *Diphyllobothrium latum,* 171
of *Monogonont Rotifer,* 175
of *Obelia* structure, 144
of schistosome fluke, 167
of *Taeniarhynchus saginatus,* 171
light microscope, 12
limbic lobe, 446
limbic system, 446
limiting factor, 87
Linné, Karl von, 107
lionfishes (Wildlife Alert), 330
Lissamphibia, 336
Lithobates sylvaticus, 341
liver, 512
fluke life cycle (*See* trematodes)
role of, 518-519
lizards, 361
locomotion. *See also* movement
Amphibia/amphibians, 341-342
Annelida/annelids, 210, 211
birds, 376, 378
Cephalopoda, 196
Clitellata, 217-218
earthworm, 217
fishes, 324
Gastropoda, 187
Insecta/insects, 269
leeches, 220
reptiles, 365
salamanders, 343, 344
Turbellaria/turbellarian, 161-162
locus, 48
logistic population growth, 90
Loligo, 197
loop of nephron, 406
Lophotrochozoa/lophotrochozoans
Annelida, 223-224
Brachiopoda, 179
Cycliophora, 177
Ectoprocta, 177-179
evolutionary perspective, 158-160
hallmarks, 159
lophophore, 158
Nemertea, 176-177
origin, 130
other lophotrochozoans, 176-179
phylogenetic considerations, 179-180
phylogeny of, 180
Platyhelminthes, 160-171
relationships, 159
smaller phyla
Acanthocephalans, 174-175
Gastrotricha, 171-172
Gnathostomulida, 172-173
Micrognathozoa, 172
Rotifera, 173-174
trochophore, 158
lophotrochozoan phyla, 467
Loricifera, 236
Lumbricus terrestris, 545
lungfish, 322
lungs
vs. tissues, gas-exchange, 498
ventilation in, 494-495, 498
vertebrate respiratory systems, 493
luteinizing hormone (LH), 473, 476, 551, 554, 555
Lyell's, Charles, 55
lymph, 489, 490
hearts, 489
lymphatic systems, 489-490
lymphocytes, 484
lysosomes, 26

M

Macroacanthorhynchus hirudinaceus, 175
macroevolution. *See* microevolution
macroevolutionary change. *See* microevolution
macronutrients, 502
magnetoreception, 459-460
Malacostraca/malacostracans, 261-265. *See also* crayfish; krill
Malpighian tubules, 532
remove nitrogenous wastes, 533
Malthus, Thomas, 57
mammalian circulatory system, 404
blood pressure, 489
heart, 488-489
mammalian digestive system
gallbladder, 518-519
gastrointestinal motility and control, 513-514
large intestine, 515-517
oral cavity, 514-515
pharynx and esophagus, 515
role of liver, 518-519

role of pancreas, 517–518
small intestine, 515
stomach, 515
mammalian gastrointestinal tract, 513
mammalian infraclasses
Ornithodelphia and Metatheria, 396
mammalian phylogeny, 394
mammalian respiratory system
air-conduction portion, 496
gas-conduction portion, 496
gas transport, 497–499
ventilation, 496–497
mammals
of adrenal gland, 477
classification of living, 395
digestive glands in, 518
diversity of, 394–397
endocrine systems
adrenal glands, 477
gonads, 478
pancreas, 477–478
parathyroid gland, 476
pineal gland, 476
pituitary gland (hypophysis), 472–476
thymus, 478
thyroid gland, 476
enzymes in, 518
with estrous cycles, 555
gas-exchange surfaces in, 495
heart rates of, 488
nutrients in, 512–513
ruminants, 511
secretions in, 518
of skin, 422
synapsid amniotes, 392–416
teeth arrangement of, 509
water retention in respiratory
system of, 537
mammary glands, 398, 399, 421, 552
evolution of, 401
mantle, 424
mantle cavity, 184
manubrium, 144
marine teleosts, 535, 536
Marsupialia, 409
marsupium, 396
mass extinctions, 60–61
mastax, 174
mating. *See also* reproduction
in hermaphroditic earthworms, 545
maxillary glands, 532
Maxillopoda, 267–268
mechanoreception, 452
mechanoreceptors, 452
meconium, 560
medulla oblongata, 443
medusa, 142
meiosis, 36–38
meiotic division I
crossing-over, 37, 38
homologous chromosomes, 36
synapsis, 36, 38
meiotic division II, 38
Meissner's (tactile) corpuscles, 455
melanocyte-stimulating hormone (MSH), 473
melatonin, 473
membrane transport
carrier-mediated transport, 16–17
non-transporter gradient exchanges, 15–16
selective permeability, 15
meninges, 443
menopause, 554
menstrual cycle, 554–556
Merostomata, 245
mesenchyme, 135
mesentery, 118, 513
mesoderm, 118
mesoglea, 118, 141
mesohyl, 136
mesonephros, 534
messenger RNA (mRNA), 24, 39
metabolic fates, of nutrients, 502–503
metacercaria, 166
metamerism, 208–209, 241
metamorphosis, 547
metanephric kidney, filtration device of, 537
metanephric nephron, 538
metanephridia, 530–531
metanephridium, 213, 530
metanephros, 534
metaphase, 35
Metatheria, 396
micelles, 519
microevolution
biogeography, 63
comparative anatomy, 64–66
developmental patterns, 66–67
homology, 64–67
molecular biology, 67
paleontology, 64
phylogeny and common descent, 67–70
microfilaments, 27
microfilariae, 233
microfossils, 122
Micrognathozoa, 106, 172
micronutrients, 502–503
micropyle, 141
Microraptor, 375
microtubule-organizing center (MTOC), 27
microtubules, 27
mictic eggs, 174
midbrain, 444
middle ear, 454
milk production, 560
Miller–Urey experiment, 122
millipedes, 253, 254
mimicry, 93
mineralization, 64
mineralocorticoids, 473, 477
minerals, 502
miracidia, 168
miracidium, 166
mitochondria, 21
chemiosmosis, 22
citric acid cycle, 20, 21
electron transport system, 22
mitochondrial DNA, 158
mitosis
continuum of, 35
M-phase, 34–36
mitotic cell division
control, 36
interphase, hereditary material replication,
34–35
mitosis, 34
mitotic spindle, 35
mobile-receptor mechanism, 465–466
modern synthesis, 68
molecular clock, 61
molecular evolution, 84
molecular genetics, 38
Mollusca/molluscs. *See also* Bivalvia/bivalves
body organization, 185
characteristics, 184–186
classification of, 185
evolutionary relationships of, 184
head-foot, 184
mantle, 184
odontophore, 186
phylogenetic considerations, 202
phylogeny, 204
radula, 186
shell and mantle, 185
visceral mass, 184
mollusc kidneys, 531
molting. *See also* ecdysis; Ecdysozoa/
ecdysozoans
crustaceans, 468
molt-inhibiting hormone (MIH), 467
monocytes, 484
monoecious sponges, 140
Monogonont Rotifer, life cycle of, 175
monohybrid cross, 46
monophyletic group, 108
Monoplacophora, 200–201
monotremes, 394
morula, 557
mosaic evolution, 84
motor nerves, 442
motor neurons, 435
mouthparts. *See* tongues

movement. *See also* locomotion
- ameboid, 428
- Amphibia/amphibians, 342-344
- cellular (*See* cellular movement)
- ciliary, 428
- nonmuscular (*See* nonmuscular movement)

M-phase
- cytokinesis, 36
- mitosis, 35-36

mRNA. *See* messenger RNA (mRNA)
mucoid mass, 188
mucous cells, 515
Müllerian mimicry, 93
Müller's larva, 164
muscle fibers, 428
muscle tissues, 28-29
muscular diaphragm, 402
muscular gizzard, 510
muscular systems, 428. *See also* nonmuscular movement
musk glands, 398, 421
mutation(s), 76-77
- pressure, 77

mutualism, 93
myelin sheath, 437
myocytes, 137
myofibrils, 429
myofilaments, 429
myoglobin, 499
myomeres, 428, 429
myosin, 429
Myriapoda, 282
- Chilopoda, 254
- Diplopoda, 253-254
- Pauropoda, 254
- Symphyla, 254

Myxini, 317-318

N

nails, structure of, 398
Nanaloricus mysticus, 236
natural selection
- evolution, 59
- modes of selection, 78-79
- neutralist/selectionist controversy, 79
- reexamined, 77-79

Neanthes, 214
Necator americanus, 232
nectophore, 145
negative feedback, 464
nematocysts, 142
- Cnidaria, 141-142
- discharge, 142

Nematoda/nematodes
- characteristics of, 228
- eutely in, 228
- reproduction and development, 230
- structure and function, 229-230

Nematomorphs, 234, 235
Nemertea, 176-177
Neognathae, 376
neopallial cortex, 445
nephridia, 212, 530, 531
nephridiopore, 162
nephrons, 537
nephrostome, 213
Nereis, 214
nerve(s), 442
- cords, 441
- net, 439
- ring, 144

nervous systems, 435-437
- amphibians, 346-347
- Annelida, 211-212
- basic functional units, 435-437
- flow of information, 436
- insects, 270-271
- invertebrate nervous systems, 439-442
- invertebrate sensory receptors, 448-451
- neuron communication, 437-439
- organization of, 442
- sense organs (*See* sensory receptors)
- sensory reception, 446-447
- vertebrate nervous systems, 442-446
- vertebrate sensory receptors, 451-460

nervous tissues, 28
nesting, hierarchical, 111
neuroendocrine system, 463
neurofibril nodes (nodes of Ranvier), 437
neurolemmocytes, 437
neuromasts, 452
neuromuscular junction, 431
neurons, 435
- communication
  - action potentials, 438-439
  - graded potentials, 438
  - resting membrane potential, 437
- structure, 436-437

neuropeptides, 463
neurosecretory cells, 463
neurotransmitters, 439, 463
neutralist/selectionist controversy, 79
neutral theory, 74
- bottleneck effect, 75-76
- founder effect, 74

neutrophils, 484
newborn/baby, 555
newts. *See* salamanders
niche concept. *See* ecological niche
niche partitioning, 94
nictitating membrane, 346
Nile perch (*Lates niloticus*), 5, 6
nociceptors, 454
nodes, 109, 111
nomenclature, 108
nonavian reptiles
- amniotic egg, 356
- Archosauria, 359-360
- chromatophores of, 364
- circulatory system, 366-368
- early amniote evolution, 356-357
- excretion and osmoregulation, 369
- external structure, 364-365
- gas exchange, 368
- interpretation of amniote phylogeny, 355-356
- Lepidosauria, 360-363
- lifestyles of, 364
- locomotion, 365
- nervous and sensory functions, 368-369
- nutrition and the digestive system, 365-366
- reproduction and development, 369-370
- skull structure, 356-358
- temperature regulation, 368
- Testudines, 358, 359

nondisjunction, 44
nonmuscular movement
- cardiac muscles, 432-433
- muscular systems, 428
- nonmuscular movement, 427-428
- skeletal muscle structure and function, 428-432
- variations in muscle structure and function, 432-433

nonrenewable resources, 103
nonruminant herbivore, 512
nonshivering thermogenesis, 528
non-transporter gradient exchanges
- aquaporins, 15
- concentration gradient, 15
- diffusion, through membrane channels, 15
- hypertonic and hypotonic solution, 16
- isotonic solution, 16
- osmosis, 15, 16
- simple diffusion, 15
- tonicity, 16

nontropic hormones, 474
norepinephrine, 439, 473
nuchal organs, 210
nuclear DNA, 158
nuclear envelope, 23, 24
nuclei, 442
nucleic acids, components of, 39
nucleolus, 23
nucleosome, 32
nucleotides, 38
nucleus, 12
nutrients
- load, 101-103
- in mammal, 512-513
- metabolic fates of, 502-503

nutrition, 501
and nutrients (*See* digestion)
nutritive-muscular cells, 142

O

Obelia, structure and life cycle, 144
occipital lobe, 446
ocean acidification, 101
ocelli, 147, 163, 449
ocellus. *See* eyes
Octacorallian Corals, 151
Odonata, 275
olfaction, 456
ommatidia, 449
omnivores, 501
oncomiracidium, 167
oncosphere, 169
Onychophora, 280, 281
oogenesis, 38
open circulatory system, 188, 483
operculum, 166, 186
Ophiuroidea, 290–291
Opiliones, 251
opisthaptor, 167
oral cavity, 508–509, 514–515
oral lobes, 147
oral sucker, 165
organelles, 12
organic evolution, 53
and pre-Darwinian theories of change, 53–54
zoology, 3
organs
food storage, 507
of human digestive system, 513
of human respiratory system, 497
Jacobson's, 369, 457
Johnston's, 272
of Ruffini, 456
system, 29
orifice, 178
Ornithodelphia, 396
oscula, 137
osmoconformer, 530
osmolarity, 530
osmoregulation, 530
amphibians, 347
osmoregulator, 530
Osphradia, 188
Osteichthyes, 315
Osteolepiformes, 331
osteon, 425
ostracoderms, 316, 318
outer ear, 454
outgroup, 109
ovarian cycle, 554
ovary, 473, 553
and testes, 477
overexploitation of resources, 103–104
oviparous, 546
ovoviviparous, 546
ovulation, 551
oxyhemoglobin, 484
dissociation curve, 498, 499
oxytocin, 473

P

Pacinian corpuscles, 456
paddlefishes, 322, 324
paleontology, 64
Panarthropoda, 280, 281
pancreas, 477–478, 512
islet cells, 473
role of, 517–518
pancreatic hormones, 478
pancreatic islets, 477
Pancrustacea
Branchiopoda, 257–268
Hexapoda, 259–260, 268–279
Pangaea, 396
papulae, 490
paracrine agents, 463
paramecium, 13
parapatric speciation, 81
paraphyletic group, 109
parapodia, 210
parasitic nematodes
Ascaris lumbricoides, 231
Enterobius vermicularis, 231, 232
filarial worms, 233–234
Necator americanus, 232
Trichinella spiralis, 232, 233
parasitism, 92
parasympathetic
division, 446
nervous system, 442
parathormone (PTH), 473, 476
parathyroid gland, 473, 476
Parayunnanolepis xitunensis, 317
parenchyma, 117, 160
parietal cells, 515
parietal lobe, 446
pars intermedia, 474
parthenogenesis, 172, 544
parturition, 557
stages of, 560
patterns of organization
symmetry, 116–117
triploblastic organization, 118–119
unicellular (cytoplasmic) level of organization, 117–118
Pauropoda, 254
pedal laceration, 148
pedipalps, 245
pelage, 397
pen/gladius, 196
pennaceous feathers, 376, 378, 379
pentaradial symmetry, 117, 286
peppered moth, 78
periostracum, 184
peripheral nervous system (PNS), 442
perisarc, 144
Perissodactyla, 402
Perissodus, 2
Perissodus microlepis, 3
peristalsis, 513, 514
peristomium, 210
peritoneum, 183, 513
peroxisomes, 26
persistent organic pollutants (POPs), 97
Petromyzon marinus, 318
Petromyzontida, 318–319
phagocytosis, 26
pharynx, 515
phenotype, 46
pheromones, 457, 463
phonoreceptors, 449
phosphorus, 502
photoreceptors, 147, 449–450
phototaxis, 87
phylogenetic species concept, 115
phylogenetic systematics, 108–111
phylogenetic trees, 67, 114. *See also* cladograms/cladistics/cladists
of hemoglobin, 68
phylogeny, 67, 107
from base sequence alignment, 113–114
cat, 69–70
cladogram vertebrate, 112
of family Felidae, 69
and phylogenetic tree (*See* cladograms/cladistics/cladists)
physoclistous, 327
physostomous, 327
pia mater, 443
pigeon, 384, 387
pigment cells, 198
pilidium larva, 177
Pilosa, 397
pinacocytes, 136
pinacoderm, 136
pineal gland, 469, 473, 476
pinocytosis, 26
pinworms, 231
pioneer community, 94
pit organs, 455
pituitary gland, 470, 473
functional links between, 475
hypophysis, 472–476

pituitary stalk, 474
placenta, 396, 474, 556, 559
placids, 234
Placoderms, 316
Placozoa, 155
Plantae, 123
planula, 143
plasma, 11, 483
  cholesterol between lipid molecules, 14
  glycocalyx, 15
  structure of, 14
platelets, 484
Platyhelminthes, 160-171
  Cestoidea, 168-170
  classification of, 160
  Monogenea, 167-168
  Trematoda, 164-167
  Turbellaria, 161-164
plerocercoid larvae, 170
plesiomorphies, 109
Pleurobranchia, 154
Pleurodira, 358
plumulaceous feathers, 378, 379
pneumatic sacs, 326, 327
pneumostome, 190, 491
polar bear, 77
polian vesicles, 287
pollution, 101-103
polychaete, 208
  circulatory system in, 212
  development of, 214
polyembryony, 165
polymerase chain reaction
    (PCR), 67, 108
polymorphic loci, proportion of, 67
polyp, 142
polypeptide formation, 41
polyphyletic groups, 108-109
polypide, 178
Polyplacophora, 199-200
polyploidy, 33
polyribosomes, 24
polysomes, 24
pons, 444
population, 72-73
  age structure of, 99
population density, 90-91
population genetics, 73
population growth, 89-90
population regulation, 90-91
Porifera. *See also* sponges
  and body forms, 137-138
  body wall, 136-137
  cell types, 136-137
  classification of, 137
  maintenance functions, 138-140
  reproduction, 140-141
  skeletons, 136-137
  water currents, 137-138
porkworm, 232
positive feedback, 465
prairie chickens, 387, 388
precocial chicks, 388
pre-Darwinian theories of change, 53-54
predation, 505-506
predatory lifestyles, 183
pregnancy. *See* prenatal development, in human
premammalian synapsids, 393
prenatal development, in human
  end and beginning, birth, 556-560
  exchange site and hormone producer, 557
  milk production and lactation, 560
  primary and secondary inductions, 557
  from zygote to newborn, 557
preoptic nuclei, 469
Priapulida/priapulids, 235-236
primates, 411
primitive brain, 163
prismatic layer, 184
proboscis worms, 176
procercoid larvae, 170
production and transport
  of egg, 551-553
  of sperm, 549-550
progesterone, 473, 474
proglottids, 168
prolactin (PRL), 473, 474
prometaphase, 35
pronephros, 534
prophase, 35
proprioceptors, 450-451
prostomium, 210
protandrous, 545
protandry, 149
protein-coding loci, 75
proteins, 502. *See also* DNA; gene(s)
  kinase, 465
prothoracicotropic hormone (PTTH), 274,
    467, 468
protists/protistology/protozoology, 123.
    See also protozoa
protogynous, 545
protonephridia, 162, 530
protonephridium, 530
protostome, 131
Protostomia, 130
protostyle, 188
protozoa, 568-572
pseudocoelom, 227
pseudocoelomate pattern, 118
pseudofeces, 192
pseudopodia, 427
pulmocutaneous circuit, 486
pulmonary circuit, 486
pulmonary ventilation, 496
pulmonate lung, 491, 492
punctuated equilibrium model, 83
Punnett square, 46
  construction, cross involving, 47
purine, 38
Purkinje fibers, 489
Pycnogonida, 252-253
pyrimidine, 38
python feeding, 505

Q

Quadrula cylindrica, 203

R

rabbitsfoot mussel, 203
radial symmetry, 116
radiation, 522
radular structure, 186
rain forests. *See* tropical rain forests
range of optimum, 87
ratfish, 321
rattlesnake (Wildlife Alert), 362-363
receptor-mediated endocytosis, 26
receptor neuron, 435
receptor potential, 447
red blood cells, 484
red-cockaded woodpecker
    (Wildlife Alert), 390
rediae, 166
refractory period, 458
regulate ions and water, 533
releaser glands, 161
renal capsule, 537
renewable resources, 103
repolarization, 438
reproduction. *See also* asexual reproduction;
    human female reproductive system;
    human male reproductive system;
    sexual reproduction
  amphibians, 347-350
  birds, 387-388
  brittle star, 291
  fishes, 330-331
  gametes, 295
  insects, 274-275
  nematodes, 230
  prenatal development and birth, in human,
    555-560
  reproductive system
    in human female, 551-555
    in human male, 549-551
  reptiles, 369-370
  sea cucumbers, 294
  sea stars, 287

sexual reproduction
in invertebrates, 544-546
in vertebrates, 546-548
reproductive isolation, 80
Reptilia/reptiles. *See also* Amniota/amniotes
median eye of, 459
nonavian (*See* nonavian reptiles)
of skin, 420
Reptiliomorpha, 338, 352
respiration. *See* circulatory systems; gas-exchange
respiratory pigments, 211, 498
respiratory systems
circulation and gas exchange
mammalian, 495-499
human organs, 497
in invertebrate, 490-492
in mammalian, 495-499
vertebrate
circulation and gas exchange, 492-495
in vertebrate, 492-495
water retention in, 537
resting membrane potential, 437
rete mirabile, 526
Reticulitermes flavipes, 510
reticulum, 509
rhabdites, 161
rhodopsin, 459
rhopalia, 147
rhopalium, 147
rhynchocoel, 176
Rhynchocoela, 176-177
ribosomal RNA (rRNA), 23, 39, 108, 123, 129, 158
ribosomes, 23
RNA, 13, 122, 502
translation, 43
types, 39
rod cells, 459
Rodentia, 406
Rotifera
classification of, 174
external features, 173-174
feeding and digestive system, 174
reproduction and development, 174
roundworms. *See* Nematoda/nematodes
rRNA. *See* ribosomal RNA (rRNA)
rumen, 509
ruminants, 402, 511

S

salamanders, 339-340, 343, 344
salmon, 327
saltatory conduction, 439
sand dollars, 291-293
sarcolemma, 429
sarcomere, 429, 430
sarcopterygians, 486
Sarcopterygii, 315, 322
Sarcoptes scabei, 252
Sauria, 360-361. *See also* lizards
sawfly, 279
scales, 419
scalids, 234
Scaphopoda, 200
scent, 421
glands, 398
schistosome fluke, life cycle, 167
sclerotization, 419, 424
Scorpionida, 247-248
scrotum, 549
scyphistoma, 148
Scyphozoa/scyphozoans, 146-147
medusa, 147
sea anemones, 149
sea cucumbers, 293-294
sea daisies, 289-290
sea lampreys, 318, 319
sea lilies, 294-295
sea stars, 287-289
sea turtles, 369, 496
sea urchins, 291-292
sea walnuts, 151
sebaceous (oil) glands, 398, 421
secondary palate, 399
of reptiles, 364, 365
secretions, in mammals, 518
Sedentaria
Clitellata, 216-221
Echiura, 216
Siboglinidae, 216
tubeworms, 216
sedimentary cycles, 98
segmentation, 513, 514
segmented worms, 130, 176
segregation
cross involving, single trait, 45
heterozygous, 46
homozygous, 46
principle of, 45
selection pressure, 77
selective permeability, 15
semen, 549
semilunar valve, 488
seminal receptacles, 219
seminal vesicles, 218, 545
sensilla, 448
sensory nerves, 442
sensory neurons, 435
sensory papillae, 220
sensory reception, 446-447
sensory receptors, 447. *See also* invertebrate sensory receptors; vertebrate sensory receptors
of skin, 454-456
sensory systems
basic functional units, 435-437
invertebrate nervous systems, 439-442
invertebrate sensory receptors, 448-451
neuron communication, 437-439
sensory reception, 446-447
vertebrate nervous systems, 442-446
vertebrate sensory receptors, 451-460
Sepiola, 198
sequential hermaphroditism, 545
seral stage, 94
sere, 94
serosa, 513
Serpentes, 361, 363
serpent stars, 290
serum, 483
Setae, 210
set point, 464
sex cells, 542
sex chromosomes, 32-33
sexes *vs.* phenotypes, 45
sex hormone, 473, 477
testosterone, 549
sexual reproduction, 544. *See also* reproduction
in invertebrates
advantages and disadvantages, 546
external fertilization, 544
internal fertilization, 545
in vertebrates
amniotes, 547-548
amphibians, 546-547
fishes, 546
reproductive strategies, 546, 547
sexual selection, 79
Seymouria baylorensis, 353
sharks, 320, 321
shivering thermogenesis, 527
Siboglinidae, 216
sickle-cell anemia, 80
Sidneyia, 129
silverfish, 275, 282
sinoatrial (SA) node, 488
Sinornis, 375-376
Sinosauropteryx, 375
sinuses, 188
siphon, 188
Siphonophora, 145
siphuncle, 196
Sipuncula, 221-222
sister chromatids, 34
sister groups, 110
skates, 321
skeletal muscle, 428, 430

skeleton/skeletal system, 422. *See also* endoskeletons; exoskeletons
appendicular, 425
axial, 425
calcium carbonate, 150
of humans and other apes, 412
hydraulic, 423
hydrostatic, 423
invertebrates, 422–425
vertebrate, 425–427
skin, 419
of amphibians, 420
of birds, 421
of bony fishes, 420
of cartilaginous fishes, 420
of jawless fishes, 419
of mammals, 422
of reptiles, 420
of sensory receptors, 454–456
sliding-filament model, of muscle contraction, 431
slit sense, 449
small intestine, 515
smell, 456–457
smooth muscle, 433
tissue, 433
snails, 166. *See also* Gastropoda
snakes, 361, 363
snake venom, 366
snow leopard *(Panthera uncia),* 50
sodium-potassium ATPase pumps, 437
soft palate, 398
Solenogastres, 201
soma, 436
somatic cells, 36
somatic nervous system, 442
somatotropin (STH), 473, 474
sonar (biosonar), 456
speciation
allopatric, 81
biological species concept, 80
parapatric, 81
reproductive isolation, 80
sympatric, 81
species
biological species concept, 80
coexistence of competing, 92
invasive, 104
keystone, 94
sperm
cells and eggs, 140
production and transport of, 549–550
spermatogenesis, 38, 551
spermatophores, 199, 545
Sphenodontia, 360
sphincters, 513
spicules, 137, 138
spiders, 239
spider silk, 248, 251
spinal cord, 442–443
of vertebrates, 443
spinal nerves, 443
spiracle, 491
sponges
body forms, 139
defense mechanisms, 139
feeding cells of, 127
larval stages, 140
morphology of, 138
sperm cells and eggs, 140
spongin, 137
spongocoel, 137
sporocyst, 166
springtails, 274
Squamata, 360, 363
squamate teeth, 366
squid, internal structure of, 197
stabilizing selection, 79
static equilibrium (balance), 455
statocysts, 144, 147, 448
statolith, 144, 448, 449
Staurozoa, 146. *See also* jellyfish
stem cells, 484
Stereospondyli, 339
sticklebacks, 331
stimulus, 447
stinging nettle, 146
stomach, 515
generalized arrangement of, 510
strobila, 147
style sac, 192
substrates, 18
sudoriferous (sweat) glands, 398, 421
summation, 438
sunfishes, 331
superior frontal lobes, brain, 199
surface nutrient absorption, 506
survivorship, 90
suspension feeding, 504
suspensory ligaments, 552
swim bladders, 322
sycon body form, 137
symbiosis, 92–93
symmetry, 116–117
sympathetic, nervous system, 442
division, 446
sympatric speciation, 81
Symphyla, 254
symplesiomorphies, 109
synapomorphies, 109
synapse, 439
chemical transmission, 440
Synapsida/synapsids, 357. *See also* mammals
diversity of mammals, 394–397
evolutionary perspective, 392–394
evolutionary pressures, 397–409
human evolution, 409–416
synaptic cleft, 439
synaptic integration, 439
Syndermata, 180
systematics, 106
systemic circuit, 486
systole, 489
systolic pressure, 489

T

tactile receptors, 451
tadpoles. *See* frogs
Taeniarhynchus saginatus, life cycle of, 171
tagmatization, 208–209, 241
Tanganicodus, 2
tapeworms, 168, 170
Tardigrada, 280, 281
target cell, 464
taste, 457
taxis, 87
taxonomy, 107
animal systematics, 108–115
categories of human and dog, 107
hierarchy, 107
methods, 108
nomenclature, 108
patterns of organization, 115–119
and phylogeny, 106–115
T cells, 484
tectum, 444
teeth
arrangement of mammals, 509
specializations of, 400
tegument, 165
telencephalon, 446
teleosts, 535
telophase, 36
temperature regulation
antelope jackrabbit, 527
control of water and electrolytes, 528–530
homeostasis and temperature regulation, 521–528
invertebrate excretory systems, 530–532
vertebrate excretory systems, 532–540
temporal lobe, 446
tendons, 429
termites, 276
testes, 474, 549
and ovaries, 477
testosterone, 551

sex hormone, 549
Testudines, 358, 359
tetrad, 36
Tetrapoda/tetrapods, 336. *See also* Amphibia/amphibians
endoskeleton, 427
Tetrapodomorpha, 322, 331, 332
thalamus, 446
thermogenesis, 527
thermoreception, 456
thermoreceptors, 451, 455
thermoregulation, 522, 523
feedback pathways, 529
in large, active fishes, 525
thermostat, 528
thin filaments, 429
threatened species, 8
thrombocytes, 484
thymopoietin, 474
thymosins, 474
thymus, 474, 478
thyroid gland, 474, 476
thyroid-stimulating hormone (TSH), 473
thyrotropin, 473
thyroxine (T_4), 474
Thysanura, 274, 275, 282
tiedemann bodies, 287
Tiktaalik, 332–334
tissue(s), 28
cardiac muscle, 432
vs. lungs, gas-exchange, 498
toads, 341, 344
golden toad (Wildlife Alert), 351
tolerance range, 87
of animal, 88
tongues, 344, 508
tonic fibers, 432
tooth shells, 200
torpor, 89, 368
torsion, 186, 187
Toxodon, 56
tracheae, 490
tracheal system, 490, 491
tracheoles, 490
tracts, 442
transducers, 447
transfer RNA (tRNA), 39, 42
transmission electron microscope (TEM), 12
transport proteins, 16
transport systems, 482. *See also* circulatory systems
blood cells, 485
formed-element fraction, 483
invertebrate characteristics, 481–483
plasma, 483
platelets, 484
red blood cells, 484
vertebrate characteristics, 483–485
white blood cells, 484
transverse (T) tubules, 429
tree of life, 124
Trematoda/trematodes. *See also* flukes
digenetic, 165
parasites of humans, 166–167
subclass digenea, 165–166
tegument, 165
trematodes, 165
Trichinella spiralis, 232, 233
trichinosis, 232
triclad Turbellarian reproductive system, 164
tricuspid valve, 488
triiodothyronine (T_3), 474
Trilobitomorpha/trilobites, 244
trimesters, 557
triploblastic, 118
acoelomate pattern, 118
body plans, 119
coelomate pattern, 119
organization, 118–119
pseudocoelomate pattern, 118
trochophore larva, 130, 189
trophic levels, 95
tropical rain forests, 7, 8
tropic hormones, 474
troponin, 430
calcium-induced changes in, 431
tubeworms, 216
tunas, of gills, 494
Turbellaria/turbellarian. *See also* flatworms
asexual reproduction in, 164
body wall, 161
digestion and nutrition, 162
digestive systems in, 163
environment exchanges, 162
locomotion, 161–162
nervous and sensory functions, 162–164
nervous organization in, 164
pharynx, 163
protonephridial excretion in a, 163, 531
reproduction and development, 164
turtle(s), 358, 359, 369
tusk shells, 200
twitch fibers, 432
tympanal organ, 449
tympanic (tympanal) organs, 449
typhlosole, 218
Tyrannosaurus, 368
tyrosine kinase receptor, 134

U

ultimobranchial glands, 470
umbilical cord, 556
under hair, 421
ungulates, 395
unicellular, level of organization, 117–118
uniformitarianism, 55
ureter, 540
urethra, 540
urinary bladder, 540
urinary (renal) system, 540
urophysis, 469
uterine tubes, 551
uterus, 551, 554

V

vagina, 552
vault organelle, 23, 24
veliger larva, 189
velum, 144
ventilation. *See also* lungs
in amphibians, 495
in lung, 494–495, 498
mammalian respiratory system, 496–497
ventricles, 443
ventricular systole, 489
venules, 485
vertebrata/vertebrates
cellular component of, 483
in cerebral cortex, 445
chromaffin tissue and adrenal glands in, 471
diversity in digestive structures, 507–512
infraphylum, 314, 318–324
lymphatic system in, 490
morphological data for five groups, 110
phylogenetic tree, 114
red blood cells variety of, 484
salt and water balance in, 535
of spinal cords, 443
vertebrate brains, 444
vertebrate circulatory systems, 485–487
vertebrate ear ossicles, evolution of, 66
vertebrate heart, evolution of, 403
vertebrate integument
skin of amphibians, 420
skin of fishes, 419–420
skin of mammals, 421–422
skin of reptiles, 420–421
vertebrate kidney, evolution of, 531
vertebrate nervous systems
autonomic nervous system, 446
brain, 443–446
cranial nerves, 446
spinal cord, 442–443

vertebrate nervous systems—*Cont.*
spinal nerves, 443
types of, 436
vertebrate respiratory systems
cutaneous respiration, 492
external gills, 493
gills, 492–493
internal gills, 493
lung, 493–495
vertebrate sensory receptors
echolocation, 456
hearing and equilibrium
in air, 452–454
in water, 454
lateral-line system
and electroreception, 451–452
and mechanoreception, 452
magnetoreception, 459–460
sensory receptors of skin, 454–456
smell, 456–457
taste, 457
vision, 458–459
vertebrate skeletons, 425–427
of fishes, 425
of tetrapods, 425–427
tissues, 425
vertical frontal lobes, brain, 199
vertical gene transfer (VGT), 123
vestigial structures, 66
vibrissae, 397, 421, 456
villi, 515
vipers, 366
vision, 458–459. *See also* eyes
vitamins, 502
functions, 503
vitreous humor fills, 458
viviparous, 546
vomeronasal organs, 457
vulva, 552, 553

W

Wallace, Alfred Russel, 60
wasps, 279
water conservation, 539
water fleas, 266, 267
water temperature control, 528–530
water-vascular system, 287, 290
Weberian apparatus, 454
whale sharks, 324
white blood cells, 484
wings. *See* birds
winter sleep, 404
woodpeckers, 387
worm(s), 235. *See also* Annelida/annelids; flatworms
human parasitic, 230–234
Wuchereria bancrofti, 233
Wuchereria malayi, 233

X

Xenoturbellida, 301
X-organs, 467
Xyloplax medusiformis, 290

Y

Y-organs, 467

Z

zebra mussel, 203, 330
zoarium, 178
zoecium, 178
zonites, 234
zooids, 145, 164
zoology, 1
animal classifications, 3–4
binomial nomenclature, 4
evolutionary process, 3
genetic relationship, animals, 4
specialization in, 2
taxonomic categories, specialization, 2
zooxanthellae, 148
zygote, 36
first trimester, 557
to newborn, 557
second trimester, 557
third trimester, 557